Physics of atom[illegible] molecules

2nd edition

We work with leading authors to develop the strongest educational materials in physics, bringing cutting-edge thinking and best learning practice to a global market.

Under a range of well-known imprints, including Prentice Hall, we craft high quality print and electronic publications which help readers to understand and apply their content, whether studying or at work.

To find out more about the complete range of our publishing, please visit us on the World Wide Web at: www.pearsoneduc.com

Physics of atoms and molecules

2nd edition

B.H. Bransden and C.J. Joachain

An imprint of Pearson Education

Harlow, England • London • New York • Boston • San Francisco • Toronto • Sydney • Singapore • Hong Kong
Tokyo • Seoul • Taipei • New Delhi • Cape Town • Madrid • Mexico City • Amsterdam • Munich • Paris • Milan

Pearson Education Limited
Edinburgh Gate
Harlow
Essex CM20 2JE
England

and Associated Companies throughout the world

Visit us on the World Wide Web at:
www.pearsoned.co.uk

First published under the Longman Scientific & Technical imprint 1983
Second edition published 2003

ISBN 978-0-582-35692-4

British Library Cataloguing-in-Publication Data
A catalogue record for this book is available from the British Library

Library of Congress Cataloging-in-Publication Data
Bransden, B.H., 1926–
Physics of atoms and molecules / B.H. Bransden and C.J. Joachain.—2nd ed.
p. cm.
Includes bibliographical references and index.
ISBN 0–582–35692–X (pbk.)
1. Atoms. 2. Molecules. I. Joachain, C.J. (Charles Jean) II. Title.

QC173 .B677 2002
539—dc21 2002029340

10 9
11

Typeset in 10/12pt TimesTen by 35
Printed in Malaysia, VVP

Contents

Preface to the second edition

In the years since the first edition of this book was published, atomic and molecular physics has remained a key component of undergraduate courses in physics, both because of its fundamental importance to the understanding of many aspects of modern physics and also because of the exciting new developments that have occurred. In this new edition we have aimed to provide, as before, a unified account of the subject within an undergraduate framework, taking the opportunity to make improvements based on the teaching experience of users of the first edition. In particular, we have revised our account of molecular structure and spectra and extended the material on electronic and atomic collisions. The last two chapters provide an introduction to the exciting advances of recent years and to some of the many important developments of the subject. The first of these chapters describes applications based on the use of the maser and the laser, including laser cooling and trapping of atoms, Bose–Einstein condensation, atom lasers and a discussion of the behaviour of atomic systems in intense laser fields. In the final chapter a selection of applications of a different nature is introduced: magnetic resonance, atom optics, atoms in cavities, ions in traps, atomic clocks and astrophysics. As in the first edition, various special topics and derivations are given in appendices together with updated tables of physical constants. Hints for the solution of selected problems are given at the end of the book.

In preparing the second edition of this book we have been greatly helped by the comments and suggestions sent to us by colleagues who have used the first edition in their teaching. We should like to take this opportunity to thank all those who have written to us. One of us (C.J.J.) would like to thank Professor H. Walther for his hospitality at the Max-Planck-Institut für Quantenoptik in Garching, where part of this work was carried out. We also wish to thank Mrs R. Lareppe for her expert and careful typing of the manuscript and Mrs H. Joachain-Bukowinski for preparing a large number of the diagrams.

B.H. Bransden, Durham
C.J. Joachain, Brussels
December 2001

Preface to the first edition

Modern undergraduate courses in physics invariably include a good deal of material on basic atomic physics, including discussions of atomic structure, the optical and X-ray spectra of atoms, the interaction of atoms with electric and magnetic fields, the theory of simple molecules and some atomic scattering theory. As a rule, part of this material is given in a course on quantum mechanics, and some separately. Correspondingly, most books on quantum mechanics deal with some of these topics, usually in a rather sketchy fashion, while texts on 'Atomic Spectra', 'Collision Theory' and the like, deal with individual topics at considerably greater length than the undergraduate requires.

The aim of this book is to present a unified account of the physics of atoms and molecules, from a modern viewpoint, in adequate detail, but keeping within the undergraduate framework. It is based on courses given by the authors at the Universities of Durham, Glasgow, California (Berkeley), Brussels and Louvain-la-Neuve, and is suitable for study at second or third year level of an undergraduate course following some study of elementary quantum theory.

Following a brief historical introduction in Chapter 1, Chapter 2 contains an outline of the ideas and approximation methods of quantum mechanics, which are used later in the book. This is in no sense intended as a substitute for a proper study of quantum mechanics, but serves to establish notation and as a convenient summary of results. In Chapters 3 to 8, the structure of atoms and the interaction of atoms with radiation are discussed, followed in Chapters 9 and 10 by an account of the structure and spectra of molecules. Selected topics dealing with the scattering of electrons by atoms, and of atoms by atoms, are given in Chapters 11 to 13 while in the final chapter, a few of the many important applications of atomic physics are considered. Various special topics and derivations are given in the appendices together with useful tables of units. For a full understanding, the reader should work through the problems given at the end of the chapters. Hints at the solutions of selected problems are given at the end of the book.

We wish to thank our colleagues and students for numerous helpful discussions and suggestions. It is also a pleasure to thank Mme E. Péan and Mrs M. Raine for their patient and careful typing of the manuscript.

B.H. Bransden, Durham
C.J. Joachain, Brussels
July 1980

Acknowledgements

We are grateful to the following for permission to reproduce copyright material:

Figure 1.30 from Demonstration of single-electron build up of an interference pattern in *American Journal of Physics,* Vol. 57, American Association of Physics Teachers (Tonomura, A. *et al.*, 1989); Figure 9.9 reprinted from *Physics Letters* A, Vol. 31, Krause, M.O., Stevie, F.A., Lewis, L.J., Carlson, T.A. and Moddeman, W.E., p. 81, 'Multiple excitation of neon by photon and electron impact', Copyright 1970, with permission from Elsevier Science; Figures 11.3, 11.5 and 11.6 reproduced with kind permission of Professor R. Colin; Figure 13.3 from Electron scattering from hydrogen atoms. II. Elastic scattering at low energies from 0.5 to 8.7 eV in *Journal of Physics B: Atomic and Molecular Physics*, Vol. 8 No. 10, Institute of Physics (Williams, J.F., 1975); Figure 13.4 from Accurate e^--He cross sections below 19 eV in *Journal of Physics B: Atomic and Molecular Physics*, Vol. 12 No. 7, Institute of Physics (Nesbet, R.K., 1979); Figure 13.6 from Electron scattering in helium at low energies: a 29-state R-matrix calculation in *Journal of Physics B: Atomic, Molecular and Optical Physics*, Vol. 23 No. 23, Institute of Physics (Sawey, P.M.J. *et al.*, 1990); Figure 13.7 from Benchmark calculations for e-H scattering between the $n = 2$ and $n = 3$ thresholds in *Journal of Physics B: Atomic, Molecular and Optical Physics*, Vol. 29, Institute of Physics (Bartschat, K. *et al.*, 1996); Figure 13.23 reprinted from *Physics Letters A*, Vol. 86, Lohmann, B. and Weigold, E., Direct Measurement of the Electron Momentum Probability Distribution in Atomic Hydrogen, p. 139, Copyright 1981 with permission from Elsevier Science; Figure 13.27 from Observation of Resonances near 11 eV in the Photodetachment Cross Section of the H^- Ion in *Physical Review Letters*, Vol. 38, American Physical Society (Bryant, H.C. *et al.*, 1977); Figure 13.28 from Photo-ionisation of the $3s^23p\ ^2P^o$ ground state of aluminium in *Journal of Physics B: Atomic and Molecular Physics*, Vol. 20, Institute of Physics (Tayal, S.S. and Burke, P.G., 1987); Figures 13.29 and 13.30 from Double Photonization of Helium by Pont, M. and Shakeshaft, R., in *Photon and Electron Collisions with Atoms and Molecules*, Plenum Press, New York and London (Burke, P.G. and Joachain, C.J. (eds), 1997); Figures 14.18 and 14.19 from Critical test of first-order theories for electron transfer in collisions between multicharged ions and atomic hydrogen: The boundary condition problem in *Physical Review A*, Vol. 36 No. 4, American Physical Society (Belkić, D. *et al.*, 1987); Figure 15.15 reproduced from www.desy.de with kind permission of Deutches Elektronen Synchrotron (DESY);

Figure 15.16 from Ultrahigh-Intensity Lasers: Physics of the Extreme on a Tabletop in *Physics Today*, January, American Institute of Physics (Mourou, G.A. *et al.*, 1998); Figure 15.18 reprinted from *Optics Communication*, Vol. 8, Lange, W., Luther, J., Nottbeck, B. and Schröder, H.W., p. 157, Copyright 1973, with permission of Elsevier Science; Figure 15.23 from Study of resonance fluorescence in cadmium: modulation effects and lifetime measurements in *Proceedings of the Physical Society,* Vol. 92, Physical Society (Dodd, J.N. *et al.*, 1967); Figure 15.24 from Nonperturbative treatment of multiphoton ionigation within the Floquet theory in *Advances in Atomic and Molecular Physics*, Sup. 1, Academic Press, San Diego (Potvliege, R.M. and Shakeshaft, R., 1992); Figure 15.26 from Multiphoton ionization of H^- and He in intense laser fields, *Physical Review Letters*, Vol. 71 No. 24, American Physical Society (Purvis, J. *et al.*, 1993); Figure 15.27 from Singly, doubly and triply resonant multiphoton processes involving autoionizing states in magnesium, *Journal of Physics B: Atomic and Molecular Physics*, Vol. 31, Institute of Physics (Kylstra, N.J. *et al.*, 1998); Figure 15.28 from *Europhysics Letters*, Vol. 26, EDP Sciences and Societa Italiana de Fisica (Latinne, O. *et al.*, 1994); Figure 15.29 from Nonsequential double ionization of helium in *Physical Review Letters*, Vol. 78 No. 10, American Physical Society (Watson, J.B. *et al.*, 1997); Figure 15.31 from Relativistic effects in the time evolution of a one-dimensional model atom in an intense laser field, *Journal of Physics B: Atomic and Molecular Physics*, Vol. 30, Institute of Physics (Kylstra, N.J. *et al.*, 1997); Figure 15.32 from Intensity dependence of non-perturbative above-threshold ionisation spectra: experimental study in *Journal of Physics B: Atomic, Molecular and Optical Physics*, Vol. 21 No. 24, Institute of Physics (Petite, G. *et al.*, 1988); Figure 15.34 from Plateau in above threshold ionization spectra in *Physical Review Letters*, Vol. 72 No. 18, American Physical Society (Paulus, G.G. *et al.*, 1994); Figures 15.35 and 15.36 from High-order harmonic generation in rare gases with a 1-ps 1053-nm laser in *Physical Review Letters*, Vol. 70 No. 6, American Physical Society (L'Huillier, A. and Balcou, P., 1993); Figure 15.37 from *Laser Physics*, Vol. 12 (Kylstra, N.J. *et al.*, 2001); Figures 15.39 and 15.40 from New mechanisms for laser cooling in *Physics Today*, October, American Institute of Physics (Cohen-Tannoudji, C. and Phillips, W.D., 1990); Figure 15.42 from Laser Cooling below the One-Photon Recoil Energy by Velocity-Selective Coherent Population Trapping in *Physical Review Letters*, Vol. 61 No. 7, American Physical Society (Aspect, A. *et al.*, 1988); Figure 15.43 from A. Aspect in *Laser Interactions with Atoms, Solids, and Plasmas,* Plenum Press, New York and London (More, R.M. (ed.), 1994); Figure 15.44 from Bose-Einstein condensation in a dilute gas: measurement of energy and ground-state occupation in *Physical Review Letters*, Vol. 77 No. 25, American Physical Society (Ensher, J.R. *et al.*, 1996); Figures 15.45 and 15.46 from Atom Lasers in *Physics World*, August, IOP Publishing Ltd., Bristol (Helmerson, K. *et al.*, 1999); Figure 16.6 reprinted from *Physics Reports*, Vol. 240, Adams, C.S., Sigel, M. and Mlynek, J., p. 143, 'Atom optics', Copyright 1994, with permission of Elsevier Science; Figure 16.7 from *Applied Physics B*, Sleator, T., Pfau, T., Balykin, V. and

Mlynek, J., Vol. 54, p. 375, Figure 16, 1992, copyright Springer-Verlag; Figures 16.8 and 16.9 from Imaging and focusing of atoms by a fresnel zone plate in *Physical Review Letters*, Vol. 67 No. 23, American Physical Society (Carnal, O. *et al.*, 1991); Figure 16.12 from Cesium atoms bouncing in a stable gravitational cavity in *Physical Review Letters*, Vol. 71 No. 19, American Physical Society (Aminoff, C.G. *et al.*, 1993); Figures 16.14 and 16.15 from Young's double-slit experiment with atoms: A simple atom interferometer in *Physical Review Letters*, Vol. 66 No. 21, American Physical Society (Carnal, O. and Mlynek, J., 1991); Figure 16.16 from Double-slit interference with ultracold metastable neon atoms in *Physical Review A*, Vol. 46 No. 1, American Physical Society (Shimizu, F. *et al.*, 1992); Figure 16.19 from Pritchard, D.E., in *Atomic Physics 13*, American Institute of Physics (Walther, H. *et al.* (eds), 1993); Figure 16.20 from Atomic interferometry in *Journal of Physics B: Atomic, Molecular and Optical Physics,* Vol. 32 No. 15, Institute of Physics (Baudon, J. *et al.*, 1999); Figure 16.22 from Atom Interferometry in *Advances in Atomic and Molecular Physics*, Vol. 34, Academic Press (Adams, C.S. *et al.*, 1994); Figures 16.30, 16.32, 16.33 reproduced with kind permission of Professor H. Walther; Figure 16.26 from The one electron maser and the generation of non-classical light in *Physica Scripta*, Vol. T34, Royal Swedish Academy of Science (Rempe, G. *et al.*, 1991); Figure 16.27 from Microlaser: A laser with One Atom in an Optical Resonator in *Physical Review Letters*, Vol. 73 No. 25, American Physical Society (An, K. *et al.*, 1994); Figure 16.29 from Berquist, J.C. *et al.,* in *Laser Specroscopy IX*, Academic Press, Boston (Feld, M.S. *et al.* (eds), 1989); Figure 16.31 from Observation of ordered structures of laser-cooled ions in a quadrupole storage ring in *Physical Review Letters*, Vol. 68 No. 13, American Physical Society (Waki, I. *et al.*, 1992); Figure 16.35 from Atomic Clocks in *Physics World*, January, IOP Publishing Ltd. (Lemonde, P., 2001); Figure 16.36 from Time Measurement at the Millennium in *Physics Today*, March, American Institute of Physics (Bergquist, J.C. *et al.*, 2001); Figure 16.37 from Sub-dekahertz Ultraviolet Spectroscopy of $^{199}Hg^{+}$ in *Physical Review Letters*, Vol. 85 No. 12, American Physical Society (Rafac, R.J. *et al.*, 2000).

In some instances we have been unable to trace the owners of copyright material, and we would appreciate any information that would enable us to do so.

1 Electrons, photons and atoms

The physics of atoms and molecules which constitutes the subject matter of this book rests on a long history of discoveries, both experimental and theoretical. A complete account of the historical development of atomic and molecular physics lies far outside the scope of this volume. Nevertheless, it is important to recognise the key steps which have occurred in this evolution. In the present chapter we shall briefly describe the major experiments and discuss the basic theoretical concepts which are at the root of modern atomic and molecular physics.

1.1 The atomic nature of matter

The first recorded speculations as to whether matter is continuous, or is composed of discrete particles, were made by the Greek philosophers. In particular, following ideas of Anaxagoras (500–428 BC) and Empedocles (484–424 BC), Leucippus (*circa* 450 BC) and his pupil Democritus (460–370 BC) argued that the universe consists of empty space and of indivisible particles, the *atoms* [1], differing from each other in form, position and arrangement. The atomic hypothesis, however, was rejected by Aristotle (384–322 BC) who strongly supported the concept of the continuity of matter.

In modern times, the question was re-opened following the experimental discovery of the gas laws by R. Boyle in 1662, and the interpretation of these laws in terms of a kinetic model by D. Bernoulli in 1738. The kinetic theory of gases developed throughout the nineteenth century, notably by R. Clausius, J.C. Maxwell and L. Boltzmann, was able to explain the physical properties of gases by assuming that:

1. A gas consists of a large number of particles called *molecules* which make elastic collisions with each other and with the walls of the container.
2. The molecules of a particular substance are all identical and are small compared with the distances that separate them.
3. The temperature of a gas is proportional to the average kinetic energy of the molecules.

[1] The Greek word '*atomos*' (atom) means 'indivisible'.

In parallel with the development of the kinetic theory, the laws of chemical combination were being discovered, which again could be interpreted by making hypotheses about the atomic nature of matter. In 1801, J.L. Proust formulated the law of definite proportions which states that when chemical elements combine to form a given compound, the proportion by weight of each element is always the same. This was followed in 1807 by J. Dalton's law of multiple proportions, according to which when two elements combine in different ways, to form different compounds, then for a fixed weight of one element, the weights of the other element are in the ratio of small integers. These laws were explained by Dalton in 1808, who made the hypothesis that the elements are composed of discrete atoms. For a given element these atoms are all identical and each atom has the same weight. Compounds are formed when atoms of different elements combine in a simple ratio.

Also in 1808, J.L. Gay-Lussac discovered that when two gases combine to form a third, the volumes are in the ratio of simple integers. This result was explained by A. Avogadro in 1811. He was the first to make a clear distinction between atoms, the discrete particles of the elements, and molecules, which are discrete particles composed of two or more atoms bound together. Avogadro was able to show that the Gay-Lussac law is satisfied if equal volumes of different gases, at the same pressure and temperature, contain equal numbers of molecules.

It is interesting that the atomic explanation of chemistry was not fully accepted until late in the nineteenth century [2], largely because chemists tended to ignore the compelling evidence from kinetic theory [3]. In addition to the properties of gases, the kinetic theory was able to explain other phenomena, for example the random motion of small particles suspended in a fluid. This motion, discovered by R. Brown in 1827, is due to the collisions of the molecules of the fluid with the suspended particles, as proved by A. Einstein in 1905.

From the chemical laws the relative weights of atoms can be established. Originally, Dalton proposed a scale in which hydrogen was given, by definition, the atomic weight 1. Later, this was superseded by a scale in which naturally occurring oxygen was assigned the atomic weight 16. On this scale, known as the *chemical scale*, atomic hydrogen has the atomic weight 1.008.

A *mole* is defined as a quantity of a substance weighing μ grams, where μ is the atomic (or molecular) weight of that substance. *Avogadro's number* N_A is the number of atoms (or molecules) in one mole of any substance. The first estimate of N_A was made by J. Loschmidt in 1865. In fact, N_A can be found in several ways, one of the most interesting being from observations of Brownian motion. The number deduced in this way in 1908 by J. Perrin, who performed experiments on the motion of suspended particles, was close to the best modern value of

$$N_A = 6.022\ 14 \times 10^{23}\ \text{mole}^{-1} \qquad \textbf{(1.1)}$$

[2] The nineteenth-century controversies are described in an interesting book by Knight (1967), while a collection of original papers covering the early history of the atomic theory, translated into English, has been compiled by Borse and Motz (1966).

[3] An account of the kinetic theory and its applications can be found in the text by Morse (1966).

1.2 The electron

The first experimental evidence that electric charge was not infinitely divisible, but existed in discrete units, was obtained by M. Faraday, who discovered the *laws of electrolysis* in 1833. In his experiments Faraday passed a current through conducting (electrolytic) solutions of chemical compounds. He found that the mass M of a substance (for example, hydrogen, oxygen or metals) liberated at an electrode during a certain time interval was proportional to the quantity of electricity, Q, passed through the solution during that time. He also found that a given quantity of electricity always liberated the same mass of a given substance, and that this mass was proportional to the equivalent weight of the substance, where the equivalent weight is defined as the atomic (or molecular) weight μ divided by the valency v. Faraday's laws of electrolysis can be summarised by the formula

$$M = \frac{Q}{F}\frac{\mu}{v} \tag{1.2}$$

where F is a constant called *Faraday's constant.* Its value in SI units is given by

$$F = 9.648\,53 \times 10^4 \text{ coulombs/mole} \tag{1.3}$$

Thus, since $M = \mu$ grams for one mole, we see from (1.2) and (1.3) that it takes 96 485.3 C (sometimes called one faraday) to liberate for example 1.008 g of hydrogen, 107.9 g of silver, 23 g of sodium, 35.5 g of chlorine (which are all monovalent), 8 g of oxygen (having a valency of 2) and so on.

Faraday interpreted his results by assuming that a given amount of electricity is carried by each atom (or group of atoms) during electrolysis. The charged atoms (or groups of atoms) he called *ions.* In electrolysis, the electric current is the result of the motion of the ions through the solution, the positively charged ions (or cations) moving towards the cathode and the negatively charged ions (or anions) moving towards the anode. At the electrodes, the ions are converted to neutral atoms (or radicals) which are liberated or which give rise to secondary reactions.

Faraday's results implied the existence of an *elementary unit* of electricity, some types of ion carrying one unit, others two units and so on. Indeed, as H. Helmholtz emphasised in 1881 during a speech in honour of Faraday: 'If we assume the existence of atoms of chemical elements, we cannot escape from drawing the further inference that electricity also, positive as well as negative, is divided into definite elementary charges that behave like atoms of electricity'. However, at the time of Faraday's experiments the idea, that electrical charge existed in discrete units, did not seem to agree with the evidence from other electrical phenomena, such as metallic conduction, and both Faraday and Maxwell were reluctant to accept it. In fact, the hypothesis that there is a 'natural unit of electricity' was only put forward by G.J. Stoney in 1874, who proposed that this unit should be taken to be the quantity of electricity which must pass through an electrolytic solution in order to liberate one atom of a monovalent substance. Since one faraday (96 485.3 C) liberates one mole of a monovalent substance and because one mole contains N_A atoms, where N_A is Avogadro's number, the 'natural unit of electricity', e, is given by

$$e = \frac{F}{N_A} \tag{1.4}$$

Stoney suggested the word 'electron' for this unit, and he obtained for e an approximate value of 10^{-20} C, using the rough estimates of N_A that were available from kinetic theory. In 1880, Helmholtz pointed out that it is apparently impossible to obtain electricity in smaller amounts than e. The first *direct* measurements of this smallest possible charge were initiated by J.J. Thomson and carried out by his student J.S. Townsend in 1897, and the first accurate value was found in the famous oil-drop experiment of R.M. Millikan in 1909, to which we shall return below.

Cathode rays and Thomson's measurement of *e/m*

When electrodes are placed in a gas at normal atmospheric pressure no current passes and the gas acts as an insulator until the electric field is increased to above 3 or 4 MV m^{-1} when sparking takes place. In contrast, at low pressures, a steady current can be maintained in a gas. At pressures of about 1 mm of mercury, the discharge is accompanied by the emission of light, but at still lower pressures a dark region forms near the cathode. The dark region, called the Crookes dark space, increases in size as the pressure falls, filling the discharge tube at pressures of 10^{-3} mm and below. If, under these low-pressure conditions, a small hole is made in the anode (see Fig. 1.1), a green glow is observed on the glass wall of the discharge tube.

The causative agents of this phenomenon were termed 'cathode rays'. The properties of these rays were studied in the latter part of the nineteenth century by W. Crookes and P. Lenard, who showed that the rays travelled in straight lines, cast 'shadows' and also carried sufficient momentum to set in motion a light paddle wheel. In 1895, J. Perrin demonstrated that the rays carried negative

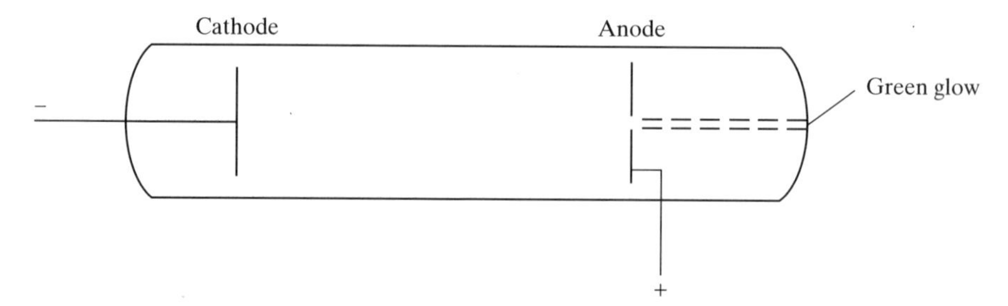

Figure 1.1 Low-pressure discharge tube. Cathode rays passing through a hole in the anode cause a green glow on the glass wall of the tube.

charge by collecting the charge on an electrometer. At that time very differing views were expressed as to the nature of the cathode rays, but J.J. Thomson set out the hypothesis that the rays consisted of a stream of particles each of mass m and charge $-e$, originating in the cathode of the discharge tube. Since the earlier investigations showed that the properties of the cathode rays were independent of the material of the cathode and of the gas in the tube, the particles could be assumed to be constituents of all matter.

In Thomson's experiments, performed in 1897, the deflection of the cathode rays by static electric and magnetic fields was investigated, which allowed the determination of the '*specific charge*', the ratio e/m, of the constituent particles. The cathode rays were passed between parallel plates, a distance D apart, to which a potential difference V could be applied, as in Fig. 1.2. The cathode rays emerging from the region of the electric field were detected on a screen S, and the deflection measured as a function of V. Neglecting end effects, the electric field strength $\mathscr{E}$ between the plates can be taken to be uniform and equal to V/D, and in this field the charged particle experiences a constant acceleration of magnitude $e\mathscr{E}/m$ in the Y direction (see Fig. 1.2). If the initial velocity of a particle is v, the time taken to traverse the region between the plates, of length x_1, is $t_1 = x_1/v$. The subsequent time to reach the screen, placed at distance x_2 from the plates, is $t_2 = x_2/v$.

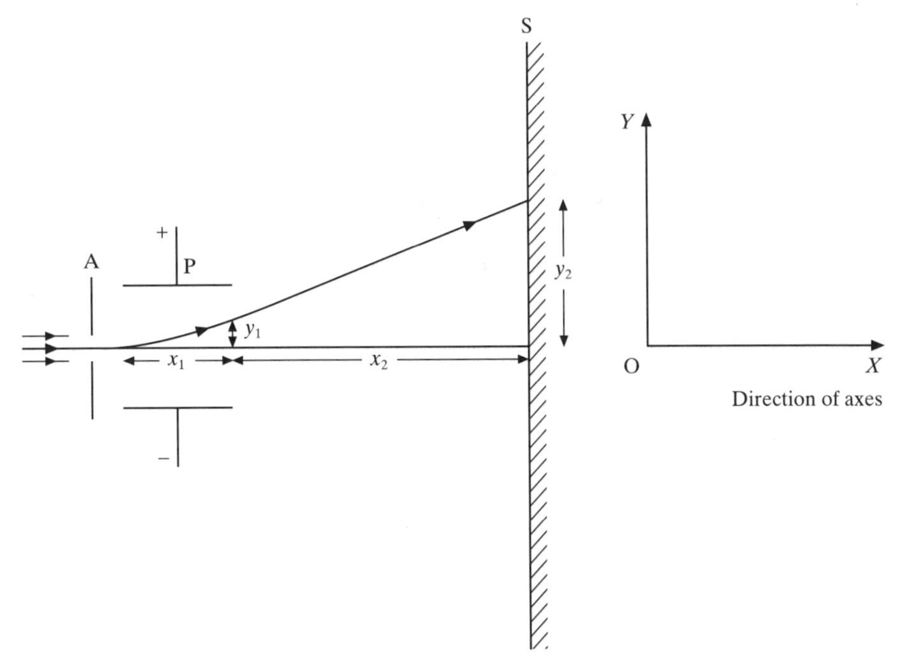

Figure 1.2 Schematic diagram of Thomson's apparatus to measure e/m. A stream of cathode rays, passing through a small hole in the anode A of a discharge tube, is deflected by being passed between the plates of a condenser P to which a potential is applied.

After the time t_1, the deflection in the Y direction is

$$y_1 = \tfrac{1}{2}\left(\frac{e\mathscr{E}}{m}\right)t_1^2 = \tfrac{1}{2}\left(\frac{e\mathscr{E}}{m}\right)\left(\frac{x_1}{v}\right)^2 \tag{1.5}$$

On leaving the region between the plates, the component of the particle velocity in the Y direction is

$$v_y = \left(\frac{e\mathscr{E}}{m}\right)t_1 = \left(\frac{e\mathscr{E}}{m}\right)\left(\frac{x_1}{v}\right) \tag{1.6}$$

from which the total deflection in reaching the screen is

$$y_2 = v_y t_2 + y_1 = \frac{e}{m}\frac{\mathscr{E}x_1}{v^2}(\tfrac{1}{2}x_1 + x_2) \tag{1.7}$$

Thus, a measurement of the deflection y_2 provides a value of the combination $e/(mv^2)$, if $\mathscr{E}$, x_1 and x_2 are known.

To determine e/m, an independent measurement is required from which v can be found. By placing the apparatus within a Helmholtz coil, Thomson could apply a constant magnetic field $\mathscr{B}$, directed in the Z direction, at right angles both to the electric field and to the undeflected path of the cathode rays. The magnetic force on the charged particles is of magnitude $ev\mathscr{B}$ and is perpendicular to the particle trajectory, being initially antiparallel to the Y direction. If both electric and magnetic fields are applied simultaneously, the net force on the particles vanishes provided $\mathscr{E}$ and $\mathscr{B}$ are adjusted so that

$$ev\mathscr{B} = e\mathscr{E} \tag{1.8}$$

Two experiments can now be performed. In the first, the values of $\mathscr{B}$ and $\mathscr{E}$ are measured for which the cathode rays are undeflected and this provides the value of v, since from (1.8) we have $v = \mathscr{E}/\mathscr{B}$. In the second, the magnetic field is switched off and the deflection due to the electric field alone is measured [4]. Knowing v, the specific charge e/m can be calculated from (1.7).

Thomson found a value for the specific charge somewhat smaller than the modern value of 1.76×10^{11} C kg^{-1}. The specific charge for the lightest known positive ion (the hydrogen ion) is smaller by a factor of approximately 1840, so either the cathode ray particles are much lighter or they carry a very large charge. Thomson assumed that the charge on a cathode ray particle was equal in magnitude (but opposite in sign) to that on the hydrogen ion, so that each particle was lighter than a hydrogen ion by a factor of about 1840. Particles with this property are now called *electrons*, thus changing the original meaning of the word electron which was applied by Stoney to the magnitude e of the charge carried by a hydrogen ion or a cathode ray particle.

---

[4] If the electric field is switched off the cathode rays move along an arc of a circle of radius R, where $(mv^2/R) = ev\mathscr{B}$. From this the deflection by a magnetic field extending over a region of length x can be calculated.

Millikan and the charge of the electron

Following his determination of an accurate value for the charge to mass ratio of the electron, J.J. Thomson, together with his student J.S. Townsend, attempted to measure the electronic charge itself. The method employed was to produce clouds of charged water droplets and to estimate the number of droplets in a cloud from a knowledge of its total mass and the rate at which the cloud settles. The total charge of the cloud could be measured with an electrometer and hence the charge on each drop estimated. In an extension of the method due to H.A. Wilson, the charge on the cloud was measured by applying an electric field in the opposite direction to gravity, and adjusting the strength of the field until the cloud ceased to settle, but remained suspended at rest. Both these methods failed to provide accurate results because of the evaporation of the droplets during the experiment. However, a brilliant modification of Wilson's method, by R.A. Millikan in 1909, gave the first accurate value for the magnitude, e, of the electronic charge.

In Millikan's experiments very small oil droplets a few microns in diameter were formed by spraying mechanically from a nozzle. The droplets became charged by friction as they were formed. They also acquired charges from the surrounding air, which could be ionised by passing X-rays through the apparatus. Some droplets were allowed to fall through a small hole into a region between two parallel plates of a condenser to which an electrostatic potential V could be applied. The motion of the drops was viewed by a microscope with a micrometer eyepiece, and the velocity of fall was measured. The whole apparatus, a schematic diagram of which is shown in Fig. 1.3, was enclosed in a thermostat to avoid convection currents of the air between the condenser plates, which were 22 cm in diameter and separated by about 15 mm.

If the condenser is uncharged ($V = 0$), a droplet of effective mass M falls under gravity, reaching a terminal velocity v_1 when the gravitational force Mg is

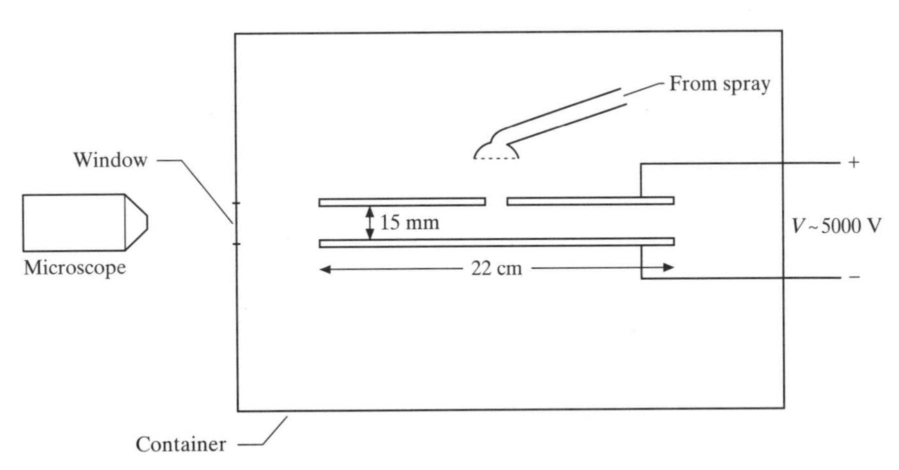

Figure 1.3 Millikan's experiment for the determination of the charge of the electron.

balanced by the viscous drag of the air. According to Stokes's law this occurs when

$$Mg = 6\pi\eta r v_1 \tag{1.9}$$

where η is the coefficient of viscosity of air and r the radius of the drop. The mass of the drop is $4\pi r^3\rho_O/3$ where ρ_O is the density of the oil, and allowing for the buoyancy of the air the effective mass is

$$M = \tfrac{4}{3}\pi r^3(\rho_O - \rho_A) \tag{1.10}$$

where ρ_A is the density of the air.

The potential V (of the order of 5 kV) can now be applied. If it is sufficiently large and in the correct direction, the drop will move upwards until a new terminal velocity v_2 is reached. If D is the distance between the plates and q is the charge on the drop

$$q\frac{V}{D} - Mg = 6\pi\eta r v_2 \tag{1.11}$$

Thus, from (1.9) and (1.11), the charge on the drop is

$$q = 6\pi\eta r\left(\frac{D}{V}\right)(v_1 + v_2) \tag{1.12}$$

which can be determined by measuring v_1, v_2 and (D/V), since the radius of the drop is given by (1.9) and (1.10), provided η, ρ_O and ρ_A are known.

The same drop could be observed for a period of some hours, during which time the charge q varied because positive or negative ions were acquired from the surrounding air. From many thousands of observations, Millikan found that as q altered, it always changed in integral units of a basic charge and in general the magnitude of q was given by

$$|q| = 1.59n \times 10^{-19}\ \text{C} \tag{1.13}$$

where n was an integer usually between 3 and 30. Thus the basic charge, which he identified with the magnitude of the electronic charge e, was found to have the value 1.59×10^{-19} C. Later measurements in which better values of the viscosity η were used gave an improved value,

$$e = 1.60 \times 10^{-19}\ \text{C} \tag{1.14}$$

Combining these results with the modern value of e/m, a value for the mass of the electron is obtained,

$$m = 9.11 \times 10^{-31}\ \text{kg} \tag{1.15}$$

which is approximately 1840 times lighter than a hydrogen ion, as postulated by Thomson.

1.3 Black body radiation

During the later part of the nineteenth century, and in the early years of the twentieth century, evidence accumulated that classical physics, represented by Newton's laws of motion and Maxwell's electromagnetic equations, is inadequate to describe atomic phenomena. The first clues to a new physics, based on the quantisation of energy, came from a study of the properties of radiation from hot bodies. It is a matter of common experience that a hot body radiates electromagnetic energy in the form of heat. In fact, at any temperature, a body emits radiation of all wavelengths, but the distribution in wavelength, the *spectral distribution*, depends on temperature. At low temperature, most of the energy is in the form of low-frequency infra-red radiation, but as the temperature increases more energy is radiated at higher frequencies, until by ~500 °C radiation of visible light is observed. At still higher temperatures, such as that of an incandescent lamp filament, the spectral distribution has shifted sufficiently to the higher frequencies for the body to be white hot. Not only the spectral distribution changes with temperature, but the total energy radiated also changes, increasing as a body becomes hotter.

In 1879, J. Stefan showed that the total power emitted per unit area, R, called the *total emissive power* (or *total emittance*) from a body at the absolute temperature T could be represented by the empirical law

$$R = e\sigma T^4 \tag{1.16}$$

where e is called the *emissivity* with $e \leqslant 1$. The emissivity varies with the nature of the surface, but the constant σ, known as Stefan's constant, is independent of the nature of the radiating surface and is given by

$$\sigma = 5.67 \times 10^{-8}\ \text{W m}^{-2}\ \text{K}^{-4} \tag{1.17}$$

When radiation falls on a body some is reflected and some is absorbed. For example, dark bodies absorb most of the radiation falling on them, while light-coloured bodies reflect most of it. The *absorptivity*, a, of a surface is defined as the fraction of the energy of the radiation falling on unit area which is absorbed, and a *black body* is defined as a body with a surface having an absorptivity equal to unity, that is a body which absorbs all the radiant energy falling upon it.

If a body is in thermal equilibrium with its surroundings, and therefore is at constant temperature, it must emit and absorb the same amount of radiant energy per unit time, for otherwise its temperature would rise or fall. The radiation emitted or absorbed under these circumstances is called *thermal radiation*. By considering the thermal equilibrium between objects made of different substances G.R. Kirchhoff in 1859 proved, using the laws of thermodynamics, that the absorptivity of a surface is equal to its emissivity, $e = a$, independently of its temperature, and that this holds for radiation of each particular wavelength. Kirchhoff's law thus shows that the emissivity of a black body is unity and that a black body is the most efficient radiator of electromagnetic energy. In 1884, L. Boltzmann derived the relation

(1.16) from thermodynamics for the case of a black body ($e = 1$). It is now known as the Stefan–Boltzmann law. It follows, from the Stefan–Boltzmann law, that the energy radiated by a black body depends only on the temperature. The spectral distribution of this radiation is of a universal nature and is of particular interest.

A perfect black body is an idealisation, but it can be very closely realised in the following way. Consider a cavity kept at a constant temperature of which the interior walls are blackened. A small hole made in the wall of such a cavity behaves like a black body, because any radiation falling on the hole from outside will pass through it and after multiple reflections will eventually be absorbed by the interior surfaces and the opening has an effective absorptivity of unity. Since the cavity is in thermal equilibrium, the radiation within it and the radiation from the small opening are characteristic of the thermal radiation from a black body. This radiation was studied experimentally as a function of the temperature of the enclosure, and the spectral distribution at each temperature was measured by O. Lummer and E. Pringsheim in 1899.

The power emitted per unit area, from a black body, at wavelengths between λ and $\lambda + d\lambda$ is denoted by $R(\lambda)\, d\lambda$, so that the total power emitted per unit area is

$$R = \int_0^\infty R(\lambda)\, d\lambda \tag{1.18}$$

and by the Stefan–Boltzmann law $R = \sigma T^4$. The observed spectral distribution function $R(\lambda)$ is shown plotted against λ, for a number of different temperatures, in Fig. 1.4. We see that, for fixed λ, $R(\lambda)$ increases with increasing T. At each temperature, there is a wavelength λ_{max}, for which $R(\lambda)$ has its maximum value. Using general thermodynamical arguments it had been predicted in 1893 by W. Wien that λ_{max} would vary inversely with T and this was confirmed by the later experiments. The relation

$$\lambda_{max} T = b \tag{1.19}$$

is known as *Wien's displacement law*, and the constant b has the value $b = 2.898 \times 10^{-3}$ mK.

The spectral distribution function $R(\lambda)$, for the power emitted, is related by a geometrical factor to the spectral distribution function $\rho(\lambda)$ for the energy density within the cavity. In fact, if $\rho(\lambda)\, d\lambda$ is defined as the energy density of the radiation with wavelengths between λ and $\lambda + d\lambda$, it can be shown [5] that $\rho(\lambda) = 4R(\lambda)/c$, where c is the velocity of light *in vacuo*. Consequently, measurements of $R(\lambda)$ determine the spectral distribution of the energy density within the cavity. It is also interesting to consider the energy density as a function of the frequency $\nu = c/\lambda$, in which case a distribution function $\rho(\nu)$ is defined so that

$$\rho(\nu) = \rho(\lambda) \left| \frac{d\lambda}{d\nu} \right| = \lambda^2 \rho(\lambda)/c \tag{1.20}$$

[5] The details of the calculation are given in the book by Richtmyer, Kennard and Cooper (1969).

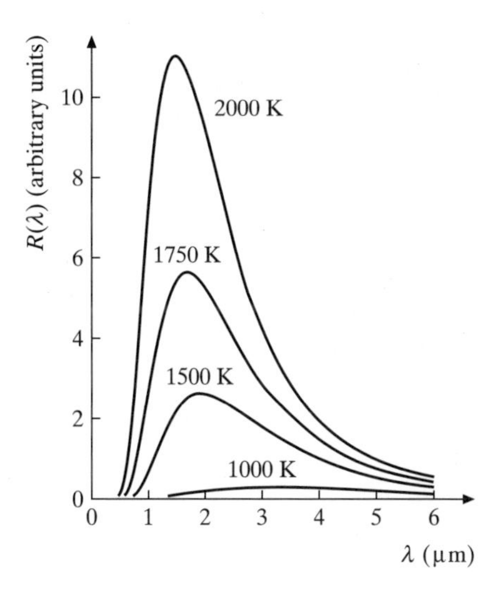

Figure 1.4 Spectral distribution of black body radiation. The spectral distribution function $R(\lambda)$ is plotted as a function of the wavelength λ for different absolute temperatures.

Again, using very general thermodynamical arguments, Wien was able to show in 1893 that $\rho(\lambda)$ had to be of the form

$$\rho(\lambda) = \lambda^{-5} f(\lambda T) \tag{1.21}$$

where $f(\lambda T)$ was some function, which could only be determined by going beyond thermodynamical reasoning to a more detailed theoretical model. After some attempts by Wien, Lord Rayleigh and J. Jeans derived a distribution function $\rho(\lambda)$ (and hence $f(\lambda T)$) in the following way.

The number of standing electromagnetic waves (modes) per unit volume within a cavity, with wavelengths within the interval λ to $\lambda + \mathrm{d}\lambda$, is given by $8\pi \,\mathrm{d}\lambda / \lambda^4$ (Problem 1.2). This number is independent of the size and shape of a sufficiently large cavity. The energy density can then be written as

$$\rho(\lambda) = \frac{8\pi}{\lambda^4} \bar{\varepsilon} \tag{1.22}$$

or alternatively as

$$\rho(\nu) = \frac{8\pi \nu^2}{c^3} \bar{\varepsilon} \tag{1.23}$$

where $\bar{\varepsilon}$ is the average energy in the mode with wavelength λ. Rayleigh and Jeans then suggested that each standing wave of radiation could be considered to be due to an assemblage of a large number of oscillating electric dipoles of frequency $\nu = c/\lambda$. The energy, ε, of each oscillator can take any value, $0 \leqslant \varepsilon < \infty$, independently of the frequency ν, but since the system is in thermal equilibrium, the average energy $\bar{\varepsilon}$ can be obtained by weighting each value of ε with the Boltzmann factor $\exp[-\varepsilon/(k_B T)]$, where k_B is Boltzmann's constant. Setting $\beta = 1/(k_B T)$, we have

$$\bar{\varepsilon} = \frac{\int_0^\infty \varepsilon \exp(-\beta\varepsilon)\, d\varepsilon}{\int_0^\infty \exp(-\beta\varepsilon)\, d\varepsilon}$$

$$= -\frac{d}{d\beta}\left[\log \int_0^\infty \exp(-\beta\varepsilon)\, d\varepsilon\right] = k_B T \quad \textbf{(1.24)}$$

Inserting this value of $\bar{\varepsilon}$ into (1.22) or (1.23) gives the Rayleigh–Jeans distribution law

$$\rho(\lambda) = \frac{8\pi}{\lambda^4}(k_B T) \quad \textbf{(1.25)}$$

from which, using (1.21), we see that $f(\lambda T) = 8\pi k_B(\lambda T)$.

In the limit of long wavelengths the Rayleigh–Jeans result (1.25) approaches the experimental results, as shown in Fig. 1.5. However, as can be seen from the figure, $\rho(\lambda)$ does not show the observed maximum, and diverges as $\lambda \to 0$. This behaviour at short wavelengths is known as the 'ultra-violet catastrophe', and as a consequence the total energy per unit volume

$$\rho_{tot} = \int_0^\infty \rho(\lambda)\, d\lambda \quad \textbf{(1.26)}$$

is seen to be infinite.

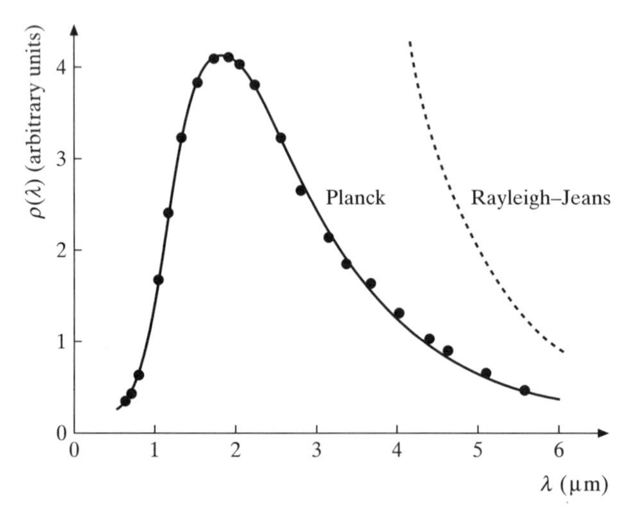

Figure 1.5 Comparison of the Rayleigh–Jeans and Planck spectral distribution laws with experiment at 1600 K. The dots represent experimental points.

Planck's quantum theory

No solution to these difficulties can be found using classical physics. However, in December 1900, M. Planck presented a new form of the black body radiation spectral distribution, based on a revolutionary hypothesis. He postulated that the energy of an oscillator of a given frequency ν cannot take arbitrary values between zero and infinity, but can only take on the discrete values $n\varepsilon_0$, where n is a positive integer or zero, and ε_0 is a finite amount, or *quantum*, of energy, which may depend on the frequency. In this case the average energy of an assemblage of oscillators, each of frequency ν, in thermal equilibrium is given by

$$\bar{\varepsilon} = \frac{\sum_{n=0}^{\infty} n\varepsilon_0 \exp(-\beta n\varepsilon_0)}{\sum_{n=0}^{\infty} \exp(-\beta n\varepsilon_0)} = -\frac{\mathrm{d}}{\mathrm{d}\beta}\left[\log \sum_{n=0}^{\infty} \exp(-\beta n\varepsilon_0)\right]$$

$$= -\frac{\mathrm{d}}{\mathrm{d}\beta}\left[\log\left(\frac{1}{1-\exp(-\beta\varepsilon_0)}\right)\right] = \frac{\varepsilon_0}{\exp(\beta\varepsilon_0)-1} \quad \textbf{(1.27)}$$

Substituting this value of $\bar{\varepsilon}$ in (1.22), we find

$$\rho(\lambda) = \frac{8\pi}{\lambda^4}\frac{\varepsilon_0}{\exp[\varepsilon_0/(k_B T)]-1} \quad \textbf{(1.28)}$$

In order to satisfy Wien's law (1.21), ε_0 must be taken to be proportional to the frequency ν:

$$\varepsilon_0 = h\nu \quad \textbf{(1.29)}$$

where h is a fundamental physical constant, called *Planck's constant*. Planck's distribution law then can be written as

$$\rho(\lambda) = \frac{8\pi hc}{\lambda^5}\frac{1}{\exp[hc/(\lambda k_B T)]-1} \quad \textbf{(1.30)}$$

so that $f(\lambda T) = 8\pi hc[\exp(hc/(\lambda k_B T)) - 1]^{-1}$. In terms of frequency,

$$\rho(\nu) = \frac{8\pi h\nu^3}{c^3}\frac{1}{\exp[h\nu/(k_B T)]-1} \quad \textbf{(1.31)}$$

By expanding the denominator in (1.30), it is easy to show that at long wavelengths $\rho(\lambda) \to 8\pi k_B T/\lambda^4$ in agreement with the Rayleigh–Jeans formula (1.25). On the other hand, for short wavelengths, the presence of the exponential in the denominator of (1.30) ensures that $\rho(\lambda) \to 0$ as $\lambda \to 0$. The value of λ for which the Planck distribution (1.30) is a maximum can also be evaluated (Problem 1.3) and it is found that

$$\lambda_{\max} T = \frac{hc}{4.965\, k_B} = b \tag{1.32}$$

where b is Wien's displacement constant.

In Planck's theory, the total energy density is finite and we find from (1.30) and (1.26) (see Problem 1.4) that

$$\rho_{\text{tot}} = aT^4 \tag{1.33}$$

where

$$a = \frac{8\pi^5}{15} \frac{k_B^4}{h^3 c^3} \tag{1.34}$$

Since ρ_{tot} is related to the total emittance R by $\rho_{\text{tot}} = 4R/c$, where R is given by the Stefan–Boltzmann law (1.16) with $e = 1$, we see that Stefan's constant σ is given by

$$\sigma = \frac{2\pi^5}{15} \frac{k_B^4}{h^3 c^2} \tag{1.35}$$

Equations (1.32) and (1.35) relate σ and b to the three fundamental physical constants c, h and k_B. In 1901, the velocity of light, c, was known accurately and the experimental values of b and σ were also known. Using this data, Planck calculated both the values of h and k_B, which he found to be $h = 6.55 \times 10^{-34}$ J s and $k_B = 1.346 \times 10^{-23}$ J K^{-1}. Not only was this the first calculation of Planck's constant, but it was, at that time, the most accurate value of Boltzmann's constant available. The modern values of h and k_B are given by (Appendix 14)

$$\begin{aligned} h &= 6.626\,07 \times 10^{-34} \text{ J s} \\ k_B &= 1.380\,65 \times 10^{-23} \text{ J K}^{-1} \end{aligned} \tag{1.36}$$

Using his values of h and k_B, Planck obtained very good agreement with the experimental data for $\rho(\lambda)$ over the entire range of wavelengths (see Fig. 1.5).

The idea of quantisation of energy, in which the energy of a system can only take certain discrete values, was totally at variance with classical physics, and Planck's theory was not accepted readily. However, it was not long before the quantum concept was used to explain other phenomena. Indeed, in 1905, A. Einstein was able to interpret the photoelectric effect by introducing the idea of *photons*, or light quanta, and in 1907 he used the Planck formula for the average energy of an oscillator (1.27) to derive the law of Dulong and Petit concerning the specific heat of solids. Subsequently N. Bohr, in 1913, was able to invoke the idea of quantisation of atomic energy levels to explain the existence of line spectra.

1.4 The photoelectric effect

In the course of experiments investigating the properties of electromagnetic waves, H. Hertz discovered in 1887 that ultra-violet light falling on metallic

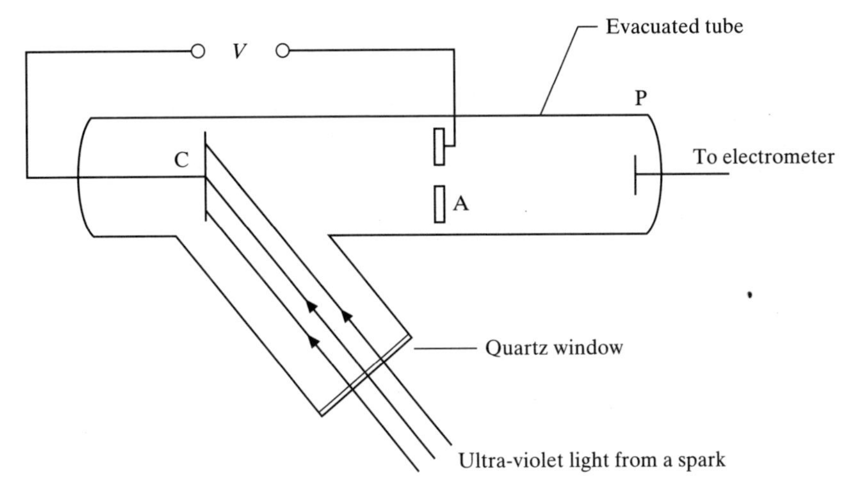

Figure 1.6 Lenard's apparatus for investigating the photoelectric effect.

electrodes facilitates the passage of a spark. Further work by W. Hallwachs, M. Stoletov, P. Lenard and others showed that charged particles are ejected from metal surfaces irradiated by high-frequency electromagnetic waves. This phenomenon is called the *photoelectric effect*. In 1900, Lenard measured the charge to mass ratio of the charged particles in experiments similar to those of J.J. Thomson, and in this way he was able to identify the particles as electrons.

In his experiments to establish the mechanism of the photoelectric effect, Lenard used the apparatus shown in schematic form in Fig. 1.6. Electrons liberated by light striking a cathode C could pass through a small hole in an anode A and be detected by an electrometer connected to a collecting plate P. The anode current was studied as a function of the voltage difference V between anode and cathode. The variation of the photoelectric current with V is shown in Fig. 1.7. As V is increased, so that the electrons are attracted towards the anode, the current increases until it saturates when V is of the order of 20 V. If V is decreased, and then reversed, so that the cathode is positive with respect to the anode, there is a definite voltage V_0 at which the current ceases and the emission of electrons from the cathode stops. From this result it follows that the electrons are emitted with velocities up to a maximum v_{max} and the stopping voltage V_0 is just enough to repel an electron with kinetic energy $\frac{1}{2}mv_{\text{max}}^2$, giving

$$eV_0 = \frac{1}{2}mv_{\text{max}}^2 \tag{1.37}$$

The most important features of the experimental data are the following:

1. There is a minimum, or threshold frequency ν_t of the radiation, below which no emission of electrons takes place, no matter what the intensity of the incident radiation, or for how long it falls on the surface.

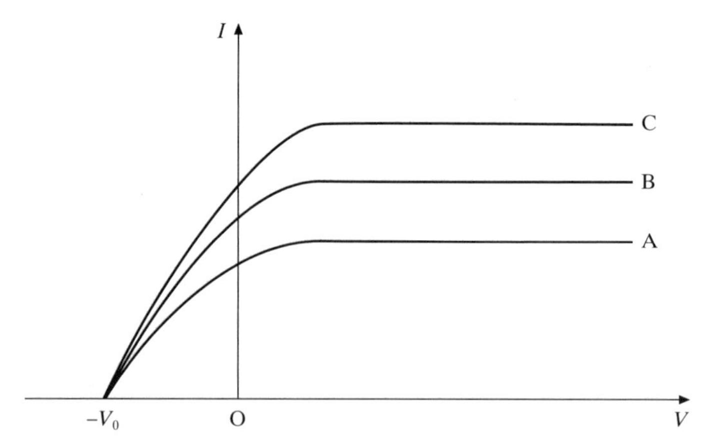

Figure 1.7 The variation of the photoelectric current with voltage, for different intensities of light, A < B < C.

2. Electrons emerge with a range of velocities from zero up to a maximum v_{max} and the maximum kinetic energy, $\frac{1}{2}mv_{\text{max}}^2$, is found to depend linearly on the frequency of the radiation and to be independent of its intensity.
3. For incident radiation of a given frequency, the number of electrons emitted per unit time is proportional to the intensity of the radiation.
4. Electron emission takes place immediately the light shines on the surface, with no detectable time delay.

According to classical, pre-quantum, physics it would be natural to suppose that the maximum kinetic energy of the emitted electrons would increase with the energy density (or intensity) of the incident radiation, independently of the frequency. In fact, this is not in accord with what is observed. Another important aspect of classical theory is that the incident energy is spread uniformly over the illuminated surface. To eject an electron from an atom, this energy would have to be concentrated over an area of atomic dimensions, and to achieve such a concentration would require a certain time delay. Experiments can be arranged for which the predicted time delay is minutes, or even hours, and yet no detectable time delay is actually observed.

In 1905, Einstein offered an explanation for these seemingly strange aspects of the photoelectric effect, based on an extension of Planck's idea of the quantisation of black body radiation. In Planck's theory, the oscillators representing the source of the electromagnetic field could only vibrate with energies given by $E = nh\nu$. In contrast, Einstein supposed that the electromagnetic field itself was quantised and that light consists of corpuscles, called *light quanta* or *photons*, and that each photon travels with the velocity of light c and carries a quantum of energy of magnitude

$$E = h\nu = hc/\lambda \qquad \textbf{(1.38)}$$

The photons are sufficiently localised, so that the whole quantum of energy can be absorbed by a single atom at one time. When a photon falls on a metallic surface, its entire energy $h\nu$ is used to eject an electron from an atom. Because of the interaction of the ejected electron with other atoms it requires a certain minimum energy to escape from the surface. The minimum energy required to escape depends on the metal and is called the *work function* W. It follows that the maximum kinetic energy of a photoelectron is given by

$$\frac{1}{2}mv^2_{\max} = h\nu - W \tag{1.39}$$

This relation is called Einstein's equation. The threshold frequency ν_t is determined by the work function since in this case $v_{\max} = 0$, from which

$$h\nu_t = W \tag{1.40}$$

The number of electrons emerging from the metal surface per unit time is proportional to the number of photons striking the surface per unit time, but the intensity of the radiation is also proportional to the number of photons falling on a certain area per unit time, since each photon carries a fixed energy $h\nu$. It follows that the photoelectric current is proportional to the intensity of the radiation and that all the experimental observations are explained by Einstein's theory.

A series of very accurate measurements carried out between 1914 and 1916 by Millikan provided further confirmation of Einstein's theory. Combining (1.37) with (1.39), we see that the stopping voltage satisfies

$$eV_0 = h\nu - W \tag{1.41}$$

Millikan measured, for a given surface, V_0 as a function of ν, and showed that his results fell on a straight line (see Fig. 1.8) of slope $\tan\alpha = h/e$. Knowing e, Millikan obtained the value 6.56×10^{-34} J s for h, which agreed well with Planck's results. It is interesting that Millikan was able to use visible, rather than ultra-violet, light for

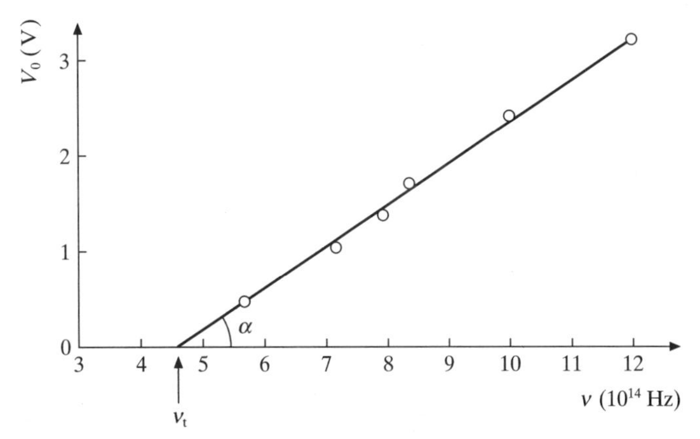

Figure 1.8 Millikan's results (circles) for the stopping potential V_0 as a function of the frequency ν. The data fall on a straight line, of slope $\tan\alpha = h/e$.

this experiment by using surfaces of lithium, sodium and potassium which have small values of the work function W.

Although the photoelectric effect provides compelling evidence for a corpuscular theory of light, it must not be forgotten that the existence of diffraction and interference phenomena demonstrate that light also exhibits a wave behaviour. The particle and the wave aspects of light are contained within modern quantum electrodynamics, which is capable of predicting both types of phenomena.

Photons and the electromagnetic spectrum

The electromagnetic spectrum extends from radio waves with low frequencies up to gamma rays of high frequency (see Fig. 1.9). At each particular frequency, the radiation consists of photons, each of energy $E = h\nu = hc/\lambda$. In Fig. 1.9 we show the photon energy corresponding to each part of the spectrum. The energies are given in electron volts (eV), which is a particularly convenient unit to use in atomic and nuclear physics. It is defined to be the energy acquired by an electron passing through a potential difference of one volt. Since the electronic charge has the absolute value $e = 1.602\,18 \times 10^{-19}$ C, we have that

$$\begin{aligned} 1\text{ eV} &= 1.602\,18 \times 10^{-19}\text{ coulomb-volts} \\ &= 1.602\,18 \times 10^{-19}\text{ J} \end{aligned} \tag{1.42}$$

1.5 X-rays and the Compton effect

The corpuscular nature of electromagnetic radiation was exhibited in a spectacular way in a quite different experiment performed in 1923 by A.H. Compton, in which a beam of X-rays was scattered through a block of material. X-rays had been discovered by W.K. Röntgen in 1895 and were known to be electromagnetic radiation of high frequency (see Fig. 1.9). The scattering of X-rays by various substances was first studied by C.G. Barkla in 1909, who interpreted his results with the help of J.J. Thomson's classical theory, developed around 1900. According to this theory, the oscillating electric field of the radiation acts on the electrons contained in the atoms of the target material. This interaction forces the atomic electrons to vibrate with the same frequency as the incident radiation. The oscillating electrons, in turn, radiate electromagnetic waves of the same frequency. The net effect is that the incident radiation is scattered with no change in wavelength and this is called *Thomson scattering*.

In general, Barkla found that the scattered intensity predicted by Thomson's theory agreed well with his experimental data. However, he found that some of his results were anomalous, particularly in the region of 'hard' X-rays, which correspond to shorter wavelengths. At the time of Barkla's work, it was not possible to measure the wavelengths of X-rays and a further advance could not be made until M. von Laue in 1912, and later W.L. Bragg, had shown that the wavelengths

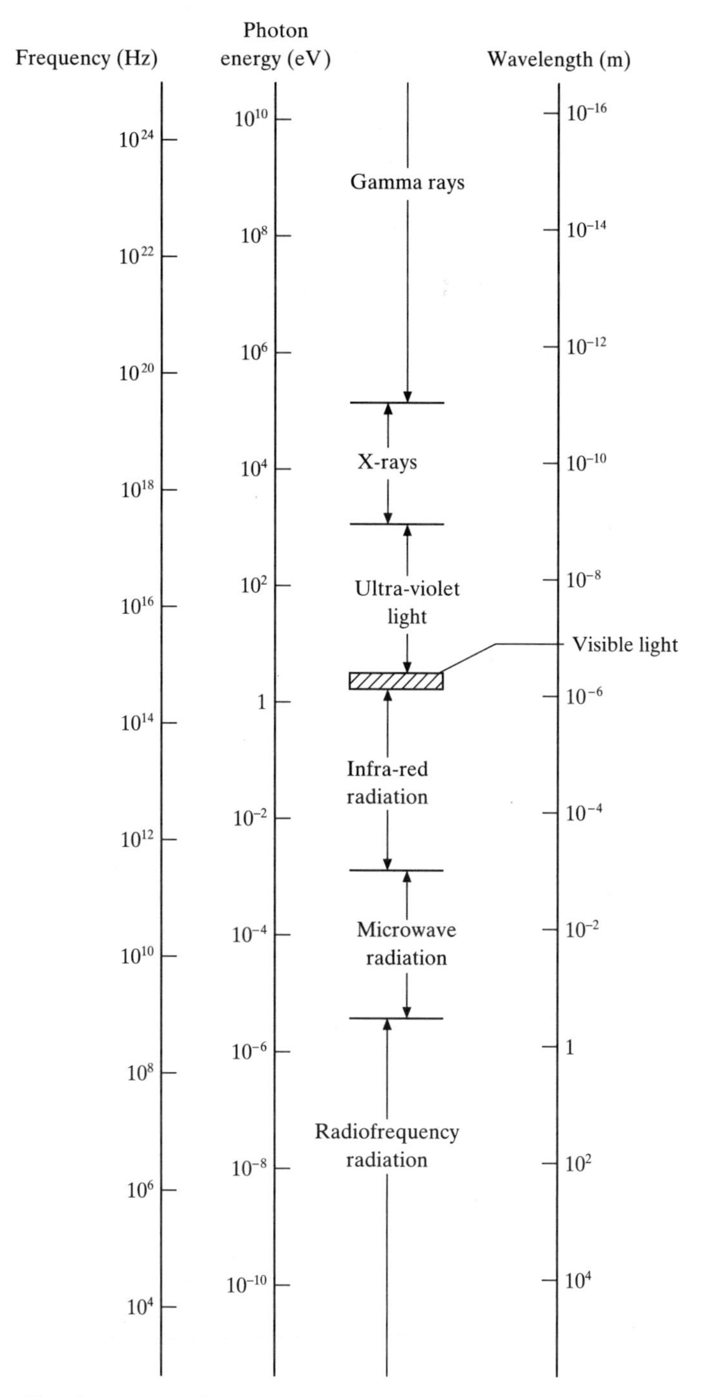

Figure 1.9 The electromagnetic spectrum.

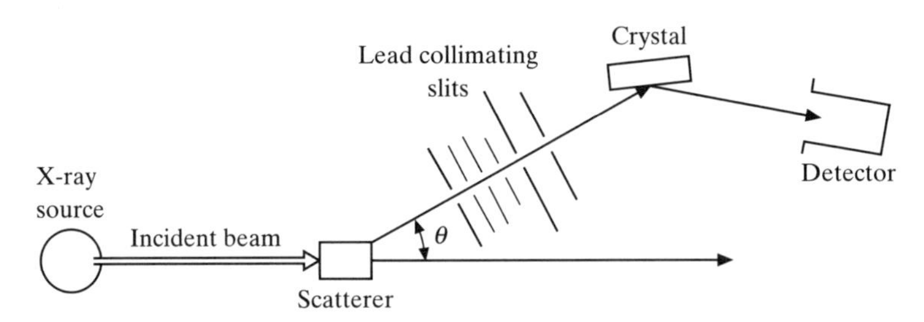

Figure 1.10 Schematic diagram of Compton's apparatus.

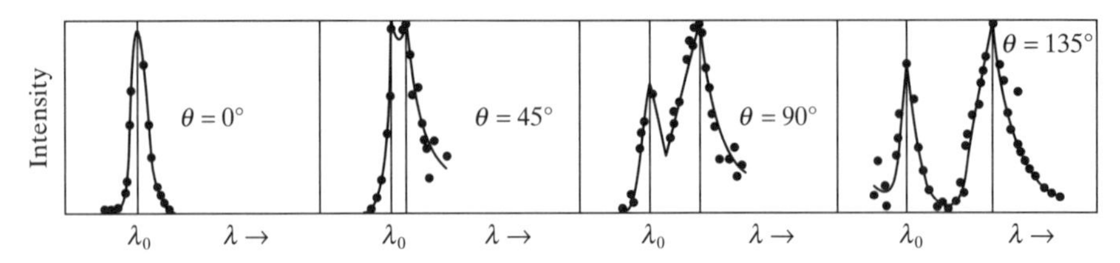

Figure 1.11 Compton's data for the scattering of X-rays by graphite.

could be determined by studying the diffraction of X-rays by crystals. The experiment of Compton, which we shall now describe, was only possible because a precise determination of wavelength could be made using a crystal spectrometer.

The experimental arrangement used by Compton is sketched in Fig. 1.10. He irradiated a graphite target with a nearly monochromatic beam of X-rays, of wavelength λ_0. He then measured the intensity of the scattered radiation as a function of wavelength. His results, illustrated in Fig. 1.11, showed that although part of the scattered radiation had the same wavelength λ_0 as the incident radiation, there was also a second component of wavelength λ_1, where $\lambda_1 > \lambda_0$. This phenomenon, called the *Compton effect*, could not be explained by the classical Thomson model. The shift in wavelength between the incident and scattered radiation, the *Compton shift* $\Delta\lambda = \lambda_1 - \lambda_0$, was found to vary with the angle of scattering (see Fig. 1.11) and to be proportional to $\sin^2(\theta/2)$ where θ is the angle between the incident and scattered beams. Further investigation showed $\Delta\lambda$ to be independent of both λ_0 and the material used as the scatterer, and that the value of the constant of proportionality was 0.048×10^{-10} m.

To interpret these results, Compton suggested that the modified line at wavelength λ_1 could be attributed to scattering of the X-ray photons by loosely bound electrons in the atoms of the target. In fact, such electrons can be treated as *free* since their binding energies of a few electron volts are small compared with the energy of an X-ray photon and this explains why the results are independent of the nature of the material used for the target.

Let us then consider the scattering of an X-ray photon by a free electron, which can be taken to be at rest initially. Since the energies involved in the collision may be large, we need to use relativisitic kinematics and we shall first outline the results required [6]. The total energy of a particle having a rest mass m and a velocity $\mathbf{v}$ is given by

$$E = \frac{mc^2}{\sqrt{1 - v^2/c^2}} \tag{1.43}$$

The kinetic energy T of the particle is defined as the difference between E and the rest mass energy mc^2, so that

$$T = E - mc^2 \tag{1.44}$$

The corresponding momentum of the particle is

$$\mathbf{p} = \frac{m\mathbf{v}}{\sqrt{1 - v^2/c^2}} \tag{1.45}$$

and from (1.43) and (1.45) we see that the energy and momentum are related by

$$E^2 = m^2c^4 + p^2c^2 \tag{1.46}$$

Since the velocity of a photon is c and its energy $E = h\nu = hc/\lambda$ is finite, we see from (1.43) that we must take the mass of a photon to be zero, in which case we observe from (1.46) that the magnitude of its momentum is

$$p = E/c = h/\lambda \tag{1.47}$$

Let us now consider the situation depicted in Fig. 1.12, where a photon of energy $E_0 = hc/\lambda_0$ and momentum $\mathbf{p}_0(p_0 = E_0/c)$ collides with an electron initially at rest. After the collision, the photon has an energy $E_1 = hc/\lambda_1$ and a momentum $\mathbf{p}_1(p_1 = E_1/c)$ in a direction making an angle θ with the direction of incidence,

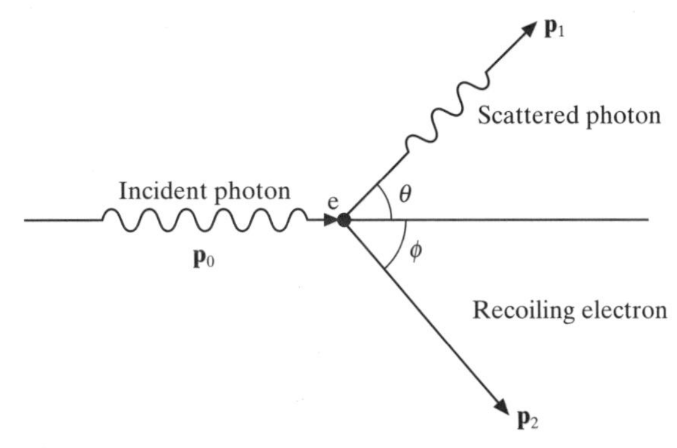

Figure 1.12 A photon of momentum $\mathbf{p}_0$ collides with a stationary electron e, and is scattered with momentum $\mathbf{p}_1$, while the electron recoils with momentum $\mathbf{p}_2$.

[6] A good account of the theory of special relativity is given in the text of Taylor and Wheeler (1966).

while the electron recoils with a momentum $\mathbf{p}_2$ making an angle ϕ with the incident direction. Conservation of momentum yields

$$\mathbf{p}_0 = \mathbf{p}_1 + \mathbf{p}_2 \tag{1.48}$$

from which

$$p_2^2 = p_0^2 + p_1^2 - 2p_0 p_1 \cos\theta \tag{1.49}$$

Conservation of energy then gives

$$E_0 + mc^2 = E_1 + (m^2c^4 + p_2^2c^2)^{1/2} \tag{1.50}$$

and defining the kinetic energy of the electron after the collision as T_2 we have

$$\begin{aligned} T_2 &= (m^2c^4 + p_2^2c^2)^{1/2} - mc^2 \\ &= E_0 - E_1 = c(p_0 - p_1) \end{aligned} \tag{1.51}$$

From (1.51) we have that

$$p_2^2 = (p_0 - p_1)^2 + 2mc(p_0 - p_1) \tag{1.52}$$

and combining (1.52) with (1.49) we find that

$$mc(p_0 - p_1) = p_0 p_1 (1 - \cos\theta) = 2p_0 p_1 \sin^2(\theta/2) \tag{1.53}$$

Since $\lambda_0 = h/p_0$ and $\lambda_1 = h/p_1$ this can be written in the form

$$\Delta\lambda = \lambda_1 - \lambda_0 = 2\lambda_c \sin^2(\theta/2) \tag{1.54}$$

where the constant λ_c is given by

$$\lambda_c = \frac{h}{mc} \tag{1.55}$$

and is called the *Compton wavelength* of the electron. Equation (1.54) is known as the *Compton equation*. The calculated value of $(2\lambda_c)$ is $0.048\,52 \times 10^{-10}$ m, and this agrees very well with the experimental data.

The existence of the unmodified component of the scattered radiation, which has the same wavelength λ_0 as the incident radiation, can be explained by assuming that it results from scattering by electrons so tightly bound that the entire atom recoils. In this case, the mass to be used in (1.55) is M, the mass of the entire atom, and since $M \gg m$, the Compton shift $\Delta\lambda$ is negligible. For the same reason, there is no Compton shift for light in the visible region, because the photon energy in this case is not large compared with the binding energy of even the loosely bound electrons. In contrast, for very energetic γ-rays only the shifted line is observed, since the photon energies are large compared with the binding energies of even the tightly bound electrons.

The recoil electrons predicted by Compton's theory were observed in 1923 by W. Bothe and also by C.T.R. Wilson. A little later, in 1925, W. Bothe and H. Geiger demonstrated that the scattered photon and the recoiling electron appear simultaneously. Finally, in 1927, A.A. Bless measured the energy of the ejected electrons, which he found to be in agreement with the prediction of Compton's theory.

1.6 The nuclear atom

By the early years of this century, the atomic nature of matter had been well established. It was known that atoms contained electrons and that an electron was much lighter than even the lightest atom. It had also been shown that electrons could be removed from atoms of a given species, producing positively charged ions, but that only a finite number of electrons could be obtained from each atom. This number is characteristic of the atoms of each element and is called the atomic number Z. In the normal state an atom is electrically neutral so that it must contain positive charge of an amount Ze, where $-e$ is the charge on the electron, and nearly all the mass of an atom must be associated with this positive charge.

Atomic sizes

Information about atomic sizes can be found from simple arguments about the nature of solids. Let us assume that in a solid the atoms are packed as closely as possible. If the diameter of each atom is D, then a length L of material contains L/D atoms, and a volume L^3 contains $(L/D)^3$ atoms. Now the number of atoms in one mole of substance is equal to Avagadro's number N_A, given by (1.1). If the density in kg m^{-3} is ρ, one mole will occupy a volume of $(10^{-3}\ \mu/\rho)$ m^3 where μ is the atomic weight. It follows that a unit volume contains $(10^3\ \rho/\mu)\ 6\times10^{23}$ atoms. This is to be equated with $1/D^3$, with the result

$$D = \left[\frac{\mu}{6\rho}10^{-26}\right]^{1/3} \quad \textbf{(1.56)}$$

In Table 1.1 some values of D obtained from this formula are shown for a number of elements. It is seen that all the atoms concerned have diameters of about 2×10^{-10} m, and this can be taken to be representative of atomic sizes, which do not vary greatly from element to element.

Table 1.1 Atomic sizes.

Element	*Atomic weight μ*	*ρ(in kg m^{-3})*	*D(in m)*
Lithium	6.94	0.53×10^3	2.8×10^{-10}
Carbon (diamond)	12.0	3.5×10^3	1.8×10^{-10}
Iron	55.8	7.9×10^3	2.3×10^{-10}
Silver	107.9	10.5×10^3	2.6×10^{-10}
Gold	197.0	19.3×10^3	2.6×10^{-10}
Lead	207.2	11.35×10^3	3.1×10^{-10}

The experiments of Geiger, Marsden and Rutherford

The question now arises as to how the mass and positive charge are distributed within the atom. The answer was provided by a series of experiments by H. Geiger, E. Marsden and E. Rutherford, carried out between 1906 and 1913, on the scattering of α particles by metallic foils of various thicknesses. Alpha particles are emitted by many radioactive substances, for example uranium. By measuring the deflection of α particles by electric and magnetic fields, Rutherford found that α particles had a charge to mass ratio q/M (that is, a specific charge) which was the same as that of the doubly ionised helium atom. Spectroscopic measurements subsequently confirmed that α particles were fully ionised helium atoms, with a mass about four times that of a hydrogen atom and a positive charge equal to $+2e$. The kinetic energies of the α particles employed in the experiments were several million electron volts (MeV).

A schematic diagram of the experiments of Geiger and Marsden is shown in Fig. 1.13. The α particles emitted from a source are collimated and then scattered from thin metallic foils, for example gold foils of thickness $\sim 10^{-6}$ m. Alpha particles produce scintillations in zinc sulphide and can be detected by observing a screen, coated with this substance, with a microscope. Most of the α particles are deflected through very small angles ($<1°$), but some are deflected through large angles, about 1 in 8000 being deflected through angles greater than 90°. The measurements established that:

1. For a fixed angle of scattering and fixed kinetic energy, the number of α particles scattered within an element of solid angle $d\Omega$ is proportional to the thickness of the foil, except at very small angles of scattering ($<1°$).
2. At a fixed angle and for a given foil, the number of particles scattered within an element of solid angle $d\Omega$ is inversely proportional to E^2, where E is the kinetic energy of the α particles.
3. For a given energy and a given foil, the number of particles scattered within an element of solid angle $d\Omega$ is proportional to $(\sin(\theta/2))^{-4}$, where θ is the angle of scattering.

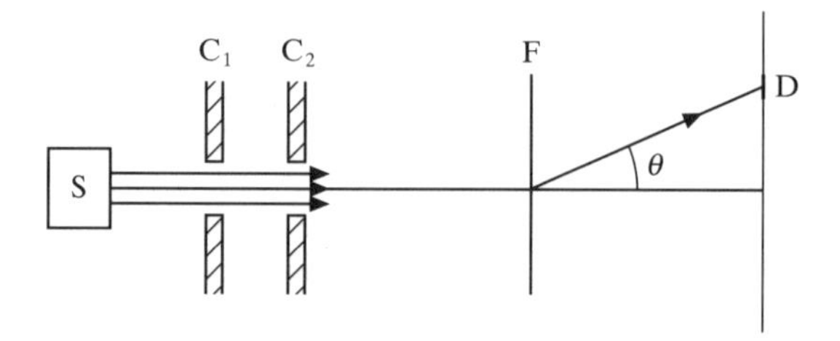

Figure 1.13 The scattering experiments of Geiger and Marsden. S represents a radioactive source emitting α particles, which are collimated by slits C_1 and C_2. The α particles fall onto a fixed foil F and those particles scattered through an angle θ are detected at D on a screen coated with a scintillating material (zinc sulphide). The apparatus is enclosed in an evacuated chamber to avoid scattering of the α particles by the air.

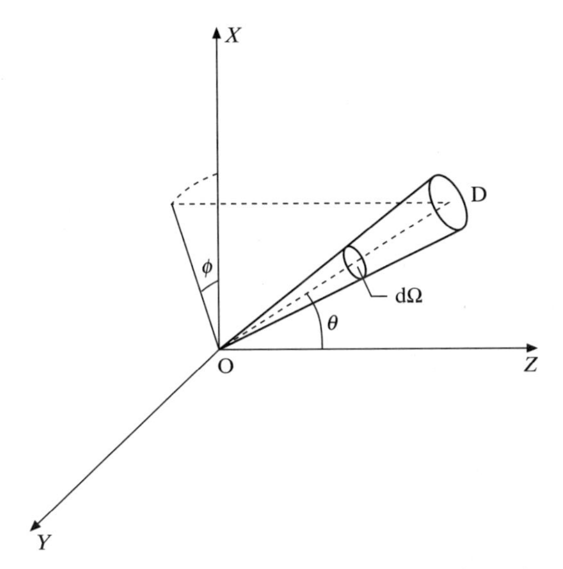

Figure 1.14 The geometry of scattering. The detector D subtends an element of solid angle dΩ at the scattering centre O. The polar coordinates of OD are (θ, ϕ). The incident beam is directed along the Z axis so that θ and ϕ are the scattering angles.

4. At a fixed energy and for a foil of given thickness, the number of particles scattered within an element of solid angle dΩ in a given direction is proportional to Z^2, where Z is the atomic number of the atoms in the foil.

To understand what is meant by scattering in a certain direction, within an element of solid angle dΩ, we refer to Fig. 1.14. Consider a particle travelling along the Z axis and being deflected along the direction OD after a collision with an atom situated at O. The direction OD is defined by the angle of scattering θ between OD and the Z axis and by the azimuthal angle ϕ, which is the angle between OX and the projection of OD on the XY plane. The element of solid angle dΩ subtended at O by a small area of the detecting screen, placed at D at right angles to OD, is given by $\mathrm{d}\Omega = \sin\theta\,\mathrm{d}\theta\,\mathrm{d}\phi$.

Except at very small angles, where multiple scattering occurs, the number of atoms, n, encountered by a beam of particles of unit cross-sectional area traversing a thin foil is proportional to the thickness of the foil. Since the intensity of scattering in a given direction is found to be proportional to the thickness of the foil, and hence to n, it can be inferred that each particle makes at most one collision with an atom within the foil. If multiple scattering occurred then the number of particles deflected through a certain angle would increase much more rapidly with n. The incident flux N is defined as the number of particles crossing a unit area normal to the direction of the beam, per unit time. The number of α particles scattered within an element of solid angle dΩ in the direction (θ, ϕ) (see Fig. 1.14) will be denoted by dN', and this quantity is proportional both to n dΩ and to the flux N. Thus

$$dN' = \left(\frac{d\sigma}{d\Omega}\right) Nn \, d\Omega \tag{1.57}$$

The constant of proportionality ($d\sigma/d\Omega$) is called the *differential cross-section* and determines the intensity of scattering from a single isolated atom. Clearly the differential cross-section is characteristic of the particular type of atom and it is a function both of the energy of the incident particles and of the scattering angles. The experimental data of Geiger and Marsden exhibit axial symmetry about the direction of incidence (the Z axis), so in this case $d\sigma/d\Omega$ depends on θ only and not on the azimuthal angle ϕ.

In explaining the experimental data, we can neglect the influence of the electrons within the atom and assume that the scattering is due to the positive charge alone. This is because an α particle is about 8000 times as heavy as an electron and the kinematics of a collision with an electron will allow a maximum deflection of much less than 0.1° (see Problem 1.10). It can also be shown that if the positive charge is spread uniformly throughout the atom, as in a model proposed in 1910 by J.J. Thomson, there would be no appreciable scattering at large angles, and even at 1° the number of scattered α particles would be negligible. The data can be fully explained by a model proposed in 1911 by Rutherford, who suggested that all the positive charge and almost all the mass of an atom are concentrated at the centre of the atom in a *nucleus* of very small dimensions. By treating the nucleus as a point charge, Rutherford obtained the differential scattering cross-section of a beam of α particles of charge $2e$ by the Coulomb force of a nucleus of charge (Ze). It is given in the centre of mass (CM) system by (see Appendices 1 and 2)

$$\frac{d\sigma}{d\Omega} = \left(\frac{2Ze^2}{4\pi\varepsilon_0}\right)^2 \frac{1}{4\mu^2 v^4} \frac{1}{\sin^4(\theta/2)} \tag{1.58}$$

Here ε_0 is the permittivity of free space, v is the initial relative velocity of the α particle and the nucleus and μ is the reduced mass $\mu = M_\alpha M_N/(M_\alpha + M_N)$ where M_α and M_N are the masses of the α particle and of the nucleus, respectively. The Geiger and Marsden experiments were performed for heavy targets, in which case the CM and laboratory system nearly coincide. Since $E = \mu v^2/2$, the formula (1.58) completely explains each of the experimental findings 2, 3 and 4.

The nucleus

The nucleus, although small compared with the atom, is finite in size. This can be demonstrated by the departure of the cross-section from the Rutherford scattering formula, at a given angle of scattering, when the energy of the incident α particle is raised to make the distance of closest approach comparable to the nuclear radius. For a head-on collision at zero impact parameter, the distance of closest approach r_0 is given by (see Appendix 1)

$$\frac{1}{2}\mu v^2 = \left(\frac{2Ze^2}{4\pi\varepsilon_0}\right)\frac{1}{r_0} \quad \textbf{(1.59)}$$

Departures from Rutherford scattering occur when r_0 becomes less than $\sim 10^{-14}$ m and this distance is characteristic of nuclear sizes. The structure of the nucleus is the subject of the text by Burcham and Jobes (1995), where it is shown that nuclei are composed of *protons*, of positive charge e, and *neutrons* which are uncharged. Both protons and neutrons have approximately the same mass (which is about 1840 times as large as the mass of the electron), and are referred to collectively as *nucleons*. The chemical properties of an atom are determined by the charge of the nucleus, Ze, where Z is the atomic number. It is found that, in general, several nuclei exist for each value of Z differing in the number of neutrons N they contain. Atoms with the same Z but with different mass numbers $A = N + Z$ are called *isotopes*. Many naturally occurring elements are mixtures of two or more different isotopes. For example, oxygen is a mixture of three stable isotopes ^{16}O, ^{17}O and ^{18}O, where the notation is such that the superscript indicates the mass number A. The isotope ^{16}O occurs with a relative abundance of 99.759 per cent. The atomic weight of an element on the chemical scale, defined in Section 1.1, is therefore based on a mixture of oxygen isotopes. Because of this a new scale has been introduced, in which the isotope of carbon ^{12}C is assigned a mass of 12 atomic mass units (a.m.u. or u). The absolute value of the atomic mass unit is

$$1 \text{ a.m.u.} = 1.660\,54 \times 10^{-27} \text{ kg} \quad \textbf{(1.60)}$$

and differs very slightly from the previous unit based on the chemical scale.

Limitations of the Rutherford model

In the Rutherford model, the electrons move in the Coulomb field of the nucleus in orbits, like a planetary system. A particle moving on a curved trajectory is accelerating and an accelerating charged particle radiates electromagnetic waves and loses energy. The laws of classical physics – Newton's laws of motion and Maxwell's electromagnetic equations – if applied to the Rutherford atom, show that in a time of the order 10^{-10} s all the energy of the atom would be radiated away and the electrons would collapse into the nucleus. This is clearly contrary to experiment and is another piece of evidence to suggest that the laws of motion need to be modified when applied to phenomena on the atomic scale. Another fact not explained by the Rutherford model is the existence of atomic line spectra, to which we now turn our attention.

1.7 Atomic spectra and the Bohr model of hydrogen

Isaac Newton was the first person to resolve white light into separate colours by dispersion with a prism, but it was not until 1752 that Th. Melvill first showed that light from an incandescent gas is composed of a large number of discrete frequencies called *emission lines*. It was subsequently discovered that atoms exposed to white light can only absorb light at certain discrete frequencies called *absorption lines*. With the development of diffraction gratings, much greater resolving powers could be obtained and towards the end of the nineteenth century much progress was made in the empirical analysis of line spectra. G.R. Kirchhoff was the first to show that only certain definite frequencies can be radiated or absorbed by a given element and that the emission frequencies coincide with the absorption frequencies. The fact that each element has its own characteristic line spectrum is of the greatest importance and is, for example, the only means by which the presence of particular elements in the Sun and stars can be determined.

The most important discovery in the search for regularities in the line spectra of atoms was made by J. Balmer in 1885, who showed that the frequencies of a series of lines in the visible part of the spectrum of atomic hydrogen were among those given by the empirical formula

$$\nu_{ab} = R\left(\frac{1}{n_a^2} - \frac{1}{n_b^2}\right); \qquad n_a = 1, 2, \ldots \quad n_b = 2, 3, \ldots \tag{1.61}$$

where ν_{ab} is the frequency of either an emission or absorption line, n_a and n_b are positive integers with $n_b > n_a$ and R is a constant, known as Rydberg's constant. It is usual in the spectroscopy of the visible and ultra-violet regions to give the frequencies in terms of inverse wavelengths, or wave numbers

$$\tilde{\nu} = \frac{1}{\lambda} = \frac{\nu}{c} \tag{1.62}$$

The corresponding Rydberg constant for hydrogen then has the value

$$\tilde{R} = 109\,677.58 \text{ cm}^{-1} \tag{1.63}$$

However, in the infra-red or microwave region, frequencies are usually expressed in megahertz (MHz). The wave number given in units of cm^{-1} is related to the frequency given in MHz by

$$\tilde{\nu}(\text{cm}^{-1}) = \frac{10^{14}}{3}\nu(\text{MHz}) \tag{1.64}$$

where we have taken the velocity of light to be 3.00×10^8 m s^{-1}.

It was subsequently discovered that Balmer's formula is not only applicable to the visible region, but in fact describes the complete spectrum of atomic hydrogen. In spectroscopy, wavelengths are generally given in ångström units, where 1 Å $\equiv 10^{-10}$ m, so that if the wave number $\tilde{\nu}_{ab}$ of a particular line is given in cm^{-1} the wavelength in ångström units is

$$\lambda_{ab} = 10^8/\tilde{\nu}_{ab} \tag{1.65}$$

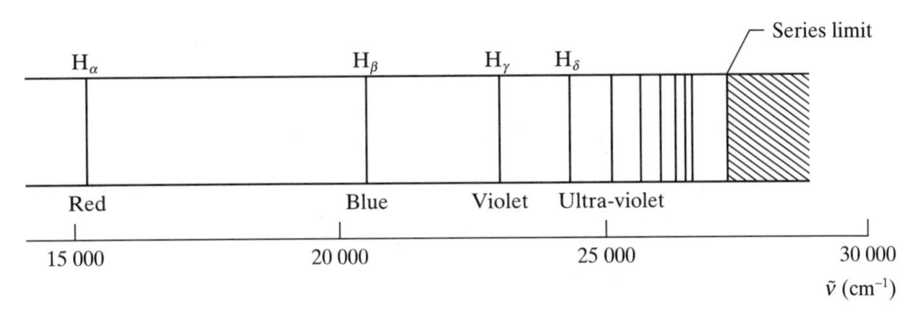

Figure 1.15 The Balmer series of atomic hydrogen.

In atomic hydrogen, the series of lines with $n_a = 1$ is known as the *Lyman series* and lies in the ultra-violet part of the spectrum. The lines are labelled $\alpha, \beta, \gamma \ldots$ in order of decreasing wavelengths; the wavelength of the Lyman α line ($n_b = 2$) is 1216 Å, while the series limit ($n_b = \infty$) is 912 Å. The Balmer series ($n_a = 2$) (the first to be discovered) lies in the visible region. The Balmer lines are denoted by H_α, H_β, H_γ. . . . The Balmer H_α line ($n_b = 3$) was first discovered by J. von Fraunhofer in the solar spectrum and was called by him the C line. It is a strong red line with a wavelength of 6563 Å. The next members of the series ($n_b = 4$ and $n_b = 5$) at 4861 and 4340 Å are blue and violet respectively and the series limit is 3646 Å (see Fig. 1.15).

Further series of lines are found in the infra-red part of the spectrum: the Paschen series ($n_a = 3$) starts with a line at 18 751 Å; the Brackett series ($n_a = 4$) starts at 40 500 Å and the Pfund series ($n_a = 5$) at 74 000 Å.

The set of quantities R/n_a^2 are called *terms*. J.R. Rydberg showed that for other atoms, particularly the alkalis, the frequencies of lines could be represented approximately as differences of terms T_n, where

$$T_n = R/(n + \alpha)^2, \qquad n = 1, 2, 3, \ldots \tag{1.66}$$

and where α was a constant for each particular series. Subsequently it was discovered by W. Ritz that the frequencies of all lines can be represented by the difference of terms, even if the terms cannot be represented by a simple formula such as (1.66). This has the consequence that if lines of frequencies ν_{12} and ν_{23} can be represented as

$$\nu_{12} = T_1 - T_2; \qquad \nu_{23} = T_2 - T_3 \tag{1.67}$$

then a line of frequency ν_{13} will usually exist, where

$$\nu_{13} = (T_1 - T_2) + (T_2 - T_3) = T_1 - T_3 \tag{1.68}$$

This is an example of the *Ritz combination principle*, which states that if lines at frequencies ν_{ij} and ν_{jk} exist in a spectrum with $j > i$ and $k > j$ then there will usually be a line at ν_{ik} where

$$\nu_{ik} = \nu_{ij} + \nu_{jk} \tag{1.69}$$

However, not all combinations of frequencies are observed because certain selection rules operate which will be discussed in Chapters 4 and 9.

Bohr's model of the hydrogen atom

A major step forward was taken by N. Bohr in 1913. Working in Rutherford's laboratory, he was able to combine the concepts of Rutherford's nuclear atom, Planck's quanta and Einstein's photons to explain the observed spectrum of atomic hydrogen.

Bohr assumed that an electron in an atom moves in an orbit about the nucleus under the influence of the electrostatic attraction of the nucleus. Circular or elliptical orbits are allowed by classical mechanics, and Bohr elected to consider circular orbits for simplicity. He then postulated that instead of the infinity of orbits which are possible in classical mechanics, only a certain set of stable orbits, which he called *stationary states*, are allowed. As a result, atoms can only exist in certain allowed energy levels, with energies E_a, E_b, E_c, ...

Bohr further postulated that an electron in a stable orbit does not radiate electromagnetic energy, and that radiation can only take place when a transition is made between the allowed energy levels. To obtain the frequency of the radiation, he made use of the idea that the energy of electromagnetic radiation is quantised and carried by photons, each photon associated with the frequency ν carrying an energy $h\nu$. Thus, if a photon of frequency ν is absorbed by an atom, conservation of energy requires that

$$h\nu = E_b - E_a \tag{1.70}$$

where E_a and E_b are the energies of the atom in the initial and final orbits, with $E_b > E_a$. Similarly, if the atom passes from a state of energy E_b to another state of lower energy E_a, the frequency of the emitted photon must be given by the *Bohr frequency relation* (1.70).

The *terms* of spectroscopy can then be interpreted as being the energies of the various allowed energy levels of an atom. It is important to note that because of the existence of the energy–frequency relationship, we can use frequency (or wave number) units of energy where convenient, and in this book we shall often use hertz or inverse centimetres as energy units. In terms of electron volts, we have

$$\begin{aligned} 1\ \text{eV} &\equiv 2.417\,94 \times 10^{14}\ \text{Hz} \\ &\equiv 8065.44\ \text{cm}^{-1} \end{aligned} \tag{1.71}$$

Other conversions of units are given in Appendix 14.

For the case of one-electron atoms Bohr was able to modify the classical planetary model to obtain the quantisation of energy levels by making the additional postulate that the angular momentum of the electron moving in a circular orbit can only take one of the values $L = nh/(2\pi) = n\hbar$, where n is a positive integer, $n = 1, 2, 3, \ldots$ and the commonly occurring quantity $h/(2\pi)$ is conventionally denoted by $\hbar$. The allowed energy levels can then be determined in the following way.

We shall make the approximation (which we shall remove later) that the nucleus is infinitely heavy compared with the bound electron and is therefore at rest. The electron will be taken to be moving with a velocity v in a circular orbit of radius r, in which case the Coulomb attractive force acting on the electron, due to its electrostatic interaction with the nucleus of charge Ze, can be equated with the electron mass m times the centripetal acceleration (v^2/r):

$$\frac{Ze^2}{(4\pi\varepsilon_0)r^2} = \frac{mv^2}{r} \tag{1.72}$$

where ε_0 is the permittivity of free space. A second equation is obtained from Bohr's postulate that the orbital angular momentum is quantised:

$$L = mvr = n\hbar, \qquad n = 1, 2, 3, \ldots \tag{1.73}$$

From (1.72) and (1.73), we obtain the possible values of v and r

$$v = \frac{Ze^2}{(4\pi\varepsilon_0)\hbar n} \tag{1.74}$$

$$r = \frac{(4\pi\varepsilon_0)\hbar^2 n^2}{Ze^2 m} \tag{1.75}$$

The kinetic energy of the electron, T, is then found to be

$$T = \frac{1}{2}mv^2 = \frac{m}{2\hbar^2}\left(\frac{Ze^2}{4\pi\varepsilon_0}\right)^2 \frac{1}{n^2} \tag{1.76}$$

and the potential energy V is, correspondingly,

$$V = -\frac{Ze^2}{(4\pi\varepsilon_0)r} = -\frac{m}{\hbar^2}\left(\frac{Ze^2}{4\pi\varepsilon_0}\right)^2 \frac{1}{n^2} \tag{1.77}$$

from which the total energy E_n of the system is

$$E_n = T + V = -\frac{m}{2\hbar^2}\left(\frac{Ze^2}{4\pi\varepsilon_0}\right)^2 \frac{1}{n^2} \tag{1.78}$$

Taking $Z = 1$ for atomic hydrogen and using the Bohr frequency relation (1.70), the frequencies of radiation emitted in a transition between two energy levels a and b, with energies $E_a < E_b$, are

$$\nu_{ab} = \frac{m}{4\pi\hbar^3}\left(\frac{e^2}{4\pi\varepsilon_0}\right)^2 \left(\frac{1}{n_a^2} - \frac{1}{n_b^2}\right) \tag{1.79}$$

where $n_a = 1, 2, 3, \ldots$ and $n_b = 2, 3, 4, \ldots$ ($n_b > n_a$), in agreement with (1.61) provided R is taken to be

$$R(\infty) = \frac{m}{4\pi\hbar^3}\left(\frac{e^2}{4\pi\varepsilon_0}\right)^2 \qquad \textbf{(1.80)}$$

Here we have written $R(\infty)$ to recall that we are using the infinite nuclear mass approximation.

If instead of frequencies we use wave numbers $\tilde{\nu}_{ab}$ we have

$$\tilde{\nu}_{ab} = \tilde{R}(\infty)\left(\frac{1}{n_a^2} - \frac{1}{n_b^2}\right), \qquad n_b > n_a \qquad \textbf{(1.81)}$$

where

$$\tilde{R}(\infty) = \frac{m}{4\pi c\hbar^3}\left(\frac{e^2}{4\pi\varepsilon_0}\right)^2 \qquad \textbf{(1.82)}$$

Evaluating $\tilde{R}(\infty)$ we find

$$\tilde{R}(\infty) = 109\,737 \text{ cm}^{-1} \qquad \textbf{(1.83)}$$

in good, but not perfect, agreement with the experimental value (1.63).

Returning to equation (1.78), we see that the energy levels of a one-electron atom are given by

$$E_n = -I_P/n^2, \qquad n = 1, 2, 3, \ldots \qquad \textbf{(1.84)}$$

where I_P is given by

$$I_P = \frac{1}{2}\frac{m}{\hbar^2}\left(\frac{Ze^2}{4\pi\varepsilon_0}\right)^2 = 13.6Z^2 \text{ eV} \qquad \textbf{(1.85)}$$

The level with the lowest energy is known as the ground state and has $n = 1$. If the atom absorbs an energy greater than I_P, the energy of the electron becomes positive and the electron is ejected from the atom. The quantity I_P is known as the *ionisation potential*, and the value (13.6 eV) can be verified by experiments in which hydrogen atoms are ionised either by absorbing photons or in collision processes. An energy level diagram for hydrogen, showing series of spectral lines, is shown in Fig. 1.16. The quantum number n is called the *principal quantum number* to distinguish it from other quantum numbers that we shall meet later. A commonly used notation is one in which an electron in a level with $n = 1$ is said to be in the K shell. Correspondingly if $n = 2$ or $n = 3$, the electron is said to be in the L or M shells respectively.

Atomic units

The radius of the orbit of the electron in the ground state of hydrogen is known as the first Bohr radius of hydrogen and is denoted by the symbol a_0:

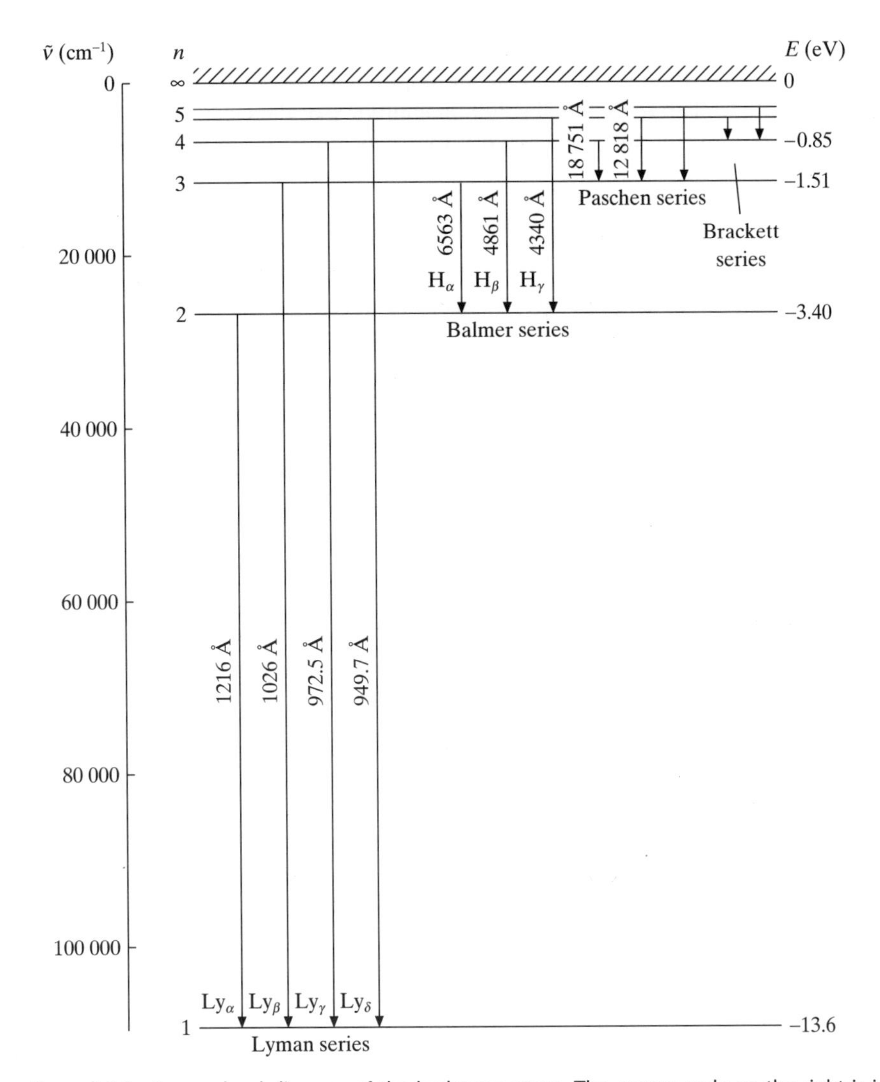

Figure 1.16 Energy level diagram of the hydrogen atom. The energy scale on the right is in electron volts; on the left another scale, increasing from top to bottom, gives the wave number $\tilde{\nu}$ in cm^{-1}. The principal quantum number n corresponding to each energy level is also indicated. A spectral line, resulting from the transition of the atom from one energy level to another, is represented by a vertical line joining the two energy levels. The numbers against the lines indicate the corresponding wavelengths in ångströms (1 Å = 10^{-10} m). For clarity, only transitions between lower-lying levels are shown.

$$a_0 = \frac{(4\pi\varepsilon_0)\hbar^2}{me^2} = 5.291\,77 \times 10^{-11}\ \text{m} \qquad \textbf{(1.86)}$$

In atomic physics, it has proved to be extremely useful to introduce a set of units, called *atomic units* (a.u.), in which a_0 is taken as the unit of length.

Correspondingly the mass of the electron is employed as the unit of mass and $\hbar$ as the unit of angular momentum. To complete the system the unit of charge is taken to be the absolute magnitude e of the electronic charge and the permittivity of free space ε_0 is $1/(4\pi)$. In atomic units ($m = \hbar = e = 1$, $4\pi\varepsilon_0 = 1$), we have

$$E_n = -\tfrac{1}{2}Z^2/n^2 \text{ a.u.} \tag{1.87}$$

The ground state energy of hydrogen ($Z = 1$, $n = 1$) is $-\frac{1}{2}$ a.u., from which we see that the atomic unit of energy is equivalent to 27.2 eV. The atomic unit of velocity is equal to the velocity v_0 of the electron in the first Bohr orbit of hydrogen. From (1.74) we see that

$$v_0 = \frac{e^2}{(4\pi\varepsilon_0)\hbar} = \alpha c \tag{1.88}$$

where we have introduced the dimensionless constant

$$\alpha = \frac{e^2}{(4\pi\varepsilon_0)\hbar c} \tag{1.89}$$

which is known as the *fine-structure constant* and has the value $\alpha \simeq 1/137$. Thus we see that in atomic units the velocity of light is $c \simeq 137$ a.u. Further details concerning atomic units can be found in Appendix 14.

Finite nuclear mass

Although the approximation, in which the nucleus is assumed to be infinitely heavy and at rest, is good enough for many purposes, a distinct improvement can be made if we allow the nucleus to move. In this case, both the nucleus A of mass M and the electron B of mass m rotate about the centre of mass of the system C as in Fig. 1.17. In the absence of forces external to the atom, the centre of mass

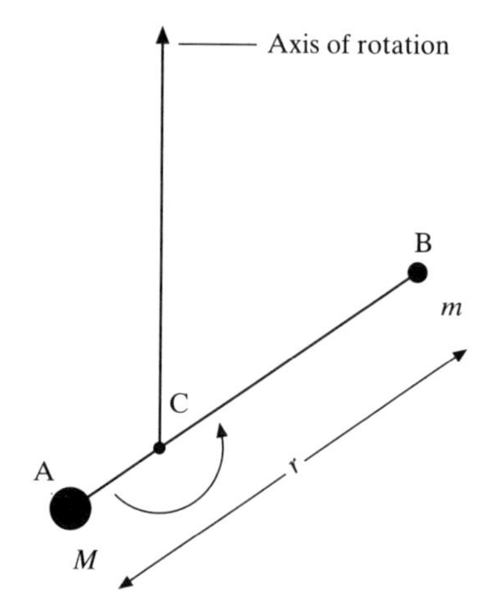

Figure 1.17 Bohr model with finite nuclear mass.

will be either at rest, or in uniform motion, according to Newton's law. The distance AB is again denoted by r and the angular velocity of the line AB about an axis through C by ω. Since C is the centre of mass

$$M \times \mathrm{AC} = m \times \mathrm{CB} \tag{1.90}$$

from which

$$\begin{aligned} \mathrm{AC} &= \left(\frac{m}{M+m}\right) r \\ \mathrm{CB} &= \left(\frac{M}{M+m}\right) r \end{aligned} \tag{1.91}$$

The total angular momentum of the system is

$$L = m\omega \mathrm{CB}^2 + M\omega \mathrm{AC}^2 = \mu\omega r^2 \tag{1.92}$$

where

$$\mu = \frac{mM}{m+M} \tag{1.93}$$

is the reduced mass. Putting $r\omega = v$, where v is the velocity of the electron with respect to the nucleus, we have

$$L = \mu v r = n\hbar, \qquad n = 1, 2, 3, \ldots \tag{1.94}$$

which is the same as equation (1.73) but with μ replacing m.

The centripetal force acting on the electron B is of magnitude

$$m\mathrm{CB}\omega^2 = \mu\omega^2 r \tag{1.95}$$

and this is to be equated with the Coulomb force

$$\mu\omega^2 r = \frac{\mu v^2}{r} = \left(\frac{Ze^2}{4\pi\varepsilon_0}\right)\frac{1}{r^2} \tag{1.96}$$

This is the same equation of motion as before, but with μ replacing m.

The kinetic energy of the system is

$$T = \tfrac{1}{2}(M+m)v_{\mathrm{CM}}^2 + \tfrac{1}{2}\mu v^2 \equiv T_{\mathrm{CM}} + T_{\mathrm{I}} \tag{1.97}$$

where v_{CM} is the velocity of the centre of mass, and T_{I} is the internal (or relative) kinetic energy. A bound state energy E_n is defined as the difference between the total energy and the kinetic energy T_{CM} of the centre of mass,

$$E_n = T_{\mathrm{I}} + V = \tfrac{1}{2}\mu v^2 - \left(\frac{Ze^2}{4\pi\varepsilon_0}\right)\frac{1}{r} \tag{1.98}$$

so that E_n is given by (1.78) with μ replacing m

$$E_n = -\frac{\mu}{2\hbar^2}\left(\frac{Ze^2}{4\pi\varepsilon_0}\right)^2\frac{1}{n^2} \tag{1.99}$$

Similarly the allowed values of r are given by (1.75) with, again, μ replacing m

$$r = \frac{(4\pi\varepsilon_0)\hbar^2 n^2}{Ze^2\mu} = \frac{n^2}{Z}\frac{m}{\mu}a_0 = \frac{n^2}{Z}a_\mu \tag{1.100}$$

where we have defined

$$a_\mu = a_0\frac{m}{\mu} \tag{1.101}$$

The Rydberg constant for a nucleus of mass M can be written immediately in terms of $R(\infty)$ and $\tilde{R}(\infty)$ by

$$R(M) = \frac{\mu}{m}R(\infty) \quad \text{and} \quad \tilde{R}(M) = \frac{\mu}{m}\tilde{R}(\infty) \tag{1.102}$$

For hydrogen $M = M_p$, where M_p is the mass of the proton and $\tilde{R}(M_p) = 109\,678\ \text{cm}^{-1}$ which agrees very well with the experimental value (1.63).

Because of the nuclear mass effect there is an *isotopic shift* between the spectral lines of different isotopes of the same atom. For example, there is such a shift between the spectrum of atomic deuterium, which has a nucleus with $Z = 1$ but containing a proton and a neutron (so that its mass $M \simeq 2M_p$), and that of atomic hydrogen. The ratio of frequencies of corresponding lines is 1.000 27, which is easily detectable, and in fact through this the discovery of the deuteron was made.

Hydrogenic ions

By taking $Z = 2, 3, 4, \ldots$, the Bohr model predicts the energy levels (and hence the line spectra) of all the ions containing one electron with a nucleus of charge Ze. The observed frequencies agree closely with the Bohr formula (1.79) until Z becomes large ($Z \gtrsim 20$). The orbital velocity of the electron in the ground state divided by the velocity of light is such that (see (1.74) and (1.89))

$$\frac{v}{c} = \alpha Z \simeq \frac{Z}{137} \tag{1.103}$$

so it is to be expected that, for large Z, relativistic effects will be significant. In Table 1.2, the values of $\tilde{R}(M)$ are shown for a few hydrogenic systems together with the wavelengths of the lines corresponding to the $n = 2$ to $n = 1$ transition, which are known as *resonance lines:*

The Franck and Hertz experiment

The Bohr model predicts that the energy levels of atoms are quantised and only certain discrete values of the total energy are allowed. This can be confirmed

Table 1.2 Rydberg constants for some hydrogenic systems and the wavelengths of the resonance lines $n = 2 \rightarrow n = 1$.

	Wavelength of the transition (Å)	$R(M)$ (cm^{-1})
H I[†]	1215.664	109 677.58
D I	1215.336	109 707.19
He II	303.779	109 722.26
Li III	134.994	109 728.72
C VI	33.734	109 732.29
O VIII	18.967	109 733.54

[†] The Roman numeral indicates the degree of ionisation. I denotes the neutral atom, II an atom from which one electron has been ionised and so on.

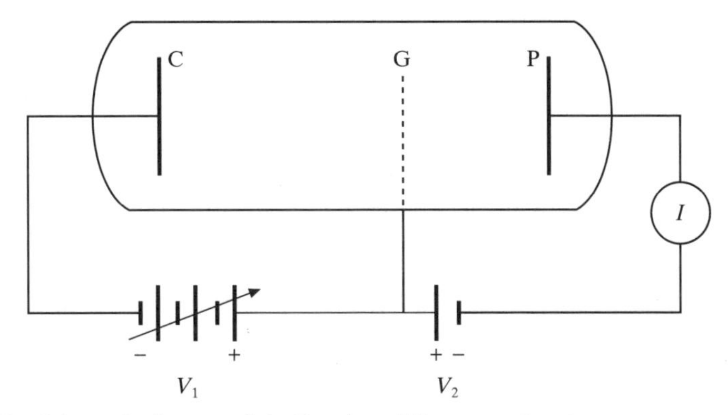

Figure 1.18 Schematic diagram of the Franck and Hertz experiment.

rather directly by an experiment originally devised by J. Franck and G. Hertz in 1914. A schematic diagram of the apparatus is shown in Fig. 1.18. In the first stage of the experiment, the apparatus is evacuated. A cathode C is heated so that it emits electrons, which are attracted to and pass through a wire grid, G, which is maintained at a positive potential V_1 with respect to the cathode. The electrons accelerated by the potential V_1 attain a kinetic energy $mv^2/2 = eV_1$. After passing through the grid they are collected by a plate P and cause a current I to flow in the circuit. The plate P is maintained at a slightly lower voltage V_2 than the grid, $V_2 = V_1 - \Delta V$, where $\Delta V \ll V_1$. The small retarding potential ΔV has the effect of reducing the kinetic energy of the electrons slightly, but not enough to stop them being collected.

The apparatus is now filled with mercury vapour. The electrons collide with the atoms of mercury, and if the collisions are elastic, so that there is no transfer of energy from the electrons to the internal structure of the atoms, the current I will be unaffected by the introduction of the gas. This follows because mercury atoms are too heavy to gain appreciable kinetic energy when struck by an

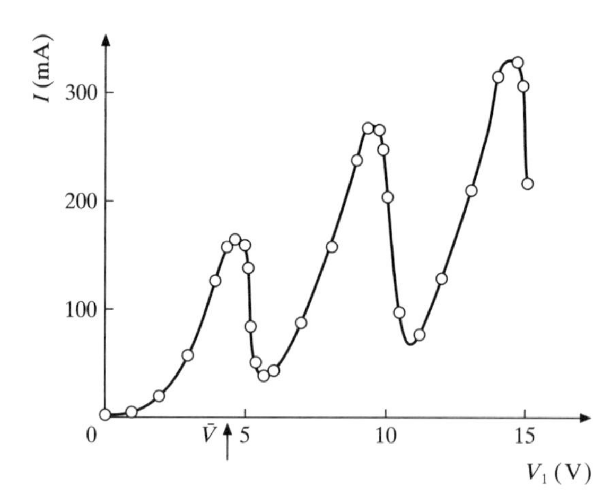

Figure 1.19 The variation of the current I as a function of the accelerating voltage V_1 in the Franck and Hertz experiment.

electron. The electrons are deflected but retain the same kinetic energy. In contrast, if an electron makes an inelastic collision with a mercury atom in which it loses an energy E, exciting the mercury atom to a level of greater internal energy, then its final kinetic energy will be $mv_1^2/2 = (eV_1 - E)$. If eV_1 is equal to E, or is only a little larger, the retarding potential ΔV will be sufficient to prevent the electron from reaching the collecting plate and it will no longer contribute to the current I.

The experiment is carried out by gradually increasing V_1 from zero and measuring the current I as a function of V_1. The result obtained is shown diagrammatically in Fig. 1.19. The current I is seen to fall sharply at a potential $\bar{V}$, which is known as the resonance potential, and which for mercury is equal to 4.9 V. The results can be interpreted by supposing that for $eV_1 < 4.9$ eV the atoms cannot absorb the energy of the electrons and the collisions are elastic, while exactly 4.9 eV above their ground state, mercury atoms possess another discrete energy level. When eV_1 reaches this value, a large number of the colliding electrons excite atoms to this level, losing their energy in the process, and reducing the current I sharply.

If the voltage V_1 is further increased the current again increases, and further sharp falls are seen. Some of these are due to electrons having sufficient energy to excite two or more atoms to the 4.9 eV level; but others are due to the excitation of higher discrete levels. We have seen that 1 eV corresponds to a wave number of 8065 cm^{-1}, so if this interpretation is correct, a line would be expected in the mercury spectrum corresponding to a transition from the first excited state at 4.9 eV to the ground state, with a wave number of 4.9×8065 cm$^{-1} \simeq 39\,500$ cm^{-1}. Franck and Hertz were indeed able to verify the existence of such a line, and also to show that radiation of this wave number was only emitted from the mercury vapour when V_1 exceeded 4.9 V.

This experiment, and corresponding experiments using other gases and vapours, provide excellent confirmation of the discrete nature of bound state energy levels. It can also be demonstrated that when sufficient energy is available to ionise an atom, the energy of the ejected electron can take any positive value, so we can say that *the energy level spectrum of an atom consists of two parts: discrete negative energies corresponding to bound states and a continuum of positive energies corresponding to unbound (ionised) states.*

Limitations of the Bohr model

Although the planetary model of the hydrogen atom is rather successful, and the idea of quantised atomic energy is correct, the model is unsatisfactory in many respects. First, it cannot be generalised to deal with systems containing two or more electrons. In addition, the assumptions made and, in particular, the hypothesis that only circular orbits are allowed, are inexplicable and arbitrary. Among other objections are the lack of any method to calculate the rate of transitions between the different energy levels when radiation is emitted or absorbed, and the inability to handle unbound systems. In later work, W. Wilson and A. Sommerfeld showed how to remove the restriction to circular orbits and Sommerfeld also obtained relativistic corrections to the Bohr model. However, the other objections still persisted, and the theory – called *the old quantum theory* – remained restricted in scope. It was eventually superseded by the quantum mechanics developed by E. Schrödinger, W. Heisenberg and others, following the ideas of L. de Broglie.

X-ray spectra and Moseley's law

Despite the general inability of the Bohr model to describe many-electron systems, it was able to provide an illuminating explanation of the regularities in X-ray spectra observed by H. Moseley in 1913. In an X-ray tube the radiation is emitted from a target bombarded by high-energy electrons. The X-ray region may be taken to be in the range of wavelengths from 0.1 to 10 Å corresponding to photon energies from a few keV to several hundred keV (see Fig. 1.9). The spectrum observed is characteristic of the material used as the target in the tube and consists of a continuous spectrum upon which is superimposed a line spectrum. Moseley studied the line spectra of 39 elements from aluminium (the lightest) to gold (the heaviest). All the spectra were remarkably similar. In most cases, the spectrum consisted of two groups of lines, the K series and the L series; for a given element the L lines were at lower frequencies than the K lines. For heavier elements other series of lines appeared at still lower frequencies.

Moseley found that the frequency ν_n of the nth line of each series varied smoothly with the atomic number of the target element, Z. By plotting $\sqrt{\nu_n}$ against Z, he established the law

$$\sqrt{\nu_n} = C_n(Z - \sigma) \tag{1.104}$$

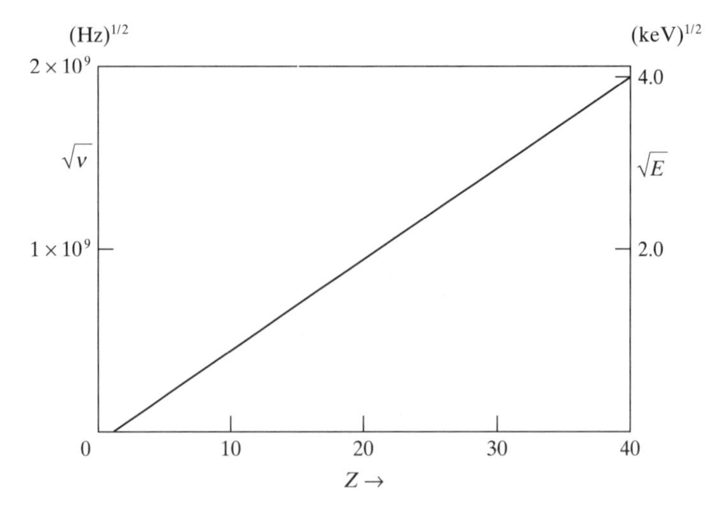

Figure 1.20 A Moseley plot of $\sqrt{\nu}$ against Z for the K_α line in the X-ray spectrum. A scale in $\sqrt{E}$, with E in keV, is shown on the right-hand side of the figure.

where C_n is independent of Z, and σ is in the range 1 to 2 for the K series and is in the range 7.4 to 9.4 for the L series. An example is the K line of lowest frequency, the K_α line, for which the *Moseley plot* is shown in Fig. 1.20. Since plots of $\sqrt{\nu_n}$ against Z were smoother than those against the atomic weight μ, this suggested that Z had a fundamental significance. At this time the atomic number had no significance other than giving the position of an element in Mendeleev's periodic table. In the main the ordering of the elements of the table was given by their atomic weights μ, although Mendeleev found that certain pairs had to be reversed in order to preserve the periodicity of the chemical properties. For example, in order of weights the 18th element is potassium ($\mu = 39.102$) and the 19th argon ($\mu = 39.948$). This arrangement puts potassium in the rare gas column and argon in the alkali metal column and to preserve the chemical periodicity argon has to be assigned the atomic number 18 and potassium 19.

An explanation of the significance of Z was given by Moseley in terms of the Bohr model, which had been published a year before. He first argued that in the high-energy electron bombardment of the target atoms, the inner (tightly bound) electrons are ejected leaving vacancies. The X-rays are emitted when a less tightly bound electron makes a transition, filling a vacancy. Since the transitions concern tightly bound electrons in orbits close to the nucleus, the effective potential experienced by these electrons is due mainly to the Coulomb field of the nucleus, screened to a small extent by the other electrons. The atomic number Z was known to be approximately equal to $\mu/2$ and the experiments of Geiger and Marsden had shown that the nuclear charge was also approximately $\mu/2$. This suggested the identification of Z with the nuclear charge and $(Z - \sigma)$ as an effective charge, σ allowing for screening of the nucleus by other electrons. Using the Bohr formula for the energy levels of one electron moving in a Coulomb field with charge $(Z - \sigma)$, the frequencies of the spectrum are given by *Moseley's law*

$$\nu_{mn} = R(Z - \sigma)^2 \left(\frac{1}{m^2} - \frac{1}{n^2} \right) \quad \textbf{(1.105)}$$

where n and m are both integers and $n > m$. The K series of lines can be attributed to transitions in which the final energy level has $m = 1$ (the K shell). The line of longest wavelength in the K series is the K_α line, and for this $n = 2$. The L series of lines are those in which the vacancy occurs in the $m = 2$ level (the L shell) and the line of longest wavelength corresponds to $n = 3$.

By interpreting the regularities he had observed in the X-ray spectra in this way, Moseley was able to establish the critical identification of Z, the atomic number, with the nuclear charge, and to show that the Bohr model could be applied to the most tightly bound inner electrons of an atom, which move in a potential dominated by the nuclear Coulomb field. It is interesting to note that in plotting his results Moseley found that to avoid breaks in his curve, he had to postulate the existence of four, hitherto unknown, elements with $Z = 43, 61, 72$ and 75. These were discovered subsequently.

A further discussion of X-ray spectra is given in Chapter 9. The optical spectra of many-electron atoms are much more complicated and cannot be interpreted so easily, since the outer electrons move in potentials which are not strongly dominated by the nuclear potential.

1.8 The Stern–Gerlach experiment – angular momentum and spin

We shall now discuss an experiment of fundamental importance, carried out by O. Stern and W. Gerlach in 1922, to measure the magnetic dipole moments of atoms. The results demonstrated, once more, the inability of classical mechanics to describe atomic phenomena and confirmed the necessity of a quantum theory of angular momentum, which had been suggested by Bohr's model.

Let us first understand how an atom comes to possess a magnetic moment. In the Bohr model of a hydrogenic atom, an electron occupies a circular orbit, rotating with an orbital angular momentum **L**. A moving charge is equivalent to an electric current, so that an electron moving in a closed orbit forms a current loop, and this in turn creates a magnetic dipole (Duffin, 1968). In fact whatever model of atomic structure we make, the electrons can be expected to possess angular momentum and accordingly atoms possess magnetic moments.

A circulating current of magnitude I enclosing a small plane area $\mathrm{d}A$ gives rise to a magnetic dipole moment

$$\mathcal{M} = I\,\mathrm{d}\mathbf{A} \quad \textbf{(1.106)}$$

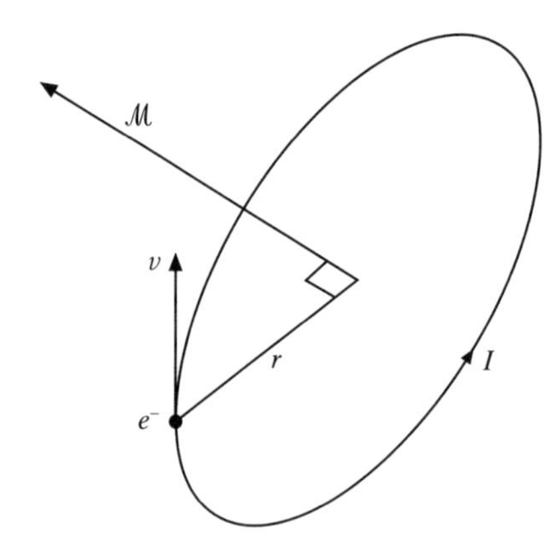

Figure 1.21 The magnetic dipole moment due to a current loop.

where the direction of d**A** is along the normal to the plane of the current loop, as shown in Fig. 1.21.

When the current is due to an electron moving with a velocity v in a circle of radius r, I is given by

$$I = \frac{ev}{2\pi r} \tag{1.107}$$

Since the area enclosed is πr^2, we have $\mathcal{M} = evr/2 = eL/2m$, and as the direction of the current is opposite to the direction of rotation of the electron, the corresponding magnetic dipole moment due to the orbital motion of the electron is

$$\boldsymbol{\mathcal{M}}_L = -\left(\frac{e}{2m}\right)\mathbf{L} \tag{1.108}$$

The Bohr quantisation rule (1.73) suggests that $\hbar$ is a natural unit of angular momentum, so we can write

$$\boldsymbol{\mathcal{M}}_L = -\mu_{\mathrm{B}}(\mathbf{L}/\hbar) \tag{1.109}$$

where

$$\mu_{\mathrm{B}} = \frac{e\hbar}{2m} \tag{1.110}$$

Because $(\mathbf{L}/\hbar)$ is dimensionless, μ_{B} has the dimensions of a magnetic moment. It is known as the *Bohr magneton* and has the numerical value

$$\mu_{\mathrm{B}} = 9.274\,01 \times 10^{-24}\ \mathrm{J\,T^{-1}} \tag{1.111}$$

Quite generally, a system of electrons possessing a total angular momentum **J** has an effective magnetic moment $\boldsymbol{\mathcal{M}}$ antiparallel to **J**, and it is usual to write

$$\mathcal{M} = -g\mu_B(\mathbf{J}/\hbar) \tag{1.112}$$

where g is a dimensionless constant called a *gyromagnetic ratio*.

Interaction with a magnetic field

If an atom with a magnetic moment $\mathcal{M}$ is placed in a magnetic field $\mathcal{B}$, the energy of interaction (potential energy) is (see Duffin, 1968)

$$W = -\mathcal{M} \cdot \mathcal{B} \tag{1.113}$$

The system experiences a torque $\boldsymbol{\Gamma}$, where

$$\boldsymbol{\Gamma} = \mathcal{M} \times \mathcal{B} \tag{1.114}$$

and a net force $\mathbf{F}$, where

$$\mathbf{F} = -\nabla W \tag{1.115}$$

Combining (1.113) with (1.115) we see that the components of $\mathbf{F}$ are

$$F_x = \mathcal{M} \cdot \frac{\partial \mathcal{B}}{\partial x}; \qquad F_y = \mathcal{M} \cdot \frac{\partial \mathcal{B}}{\partial y}; \qquad F_z = \mathcal{M} \cdot \frac{\partial \mathcal{B}}{\partial z} \tag{1.116}$$

In a magnetic field that is uniform, no net force is experienced by a magnetic dipole, which precesses with a constant angular frequency. For an orbiting electron this angular frequency is

$$\omega_L = \frac{\mu_B}{\hbar} \mathcal{B} \tag{1.117}$$

It is called the *Larmor angular frequency*. On the other hand, in an inhomogeneous field an atom experiences a net force proportional to the magnitude of the magnetic moment.

The Stern–Gerlach experiment

In 1921, Stern suggested that magnetic moments of atoms could be measured by detecting the deflection of an atomic beam by such an inhomogeneous field. The experiment was carried out a year later by Stern and Gerlach. The apparatus is shown in schematic form in Fig. 1.22.

The first experiments were made using atoms of silver. A beam is produced by heating the metallic vapour in an enclosure, which is situated in an evacuated region into which the silver atoms stream through a small hole. The beam can be collimated with a system of slits and passed between the poles of a magnet shaped to produce an inhomogeneous field, as shown in the figure. The beam is then detected by allowing it to fall onto a cool plate. The density of the deposit is proportional to the intensity of the beam and to the length of time for which the beam falls on the plate.

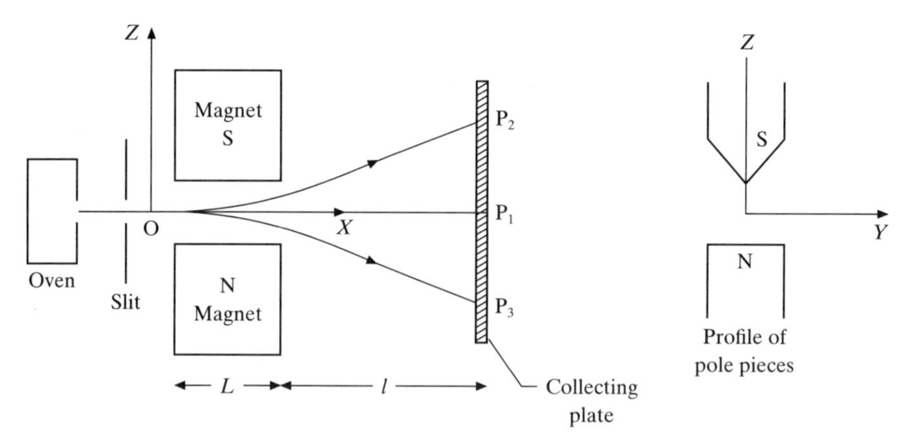

Figure 1.22 The Stern-Gerlach apparatus.

Taking the shape of the magnets as shown in Fig. 1.22, the force on each atom is given by (see (1.116))

$$F_x = \mathcal{M}_z \frac{\partial \mathcal{B}_z}{\partial x}; \qquad F_y = \mathcal{M}_z \frac{\partial \mathcal{B}_z}{\partial y}; \qquad F_z = \mathcal{M}_z \frac{\partial \mathcal{B}_z}{\partial z} \tag{1.118}$$

The magnet is symmetrical about the XZ plane and the beam is confined to this plane. It follows that $\partial \mathcal{B}_z/\partial y = 0$. Also, apart from edge effects, $\partial \mathcal{B}_z/\partial x = 0$, so that the only force on the atoms in the beam is in the Z direction.

In the incident beam, the direction of the magnetic moment $\boldsymbol{\mathcal{M}}$ of the atoms will be completely at random and in the Z direction it would be expected that every value of $\mathcal{M}_z$ would occur in the interval $-\mathcal{M} \leqslant \mathcal{M}_z \leqslant \mathcal{M}$, with the consequence that the deposit on the collecting plate would be spread over a region symmetrically disposed about the point of no deflection. The surprising result that Stern and Gerlach obtained in their experiments on silver was that two distinct and separate lines were formed on the plate (see Fig. 1.23), symmetrically about the point

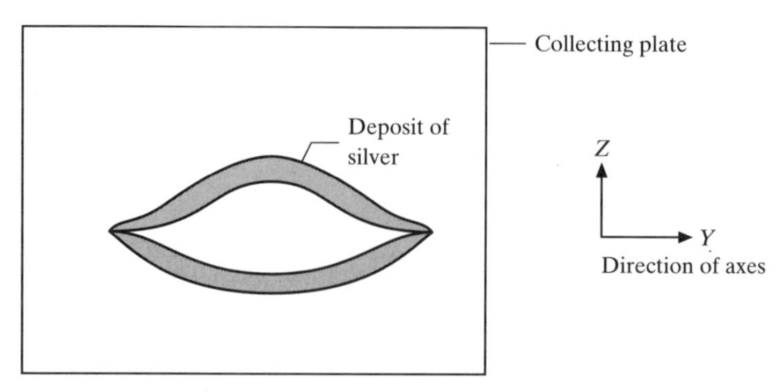

Figure 1.23 Results of the Stern–Gerlach experiment for silver.

of no deflection. Similar results were found for atoms of copper and gold, and in later work, for sodium, potassium and hydrogen.

The quantisation of the component of the magnetic moment along the direction defined by the magnetic field is termed *space quantisation*. This implies that the component of the angular momentum in a certain direction is quantised so that it can only take certain values. In general for each type of atom, the values of $\mathcal{M}_z$ will range from $(\mathcal{M}_z)_{\max}$ to $-(\mathcal{M}_z)_{\max}$ and correspondingly L_z will range from $-(L_z)_{\max}$ to $(L_z)_{\max}$. If we denote the observed multiplicity of values of $\mathcal{M}_z$ (and hence L_z) by α we can try to interpret α and to deduce the allowed values of L using the Bohr model. Indeed, the Bohr quantisation of angular momentum suggests that orbital angular momentum only occurs in integral units of $\hbar$. We may postulate that the magnitude of orbital angular momentum can only take values $L = l\hbar$, where l is a positive integer or zero. Thus the maximum value of L_z is $+l\hbar$ and its minimum value $-l\hbar$. If L_z is also quantised in the form

$$L_z = m\hbar \qquad \textbf{(1.119)}$$

where m is a positive or negative integer or zero, then m must take on the values $-l, -l+1, \dots, l-1, l$ and the multiplicity α must be equal to $(2l+1)$. The number m is known as a *magnetic quantum number*. In fact, as we shall see in the next chapter, this result turns out to be correct in quantum mechanics, with the difference that the possible values of the total orbital angular momentum are of the form $L = \sqrt{l(l+1)}\hbar$ with $l = 0, 1, 2, \dots$ rather than of the form $L = l\hbar$ suggested by the Bohr model. However, the results of Stern and Gerlach for silver do not fit with this scheme, since the multiplicity of values of the Z component of the angular momentum for silver is 2. This implies that $(2l+1) = 2$ giving $l = \frac{1}{2}$, a non-integer value.

Electron spin

The explanation of this result for silver came in 1925, when S. Goudsmit and G.E. Uhlenbeck showed that the splitting of spectral lines occurring when atoms are placed in a magnetic field (the Zeeman effect) could be explained if electrons possess an *intrinsic magnetic moment* $\boldsymbol{\mathcal{M}}_s$ in addition to the magnetic moment produced by orbital motion, where the component of $\boldsymbol{\mathcal{M}}_s$ in a given direction can take the two values $\pm\mathcal{M}_s$ only. We can postulate that this magnetic moment is due to an *intrinsic angular momentum*, or *spin*, of the electron, which we denote by $\mathbf{S}$. By analogy with (1.112), we then have

$$\boldsymbol{\mathcal{M}}_s = -g_s\mu_B(\mathbf{S}/\hbar) \qquad \textbf{(1.120)}$$

where g_s is the *spin gyromagnetic ratio*. If we introduce a spin quantum number s, analogous to l, so that the multiplicity of the spin component in a given direction is $(2s+1)$, we must have $s = \frac{1}{2}$ and the possible values of the component of the spin $\mathbf{S}$ in the Z direction are $\pm\hbar/2$, while the magnitude of the spin is $\sqrt{s(s+1)}\hbar = \sqrt{\frac{3}{4}}\hbar$. The Stern–Gerlach results are then explained if we assume that the orbital angular momentum of a silver atom is zero, but its spin angular momentum is given by $s = \frac{1}{2}$.

We have seen that the magnetic moment of an electron is due partly to its orbital angular momentum and partly to its spin angular momentum, and we can write

$$\mathcal{M} = -\mu_B(\mathbf{L} + g_s\mathbf{S})/\hbar \tag{1.121}$$

Measurements of the magnitude of $\mathcal{M}$ for atomic hydrogen have shown that an accurate value of the electron spin gyromagnetic ratio is $g_s = 2$. The discovery of this intrinsic property of the electron is of fundamental importance. In fact, it is now known that all particles can be assigned an intrinsic angular momentum (spin), and hence a spin quantum number s. In some cases, such as the pion (π meson), it is zero, but in others, such as the electron, the proton, the neutron, it is one-half (that is, $s = \frac{1}{2}$) and for other elementary particles it may be $s = 1$, $s = \frac{3}{2}, \ldots$.

Angular momentum

The total angular momentum $\mathbf{J}$ of an atom containing N electrons is given by

$$\mathbf{J} = \mathbf{L} + \mathbf{S} \tag{1.122}$$

where

$$\mathbf{L} = \sum_{i=1}^{N} \mathbf{L}_i \tag{1.123}$$

is the total orbital angular momentum, which is the vector sum of the orbital angular momenta $\mathbf{L}_i$ of the N electrons and

$$\mathbf{S} = \sum_{i=1}^{N} \mathbf{S}_i \tag{1.124}$$

is the total spin angular momentum, which is the vector sum of the spin angular momenta $\mathbf{S}_i$ of the N electrons. A measurement of the component of $\mathbf{J}$ in the Z direction can only provide $(2j + 1)$ values, given by $m_j\hbar$, where the *magnetic quantum number* m_j can take the values $-j, -j + 1, \ldots, j - 1, j$. It is found that j can take integer (including zero) or half-odd integer values only: $j = 0, \frac{1}{2}, 1, \frac{3}{2}, \ldots$. For an angular momentum whose component in a given direction has the multiplicity $(2j + 1)$, a measurement of the magnitude of the angular momentum produces the value $\sqrt{j(j+1)}\hbar$. Thus, in a Stern–Gerlach experiment, a beam of atoms with angular momentum of magnitude $\sqrt{j(j+1)}\hbar$ will produce $(2j + 1)$ spots on the detecting screen, symmetrically disposed about the point of zero deflection.

Another property of angular momentum in quantum mechanics (which will be discussed in Chapter 2) is that there is a limitation on the precision with which simultaneous measurements of two (or three) components of an angular momentum can be measured. In fact, if the value of the z component is known precisely, the values of the x and y components are indefinite, but on the *average*

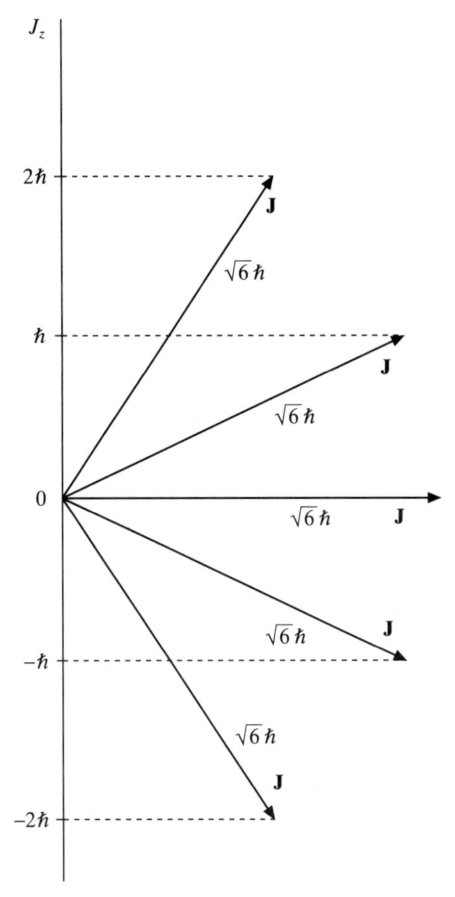

Figure 1.24 Precession of the angular momentum **J** about the axis of quantisation for the case $j = 2$.

are zero. This situation is often represented by a *vector model* in which the angular momentum vector **J** of length $\sqrt{j(j+1)}\hbar$ precesses about the Z axis such that J_z has one of the $(2j + 1)$ values $m_j\hbar$. This is shown in Fig. 1.24 for the particular case $j = 2$.

In the vector model, the total orbital angular momentum **L** and the total spin **S** precess about the total angular momentum **J**, and the effective magnetic moment of an atom is given by

$$\mathcal{M} = -\mu_B \langle \mathbf{L} + g_s \mathbf{S} \rangle / \hbar \qquad \textbf{(1.125)}$$

where we have used (1.121) and the symbol $\langle\ \rangle$ indicates an average over the precession. It can be shown that this averaging yields

$$\mathcal{M} = -g\mu_B(\mathbf{J}/\hbar) \qquad \textbf{(1.126)}$$

where g is a gyromagnetic ratio, in conformity with (1.112). The corresponding quantum mechanical result will be discussed in Chapter 6.

1.9 de Broglie's hypothesis and the genesis of wave mechanics

In our brief historical survey, we have seen how, as knowledge of atomic structure increased, evidence accumulated that a description in terms of classical physics – Newton's laws of mechanics and Maxwell's electromagnetic equations – was inadequate. Electromagnetic radiation displays particle as well as wave characteristics; the energy of the field is quantised, each packet of energy $h\nu$ being carried by a photon. On the other hand, the energy levels and angular momenta of bound electrons in atoms are also quantised. In contrast, beams of electrons moving under electric and magnetic fields, as in J.J. Thomson's experiments, behave like classical charged particles.

In 1924, L. de Broglie made a great unifying, but speculative, hypothesis, that just as radiation has particle-like properties, electrons and other material particles possess wave-like properties. The energy of a photon is given by $E = h\nu$, where ν is the frequency and the corresponding magnitude of the momentum is $p = h\nu/c = h/\lambda$, where λ is the wavelength. For free material particles, de Broglie assumed that the associated wave also has a frequency ν and a wavelength λ, related to the energy E and the magnitude of the momentum p of the particle by

$$\nu = \frac{E}{h} \tag{1.127}$$

and

$$\lambda = \frac{h}{p} \tag{1.128}$$

In particular, for a particle of mass m moving at a non-relativistic speed v, one has $p = mv$ so that $\lambda = h/(mv)$.

This idea immediately gives a *qualitative* explanation of the quantum condition (1.73), used in the Bohr model of the hydrogen atom. Indeed, let us suppose that an electron in a hydrogenic atom moves in a circular orbit of radius r, with speed v. If this is to be a stable stationary state, the wave associated with the electron must be a standing wave and a whole number of wavelengths must fit into the circumference $2\pi r$. Thus

$$n\lambda = 2\pi r \qquad n = 1, 2, 3, \dots \tag{1.129}$$

Since $\lambda = h/p$ and $L = rp$, we immediately find the condition

$$L = nh/(2\pi) = n\hbar \tag{1.130}$$

which is identical with (1.73). Later, in 1925, these qualitative ideas were incorporated into the systematic theory of quantum mechanics developed by E. Schrödinger, which will be discussed in the next chapter.

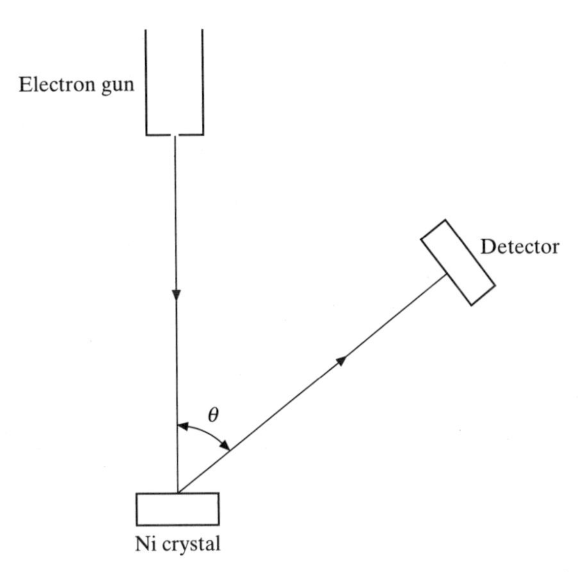

Figure 1.25 Schematic diagram of the Davisson–Germer experiment. Electrons strike at normal incidence the surface of a single nickel crystal. The number of electrons scattered at an angle θ to the incident direction is measured by means of a detector.

When waves are scattered or pass through slits which have dimensions comparable to their wavelength, interference and diffraction effects are observed. Now, as seen from (1.128), the *de Broglie wavelengths* associated with electrons of energy 1, 10 and 100 eV are, respectively, 12, 3.9 and 1.2 Å (where 1 Å = 10^{-10} m). Thus, in macroscopic situations, as in Thomson's experiments, the de Broglie electron wavelengths are exceedingly small compared with the dimensions of any obstacles or slits in the apparatus and no interference or diffraction effects can be observed. However, the spacing of atoms in a crystal lattice is of the order of an ångström and therefore, just as in the case of X-rays, a crystal could be used as a grating to observe the diffraction and interference effects due to the electron matter waves. Experiments of this type were performed in 1927 by C.J. Davisson and L.H. Germer, and independently by G.P. Thomson.

In the Davisson–Germer experiment, the *reflection* of electrons from the face of a crystal was investigated. Electrons from a heated filament were accelerated through a potential difference V_0 and emerged from the 'electron gun' with kinetic energy eV_0. This beam of monoenergetic electrons was directed to strike at normal incidence the surface of a single nickel crystal, and the number $N(\theta)$ of electrons scattered at an angle θ to the incident direction was measured by means of a detector (see Fig. 1.25). The data obtained by Davisson and Germer for 54 eV electrons are shown in Fig. 1.26. The scattered intensity is seen to fall from a maximum at $\theta = 0°$ to a minimum near 35°, and then to rise to a peak near 50°. The strong scattering at $\theta = 0°$ is expected from either a particle or a wave

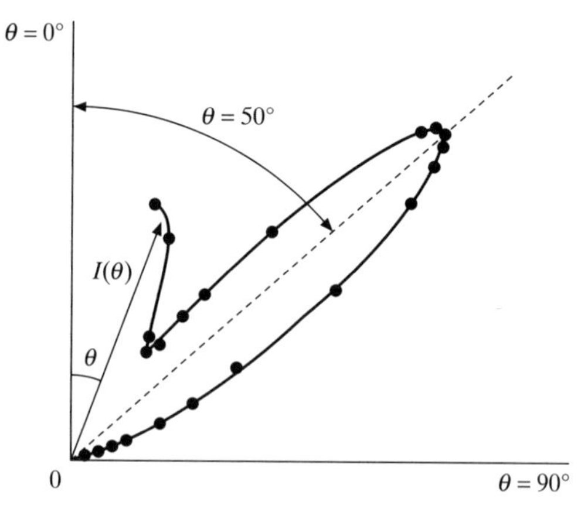

Figure 1.26 Polar plot of the scattered intensity as a function of the scattering angle θ for 54 eV electrons in the Davisson–Germer experiment. At each angle the intensity $I(\theta)$ is given by the distance of the point from the origin. A maximum is observed at $\theta = 50°$, which can only be explained by constructive interference of the waves scattered by the crystal lattice.

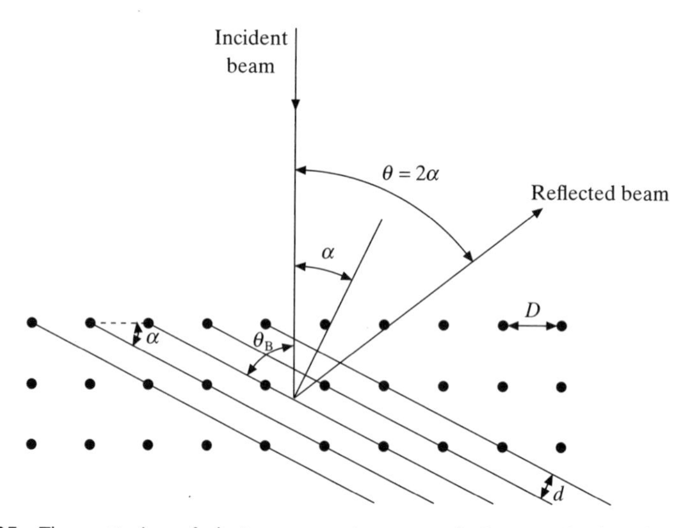

Figure 1.27 The scattering of electron waves by a crystal. Constructive interference occurs when the Bragg condition $n\lambda = 2d \sin \theta_B$ is satisfied.

theory, but the peak at 50° can only be explained by constructive interference of the waves scattered by the regular crystal lattice.

Referring to Fig. 1.27, the Bragg condition for constructive interference is

$$n\lambda = 2d \sin \theta_B \tag{1.131}$$

where d is the spacing of the Bragg planes and n is an integer. If D denotes the spacing of the atoms in the crystal, we see from Fig. 1.27 that $d = D \sin \alpha$, with

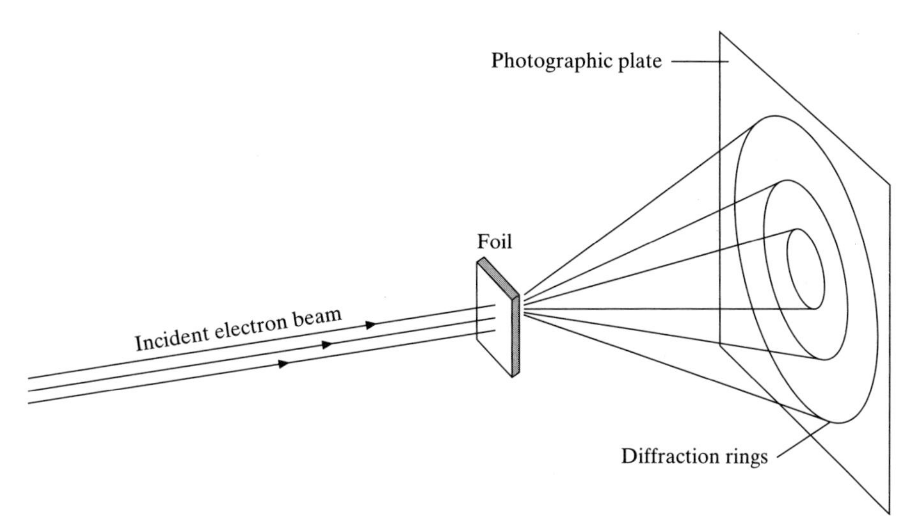

Figure 1.28 The experimental arrangement of G.P. Thomson for observing electron diffraction through a thin film of polycrystalline material.

$\alpha = \pi/2 - \theta_B$. Moreover, the scattering angle $\theta = 2\alpha$, so that the Bragg condition (1.131) becomes

$$n\lambda = D \sin \theta \tag{1.132}$$

Experiments in which X-rays were diffracted established that for a nickel crystal the atomic spacing is $D = 2.15$ Å. Assuming that the peak at $\theta = 50°$ corresponds to first-order diffraction ($n = 1$) we see from (1.132) that the experimental electron wavelength is given by $\lambda = (2.15 \sin 50°)$ Å $= 1.65$ Å. On the other hand, the de Broglie wavelength of a 54 eV electron is 1.67 Å, which agrees with the value of 1.65 Å, within the experimental error. By varying the voltage V_0, measurements were also performed at other incident electron energies, which confirmed the variation of λ with the magnitude of the momentum as predicted by the de Broglie relation (1.128). Higher-order maxima, corresponding to $n > 1$ in (1.132), were also observed and found to be in agreement with the theoretical predictions.

In the experiment of G.P. Thomson, the *transmission* of electrons through a thin foil of polycrystalline material was analysed. A beam of monoenergetic electrons was directed towards the foil, and after passing through it the scattered electrons struck a photographic plate (see Fig. 1.28). This method is analogous to the Debye–Scherrer method used in the study of X-ray diffraction. Because the foil consists of many small randomly oriented microcrystals, 'classical' electrons behaving only as particles would yield a blurred image. However, the result obtained by G.P. Thomson was similar to the X-ray Debye–Scherrer diffraction pattern, which consists of a series of concentric rings. In the same way, when an electron beam passes through a single crystal, Laue spot patterns are observed as in the case of the diffraction of X-rays.

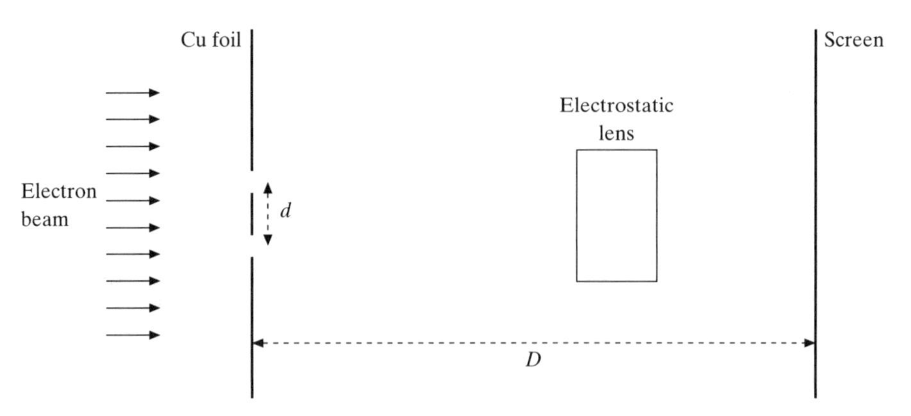

Figure 1.29 Schematic diagram illustrating the Jönsson double slit electron diffraction experiment.

The wave behaviour of electrons has also been demonstrated by observing the diffraction of electrons from edges and slits. In particular, electron diffraction experiments analogous to T. Young's famous double slit experiment (which in 1803 gave conclusive evidence of the wave properties of light) were performed by G. Möllenstedt and H. Dücker in 1956, and by C. Jönsson in 1961. The principle of the experiment carried out by Jönsson is illustrated in Fig. 1.29. He used 40 keV electrons, having a de Broglie wavelength $\lambda = 5 \times 10^{-2}$ Å. The slits, formed in copper foil, were very small (about 0.5×10^{-6} m wide), the slit separation being typically $d \simeq 2 \times 10^{-6}$ m. The interference fringes were observed on a screen at a distance $D = 0.35$ m from the slits. The spacing s of adjacent fringes is given by $s \simeq D\lambda/d \simeq 10^{-6}$ m. Because this spacing is very small, the fringes were magnified by placing electrostatic lenses between the slits and the screen. More recently, in 1989, A. Tonomura, J. Endo, T. Matsuda, T. Kawasaki and H. Ezawa performed two-slit experiments in which the accumulation of the interference pattern due to incoming single electrons was observed (see Fig. 1.30).

According to de Broglie, not only electrons but *all material particles* possess wave-like characteristics. This universality of matter waves has been confirmed by many experiments. In all cases the measured wavelength was found to agree with the de Broglie formula (1.128). For example, in 1929, O. Stern and co-workers demonstrated the diffraction of atoms from metallic and crystalline surfaces. In 1931, I. Esterman, R. Frisch and O. Stern observed the diffraction of hydrogen molecules by a crystal. The diffraction of neutrons was studied in 1938 by R. Frisch, H. von Halban and J. Koch, and an interferometer for neutrons based on Bragg diffraction from crystals was realised in 1974 by H. Rauch, W. Treimar and U. Bonse. In 1991, O. Carnal and J. Mlynek used a Young double slit arrangement to observe the interference pattern of helium atoms. Their atom interferometer, as well as other ones, will be discussed in Section 16.2, which is devoted to atom optics.

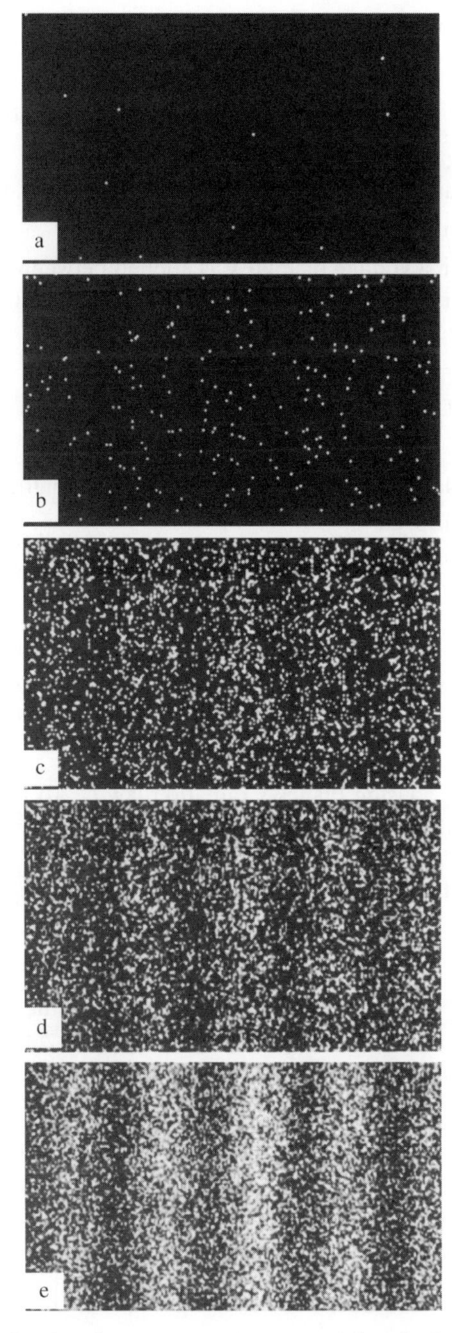

Figure 1.30 Buildup of an interference pattern by accumulating single electrons in the two-slit experiment of Tonomura *et al.*
(a) Number of electrons = 10; (b) number of electrons = 100;
(c) number of electrons = 3000; (d) number of electrons = 20 000; and
(e) number of electrons = 70 000.
(From A. Tonomura, J. Endo, T. Matsuda, T. Kawasaki and H. Ezawa, *American Journal of Physics* **57**, 117, 1989.)

Problems

1.1 Consider a Thomson apparatus with the following dimensions

Length of the condenser plates: $x_1 = 0.1$ m
Distance from the plates to the screen: $x_2 = 0.45$ m
Distance between the plates: $d = 0.03$ m

When a voltage of 1500 V is applied between the plates, a deflection $y_2 = 0.2$ m is observed. This deflection is reduced to zero if a magnetic field $\mathcal{B} = 1.1 \times 10^{-3}$ T is applied.

(a) Find the velocity v of the electrons.
(b) Determine the value of e/m given by this experiment.

1.2 Show that the number $N(\lambda)\,\mathrm{d}\lambda$ of standing electromagnetic waves (modes) in a large cube of volume V with wavelengths within the interval λ to $\lambda + \mathrm{d}\lambda$ is given by $N(\lambda)\,\mathrm{d}\lambda = 8\pi V\,\mathrm{d}\lambda/\lambda^4$.

1.3 Using Planck's radiation law (1.30) for $\rho(\lambda)$, prove that

$$\lambda_{\max} T = \frac{hc}{4.965 k_B}$$

where $\lambda_{\max}$ is the wavelength at which $\rho(\lambda)$ is a maximum. From this result and the values of h, c and k_B given in Appendix 14, obtain the constant b which occurs in Wien's displacement law (1.19).

1.4 Using Planck's radiation law (1.30) for $\rho(\lambda)$, prove that the total energy density ρ_{tot} is given by $\rho_{tot} = aT^4$, where $a = 8\pi^5 k_B^4/(15h^3c^3)$.

$$\left(\textbf{Hint: } \int_0^\infty \frac{x^3}{e^x - 1}\,\mathrm{d}x = \frac{\pi^4}{15}\right)$$

1.5 The photoelectric work function W for lithium is 2.3 eV.

(a) Find the threshold wavelength λ_t for the photoelectric effect.
(b) If ultra-violet light of wavelength $\lambda = 2000$ Å is incident on a lithium surface, obtain the maximum kinetic energy of the photoelectrons and the value of the stopping voltage V_0.

1.6 Consider black body radiation at absolute temperature T. Show that:

(a) The number of photons per unit volume is $N = 2.029 \times 10^7 T^3$ photons/m^3. (**Hint:** $\int_0^\infty x^2(e^x - 1)^{-1}\,\mathrm{d}x = 2.404\,11$)
(b) The average energy per photon is $\bar{E} = 3.73 \times 10^{-23} T$ joules $= 2.33 \times 10^{-4} T$ eV.

1.7 A photon of wavelength $\lambda_0 = 0.708$ Å is incident on an electron which is initially at rest.

(a) What is the wavelength shift $\Delta\lambda$ at the photon scattering angle $\theta = 30°$?
(b) What is the angle ϕ (measured from the incident photon direction) at which the electron recoils?
(c) What is the kinetic energy T_2 of the recoiling electron?

1.8 Consider the Compton scattering of a photon of wavelength λ_0 by a free electron moving with a momentum of magnitude P in the same direction as that of the incident photon.

(a) Show that in this case the Compton equation (1.54) becomes

$$\Delta\lambda = 2\lambda_0 \frac{(p_0 + P)c}{E - Pc} \sin^2\frac{\theta}{2}$$

where $p_0 = h/\lambda_0$ is the magnitude of the incident photon momentum, θ is the photon scattering angle and $E = (m^2c^4 + P^2c^2)^{1/2}$ is the initial electron energy.

(b) What is the maximum value of the final electron momentum? Compare with the case $P = 0$ discussed in the text.

(c) Show that if the free electron initially moves with a momentum of magnitude P in a direction opposite to that of the incident photon, the wavelength shift is given by

$$\Delta\lambda = 2\lambda_0 \frac{(p_0 - P)c}{E + Pc} \sin^2\frac{\theta}{2}$$

1.9 Consider a photon of energy $E_0 = 2$ eV which is scattered through an angle $\theta = \pi$ from an electron having a kinetic energy $T = 20$ GeV (1 GeV = 10^9 eV) and moving initially in a direction opposite to that of the photon. What is the energy E_1 of the scattered photon? (**Hint**: use the result (c) of Problem 1.8.)

1.10 Consider the scattering of an α particle of energy 10 MeV by an atomic electron, which we assume for simplicity to be initially at rest.

(a) What is the maximum momentum that can be transferred to the atomic electron?

(b) What is the maximum angle of deflection of the α particle?

1.11 Consider a Rutherford scattering experiment in which an α particle of laboratory energy 10 MeV scatters from a gold nucleus ($Z = 79$).

(a) Find the distance of closest approach for a head-on collision.

(b) Using the results of Appendices 1 and 2, find the centre of mass scattering angle θ of the α particle if its impact parameter is $b = 8 \times 10^{-14}$ m.

(The mass of the α particle is $M_\alpha \simeq 4$ a.m.u. and the mass of the gold nucleus is $M_N \simeq 197$ a.m.u.)

1.12 Alpha particles of laboratory energy E are scattered by a copper nucleus ($Z = 29$) of mass $M_N \simeq 63$ a.m.u. and radius $R \simeq 5 \times 10^{-15}$ m. Find the value of E for which departures from Rutherford scattering occur for head-on collisions ($\theta = 180°$).

1.13 Using the infinite nuclear mass approximation in the Bohr model, obtain the wavelengths of the first four lines of the Lyman, Balmer and Paschen series for H, He^+, Li^{2+} and C^{5+}.

1.14 Consider a one-electron atom (or ion), the nucleus of which contains A nucleons (Z protons and $N = A - Z$ neutrons). The mass of that nucleus is given approximately by $M \simeq AM_p$, where $M_p \simeq 1.67 \times 10^{-27}$ kg is the proton mass. Using this value of M, obtain the relative correction $\Delta E/E$ to the Bohr energy levels due to the finite nuclear mass for the case of hydrogen ($A = 1$, $Z = 1$), deuterium ($A = 2$, $Z = 1$), tritium ($A = 3$, $Z = 1$), $^4He^+$ ($A = 4$, $Z = 2$) and $^7Li^{2+}$ ($A = 7$, $Z = 3$).

1.15 Calculate the difference in wavelengths $\Delta\lambda$ between the Balmer H_α lines in atomic hydrogen and deuterium.

1.16 (a) Consider the absorption or emission of a photon of energy $h\nu$ by an atom initially at rest. After the transition, the atom recoils with a momentum $\mathbf{P}_R$ whose magnitude is equal to the magnitude $h\nu/c$ of the photon momentum, and with an energy $P_R^2/(2M)$, where M is the mass of the atom. If ν_0 denotes the frequency of the transition uncorrected for the recoil effect, show that the fractional change in the frequency due to recoil is

$$\frac{\nu - \nu_0}{\nu} = \pm\frac{h\nu}{2Mc^2}$$

where the plus sign corresponds to the absorption and the minus sign to the emission of a photon.

(b) Calculate the energy E_R with which an hydrogen atom initially at rest recoils when absorbing a photon in a transition from the ground state ($n = 1$) to the $n = 4$ level.

1.17 Use Moseley's law (1.105) with $\sigma = 1$ for K lines and $\sigma = 7.4$ for L lines to

(a) Calculate the two longest wavelengths in the K series of Cu ($Z = 29$).
(b) Calculate the three longest wavelengths in the L series of gold ($Z = 79$).
(c) Find the element whose K_α line has the wavelength 0.723 Å.

1.18 With reference to Fig. 1.22 showing a Stern–Gerlach apparatus, calculate the distance P_2P_3 from the following data:

Field gradient: $\dfrac{\partial \mathcal{B}_z}{\partial z} = 10^3$ T m^{-1}

Length of pole piece: $L = 0.1$ m
Distance to screen: $l = 1$ m
Atomic beam composed of silver atoms, for which $\mathcal{M}_z = \pm\mu_B$
Temperature of oven: 600 K

Assume that the velocity of the silver atoms is equal to the root mean square velocity of $(3k_BT/M)^{1/2}$, where k_B is Boltzmann's constant and M is the mass of a silver atom.

1.19 Calculate the de Broglie wavelength of:

(a) A mass of 1 kg moving at a velocity of 1 m s^{-1}.
(b) A free electron of energy $E = 200$ eV.
(c) A free α particle of energy $E = 5$ MeV.
(d) A free neutron of energy $E = 0.02$ eV.
(e) A free electron of kinetic energy $E = 1$ MeV. (Consider whether you need to use relativistic kinematics.)

2 The elements of quantum mechanics

In Chapter 1, we discussed some of the evidence for the atomic nature of matter. We also learned that the classical Newtonian form of mechanics could not describe phenomena on the atomic scale. In particular, we saw that experiments involving the diffraction of electrons or atoms by crystals demonstrate that particles exhibit wave properties. Equally, experimental evidence for the quantisation of the radiation field, from phenomena as different as black body radiation, the photoelectric effect and the Compton effect, points to a fundamental wave–particle duality which must be taken into account in satisfactory theories of both matter and radiation. For material particles, such a theory was developed in the years 1925 and 1926 by W. Heisenberg, M. Born and P. Jordan in a form known as *matrix mechanics*. An equivalent form of the theory, called *wave mechanics*, was proposed at the same time by E. Schrödinger, following the ideas put forward in 1924 by L. de Broglie. A more abstract form of quantum mechanics, which includes both matrix mechanics and wave mechanics, was published by P.A.M. Dirac in 1930.

In this chapter, we shall outline the main results and approximation methods of quantum mechanics, which will be used in the detailed discussions of atomic and molecular phenomena following in later chapters. In the space available, only those aspects of the subject which find immediate application will be discussed. For detailed accounts of quantum theory, we refer the reader to the texts of Dirac (1958), Landau and Lifshitz (1965), Messiah (1968), Schiff (1968), Merzbacher (1998) and Bransden and Joachain (2000).

2.1 Waves and particles, wave packets and the uncertainty principle

The experiments on the corpuscular nature of the electromagnetic radiation, which we discussed in the previous chapter, require that with the electromagnetic field we must associate a particle, the photon, whose energy E and magnitude of momentum p are related to the frequency ν and wavelength λ of the electromagnetic radiation by

$$E = h\nu, \qquad p = h/\lambda \tag{2.1}$$

On the other hand, the experiments on the wave-like properties of particles, also discussed in Chapter 1, imply that we associate with each particle a wave or matter field, the de Broglie relations which link the frequency ν and the wavelength λ of the wave with the particle energy E and magnitude of momentum p being also given by (2.1). As a consequence, we can assume that the relations (2.1) hold for all types of particles and field quanta. Introducing the angular frequency $\omega = 2\pi\nu$, the wave number $k = 2\pi/\lambda$ and the reduced Planck constant $\hbar = h/(2\pi)$, we may write the relations (2.1) in the more symmetric form

$$E = \hbar\omega, \qquad p = \hbar k \tag{2.2}$$

As a first step in formulating non-relativistic wave mechanics for material particles, let us consider a free particle of mass m, having a well-defined momentum $\mathbf{p} = p_x\hat{\mathbf{x}}$ directed along the positive X direction and a non-relativistic energy $E = p_x^2/(2m)$. Guided by (2.2), we associate with this particle a plane wave

$$\Psi(x, t) = A \exp[\mathrm{i}(kx - \omega t)] \tag{2.3}$$

where A is a constant. This plane wave travels in the positive X direction, has a wave number $k = 2\pi/\lambda = p_x/\hbar$ and an angular frequency $\omega = E/\hbar$ which we may also write as

$$\omega = \frac{\hbar k^2}{2m} \tag{2.4}$$

The function $\Psi(x, t)$ is known as a *wave function*, and we shall discuss its significance shortly. For the moment, we note that for a free particle, represented by the plane wave (2.3), we have

$$-\mathrm{i}\hbar \frac{\partial}{\partial x}\Psi = p_x\Psi \tag{2.5}$$

and

$$\mathrm{i}\hbar \frac{\partial}{\partial t}\Psi = E\Psi \tag{2.6}$$

This one-dimensional treatment is easily extended to three dimensions. To a free particle of mass m, having a well-defined momentum $\mathbf{p}$ and an energy $E = p^2/(2m)$, we now associate a plane wave

$$\Psi(\mathbf{r}, t) = A \exp[\mathrm{i}(\mathbf{k}\cdot\mathbf{r} - \omega t)] \tag{2.7}$$

where the *propagation vector* $\mathbf{k}$ is related to the momentum $\mathbf{p}$ by

$$\mathbf{p} = \hbar\mathbf{k} \tag{2.8}$$

with

$$k = |\mathbf{k}| = \frac{|\mathbf{p}|}{\hbar} = \frac{2\pi}{\lambda} \tag{2.9}$$

The vector **r** is the position vector of the particle and the angular frequency ω is still given by $\omega = E/\hbar = \hbar k^2/(2m)$. The equation (2.6) remains unchanged for the plane wave (2.7), while (2.5) is now replaced by its obvious generalisation

$$-i\hbar\nabla\Psi = \mathbf{p}\Psi \tag{2.10}$$

where ∇ is the gradient operator, having Cartesian components $(\partial/\partial x, \partial/\partial y, \partial/\partial z)$. The relations (2.6) and (2.10) show that for a free particle the energy and momentum can be represented by the differential operators

$$E_{op} = i\hbar\frac{\partial}{\partial t}, \qquad \mathbf{p}_{op} = -i\hbar\nabla \tag{2.11}$$

acting on the wave function Ψ. It is a *postulate* of wave mechanics that when the particle is not free the dynamical variables E and $\mathbf{p}$ are still represented by these differential operators.

The plane waves (2.3) or (2.7) represent particles having a definite momentum, but since their amplitude is constant, these plane waves correspond to a complete absence of localisation of the particle in space. To describe a particle which is confined in a certain spatial region, a *wave packet* can be formed by superposing plane waves of different wave numbers. For example, in order to describe a free particle which is confined in a region of the X axis, we shall superpose plane waves of the form (2.3) to obtain the wave packet

$$\Psi(x, t) = (2\pi\hbar)^{-1/2}\int_{-\infty}^{+\infty} \exp[i(p_x x - Et)/\hbar]\phi(p_x)\,dp_x \tag{2.12}$$

where the factor in front of the integral has been chosen for future convenience. Writing $\psi(x) \equiv \Psi(x, t = 0)$ we see that the functions

$$\psi(x) = (2\pi\hbar)^{-1/2}\int_{-\infty}^{+\infty} \exp(ip_x x/\hbar)\phi(p_x)\,dp_x \tag{2.13a}$$

and

$$\phi(p_x) = (2\pi\hbar)^{-1/2}\int_{-\infty}^{+\infty} \exp(-ip_x x/\hbar)\psi(x)\,dx \tag{2.13b}$$

are just *Fourier transforms* of each other. More generally, we write at time t

$$\Psi(x, t) = (2\pi\hbar)^{-1/2}\int_{-\infty}^{+\infty} \exp(ip_x x/\hbar)\Phi(p_x, t)\,dp_x \tag{2.14a}$$

and

$$\Phi(p_x, t) = (2\pi\hbar)^{-1/2}\int_{-\infty}^{+\infty} \exp(-ip_x x/\hbar)\Psi(x, t)\,dx \tag{2.14b}$$

so that the functions $\Psi(x, t)$ and $\Phi(p_x, t)$ are also mutual Fourier transforms. The functions $\Phi(p_x, t)$ and $\phi(p_x) \equiv \Phi(p_x, t = 0)$ are called *wave functions in momentum space*.

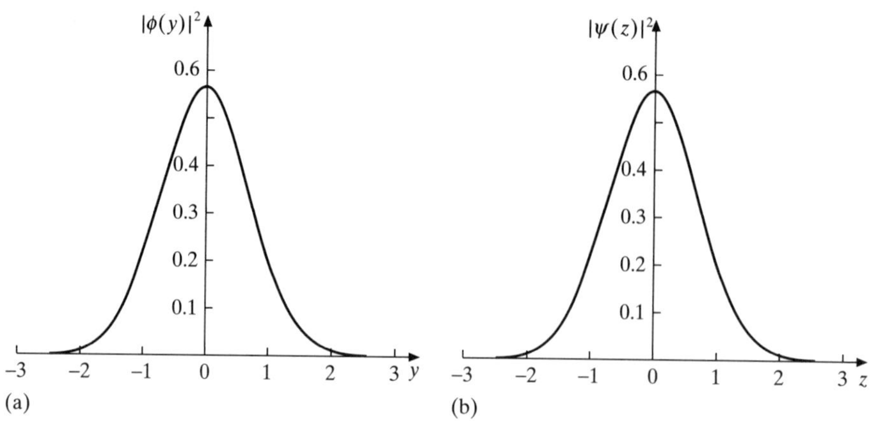

Figure 2.1 (a) The function $|\phi(y)|^2 = \pi^{-1/2}\exp(-y^2)$, where $y = (p_x - p_0)/\Delta p_x$. (b) The function $|\psi(z)|^2 = \pi^{-1/2}\exp(-z^2)$, where $z = (\Delta p_x/\hbar)x$. Note that the functions $\phi(y)$ and $\psi(z)$ have been normalised so that $\int_{-\infty}^{+\infty}|\phi(y)|^2\,dy = \int_{-\infty}^{+\infty}|\psi(z)|^2\,dz = 1$.

The Heisenberg uncertainty principle

Consider the case for which the wave function $\phi(p_x)$ is localised in a certain region of the p_x variable. As a simple example, we shall assume that $\phi(p_x)$ is the Gaussian function

$$\phi(p_x) = \exp\left[-\frac{(p_x - p_0)^2}{2(\Delta p_x)^2}\right] \tag{2.15}$$

where Δp_x, which we shall call the 'width' of the distribution in p_x, is a constant such that $|\phi(p_x)|^2$ drops to 1/e of its maximum value at $p_x = p_0 \pm \Delta p_x$ (see Fig. 2.1(a)). Using the known result

$$\int_{-\infty}^{+\infty}\exp(-\alpha u^2)\exp(-\beta u)\,du = \left(\frac{\pi}{\alpha}\right)^{1/2}\exp[\beta^2/(4\alpha)] \tag{2.16}$$

we find from (2.13) that

$$\psi(x) = \hbar^{-1/2}\,\Delta p_x\exp(ip_0x/\hbar)\exp[-(\Delta p_x)^2x^2/(2\hbar^2)]. \tag{2.17}$$

Apart from the phase factor $\exp(ip_0x/\hbar)$, this function is again a Gaussian. We remark that $|\psi(x)|^2$ has a maximum value at $x = 0$ and falls to 1/e of its maximum value at $x = \pm\Delta x$, where $\Delta x = \hbar/\Delta p_x$ is the 'width' of the distribution in the x variable (see Fig. 2.1(b)). Given the above definitions of the 'widths' Δx and Δp_x, we see that for a Gaussian wave packet $\Delta x\Delta p_x = \hbar$. Thus, if we decrease Δp_x so that the wave function in momentum space, $\phi(p_x)$, is more sharply peaked about $p_x = p_0$, then Δx will increase and $\psi(x)$ becomes increasingly 'delocalised'. Conversely, if Δp_x is increased, so that $\phi(p_x)$ is 'delocalised' in momentum space, then $\psi(x)$ will become more strongly localised about $x = 0$. Similar conclusions can be derived for functions $\phi(p_x)$ having different shapes (Problem 2.1). In fact, it is a general result of Fourier analysis that if $\psi(x)$ and $\phi(p_x)$ are mutual Fourier transforms as in (2.13)–(2.14), then

$$\Delta x \, \Delta p_x \gtrsim \hbar \tag{2.18}$$

This is the *Heisenberg uncertainty relation for position and momentum*, according to which the uncertainty Δx in measuring the x coordinate of a particle is related to the uncertainty Δp_x in measuring the x component of the momentum, the product of the uncertainties being larger than a quantity of order $\hbar$. A precise definition of the uncertainties Δx and Δp_x will be given below.

The foregoing discussion is readily generalised to more than one dimension. By superposing plane waves of the form (2.7) we obtain the wave packet

$$\begin{aligned}\Psi(\mathbf{r}, t) &= (2\pi\hbar)^{-3/2} \int \exp[\mathrm{i}(\mathbf{p}\cdot\mathbf{r} - Et)/\hbar]\phi(\mathbf{p}) \, \mathrm{d}\mathbf{p} \\ &= (2\pi\hbar)^{-3/2} \int \exp(\mathrm{i}\mathbf{p}\cdot\mathbf{r}/\hbar)\Phi(\mathbf{p}, t) \, \mathrm{d}\mathbf{p}\end{aligned} \tag{2.19}$$

where the wave functions in momentum space $\Phi(\mathbf{p}, t)$ and $\phi(\mathbf{p}) \equiv \Phi(\mathbf{p}, t = 0)$ are the Fourier transforms of $\Psi(\mathbf{r}, t)$ and $\psi(\mathbf{r}) \equiv \Psi(\mathbf{r}, t = 0)$, respectively. That is,

$$\Phi(\mathbf{p}, t) = (2\pi\hbar)^{-3/2} \int \exp(-\mathrm{i}\mathbf{p}\cdot\mathbf{r}/\hbar)\Psi(\mathbf{r}, t) \, \mathrm{d}\mathbf{r} \tag{2.20}$$

and

$$\phi(\mathbf{p}) = (2\pi\hbar)^{-3/2} \int \exp(-\mathrm{i}\mathbf{p}\cdot\mathbf{r}/\hbar)\psi(\mathbf{r}) \, \mathrm{d}\mathbf{r} \tag{2.21}$$

The three-dimensional form of the Heisenberg uncertainty relations for position and momentum is now

$$\Delta x \, \Delta p_x \gtrsim \hbar, \qquad \Delta y \, \Delta p_y \gtrsim \hbar, \qquad \Delta z \, \Delta p_z \gtrsim \hbar \tag{2.22}$$

The position–momentum uncertainty relations (2.22) have been derived by using the theory of Fourier analysis. This theory may also be invoked to obtain a *time–energy uncertainty relation*. Indeed, according to Fourier analysis, a wave packet of duration Δt must be composed of plane-wave components whose angular frequencies extend over a range $\Delta\omega$ such that $\Delta t \, \Delta\omega \gtrsim 1$. Since $E = \hbar\omega$, we therefore have

$$\Delta t \, \Delta E \gtrsim \hbar \tag{2.23}$$

which is the *Heisenberg uncertainty relation for time and energy*. It connects the uncertainty ΔE in the determination of the energy of a system with the time interval Δt available for this energy determination. Thus, if a system does not stay longer than a time Δt in a given state of motion, its energy in that state will be uncertain by an amount $\Delta E \gtrsim \hbar/\Delta t$.

The Heisenberg uncertainty relations (2.22) and (2.23), which are fundamental to any wave theory of matter, are particular cases of the *uncertainty principle* formulated by W. Heisenberg in 1927. This principle states that it is impossible to specify precisely and simultaneously the values of 'complementary' dynamical variables such as x and p_x, y and p_y, z and p_z, or t and E.

The uncertainty relations (2.22) and (2.23) may be used to estimate various basic quantities which occur in quantum physics. For example, it is possible by using the position–momentum uncertainty relations (2.22) to obtain an estimate of the size and of the energy of the hydrogen atom in its ground state (Problem 2.2). On the other hand, the time–energy uncertainty relation (2.23) is very useful in studying the lifetime of unstable systems.

Interpretation of the wave function

To make progress in the theory, we need to interpret the meaning of the wave function $\Psi(\mathbf{r}, t)$. In searching for this interpretation, we must remember that $\Psi(\mathbf{r}, t)$ is in general a complex function, and that $|\Psi(\mathbf{r}, t)|$ is large where the particle is likely to be found, and small everywhere else. We also recall that diffraction patterns, made by light, depend on the intensity of the radiation determined by the Poynting vector, which can be interpreted as measuring the flux of photons, and which depends on the square of the vector potential. In a similar way, M. Born in 1926 made the fundamental postulate that if a particle is described by a wave function $\Psi(\mathbf{r}, t)$, the probability of finding the particle within the volume element $\mathrm{d}\mathbf{r} = \mathrm{d}x\,\mathrm{d}y\,\mathrm{d}z$ about the point $\mathbf{r}$ at the time t is

$$P(\mathbf{r}, t)\,\mathrm{d}\mathbf{r} = |\Psi(\mathbf{r}, t)|^2\,\mathrm{d}\mathbf{r} \qquad \textbf{(2.24)}$$

so that

$$P(\mathbf{r}, t) = |\Psi(\mathbf{r}, t)|^2 = \Psi^*(\mathbf{r}, t)\Psi(\mathbf{r}, t) \qquad \textbf{(2.25)}$$

is the (position) *probability density*. Born's statistical interpretation of the wave function has to be justified by the success of the theory built on it. For $P(\mathbf{r}, t)$ to be unique everywhere, one must require that the wave function $\Psi(\mathbf{r}, t)$ should be continuous and single-valued. It is also worth noting that since $|\Psi(\mathbf{r}, t)|^2$ is the physically significant quantity, two wave functions which differ from each other by a constant multiplicative factor of modulus one (that is, a constant phase factor of the form $\exp(\mathrm{i}\alpha)$, where α is a real constant) are equivalent.

Since the probability of finding the particle somewhere must be unity, we deduce from (2.24) that the wave function $\Psi(\mathbf{r}, t)$ should be *normalised* so that

$$\int |\Psi(\mathbf{r}, t)|^2\,\mathrm{d}\mathbf{r} = 1 \qquad \textbf{(2.26)}$$

where the integral extends over all space. Wave functions satisfying this condition are said to be *square integrable*. However, not every wave function can be normalised like this, for example the plane wave (2.7). In this case $|\Psi(\mathbf{r}, t)|^2\,\mathrm{d}\mathbf{r}$ can be interpreted as the *relative* probability of finding the particle at time t in a volume element $d\mathbf{r}$ centred about $\mathbf{r}$, so that the ratio $|\Psi(\mathbf{r}_1, t)|^2/|\Psi(\mathbf{r}_2, t)|^2$ gives the probability of finding the particle within a volume element centred about $\mathbf{r} = \mathbf{r}_1$, compared with that of finding it within the same volume element at $\mathbf{r} = \mathbf{r}_2$.

There are alternative methods of normalising wave functions which are not square integrable. For example, we may assume that the system is confined to a large box of volume V, in which case the normalisation condition (2.26) reads

$$\int_V |\Psi(\mathbf{r}, t)|^2 \, d\mathbf{r} = 1 \tag{2.27}$$

where the integration now extends over the volume V. It can be shown that calculated physical quantities are independent of V, provided the box is sufficiently large. Plane waves such as (2.7) can also be 'normalised' with the help of the *Dirac delta function* $\delta(x)$, which is defined by the relation

$$f(0) = \int_{-\infty}^{+\infty} f(x)\, \delta(x)\, dx \tag{2.28}$$

where $f(x)$ is an arbitrary well-behaved function. There are several ways of representing the delta function, one of which is

$$\delta(x) = \lim_{\alpha \to \infty} \frac{\sin \alpha x}{\pi x} \tag{2.29}$$

When used in (2.28) the infinitely rapid oscillations of this function when $x \neq 0$ imply that the only contribution to the integral arises from the neighbourhood of $x = 0$, and

$$\int_{-\infty}^{+\infty} f(x)\, \delta(x)\, dx = f(0) \lim_{\alpha \to \infty} \int_{-\infty}^{+\infty} \frac{\sin \alpha x}{\pi x}\, dx = f(0) \tag{2.30}$$

Other useful representations of the Dirac delta function are

$$\begin{aligned} \delta(x) &= (2\pi)^{-1} \lim_{K \to \infty} \int_{-K}^{+K} \exp(ikx)\, dk \\ &= (2\pi)^{-1} \int_{-\infty}^{+\infty} \exp(ikx)\, dk \end{aligned} \tag{2.31a}$$

$$\begin{aligned} \delta(x) &= \lim_{\varepsilon \to 0^+} (2\pi)^{-1} \int_{-\infty}^{+\infty} \exp(ikx) \exp(-\varepsilon |k|)\, dk \\ &= \lim_{\varepsilon \to 0^+} \frac{\varepsilon}{\pi(x^2 + \varepsilon^2)} \end{aligned} \tag{2.31b}$$

$$\delta(x) = \lim_{\beta \to \infty} \frac{1 - \cos \beta x}{\pi \beta x^2} \tag{2.31c}$$

$$\delta(x) = \lim_{\varepsilon \to 0} \frac{\theta(x + \varepsilon) - \theta(x)}{\varepsilon} \tag{2.31d}$$

where in the last equation $\theta(x)$ denotes the step function such that $\theta(x) = 1$ for $x > 0$, and $\theta(x) = 0$ for $x < 0$.

We see that the Dirac delta function is an unusual function (it is in fact a 'distribution') which is such that, in effect,

$$\delta(x) = \begin{cases} 0 & \text{if } x \neq 0 \\ +\infty & \text{if } x = 0 \end{cases} \tag{2.32a}$$

and

$$\int_{-\infty}^{+\infty} \delta(x)\,\mathrm{d}x = 1. \tag{2.32b}$$

The main properties of the Dirac delta functions are the following:

$$\begin{aligned} \int_a^b f(x)\delta(x - x_0)\,\mathrm{d}x &= f(x_0) && \text{if } a < x_0 < b \\ &= 0 && \text{if } x_0 < a \text{ or } x_0 > b \end{aligned} \tag{2.33a}$$

$$\delta(x) = \delta(-x) \tag{2.33b}$$

$$x\delta(x) = 0 \tag{2.33c}$$

$$\delta(ax) = \frac{1}{|a|}\delta(x), \qquad a \neq 0 \tag{2.33d}$$

$$f(x)\delta(x - a) = f(a)\delta(x - a) \tag{2.33e}$$

$$\int \delta(a - x)\delta(x - b)\,\mathrm{d}x = \delta(a - b) \tag{2.33f}$$

$$\delta[g(x)] = \sum_i \frac{1}{|g'(x_i)|}\delta(x - x_i) \tag{2.33g}$$

where in the last relation x_i is a zero of $g(x)$ and $g'(x_i) \neq 0$. Particular cases of (2.33g) are

$$\delta[(x - a)(x - b)] = \frac{1}{|a - b|}[\delta(x - a) + \delta(x - b)], \quad a \neq b \tag{2.33h}$$

and

$$\delta(x^2 - a^2) = \frac{1}{2|a|}[\delta(x - a) + \delta(x + a)] \tag{2.33i}$$

Using the representation (2.31a) of the Dirac delta function, we find that the plane waves

$$\psi_{\mathbf{k}}(\mathbf{r}) = (2\pi)^{-3}\exp(\mathrm{i}\mathbf{k}\cdot\mathbf{r}) \tag{2.34}$$

satisfy the relation

$$\int \psi^*_{\mathbf{k}'}(\mathbf{r})\psi_{\mathbf{k}}(\mathbf{r})\, \mathrm{d}\mathbf{r} = (2\pi)^{-3} \int_{-\infty}^{+\infty} \exp[\mathrm{i}(k_x - k'_x)x]\, \mathrm{d}x$$

$$\times \int_{-\infty}^{+\infty} \exp[\mathrm{i}(k_y - k'_y)y]\, \mathrm{d}y \int_{-\infty}^{+\infty} \exp[\mathrm{i}(k_z - k'_z)z]\, \mathrm{d}z$$

$$= \delta(k_x - k'_x)\delta(k_y - k'_y)\delta(k_z - k'_z)$$

$$= \delta(\mathbf{k} - \mathbf{k}') \tag{2.35}$$

which is known as an *orthonormality relation*. In addition, the plane waves (2.34) satisfy the *closure relation*

$$\int \psi^*_{\mathbf{k}}(\mathbf{r}')\psi_{\mathbf{k}}(\mathbf{r})\, \mathrm{d}\mathbf{k} = \delta(\mathbf{r} - \mathbf{r}') \tag{2.36}$$

The normalisation condition expressed by (2.35) is often referred to as *k-normalisation*. Other normalisations for wave functions which are not square integrable may be used, for example the *p-normalisation* and the *energy normalisation*, which are the subject of Problem 2.3.

Expectation values

Consider a one-dimensional system in which the normalised wave function is $\Psi(x, t)$. Since $|\Psi(x, t)|^2\, \mathrm{d}x$ is the probability of finding the particle at the position x in the interval $\mathrm{d}x$, the *average* or *expectation value* of x may be defined as

$$\langle x \rangle = \int_{-\infty}^{+\infty} x\,|\Psi(x, t)|^2\, \mathrm{d}x \tag{2.37}$$

Likewise, the expectation value of any function $f(\mathbf{r})$ of the coordinates x, y, z, in a state represented by the normalised wave function $\Psi(\mathbf{r}, t)$ is

$$\langle f \rangle = \int f(\mathbf{r})\,|\Psi(\mathbf{r}, t)|^2\, \mathrm{d}\mathbf{r} \tag{2.38}$$

where the integral is taken over all space.

How should we calculate the average value of the momentum of the particle? The wavelength, which determines the magnitude of the momentum of a particle through (2.1), has only a precise value for a plane wave such as (2.7). For such a wave the momentum is obtained by operating on the wave function with the differential operator $-\mathrm{i}\hbar\nabla$, as in (2.10). When the wave function is not a plane wave, we shall still assume that the momentum is represented by this operator (see (2.11)) and the average value of the momentum in a state represented by a wave function $\Psi(\mathbf{r}, t)$ normalised to unity is

$$\langle \mathbf{p} \rangle = \int \Psi^*(\mathbf{r}, t)\mathbf{p}_{\mathrm{op}}\Psi(\mathbf{r}, t)\, \mathrm{d}\mathbf{r} = \int \Psi^*(\mathbf{r}, t)(-\mathrm{i}\hbar\nabla)\Psi(\mathbf{r}, t)\, \mathrm{d}\mathbf{r} \tag{2.39}$$

In general, if an operator A depends on $\mathbf{r}$, $\mathbf{p}$ and t, its expectation value is defined as

$$\langle A\rangle = \int \Psi^*(\mathbf{r}, t)A(\mathbf{r}, -\mathrm{i}\hbar\nabla, t)\Psi(\mathbf{r}, t)\,\mathrm{d}\mathbf{r} \tag{2.40}$$

The uncertainties Δx, Δp_x and so on which appeared in the uncertainty relations (2.22) can now be defined as standard deviations by the formulae

$$\Delta x = \langle(x - \langle x\rangle)^2\rangle^{1/2}, \qquad \Delta p_x = \langle(p_x - \langle p_x\rangle)^2\rangle^{1/2} \tag{2.41}$$

2.2 The Schrödinger equation

In order to compute the average values of functions of the position and momentum of a particle, which are the quantities determined by a series of measurements, some way is needed of calculating wave functions. We first notice that the wave packet (2.19) for a free particle satisfies the differential equation

$$\mathrm{i}\hbar\frac{\partial}{\partial t}\Psi = -\frac{\hbar^2}{2m}\nabla^2\Psi \tag{2.42}$$

because $E(p) = p^2/(2m)$, and this is called the Schrödinger equation for a free particle. What is the Schrödinger equation for a particle moving in a potential $V(\mathbf{r}, t)$? Using (2.11), the Schrödinger equation for a free particle can be written

$$E_{\mathrm{op}}\Psi = \frac{1}{2m}(\mathbf{p}_{\mathrm{op}})^2\,\Psi \tag{2.43}$$

The operator $(\mathbf{p}_{\mathrm{op}})^2/(2m)$ represents the kinetic energy T of the particle. The total energy or Hamiltonian operator of a particle is just

$$H = T + V = -\frac{\hbar^2}{2m}\nabla^2 + V \tag{2.44}$$

and the generalisation of the Schrödinger equation for a particle in a potential is

$$E_{\mathrm{op}}\Psi = (T + V)\Psi \tag{2.45}$$

or

$$\mathrm{i}\hbar\frac{\partial}{\partial t}\Psi = H\Psi \tag{2.46}$$

which explicitly reads

$$\mathrm{i}\hbar\frac{\partial}{\partial t}\Psi(\mathbf{r}, t) = \left[-\frac{\hbar^2}{2m}\nabla^2 + V(\mathbf{r}, t)\right]\Psi(\mathbf{r}, t) \tag{2.47}$$

This is the *time-dependent Schrödinger equation* for a particle moving in a potential. We remark that this equation is *linear* in $\Psi(\mathbf{r}, \mathrm{t})$. As a result, it satisfies the *superposition principle*, according to which a linear superposition of possible wave functions is also a possible wave function. The fact that the superposition principle applies is directly related to the wave nature of matter, and in particular to the existence of interference effects for de Broglie waves. We also note that the Schrödinger equation (2.47) is of first order in the time derivative $\partial/\partial t$, so that once the initial value of the wave function Ψ is given at some time t_0, namely $\Psi(\mathbf{r}, t_0)$, its values at all other times can be found.

The time-dependent Schrödinger equation (2.47) will now be used to obtain two important results concerning probability conservation and the time variation of expectation values.

Probability conservation

The probability of finding a particle within a fixed volume V is seen from (2.24) to be

$$\int_V P(\mathbf{r}, t)\,\mathrm{d}\mathbf{r} = \int_V |\Psi(\mathbf{r}, t)|^2\,\mathrm{d}\mathbf{r} \tag{2.48}$$

The rate of change of this probability can be obtained with the help of (2.47):

$$\begin{aligned}\frac{\partial}{\partial t}\int_V P(\mathbf{r}, t)\,\mathrm{d}\mathbf{r} &= \int_V \left(\Psi^*\frac{\partial \Psi}{\partial t} + \frac{\partial \Psi^*}{\partial t}\Psi\right)\mathrm{d}\mathbf{r} \\ &= \frac{\mathrm{i}\hbar}{2m}\int_V [\Psi^*\nabla^2\Psi - (\nabla^2\Psi^*)\Psi]\,\mathrm{d}\mathbf{r} \\ &= -\int_V \nabla\cdot\mathbf{j}\,\mathrm{d}\mathbf{r}\end{aligned} \tag{2.49}$$

where we have introduced the vector

$$\mathbf{j}(\mathbf{r}, t) = \frac{\hbar}{2m\mathrm{i}}[\Psi^*\nabla\Psi - (\nabla\Psi^*)\Psi] \tag{2.50}$$

Using Green's theorem, we find that

$$\frac{\partial}{\partial t}\int_V P(\mathbf{r}, t)\,\mathrm{d}\mathbf{r} = -\int_S \mathbf{j}\cdot\mathrm{d}\mathbf{S} \tag{2.51}$$

where the integral on the right-hand side is over the surface S bounding the volume V. Since the rate of change of the probability of finding the particle in the region V is equal to the probability flux passing through the surface S, the vector $\mathbf{j}$ can be interpreted as a *probability current density*. The equation

$$\nabla\cdot\mathbf{j} + \frac{\partial P}{\partial t} = 0 \tag{2.52}$$

which follows from (2.49) is analogous to the continuity equation of electromagnetism, expressing charge conservation.

Time variation of expectation values

The rate of change of the expectation value (2.40) of an operator A is given by

$$\begin{aligned}\frac{\mathrm{d}}{\mathrm{d}t}\langle A\rangle &= \frac{\mathrm{d}}{\mathrm{d}t}\int (\Psi^* A\Psi)\,\mathrm{d}\mathbf{r}\\ &= \int\left(\frac{\partial\Psi^*}{\partial t}A\Psi + \Psi^*\frac{\partial A}{\partial t}\Psi + \Psi^* A\frac{\partial\Psi}{\partial t}\right)\mathrm{d}\mathbf{r}\\ &= \left\langle\frac{\partial A}{\partial t}\right\rangle + \frac{1}{\mathrm{i}\hbar}\int \Psi^*(AH - HA)\Psi\,\mathrm{d}\mathbf{r} \qquad \textbf{(2.53)}\end{aligned}$$

where we have used the Schrödinger equation (2.46) and the fact that H is Hermitian (see below). The *commutator* of two operators A and B is defined as

$$[A, B] = AB - BA \qquad \textbf{(2.54)}$$

and two operators are said to commute if their commutator vanishes. Using this definition (2.53) can be written as

$$\frac{\mathrm{d}}{\mathrm{d}t}\langle A\rangle = \left\langle\frac{\partial A}{\partial t}\right\rangle + \frac{1}{\mathrm{i}\hbar}\langle[A, H]\rangle \qquad \textbf{(2.55)}$$

In particular, if the operator A does not depend explicitly on time (that is, $\partial A/\partial t = 0$), we have

$$\mathrm{i}\hbar\frac{\mathrm{d}}{\mathrm{d}t}\langle A\rangle = \langle[A, H]\rangle \qquad \textbf{(2.56)}$$

and if A commutes with H, A is a *constant of the motion* ($\mathrm{d}\langle A\rangle/\mathrm{d}t = 0$).

Using the result (2.56) one can prove (Problem 2.6) *Ehrenfest's theorem*, according to which the expectation values of the variables $\mathbf{r}$ and $\mathbf{p}$ obey equations corresponding to Newton's equations of motion. That is,

$$m\frac{\mathrm{d}\langle\mathbf{r}\rangle}{\mathrm{d}t} = \langle\mathbf{p}\rangle, \qquad \frac{\mathrm{d}\langle\mathbf{p}\rangle}{\mathrm{d}t} = -\langle\boldsymbol{\nabla}V\rangle \qquad \textbf{(2.57)}$$

Time-independent Schrödinger equation and energy eigenfunctions

When the potential does not depend on the time, the Schrödinger equation (2.47) admits *stationary state solutions* of the form

$$\Psi(\mathbf{r}, t) = \psi_E(\mathbf{r})\exp(-\mathrm{i}Et/\hbar) \qquad \textbf{(2.58)}$$

where E is a constant and where $\psi_E(\mathbf{r})$ satisfies the *time-independent Schrödinger equation*

$$\left[-\frac{\hbar^2}{2m}\nabla^2 + V(\mathbf{r})\right]\psi_E(\mathbf{r}) = E\psi_E(\mathbf{r}) \tag{2.59}$$

or

$$H\psi_E = E\psi_E \tag{2.60}$$

The significance of the constant E is seen by recognising that E is the expectation value of the total energy $H = T + V$. Indeed, using (2.58) and (2.60), we have

$$E = \langle H \rangle = \int \Psi^*(\mathbf{r}, t) H \Psi(\mathbf{r}, t)\, d\mathbf{r} \tag{2.61}$$

In writing (2.61) we have assumed that the wave function Ψ given by (2.58) is normalised to unity, so that

$$\int \psi_E^*(\mathbf{r})\psi_E(\mathbf{r})\, d\mathbf{r} = 1 \tag{2.62}$$

where the integral can be taken over all space for square integrable functions, or over a large volume V in the other cases. We shall find that only certain values of E are compatible with normalisable solutions. These values are called the *energy eigenvalues* and the corresponding solutions ψ_E are the *eigenfunctions* of the total energy or Hamiltonian operator H. Equations of the type

$$A\psi_n = a_n\psi_n \tag{2.63}$$

– where A is an operator and a_n is a number – are called *eigenvalue equations*, a_n being called the *eigenvalue* and ψ_n the *eigenfunction* of the operator A. If more than one eigenfunction corresponds to a given eigenvalue, this eigenvalue is said to be *degenerate*.

It is a general postulate of quantum mechanics that each dynamical variable (such as the position, momentum, energy, . . .) can be represented by a linear operator [1] and that the result of a precise measurement of the variable can only be one of the eigenvalues of this operator. For example, the only possible result of a measurement of the energy of the system is one of the eigenvalues E of the Hamiltonian operator H.

The spectrum of the Hamiltonian

The set of all eigenvalues of the Hamiltonian is called the *energy spectrum*. It may consist of *discrete* values or a *continuous* range, or both. In general the discrete eigenvalues are associated with *bound states* (analogous to the closed orbits of classical mechanics) and the continuum with *scattering states* (corresponding to open classical orbits).

---

[1] For a definition of linear operators, see Byron and Fuller (1969) or Mathews and Walker (1973).

It is easy to show by integrating by parts that for any normalised eigenfunction ψ_E of H one has

$$E^* = \langle H \rangle^* = \left[\int \psi_E^* \left[-\frac{\hbar^2}{2m} \nabla^2 + V \right] \psi_E \, d\mathbf{r} \right]^*$$

$$= \int \psi_E \left[-\frac{\hbar^2}{2m} \nabla^2 + V \right] \psi_E^* \, d\mathbf{r} = \int \psi_E^* \left[-\frac{\hbar^2}{2m} \nabla^2 + V \right] \psi_E \, d\mathbf{r}$$

$$= \langle H \rangle = E \quad \textbf{(2.64)}$$

so that the eigenvalues E are *real*.

In general, if A is a linear operator and Φ and Ψ are any two wave functions, the adjoint operator $A^\dagger$ is defined by the relation

$$\int \Phi^* A^\dagger \Psi \, d\mathbf{r} = \int (A\Phi)^* \Psi \, d\mathbf{r} \quad \textbf{(2.65)}$$

If $A^\dagger = A$, then A is said to be *self-adjoint* or *Hermitian*, in which case $\langle A \rangle = \langle A \rangle^*$ and the eigenvalues of A are real. Operators representing physical quantities (such as H, $\mathbf{r}$ or $\mathbf{p}$) are called *observables*, and it is a postulate of quantum mechanics that observables must be Hermitian.

We shall now prove that if ψ_E and $\psi_{E'}$ are two energy eigenfunctions corresponding to unequal eigenvalues E and E', then

$$\int \psi_E^*(\mathbf{r}) \psi_{E'}(\mathbf{r}) \, d\mathbf{r} = 0 \quad \textbf{(2.66)}$$

in which case ψ_E and $\psi_{E'}$ are said to be *orthogonal*. We first note that

$$\psi_{E'}^*(H - E)\psi_E = 0 \quad \textbf{(2.67a)}$$

and

$$\psi_E[(H - E')\psi_{E'}]^* = 0 \quad \textbf{(2.67b)}$$

Subtracting these two equations and integrating, we have

$$(E - E') \int \psi_{E'}^* \psi_E \, d\mathbf{r} = \int [\psi_{E'}^* H \psi_E - (H\psi_{E'})^* \psi_E] \, d\mathbf{r} = 0 \quad \textbf{(2.68)}$$

where in the last step we have used the fact that H is Hermitian. Since $E \neq E'$, we see that (2.66) follows from (2.68).

When several eigenfunctions $\psi_{E,r}$, $r = 1, 2, \ldots, \alpha$, correspond to a given eigenvalue E (that is, if the eigenvalue E is degenerate) linear combinations of the eigenfunctions $\psi_{E,r}$ can always be constructed so that the resulting α eigenfunctions belonging to the eigenvalue E are mutually orthogonal [see for example Bransden and Joachain (2000)]. Each of these eigenfunctions is evidently orthogonal to every eigenfunction belonging to a different eigenvalue $E' \neq E$. Thus all the energy eigenfunctions can be made orthogonal to each other.

The eigenfunctions ψ_n of an operator A are said to constitute an *orthonormal set* if they are both normalised and mutually orthogonal. That is,

$$\int \psi_{n'}^*(\mathbf{r})\psi_n(\mathbf{r})\, \mathrm{d}\mathbf{r} = \delta_{nn'} \tag{2.69}$$

where $\delta_{nn'}$ is the Kronecker symbol

$$\delta_{nn'} = \begin{cases} 1, & n = n' \\ 0, & n \neq n' \end{cases} \tag{2.70}$$

The energy eigenfunctions ψ_E constitute such a set, and in the same way so do the eigenfunctions of an Hermitian operator representing any observable [2].

2.3 Expansions, operators and observables

We shall assume that all the orthonormal eigenfunctions $\psi_n(\mathbf{r})$ of a given Hermitian operator A form a *complete set*, in the mathematical sense that an arbitrary normalised wave function $\Psi(\mathbf{r}, t)$ can be expanded in terms of them:

$$\Psi(\mathbf{r}, t) = \sum_n c_n(t)\psi_n(\mathbf{r}) \tag{2.71}$$

Suppose now that at time t a measurement is made of the observable represented by A. Then, as we have seen above, the value obtained must be one of the eigenvalues a_n. A further postulate, due to Born, is that the probability $P_n(t)$ of obtaining the value a_n is given by

$$P_n(t) = |c_n(t)|^2 \tag{2.72}$$

As a consequence, we are certain to obtain a particular eigenvalue a_n only when the wave function that describes the particle at time t is the corresponding eigenfunction ψ_n (apart from a constant phase factor). Immediately after a measurement in which the result a_n is obtained, the system will be found in the state represented by the wave function ψ_n.

We can calculate the *probability amplitudes* c_n from (2.71) using the orthonormal property (2.69) of the expansion functions ψ_n. Multiplying (2.71) by $\psi_{n'}^*$ and integrating, we obtain

$$\begin{aligned} \int \psi_{n'}^*(\mathbf{r})\Psi(\mathbf{r}, t)\, \mathrm{d}\mathbf{r} &= \sum_n c_n(t) \int \psi_{n'}^*(\mathbf{r})\psi_n(\mathbf{r})\, \mathrm{d}\mathbf{r} \\ &= \sum_n c_n(t)\, \delta_{nn'} = c_{n'}(t) \end{aligned} \tag{2.73}$$

[2] If n is a continuous index, then a 'delta function' normalisation analogous to (2.35) can be used. Alternatively, a 'box' normalisation of the type (2.27) can be introduced, in which case n remains discrete and the relations (2.69)–(2.70) apply without modification.

It is convenient at this point to introduce the Dirac notation

$$\langle f|g\rangle = \int f^*(\mathbf{r})g(\mathbf{r})\,\mathrm{d}\mathbf{r} \tag{2.74}$$

so that the result (2.73) reads $c_{n'} = \langle \psi_{n'}|\Psi\rangle$. More generally, we define the *matrix element* of an operator A as

$$\langle f|A|g\rangle = \int f^*(\mathbf{r})Ag(\mathbf{r})\,\mathrm{d}\mathbf{r} \tag{2.75}$$

so that the expectation, or average, value of A can be written as

$$\begin{aligned}\langle A\rangle &= \langle \Psi|A|\Psi\rangle \\ &= \sum_n \sum_{n'} c_{n'}^* c_n \langle \psi_{n'}|A|\psi_n\rangle \\ &= \sum_n \sum_{n'} c_{n'}^* c_n a_n \langle \psi_{n'}|\psi_n\rangle \\ &= \sum_n |c_n|^2 a_n \end{aligned} \tag{2.76}$$

where we have used (2.63) and the orthonormality relation (2.69), which reads $\langle \psi_{n'}|\psi_n\rangle = \delta_{nn'}$ in Dirac's notation.

Another important relation can be proved from (2.71) and (2.73). Indeed, since

$$\begin{aligned}\Psi(\mathbf{r}, t) &= \sum_n c_n(t)\psi_n(\mathbf{r}) \\ &= \sum_n \left[\int \psi_n^*(\mathbf{r}')\Psi(\mathbf{r}', t)\,\mathrm{d}\mathbf{r}'\right]\psi_n(\mathbf{r})\end{aligned} \tag{2.77}$$

we have

$$\sum_n \psi_n^*(\mathbf{r}')\psi_n(\mathbf{r}) = \delta(\mathbf{r} - \mathbf{r}') \tag{2.78}$$

where $\delta(\mathbf{r} - \mathbf{r}')$ is the Dirac delta function discussed in Section 2.1. The result (2.78) (see also (2.36) for the special case of the plane waves) is known as the *closure* property of the orthonormal set of functions ψ_n.

General solution of the Schrödinger equation for a time-independent potential

Let us assume that the potential is independent of time, so that the time-dependent Schrödinger equation (2.47) admits stationary state solutions of the form (2.58). Expanding the general solution of (2.47) in terms of energy eigenfunctions, we then write (see (2.71) and (2.73))

$$\Psi(\mathbf{r}, t) = \sum_E c_E(t)\psi_E(\mathbf{r}) \tag{2.79}$$

where

$$c_E(t) = \int \psi_E^*(\mathbf{r})\Psi(\mathbf{r}, t)\, \mathrm{d}\mathbf{r} \tag{2.80}$$

Substituting (2.79) into the Schrödinger equation (2.47), and using the orthonormality of the eigenfunctions ψ_E, together with the fact that they satisfy the time-independent Schrödinger equation (2.59), we find that the coefficients $c_E(t)$ satisfy the equation

$$\mathrm{i}\hbar \frac{\mathrm{d}}{\mathrm{d}t} c_E(t) = E c_E(t) \tag{2.81}$$

which is readily integrated to yield

$$c_E(t) = c_E(t_0) \exp\left[-\frac{\mathrm{i}}{\hbar} E(t - t_0)\right] \tag{2.82}$$

We note that the probability P_E of obtaining the energy value E is constant, since from (2.72) and (2.82) we have

$$P_E = |c_E(t)|^2 = |c_E(t_0)|^2 \tag{2.83}$$

Using (2.79) and (2.82), we find that

$$\Psi(\mathbf{r}, t) = \sum_E c_E(t_0) \exp\left[-\frac{\mathrm{i}}{\hbar} E(t - t_0)\right] \psi_E(\mathbf{r}) \tag{2.84}$$

or

$$\Psi(\mathbf{r}, t) = \sum_E \tilde{c}_E(t_0) \psi_E(\mathbf{r}) \exp\left(-\frac{\mathrm{i}}{\hbar} Et\right); \qquad \tilde{c}_E(t_0) = c_E(t_0) \exp\left(\frac{\mathrm{i}}{\hbar} Et_0\right) \tag{2.85}$$

so that the general solution $\Psi(\mathbf{r}, t)$ is a linear superposition of stationary state solutions (2.58), as we expect from the linearity of the Schrödinger equation (2.47).

These results may also be used to write down an expression for $\Psi(\mathbf{r}, t)$ at any time, once it is known at the time $t = t_0$. Indeed, from (2.80) and (2.84), we have

$$\begin{aligned}\Psi(\mathbf{r}, t) &= \sum_E \left[\int \psi_E^*(\mathbf{r}')\Psi(\mathbf{r}', t_0)\, \mathrm{d}\mathbf{r}'\right] \exp\left[-\frac{\mathrm{i}}{\hbar} E(t - t_0)\right] \psi_E(\mathbf{r}) \\ &= \int K(\mathbf{r}, t; \mathbf{r}', t_0)\Psi(\mathbf{r}', t_0)\, \mathrm{d}\mathbf{r}' \end{aligned} \tag{2.86}$$

where

$$K(\mathbf{r}, t; \mathbf{r}', t_0) = \sum_E \psi_E^*(\mathbf{r}')\psi_E(\mathbf{r}) \exp\left[-\frac{\mathrm{i}}{\hbar} E(t - t_0)\right] \tag{2.87}$$

We see from (2.86) that K may be interpreted as the probability amplitude that a particle originally at $\mathbf{r}'$ will propagate to the point $\mathbf{r}$ in the time interval $t - t_0$. The function K is therefore called a *propagator*.

Matrix representations of wave functions and operators

Let us consider a complete orthonormal set of functions $\{\psi_j\}$. Any physically admissible wave function Ψ can be expanded in terms of them as (see (2.71) and (2.73))

$$\Psi = \sum_j c_j \psi_j, \qquad c_j = \langle \psi_j | \Psi \rangle \tag{2.88}$$

The set $\{\psi_j\}$ is called a *basis* and the coefficients c_j are the components of Ψ in that basis. Once a given basis – also called a *representation* – has been chosen, Ψ is completely specified by its components c_j. Thus the set $\{\psi_j\}$ may be thought of as a basis in a vector space [3], and Ψ as being represented by a vector in that vector space, whose components are the coefficients c_j.

The action of a linear operator A on Ψ may be specified in terms of its effect on the components c_j in a given representation. Indeed, starting from (2.88) we have

$$A\Psi = A \sum_j c_j \psi_j = \sum_j c_j A \psi_j \tag{2.89}$$

Let us express the new vectors $\psi_j' = A\psi_j$ in our basis as

$$\psi_j' = A\psi_j = \sum_i A_{ij} \psi_i \tag{2.90}$$

where

$$A_{ij} = \langle \psi_i | A | \psi_j \rangle \tag{2.91}$$

Then, if we write

$$\Psi' = A\Psi = \sum_i c_i' \psi_i \tag{2.92}$$

so that the coefficients c_i' are the components of Ψ' in our basis, we have from (2.89)–(2.92)

$$\Psi' = A\Psi = \sum_i \sum_j A_{ij} c_j \psi_i = \sum_i c_i' \psi_i \tag{2.93}$$

so that

$$c_i' = \sum_j A_{ij} c_j \tag{2.94}$$

This is a *matrix equation* which relates the components c_j of Ψ and c_i' of $\Psi' = A\Psi$ by means of the *matrix elements* A_{ij} of the linear operator A.

Thus far we have adopted a given representation (that is, a given complete orthonormal set of basis vectors $\{\psi_j\}$) in which Ψ is represented by a vector having

[3] Vector spaces are discussed by Byron and Fuller (1969) and Mathews and Walker (1973).

components c_j and the operator A by a matrix whose matrix elements are given by (2.91). This representation, however, is not unique. Indeed, we may develop Ψ in terms of another complete orthonormal set of basis vectors $\{\varphi_k\}$ as

$$\Psi = \sum_k d_k \varphi_k, \qquad d_k = \langle \varphi_k | \Psi \rangle \tag{2.95}$$

and the coefficients d_k are the components of a vector representing Ψ in the 'new' basis $\{\varphi_k\}$. In particular, any 'old' basis vector may be expressed in the 'new' basis as

$$\psi_j = \sum_k U_{kj} \varphi_k \tag{2.96}$$

where

$$U_{kj} = \langle \varphi_k | \psi_j \rangle \tag{2.97}$$

In the same way, the reverse expansion is

$$\varphi_k = \sum_j \langle \psi_j | \varphi_k \rangle \psi_j = \sum_j U^*_{kj} \psi_j \tag{2.98}$$

Since

$$\Psi = \sum_j c_j \psi_j = \sum_j \sum_k c_j U_{kj} \varphi_k = \sum_k d_k \varphi_k \tag{2.99}$$

we also see that

$$d_k = \sum_j U_{kj} c_j \tag{2.100}$$

The set of numbers U_{kj} can be regarded as the elements of a matrix U and the equation (2.100) written in matrix form as

$$d = Uc \tag{2.101}$$

The matrix U is readily shown to be *unitary*, namely such that

$$UU^\dagger = U^\dagger U = I \tag{2.102}$$

where I is the unit matrix. Indeed, we have

$$\begin{aligned}
(UU^\dagger)_{kn} &= \sum_j U_{kj} U^\dagger_{jn} = \sum_j U_{kj} U^*_{nj} \\
&= \sum_j \langle \varphi_k | \psi_j \rangle \langle \varphi_n | \psi_j \rangle^* \\
&= \int \mathrm{d}\mathbf{r}\, \varphi^*_k(\mathbf{r}) \int \mathrm{d}\mathbf{r}'\, \varphi_n(\mathbf{r}') \sum_j \psi^*_j(\mathbf{r}') \psi_j(\mathbf{r}) \\
&= \int \mathrm{d}\mathbf{r}\, \varphi^*_k(\mathbf{r}) \int \mathrm{d}\mathbf{r}'\, \varphi_n(\mathbf{r}') \delta(\mathbf{r} - \mathbf{r}') \\
&= \int \mathrm{d}\mathbf{r}\, \varphi^*_k(\mathbf{r}) \varphi_n(\mathbf{r}) = \delta_{kn}
\end{aligned} \tag{2.103}$$

where we have used the closure relation for the functions ψ_j. Similarly, one has

$$(U^\dagger U)_{kn} = \delta_{kn} \tag{2.104}$$

so that the unitarity property (2.102) follows from (2.103) and (2.104). Therefore the passage from one representation of quantum mechanics to another is effected by *unitary transformations.*

Instead of changing the representation by means of a unitary transformation, we may also apply unitary transformations directly to the wave functions and operators. For example, let Ψ and Φ be two wave functions and A an operator such that

$$A\Psi = \Phi \tag{2.105}$$

Let us now apply a unitary transformation such that

$$\Psi' = U\Psi, \qquad \Phi' = U\Phi \tag{2.106}$$

where U is a unitary operator, namely a linear operator satisfying the equations (2.102), where I is the unit operator. Writing

$$A'\Psi' = \Phi' \tag{2.107}$$

we have

$$A'U\Psi = U\Phi = UA\Psi \tag{2.108}$$

Hence, using (2.102), we find that

$$A = U^\dagger A'U, \qquad A' = UAU^\dagger \tag{2.109}$$

These equations imply (Problem 2.9) that if A and A' are two operators connected by a unitarity transformation:

1. If A is Hermitian, then A' is also Hermitian.
2. The eigenvalues of A' are the same as those of A.
3. One has

$$\langle \Phi' | A' | \Psi' \rangle = \langle \Phi | A | \Psi \rangle \tag{2.110}$$

where Ψ' and Φ' are defined by (2.106). In particular, the expectation value $\langle \Psi | A | \Psi \rangle$ remains unchanged. By choosing $A = I$ (the unit operator) we also see that the scalar (inner) product $\langle \Phi | \Psi \rangle$ is invariant under a unitary transformation. As a result, the normalisation is also preserved, since $\langle \Psi' | \Psi' \rangle = \langle \Psi | \Psi \rangle$.

Let us assume that the orthonormal functions $\{\psi_j\}$ which we use as a basis set for a matrix representation are the eigenfunctions of some quantum mechanical operator A. Then, using (2.63), we find that

$$A_{ij} = \langle \psi_i | A | \psi_j \rangle = \langle \psi_i | a_j | \psi_j \rangle = a_j \delta_{ij} \tag{2.111}$$

so that the matrix which represents the operator A in this basis is *diagonal*. Solving the eigenvalue equation (2.63) for an observable A is therefore equivalent to finding a unitary transformation which diagonalises the corresponding matrix.

The Heisenberg equations of motion

We have learned in Section 2.2 how to calculate the time rate of change of the expectation value of an operator A (see (2.55)). In a similar way, one can show that

$$\frac{\mathrm{d}}{\mathrm{d}t}\int \Phi^* A \Psi \,\mathrm{d}\mathbf{r} = \int \Phi^* \frac{\partial A}{\partial t} \Psi \,\mathrm{d}\mathbf{r} + \frac{1}{\mathrm{i}\hbar}\int \Phi^* [A, H] \Psi \,\mathrm{d}\mathbf{r} \tag{2.112}$$

where $\Phi(\mathbf{r}, t)$ and $\Psi(\mathbf{r}, t)$ are two arbitrary wave functions. Thus we have the operator (or matrix) equation

$$\frac{\mathrm{d}A}{\mathrm{d}t} = \frac{\partial A}{\partial t} + \frac{1}{\mathrm{i}\hbar}[A, H] \tag{2.113}$$

where $\mathrm{d}A/\mathrm{d}t$ is the operator whose matrix elements are the time rate of change of the matrix elements of A. The equation (2.113) is known as the *Heisenberg equation of motion* of a dynamical variable.

Commuting observables

Let two different observables be represented by the operators A and B. From Born's postulate, it follows that if there is a state in which a simultaneous measurement of the two observables is certain to yield the values a_i and b_j, then a_i and b_j must be eigenvalues of A and B, respectively. The corresponding wave function, ψ_{ij}, must be an eigenfunction of both A and B. That is,

$$A\psi_{ij} = a_i\psi_{ij}; \qquad B\psi_{ij} = b_j\psi_{ij} \tag{2.114}$$

If such eigenfunctions of A and B can be found for all the eigenvalues a_i and b_j, then the two observables are said to be *compatible*, and the ψ_{ij} form a complete set. By taking suitable linear combinations of any degenerate eigenfunctions, this set can be made orthonormal,

$$\langle \psi_{i'j'} | \psi_{ij} \rangle = \delta_{ii'}\delta_{jj'} \tag{2.115}$$

From (2.114), we see that

$$[A, B]\psi_{ij} = (AB - BA)\psi_{ij} = 0 \tag{2.116}$$

and since any wave function Ψ can be expanded in terms of the orthonormal set $\{\psi_{ij}\}$, we must have

$$[A, B] = 0 \tag{2.117}$$

so that the two operators A and B commute with each other. Conversely, if A and B are two operators which commute with each other, there exists a complete set of eigenfunctions which are simultaneously eigenstates of both A and B. On the other hand, if $[A, B] \neq 0$, a precise simultaneous measurement of both the observables is impossible.

As a first example, let us consider the Cartesian components x, y, z of the position operator and the Cartesian components of the momentum operator, $p_x = -i\hbar\, \partial/\partial x$, $p_y = -i\hbar\, \partial/\partial y$ and $p_z = -i\hbar\, \partial/\partial z$. If $f(x)$ is a function of x, we have

$$[x, p_x]f(x) = -i\hbar\left\{x\frac{\partial f}{\partial x} - \frac{\partial}{\partial x}(xf)\right\} = i\hbar\, f(x) \tag{2.118}$$

so that we may write the relation $[x, p_x] = i\hbar$. More generally, we have

$$[x, p_x] = [y, p_y] = [z, p_z] = i\hbar \tag{2.119}$$

while all other commutators – such as $[x, p_y]$ – vanish. It follows that, for example, x and p_y have common eigenfunctions and can be measured simultaneously with arbitrary accuracy. In contrast, since x and p_x do not commute with each other, a precise simultaneous measurement of both of these observables is impossible. Indeed, as we saw in Section 2.1, the accuracy to which both x and p_x can be measured is limited by the uncertainty principle.

As a second example, let us consider a system described by a *time-independent* Hamiltonian H, so that stationary states corresponding to a definite value E of the total energy exist. Now, as we have seen above, an operator A corresponding to a constant of the motion commutes with the Hamiltonian. Hence, in this case a constant of the motion can be measured simultaneously with the energy E. It can also be shown (Problem 2.10) that if two non-commuting operators A and B are constants of the motion, then the energy levels of the system are, in general, degenerate.

The foregoing discussion can be extended to any number of observables. In particular if $A, B, C, \ldots$ are a set of commuting observables, then a complete set of simultaneous eigenfunctions of these observables exists. The largest set of commuting observables which can be found for a given system is called a *complete set of commuting observables.*

Commutator algebra

It is convenient to list here some elementary rules for the calculation of commutators. These rules are easily verified from the basic definition (2.54). If A, B and C are three linear operators

$$[A, B] = -[B, A] \tag{2.120a}$$

$$[A, B + C] = [A, B] + [A, C] \tag{2.120b}$$

$$[A, BC] = [A, B]C + B[A, C] \tag{2.120c}$$

$$[A, [B, C]] + [B, [C, A]] + [C, [A, B]] = 0 \tag{2.120d}$$

2.4 One-dimensional examples

In this section we shall analyse the time-independent Schrödinger equation

$$-\frac{\hbar^2}{2m}\frac{\mathrm{d}^2\psi(x)}{\mathrm{d}x^2} + V(x)\psi(x) = E\psi(x) \qquad \textbf{(2.121)}$$

for two simple one-dimensional potentials: the infinite square well and the linear harmonic oscillator. This will allow us to illustrate the theory and to obtain several results which will be useful in further chapters.

The infinite square well

Let us consider a particle of mass m which is constrained by impenetrable walls to move in a region of width L, where the potential energy is constant. Taking this constant to be zero, and setting $a = L/2$, the potential energy for this problem is

$$V(x) = \begin{cases} 0, & |x| < a \\ \infty, & x < -a, \quad x > a \end{cases} \qquad \textbf{(2.122)}$$

and is illustrated in Fig. 2.2(a).

Because the potential energy is infinite at $x = \pm a$, the probability of finding the particle outside the well is zero. Hence the wave function $\psi(x)$ must vanish for $|x| > a$, and we only need to solve the Schrödinger eigenvalue equation (2.121) inside the well. Moreover, since the wave function must be continuous, $\psi(x)$ must vanish at the constraining walls, namely

$$\psi(x) = 0 \quad \text{at} \quad x = \pm a \qquad \textbf{(2.123)}$$

We shall see shortly that it is precisely this *boundary condition* which leads to the quantisation of the energy.

The time-independent Schrödinger equation for $|x| < a$ reads

$$-\frac{\hbar^2}{2m}\frac{\mathrm{d}^2\psi(x)}{\mathrm{d}x^2} = E\psi(x) \qquad \textbf{(2.124)}$$

and has the general solution

$$\psi(x) = A\sin kx + B\cos kx, \qquad k = \left(\frac{2m}{\hbar^2}E\right)^{1/2} \qquad \textbf{(2.125)}$$

Applying the boundary condition (2.123) we find that

$$A\sin ka = 0, \qquad B\cos ka = 0 \qquad \textbf{(2.126)}$$

As a result, there are two possible classes of solutions. For the *first class* $A = 0$ and $\cos ka = 0$, so that the only allowed values of k are

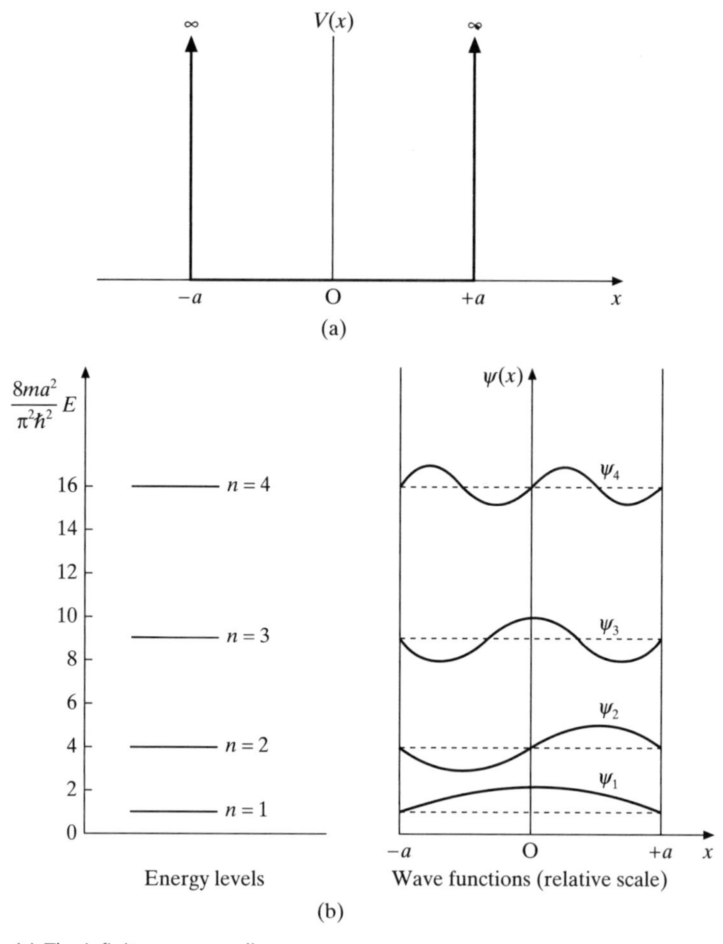

Figure 2.2 (a) The infinite square well
(b) The first four energy eigenvalues and eigenfunctions for the infinite square well.

$$k_n = \frac{n\pi}{2a} = \frac{n\pi}{L} \tag{2.127}$$

with $n = 1, 3, 5, \ldots$. The corresponding eigenfunctions $\psi_n(x) = B_n \cos k_n x$ can be normalised so that

$$\int_{-a}^{+a} \psi_n^*(x)\psi_n(x)\,\mathrm{d}x = 1 \tag{2.128}$$

from which the normalisation constants B_n are found (within an arbitrary multiplicative factor of modulus one) to be $B_n = a^{-1/2}$. The normalised eigenfunctions of the first class may therefore be written as

$$\psi_n(x) = \frac{1}{\sqrt{a}} \cos \frac{n\pi}{2a} x, \qquad n = 1, 3, 5, \ldots \tag{2.129}$$

Similarly, for the *second class* of solutions, such that $B = 0$ and $\sin ka = 0$, the allowed values of k are given by (2.127) with $n = 2, 4, 6, \ldots$ and the corresponding normalised eigenfunctions are

$$\psi_n(x) = \frac{1}{\sqrt{a}} \sin \frac{n\pi}{2a} x, \qquad n = 2, 4, 6, \ldots \tag{2.130}$$

For both classes of eigenfunctions it is unnecessary to consider negative values of n, since these lead to solutions which are not linearly independent of those corresponding to positive n. The *energy eigenvalues* E_n are given by

$$E_n = \frac{\hbar^2 k_n^2}{2m} = \frac{\hbar^2}{8m} \frac{\pi^2 n^2}{a^2} = \frac{\hbar^2}{2m} \frac{\pi^2 n^2}{L^2}, \qquad n = 1, 2, 3, \ldots \tag{2.131}$$

We see that there is an infinite number of discrete energy levels, with one eigenfunction corresponding to each energy level, so that the energy levels are *non-degenerate*. The energy eigenvalues and the corresponding eigenfunctions are shown in Fig. 2.2(b) for the first few states. It is interesting to note that the lowest energy or *zero-point energy* is $E_1 = \hbar^2\pi^2/(8ma^2)$ so that there is no state of zero energy. This is in agreement with the requirements of the uncertainty principle. Indeed, the position uncertainty is roughly given by $\Delta x \simeq a$. The corresponding momentum uncertainty is therefore $\Delta p_x \gtrsim \hbar/a$, leading to a minimum kinetic energy of order $\hbar^2/(ma^2)$, in qualitative agreement with the value of E_1.

There is an important difference between the two classes of eigenfunctions which we have obtained. That is, the eigenfunctions belonging to the first class are such that $\psi_n(-x) = \psi_n(x)$, and are therefore *even* functions of x, while those of the second class are such that $\psi_n(-x) = -\psi_n(x)$, and hence are *odd*. This division of the eigenfunctions into even and odd types is a consequence of the fact that the potential $V(x)$ is symmetric about $x = 0$, $V(-x) = V(x)$, so that the Hamiltonian is invariant under the *parity* or reflection operation $x \to -x$.

Finally, we observe that a general solution of the time-dependent Schrödinger equation (2.46) for the present problem can be written as a linear superposition of stationary solutions (see (2.85)), namely

$$\Psi(x, t) = \sum_{n=1}^{\infty} c_n \psi_n(x) \exp(-\mathrm{i} E_n t/\hbar) \tag{2.132}$$

where the coefficients c_n can be determined from the knowledge of the wave function Ψ at some particular time, say $t = 0$. Thus, using (2.80), we find that

$$c_n = \int_{-a}^{+a} \psi_n^*(x) \Psi(x, t = 0)\, \mathrm{d}x \tag{2.133}$$

The linear harmonic oscillator

We now consider the one-dimensional motion of a particle of mass m which is attracted to a fixed centre by a force proportional to the displacement from that centre. Thus, choosing the origin at the centre of force, the restoring force is given by $F = -kx$ (Hooke's law), where k is the force constant. This force can thus be represented by the potential energy

$$V(x) = \frac{1}{2}kx^2 \tag{2.134}$$

which is shown in Fig. 2.3(a). Such a parabolic potential is of great importance in quantum physics as well as in classical physics, since it can be used to approximate an arbitrary continuous potential $W(x)$ in the neighbourhood of a stable equilibrium position $x = a$ (see Fig. 2.3(b)). Indeed, if we expand $W(x)$ in a Taylor series about $x = a$, we find that

$$W(x) = W(a) + (x - a)W'(a) + \frac{1}{2}(x - a)^2W''(a) + \dots \tag{2.135}$$

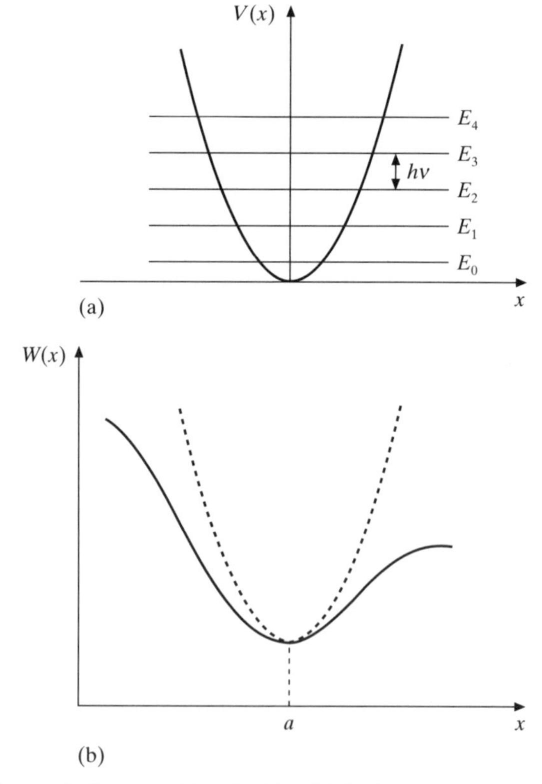

Figure 2.3 (a) The parabolic potential well $V(x) = \frac{1}{2}kx^2$. This is the potential of the linear harmonic oscillator. Also shown are the first few energy eigenvalues (2.143). (b) A continuous potential well $W(x)$, represented by the solid line, can be approximated in the vicinity of a stable equilibrium position at $x = a$ by a linear harmonic oscillator potential, shown as the dashed line.

where $W'(a) = (\mathrm{d}W/\mathrm{d}x)_{x=a}$ and $W''(a) = (\mathrm{d}^2W/\mathrm{d}x^2)_{x=a}$. Since $W(x)$ has a minimum at $x = a$, we have $W'(a) = 0$ and $W''(a) > 0$. Choosing a as the coordinate origin and $W(a)$ as the origin of the energy scale, we see that the harmonic oscillator potential (2.134) is the first approximation to $W(x)$. Hence the linear harmonic oscillator is the prototype for systems in which there exist small vibrations about a point of stable equilibrium. This will be illustrated in Chapters 10 and 11, where we shall study the vibrational motion of nuclei in molecules.

The one-dimensional Schrödinger equation (2.121) becomes, for the linear harmonic oscillator,

$$-\frac{\hbar^2}{2m}\frac{\mathrm{d}^2\psi(x)}{\mathrm{d}x^2} + \frac{1}{2}kx^2\psi(x) = E\psi(x) \tag{2.136}$$

It is convenient to rewrite this equation in terms of dimensionless quantities. To this end we introduce the dimensionless variable

$$\xi = \alpha x, \qquad \alpha = \left(\frac{mk}{\hbar^2}\right)^{1/4} \tag{2.137}$$

and the dimensionless eigenvalue

$$\lambda = \frac{2E}{\hbar\omega} \tag{2.138}$$

where

$$\omega = \left(\frac{k}{m}\right)^{1/2} \tag{2.139}$$

is the angular frequency of the corresponding classical oscillator. The Schrödinger equation (2.136) then becomes

$$\frac{\mathrm{d}^2\psi}{\mathrm{d}\xi^2} + (\lambda - \xi^2)\psi = 0 \tag{2.140}$$

Let us first analyse the behaviour of ψ in the asymptotic region $|\xi| \to \infty$. We may then neglect the term λ compared to ξ^2, and the resulting equation is readily seen to have solutions of the form $\xi^n \exp(\pm\xi^2/2)$, where n is finite. The physically acceptable solution must contain only the minus sign in the exponent. This suggests looking for solutions to (2.140) of the form

$$\psi(\xi) = \mathrm{e}^{-\xi^2/2}H(\xi) \tag{2.141}$$

where $H(\xi)$ are functions which must not affect the asymptotic behaviour. Substituting (2.141) into (2.140) we obtain for $H(\xi)$ the equation

$$\frac{\mathrm{d}^2H}{\mathrm{d}\xi^2} - 2\xi\frac{\mathrm{d}H}{\mathrm{d}\xi} + (\lambda - 1)H = 0 \tag{2.142}$$

Energy levels

Solutions of (2.142) which satisfy the above requirement and are finite everywhere are only found for $\lambda = 2n + 1$, where $n = 0, 1, 2, \ldots$. They are called *Hermite polynomials* $H_n(\xi)$. The energy spectrum of the linear harmonic oscillator is therefore given by (see (2.138))

$$E_n = \hbar\omega\left(n + \frac{1}{2}\right) = h\nu\left(n + \frac{1}{2}\right), \qquad n = 0, 1, 2, \ldots \tag{2.143}$$

where $\nu = \omega/(2\pi)$.

The infinite sequence of discrete, equally spaced energy levels (2.143) is similar to that discovered in 1900 by Planck for the radiation field modes (see Section 1.3). This is due to the fact that a decomposition of the electromagnetic field into normal modes is essentially a decomposition into uncoupled harmonic oscillators. We note, however, that according to (2.143) the linear harmonic oscillator, even in its lowest energy state, $n = 0$, has the energy $\hbar\omega/2$. On the other hand the lowest energy of a classical harmonic oscillator is zero. The finite value $\hbar\omega/2$ of the ground state energy level, which is called the *zero-point energy*, is therefore a purely quantum mechanical effect, and is directly related to the uncertainty principle (Problem 2.12). The eigenvalues (2.143) of the linear harmonic oscillator are *non-degenerate*, since for each eigenvalue there exists only *one* eigenfunction, apart from an arbitrary multiplicative factor.

Hermite polynomials

The Hermite polynomials $H_n(\xi)$, which are the physically acceptable solutions of (2.142) corresponding to the eigenvalues $\lambda = 2n + 1$, are polynomials of order n, having the parity of n, which are uniquely defined except for an arbitrary multiplicative constant. This constant is traditionally chosen so that the highest power of ξ appears with the coefficient 2^n in $H_n(\xi)$. This is consistent with the following definition of the Hermite polynomials:

$$H_n(\xi) = (-1)^n e^{\xi^2} \frac{d^n e^{-\xi^2}}{d\xi^n} \tag{2.144}$$

The first few Hermite polynomials, obtained from (2.144), are

$$\begin{aligned} H_0(\xi) &= 1 \\ H_1(\xi) &= 2\xi \\ H_2(\xi) &= 4\xi^2 - 2 \\ H_3(\xi) &= 8\xi^3 - 12\xi \\ H_4(\xi) &= 16\xi^4 - 48\xi^2 + 12 \\ H_5(\xi) &= 32\xi^5 - 160\xi^3 + 120\xi \end{aligned} \tag{2.145}$$

Another definition of the Hermite polynomials $H_n(\xi)$, which is equivalent to (2.144), involves the use of a *generating function* $G(\xi, s)$. That is,

$$\begin{aligned} G(\xi, s) &= e^{-s^2+2s\xi} \\ &= \sum_{n=0}^{\infty} \frac{H_n(\xi)}{n!} s^n \end{aligned} \tag{2.146}$$

The relation (2.146) means that if the function $\exp(-s^2 + 2s\xi)$ is expanded in a power series in s, the coefficients of successive powers of s are just $1/n!$ times the Hermite polynomials $H_n(\xi)$. Using the generating function (2.146) it may be shown that the Hermite polynomials satisfy the recurrence relation

$$H_{n+1}(\xi) - 2\xi H_n(\xi) + 2nH_{n-1}(\xi) = 0 \tag{2.147}$$

The wave functions for the linear harmonic oscillator

Using (2.141), we see that to each of the discrete values E_n of the energy, given by (2.143), there corresponds one, and only one, physically acceptable eigenfunction, namely

$$\psi_n(x) = N_n e^{-\alpha^2 x^2/2} H_n(\alpha x) \tag{2.148}$$

where we have returned to our original variable x, and N_n is a constant which (apart from an arbitrary phase factor) is determined by requiring that the wave function (2.148) be normalised to unity. That is,

$$\int_{-\infty}^{+\infty} |\psi_n(x)|^2 \, dx = \frac{|N_n|^2}{\alpha} \int_{-\infty}^{+\infty} e^{-\xi^2} H_n^2(\xi) \, d\xi = 1 \tag{2.149}$$

The integral on the right is evaluated in Appendix 3 by using the generating function (2.146). It is found (see equation (A3.10)) that the normalisation constant N_n can be chosen to be

$$N_n = \left(\frac{\alpha}{\sqrt{\pi} 2^n n!} \right)^{1/2} \tag{2.150}$$

so that the normalised linear harmonic oscillator eigenfunctions are given by

$$\psi_n(x) = \left(\frac{\alpha}{\sqrt{\pi} 2^n n!} \right)^{1/2} e^{-\alpha^2 x^2/2} H_n(\alpha x) \tag{2.151}$$

It is also shown in Appendix 3 that the wave functions $\psi_n(x)$ and $\psi_m(x)$ are orthogonal if $n \neq m$, in agreement with the fact that they correspond to non-degenerate energy eigenvalues.

Because the Hamiltonian of the linear harmonic oscillator,

$$H = -\frac{\hbar^2}{2m} \frac{d^2}{dx^2} + \frac{1}{2} kx^2 \tag{2.152}$$

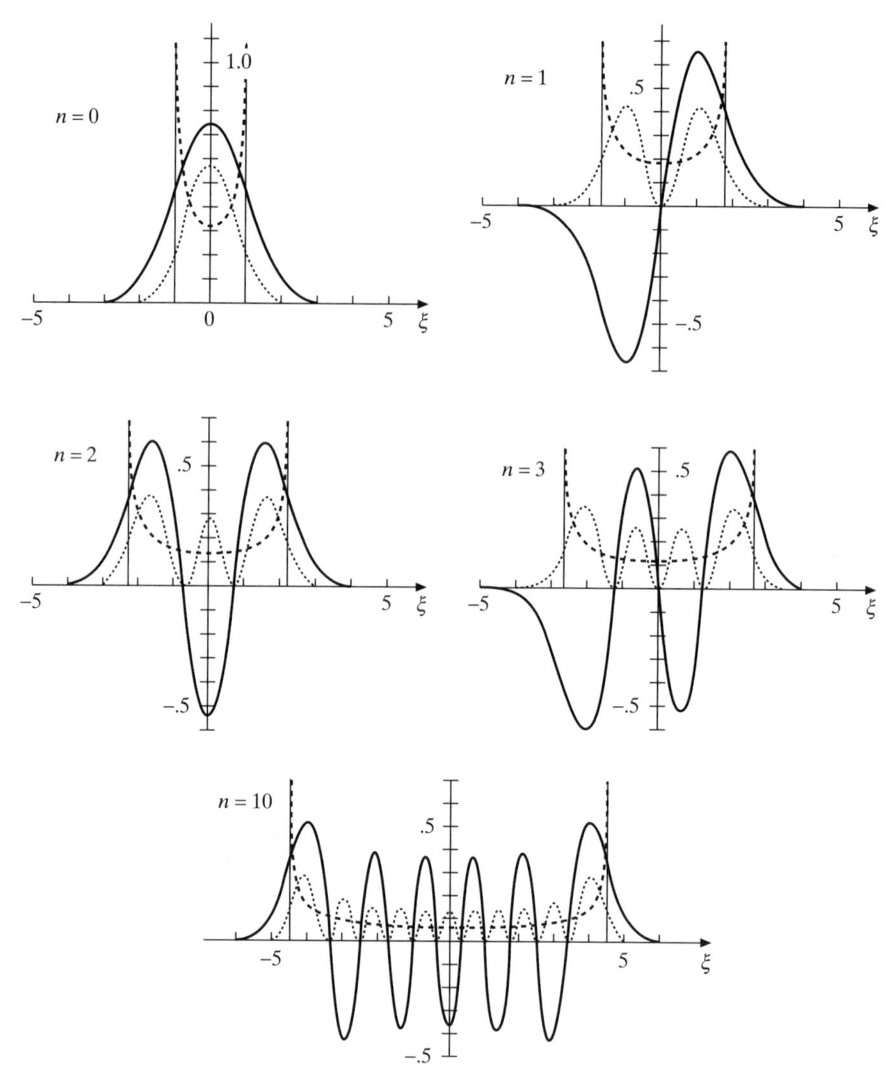

Figure 2.4 Wave functions for the linear harmonic oscillator
—— $\psi_n(\xi)/\sqrt{\alpha}$ with $\xi = \alpha x$, see (2.148) and (2.137)
······ $|\psi_n(\xi)|^2/\alpha$
------ The probability distribution for a classical oscillator with energy E_n. The classical motion is confined to the region between the vertical lines.

is invariant under the *parity* operation $x \to -x$, the eigenfunctions $\psi_n(x)$ are either *even* functions of x (for $n = 0, 2, 4, \dots$) or *odd* functions of x (for $n = 1, 3, 5, \dots$). This is apparent in Fig. 2.4 where plots of the first few harmonic oscillator wave functions are shown, together with the corresponding probability densities $|\psi_n|^2$. It is also clear from this figure that for small values of n the quantum mechanical probability densities are very different from the corresponding densities for the classical harmonic oscillator. However, the agreement between the classical and quantum mechanical probability densities quickly improves with increasing values of n. This is an illustration of Bohr's *correspondence principle*, according to which the predictions of quantum physics must correspond to those of classical physics in the limit in which the quantum numbers specifying the state of the system become large.

2.5 Angular momentum

The classical orbital angular momentum of a particle is

$$\mathbf{L} = \mathbf{r} \times \mathbf{p} \qquad \textbf{(2.153)}$$

where $\mathbf{r}$ and $\mathbf{p}$ are the position and momentum vectors of the particle, respectively. Using the fact that $\mathbf{p}$ is represented in wave mechanics by the vector operator $-\mathrm{i}\hbar\boldsymbol{\nabla}$, we see that $\mathbf{L}$ is represented by the vector operator $-\mathrm{i}\hbar(\mathbf{r} \times \boldsymbol{\nabla})$. Its Cartesian components are given by

$$\begin{aligned}
L_x &= yp_z - zp_y = -\mathrm{i}\hbar\left(y\frac{\partial}{\partial z} - z\frac{\partial}{\partial y}\right) \\
L_y &= zp_x - xp_z = -\mathrm{i}\hbar\left(z\frac{\partial}{\partial x} - x\frac{\partial}{\partial z}\right) \\
L_z &= xp_y - yp_x = -\mathrm{i}\hbar\left(x\frac{\partial}{\partial y} - y\frac{\partial}{\partial x}\right)
\end{aligned} \qquad \textbf{(2.154)}$$

Using the rules (2.120) of commutator algebra and the basic commutation relations (2.119) one finds that

$$[L_x, L_y] = \mathrm{i}\hbar L_z, \qquad [L_y, L_z] = \mathrm{i}\hbar L_x, \qquad [L_z, L_x] = \mathrm{i}\hbar L_y \qquad \textbf{(2.155)}$$

so that the operators L_x, L_y and L_z do not mutually commute. As a consequence, it is impossible to find a representation that diagonalises more than one of them, and the components of the orbital angular momentum cannot in general [4] be assigned definite values simultaneously. However, each of the three components of $\mathbf{L}$ is easily seen to commute with the operator $\mathbf{L}^2 = L_x^2 + L_y^2 + L_z^2$. For example,

$$\begin{aligned}
[L_z, \mathbf{L}^2] &= [L_z, L_x^2] + [L_z, L_y^2] \\
&= [L_z, L_x]L_x + L_x[L_z, L_x] + [L_z, L_y]L_y + L_y[L_z, L_y] \\
&= \mathrm{i}\hbar(L_yL_x + L_xL_y) - \mathrm{i}\hbar(L_xL_y + L_yL_x) = 0
\end{aligned} \qquad \textbf{(2.156)}$$

[4] An exception occurs when the angular momentum is zero ($L_x = L_y = L_z = 0$). In this case any function which only depends on the magnitude r of the position vector $\mathbf{r}$ is a simultaneous eigenfunction of L_x, L_y and L_z.

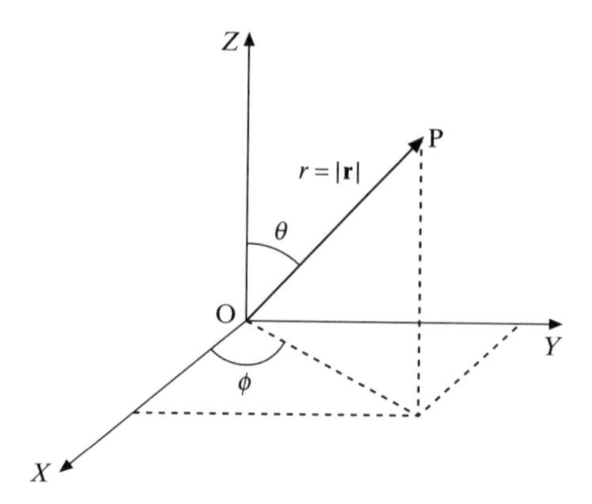

Figure 2.5 Spherical polar coordinates (r, θ, ϕ) of a point P.

Thus it is possible to construct simultaneous eigenfunctions of $\mathbf{L}^2$ and one component of $\mathbf{L}$, so that $\mathbf{L}^2$ and this component can be simultaneously defined precisely. In what follows we shall choose this component to be L_z.

It is convenient to express L_z and $\mathbf{L}^2$ in spherical polar coordinates (r, θ, ϕ), which are related to the Cartesian coordinates (x, y, z) of the vector $\mathbf{r}$ by

$$\begin{aligned} x &= r \sin \theta \cos \phi \\ y &= r \sin \theta \sin \phi \\ z &= r \cos \theta \end{aligned} \tag{2.157}$$

with $0 \leqslant r \leqslant \infty$, $0 \leqslant \theta \leqslant \pi$, $0 \leqslant \phi \leqslant 2\pi$ (see Fig. 2.5). It is found that (Problem 2.13)

$$L_z = -i\hbar \frac{\partial}{\partial \phi} \tag{2.158}$$

and

$$\mathbf{L}^2 = -\hbar^2 \left[\frac{1}{\sin \theta} \frac{\partial}{\partial \theta} \left(\sin \theta \frac{\partial}{\partial \theta} \right) + \frac{1}{\sin^2 \theta} \frac{\partial^2}{\partial \phi^2} \right] \tag{2.159}$$

The eigenfunctions $\Phi(\phi)$ of L_z satisfy the eigenvalue equation

$$L_z \Phi(\phi) = m\hbar \Phi(\phi) \tag{2.160}$$

where the eigenvalues have been written as $m\hbar$ for convenience. The normalised solutions of (2.160) are

$$\Phi_m(\phi) = \frac{1}{\sqrt{2\pi}} e^{im\phi} \tag{2.161}$$

Since the functions Φ_m must be single-valued, we have $\Phi_m(2\pi) = \Phi_m(0)$, so that m, which is called the *magnetic quantum number*, is restricted to positive or negative integers, or zero ($m = 0, \pm 1, \pm 2, \ldots$). We also note that the functions (2.161) are orthonormal,

$$\int_0^{2\pi} \Phi^*_{m'}(\phi)\Phi_m(\phi)\,\mathrm{d}\phi = \delta_{mm'} \tag{2.162}$$

The simultaneous eigenfunctions of $\mathbf{L}^2$ and L_z are called the *spherical harmonics* and are denoted by $Y_{lm}(\theta, \phi)$. They satisfy the eigenvalue equations

$$\mathbf{L}^2 Y_{lm}(\theta, \phi) = l(l+1)\hbar^2 Y_{lm}(\theta, \phi) \tag{2.163}$$

and

$$L_z Y_{lm}(\theta, \phi) = m\hbar Y_{lm}(\theta, \phi) \tag{2.164}$$

where the eigenvalues of $\mathbf{L}^2$ have been written as $l(l+1)\hbar^2$. The quantum number l is known as the *orbital angular momentum quantum number*. Setting

$$Y_{lm}(\theta, \phi) = \Theta_{lm}(\theta)\Phi_m(\phi) \tag{2.165}$$

substituting into (2.163) and using (2.159), we find that the functions $\Theta_{lm}(\theta)$ satisfy the equation

$$\left[-\frac{1}{\sin\theta}\frac{\partial}{\partial\theta}\left(\sin\theta\frac{\partial}{\partial\theta}\right) + \frac{m^2}{\sin^2\theta}\right]\Theta_{lm}(\theta) = l(l+1)\,\Theta_{lm}(\theta) \tag{2.166}$$

The physically acceptable solutions of this equation that remain finite over the range $0 \leqslant \theta \leqslant \pi$ exist only when $l = 0, 1, 2, \ldots$ and $m = -l, -l+1, \ldots, +l$. They can be expressed in terms of *associated Legendre functions* $P_l^m(\cos\theta)$, which are defined in the following way. Introducing the variable $w = \cos\theta$, we first define the *Legendre polynomials* $P_l(w)$ of degree l by the relation

$$P_l(w) = \frac{1}{2^l l!}\frac{\mathrm{d}^l}{\mathrm{d}w^l}(w^2 - 1)^l \tag{2.167}$$

An equivalent definition of $P_l(w)$ can be given in terms of a generating function, namely

$$\begin{aligned} T(w, s) &= (1 - 2sw + s^2)^{-1/2} \\ &= \sum_{l=0}^{\infty} P_l(w)s^l, \qquad |s| < 1 \end{aligned} \tag{2.168}$$

The Legendre polynomials satisfy the differential equation

$$\left[(1 - w^2)\frac{\mathrm{d}^2}{\mathrm{d}w^2} - 2w\frac{\mathrm{d}}{\mathrm{d}w} + l(l+1)\right]P_l(w) = 0 \tag{2.169}$$

which is readily shown to be equivalent to (2.166), with $m = 0$. One has the recurrence relation

$$(2l+1)wP_l - (l+1)P_{l+1} - lP_{l-1} = 0 \tag{2.170}$$

which is also valid for $l = 0$ if one defines $P_{-1} = 0$. The orthogonality relations read

$$\int_{-1}^{+1} P_l(w)P_{l'}(w)\,\mathrm{d}w = \frac{2}{2l+1}\delta_{ll'} \tag{2.171}$$

One also has the closure relation

$$\frac{1}{2}\sum_{l=0}^{\infty}(2l+1)P_l(w)P_l(w') = \delta(w-w') \tag{2.172}$$

Important particular values of the Legendre polynomials are

$$P_l(1) = 1, \qquad P_l(-1) = (-1)^l \tag{2.173}$$

For the lowest values of l one has explicitly

$$\begin{aligned}
P_0(w) &= 1\\
P_1(w) &= w\\
P_2(w) &= \tfrac{1}{2}(3w^2-1)\\
P_3(w) &= \tfrac{1}{2}(5w^3-3w)\\
P_4(w) &= \tfrac{1}{8}(35w^4-30w^2+3)\\
P_5(w) &= \tfrac{1}{8}(63w^5-70w^3+15w)
\end{aligned} \tag{2.174}$$

The associated Legendre functions $P_l^m(w)$ are now defined by the relations

$$P_l^m(w) = (1-w^2)^{m/2}\frac{\mathrm{d}^m}{\mathrm{d}w^m}P_l(w), \qquad m = 0, 1, 2, \ldots, l \tag{2.175}$$

They satisfy the recurrence relations

$$(2l+1)wP_l^m = (l+1-m)P_{l+1}^m + (l+m)P_{l-1}^m \tag{2.176a}$$

$$(2l+1)(1-w^2)^{1/2}P_l^{m-1} = P_{l+1}^m - P_{l-1}^m \tag{2.176b}$$

and the orthogonality relations

$$\int_{-1}^{+1} P_l^m(w)P_{l'}^m(w)\,\mathrm{d}w = \frac{2}{2l+1}\frac{(l+m)!}{(l-m)!}\delta_{ll'} \tag{2.177}$$

The first few associated Legendre functions are given explicitly by

$$\begin{aligned}
P_1^1(w) &= (1-w^2)^{1/2}\\
P_2^1(w) &= 3(1-w^2)^{1/2}w\\
P_2^2(w) &= 3(1-w^2)\\
P_3^1(w) &= \tfrac{3}{2}(1-w^2)^{1/2}(5w^2-1)\\
P_3^2(w) &= 15w(1-w^2)\\
P_3^3(w) &= 15(1-w^2)^{3/2}
\end{aligned} \tag{2.178}$$

The functions $\Theta_{lm}(\theta)$, normalised so that

$$\int_0^{\pi} \Theta^*_{l'm}(\theta)\Theta_{lm}(\theta) \sin\theta \, d\theta = \delta_{ll'} \tag{2.179}$$

are given in terms of the associated Legendre functions P_l^m by

$$\begin{aligned}\Theta_{lm}(\theta) &= (-1)^m \left[\frac{(2l+1)(l-m)!}{2(l+m)!}\right]^{1/2} P_l^m(\cos\theta), \qquad m \geqslant 0 \\ &= (-1)^{|m|}\Theta_{l|m|}(\theta), \qquad m < 0\end{aligned} \tag{2.180}$$

Using (2.165), (2.161) and (2.180), the spherical harmonics are given by

$$Y_{lm}(\theta, \phi) = (-1)^m \left[\frac{(2l+1)(l-m)!}{4\pi(l+m)!}\right]^{1/2} P_l^m(\cos\theta)e^{im\phi}, \qquad m \geqslant 0 \tag{2.181a}$$

$$Y_{l,-m}(\theta, \phi) = (-1)^m Y^*_{lm}(\theta, \phi) \tag{2.181b}$$

They satisfy the orthonormality relations

$$\begin{aligned}\int Y^*_{l'm'}(\theta, \phi)Y_{lm}(\theta, \phi)\, d\Omega &= \int_0^{2\pi} d\phi \int_0^{\pi} d\theta \sin\theta \, Y^*_{l'm'}(\theta, \phi)Y_{lm}(\theta, \phi) \\ &= \delta_{ll'}\delta_{mm'}\end{aligned} \tag{2.182}$$

where we have written $d\Omega = \sin\theta \, d\theta \, d\phi$. The closure relation reads

$$\sum_{l=0}^{\infty} \sum_{m=-l}^{+l} Y^*_{lm}(\theta, \phi)Y_{lm}(\theta', \phi') = \delta(\Omega - \Omega'),$$

$$\delta(\Omega - \Omega') = \frac{\delta(\theta - \theta')\,\delta(\phi - \phi')}{\sin\theta} \tag{2.183}$$

The first few spherical harmonics are listed in Table 2.1. Polar plots of the probability distributions

$$|Y_{lm}(\theta, \phi)|^2 = (2\pi)^{-1}|\Theta_{lm}(\theta)|^2 \tag{2.184}$$

are shown in Fig. 2.6. Additional useful formulae involving the Legendre polynomials and the spherical harmonics are given in Appendix 4. In that appendix we also discuss *matrix representations* of angular momentum operators, and define the *raising* and *lowering* operators

$$L_\pm = L_x \pm iL_y \tag{2.185}$$

These operators are such that (see (A4.21))

$$L_\pm Y_{lm}(\theta, \phi) = \hbar[l(l+1) - m(m \pm 1)]^{1/2} Y_{l,m\pm1}(\theta, \phi) \tag{2.186a}$$

or

$$L_\pm|lm\rangle = \hbar[l(l+1) - m(m \pm 1)]^{1/2}|lm \pm 1\rangle \tag{2.186b}$$

Table 2.1 The first few spherical harmonics $Y_{lm}(\theta, \phi)$.

l	m	Spherical harmonic $Y_{lm}(\theta, \phi)$
0	0	$Y_{0,0} = \frac{1}{(4\pi)^{1/2}}$
1	0	$Y_{1,0} = \left(\frac{3}{4\pi}\right)^{1/2} \cos\theta$
	±1	$Y_{1,\pm1} = \mp\left(\frac{3}{8\pi}\right)^{1/2} \sin\theta\, e^{\pm i\phi}$
2	0	$Y_{2,0} = \left(\frac{5}{16\pi}\right)^{1/2} (3\cos^2\theta - 1)$
	±1	$Y_{2,\pm1} = \mp\left(\frac{15}{8\pi}\right)^{1/2} \sin\theta\cos\theta\, e^{\pm i\phi}$
	±2	$Y_{2,\pm2} = \left(\frac{15}{32\pi}\right)^{1/2} \sin^2\theta\, e^{\pm 2i\phi}$
3	0	$Y_{3,0} = \left(\frac{7}{16\pi}\right)^{1/2} (5\cos^3\theta - 3\cos\theta)$
	±1	$Y_{3,\pm1} = \mp\left(\frac{21}{64\pi}\right)^{1/2} \sin\theta\,(5\cos^2\theta - 1)\, e^{\pm i\phi}$
	±2	$Y_{3,\pm2} = \left(\frac{105}{32\pi}\right)^{1/2} \sin^2\theta\cos\theta\, e^{\pm 2i\phi}$
	±3	$Y_{3,\pm3} = \mp\left(\frac{35}{64\pi}\right)^{1/2} \sin^3\theta\, e^{\pm 3i\phi}$

where in the last line we have used the *Dirac ket notation*, in which the eigenfunctions Y_{lm} are written in the form $|lm\rangle$.

Let us assume that the particle is in the orbital angular momentum state $|lm\rangle$ such that $\mathbf{L}^2|lm\rangle = l(l+1)\hbar^2|lm\rangle$ and $L_z|lm\rangle = m\hbar|lm\rangle$ (see (2.163) and (2.164)). Although two components of the orbital angular momentum cannot in general be assigned precise values simultaneously, it is nevertheless possible to say something about the components L_x and L_y. Indeed, one can readily show (Problem 2.14) that

$$\langle lm|L_x|lm\rangle = \langle lm|L_y|lm\rangle = 0 \tag{2.187}$$

and

$$\langle lm|L_x^2|lm\rangle = \langle lm|L_y^2|lm\rangle = \tfrac{1}{2}[l(l+1) - m^2]\hbar^2 \tag{2.188}$$

We note that when $m = +l$ or $m = -l$, so that the orbital angular momentum is respectively 'parallel' or 'antiparallel' to the Z axis, its x and y components are

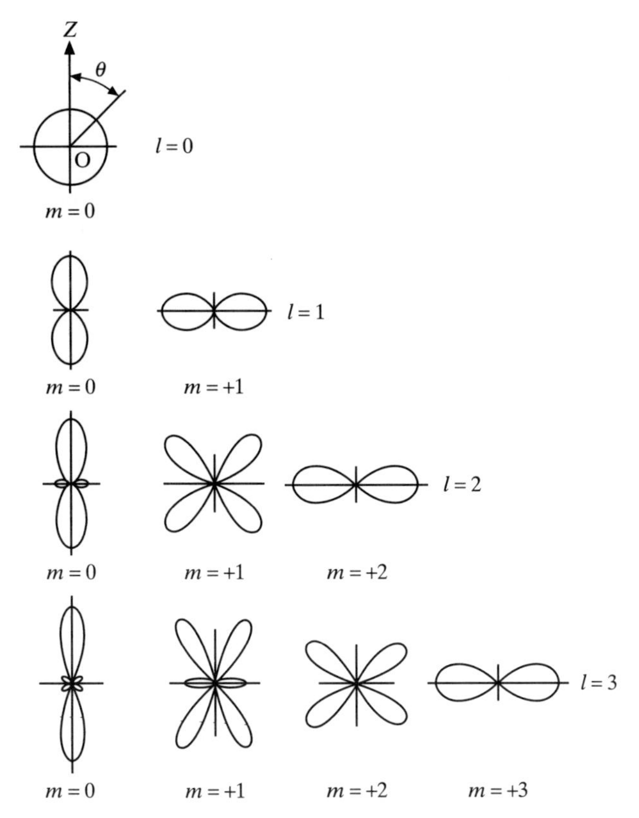

Figure 2.6 Polar plots of the probability distributions $|Y_{lm}(\theta, \phi)|^2 = (2\pi)^{-1}|\Theta_{lm}(\theta)|^2$.

still not zero. It is helpful to visualise these results in terms of the *vector model* introduced in Section 1.8 (see Fig. 1.24). According to this model, the orbital angular momentum vector $\mathbf{L}$, of length $\sqrt{l(l+1)}\hbar$, precesses about the Z axis, the $(2l + 1)$ allowed projections of $\mathbf{L}$ on the Z axis being given by $m\hbar$, with $m = -l, -l+1, \ldots, +l$.

Spherical harmonics in real form

In some applications it is convenient to use an alternative set of eigenfunctions of $\mathbf{L}^2$, which are the *real forms of the spherical harmonics*. That is,

$$\begin{aligned} Y_{l,\cos}(\theta, \phi) &= N\Theta_{l|m|}(\theta) \cos|m|\phi \\ Y_{l,\sin}(\theta, \phi) &= N\Theta_{l|m|}(\theta) \sin|m|\phi \end{aligned} \tag{2.189}$$

where the normalisation constant N is equal to $\pi^{-1/2}$, except for $m = 0$ for which $N = (2\pi)^{-1/2}$. For $m = 0$, the function $Y_{l,\cos}$ is clearly identical to the spherical harmonic $Y_{l,0}$. On the other hand, for $m \neq 0$ we have

$$Y_{l,\cos} = \frac{1}{\sqrt{2}}(Y_{l|m|} + Y^*_{l|m|})$$
$$Y_{l,\sin} = -\frac{\mathrm{i}}{\sqrt{2}}(Y_{l|m|} - Y^*_{l|m|}) \tag{2.190}$$

The spherical harmonics in real form are eigenfunctions of $\mathbf{L}^2$ and L_z^2, but not of L_z (except, of course, when $m = 0$). They behave like simple functions of the Cartesian coordinates, and for this reason are well suited for describing the directional properties of chemical bonds (see Chapter 10).

The first few spherical harmonics in real form are listed in Table 2.2. We have used in this table the so-called 'spectroscopic' notation, in which the value of the orbital angular momentum quantum number l is indicated by a letter, according to the correspondence

Value of l	0	1	2	3	4	5
	↕	↕	↕	↕	↕	↕
Code letter	s	p	d	f	g	h, . . .

Table 2.2 The first few spherical harmonics in real form.

l	$\lvert m \rvert$	*Spherical harmonic in real form*
0	0	$\mathrm{s} = \dfrac{1}{(4\pi)^{1/2}}$
1	0	$\mathrm{p}_z = \left(\dfrac{3}{4\pi}\right)^{1/2} \cos\theta$
	1	$\mathrm{p}_x = \left(\dfrac{3}{4\pi}\right)^{1/2} \sin\theta\cos\phi$
		$\mathrm{p}_y = \left(\dfrac{3}{4\pi}\right)^{1/2} \sin\theta\sin\phi$
2	0	$\mathrm{d}_{3z^2-r^2} = \left(\dfrac{5}{16\pi}\right)^{1/2} (3\cos^2\theta - 1)$
	1	$\mathrm{d}_{xz} = \left(\dfrac{15}{4\pi}\right)^{1/2} \sin\theta\cos\theta\cos\phi$
		$\mathrm{d}_{yz} = \left(\dfrac{15}{4\pi}\right)^{1/2} \sin\theta\cos\theta\sin\phi$
	2	$\mathrm{d}_{x^2-y^2} = \left(\dfrac{15}{16\pi}\right)^{1/2} \sin^2\theta\cos 2\phi$
		$\mathrm{d}_{xy} = \left(\dfrac{15}{16\pi}\right)^{1/2} \sin^2\theta\sin 2\phi$

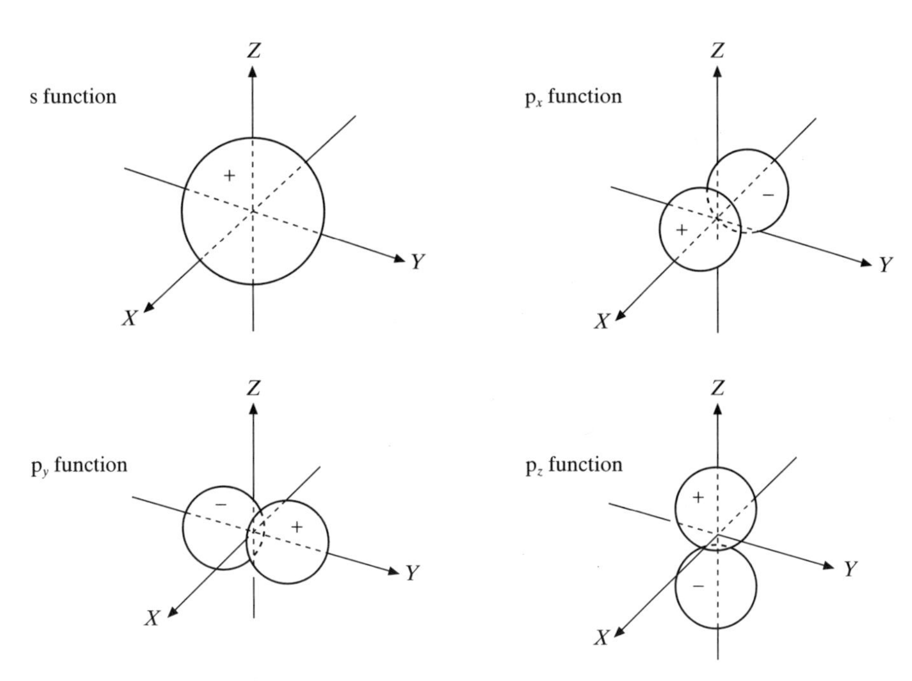

Figure 2.7 Polar diagrams of the real spherical harmonics s, p_x, p_y and p_z.

These code letters are remnants of the spectroscopist's description of various series of spectral lines, the letters s, p, d and f being the first letters of the adjectives 'sharp', 'principal', 'diffuse' and 'fundamental', respectively. For values of l greater than three the letters follow in alphabetical order (g for $l = 4$, h for $l = 5$, and so on). The subscripts z, x, y, xz, yz, etc. used in Table 2.2 indicate the behaviour of the real spherical harmonic in terms of Cartesian coordinates. Polar graphs of the s, p_x, p_y and p_z functions are given in Fig. 2.7. We also remark that while the probability distributions (2.184) corresponding to the 'genuine' spherical harmonics are independent of the azimuthal angle ϕ, those corresponding to the spherical harmonics in real form (that is, $|Y_{l,\cos}|^2$ and $|Y_{l,\sin}|^2$) depend on ϕ in the same way as the functions $\cos^2|m|\phi$ and $\sin^2|m|\phi$, as seen from (2.189).

The rigid rotator

As an example of the use of the operator $\mathbf{L}^2$ and of its eigenfunctions, consider the motion of a particle of mass m, constrained to remain at a given distance R_0 from a fixed point, which we choose as the origin of coordinates. Denoting by $I = mR_0^2$ the moment of inertia, the Hamiltonian of this system, which is known as a (three-dimensional) *rigid rotator*, is

$$H = \frac{\mathbf{L}^2}{2I} \tag{2.191}$$

Using the expression (2.159) of $\mathbf{L}^2$ in spherical polar coordinates, the Schrödinger eigenvalue equation reads

$$\frac{\mathbf{L}^2}{2I}\psi(\theta, \phi) \equiv -\frac{\hbar^2}{2I}\left[\frac{1}{\sin\theta}\frac{\partial}{\partial\theta}\left(\sin\theta\frac{\partial}{\partial\theta}\right)+\frac{1}{\sin^2\theta}\frac{\partial^2}{\partial\phi^2}\right]\psi(\theta, \phi)$$
$$= E\psi(\theta, \phi) \qquad \textbf{(2.192)}$$

Thus, we see from (2.163) that the eigenfunctions are just the spherical harmonics $Y_{lm}(\theta, \phi)$, and the energy eigenvalues are

$$E = \frac{\hbar^2}{2I}l(l+1), \qquad l = 0, 1, 2, \ldots \qquad \textbf{(2.193)}$$

Spin angular momentum

We saw in Chapter 1 that the Stern–Gerlach experiment can be interpreted as showing that the electron possesses an intrinsic degree of freedom, the *spin*, which behaves like an angular momentum in the way it couples with a magnetic field. The z component of this spin angular momentum can only take on two values $m_s\hbar$, where $m_s = \pm 1/2$. Therefore, the electron spin cannot be described by the orbital angular momentum operator $\mathbf{L}$ we have considered thus far, since the z component of $\mathbf{L}$ only takes on the values $m\hbar$, with $m = -l, -l+1, \ldots, +l$, and $l = 0, 1, 2, \ldots$. We shall assume, however, that *all* angular momentum operators, whether orbital or spin, satisfy commutation relations of the form (2.155). Thus, if S_x, S_y and S_z are the three Cartesian components of the electron spin angular momentum operator $\mathbf{S}$, they must satisfy the commutation relations

$$[S_x, S_y] = i\hbar S_z, \qquad [S_y, S_z] = i\hbar S_x, \qquad [S_z, S_x] = i\hbar S_y \qquad \textbf{(2.194)}$$

Using the results of Appendix 4, we may readily obtain the properties of the spin angular momentum and of the spin eigenfunctions. Denoting by χ_{s,m_s} the simultaneous eigenfunctions of $\mathbf{S}^2$ and S_z, we have

$$\mathbf{S}^2\chi_{s,m_s} = s(s+1)\hbar^2\chi_{s,m_s} \qquad \textbf{(2.195)}$$

and

$$S_z\chi_{s,m_s} = m_s\hbar\chi_{s,m_s} \qquad \textbf{(2.196)}$$

Since $m_s = \pm 1/2$ for an electron, we must have $s = 1/2$, and we say that the electron has spin one-half. There are only two different normalised spin eigenfunctions χ_{s,m_s}, namely

$$\alpha \equiv \chi_{1/2,1/2}; \qquad \beta \equiv \chi_{1/2,-1/2} \qquad \textbf{(2.197)}$$

and we see from (2.195) and (2.196) that

$$\mathbf{S}^2\alpha = \tfrac{3}{4}\hbar^2\alpha; \qquad \mathbf{S}^2\beta = \tfrac{3}{4}\hbar^2\beta \qquad \textbf{(2.198)}$$

and

$$S_z\alpha = \frac{\hbar}{2}\alpha; \qquad S_z\beta = -\frac{\hbar}{2}\beta \tag{2.199}$$

The spin eigenfunctions α and β are said to correspond respectively to *spin up* (↑) and *spin down* (↓) states. A general spin-1/2 function χ is an arbitrary linear superposition of the two basic spin states α and β. That is,

$$\chi = \chi_+\alpha + \chi_-\beta \tag{2.200}$$

where χ_+ and χ_- are complex coefficients such that $|\chi_+|^2$ is the probability of finding the electron in the 'spin up' state α, while $|\chi_-|^2$ is the probability of finding it in the 'spin down' state β. The normalisation condition $\langle\chi|\chi\rangle = 1$ gives

$$|\chi_+|^2 + |\chi_-|^2 = 1 \tag{2.201}$$

provided that the basic spin states α and β are orthonormal, namely

$$\begin{aligned} \langle\alpha|\alpha\rangle &= \langle\beta|\beta\rangle = 1 \\ \langle\alpha|\beta\rangle &= \langle\beta|\alpha\rangle = 0 \end{aligned} \tag{2.202}$$

We note from (2.198) and (2.200) that for an arbitrary spin-1/2 function χ we have $\mathbf{S}^2\chi = (3\hbar^2/4)\chi$, so that $\mathbf{S}^2$ is the purely numerical operator

$$\mathbf{S}^2 = \tfrac{3}{4}\hbar^2 \tag{2.203}$$

Introducing the raising and lowering operators

$$S_\pm = S_x \pm \mathrm{i}S_y \tag{2.204}$$

and using the general relations (A4.15)–(A4.16) of Appendix 4, with $j = 1/2$ and $m = \pm 1/2$, we have

$$\begin{aligned} S_+\alpha &= 0; \qquad S_+\beta = \hbar\alpha \\ S_-\alpha &= \hbar\beta; \qquad S_-\beta = 0 \end{aligned} \tag{2.205}$$

From (2.199), (2.204) and (2.205) we may construct a table which tells us how the components of $\mathbf{S}$ act on α and β. That is,

$$\begin{aligned} S_x\alpha &= \frac{\hbar}{2}\beta; \qquad S_x\beta = \frac{\hbar}{2}\alpha \\ S_y\alpha &= \frac{\mathrm{i}\hbar}{2}\beta; \qquad S_y\beta = -\frac{\mathrm{i}\hbar}{2}\alpha \\ S_z\alpha &= \frac{\hbar}{2}\alpha; \qquad S_z\beta = -\frac{\hbar}{2}\beta \end{aligned} \tag{2.206}$$

Using (2.200) and the results of this table, we remark that if χ is an arbitrary spin-1/2 function one has $S_x^2\chi = (\hbar^2/4)\chi$, with a similar result for S_y^2 and S_z^2.

Thus we have

$$S_x^2 = S_y^2 = S_z^2 = \frac{\hbar^2}{4} \tag{2.207}$$

Since there are only two basic spin states α and β, a *matrix representation* of the spin operators will only require two-by-two matrices. It is apparent from (2.195)–(2.199) that the matrices representing the operators $\mathbf{S}^2$ and S_z may be taken to be the two-by-two diagonal matrices

$$\mathbf{S}^2 = \frac{3}{4}\hbar^2\begin{pmatrix} 1 & 0 \\ 0 & 1 \end{pmatrix}; \qquad S_z = \frac{\hbar}{2}\begin{pmatrix} 1 & 0 \\ 0 & -1 \end{pmatrix} \tag{2.208}$$

The normalised spin-1/2 eigenfunctions α and β are given by the two-component column vectors (also called spinors)

$$\alpha = \begin{pmatrix} 1 \\ 0 \end{pmatrix}; \qquad \beta = \begin{pmatrix} 0 \\ 1 \end{pmatrix} \tag{2.209}$$

and may be considered as the basis vectors of a two-dimensional 'spin space'. The orthonormality relations (2.202) can then be written in the form

$$\begin{aligned} \alpha^\dagger\alpha &= \beta^\dagger\beta = 1 \\ \alpha^\dagger\beta &= \beta^\dagger\alpha = 0 \end{aligned} \tag{2.210}$$

where the dagger denotes the adjoint. Thus $\alpha^\dagger$ and $\beta^\dagger$ are the row vectors

$$\alpha^\dagger = (1 \quad 0); \qquad \beta^\dagger = (0 \quad 1) \tag{2.211}$$

and according to the rules of matrix multiplication, we have explicitly

$$\begin{aligned} \alpha^\dagger\alpha &= (1 \quad 0)\begin{pmatrix} 1 \\ 0 \end{pmatrix} = 1 \\ \alpha^\dagger\beta &= (1 \quad 0)\begin{pmatrix} 0 \\ 1 \end{pmatrix} = 0, \quad \text{etc.} \end{aligned} \tag{2.212}$$

It is also readily verified that the equations (2.205) are satisfied if

$$S_+ = \hbar\begin{pmatrix} 0 & 1 \\ 0 & 0 \end{pmatrix} \qquad S_- = \hbar\begin{pmatrix} 0 & 0 \\ 1 & 0 \end{pmatrix} \tag{2.213}$$

whence

$$S_x = \frac{\hbar}{2}\begin{pmatrix} 0 & 1 \\ 1 & 0 \end{pmatrix} \qquad S_y = \frac{\hbar}{2}\begin{pmatrix} 0 & -\mathrm{i} \\ \mathrm{i} & 0 \end{pmatrix} \tag{2.214}$$

The results (2.208) and (2.214) can also be written as

$$\mathbf{S} = \frac{\hbar}{2}\boldsymbol{\sigma} \tag{2.215}$$

where

$$\sigma_x = \begin{pmatrix} 0 & 1 \\ 1 & 0 \end{pmatrix} \qquad \sigma_y = \begin{pmatrix} 0 & -\mathrm{i} \\ \mathrm{i} & 0 \end{pmatrix} \qquad \sigma_z = \begin{pmatrix} 1 & 0 \\ 0 & -1 \end{pmatrix} \tag{2.216}$$

are called the *Pauli spin matrices*. Their principal properties, which can readily be verified (Problem 2.15), are summarised by the equations

$$\sigma_x^2 = \sigma_y^2 = \sigma_z^2 = 1 \tag{2.217a}$$

$$\sigma_x\sigma_y = -\sigma_y\sigma_x = \mathrm{i}\sigma_z; \quad \sigma_y\sigma_z = -\sigma_z\sigma_y = \mathrm{i}\sigma_x; \quad \sigma_z\sigma_x = -\sigma_x\sigma_z = \mathrm{i}\sigma_y \tag{2.217b}$$

$$\mathrm{Tr}\ \sigma_x = \mathrm{Tr}\ \sigma_y = \mathrm{Tr}\ \sigma_z = 0 \tag{2.217c}$$

$$\det \sigma_x = \det \sigma_y = \det \sigma_z = -1 \tag{2.217d}$$

where Tr means the trace and det the determinant. Moreover, the three Pauli matrices $\sigma_x, \sigma_y, \sigma_z$ and the unit two-by-two matrix form a complete set of 2×2 matrices, in the sense that an arbitrary 2×2 matrix can be expressed in terms of them. Finally, one can prove (Problem 2.15) the identity

$$(\boldsymbol{\sigma} \cdot \mathbf{A})(\boldsymbol{\sigma} \cdot \mathbf{B}) = \mathbf{A} \cdot \mathbf{B} + \mathrm{i}\boldsymbol{\sigma} \cdot (\mathbf{A} \times \mathbf{B}) \tag{2.218}$$

where **A** and **B** are any two vectors, or two vector operators whose components commute with those of the spin **S** (that is, with those of $\boldsymbol{\sigma}$). In the latter case the order of **A** and **B** on both sides of (2.218) must be respected.

Using the explicit form (2.209) of the basic spinors α and β, an arbitrary spin-1/2 function (2.200) may be written as the spinor

$$\chi = \begin{pmatrix} \chi_+ \\ \chi_- \end{pmatrix} \tag{2.219}$$

and the normalisation condition (2.201) becomes

$$\chi^\dagger \chi = 1 \tag{2.220}$$

where $\chi^\dagger$ denotes the adjoint of the spinor χ, namely

$$\chi^\dagger = (\chi_+^* \quad \chi_-^*) \tag{2.221}$$

It is worth noting that if the electron is in a pure 'spin up' state α or a pure 'spin down' state β, the expectation values of S_x and S_y vanish, $\langle S_x \rangle = \langle S_y \rangle = 0$, while $\langle S_x^2 \rangle = \langle S_y^2 \rangle = \hbar^2/4$ (see Problem 2.14). Thus, even when the spin angular momentum **S** is said to be 'up' ($m_s = +1/2$) or 'down' ($m_s = -1/2$), its x and y components are still not zero. As in the case of the orbital angular momentum **L**, these results can be visualised with the help of a vector model. According to this model, the spin vector **S**, of length $\sqrt{3/4}\,\hbar$, precesses about the Z axis, the only allowed projections of **S** on the Z axis being $m_s\hbar$, with $m_s = \pm 1/2$. This is illustrated in Fig. 2.8.

Thus far we have only considered the spin of the electron. Other particles, such as the atomic nuclei, may also possess a spin angular momentum **S**, for which the quantum number s (see (2.195)) can be either integer or half-odd integer.

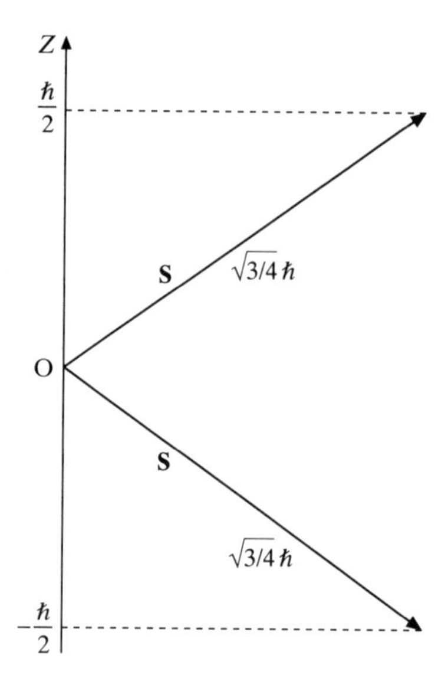

Figure 2.8 The vector model of the spin, for a spin-1/2 particle.

From the results of Appendix 4, with $j = s$ and $m = m_s$, we see that the quantum number m_s takes on the $2s + 1$ values $-s, -s + 1, \dots, s - 1, s$. The matrices representing the spin operators S_x, S_y, S_z and $\mathbf{S}^2$ then have dimensions $2s + 1$; they can be written explicitly by using the methods of Appendix 4 (see Problem 2.16).

Total angular momentum

The total angular momentum of a particle can be written as

$$\mathbf{J} = \mathbf{L} + \mathbf{S} \tag{2.222}$$

The orbital angular momentum $\mathbf{L} = \mathbf{r} \times \mathbf{p}$ operates only in 'ordinary' space and satisfies the commutation relations (2.155). On the other hand, the spin angular momentum $\mathbf{S}$ satisfies the commutation relations (2.194) and operates only in 'spin space'. All its components therefore commute with those of $\mathbf{r}$ and $\mathbf{p}$, and hence with all those of $\mathbf{L}$. As a result, the total angular momentum $\mathbf{J}$ satisfies the commutation relations

$$[J_x, J_y] = \mathrm{i}\hbar J_z, \qquad [J_y, J_z] = \mathrm{i}\hbar J_x, \qquad [J_z, J_x] = \mathrm{i}\hbar J_y \tag{2.223}$$

which characterise an angular momentum operator.

It is shown in Appendix 4 that the simultaneous eigenfunctions of $\mathbf{J}^2 = J_x^2 + J_y^2 + J_z^2$ and J_z^2 satisfy the eigenvalue equations

$$\mathbf{J}^2 \psi_{jm_j} = j(j+1)\hbar^2 \psi_{jm_j} \tag{2.224}$$

and

$$J_z \psi_{jm_j} = m_j \hbar \psi_{jm_j} \tag{2.225}$$

where j is a an integer (including zero) or a half-odd integer ($j = 0, 1/2, 1, 3/2, \ldots$) and $m_j = -j, -j+1, \ldots, j-1, j$.

Since all the components of $\mathbf{L}$ commute with all those of $\mathbf{S}$, the operators $\mathbf{L}^2$, L_z, $\mathbf{S}^2$ and S_z mutually commute, and have the simultaneous eigenfunctions

$$\psi_{lsm_lm_s} = Y_{lm_l}(\theta, \phi)\chi_{s,m_s} \tag{2.226}$$

where we have written $m_l \equiv m$ [5]. It can be shown (Problem 2.17) that the four operators $\mathbf{L}^2$, $\mathbf{S}^2$, $\mathbf{J}^2$ and J_z also form a commuting set of operators. Their simultaneous eigenfunctions are linear combinations of the functions $\psi_{lsm_lm_s}$ and are often denoted by the symbol $\mathcal{Y}_{ls}^{jm_j}$ (see Appendix 4). For a given value of l and s the possible values of j are given by

$$|l-s|, |l-s|+1, \ldots, l+s \tag{2.227}$$

For a given j the quantum number m_j can take on the $2j+1$ values $-j, -j+1, \ldots, j-1, j$, as we have seen above.

Wave functions for a spin-1/2 particle

Thus far in this section we have focused our attention on the angular and spin parts of the wave functions. In general, the wave functions also depend on the radial coordinate r and on the time t. For example, in the case of a spin-1/2 particle (for example, an electron), a general expression for the wave function is

$$\Psi(q, t) = \Psi_+(\mathbf{r}, t)\alpha + \Psi_-(\mathbf{r}, t)\beta \tag{2.228}$$

where q denotes the ensemble of the (continuous) spatial variables $\mathbf{r}$ *and* the (discrete) spin variable ($m_s = \pm 1/2$) of the particle. The probability density for finding at time t the particle at $\mathbf{r}$ with 'spin up' is $|\Psi_+(\mathbf{r}, t)|^2$, and with 'spin down' it is $|\Psi_-(\mathbf{r}, t)|^2$. Using (2.209) we may also write the wave function Ψ as a two-component or spinor wave function, namely

$$\Psi = \begin{pmatrix} \Psi_+ \\ \Psi_- \end{pmatrix} \tag{2.229}$$

2.6 Central forces

Let us now return to the Schrödinger equation (2.59) describing the motion of a spinless particle of mass m in a time-independent potential $V(\mathbf{r})$. We shall consider in this section the important case of *central* potentials, that is potentials $V(r)$ which depend only upon the magnitude $r = |\mathbf{r}|$ of the vector $\mathbf{r}$. Since $V(r)$ is spherically symmetric, it is natural to use the spherical polar coordinates defined in (2.157). The Hamiltonian of the system is then given by

[5] In general we shall use the notation m for a magnetic quantum number, but where it is important to distinguish between different kinds of angular momenta, we shall use the notation m_l, m_s, m_j, as necessary.

$$H = -\frac{\hbar^2}{2m}\nabla^2 + V(r)$$

$$= -\frac{\hbar^2}{2m}\left[\frac{1}{r^2}\frac{\partial}{\partial r}\left(r^2\frac{\partial}{\partial r}\right) + \frac{1}{r^2 \sin\theta}\frac{\partial}{\partial\theta}\left(\sin\theta\frac{\partial}{\partial\theta}\right) + \frac{1}{r^2\sin^2\theta}\frac{\partial^2}{\partial\phi^2}\right] + V(r) \tag{2.230}$$

Using the expression (2.159) of $\mathbf{L}^2$, we may also write

$$H = -\frac{\hbar^2}{2m}\left[\frac{1}{r^2}\frac{\partial}{\partial r}\left(r^2\frac{\partial}{\partial r}\right) - \frac{\mathbf{L}^2}{\hbar^2 r^2}\right] + V(r) \tag{2.231}$$

so that the Schrödinger equation (2.59) reads

$$\left\{-\frac{\hbar^2}{2m}\left[\frac{1}{r^2}\frac{\partial}{\partial r}\left(r^2\frac{\partial}{\partial r}\right) - \frac{\mathbf{L}^2}{\hbar^2 r^2}\right] + V(r)\right\}\psi(\mathbf{r}) = E\psi(\mathbf{r}) \tag{2.232}$$

Now $\mathbf{L}^2$ and L_z only operate on angular variables, and in addition $[\mathbf{L}^2, L_z] = 0$. Therefore, we see from (2.231) that

$$[H, \mathbf{L}^2] = [H, L_z] = 0 \tag{2.233}$$

We may thus look for solutions of the Schrödinger equation (2.232) which are simultaneous eigenfunctions of the operators H, $\mathbf{L}^2$ and L_z. Since the spherical harmonics $Y_{lm}(\theta, \phi)$ are simultaneous eigenfunctions of $\mathbf{L}^2$ and L_z (see (2.163) and (2.164)) we can write a particular solution as

$$\psi_{E,l,m}(r, \theta, \phi) = R_{E,l}(r)Y_{lm}(\theta, \phi) \tag{2.234}$$

Substituting (2.234) into (2.232), and using (2.163), we obtain for the radial function $R_{E,l}(r)$ the equation

$$\left\{-\frac{\hbar^2}{2m}\left[\frac{1}{r^2}\frac{\mathrm{d}}{\mathrm{d}r}\left(r^2\frac{\mathrm{d}}{\mathrm{d}r}\right) - \frac{l(l+1)}{r^2}\right] + V(r)\right\}R_{E,l}(r) = ER_{E,l}(r) \tag{2.235}$$

which shows that $R_{E,l}$ does not depend on the magnetic quantum number m.

The radial equation (2.235) can be simplified by introducing the new radial function

$$u_{E,l}(r) = rR_{E,l}(r) \tag{2.236}$$

The new radial equation which we obtain for $u_{E,l}(r)$ is then

$$\left[-\frac{\hbar^2}{2m}\frac{\mathrm{d}^2}{\mathrm{d}r^2} + \frac{l(l+1)\hbar^2}{2mr^2} + V(r)\right]u_{E,l}(r) = Eu_{E,l}(r) \tag{2.237}$$

For potentials which are less singular than r^{-2} at the origin, a power series expansion of $u_{E,l}(r)$ can be made for small r, and the examination of the indicial equation shows that

$$u_{E,l}(r) \underset{r\to 0}{\sim} r^{l+1} \tag{2.238}$$

The equation (2.237) is similar to the one-dimensional Schrödinger equation (2.121) for an *effective potential*

$$V_{\text{eff}}(r) = V(r) + \frac{l(l+1)\hbar^2}{2mr^2} \tag{2.239}$$

which contains the repulsive *centrifugal barrier* term $l(l+1)\hbar^2/(2mr^2)$ in addition to $V(r)$. We remark, however, that the variable r is confined to positive values $0 \leqslant r \leqslant \infty$, in contrast to the variable x in (2.121).

Parity

The *parity* operator $\mathcal{P}$ is defined by the relation

$$\mathcal{P}f(\mathbf{r}) = f(-\mathbf{r}) \tag{2.240}$$

where $f(\mathbf{r})$ is an arbitrary function. Thus the parity operator corresponds to an inversion of the position coordinate $\mathbf{r}$ through the origin. It is a Hermitian operator since, for any two wave functions $\phi(\mathbf{r})$ and $\psi(\mathbf{r})$, we have

$$\begin{aligned}\int \phi^*(\mathbf{r})\mathcal{P}\psi(\mathbf{r})\,\mathrm{d}\mathbf{r} &= \int \phi^*(\mathbf{r})\psi(-\mathbf{r})\,\mathrm{d}\mathbf{r} \\ &= \int \phi^*(-\mathbf{r})\psi(\mathbf{r})\,\mathrm{d}\mathbf{r} \\ &= \int [\mathcal{P}\phi(\mathbf{r})]^*\psi(\mathbf{r})\,\mathrm{d}\mathbf{r}\end{aligned} \tag{2.241}$$

Let us now consider the eigenvalue equation for $\mathcal{P}$, which we write as

$$\mathcal{P}\psi_\alpha(\mathbf{r}) = \alpha\psi_\alpha(\mathbf{r}) \tag{2.242}$$

From the definition (2.240) of $\mathcal{P}$ and the fact that $\mathcal{P}f(-\mathbf{r}) = f(\mathbf{r})$, we deduce that

$$\mathcal{P}^2 = I \tag{2.243}$$

where I is the unit operator. Hence

$$\mathcal{P}^2\psi_\alpha(\mathbf{r}) = \alpha\mathcal{P}\psi_\alpha(\mathbf{r}) = \alpha^2\psi_\alpha(\mathbf{r}) = \psi_\alpha(\mathbf{r}) \tag{2.244}$$

so that $\alpha^2 = 1$ and the eigenvalues of $\mathcal{P}$ are $\alpha = \pm 1$. Denoting the corresponding eigenfunctions by ψ_+ and ψ_-, we have

$$\mathcal{P}\psi_+(\mathbf{r}) = \psi_+(\mathbf{r}), \qquad \mathcal{P}\psi_-(\mathbf{r}) = -\psi_-(\mathbf{r}) \tag{2.245}$$

or

$$\psi_+(-\mathbf{r}) = \psi_+(\mathbf{r}), \qquad \psi_-(-\mathbf{r}) = -\psi_-(\mathbf{r}) \tag{2.246}$$

Thus $\psi_+(\mathbf{r})$ is an *even* function of $\mathbf{r}$, and $\psi_-(\mathbf{r})$ is an *odd* function of $\mathbf{r}$. The eigenfunctions ψ_+ are said to have *even parity*, while the eigenfunctions ψ_- have *odd parity*. We note that ψ_+ and ψ_- are orthogonal. They also form a complete set, since any function $\psi(\mathbf{r})$ can always be written as

$$\psi(\mathbf{r}) = \psi_+(\mathbf{r}) + \psi_-(\mathbf{r}) \tag{2.247}$$

where

$$\psi_+(\mathbf{r}) = \tfrac{1}{2}[\psi(\mathbf{r}) + \psi(-\mathbf{r})] \tag{2.248}$$

has obviously even parity, while

$$\psi_-(\mathbf{r}) = \tfrac{1}{2}[\psi(\mathbf{r}) - \psi(-\mathbf{r})] \tag{2.249}$$

has odd parity.

Let us now return to the central force problem. Under the parity operation $\mathbf{r} \to -\mathbf{r}$, the spherical polar coordinates (r, θ, ϕ) become $(r, \pi - \theta, \phi + \pi)$. The central force Hamiltonian (2.230) is clearly unaffected by this operation, or in other words the parity operator $\mathcal{P}$ commutes with the Hamiltonian (2.230)

$$[\mathcal{P}, H] = 0 \tag{2.250}$$

As a result, simultaneous eigenfunctions of the operators $\mathcal{P}$ and H can be found. Applying the parity operator on the wave function $\psi_{E,l,m}(r, \theta, \phi) = R_{E,l}(r)Y_{lm}(\theta, \phi)$ (see (2.234)), we have

$$\mathcal{P}[R_{E,l}(r)Y_{lm}(\theta, \phi)] = R_{E,l}(r)Y_{lm}(\pi - \theta, \phi + \pi) \tag{2.251}$$

Now, from the definition (2.181) of the spherical harmonics, it can be shown that

$$Y_{lm}(\pi - \theta, \phi + \pi) = (-1)^l Y_{lm}(\theta, \phi) \tag{2.252}$$

so that Y_{lm} has the parity of l. Thus

$$\mathcal{P}[R_{E,l}(r)Y_{lm}(\theta, \phi)] = R_{E,l}(r)(-1)^l Y_{lm}(\theta, \phi) \tag{2.253}$$

and the wave function $\psi_{E,l,m}$ itself has the parity of l (even for even l, odd for odd l).

The free particle

We shall discuss at length various applications involving central potentials in subsequent chapters. For further reference, we consider here the (very) special case of the free particle, for which $V(r) = 0$. Writing $k = \sqrt{2mE}/\hbar$, the radial equation (2.235) becomes

$$\left[\frac{d^2}{dr^2} + \frac{2}{r}\frac{d}{dr} - \frac{l(l+1)}{r^2} + k^2\right]R_{E,l}(r) = 0 \tag{2.254}$$

It is convenient to set $\rho = kr$ and to write $R_l(\rho) \equiv R_{E,l}(r)$, so that the above equation reads

$$\left[\frac{\mathrm{d}^2}{\mathrm{d}\rho^2} + \frac{2}{\rho}\frac{\mathrm{d}}{\mathrm{d}\rho} + \left(1 - \frac{l(l+1)}{\rho^2}\right)\right] R_l(\rho) = 0 \tag{2.255}$$

This equation is known as the *spherical Bessel differential equation.* The solutions of (2.255) which are regular and finite everywhere are given (up to a multiplicative constant) by the spherical Bessel functions $j_l(\rho)$, defined by

$$j_l(\rho) = \left(\frac{\pi}{2\rho}\right)^{1/2} J_{l+\frac{1}{2}}(\rho) \tag{2.256}$$

where $J_\upsilon(\rho)$ is a Bessel function of order υ. The first few spherical Bessel functions are given explicitly by

$$\begin{aligned} j_0(\rho) &= \frac{\sin\rho}{\rho} \\ j_1(\rho) &= \frac{\sin\rho}{\rho^2} - \frac{\cos\rho}{\rho} \\ j_2(\rho) &= \left(\frac{3}{\rho^3} - \frac{1}{\rho}\right)\sin\rho - \frac{3}{\rho^2}\cos\rho \end{aligned} \tag{2.257}$$

Additional properties of the functions $j_l(\rho)$ will be discussed in Chapter 12.

The eigenvalues k^2 of (2.254) can take on any positive real value. As a result, the energy $E = \hbar^2k^2/(2m)$ can assume any value in the interval $(0, \infty)$, giving an example of a *continuous spectrum.* Thus, returning to (2.234) and using the foregoing results, we see that for every positive value of E there exist eigenfunctions of the free particle Schrödinger equation labelled by the orbital angular momentum quantum numbers (l, m), namely

$$\psi_{E,l,m}(r, \theta, \phi) = Cj_l(kr)Y_{lm}(\theta, \phi) \tag{2.258}$$

where C is a constant. It can be shown that the ensemble of *spherical waves* (2.258) forms a complete set.

On the other hand, we have seen in Section 2.1 that a free particle of momentum $\hbar\mathbf{k}$ and energy $E = \hbar^2k^2/(2m)$ is represented by the plane wave $\exp(\mathrm{i}\mathbf{k}\cdot\mathbf{r})$. Since the spherical waves (2.258) form a complete set, we may expand the plane wave $\exp(\mathrm{i}\mathbf{k}\cdot\mathbf{r})$ in terms of them. That is,

$$\exp(\mathrm{i}\mathbf{k}\cdot\mathbf{r}) = \sum_{l=0}^{\infty}\sum_{m=-l}^{+l} c_{lm}(\mathbf{k})j_l(kr)Y_{lm}(\theta, \phi) \tag{2.259}$$

If we choose the Z axis to be along the wave vector $\mathbf{k}$, then the left-hand side of (2.259) reads $\exp(\mathrm{i}\mathbf{k}\cdot\mathbf{r}) = \exp(\mathrm{i}kr\cos\theta)$, which is independent of ϕ. Setting $w = \cos\theta$, the expansion of $\exp(\mathrm{i}\mathbf{k}\cdot\mathbf{r})$ reduces to an expansion in terms of

Legendre polynomials $P_l(w)$. A straightfoward calculation (Problem 2.18) then yields

$$\exp(\mathrm{i}\mathbf{k}\cdot\mathbf{r}) = \sum_{l=0}^{\infty}(2l+1)\mathrm{i}^l j_l(kr)P_l(\cos\theta) \tag{2.260}$$

Using the addition theorem of the spherical harmonics (see the equation (A4.23) of Appendix 4) we can also write the above formula in the form of equation (2.259). That is,

$$\exp(\mathrm{i}\mathbf{k}\cdot\mathbf{r}) = 4\pi\sum_{l=0}^{\infty}\sum_{m=-l}^{+l}\mathrm{i}^l j_l(kr)Y^*_{lm}(\hat{\mathbf{k}})\ Y_{lm}(\hat{\mathbf{r}}) \tag{2.261}$$

where $\hat{\mathbf{x}}$ denotes the polar angles of a vector $\mathbf{x}$. Upon comparison of (2.259) and (2.261) we see that the coefficients $c_{lm}(\mathbf{k})$ are given by $c_{lm}(\mathbf{k}) = 4\pi\mathrm{i}^l Y^*_{lm}(\hat{\mathbf{k}})$.

2.7 Several-particle systems

Until now we have only discussed the motion of a single particle. In this section we shall generalise our results to N-particle systems. We begin by considering a system of N non-relativistic *spinless* particles having masses m_i, position coordinates $\mathbf{r}_i$ and momenta $\mathbf{p}_i$ ($i = 1, 2, \ldots, N$). The classical Hamiltonian function of this system is

$$H(\mathbf{r}_1, \ldots, \mathbf{r}_N, \mathbf{p}_1, \ldots, \mathbf{p}_N, t) = \sum_{i=1}^{N}\frac{\mathbf{p}_i^2}{2m_i} + V(\mathbf{r}_1, \ldots, \mathbf{r}_N, t) \tag{2.262}$$

where V is the potential energy. The total classical energy E_{tot} of the system is

$$E_{\text{tot}} = H(\mathbf{r}_1, \ldots, \mathbf{r}_N, \mathbf{p}_1, \ldots, \mathbf{p}_N, t) \tag{2.263}$$

The Schrödinger equation for the wave function $\Psi(\mathbf{r}_1, \ldots, \mathbf{r}_N, t)$ which describes the dynamical state of the system is obtained by considering E_{tot} and $\mathbf{p}_i$ to be the differential operators (compare with (2.11))

$$E_{\text{op}} = \mathrm{i}\hbar\frac{\partial}{\partial t}, \qquad (\mathbf{p}_i)_{\text{op}} = -\mathrm{i}\hbar\nabla_{\mathbf{r}_i} \tag{2.264}$$

and writing that E_{op} and $H_{\text{op}} \equiv H(\mathbf{r}_1, \ldots, \mathbf{r}_N, -\mathrm{i}\hbar\nabla_{\mathbf{r}_1}, \ldots, -\mathrm{i}\hbar\nabla_{\mathbf{r}_N}, t)$ give identical results when acting on Ψ. That is,

$$\mathrm{i}\hbar\frac{\partial}{\partial t}\Psi = H_{\text{op}}\Psi = \left[\sum_{i=1}^{N}\left(-\frac{\hbar^2}{2m_i}\nabla^2_{\mathbf{r}_i}\right) + V(\mathbf{r}_1, \ldots, \mathbf{r}_N, t)\right]\Psi \tag{2.265}$$

The normalisation condition (2.26) now takes the form

$$\int|\Psi(\mathbf{r}_1, \ldots, \mathbf{r}_N, t)|^2\,\mathrm{d}\mathbf{r}_1\ldots\mathrm{d}\mathbf{r}_N = 1 \tag{2.266}$$

The total orbital angular momentum $\mathbf{L}$ of the system is the sum of the individual orbital angular momenta $\mathbf{L}_i = \mathbf{r}_i \times \mathbf{p}_i$:

$$\mathbf{L} = \sum_{i=1}^{N} \mathbf{L}_i \tag{2.267}$$

It is also worth noting that all the components of the position $\mathbf{r}_i$ and momentum $\mathbf{p}_i$ of particle i commute with all those pertaining to particle j provided that $i \neq j$, so that the fundamental commutation relations read

$$[x_i, p_{x_j}] = \mathrm{i}\hbar\delta_{ij}, \qquad [y_i, p_{x_j}] = 0, \text{ etc.} \tag{2.268}$$

Two-body systems

Of particular interest is a system of two particles, of masses m_1 and m_2, interacting via a time-independent potential $V(\mathbf{r}_1 - \mathbf{r}_2)$ which depends only upon the relative coordinate $\mathbf{r}_1 - \mathbf{r}_2$. The classical Hamiltonian of the system is therefore given by

$$H = \frac{\mathbf{p}_1^2}{2m_1} + \frac{\mathbf{p}_2^2}{2m_2} + V(\mathbf{r}_1 - \mathbf{r}_2) \tag{2.269}$$

Making the substitutions $\mathbf{p}_1 \to -\mathrm{i}\hbar\nabla_{\mathbf{r}_1}$ and $\mathbf{p}_2 \to -\mathrm{i}\hbar\nabla_{\mathbf{r}_2}$ in (2.269), we obtain the quantum mechanical Hamiltonian operator, and the corresponding Schrödinger equation reads

$$\mathrm{i}\hbar\frac{\partial}{\partial t}\Psi(\mathbf{r}_1, \mathbf{r}_2, t) = \left[-\frac{\hbar^2}{2m_1}\nabla_{\mathbf{r}_1}^2 - \frac{\hbar^2}{2m_2}\nabla_{\mathbf{r}_2}^2 + V(\mathbf{r}_1 - \mathbf{r}_2)\right]\Psi(\mathbf{r}_1, \mathbf{r}_2, t) \tag{2.270}$$

Because the potential V only depends on the difference of coordinates $\mathbf{r}_1 - \mathbf{r}_2$, an important simplification can now be made. We introduce the relative coordinate

$$\mathbf{r} = \mathbf{r}_1 - \mathbf{r}_2 \tag{2.271}$$

together with the vector

$$\mathbf{R} = \frac{m_1\mathbf{r}_1 + m_2\mathbf{r}_2}{m_1 + m_2} \tag{2.272}$$

which determines the position of the centre of mass (CM) of the system. Changing variables from the coordinates $(\mathbf{r}_1, \mathbf{r}_2)$ to the new coordinates $(\mathbf{r}, \mathbf{R})$ one finds that the Schrödinger equation (2.270) becomes

$$\mathrm{i}\hbar\frac{\partial}{\partial t}\Psi(\mathbf{R}, \mathbf{r}, t) = \left[-\frac{\hbar^2}{2M}\nabla_{\mathbf{R}}^2 - \frac{\hbar^2}{2\mu}\nabla_{\mathbf{r}}^2 + V(\mathbf{r})\right]\Psi(\mathbf{R}, \mathbf{r}, t) \tag{2.273}$$

where

$$M = m_1 + m_2 \tag{2.274}$$

is the total mass of the system and

$$\mu = \frac{m_1 m_2}{m_1 + m_2} \tag{2.275}$$

is the reduced mass of the two particles. The Schrödinger equation (2.273) may also be obtained by introducing the relative momentum

$$\mathbf{p} = \frac{m_2 \mathbf{p}_1 - m_1 \mathbf{p}_2}{m_1 + m_2} \tag{2.276}$$

together with the total momentum

$$\mathbf{P} = \mathbf{p}_1 + \mathbf{p}_2 \tag{2.277}$$

Since

$$\frac{\mathbf{p}_1^2}{2m_1} + \frac{\mathbf{p}_2^2}{2m_2} = \frac{\mathbf{P}^2}{2M} + \frac{\mathbf{p}^2}{2\mu} \tag{2.278}$$

the classical Hamiltonian (2.269) can be written as

$$H = \frac{\mathbf{P}^2}{2M} + \frac{\mathbf{p}^2}{2\mu} + V(\mathbf{r}) \tag{2.279}$$

Performing the substitutions $\mathbf{P} \rightarrow -i\hbar\nabla_{\mathbf{R}}$ and $\mathbf{p} \rightarrow -i\hbar\nabla_{\mathbf{r}}$ in (2.279), we then obtain the quantum mechanical Hamiltonian operator leading to the Schrödinger equation (2.273).

Two separations of the equation (2.273) can now be made. The time dependence can first be separated as in (2.58) since the potential is time-independent. Secondly, the spatial part of the wave function can be separated into a product of functions of the centre of mass coordinate $\mathbf{R}$ and of the relative coordinate $\mathbf{r}$. Thus the Schrödinger equation (2.273) admits solutions of the form

$$\Psi(\mathbf{R}, \mathbf{r}, t) = \Phi(\mathbf{R})\psi(\mathbf{r}) \exp[-i(E_{CM} + E)t/\hbar] \tag{2.280}$$

where the functions $\Phi(\mathbf{R})$ and $\psi(\mathbf{r})$ satisfy respectively the equations

$$-\frac{\hbar^2}{2M}\nabla_{\mathbf{R}}^2\Phi(\mathbf{R}) = E_{CM}\Phi(\mathbf{R}) \tag{2.281}$$

and

$$\left[-\frac{\hbar^2}{2\mu}\nabla_{\mathbf{r}}^2 + V(\mathbf{r})\right]\psi(\mathbf{r}) = E\psi(\mathbf{r}) \tag{2.282}$$

We see that the equation (2.281) is a time-independent Schrödinger equation describing the centre of mass as a free particle of mass M and energy E_{CM}. The second time-independent Schrödinger equation (2.282) describes the relative motion of the two particles; it is the same as the equation corresponding to the motion of a particle having the reduced mass μ in the potential $V(\mathbf{r})$. The total energy of the system is clearly

$$E_{\text{tot}} = E_{\text{CM}} + E \tag{2.283}$$

We have therefore 'decoupled' the original two-body problem into two one-body problems, that of a free particle (the centre of mass) and that of a single particle of reduced mass μ in a potential $V(\mathbf{r})$. We remark that if we elect to work in the centre of mass system of the two particles, we need not be concerned with the motion of the centre of mass, the coordinates of which are eliminated.

Systems of particles with spin. Addition of angular momenta

Let us now consider a system of particles which may possess spin. We shall denote by q_i the ensemble of space and spin coordinates of the particle i. The angular momentum $\mathbf{J}_i$ of that particle is the sum of its orbital angular momentum $\mathbf{L}_i$ and its spin angular momentum $\mathbf{S}_i$,

$$\mathbf{J}_i = \mathbf{L}_i + \mathbf{S}_i \tag{2.284}$$

The total orbital angular momentum $\mathbf{L}$ of the system is given by (2.267). The total spin angular momentum is

$$\mathbf{S} = \sum_{i=1}^{N} \mathbf{S}_i \tag{2.285}$$

Finally, the total angular momentum of the system is the sum of the individual angular momenta $\mathbf{J}_i$,

$$\mathbf{J} = \sum_{i=1}^{N} \mathbf{J}_i \tag{2.286}$$

In many problems of quantum physics the Hamiltonian H of the system is rotationally invariant. As a consequence, it may be shown [6] that H commutes with the components of the total angular momentum $\mathbf{J}$, so that we can look for the eigenfunctions of H among the simultaneous eigenfunctions of $\mathbf{J}^2$ and J_z. On the other hand, we know in general how to obtain the eigenfunctions of the individual angular momentum operators $\mathbf{J}_i^2$, J_{iz} (see Appendix 4). The problem of the *addition of angular momenta* consists in obtaining the eigenvalues and eigenfunctions of $\mathbf{J}^2$ and J_z in terms of those of $\mathbf{J}_i^2$ and J_{iz}. The simplest addition problem, namely that of adding *two* angular momenta, is discussed in Appendix 4.

Indistinguishable particles

Two particles are said to be *identical* when they cannot be distinguished by means of any intrinsic property. Many atomic or molecular systems contain a number of

[6] See Bransden and Joachain (2000).

particles (notably electrons) which are all identical. While in classical physics the existence of sharp trajectories makes it possible, in principle, to distinguish 'classical' particles by their paths, in quantum mechanics there is no way of keeping track of individual particles when the wave functions of identical particles overlap. Thus, it is impossible in quantum mechanics to distinguish between identical particles in regions of space where they may be found simultaneously, such as their interaction region. As we shall now see, this quantum mechanical indistinguishability of identical particles has profound consequences.

Let us consider a quantum mechanical system of N identical particles. Since the N particles are identical, all observables corresponding to this system must be symmetric functions of the basic dynamical variables. In particular, the Hamiltonian H of the system must be symmetric with respect to any interchange of the space *and* spin coordinates of the particles. Thus an interchange operator P_{ij} that permutes the variables q_i and q_j of the particles i and j commutes with the Hamiltonian:

$$[P_{ij}, H] = 0 \tag{2.287}$$

In general, an exact eigenfunction $\psi(q_1, \ldots, q_N)$ of H has no particular symmetry property under the interchange of the variables q_i. However, it is important to recognise that if $\psi(q_1, \ldots, q_i, \ldots, q_j, \ldots, q_N)$ is an eigenfunction of H corresponding to the eigenvalue E, then so is $P_{ij}\psi$, where

$$P_{ij}\psi(q_1, \ldots, q_i, \ldots, q_j, \ldots, q_N) = \psi(q_1, \ldots, q_j, \ldots, q_i, \ldots, q_N) \tag{2.288}$$

Since two successive interchanges of q_i and q_j bring the particles back to their initial configurations, we have

$$P_{ij}^2 = I \tag{2.289}$$

so that the eigenvalues of the operator P_{ij} are $\varepsilon = \pm 1$. Wave functions corresponding to the eigenvalue $\varepsilon = +1$ are such that

$$\begin{aligned} P_{ij}\psi(q_1, \ldots, q_i, \ldots, q_j, \ldots, q_N) &= \psi(q_1, \ldots, q_j, \ldots, q_i, \ldots, q_N) \\ &= \psi(q_1, \ldots, q_i, \ldots, q_j, \ldots, q_N) \end{aligned} \tag{2.290}$$

and are said to be *symmetric* under the interchange P_{ij}. On the other hand, wave functions which correspond to the eigenvalue $\varepsilon = -1$ are such that

$$\begin{aligned} P_{ij}\psi(q_1, \ldots, q_i, \ldots, q_j, \ldots, q_N) &= \psi(q_1, \ldots, q_j, \ldots, q_i, \ldots, q_N) \\ &= -\psi(q_1, \ldots, q_i, \ldots, q_j, \ldots, q_N) \end{aligned} \tag{2.291}$$

and are said to be *antisymmetric* under the interchange P_{ij}.

More generally, there are $N!$ different permutations of the variables $q_1, \ldots, q_N$. Defining P as the permutation that replaces q_n by q_{Pn}, with $n = 1, 2, \ldots, N$ and noting that P can be obtained as a succession of interchanges, we have

$$[P, H] = 0 \tag{2.292}$$

A permutation P is said to be *even* or *odd* depending on whether the number of interchanges leading to it is even or odd. If we let the operator P act on a wave function $\psi(q_1, \ldots, q_N)$, we have

$$P\psi(q_1, \ldots, q_N) = \psi(q_{P1}, \ldots, q_{PN}) \quad \textbf{(2.293)}$$

It is worth stressing that except for the case $N = 2$ the $N!$ permutations P do *not* commute among themselves. This is due to the fact that the interchange operators P_{ij} and P_{ik} $(k \neq j)$ do not mutually commute. Therefore, the eigenfunctions $\psi(q_1, \ldots, q_N)$ of H are not *in general* eigenfunctions of all the $N!$ permutation operators P. However, there are *two* exceptional states which *are* eigenstates of H and of the $N!$ permutation operators P. The first one is the *totally symmetric state* $\psi_S(q_1, \ldots, q_N)$ satisfying (2.290) for *any* interchange P_{ij}, so that for all P

$$\begin{aligned} P\psi_S(q_1, \ldots, q_N) &= \psi_S(q_{P1}, \ldots, q_{PN}) \\ &= \psi_S(q_1, \ldots, q_N) \end{aligned} \quad \textbf{(2.294)}$$

The other one is the *totally antisymmetric state* $\psi_A(q_1, \ldots, q_N)$, which satisfies (2.291) for *any* interchange P_{ij}. Hence, for all P

$$\begin{aligned} P\psi_A(q_1, \ldots, q_N) &= \psi_A(q_{P1}, \ldots, q_{PN}) \\ &= \begin{cases} \psi_A(q_1, \ldots, q_N) & \text{for an even permutation} \\ -\psi_A(q_1, \ldots, q_N) & \text{for an odd permutation} \end{cases} \end{aligned} \quad \textbf{(2.295)}$$

We also remark that the equation (2.292) implies that P is a constant of the motion, so that a system of identical particles represented by a wave function of a given symmetry (S or A) will keep that symmetry at all times.

Bosons and fermions

According to our present knowledge of particles occurring in nature, the two types of states ψ_S and ψ_A are thought to be sufficient to describe all systems of identical particles. This is called the *symmetrisation postulate*. Particles having states described by totally symmetric wave functions are called *bosons*; they obey Bose–Einstein statistics [7]. Experiment shows that particles of *zero* or *integer* spin ($s = 0, 1, 2, \ldots$) are bosons. For example, all the *mesons* (such as the π and K mesons, which have spin $s = 0$, the ρ meson, which has spin $s = 1$, etc.) are bosons. The photon (which has spin $s = 1$), is also a boson, as are the newly discovered W and Z particles, which also have spin $s = 1$, and are called 'intermediate vector bosons'. On the other hand, particles having states described by totally antisymmetric wave functions are called *fermions*; they satisfy Fermi–Dirac statistics [7]. Experiment shows that particles having half-odd integer spin ($s = 1/2, 3/2, \ldots$) are fermions. For example, all the *leptons* (such as the electron, the muon and the

[7] See Bransden and Joachain (2000).

neutrinos, which have spin $s = 1/2$) are fermions, as are the *baryons* (such as the proton and the neutron, which have spin $s = 1/2$, and the hyperons). A composite particle, such as an atom, is a boson if the sum of its protons, neutrons and electrons is an even number; it is a fermion if this sum is an odd number.

If a system is composed of different kinds of bosons (fermions), then its wave function must be separately totally symmetric (antisymmetric) with respect to permutations of each kind of identical particle. For example, the wave function of an alpha particle (that is, a ^{4}He nucleus which contains two protons and two neutrons) must be antisymmetric under the interchange of the two protons and also antisymmetric under the interchange of the two neutrons. It should be noted, however, that in some physical situations one can ignore the possible composite structure of particles. For example, in studying the scattering of two alpha particles at low energies (where the effect of nuclear forces is negligible), the symmetrisation process must be accomplished only with respect to the permutation of the two alpha particles, each of them being considered as a spin zero boson [8]. Similarly, in studying the properties of atomic gases at low temperature, one can treat the atoms as 'elementary' bosons or fermions.

2.8 Approximation methods

As in the case of classical mechanics, there are relatively few physically interesting problems in quantum mechanics which can be solved exactly. Approximation methods are therefore of great importance in discussing the application of quantum theory to specific systems, such as the atomic and molecular ones considered in this book. In this section we shall review several approximation methods which will be used in further chapters.

Time-independent perturbation theory

Perturbation theory deals with the changes induced in a system by a 'small' disturbance. Although we shall also apply perturbation methods to scattering problems at a later stage (see Chapters 12–14) we shall start here by discussing the *Rayleigh–Schrödinger perturbation theory*, which is concerned with the modifications in the *discrete* energy levels and corresponding eigenfunctions of a system when a perturbation is applied.

Let us assume that the time-independent Hamiltonian H of a system may be separated into two parts,

$$H = H_0 + \lambda H' \tag{2.296}$$

[8] On the other hand, if one wants to study the scattering of two alpha particles at high energies (where nuclear forces are important), the symmetrisation of the total wave function must be performed separately with respect to the constituent protons and neutrons.

where the 'unperturbed' Hamiltonian H_0 is sufficiently simple so that the corresponding Schrödinger eigenvalue equation

$$H_0\psi_k^{(0)} = E_k^{(0)}\psi_k^{(0)} \quad \textbf{(2.297)}$$

may be solved, and the term $\lambda H'$ is a small perturbation. The parameter λ will be used below to distinguish between the various orders of the perturbation calculation; we shall take $\lambda = 1$ for the actual physical problem. We assume that the (known) eigenfunctions $\psi_k^{(0)}$ corresponding to the (known) eigenvalues $E_k^{(0)}$ of H_0 form a complete orthonormal set (which may be partly continuous). Thus, if $\psi_i^{(0)}$ and $\psi_j^{(0)}$ are two members of that set, we have

$$\langle\psi_i^{(0)}|\psi_j^{(0)}\rangle = \delta_{ij} \quad \text{or} \quad \delta(i-j) \quad \textbf{(2.298)}$$

where the symbol $\delta(i-j)$ should be used when both $\psi_i^{(0)}$ and $\psi_j^{(0)}$ correspond to continuous states. In what follows we shall simplify the notation by extending the meaning of δ_{ij} to cover both possibilities in (2.298). The eigenvalue problem which we want to solve is

$$H\psi_k = E_k\psi_k \quad \textbf{(2.299)}$$

where we have used the notation E_k and ψ_k to denote the perturbed energy levels and eigenfunctions, respectively.

Non-degenerate case

Let us focus our attention on a particular unperturbed, discrete energy level $E_k^{(0)}$, which we assume to be *non-degenerate*. We suppose that the effect of the perturbation $\lambda H'$ is small enough so that the perturbed energy level E_k is much closer to $E_k^{(0)}$ than to any other unperturbed level. It is then reasonable to expand both ψ_k and E_k in powers of λ, namely

$$\psi_k = \sum_{n=0}^{\infty} \lambda^n \psi_k^{(n)} \quad \textbf{(2.300)}$$

and

$$E_k = \sum_{n=0}^{\infty} \lambda^n E_k^{(n)} \quad \textbf{(2.301)}$$

where the index n refers to the order of the perturbation. Substituting the expansions (2.300) and (2.301) in (2.299), and using (2.296), we have

$$\begin{aligned}&(H_0 + \lambda H')(\psi_k^{(0)} + \lambda\psi_k^{(1)} + \lambda^2\psi_k^{(2)} + \dots)\\ &= (E_k^{(0)} + \lambda E_k^{(1)} + \lambda^2 E_k^{(2)} + \dots)(\psi_k^{(0)} + \lambda\psi_k^{(1)} + \lambda^2\psi_k^{(2)} + \dots)\end{aligned} \quad \textbf{(2.302)}$$

Let us now equate the coefficients of equal powers of λ. Beginning with λ^0, we see that

$$H_0\psi_k^{(0)} = E_k^{(0)}\psi_k^{(0)} \quad \textbf{(2.303)}$$

as expected. The coefficient of λ then gives

$$H_0\psi_k^{(1)} + H'\psi_k^{(0)} = E_k^{(0)}\psi_k^{(1)} + E_k^{(1)}\psi_k^{(0)} \tag{2.304}$$

while that of λ^2 yields

$$H_0\psi_k^{(2)} + H'\psi_k^{(1)} = E_k^{(0)}\psi_k^{(2)} + E_k^{(1)}\psi_k^{(1)} + E_k^{(2)}\psi_k^{(0)} \tag{2.305}$$

and so on.

In order to obtain the first energy correction $E_k^{(1)}$, we premultiply (2.304) by $\psi_k^{(0)*}$ and integrate over all space. This gives

$$\langle\psi_k^{(0)}|H_0 - E_k^{(0)}|\psi_k^{(1)}\rangle + \langle\psi_k^{(0)}|H' - E_k^{(1)}|\psi_k^{(0)}\rangle = 0 \tag{2.306}$$

Now, using (2.297) and the fact that the operator H_0 is Hermitian, we have

$$\begin{aligned}\langle\psi_k^{(0)}|H_0|\psi_k^{(1)}\rangle &= \langle H_0\psi_k^{(0)}|\psi_k^{(1)}\rangle \\ &= E_k^{(0)}\langle\psi_k^{(0)}|\psi_k^{(1)}\rangle\end{aligned} \tag{2.307}$$

so that the first term on the left of (2.306) vanishes. Moreover, since $\langle\psi_k^{(0)}|\psi_k^{(0)}\rangle = 1$, we see that (2.306) reduces to

$$E_k^{(1)} = \langle\psi_k^{(0)}|H'|\psi_k^{(0)}\rangle \equiv H'_{kk} \tag{2.308}$$

This important result tells us that the first-order correction to the energy for a non-degenerate level is just the expectation value of the perturbation H' with respect to the corresponding unperturbed state of the system.

Proceeding in a similar way with equation (2.305), we have

$$\langle\psi_k^{(0)}|H_0 - E_k^{(0)}|\psi_k^{(2)}\rangle + \langle\psi_k^{(0)}|H' - E_k^{(1)}|\psi_k^{(1)}\rangle - E_k^{(2)}\langle\psi_k^{(0)}|\psi_k^{(0)}\rangle = 0 \tag{2.309}$$

and therefore

$$E_k^{(2)} = \langle\psi_k^{(0)}|H' - E_k^{(1)}|\psi_k^{(1)}\rangle \tag{2.310}$$

An equivalent expression for $E_k^{(2)}$ may be obtained by starting from (2.304), and is given (Problem 2.19) by

$$E_k^{(2)} = -\langle\psi_k^{(1)}|H_0 - E_k^{(0)}|\psi_k^{(1)}\rangle \tag{2.311}$$

Expressions for higher order corrections $E_k^{(n)}$, $n \geqslant 3$, can be obtained in a similar way. For example, one has (Problem 2.19)

$$E_k^{(3)} = \langle\psi_k^{(1)}|H' - E_k^{(1)}|\psi_k^{(1)}\rangle - 2E_k^{(2)}\langle\psi_k^{(0)}|\psi_k^{(1)}\rangle \tag{2.312}$$

Let us now return to (2.304). The Rayleigh–Schrödinger method attempts to obtain the solution $\psi_k^{(1)}$ of that equation in the following way. First, the 'unperturbed' equation (2.297) is solved for all eigenvalues and eigenfunctions (including those belonging to the continuous part of the spectrum, if one exists). The unknown function $\psi_k^{(1)}$ is then expanded in the basis set of the unperturbed eigenfunctions. That is,

$$\psi_k^{(1)} = \sum_m a_m^{(1)}\psi_m^{(0)} \tag{2.313}$$

where the sum over m means a summation over the discrete part of the set and an integration over its continuous part. Substituting (2.313) into (2.304), we obtain

$$(H_0 - E_k^{(0)}) \sum_m a_m^{(1)} \psi_m^{(0)} + (H' - E_k^{(1)}) \psi_k^{(0)} = 0 \tag{2.314}$$

Premultiplying by $\psi_l^{(0)*}$, integrating over all space and using the fact that $H_0 \psi_l^{(0)} = E_l^{(0)} \psi_l^{(0)}$ and $\langle \psi_l^{(0)} | \psi_k^{(0)} \rangle = \delta_{kl}$, we find that

$$a_l^{(1)}(E_l^{(0)} - E_k^{(0)}) + \langle \psi_l^{(0)} | H' | \psi_k^{(0)} \rangle - E_k^{(1)} \delta_{kl} = 0 \tag{2.315}$$

For $l = k$ this reduces to our basic result (2.308). On the other hand, for $l \neq k$ we have

$$a_l^{(1)} = \frac{H'_{lk}}{E_k^{(0)} - E_l^{(0)}}, \qquad l \neq k \tag{2.316}$$

where we have set $H'_{lk} \equiv \langle \psi_l^{(0)} | H' | \psi_k^{(0)} \rangle$. We note that the equation (2.304) does not determine the coefficient $a_k^{(1)}$, which is the 'component' of $\psi_k^{(1)}$ along $\psi_k^{(0)}$. We can thus require without loss of generality that

$$a_k^{(1)} = \langle \psi_k^{(0)} | \psi_k^{(1)} \rangle = 0 \tag{2.317}$$

and rewrite (2.313) as

$$\begin{aligned} \psi_k^{(1)} &= \sum_{m \neq k} a_m^{(1)} \psi_m^{(0)} \\ &= \sum_{m \neq k} \frac{H'_{mk}}{E_k^{(0)} - E_m^{(0)}} \psi_m^{(0)} \end{aligned} \tag{2.318}$$

Substituting this result in (2.310) we obtain

$$E_k^{(2)} = \sum_{m \neq k} \frac{H'_{km} H'_{mk}}{E_k^{(0)} - E_m^{(0)}} = \sum_{m \neq k} \frac{|H'_{mk}|^2}{E_k^{(0)} - E_m^{(0)}} \tag{2.319}$$

The third-order correction $E_k^{(3)}$ may be obtained in a similar way from (2.312) and (2.318).

Degenerate case

Thus far we have assumed that the perturbed eigenfunction ψ_k differs slightly from a given function $\psi_k^{(0)}$, the solution of the 'unperturbed' equation (2.297). When the level $E_k^{(0)}$ is α-fold degenerate, there are *several* 'unperturbed' wave functions $\psi_{kr}^{(0)} (r = 1, 2, \ldots, \alpha)$ corresponding to this level and we do not know *a priori* to which functions the perturbed eigenfunctions tend when $\lambda \to 0$. This means that the above treatment – and in particular the basic expansion (2.300) – must be modified to deal with the degenerate case.

The α unperturbed wave functions $\psi_{kr}^{(0)}$ corresponding to the level $E_k^{(0)}$ are orthogonal to the unperturbed wave functions $\psi_l^{(0)}$ corresponding to *other* energy levels $E_l^{(0)} \neq E_k^{(0)}$. Although they need not be orthogonal among themselves, it is always possible to construct from linear combinations of them a new set of α unperturbed wave functions which are mutually orthogonal and normalised to unity. We may therefore assume without loss of generality that this has already been done, so that

$$\langle \psi_{kr}^{(0)} | \psi_{ks}^{(0)} \rangle = \delta_{rs} \qquad (r, s = 1, 2, \ldots, \alpha) \tag{2.320}$$

Let us now introduce the correct zero-order functions $\chi_{kr}^{(0)}$ which yield the first term in the expansion of the exact wave function ψ_{kr} in powers of λ. That is,

$$\psi_{kr} = \chi_{kr}^{(0)} + \lambda \psi_{kr}^{(1)} + \lambda^2 \psi_{kr}^{(2)} + \ldots \tag{2.321}$$

We shall also write the perturbed energy E_{kr} as

$$E_{kr} = E_k^{(0)} + \lambda E_{kr}^{(1)} + \lambda^2 E_{kr}^{(2)} + \ldots \tag{2.322}$$

with $E_k^{(0)} \equiv E_{kr}^{(0)}$ $(r = 1, 2, \ldots, \alpha)$ since the level $E_k^{(0)}$ is α-fold degenerate. Using the above expansions in (2.299) and equating the coefficients of λ we find that

$$H_0 \psi_{kr}^{(1)} + H' \chi_{kr}^{(0)} = E_k^{(0)} \psi_{kr}^{(1)} + E_{kr}^{(1)} \chi_{kr}^{(0)} \tag{2.323}$$

Since the functions $\chi_{kr}^{(0)}$ are linear combinations of the unperturbed wave functions $\psi_{kr}^{(0)}$, we may write

$$\chi_{kr}^{(0)} = \sum_{s=1}^{\alpha} c_{rs} \psi_{ks}^{(0)} \qquad (r = 1, 2, \ldots, \alpha) \tag{2.324}$$

where the coefficients c_{rs} are to be determined. Similarly, expanding $\psi_{kr}^{(1)}$ in the basis set of the unperturbed wave functions, we have

$$\psi_{kr}^{(1)} = \sum_m \sum_s a_{kr,ms}^{(1)} \psi_{ms}^{(0)} \tag{2.325}$$

where the indices r and s refer explicitly to the degeneracy. Substituting the above expressions of $\chi_{kr}^{(0)}$ and $\psi_{kr}^{(1)}$ in (2.323) and using the fact that $H_0 \psi_{ms}^{(0)} = E_m^{(0)} \psi_{ms}^{(0)}$, we find that

$$\sum_m \sum_s a_{kr,ms}^{(1)} (E_m^{(0)} - E_k^{(0)}) \psi_{ms}^{(0)} + \sum_s c_{rs} (H' - E_{kr}^{(1)}) \psi_{ks}^{(0)} = 0 \tag{2.326}$$

Premultiplying by $\psi_{ku}^{(0)*}$ and integrating over all space, we obtain

$$\sum_m \sum_s a_{kr,ms}^{(1)} (E_m^{(0)} - E_k^{(0)}) \langle \psi_{ku}^{(0)} | \psi_{ms}^{(0)} \rangle + \sum_s c_{rs} [\langle \psi_{ku}^{(0)} | H' | \psi_{ks}^{(0)} \rangle - E_{kr}^{(1)} \delta_{us}] = 0$$

$$(u = 1, 2, \ldots, \alpha) \tag{2.327}$$

where we have used (2.320). Since $\langle \psi_{ku}^{(0)} | \psi_{ms}^{(0)} \rangle = 0$ when $k \neq m$ and $E_k^{(0)} = E_m^{(0)}$ if $k = m$, we see that (2.327) reduces to

$$\sum_{s=1}^{\alpha} c_{rs}[\langle \psi_{ku}^{(0)} | H' | \psi_{ks}^{(0)} \rangle - E_{kr}^{(1)} \delta_{us}] = 0 \qquad (u = 1, 2, \ldots, \alpha) \tag{2.328}$$

This is a linear, homogeneous system of equations for the α unknown quantities $c_{r1}, c_{r2}, \ldots, c_{r\alpha}$. A non-trivial solution is obtained if the determinant of the quantity in square brackets vanishes,

$$\det |\langle \psi_{ku}^{(0)} | H' | \psi_{ks}^{(0)} \rangle - E_{kr}^{(1)} \delta_{us}| = 0 \qquad (s, u = 1, 2, \ldots, \alpha) \tag{2.329}$$

This equation yields α real roots $E_{k1}^{(1)}, E_{k2}^{(1)}, \ldots, E_{k\alpha}^{(1)}$. If all of these roots are *distinct* the degeneracy is *completely removed* to first order in the perturbation. On the other hand, if some or all roots of (2.329) are identical the degeneracy is only *partially* (or *not at all*) *removed*. The residual degeneracy may then either be removed in higher order of perturbation theory, or persist to all orders. The latter case occurs when the operators H_0 and H share symmetry properties.

For a given value of r, the coefficients c_{rs} $(s = 1, 2, \ldots, \alpha)$ which determine the 'correct' unperturbed zero-order wave functions $\chi_{kr}^{(0)}$ via (2.324) may be obtained by substituting the value of $E_{kr}^{(1)}$ in the system (2.328) and solving for the coefficients $c_{r1}, c_{r2}, \ldots, c_{r\alpha}$ in terms of one of them. The last coefficient is then obtained (up to a phase factor) by requiring the function $\chi_{kr}^{(0)}$ to be normalised to unity. It is clear that this procedure does not lead to a unique result when two or more roots $E_{kr}^{(1)}$ of equation (2.329) coincide, since in this case the degeneracy is not fully removed.

Time-dependent perturbation theory

We shall now discuss the perturbation theory for a system whose total Hamiltonian H may be split as

$$H = H_0 + \lambda H'(t) \tag{2.330}$$

where the unperturbed Hamiltonian H_0 is time-independent and $\lambda H'(t)$ is a small *time-dependent* perturbation. The method which we outline below is known as *Dirac's method of variation of constants*.

Let us suppose that we know the eigenvalues $E_k^{(0)}$ of the unperturbed Hamiltonian H_0, together with the corresponding stationary eigenfunctions $\psi_k^{(0)}$, which we assume to be orthonormal and to form a complete set. Thus, since $H_0 \psi_k^{(0)} = E_k^{(0)} \psi_k^{(0)}$, the general solution of the time-dependent Schrödinger equation

$$\mathrm{i}\hbar \frac{\partial \Psi_0}{\partial t} = H_0 \Psi_0 \tag{2.331}$$

is given by

$$\Psi_0 = \sum_k c_k^{(0)} \psi_k^{(0)} \exp(-\mathrm{i}E_k^{(0)} t/\hbar) \tag{2.332}$$

where the coefficients $c_k^{(0)}$ are *constants* and the sum is over the entire set of eigenfunctions $\psi_k^{(0)}$. Because the functions $\psi_k^{(0)}$ form a complete set, the general solution Ψ of the time-dependent Schrödinger equation

$$\mathrm{i}\hbar\frac{\partial\Psi}{\partial t} = H\Psi \tag{2.333}$$

can be expanded as

$$\Psi = \sum_k c_k(t)\psi_k^{(0)}\exp(-\mathrm{i}E_k^{(0)}t/\hbar) \tag{2.334}$$

where the unknown coefficients $c_k(t)$ clearly *depend on the time*. Since the wave functions $\psi_k^{(0)}$ are orthonormal, and provided Ψ is normalised to unity, we can interpret the quantity $|c_k(t)|^2$ as the probability of finding the system in the state labelled k at the time t, and $c_k(t)$ as the corresponding probability amplitude. Upon comparison of (2.332) and (2.334) we see that if $H'(t) = 0$ the coefficients $c_k(t)$ reduce to the constants $c_k^{(0)}$ which are therefore the *initial values* of the c_k. Thus, as we expect from (2.332) the quantity $|c_k^{(0)}|^2$ gives the probability of finding the system in the stationary state $\psi_k^{(0)}$ before the perturbation is applied.

To find equations for the coefficients $c_k(t)$ the expansion (2.334) is inserted into the Schrödinger equation (2.333). From (2.330) and the fact that $H_0\psi_k^{(0)} = E_k^{(0)}\psi_k^{(0)}$, we then have

$$\mathrm{i}\hbar\sum_k \dot{c}_k(t)\psi_k^{(0)}\exp(-\mathrm{i}E_k^{(0)}t/\hbar) = \sum_k c_k(t)\lambda H'(t)\psi_k^{(0)}\exp(-\mathrm{i}E_k^{(0)}t/\hbar) \tag{2.335}$$

where the dot indicates a derivative with respect to the time. Taking the scalar product with a particular function $\psi_b^{(0)}$ belonging to the set $\{\psi_k^{(0)}\}$ and using the fact that $\langle\psi_b^{(0)}|\psi_k^{(0)}\rangle = \delta_{bk}$, we then find from (2.335) the set of coupled equations

$$\dot{c}_b(t) = (\mathrm{i}\hbar)^{-1}\sum_k \lambda H'_{bk}(t)c_k(t)\exp(\mathrm{i}\omega_{bk}t) \tag{2.336}$$

where

$$H'_{bk}(t) = \langle\psi_b^{(0)}|H'(t)|\psi_k^{(0)}\rangle \tag{2.337}$$

and where the *Bohr angular frequency* ω_{bk} is defined by

$$\omega_{bk} = \frac{E_b^{(0)} - E_k^{(0)}}{\hbar} \tag{2.338}$$

The system of coupled differential equations (2.336) is completely equivalent to the original time-dependent Schrödinger equation (2.333), and no approximation has been made thus far. However, if the perturbation $\lambda H'$ is weak, we can expand the coefficients c_k in powers of the parameter λ as

$$c_k = c_k^{(0)} + \lambda c_k^{(1)} + \lambda^2 c_k^{(2)} + \dots \tag{2.339}$$

Substituting this expansion into the system (2.336) and equating the coefficients of equal powers of λ, we find that

$$\dot{c}_b^{(0)} = 0 \tag{2.340a}$$

$$\dot{c}_b^{(1)} = (\mathrm{i}\hbar)^{-1} \sum_k H'_{bk}(t) \exp(\mathrm{i}\omega_{bk}t) c_k^{(0)} \tag{2.340b}$$

$$\vdots \quad \vdots \quad \vdots$$

$$\dot{c}_b^{(s+1)} = (\mathrm{i}\hbar)^{-1} \sum_k H'_{bk}(t) \exp(\mathrm{i}\omega_{bk}t) c_k^{(s)}, \qquad s = 0, 1, \ldots \tag{2.340c}$$

Thus the original system (2.336) has been decoupled in such a way that the equations (2.340) can now in principle be integrated successively to any given order.

The first equation (2.340a) simply confirms that the coefficients $c_k^{(0)}$ are time-independent. As we have seen above, the constants $c_k^{(0)}$ define the initial conditions of the problem. In what follows, we shall assume for the sake of simplicity that the system is initially (that is, for $t \leqslant t_0$) in a well-defined stationary state $\psi_a^{(0)}$ of energy $E_a^{(0)}$. Thus

$$c_k^{(0)} = \begin{cases} \delta_{ka} & \text{for discrete states} \\ \delta(k-a) & \text{for continuous states.} \end{cases} \tag{2.341}$$

We note that this statement is not in contradiction with the uncertainty relation $\Delta E\, \Delta t \gtrsim \hbar$ since we have essentially an 'infinite' amount of time available to prepare our initial state. Upon substitution of (2.341) into (2.340b) we then have

$$\dot{c}_b^{(1)}(t) = (\mathrm{i}\hbar)^{-1} H'_{ba}(t) \exp(\mathrm{i}\omega_{ba}t) \tag{2.342}$$

where $\omega_{ba} = (E_b^{(0)} - E_a^{(0)})/\hbar$. The solution of these first-order equations is

$$c_a^{(1)}(t) = (\mathrm{i}\hbar)^{-1} \int_{t_0}^{t} H'_{aa}(t')\mathrm{d}t' \tag{2.343a}$$

for $b = a$, while for $b \neq a$ one has

$$c_b^{(1)}(t) = (\mathrm{i}\hbar)^{-1} \int_{t_0}^{t} H'_{ba}(t') \exp(\mathrm{i}\omega_{ba}t')\mathrm{d}t', \qquad b \neq a \tag{2.343b}$$

where the integration constant has been chosen in such a way that $c_a^{(1)}(t)$ and $c_b^{(1)}(t)$ vanish at $t = t_0$, before the perturbation is applied. To first order in the perturbation, the *transition probability* corresponding to the transition $a \rightarrow b$ is therefore given by

$$P_{ba}^{(1)}(t) = |c_b^{(1)}(t)|^2 = \hbar^{-2} \left| \int_{t_0}^{t} H'_{ba}(t') \exp(\mathrm{i}\omega_{ba}t')\, \mathrm{d}t' \right|^2, \qquad b \neq a \tag{2.344}$$

It is also worth noting that for $t > t_0$ the coefficient c_a of the state a is given to first order in the perturbation by

$$\begin{aligned} c_a(t) &\simeq c_a^{(0)} + c_a^{(1)}(t) \\ &\simeq 1 + (\mathrm{i}\hbar)^{-1} \int_{t_0}^{t} H'_{aa}(t')\mathrm{d}t' \\ &\simeq \exp\left[-\frac{\mathrm{i}}{\hbar} \int_{t_0}^{t} H'_{aa}(t')\,\mathrm{d}t' \right] \end{aligned} \quad \textbf{(2.345)}$$

so that $|c_a(t)|^2 \simeq 1$ and the main effect of the perturbation on the initial state is to change its phase.

Time-independent perturbation

The results (2.343) take a particularly simple form if the perturbation H' is independent of time, except for being switched on suddenly at a given time (say $t_0 = 0$). We then have

$$c_a^{(1)}(t) = (\mathrm{i}\hbar)^{-1} H'_{aa} t \quad \textbf{(2.346a)}$$

and

$$c_b^{(1)}(t) = \frac{H'_{ba}}{\hbar\omega_{ba}} [1 - \exp(\mathrm{i}\omega_{ba}t)], \qquad b \neq a \quad \textbf{(2.346b)}$$

Note that if the perturbation is switched off at time t, the above amplitudes are also those at any subsequent time $t_1 > t$. The first-order transition probability from state a to state $b \neq a$ is given by

$$P_{ba}^{(1)}(t) = |c_b^{(1)}(t)|^2 = \frac{2}{\hbar^2} |H'_{ba}|^2 F(t, \omega_{ba}) \quad \textbf{(2.347)}$$

where

$$F(t, \omega) = \frac{1 - \cos \omega t}{\omega^2} = \frac{2 \sin^2(\omega t/2)}{\omega^2} \quad \textbf{(2.348)}$$

The function $F(t, \omega)$ is shown in Fig. 2.9 for fixed t. We see that it exhibits a sharp peak about the value $\omega = 0$. The height of this peak is proportional to t^2, while its width is approximately $2\pi/t$. Setting $x = \omega t/2$, we also note that

$$\int_{-\infty}^{+\infty} F(t, \omega)\,\mathrm{d}\omega = t \int_{-\infty}^{+\infty} \frac{\sin^2 x}{x^2}\,\mathrm{d}x = \pi t \quad \textbf{(2.349)}$$

where we have used a standard integral. Using (2.31c) we also deduce that in the limit $t \to \infty$

$$F(t, \omega) \underset{t\to\infty}{\sim} \pi t\, \delta(\omega) \quad \textbf{(2.350)}$$

Let us first analyse (2.347) for a fixed value of t. Since the function $F(t, \omega_{ba})$ has a sharp peak of width $2\pi/t$ about the value $\omega_{ba} = 0$, it is clear from (2.347) that

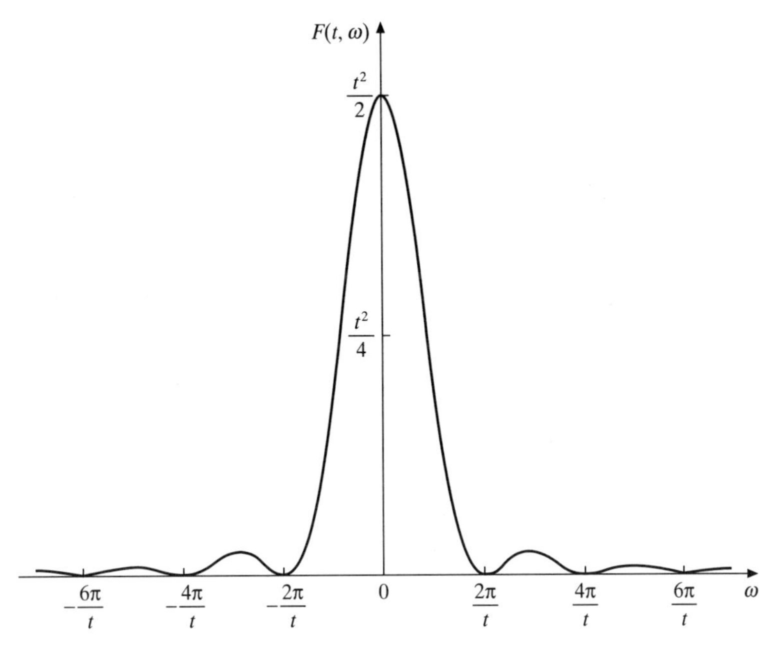

Figure 2.9 The function $F(t, \omega)$ of equation (2.348), for fixed t.

transitions to those final states b for which ω_{ba} does not deviate from zero by more than $\delta\omega_{ba} \simeq 2\pi/t$ will be strongly favoured. Therefore the transitions $a \to b$ will occur mainly towards those final states whose energy $E_b^{(0)}$ is located in a band of width

$$\delta E \simeq 2\pi\hbar/t \tag{2.351}$$

about the initial energy $E_a^{(0)}$, so that the unperturbed energy is conserved to within $2\pi\hbar/t$. This result may easily be related to the time–energy uncertainty relation $\Delta E\ \Delta t \gtrsim \hbar$. Indeed, since the perturbation gives a way of measuring the energy of the system by inducing transitions $a \to b$, and because this perturbation acts during a time t, the uncertainty related to this energy measurement should be approximately $\hbar/t$, in qualitative agreement with (2.351).

We now study the transition probability (2.347) as a function of t. For a transition to a *given state* b we must distinguish two cases:

1. If the transition is such that $\omega_{ba} = 0$ (so that $E_b^{(0)} = E_a^{(0)}$ and the states a and b are degenerate), then we see from (2.347) that

$$P_{ba}^{(1)}(t) = \frac{|H'_{ba}|^2}{\hbar^2}\, t^2 \tag{2.352}$$

so that the transition probability increases as t^2. Thus, after a sufficient length of time the quantity $P_{ba}^{(1)}(t)$ will no longer satisfy the inequality

$$P_{ba}^{(1)}(t) \ll 1 \qquad \textbf{(2.353)}$$

required by a perturbative approach. Hence, the present perturbation method cannot be applied to degenerate systems which are perturbed over long periods of time.

2. If on the contrary $\omega_{ba} \neq 0$ (so that the state b is not degenerate with the initial state a), we have

$$P_{ba}^{(1)}(t) = \frac{4|H'_{ba}|^2}{\hbar^2\omega_{ba}^2}\sin^2(\omega_{ba}t/2) \qquad \textbf{(2.354)}$$

and we note that $P_{ba}^{(1)}(t)$ oscillates with a period $2\pi/|\omega_{ba}|$ about the average value $2|H'_{ba}|^2/(\hbar^2\omega_{ba}^2)$. Since

$$P_{ba}^{(1)}(t) \leqslant \frac{4|H'_{ba}|^2}{\hbar^2\omega_{ba}^2} \qquad \textbf{(2.355)}$$

the condition (2.353) can always be satisfied if the perturbation H' is sufficiently weak. It is also worth noting that for times t small with respect to the period of oscillation $2\pi/|\omega_{ba}|$ one has $\sin(\omega_{ba}t/2) \simeq \omega_{ba}t/2$ so that $P_{ba}^{(1)}(t) \simeq |H'_{ba}|^2t^2/\hbar^2$ increases quadratically with time.

Instead of considering transitions to a particular state b, it is often necessary to deal with transitions involving a *group of states* n whose energy E_n lies within a given interval $(E_b^{(0)} - \eta, E_b^{(0)} + \eta)$ centred about the value $E_b^{(0)}$. Let us denote by $\rho_n(E_n)$ the density of levels on the energy scale, so that $\rho_n(E_n)\,\mathrm{d}E_n$ is the number of final states n in an interval $\mathrm{d}E_n$ containing the energy E_n. We shall continue to assume that the perturbation H' is constant in time, except that it is switched on at $t = 0$. The first-order transition probability $P_{ba}^{(1)}(t)$ from the initial state a to the group of final states n is then given by

$$P_{ba}^{(1)}(t) = \frac{2}{\hbar^2}\int_{E_b^{(0)}-\eta}^{E_b^{(0)}+\eta} |H'_{na}|^2 F(t, \omega_{na})\rho_n(E_n)\,\mathrm{d}E_n \qquad \textbf{(2.356)}$$

where $\omega_{na} = (E_n - E_a^{(0)})/\hbar$. Assuming that η is small enough so that H'_{na} and $\rho_n(E_n)$ are nearly constant within the integration range, we have

$$P_{ba}^{(1)}(t) = \frac{2}{\hbar^2}|H'_{ba}|^2\rho_b(E_b^{(0)})\int_{E_b^{(0)}-\eta}^{E_b^{(0)}+\eta} F(t, \omega_{na})\,\mathrm{d}E_n \qquad \textbf{(2.357)}$$

We shall also assume that t is large enough so that the quantity η satisfies the condition

$$\eta \gg 2\pi\hbar/t. \qquad \textbf{(2.358)}$$

It is clear from the examination of the function $F(t, \omega)$ (see (2.348) and Fig. 2.9) that the overwhelming part of the integral on the right of (2.357) arises from transitions which conserve the energy (within $\delta E = 2\pi\hbar/t$). Since $\eta \gg 2\pi\hbar/t$, we may write

$$\int_{E_b^{(0)}-\eta}^{E_b^{(0)}+\eta} F(t, \omega_{na})\, \mathrm{d}E_n \simeq \hbar \int_{-\infty}^{+\infty} F(t, \omega_{na})\, \mathrm{d}\omega_{na} = \pi\hbar t \tag{2.359}$$

where we have used (2.349). Thus, (2.357) reduces to

$$P_{ba}^{(1)}(t) = \frac{2\pi}{\hbar} |H'_{ba}|^2 \rho_b(E) t \tag{2.360}$$

with $E = E_a^{(0)} = E_b^{(0)}$. Thus the transition probability increases linearly with time for energy-conserving transitions to a group of states. It is worth stressing that the result (2.360) is soundly based only if the condition (2.353) required by a perturbative approach is obeyed.

Introducing the *transition probability per unit time* or *transition rate*

$$W_{ba} = \frac{\mathrm{d}P_{ba}}{\mathrm{d}t} \tag{2.361}$$

we see from (2.360) that to first order in perturbation theory, we have

$$W_{ba} = \frac{2\pi}{\hbar} |H'_{ba}|^2 \rho_b(E) \tag{2.362}$$

This formula, first obtained by P.A.M. Dirac, was later called by E. Fermi 'The Golden Rule' of perturbation theory. Although we have derived it here for a perturbation H' which is constant in time (except for being switched on at $t = 0$), it can be generalised to other perturbations, as we shall now illustrate.

Periodic perturbation

Let us now consider a perturbation $H'(t)$ which is a *periodic* function of time, except for being turned on at $t = 0$. This case is of particular importance for studying the interaction of atoms and molecules with electromagnetic fields, as we shall see in further chapters. We shall assume first that the perturbation is *harmonic* in time, with angular frequency ω. That is,

$$H'(t) = A \exp(\mathrm{i}\omega t) + A^\dagger \exp(-\mathrm{i}\omega t) \tag{2.363}$$

where A is a time-independent operator and $A^\dagger$ is its adjoint so that the operator H' is Hermitian. It is supposed that the system is initially (for $t \leqslant 0$) in the unperturbed bound state $\psi_a^{(0)}$, of energy $E_a^{(0)}$, so that the initial conditions are $c_a(t \leqslant 0) = 1$ and $c_b(t \leqslant 0) = 0$ for $b \neq a$.

According to (2.345), we have $|c_a(t)|^2 \simeq 1$ for $t > 0$. In order to find the coefficient $c_b^{(1)}(t)$ for $t > 0$ and $b \neq a$, we substitute (2.363) into (2.343b) and use the fact that $t_0 = 0$. This gives

$$c_b^{(1)}(t) = (\mathrm{i}\hbar)^{-1}\left\{A_{ba}\int_0^t \exp[\mathrm{i}(\omega_{ba}+\omega)t']\,dt' + A_{ba}^{\dagger}\int_0^t \exp[\mathrm{i}(\omega_{ba}-\omega)t']\,dt'\right\} \quad \textbf{(2.364)}$$

where $A_{ba} = \langle\psi_b^{(0)}|A|\psi_a^{(0)}\rangle$ and $A_{ba}^{\dagger} = A_{ab}^{*}$. Performing the integrals, and remembering that $\hbar\omega_{ba} = E_b^{(0)} - E_a^{(0)}$, we find that the corresponding first-order transition probability is given by

$$P_{ba}^{(1)}(t) = \left| A_{ba}\frac{1-\exp[\mathrm{i}(E_b^{(0)} - E_a^{(0)} + \hbar\omega)t/\hbar]}{E_b^{(0)} - E_a^{(0)} + \hbar\omega} + A_{ba}^{\dagger}\frac{1-\exp[\mathrm{i}(E_b^{(0)} - E_a^{(0)} - \hbar\omega)t/\hbar]}{E_b^{(0)} - E_a^{(0)} - \hbar\omega}\right|^2 \quad \textbf{(2.365)}$$

It is clear from this equation that if t is large enough the probability of finding the system in the state b will only be appreciable if the denominator of one or the other of the two terms on the right-hand side is close to zero. Moreover, assuming that $E_b^{(0)} \neq E_a^{(0)}$ (so that the levels $E_a^{(0)}$ and $E_b^{(0)}$ are not degenerate), both denominators cannot simultaneously be close to zero. A good approximation is therefore to neglect the interference between the two terms in calculating the transition probability. Thus, if the energy $E_b^{(0)}$ lies in a small band about the value

$$E = E_a^{(0)} + \hbar\omega \quad \textbf{(2.366)}$$

only the second term in (2.365) will have an appreciable magnitude, the corresponding transition probability being given by

$$P_{ba}^{(1)}(t) = \frac{2}{\hbar^2}|A_{ba}^{\dagger}|^2 F(t, \omega_{ba} - \omega) \quad \textbf{(2.367)}$$

The main difference with the expression (2.347), obtained for a time-independent perturbation, is that the angular frequency ω_{ba} has now been replaced by $\omega_{ba} - \omega$. From the properties of the function $F(t, \omega)$ it is apparent that the transition probability (2.367) will only be significant if $E_b^{(0)}$ is located in an interval of width $2\pi\hbar/t$ about the value $E_a^{(0)} + \hbar\omega$. Hence the first-order transition probability (2.367) will be appreciable if the system has *absorbed* an amount of energy given (to within $2\pi\hbar/t$) by $\hbar\omega = E_b^{(0)} - E_a^{(0)}$. This, of course, is nothing but the Bohr frequency rule (1.70). When it is exactly satisfied, so that $\omega = \omega_{ba}$, *resonance* is said to occur and we see from (2.367) and (2.348) that the first-order transition probability increases quadratically with time according to the formula

$$P_{ba}^{(1)}(t) = \frac{|A_{ba}^{\dagger}|^2}{\hbar^2}t^2 \quad \textbf{(2.368)}$$

In the same way, if the energy $E_b^{(0)}$ lies in a small interval in the neighbourhood of the value

$$E = E_a^{(0)} - \hbar\omega \quad \textbf{(2.369)}$$

only the first term on the right of (2.365) will be significant. The corresponding first-order transition probability will be given by

$$P_{ba}^{(1)}(t) = \frac{2}{\hbar^2} |A_{ba}|^2 F(t, \omega_{ba} + \omega) \tag{2.370}$$

and will only be significant if the system has *emitted* an amount of energy given (to within $2\pi\hbar/t$) by $\hbar\omega = E_a^{(0)} - E_b^{(0)}$. Again *resonance* occurs if this condition is exactly satisfied, in which case $\omega = -\omega_{ba}$ and the transition probability (2.370) increases quadratically with time. It should be noted that in practice t is large enough ($t \gg 2\pi/\omega$) so that the two bands of width $2\pi\hbar/t$ about the values (2.366) and (2.369) do not overlap. Thus our neglect of the interference between the two terms on the right of (2.365) in calculating $P_{ba}^{(1)}$ is indeed justified.

As in the case of a time-independent perturbation, one can consider transitions to a *group* of final states n whose energy E_n lies within an interval $(E_b^{(0)} - \eta, E_b^{(0)} + \eta)$ about the value $E_b^{(0)} = E_a^{(0)} + \hbar\omega$ (for absorption) or $E_b^{(0)} = E_a^{(0)} - \hbar\omega$ (for emission) with $\eta \gg 2\pi\hbar/t$. Let $\rho_n(E_n)$ be the density of levels E_n on the energy scale. Under conditions similar to those discussed above for a time-independent perturbation, a transition probability per unit time (transition rate) can be defined. For transitions in which the system *absorbs* an energy $\hbar\omega \simeq E_b^{(0)} - E_a^{(0)}$, this transition rate is given to first order by

$$W_{ba} = \frac{2\pi}{\hbar} |A_{ba}^{\dagger}|^2 \rho_b(E) \tag{2.371}$$

where $E = E_a^{(0)} + \hbar\omega$. The above expression is clearly the direct generalisation of the Golden Rule (2.362). Similarly, for transitions in which the system *emits* an energy $\hbar\omega \simeq E_a^{(0)} - E_b^{(0)}$ by making transitions to a group of final states, the corresponding transition rate is given (under the same conditions) by

$$W_{ba} = \frac{2\pi}{\hbar} |A_{ba}|^2 \rho_b(E) \tag{2.372}$$

with $E = E_a^{(0)} - \hbar\omega$.

Until now we have considered a perturbation $H'(t)$ which is a harmonic function of time. However, the generalisation of the above results to a perturbation $H'(t)$ which is a general periodic function of time is straightforward. Indeed, we can then develop $H'(t)$ in the Fourier series

$$H'(t) = \sum_{n=1}^{\infty} [A_n \exp(in\omega t) + A_n^{\dagger} \exp(-in\omega t)] \tag{2.373}$$

where the operators A_n are time-independent. For large enough times ($t \gg 2\pi/\omega$), there is no interference between the contributions of the various terms of this development to the transition probability, because each term corresponds to a different energy transfer. A term of the type $A_n \exp(in\omega t)$ will therefore lead to the *emission* by the system of an energy given (to within $2\pi\hbar/t$) by $n\hbar\omega = E_a^{(0)} - E_b^{(0)}$

while a term of the type $A_n^\dagger \exp(-\mathrm{i}n\omega t)$ corresponds to the *absorption* by the system of the energy $n\hbar\omega = E_b^{(0)} - E_a^{(0)}$ (to within $2\pi\hbar/t$). The corresponding first-order transition probabilities are readily obtained, either for transitions to a given state b or to a group of final states.

Two–level system with harmonic perturbation

As a simple application of some of the concepts developed above, let us consider a two-level system, with unperturbed energies $E_a^{(0)} < E_b^{(0)}$ and corresponding eigenfunctions $\psi_a^{(0)}$ and $\psi_b^{(0)}$, respectively. The system being initially in the state a, a perturbation of the form (2.363) is switched on at time $t = 0$. Setting $\lambda = 1$ in (2.336), we obtain the two coupled equations

$$\mathrm{i}\hbar\dot{c}_a(t) = \{A_{aa}\exp(\mathrm{i}\omega t) + A_{aa}^\dagger \exp(-\mathrm{i}\omega t)\}c_a + \{A_{ab}\exp[\mathrm{i}(\Delta\omega)t] + A_{ab}^\dagger \exp[-\mathrm{i}(\omega + \omega_{ba})t]\}c_b \tag{2.374a}$$

and

$$\mathrm{i}\hbar\dot{c}_b(t) = \{A_{ba}\exp[\mathrm{i}(\omega_{ba} + \omega)t] + A_{ba}^\dagger \exp[-\mathrm{i}(\Delta\omega)t]\}c_a + \{A_{bb}\exp(\mathrm{i}\omega t) + A_{bb}^\dagger \exp(-\mathrm{i}\omega t)\}c_b \tag{2.374b}$$

where $\omega_{ba} = (E_b^{(0)} - E_a^{(0)})/\hbar$ and we have introduced the 'detuning' angular frequency

$$\Delta\omega = \omega - \omega_{ba} \tag{2.375}$$

The system (2.374) must be solved subject to the initial conditions

$$c_a(t \leqslant 0) = 1, \qquad c_b(t \leqslant 0) = 0 \tag{2.376}$$

The equations (2.374) cannot be solved exactly, but if it is assumed that $|\Delta\omega| \ll \omega$ (so that the angular frequency ω is always close to its resonant value $\omega = \omega_{ba}$) then the terms in $\exp[\pm\mathrm{i}(\Delta\omega)t]$ will be much more important than those in $\exp[\pm\mathrm{i}(\omega + \omega_{ba})t]$ and $\exp(\pm\mathrm{i}\omega t)$. This is because the latter terms oscillate much more rapidly and on the average make little contribution to $\dot{c}_a$ or $\dot{c}_b$. It is therefore reasonable to neglect the higher frequency terms. This is known as the *rotating wave approximation* because the only terms which are kept are those in which the time dependence of the system and of the perturbation are in phase. In this approximation, the pair of equations (2.374) reduces to

$$\mathrm{i}\hbar\dot{c}_a(t) = A_{ab}\exp[\mathrm{i}(\Delta\omega)t]c_b \tag{2.377 a}$$

and

$$\mathrm{i}\hbar\dot{c}_b(t) = A_{ba}^\dagger \exp[-\mathrm{i}(\Delta\omega)t]c_a \tag{2.377b}$$

This system, which is much simpler than (2.374), can be solved exactly. The solutions $c_a(t)$ and $c_b(t)$ satisfying the initial conditions (2.376) are

$$c_a(t) = \exp[\mathrm{i}(\Delta\omega)t/2]\left[\cos(\omega_\mathrm{R}t/2) - \mathrm{i}\left(\frac{\Delta\omega}{\omega_\mathrm{R}}\right)\sin(\omega_\mathrm{R}t/2)\right] \tag{2.378a}$$

and

$$c_b(t) = \frac{2A^{\dagger}_{ba}}{\mathrm{i}\hbar\omega_{\mathrm{R}}} \exp[-\mathrm{i}(\Delta\omega)t/2] \sin(\omega_{\mathrm{R}}t/2) \tag{2.378b}$$

where

$$\omega_{\mathrm{R}} = \left[(\Delta\omega)^2 + \frac{4|A^{\dagger}_{ba}|^2}{\hbar^2}\right]^{1/2} \tag{2.379}$$

is called the Rabi '*flopping frequency*'.

The probability of finding the system at time $t > 0$ in the state a is therefore given by

$$|c_a(t)|^2 = \cos^2(\omega_{\mathrm{R}}t/2) + \frac{(\Delta\omega)^2}{\omega_{\mathrm{R}}^2} \sin^2(\omega_{\mathrm{R}}t/2) \tag{2.380}$$

while that of finding it in the state b (that is, the probability that the transition $a \to b$ will take place) is

$$P_{ba}(t) = |c_b(t)|^2 = \frac{4|A^{\dagger}_{ba}|^2}{\hbar^2\omega_{\mathrm{R}}^2} \sin^2(\omega_{\mathrm{R}}t/2) \tag{2.381}$$

As expected, the excitation is a typical *resonance* process, since the probability (2.381) rapidly decreases when the absolute value $|\Delta\omega|$ of the detuning angular frequency increases. It is also readily verified from (2.378)–(2.381) that $|c_a(t)|^2 + |c_b(t)|^2 = 1$, and that the system oscillates between the two levels with a period $T = 2\pi/\omega_{\mathrm{R}}$.

Having obtained 'exact' results for this problem (within the framework of the rotating wave approximation), we can compare them with those arising from first-order perturbation theory. Using (2.380), we see that when $\omega_{\mathrm{R}}t \ll 1$ we have $|c_a(t)|^2 \simeq 1$, which is in agreement with the perturbative result following from (2.345). From (2.367) and (2.348) we also note that the first-order transition probability is given by

$$\begin{aligned} P^{(1)}_{ba}(t) &= \frac{2}{\hbar^2} |A^{\dagger}_{ba}|^2 F(t, \Delta\omega) \\ &= \frac{4|A^{\dagger}_{ba}|^2}{\hbar^2(\Delta\omega)^2} \sin^2[(\Delta\omega)t/2] \end{aligned} \tag{2.382}$$

If $\Delta\omega \neq 0$ this result agrees with (2.381) provided that the perturbation is weak enough so that one can write $\omega_{\mathrm{R}} \simeq \Delta\omega$ (see (2.379)). At resonance ($\Delta\omega = 0$), we see that $P^{(1)}_{ba}(t)$ increases quadratically with time according to (2.368). As expected, this result is only in agreement with the 'exact' expression (2.381) for small enough times and small enough perturbations.

The variational method

We shall now discuss an approximation method, known as the *variational method*, which is very useful in obtaining the bound state energies and wave functions of a time-independent Hamiltonian H. We denote by E_n the eigenvalues of this Hamiltonian and by ψ_n the corresponding orthonormal eigenfunctions, and assume that H has at least one discrete eigenvalue. Let ϕ be an arbitrary normalisable function, and let $E[\phi]$ be the functional

$$E[\phi] = \frac{\langle\phi|H|\phi\rangle}{\langle\phi|\phi\rangle} = \frac{\int \phi^* H\phi \, d\tau}{\int \phi^*\phi \, d\tau} \tag{2.383}$$

where the integration is extended over the full range of all the coordinates of the system.

It is clear that if the function ϕ is identical to one of the exact eigenfunctions ψ_n of H, then $E[\phi]$ will be identical to the corresponding exact eigenvalue E_n. Moreover, it will now be shown that any function ϕ for which the functional $E[\phi]$ is stationary is an eigenfunction of the discrete spectrum of H. Thus, if ϕ and ψ_n differ by an arbitrary infinitesimal variation $\delta\phi$,

$$\phi = \psi_n + \delta\phi \tag{2.384}$$

then the corresponding first-order variation of $E[\phi]$ vanishes:

$$\delta E = 0 \tag{2.385}$$

and the eigenfunctions of H are solutions of the variational equation (2.385).

To prove this statement, we note that upon clearing the fractions and varying, we have from (2.383)

$$\delta E \int \phi^*\phi \, d\tau + E \int \delta\phi^*\phi \, d\tau + E \int \phi^* \, \delta\phi \, d\tau$$
$$= \int \delta\phi^* H\phi \, d\tau + \int \phi^* H \, \delta\phi \, d\tau \tag{2.386}$$

Since $\langle\phi|\phi\rangle$ is assumed to be finite and non-vanishing, we see that the equation (2.385) is equivalent to

$$\int \delta\phi^*(H - E)\phi \, d\tau + \int \phi^*(H - E) \, \delta\phi \, d\tau = 0 \tag{2.387}$$

Although the variations $\delta\phi$ and $\delta\phi^*$ are not independent, they may in fact be treated as such, so that the individual terms in (2.387) can be set equal to zero. To see how this comes about, we replace the arbitrary variation $\delta\phi$ by $\mathrm{i}\,\delta\phi$ in (2.387) so that we have

$$-\mathrm{i}\int \delta\phi^*(H-E)\phi\,\mathrm{d}\tau + \mathrm{i}\int \phi^*(H-E)\,\delta\phi\,\mathrm{d}\tau = 0 \tag{2.388}$$

Upon combining (2.388) with (2.387) we then obtain the two equations

$$\int \delta\phi^*(H-E)\phi\,\mathrm{d}\tau = 0 \tag{2.389a}$$

$$\int \phi^*(H-E)\delta\phi\,\mathrm{d}\tau = 0 \tag{2.389b}$$

which is the desired result. Using the fact that H is Hermitian, we see that the two equations (2.389) are equivalent to the Schrödinger equation

$$(H - E[\phi])\phi = 0 \tag{2.390}$$

Thus any function $\phi = \psi_n$ for which the functional (2.383) is stationary is an eigenfunction of H corresponding to the eigenvalue $E_n \equiv E[\psi_n]$. Conversely, if ψ_n is an eigenfunction of H and E_n the corresponding energy, we have $E_n = E[\psi_n]$ and the functional $E[\psi_n]$ is stationary because ψ_n satisfies the equations (2.389). It is worth stressing that if ϕ and ψ_n differ by $\delta\phi$, the variational principle (2.385) implies that the leading term of the difference between $E[\phi]$ and the true eigenvalue E_n is *quadratic* in $\delta\phi$. As a result, errors in the approximate energy are of *second order* in $\delta\phi$ when the energy is calculated from the functional (2.383).

We also remark that the functional (2.383) is independent of the normalisation and of the phase of ϕ. In particular, it is often convenient to impose the condition $\langle\phi|\phi\rangle = 1$. The above results may then be retrieved by varying the functional $\langle\phi|H|\phi\rangle$ subject to the condition $\langle\phi|\phi\rangle = 1$, namely

$$\delta\int \phi^* H\phi\,\mathrm{d}\tau = 0, \qquad \int \phi^*\phi\,\mathrm{d}\tau = 1 \tag{2.391}$$

The constraint $\langle\phi|\phi\rangle = 1$ may be taken care of by introducing a Lagrange multiplier [9] which we denote by E, so that the variational equation reads

$$\delta\left[\int \phi^* H\phi\,\mathrm{d}\tau - E\int \phi^*\phi\,\mathrm{d}\tau\right] = 0 \tag{2.392}$$

or

$$\int \delta\phi^*(H-E)\phi\,\mathrm{d}\tau + \int \phi^*(H-E)\delta\phi\,\mathrm{d}\tau = 0 \tag{2.393}$$

This equation is identical to (2.387), and we see that the Lagrange multiplier E has the significance of an energy eigenvalue.

---

[9] Lagrange multipliers are discussed for example in Byron and Fuller (1969).

An important additional property of the functional (2.383) is that it provides an *upper bound* to the exact ground state energy E_0. To prove this result, we expand the arbitrary, normalisable function ϕ in the complete set of orthonormal eigenfunctions ψ_n of H. That is,

$$\phi = \sum_n a_n \psi_n \tag{2.394}$$

Substituting (2.394) into (2.383), we find that

$$E[\phi] = \frac{\sum_n |a_n|^2 E_n}{\sum_n |a_n|^2} \tag{2.395}$$

where we have used the fact that $H\psi_n = E_n\psi_n$ and $\langle\phi|\phi\rangle = \Sigma_n|a_n|^2$. If we now subtract E_0, the lowest energy eigenvalue, from both sides of (2.395) we have

$$E[\phi] - E_0 = \frac{\sum_n |a_n|^2(E_n - E_0)}{\sum_n |a_n|^2} \tag{2.396}$$

Since $E_n \geqslant E_0$, the right-hand side of (2.396) is non-negative, so that

$$E_0 \leqslant E[\phi] \tag{2.397}$$

and the functional $E[\phi]$ gives an upper bound – or in other words a *minimum principle* for the ground state energy.

The property (2.397) constitutes the basis of the *Rayleigh–Ritz variational method* for the approximate calculation of E_0. This method consists in evaluating the quantity $E[\phi]$ by using *trial functions* ϕ which depend on a certain number of *variational parameters*, and then to minimise $E[\phi]$ with respect to these parameters in order to obtain the best approximation of E_0 allowed by the form chosen for ϕ.

The Rayleigh–Ritz variational method can also be used to obtain an upper bound for the energy of an excited state, provided that the trial function ϕ is made orthogonal to all the energy eigenfunctions corresponding to states having a lower energy than the energy level considered. Indeed, let us arrange the energy levels in an ascending sequence: $E_0, E_1, E_2, \ldots$ and let the trial function ϕ be orthogonal to the energy eigenfunctions ψ_n $(n = 0, 1, \ldots, i)$, namely

$$\langle\psi_n|\phi\rangle = 0, \qquad n = 0, 1, \ldots, i \tag{2.398}$$

Then, if we expand ϕ in the orthonormal set $\{\psi_n\}$ as in (2.394) we have $a_n = \langle\psi_n|\phi\rangle = 0$ $(n = 0, 1, \ldots, i)$ and the functional $E[\phi]$ becomes

$$E[\phi] = \frac{\sum_{n=i+1} |a_n|^2 E_n}{\sum_{n=i+1} |a_n|^2} \tag{2.399}$$

so that

$$E_{i+1} \leqslant E[\phi] \tag{2.400}$$

As an example, suppose that the lowest energy eigenfunction ψ_0 is known, and let ϕ be a trial function. The function

$$\tilde{\phi} = \phi - \psi_0\langle\psi_0|\phi\rangle \tag{2.401}$$

is orthogonal to ψ_0 (that is, $\langle\psi_0|\tilde{\phi}\rangle = 0$) and can therefore be used to obtain an upper limit of E_1, the exact energy of the first excited state.

In many practical situations the lower energy eigenfunctions ψ_n $(n = 0, 1, \ldots, i)$ are not known exactly and one only has approximations (obtained for example from a variational calculation) of these functions. In this case the orthogonality conditions (2.398) cannot be achieved exactly, and the relation (2.400) does not hold. For example, let us suppose that the function ϕ_0 (normalised to unity) is an approximation to the true ground state eigenfunction ψ_0. If ϕ_1 is a trial function orthogonal to ϕ_0 (that is, if $\langle\phi_0|\phi_1\rangle = 0$) it may be shown (Problem 2.24) that

$$E_1 - \varepsilon_0(E_1 - E_0) \leqslant E[\phi_1] \tag{2.402}$$

where ε_0 is the positive quantity

$$\varepsilon_0 = 1 - |\langle\psi_0|\phi_0\rangle|^2 \tag{2.403}$$

Thus $E[\phi_1]$ does not provide a rigorous upper bound to E_1. However, if ϕ_0 is a good approximation to ψ_0, then ε_0 will be small and the violation of the relation $E_1 \leqslant E[\phi_1]$ will be mild.

The application of the variational method to excited states is greatly facilitated if the Hamiltonian of the system has certain *symmetry properties*, since in this case the orthogonality conditions (2.398) can be satisfied exactly for certain states. For example, if the excited state in question is of different parity or angular momentum than the lower states, then the orthogonality conditions are automatically satisfied.

Particularly useful trial functions ϕ can be constructed by choosing a certain number (N) of linearly independent functions $\chi_1, \chi_2, \ldots, \chi_N$ and forming the linear combination

$$\phi = \sum_{n=1}^{N} c_n \chi_n \tag{2.404}$$

where the coefficients $c_1, c_2, \ldots, c_N$ are linear variational parameters which must be determined by minimising the functional $E[\phi]$ in order to obtain the best approximation to E_0. Substituting (2.404) in (2.383) we find that

$$E[\phi] = \frac{\displaystyle\sum_{n=1}^{N}\sum_{n'=1}^{N} c_{n'}^* c_n H_{n'n}}{\displaystyle\sum_{n=1}^{N}\sum_{n'=1}^{N} c_{n'}^* c_n \Delta_{n'n}} \tag{2.405}$$

where we have set

$$\begin{aligned} H_{n'n} &= \langle \chi_{n'} | H | \chi_n \rangle \\ \Delta_{n'n} &= \langle \chi_{n'} | \chi_n \rangle \end{aligned} \qquad \textbf{(2.406)}$$

We remark that if the functions χ_n are orthonormal, then $\Delta_{n'n} = \delta_{n'n}$.

In order to find the values of the variational parameters $c_1, c_2, \ldots, c_N$ which minimise $E[\phi]$, we first rewrite (2.405) as

$$E[\phi] \sum_{n=1}^{N} \sum_{n'=1}^{N} c_{n'}^* c_n \Delta_{n'n} = \sum_{n=1}^{N} \sum_{n'=1}^{N} c_{n'}^* c_n H_{n'n} \qquad \textbf{(2.407)}$$

Differentiating with respect to each c_n or $c_{n'}^*$, expressing that $\partial E/\partial c_n = 0$ (or $\partial E/\partial c_{n'}^* = 0$), we obtain a system of N linear and homogeneous equations in the variables $c_1, c_2, \ldots, c_N$, namely

$$\sum_{n=1}^{N} c_n (H_{n'n} - \Delta_{n'n} E) = 0; \qquad n' = 1, 2, \ldots, N \qquad \textbf{(2.408)}$$

The necessary and sufficient condition for this system to have a non-trivial solution is that the determinant of the coefficients vanishes. That is,

$$\det | H_{n'n} - \Delta_{n'n} E | = 0 \qquad \textbf{(2.409)}$$

Let $E_0^{(N)}, E_1^{(N)}, \ldots, E_{n-1}^{(N)}$ be the N roots of this equation, arranged in an ascending sequence; the superscript (N) indicates that we are dealing with an $N \times N$ determinant. The lowest root $E_0^{(N)}$ is of course an upper bound to the ground state energy E_0. Upon substituting $E_0^{(N)}$ in the system of equations (2.408) and solving for the coefficients c_n in terms of one of them (for example, c_1, which may be used as a normalisation factor), we then obtain the corresponding 'optimum' approximation ϕ_0 to the ground state wave function ψ_0. It may also be shown that the other roots $E_i^{(N)}$ of (2.409), with $i = 1, 2, \ldots, N-1$, are upper bounds to excited state energies of the system. In particular, if the Hamiltonian commutes with a Hermitian operator A and the trial function (2.404) has been constructed from eigenfunctions corresponding to a given eigenvalue α of A, then the roots $E_i^{(N)}$, with $i = 1, 2, \ldots, N-1$, are upper bounds to the energies E_i associated with excited states belonging to the eigenvalue α of A (for example, a given value of the angular momentum or the parity). The 'optimum' approximations $\phi_0, \phi_1, \ldots, \phi_{N-1}$ to the true wave functions, obtained by the above method, may also be shown to be mutually orthogonal. Moreover, if we construct a new trial function $\hat{\phi}$ containing an additional basis function χ_{N+1}, namely

$$\hat{\phi} = \sum_{n=1}^{N+1} c_n \chi_n \qquad \textbf{(2.410)}$$

it can be proved that the 'new' $(N+1)$ roots $E_0^{(N+1)}, E_1^{(N+1)}, \ldots, E_N^{(N+1)}$ of the determinantal equation (2.409) are separated by the 'old' roots $E_0^{(N)}, E_1^{(N)}, \ldots, E_{N-1}^{(N)}$. This property, which is illustrated in Fig. 2.10, is known as the *Hylleraas–Undheim theorem*.

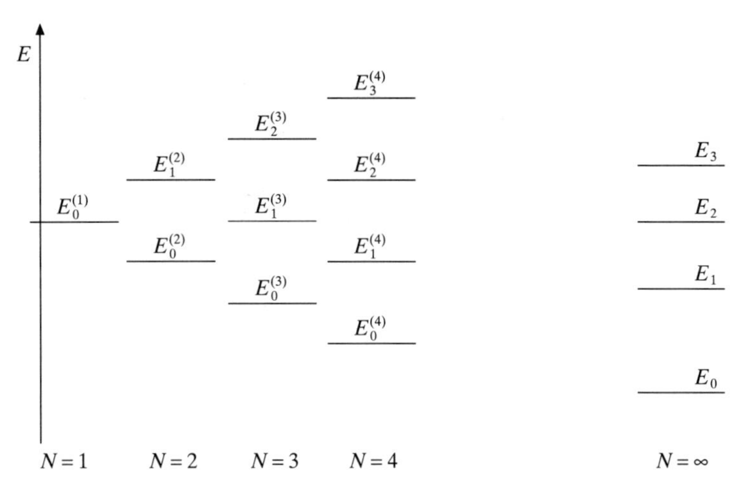

Figure 2.10 Approximate energy eigenvalues given by the Rayleigh–Ritz variational method with linear trial functions. Each root $E_i^{(N)}$ of the determinantal equation (2.409) is an upper bound to the corresponding exact eigenvalue E_i.

The variation–perturbation method

We have seen in our discussion of perturbation theory that according to the Rayleigh–Schrödinger method the solution $\psi_k^{(1)}$ of (2.304) (for the non-degenerate case) or the solution $\psi_{kr}^{(1)}$ of (2.323) (for the degenerate case) is obtained by expanding these functions in the basis set $\{\psi_m^{(0)}\}$ of the unperturbed eigenfunctions. The coefficients of this expansion and the second- and third-order corrections to the energy are then given in terms of matrix elements of H' between unperturbed eigenfunctions (see for example (2.316) and (2.319)). Unfortunately, in many cases the evaluation of all the necessary matrix elements of H' and of the required summations is very difficult. In these cases, however, it is possible to obtain *approximations* to the perturbed energy levels and eigenfunctions by using the *variation–perturbation method*, which we shall now discuss for the non-degenerate case.

Let us assume that the unperturbed eigenfunction $\psi_k^{(0)}$ as well as the corresponding energy $E_k^{(0)}$ and the first-order energy correction $E_k^{(1)} = \langle\psi_k^{(0)}|H'|\psi_k^{(0)}\rangle$ are known for a given state k. Let $\phi_k^{(1)}$ be an arbitrary trial function and $F_1[\phi_k^{(1)}]$ the functional

$$F_1[\phi_k^{(1)}] = \langle\phi_k^{(1)}|H_0 - E_k^{(0)}|\phi_k^{(1)}\rangle + 2\langle\phi_k^{(1)}|H' - E_k^{(1)}|\psi_k^{(0)}\rangle \tag{2.411}$$

We now express that this functional is *stationary* for variations of $\phi_k^{(1)}$ around the correct first-order wave function $\psi_k^{(1)}$. Proceeding as in the case of the variational method studied above, we find that the variational equation

$$\delta F_1 = 0 \tag{2.412}$$

implies that

$$(H_0 - E_k^{(0)})\phi_k^{(1)} + (H' - E_k^{(1)})\psi_k^{(0)} = 0 \qquad \textbf{(2.413)}$$

so that the function $\phi_k^{(1)}$ which makes the functional F_1 stationary is a solution $\psi_k^{(1)}$ of the equation (2.304). Moreover, by comparing (2.304), (2.310) and (2.411) we see that $F_1[\phi_k^{(1)}]$ reduces to the correct value of $E_k^{(2)}$ when $\phi_k^{(1)} \equiv \psi_k^{(1)}$.

Let us now consider the particular case of a state k for which $E_k^{(0)}$ is the *lowest* eigenvalue of H_0 corresponding to a given symmetry. It is then straightforward to show that (compare with (2.397))

$$E_k^{(2)} \leqslant F_1[\phi_k^{(1)}] \qquad \textbf{(2.414)}$$

One can then proceed in a way similar to that followed in the Rayleigh–Ritz method. First, a trial function $\phi_k^{(1)}$ is chosen containing a certain number of variational parameters. One then inserts that trial function in the functional $F_1[\phi_k^{(1)}]$, which is minimised with respect to the variational parameters. Because of (2.414) this minimum value of $F_1[\phi_k^{(1)}]$ gives an *upper bound* to $E_k^{(2)}$. The corresponding 'optimum' function $\phi_k^{(1)}$ can also be used to calculate an approximate value of $E_k^{(3)}$ by replacing $\psi_k^{(1)}$ by $\phi_k^{(1)}$ in (2.312).

The method can be extended to calculate approximate values of $\psi_k^{(2)}$, $E_k^{(4)}$ and $E_k^{(5)}$. To this end, we consider a trial function $\phi_k^{(2)}$ and the functional

$$F_2[\phi_k^{(2)}] = \langle \phi_k^{(2)} | H_0 - E_k^{(0)} | \phi_k^{(2)} \rangle + 2\langle \phi_k^{(2)} | H' - E_k^{(1)} | \psi_k^{(1)} \rangle - 2E_k^{(2)} \langle \phi_k^{(2)} | \psi_k^{(0)} \rangle \qquad \textbf{(2.415)}$$

where $\psi_k^{(1)}$ is now assumed to be known. Let us express that this functional is stationary with respect to variations of $\phi_k^{(2)}$. The condition

$$\delta F_2 = 0 \qquad \textbf{(2.416)}$$

yields the equation

$$(H_0 - E_k^{(0)})\phi_k^{(2)} + (H' - E_k^{(1)})\psi_k^{(1)} - E_k^{(2)}\psi_k^{(0)} = 0 \qquad \textbf{(2.417)}$$

so that the function $\phi_k^{(2)}$ which makes the functional F_2 stationary is a solution $\psi_k^{(2)}$ of (2.305). In addition, it is easily shown that when $\phi_k^{(2)} \equiv \psi_k^{(2)}$ the functional F_2 gives the exact value of $E_k^{(4)}$, apart from terms which are independent of $\phi_k^{(2)}$. Moreover, F_2 yields a minimum principle (upper bound) for $E_k^{(4)}$ when the state k is such that $E_k^{(0)}$ is the *lowest* eigenvalue of H_0 corresponding to a given symmetry. The function $\phi_k^{(2)}$ determined from the variational principle (2.416) can also be used to calculate the energy correction $E_k^{(5)}$. By constructing functionals $F_3[\phi_k^{(3)}]$, $F_4[\phi_k^{(4)}]$, and so on, the variation–perturbation method can be used to calculate higher order perturbation corrections to the wave functions and the energy eigenvalues.

The WKB approximation

We shall conclude this chapter with a discussion of an approximation method which can be used when the potential energy is a slowly varying function of position.

The method we shall study is called the *WKB approximation*, after G. Wentzel, H.A. Kramers and L. Brillouin, who independently introduced it in quantum mechanics in 1926. It can only be readily applied to one-dimensional problems, and to problems of higher dimensionality that can be reduced to the solution of a one-dimensional Schrödinger equation. We shall therefore focus our attention on the one-dimensional motion of a particle of mass m in a potential $V(x)$, the corresponding time-independent Schrödinger equation being the equation (2.121).

If the potential were constant, $V(x) = V_0$, solutions of (2.121) would be written as linear combinations of the plane waves

$$\psi(x) = A \exp\left(\pm\frac{\mathrm{i}}{\hbar} p_0 x\right) \tag{2.418}$$

where

$$p_0 = [2m(E - V_0)]^{1/2} \tag{2.419}$$

and A is a constant. The fact that the potential is 'slowly varying' means that $V(x)$ changes only slightly over the de Broglie wavelength

$$\lambda(x) = \frac{h}{p(x)} \tag{2.420}$$

associated with the particle (having an energy $E > V$) in this potential, where

$$p(x) = \{2m[E - V(x)]\}^{1/2} \tag{2.421}$$

is the classical momentum at the point x. Now, in the classical limit ($h \to 0$) the de Broglie wavelength λ tends to zero, so that the slow variation of the potential in a de Broglie wavelength is always satisfied for physically realizable potentials. In the quantum case, the fact that the potential varies slowly over a de Broglie wavelength means that we are dealing with a quasi-classical (or semi-classical) situation. For this reason, the WKB method is said to belong to the category of *semi-classical approximations*.

The result (2.418) obtained for constant potentials suggests that in the case of slowly varying potentials, solutions of the Schrödinger equation (2.121) should be sought which have the form

$$\psi(x) = A \exp\left[\frac{\mathrm{i}}{\hbar} S(x)\right] \tag{2.422}$$

Substituting (2.422) into (2.121), we obtain for $S(x)$ the equation

$$-\frac{\mathrm{i}\hbar}{2m}\frac{\mathrm{d}^2 S(x)}{\mathrm{d}x^2} + \frac{1}{2m}\left[\frac{\mathrm{d}S(x)}{\mathrm{d}x}\right]^2 + V(x) - E = 0 \tag{2.423}$$

So far, no approximation has been made, the above equation being strictly equivalent to the original Schrödinger equation (2.121). Unfortunately, equation (2.423) is a non-linear equation which is in fact more complicated than the Schrödinger equation (2.121) itself! We must therefore try to solve (2.423) approximately. To this end, we first remark that if the potential is constant, then $S(x) = \pm p_0 x$ (see (2.418)) and the first term on the left of (2.423) vanishes. Moreover, this term is proportional to $\hbar$, and hence vanishes in the classical limit. This suggest that we treat $\hbar$ as a parameter of smallness and expand the function $S(x)$ in the power series

$$S(x) = S_0(x) + \hbar S_1(x) + \frac{\hbar^2}{2} S_2(x) + \ldots \tag{2.424}$$

Inserting this expansion into (2.423), and equating to zero the coefficients of each power of $\hbar$ separately, we find the set of equations (Problem 2.26)

$$\frac{1}{2m}\left[\frac{\mathrm{d}S_0(x)}{\mathrm{d}x}\right]^2 + V(x) - E = 0 \tag{2.425a}$$

$$\frac{\mathrm{d}S_0(x)}{\mathrm{d}x}\frac{\mathrm{d}S_1(x)}{\mathrm{d}x} - \frac{\mathrm{i}}{2}\frac{\mathrm{d}^2 S_0(x)}{\mathrm{d}x^2} = 0 \tag{2.425b}$$

$$\frac{\mathrm{d}S_0(x)}{\mathrm{d}x}\frac{\mathrm{d}S_2(x)}{\mathrm{d}x} + \left[\frac{\mathrm{d}S_1(x)}{\mathrm{d}x}\right]^2 - \mathrm{i}\frac{\mathrm{d}^2 S_1(x)}{\mathrm{d}x^2} = 0 \tag{2.425c}$$

which must be solved successively to find $S_0(x)$, $S_1(x)$, $S_2(x)$, and so on. As we shall see below, the WKB method yields the first *two* terms [10] of the expansion (2.424).

Let us begin by solving the equation (2.425a). Assuming first that $E > V(x)$, so that we are in a classically allowed region of positive kinetic energy, we find that

$$S_0(x) = \pm \int^x p(x')\,\mathrm{d}x' \tag{2.426}$$

where $p(x)$ is the classical momentum given by (2.421) and an arbitrary integration constant, which can be absorbed in the coefficient A on the right of (2.422), has been omitted. Substituting (2.426) into (2.422), we obtain the simple semi-classical wave function

$$\psi(x) = A \exp\left[\pm\mathrm{i}\int^x \kappa(x')\,\mathrm{d}x'\right] \tag{2.427}$$

[10] It can be shown that the series (2.424) does not converge, but is an asymptotic series for the function $S(x)$. As a result, the best approximation to $S(x)$ is obtained by keeping a finite number of terms on the right-hand side of (2.424).

where

$$\kappa(x) = \hbar^{-1}p(x) = \hbar^{-1}\{2m[E - V(x)]\}^{1/2} \tag{2.428}$$

is the 'effective wave number'. If we assume that $E \gg |V(x)|$, we may expand the effective wave number as

$$\kappa(x) = k\left[1 - \frac{1}{2}\frac{V(x)}{E} + \ldots\right] = k - \frac{U(x)}{2k} + \ldots \tag{2.429}$$

where $k = (2mE)^{1/2}/\hbar$ and $U(x) = 2mV(x)/\hbar^2$. The semi-classical wave function (2.427) then becomes

$$\psi(x) = A \exp\left\{\pm\mathrm{i}\left[kx - \frac{1}{2k}\int^x U(x')\,\mathrm{d}x'\right]\right\} \tag{2.430}$$

which is called a (one-dimensional) *eikonal* wave function.

Using the result (2.426) in equation (2.425b), the function $S_1(x)$ is found to be

$$S_1(x) = \frac{\mathrm{i}}{2}\log p(x) \tag{2.431}$$

where an arbitrary constant of integration has again been omitted. Next, by using the expressions for $S_0(x)$ and $S_1(x)$ in (2.425c) one finds (Problem 2.27) that

$$S_2(x) = \frac{1}{2}m[p(x)]^{-3}\frac{\mathrm{d}V(x)}{\mathrm{d}x} - \frac{1}{4}m^2\int^x [p(x')]^{-5}\left[\frac{\mathrm{d}V(x')}{\mathrm{d}x'}\right]^2 \mathrm{d}x' \tag{2.432}$$

From this expression and that of $p(x)$ (see (2.421)), it is clear that S_2 will be small whenever $\mathrm{d}V(x)/\mathrm{d}x$ is small, and provided that $E - V$ is not too close to zero. If, in addition, all the higher derivatives of $V(x)$ are small, it can be verified that $S_3, S_4, \ldots$ will also be small. Note that in the case of a constant potential $S_0 = \pm p_0 x$ (apart from an irrelevant additive constant) and $S_1, S_2, \ldots$ are all zero.

These considerations show that if $V(x)$ is a slowly varying function of position, and provided $E - V$ is not too small, one can retain only the first two terms on the right of the expansion (2.424). Using (2.426) and (2.431), we then obtain the *WKB approximation* to the wave function (2.422) in a classically allowed region, namely

$$\psi(x) = A[p(x)]^{-1/2}\exp\left[\pm\frac{\mathrm{i}}{\hbar}\int^x p(x')\,\mathrm{d}x'\right], \qquad E > V \tag{2.433}$$

Of course, the general WKB solution in this region is a linear combination of such solutions. That is,

$$\psi(x) = [p(x)]^{-1/2} \left\{ A \exp\left[\frac{\mathrm{i}}{\hbar} \int^{x} p(x')\,\mathrm{d}x' \right] + B \exp\left[-\frac{\mathrm{i}}{\hbar} \int^{x} p(x')\,\mathrm{d}x' \right] \right\}, \qquad E > V \tag{2.434}$$

where A and B are arbitrary constants. The first exponential corresponds to a wave moving in the positive direction, the second exponential to a wave moving in the opposite direction. We also remark that if the potential is constant ($V = V_0$) these exponentials reduce to the plane waves $\exp(\mathrm{i}p_0 x/\hbar)$ and $\exp(-\mathrm{i}p_0 x/\hbar)$, respectively, where p_0 is given by (2.419).

A criterion of validity for the WKB approximation may be obtained by requiring that the contribution to the wave function (2.422) arising from the third term on the right of (2.424) be negligible. This will be the case if

$$|(\hbar/2)S_2(x)| \ll 1 \tag{2.435}$$

Using the expression (2.432) of $S_2(x)$ and taking into account that both terms on the right of (2.432) are of the same order of magnitude, we may rewrite the condition (2.435) more explicitly as

$$\left| \frac{\hbar m\,\mathrm{d}V(x)/\mathrm{d}x}{[2m(E - V(x))]^{3/2}} \right| \ll 1 \tag{2.436}$$

where we have used the expression (2.421) of $p(x)$. This criterion is seen to be satisfied if the potential energy varies slowly enough, provided that the kinetic energy $E - V$ is sufficiently large. An alternative way of writing the condition (2.436) is

$$\left| \frac{\hbar}{[p(x)]^2} \frac{\mathrm{d}p(x)}{\mathrm{d}x} \right| = \left| \frac{1}{p(x)} \frac{\mathrm{d}p(x)}{\mathrm{d}x} ƛ(x) \right| \ll 1 \tag{2.437}$$

where $ƛ(x) = \hbar/p(x) = \lambda(x)/(2\pi)$ is the reduced de Broglie wavelength of the particle. The inequality (2.437) means that the fractional change in the momentum must be small in a wavelength. In other words, the potential must change so slowly that the momentum of the particle is nearly constant over many wavelengths. We remark that the condition (2.437) may equally be written in the form

$$\left| \frac{\mathrm{d}ƛ(x)}{\mathrm{d}x} \right| \ll 1 \tag{2.438}$$

Now $\delta\lambda\!\!\!^{-} = (\mathrm{d}\lambda\!\!\!^{-}/\mathrm{d}x)\delta x$ is the change occurring in the reduced wavelength $\lambda\!\!\!^{-}$ in the distance δx. Upon setting $\delta x = \lambda\!\!\!^{-}$, we see that the condition (2.438) is equivalent to

$$|\delta\lambda\!\!\!^{-}(x)| = \left|\frac{\mathrm{d}\lambda\!\!\!^{-}(x)}{\mathrm{d}x}\lambda\!\!\!^{-}(x)\right| \ll \lambda\!\!\!^{-}(x) \tag{2.439}$$

showing that $\lambda\!\!\!^{-}$ must only change by a small fraction of itself over a distance of the order of $\lambda\!\!\!^{-}$. This condition is well known in wave optics as the one to be satisfied if a wave is to propagate in a medium of varying index of refraction without giving rise to significant reflection.

Until now we have only considered classically allowed regions for which $E > V$. However, we can equally well obtain the WKB approximation to the wave function (2.422) in classically forbidden regions of negative kinetic energy ($E < V$), where $p(x)$ becomes purely imaginary (see (2.421)). Solving equations (2.425a) and (2.425b) in this case, we obtain the WKB solutions (Problem 2.28)

$$\psi(x) = A|p(x)|^{-1/2} \exp\left[\pm\frac{1}{\hbar}\int^x |p(x')|\,\mathrm{d}x'\right], \qquad E < V \tag{2.440}$$

where A is a constant. The general WKB solution for $E < V$ is a linear combination of such solutions, namely

$$\psi(x) = |p(x)|^{-1/2}\left\{C\exp\left[-\frac{1}{\hbar}\int^x |p(x')|\,\mathrm{d}x'\right]\right.$$
$$\left. + D\exp\left[\frac{1}{\hbar}\int^x |p(x')|\,\mathrm{d}x'\right]\right\}, \qquad E < V \tag{2.441}$$

where C and D are arbitrary constants. These WKB solutions are accurate provided the criterion (2.436) is satisfied, which requires that $V(x)$ varies slowly enough and that $V - E$ be sufficiently large.

It is apparent from (2.436) that the WKB approximation breaks down in the vicinity of a *classical turning point*, for which $E = V$. The two kinds of WKB wave functions (2.434) and (2.441) we have obtained are therefore 'asymptotically' valid, because they may only be used sufficiently far away from the nearest turning point. Since the exact wave function is continuous and smooth for all x, it is possible to obtain *connection formulae* which allow one to join smoothly the two types of WKB solutions across a turning point. This interpolation is based on the fact that the original Schrödinger equation (2.121) can be modified slightly so that an 'exact' solution at and near a turning point can be written down. A detailed discussion of the WKB connection formulae may be found in Bransden and Joachain (2000).

Problems

2.1 Consider the momentum space wave function

$$\begin{aligned}\phi(p_x) &= 0, \quad |p_x - p_0| > \gamma \\ &= C, \quad |p_x - p_0| \leqslant \gamma\end{aligned}$$

where p_0, C and γ are constants. Find the corresponding wave function $\psi(x)$ in configuration space and determine the constant C, so that $\psi(x)$ satisfies the normalisation condition

$$\int_{-\infty}^{\infty} \mathrm{d}x |\psi(x)|^2 = 1$$

Using a reasonable definition of the 'width' Δx of $|\psi(x)|^2$, show that $\Delta x \, \Delta p_x \gtrsim \hbar$.

2.2 Consider an electron of momentum $\mathbf{p}$ in the Coulomb field of a proton. The total energy is

$$E = \frac{p^2}{2m} - \frac{e^2}{(4\pi\varepsilon_0)r}$$

where r is the distance of the electron from the proton. Assuming that the uncertainty Δr of the radial coordinate is $\Delta r \simeq r$ and that $\Delta p \simeq p$, use Heisenberg's uncertainty principle to obtain estimates of the size and of the energy of the hydrogen atom in the ground state.

2.3 Consider the one-dimensional wave function

$$\psi_{p_x}(x) = C \exp(\mathrm{i}p_x x/\hbar)$$

which is an eigenfunction of the operator $(p_x)_{\mathrm{op}} = -\mathrm{i}\hbar\partial/\partial x$ corresponding to the eigenvalue $p_x = \hbar k$, and is therefore called a *momentum eigenfunction*.

(a) Determine the constant C (independent of x) so that $\psi_{p_x}(x)$ is 'normalised' according to the *p-normalisation* condition

$$\int \psi^*_{p'_x}(x)\psi_{p_x}(x)\, \mathrm{d}x = \delta(p_x - p'_x)$$

(b) Determine the constant C so that $\psi_{p_x}(x)$ is 'normalised' *on the energy scale* (*energy normalisation*) according to

$$\int \psi^*_{p'_x}(x)\psi_{p_x}(x)\, \mathrm{d}x = \delta(E - E')$$

where $E = p_x^2/(2m)$ and $E' = p_x'^2/(2m)$.

(**Hint**: Use the representation (2.31a) and the properties (2.33d) and (2.33g) of the Dirac delta function.)

2.4 Consider the normalised Gaussian wave packet $\psi(x) = N\exp(-\lambda^2 x^2)$ where $N = (2\lambda^2/\pi)^{1/4}$. Using (2.16), calculate the uncertainties Δx and Δp_x from (2.41) and show that $\Delta x \, \Delta p_x = \hbar/2$.

2.5 Starting from (2.39) show that

$$\langle \mathbf{p} \rangle = \int \Phi^*(\mathbf{p}, t)\mathbf{p}\Phi(\mathbf{p}, t)\, d\mathbf{p}$$

where $\Phi(\mathbf{p}, t)$ is the momentum space wave function (normalised to unity) corresponding to $\Psi(\mathbf{r}, t)$ (see (2.20)).

2.6 A particle of mass m moving in a potential $V(\mathbf{r})$ has the Hamiltonian $H = -(\hbar^2/2m)\nabla^2 + V(\mathbf{r})$. Using the result (2.56) prove Ehrenfest's theorem (2.57).

2.7 (a) Using the definition of an adjoint operator (2.65) prove that if A and B are two operators then $(AB)^\dagger = B^\dagger A^\dagger$.

(b) Suppose that A and B are two non-commuting Hermitian operators. Determine which of the following operators are Hermitian: (i) AB (ii) $[A, B]$ (iii) $AB + BA$ (iv) ABA (v) A^n where n is a positive integer.

2.8 Let E_n denote the energy eigenvalues of a one-dimensional system and $\psi_n(x)$ the corresponding energy eigenfunctions. Suppose that the normalised wave function of the system at $t = 0$ is given by

$$\Psi(x, t=0) = \frac{1}{\sqrt{2}}\exp(i\alpha_1)\psi_1(x) + \frac{1}{\sqrt{3}}\exp(i\alpha_2)\psi_2(x) + \frac{1}{\sqrt{6}}\exp(i\alpha_3)\psi_3(x)$$

where the α_i are constants.

(a) Write down the wave function $\Psi(x, t)$ at time t.

(b) Find the probability that at time t a measurement of the energy of the system gives the value E_2.

(c) Does $\langle x \rangle$ vary with time? Does $\langle p_x \rangle$ vary with time? Does $E = \langle H \rangle$ vary with time?

2.9 If U is a unitary operator and $A' = UAU^\dagger$ where A is a Hermitian operator, show that:

(a) A' is Hermitian.

(b) The eigenvalues of A' are the same as those of A.

(c) $\langle \Phi' | A' | \Psi' \rangle = \langle \Phi | A | \Psi \rangle$ where Ψ' and Φ' are related to Ψ and Φ by (2.106).

2.10 Consider a system described by a time-independent Hamiltonian H. Let A and B be two non-commuting operators ($[A, B] \neq 0$) which are constants of the motion, so that they commute with H. Prove that in this case the energy levels of the system are, in general, degenerate.

2.11 Consider a particle of mass m moving in one dimension in the infinite square well (2.122). Suppose that at time $t = 0$ its wave function is given by

$$\Psi(x, t=0) = A(a^2 - x^2)$$

(a) Find the probability P_n that a measurement of the energy will give the value E_n, where E_n is given by (2.131).

(b) Obtain the average value of the energy.

2.12 Classically, the linear harmonic oscillator having the total energy $E = p^2/2m + m\omega^2x^2/2$ could have $x = p = 0$ and hence $E = 0$. In quantum mechanics, there will be uncertainties $x = 0 + \Delta x$, $p_x = 0 + \Delta p_x$ such that $\Delta x \Delta p_x \gtrsim \hbar$. Show that because of this condition the minimum total energy cannot be zero and must be of the order of $\hbar\omega$.

2.13 Starting from the definitions (2.154) show that L_x, L_y and L_z can be expressed in spherical polar coordinates (see (2.157)) as

$$L_x = \mathrm{i}\hbar\left(\sin\phi\frac{\partial}{\partial\theta} + \cot\theta\cos\phi\frac{\partial}{\partial\phi}\right)$$

$$L_y = \mathrm{i}\hbar\left(-\cos\phi\frac{\partial}{\partial\theta} + \cot\theta\sin\phi\frac{\partial}{\partial\phi}\right)$$

$$L_z = -\mathrm{i}\hbar\frac{\partial}{\partial\phi}$$

and verify that $\mathbf{L}^2 = L_x^2 + L_y^2 + L_z^2$ is given by (2.159). Using these results show that $[\mathbf{L}, f(r)] = 0$ and $[\mathbf{L}^2, f(r)] = 0$ where $f(r)$ is an arbitrary function of the radial coordinate r. Show also that $[\mathbf{L}, \mathbf{p}^2] = 0$, where $\mathbf{p}^2 = -\hbar^2\nabla^2$.

2.14 Obtain the results (2.187) and (2.188). Prove also that for a spin-1/2 particle, one has

$$\langle S_x\rangle = \langle S_y\rangle = 0 \text{ while } \langle S_x^2\rangle = \langle S_y^2\rangle = \hbar^2/4.$$

2.15 Verify the equations (2.217) and (2.218).

2.16 Write down explicitly the matrices representing the operators S_x, S_y, S_z, S_+, S_- and $\mathbf{S}^2$ for $s = 1$ and $s = 3/2$.

2.17 Prove that $\mathbf{L}^2$, $\mathbf{S}^2$, $\mathbf{J}^2$ and J_z form a commuting set of operators, and that $[(\mathbf{L}\cdot\mathbf{S}), J_z] = 0$.

2.18 Assuming that the expansion of a plane wave in a series of Legendre polynomials is of the form

$$\exp(\mathrm{i}kr\cos\theta) = \sum_{l=0}^{\infty} c_l j_l(kr) P_l(\cos\theta)$$

find the coefficients c_l.
(**Hint**: Use the fact that $j_l(kr) \underset{r\to\infty}{\to} (kr)^{-1}\sin(kr - l\pi/2)$.)

2.19 Show that $E_k^{(2)}$ and $E_k^{(3)}$ are given by (2.311) and (2.312) respectively.

2.20 Consider a linear harmonic oscillator for which the Hamiltonian is

$$H_0 = -\frac{\hbar^2}{2m}\frac{\mathrm{d}^2}{\mathrm{d}x^2} + \frac{1}{2}kx^2$$

If this oscillator is perturbed by an additional potential of the form $H' = \frac{1}{2}k'x^2$, find the first- and second-order corrections to the energy levels using perturbation theory. Also find the first-order correction to the wave function. Compare your results with the exact solution

$$E_n = \hbar\omega\left(n + \frac{1}{2}\right)\left(1 + \frac{k'}{k}\right)^{1/2}$$

where $\omega = (k/m)^{1/2}$. (*Hint*: Obtain first the matrix elements:

$$x_{ij}^2 = \langle\psi_i|x^2|\psi_j\rangle$$

$$= \begin{cases} (2\alpha^2)^{-1}[(j+1)(j+2)]^{1/2} & i = j+2 \\ (2\alpha^2)^{-1}(2j+1) & i = j \\ (2\alpha^2)^{-1}[j(j-1)]^{1/2} & i = j-2 \end{cases}$$

where $\alpha = \left(\dfrac{mk}{\hbar^2}\right)^{1/4}$.)

2.21 A two-dimensional isotropic harmonic oscillator has the Hamiltonian

$$H = -\frac{\hbar^2}{2m}\left(\frac{\partial^2}{\partial x^2} + \frac{\partial^2}{\partial y^2}\right) + \frac{1}{2}k(x^2 + y^2)$$

(a) Show that the energy levels are given by

$$E_{n_x n_y} = \hbar\omega(n_x + n_y + 1); \qquad n_x = 0, 1, 2, \ldots$$
$$n_y = 0, 1, 2, \ldots, \qquad \omega = (k/m)^{1/2}$$

What is the degeneracy of each level?

(b) If this oscillator is perturbed by an interaction of the form $H' = \lambda xy$, where λ is a constant, find the first-order modification of the energy of the first excited level.

2.22 Consider a particle of charge q and mass m, which is in simple harmonic motion along the X axis with a force constant k. An electric field $\mathscr{E}(t)$, directed along the X axis, is switched on at time $t = 0$ so that the system is perturbed by an interaction

$$H'(t) = -qx\mathscr{E}(t)$$

If $\mathscr{E}(t)$ has the form

$$\mathscr{E}(t) = \mathscr{E}_0 \exp(-t/\tau)$$

where $\mathscr{E}_0$ and τ are constants, and if the oscillator is in the ground state for $t \leqslant 0$, find the probability that the oscillator will be found in an excited state as $t \to \infty$.

2.23 (a) By varying the parameter a in the trial function

$$\phi_0(x) = (a^2 - x^2)^2, \qquad |x| < a$$
$$= 0 \qquad |x| \geqslant a$$

obtain an upper bound for the ground state energy of a linear harmonic oscillator having the Hamiltonian

$$H = -\frac{\hbar^2}{2m}\frac{d^2}{dx^2} + \frac{1}{2}m\omega^2 x^2$$

(b) Show that the function $\phi_1(x) = x\phi_0(x)$ is a suitable trial function for the first excited state and obtain a variational estimate of the energy of this level.

2.24 Prove the result given in equation (2.402).

2.25 Let $E_{n,\bar{l}}$ denote the discrete energy levels of a particle of mass m in a central potential $V(r)$, corresponding to a given orbital angular momentum quantum number $\bar{l}$, and let $E_{n,\bar{l}}^{\min}$ be their minimum value. Prove that $E_{n,\bar{l}}^{\min} < E_{n,\bar{l}+1}^{\min}$ (**Hint**: Write the Hamiltonian of the particle as

$$H = H_r + \frac{\mathbf{L}^2}{2mr^2}, \qquad H_r = -\frac{\hbar^2}{2m}\frac{1}{r^2}\frac{\partial}{\partial r}\left(r^2\frac{\partial}{\partial r}\right) + V(r)$$

and note that H_r is a purely radial operator.)

2.26 Derive the equations (2.425) for $S_0(x)$, $S_1(x)$ and $S_2(x)$.

2.27 Derive the expression (2.432) for $S_2(x)$.

2.28 Solve the equations (2.425a) and (2.425b) in the case $E < V$ to obtain the WKB solutions (2.440).

3 One-electron atoms

In this chapter we begin our quantum mechanical study of atomic structure by considering the simplest atom, namely the hydrogen atom, which consists of a proton and an electron. Apart from small corrections, which we shall discuss in Chapter 5, the hydrogen atom may be considered as a non-relativistic system of two particles interacting by means of an attractive Coulomb potential. Other similar one-electron systems, called hydrogenic atoms, include the isotopes of hydrogen (deuterium, tritium) and the hydrogenic ions (He^+, Li^{++}, etc.) which we have already encountered in our study of the Bohr model. These hydrogenic systems will also be discussed in the present chapter.

Our starting point is the Schrödinger equation for one-electron atoms. After separating the centre of mass motion, we solve the eigenvalue equation for the relative motion in spherical polar coordinates, and obtain the energy levels and wave functions of the discrete spectrum. We then consider the expectation values of various operators, and prove the virial theorem. This is followed by a study of the bound state wave functions of one-electron atoms in parabolic coordinates. We conclude this chapter with a brief discussion of 'special' hydrogenic systems such as positronium, muonium, antihydrogen, muonic and hadronic atoms, and Rydberg atoms.

3.1 The Schrödinger equation for one-electron atoms

Let us consider a hydrogenic atom containing an atomic nucleus of charge Ze and an electron of charge $-e$ interacting by means of the Coulomb potential

$$V(r) = -\frac{Ze^2}{(4\pi\varepsilon_0)r} \tag{3.1}$$

where r is the distance between the two particles. We denote by m the mass of the electron and M the mass of the nucleus. Since the interaction potential (3.1) only depends on the relative coordinate of the two particles, we may use the results of Section 2.7 to separate the motion of the centre of mass. Remembering that $\mathbf{P}$ is the total momentum associated with the motion of the centre of mass and $\mathbf{p}$ the relative momentum, the total energy of the atom can be split into two parts. The first one is the kinetic energy $P^2/[2(M + m)]$ corresponding to the motion of the centre of mass, and the second one is the (internal) energy of the relative motion, governed by the Hamiltonian

$$H = \frac{p^2}{2\mu} - \frac{Ze^2}{(4\pi\varepsilon_0)r} \tag{3.2}$$

where

$$\mu = \frac{mM}{m + M} \tag{3.3}$$

is the reduced mass of the two particles. Thus, working in the centre of mass system (where $P = 0$) and in the position representation, we must solve the one-body time-independent Schrödinger equation

$$\left[-\frac{\hbar^2}{2\mu}\nabla^2 - \frac{Ze^2}{(4\pi\varepsilon_0)r}\right]\psi(\mathbf{r}) = E\psi(\mathbf{r}) \tag{3.4}$$

Instead of using the position representation, as we shall do below, it is also possible to solve the Schrödinger equation for one-electron atoms in momentum space. This is carried out in Appendix 5, where the hydrogenic wave functions in momentum space are obtained.

Solution in spherical polar coordinates

Because the interaction potential (3.1) is central, the wave equation (3.4) may be separated in spherical polar coordinates (see Section 2.6). Thus we write a particular solution of this equation as

$$\psi_{E,l,m}(r, \theta, \phi) = R_{E,l}(r)Y_{lm}(\theta, \phi) \tag{3.5}$$

where $Y_{lm}(\theta, \phi)$ is a spherical harmonic corresponding to the orbital angular momentum quantum number l and to the magnetic quantum number m (with $m = -l, -l + 1, \ldots, +l$). The function $R_{E,l}(r)$ satisfies the radial Schrödinger equation

$$\left\{-\frac{\hbar^2}{2\mu}\left[\frac{1}{r^2}\frac{\mathrm{d}}{\mathrm{d}r}\left(r^2\frac{\mathrm{d}}{\mathrm{d}r}\right) - \frac{l(l+1)}{r^2}\right] - \frac{Ze^2}{(4\pi\varepsilon_0)r}\right\}R_{E,l}(r) = ER_{E,l}(r) \tag{3.6}$$

which we have obtained by substituting into (2.235) the expression (3.1) of the Coulomb potential, and making the change $m \to \mu$. As we remarked in Section 2.6, we can simplify (3.6) by introducing the new unknown function

$$u_{E,l}(r) = rR_{E,l}(r) \tag{3.7}$$

Thus, using (2.239) we find that $u_{E,l}(r)$ satisfies the equation

$$\frac{\mathrm{d}^2 u_{E,l}}{\mathrm{d}r^2} + \frac{2\mu}{\hbar^2}[E - V_{\text{eff}}(r)]u_{E,l}(r) = 0 \tag{3.8}$$

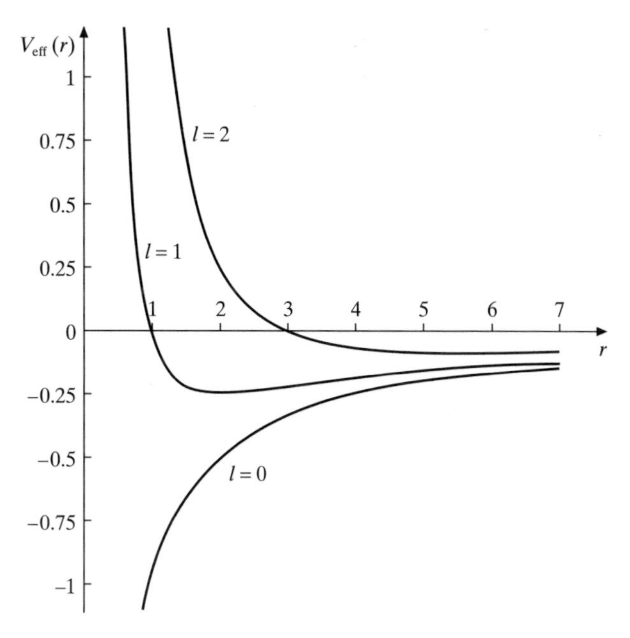

Figure 3.1 The effective potential $V_{eff}(r)$ given by (3.9) for the case $Z = 1$ and for the values $l = 0, 1, 2$. The unit of length is $a_\mu = (m/\mu)a_0$ where a_0 is the Bohr radius (1.86). The unit of energy is $e^2/(4\pi\varepsilon_0 a_\mu)$.

where

$$V_{eff}(r) = -\frac{Ze^2}{(4\pi\varepsilon_0)r} + \frac{l(l+1)\hbar^2}{2\mu r^2} \tag{3.9}$$

is the effective potential. Figure 3.1 shows $V_{eff}(r)$ for the case $Z = 1$ and for the values $l = 0, 1, 2$.

The problem of solving the Schrödinger equation (3.4) therefore reduces to that of solving the radial one-dimensional equation (3.8) corresponding to a particle of mass μ moving in an effective potential V_{eff} made up of the Coulomb potential (3.1) plus the 'centrifugal barrier' potential $l(l+1)\hbar^2/(2\mu r^2)$. It is clear that since $V_{eff}(r)$ tends to zero for large r, the solution $u_{E,l}(r)$ for $E > 0$ will have an oscillatory behaviour at infinity and will be an acceptable eigenfunction for any positive value of E. We therefore have a continuum spectrum for $E > 0$. The corresponding unbound states play an important role in the analysis of the scattering of electrons by ions; we shall return to such collision phenomena in Chapters 12 and 13. For the moment, however, we will only be concerned with the bound states of hydrogenic atoms, corresponding to the case $E < 0$.

Let us return to the radial equation (3.8). We shall look for solutions which vanish at $r = 0$, that is

$$u_{E,l}(0) = 0 \tag{3.10}$$

so that the radial function $R_{E,l}(r)$ – and hence the full wave function $\psi_{E,l,m}(\mathbf{r})$ given by (3.5) – remains finite at the origin [1]. It is convenient to introduce the dimensionless quantities

$$\rho = \left(-\frac{8\mu E}{\hbar^2} \right)^{1/2} r \tag{3.11}$$

and

$$\lambda = \frac{Ze^2}{4\pi\varepsilon_0 \hbar} \left(-\frac{\mu}{2E} \right)^{1/2} = Z\alpha \left(-\frac{\mu c^2}{2E} \right)^{1/2} \tag{3.12}$$

where $\alpha = e^2/(4\pi\varepsilon_0\hbar c) \simeq 1/137$ is the fine structure constant and we recall that $E < 0$. In terms of the new quantities ρ and λ, the equation (3.8) now reads

$$\left[\frac{\mathrm{d}^2}{\mathrm{d}\rho^2} - \frac{l(l+1)}{\rho^2} + \frac{\lambda}{\rho} - \frac{1}{4} \right] u_{E,l}(\rho) = 0 \tag{3.13}$$

As in the case of the harmonic oscillator (see Section 2.4) we first analyse the asymptotic behaviour of $u_{E,l}(\rho)$. To this end, we remark that when $\rho \to \infty$ the terms in $1/\rho$ and $1/\rho^2$ become negligible with respect to the constant term $(-1/4)$, so that (3.13) reduces to the equation

$$\left[\frac{\mathrm{d}^2}{\mathrm{d}\rho^2} - \frac{1}{4} \right] u_{E,l}(\rho) = 0 \tag{3.14}$$

whose solutions are $\exp(\pm\rho/2)$. Therefore, using the fact that the function $u_{E,l}(r)$ must be bounded everywhere, including infinity, we keep only the exponentially decreasing function and we have

$$u_{E,l}(\rho) \underset{\rho\to\infty}{\sim} \mathrm{e}^{-\rho/2} \tag{3.15}$$

This result suggests that we should search for a solution of the radial equation (3.13) having the form

$$u_{E,l}(\rho) = \mathrm{e}^{-\rho/2} f(\rho) \tag{3.16}$$

where we have written $f(\rho) \equiv f_{E,l}(\rho)$ to simplify the notation. Substituting (3.16) in (3.13), we obtain for $f(\rho)$ the equation

$$\left[\frac{\mathrm{d}^2}{\mathrm{d}\rho^2} - \frac{\mathrm{d}}{\mathrm{d}\rho} - \frac{l(l+1)}{\rho^2} + \frac{\lambda}{\rho} \right] f(\rho) = 0 \tag{3.17}$$

[1] The finiteness of the wave function $\psi_{E,l,m}(\mathbf{r})$ is not really a necessary requirement. In fact, mildly singular wave functions are encountered in some cases, for example in the Dirac relativistic theory of one-electron atoms. The correct boundary condition is obtained by requiring that all the possible physical states are described by a complete, orthogonal set of wave functions. However, for a large class of potentials including the Coulomb potential (3.1), this condition may be shown to lead to the simpler requirement (3.10).

We now write a power series expansion for $f(\rho)$ in the form

$$f(\rho) = \rho^{l+1} g(\rho) \tag{3.18}$$

where

$$g(\rho) = \sum_{k=0}^{\infty} c_k \rho^k, \qquad c_0 \neq 0 \tag{3.19}$$

and we have used the fact (see Section 2.6) that $u_{E,l}(\rho)$ – and therefore $f(\rho)$ – behaves like ρ^{l+1} for all central potentials $V(r)$ which are less singular than r^{-2} at the origin. Upon substitution of (3.18) into (3.17) we find that the function $g(\rho)$ satisfies the differential equation

$$\left[\rho \frac{\mathrm{d}^2}{\mathrm{d}\rho^2} + (2l + 2 - \rho)\frac{\mathrm{d}}{\mathrm{d}\rho} + (\lambda - l - 1)\right] g(\rho) = 0 \tag{3.20}$$

Moreover, using the expansion (3.19) to solve the equation (3.20), we find that

$$\sum_{k=0}^{\infty} [k(k-1)c_k\rho^{k-1} + (2l+2-\rho)kc_k\rho^{k-1} + (\lambda - l - 1)c_k\rho^k] = 0 \tag{3.21}$$

or

$$\sum_{k=0}^{\infty} \{[k(k+1) + (2l+2)(k+1)]c_{k+1} + (\lambda - l - 1 - k)c_k\}\rho^k = 0 \tag{3.22}$$

so that the coefficients c_k must satisfy the recursion relation

$$c_{k+1} = \frac{k + l + 1 - \lambda}{(k+1)(k+2l+2)} c_k \tag{3.23}$$

If the series (3.19) does not terminate, we see from (3.23) that for large k

$$\frac{c_{k+1}}{c_k} \sim \frac{1}{k} \tag{3.24}$$

a ratio which is the same as that of the series for $\rho^p \exp(\rho)$, where p has a finite value. Thus in this case we deduce from (3.16) and (3.18) that the function $u_{E,l}(r)$ has an asymptotic behaviour of the type

$$u_{E,l}(\rho) \underset{\rho\to\infty}{\sim} \rho^{l+1+p}\, \mathrm{e}^{\rho/2} \tag{3.25}$$

which is clearly unacceptable.

The series (3.19) must therefore terminate, or in other words $g(\rho)$ must be a polynomial in ρ. Let us assume that the highest power of ρ in $g(\rho)$ is ρ^{n_r}, where the *radial quantum number* $n_r = 0, 1, 2, \ldots$ is a positive integer or zero. Then the coefficient $c_{n_r+1} = 0$, and from the recursion relation (3.23) we deduce that

$$\lambda = n_r + l + 1 \tag{3.26}$$

Let us introduce the *principal quantum number*

$$n = n_r + l + 1 \tag{3.27}$$

which is a positive integer ($n = 1, 2, \ldots$) since n_r and l can take on positive integer or zero values. Thus, from (3.26) and (3.27), we see that the eigenvalues of (3.13) are given by

$$\lambda = n, \qquad n = 1, 2, \ldots \tag{3.28}$$

3.2 Energy levels

Replacing in (3.12) the parameter λ by its value (3.28) we obtain the energy eigenvalues

$$\begin{aligned} E_n &= -\frac{1}{2n^2}\left(\frac{Ze^2}{4\pi\varepsilon_0}\right)^2\frac{\mu}{\hbar^2} \\ &= -\frac{e^2}{(4\pi\varepsilon_0)a_0}\frac{\mu}{m}\frac{Z^2}{2n^2} \\ &= -\frac{e^2}{(4\pi\varepsilon_0)a_\mu}\frac{Z^2}{2n^2} \end{aligned} \tag{3.29}$$

where $a_0 = 4\pi\varepsilon_0\hbar^2/(me^2)$ is the Bohr radius (1.86) and $a_\mu = 4\pi\varepsilon_0\hbar^2/(\mu e^2) = a_0 m/\mu$ is the modified Bohr radius (1.101). We may also write

$$E_n = -\frac{1}{2}\mu c^2\frac{(Z\alpha)^2}{n^2} \tag{3.30}$$

where $\alpha = e^2/(4\pi\varepsilon_0\hbar c)$ is the fine structure constant. The first form (3.29) which does not contain the velocity of light c clearly shows that the energy levels E_n have been obtained by solving a non-relativistic equation. The second form (3.30), in which the energies E_n are expressed in terms of the rest mass energy μc^2, will be useful at a later stage when we shall discuss the relativistic corrections to the energy levels of one-electron atoms (see Chapter 5). Using atomic units (a.u.) defined in Appendix 14, we also have

$$E_n = -\frac{Z^2}{2n^2}\left(\frac{\mu}{m}\right) \tag{3.31}$$

where we have written explicitly the electron mass m (which is equal to unity in a.u.).

The energy values E_n, which we have obtained here by solving the Schrödinger equation for one-electron atoms, agree exactly with those found in Section 1.7 from the Bohr model. The agreement of this energy spectrum with the main features of the experimental spectrum was pointed out when we analysed the Bohr results. This agreement, however, is not perfect and we shall discuss in Chapter 5 various corrections (such as the fine structure arising from relativistic effects and the electron spin, the Lamb shift and the hyperfine structure due to nuclear effects) which must be taken into account in order to explain the details of the experimental spectrum.

We note from (3.31) that since n may take on all integral values from 1 to $+\infty$, the energy spectrum corresponding to the Coulomb potential (3.1) contains an infinite number of discrete energy levels extending from $-(Z^2/2)(\mu/m)$ to zero. This is due to the fact that the magnitude of the Coulomb potential falls off slowly at large r. On the contrary, short-range potentials such as the square well have a finite (sometimes zero) number of bound states.

Another striking feature of the result (3.31) is that the energy eigenvalues depend only on the principal quantum number n, and are therefore degenerate with respect to l and m. Indeed, for each value of n the orbital quantum number l may take on the values $0, 1, \ldots, n-1$ and for each value of l the $(2l+1)$ possible values of the magnetic quantum number m are $-l, -l+1, \ldots, +l$. The total degeneracy of the energy level E_n is therefore given by

$$\sum_{l=0}^{n-1} (2l+1) = 2\frac{n(n-1)}{2} + n = n^2 \qquad \textbf{(3.32)}$$

As we have already pointed out in Section 2.6, the degeneracy with respect to m is present for any central potential $V(r)$. On the other hand, the degeneracy with respect to l is characteristic of the Coulomb potential. This 'accidental' degeneracy can be traced to the existence of an additional constant of the motion, as we shall see in Section 3.5. The degeneracy with respect to l is removed if the dependence of the potential on r is modified. For example, we shall see in Chapters 8 and 9 that many properties of the alkali atoms can be understood in terms of the motion of a single 'valence' electron in a potential which is central, but which deviates from the $1/r$ Coulomb behaviour because of the presence of the 'inner' electrons. As a result, the energy of this valence electron does depend on l and the degeneracy with respect to l is removed, leading to n distinct levels E_{nl} for a given principal quantum number n. Finally, if an external magnetic field is applied to the atom, we shall see in Chapter 6 that the $(2l+1)$ degeneracy with respect to the magnetic quantum number m is removed.

Figure 3.2 shows an energy-level diagram for the hydrogen atom; it is similar to that displayed in Fig. 1.16 except that the degenerate energy levels with the same n but different l are shown separately. Following the usual spectroscopic notation, these levels are labelled by two symbols. The first one gives the value of the principal quantum number n and the second one indicates the orbital quantum number l according to the correspondence discussed in Section 2.5, namely

Value of l	0	1	2	3	4	5
	↕	↕	↕	↕	↕	↕
Code letter	s	p	d	f	g	h

Looking at the hydrogen atom spectrum (Fig. 3.2), we see that the ground state is a 1s state, the first excited state is fourfold degenerate and contains a 2s state and three 2p states (with $m = -1, 0, +1$), etc.

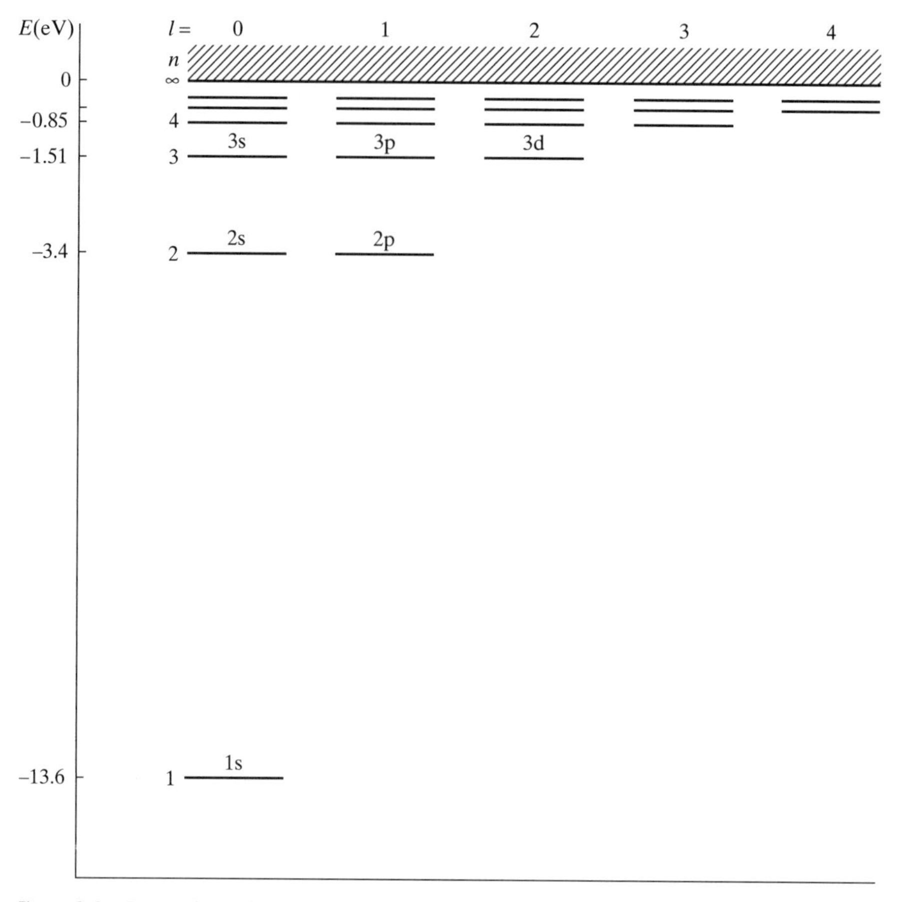

Figure 3.2 Energy-level diagram for atomic hydrogen.

Having obtained the energy levels of one-electron atoms within the framework of the Schrödinger non-relativistic theory, we may now ask about the spectral lines corresponding to transitions from one level to another. This problem will be discussed in detail in the next chapter, where we shall study the interaction of one-electron atoms with electromagnetic radiation. In particular, we shall calculate the transition rates for the most common transitions, the so-called electric dipole transitions, and we shall prove that these transitions obey the selection rules

$$\begin{aligned}\Delta l &= l - l' = \pm 1 \\ \Delta m &= m - m' = 0, \pm 1\end{aligned} \tag{3.33}$$

while $\Delta n = n - n'$ is arbitrary. Here the symbols n, l, m refer to the quantum numbers of the upper state and n', l', m' to those of the lower state of the transition. Since the bound state energies E_n depend only on n, and because transitions can occur between states with any two values of n, it is clear that the Bohr frequency rule (1.70) can still be applied to obtain the frequencies of the spectral lines corresponding to transitions between the energy levels. Thus, for a transition between two energy levels a and b, with energies $E_a < E_b$, we have

$$\nu_{ab} = Z^2 R(M)\left(\frac{1}{n_a^2} - \frac{1}{n_b^2}\right) \qquad \textbf{(3.34)}$$

where $R(M)$ is given by (1.102) and where $n_a = 1, 2, 3, \dots, n_b = 2, 3, 4, \dots$ with $n_b > n_a$.

3.3 The eigenfunctions of the bound states

Until now we have seen that the energy levels predicted by the Schrödinger theory for one-electron atoms agree with those already obtained in Section 1.7 using the Bohr model. However, the Schrödinger theory has much more predictive power than the old quantum theory since it also yields the *eigenfunctions* which enable one to calculate probability densities, expectation values of operators, transition rates, etc.

The radial eigenfunctions of the bound states

In order to obtain these eigenfunctions explicitly, let us return to (3.20). This equation can be identified with the *Kummer–Laplace differential equation*

$$z\frac{d^2w}{dz^2} + (c - z)\frac{dw}{dz} - aw = 0 \qquad \textbf{(3.35)}$$

with $z = \rho$, $w = g$, $a = l + 1 - \lambda$ ar[illegible] Within a multiplicative constant, the solution of (3.35), regular at the origin is the *confluent hypergeometric function*

$$\begin{aligned} {}_1F_1(a, c, z) &= 1 + \frac{az}{c\,1!} + \frac{a(a+1)z^2}{c(c+1)\,2!} + \\ &= \sum_{k=0}^{\infty} \frac{(a)_k}{(c)_k}\frac{z^k}{k!} \end{aligned} \qquad \textbf{(3.36a)}$$

where

$$\alpha_k = \alpha(\alpha+1)\dots(\alpha+k-1), \qquad (\alpha)_0 = 1 \qquad \textbf{(3.36b)}$$

In general, for large positive values of its argument the confluent hypergeometric series (3.36a) behaves asymptotically as

$${}_1F_1(a, c, z) \to \frac{\Gamma(c)}{\Gamma(a)} e^z z^{a-c} \qquad \textbf{(3.37)}$$

where Γ is Euler's gamma function. Thus, in the present case the series (3.36a) for ${}_1F_1(l + 1 - \lambda, 2l + 2, \rho)$ is in general proportional to $\rho^{-l-1-\lambda}\exp(\rho)$ for large ρ, leading to a function $u_{El}(\rho)$ having the unacceptable asymptotic behaviour $u_{El}(\rho) \sim \rho^{-\lambda} \exp(\rho/2)$ (see (3.16) and (3.18)). The only way to obtain physically acceptable solutions of (3.20) is to require that the hypergeometric series for ${}_1F_1(l+1-\lambda, 2l+2, \rho)$ terminates, which implies that $l+1-\lambda = -n_r$ ($n_r = 0, 1, 2, \ldots$), and hence $\lambda = n_r + l + 1 = n$. The confluent hypergeometric function ${}_1F_1(l+1-\lambda, 2l+2, \rho) \equiv {}_1F_1(-n_r, 2l+2, \rho)$ then reduces to a *polynomial* of degree n_r, namely

$$
\begin{aligned}
&{}_1F_1(l+1-n, 2l+2, \rho) \\
&= \sum_{k=0}^{n-l-1} \frac{(k+l-n)(k-1+l-n)\ldots(1+l-n)}{(k+2l+1)(k-1+2l+1)\ldots(1+2l+1)} \frac{\rho^k}{k!}
\end{aligned} \tag{3.38}
$$

in accordance with our foregoing discussion. We may readily verify the correctness of this result by using the recursion relation (3.23) which we derived above for the coefficients c_k of the function $g(\rho)$. Thus, setting $c_0 = 1$ and using the fact that $\lambda = n$, we find from (3.23) that

$$
c_k = \frac{(k+l-n)(k-1+l-n)\ldots(1+l-n)}{(k+2l+1)(k-1+2l+1)\ldots(1+2l+1)} \frac{1}{k!} \tag{3.39}
$$

in agreement with (3.38). We also remark from (3.11) and (3.29) that

$$
\rho = \frac{2Z}{na_\mu} r = \frac{2Z}{na_0} \frac{\mu}{m} r \tag{3.40}
$$

where we recall that $a_\mu = a_0 m/\mu$ is the modified Bohr radius. In atomic units (a.u.) such that $a_0 = 1$ and $m = 1$, we have

$$
\rho = \frac{2Z}{n} \mu r \tag{3.41}
$$

The physically admissible solutions $g(\rho)$ of (3.20), corresponding to $\lambda = n$, may also be expressed in terms of *associated Laguerre polynomials*. To see how this comes about, we first define the *Laguerre polynomials* $L_q(\rho)$ by the relation

$$
L_q(\rho) = e^{\rho} \frac{d^q}{d\rho^q} (\rho^q e^{-\rho}) \tag{3.42}
$$

and we note that these Laguerre polynomials may also be obtained from the generating function

$$
\begin{aligned}
U(\rho, s) &= \frac{\exp[-\rho s/(1-s)]}{1-s} \\
&= \sum_{q=0}^{\infty} \frac{L_q(\rho)}{q!} s^q, \qquad |s| < 1
\end{aligned} \tag{3.43}
$$

Differentiation of this generating function with respect to s yields the recurrence formula

$$
L_{q+1}(\rho) + (\rho - 1 - 2q)L_q(\rho) + q^2 L_{q-1}(\rho) = 0 \tag{3.44}
$$

Similarly, upon differentiation of $U(\rho, s)$ with respect to ρ, we find that

$$\frac{\mathrm{d}}{\mathrm{d}\rho}L_q(\rho) - q\frac{\mathrm{d}}{\mathrm{d}\rho}L_{q-1}(\rho) + qL_{q-1}(\rho) = 0 \tag{3.45}$$

Using (3.44) and (3.45), it is readily shown that the lowest order differential equation involving only $L_q(\rho)$ is

$$\left[\rho\frac{\mathrm{d}^2}{\mathrm{d}\rho^2} + (1-\rho)\frac{\mathrm{d}}{\mathrm{d}\rho} + q\right]L_q(\rho) = 0 \tag{3.46}$$

Next, we define the *associated Laguerre polynomials* $L_q^p(\rho)$ by the relation

$$L_q^p(\rho) = \frac{\mathrm{d}^p}{\mathrm{d}\rho^p}L_q(\rho) \tag{3.47}$$

Differentiating (3.46) p times, we find that $L_q^p(\rho)$ satisfies the differential equation

$$\left[\rho\frac{\mathrm{d}^2}{\mathrm{d}\rho^2} + (p+1-\rho)\frac{\mathrm{d}}{\mathrm{d}\rho} + (q-p)\right]L_q^p(\rho) = 0 \tag{3.48}$$

Setting $\lambda = n$ in (3.20) and comparing with (3.48), we see that the physically acceptable solution $g(\rho)$ of (3.20) is given (up to a multiplicative constant) by the associated Laguerre polynomial $L_{n+l}^{2l+1}(\rho)$. Note that this polynomial is of order $(n+l)-(2l+1) = n-l-1 = n_r$, in accordance with the discussion following (3.25).

The generating function for the associated Laguerre polynomials may be obtained by differentiating (3.43) p times with respect to ρ. That is,

$$\begin{aligned} U_p(\rho, s) &= \frac{(-s)^p\exp[-\rho s/(1-s)]}{(1-s)^{p+1}} \\ &= \sum_{q=p}^{\infty}\frac{L_q^p(\rho)}{q!}s^q, \qquad |s| < 1 \end{aligned} \tag{3.49}$$

An explicit expression for $L_{n+l}^{2l+1}(\rho)$ is given by

$$L_{n+l}^{2l+1}(\rho) = \sum_{k=0}^{n-l-1}(-1)^{k+1}\frac{[(n+l)!]^2}{(n-l-1-k)!(2l+1+k)!}\frac{\rho^k}{k!} \tag{3.50}$$

and is easily verified by substitution into (3.49), with $q = n+l$ and $p = 2l+1$.

Since the physically admissible solutions $g(\rho)$ of (3.20) are given within a multiplicative constant either by the confluent hypergeometric function ${}_1F_1(l+1-n, 2l+2, \rho)$ or by the associated Laguerre polynomial $L_{n+l}^{2l+1}(\rho)$, it is clear that these two functions differ only by a constant factor. This factor is readily found by comparing (3.38) and (3.50) at $\rho = 0$ and remembering that the confluent hypergeometric function is equal to unity at the origin (see (3.36)). Thus

$$L_{n+l}^{2l+1}(\rho) = -\frac{[(n+l)!]^2}{(n-l-1)!(2l+1)!}\,{}_1F_1(l+1-n, 2l+2, \rho) \tag{3.51}$$

Using (3.7), (3.16), (3.18) and the foregoing results, we may now write the full hydrogenic radial functions as

$$R_{nl}(r) = N_{nl}e^{-\rho/2}\rho^l L_{n+l}^{2l+1}(\rho) \tag{3.52a}$$

$$= \tilde{N}_{nl}e^{-\rho/2}\rho^l {}_1F_1(l+1-n, 2l+2, \rho) \tag{3.52b}$$

where N_{nl} and $\tilde{N}_{nl}$ are constants which will be determined below (apart from an arbitrary phase factor) by the normalisation condition. In (3.52) we have used the notation R_{nl} (which displays explicitly the quantum numbers n and l) instead of the symbol R_{El}, and we recall that $\rho = 2Zr/(na_\mu)$.

The hydrogenic wave functions of the discrete spectrum

Using (3.5), we see that the full eigenfunctions of the discrete spectrum for hydrogenic atoms may be written as

$$\psi_{nlm}(r, \theta, \phi) = R_{nl}(r)Y_{lm}(\theta, \phi) \tag{3.53}$$

where the radial functions are given by (3.52) and the spherical harmonics provide the angular part of the wave functions. We require that the eigenfunctions (3.53) be normalised to unity, so that

$$\int_0^\infty \mathrm{d}r r^2 \int_0^\pi \mathrm{d}\theta \sin\theta \int_0^{2\pi} \mathrm{d}\phi |\psi_{nlm}(r, \theta, \phi)|^2 = 1 \tag{3.54}$$

Because the spherical harmonics are normalisd on the unit sphere (see (2.182)), thc normalisation condition (3.54) implies that

$$\int_0^\infty |R_{nl}(r)|^2 r^2 \,\mathrm{d}r = 1 \tag{3.55}$$

or

$$|N_{nl}|^2 \left(\frac{na_\mu}{2Z}\right)^3 \int_0^\infty e^{-\rho}\rho^{2l}[L_{n+l}^{2l+1}(\rho)]^2 \rho^2 \,\mathrm{d}\rho = 1 \tag{3.56}$$

where we have used (3.52a). The integral over ρ can be evaluated by using the generating function (3.49) for the associated Laguerre polynomials (see Appendix 3). The result is

$$\int_0^\infty e^{-\rho}\rho^{2l}[L_{n+l}^{2l+1}(\rho)]^2 \rho^2 \,\mathrm{d}\rho = \frac{2n[(n+l)!]^3}{(n-l-1)!} \tag{3.57}$$

so that the normalised radial functions for the bound states of hydrogenic atoms may be written as [2]

[2] In writing (3.58) we have used the fact that the radial eigenfunctions $R_{nl}(r)$ may be taken to be real without loss of generality.

$$R_{nl}(r) = -\left\{\left(\frac{2Z}{na_\mu}\right)^3 \frac{(n-l-1)!}{2n[(n+l)!]^3}\right\}^{1/2} e^{-\rho/2}\rho^l L_{n+l}^{2l+1}(\rho) \tag{3.58a}$$

or

$$R_{nl}(r) = \frac{1}{(2l+1)!}\left\{\left(\frac{2Z}{na_\mu}\right)^3 \frac{(n+l)!}{2n(n-l-1)!}\right\}^{1/2}$$

$$\times e^{-\rho/2}\rho^l {}_1F_1(l+1-n, 2l+2, \rho) \tag{3.58b}$$

with

$$\rho = \frac{2Z}{na_\mu}r, \qquad a_\mu = \frac{(4\pi\varepsilon_0)\hbar^2}{\mu e^2} = a_0\frac{m}{\mu} \tag{3.58c}$$

where a constant multiplicative factor of modulus one is still arbitrary. In writing (3.58b) we have used equation (3.51) which relates the associated Laguerre polynomial $L_{n+l}^{2l+1}(\rho)$ to the confluent hypergeometric function ${}_1F_1(l+1-n, 2l+2, \rho)$.

The first few radial eigenfunctions (3.58) are given explicitly by

$$\begin{aligned}
R_{10}(r) &= 2(Z/a_\mu)^{3/2}\exp(-Zr/a_\mu)\\
R_{20}(r) &= 2(Z/2a_\mu)^{3/2}(1 - Zr/2a_\mu)\exp(-Zr/2a_\mu)\\
R_{21}(r) &= \frac{1}{\sqrt{3}}(Z/2a_\mu)^{3/2}(Zr/a_\mu)\exp(-Zr/2a_\mu)\\
R_{30}(r) &= 2(Z/3a_\mu)^{3/2}(1 - 2Zr/3a_\mu + 2Z^2r^2/27a_\mu^2)\exp(-Zr/3a_\mu)\\
R_{31}(r) &= \frac{4\sqrt{2}}{9}(Z/3a_\mu)^{3/2}(1 - Zr/6a_\mu)(Zr/a_\mu)\exp(-Zr/3a_\mu)\\
R_{32}(r) &= \frac{4}{27\sqrt{10}}(Z/3a_\mu)^{3/2}(Zr/a_\mu)^2\exp(-Zr/3a_\mu)
\end{aligned} \tag{3.59}$$

and are illustrated in Fig. 3.3 below. We recall that for an 'infinitely heavy' nucleus, a_μ reduces to the first Bohr radius a_0. We also note that in atomic units (a.u.) one has $a_0 = 1$ and $a_\mu = 1/\mu$.

Using the radial wave functions (3.58) together with the explicit expressions of the spherical harmonics given in Table 2.1, we display in Table 3.1 the complete normalised bound state hydrogenic eigenfunctions $\psi_{nlm}(r, \theta, \phi)$ for the first three shells (that is, the K, L and M shells corresponding respectively to the values $n = 1, 2$ and 3 of the principal quantum number). We have also indicated in Table 3.1 the spectroscopic notation, introduced in our discussion of the energy levels. We shall also refer to one-electron orbital wave functions such as the hydrogenic wave functions ψ_{nlm} as *orbitals*. In accordance with the spectroscopic notation, orbitals corresponding to $l = 0$ will be called s orbitals, those with $l = 1$ will be denoted as p orbitals, and so on.

Table 3.1 The complete normalised hydrogenic wave functions corresponding to the first three shells.

Shell	*Quantum numbers* *n*	*l*	*m*	*Spectroscopic notation*	*Wave function* $\psi_{nlm}(r, \theta, \phi)$
K	1	0	0	1s	$\frac{1}{\sqrt{\pi}}(Z/a_\mu)^{3/2}\exp(-Zr/a_\mu)$
L	2	0	0	2s	$\frac{1}{2\sqrt{2\pi}}(Z/a_\mu)^{3/2}(1 - Zr/2a_\mu)\exp(-Zr/2a_\mu)$
	2	1	0	$2p_0$	$\frac{1}{4\sqrt{2\pi}}(Z/a_\mu)^{3/2}(Zr/a_\mu)\exp(-Zr/2a_\mu)\cos\theta$
	2	1	±1	$2p_{\pm1}$	$\mp\frac{1}{8\sqrt{\pi}}(Z/a_\mu)^{3/2}(Zr/a_\mu)\exp(-Zr/2a_\mu)\sin\theta\exp(\pm i\phi)$
M	3	0	0	3s	$\frac{1}{3\sqrt{3\pi}}(Z/a_\mu)^{3/2}(1 - 2Zr/3a_\mu + 2Z^2r^2/27a_\mu^2)\exp(-Zr/3a_\mu)$
	3	1	0	$3p_0$	$\frac{2\sqrt{2}}{27\sqrt{\pi}}(Z/a_\mu)^{3/2}(1 - Zr/6a_\mu)(Zr/a_\mu)\exp(-Zr/3a_\mu)\cos\theta$
	3	1	±1	$3p_{\pm1}$	$\mp\frac{2}{27\sqrt{\pi}}(Z/a_\mu)^{3/2}(1 - Zr/6a_\mu)(Zr/a_\mu)\exp(-Zr/3a_\mu)\sin\theta\exp(\pm i\phi)$
	3	2	0	$3d_0$	$\frac{1}{81\sqrt{6\pi}}(Z/a_\mu)^{3/2}(Z^2r^2/a_\mu^2)\exp(-Zr/3a_\mu)(3\cos^2\theta - 1)$
	3	2	±1	$3d_{\pm1}$	$\mp\frac{1}{81\sqrt{\pi}}(Z/a_\mu)^{3/2}(Z^2r^2/a_\mu^2)\exp(-Zr/3a_\mu)\sin\theta\cos\theta\exp(\pm i\phi)$
	3	2	±2	$3d_{\pm2}$	$\frac{1}{162\sqrt{\pi}}(Z/a_\mu)^{3/2}(Z^2r^2/a_\mu^2)\exp(-Zr/3a_\mu)\sin^2\theta\exp(\pm 2i\phi)$

In some applications it is convenient to consider a different set of hydrogenic wave functions, in which the *real form* of the spherical harmonics is used for the angular part. As we saw in Section 2.5, the spherical harmonics in real form exhibit a directional dependence and behave like simple functions of Cartesian coordinates. Orbitals using the real form of the spherical harmonics for their angular part are therefore particularly convenient for discussion of some properties such as the directed valence characteristic of chemical bonds. We recall that the spherical harmonics in real form are not eigenfunctions of L_z (except, of course, for $m = 0$ where they coincide with the usual spherical harmonics). For a given n and l the hydrogenic wave functions obtained by using the real form of the spherical harmonics are distinguished by the symbols x, y, z, $3z^2 - r^2$, xz, yz, xy, etc. which have been introduced in Section 2.5. As an example, let us consider the three 2p wave functions (for which $n = 2$, $l = 1$ and $m = 0, \pm1$). Using the real forms of the spherical harmonics given in Table 2.2, together with the radial function $R_{21}(r)$ from (3.59), we see that the corresponding normalised hydrogenic wave functions are given by

$$\psi_{2p_x} = \psi_{2,1,\cos\phi} = \frac{1}{4\sqrt{2\pi}}(Z/a_\mu)^{3/2}(Zr/a_\mu)\exp(-Zr/2a_\mu)\sin\theta\cos\phi \tag{3.60a}$$

$$\psi_{2p_y} = \psi_{2,1,\sin\phi} = \frac{1}{4\sqrt{2\pi}}(Z/a_\mu)^{3/2}(Zr/a_\mu)\exp(-Zr/2a_\mu)\sin\theta\sin\phi \tag{3.60b}$$

$$\psi_{2p_z} = \psi_{2p_0} = \psi_{2,1,0} = \frac{1}{4\sqrt{2\pi}}(Z/a_\mu)^{3/2}(Zr/a_\mu)\exp(-Zr/2a_\mu)\cos\theta \tag{3.60c}$$

In what follows, unless otherwise stated, we shall always use the usual (complex) form of the spherical harmonics.

Discussion of the hydrogenic bound state wave functions. Probability density. Parity

Let us return to the hydrogenic wave functions (3.53). First of all, we note that

$$|\psi_{nlm}(r, \theta, \phi)|^2 \, d\mathbf{r} = \psi^*_{nlm}(r, \theta, \phi)\psi_{nlm}(r, \theta, \phi)r^2 \, dr \sin\theta \, d\theta \, d\phi \tag{3.61}$$

represents the probability of finding the electron in the volume element $d\mathbf{r}$ (given in spherical polar coordinates by $d\mathbf{r} = r^2 \, dr \sin\theta \, d\theta \, d\phi$) when the system is in the stationary state specified by the quantum numbers (n, l, m). The quantity $|\psi_{nlm}|^2 = \psi^*_{nlm}\psi_{nlm}$ is the *probability density*. Using (3.53) and (2.184), we see that

$$\begin{aligned} |\psi_{nlm}(r, \theta, \phi)|^2 &= |R_{nl}(r)|^2 |Y_{lm}(\theta, \phi)|^2 \\ &= |R_{nl}(r)|^2 (2\pi)^{-1} |\Theta_{lm}(\theta)|^2 \end{aligned} \tag{3.62}$$

so that the probability density does not depend on the coordinate ϕ. In fact, we see from (3.62) that the behaviour of $|\psi_{nlm}|^2$ is completely specified by the product of the quantity $|R_{nl}(r)|^2$, which gives the *electron density* as a function of r *along a given direction*, and the *angular factor* $(2\pi)^{-1}|\Theta_{lm}(\theta)|^2$.

The spherical harmonics $Y_{lm}(\theta, \phi)$ and the angular factors $(2\pi)^{-1}|\Theta_{lm}(\theta)|^2$ have been studied in detail in Section 2.5. In particular, we refer the reader to polar plots shown in Fig. 2.6. We also recall that if the real form of the spherical harmonics is used (that is, if the sine and cosine functions of ϕ are used), then the probability density will depend on ϕ, the dependence being through the functions $\sin^2 m\phi$ and $\cos^2 m\phi$. Polar representations of the angular dependence of the probability density for s and p orbitals (in the real form) are given by Fig. 2.7.

We now turn our attention to the properties of the radial eigenfunctions $R_{nl}(r)$. We have already seen that the quantity $|R_{nl}(r)|^2$ represents the electron density as a function of r along a given direction. On the other hand, the *radial distribution function*

$$D_{nl}(r) = r^2 |R_{nl}(r)|^2 \tag{3.63}$$

gives the probability per unit length that the electron is to be found at a distance r from the nucleus. Indeed, by integrating (3.61) over the polar angles θ and ϕ and using (3.53) we see that

$$D_{nl}(r)\,\mathrm{d}r = r^2|R_{nl}(r)|^2\,\mathrm{d}r \int_0^{\pi} \mathrm{d}\theta \sin\theta \int_0^{2\pi} \mathrm{d}\phi |Y_{lm}(\theta, \phi)|^2$$

$$= r^2|R_{nl}(r)|^2\,\mathrm{d}r \tag{3.64}$$

represents the probability of finding the electron between the distances r and $r + \mathrm{d}r$ from the nucleus, regardless of direction. The appearance of the factor of r^2 on the right side of (3.64) is because the volume enclosed between two spheres of radii r and $r + \mathrm{d}r$ is proportional to that factor. The radial distribution functions $D_{nl}(r)$ corresponding to the first few radial eigenfunctions are plotted in Fig. 3.3.

Several interesting features emerge from the examination of the eigenfunctions $R_{nl}(r)$ and the radial distribution functions $D_{nl}(r)$.

1. Only for s states ($l = 0$) are the radial wave functions different from zero at $r = 0$. We also note that since $Y_{00} = (4\pi)^{-1/2}$ is independent of θ and ϕ, one has from (3.58)

$$|\psi_{n00}(0)|^2 = \frac{1}{4\pi}|R_{n0}(0)|^2 = \frac{Z^3}{\pi a_\mu^3 n^3} \tag{3.65}$$

   a result which plays an important role in the theory of hyperfine structure (see Chapter 5). Moreover, each of the s-state radial eigenfunctions R_{n0} is such that $\mathrm{d}R_{n0}/\mathrm{d}r \neq 0$ at $r = 0$. This peculiar behaviour is due to the fact that the potential energy (3.1) is infinite at the origin.
2. For $l \neq 0$ the fact that R_{nl} is proportional to r^l for small r forces the wave function to remain small over distances from the nucleus which increase with l. This is because the effective potential (3.9) contains the centrifugal barrier term $l(l + 1)\hbar^2/(2\mu r^2)$ which prevents the electron from approaching the nucleus. Among the eigenfunctions which have the same n, the one with the lowest value of l has the largest amplitude in the vicinity of the nucleus.
3. The associated Laguerre polynomial $L_{n+l}^{2l+1}(\rho)$ is a polynomial of degree $n_r = n - l - 1$ having n_r radial nodes (zeros). Thus the radial distribution function $D_{nl}(r)$ will exhibit $n - l$ maxima. We note that there is only one maximum when, for a given n, the orbital quantum number l has its largest value $l = n - 1$. In this case $n_r = 0$, and we see from (3.50) and (3.58) that

$$R_{n,n-1}(r) \sim r^{n-1}\exp[-Zr/(na_\mu)] \tag{3.66}$$

   Hence, $D_{n,n-1}(r) = r^2R_{n,n-1}^2(r)$ will have a maximum at a value of r obtained by solving the equation

$$\frac{\mathrm{d}D_{n,n-1}}{\mathrm{d}r} = \left(2nr^{2n-1} - \frac{2Z}{na_\mu}r^{2n}\right)\exp[-2Zr/(na_\mu)] = 0 \tag{3.67}$$

   that is at

$$r = \frac{n^2 a_\mu}{Z} \tag{3.68}$$

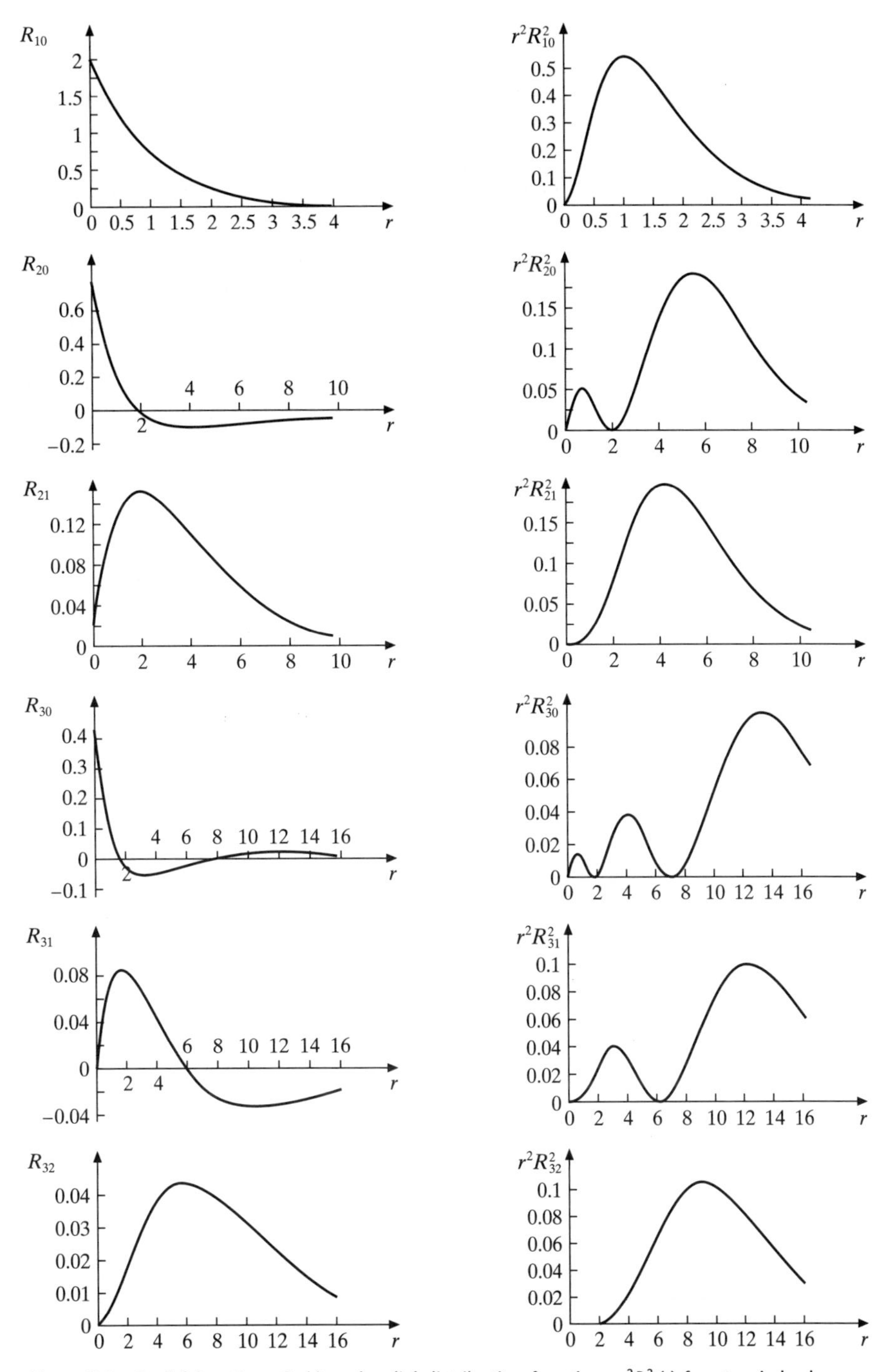

Figure 3.3 Radial functions $R_{nl}(r)$ and radial distribution functions $r^2R_{nl}^2(r)$ for atomic hydrogen.

which is precisely the value (1.100) appearing in the Bohr model. However, in contrast to the Bohr model, the diffuseness of the electron cloud implies that the concept of size is less precise in the quantum mechanical theory, so that the value (3.68) should be interpreted as 'a most probable distance'. We see from (3.68) that this most probable distance is proportional to n^2 and inversely proportional to Z. More generally, the maximum value of $D_{nl}(r)$ recedes from the nucleus with increasing values of n (see Fig. 3.3) and becomes closer to the nucleus by a factor of Z^{-1} when Z increases.

To conclude our study of the hydrogenic bound state wave functions (3.53) we now discuss their *parity*. Since we have reduced the study of hydrogenic atoms to the problem of a particle of mass μ in a central field, we may use directly the results of Section 2.6. Thus the parity operation $\mathbf{r} \to -\mathbf{r}$ (that is, $(r, \theta, \phi) \to (r, \pi - \theta, \phi + \pi)$ in spherical polar coordinates) leaves the radial part $R_{nl}(r)$ of the hydrogenic wave function unaffected, while the angular part $Y_{lm}(\theta, \phi)$ has the parity of l, as shown by (2.252). As a result, under the parity operation $\mathcal{P}\mathbf{r} = -\mathbf{r}$ the hydrogenic wave functions (3.53) transform according to

$$\mathcal{P}[R_{nl}(r)Y_{lm}(\theta, \phi)] = R_{nl}(r)Y_{lm}(\pi - \theta, \phi + \pi) = R_{nl}(r)(-1)^l Y_{lm}(\theta, \phi) \qquad \mathbf{(3.69)}$$

We see that for *l even* the hydrogenic wave functions $\psi_{nlm}(r, \theta, \phi) = R_{nl}(r)Y_{lm}(\theta, \phi)$ are unaffected by the parity operation: they are said to be of *even parity*. For *l odd* the wave functions ψ_{nlm} change sign under the parity operation and are said to be of *odd parity*.

3.4 Expectation values. The virial theorem

Using hydrogenic wave functions $\psi_{nlm}(\mathbf{r})$ normalised to unity, we can calculate the expectation (or average) values of various operators. As a simple example, let us consider the average value of the distance r when the hydrogenic atom is in the ground state, that is the quantity $\langle \psi_{100}|r|\psi_{100}\rangle \equiv \langle r\rangle_{100}$. We have $\psi_{100} = \pi^{-1/2}(Z/a_\mu)^{3/2}$ $\exp(-Zr/a_\mu)$ and therefore

$$\begin{aligned}\langle r\rangle_{100} &= \int \psi^*_{100}(r) r \psi_{100}(r)\, \mathrm{d}\mathbf{r} \\ &= \int |\psi_{100}(r)|^2 r\, \mathrm{d}\mathbf{r} \\ &= \frac{Z^3}{\pi a_\mu^3}\int_0^\infty \mathrm{d}r\, r^3 \exp(-2Zr/a_\mu)\int_0^\pi \mathrm{d}\theta \sin\theta \int_0^{2\pi} \mathrm{d}\phi \end{aligned} \qquad \mathbf{(3.70)}$$

The angular integral has the value 4π and the radial integral is readily performed to give

$$\langle r\rangle_{100} = \frac{3a_\mu}{2Z} \qquad \mathbf{(3.71)}$$

For a general hydrogenic eigenstate $\psi_{nlm}(\mathbf{r})$ we may calculate in the same way

$$\langle r \rangle_{nlm} = \int \psi^*_{nlm}(\mathbf{r}) r \psi_{nlm}(\mathbf{r}) \, d\mathbf{r}$$

$$= \int_0^\infty |R_{nl}(r)|^2 r^3 \, dr \tag{3.72}$$

Using the normalised radial eigenfunctions (3.58) it is found that

$$\langle r \rangle_{nlm} = a_\mu \frac{n^2}{Z} \left\{ 1 + \frac{1}{2} \left[1 - \frac{l(l+1)}{n^2} \right] \right\} \tag{3.73}$$

which is seen to agree with (3.71) when $n = 1$ and $l = 0$. We remark from (3.73) that $\langle r \rangle_{nlm}$, which we may interpret as the 'size' of the atom, is inversely proportional to Z and roughly proportional to n^2, in agreement with the discussion following (3.68). In fact, we see that for s states ($l = 0$) $\langle r \rangle_{nlm}$ is directly proportional to n^2; for states with $l \neq 0$ the deviations from this proportionality are small.

It is also interesting to evaluate the average values of r^k, where k is a positive or a negative integer, since the quantities $\langle r^k \rangle_{nlm}$ clearly exhibit the differences between the radial eigenfunctions. For example, one has

$$\langle r^2 \rangle_{nlm} = a_\mu^2 \frac{n^4}{Z^2} \left\{ 1 + \frac{3}{2} \left[1 - \frac{l(l+1) - 1/3}{n^2} \right] \right\} \tag{3.74}$$

$$\langle r^3 \rangle_{nlm} = a_\mu^3 \frac{n^6}{Z^3} \left\{ 1 + \frac{27}{8} \left[1 - \frac{1}{n^2} \left(\frac{35}{27} + \frac{10}{9}(l+2)(l-1) \right) + \frac{1}{9n^4}(l+2)(l+1)l(l-1) \right] \right\} \tag{3.75}$$

$$\left\langle \frac{1}{r} \right\rangle_{nlm} = \frac{Z}{a_\mu n^2} \tag{3.76}$$

$$\left\langle \frac{1}{r^2} \right\rangle_{nlm} = \frac{Z^2}{a_\mu^2 n^3 (l + 1/2)} \tag{3.77}$$

$$\left\langle \frac{1}{r^3} \right\rangle_{nlm} = \frac{Z^3}{a_\mu^3 n^3 l(l + 1/2)(l + 1)} \tag{3.78}$$

We note from (3.73)–(3.78) that $\langle r^k \rangle_{nlm} \sim (a_\mu/Z)^k$, a result which is easily explained since the Z dependence can be 'factored out' of the problem by defining a new reference length $\tilde{a}_\mu = a_\mu/Z$. We also remark that the expectation values of positive powers of r ($k > 0$) are mainly controlled by the principal quantum number n, while those corresponding to negative powers of r are strongly dependent on l

for $k < -1$. This is due to the fact that for positive powers ($k > 0$) the important contributions to the integral

$$\langle r^k \rangle_{nlm} = \int_0^\infty |R_{nl}(r)|^2 r^{k+2}\, dr \tag{3.79}$$

arise from large values of r, for which R_{nl} behaves essentially like $r^{n-1} \exp[-Zr/(na_\mu)]$. On the other hand, for negative powers such that $k < -1$ the main contributions to the integral (3.79) come from the region of small r, where R_{nl} is proportional to r^l.

Using the result (3.76) we may immediately obtain the average value of the potential energy $V(r) = -Ze^2/[(4\pi\varepsilon_0)r]$. The result is

$$\langle V \rangle_{nlm} = -\frac{Ze^2}{4\pi\varepsilon_0}\left\langle \frac{1}{r} \right\rangle_{nlm} = -\frac{e^2}{(4\pi\varepsilon_0)a_\mu}\frac{Z^2}{n^2} = 2E_n \tag{3.80}$$

where we have also used the equation (3.29) which gives the energy eigenvalues E_n. From (3.80) we can also deduce the average value of the kinetic energy operator $T = -(\hbar^2/2\mu)\nabla^2$. That is,

$$\langle T \rangle_{nlm} = E_n - \langle V \rangle_{nlm} = -E_n \tag{3.81}$$

so that

$$2\langle T \rangle = -\langle V \rangle \tag{3.82}$$

where we have dropped the subscripts. The result (3.82) is a particular case of the *virial theorem*, which we shall now prove.

The virial theorem

Let us denote by H the Hamiltonian of a physical system and by Ψ its state vector, solution of the time-dependent Schrödinger equation (2.46). We have proved in Section 2.2 that the time rate of change of the expectation value $\langle A \rangle \equiv \langle \Psi | A | \Psi \rangle$ of an operator which does not depend explicitly on the time t satisfies the equation (2.56), namely

$$i\hbar \frac{d}{dt}\langle \Psi | A | \Psi \rangle = \langle \Psi | [A, H] | \Psi \rangle \tag{3.83}$$

where $[A, H] = AH - HA$ is the commutator of the operators A and H.

Let us further assume that H is time-independent and denote respectively by E_n and ψ_n its eigenenergies and eigenfunctions. For a *stationary state* $\Psi_n = \psi_n \exp(-iE_n t/\hbar)$ and a time-independent operator A it is clear that the expectation value $\langle \Psi_n | A | \Psi_n \rangle = \langle \psi_n | A | \psi_n \rangle$ does not depend on t so that (3.83) reduces in this case to

$$\langle \psi_n | [A, H] | \psi_n \rangle = 0 \tag{3.84}$$

We now apply this result to the particular case of the non-relativistic motion of a particle of mass μ in a potential $V(\mathbf{r})$, the corresponding time-independent Hamiltonian being

$$H = \frac{p^2}{2\mu} + V = T + V \quad \textbf{(3.85)}$$

where $T = p^2/(2\mu) = -\hbar^2\nabla^2/(2\mu)$ is the kinetic energy operator. Moreover, we choose A to be the time-independent operator $\mathbf{r}\cdot\mathbf{p}$. We then have from (3.84)

$$\langle\psi_n|[(\mathbf{r}\cdot\mathbf{p}), H]|\psi_n\rangle \equiv \langle[(\mathbf{r}\cdot\mathbf{p}), H]\rangle = 0 \quad \textbf{(3.86)}$$

Using the algebraic properties (2.120) of the commutators, together with the fundamental commutation relations (2.119) and the fact that $\mathbf{p} = -\mathrm{i}\hbar\boldsymbol{\nabla}$, we find that

$$\begin{aligned}[(\mathbf{r}\cdot\mathbf{p}), H] &= \left[(xp_x + yp_y + zp_z), \frac{1}{2\mu}(p_x^2 + p_y^2 + p_z^2) + V(x, y, z)\right] \\ &= \frac{\mathrm{i}\hbar}{\mu}(p_x^2 + p_y^2 + p_z^2) - \mathrm{i}\hbar\left(x\frac{\partial V}{\partial x} + y\frac{\partial V}{\partial y} + z\frac{\partial V}{\partial z}\right) \\ &= 2\mathrm{i}\hbar T - \mathrm{i}\hbar(\mathbf{r}\cdot\boldsymbol{\nabla}V)\end{aligned} \quad \textbf{(3.87)}$$

From (3.86) and (3.87) we therefore obtain the relation

$$2\langle T\rangle = \langle\mathbf{r}\cdot\boldsymbol{\nabla}V\rangle \quad \textbf{(3.88)}$$

which is known as the virial theorem [3]. It is worth noting that this result may also be obtained by choosing the operator A to be $\mathbf{p}\cdot\mathbf{r}$ instead of $\mathbf{r}\cdot\mathbf{p}$. Indeed, the difference between $\mathbf{r}\cdot\mathbf{p}$ and $\mathbf{p}\cdot\mathbf{r}$ is a constant, and therefore commutes with H.

If the interaction potential is spherically symmetric and proportional to r^s, we deduce from (3.88) that

$$\begin{aligned}2\langle T\rangle &= \left\langle r\frac{\partial V}{\partial r}\right\rangle \\ &= s\langle V\rangle\end{aligned} \quad \textbf{(3.89)}$$

---

[3] We recall that in classical mechanics the *virial* of a particle is defined as the quantity $-(1/2)\overline{\mathbf{F}\cdot\mathbf{r}}$, where $\mathbf{F}$ is the force acting on the particle and the bar denotes a *time average*. If the motion is periodic (or even if the motion is not periodic, but the coordinate and velocity of the particle remain finite) and $\bar{T}$ denotes the time average of the kinetic energy of the particle, one has (Goldstein, 1980)

$$\bar{T} = -\tfrac{1}{2}\overline{\mathbf{F}\cdot\mathbf{r}}$$

and this relation is known as the *virial theorem*. If the force is derivable from a potential V the virial theorem becomes

$$2\bar{T} = \overline{\mathbf{r}\cdot\boldsymbol{\nabla}V}$$

which is the classical analogue of (3.88).

For example the case $s = 2$ corresponds to the isotropic three dimensional harmonic oscillator, for which $\langle T \rangle = \langle V \rangle$. On the other hand the case $s = -1$, corresponding to the hydrogenic atom, yields the relation $2\langle T \rangle = -\langle V \rangle$, in agreement with our result (3.82).

3.5 One-electron atoms in parabolic coordinates

We have seen in Section 2.6 that the time-independent Schrödinger equation describing the motion of a particle in *any central potential* can always be separated in spherical polar coordinates. This property has been employed above to obtain the energy levels and wave functions of the discrete spectrum of one-electron atoms. In this section, we shall use the fact that if the central potential is the *Coulomb potential* (3.1), then a separation of the Schrödinger equation (3.4) can also be carried out in *parabolic coordinates*. Such a separation turns out to be useful in the treatment of problems in which a particular direction of space is singled out, for example by the existence of an additional external electric field as in the Stark effect (see Section 6.1) or in Coulomb scattering (see Section 12.5).

The parabolic coordinates (ξ, η, ϕ) are related to the Cartesian coordinates (x, y, z) and the spherical polar coordinates (r, θ, ϕ) of the vector $\mathbf{r}$ by

$$
\begin{aligned}
x &= \sqrt{\xi\eta}\cos\phi \\
y &= \sqrt{\xi\eta}\sin\phi \\
z &= \tfrac{1}{2}(\xi - \eta) \\
r &= \tfrac{1}{2}(\xi + \eta) \\
\xi &= r + z = r(1 + \cos\theta) \\
\eta &= r - z = r(1 - \cos\theta) \\
\phi &= \tan^{-1}(y/x)
\end{aligned}
\tag{3.90}
$$

with $0 \leqslant \xi \leqslant \infty$, $0 \leqslant \eta \leqslant \infty$, $0 \leqslant \phi \leqslant 2\pi$. The surfaces $\xi =$ constant and $\eta =$ constant are paraboloids of revolution about the Z axis with the origin as focus. In parabolic coordinates, the volume element is given by

$$\mathrm{d}\mathbf{r} = \tfrac{1}{4}(\xi + \eta)\,\mathrm{d}\xi\,\mathrm{d}\eta\,\mathrm{d}\phi \tag{3.91}$$

the expression for the operator ∇^2 is

$$\nabla^2 = \frac{4}{\xi + \eta}\left[\frac{\partial}{\partial\xi}\left(\xi\frac{\partial}{\partial\xi}\right) + \frac{\partial}{\partial\eta}\left(\eta\frac{\partial}{\partial\eta}\right)\right] + \frac{1}{\xi\eta}\frac{\partial^2}{\partial\phi^2} \tag{3.92}$$

and the Coulomb potential (3.1) is given by

$$V(r) = -\frac{Ze^2}{(4\pi\varepsilon_0)r} = -\frac{2Ze^2}{(4\pi\varepsilon_0)(\xi + \eta)} \tag{3.93}$$

Using (3.92) and (3.93), the Schrödinger equation (3.4) reads in parabolic coordinates

$$-\frac{\hbar^2}{2\mu}\left\{\frac{4}{\xi+\eta}\left[\frac{\partial}{\partial\xi}\left(\xi\frac{\partial}{\partial\xi}\right)+\frac{\partial}{\partial\eta}\left(\eta\frac{\partial}{\partial\eta}\right)\right]+\frac{1}{\xi\eta}\frac{\partial^2}{\partial\phi^2}\right\}\psi-\frac{2Ze^2}{(4\pi\varepsilon_0)(\xi+\eta)}\psi$$
$$= E\Psi \tag{3.94}$$

Let us look for eigenfunctions having the form

$$\psi(\xi,\eta,\phi)=f(\xi)g(\eta)\Phi(\phi) \tag{3.95}$$

Upon substitution of (3.95) into (3.94), and dividing through by ψ, we find that

$$\frac{4\xi\eta}{\xi+\eta}\left[\frac{1}{f}\frac{\mathrm{d}}{\mathrm{d}\xi}\left(\xi\frac{\mathrm{d}f}{\mathrm{d}\xi}\right)+\frac{1}{g}\frac{\mathrm{d}}{\mathrm{d}\eta}\left(\eta\frac{\mathrm{d}g}{\mathrm{d}\eta}\right)\right]+\frac{4\mu Ze^2\xi\eta}{\hbar^2(4\pi\varepsilon_0)(\xi+\eta)}+\frac{2\mu E\xi\eta}{\hbar^2}=-\frac{1}{\Phi}\frac{\mathrm{d}^2\Phi}{\mathrm{d}\phi^2} \tag{3.96}$$

so that the ϕ part of the equation separates at once. Since the left-hand side of (3.96) depends only on ξ and η, and the right-hand side depends only on ϕ, both sides must be equal to a constant that we call m^2. The equation

$$\frac{1}{\Phi}\frac{\mathrm{d}^2\Phi}{\mathrm{d}\phi^2}=-m^2 \tag{3.97}$$

is then readily solved. The physically acceptable normalised solutions are the same as (2.161),

$$\Phi_m(\phi)=\frac{1}{\sqrt{2\pi}}\mathrm{e}^{im\phi} \tag{3.98}$$

where m is the *magnetic quantum number* such that $m = 0, \pm 1, \pm 2, \ldots$.

The remaining part of the equation (3.96) can be separated into its ξ and η parts as follows. Let us introduce the 'separation constants' ν_1 and ν_2 such that

$$\nu_1+\nu_2=\frac{\mu Ze^2}{\hbar^2(4\pi\varepsilon_0)}=\frac{Z}{a_\mu} \tag{3.99}$$

where we recall that $a_\mu = 4\pi\varepsilon_0\hbar^2/(\mu e^2)$. Multiplying the equation (3.96) by $(\xi+\eta)/(4\xi\eta)$ and using (3.98), we find that the equations for $f(\xi)$ and $g(\eta)$ are

$$\frac{\mathrm{d}}{\mathrm{d}\xi}\left(\xi\frac{\mathrm{d}f}{\mathrm{d}\xi}\right)+\left(\frac{\mu E\xi}{2\hbar^2}-\frac{m^2}{4\xi}+\nu_1\right)f=0 \tag{3.100a}$$

and

$$\frac{\mathrm{d}}{\mathrm{d}\eta}\left(\eta\frac{\mathrm{d}g}{\mathrm{d}\eta}\right)+\left(\frac{\mu E\eta}{2\hbar^2}-\frac{m^2}{4\eta}+\nu_2\right)g=0 \tag{3.100b}$$

These two equations are of the same form, differing only in their constant terms.

Energy levels

Let us restrict our attention on the bound states ($E < 0$). The equations (3.100) may be solved by a method similar to that used to solve (3.6). Setting

$$\beta = \left(-\frac{2\mu E}{\hbar^2}\right)^{1/2}, \qquad \rho_1 = \beta\xi, \qquad \lambda_1 = \frac{\nu_1}{\beta} \tag{3.101}$$

we find that equation (3.100a) reduces to

$$\frac{d^2 f}{d\rho_1^2} + \frac{1}{\rho_1}\frac{df}{d\rho_1} + \left(-\frac{1}{4} + \frac{\lambda_1}{\rho_1} - \frac{m^2}{4\rho_1^2}\right) f = 0 \tag{3.102}$$

Similarly, starting from (3.100b) and setting

$$\rho_2 = \beta\eta, \qquad \lambda_2 = \frac{\nu_2}{\beta} \tag{3.103}$$

one obtains for g the equation

$$\frac{d^2 g}{d\rho_2^2} + \frac{1}{\rho_2}\frac{dg}{d\rho_2} + \left(-\frac{1}{4} + \frac{\lambda_2}{\rho_2} - \frac{m^2}{4\rho_2^2}\right) g = 0 \tag{3.104}$$

Following the method used in Section 3.1, one finds that the asymptotic behaviour of f is given by $\exp(-\rho_1/2)$, and that f behaves like $\rho_1^{|m|/2}$ for small values of ρ_1. We therefore look for a solution of the equation (3.102) having the form

$$f(\rho_1) = e^{-\rho_1/2}\rho_1^{|m|/2} v(\rho_1) \tag{3.105}$$

and we obtain for $v(\rho_1)$ the equation

$$\left[\rho_1 \frac{d^2}{d\rho_1^2} + (|m| + 1 - \rho_1)\frac{d}{d\rho_1} + \left(\lambda_1 - \frac{1}{2}(|m| + 1)\right)\right] v = 0 \tag{3.106}$$

This equation can be identified with the Kummer–Laplace differential equation (3.35). The physically acceptable solutions are given by

$$v = C\, L_{n_1+|m|}^{|m|}(\rho_1) = \tilde{C}\, {}_1F_1(-n_1, |m| + 1, \rho_1) \tag{3.107}$$

where C and $\tilde{C}$ are constants, and the number

$$n_1 = \lambda_1 - \tfrac{1}{2}(|m| + 1) \tag{3.108}$$

is a positive integer or zero.

In a similar way, the solution of equation (3.104) shows that the number

$$n_2 = \lambda_2 - \tfrac{1}{2}(|m| + 1) \tag{3.109}$$

must be a positive integer or zero. From (3.108) and (3.109), it follows that

$$\lambda_1 + \lambda_2 = n_1 + n_2 + |m| + 1 = n \tag{3.110}$$

where n is a positive integer ($n = 1, 2, \dots$). By combining (3.99), (3.101), (3.103) and (3.110), we find that

$$\beta = \frac{\nu_1 + \nu_2}{n} = \frac{Z}{na_\mu} \tag{3.111}$$

so that the energy eigenvalues are given by

$$E_n = -\frac{\hbar^2\beta^2}{2\mu} = -\frac{e^2}{(4\pi\varepsilon_0)a_\mu}\frac{Z^2}{2n^2} \tag{3.112}$$

in agreement with (3.29).

The energy levels E_n are degenerate, since they depend only on the principal quantum number n, and we see from (3.110) that there are various ways in which the two *parabolic quantum numbers* n_1 and n_2 and the magnetic quantum number m can be combined to obtain n. Indeed, for a given n, the number $|m|$ can take on the n distinct values $|m| = 0, 1, \dots, n-1$. If $m = 0$, there are n ways of choosing the pair (n_1, n_2). If $|m| > 0$, there are $n - |m|$ ways of choosing the pair (n_1, n_2), and two ways of choosing m ($m = \pm|m|$). The total degeneracy of the energy level E_n is therefore

$$n + 2\sum_{|m|=1}^{n-1}(n - |m|) = n + 2\left[n(n-1) - \frac{n(n-1)}{2}\right] = n^2 \tag{3.113}$$

in agreement with our previous result (3.32).

Eigenfunctions of the bound states

It follows from the above discussion that the hydrogenic eigenfunctions of the discrete spectrum can be labelled in parabolic coordinates by the three quantum numbers n_1, n_2 and m, and are given by

$$\psi_{n_1n_2m}(\xi, \eta, \phi) = f_{n_1|m|}(\rho_1)g_{n_2|m|}(\rho_2)\frac{e^{im\phi}}{\sqrt{2\pi}} \tag{3.114}$$

where $\rho_1 = \beta\xi$, $\rho_2 = \beta\eta$ and $\beta = Z/(na_\mu)$. From (3.105) and (3.107), we have

$$f_{n_1|m|}(\rho_1) = Ce^{-\rho_1/2}\rho_1^{|m|/2}L_{n_1+|m|}^{|m|}(\rho_1) \tag{3.115a}$$

$$= \tilde{C}e^{-\rho_1/2}\rho_1^{|m|/2}{}_1F_1(-n_1, |m|+1, \rho_1) \tag{3.115b}$$

Equations similar to (3.115) may be written down for $g_{n_2|m|}(\rho_2)$, with n_2 replacing n_1 and ρ_2 replacing ρ_1. The constants C and $\tilde{C}$ can be determined (apart from an arbitrary multiplicative factor of modulus one) by requiring that the eigenfunctions $\psi_{n_1n_2m}$ be normalised to unity. That is,

$$\int |\psi_{n_1n_2m}|^2\,d\mathbf{r} = \frac{1}{4}\int_0^\infty d\xi\int_0^\infty d\eta\int_0^{2\pi} d\phi\,|\psi_{n_1n_2m}(\xi, \eta, \phi)|^2(\xi + \eta) = 1 \tag{3.116}$$

where we have used the expression (3.91) of the volume element in parabolic coordinates. It is then found that

$$f_{n_1|m|}(\rho_1) = \frac{2^{1/4}}{n}\left(\frac{Z}{a_\mu}\right)^{3/4}\left\{\frac{n_1!}{[(n_1+|m|)!]^3}\right\}^{1/2} e^{-\rho_1/2}\rho_1^{|m|/2}L_{n_1+|m|}^{|m|}(\rho_1) \quad \textbf{(3.117a)}$$

and

$$g_{n_2|m|}(\rho_2) = \frac{2^{1/4}}{n}\left(\frac{Z}{a_\mu}\right)^{3/4}\left\{\frac{n_2!}{[(n_2+|m|)!]^3}\right\}^{1/2} e^{-\rho_2/2}\rho_2^{|m|/2}L_{n_2+|m|}^{|m|}(\rho_2) \quad \textbf{(3.117b)}$$

Thus, the normalised eigenfunctions of the bound states of one-electron atoms are given in parabolic coordinates by

$$\psi_{n_1n_2m}(\xi, \eta, \phi) = \frac{\sqrt{2}}{n^2}\left(\frac{Z}{a_\mu}\right)^{3/2}\left\{\frac{n_1!n_2!}{[(n_1+|m|)!(n_2+|m|)!]^3}\right\}^{1/2}$$

$$\times\, e^{-(\rho_1+\rho_2)/2}(\rho_1\rho_2)^{|m|/2}L_{n_1+|m|}^{|m|}(\rho_1)L_{n_2+|m|}^{|m|}(\rho_2)\,\frac{e^{im\phi}}{\sqrt{2\pi}} \quad \textbf{(3.118a)}$$

with

$$\rho_1 = \beta\xi,\ \rho_2 = \beta\eta,\ \beta = \frac{Z}{na_\mu},\ n = n_1 + n_2 + |m| + 1 \quad \textbf{(3.118b)}$$

These eigenfunctions, in contrast to the eigenfunctions $\psi_{nlm}(r, \theta, \phi)$ in spherical polar coordinates, are asymmetrical with respect to the plane $z = 0$. For $n_1 > n_2$, the probability of finding the electron with $z > 0$ is larger than that of finding it with $z < 0$; the opposite happens when $n_1 < n_2$.

For a given energy level E_n, that is for a fixed value of the principal quantum number n, and for a fixed value of the magnetic quantum number m (with $n > |m|$), the parabolic quantum numbers n_1 and n_2 can be chosen in $n - |m|$ different ways so that $n_1 + n_2 = n - |m| - 1$. Similarly, in the treatment using spherical polar coordinates, for fixed values of n and m the orbital angular momentum quantum number l can be chosen in $n - |m|$ different ways such that $|m| \leqslant l \leqslant n - 1$. Hence, for fixed n and m, the $n - |m|$ products of functions $f_{n_1|m|}(\beta\xi)g_{n_2|m|}(\beta\eta)$ are linear combinations of the $n - |m|$ products of radial functions $R_{nl}(r)$ and angular functions $\Theta_{lm}(\theta)$. It follows that for fixed values of n and m, any eigenfunction $\psi_{n_1n_2m}(\xi, \eta, \phi)$ in parabolic coordinates is a linear combination of eigenfunctions $\psi_{nlm}(r, \theta, \phi)$ in spherical polar coordinates. For the non-degenerate ground state with $n = 1$ ($n_1 = n_2 = m = 0$), we have

$$\psi_{000}(\xi, \eta, \phi) = \frac{1}{\sqrt{\pi}}\left(\frac{Z}{a_\mu}\right)^{3/2} \exp\left[-\frac{Z}{2a_\mu}(\xi + \eta)\right]$$

$$= \frac{1}{\sqrt{\pi}}\left(\frac{Z}{a_\mu}\right)^{3/2} \exp(-Zr/a_\mu) \quad \textbf{(3.119)}$$

which is identical to the ground state wave function ψ_{100} in spherical polar coordinates.

We now examine briefly the cause of the separability of the Schrödinger equation (3.4) for one-electron atoms in parabolic coordinates. In classical mechanics, it can be shown that the *constants of the motion* for the two-body Coulomb problem are the energy E, the orbital angular momentum $\mathbf{L} = \mathbf{r} \times \mathbf{p}$ and the (classical) *Runge–Lenz vector* [4]

$$\mathbf{A}_c = \frac{4\pi\varepsilon_0}{\mu Z e^2}(\mathbf{L} \times \mathbf{p}) + \frac{\mathbf{r}}{r} \qquad \textbf{(3.120)}$$

which is such that $\mathbf{A}_c \cdot \mathbf{L} = 0$ and is a symmetry axis for the trajectory.

In quantum mechanics, the analogue of the classical Runge–Lenz vector (3.120) is the *Runge–Lenz operator*

$$\mathbf{A} = \frac{4\pi\varepsilon_0}{2\mu Z e^2}[\mathbf{L} \times \mathbf{p} - \mathbf{p} \times \mathbf{L}] + \frac{\mathbf{r}}{r} \qquad \textbf{(3.121)}$$

where $\mathbf{p} = -i\hbar\boldsymbol{\nabla}$ and the term in brackets has been symmetrized so that the operator $\mathbf{A}$ is Hermitian. It can be proved (Problem 3.8) that

$$[H, \mathbf{A}] = 0 \qquad \textbf{(3.122)}$$

where H is the hydrogenic Hamiltonian (3.2). In addition, it can be shown (Problem 3.8) that $\mathbf{A} \cdot \mathbf{L} = \mathbf{L} \cdot \mathbf{A} = 0$ and that the three operators A_x, A_y and A_z do not mutually commute. Each of these three operators commutes with the corresponding component of $\mathbf{L}$ (for example A_z commutes with L_z) but not with $\mathbf{L}^2$. The fact that there exists a new constant of the motion (A_z) which does not commute with another conserved quantity ($\mathbf{L}^2$) leads to an additional degeneracy of the energy levels, which in the present case is the 'accidental' Coulomb degeneracy mentioned in Section 3.2. From our study of central forces in Section 2.6 and the foregoing discussion, we note that one can construct *two independent complete sets of commuting observables* for the two-body Coulomb problem. The first set consists of the operators H, $\mathbf{L}^2$, L_z and leads to the separability of the Schrödinger equation in spherical polar coordinates for all central potentials. The second set is made of the operators H, L_z, A_z, and leads to the separability of the Schrödinger equation for the two-body Coulomb problem in parabolic coordinates.

3.6 Special hydrogenic systems: positronium; muonium; antihydrogen; muonic and hadronic atoms; Rydberg atoms

Let us recall some of the key results we have obtained for hydrogenic systems. The energy eigenvalues are given by (3.29) and the frequencies of the transitions by (3.34). In particular, the ionisation potential $I_P = |E_{n=1}|$ is just

[4] See Goldstein (1980).

$$I_P = \frac{e^2}{(4\pi\varepsilon_0)a_\mu}\frac{Z^2}{2} \tag{3.123}$$

and the 'extension' a of the wave function describing the relative motion of the system is roughly given in the ground state (see (3.68)) by

$$a = \frac{a_\mu}{Z} = \frac{(4\pi\varepsilon_0)\hbar^2}{Z\mu e^2} \tag{3.124}$$

where $\mu = mM/(m + M)$ is the reduced mass.

The hydrogenic systems we have considered so far correspond to an atomic nucleus of mass M and charge Ze and an electron of mass m and charge $-e$ interacting by means of the Coulomb potential (3.1). The 'normal' hydrogen atom, containing a proton and an electron is the prototype of these hydrogenic systems. The hydrogenic ions He^+ ($Z = 2$), Li^{++} ($Z = 3$), Be^{3+} ($Z = 4$), etc. are also examples of such systems. As we have already noted in Chapter 1, and as we can see again from the above formulae, the value of a for these ions is reduced with respect to that of the hydrogen atom by a factor of Z and their ionisation potential is increased by a factor of Z^2 (neglecting small reduced mass effects).

The (neutral) isotopes of atomic hydrogen, deuterium and tritium, also provide examples of hydrogenic systems. Here the proton is replaced by a nucleus having the same charge $+e$, namely a deuteron (containing one proton and one neutron) in the case of deuterium or a triton (containing one proton and two neutrons) in the case of tritium. Since $M_d \simeq 2M_p$ and $M_t \simeq 3M_p$, where M_p is the mass of the proton, M_d the mass of the deuteron and M_t the mass of the triton, we see that the reduced mass μ is slightly different for hydrogen, deuterium and tritium, the relative differences being of the order of 10^{-3}. Thus the quantities I_P and a are nearly identical for the three atoms, the small differences in the value of μ giving rise to *isotopic shifts* of the spectral lines (of the order of 10^{-3}) which we have already discussed in Chapter 1.

Positronium, muonium

In addition to deuterium and tritium, there also exist other 'less conventional' isotopes of hydrogen, in which the role of the nucleus is played by another particle. For example, *positronium* (e^+e^-) is a bound hydrogenic system made of a positron e^+ (the antiparticle of the electron, having the same mass as the electron, but the opposite charge) and an electron e^-. *Muonium* (μ^+e^-) is another non-conventional isotope of hydrogen, in which the proton has been replaced by a positive muon μ^+, a particle which is very similar to the positron e^+, except that it has a mass $M_{\mu^+} \simeq 207m$ and that it is unstable, with a lifetime of about 2.2×10^{-6} s. Both positronium and muonium may thus be considered as light isotopes of hydrogen. Positronium was first observed in 1951 by M. Deutsch *et al.* and muonium in 1960 by V.W. Hughes *et al.* Table 3.2 gives the values of the reduced mass μ, the 'radius' a and the ionisation potential I_P for positronium and muonium, compared with those of the hydrogen atom.

Positronium and muonium have attracted a great deal of interest because they only contain *leptons* (that is, particles which are not affected by the strong

Table 3.2 The reduced mass μ, ground state 'radius' a and ionisation potential I_P of some 'unconventional' hydrogenic systems, compared with the corresponding quantities for the hydrogen atom (pe^-). Atomic units are used. The masses of the particles considered are $M_p = M_{\bar{p}} \simeq 1836$, $M_{\mu^-} = M_{\mu^+} \simeq 207$, $M_{\pi^-} \simeq 273$, $M_{K^-} \simeq 966$, $M_{\Sigma^-} \simeq 2343$, the unit of mass being the electron mass m.

System	*Reduced mass* μ	*'Radius'* a	*Ionisation potential* I_P
(pe^-), $(\bar{p}e^+)$	$\frac{1836}{1837} \simeq 1$	$\simeq a_0 = 1$	$\simeq \frac{e^2}{(4\pi\varepsilon_0)2a_0} = 0.5$
(e^+e^-)	0.5	2	0.25
(μ^+e^-)	$\frac{207}{208} \simeq 1$	$\simeq a_0 = 1$	$\simeq 0.5$
$(p\mu^-)$	$\simeq 186$	$\simeq 5.4 \times 10^{-3}$	$\simeq 93$
$(p\pi^-)$	$\simeq 238$	$\simeq 4.2 \times 10^{-3}$	$\simeq 119$
(pK^-)	$\simeq 633$	$\simeq 1.6 \times 10^{-3}$	$\simeq 317$
$(p\bar{p})$	$\simeq 918$	$\simeq 1.1 \times 10^{-3}$	$\simeq 459$
$(p\Sigma^-)$	$\simeq 1029$	$\simeq 9.7 \times 10^{-4}$	$\simeq 515$

interactions) and hence are particularly suitable systems to verify the predictions of quantum electrodynamics (QED). We remark that both positronium and muonium are *unstable*. Indeed, muonium has a lifetime of 2.2×10^{-6} s (which is the lifetime of the muon μ^+ itself) while in positronium the electron and the positron may annihilate, their total energy including their rest mass energy being completely converted into electromagnetic radiation (photons).

Antihydrogen

In 1996, G. Baur *et al.* reported the first observation of atoms of *antihydrogen* ($\bar{p}e^+$), a bound state system made of an antiproton $\bar{p}$ and a positron e^+, in an experiment performed at CERN (the European laboratory for particle physics in Geneva) using a 'flight' method. The basic idea is that an antiproton $\bar{p}$ passing through the Coulomb field of a nucleus with charge Ze will create an electron–positron pair. Occasionally, the antiproton will capture a positron from the produced pair and form a fast-moving antihydrogen atom. In the experiment of Baur *et al.*, a jet of xenon atoms ($Z = 54$) was fired across the antiproton beam from CERN's Low Energy Antiproton Ring (LEAR). Nine atoms of antihydrogen were detected, which lasted about 40 nanoseconds (1 ns = 10^{-9} s) before the positrons were annihilated by collisions with electrons in the detector. That time was too short and the number of antihydrogen atoms formed was too few to make high-precision comparisons between hydrogen and antihydrogen, its antimatter counterpart. The new challenge is to trap antihydrogen atoms, and to use high-resolution laser spectroscopy (see Section 15.2) to compare the spectral lines of antihydrogen with those of hydrogen. Such studies would provide a stringent test of the charge–parity–time ($\mathscr{CPT}$) theorem [5] and might reveal conceivable small differences between the gravitational forces acting on matter and antimatter.

Muonic atoms

In 1947, J.A. Wheeler suggested that negatively charged particles other than the electron could form a bound system with a nucleus. These negative particles can be *leptons* such as the negative muon μ^- (which is a kind of 'heavy electron' having the same mass and lifetime as the positive muon μ^+, but a negative charge $-e$) or *hadrons* (particles which can have strong interactions). The unusual 'atoms' formed in this way are sometimes called 'exotic atoms'. We shall return shortly to hadronic atoms and examine for the moment the muonic atoms which are formed when a negative muon μ^- is captured by the Coulomb attraction of a nucleus of charge Ze as the muon is slowing down in bulk matter.

As a first example, let us consider the simplest muonic atom $(p\mu^-)$ which contains a proton p and a muon μ^-. Since the muon has a mass which is about 207 times that of the electron, the reduced mass of the muon with respect to the proton is approximately 186 times the electron mass. As a result, the 'radius' a of the muonic atom $(p\mu^-)$ is 186 times smaller than that of the hydrogen atom, the ionisation potential I_P of $(p\mu^-)$ being 186 times larger than the corresponding quantity for atomic hydrogen (see Table 3.2). The frequencies of the spectral lines corresponding to transitions between the energy levels of $(p\mu^-)$ may thus be obtained from those of the hydrogen atom by multiplying the latter by a factor of 186. For transitions between the lowest energy levels of $(p\mu^-)$ the spectral lines are therefore lying in the X-ray region.

Let us now assume that the negative muon μ^- is captured by the Coulomb field of a nucleus $\mathcal{N}$ of charge Ze. Assuming that we are dealing with a heavy nucleus, so that we may neglect the reduced mass effect, we see that equation (3.124) would then yield for the bound system $(\mathcal{N}\mu^-)$ the value $a \simeq a_0/(207Z)$, while (3.123) would give an ionisation potential I_P larger than that of hydrogen by a factor of $207Z^2$. Thus for the case of muonic lead (corresponding to a nucleus with $Z = 82$) we would have $I_P \simeq 19$ MeV (1 MeV = 10^6 eV) and $a \simeq 3 \times 10^{-15}$ m = 3 fermi (1 fermi or femtometer = 10^{-15} m). In fact this value of a is smaller than the radius R of the lead nucleus, which is given by $R \simeq 6.7$ fermi, so that the expressions (3.123)–(3.124) cannot be used any more! Indeed, they have been derived on the assumption that the two particles of the hydrogenic system interact by means of the Coulomb potential (3.1) for *all values* of the relative distance r, that is both particles are considered to be *point-like*. This assumption is an excellent one for 'usual' (electronic) atoms or ions such as hydrogen, deuterium, tritium, He^+, Li^{2+}, etc., where the finite extension of the nucleus gives rise to very small effects such

[5] The $\mathscr{CPT}$ theorem says that if (mathematically) a particle is replaced by its antiparticle (charge conjugation operation $\mathscr{C}$), its position in space is reflected (parity operation $\mathscr{P}$) and the direction of time is reversed (time reversal operation $\mathscr{T}$), then the equations governing the mass and interactions of the particle are unchanged.

as the *volume effect*, which will be shown in Chapter 5 to yield tiny shifts of the energies associated with low-lying s states. However, for muonic atoms with large values of Z (such as muonic lead) the volume effect may lead to important shifts of the low-lying levels (in particular of the 1s and 2s states). Nevertheless the main *qualitative* features predicted by (3.123)–(3.124) are correct: the ionisation potential I_P is of the order of several MeV, the spectral lines corresponding to transitions between the lowest energy levels lie at the limit of the X-ray and γ-ray regions, and in the 1s state the muon spends a significant fraction of its time within the nucleus. The fact that the muon acts as a probe of the nucleus and that the energy spectrum of muonic atoms is therefore sensitive to the internal structure of the nucleus constitutes one of the major interests of the study of muonic atoms.

We conclude this brief discussion of muonic atoms by three remarks. First, muonic atoms are unstable, since the negative muon μ^- has a finite lifetime $\tau \simeq 2.2 \times 10^{-6}$ s (which is the same as that of its antiparticle, the positive muon μ^+). Secondly, muonic atoms usually keep an electron cloud, but the influence of the electrons on the hydrogenic system (nucleus + muon μ^-) is almost always negligible, since the muon μ^- remains on the average much closer to the nucleus than the electrons. Thirdly, the muon is often captured into an excited state of the (nucleus + muon) system. It will then 'cascade' down to the ground state, either by emitting radiation in the form of X-rays or by means of radiationless transitions known as *Auger transitions* (see Chapter 9), in which electrons from the cloud are ejected.

Hadronic atoms

In contrast with the leptons (such as the electron e^-, the positron e^+ and the muons μ^- and μ^+) which participate only in the electromagnetic and weak interactions, the hadrons participate in the *strong* (nuclear-type) interactions in addition to the electromagnetic and weak interactions. There are two kinds of hadrons, the *baryons* (such as the proton p, the neutron n, the antiproton $\bar{p}$, the antineutron $\bar{n}$, the hyperons $\Sigma, \Xi, \ldots$) which have half-odd integer spin ($\frac{1}{2}, \frac{3}{2}, \ldots$) and are therefore fermions, and the *mesons* (such as the π-mesons, the K-mesons, etc.) which have zero or integer spin (0, 1, ...) and hence are bosons. Among the hadrons, those having a negative charge can form with a nucleus $\mathcal{N}$ a 'hydrogenic-type' system which is referred to as a *hadronic atom*. In particular, the system $(\mathcal{N}\pi^-)$ is called a *pionic atom*, $(\mathcal{N}K^-)$ is known as a *kaonic atom* while $(\mathcal{N}\bar{p})$ is called an *antiprotonic atom*. Hydrogenic-type systems containing a nucleus and a negative hyperon – for example, $(\mathcal{N}\Sigma^-)$ – are known as *hyperonic atoms*. All these hadronic atoms are unstable, but their lifetime is long enough so that some of their spectral lines have actually been observed.

Since hadrons interact strongly with nuclei, it is clear that the theory of hydrogenic systems which we have developed in this chapter – and which only takes into account the Coulomb interaction (3.1) – cannot be directly applied to hadronic atoms. Thus the values of a and I_P listed in Table 3.2 only give a rough estimate of the 'radius' and of the ionisation potentials of the hadronic atoms $(p\pi^-)$, (pK^-), $(p\bar{p})$

and $(p\Sigma^-)$. However, because the strong interactions have a short range, *excited states* of hadronic atoms and in particular those with $l \neq 0$ for which the wave function is very small in the vicinity of the origin can essentially be studied by using the theory of this chapter. The energies of these excited states are thus given by

$$E_n \simeq -\frac{1}{n^2} I_P \tag{3.125}$$

and their 'radii' by

$$a_n \simeq n^2 a \tag{3.126}$$

where I_P and a are given respectively by (3.123) and (3.124).

Rydberg atoms

A highly excited atom (or ion) has an electron with a large principal quantum number n. The electron (or the atom) is said to be in a 'high Rydberg state' and the highly excited atom is also referred to more simply as a 'Rydberg atom'.

Several characteristic quantities of the hydrogen atom are compared in Table 3.3 for $n = 1$, arbitrary n and $n = 100$. It is clear from the examination of this table that highly excited hydrogen atoms with $n \simeq 100$ exhibit some remarkable properties. For example, *their size is enormous* on the atomic scale. Indeed, with electron orbital radii of the order of 10^{-7} m, such atoms are as big as simple bacteria! Also, their geometrical cross-section being proportional to n^4 is therefore 10^8 larger when $n = 100$ than in the ground state $n = 1$. On the other hand, the electron in a high Rydberg state is *very weakly bound*, its binding energy being smaller than the binding energy of the ground state by a factor n^2. For example, the energy required to ionise a hydrogen atom with $n = 100$ is only 1.36×10^{-3} eV. We also remark that the energy separation ΔE between adjacent levels is given for large n by

$$\Delta E = E_{n+1} - E_n = I_P^H \left(\frac{1}{n^2} - \frac{1}{(n+1)^2} \right) \simeq 2I_P^H/n^3 \tag{3.127}$$

Table 3.3 Comparison of some characteristic quantities of the hydrogen atom for different values of the principal quantum number n.

Quantity	$n = 1$	*Arbitrary* n	$n = 100$
Radius a_n of Bohr orbit (in m)	$a_0 \simeq 5.3 \times 10^{-11}$	$\simeq n^2 a_0$	5.3×10^{-7}
Geometric cross-section πa_n^2 (in m^2)	$\pi a_0^2 \simeq 8.8 \times 10^{-21}$	$\simeq n^4 \pi a_0^2$	8.8×10^{-13}
Binding energy $\lvert E_n \rvert$ (in eV)	$I_P^H \simeq 13.6$	I_P^H/n^2	1.36×10^{-3}
Energy separation ΔE between adjacent levels (in eV)		$\simeq 2I_P^H/n^3$ (n large)	2.7×10^{-5}
Root-mean-square velocity of electron v_n (in m s^{-1})	$v_0 \simeq c\alpha \simeq 2.2 \times 10^6$	$\simeq v_0/n$	2.2×10^4
Period T_n (in s)	$T_0 = 1.5 \times 10^{-16}$	$\simeq n^3 T_0$	1.5×10^{-10}

where $I_P^H \simeq 13.6$ eV is the ionisation potential of the hydrogen atom. Thus for $n = 100$ this energy separation is given by $\Delta E \simeq 2.7 \times 10^{-5}$ eV ($= 0.22$ cm^{-1} in units of reciprocal centimetres), so that the selective excitation of atoms in highly excited states requires experimental techniques with extremely high resolution. A highly excited hydrogen atom left to itself has a relatively long lifetime, increasing roughly as n^3 for a fixed angular momentum quantum number l. However, even thermal collisions can transfer enough energy to the atom to ionise it, although it is possible that a neutral system passing through the (very large) Rydberg atom will leave it undisturbed.

Many of the studies concerning Rydberg atoms also deal with excited states of atoms other than hydrogen. However, for a large enough n, a Rydberg atom of any kind may be considered as an 'ionic core' plus a single highly excited electron. If this electron has enough angular momentum, so that it does not significantly penetrate the core, it will essentially move in a Coulomb field corresponding to an effective charge $Z_{\text{eff}} = 1$ (in atomic units). Such Rydberg atoms are therefore very similar to highly excited (Rydberg) hydrogen atoms.

Rydberg atoms have been studied intensively [6], since they are important in such diverse areas as astrophysics, plasma physics and quantum optics, as well as in tests of quantum electrodynamics and in studies of the classical limit of quantum mechanics. For instance, when making radiative transitions from an initial (high n) energy level to a neighbouring one of lower energy, Rydberg atoms in interstellar space emit radio waves which can be detected by radio telescopes (see Chapter 16). In astrophysical and laboratory plasmas, the recombination of low-energy electrons and ions accompanied with the emission of radiation (radiative recombination) results in Rydberg atoms.

The development of widely tunable and monochromatic dye lasers [7] around 1970 made it possible to excite atoms to single, well-defined Rydberg states, and to study the properties of these Rydberg states in great detail [8]. Of particular interest is the behaviour of Rydberg atoms in the presence of external electric and magnetic fields, which we shall study in Chapter 6.

The fact that transitions between neighbouring levels of Rydberg atoms occur at long wavelengths (corresponding to radio-frequency or microwave radiation) makes it possible to construct resonant cavities large enough to study the coupling of Rydberg states to radiation during a relatively long time. If the cavity is tuned to a particular transition frequency, the properties of the atoms can be altered. For instance the rate of spontaneous emission of radiation can be changed. By studying these changes, it is possible to make fundamental tests of low-energy QED. More generally, the study of Rydberg atoms interacting with radiation in a cavity is the subject of *cavity quantum electrodynamics*. Of particular interest is the *one-atom maser* (or micromaser) demonstrated in 1985 by D. Meschede,

[6] For a detailed account, see Gallagher (1994).

[7] Dye lasers are discussed in Chapter 15.

[8] See Connerade (1998).

H. Walther and G. Müller, which will be discussed in Chapter 16. With this device, the interaction of a single Rydberg atom with a single mode of an electromagnetic field in a cavity could be studied. At the other intensity extreme, J.E. Bayfield and P.M. Koch showed in 1974 that Rydberg atoms in microwave fields provide an excellent testing ground to study the properties of atoms in strong radiation fields.

From Bohr's correspondence principle, we expect that Rydberg atoms, which are in states with a large quantum number n, will behave quasi-classically, and hence are ideal systems for studying the quantum–classical interface. Of special interest is the transition from quantum systems to corresponding classical systems which exhibit 'chaotic' behaviour resulting in the exponential increase with time in sensitivity to the initial state of the system [9]. It follows that with the passage of time the state of such classical systems cannot be predicted from the initial values of the dynamical variables. In contrast, the evolution of a quantum system depends on the first derivative of the wave function with respect to time, so that it cannot display chaotic behaviour in the classical sense [9]. Nevertheless, there are characteristic signs in the distribution of energy levels of quantum systems with large quantum numbers which indicate that the corresponding classical system is chaotic. Rydberg atoms are an ideal test-bed for studying these connections.

HW 3

Problems

3.1 Using the explicit expressions of the hydrogenic wave functions given in Table 3.1, calculate the expectation values $\langle r \rangle$, $\langle r^2 \rangle$, $\langle 1/r \rangle$, $\langle p \rangle$ and $\langle p^2 \rangle$ for the following states: (i) 1s, (ii) 2s, (iii) 2p. Verify that the virial theorem (3.82) is satisfied.

3.2 Any region of space in which the kinetic energy T of a particle would become negative is forbidden for classical motion. For a hydrogen atom in the ground state, of total energy $E_{1s} = -\frac{1}{2}$ a.u.

(a) Find the classically forbidden region.
(b) Using the ground state wave function $\psi_{1s}(r)$, calculate the probability of finding the electron in this region.

3.3 Consider a hydrogen atom of which the wave function at $t = 0$ is the following superposition of energy eigenfunctions $\psi_{nlm}(\mathbf{r})$

$$\Psi(\mathbf{r}, t = 0) = \frac{1}{\sqrt{14}}[2\psi_{100}(r) - 3\psi_{200}(r) + \psi_{322}(\mathbf{r})]$$

(a) Is this wave function an eigenfunction of the parity operator?
(b) What is the probability of finding the system in the ground state (100)? In the state (200)? In the state (322)? In another eigenstate?
(c) What is the expectation value of the energy? Of the operator $\mathbf{L}^2$? Of the operator L_z?

[9] See Blümel and Reinhardt (1997).

3.4 Consider a tritium atom, containing a nucleus, ^{3}H (the triton) and an electron. The triton nucleus, which consists of one proton ($Z = 1$) and two neutrons, is unstable, since by beta emission it decays to ^{3}He, which contains two protons ($Z = 2$) and one neutron. This decay process occurs very rapidly with respect to characteristic atomic times, and will be assumed here to take place instantaneously. As a result, there is a sudden doubling of the Coulomb attraction between the atomic electron and the nucleus when the tritium nucleus ^{3}H decays by beta emission into ^{3}He. Assuming that the tritium atom is in the ground state when the decay takes place, and neglecting recoil effects ($M = \infty$), find the probability that immediately after the decay the He^+ ion can be found:

(a) In its ground state 1s.
(b) In any state other than the ground state. (Total probability for excitation or ionisation.)
(c) In the 2s state.
(d) In a state with $l \neq 0$.

3.5 The electron and the proton of an hydrogen atom interact not only through the electrostatic potential (3.1), but also by means of the gravitational interaction. Using perturbation theory (Section 2.8), obtain the relative energy shift $\Delta E/E_{1s}$, where ΔE is the energy change due to the gravitational force and E_{1s} is the ground state energy of hydrogen, as given by (3.29) with $Z = 1$ and $n = 1$. (**Note**: The gravitational constant is $G = 6.673 \times 10^{-11}$ N m^2 kg^{-2}.)

3.6 Using the definition (2.21), obtain the momentum space wave functions of the hydrogen atom for the 1s, 2s and 2p states. Compare your results with those of Appendix 5.
(**Hints**: Use the expansion (2.261) of a plane wave in spherical harmonics, and the known integral

$$\int_0^{\infty} e^{-ax} j_l(bx) x^{l+1} \, dx = \frac{(2b)^l l!}{(a^2 + b^2)^{l+1}}$$

Note that integrals involving higher powers of x can be obtained by differentiating this result with respect to the parameter a.)

3.7 Consider a one-electron atom bound state labelled by the parabolic quantum numbers $n_1 = 1$, $n_2 = 0$ and the magnetic quantum number $m = 0$. Show that the corresponding eigenfunction in parabolic coordinates, $\psi_{100}(\xi, \eta, \phi)$, can be written as the following linear combination of eigenfunctions $\psi_{nlm}(r, \theta, \phi)$ in spherical polar coordinates:

$$\psi_{100}(\xi, \eta, \phi) = -\frac{1}{\sqrt{2}} \psi_{200}(r, \theta, \phi) + \frac{1}{\sqrt{2}} \psi_{210}(r, \theta, \phi)$$

3.8 The Runge–Lenz operator **A** is defined by the relation (3.121). Prove that

(a)

$$[H, \mathbf{A}] = 0$$

where H is the hydrogenic Hamiltonian (3.2).

(b)

$$\mathbf{L}\cdot\mathbf{A} = 0$$

(c)

$$[A_x, A_y] = \mathrm{i}\hbar\left(-\frac{2(4\pi\varepsilon_0)^2}{\mu Z^2 e^4}H\right)L_z$$

$$[A_y, A_z] = \mathrm{i}\hbar\left(-\frac{2(4\pi\varepsilon_0)^2}{\mu Z^2 e^4}H\right)L_x$$

$$[A_z, A_x] = \mathrm{i}\hbar\left(-\frac{2(4\pi\varepsilon_0)^2}{\mu Z^2 e^4}H\right)L_y$$

3.9 Consider the Hamiltonian

$$H = H_0 + H'$$

where

$$H_0 = \frac{p^2}{2\mu} - \frac{Ze^2}{(4\pi\varepsilon_0)r}$$

is the hydrogenic Hamiltonian, and

$$H' = -\frac{\varepsilon}{r^2}$$

is a small perturbation. Show that this perturbation removes the 'accidental' Coulomb degeneracy with respect to l, the energy eigenvalues E_{nl} depending now on both quantum numbers n and l.

4 Interaction of one-electron atoms with electromagnetic radiation

In this chapter, we shall discuss the interaction of hydrogenic atoms with electromagnetic radiation. We shall first show how spectral lines arise, and at a later stage we shall study the photoelectric effect and the scattering of radiation by atomic systems. In considering the interaction of an atom with radiation, there are three basic processes to analyse. First, just as a classical oscillating charge will radiate spontaneously, an atom can make a spontaneous transition from an excited state to a state of lower energy, emitting a photon which is the quantum of the electromagnetic field. This process is called *spontaneous emission*. Secondly, an atom can *absorb* a photon from an external radiation field, making a transition from a state of lower to a state of higher energy. Finally, an atom can also emit a photon under the influence of an external radiation field. This process is called *stimulated emission*; it is distinct from spontaneous emission because it requires (like absorption) the presence of an external radiation field. Stimulated emission has important applications in the LASER (an acronym for Light Amplification by Stimulated Emission of Radiation) and the MASER (Microwave Amplification by Stimulated Emission of Radiation), which produce intense beams of coherent radiation, and which will be discussed in Chapter 15.

In a rigorous treatment, we would have to start by studying quantum electrodynamics, in which the electromagnetic field is expressed in terms of its quanta – the photons. Each photon corresponding to a field of frequency ν carries an amount of energy $h\nu$. Even in comparatively weak fields the photon density can be very high (see Problem 4.1). Under these circumstances the number of photons can be treated as a continuous variable and the field can be described classically by using Maxwell's equations. We shall proceed by using a *semi-classical* theory in which the radiation field is treated classically, but the atomic system is described by using quantum mechanics. The approximation will also be made that the influence of the atom on the external field can be neglected. Clearly these assumptions do not hold in the case of spontaneous emission, because only one photon is concerned – and one is not a large number! The proper treatment of spontaneous emission is well understood, but is beyond the scope of this book. Nevertheless, we shall be able to find the transition rate for spontaneous emission indirectly using a statistical argument due to Einstein.

4.1 The electromagnetic field and its interaction with charged particles

The classical electromagnetic field is described by electric and magnetic field vectors $\mathcal{E}$ and $\mathcal{B}$, which satisfy Maxwell's equations [1]. We shall express these and other electromagnetic quantities in rationalised MKS units, which form part of the standard SI system. The electric field $\mathcal{E}$ and magnetic field $\mathcal{B}$ can be generated from scalar and vector potentials ϕ and $\mathbf{A}$ by

$$\mathcal{E}(\mathbf{r}, t) = -\boldsymbol{\nabla}\phi(\mathbf{r}, t) - \frac{\partial}{\partial t}\mathbf{A}(\mathbf{r}, t) \qquad \textbf{(4.1)}$$

and

$$\mathcal{B}(\mathbf{r}, t) = \boldsymbol{\nabla} \times \mathbf{A}(\mathbf{r}, t) \qquad \textbf{(4.2)}$$

The potentials are not completely defined by (4.1) and (4.2), since the fields, $\mathcal{E}$ and $\mathcal{B}$, are invariant under the (classical) *gauge transformation* $\mathbf{A} \to \mathbf{A} + \boldsymbol{\nabla}\chi$, $\phi \to \phi - \partial\chi/\partial t$, where χ is any real, differentiable function of $\mathbf{r}$ and t. The freedom implied by this *gauge invariance* allows us to impose a further condition on the vector potential $\mathbf{A}$, which we shall choose to be

$$\boldsymbol{\nabla}\cdot\mathbf{A} = 0 \qquad \textbf{(4.3)}$$

When $\mathbf{A}$ satisfies this condition, we are said to be using the *Coulomb gauge*. This choice of gauge is convenient when no sources are present, which is the case considered here. One may then take $\phi = 0$, and $\mathbf{A}$ satisfies the wave equation

$$\nabla^2\mathbf{A} - \frac{1}{c^2}\frac{\partial^2\mathbf{A}}{\partial t^2} = 0 \qquad \textbf{(4.4)}$$

where c is the velocity of light *in vacuo*.

A monochromatic plane wave solution of (4.4) corresponding to the angular frequency ω (that is, to the frequency $\nu = \omega/(2\pi)$) is

$$\mathbf{A}(\mathbf{r}, t) = A_0(\omega)\hat{\boldsymbol{\varepsilon}}\cos(\mathbf{k}\cdot\mathbf{r} - \omega t + \delta_\omega) \qquad \textbf{(4.5)}$$

where $\mathbf{k}$ is the wave (or propagation) vector and δ_ω is a real constant phase. Substituting (4.5) in (4.4), it is found that the angular frequency ω and the wave number k (the magnitude of the wave vector $\mathbf{k}$) are related by

$$\omega = kc \qquad \textbf{(4.6)}$$

The vector potential $\mathbf{A}$ has an amplitude $|A_0(\omega)|$ and is in the direction specified by the unit vector $\hat{\boldsymbol{\varepsilon}}$, called the *polarisation vector*. In addition, equation (4.3) is satisfied if

[1] Useful texts on electromagnetism are those by Duffin (1968) and Jackson (1998).

$$\mathbf{k} \cdot \hat{\boldsymbol{\varepsilon}} = 0 \tag{4.7}$$

so that $\hat{\boldsymbol{\varepsilon}}$ is perpendicular to **k** and the wave is *transverse*.

Using the Coulomb gauge, with $\phi = 0$, the electric and magnetic fields are given from (4.1), (4.2) and (4.5) by

$$\mathscr{E}(\mathbf{r}, t) = \mathscr{E}_0(\omega)\hat{\boldsymbol{\varepsilon}} \sin(\mathbf{k}\cdot\mathbf{r} - \omega t + \delta_\omega) \tag{4.8a}$$

and

$$\mathscr{B}(\mathbf{r}, t) = \mathscr{E}_0(\omega)\omega^{-1}(\mathbf{k} \times \hat{\boldsymbol{\varepsilon}}) \sin(\mathbf{k}\cdot\mathbf{r} - \omega t + \delta_\omega) \tag{4.8b}$$

where $\mathscr{E}_0(\omega) = -\omega A_0(\omega)$. The electric field vector $\mathscr{E}$ has an amplitude $|\mathscr{E}_0(\omega)|$ and is in the direction of the polarisation vector $\hat{\boldsymbol{\varepsilon}}$. From (4.6) and (4.8), we see that for a radiation field of a given frequency, $|\mathscr{B}|/|\mathscr{E}| = 1/c$. We also note that the vectors $\mathscr{E}$, $\mathscr{B}$ and **k** are mutually perpendicular. An electromagnetic plane wave such as (4.8), for which the electric field vector points in a fixed direction $\hat{\boldsymbol{\varepsilon}}$, is said to be *linearly polarised*. A general state of polarisation for a plane wave propagating in the direction $\hat{\mathbf{k}}$ can be described by combining two independent linearly polarised plane waves with polarisation vectors $\hat{\boldsymbol{\varepsilon}}_\lambda$ ($\lambda = 1, 2$) perpendicular to $\hat{\mathbf{k}}$, where the phases of the two component waves are, in general, different. Any radiation field can be expressed as a superposition of monochromatic fields.

It is useful to relate the energy density of the field to the photon density, keeping in mind that each photon at a frequency ν carries a quantum of energy of magnitude $h\nu = \hbar\omega$. The energy density of the field is given by

$$\frac{1}{2}(\varepsilon_0|\mathscr{E}|^2 + |\mathscr{B}|^2/\mu_0) = \varepsilon_0\mathscr{E}_0^2(\omega) \sin^2(\mathbf{k}\cdot\mathbf{r} - \omega t + \delta_\omega) \tag{4.9}$$

where ε_0 and μ_0 are the permittivity and permeability of free space, and $\varepsilon_0\mu_0 = c^{-2}$. The average of $\sin^2(\mathbf{k}\cdot\mathbf{r} - \omega t + \delta_\omega)$ over a period $T = 2\pi/\omega$ is given by

$$\frac{1}{T}\int_0^T \sin^2(\mathbf{k}\cdot\mathbf{r} - \omega t + \delta_\omega)\, \mathrm{d}t = \frac{1}{2} \tag{4.10}$$

Using this result the average energy density $\rho(\omega)$ is

$$\rho(\omega) = \frac{1}{2}\varepsilon_0\mathscr{E}_0^2(\omega) = \frac{1}{2}\varepsilon_0\omega^2 A_0^2(\omega) \tag{4.11}$$

If the number of photons of angular frequency ω within a volume V is $N(\omega)$, the energy density is $\hbar\omega N(\omega)/V$, and equating this with (4.11) the amplitude of the electric field is found to be

$$|\mathscr{E}_0(\omega)| = [2\rho(\omega)/\varepsilon_0]^{1/2} = [2\hbar\omega N(\omega)/(\varepsilon_0 V)]^{1/2} \tag{4.12}$$

The average rate of energy flow through a unit cross-sectional area, normal to the direction of propagation of the radiation, defines the intensity $I(\omega)$. Since the velocity of electromagnetic waves in free space is c, we have

$$I(\omega) = \rho(\omega)c = \frac{1}{2}\varepsilon_0 c\mathscr{E}_0^2(\omega) = \frac{1}{2}\varepsilon_0 c\omega^2 A_0^2(\omega) = \hbar\omega N(\omega)c/V \quad \textbf{(4.13)}$$

A general pulse of radiation can be described by representing the vector potential $\mathbf{A}(\mathbf{r}, t)$ as a superposition of plane waves of the form (4.5). Taking each plane wave component to have the same direction of propagation $\hat{\mathbf{k}}$ and to be linearly polarised in the direction $\hat{\boldsymbol{\varepsilon}}$, we have

$$\mathbf{A}(\mathbf{r}, t) = \hat{\boldsymbol{\varepsilon}} \int_0^\infty A_0(\omega) \cos(\mathbf{k}\cdot\mathbf{r} - \omega t + \delta_\omega)\, \mathrm{d}\omega \quad \textbf{(4.14)}$$

When the radiation is nearly monochromatic, the amplitude $A_0(\omega)$ is sharply peaked about some value ω_0 of ω. The radiation from a hot gas or glowing filament arises from many atoms each emitting photons independently. As a result, within the integral over ω the phases δ_ω are distributed completely at random, and the radiation is said to be *incoherent*. This is characteristic of light from all sources with the exception of lasers. Because of the random phase distribution, when the average energy density is calculated from the squares of $\mathscr{E}$ and $\mathscr{B}$, the contribution to the energy density from each frequency can be added together, the cross terms averaging to zero [2]. The average energy density ρ and intensity I for radiation composed of a range of frequencies can then be expressed as

$$\rho = \int_0^\infty \rho(\omega)\, \mathrm{d}\omega, \qquad I = \int_0^\infty I(\omega)\, \mathrm{d}\omega \quad \textbf{(4.15)}$$

where $\rho(\omega)$ and $I(\omega)$ are the *energy density* and *intensity per unit angular frequency range*, given by (4.11) and (4.13), respectively.

In Chapter 15, the radiation from a single mode laser, which exhibits a high degree of coherence and is nearly monochromatic, is discussed. In this case, the phase δ_ω is constant in a small region of width $\Delta\omega$ centred about an angular frequency ω_0, so that δ_ω can be eliminated from (4.14) by changing the (arbitrary) zero of time. It follows that the expressions (4.15) remain valid for nearly monochromatic coherent radiation.

Charged particles in an electromagnetic field

The Hamiltonian of a spinless particle of charge q and mass m in an electromagnetic field is

$$H = \frac{1}{2m}(\mathbf{p} - q\mathbf{A})^2 + q\phi \quad \textbf{(4.16)}$$

where $\mathbf{p}$ is the generalised momentum of the particle. The steps leading to (4.16) are given in Appendix 6. Ignoring for the present small spin-dependent terms, the

[2] While this is intuitively clear, a detailed proof is too long to be given here. This point is discussed by Marion and Heals (1980).

Hamiltonian of an electron of mass m in an electromagnetic field is given by (4.16), with $q = -e$.

Since the Hamiltonian H must be Hermitian, we shall write (4.16) in the form

$$H = \frac{\mathbf{p}^2}{2m} - \frac{q}{2m}(\mathbf{A}\cdot\mathbf{p} + \mathbf{p}\cdot\mathbf{A}) + \frac{q^2}{2m}\mathbf{A}^2 + q\phi \tag{4.17}$$

In the position representation, $\mathbf{p}$ is the operator $-\mathrm{i}\hbar\boldsymbol{\nabla}$ and the time-dependent Schrödinger equation is

$$\mathrm{i}\hbar\frac{\partial}{\partial t}\Psi(\mathbf{r}, t) = \left[-\frac{\hbar^2}{2m}\nabla^2 + \mathrm{i}\hbar\frac{q}{2m}(\mathbf{A}\cdot\boldsymbol{\nabla} + \boldsymbol{\nabla}\cdot\mathbf{A}) + \frac{q^2}{2m}\mathbf{A}^2 + q\phi\right]\Psi(\mathbf{r}, t) \tag{4.18}$$

An important property of equation (4.18) is that its form is unchanged under the *gauge transformation*

$$\mathbf{A}(\mathbf{r}, t) = \mathbf{A}'(\mathbf{r}, t) + \boldsymbol{\nabla}\chi(\mathbf{r}, t) \tag{4.19a}$$

$$\phi(\mathbf{r}, t) = \phi'(\mathbf{r}, t) - \frac{\partial}{\partial t}\chi(\mathbf{r}, t) \tag{4.19b}$$

$$\Psi(\mathbf{r}, t) = \Psi'(\mathbf{r}, t)\exp[\mathrm{i}q\chi(\mathbf{r}, t)/\hbar] \tag{4.19c}$$

where χ is an arbitrary real, differentiable function of $\mathbf{r}$ and t. That is, the wave function $\Psi'(\mathbf{r}, t)$ satisfies the equation

$$\mathrm{i}\hbar\frac{\partial}{\partial t}\Psi'(\mathbf{r}, t) = \left[-\frac{\hbar^2}{2m}\nabla^2 + \mathrm{i}\hbar\frac{q}{2m}(\mathbf{A}'\cdot\boldsymbol{\nabla} + \boldsymbol{\nabla}\cdot\mathbf{A}') + \frac{q^2}{2m}\mathbf{A}'^2 + q\phi'\right]\Psi'(\mathbf{r}, t) \tag{4.20}$$

Since, as seen from (4.19c), a gauge transformation is a particular case of a unitary transformation, measurable quantities (such as expectation values or transition probabilities) calculated in different gauges must be the same. The property of gauge invariance allows us to adopt the Coulomb gauge defined by (4.3) and to take $\phi = 0$, as we have seen above. The time-dependent Schrödinger equation (4.18) then reduces to

$$\mathrm{i}\hbar\frac{\partial}{\partial t}\Psi(\mathbf{r}, t) = \left[-\frac{\hbar^2}{2m}\nabla^2 + \mathrm{i}\hbar\frac{q}{m}\mathbf{A}\cdot\boldsymbol{\nabla} + \frac{q^2}{2m}\mathbf{A}^2\right]\Psi(\mathbf{r}, t) \tag{4.21}$$

where we have used the fact that in the Coulomb gauge

$$\begin{aligned}\boldsymbol{\nabla}\cdot(\mathbf{A}\Psi) &= \mathbf{A}\cdot(\boldsymbol{\nabla}\Psi) + (\boldsymbol{\nabla}\cdot\mathbf{A})\Psi \\ &= \mathbf{A}\cdot(\boldsymbol{\nabla}\Psi)\end{aligned} \tag{4.22}$$

Interaction of one-electron atoms with an electromagnetic field

Let us now consider the interaction of the electromagnetic field (4.8) with a one-electron atom (ion), containing a nucleus of charge Ze and mass M and an electron of charge $q = -e$ and mass m. We shall restrict ourselves here to 'ordinary' hydrogenic systems such as H, He^+, ... In that case the nuclear mass M is very large compared to the electronic mass m. In fact, we shall make the infinite nuclear mass approximation, thus neglecting recoil effects (see Problem 1.16) and reduced mass effects. The interaction between the radiation field and the nucleus can be ignored to a high degree of accuracy. However, we must include in the Hamiltonian the electrostatic Coulomb potential $-Ze^2/(4\pi\varepsilon_0 r)$ between the electron and the nucleus. It is convenient to regard this electrostatic interaction as an additional potential energy term, while the radiation field is described in terms of a vector potential alone satisfying the Coulomb gauge condition (4.3), as discussed above. The time-dependent Schrödinger equation for a one-electron atom in an electromagnetic field then reads

$$\mathrm{i}\hbar\frac{\partial}{\partial t}\Psi(\mathbf{r}, t) = H(t)\Psi(\mathbf{r}, t) \tag{4.23}$$

where

$$H(t) = -\frac{\hbar^2}{2m}\nabla^2 - \frac{Ze^2}{(4\pi\varepsilon_0)r} - \mathrm{i}\hbar\frac{e}{m}\mathbf{A}\cdot\nabla + \frac{e^2}{2m}\mathbf{A}^2 \tag{4.24}$$

We can also write (4.23) in the form

$$\mathrm{i}\hbar\frac{\partial}{\partial t}\Psi(\mathbf{r}, t) = [H_0 + H_{\mathrm{int}}(t)]\,\Psi(\mathbf{r}, t) \tag{4.25}$$

where

$$H_0 = -\frac{\hbar^2}{2m}\nabla^2 - \frac{Ze^2}{(4\pi\varepsilon_0)r} \tag{4.26}$$

is the time-independent Hamiltonian of the one-electron atom (ion) in the absence of the electromagnetic field, and

$$\begin{aligned} H_{\mathrm{int}}(t) &= \frac{e}{m}\mathbf{A}\cdot\mathbf{p} + \frac{e^2}{2m}\mathbf{A}^2 \\ &= -\mathrm{i}\hbar\frac{e}{m}\mathbf{A}\cdot\nabla + \frac{e^2}{2m}\mathbf{A}^2 \end{aligned} \tag{4.27}$$

is the Hamiltonian describing the interaction of the hydrogenic atom with the radiation field.

In this chapter, we shall treat only the weak field case [3] in which the term in $\mathbf{A}^2$ is negligible [4] compared with the term linear in $\mathbf{A}$. Accordingly, we shall write $H_{\mathrm{int}}(t) = H'(t)$, where

[3] Atoms in intense electromagnetic fields will be discussed in Chapter 15.

[4] Although we are treating the case for which $\mathbf{A}^2$ is very small compared with $\mathbf{A}$, the photon density is assumed to be high enough for the radiation field to be treated classically. Both conditions are well satisfied in the emission and absorption processes we shall describe.

$$H'(t) = -i\hbar\frac{e}{m}\mathbf{A}\cdot\nabla = \frac{e}{m}\mathbf{A}\cdot\mathbf{p} \tag{4.28}$$

will be treated as a small perturbation.

4.2 Transition rates

Having neglected the term in $\mathbf{A}^2$, we see that the time-dependent Schrödinger equation (4.25) may be written as

$$i\hbar\frac{\partial}{\partial t}\Psi(\mathbf{r}, t) = [H_0 + H'(t)]\Psi(\mathbf{r}, t) \tag{4.29}$$

where H_0 is given by (4.26) and $H'(t)$ by (4.28).

We shall study this problem by using the time-dependent perturbation theory given in Chapter 2. Referring to equation (2.297), and dropping the superscript (0) for notational simplicity, we denote by E_k the eigenvalues and by ψ_k the corresponding normalised eigenfunctions of the hydrogenic Hamiltonian (4.26), so that

$$H_0\psi_k = E_k\psi_k \tag{4.30}$$

Because the set of functions ψ_k (including both the discrete set studied in Chapter 3 and the continuous set corresponding to unbound states) is complete, the general solution Ψ of the time-dependent Schrödinger equation (4.29), which we assume to be normalised to unity, can be expanded as

$$\Psi(\mathbf{r}, t) = \sum_k c_k(t)\psi_k(\mathbf{r})\exp(-iE_kt/\hbar) \tag{4.31}$$

where the sum is over both the discrete set and the continuous set of hydrogenic eigenfunctions ψ_k. The coefficients $c_k(t)$ satisfy the coupled equations (2.336) with $\lambda = 1$, namely

$$\dot{c}_b(t) = (i\hbar)^{-1}\sum_k H'_{bk}(t)c_k(t)\exp(i\omega_{bk}t) \tag{4.32}$$

where

$$H'_{bk}(t) = \langle\psi_b|H'(t)|\psi_k\rangle \tag{4.33}$$

and

$$\omega_{bk} = (E_b - E_k)/\hbar \tag{4.34}$$

Let us suppose that the system is initially in a well-defined stationary bound state of energy E_a described by the wave function ψ_a and that the pulse of radiation is switched on at the time $t = 0$. Thus the initial conditions are given by

$$c_k(t \leqslant 0) = \delta_{ka} \tag{4.35}$$

and, to first order in the perturbation H', we have (see (2.343b))

$$c_b^{(1)}(t) = (\mathrm{i}\hbar)^{-1}\int_0^t H'_{ba}(t')\exp(\mathrm{i}\omega_{ba}t')\,\mathrm{d}t'$$

$$= -\frac{e}{m}\int_0^t \langle\psi_b|\mathbf{A}\cdot\nabla|\psi_a\rangle\exp(\mathrm{i}\omega_{ba}t')\,\mathrm{d}t' \tag{4.36}$$

where $\omega_{ba} = (E_b - E_a)/\hbar$ and

$$\langle\psi_b|\mathbf{A}\cdot\nabla|\psi_a\rangle = \int \psi_b^*(\mathbf{r})\mathbf{A}\cdot\nabla\psi_a(\mathbf{r})\,\mathrm{d}\mathbf{r} \tag{4.37}$$

To proceed further, we use the vector potential $\mathbf{A}(\mathbf{r}, t)$ given by (4.14) to obtain

$$c_b^{(1)}(t) = -\frac{e}{2m}\int_0^\infty \mathrm{d}\omega A_0(\omega)\Bigg[\exp(\mathrm{i}\delta_\omega)\langle\psi_b|\exp(\mathrm{i}\mathbf{k}\cdot\mathbf{r})\hat{\boldsymbol{\varepsilon}}\cdot\nabla|\psi_a\rangle\int_0^t \mathrm{d}t'\exp[\mathrm{i}(\omega_{ba}-\omega)t']$$

$$+\exp(-\mathrm{i}\delta_\omega)\langle\psi_b|\exp(-\mathrm{i}\mathbf{k}\cdot\mathbf{r})\hat{\boldsymbol{\varepsilon}}\cdot\nabla|\psi_a\rangle\int_0^t \mathrm{d}t'\exp[\mathrm{i}(\omega_{ba}+\omega)t']\Bigg] \tag{4.38}$$

In general, the duration of the pulse is much larger than the periodic time $(2\pi/\omega_{ba})$ which is for example about 2×10^{-15} s for the yellow sodium D line at 5890 Å. It follows that the first integral over t' will be negligible unless $\omega_{ba} \simeq \omega$, that is unless $E_b \simeq E_a + \hbar\omega$. Thus we see that in this case the final state of the atom has greater energy than the initial state and one photon of energy $\hbar\omega$ is *absorbed* from the radiation field. On the other hand, the second integral over t' in (4.38) will be negligible unless $\omega_{ba} \simeq -\omega$, that is unless $E_b \simeq E_a - \hbar\omega$. In this case the initial state of the atom has greater energy than the final state and one photon of energy $\hbar\omega$ is *emitted*. Since only one of these conditions can be satisfied for a pair of states a and b, we can deal with the two terms separately.

We shall assume for the moment that both the initial and final atomic states are *discrete* ('bound–bound' transitions). The photoelectric effect, which corresponds to transitions from a discrete initial state to final states lying in the continuum ('bound–free' transitions) will be studied in Section 4.8.

Absorption

We start with the first term of (4.38), describing absorption. Using the fact that the radiation is incoherent – so that no interference terms occur – we find that the probability for the system to be in the state b at time t is

$$|c_b^{(1)}(t)|^2 = \frac{1}{2}\left(\frac{e}{m}\right)^2\int_0^\infty A_0^2(\omega)|M_{ba}(\omega)|^2 F(t, \omega - \omega_{ba})\,\mathrm{d}\omega \tag{4.39}$$

where we have defined the matrix element M_{ba} as

$$M_{ba} = \langle\psi_b|\exp(\mathrm{i}\mathbf{k}\cdot\mathbf{r})\hat{\boldsymbol{\varepsilon}}\cdot\nabla|\psi_a\rangle = \int \psi_b^*(\mathbf{r})\exp(\mathrm{i}\mathbf{k}\cdot\mathbf{r})\hat{\boldsymbol{\varepsilon}}\cdot\nabla\psi_a(\mathbf{r})\,\mathrm{d}\mathbf{r} \tag{4.40}$$

and we recall that $\omega = kc$. Upon setting $\bar{\omega} = \omega - \omega_{ba}$, the function $F(t, \bar{\omega})$ which appears in (4.39) is seen to be the same as the function $F(t, \omega)$ introduced in Chapter 2. That is (see (2.348)),

$$F(t, \bar{\omega}) = \frac{1 - \cos \bar{\omega} t}{\bar{\omega}^2}, \qquad \bar{\omega} = \omega - \omega_{ba} \tag{4.41}$$

The properties of $F(t, \omega)$ discussed in Section 2.8 may therefore be used directly here, provided that we make the substitution $\omega \to \bar{\omega}$. In particular, since for large t the function $F(t, \bar{\omega})$ has a sharp maximum at $\bar{\omega} = 0$, namely at $\omega = \omega_{ba}$, we can set $\omega = \omega_{ba}$ in the slowly varying quantities $A_0^2(\omega)$ and $|M_{ba}(\omega)|^2$, take these factors outside the integral in (4.39) and extend the limits of integration to $\pm\infty$. Hence we have

$$|c_b^{(1)}(t)|^2 = \frac{1}{2}\left(\frac{e}{m}\right)^2 A_0^2(\omega_{ba})|M_{ba}(\omega_{ba})|^2 \int_{-\infty}^{+\infty} F(t, \bar{\omega})\, \mathrm{d}\omega \tag{4.42}$$

and using the result (2.349), we obtain

$$|c_b^{(1)}(t)|^2 = \frac{\pi}{2}\left(\frac{e}{m}\right)^2 A_0^2(\omega_{ba})|M_{ba}(\omega_{ba})|^2 t \tag{4.43}$$

Thus the probability $|c_b^{(1)}(t)|^2$ increases linearly with time and a *transition rate for absorption* (integrated over ω) can be defined in first-order perturbation theory as

$$W_{ba} = \frac{\mathrm{d}}{\mathrm{d}t}|c_b^{(1)}(t)|^2 = \frac{\pi}{2}\left(\frac{e}{m}\right)^2 A_0^2(\omega_{ba})|M_{ba}(\omega_{ba})|^2 \tag{4.44}$$

In terms of the intensity per unit angular frequency range, $I(\omega)$, given by (4.13), the integrated transition rate for absorption is given by

$$W_{ba} = \frac{4\pi^2}{m^2 c}\left(\frac{e^2}{4\pi\varepsilon_0}\right)\frac{I(\omega_{ba})}{\omega_{ba}^2}|M_{ba}(\omega_{ba})|^2 \tag{4.45}$$

and is seen to be proportional to $I(\omega_{ba})$.

The rate of absorption of energy from the radiation field, per atom, is $(\hbar\omega_{ba})W_{ba}$. It is convenient to define an integrated *absorption cross-section* σ_{ba} which is the rate of absorption of energy (per atom) divided by $I(\omega_{ba})$. That is,

$$\sigma_{ba} = \frac{4\pi^2 \alpha \hbar^2}{m^2 \omega_{ba}}|M_{ba}(\omega_{ba})|^2 \tag{4.46}$$

where $\alpha = e^2/(4\pi\varepsilon_0\hbar c) \simeq 1/137$ is the fine structure constant. Since the incident flux of photons of angular frequency ω_{ba} is given by $I(\omega_{ba})/(\hbar\omega_{ba})$, we see that the integrated cross section σ_{ba} may also be defined as the transition probability per unit time per atom (integrated over ω), W_{ba}, divided by the incident photon flux. It should be noted that the integrated absorption cross section σ_{ba} has the dimensions of area divided by time ($[\mathrm{L}]^2\,[\mathrm{T}]^{-1}$).

Stimulated emission

To calculate the transition rate for stimulated emission, we return to (4.38) and in particular to the second term in the expression for $c_b^{(1)}(t)$, which corresponds to a downward transition ($E_b \simeq E_a - \hbar\omega$) in which a photon of energy $\hbar\omega$ is emitted. It is convenient to interchange the labels of the states a and b so that the state b is again the one with higher energy. The transition $b \to a$ corresponding to stimulated emission may then be viewed as the reverse transition of the absorption process $a \to b$ which we have just studied. Carrying out the same manipulations as we did for absorption, we find that the *transition rate* for stimulated emission (integrated over ω) is given by

$$\bar{W}_{ab} = \frac{4\pi^2}{m^2 c}\left(\frac{e^2}{4\pi\varepsilon_0}\right)\frac{I(\omega_{ba})}{\omega_{ba}^2}|\bar{M}_{ab}(\omega_{ba})|^2 \tag{4.47}$$

where

$$\begin{aligned}\bar{M}_{ab} &= \langle\psi_a|\exp(-\mathrm{i}\mathbf{k}\cdot\mathbf{r})\hat{\boldsymbol{\varepsilon}}\cdot\boldsymbol{\nabla}|\psi_b\rangle \\ &= \int \psi_a^*(\mathbf{r})\exp(-\mathrm{i}\mathbf{k}\cdot\mathbf{r})\hat{\boldsymbol{\varepsilon}}\cdot\boldsymbol{\nabla}\psi_b(\mathbf{r})\,\mathrm{d}\mathbf{r}\end{aligned} \tag{4.48}$$

Integrating by parts, and using the fact that $\hat{\boldsymbol{\varepsilon}}\cdot\mathbf{k} = 0$, we have

$$\bar{M}_{ab} = -M_{ba}^* \tag{4.49}$$

and comparing (4.45) and (4.47), we find that

$$\bar{W}_{ab} = W_{ba} \tag{4.50}$$

Thus we see that under the same radiation field the number of transitions per second exciting the atom from the state a to the state b is the same as the number de-exciting the atom from the state b to the state a. This is consistent with the *principle of detailed balancing*, which says that in an enclosure containing atoms and radiation in equilibrium, the transition probability from a to b is the same as that from b to a, where a and b are any pair of states.

An integrated *stimulated emission cross-section* $\bar{\sigma}_{ab}$ can be defined in analogy with the absorption cross-section (4.46) by dividing the rate at which energy is radiated by the atom, $(\hbar\omega_{ba})\bar{W}_{ab}$, by the intensity $I(\omega_{ba})$. From (4.50) we have

$$\bar{\sigma}_{ab} = \sigma_{ba} \tag{4.51}$$

Despite the fact that the transition rates W_{ba} and $\bar{W}_{ab}$ are equal, stimulated emission is usually much less intense than absorption. Indeed, under equilibrium conditions the initial population of the upper level b is smaller than that of the lower level a because of the Boltzmann factor $\exp(-\hbar\omega_{ba}/(k_B T))$. However, if a *population inversion* is achieved between the two levels a and b, then stimulated emission becomes the dominant process. This is the case in the *maser* and the *laser* where stimulated emission enables atomic or molecular systems to amplify incident radiation, as we shall see in Chapter 15.

Spontaneous emission

In quantum electrodynamics (QED), the part of the vector potential describing the *absorption* of a linearly polarised photon with wave vector **k** from an N-photon state has the form [5]

$$\mathbf{A}_1 = \hat{\boldsymbol{\varepsilon}}\left[\frac{2N(\omega)\hbar}{V\varepsilon_0\omega}\right]^{1/2}\frac{1}{2}\exp[\mathrm{i}(\mathbf{k}\cdot\mathbf{r} - \omega t + \delta_\omega)] \tag{4.52}$$

and it can be shown that the QED transition rate for absorption is given in first-order perturbation theory by

$$W_{ba} = \frac{4\pi^2}{m^2}\left(\frac{e^2}{4\pi\varepsilon_0}\right)\frac{N(\omega_{ba})\hbar}{V\omega_{ba}}|M_{ba}|^2\delta(\omega - \omega_{ba}) \tag{4.53}$$

Using (4.13) and integrating over a range of angular frequencies about ω_{ba}, this result is seen to be identical to (4.45).

The corresponding part of the vector potential describing the *creation* of a photon, adding a single photon to an N-photon state, is

$$\mathbf{A}_2 = \hat{\boldsymbol{\varepsilon}}\left[\frac{2[N(\omega)+1]\hbar}{V\varepsilon_0\omega}\right]^{1/2}\frac{1}{2}\exp[-\mathrm{i}(\mathbf{k}\cdot\mathbf{r} - \omega t + \delta_\omega)] \tag{4.54}$$

and the first-order QED transition rate for emission is given by

$$\bar{W}_{ab} = \frac{4\pi^2}{m^2}\left(\frac{e^2}{4\pi\varepsilon_0}\right)\frac{[N(\omega_{ba})+1]\hbar}{V\omega_{ba}}|M_{ba}|^2\delta(\omega - \omega_{ba}) \tag{4.55}$$

After integrating over ω, this expression is seen to be identical to the semi-classical result (4.47), *provided* $N(\omega_{ba}) + 1$ *is replaced by* $N(\omega_{ba})$. The semi-classical approximation amounts to the neglect of 1 compared with $N(\omega_{ba})$, which is the same as neglecting the possibility of spontaneous emission. In the absence of an external field one has $N = 0$ and the *transition rate* for the spontaneous emission of a photon, W^{s}_{ab}, is given from (4.55) by

$$W^{\mathrm{s}}_{ab} = \frac{4\pi^2}{m^2}\left(\frac{e^2}{4\pi\varepsilon_0}\right)\frac{\hbar}{V\omega_{ba}}|M_{ba}|^2\delta(\omega - \omega_{ba}) \tag{4.56}$$

What can be observed is the emission of a photon within an element of solid angle $\mathrm{d}\Omega$ about the direction $\hat{\mathbf{k}}$ specified by the polar angles (θ, ϕ). In order to obtain the physical transition rate, we must sum (4.56) over the number of allowed photon states in this interval. To do this, we need to calculate the density $\rho_a(\omega)$ of the final photon states, in accordance with the Golden Rule (2.362).

[5] A detailed discussion can be found in Sakurai (1967).

Density of states

Let the volume V be a cube of side L. (In fact the shape does not matter provided V is large.) We can impose periodic boundary conditions [6] on the function $\exp(-i\mathbf{k}\cdot\mathbf{r})$ which is contained in the expression (4.54) representing the wave function of the emitted photon. That is,

$$k_x = \frac{2\pi}{L} n_x, \qquad k_y = \frac{2\pi}{L} n_y, \qquad k_z = \frac{2\pi}{L} n_z \tag{4.57}$$

where n_x, n_y, n_z are positive or negative integers, or zero. Since L is very large, we can treat n_x, n_y and n_z as continuous variables, and the number of states in the range $\mathrm{d}\mathbf{k} = \mathrm{d}k_x\,\mathrm{d}k_y\,\mathrm{d}k_z$ is

$$\begin{aligned} \mathrm{d}n_x\,\mathrm{d}n_y\,\mathrm{d}n_z &= \left(\frac{L}{2\pi}\right)^3 \mathrm{d}k_x\,\mathrm{d}k_y\,\mathrm{d}k_z \\ &= \left(\frac{L}{2\pi}\right)^3 k^2\,\mathrm{d}k\,\mathrm{d}\Omega \end{aligned} \tag{4.58}$$

Expressed in terms of $V = L^3$ and $\omega = kc$, the number of states in the angular frequency interval $\mathrm{d}\omega$ with directions of propagation within $\mathrm{d}\Omega$ is

$$\rho_a(\omega)\,\mathrm{d}\omega\,\mathrm{d}\Omega = \frac{V}{(2\pi)^3}\frac{\omega^2}{c^3}\,\mathrm{d}\omega\,\mathrm{d}\Omega \tag{4.59}$$

Using (4.56) and integrating over the angular frequency ω, the transition rate for the emission of a linearly polarised photon into the solid angle $\mathrm{d}\Omega$ in the direction (θ, ϕ) is then given by

$$W_{ab}^{s}(\theta, \phi)\,\mathrm{d}\Omega = \frac{\hbar}{2\pi m^2 c^3}\left(\frac{e^2}{4\pi\varepsilon_0}\right)\omega_{ba}|M_{ba}(\omega_{ba})|^2\,\mathrm{d}\Omega \tag{4.60}$$

The total transition rate is found by summing over each of the two independent polarisations of the photon, corresponding to polarisation vectors $\hat{\boldsymbol{\varepsilon}}_\lambda$ $(\lambda = 1, 2)$ and integrating over all angles of emission. That is,

$$W_{ab}^{s} = \frac{\hbar}{2\pi m^2 c^3}\left(\frac{e^2}{4\pi\varepsilon_0}\right)\int \mathrm{d}\Omega \sum_{\lambda=1}^{2} \omega_{ba}|M_{ba}^{\lambda}(\omega_{ba})|^2 \tag{4.61}$$

where M_{ba}^{λ} is given by (4.40), with $\hat{\boldsymbol{\varepsilon}}$ replaced by $\hat{\boldsymbol{\varepsilon}}_\lambda$.

---

[6] The imposition of periodic boundary conditions amounts to assuming that all space can be divided into identical large cubes of volume L^3, each containing an identical physical system. The vector potential $\mathbf{A}$ must then be periodic with period L along each of the three Cartesian axes.

4.3 The dipole approximation

In many cases of practical interest the matrix element M_{ba} defined in (4.40) can be simplified by expanding the exponential $\exp(\mathrm{i}\mathbf{k}\cdot\mathbf{r})$ as

$$\exp(\mathrm{i}\mathbf{k}\cdot\mathbf{r}) = 1 + (\mathrm{i}\mathbf{k}\cdot\mathbf{r}) + \frac{1}{2!}(\mathrm{i}\mathbf{k}\cdot\mathbf{r})^2 + \dots \tag{4.62}$$

Consider for example the case of optical transitions. The atomic wave functions extend over distances of the order of the first Bohr radius of the atom, that is about 1 Å (= 10^{-8} cm). On the other hand, the wavelengths associated with optical transitions are of the order of several thousand ångströms, so that the corresponding wave number $k = 2\pi/\lambda$ is of the order of 10^5 cm^{-1}. Thus the quantity (kr) is small for $r < 1$ Å and we can replace $\exp(\mathrm{i}\mathbf{k}\cdot\mathbf{r})$ by unity in (4.40). More generally, if a is a distance characteristic of the linear dimensions of the atomic wave functions, and if $ka \ll 1$, we can replace $\exp(\mathrm{i}\mathbf{k}\cdot\mathbf{r})$ by unity in (4.40). This is known as the *dipole approximation*. We note that since $\lambda \gg a$ this is a *long-wavelength approximation*, which amounts to *neglecting the spatial variation of the radiation field* (that is, neglecting *retardation effects*) across the atom.

As the wavelength of the radiation decreases (that is, as the frequency increases), the dipole approximation becomes less accurate. For example, it is a poor approximation for 'bound–bound' X-ray transitions. Retardation effects must also be taken into account for the continuous spectrum when photons of high frequency (whose wavelength λ does not satisfy the condition $\lambda \gg a$) are absorbed or emitted. This will be illustrated in Section 4.8, where we shall study the photoelectric effect.

It is important to note that in the dipole approximation both the vector potential $\mathbf{A}(t)$ and the electric field $\mathcal{E}(t)$ depend only on the time t, and it follows from (4.2) that the magnetic field $\mathcal{B}$ vanishes. In addition, the matrix element M_{ba} of equation (4.40) is replaced by M_{ba}^{D}, where

$$M_{ba}^{\mathrm{D}} = \hat{\boldsymbol{\varepsilon}}\cdot\langle\psi_b|\boldsymbol{\nabla}|\psi_a\rangle \tag{4.63}$$

In terms of the momentum operator $\mathbf{p} = -\mathrm{i}\hbar\boldsymbol{\nabla} = m\dot{\mathbf{r}}$, we can also write (4.63) in the form

$$M_{ba}^{\mathrm{D}} = \frac{\mathrm{i}}{\hbar}\hat{\boldsymbol{\varepsilon}}\cdot\langle\psi_b|\mathbf{p}|\psi_a\rangle = \frac{\mathrm{i}m}{\hbar}\hat{\boldsymbol{\varepsilon}}\cdot\langle\psi_b|\dot{\mathbf{r}}|\psi_a\rangle \tag{4.64}$$

Now, applying the Heisenberg equation of motion (2.113) to the operator $\mathbf{r}$, we have

$$\dot{\mathbf{r}} = (\mathrm{i}\hbar)^{-1}[\mathbf{r}, H_0] \tag{4.65}$$

where we have replaced H by H_0 since we are working in perturbation theory. Therefore

$$\begin{aligned}\langle\psi_b|\dot{\mathbf{r}}|\psi_a\rangle &= (\mathrm{i}\hbar)^{-1}\langle\psi_b|\mathbf{r}H_0 - H_0\mathbf{r}|\psi_a\rangle \\ &= (\mathrm{i}\hbar)^{-1}(E_a - E_b)\,\langle\psi_b|\mathbf{r}|\psi_a\rangle\end{aligned} \tag{4.66}$$

or, in a more compact notation,

$$\mathbf{p}_{ba} = \mathrm{i}m\omega_{ba}\mathbf{r}_{ba} \tag{4.67}$$

where

$$\mathbf{p}_{ba} = \langle\psi_b|\mathbf{p}|\psi_a\rangle = m\langle\psi_b|\dot{\mathbf{r}}|\psi_a\rangle \tag{4.68}$$

and

$$\mathbf{r}_{ba} = \langle\psi_b|\mathbf{r}|\psi_a\rangle \tag{4.69}$$

This allows us to express M_{ba}^{D} in the form

$$M_{ba}^{\mathrm{D}} = -\frac{m\omega_{ba}}{\hbar}\,\hat{\boldsymbol{\varepsilon}}\cdot\mathbf{r}_{ba} \tag{4.70}$$

The transition rate for absorption in the electric dipole approximation may now be obtained by substituting M_{ba}^{D} for M_{ba} in (4.45). That is,

$$W_{ba}^{\mathrm{D}} = \frac{4\pi^2}{c\hbar^2}\left(\frac{e^2}{4\pi\varepsilon_0}\right)I(\omega_{ba})|\hat{\boldsymbol{\varepsilon}}\cdot\mathbf{r}_{ba}|^2 \tag{4.71}$$

It is convenient at this point to introduce the *electric dipole moment* operator

$$\mathbf{D} = -e\mathbf{r} \tag{4.72}$$

and its matrix element

$$\mathbf{D}_{ba} = -e\mathbf{r}_{ba} \tag{4.73}$$

so that

$$M_{ba}^{\mathrm{D}} = \frac{m\omega_{ba}}{\hbar e}\,\hat{\boldsymbol{\varepsilon}}\cdot\mathbf{D}_{ba} \tag{4.74}$$

and the transition rate W_{ba}^{D} becomes

$$W_{ba}^{\mathrm{D}} = \frac{4\pi^2}{c\hbar^2}\left(\frac{1}{4\pi\varepsilon_0}\right)I(\omega_{ba})|\hat{\boldsymbol{\varepsilon}}\cdot\mathbf{D}_{ba}|^2 \tag{4.75}$$

The quantity $\hat{\boldsymbol{\varepsilon}}\cdot\mathbf{D}_{ba}$ is the matrix element of the component of the electric dipole moment in the direction of polarisation $\hat{\boldsymbol{\varepsilon}}$, between the states b and a. If $\mathbf{D}_{ba}$ is non-vanishing, the transition is said to be an *allowed* or *electric dipole* (E1) transition. If $\mathbf{D}_{ba}$ vanishes, the transition is said to be *forbidden*. When the transition is forbidden, higher terms in the series (4.62) may give rise to non-vanishing transition rates, but these are much smaller than for allowed transitions. These higher terms lead to transitions which are somewhat similar to the types of radiation arising from a multipole expansion in classical radiation theory [7]. For example, the second term ($\mathrm{i}\mathbf{k}\cdot\mathbf{r}$) in the expansion (4.62) gives rise to *magnetic dipole* (M1)

[7] See for example Jackson (1998).

and *electric quadrupole* (E2) transitions. These transitions will be studied in Chapter 9 for a general atom.

If the matrix element M_{ba} in its unapproximated form (4.40) vanishes, the transition is said to be *strictly forbidden* (in first order of perturbation theory). In this case, higher orders of perturbation theory must be considered, and the transition may occur through the simultaneous absorption or emission of two or more photons. It should be noted that the quadratic term $e^2\mathbf{A}^2/(2m)$ of the interaction Hamiltonian (4.27) must then be included in addition to the linear term (e/m) $\mathbf{A}\cdot\mathbf{p}$ considered in the above treatment.

Let us now return to (4.71). Defining Θ as the angle between the vectors $\hat{\boldsymbol{\varepsilon}}$ and $\mathbf{r}_{ba}$, we may write

$$W_{ba}^{\mathrm{D}} = \frac{4\pi^2}{c\hbar^2}\left(\frac{e^2}{4\pi\varepsilon_0}\right) I(\omega_{ba})|\mathbf{r}_{ba}|^2 \cos^2\Theta \tag{4.76}$$

where

$$|\mathbf{r}_{ba}|^2 = |x_{ba}|^2 + |y_{ba}|^2 + |z_{ba}|^2 \tag{4.77}$$

For unpolarised radiation, the orientation of $\hat{\boldsymbol{\varepsilon}}$ is at random, and $\cos^2\Theta$ can be replaced by its average over all solid angles of 1/3 (Problem 4.2), giving

$$W_{ba}^{\mathrm{D}} = \frac{4\pi^2}{3c\hbar^2}\left(\frac{1}{4\pi\varepsilon_0}\right) I(\omega_{ba})|\mathbf{D}_{ba}|^2 \tag{4.78}$$

It is worth noting that because of (4.50) the expression (4.78) also represents the transition rate for *stimulated emission in the dipole approximation* corresponding to the transition $b \to a$, namely the dipole approximation to $\bar{W}_{ab}$. On the other hand, the transition rate for *spontaneous emission* of a photon into the solid angle $\mathrm{d}\Omega$ is given in the *dipole approximation* by substituting (4.74) into (4.60). That is,

$$\begin{aligned} W_{ab}^{\mathrm{s,D}}(\theta,\phi)\,\mathrm{d}\Omega &= \frac{1}{2\pi\hbar c^3}\left(\frac{1}{4\pi\varepsilon_0}\right)\omega_{ba}^3|\hat{\boldsymbol{\varepsilon}}\cdot\mathbf{D}_{ba}|^2\,\mathrm{d}\Omega \\ &= \frac{1}{2\pi\hbar c^3}\left(\frac{e^2}{4\pi\varepsilon_0}\right)\omega_{ba}^3|\hat{\boldsymbol{\varepsilon}}\cdot\mathbf{r}_{ba}|^2\,\mathrm{d}\Omega \end{aligned} \tag{4.79}$$

By summing this expression with respect to the two polarisation directions of the photon and integrating over the angles one obtains the full transition rate for spontaneous emission of a photon in the dipole approximation, namely (Problem 4.2)

$$W_{ab}^{\mathrm{s,D}} = \frac{4}{3\hbar c^3}\left(\frac{1}{4\pi\varepsilon_0}\right)\omega_{ba}^3|\mathbf{D}_{ba}|^2 \tag{4.80a}$$

$$= \frac{4\alpha}{3c^2}\,\omega_{ba}^3|\mathbf{r}_{ba}|^2 \tag{4.80b}$$

where α is the fine structure constant.

The velocity and acceleration forms of the dipole matrix elements

The matrix elements $\mathbf{D}_{ba}$ of the electric dipole operator have been written in terms of what is often called the length matrix elements $\mathbf{r}_{ba}$ given by (4.69). They can also be expressed in terms of the momentum operator $\mathbf{p} = -i\hbar\boldsymbol{\nabla}$ or the gradient of the potential energy

$$V(r) = -\frac{Ze^2}{(4\pi\varepsilon_0)r} \tag{4.81}$$

in which case the matrix elements are said to be expressed in *velocity* or *acceleration* forms, respectively. To see this, we first remark that by using (4.69) and (4.30) we can write

$$\mathbf{r}_{ba} = \frac{1}{E_b - E_a}\langle\psi_b|H_0\mathbf{r} - \mathbf{r}H_0|\psi_a\rangle \tag{4.82}$$

Taking $H_0 = -(\hbar^2/2m)\nabla^2 + V$, and using the fact that V commutes with $\mathbf{r}$, we have

$$\mathbf{r}_{ba} = -\frac{\hbar^2}{2m}\frac{1}{E_b - E_a}\langle\psi_b|\nabla^2\mathbf{r} - \mathbf{r}\nabla^2|\psi_a\rangle \tag{4.83}$$

Noting that

$$\nabla^2[\mathbf{r}\psi_a(\mathbf{r})] = 2\boldsymbol{\nabla}\psi_a(\mathbf{r}) + \mathbf{r}\nabla^2\psi_a(\mathbf{r}) \tag{4.84}$$

and that $\mathbf{p} = -i\hbar\boldsymbol{\nabla}$, we retrieve the result (4.67). That is,

$$\mathbf{r}_{ba} = -\frac{i\hbar}{m}\frac{1}{E_b - E_a}\langle\psi_b|\mathbf{p}|\psi_a\rangle = -\frac{i}{m\omega_{ba}}\mathbf{p}_{ba} \tag{4.85}$$

In a similar way, we can write the matrix element $\mathbf{p}_{ba}$ in the form

$$\begin{aligned}\mathbf{p}_{ba} &= -\frac{i\hbar}{E_b - E_a}\langle\psi_b|H_0\boldsymbol{\nabla} - \boldsymbol{\nabla}H_0|\psi_a\rangle \\ &= -\frac{i\hbar}{E_b - E_a}\langle\psi_b|V\boldsymbol{\nabla} - \boldsymbol{\nabla}V|\psi_a\rangle \\ &= \frac{i\hbar}{E_b - E_a}\langle\psi_b|(\boldsymbol{\nabla}V)|\psi_a\rangle = \frac{i}{\omega_{ba}}(\boldsymbol{\nabla}V)_{ba}\end{aligned} \tag{4.86}$$

Using these results the length, velocity and acceleration forms of $\mathbf{D}_{ba}$ are, respectively,

$$\mathbf{D}^{\mathrm{L}}_{ba} = -e\mathbf{r}_{ba} \tag{4.87a}$$

$$\mathbf{D}^{\mathrm{V}}_{ba} = \frac{ie}{m\omega_{ba}}\mathbf{p}_{ba} \tag{4.87b}$$

and

$$\mathbf{D}^{\mathrm{A}}_{ba} = -\frac{e}{m\omega^2_{ba}}(\boldsymbol{\nabla}V)_{ba} \tag{4.87c}$$

Provided that the wave functions ψ_a and ψ_b are exact eigenfunctions of H_0, the three forms of $\mathbf{D}_{ba}$ give identical results. However, if approximate wave functions are employed, the three forms of $\mathbf{D}_{ba}$ will yield different numerical values.

The Schrödinger equation in the velocity and length gauges

Let us rewrite the time-dependent Schrödinger equation (4.25) in the form

$$\mathrm{i}\hbar\frac{\partial}{\partial t}\Psi(\mathbf{r},t) = \left[H_0 + \frac{e}{m}\mathbf{A}\cdot\mathbf{p} + \frac{e^2}{2m}\mathbf{A}^2\right]\Psi(\mathbf{r},t) \tag{4.88}$$

where H_0 is given by (4.26) and we have used (4.27). It is interesting to note that *within the dipole approximation* (so that the vector potential $\mathbf{A}$ depends only on the time t), the term in $\mathbf{A}^2$ in (4.88) can be eliminated by performing the gauge transformation

$$\Psi(\mathbf{r},t) = \exp\left[-\frac{\mathrm{i}}{\hbar}\frac{e^2}{2m}\int^t \mathbf{A}^2(t')\,\mathrm{d}t'\right]\Psi^{\mathrm{V}}(\mathbf{r},t) \tag{4.89}$$

This gives for $\Psi^{\mathrm{V}}(\mathbf{r},t)$ the new time-dependent Schrödinger equation (Problem 4.3)

$$\mathrm{i}\hbar\frac{\partial}{\partial t}\Psi^{\mathrm{V}}(\mathbf{r},t) = \left[H_0 + \frac{e}{m}\mathbf{A}(t)\cdot\mathbf{p}\right]\Psi^{\mathrm{V}}(\mathbf{r},t) \tag{4.90}$$

which is said to be in the *velocity gauge* since the interaction term

$$H^{\mathrm{V}}_{\mathrm{int}}(t) = \frac{e}{m}\mathbf{A}(t)\cdot\mathbf{p} = -\frac{\mathrm{i}\hbar e}{m}\mathbf{A}(t)\cdot\boldsymbol{\nabla} \tag{4.91}$$

couples the vector potential $\mathbf{A}(t)$ to the velocity operator $\mathbf{p}/m$.

Another form of the time-dependent Schrödinger equation in the dipole approximation can be obtained by returning to equation (4.88) and performing a gauge transformation specified by taking $\chi(\mathbf{r},t) = \mathbf{A}(t)\cdot\mathbf{r}$. Using the fact that in the dipole approximation the electric field is given by $\mathcal{E}(t) = -\mathrm{d}\mathbf{A}(t)/\mathrm{d}t$, we see from equations (4.19) with $q = -e$ that

$$\mathbf{A}' = 0 \tag{4.92a}$$

$$\phi' = \frac{\mathrm{d}}{\mathrm{d}t}\mathbf{A}(t)\cdot\mathbf{r} = -\mathcal{E}(t)\cdot\mathbf{r} \tag{4.92b}$$

$$\Psi'(\mathbf{r},t) = \exp\left[\frac{\mathrm{i}e}{\hbar}\mathbf{A}(t)\cdot\mathbf{r}\right]\Psi(\mathbf{r},t) \tag{4.92c}$$

The new time-dependent Schrödinger equation for $\Psi'(\mathbf{r},t) \equiv \Psi^{\mathrm{L}}(\mathbf{r},t)$ is (Problem 4.4)

$$\mathrm{i}\hbar\frac{\partial}{\partial t}\Psi^{\mathrm{L}}(\mathbf{r},t) = [H_0 + e\mathcal{E}(t)\cdot\mathbf{r}]\Psi^{\mathrm{L}}(\mathbf{r},t) \tag{4.93}$$

which is said to be in the *length gauge* because the interaction term

$$H^{\rm L}_{\rm int}(t) = e\mathcal{E}(t)\cdot\mathbf{r} \tag{4.94}$$

couples the electric field $\mathcal{E}(t)$ to the position operator $\mathbf{r}$. Since the electric dipole operator is given by $\mathbf{D} = -e\mathbf{r}$, we see that $H^{\rm L}_{\rm int} = -\mathcal{E}(t)\cdot\mathbf{D}$. This is the reason why the approximation in which the electromagnetic field is taken to be uniform over the atom is called the *dipole approximation*.

Spontaneous emission from the 2p level of hydrogenic atoms

As an example of the calculation of a transition rate for spontaneous emission in the dipole approximation, we shall consider the transition from the 2p level of a hydrogenic atom with nuclear charge Ze to the ground state, starting from (4.80b). In this equation, b represents the 2p state of the atom with magnetic quantum number m while a is the 1s ground state. The angular frequency ω_{ba} can be found from (4.34) and (3.30). It is

$$\begin{aligned}\omega_{ba} &= (E_b - E_a)/\hbar \\ &= \frac{3}{8}\frac{mc^2}{\hbar}(Z\alpha)^2\end{aligned} \tag{4.95}$$

From (4.69) and (3.53) the matrix element $\mathbf{r}_{ba}$ is seen to be

$$\mathbf{r}_{ba} = \int_0^\infty R_{21}(r)R_{10}(r)r^3\,\mathrm{d}r \int Y^*_{1m}(\theta,\phi)\,\hat{\mathbf{r}}\,Y_{00}(\theta,\phi)\,\mathrm{d}\Omega \tag{4.96}$$

where $\hat{\mathbf{r}}$ is a unit vector in the direction of $\mathbf{r}$. Using the expressions for $R_{21}(r)$ and $R_{10}(r)$ given in (3.59) the radial integral in (4.96) can be evaluated:

$$\begin{aligned}\int_0^\infty R_{21}(r)R_{10}(r)r^3\,\mathrm{d}r &= \left(\frac{Z}{a_0}\right)^4 \frac{1}{\sqrt{6}}\int_0^\infty r^4 \exp[-3Zr/(2a_0)]\,\mathrm{d}r \\ &= \frac{a_0}{Z}\frac{24}{\sqrt{6}}\left(\frac{2}{3}\right)^5\end{aligned} \tag{4.97}$$

To perform the angular integration in (4.96), the x, y and z components of $\hat{\mathbf{r}}$ are expressed as

$$\begin{aligned}(\hat{\mathbf{r}})_x &= \sin\theta\cos\phi = \sqrt{\frac{2\pi}{3}}\,[-Y_{1,1}(\theta,\phi) + Y_{1,-1}(\theta,\phi)] \\ (\hat{\mathbf{r}})_y &= \sin\theta\sin\phi = \sqrt{\frac{2\pi}{3}}\,\mathrm{i}\,[Y_{1,1}(\theta,\phi) + Y_{1,-1}(\theta,\phi)] \\ (\hat{\mathbf{r}})_z &= \cos\theta = \sqrt{\frac{4\pi}{3}}\,Y_{1,0}(\theta,\phi)\end{aligned} \tag{4.98}$$

From (4.96)–(4.98), we have

$$|\mathbf{r}_{ba}|^2 = \left(\frac{a_0}{Z}\right)^2 \frac{2^{15}}{3^{10}} [\delta_{m,1} + \delta_{m,-1} + \delta_{m,0}]$$

$$= \frac{1}{Z^2}\left(\frac{\hbar}{mc\alpha}\right)^2 \frac{2^{15}}{3^{10}} [\delta_{m,1} + \delta_{m,-1} + \delta_{m,0}] \qquad \textbf{(4.99)}$$

The transition rate from each magnetic substate is the same, and if each state is equally populated the full transition rate in the dipole approximation is, from (4.80b), (4.95) and (4.99),

$$W_{ab}^{\mathrm{s,D}} = \frac{1}{3}\sum_{m=-1}^{1} W_{1\mathrm{s},2pm}^{\mathrm{s,D}}$$

$$= \left(\frac{2}{3}\right)^8 \frac{m\alpha^5 Z^4 c^2}{\hbar}$$

$$\simeq 6.27 \times 10^8 Z^4 \ \mathrm{s}^{-1} \qquad \textbf{(4.100)}$$

4.4 The Einstein coefficients

We shall verify that (4.80) is the correct expression for the rate of spontaneous emission by using the treatment of emission and absorption of radiation given by Einstein in 1916. Consider an enclosure containing atoms (of a single kind) and radiation in equilibrium at absolute temperature T, and let a and b denote two non-degenerate atomic levels, with energy values E_a and E_b such that $E_b > E_a$. We assume that the radiation field is weak enough for a first-order perturbative treatment of absorption and stimulated emission to be valid, and denote by $\rho(\omega_{ba})$ the energy density of the radiation at the angular frequency $\omega_{ba} = (E_b - E_a)/\hbar$. The number of atoms making the transition from a to b per unit time by absorbing radiation, $\dot{N}_{ba}$, is proportional to the total number N_a of atoms in the state a and to the energy density $\rho(\omega_{ba})$. That is,

$$\dot{N}_{ba} = B_{ba} N_a \rho(\omega_{ba}) \qquad \textbf{(4.101)}$$

where B_{ba} is called the *Einstein coefficient for absorption*. Since $\rho = I/c$ (see (4.13)) and the transition rate for absorption (per atom) is W_{ba}, we have

$$B_{ba} = \frac{W_{ba}}{\rho} = \frac{4\pi^2}{3\hbar^2}\left(\frac{1}{4\pi\varepsilon_0}\right)|\mathbf{D}_{ba}|^2 \qquad \textbf{(4.102)}$$

where in the last step we have used the dipole approximation (4.78) for W_{ba}.

On the other hand, the number of atoms making the transition $b \to a$ per unit time, $\dot{N}_{ab}$, is the sum of the number of spontaneous transitions per unit time, which is independent of ρ, and the number of stimulated transitions per unit time, which is proportional to ρ. Thus

$$\dot{N}_{ab} = A_{ab}N_b + B_{ab}N_b\rho(\omega_{ba}) \tag{4.103}$$

where N_b is the total number of atoms in the state b, A_{ab} is the *Einstein coefficient for spontaneous emission* and B_{ab} is the *Einstein coefficient for stimulated emission.* In our notation $A_{ab} \equiv W^{s}_{ab}$. At equilibrium we have $\dot{N}_{ba} = \dot{N}_{ab}$, so that from (4.101) and (4.103) we deduce that

$$\frac{N_a}{N_b} = \frac{A_{ab} + B_{ab}\rho(\omega_{ba})}{B_{ba}\rho(\omega_{ba})} \tag{4.104}$$

We also know that at thermal equilibrium the ratio N_a/N_b is given by [8]

$$\frac{N_a}{N_b} = \exp[-(E_a - E_b)/(k_B T)] = \exp[\hbar\omega_{ba}/(k_B T)] \tag{4.105}$$

where k_B is Boltzmann's constant. From (4.104) and (4.105) we find for $\rho(\omega_{ba})$ the expression

$$\rho(\omega_{ba}) = \frac{A_{ab}}{B_{ba}\exp[\hbar\omega_{ba}/(k_B T)] - B_{ab}} \tag{4.106}$$

Since the atoms are in equilibrium with the radiation at temperature T, the energy density $\rho(\omega)$ is given by the Planck distribution law discussed in Section 1.3. Using (1.31) together with the fact that $\rho(\omega)\,d\omega = \rho(\nu)\,d\nu$, with $\omega = 2\pi\nu$, the energy density at the particular angular frequency ω_{ba} is

$$\rho(\omega_{ba}) = \frac{\hbar\omega_{ba}^3}{\pi^2 c^3}\frac{1}{\exp[\hbar\omega_{ba}/(k_B T)] - 1} \tag{4.107}$$

In order for (4.106) and (1.107) to be identical, the three Einstein coefficients must be related by the two equations

$$B_{ba} = B_{ab} \tag{4.108a}$$

$$A_{ab} = \frac{\hbar\omega_{ba}^3}{\pi^2 c^3}B_{ab} \tag{4.108b}$$

The relation (4.108a) expresses the principle of detailed balancing discussed previously. Using (4.102) and (4.108), we verify that W^{s}_{ab} ($\equiv A_{ab}$) is indeed given in the dipole approximation by the expression (4.80). It is a simple matter to generalise the above results to the case in which the energy levels E_a and (or) E_b are degenerate. Denoting by g_a and g_b the degeneracy of these levels, one finds (Problem 4.5) that (4.108a) becomes

[8] See for instance the text by Kittel and Kroemer (1980).

$$g_a B_{ba} = g_b B_{ab} \tag{4.109}$$

while the relation (4.108b) remains unchanged.

4.5 Selection rules and the spectrum of one-electron atoms

In the last section, we obtained the probability of a radiative transition between two levels a and b, in the electric dipole approximation. For stimulated emission or absorption of linearly polarised radiation in the direction $\hat{\boldsymbol{\varepsilon}}$, the basic expression is given by (4.71) and for spontaneous emission by (4.79). In each case the transition rate depends on the key quantity $|\hat{\boldsymbol{\varepsilon}} \cdot \mathbf{r}_{ba}|^2$. In order to study this expression it is convenient to introduce the *spherical components* of the vectors $\hat{\boldsymbol{\varepsilon}}$ and $\mathbf{r}$. According to the definition (A4.48) of Appendix 4 the spherical components ε_q ($q = 0, \pm 1$) of $\hat{\boldsymbol{\varepsilon}}$ are given in terms of its Cartesian components $(\hat{\varepsilon}_x, \hat{\varepsilon}_y, \hat{\varepsilon}_z)$ by

$$\varepsilon_1 = -\frac{1}{\sqrt{2}}(\hat{\varepsilon}_x + \mathrm{i}\hat{\varepsilon}_y), \qquad \varepsilon_0 = \hat{\varepsilon}_z, \qquad \varepsilon_{-1} = \frac{1}{\sqrt{2}}(\hat{\varepsilon}_x - \mathrm{i}\hat{\varepsilon}_y) \tag{4.110}$$

As we shall see later, if the direction of propagation of the radiation is along the Z axis ($\hat{\varepsilon}_z = 0$), ε_1 and ε_{-1} describe states of circular polarisation.

Similarly, the spherical components r_q ($q = 0, \pm 1$) of the vector $\mathbf{r}$ are given by

$$r_1 = -\frac{1}{\sqrt{2}}(x + \mathrm{i}y) = -\frac{1}{\sqrt{2}} r \sin\theta\, \mathrm{e}^{\mathrm{i}\phi} = r\left(\frac{4\pi}{3}\right)^{1/2} Y_{1,1}(\theta, \phi)$$

$$r_0 = z = r\cos\theta = r\left(\frac{4\pi}{3}\right)^{1/2} Y_{1,0}(\theta, \phi) \tag{4.111}$$

$$r_{-1} = \frac{1}{\sqrt{2}}(x - \mathrm{i}y) = \frac{1}{\sqrt{2}} r \sin\theta\, \mathrm{e}^{-\mathrm{i}\phi} = r\left(\frac{4\pi}{3}\right)^{1/2} Y_{1,-1}(\theta, \phi)$$

The scalar product $\hat{\boldsymbol{\varepsilon}} \cdot \mathbf{r}_{ba}$ can be expressed in terms of spherical components as

$$\begin{aligned}\hat{\boldsymbol{\varepsilon}} \cdot \mathbf{r}_{ba} &= \sum_{q=0,\pm1} \varepsilon_q^*(\mathbf{r}_{ba})_q \\ &= \sum_{q=0,\pm1} \varepsilon_q^* I^q_{n'l'm',nlm}\end{aligned} \tag{4.112}$$

where

$$\begin{aligned}I^q_{n'l'm',nlm} &= \left(\frac{4\pi}{3}\right)^{1/2} \int_0^\infty \mathrm{d}r\, r^3 R_{n'l'}(r) R_{nl}(r) \\ &\quad \times \int \mathrm{d}\Omega\, Y^*_{l'm'}(\theta, \phi) Y_{1,q}(\theta, \phi) Y_{lm}(\theta, \phi)\end{aligned} \tag{4.113}$$

and where we have written the quantum numbers of the levels a and b of the hydrogenic atom as (nlm) and $(n'l'm')$, respectively. The radial integral in (4.113) is always non-zero, but the angular integrals are only non-zero for certain values of (l, m) and (l', m'), giving rise to *selection rules* which we shall now investigate.

Parity

Under the reflection $\mathbf{r} \to -\mathbf{r}$ we have shown in Section 3.3 that the hydrogenic wave functions behave like (see (3.69))

$$\begin{aligned} R_{nl}(r)Y_{lm}(\theta, \phi) &\to R_{nl}(r)Y_{lm}(\pi - \theta, \phi + \pi) \\ &= R_{nl}(r)(-1)^l Y_{lm}(\theta, \phi) \end{aligned} \tag{4.114}$$

and the parity of the wave function is even or odd according to whether l is even or odd. By making the coordinate transformation $\mathbf{r} \to -\mathbf{r}$ in (4.113) we see that

$$I^q_{n'l'm',nlm} = (-1)^{l+l'+1} I^q_{n'l'm',nlm} \tag{4.115}$$

Hence the quantity $I^q_{n'l'm',nlm}$ is only non-vanishing if $(l + l' + 1)$ is even. In other words, *the electric dipole operator only connects states of opposite parity.*

Magnetic quantum numbers

The integral over ϕ which must be performed in (4.113) is of the form

$$J(m, m', q) = \int_0^{2\pi} \exp[\mathrm{i}(m + q - m')\phi]\, \mathrm{d}\phi \tag{4.116}$$

We shall consider separately the two cases $q = 0$ and $q = \pm 1$, which correspond respectively to radiation polarised *parallel* to the Z axis and *perpendicular* to the Z axis, the Z direction being the quantisation direction to which the magnetic quantum number m refers.

1. $q = 0$ (polarisation vector $\hat{\varepsilon}$ in the Z direction).

In this case the integral (4.116) – and therefore the matrix element (4.113) – vanishes unless

$$m' = m, \quad \text{i.e.} \quad \Delta m = 0 \tag{4.117a}$$

2. $q = \pm 1$ (propagation vector $\mathbf{k}$ in the Z direction).

Here the ϕ integration in (4.116) yields for the matrix element (4.113) the selection rule

$$m' = m \pm 1, \quad \text{i.e.} \quad \Delta m = \pm 1 \tag{4.117b}$$

In a given transition, only one of the conditions $\Delta m = 0$ or $\Delta m = \pm 1$ can be satisfied, and hence only one of the matrix elements z_{ba} and $(x \pm iy)_{ba}$ will be non-zero.

Orbital angular momentum

The integral over the angles in (4.113), which we call $\mathscr{A}(l, m; l', m'; q)$, can be evaluated by the methods of Appendix 4. The result, expressed in terms of Clebsch–Gordan coefficients, is (see (A4.40))

$$\mathscr{A}(l, m; l', m'; q) = \int d\Omega\, Y^*_{l'm'}(\theta, \phi) Y_{1,q}(\theta, \phi) Y_{lm}(\theta, \phi)$$
$$= \left(\frac{3}{4\pi}\frac{2l+1}{2l'+1}\right)^{1/2} \langle l100|l'0\rangle\langle l1mq|l'm'\rangle \quad \textbf{(4.118)}$$

From the properties of the Clebsch–Gordan coefficients we note that $\mathscr{A}(l, m; l', m'; q)$ vanishes unless $m' = m + q$, which is in agreement with the selection rules (4.117) we have just obtained for the magnetic quantum numbers. In addition, the properties of the Clebsch–Gordan coefficients also imply that $\mathscr{A}(l, m; l', m'; q)$ vanishes unless

$$l' = l \pm 1, \quad \text{i.e.} \quad \Delta l = \pm 1 \quad \textbf{(4.119)}$$

This is the orbital angular momentum selection rule for electric dipole transitions. This rule can also be deduced in a more elementary fashion by using the recurrence relations satisfied by the associated Legendre functions $P_l^m(\cos\theta)$. Indeed, using the expressions (2.181) for the spherical harmonics, together with the selection rules (4.117) for the magnetic quantum numbers, we see that the θ integration in (4.118) can be written, apart from numerical factors, as

$$\int_{-1}^{+1} d(\cos\theta) P_l^m(\cos\theta) P_{l'}^m(\cos\theta)\cos\theta \quad \text{for} \quad q = 0 \quad \textbf{(4.120a)}$$

$$\int_{-1}^{+1} d(\cos\theta) P_l^m(\cos\theta) P_{l'}^{m\pm1}(\cos\theta)\sin\theta \quad \text{for} \quad q = \pm 1 \quad \textbf{(4.120b)}$$

where, in view of (2.181b), we need only deal with magnetic quantum numbers which are positive or zero. From the recurrence relations (2.176),

$$(2l+1)\cos\theta\, P_l^m(\cos\theta) = (l+1-m)P_{l+1}^m(\cos\theta) + (l+m)P_{l-1}^m(\cos\theta) \quad \textbf{(4.121a)}$$

and

$$(2l+1)\sin\theta\, P_l^{m-1}(\cos\theta) = P_{l+1}^m(\cos\theta) - P_{l-1}^m(\cos\theta) \quad \textbf{(4.121b)}$$

together with the orthogonality relation (2.177)

$$\int_{-1}^{+1} d(\cos\theta) P_l^m(\cos\theta) P_{l'}^m(\cos\theta) = \frac{2}{2l+1}\frac{(l+m)!}{(l-m)!}\delta_{ll'} \quad \textbf{(4.122)}$$

we find that $l' = l \pm 1$, in accordance with (4.119). Using either of the above methods one can also obtain the explicit forms of the quantities $\mathscr{A}(l, m; l', m'; q)$, which is left as an exercise for the reader (Problem 4.6).

Electron spin

We note that the electric dipole operator does not act on the spin of the electron. It follows that the component of the electron spin in the direction of quantisation remains unaltered by the absorption or emission of dipole radiation.

The spin of the photon

The selection rules for electric dipole transitions have a simple interpretation in terms of the spin of the photon. To discuss this point, we must first explore in more detail the possible states of polarisation of an electromagnetic wave. In Section 4.1 we saw that a general state of polarisation for a plane wave propagating in the direction $\hat{\mathbf{k}}$ can be described by combining two independent *linearly polarised* plane waves (having in general different phases) with polarisation vectors $\hat{\varepsilon}_\lambda$ ($\lambda = 1, 2$) orthogonal to $\hat{\mathbf{k}}$. The resulting polarisation vector $\hat{\varepsilon}$ lies in a plane perpendicular to $\hat{\mathbf{k}}$, so that a state of arbitrary polarisation can always be represented as

$$\hat{\varepsilon} = a_1\hat{\mathbf{e}}_1 + a_2\hat{\mathbf{e}}_2; \qquad a_1^2 + a_2^2 = 1 \tag{4.123}$$

where $\hat{\mathbf{e}}_i$ ($i = 1, 2$) are fixed mutually orthogonal real unit vectors in a plane perpendicular to $\hat{\mathbf{k}}$. We shall take $\hat{\mathbf{e}}_1$, $\hat{\mathbf{e}}_2$ and $\hat{\mathbf{k}}$ to form a right-handed system, so that (see Fig. 4.1)

$$\hat{\mathbf{k}} = \hat{\mathbf{e}}_1 \times \hat{\mathbf{e}}_2; \qquad \hat{\mathbf{e}}_1 \cdot \hat{\mathbf{e}}_2 = 0 \tag{4.124}$$

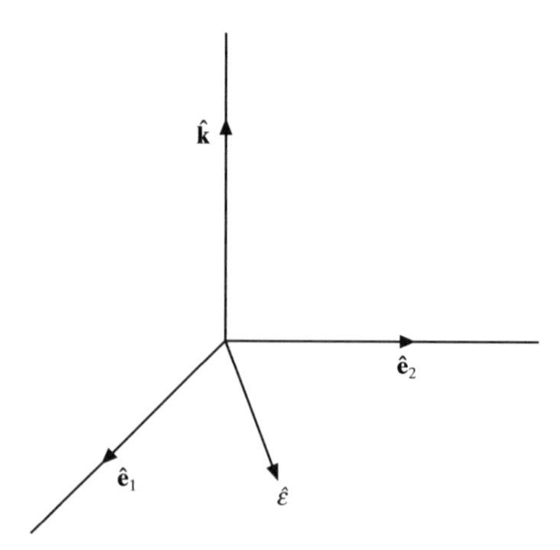

Figure 4.1 The vectors $\hat{\mathbf{e}}_1$, $\hat{\mathbf{e}}_2$ and $\hat{\mathbf{k}}$, forming a right-handed set of mutually orthogonal unit vectors. Also shown is the polarisation vector $\hat{\varepsilon}$ which lies in the plane of $\hat{\mathbf{e}}_1$ and $\hat{\mathbf{e}}_2$.

The foregoing discussion may of course be directly generalised to pulses of the form (4.14), for which the polarisation vector $\hat{\boldsymbol{\varepsilon}}$ is independent of ω.

An alternative description of a general state of polarisation of an electromagnetic wave may be given in terms of two *circularly polarised waves*. Let us first consider the particular case in which the direction of propagation is along the Z axis. In place of the vector potential (4.5), we consider the potentials $\mathbf{A}^{\mathrm{L}}(\mathbf{r}, t)$ and $\mathbf{A}^{\mathrm{R}}(\mathbf{r}, t)$ defined by

$$\begin{aligned} A_x^{\mathrm{L}} &= A_x^{\mathrm{R}} = (2)^{-1/2} A_0(\omega) \cos(kz - \omega t + \delta_\omega) \\ A_y^{\mathrm{L}} &= -A_y^{\mathrm{R}} = -(2)^{-1/2} A_0(\omega) \sin(kz - \omega t + \delta_\omega) \\ A_z^{\mathrm{L}} &= A_z^{\mathrm{R}} = 0 \end{aligned} \tag{4.125}$$

The corresponding electric field vectors $\mathscr{E}^{\mathrm{L}}$ and $\mathscr{E}^{\mathrm{R}}$ are such that

$$\begin{aligned} \mathscr{E}_x^{\mathrm{L}} &= \mathscr{E}_x^{\mathrm{R}} = -(2)^{-1/2} \omega A_0(\omega) \sin(kz - \omega t + \delta_\omega) \\ \mathscr{E}_y^{\mathrm{L}} &= -\mathscr{E}_y^{\mathrm{R}} = -(2)^{-1/2} \omega A_0(\omega) \cos(kz - \omega t + \delta_\omega) \\ \mathscr{E}_z^{\mathrm{L}} &= \mathscr{E}_z^{\mathrm{R}} = 0 \end{aligned} \tag{4.126}$$

On facing into the oncoming wave, the vector $\mathscr{E}^{\mathrm{L}}$ is seen to be of constant magnitude and to be rotating in an anticlockwise way in the (X, Y) plane at a frequency ω (see Fig. 4.2(a)), while the vector $\mathscr{E}^{\mathrm{R}}$ is of the same magnitude but rotates at a frequency ω in a clockwise way (see Fig. 4.2(b)). The radiation described by $\mathscr{E}^{\mathrm{L}}$ is said to be left-hand circularly polarised and that corresponding to $\mathscr{E}^{\mathrm{R}}$ is right-hand circularly polarised. By forming the combination

$$\mathscr{E} = a_{\mathrm{L}} \mathscr{E}^{\mathrm{L}} + a_{\mathrm{R}} \mathscr{E}^{\mathrm{R}} \tag{4.127}$$

where a_{L} and a_{R} are complex coefficients, radiation in any state of polarisation can be produced. For example, if $a_{\mathrm{L}} = a_{\mathrm{R}} = 1$, we obtain linearly polarised radiation, with the electric field vector oriented along the X axis.

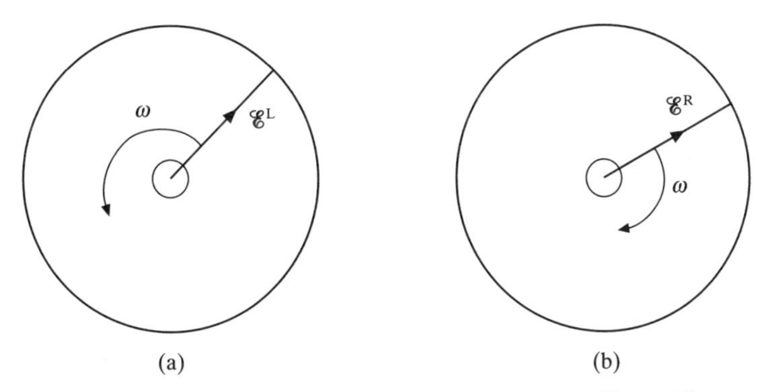

Figure 4.2 In circularly polarised radiation the electric field vectors $\mathscr{E}^{\mathrm{L}}$ and $\mathscr{E}^{\mathrm{R}}$ rotate in anticlockwise and clockwise directions when facing into the oncoming wave.

In terms of complex exponentials, $\mathbf{A}^{\mathrm{L}}$ and $\mathbf{A}^{\mathrm{R}}$ can be written as

$$\mathbf{A}^{\mathrm{L}} = \frac{1}{2} A_0(\omega)[\hat{\mathbf{e}}^{\mathrm{L}}\, \mathrm{e}^{\mathrm{i}(kz-\omega t+\delta_\omega)} + \text{c.c.}]$$
$$\mathbf{A}^{\mathrm{R}} = \frac{1}{2} A_0(\omega)[\hat{\mathbf{e}}^{\mathrm{R}}\, \mathrm{e}^{\mathrm{i}(kz-\omega t+\delta_\omega)} + \text{c.c.}] \qquad \textbf{(4.128)}$$

where c.c. denotes the complex conjugate. In (4.128), $\hat{\mathbf{e}}^{\mathrm{L}}$ and $\hat{\mathbf{e}}^{\mathrm{R}}$ are two complex orthogonal unit vectors such that

$$\hat{\mathbf{e}}^{\mathrm{L}} = \frac{1}{\sqrt{2}}(\hat{\mathbf{x}} + \mathrm{i}\hat{\mathbf{y}}), \qquad \hat{\mathbf{e}}^{\mathrm{R}} = \frac{1}{\sqrt{2}}(\hat{\mathbf{x}} - \mathrm{i}\hat{\mathbf{y}}) \qquad \textbf{(4.129)}$$

describing respectively the states of left-hand and right-hand circularly polarised radiation. An arbitrary state of polarisation can be specified by a complex vector $\hat{\mathbf{n}}$ such that

$$\hat{\mathbf{n}} = a^{\mathrm{L}}\hat{\mathbf{e}}^{\mathrm{L}} + a^{\mathrm{R}}\hat{\mathbf{e}}^{\mathrm{R}}, \qquad |a^{\mathrm{L}}|^2 + |a^{\mathrm{R}}|^2 = 1 \qquad \textbf{(4.130)}$$

and this description is as general as the one given by (4.123). We note that if the direction of propagation is not along the Z axis, then we can write more generally

$$\hat{\mathbf{e}}^{\mathrm{L}} = \frac{1}{\sqrt{2}}(\hat{\mathbf{e}}_1 + \mathrm{i}\hat{\mathbf{e}}_2), \qquad \hat{\mathbf{e}}^{\mathrm{R}} = \frac{1}{\sqrt{2}}(\hat{\mathbf{e}}_1 - \mathrm{i}\hat{\mathbf{e}}_2) \qquad \textbf{(4.131)}$$

where $\hat{\mathbf{e}}_1$ and $\hat{\mathbf{e}}_2$ are the unit vectors introduced in (4.123).

We have already seen in Section 4.2 that the terms in $\mathbf{A}$ associated with $\exp[\mathrm{i}(\mathbf{k}\cdot\mathbf{r} - \omega t)]$ give rise to the absorption of photons (see (4.52)) and those with $\exp[-\mathrm{i}(\mathbf{k}\cdot\mathbf{r} - \omega t)]$ to the emission of photons (see (4.54)). From (4.56), (4.49), (4.71) and (4.79) we see that, in the electric dipole approximation, we should use the expressions $\hat{\mathbf{e}}^{\mathrm{L}}\cdot\mathbf{r}_{ba}$ or $\hat{\mathbf{e}}^{\mathrm{R}}\cdot\mathbf{r}_{ba}$ to describe the *absorption* of a left-hand or a right-hand circularly polarised photon, respectively, while the expressions $\hat{\mathbf{e}}^{\mathrm{L}*}\cdot\mathbf{r}^*_{ba}$ $(= \hat{\mathbf{e}}^{\mathrm{L}*}\cdot\mathbf{r}_{ab})$ or $\hat{\mathbf{e}}^{\mathrm{R}*}\cdot\mathbf{r}^*_{ba}$ $(= \hat{\mathbf{e}}^{\mathrm{R}*}\cdot\mathbf{r}_{ab})$ must be used to describe the *emission* of the corresponding circularly polarised photons. Thus, if a left-hand circularly polarised photon is emitted in the Z direction, the appropriate expression is

$$\hat{\mathbf{e}}^{\mathrm{L}*}\cdot\mathbf{r}_{ab} = \frac{1}{\sqrt{2}}(\hat{\mathbf{x}} - \mathrm{i}\hat{\mathbf{y}})\cdot\mathbf{r}_{ab}$$
$$= \frac{1}{\sqrt{2}}(x_{ab} - \mathrm{i}y_{ab}) \qquad \textbf{(4.132)}$$

If we denote by $m\hbar$ the Z component of the angular momentum for the initial (upper) state b, we see from (4.111) and (4.117) that the matrix element (4.132) vanishes unless the final (lower) state a of the atom has a component $(m - 1)\hbar$ of angular momentum in the Z direction. A similar reasoning leads to the conclusion that the emission of a right-hand circularly polarised photon increases the component of the angular momentum of the atom along the Z axis by $\hbar$.

By conservation of angular momentum, each photon must have a component of angular momentum parallel to the Z axis (the direction of propagation) of $\pm\hbar$. Since photons travelling parallel to the Z axis cannot have a component of *orbital* angular momentum in the Z direction, the angular momentum carried by the photons in this case can only be due to their intrinsic spin. Further, for electric dipole radiation, the orbital angular momentum of the photon must be zero, since the wave function (the vector potential) is spherically symmetrical (we have replaced $\exp(i\mathbf{k}\cdot\mathbf{r})$ by 1). From these remarks, and from the selection rule $\Delta l = \pm 1$, we infer that the photon has spin of unit magnitude, that is $\mathbf{S}^2 = s(s+1)\hbar^2$ with $s = 1$. The components of the spin in the direction of propagation are $S_z = m_s\hbar$ with $m_s = \pm 1$. The component of the spin along the direction of motion is called the *helicity* of a particle. For the photon, only two helicity states are possible, because the electromagnetic wave is transverse and the case $m_s = 0$ is excluded. From the definition of helicity it is clear that a photon with helicity $+\hbar$ is always left-hand circularly polarised and one with helicity $-\hbar$ is always right-hand circularly polarised, and this is independent of any particular choice of axes.

Beth's experiment

If a beam of light, propagating parallel to the positive Z axis, is left circularly polarised, each photon in the beam will have a positive angular momentum $+\hbar$ along the Z axis. If the beam contains N photons per unit volume, the energy density of the beam, ρ, will be $\rho = N\hbar\omega$, where ω is the angular frequency of the radiation, and a unit volume will possess an angular momentum $L_z = N\hbar$. The ratio $\rho/\omega = N\hbar$ is independent of frequency and is equal in magnitude to the angular momentum L_z. Similarly, for a right-hand circularly polarised beam $\rho/\omega = -L_z$ (a plane polarised beam carries no angular momentum). These facts are consistent with the results of a remarkable experiment carried out in 1936 by R.A. Beth.

In Beth's experiment an anisotropic crystalline plate is prepared, which has the property that (at a certain wavelength) it converts left-hand circularly polarised light passing through it to right-hand circularly polarised light, acting as a half-wave plate. Because of this, the plate must be subjected to a couple of magnitude Γ_z per unit area, where

$$\Gamma_z = 2cL_z = 2\rho c/\omega \qquad \textbf{(4.133)}$$

The couple is measured by suspending the plate from a quartz fibre and measuring the angle through which the plate rotates. The effect can be doubled by reflecting the light which has passed through the plate, so it passes through the plate a second time. A fixed quarter-wave plate must be inserted, so that the polarisation of the light is in the correct sense to reinforce the couple (see Fig. 4.3). The angle of deflection is extremely small, but the constancy of the ratio ρ/ω can be observed, and a result of the expected order of magnitude obtained

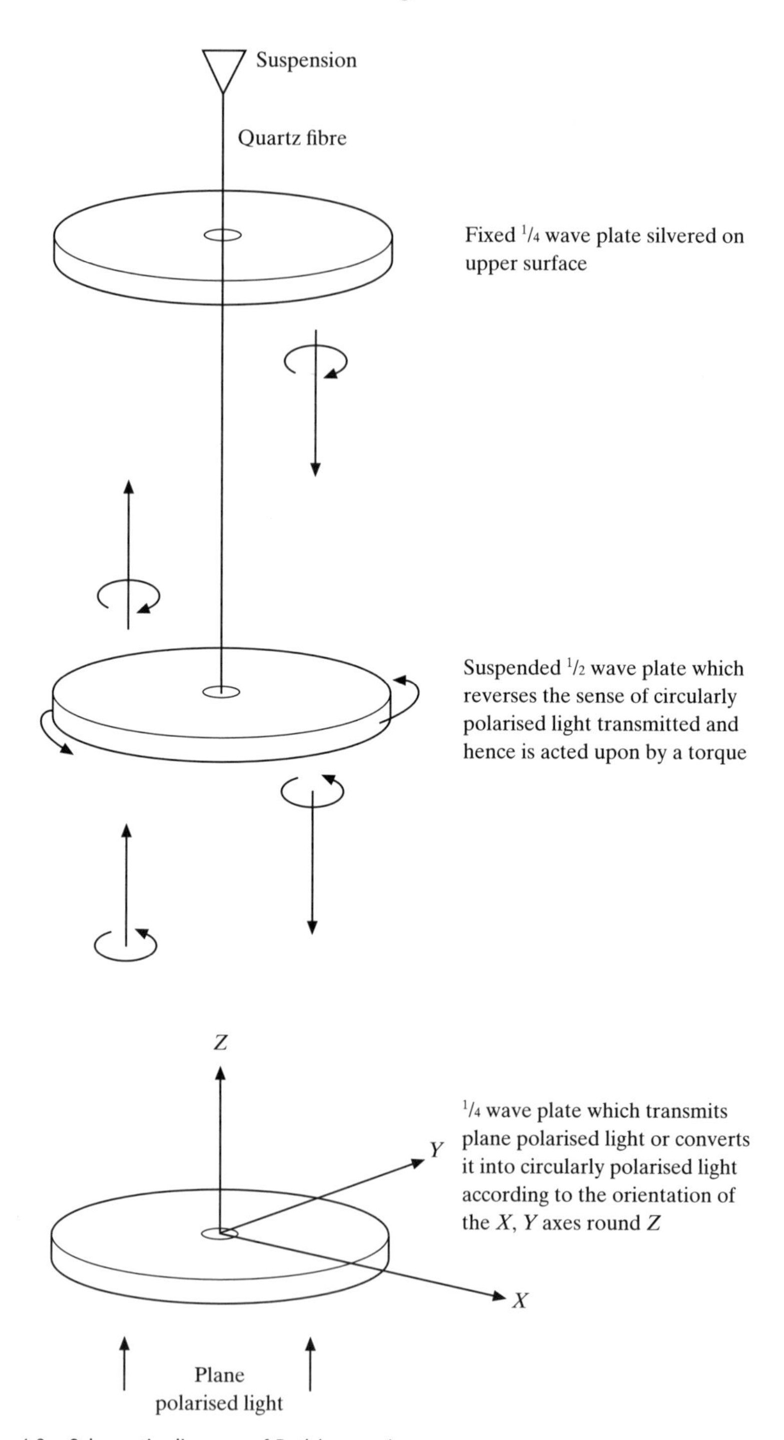

Figure 4.3 Schematic diagram of Beth's experiment.

provided the experiment is carried out with due precautions. For example, the whole apparatus must be in a vacuum to avoid the effect of currents in the air, and the power of the source of illumination must be known accurately.

Parity of the photon

Provided tiny effects due to the weak nuclear interactions are neglected, a system of electrons interacting with the electromagnetic field conserves parity [9]. From the behaviour of the vector potential **A** under reflections, it can be inferred that the photon has negative parity, which is consistent with the selection rule (4.115) showing that an electric dipole transition causes a change in parity of the atom.

The spectrum of one-electron atoms

In Chapter 3 it was shown that in the non-relativistic approximation, and neglecting spin–orbit coupling, the bound states of a one-electron atom were degenerate in the quantum numbers l and m and the energy of a level depended only on the principal quantum number n. That is (see (3.29) and (3.31)),

$$E_n = -\frac{1}{2n^2}\left(\frac{Ze^2}{4\pi\varepsilon_0}\right)^2 \frac{\mu}{\hbar^2}$$

$$= -\frac{Z^2}{2n^2}\left(\frac{\mu}{m}\right) \quad \text{in a.u.} \tag{4.134}$$

where μ is the reduced mass, given in terms of the mass of the nucleus M and the mass of the electron m by $\mu = mM/(m + M)$ (see (3.3)). Since there is no selection rule limiting n, the hydrogenic spectrum contains all frequencies given by the expression (3.34) which we recall here, namely

$$\nu_{ab} = Z^2 R(M)\left(\frac{1}{n_a^2} - \frac{1}{n_b^2}\right) \tag{4.135}$$

where $n_a = 1, 2, 3, \ldots, n_b = 2, 3, 4, \ldots$ with $n_b > n_a$ and $R(M)$ is given by (1.102). The gross structure of the spectra of one-electron atoms, described in Chapter 1 within the framework of the Bohr model, agrees with this formula. The foregoing discussion in this chapter has led to a consistent derivation of the result (4.135). However, it is important to note that the selection rules (4.117) and (4.119) limit the values of the quantum numbers m and l of the levels concerned. This is illustrated in Fig. 4.4 for the case of the orbital angular momentum quantum number l.

[9] Very small parity-violating terms in the electromagnetic interaction, involving the so-called 'neutral currents', were discovered in 1978 in the deep inelastic scattering of polarised electrons by protons and deuterons. These neutral currents, which are a consequence of a unified description of electromagnetic and weak interactions, imply very small parity-violating effects in atomic transitions.

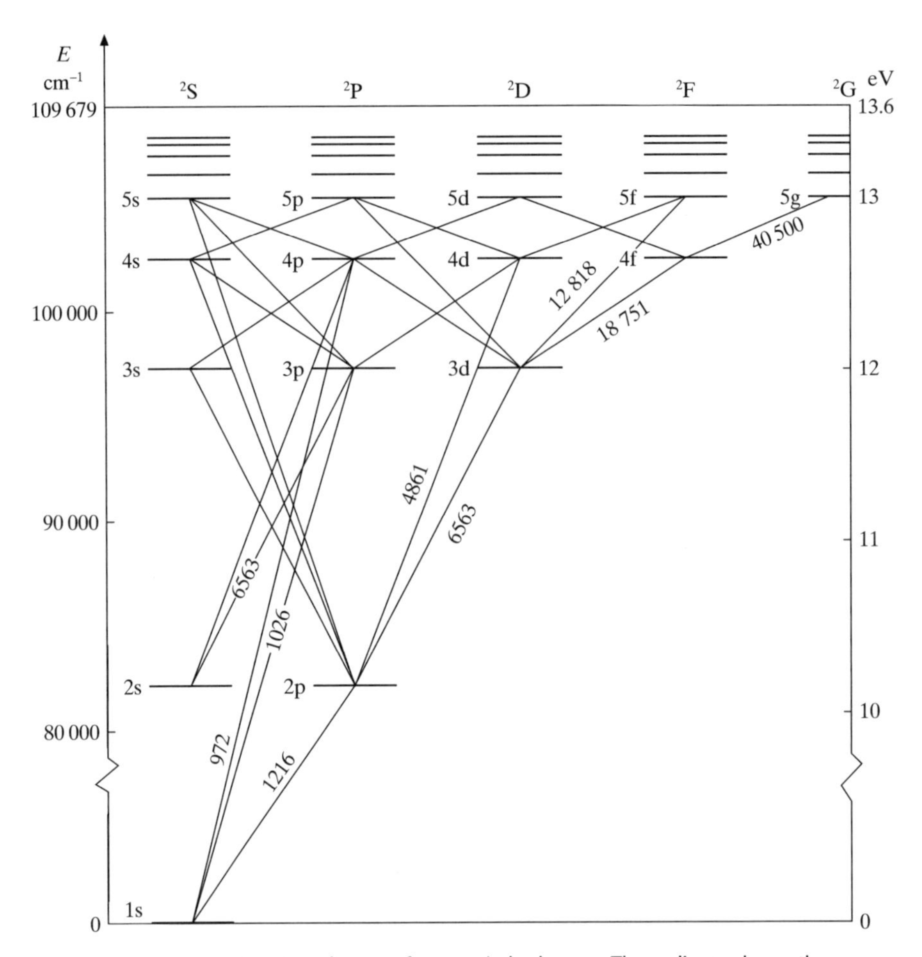

Figure 4.4 Term, or Grotrian, diagram for atomic hydrogen. The ordinate shows the energy above the 1s ground state in cm^{-1} (8065 cm^{-1} ≡ 1 eV) on the left and in eV on the right and the energy levels are shown plotted against the orbital angular momentum. Transitions obeying the $\Delta l = \pm 1$ selection rule are indicated by solid lines. The numbers against the lines indicate the wavelength in ångström units (1 Å ≡ 10^{-8} cm). For clarity, only transitions between the lower-lying levels are shown, and the wavelengths are shown only for a selection of lines. The splitting due to fine structure is too small to be shown on a diagram of this scale.

4.6 Line intensities and the lifetimes of excited states

As we have seen in Section 4.3, the intensity of a transition between a pair of states a and b is proportional, in the dipole approximation, to the quantity $|\mathbf{r}_{ba}|^2$. Thus the relative intensities of a series of transitions from a given initial state a to various final states k are determined by the quantities $|\mathbf{r}_{ka}|^2$.

Oscillator strengths and the Thomas–Reiche–Kuhn sum rule

In discussions of intensities it is customary to introduce a related dimensionless quantity f_{ka}, called the *oscillator strength*. It is defined as

$$f_{ka} = \frac{2m\omega_{ka}}{3\hbar} |\mathbf{r}_{ka}|^2 \tag{4.136}$$

with $\omega_{ka} = (E_k - E_a)/\hbar$. We note that this definition implies that $f_{ka} > 0$ for absorption, where $E_k > E_a$. On the other hand we have $f_{ka} < 0$ for emission processes.

The oscillator strengths (4.136) obey the sum rule, due to Thomas, Reiche and Kuhn,

$$\sum_k f_{ka} = 1 \tag{4.137}$$

where the sum is over all states, including the continuum. This sum rule can be proved as follows. Let f^x_{ka} be defined as

$$\begin{aligned} f^x_{ka} &= \frac{2m\omega_{ka}}{3\hbar} |x_{ka}|^2 \\ &= \frac{2m\omega_{ka}}{3\hbar} \langle a|x|k\rangle\langle k|x|a\rangle \end{aligned} \tag{4.138}$$

where we have used the simplified notation $\langle a|x|k\rangle \equiv \langle \psi_a|x|\psi_k\rangle$. From (4.67), we have

$$x_{ka} = \langle k|x|a\rangle = -\frac{\mathrm{i}}{m\omega_{ka}}\langle k|p_x|a\rangle \tag{4.139a}$$

$$x_{ak} = \langle a|x|k\rangle = \frac{\mathrm{i}}{m\omega_{ka}}\langle a|p_x|k\rangle \tag{4.139b}$$

and hence

$$f^x_{ka} = \frac{2\mathrm{i}}{3\hbar}\langle a|p_x|k\rangle\langle k|x|a\rangle \tag{4.140a}$$

$$= -\frac{2\mathrm{i}}{3\hbar}\langle a|x|k\rangle\langle k|p_x|a\rangle \tag{4.140b}$$

$$= \frac{\mathrm{i}}{3\hbar}\{\langle a|p_x|k\rangle\langle k|x|a\rangle - \langle a|x|k\rangle\langle k|p_x|a\rangle\} \tag{4.140c}$$

where the last line has been obtained by taking half the sum of the two expressions (4.140a) and (4.140b).

We can now use the closure property of the hydrogenic wave functions which form a complete set, namely $\sum_k |k\rangle\langle k| = I$ to find from (4.140c) that

$$\sum_k f^x_{ka} = \frac{\mathrm{i}}{3\hbar}\langle a|p_x x - x p_x|a\rangle \tag{4.141}$$

But since $[x, p_x] = i\hbar$, we have the sum rule

$$\sum_k f_{ka}^x = \frac{1}{3} \tag{4.142}$$

The same argument holds for f_{ka}^y and f_{ka}^z, which proves the sum rule (4.137).

The oscillator strengths and transition probabilities can be easily calculated for one-electron atoms and ions, because the hydrogenic wave functions are known exactly. The labels a and k in f_{ka} include all the quantum numbers of the initial and final states, and in particular f_{ka} depends on the magnetic quantum numbers. It is convenient to define an *average oscillator strength* for the transition $nl \to n'l'$, which is independent of the magnetic quantum numbers and hence of the polarisation of the radiation, by

$$\bar{f}_{n'l',nl} = \frac{1}{2l+1} \sum_{m'=-l'}^{l'} \sum_{m=-l}^{l} f_{n'l'm',nlm} \tag{4.143}$$

It can be shown (Problem 4.8) that the average oscillator strengths also obey the sum rule (4.137). Some calculated values of $\bar{f}_{n'l',nl}$ for hydrogenic atoms (or ions) are given in Table 4.1.

The transition rates for spontaneous emission in the dipole approximation are given in terms of oscillator strengths (see (4.80) and (4.136)) by

$$W_{ka}^{s,D} = \frac{2\hbar\alpha}{mc^2} \omega_{ka}^2 |f_{ka}| \tag{4.144}$$

For hydrogenic atoms the oscillator strengths and transition probabilities decrease as the principal quantum number n of the upper level increases, W_{ka}^s decreasing like n^{-3} for large n.

Atomic lifetimes

If $N(t)$ atoms are in an excited state b at a particular time t, the rate of change of $N(t)$ is

$$\dot{N}(t) = -N(t) \sum_k W_{kb}^s \tag{4.145}$$

Table 4.1 Average oscillator strengths for some transitions in hydrogenic atoms and ions†.

Initial level	*Final level*	Discrete spectrum $n = 1$	$n = 2$	$n = 3$	$n = 4$	$\sum_{n=5}^{\infty}$	*Conrinuum spectrum*
1s	*np*	–	0.416	0.079	0.029	0.041	0.435
2s	*np*	–	–	0.435	0.103	0.111	0.351
2p	*ns*	−0.139	–	0.014	0.003	0.003	0.008
2p	*nd*	–	–	0.696	0.122	0.109	0.183

† More complete tabulations can be found in Bethe and Salpeter (1957).

Table 4.2 Lifetime of some levels of atomic hydrogen (in 10^{-8} s).

Level	2p	3s	3p	3d	4s	4p	4d	4f
Lifetime	0.16	16	0.54	1.56	23	1.24	3.65	7.3

where W^{s}_{kb} is the transition rate for spontaneous emission and the sum is over all states k, of lower energy, to which decay is allowed by the selection rules. On integration, $N(t)$ can be expressed in terms of $N(t = 0)$ by

$$N(t) = N(t = 0) \exp(-t/\tau_b) \quad \textbf{(4.146)}$$

where τ_b is called the *lifetime* or *half-life* of the level b, and

$$\tau_b^{-1} = \sum_k W^{s}_{kb} \quad \textbf{(4.147)}$$

For example, using the result (4.100), the lifetime of the 2p level of a hydrogenic atom is seen to be $\tau = (0.16 \times 10^{-8}/Z^4)$ s.

In the absence of external fields, the lifetime of an atomic level cannot depend on the orientation of the atom, and hence cannot depend on the magnetic quantum number m of the level b. This property can also be verified by evaluating (4.147) explicitly in the dipole approximation, and remembering that the sum over k includes all the magnetic substates of the final levels k to which the atom can decay. The lifetimes τ of some of the lower levels of atomic hydrogen, calculated in the dipole approximation, are shown in Table 4.2. The corresponding lifetimes of hydrogenic ions, with nuclear charge Z, are shorter and are given by the scaling law (Problem 4.9)

$$\tau(Z) = Z^{-4}\tau(Z = 1) \quad \textbf{(4.148)}$$

In general the lifetime of a highly excited state is longer than that of a low-lying level. It is also interesting to note that the 2s level has an infinite lifetime in the dipole approximation. In fact the 2s level of atomic hydrogen has a lifetime of 1/7 s, the dominant decay process, 2s $\to$ 1s, occurring by the emission of two photons (that is, through higher order in the interaction between the atom and the electromagnetic field). The lifetime of 1/7 s is very long on the atomic time-scale, and the 2s level is said to be *metastable*.

4.7 Line shapes and widths

In the approximation used in Section 4.2 to calculate transition rates, we found that the angular frequency of the radiation emitted or absorbed between two atomic levels of energies E_a and E_b (with $E_b > E_a$) was exactly $\omega_{ba} = (E_b - E_a)/\hbar$, so that the spectral line was infinitely sharp. This cannot be completely accurate for the following reason. Every atomic level, except the ground state, decays with a finite lifetime τ. By the uncertainty principle, the energy of such a level cannot be precisely determined, but must be uncertain by an amount of order $\hbar/\tau$, which is called the *natural width* of the level. Therefore there is a finite probability that photons will be emitted with energies in an interval about $(E_b - E_a)$ of width $(\hbar/\tau_a + \hbar/\tau_b)$, where τ_a and τ_b are the lifetimes of the states a and b, respectively.

Let us consider for example the spontaneous decay of an excited state b of the atom to the state a which we choose to be the ground state. We return to the coupled equations (4.32) – with the perturbation H' given by (4.28) – and retain only those terms which contain the two atomic states a and b. The initial state of the system is characterised by an amplitude $c_b(t)$, while the final state consists of a photon of angular frequency ω, emitted in a direction (θ, ϕ) with a polarisation $\hat{\boldsymbol{\varepsilon}}_\lambda$, in addition to the ground state atom. The corresponding amplitude depends on ω, (θ, ϕ) and λ, but we will write it in shortened form as $c_a(\omega, t)$. When summing over the possible final states, we must make use of the density of states factor (4.59). Using (4.28), the expression (4.14) of $\mathbf{A}(\mathbf{r}, t)$ and remembering that M_{ba}^{λ} is given by (4.40), with $\hat{\boldsymbol{\varepsilon}}$ replaced by $\hat{\boldsymbol{\varepsilon}}_\lambda$, the equation (4.32) for $\dot{c}_b(t)$ can be written in explicit form as

$$\dot{c}_b(t) = -\frac{e}{m}\frac{V}{(2\pi)^3 c^3}\sum_{\lambda=1}^{2}\int \mathrm{d}\omega\,\omega^2 \int \mathrm{d}\Omega A_0(\omega) M_{ba}^{\lambda}(\omega)\exp[\mathrm{i}(\omega_{ba} - \omega)t + \mathrm{i}\delta_\omega]c_a(\omega, t) \tag{4.149}$$

where we have only retained the part of $\mathbf{A}(\mathbf{r}, t)$ that corresponds to the emission of a photon. Since a single photon is emitted $A_0(\omega)$ is found from (4.54) with $N(\omega) = 0$, namely

$$A_0(\omega) = \left(\frac{2\hbar}{V\varepsilon_0\omega}\right)^{1/2} \tag{4.150}$$

The equation for the time derivative of the amplitude $c_a(\omega, t)$ is again given by (4.32), with the same value of $A_0(\omega)$. We find that

$$\dot{c}_a(\omega, t) = -\frac{e}{m}A_0(\omega)\bar{M}_{ab}^{\lambda}(\omega)\exp[\mathrm{i}(\omega - \omega_{ba})t - \mathrm{i}\delta_\omega]\,c_b(t) \tag{4.151}$$

where (see (4.49))

$$\bar{M}_{ab}^{\lambda} = -M_{ba}^{\lambda *} \tag{4.152}$$

Since there is only a single amplitude $c_b(t)$, there is no sum over states on the right-hand side of (4.151).

In our previous treatment we solved the coupled equations by making the approximation $c_b(t) = 1$ on the right-hand side of (4.151), but we now allow for the decay of the upper level by writing

$$\begin{aligned} c_b(t) &= 1, & t &< 0 \\ c_b(t) &= \exp[-t/(2\tau_b)], & t &\geqslant 0 \end{aligned} \tag{4.153}$$

so that for $t \geqslant 0$ the component of the total wave function which describes the initial state b can be expressed as

$$\Psi_b(\mathbf{r}, t) = c_b(t)\psi_b(\mathbf{r}) \exp(-\mathrm{i}E_b t/\hbar) = \psi_b(\mathbf{r}) \exp\{-\mathrm{i}[E_b - \mathrm{i}\hbar/(2\tau_b)]t/\hbar\} \tag{4.154}$$

where ψ_b is a time-independent atomic wave function satisfying (4.30). For later convenience, since we are not interested in $\Psi_b(\mathbf{r}, t)$ for $t < 0$, we set $\Psi_b = 0$ for $t < 0$.

In the absence of any coupling to the radiation field, an excited atomic state b would be stable and the wave function would be $\Phi_b(\mathbf{r}, t) = \psi_b(\mathbf{r}) \exp(-\mathrm{i}E_b t/\hbar)$. This is a stationary state which is an eigenstate of the energy operator since

$$\mathrm{i}\hbar \frac{\partial}{\partial t} \Phi_b(\mathbf{r}, t) = E_b \Phi_b(\mathbf{r}, t) \tag{4.155}$$

and the system possesses a well-defined real energy E_b. In contrast, when the coupling to the radiation field is taken into account, the time variation of the initial wave function is given by (4.154) and

$$\mathrm{i}\hbar \frac{\partial}{\partial t} \Psi_b(\mathbf{r}, t) = \left(E_b - \mathrm{i}\frac{\hbar}{2\tau_b} \right) \Psi_b(\mathbf{r}, t) \tag{4.156}$$

showing that $\Psi_b(\mathbf{r}, t)$ does not describe a state with a well-defined *real* energy. This is a general result: a decaying state is never a state with a definite real energy.

Inserting $c_b(t)$ given by (4.153) into (4.151), and integrating over t, we find that

$$\begin{aligned} c_a(\omega, t) &= -\frac{e}{m} A_0(\omega) \bar{M}^{\lambda}_{ab}(\omega)\, \mathrm{e}^{-\mathrm{i}\delta_\omega} \int_0^t \exp[\mathrm{i}(\omega - \omega_{ba})t' - t'/(2\tau_b)]\, \mathrm{d}t' \\ &= -\frac{e}{m} A_0(\omega) \bar{M}^{\lambda}_{ab}(\omega)\, \mathrm{e}^{-\mathrm{i}\delta_\omega} \frac{\exp[\mathrm{i}(\omega - \omega_{ba})t - t/(2\tau_b)] - 1}{\mathrm{i}(\omega - \omega_{ba}) - 1/(2\tau_b)} \end{aligned} \tag{4.157}$$

At times $t \gg \tau_b$ the probability that a photon has been emitted is given by $|c_a(\omega, t)|^2$, which is proportional to

$$\left| \frac{1}{\mathrm{i}(\omega - \omega_{ba}) - 1/(2\tau_b)} \right|^2 = \frac{1}{(\omega - \omega_{ba})^2 + 1/(4\tau_b^2)} \tag{4.158}$$

The intensity of the emitted radiation therefore reaches a maximum when $\omega = \omega_{ba} = (E_b - E_a)/\hbar$, and decreases to one-half of the maximum value at

$$\begin{aligned} \omega &= \omega_{ba} \pm 1/(2\tau_b) \\ &= (E_b - E_a \pm \Gamma_b/2)/\hbar \end{aligned} \tag{4.159}$$

where

$$\Gamma_b = \frac{\hbar}{\tau_b} \tag{4.160}$$

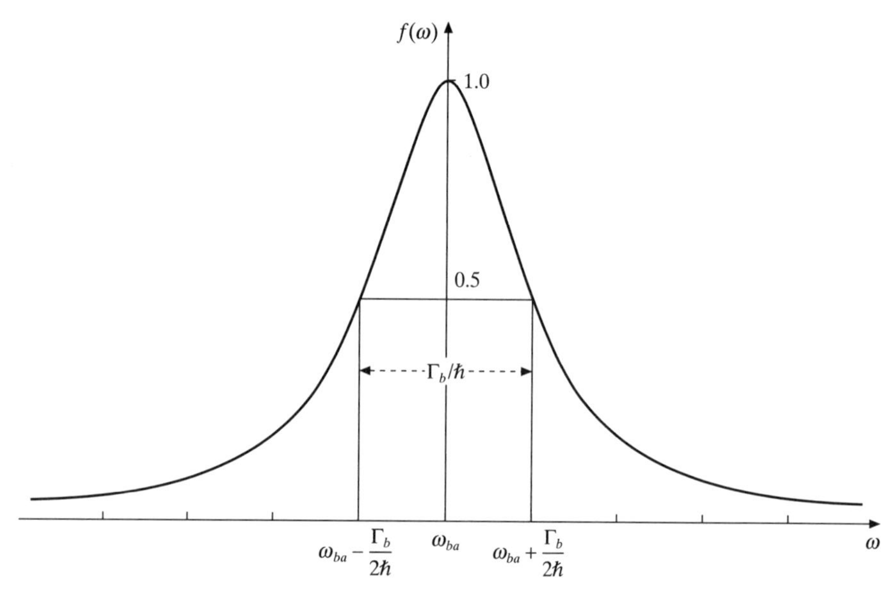

Figure 4.5 A plot of the Lorentzian intensity distribution.

is the *natural width* of the line. The intensity distribution given by (4.158) is said to be *Lorentzian* in shape. It is proportional to the function

$$f(\omega) = \frac{\Gamma_b^2/(4\hbar^2)}{(\omega - \omega_{ba})^2 + \Gamma_b^2/(4\hbar^2)} \tag{4.161}$$

which is plotted in Fig. 4.5.

To calculate the lifetime τ_b, we insert the expression (4.157) for $c_a(\omega, t)$ into (4.149). Using (4.150) for $A_0(\omega)$ we find that

$$\dot{c}_b(t) = -\left(\frac{e^2}{4\pi\varepsilon_0}\right)\frac{\hbar}{4\pi^2 m^2 c^3} \times \sum_{\lambda=1}^{2}\int d\Omega \int d\omega\, \omega\, |M_{ba}^{\lambda}(\omega)|^2 \left\{\frac{\exp[-t/(2\tau_b)] - \exp[i(\omega_{ba} - \omega)t]}{i(\omega - \omega_{ba}) - 1/(2\tau_b)}\right\} \tag{4.162}$$

The function in curly brackets is sharply peaked about $\omega = \omega_{ba}$. The term $\omega|M_{ba}^{\lambda}(\omega)|^2$ is slowly varying, and so can be evaluated at $\omega = \omega_{ba}$. The range of integration can then be taken from $\omega_{ba} - \eta$ to $\omega_{ba} + \eta$, with $\eta \to +\infty$. Using [10]

$$\lim_{\eta\to+\infty}\int_{-\eta}^{+\eta}\frac{1}{x - \alpha + i\beta}\,dx = -i\pi \tag{4.163a}$$

[10] The second integral can be evaluated by using contour integration.

$$\int_{-\infty}^{+\infty} \frac{\exp(-ixt)}{x - \alpha + i\beta} dx = -2\pi i \exp[-i(\alpha - i\beta)t] \tag{4.163b}$$

with $x = \omega$, $\alpha = \omega_{ba}$ and $\beta = 1/(2\tau_b)$, we find that

$$\dot{c}_b(t) = -\left(\frac{e^2}{4\pi\varepsilon_0}\right) \frac{\hbar}{4\pi m^2 c^3} \int d\Omega \sum_{\lambda=1}^{2} \omega_{ba} |M_{ba}^{\lambda}(\omega_{ba})|^2 \exp[-t/(2\tau_b)] \tag{4.164}$$

Now, from (4.153), we see that

$$\dot{c}_b(t) = -\frac{1}{2\tau_b} \exp[-t/(2\tau_b)], \quad t \geqslant 0 \tag{4.165}$$

Comparing (4.164) with (4.165), and remembering that the transition rate is given by $W_{ab}^{s} = \tau_b^{-1}$, we have

$$W_{ab}^{s} = \frac{1}{\tau_b} = \frac{\hbar}{2\pi m^2 c^3} \left(\frac{e^2}{4\pi\varepsilon_0}\right) \int d\Omega \sum_{\lambda=1}^{2} \omega_{ba} |M_{ba}^{\lambda}(\omega_{ba})|^2 \tag{4.166}$$

which agrees with our previous result (4.61).

In the dipole approximation, after integration over the angles and summing over the two directions of polarisation, the expressions (4.80) for $W_{ab}^{s,D}$ are regained. In particular, we have that

$$W_{ab}^{s,D} = \frac{1}{\tau_b} = \frac{\Gamma_b}{\hbar} = \frac{4}{3\hbar c^3} \left(\frac{1}{4\pi\varepsilon_0}\right) \omega_{ba}^3 |\mathbf{D}_{ba}|^2 \tag{4.167}$$

For a one-electron atom in which the quantum numbers of the initial (decaying) and final states are (nlm) and $(n'l'm')$, respectively, the total transition rate from the level b to the level a is given by summing over the final and averaging over the initial magnetic substates. We have that

$$W_{ab}^{s,D} = \frac{1}{\tau_b} = \frac{4}{3\hbar c^3} \left(\frac{1}{4\pi\varepsilon_0}\right) \omega_{ba}^3 \frac{1}{g_b} \sum_{m'=-l'}^{l'} \sum_{m=-l}^{l} |\langle nlm|\mathbf{D}|n'l'm'\rangle|^2 \tag{4.168}$$

where $g_b = (2l + 1)$ is the degeneracy of the initial state b.

The distribution in angular frequency $a(\omega)$ of the spectral lines corresponding to absorption and stimulated emission can be shown, in the same way, to be proportional to the expression (4.161). To determine the constant of proportionality we note that for a line of zero width the distribution in ω is the delta function $\delta(\omega - \omega_{ba})$. Using the result

$$\lim_{\varepsilon \to 0^+} \frac{\varepsilon}{\pi(x^2 + \varepsilon^2)} = \delta(x) \tag{4.169}$$

we obtain, setting $\varepsilon = \Gamma_b/(2\hbar)$ and $x = \omega - \omega_{ba}$,

$$a(\omega) = \frac{1}{\pi} \frac{\Gamma_b/(2\hbar)}{(\omega - \omega_{ba})^2 + \Gamma_b^2/(4\hbar^2)} \tag{4.170}$$

In the limit of zero line width, an absorption cross-section $\sigma_{ba}(\omega)$ can be defined as

$$\sigma_{ba}(\omega) = \frac{\hbar\omega_{ba}}{I(\omega_{ba})} W_{ba}\ \delta(\omega - \omega_{ba}) \tag{4.171}$$

where W_{ba} is the integrated transition rate for absorption. Using the expression (4.45) for W_{ba}, we find that, for linearly polarised radiation,

$$\sigma_{ba}(\omega) = \frac{4\pi^2\alpha\hbar^2}{m^2\omega_{ba}} |M_{ba}(\omega_{ba})|^2\delta(\omega - \omega_{ba}) \tag{4.172}$$

To allow for the finite width of the line, we replace $\delta(\omega - \omega_{ba})$ in the above expression by the spectral distribution $a(\omega)$ given by (4.170), so that

$$\sigma_{ba}(\omega) = \frac{4\pi^2\alpha\hbar^2}{m^2\omega_{ba}} |M_{ba}(\omega_{ba})|^2 a(\omega) \tag{4.173}$$

Since the natural width Γ_b is extremely small compared to $(E_b - E_a)$, we have, to a high degree of accuracy,

$$\int_0^\infty a(\omega)\ \mathrm{d}\omega \simeq 1 \tag{4.174}$$

The integrated cross-section for absorption can be defined as

$$\sigma_{ba} - \int_0^\infty \sigma_{ba}(\omega)\ \mathrm{d}\omega \tag{4.175}$$

Using (4.173) and (4.174), we find that

$$\sigma_{ba} = \frac{4\pi^2\alpha\hbar^2}{m^2\omega_{ba}} |M_{ba}(\omega_{ba})|^2 \tag{4.176}$$

This result, which also follows by substituting the zero line width absorption cross-section (4.172) in (4.175), is in agreement with the expression (4.46) obtained in Section 4.2. It is worth noting that the absorption cross-section $\sigma_{ba}(\omega)$ has the dimensions of an area, whereas the integrated cross-section σ_{ba} has the dimensions of area divided by time, as we saw earlier.

In the dipole approximation, the absorption cross-section (4.173) for linearly polarised radiation becomes, using (4.74),

$$\sigma_{ba}^{\mathrm{D}}(\omega) = \frac{4\pi^2}{c\hbar}\left(\frac{1}{4\pi\varepsilon_0}\right)\omega_{ba}\,|\hat{\boldsymbol{\varepsilon}}\cdot\mathbf{D}_{ba}|^2 a(\omega) \tag{4.177}$$

For unpolarised radiation, we have (see (4.78))

$$\sigma_{ba}^{\mathrm{D}}(\omega) = \frac{4\pi^2}{3c\hbar}\left(\frac{1}{4\pi\varepsilon_0}\right)\omega_{ba}|\mathbf{D}_{ba}|^2 a(\omega) \tag{4.178}$$

Comparing this result with (4.167), we see that $\sigma_{ba}^{\rm D}(\omega)$ can be written in the form

$$\sigma_{ba}^{\rm D}(\omega) = \left(\frac{c\pi}{\omega_{ba}}\right)^2 \frac{\Gamma_b}{\hbar} a(\omega) \tag{4.179}$$

If the ground state level E_a is non-degenerate, and the excited level E_b is degenerate, with degeneracy g_b, the generalisation of (4.179) is

$$\sigma_{ba}^{\rm D}(\omega) = g_b \left(\frac{c\pi}{\omega_{ba}}\right)^2 \frac{\Gamma_b}{\hbar} a(\omega) \tag{4.180}$$

The maximum of this cross-section occurs when $\omega = \omega_{ba}$, and from (4.170) is given by

$$\sigma_{ba}^{\rm D,max} = 2\pi g_b \left(\frac{c}{\omega_{ba}}\right)^2 \tag{4.181}$$

This interesting result which is independent of the value of the matrix element $\mathbf{D}_{ba}$ will be used in Section 4.9 to make a comparison with the magnitude of the cross-section for scattering of radiation.

As the uncertainty principle suggests, a more complete treatment demonstrates that the natural width of a line from one excited level b to another excited level a is given by

$$\Gamma = \hbar\left(\frac{1}{\tau_a} + \frac{1}{\tau_b}\right) \tag{4.182}$$

where τ_a and τ_b are the lifetimes of each of the levels, taking into account all the possible ways in which the levels can decay. The line intensity is proportional to the function $f(\omega)$ given by (4.161), with Γ_b replaced by Γ.

The natural width of atomic energy levels is very small. For example, the width of the 2p level of atomic hydrogen, which has an unperturbed energy of -3.40 eV and a lifetime of 0.16×10^{-8} s, is $\Gamma = 4.11 \times 10^{-7}$ eV. The profile of a spectral line can be measured, either by recording the spectrum on a photographic plate and subsequently measuring the density of the image as a function of wavelength, or by scanning the spectrum with a photoelectric detector. It is found, after allowing for the finite resolving power of the spectrograph employed, that observed spectral lines usually have much greater widths than the natural width. The reasons for this will now be examined.

Pressure broadening

In deriving the exponential law (4.146), we assumed that the transition rate between the atomic state b of a higher energy E_b and the states k of lower energy was entirely due to spontaneous emission. However, the population of the state b must also decrease if there are other mechanisms which lead to transitions out of b. Thus in (4.147) each of the spontaneous transition rates $W_{kb}^{\rm s}$ should be replaced by the sum of the transition rates for all processes depleting the level b. If W_{tot} is this sum, then the lifetime of the state b is now reduced to

$$\tilde{\tau}_b = \frac{1}{W_{\text{tot}}} \tag{4.183}$$

Similarly, if the level a is unstable, and there are other mechanisms than spontaneous emission leading out of a, the natural lifetime τ_a of this level is reduced to $\tilde{\tau}_a$.

The observed width $\tilde{\Gamma}$ of a spectral line from b to a is now given by

$$\tilde{\Gamma} = \hbar\left(\frac{1}{\tilde{\tau}_a} + \frac{1}{\tilde{\tau}_b}\right) \tag{4.184}$$

and the line intensity is proportional to the Lorentzian function $f(\omega)$ given by (4.161), with Γ_b replaced by $\tilde{\Gamma}$. The principal mechanism of this type, broadening lines of radiation from atoms in a gas, arises from *collisions* between the atoms. In each collision there is a certain probability that an atom initially in a state b will make a radiationless transition to some other state. The corresponding transition rate W_c is proportional to the number density of the atoms concerned, n, and to the relative velocity between pairs of atoms, v, so that

$$W_c = nv\sigma \tag{4.185}$$

where σ is a quantity with dimensions of area called the *collision cross-section*. This cross-section depends on the species of atom and on v. Since n depends on the pressure, the broadening of a spectral line due to this cause is called *pressure broadening* or, alternatively, *collisional broadening*. Both n and v also depend on the temperature, so that information about the temperature and the pressure of a gas can be obtained by measuring the profiles of spectral lines. This is a major source of knowledge about physical conditions in stellar atmospheres, which in turn provides most of our evidence about stellar structure.

When the observed width $\tilde{\Gamma}_b = \hbar/\tilde{\tau}_b$ of a spectral line between an excited state b and the ground state a is greater than the natural width Γ_b, the absorption cross-section $\sigma_{ba}(\omega)$ is still given by the expression (4.180), provided $a(\omega)$ represents the observed line profile. When the line profile is Lorentzian as in the case of pressure broadening, the maximum of the absorption cross-section (see (4.181)) becomes

$$\sigma_{ba}^{\text{D, max}} = 2\pi g_b \left(\frac{c}{\omega_{ba}}\right)^2 \frac{\Gamma_b}{\tilde{\Gamma}_b} \tag{4.186}$$

where the initial state a is again taken to be non-degenerate.

Doppler broadening

The wavelength of the light emitted by a moving source is shifted by the Doppler effect. If the emitting source is moving at a non-relativistic velocity, and v is the component of the velocity of the source along the line of sight, the wavelength λ of the emitted light is, to first order in v/c,

$$\lambda = \lambda_0 \left(1 \pm \frac{v}{c}\right) \tag{4.187}$$

where λ_0 is the wavelength emitted by a stationary source. The plus sign corresponds to a source receding from the observer, and the minus sign to an approaching source. To first order in v/c, the angular frequency $\omega = 2\pi c/\lambda$ of the light emitted by a moving source is therefore related to the angular frequency $\omega_0 = 2\pi c/\lambda_0$ emitted by a stationary source as

$$\omega = \omega_0 \left(1 \mp \frac{v}{c}\right) \tag{4.188}$$

The formulae (4.187) and (4.188) describe the *first-order* Doppler effect [11].

If the light is emitted from a gas at absolute temperature T, the number of atoms, $\mathrm{d}N$, with velocities between v and $v + \mathrm{d}v$ is given by Maxwell's distribution

$$\mathrm{d}N = N_0 \exp[-Mv^2/(2k_B T)]\, \mathrm{d}v \tag{4.189}$$

where k_B is Boltzmann's constant, M is the atomic mass and N_0 is a constant. The intensity $\mathcal{I}(\omega)$ of light emitted in an angular frequency interval ω to $\omega + \mathrm{d}\omega$ is proportional to the number of atoms with velocities between v and $v + \mathrm{d}v$. Hence, using (4.188) and (4.189), we obtain the *Gaussian distribution*

$$\mathcal{I}(\omega) = \mathcal{I}(\omega_0) \exp\left[-\frac{Mc^2}{2k_B T}\left(\frac{\omega - \omega_0}{\omega_0}\right)^2\right] \tag{4.190}$$

If ω_1 is the angular frequency at half-maximum, then

$$(\omega_1 - \omega_0)^2 = \frac{2k_B T}{Mc^2}\omega_0^2 \log 2 \tag{4.191}$$

---

[11] The formulae (4.187) and (4.188) are only accurate to first order in v/c. The *relativistic* Doppler formula replacing (4.188) is

$$\omega = \left(\frac{1 \mp v/c}{1 \pm v/c}\right)^{1/2} \omega_0$$

Expanding the square root as a power series in v/c, we find that

$$\omega - \omega_0 = \mp\frac{v}{c}\omega_0 + \frac{1}{2}\frac{v^2}{c^2}\omega_0 + \dots$$

The first term on the right, which involves v/c to the first power, yields the *linear Doppler effect.* The second term, involving $(v/c)^2$, yields a *second-order Doppler* shift which does not change if the sign of the velocity is changed, that is, whether the source and observer are approaching or receding from each other.

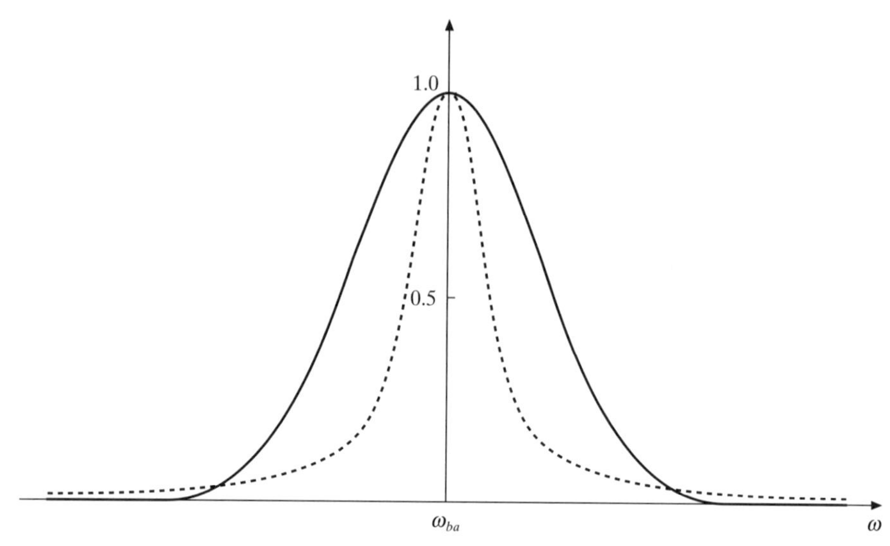

Figure 4.6 A comparison of a Gaussian distribution (solid line) of the form

$$F(\omega) = \exp[-\alpha(\omega - \omega_{ba})^2]$$

with a Lorentzian distribution (dashed line) of the form

$$f(\omega) = \frac{\Gamma^2/(4\hbar^2)}{(\omega - \omega_{ba})^2 + \Gamma^2/(4\hbar^2)}$$

The total Doppler width at half-maximum $\Delta\omega^{\mathrm{D}}$ is $2|\omega_1 - \omega_0|$ and hence

$$\Delta\omega^{\mathrm{D}} = \frac{2\omega_0}{c}\left[\frac{2k_{\mathrm{B}}T}{M}\log 2\right]^{1/2} \quad \textbf{(4.192)}$$

This width increases with temperature and with the frequency of the line, and decreases as the atomic mass increases. Several spectroscopic methods have been developed in which Doppler broadening is reduced or completely eliminated. One of these methods, *level crossing spectroscopy*, will be discussed in Chapter 9 where we consider the effect of external magnetic fields on atomic energy levels. Another method, *saturation spectroscopy*, will be described in Section 15.2, devoted to laser spectroscopy.

While pressure broadening increases the width of a spectral line, but preserves the Lorentzian profile, the Gaussian profile produced by Doppler broadening is quite different. The two shapes are compared in Fig. 4.6. In general, both pressure and Doppler broadening are present, and the observed profile, called a *Voigt profile*, is due to a combination of both effects. The decrease of the Gaussian distribution away from ω_0 ($= \omega_{ba}$) is so rapid that the 'wings' of spectral lines are determined by the residual Lorentzian distribution. Thus, if both Doppler and pressure broadening are present, the characteristics of each effect can be distinguished provided sufficiently accurate experimental profiles can be obtained.

4.8 The photoelectric effect

If electromagnetic radiation of sufficiently high frequency is absorbed by an atomic system A the final state may lie in the continuum and one or several electrons will be ejected from A. This is known as *photoionisation* and is the process responsible for the photoelectric effect (see Section 1.4).

In this section we shall obtain the cross-section for a particular photoionisation process, in which the electron is ejected from a hydrogenic atom (ion). We assume that this atom (ion) is initially in the ground state (1s), described by the wave function $\psi_a(\mathbf{r}) \equiv \psi_{1s}(r)$ and having the energy E_{1s}. We denote by $E = h\nu = \hbar\omega$ the energy of the absorbed photon, and by $\mathbf{k}$ its wave vector. Let $\mathbf{k}_f$ be the wave vector of the electron in the final state and $\mathbf{p}_f = \hbar\mathbf{k}_f$ its momentum. Assuming that the ejected electron is non-relativistic, its kinetic energy in the final state is given by

$$E_f = \frac{\hbar^2 k_f^2}{2m} \tag{4.193}$$

with $E_f \ll mc^2$.

The final state $\psi_b(\mathbf{k}_f, \mathbf{r})$ represents a continuum state corresponding to an electron with a wave vector $\mathbf{k}_f$ and an energy E_f moving in the Coulomb field of a nucleus of charge Ze, which we assume to be infinitely heavy. Thus $\psi_b(\mathbf{k}_f, \mathbf{r})$ is a positive energy Coulomb wave function [12] satisfying the equation

$$\left(-\frac{\hbar^2}{2m}\nabla^2 - \frac{Ze^2}{(4\pi\varepsilon_0)r} - E_f\right)\psi_b(\mathbf{k}_f, \mathbf{r}) = 0 \tag{4.194}$$

For sufficiently high energies of the ejected electron (that is, when $E_f \gg |E_{1s}|$), and when the nuclear charge Ze is relatively small [13] ($Z\alpha \ll 1$), the interaction of the outgoing electron with the nucleus can be neglected in the first approximation [14] and $\psi_b(\mathbf{k}_f, \mathbf{r})$ can be represented by a plane wave

$$\psi_b(\mathbf{k}_f, \mathbf{r}) = (2\pi)^{-3/2} \exp(\mathrm{i}\mathbf{k}_f \cdot \mathbf{r}) \tag{4.195}$$

where we have chosen the Dirac delta function normalisation (2.35).

The photoelectric total cross-section can be obtained by using (4.172) and integrating over all final states of the ejected electron. That is,

$$\sigma_{\text{tot}} = \frac{4\pi^2\alpha\hbar^2}{m^2} \int \mathrm{d}\mathbf{k}_f \, \frac{1}{\omega_{ba}} |M_{ba}(\omega_{ba})|^2 \, \delta(\omega - \omega_{ba}) \tag{4.196}$$

where $\omega_{ba} = (E_f - E_{1s})/\hbar$. The delta function in (4.196) ensures energy conservation:

$$E_f = \hbar\omega + E_{1s} \tag{4.197}$$

[12] Positive energy Coulomb wave functions will be discussed in Section 12.5.

[13] If $Z\alpha \ll 1$, the bound state of the electron can be treated non-relativistically (see (1.103)).

[14] A more detailed treatment of the photoionisation of one-electron atoms, including the Coulomb interaction of the ejected electron with the nucleus, will be given in Section 13.5.

Let us write

$$\mathrm{d}\mathbf{k}_f = k_f^2 \,\mathrm{d}k_f \,\mathrm{d}\Omega \tag{4.198}$$

where $\mathrm{d}\Omega \equiv \sin\theta \,\mathrm{d}\theta \,\mathrm{d}\phi$ and (θ, ϕ) are the polar angles of $\mathbf{k}_f$. Using (4.193), we also have

$$\mathrm{d}\mathbf{k}_f = \frac{k_f m}{\hbar^2} \,\mathrm{d}E_f \,\mathrm{d}\Omega \tag{4.199}$$

so that we may rewrite (4.196) in the form

$$\sigma_{\text{tot}} = \frac{4\pi^2\alpha}{m} \int_0^\infty \mathrm{d}E_f \int \mathrm{d}\Omega \frac{k_f}{\omega_{ba}} |M_{ba}(\omega_{ba})|^2 \,\delta\left(\omega - \frac{E_f - E_{1\text{s}}}{\hbar}\right) \tag{4.200}$$

Performing the integration over E_f with the help of the delta function, we find that

$$\sigma_{\text{tot}} = \frac{4\pi^2\alpha\hbar}{m}\left(\frac{k_f}{\omega}\right) \int |M_{ba}(\omega)|^2 \,\mathrm{d}\Omega \tag{4.201}$$

The differential cross-section for an electron to be ejected within the solid angle $\mathrm{d}\Omega$ in the direction (θ, ϕ) is therefore

$$\frac{\mathrm{d}\sigma}{\mathrm{d}\Omega} = \frac{4\pi^2\alpha\hbar}{m}\left(\frac{k_f}{m}\right) |M_{ba}(\omega)|^2 \tag{4.202}$$

From (4.40) and (4.195) the matrix element M_{ba} is given by

$$M_{ba} = (2\pi)^{-3/2} \int \exp(-\mathrm{i}\mathbf{k}_f \cdot \mathbf{r}) \exp(\mathrm{i}\mathbf{k} \cdot \mathbf{r}) \,\hat{\boldsymbol{\varepsilon}} \cdot \nabla \psi_{1\text{s}}(r) \,\mathrm{d}\mathbf{r} \tag{4.203}$$

Upon integration by parts, we find that

$$M_{ba} = (2\pi)^{-3/2}[-\mathrm{i}\hat{\boldsymbol{\varepsilon}} \cdot (\mathbf{k} - \mathbf{k}_f)] \int \exp[\mathrm{i}(\mathbf{k} - \mathbf{k}_f) \cdot \mathbf{r}] \,\psi_{1\text{s}}(r) \,\mathrm{d}\mathbf{r} \tag{4.204}$$

Since $\mathbf{k} \cdot \hat{\boldsymbol{\varepsilon}} = 0$, we have

$$\hat{\boldsymbol{\varepsilon}} \cdot (\mathbf{k} - \mathbf{k}_f) = -k_f \cos\gamma \tag{4.205}$$

where γ is the angle between the direction of ejection and the direction of polarisation (see Fig. 4.7). The integral appearing in (4.204) is proportional to the Fourier transform of the ground state wave function $\psi_{1\text{s}}(r)$, namely

$$\int \exp(\mathrm{i}\mathbf{K} \cdot \mathbf{r}) \,\psi_{1\text{s}}(r) \,\mathrm{d}\mathbf{r} = 8\pi \left(\frac{Z^3}{\pi a_0^3}\right)^{1/2} \frac{Z/a_0}{[(Z/a_0)^2 + K^2]^2} \tag{4.206}$$

where we have introduced the vector

$$\mathbf{K} = \mathbf{k} - \mathbf{k}_f \tag{4.207}$$

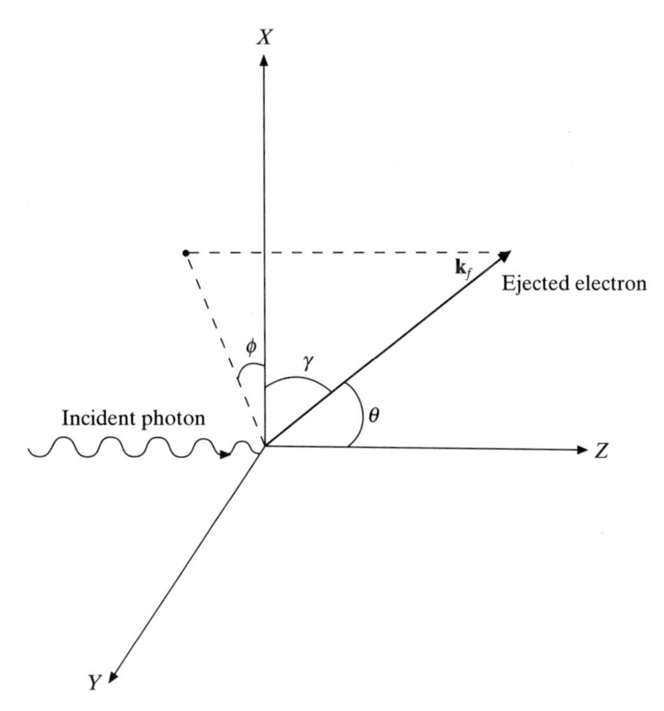

Figure 4.7 The angles employed in the discussion of the photoelectric effect. The incident radiation is in the Z direction, with polarisation vector in the X direction. The momentum of the ejected electron is $\hbar\mathbf{k}_f$.

From (4.202) and (4.204)–(4.207), the differential cross-section is seen to be given by

$$\frac{\mathrm{d}\sigma}{\mathrm{d}\Omega} = 32\alpha\left(\frac{\hbar}{m}\right)\left(\frac{k_f^3}{\omega}\right)\frac{Z^5 a_0^3 \cos^2\gamma}{[Z^2 + K^2 a_0^2]^4} \quad \textbf{(4.208)}$$

Without loss of generality we can take the direction of propagation $\hat{\mathbf{k}}$ of the radiation to be along the Z axis and the direction of the polarisation vector $\hat{\boldsymbol{\varepsilon}}$ to be the X direction. Then, as $\mathbf{k}_f$ is in the direction $\Omega \equiv (\theta, \phi)$, we have (see Fig. 4.7)

$$\cos\gamma = \sin\theta\cos\phi \quad \textbf{(4.209)}$$

and

$$K^2 = k^2 + k_f^2 - 2k\,k_f\cos\theta \quad \textbf{(4.210)}$$

At incident photon energies well in excess of the ionisation threshold ($\simeq$13.6 eV for atomic hydrogen), when $\hbar\omega \gg |E_{1s}|$, we have from (4.193) and (4.197)

$$\hbar\omega \simeq \frac{\hbar^2 k_f^2}{2m} \quad \textbf{(4.211)}$$

and since $\omega = kc$ it follows that

$$\frac{k}{k_f} \simeq \frac{\hbar k_f}{2mc} = \frac{v_f}{2c} \qquad \textbf{(4.212)}$$

where v_f is the velocity of the ejected electron. In the non-relativistic regime, for which $v_f/c \ll 1$, we can use this result to write

$$K^2 \simeq k_f^2\left(1 - \frac{v_f}{c}\cos\theta\right) \qquad \textbf{(4.213)}$$

Moreover, since $\hbar\omega \simeq \hbar^2 k_f^2/(2m) \gg |E_{1s}|$ and $E_{1s} = -Z^2[e^2/(4\pi\varepsilon_0)]/(2a_0)$, we have

$$k_f^2 a_0^2 \gg \frac{2m}{\hbar^2}|E_{1s}|a_0^2 = Z^2 \qquad \textbf{(4.214)}$$

so that

$$Z^2 + K^2 a_0^2 \simeq k_f^2 a_0^2\left(1 - \frac{v_f}{c}\cos\theta\right) \qquad \textbf{(4.215)}$$

and the differential cross-section (4.208) becomes

$$\frac{d\sigma}{d\Omega} = 32\alpha\left(\frac{\hbar}{m}\right)\frac{Z^5}{\omega(k_f a_0)^5}\frac{\sin^2\theta\cos^2\phi}{\left(1 - \frac{v_f}{c}\cos\theta\right)^4} \qquad \textbf{(4.216)}$$

We note that the ejected electrons have a cosine-squared distribution with respect to the polarisation vector $\hat{\boldsymbol{\varepsilon}}$ of the incident radiation. For an *unpolarised* photon beam an average must be made over the polarisations of the photon, so that we have in that case

$$\frac{d\sigma}{d\Omega} = 16\alpha\left(\frac{\hbar}{m}\right)\frac{Z^5}{\omega(k_f a_0)^5}\frac{\sin^2\theta}{\left(1 - \frac{v_f}{c}\cos\theta\right)^4} \qquad \textbf{(4.217)}$$

We remark that both cross-sections (4.216) and (4.217) exhibit a sine-squared distribution in the angle θ, which favours the ejection of the electrons at right angles to the incident photon beam. The quantity $(1 - v_f\cos\theta/c)^{-4}$ also affects the angular distribution by enhancing the ejection of electrons at small angles, but since $v_f/c \ll 1$ it only yields a small correction to the $\sin^2\theta$ distribution. In fact, because

$$\left(1 - \frac{v_f}{c}\cos\theta\right)^{-4} = 1 + 4\frac{v_f}{c}\cos\theta + \ldots \qquad \textbf{(4.218)}$$

we see that if we drop terms of order $(v_f/c)^2$ we may write the differential cross-section (4.217) as

$$\frac{d\sigma}{d\Omega} = 16\alpha\left(\frac{\hbar}{m}\right)\frac{Z^5}{\omega(k_f a_0)^5}\sin^2\theta\left(1 + 4\frac{v_f}{c}\cos\theta\right) \tag{4.219}$$

Upon integration over the angles (θ, ϕ) of the ejected electron, the total cross-section (for an unpolarised incident photon beam) is given by

$$\sigma_{\text{tot}} = \frac{128\pi}{3}\alpha\left(\frac{\hbar}{m}\right)\frac{Z^5}{\omega(k_f a_0)^5} \tag{4.220a}$$

Using (4.211) and the fact that $E_{1s} = -Z^2[e^2/(4\pi\varepsilon_0)]/(2a_0) = -(1/2)mc^2(Z\alpha)^2$, we may also write (4.220a) as

$$\sigma_{\text{tot}} = \frac{256\pi}{3}\alpha Z^{-2}\left(\frac{|E_{1s}|}{\hbar\omega}\right)^{7/2} a_0^2 \tag{4.220b}$$

or

$$\sigma_{\text{tot}} = \frac{16\sqrt{2}\pi}{3}\alpha^8 Z^5\left(\frac{mc^2}{\hbar\omega}\right)^{7/2} a_0^2 \tag{4.220c}$$

We note from (4.220) that the photoelectric total cross-section decreases like $(\hbar\omega)^{-7/2}$ with increasing photon energy and increases like Z^5 with increasing nuclear charge. We also remark that the *dipole approximation* (obtained by setting $\exp(i\mathbf{k}\cdot\mathbf{r}) = 1$ in the matrix element (4.203)) yields the leading term in the differential cross-section (4.219), such that retardation effects of order v_f/c are neglected. As seen from (4.212), these retardation effects become more important if the wave number k (and hence the frequency ν) of the photon increases, as expected from our discussion of the dipole approximation in Section 4.3.

Let us call $d\sigma^{(0)}/d\Omega$ the leading (dipole) term in (4.219). We then have

$$\frac{d\sigma^{(0)}}{d\Omega} = 16\alpha\left(\frac{\hbar}{m}\right)\frac{Z^5}{\omega(k_f a_0)^5}\sin^2\theta \tag{4.221a}$$

or

$$\frac{d\sigma^{(0)}}{d\Omega} = 32\alpha Z^{-2}\left(\frac{|E_{1s}|}{\hbar\omega}\right)^{7/2}\sin^2\theta\, a_0^2 \tag{4.221b}$$

It is also worth noting that the dipole approximation yields the *complete* result (4.220) for the total cross-section σ_{tot}.

Finally, we recall that the above formulae are applicable for sufficiently high, but non-relativistic energies E_f of the ejected electron, such that $|E_{1s}| \ll E_f \ll mc^2$ and for small enough values of Z. These formulae can be applied not only to photoionisation from the ground state of hydrogenic atoms and ions, but also approximately to the photoionisation of electrons from the K shell of complex atoms.

4.9 The scattering of radiation by atomic systems

So far in this chapter, the processes considered for obtaining information about the properties of atoms are the emission or absorption of radiation in the case that the electromagnetic field is weak enough for first-order perturbation theory to apply. Additional information can be obtained by using strong fields, and this will be considered in Chapter 15. Here we shall discuss briefly the *scattering* of radiation by atomic systems again in the weak field case.

The scattering of radiation by an atom or molecule must be at least a second-order process. In step 1, a photon of energy $\hbar\omega$ and wave vector $\mathbf{k}$ is absorbed exciting the atomic system from a state a to a state n. In step 2 the atomic system emits a photon of energy $\hbar\omega'$ and wave vector $\mathbf{k}'$, and is de-excited from the state n to a final state b. Alternatively, the two steps can occur in reverse order, the photon of energy $\hbar\omega'$ being emitted first and the photon of energy $\hbar\omega$ being absorbed subsequently. In what follows, it will be assumed that a represents a non-degenerate ground state.

If the final state b of the system is the same as the initial state a, then, neglecting very small recoil effects (see Problem 1.16), the emitted radiation has the same frequency as the incident radiation ($\omega' = \omega$). This elastic scattering process is called *Rayleigh scattering*; it was first discussed by Lord Rayleigh in connection with the scattering of light by molecules in the atmosphere. If the final state of the atom is different from the initial state, the scattering is inelastic, and by conservation of energy the angular frequency ω' of the emitted radiation is given by

$$\begin{aligned}\omega' &= \omega + (E_a - E_b)/\hbar \\ &= \omega - \omega_{ba}\end{aligned} \qquad \textbf{(4.222)}$$

This inelastic scattering process is called *Raman scattering* (or the Raman effect) after C.V. Raman who discovered this effect experimentally in 1927. The theory of both Rayleigh and Raman scattering may be developed in a straightforward way by solving the coupled equations (4.32) to second order. If the dipole approximation is made, the differential cross-section for the emission of a photon in the direction (θ, ϕ) when the incident photon direction is along the Z axis is found to be

$$\frac{\mathrm{d}\sigma}{\mathrm{d}\Omega}(\theta, \phi) = r_0^2 \omega\omega'^3 \frac{m^2}{\hbar^2 e^4} \left| \sum_n \left[\frac{(\hat{\varepsilon}' \cdot \mathbf{D}_{bn})(\hat{\varepsilon} \cdot \mathbf{D}_{na})}{\omega_{na} - \omega} + \frac{(\hat{\varepsilon} \cdot \mathbf{D}_{bn})(\hat{\varepsilon}' \cdot \mathbf{D}_{na})}{\omega_{na} + \omega'} \right] \right|^2 \qquad \textbf{(4.223)}$$

where

$$r_0 = \frac{e^2}{(4\pi\varepsilon_0)mc^2} = 2.82 \times 10^{-15} \text{ m} \qquad \textbf{(4.224)}$$

is the *classical radius of the electron*. In (4.223), $\mathbf{D}_{ij}$ denotes the matrix element of the electric dipole moment operator of the atom between the states i and j, while $\hat{\boldsymbol{\varepsilon}}$ and $\hat{\boldsymbol{\varepsilon}}'$ are the polarisation vectors of the photons in the initial and final states, respectively. The sum over n is over all possible intermediate states.

By the usual electric dipole selection rules the intermediate state n must be of opposite parity to each of the states a and b. Thus, in second order, scattering does not change the parity of the atomic system, and the selection rules are, for a one-electron atom [15],

$$\Delta l = 0, \pm 2 \tag{4.225}$$

If the angular frequency of the incident radiation is such that $\omega = \omega_{na}$, where $\omega_{na} = (E_n - E_a)/\hbar$ and E_n is the energy of one of the intermediate states, then the first term in the cross-section formula (4.223) becomes infinite. This is due to the fact that in deriving the result (4.223) the width of the atomic energy levels has been neglected. Taking the state a to be a stable ground state of zero width and replacing E_n by $E_n - \mathrm{i}\Gamma_n/2$ where Γ_n is the width of the level n, the correct behaviour of the cross-section in the *resonance region*, in which ω is close to ω_{na}, is found to be

$$\frac{\mathrm{d}\sigma_{ba}}{\mathrm{d}\Omega}(\theta, \phi) = r_0^2 \omega\omega'^3 \frac{m^2}{\hbar^2 e^4} \left| \frac{(\hat{\boldsymbol{\varepsilon}}' \cdot \mathbf{D}_{bn})(\hat{\boldsymbol{\varepsilon}} \cdot \mathbf{D}_{na})}{\omega_{na} - \mathrm{i}\Gamma_n/(2\hbar) - \omega} \right|^2 \tag{4.226}$$

The corresponding total cross-section $\sigma_{ba}(\omega)$ is of Lorentzian form, being given by

$$\sigma_{ba}(\omega) = r_0^2 \omega\omega'^3 \frac{m^2}{\hbar e^4} \frac{1}{(\omega - \omega_{na})^2 + \Gamma_n^2/(4\hbar^2)} \int |(\hat{\boldsymbol{\varepsilon}}' \cdot \mathbf{D}_{bn})\,(\hat{\boldsymbol{\varepsilon}} \cdot \mathbf{D}_{na})|^2 \,\mathrm{d}\Omega \tag{4.227}$$

In the elastic case, with $\omega = \omega'$ and $E_a = E_b$, the scattered light is known as *resonance radiation*, while in the inelastic case (resonant Raman scattering), with $\omega \neq \omega'$ and $E_a \neq E_b$, the scattered radiation is called *resonance fluorescence*. As we have taken a to be the ground state, E_n is necessarily greater than E_a, so that $\omega' < \omega$ and the fluorescence is of longer wavelength than the incident radiation. Resonance radiation and resonance fluorescence are easily observed, for example by shining a collimated beam from a sodium vapour lamp on sodium vapour confined in a glass vessel.

Let us define τ_{an} as the lifetime of the state n for decay into the state a and τ_{bn} as the corresponding lifetime for decay from the state n to the state b. *Partial widths* Γ_{an} and Γ_{bn} can then be defined as $\Gamma_{an} = \hbar/\tau_{an}$ and $\Gamma_{bn} = \hbar/\tau_{bn}$. If decay to the states a and b are the only decay modes of the state n, then the full natural width Γ_n is given by the sum of the partial widths:

$$\Gamma_n = \Gamma_{an} + \Gamma_{bn} \tag{4.228}$$

In general, the state n will decay in additional ways, for example by collisions, and the observed width $\tilde{\Gamma}_n$ will be larger than $(\Gamma_{an} + \Gamma_{bn})$.

[15] For a general atom the selection rules for second-order scattering are $\Delta J = 0, \pm 2$, where J is the total angular momentum quantum number, and the parity does not change.

Making use of (4.170) and (4.180), the total cross-section in the resonance region $\sigma_{ba}(\omega)$ can be written in the form

$$\sigma_{ba}(\omega) = \frac{1}{2} g_n \frac{\pi c^2}{\hbar^2 \omega^2} \frac{\Gamma_{bn}\Gamma_{an}}{(\omega - \omega_{na})^2 + \Gamma_n^2/(4\hbar^2)} \tag{4.229}$$

where g_n is the degeneracy of the intermediate state n. The maximum of the cross-section occurs at exact resonance ($\omega = \omega_{na}$). For Rayleigh (elastic) scattering ($\omega' = \omega$), it has the value

$$\sigma_{\text{el}}^{\max} = 2\pi g_n \left(\frac{c}{\omega_{na}}\right)^2 \left(\frac{\Gamma_{an}}{\Gamma_n}\right)^2 \tag{4.230}$$

while for resonant Raman scattering ($\omega' \neq \omega$), it is given by

$$\sigma_{ba}^{\max} = 2\pi g_n \left(\frac{c}{\omega_{na}}\right)^2 \frac{\Gamma_{bn}\Gamma_{an}}{\Gamma_n^2} \tag{4.231}$$

By setting $b = n$ in (4.181), and assuming that Γ_{an} and Γ_{bn} are similar in magnitude, it is found that the absorption cross-section from the state a to the state n is of the same order of magnitude as the resonant scattering cross-sections (4.230) and (4.231). The cross-section for non-resonant Rayleigh scattering is smaller than that for resonant Rayleigh scattering by a factor of up to 10^{10} while that for non-resonant Raman scattering is even smaller by a further factor of about 10^3. Nevertheless, Raman scattering is very important in the theory of rotational molecular spectra; this will be discussed in Chapter 11.

Non-resonant elastic scattering in the low- and high-frequency limits

The cross-section for non-resonant Rayleigh (elastic) scattering given by (4.223) can be written as

$$\frac{d\sigma}{d\Omega}(\theta, \phi) = r_0^2 \omega^4 \frac{m^2}{\hbar^2} \left| \sum_n (\hat{\boldsymbol{\varepsilon}}' \cdot \mathbf{r}_{an})(\hat{\boldsymbol{\varepsilon}} \cdot \mathbf{r}_{na}) \frac{2\omega_{na}}{\omega_{na}^2 - \omega^2} \right|^2 \tag{4.232}$$

where, for a one-electron atom, $\mathbf{D}$ has been set equal to $-e\mathbf{r}$ (see (4.72)).

We shall start by looking at the low-frequency limit of the cross-section for scattering by the ground state of the atom, where the initial state a is an s state with $l_a = 0$. By the dipole selection rules all the intermediate states n must be p states ($l_n = 1$) with magnetic quantum numbers $m_n = 0, \pm 1$. It is convenient to take the Z axis parallel to the unit vector $\hat{\boldsymbol{\varepsilon}}$, which is in the direction of polarisation of the incident radiation, so that

$$\hat{\boldsymbol{\varepsilon}} \cdot \mathbf{r}_{na} = z_{na} \tag{4.233}$$

From (4.117a) it is seen that z_{na} vanishes unless $m_n = 0$. It follows that the only non-vanishing component of $\mathbf{r}_{an}$ is z_{an} so that $\hat{\boldsymbol{\varepsilon}}' \cdot \mathbf{r}_{an} = \hat{\varepsilon}'_z z_{an}$. Since $\hat{\boldsymbol{\varepsilon}}$ is a unit vector parallel to the Z axis, $\hat{\varepsilon}'_z = \hat{\boldsymbol{\varepsilon}}' \cdot \hat{\boldsymbol{\varepsilon}}$ and the cross-section (4.232) reduces to

$$\frac{d\sigma}{d\Omega}(\theta, \phi) = r_0^2 \omega^4 \frac{m^2}{\hbar^2} (\hat{\boldsymbol{\varepsilon}}' \cdot \hat{\boldsymbol{\varepsilon}})^2 \left| \sum_n |z_{an}|^2 \frac{2\omega_{na}}{\omega_{na}^2 - \omega^2} \right|^2 \tag{4.234}$$

where the sum is over all p states of the atom, including both discrete and continuum states.

In the low-frequency limit ($\omega \ll \omega_{\bar{n}a}^2$) where $\bar{n}$ is the p state of lowest energy, the sum over n reduces to

$$\sum_n \frac{2}{\omega_{na}} |z_{an}|^2 = \frac{\hbar}{e^2} \bar{\alpha} \tag{4.235}$$

where

$$\bar{\alpha} = 2e^2 \sum_{n \neq a} \frac{|z_{an}|^2}{E_n - E_a} \tag{4.236}$$

is a quantity known as the *static dipole polarisability* of the atom in the ground state a.

From (4.234) and (4.235) the total cross-section for low-frequency scattering of radiation by the ground state of a one-electron atom is

$$\sigma_{\text{tot}} = \left(\frac{r_0 m \bar{\alpha}}{e^2} \right)^2 \omega^4 \int (\hat{\boldsymbol{\varepsilon}}' \cdot \hat{\boldsymbol{\varepsilon}}) \, d\Omega \tag{4.237}$$

To evaluate (4.237) it is easiest to take the polar axis parallel to $\hat{\boldsymbol{\varepsilon}}$. Two cases need to be considered, corresponding to the two independent possible polarisation directions of the scattered radiation: either $\hat{\boldsymbol{\varepsilon}}' = \hat{\boldsymbol{\varepsilon}}'(1)$ where $\hat{\boldsymbol{\varepsilon}}'(1)$ is in the plane containing $\hat{\boldsymbol{\varepsilon}}$ and $\mathbf{k}'$, or $\hat{\boldsymbol{\varepsilon}}' = \hat{\boldsymbol{\varepsilon}}'(2)$ where $\hat{\boldsymbol{\varepsilon}}'(2)$ is normal to the $(\hat{\boldsymbol{\varepsilon}}, \mathbf{k}')$ plane. In the second case the cross-section vanishes since $\hat{\boldsymbol{\varepsilon}} \cdot \hat{\boldsymbol{\varepsilon}}'(2) = 0$. In the first case, since $\mathbf{k}'$ and $\hat{\boldsymbol{\varepsilon}}'(1)$ must be at right angles to each other, we have

$$\hat{\boldsymbol{\varepsilon}}'(1) \cdot \hat{\boldsymbol{\varepsilon}} = \cos \alpha = \sin \theta \tag{4.238}$$

where α is the angle between the vectors $\hat{\boldsymbol{\varepsilon}}'(1)$ and $\hat{\boldsymbol{\varepsilon}}$ and θ, the angle of scattering, is the angle between $\mathbf{k}$ and $\mathbf{k}'$. From (4.237) and (4.238), we have

$$\begin{aligned} \sigma_{\text{tot}} &= \left(\frac{r_0 m \bar{\alpha}}{e^2} \right)^2 \omega^4 \int \sin^2\theta \, d\Omega \\ &= \frac{8\pi}{3} \left(\frac{r_0 m \bar{\alpha}}{e^2} \right)^2 \omega^4 \end{aligned} \tag{4.239}$$

The rapid increase of the scattering cross-section with ω at angular frequencies $\omega < \omega_{\bar{n}a}$, where $\bar{n}$ is the first excited p state, is a general feature, although the formula (4.239) is specific to hydrogenic atoms. As a consequence, the scattering of sunlight by atmospheric atoms increases towards the high-frequency end of the visible spectrum and is responsible for the blue appearance of the sky.

The second special case of the cross-section formula (4.232) is when ω is large compared with ω_{na} for all important intermediate states n. In this limit

$$\frac{\mathrm{d}\sigma}{\mathrm{d}\Omega}(\theta,\phi) = 4r_0^2\left(\frac{m}{\hbar}\right)^2\left|\sum_n(\hat{\boldsymbol{\varepsilon}}'\cdot\mathbf{r}_{an})(\hat{\boldsymbol{\varepsilon}}\cdot\mathbf{r}_{na})\,\omega_{na}\right|^2 \tag{4.240}$$

Let us introduce axes so that $\hat{\boldsymbol{\varepsilon}}$ is parallel to the Z axis, with $\hat{\boldsymbol{\varepsilon}}'$ lying in the XZ plane. The sum over n is then given by

$$\sum_n(\hat{\boldsymbol{\varepsilon}}'\cdot\mathbf{r}_{an})(\hat{\boldsymbol{\varepsilon}}\cdot\mathbf{r}_{na})\,\omega_{na} = \hat{\varepsilon}_x'\sum_n x_{an}z_{na}\omega_{na} + \hat{\varepsilon}_z'\sum_n|z_{na}|^2\omega_{na} \tag{4.241}$$

Now z_{na} vanishes unless $m_n = m_a$, where m_n and m_a are the magnetic quantum numbers of the states n and a respectively. Similarly x_{an} vanishes unless $m_n = m_a \pm 1$. It follows that the sum $\Sigma_n\, x_{an}z_{na}\omega_{na}$ vanishes. The second sum in (4.241) can be seen from (4.138) and (4.142) to be given by $\Sigma_n|z_{na}|^2\omega_{na} = \hbar/(2m)$. From (4.240) the total scattering cross-section becomes

$$\sigma_{\text{tot}} = r_0^2\int\hat{\varepsilon}_z'^2\,\mathrm{d}\Omega = r_0^2\int(\hat{\boldsymbol{\varepsilon}}'\cdot\hat{\boldsymbol{\varepsilon}})^2\,\mathrm{d}\Omega \tag{4.242}$$

The integration over the angles in (4.242) is the same as in the previous case of low-frequency scattering, so that

$$\sigma_{\text{tot}} = \frac{8\pi}{3}r_0^2 \tag{4.243}$$

This result is independent of the initial state a, and hence can be applied to the scattering of radiation by a free electron. This is called *Thomson scattering*. The derivation which has been given here is non-relativistic and the cross-section formula (4.243) ceases to be accurate when the energy of the incident photon $\hbar\omega$ becomes comparable with the rest energy of the target electron mc^2. Both the low-frequency limit of Rayleigh scattering and the high-frequency Thomson scattering do not depend on Planck's constant h; these cross-sections were originally derived using classical electromagnetic theory (see Jackson, 1998).

Problems

4.1 Calculate how many photons per second are radiated from a monochromatic source, 1 watt in power, for the following wavelengths: (a) 10 m (radio wave) (b) 10 cm (microwave) (c) 5890 Å (yellow sodium light) (d) 1 Å (soft X-ray). At a distance of 10 m from the source, calculate the number of photons passing through unit area, normal to the direction of propagation, per unit time and the density of photons, in each case.

4.2 (a) Establish the result (4.78) starting from equation (4.76).

(b) The full transition rate for spontaneous emission of a photon from an atom is obtained in the dipole approximation from (4.79), by integrating over all angles of emission and summing over two independent polarisation directions $\hat{\varepsilon}_1, \hat{\varepsilon}_2$:

$$W_{ab}^{s,D} = \frac{1}{2\pi\hbar c^3}\left(\frac{1}{4\pi\varepsilon_0}\right)\omega_{ba}^3 \int d\Omega\{|\hat{\varepsilon}_1\cdot\mathbf{D}_{ba}|^2 + |\hat{\varepsilon}_2\cdot\mathbf{D}_{ba}|^2\}$$

Starting from this expression obtain the result (4.80).

4.3 Show that by performing the gauge transformation (4.89), one obtains for the new wave function $\Psi^{V}(\mathbf{r}, t)$ the Schrödinger equation in the velocity gauge (4.90).

4.4 Show that by performing the gauge transformation (4.92), one obtains for the new wave function $\Psi^{L}(\mathbf{r}, t)$ the Schrödinger equation in the length gauge (4.93).

4.5 Generalise the results of Section 4.4 to the case in which the level E_a is g_a times degenerate and the level E_b is g_b times degenerate, and show that the Einstein coefficients satisfy the relations

$$g_a B_{ba} = g_b B_{ab}; \qquad A_{ab} = \frac{\hbar\omega_{ba}^3}{\pi^2 c^3} B_{ab}$$

4.6 Obtain the explicit forms of the coefficients $\mathcal{A}(l, m; l', m'; q)$, either by using a table of Clebsch–Gordan coefficients, or by using the recurrence relations (4.121).

4.7 Show that

$$\sum_{m'} |\mathcal{A}(l, m; l', m'; m' - m)|^2 = \frac{3}{4\pi}\frac{l+1}{2l+1}$$

if $l' = l + 1$ and

$$\sum_{m'} |\mathcal{A}(l, m; l', m'; m' - m)|^2 = \frac{3}{4\pi}\left(\frac{l}{2l+1}\right)$$

if $l' = l - 1$. Find the transition rate for absorption of unpolarised isotropic radiation by a hydrogenic atom from a level (nl) to the $(2l' + 1)$ degenerate levels $(n'l')$ and show that it is independent of the magnetic quantum number of the initial sublevel.

4.8 Prove that the average oscillator strengths defined by (4.143) obey the Thomas–Reiche–Kuhn sum rule (4.137).

4.9 Prove the scaling law (4.148) for the lifetimes of the hydrogenic ions in the dipole approximation.

4.10 For a given initial level a of a hydrogen atom, show that

$$\sum_b \sigma_{ba} - \sum_b{}' \bar{\sigma}_{ba} = 2\pi^2 r_0 c$$

where σ_{ba} is the integrated absorption cross-section, in the electric dipole approximation, the sum Σ_b being over all states (including continuum states) with $E_b > E_a$, and $\bar{\sigma}_{ba}$ is the corresponding cross-section for stimulated emission, the sum Σ_b' being over all states with $E_b < E_a$. The quantity r_0 is the classical electron radius, which is given by $r_0 = e^2/(4\pi\varepsilon_0 mc^2)$.

5 One-electron atoms: fine structure and hyperfine structure

Our discussion of the energy levels and wave functions of one-electron atoms in Chapter 3 was based on the simple, non-relativistic Hamiltonian

$$H = \frac{p^2}{2\mu} - \frac{Ze^2}{(4\pi\varepsilon_0)r} \quad \textbf{(5.1)}$$

where the first term represents the (non-relativistic) kinetic energy of the atom in the centre of mass system, and the second term is the electrostatic (Coulomb) interaction between the electron and the nucleus. Although the energy levels obtained in Chapter 3 from the Hamiltonian (5.1) are in good agreement with experiment, the very precise measurements carried out in atomic physics demonstrate the existence of several effects which cannot be derived from the Hamiltonian (5.1) and require the addition of correction terms to (5.1). In this chapter we shall discuss these corrections.

We begin by analysing the relativistic corrections to (5.1), which give rise to a splitting of the energy levels known as fine structure. We then describe a subtle effect called the Lamb shift, which displaces the fine structure components and is therefore responsible for additional splittings of the energy levels. Finally, we consider various small corrections such as the hyperfine structure splitting and the volume effect, which take into account the fact that the nucleus is not simply a point charge, but has a finite size, and may possess an intrinsic angular momentum (spin), a magnetic dipole moment, an electric quadrupole moment, and higher moments.

5.1 Fine structure of hydrogenic atoms

The fine structure of the energy levels of hydrogenic atoms is due to relativistic effects. In order to analyse these effects we therefore need for the electron a basic wave equation which satisfies the requirements of special relativity as well as those of quantum mechanics. This is the Dirac equation, which is discussed briefly in Appendix 7, and which provides the correct relativistic wave equation for electrons.

The most rigorous way of obtaining the relativistic corrections to the Schrödinger (Bohr) energy levels of one-electron atoms is to solve the Dirac equation for an electron in the central field $V(r) = -Ze^2/(4\pi\varepsilon_0 r)$ of the nucleus which is assumed to be of infinite mass and at the origin of the coordinates. It turns out that the Dirac equation for a central field can be separated in spherical polar coordinates and that the resulting radial equations can be solved exactly for the Coulomb potential $V(r) = -Ze^2/(4\pi\varepsilon_0 r)$ [1]. However, these calculations are rather lengthy and since the relativistic corrections are very small (provided that Z is not too large), it is convenient to use perturbation theory, keeping terms up to order v^2/c^2 in the Dirac Hamiltonian. We shall therefore start from the Hamiltonian (A7.65) of Appendix 7 which we rewrite as

$$H = H_0 + H' \tag{5.2}$$

where

$$H_0 = \frac{p^2}{2m} - \frac{Ze^2}{(4\pi\varepsilon_0)r} \tag{5.3}$$

is simply the Hamiltonian (5.1) with $\mu = m$ [2] and

$$H' = H_1' + H_2' + H_3' \tag{5.4}$$

with

$$H_1' = -\frac{p^4}{8m^3c^2} \tag{5.5}$$

$$H_2' = \frac{1}{2m^2c^2}\frac{1}{r}\frac{\mathrm{d}V}{\mathrm{d}r}\mathbf{L}\cdot\mathbf{S} \tag{5.6}$$

and

$$H_3' = \frac{\pi\hbar^2}{2m^2c^2}\left(\frac{Ze^2}{4\pi\varepsilon_0}\right)\delta(\mathbf{r}) \tag{5.7}$$

The physical interpretation of the three terms which constitute H' is discussed in Appendix 7. We simply note here that H_1' is a relativistic correction to the kinetic energy, H_2' represents the spin–orbit interaction and H_3' is the Darwin term.

Before we proceed to the evaluation of the energy shifts due to these three terms by using perturbation theory, we remark that the Schrödinger theory discussed in Chapter 3 does not include the spin of the electron. In order to calculate

[1] See Bransden and Joachain (2000).

[2] For the sake of simplicity we shall ignore all reduced mass effects in discussing the fine structure calculations. It is of course straightforward to incorporate the reduced mass effect in H_0 and in the corresponding unperturbed energy levels E_n by replacing the electron mass m by its reduced mass μ. On the other hand, the reduced mass effects arising in H' cannot be obtained by just replacing m by μ in the results of the perturbation calculation. Fortunately, these latter reduced mass effects are very small since H' is already a correction to H_0.

corrections involving the spin operator – such as those arising from H_2' – we start from the 'unperturbed' equation

$$H_0\psi_{nlm_lm_s} = E_n\psi_{nlm_lm_s} \tag{5.8}$$

where E_n are the Schrödinger eigenvalues (3.29) (with $\mu = m$) and the zero-order wave functions $\psi_{nlm_lm_s}$ are modified (two-component) Schrödinger wave functions (also referred to as Pauli wave functions or 'spin-orbitals') given by

$$\psi_{nlm_lm_s}(q) = \psi_{nlm_l}(\mathbf{r})\chi_{1/2,m_s} \tag{5.9}$$

where q denotes the space and spin variables collectively. The quantum number m_l which can take the values $-l, -l+1, \ldots, +l$ is the magnetic quantum number previously denoted by m [3], $\psi_{nlm_l}(\mathbf{r})$ is a one-electron Schrödinger wave function (see (3.53)) such that

$$H_0\psi_{nlm_l}(\mathbf{r}) = E_n\psi_{nlm_l}(\mathbf{r}) \tag{5.10}$$

and $\chi_{1/2,m_s}$ are the spin eigenfunctions for spin one-half ($s = 1/2$) introduced in Section 2.5, with $m_s = \pm 1/2$. We recall that $\chi_{1/2,m_s}$ is a two-component spinor and that the normalised spinors corresponding respectively to 'spin up' ($m_s = +1/2$) and 'spin down' ($m_s = -1/2$) are conveniently denoted by

$$\alpha = \begin{pmatrix} 1 \\ 0 \end{pmatrix} \quad \text{and} \quad \beta = \begin{pmatrix} 0 \\ 1 \end{pmatrix} \tag{5.11}$$

Since H_0 does not act on the spin variable, the two-component wave functions (5.9) are separable in space and spin variables. It is also worth noting that we now have *four* quantum numbers (n, l, m_l, m_s) to describe a one-electron atom, the effect of the spin on the 'unperturbed' solutions being to double the degeneracy, so that each Schrödinger energy level E_n is now $2n^2$ degenerate.

Energy shifts

We now calculate the energy corrections due to the three terms (5.5)–(5.7), using the Pauli wave functions (5.9) as our zero-order wave functions.

1. $H_1' = -\dfrac{p^4}{8m^3c^2}$ (relativistic correction to the kinetic energy)

Since the unperturbed energy level E_n is $2n^2$ degenerate, we should use the degenerate perturbation theory discussed in Section 2.8. However, we first note that H_1' does not act on the spin variable. Moreover, it commutes with the components of the orbital angular momentum (see Problem 2.13) so that the perturbation

[3] When no confusion is possible, we shall continue to write m instead of m_l for the magnetic quantum number associated with the operator L_z.

H_1' is already 'diagonal' in l, m_l and m_s. The energy correction ΔE_1 due to H_1' is therefore given in first-order perturbation theory by

$$\Delta E_1 = \left\langle \psi_{nlm_lm_s} \left| -\frac{p^4}{8m^3c^2} \right| \psi_{nlm_lm_s} \right\rangle$$

$$= \left\langle \psi_{nlm_l} \left| -\frac{p^4}{8m^3c^2} \right| \psi_{nlm_l} \right\rangle$$

$$= -\frac{1}{2mc^2} \langle \psi_{nlm_l} | T^2 | \psi_{nlm_l} \rangle \tag{5.12}$$

where $T = p^2/(2m)$ is the kinetic energy operator. From (5.3) we have

$$T = H_0 + \frac{Ze^2}{(4\pi\varepsilon_0)r} \tag{5.13}$$

and therefore

$$\Delta E_1 = -\frac{1}{2mc^2} \left\langle \psi_{nlm_l} \left| \left(H_0 + \frac{Ze^2}{(4\pi\varepsilon_0)r} \right) \left(H_0 + \frac{Ze^2}{(4\pi\varepsilon_0)r} \right) \right| \psi_{nlm_l} \right\rangle$$

$$= -\frac{1}{2mc^2} \left[E_n^2 + 2E_n \left(\frac{Ze^2}{4\pi\varepsilon_0} \right) \left\langle \frac{1}{r} \right\rangle_{nlm_l} + \left(\frac{Ze^2}{4\pi\varepsilon_0} \right)^2 \left\langle \frac{1}{r^2} \right\rangle_{nlm_l} \right] \tag{5.14}$$

where we have used (5.10). From the results (3.30), (3.76) and (3.77) (with $\mu = m$) one obtains (Problem 5.1)

$$\Delta E_1 = \frac{1}{2} mc^2 \frac{(Z\alpha)^2}{n^2} \frac{(Z\alpha)^2}{n^2} \left[\frac{3}{4} - \frac{n}{l + 1/2} \right]$$

$$= -E_n \frac{(Z\alpha)^2}{n^2} \left[\frac{3}{4} - \frac{n}{l + 1/2} \right] \tag{5.15}$$

2. $H_2' = \dfrac{1}{2m^2c^2} \dfrac{1}{r} \dfrac{dV}{dr} L \cdot S$ **(spin–orbit term)**

We shall first rewrite this term more simply as

$$H_2' = \xi(r) \mathbf{L} \cdot \mathbf{S} \tag{5.16}$$

where we have introduced the quantity

$$\xi(r) = \frac{1}{2m^2c^2} \frac{1}{r} \frac{dV}{dr} \tag{5.17}$$

In our case $V(r) = -Ze^2/(4\pi\varepsilon_0 r)$, so that

$$\xi(r) = \frac{1}{2m^2c^2}\frac{Ze^2}{4\pi\varepsilon_0}\frac{1}{r^3} \tag{5.18}$$

Since the operator $\mathbf{L}^2$ does not act on the radial variable r nor on the spin variable, and commutes with the components of $\mathbf{L}$, we see from (5.16) that $\mathbf{L}^2$ commutes with H'_2. It follows that the perturbation H'_2 does not connect states with different values of the orbital angular momentum l. For a given value of n and l there are $2(2l+1)$ degenerate eigenstates of H_0 (the factor of 2 arising from the two spin states), so that the calculation of the energy shift due to H'_2 requires the diagonalisation of $2(2l+1) \times 2(2l+1)$ submatrices.

This diagonalisation is greatly simplified by using for the zero-order wave functions a representation in which $\mathbf{L}\cdot\mathbf{S}$ is diagonal. It is clear that the functions $\psi_{nlm_lm_s}$ given by (5.9), which are simultaneous eigenfunctions of the operators H_0, $\mathbf{L}^2$, $\mathbf{S}^2$, L_z and S_z, are not adequate because $\mathbf{L}\cdot\mathbf{S}$ does not commute with L_z or S_z. However, we shall now show that satisfactory zero-order wave functions may be obtained by forming certain linear combinations of the functions $\psi_{nlm_lm_s}$. To this end, we introduce the total angular momentum of the electron

$$\mathbf{J} = \mathbf{L} + \mathbf{S} \tag{5.19}$$

and we note that

$$\mathbf{J}^2 = \mathbf{L}^2 + 2\mathbf{L}\cdot\mathbf{S} + \mathbf{S}^2 \tag{5.20}$$

so that

$$\mathbf{L}\cdot\mathbf{S} = \tfrac{1}{2}(\mathbf{J}^2 - \mathbf{L}^2 - \mathbf{S}^2) \tag{5.21}$$

Consider now wave functions ψ_{nljm_j} which are eigenstates of the operators H_0, $\mathbf{L}^2$, $\mathbf{S}^2$, $\mathbf{J}^2$ and J_z, the corresponding eigenvalues being E_n, $l(l+1)\hbar^2$, $s(s+1)\hbar^2$, $j(j+1)\hbar^2$ and $m_j\hbar$. In this particular case we have $s = 1/2$ and therefore (see Section 2.5)

$$\begin{aligned} j &= l \pm 1/2, \qquad l \neq 0 \\ j &= 1/2, \qquad l = 0 \end{aligned} \tag{5.22}$$

and

$$m_j = -j, -j+1, \ldots, +j \tag{5.23}$$

By using the methods of Section 2.5 and Appendix 4, we can form the functions ψ_{nljm_j} from linear combinations of the functions $\psi_{nlm_lm_s}$ [4]. Since $\mathbf{L}\cdot\mathbf{S}$ commutes

[4] Specifically, if we use the Dirac notation so that the ket $|nlsm_lm_s\rangle$ corresponds to the wave function $\psi_{nlm_lm_s}$ and the ket $|nlsjm_j\rangle$ to the wave function ψ_{nljm_j} (with $s = 1/2$), we have

$$|nlsjm_j\rangle = \sum_{m_lm_s} \langle lsm_lm_s|jm_j\rangle\, |nlsm_lm_s\rangle$$

The Clebsch–Gordan coefficients $\langle lsm_lm_s|jm_j\rangle$ are not needed in the present calculation since we are only interested in expectation values.

with $\mathbf{L}^2$, $\mathbf{S}^2$, $\mathbf{J}^2$ and J_z it is apparent that the new zero-order wave functions ψ_{nljm_j} form a satisfactory basis set in which the operator $\mathbf{L}\cdot\mathbf{S}$ (and hence the perturbation H'_2) is diagonal. Using (5.16) and (5.21), we see that for $l \neq 0$ the energy shift due to the term H'_2 is given by

$$\Delta E_2 = \left\langle \psi_{nljm_j} \left| \frac{1}{2}\xi(r)[\mathbf{J}^2 - \mathbf{L}^2 - \mathbf{S}^2] \right| \psi_{nljm_j} \right\rangle$$

$$= \frac{\hbar^2}{2}\langle\xi(r)\rangle\left[j(j+1) - l(l+1) - \frac{3}{4}\right] \tag{5.24}$$

where $\langle\xi(r)\rangle$ denotes the average value of $\xi(r)$ in the state ψ_{nljm_j}. From (5.17) and (3.78), we have

$$\langle\xi(r)\rangle = \frac{1}{2m^2c^2}\left(\frac{Ze^2}{4\pi\varepsilon_0}\right)\left\langle\frac{1}{r^3}\right\rangle = \frac{1}{2m^2c^2}\left(\frac{Ze^2}{4\pi\varepsilon_0}\right)\frac{Z^3}{a_0^3 n^3 l(l+1/2)(l+1)} \tag{5.25}$$

Thus, for $l \neq 0$, one obtains from (5.24) and (5.25) (Problem 5.2)

$$\Delta E_2 = \frac{mc^2(Z\alpha)^4}{4n^3 l(l+1/2)(l+1)} \times \begin{cases} l & \text{for } j = l + 1/2 \\ -l-1 & \text{for } j = l - 1/2 \end{cases}$$

$$= -E_n \frac{(Z\alpha)^2}{2nl(l+1/2)(l+1)} \times \begin{cases} l & \text{for } j = l + 1/2 \\ -l-1 & \text{for } j = l - 1/2 \end{cases} \tag{5.26}$$

For $l = 0$ the spin–orbit interaction (5.16) vanishes and therefore $\Delta E_2 = 0$ in that case.

3. $H'_3 = \frac{\pi\hbar^2}{2m^2c^2}\left(\frac{Ze^2}{4\pi\varepsilon_0}\right)\delta(\mathbf{r})$ (Darwin term)

This term does not act on the spin variable, is diagonal in l, m_l and m_s and applies only to the case $l = 0$. Calling ΔE_3 the corresponding energy correction and using the result (3.65), we have

$$\Delta E_3 = \frac{\pi\hbar^2}{2m^2c^2}\frac{Ze^2}{4\pi\varepsilon_0}\langle\psi_{n00}|\delta(\mathbf{r})|\psi_{n00}\rangle$$

$$= \frac{\pi\hbar^2}{2m^2c^2}\frac{Ze^2}{4\pi\varepsilon_0}|\psi_{n00}(0)|^2$$

$$= \frac{1}{2}mc^2\frac{(Z\alpha)^2}{n^2}\frac{(Z\alpha)^2}{n}$$

$$= -E_n\frac{(Z\alpha)^2}{n}, \qquad l = 0 \tag{5.27}$$

We may now combine the effects of H_1', H_2' and H_3' to obtain the total energy shift $\Delta E = \Delta E_1 + \Delta E_2 + \Delta E_3$ due to relativistic corrections. From (5.15), (5.26) and (5.27) we have for all l (Problem 5.3)

$$\begin{aligned}\Delta E_{nj} &= -\frac{1}{2}mc^2\frac{(Z\alpha)^2}{n^2}\frac{(Z\alpha)^2}{n^2}\left(\frac{n}{j+1/2}-\frac{3}{4}\right)\\ &= E_n\frac{(Z\alpha)^2}{n^2}\left(\frac{n}{j+1/2}-\frac{3}{4}\right)\end{aligned} \tag{5.28}$$

where the subscripts nj indicate that the correction depends on both the principal quantum number n and the total angular momentum quantum number j, with $j = 1/2, 3/2, \ldots, n - 1/2$. To each value of j correspond two possible values of l given by $l = j \pm 1/2$, except for $j = n - 1/2$ where one can only have $l = j - 1/2 = n - 1$.

Adding the relativistic correction ΔE_{nj} to the non-relativistic energy E_n, we find that the energy levels of one-electron atoms are now given by

$$E_{nj} = E_n\left[1+\frac{(Z\alpha)^2}{n^2}\left(\frac{n}{j+1/2}-\frac{3}{4}\right)\right] \tag{5.29}$$

so that the binding energy $|E_{nj}|$ of the electron is slightly increased with respect to the non-relativistic value $|E_n|$, the absolute value $|\Delta E_{nj}|$ of the energy shift becoming smaller as n or j increases, and larger as Z increases. The formula (5.29) can be shown (Problem 5.4) to agree up to order $(Z\alpha)^2$ with the result

$$E_{nj}^{\text{exact}} = mc^2\left\{\left[1+\left(\frac{Z\alpha}{n-j-1/2+[(j+1/2)^2-Z^2\alpha^2]^{1/2}}\right)^2\right]^{-1/2}-1\right\} \tag{5.30}$$

obtained by solving the Dirac equation for the potential $V(r) = -Ze^2/(4\pi\varepsilon_0 r)$ [1].

Fine structure splitting

Starting from non-relativistic energy levels E_n which are $2n^2$ times degenerate (the factor of two arising from the spin) we see that in the Dirac theory this degeneracy is partly removed. In fact, a non-relativistic energy level E_n depending only on the principal quantum number n splits into n different levels in the Dirac theory, one for each value $j = 1/2, 3/2, \ldots, n - 1/2$ of the total angular momentum quantum number j. This splitting is called *fine structure splitting*, and the n levels $j = 1/2, 3/2, \ldots, n - 1/2$ are said to form a *fine structure multiplet*. We note that the dimensionless constant $\alpha \simeq 1/137$ controls the scale of the splitting, and it is for this reason that it has been called the fine structure constant.

The fine structure splitting of the energy levels corresponding to $n = 1, 2, 3$ is illustrated in Fig. 5.1. We have used in that figure the spectroscopic notation nl_j (with the usual association of the letters s, p, d, ... with the values $l = 0, 1, 2, \ldots$

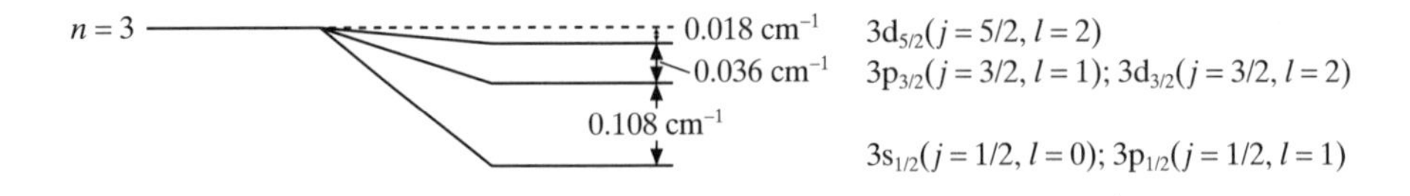

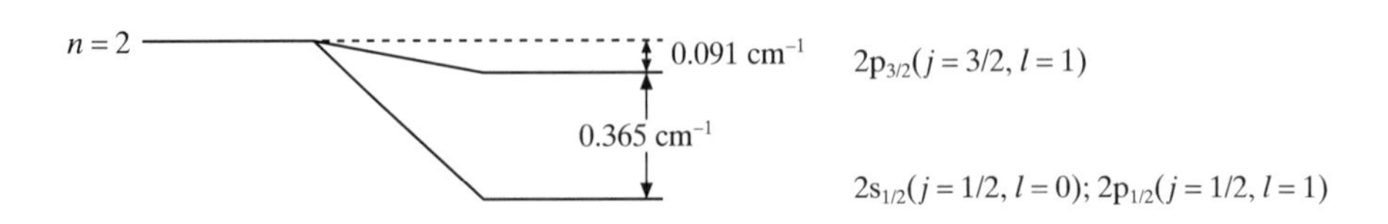

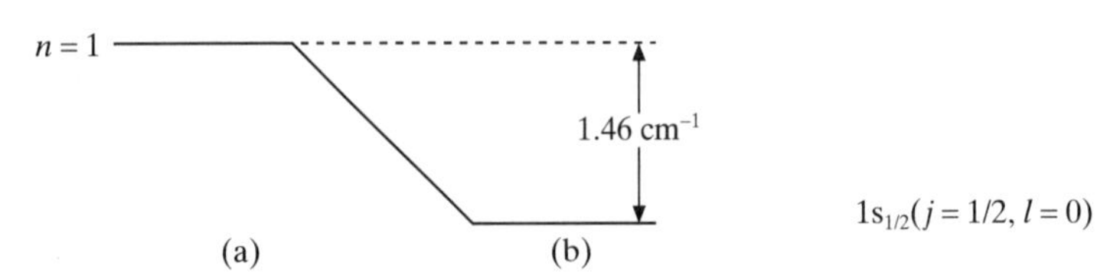

Figure 5.1 Fine structure of the hydrogen atom. The non-relativistic levels are shown on the left in column (a) and the split levels on the right in column (b), for n = 1, 2 and 3. For clarity, the scale in each diagram is different.

and an additional subscript for the value of j) to distinguish the various spectral terms corresponding to the Dirac theory [5].

It is important to emphasise that in Dirac's theory two states having the same value of the quantum number n and j but with values of l such that $l = j \pm 1/2$ have the same energy. The *parity* of the solutions is still given by $(-)^l$. Thus to each value of j correspond two series of $(2j + 1)$ solutions of opposite parity, except for $j = n - 1/2$ where there is only one series of solutions of parity $(-)^{n-1}$. It is also worth remarking that although the three separate contributions ΔE_1, ΔE_2 and ΔE_3 depend on l (see (5.15), (5.26) and (5.27)), the total energy shift ΔE_{nj} (given by (5.28)) does not! This is illustrated in Fig. 5.2, where we show the splitting of the $n = 2$ levels of atomic hydrogen due to each of the three terms H'_1, H'_2 and H'_3, as well as the resulting degeneracy of the $2s_{1/2}$ and $2p_{1/2}$ levels. We shall see in Section 5.2 that this degeneracy of the levels with $l = j \pm 1/2$ is actually removed by small quantum electrodynamics effects, known as *radiative corrections*, which are responsible for additional energy shifts called *Lamb shifts*.

[5] A similar notation with capital letters, such as $1S_{1/2}$, $2S_{1/2}$, $2P_{1/2}$, $2P_{3/2}$, etc., is also frequently used.

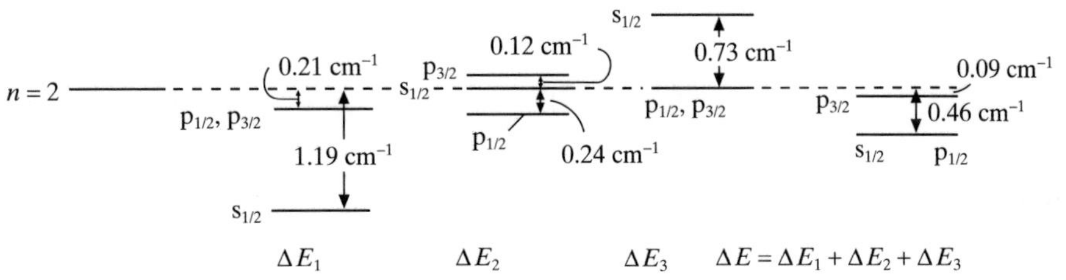

Figure 5.2 The contributions ΔE_1, ΔE_2, ΔE_3 to the splitting of the $n = 2$ level of the hydrogen atom.

Another interesting point is that the three relativistic energy shifts ΔE_1, ΔE_2 and ΔE_3 we have obtained above have the same order of magnitude, and must therefore be treated together. This is a special feature of hydrogenic atoms. For many-electron atoms (and in particular for alkali atoms) we shall see in Chapter 8 that it is the spin–orbit effect (due here to the term H'_2) which is mainly responsible for the fine structure splitting.

According to (5.28), for any Z and $n \neq 1$, the energy difference between the two extreme components of a fine structure multiplet (corresponding respectively to the values $j_1 = n - 1/2$ and $j_2 = 1/2$) is given by

$$\begin{aligned}\delta E(j_1 = n - 1/2, j_2 = 1/2) &= |E_n|(Z\alpha)^2 \frac{n-1}{n^2} \\ &= \frac{\alpha^2 Z^4 (n-1)}{2n^4} \quad \text{a.u.,} \quad n \neq 1 \end{aligned} \tag{5.31}$$

We may also use (5.28) to obtain for any Z, $n \neq 1$ and $l \neq 0$ the energy separation between two levels corresponding respectively to $j_1 = l + 1/2$ and $j_2 = l - 1/2$. The result is

$$\begin{aligned}\delta E(j_1 = l + 1/2, j_2 = l - 1/2) &= |E_n| \frac{(Z\alpha)^2}{nl(l+1)} \\ &= \frac{\alpha^2 Z^4}{2n^3 l(l+1)} \quad \text{a.u.}\end{aligned} \tag{5.32}$$

For example, in the case of atomic hydrogen the splitting of the levels $j = 3/2$ and $j = 1/2$ for $n = 2$ and $n = 3$ is, respectively, 0.365 cm^{-1} (4.52×10^{-5} eV) and 0.108 cm^{-1} (1.34×10^{-5} eV), while the splitting of the levels $j = 5/2$ and $j = 3/2$ for $n = 3$ is 0.036 cm^{-1} (4.48×10^{-6} eV) as shown in Fig. 5.1.

Fine structure of spectral lines

The set of *spectral lines* due to the transitions $nlj \rightarrow n'l'j'$ between the fine structure components of the levels nl and $n'l'$ is known as a *multiplet* of lines. Since the

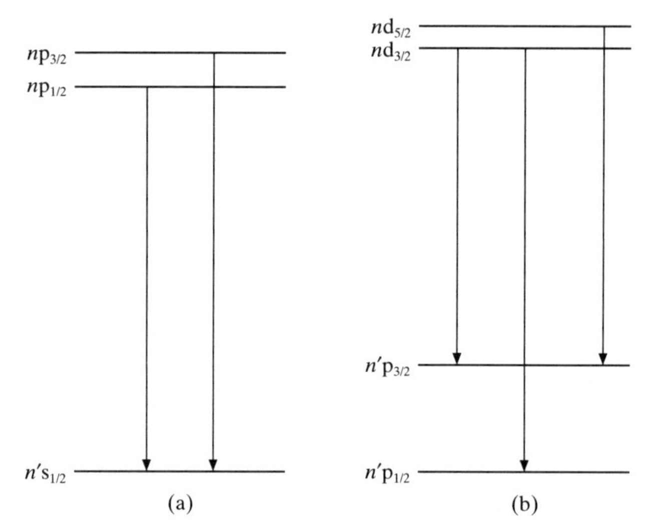

Figure 5.3 Allowed transitions in the multiplets (a) $np - n's$ and (b) $nd - n'p$.

electric dipole operator $\mathbf{D} = -e\mathbf{r}$ does not depend on the spin, the selection rule derived in Chapter 4 for the quantum number l (in the dipole approximation) remains

$$\Delta l = \pm 1 \tag{5.33}$$

from which it follows that the selection rule with respect to the quantum number j is

$$\Delta j = 0, \pm 1 \tag{5.34}$$

Using (5.33) and (5.34), it is a simple matter to establish the character of the fine structure splitting of the hydrogenic atom spectral lines. For example, we see from Fig. 5.3 that the multiplet np–$n's$ has two components. Thus each line of the Lyman series (lower state $n = 1$) is split by the fine structure into a pair of lines called a *doublet*, corresponding to the transitions

$$np_{1/2}\text{–}1s_{1/2}, \qquad np_{3/2}\text{–}1s_{1/2}$$

This is illustrated in Fig. 5.4 for the Lyman α line (upper state $n = 2$).

Referring to Fig. 5.3, we see that the multiplet np–$n's$ has two components, while the multiplet nd–$n'p$ has three components. Thus, in the case of the Balmer series (lower state $n = 2$) the following seven transitions are allowed:

$$np_{1/2}\text{–}2s_{1/2}, \qquad np_{3/2}\text{–}2s_{1/2}$$
$$ns_{1/2}\text{–}2p_{1/2}, \qquad ns_{1/2}\text{–}2p_{3/2}$$
$$nd_{3/2}\text{–}2p_{1/2}, \qquad nd_{3/2}\text{–}2p_{3/2}$$
$$nd_{5/2}\text{–}2p_{3/2}$$

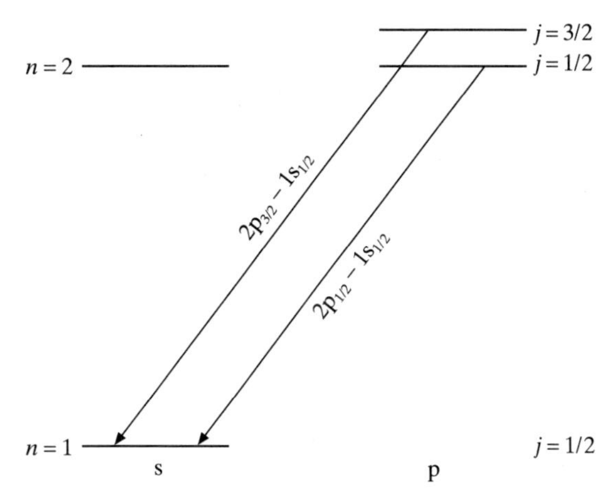

Figure 5.4 Allowed transitions between the n = 2 and n = 1 levels of atomic hydrogen giving rise to the Lyman alpha doublet (L_α).

However, since the levels $n\mathrm{s}_{1/2}$ and $n\mathrm{p}_{1/2}$ coincide, as well as the levels $n\mathrm{p}_{3/2}$ and $n\mathrm{d}_{3/2}$, each Balmer line only contains five distinct components. This is illustrated in Fig. 5.5 for the case of the fine structure of the H_α line, that is the red line of the Balmer series at 6563 Å, corresponding to the transition between the upper state $n = 3$ and the lower state $n = 2$.

Because the energy differences (5.31) or (5.32) rapidly decrease with increasing n, the fine structure splitting of a spectral line corresponding to a transition between two levels of different n is mainly due to the fine structure of the *lower* level, with additional (finer) fine structure arising from the smaller splitting of the upper level. For example, each line of the Balmer series essentially consists of a *doublet*, or more precisely of *two groups* of closely spaced lines. The distance between these two groups is approximately given by the fine structure splitting of the lower ($n = 2$) level (that is, about 0.365 cm^{-1}) and this distance is constant for all the lines of the series. Within each of the two groups the magnitude of the (small) residual splitting due to the fine structure of the upper level rapidly falls off as n increases, that is as one goes to higher lines of the series. Similarly, each line of the Paschen series (lower state $n = 3$) consists of *three groups* of closely spaced lines, etc. Finally, we remark that for hydrogenic ions the fine structure splitting is more important than for hydrogen since the energy shift ΔE_{nj} given by (5.28) is proportional to Z^4.

Intensities of fine structure lines

Since the radial integrals in (4.113) are the same for both the transitions $n\mathrm{p}_{3/2}$–$n'\mathrm{s}_{1/2}$ and $n\mathrm{p}_{1/2}$–$n'\mathrm{s}_{1/2}$, it is easy to obtain from the angular parts of those integrals (that

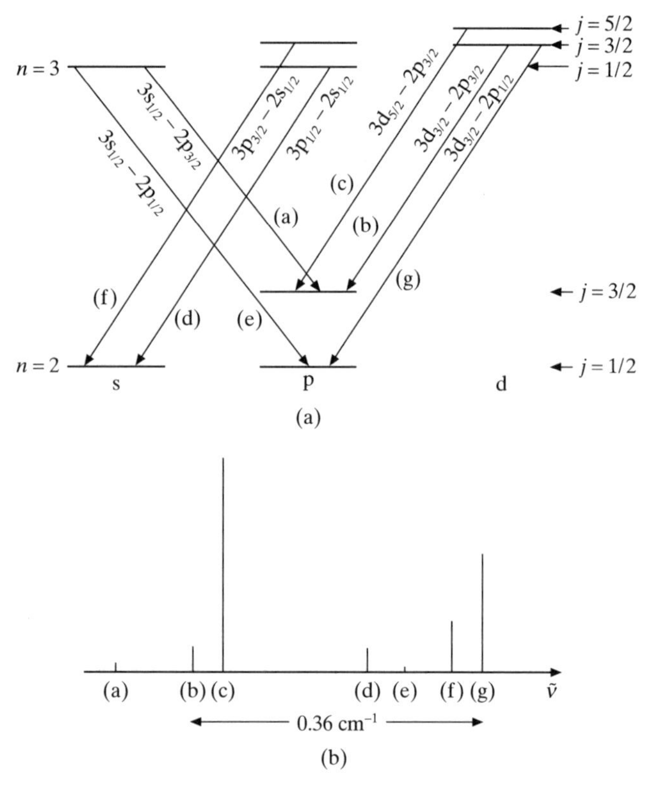

Figure 5.5 (a) Transitions contributing to the Balmer alpha (H_α) line between the n = 3 and n = 2 levels of atomic hydrogen.
(b) The observed relative intensities of the lines (a), (b)–(g). Note that (b) and (g) have the same upper level, so that the wave number difference between the lines is determined by the $2p_{1/2} - 2p_{3/2}$ energy difference and is 0.36 cm^{-1}. In the same way, the wave number difference between lines (a) and (e) is also 0.36 cm^{-1}. The lines (d) and (e) should coincide according to Dirac theory, as well as the lines (f) and (g); the differences are due to the Lamb shift.

is, from angular momentum considerations) the ratio of the two transition probabilities, which is found to be equal to 2 (Problem 5.5). More generally, the ratios of the transition probabilities for the most important special cases are (Bethe and Salpeter, 1957)

$$
\begin{aligned}
\text{for}\quad & \text{sp transitions: } s_{1/2}\text{–}p_{3/2} : s_{1/2}\text{–}p_{1/2} = 2:1 \\
& \text{pd transitions: } p_{3/2}\text{–}d_{5/2} : p_{3/2}\text{–}d_{3/2} : p_{1/2}\text{–}d_{3/2} = 9:1:5 \\
& \text{df transitions: } d_{5/2}\text{–}f_{7/2} : d_{5/2}\text{–}f_{5/2} : d_{3/2}\text{–}f_{5/2} = 20:1:14
\end{aligned} \tag{5.35}
$$

Under most circumstances the initial states are excited in proportion to their statistical weights, that is the $(2j + 1)$ degenerate levels corresponding to an initial state with a given value of j (but differing in $m_j = -j, -j + 1, \ldots, +j$) are equally populated. In this case the ratios of line intensities are the same as those of the corresponding transition probabilities. The relative intensities of the fine structure components of the H_α line are shown in Fig. 5.5.

Comparison with experiment

Many spectroscopic studies of the fine structure of atomic hydrogen and hydrogenic ions (in particular He^+) were made to test the Dirac theory, but no definite conclusion had been reached by 1940. Although there was some evidence strongly supporting the theory, the measurements performed by W.V. Houston in 1937 and R.C. Williams in 1938 were interpreted in 1938 by S. Pasternack as indicating that the $2s_{1/2}$ and $2p_{1/2}$ levels did not coincide exactly, but that there existed a slight upward shift of the $2s_{1/2}$ level with respect to the $2p_{1/2}$ level of about 0.03 cm^{-1}. However, the experimental attempts to obtain accurate information about the fine structure of hydrogenic atoms were frustrated by the broadening of the spectral lines, due mainly to the Doppler effect. In fact, other spectroscopists disagreed with the results of Houston and Williams, and found no discrepancy with the Dirac theory.

The question was settled in 1947 by W.E. Lamb and R.C. Retherford, who demonstrated in a decisive way the existence of an energy difference between the two levels $2s_{1/2}$ and $2p_{1/2}$. This 'Lamb shift', to which we have already alluded in the discussion following (5.30), will now be considered.

5.2 The Lamb shift

Instead of attempting to resolve the fine structure of hydrogen by investigating its optical spectrum, Lamb and Retherford used *microwave techniques* [6] to stimulate a direct *radio-frequency* transition between the $2s_{1/2}$ and $2p_{1/2}$ levels. As we noted in Section 4.5 there is no selection rule on the principal quantum number n for electric dipole transitions. In particular, these transitions can occur between levels having the *same* principal quantum number. This fact was pointed out as early as 1928 by W. Grotrian, who suggested that it should be possible with radio waves to induce such transitions among the excited states of the hydrogen atom. For example, in the case of the transition $2s_{1/2}$–$2p_{3/2}$, the energy separation $\delta E = 4.52 \times 10^{-5}$ eV = 0.365 cm^{-1} which we obtained in (5.32) corresponds to a wavelength of 2.74 cm or a frequency of 10 949 MHz. Because the frequencies of radio waves are much smaller than those corresponding to optical lines (such as the H_α line), the Doppler broadening, which is proportional to the frequency (see (4.192)), is considerably reduced in radio-frequency experiments, and could in fact be neglected in the experiment of Lamb and Retherford. Of course, since

[6] A detailed account of microwave spectroscopy may be found in Townes and Schawlow (1975).

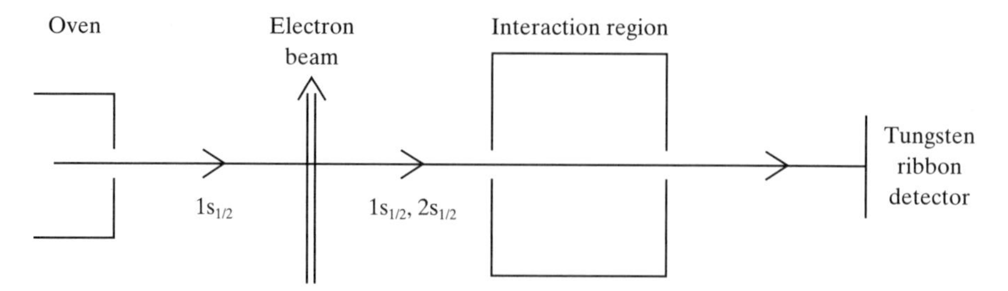

Figure 5.6 Schematic diagram of the Lamb–Retherford experiment. A collimated beam of hydrogen atoms emerges from an oven. A fraction of the atoms is excited to the $n = 2$ level by electron bombardment. The beam then passes through a region of radio-frequency electric field and a variable magnetic field, and is detected by an apparatus which records only atoms in the $n = 2$ level.

the frequencies of radio waves are small, the transition rates for spontaneous emission, which are proportional to ν^3 (see (4.80)) are very small. However, *stimulated* (induced) transitions can occur if the atoms are sent through a region where there is an electric field oscillating at the appropriate frequency corresponding to the transition to be studied. In the experiment of Lamb and Retherford such stimulated transitions are observed between the levels $2s_{1/2}$–$2p_{1/2}$ and $2s_{1/2}$–$2p_{3/2}$. Since the transition rates for stimulated absorption and emission are equal (see (4.50)), it is necessary that the two states between which the transitions are studied should be unequally populated.

The experimental method of Lamb and Retherford is based on the fact that the $2s_{1/2}$ level is *metastable*. Indeed, as we have seen in Chapter 4, the electric dipole transition from the state $2s_{1/2}$ to the ground state $1s_{1/2}$ is forbidden by the selection rule $\Delta l = \pm 1$. The most probable decay mechanism of the $2s_{1/2}$ state is two-photon emission, with a lifetime of 1/7 s. Thus, in the absence of perturbations, the lifetime of the $2s_{1/2}$ state is very long compared to that of the 2p states, which is about 1.6×10^{-9} s. In the apparatus of Lamb and Retherford, shown in Fig. 5.6, a beam of atomic hydrogen containing atoms in the metastable $2s_{1/2}$ state is produced by first dissociating molecular hydrogen in a tungsten oven (at a temperature of 2500 K where the dissociation is about 64 per cent complete), selecting a jet of atoms by means of slits, and bombarding this jet with a beam of electrons having a kinetic energy somewhat larger than 10.2 eV, which is the threshold energy for excitation of the $n = 2$ levels of atomic hydrogen. In this way a small fraction of hydrogen atoms (about one in 10^8) is excited to the $2s_{1/2}$, $2p_{1/2}$ and $2p_{3/2}$ states. The average velocity of the atomic beam is about 8×10^5 cm s^{-1}. Because of their long lifetime, the atoms in the metastable $2s_{1/2}$ state can easily reach a detector placed at a distance of about 10 cm from the region where they are produced. On the other hand the atoms which are excited in the $2p_{1/2}$ or $2p_{3/2}$ states quickly decay to the ground state $1s_{1/2}$ in 1.6×10^{-9} s, moving only about 1.3×10^{-3} cm in that time, so that they cannot reach the detector. This detector is a metallic surface (a tungsten ribbon), from which the atoms in the metastable state $2s_{1/2}$ can eject electrons by giving up their excitation energy. Atoms in the ground state are not detected, the measured electronic current being proportional to the number of metastable atoms reaching the detector. Now, if the beam containing the metastable $2s_{1/2}$ atoms passes through an 'interaction region' in which a radio-frequency field of the proper frequency is applied, the metastable atoms will undergo induced transitions to the $2p_{1/2}$ and $2p_{3/2}$ states, and decay to the ground state $1s_{1/2}$ in which they are not detected. As a result, there is a reduction of the number of metastable ($2s_{1/2}$) atoms registered by the detector at the (resonant) radio frequencies corresponding to the frequencies of the $2s_{1/2}$–$2p_{1/2}$ and $2s_{1/2}$–$2p_{3/2}$ transitions. In the 'interaction region' the atomic beam also passes in a variable magnetic field. In this way Lamb and Retherford could not only separate the Zeeman components of the $2s_{1/2}$, $2p_{1/2}$ and $2p_{3/2}$ levels, but also reduce the probability of fortuitous depletion of the $2s_{1/2}$ state due to Stark effect mixing of the $2s_{1/2}$ and the 2p levels caused by perturbing electric fields, as we shall see in Chapter 6. Moreover, the use of a variable external magnetic field avoids the difficulty of producing a radio-frequency field with a variable frequency but a constant radio-frequency power. Instead, Lamb and Retherford could operate at a fixed frequency of the radio-frequency field and obtain the passage through the resonance by varying the magnetic field. The resonance frequency for zero magnetic field was found by extrapolation. In this way Lamb and Retherford found in 1947 that the $2s_{1/2}$ level lies above the $2p_{1/2}$ level by an amount of about 1000 MHz. Further experiments carried out in 1953 by S. Triebwasser, E.S. Dayhoff and W.E. Lamb gave the very precise value (1057.77 ± 0.10) MHz for this energy difference, which is now called a 'Lamb shift'. We note that this value, which corresponds to $4.374\,62 \times 10^{-6}$ eV or $0.035\,283\,4$ cm^{-1}, is about one-tenth of the fine structure splitting of the $n = 2$ term.

The need to explain the Lamb shift stimulated numerous theoretical developments which led H.A. Bethe, S. Tomonaga, J. Schwinger, R.P. Feynman and F.J. Dyson to fundamental revisions of physical concepts (such as the renormalisation of mass) and to the formulation of the theory of *quantum electrodynamics* (QED). In this theory, 'radiative corrections' to the Dirac theory are obtained by taking into account the interaction of the electron with the quantised electromagnetic field. These calculations are outside the scope of this book, and we only mention the following qualitative explanation of the Lamb shift given in 1948 by T.A. Welton. A quantised radiation field in its lowest energy state is not one with zero electromagnetic fields, but there exist zero-point oscillations similar to those we discussed for the case of the harmonic oscillator in Section 2.4. This means that even in the vacuum there are fluctuations in this zero-point radiation field which can act on the electron, causing it to execute rapid oscillatory motions so that its charge is 'smeared out' and the point electron effectively becomes a sphere of a certain radius. If the electron is bound by a non-uniform electric field, as in atomic systems, it will therefore experience a potential which is slightly different from that corresponding to its mean position. In particular, the electron in a

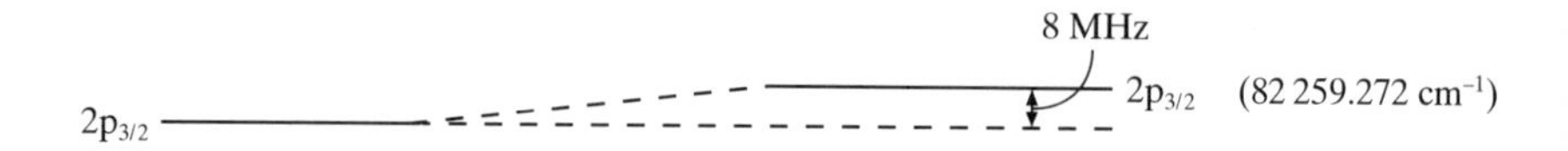

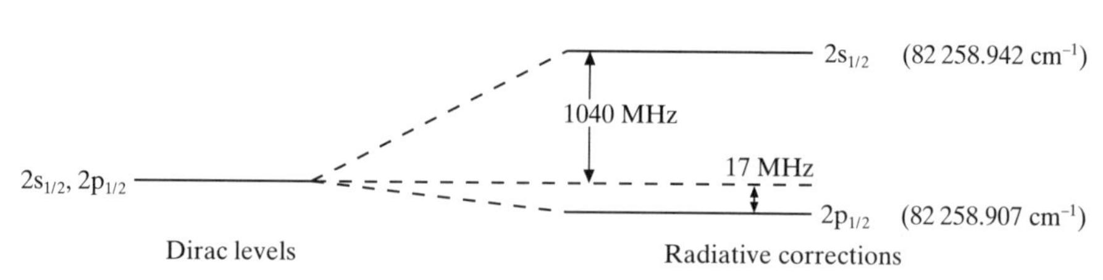

Figure 5.7 Diagram (not to scale) showing the calculated energy shifts due to radiative corrections to the Dirac theory for the $n = 2$ levels of atomic hydrogen.

one-electron atom is not so strongly attracted to the nucleus at short distances. As a result, s states (which are most sensitive to short-distance modifications because $|\psi(0)|^2 \neq 0$ for these states) are raised in energy with respect to other states, for which the corresponding modifications are much smaller.

The calculated energy shifts for the $2s_{1/2}$, $2p_{1/2}$ and $2p_{3/2}$ levels with respect to the Dirac theory are illustrated in Fig. 5.7 for the case of atomic hydrogen. The theoretical value for the Lamb shift (the energy difference between the $2s_{1/2}$ and $2p_{1/2}$ levels) calculated in 1971 by G.W. Erickson is (1057.916 ± 0.010) MHz and that obtained in 1975 by P.J. Mohr is (1057.864 ± 0.014) MHz. Both calculations are in excellent agreement with the experimental value of (1057.77 ± 0.10) MHz measured in 1953 by Lamb and his co-workers, and with more recent experiments [7]. In particular, R.T. Robiscoe and T.W. Shyn measured in 1970 the value (1057.90 ± 0.06) MHz, S.R. Lundeen and F.M. Pipkin found the result (1057.893 ± 0.020) MHz and D.A. Andrews and G. Newton found the value (1057.862 ± 0.020) MHz.

It is also possible to measure the Lamb shift by resolving the Balmer alpha (H_α) line (see Fig. 5.5) using the method of *saturation (Doppler-free) spectroscopy*, which will be described in Section 15.2. We show in Fig. 5.8 the results obtained in this way by T.W. Hänsch, I.S. Shahin and A.L. Schawlow in 1972, compared with the Doppler-broadened H_α line, as it can be observed at room temperature using conventional spectroscopy.

The Lamb shift has also been measured and calculated for other levels of atomic hydrogen and for other hydrogenic systems such as deuterium, He^+ and hydrogen-like multiply charged ions. As an example, we show in Fig. 5.9 the

[7] A detailed account of Lamb shift experiments and calculations can be found in Series (1988).

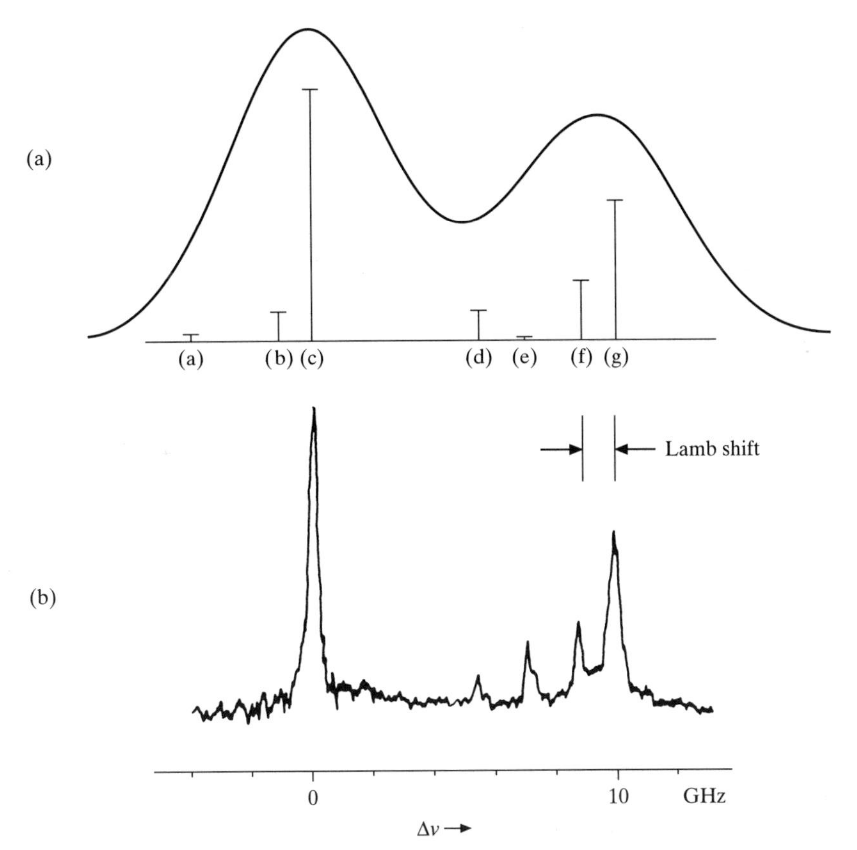

Figure 5.8 (a) The structure of the Balmer alpha (H_α) line, as observed by conventional spectroscopy with Doppler broadening at room temperature (300 K). The lines marked (a), (b)–(g) refer to those shown in Fig. 5.5.
(b) The fully resolved structure of the H_α line obtained by saturation spectroscopy (after T.W. Hänsch *et al.*).

$n = 1$ and $n = 2$ energy levels of hydrogen-like U^{91+} according to the Dirac theory, and with QED corrections. The size of these corrections increases rapidly with increasing Z. In particular, the value of the $2s_{1/2}$–$2p_{1/2}$ Lamb shift is 75 eV for U^{91+}. It is interesting to note that for one-electron ions with high Z the dominant decay mechanism of the $2s_{1/2}$ state is not two-photon emission, but single photon emission through a magnetic dipole (M1) transition. The measured values of the $2p_{3/2} \to 1s_{1/2}$ and $2s_{1/2}$, $2p_{1/2} \to 1s_{1/2}$ energy differences in U^{91+}, obtained in 1993 by Th. Stöhlker *et al.* (who performed X-ray experiments using a heavy ion storage ring), are given by 102.209 keV and 97.706 keV, respectively. The corresponding theoretical values, calculated in 1985 by W.R. Johnson and G. Soff, are 102.180 keV and 97.673 keV.

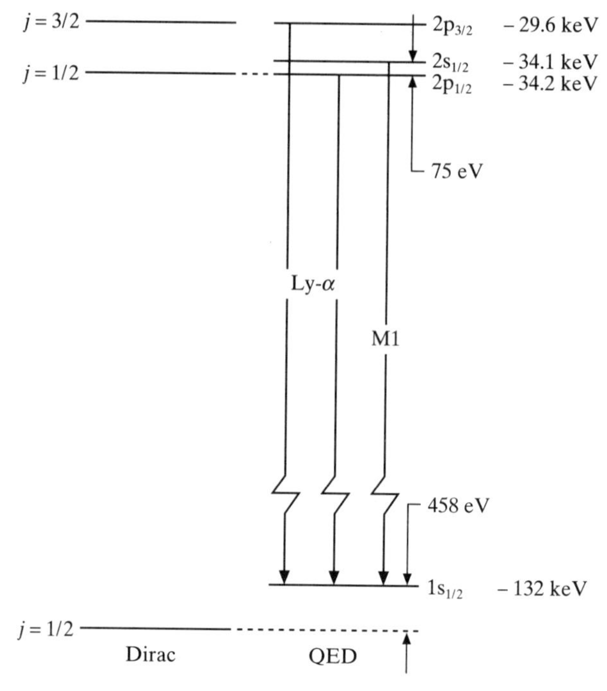

Figure 5.9 The $n = 1$ and $n = 2$ energy levels of U^{91+} according to the Dirac theory and with QED corrections.

5.3 Hyperfine structure and isotope shifts

Atomic nuclei have radii of the order of 10^{-4} Å (10^{-14} m) which are very small compared with typical distances of an electron from the nucleus (~1 Å). The nuclei are also much heavier (about 10^4 times) than electrons. It is therefore a very good approximation to consider the nuclei to be positive point charges of infinite mass. However, the high-precision experiments which can be carried out in atomic physics reveal the existence of tiny effects on the electronic energy levels, which cannot be explained if the nuclei are considered to be point charges of infinite mass. These effects, first observed by A. Michelson in 1891 and C. Fabry and A. Perot in 1897, are called *hyperfine effects*, because they produce shifts of the electronic energy levels which are usually much smaller than those corresponding to the fine structure studied in Section 5.1.

It is convenient to classify the hyperfine effects into those which give rise to *splittings* of the electronic energy levels, and those which slightly shift the energy levels, but without giving rise to splittings. The former are called *hyperfine structure* effects while the latter are known as *isotope shifts* (or isotope effects) since they can usually be detected only by examining their variation between two or more isotopes. We have already encountered examples of isotopic shifts in

Chapters 1 and 3, when we studied the modification of the energy levels of hydrogenic atoms due to the fact that *the nuclear mass is finite* (reduced mass effect). In particular, we saw that the introduction of the reduced mass gives a very good account of the frequency difference between the spectral lines of 'ordinary' atomic hydrogen (proton + electron) and its heavy isotope, deuterium (deuteron + electron). Another isotope shift is the *volume effect*, which arises because the nuclear charge is distributed within a finite volume, so that the potential felt by the electron is modified at short distances. We shall briefly consider this effect at the end of this section.

Let us now turn our attention to the *hyperfine structure effects*, which are responsible for splittings (extending over the range from 10^{-3} to 1 cm^{-1}) of the energy levels of the atoms. These effects result from the fact that a nucleus may possess *electromagnetic multipole moments* (of higher order than the electric monopole) which can interact with the electromagnetic field produced at the nucleus by the electrons. By using general symmetry arguments of parity and time-reversal invariance it may be shown [8] that the number of possible multipole (2^k pole) nuclear moments is severely restricted. Indeed, the only non-vanishing nuclear multipole moments are the *magnetic moments* for *odd* k and the *electric moments* for *even* k, namely the magnetic dipole ($k = 1$), electric quadrupole ($k = 2$), magnetic octupole ($k = 3$) and so on. The most important of these moments are the *magnetic dipole moment* (associated with the nuclear spin) and the *electric quadrupole moment* (caused by the departure from a spherical charge distribution in the nucleus). We shall first examine the hyperfine structure due to the magnetic dipole interaction and then discuss briefly the electric quadrupole interaction.

Magnetic dipole hyperfine structure

In 1924 W. Pauli suggested that a nucleus has a total angular momentum **I** (called 'nuclear spin') and that hyperfine structure effects might be due to magnetic interactions between the nucleus and the moving electrons of the atom, dependent upon the orientation of this nuclear spin. The eigenvalues of the operator $\mathbf{I}^2$ will be written as $I(I + 1)\hbar^2$, where I is the *nuclear spin quantum number* (also often called the spin of the nucleus) or in other words the maximum possible component of **I** (measured in units of $\hbar$) in any given direction. Now the nucleus is a compound structure of nucleons (protons and neutrons) which have an intrinsic spin 1/2 and may participate in orbital motion within the nucleus. Thus the nuclear spin is compounded from the spins of the nucleons, and can also contain an orbital component. The corresponding spin quantum number I may have integer (including zero) or half-odd integer values. In the former case the nucleus is a *boson* (obeying Bose–Elnstein statistics) while in the latter case it is a *fermion* (obeying Fermi–Dirac statistics). We shall also denote by $M_I\hbar$ the eigenvalues of the operator I_z, so that the possible values of M_I are $M_I = -I, -I + 1, \ldots, I$.

---

[8] See for example Ramsey (1953).

As we pointed out above, a nucleus may possess 2^k-pole moments, with k odd for magnetic moments and k even for electric moments. Furthermore, it may be shown [8] that a nucleus of spin quantum number I cannot have a multipole moment of order 2^n, where n is greater than $2I$. We shall begin by considering the nucleus as a point dipole with a magnetic dipole moment $\mathcal{M}_N$ proportional to the nuclear spin $\mathbf{I}$. That is,

$$\mathcal{M}_N = g_I \mu_N \mathbf{I}/\hbar \tag{5.36}$$

where g_I is a dimensionless number (whose order of magnitude is unity) called the *nuclear g factor* or *nuclear Landé factor*. We note that g_I is positive if $\mathcal{M}_N$ lies along $\mathbf{I}$. The quantity μ_N which appears in (5.36) is called the *nuclear magneton*; it is defined by

$$\mu_N = \frac{e\hbar}{2M_p} = \frac{m}{M_p}\mu_B \tag{5.37}$$

where m is the mass of the electron, M_p the mass of the proton and μ_B the Bohr magneton. Thus the nuclear magneton μ_N is smaller than the Bohr magneton μ_B by the factor $m/M_p = 1/1836.15$. The numerical value of the nuclear magneton is

$$\mu_N = 5.050\,78 \times 10^{-27} \text{ joule/tesla} \tag{5.38}$$

It is worth noting that (5.36) is sometimes written in units of Bohr magnetons as

$$\mathcal{M}_N = g'_I \mu_B \mathbf{I}/\hbar \tag{5.39}$$

in which case

$$g'_I = \frac{\mu_N}{\mu_B} g_I = \frac{m}{M_p} g_I = \frac{g_I}{1836.15} \tag{5.40}$$

is a very small number. Since $I\hbar$ is the maximum component of $\mathbf{I}$ in a given direction, we may also write (5.36) as

$$\mathcal{M}_N = \left(\frac{\mathcal{M}_N}{I}\right)\mathbf{I}/\hbar \tag{5.41}$$

where $\mathcal{M}_N$ is the value of the nuclear magnetic moment. In units of nuclear magnetons, we have

$$\mathcal{M}_N = g_I I \tag{5.42}$$

The values of the spin quantum number I, the nuclear Landé factor g_I and the nuclear magnetic moment $\mathcal{M}_N$ are given in Table 5.1 for the nucleons and a few nuclei.

Let us consider a hydrogenic atom with a nucleus of charge Ze such that $Z\alpha \ll 1$, and a magnetic dipole moment $\mathcal{M}_N$. We shall write the Hamiltonian of this system as

$$H = H_0 + H'_{MD} \tag{5.43}$$

where the zero-order Hamiltonian H_0 now includes the Coulomb interaction $-Ze^2/(4\pi\varepsilon_0 r)$ and the relativistic (fine structure) corrections discussed in Section 5.1 (which are of order $(Z\alpha)^2$, as seen from (5.29)) while H'_{MD} is a perturbative

Table 5.1 Values of the spin, Landé factor and magnetic moment of the nucleons and some nuclei. The notation is such that ${}^a_bX$ represents a nucleus with a total of *a* nucleons, *b* of which are protons.

Nucleus	*Spin I*	*Landé factor* g_I	*Magnetic moment* $\mathcal{M}_N$ *(in nuclear magnetons)*
proton p	1/2	5.588 3	2.792 78
neutron n	1/2	−3.826 3	−1.913 15
deuteron ${}^2_1D$	1	0.857 42	0.857 42
${}^3_2He$	1/2	−4.255	−2.127 6
${}^4_2He$	0	–	0
${}^{12}_6C$	0	–	0
${}^{13}_6C$	1/2	1.404 82	0.702 41
${}^{16}_8O$	0	–	0
${}^{19}_9F$	1/2	5.257 732	2.628 866
${}^{31}_{15}P$	1/2	2.263 20	1.131 6
${}^{39}_{19}K$	3/2	0.260 9	0.391 4
${}^{67}_{30}Zn$	5/2	0.350 28	0.875 7
${}^{85}_{37}Rb$	5/2	0.541 08	1.352 7
${}^{129}_{54}Xe$	1/2	−1.553 6	−0.776 8
${}^{133}_{55}Cs$	7/2	0.736 9	2.579
${}^{199}_{80}Hg$	1/2	1.005 4	0.502 7
${}^{201}_{80}Hg$	3/2	−0.371 13	−0.556 7

term due to the presence of the magnetic dipole moment $\mathcal{M}_N$. This term will clearly lead to even smaller corrections than those corresponding to the fine structure, since the magnetic moment of the nucleus is much smaller than that of the electron. We may therefore assume that we can deal with an isolated electronic level labelled by the total electronic angular momentum quantum number j. The zero-order wave functions (eigenfunctions of H_0) are separable in the electronic and nuclear variables and are eigenfunctions of $\mathbf{J}^2$, J_z, $\mathbf{I}^2$ and I_z (where $\mathbf{J} = \mathbf{L} + \mathbf{S}$ is the total electronic angular momentum operator). Using the Dirac notation, we shall write them as $|\gamma j m_j I M_I\rangle$, where the symbol γ represents additional quantum numbers. These zero-order wave functions are $(2j+1)(2I+1)$-fold degenerate in m_j and M_I. We also remark that in the Pauli approximation (see Appendix 7) – which we shall adopt here – the zero-order wave functions are also eigenfunctions of $\mathbf{L}^2$ and $\mathbf{S}^2$, and will thus be written more explicitly as $|lsjm_jIM_I\rangle$.

We now examine the perturbation $H'_{\rm MD}$ due to the magnetic dipole moment $\mathcal{M}_N$ of the nucleus. The magnetic field due to this dipole moment will interact with both the orbital angular momentum $\mathbf{L}$ and the spin $\mathbf{S}$ of the atomic electron. We shall denote the former interaction by H'_1 and the second by H'_2, so that

$$H'_{\rm MD} = H'_1 + H'_2 \tag{5.44}$$

The term H_1' is readily evaluated as follows. The vector potential $\mathbf{A}(\mathbf{r})$ due to a point dipole located at the origin is (Jackson, 1998)

$$\mathbf{A}(\mathbf{r}) = -\frac{\mu_0}{4\pi}\left[\mathcal{M}_\mathrm{N} \times \nabla\left(\frac{1}{r}\right)\right]$$

$$= \frac{\mu_0}{4\pi}(\mathcal{M}_\mathrm{N} \times \mathbf{r})\frac{1}{r^3} \tag{5.45}$$

Neglecting for a moment the spin of the electron, the interaction term due to the presence of the vector potential $\mathbf{A}(\mathbf{r})$ is (see (4.28))

$$H_1' = -\frac{i\hbar e}{m}\mathbf{A}\cdot\nabla \tag{5.46}$$

Inserting (5.45) into (5.46) one obtains (Problem 5.6)

$$H_1' = \frac{\mu_0}{4\pi}\frac{2}{\hbar}\mu_\mathrm{B}\frac{1}{r^3}\mathbf{L}\cdot\mathcal{M}_\mathrm{N}$$

$$= \frac{\mu_0}{4\pi}\frac{2}{\hbar^2}g_I\mu_\mathrm{B}\mu_\mathrm{N}\frac{1}{r^3}\mathbf{L}\cdot\mathbf{I} \tag{5.47}$$

where we have used (5.36) and we recall that $\mathbf{L} = \mathbf{r} \times \mathbf{p}$. We remark that the term H_1' may be interpreted as the interaction of the nuclear dipole moment $\mathcal{M}_\mathrm{N}$ with the magnetic field $-[\mu_0/(4\pi)]e\mathbf{L}/(mr^3)$ created at the nucleus by the rotation of the electronic charge. We also note that H_1' has non-zero matrix elements only between states for which $l \neq 0$.

Next, we find the contribution H_2' arising from the electron spin $\mathbf{S}$. The magnetic field associated with the vector potential (5.45) is

$$\mathcal{B} = \nabla \times \mathbf{A} = -\frac{\mu_0}{4\pi}\left[\mathcal{M}_\mathrm{N}\nabla^2\left(\frac{1}{r}\right) - \nabla(\mathcal{M}_\mathrm{N}\cdot\nabla)\frac{1}{r}\right] \tag{5.48}$$

The spin magnetic moment of the electron is $\mathcal{M}_s = -g_s\mu_\mathrm{B}\mathbf{S}/\hbar$ so that the corresponding interaction energy is (with $g_s = 2$)

$$H_2' = -\mathcal{M}_s\cdot\mathcal{B} = 2\mu_\mathrm{B}\mathbf{S}\cdot\mathcal{B}/\hbar \tag{5.49}$$

or

$$H_2' = \frac{\mu_0}{4\pi}\left[\mathcal{M}_s\cdot\mathcal{M}_\mathrm{N}\nabla^2\left(\frac{1}{r}\right) - (\mathcal{M}_s\cdot\nabla)(\mathcal{M}_\mathrm{N}\cdot\nabla)\frac{1}{r}\right]$$

$$= -\frac{\mu_0}{4\pi}\frac{2}{\hbar^2}g_I\mu_\mathrm{B}\mu_\mathrm{N}\left[\mathbf{S}\cdot\mathbf{I}\,\nabla^2\left(\frac{1}{r}\right) - (\mathbf{S}\cdot\nabla)(\mathbf{I}\cdot\nabla)\frac{1}{r}\right] \tag{5.50}$$

where we have used (5.36).

It is convenient to examine the term H'_2 separately for the two cases $r \neq 0$ and $r = 0$. Since the hydrogenic wave functions behave like r^l at the origin, the expression of H'_2 at $r = 0$ will only be relevant for states with $l = 0$ (s states). We first note that since

$$\nabla^2 \left(\frac{1}{r} \right) = -4\pi\, \delta(\mathbf{r}) \tag{5.51}$$

the first term in square brackets in (5.50) vanishes for $r \neq 0$. It can also be shown (Problem 5.7) that for $r \neq 0$

$$(\mathbf{S}\cdot\nabla)(\mathbf{I}\cdot\nabla)\frac{1}{r} = -\frac{1}{r^3}\left[\mathbf{S}\cdot\mathbf{I} - 3\frac{(\mathbf{S}\cdot\mathbf{r})(\mathbf{I}\cdot\mathbf{r})}{r^2}\right], \qquad r \neq 0 \tag{5.52}$$

Hence, using (5.50) and (5.52), we have

$$\begin{aligned} H'_2 &= -\frac{\mu_0}{4\pi}\frac{2}{\hbar^2} g_I \mu_B \mu_N \frac{1}{r^3}\left[\mathbf{S}\cdot\mathbf{I} - 3\frac{(\mathbf{S}\cdot\mathbf{r})(\mathbf{I}\cdot\mathbf{r})}{r^2}\right] \\ &= \frac{\mu_0}{4\pi}\frac{1}{r^3}\left[\mathcal{M}_s\cdot\mathcal{M}_N - 3\frac{(\mathcal{M}_s\cdot\mathbf{r})(\mathcal{M}_N\cdot\mathbf{r})}{r^2}\right], \qquad r \neq 0 \end{aligned} \tag{5.53}$$

which represents the dipole–dipole interaction between the magnetic moments of the electron and the nucleus. Adding the results (5.47) and (5.53), the interaction between the nuclear magnetic dipole moment and an electron for which $l \neq 0$ is seen to be

$$H'_{MD} = \frac{\mu_0}{4\pi}\frac{2}{\hbar^2} g_I \mu_B \mu_N \frac{1}{r^3}\left[\mathbf{L}\cdot\mathbf{I} - \mathbf{S}\cdot\mathbf{I} + 3\frac{(\mathbf{S}\cdot\mathbf{r})(\mathbf{I}\cdot\mathbf{r})}{r^2}\right], \qquad r \neq 0 \tag{5.54}$$

Let us now return to the expression (5.50) of H'_2 and consider the case $r = 0$, which is important for s states ($l = 0$). We have already seen that the first term in square brackets in (5.50) is proportional to $\delta(\mathbf{r})$. The second term in square brackets contains a similar term proportional to $\delta(\mathbf{r})$, as we now show. Indeed, for matrix elements involving spherically symmetric states (with $l = 0$) we remark that out of the expression (with $x_1 = x$, $x_2 = y$, $x_3 = z$)

$$(\mathbf{S}\cdot\nabla)(\mathbf{I}\cdot\nabla)\frac{1}{r} = \sum_{i=1}^{3}\sum_{j=1}^{3} S_i I_j \frac{\partial^2}{\partial x_i \partial x_j}\left(\frac{1}{r}\right) \tag{5.55}$$

all terms will vanish except those with $i = j$. Each of the matrix elements of

$$\frac{\partial^2}{\partial x_1^2}\left(\frac{1}{r}\right), \qquad \frac{\partial^2}{\partial x_2^2}\left(\frac{1}{r}\right), \qquad \frac{\partial^2}{\partial x_3^2}\left(\frac{1}{r}\right)$$

must have the same value, so that for $l = 0$

$$(\mathbf{S}\cdot\nabla)(\mathbf{I}\cdot\nabla)\frac{1}{r}=\frac{1}{3}(\mathbf{S}\cdot\mathbf{I})\nabla^2\left(\frac{1}{r}\right)=-\frac{4\pi}{3}\mathbf{S}\cdot\mathbf{I}\,\delta(\mathbf{r}) \tag{5.56}$$

From these equations and the fact that the term H_1' does not contribute for states with $l = 0$, we deduce that the interaction between the nuclear magnetic dipole moment and an s electron is given by

$$\begin{aligned} H'_{\text{MD}} &= \frac{\mu_0}{4\pi}\frac{2}{\hbar^2}g_I\mu_B\mu_N\frac{8\pi}{3}\delta(\mathbf{r})\mathbf{S}\cdot\mathbf{I} \\ &= -\frac{\mu_0}{4\pi}\frac{8\pi}{3}\mathcal{M}_s\cdot\mathcal{M}_N\,\delta(\mathbf{r}), \qquad l=0 \end{aligned} \tag{5.57}$$

This expression, which is proportional to $\delta(\mathbf{r})$, is called the *Fermi contact interaction.*

We now proceed to the calculation of the *first-order energy shifts* due to the perturbations (5.54) and (5.57). We begin by considering the case $l \neq 0$, and write (5.54) more simply as

$$H'_{\text{MD}} = \frac{\mu_0}{4\pi}\frac{2}{\hbar^2}g_I\mu_B\mu_N\frac{1}{r^3}\mathbf{G}\cdot\mathbf{I} \tag{5.58}$$

where

$$\mathbf{G} = \mathbf{L}-\mathbf{S}+3\frac{(\mathbf{S}\cdot\mathbf{r})\mathbf{r}}{r^2}. \tag{5.59}$$

We have seen above that the zero-order wave functions $|lsjm_jIM_I\rangle$ are $(2j + 1)(2I + 1)$-fold degenerate in m_j and M_I. By analogy with the spin–orbit coupling discussed in Section 5.1, the diagonalisation of the perturbation is greatly simplified by introducing the total angular momentum of the atom (nucleus + electron)

$$\mathbf{F} = \mathbf{I}+\mathbf{J} \tag{5.60}$$

We shall denote by $F(F + 1)\hbar^2$ the eigenvalues of the operator $\mathbf{F}^2$ and by $M_F\hbar$ those of F_z, with $M_F = -F, -F + 1, \ldots, +F$. From the rules concerning the addition of angular momenta, the possible values of the quantum number F are given by

$$F = |I-j|, |I-j|+1, \ldots, I+j-1, I+j \tag{5.61}$$

Since F and M_F remain good quantum numbers under the application of the perturbation H'_{MD}, it is convenient to form new zero-order functions $|lsjIFM_F\rangle$ which are linear combinations of the functions $|lsjm_jIM_I\rangle$. The energy shift due to the perturbation (5.58) is then

$$\Delta E = \frac{\mu_0}{4\pi}\frac{2}{\hbar^2}g_I\mu_B\mu_N\left\langle lsjIFM_F\left|\frac{1}{r^3}\mathbf{G}\cdot\mathbf{I}\right|lsjIFM_F\right\rangle, \qquad l\neq 0 \tag{5.62}$$

Using the identity (A4.58) of Appendix 4 we can replace the matrix element of $\mathbf{G}\cdot\mathbf{I}$ taken between states with equal j by that of $[j(j+1)\hbar^2]^{-1}(\mathbf{G}\cdot\mathbf{J})(\mathbf{I}\cdot\mathbf{J})$. Moreover, since

$$\mathbf{F}^2 = \mathbf{I}^2 + 2\mathbf{I}\cdot\mathbf{J} + \mathbf{J}^2 \tag{5.63}$$

so that

$$\mathbf{I}\cdot\mathbf{J} = \tfrac{1}{2}(\mathbf{F}^2 - \mathbf{I}^2 - \mathbf{J}^2) \tag{5.64}$$

we have

$$\Delta E = \frac{C}{2}[F(F+1) - I(I+1) - j(j+1)] \tag{5.65}$$

with

$$C = \frac{\mu_0}{4\pi} 2g_I \mu_B \mu_N \frac{1}{j(j+1)\hbar^2}\left\langle \frac{1}{r^3}\mathbf{G}\cdot\mathbf{J}\right\rangle, \qquad l \neq 0 \tag{5.66}$$

and we have used the simplified notation $\langle\,\rangle$ for the expectation value.

The quantity $\langle r^{-3}\mathbf{G}\cdot\mathbf{J}\rangle$ is readily obtained as follows. We first note that since $\mathbf{L}\cdot\mathbf{r} = 0$, we may write

$$\begin{aligned}\mathbf{G}\cdot\mathbf{J} &= \left(\mathbf{L} - \mathbf{S} + 3\frac{(\mathbf{S}\cdot\mathbf{r})\mathbf{r}}{r^2}\right)\cdot(\mathbf{L}+\mathbf{S}) \\ &= \mathbf{L}^2 - \mathbf{S}^2 + 3\frac{(\mathbf{S}\cdot\mathbf{r})^2}{r^2}\end{aligned} \tag{5.67}$$

It is easily shown (Problem 5.7) that

$$\mathbf{S}^2 - 3\frac{(\mathbf{S}\cdot\mathbf{r})^2}{r^2} = 0 \tag{5.68}$$

so that $\mathbf{G}\cdot\mathbf{J} = \mathbf{L}^2$ and

$$\left\langle \frac{1}{r^3}\mathbf{G}\cdot\mathbf{J}\right\rangle = l(l+1)\hbar^2\left\langle\frac{1}{r^3}\right\rangle \tag{5.69}$$

Thus we have

$$\begin{aligned}C &= \frac{\mu_0}{4\pi}2g_I\mu_B\mu_N\frac{l(l+1)}{j(j+1)}\left\langle\frac{1}{r^3}\right\rangle \\ &= \frac{\mu_0}{4\pi}2g_I\mu_B\mu_N\frac{l(l+1)}{j(j+1)}\frac{Z^3}{a_\mu^3 n^3 l(l+1/2)(l+1)}, \qquad l\neq 0\end{aligned} \tag{5.70}$$

where we have used the expectation value of r^{-3} given by (3.78), and we recall that $a_\mu = a_0(m/\mu)$, μ being the reduced mass of the electron with respect to the nucleus.

Turning now to the case of s states ($l = 0$), the first-order energy shift due to the perturbation (5.57) is

$$\Delta E = \frac{\mu_0}{4\pi}\frac{2}{\hbar^2} g_I \mu_B \mu_N \frac{8\pi}{3}\langle \delta(\mathbf{r})\, \mathbf{S}\cdot\mathbf{I}\rangle, \qquad l = 0 \tag{5.71}$$

As $\mathbf{L} = 0$, we have $\mathbf{F} = \mathbf{I} + \mathbf{S}$, from which

$$\mathbf{S}\cdot\mathbf{I} = \tfrac{1}{2}(\mathbf{F}^2 - \mathbf{I}^2 - \mathbf{S}^2) \tag{5.72}$$

and therefore

$$\Delta E = \frac{C_0}{2}[F(F+1) - I(I+1) - s(s+1)], \; s = \frac{1}{2} \tag{5.73}$$

with

$$C_0 = \frac{\mu_0}{4\pi} 2 g_I \mu_B \mu_N \frac{8\pi}{3}\langle \delta(\mathbf{r})\rangle, \qquad l = 0 \tag{5.74}$$

Now

$$\langle \delta(\mathbf{r})\rangle = \int |\psi_{n00}(r)|^2 \delta(\mathbf{r})\, \mathrm{d}\mathbf{r} = |\psi_{n00}(0)|^2 = \frac{Z^3}{\pi a_\mu^3 n^3} \tag{5.75}$$

where we have used the result (3.65). Thus

$$C_0 = \frac{\mu_0}{4\pi}\frac{16}{3} g_I \mu_B \mu_N \frac{Z^3}{a_\mu^3 n^3} \tag{5.76}$$

Comparing (5.65) and (5.73), and recalling that $j = s = 1/2$ for s states, we see that for both cases $l \neq 0$ and $l = 0$ we have

$$\Delta E = \frac{C}{2}[F(F+1) - I(I+1) - j(j+1)] \tag{5.77}$$

with

$$C = \frac{\mu_0}{4\pi} 4 g_I \mu_B \mu_N \frac{1}{j(j+1)(2l+1)}\frac{Z^3}{a_\mu^3 n^3} \tag{5.78}$$

Using atomic units and introducing the fine structure constant α, we may also display this result (writing explicitly the electron mass m) as

$$\Delta E = \frac{1}{2}\frac{m}{M_p} g_I \frac{Z^3\alpha^2}{n^3}\left(\frac{\mu}{m}\right)^3 \frac{F(F+1) - I(I+1) - j(j+1)}{j(j+1)(2l+1)} \quad \text{a.u.} \tag{5.79}$$

For a given nucleus having a spin quantum number I, a fine structure atomic energy level corresponding to fixed values of l and j is therefore split further into hyperfine components labelled by F. Since the energy correction does not depend on M_F, each of these hyperfine energy levels is $(2F + 1)$-fold degenerate. The possible values of F being $|I - j|, |I - j| + 1, \ldots, I + j$ (see (5.61)), the number of hyperfine structure components corresponding to a fine structure energy level is the smaller of the two numbers $(2j + 1)$ and $(2I + 1)$. These components are said to form a *hyperfine structure multiplet*. As an example, we show in Fig. 5.10 a schematic drawing of the hyperfine structure splitting of the $n = 1$ and $n = 2$ levels of 'ordinary' hydrogen (H) and deuterium (D). For 'ordinary' hydrogen the spin of the nucleus is just the spin of the proton, $I = 1/2$, and since $j = 1/2, 3/2, \ldots$ we always have hyperfine *doublets*. On the other hand, for deuterium the spin of the nucleus is $I = 1$, so that we have doublets for $j = 1/2$ and triplets for the other values of j.

We remark from (5.78) that since the quantity C is independent of F the *energy difference* between two neighbouring hyperfine levels – called *hyperfine separation* – is just

$$\Delta E(F) - \Delta E(F - 1) = CF \tag{5.80}$$

and is thus proportional to F. This is an example of an *interval rule*. From (5.79) we also see that the energy separation δE between the two outermost components of the hyperfine multiplet (corresponding to the values $F_1 = I + j$ and $F_2 = |I - j|$ of the quantum number F) is given in atomic units by

$$\delta E = \frac{m}{M_\text{p}}\left(\frac{\mu}{m}\right)^3 g_I \frac{2Z^3\alpha^2}{n^3(j+1)(2l+1)} \times \begin{bmatrix} I + 1/2 & \text{for} \quad j \leqslant I \\ \dfrac{I(j+1/2)}{j} & \text{for} \quad j \geqslant I \end{bmatrix} \tag{5.81}$$

The *hyperfine structure of spectral lines* resulting from the magnetic dipole interaction may be obtained (in a way similar to the fine structure discussed in Section 5.1) by combining the above results with the *selection rules* for electromagnetic transitions between energy levels. For *electric dipole transitions* the selection rules obtained in Section 5.1 ($\Delta l = \pm 1$ and $\Delta j = 0, \pm 1$) remain valid, and in addition it may be shown that the quantum number F obeys the selection rule

$$\Delta F = 0, \pm 1 \tag{5.82}$$

the transition $F = 0 \rightarrow F = 0$ being excluded. Examples of allowed hyperfine transitions are shown in Fig. 5.11. We note that transitions between levels having the same value of j but different values of F can also take place. These transitions are in the microwave region and are generally weak, so that they are best observed by using stimulated emission techniques.

The hyperfine transitions, observed by optical or microwave spectroscopy, can be used to determine the spin I and magnetic dipole moment $\mathcal{M}_\text{N} = g_I I$ of the nucleus. Indeed, the maximum hyperfine multiplicity of levels with large enough j gives $(2I + 1)$ and the hyperfine separation allows the determination of the nuclear Landé factor g_I. Using the generalisation of the above equations for complex atoms (see Chapter 9), the dipole magnetic moments of many nuclei have been obtained.

The hyperfine structure of the $1s_{1/2}$ ground state of 'ordinary' hydrogen (H) is of particular interest, because in this case very elaborate calculations can be carried

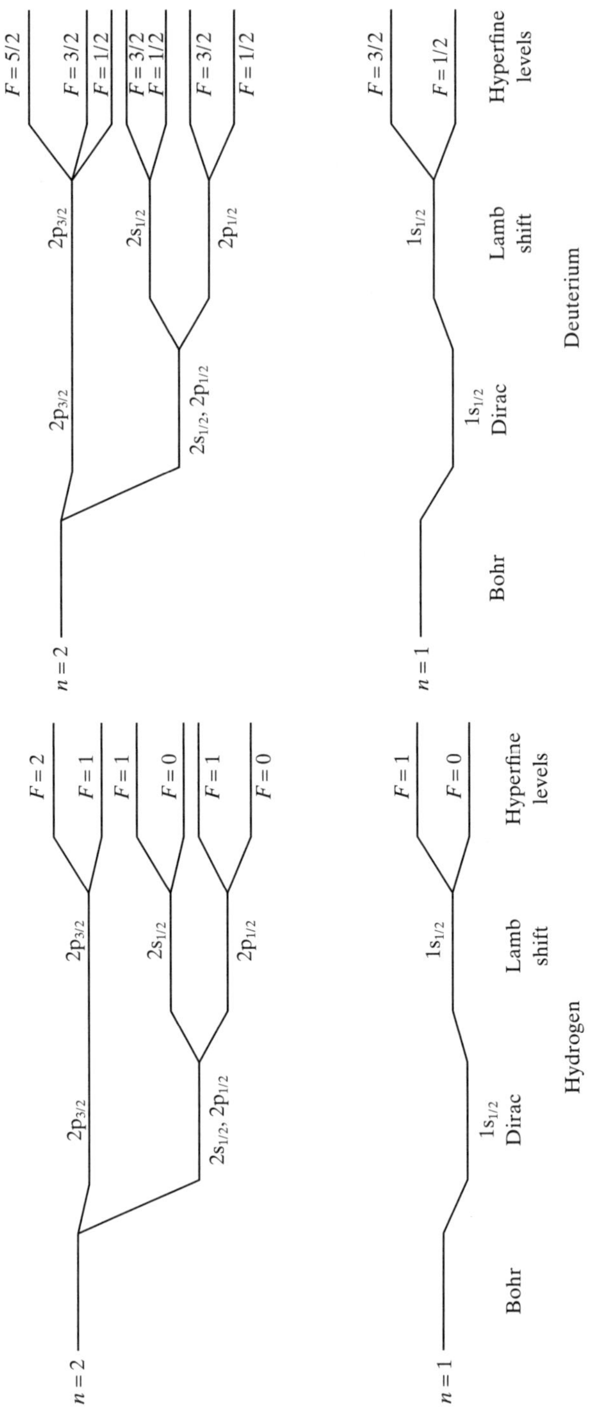

Figure 5.10 The splitting of the n = 1 and n = 2 levels of hydrogen and deuterium. The splittings are not to scale and are magnified from the left to the right of the diagram.

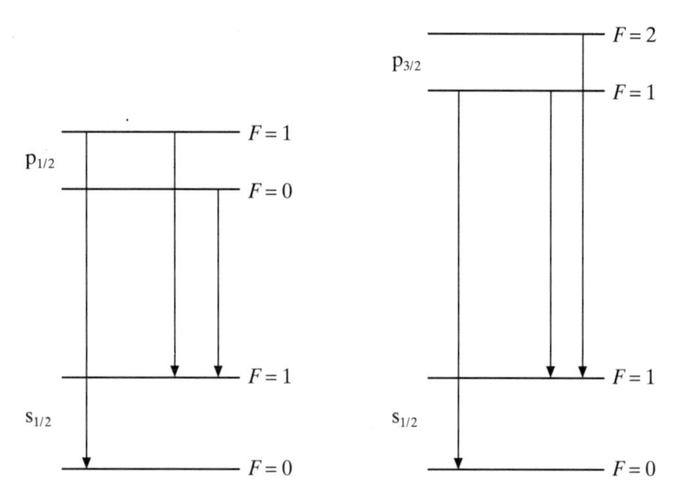

Figure 5.11 Allowed dipole transitions between *n*′p and *ns* levels of hydrogen. There is no restriction on *n*′ and *n*, and the case *n*′ = *n* is allowed.

out and compared with extremely high-precision measurements, performed by using atomic beam magnetic resonance methods [9]. Since the proton spin is $I = 1/2$ and the level $1s_{1/2}$ has a total electronic angular momentum quantum number $j = 1/2$, this level splits into two hyperfine components corresponding to the values $F = 0$ and $F = 1$, the state with $F = 0$ being the ground state (see Fig. 5.10). Using (5.81) we see that the energy difference between the two hyperfine levels is given in atomic units by

$$\delta E = \frac{4}{3}\frac{m}{M_p}\left(\frac{\mu}{m}\right)^3 g_p \alpha^2 \tag{5.83}$$

where $g_p = 5.5883$ is the Landé factor of the proton. From this result we find that the frequency $\nu = \delta E/h$ of the transition between the two hyperfine levels (which is a magnetic dipole transition) is $\nu \simeq 1420$ MHz, the corresponding wavelength being $\lambda \simeq 21$ cm. The 'hydrogen maser', invented in 1960 by H.M. Goldenberg, D. Kleppner and N.F. Ramsey (see Section 15.1), has given very accurate data for the hyperfine frequency ν. By comparing with the frequency of a caesium atomic clock (see Section 16.5) used as a reference, the best value of ν obtained by using the hydrogen maser is

$$\nu = (1\,420\,405\,751.766\,7 \pm 0.000\,9)\ \text{Hz} \tag{5.84}$$

which is one of the most accurately measured quantities in physics.

It is gratifying to note that the simple theory presented above agrees with this result to within about 0.1 per cent. Much better agreement between theory and

[9] Magnetic resonance methods are discussed in Chapter 16. See also Ramsey (1953).

experiment can be obtained by including various corrections in the theoretical calculations. The most important of these is the introduction of the *anomalous magnetic moment of the electron*, according to which the spin gyromagnetic ratio g_s of the electron is slightly different from the value $g_s = 2$ predicted by the Dirac theory [10].

We also remark that the transition between the two hyperfine levels $F = 1$ and $F = 0$ of the ground state of hydrogen plays a very important role in *radio-astronomy*. Indeed, from the analysis of the intensity of the 21 cm radio-frequency radiation received, the astronomers have been able to learn a great deal about the distribution of neutral hydrogen atoms in interstellar space, as we shall see in Chapter 16.

Electric quadrupole hyperfine structure

A second important characteristic of the structure of a nucleus is the *electric quadrupole moment*. It is a symmetric, second-order tensor whose components Q_{ij} are defined in the following way. Let $\mathbf{R}_\mathrm{p}$ be the coordinate of a proton with respect to the centre of mass of the nucleus, and let $X_{\mathrm{p}1} = X_\mathrm{p}$, $X_{\mathrm{p}2} = Y_\mathrm{p}$, $X_{\mathrm{p}3} = Z_\mathrm{p}$ be its Cartesian components. Then

$$Q_{ij} = \sum_\mathrm{p} 3X_{\mathrm{p}i}X_{\mathrm{p}j} - \delta_{ij}R_\mathrm{p}^2 \qquad (i, j = 1, 2, 3) \tag{5.85}$$

where the sum is over all the protons in the nucleus. It is customary to define the *magnitude Q of the electric quadrupole moment* as the average value of the component $Q_{zz} = Q_{33}$ in the state $|I, M_I = I\rangle$. That is,

$$\begin{aligned} Q &= \langle I, M_I = I | Q_{zz} | I, M_I = I \rangle \\ &= \left\langle I, M_I = I \left| \sum_\mathrm{p} 3Z_\mathrm{p}^2 - R_\mathrm{p}^2 \right| I, M_I = I \right\rangle \end{aligned} \tag{5.86}$$

The quantity Q has the dimensions of an area and is often measured in barns (10^{-24} cm^2). For example, the deuteron has an electric quadrupole moment of magnitude $Q = 0.0028$ barns. It is clear from (5.86) that a nucleus whose charge distribution is spherically symmetric has no electric quadrupole moment, since then the average value of $3Z_\mathrm{p}^2$ is equal to that of $R_\mathrm{p}^2 = X_\mathrm{p}^2 + Y_\mathrm{p}^2 + Z_\mathrm{p}^2$. In fact the value of Q gives a measure of the deviation from a spherical charge distribution in the nucleus. If the nuclear charge distribution is elongated along the direction of $\mathbf{I}$ (prolate), then $Q > 0$; on the other hand $Q < 0$ if the charge distribution is flattened (oblate).

[10] The quantity $a = (g_s - 2)$ has been measured with very high accuracy by R.S. Van Dyck, Jr, P.B. Schwinberg and H.G. Dehmelt in 1987 to give the value $a = (1.159\,652\,188\,4 \pm 0.000\,000\,004\,3) \times 10^{-3}$. The current theoretical result, calculated by using quantum electrodynamics, is $(1.159\,652\,2 \pm 0.000\,000\,2) \times 10^{-3}$, the main source of inaccuracy being uncertainties of the order of 10^{-7} in the value of the fine structure constant α.

The interaction energy H'_{EQ} between the electric quadrupole moment of the nucleus and the electrostatic potential V_{e} created by an electron at the nucleus was first obtained by H. Casimir. Provided I and j are both good quantum numbers, it is given in atomic units by [11]

$$H'_{\mathrm{EQ}} = B \frac{\frac{3}{2}\mathbf{I}\cdot\mathbf{J}(2\mathbf{I}\cdot\mathbf{J}+1) - \mathbf{I}^2\mathbf{J}^2}{2I(2I-1)j(2j-1)} \tag{5.87}$$

where the *quadrupole coupling constant* B is given by

$$B = Q\left\langle \frac{\partial^2 V_{\mathrm{e}}}{\partial z^2} \right\rangle \tag{5.88}$$

Here

$$\left\langle \frac{\partial^2 V_{\mathrm{e}}}{\partial z^2} \right\rangle = \left\langle j, m_j = j \left| \frac{\partial^2 V_{\mathrm{e}}}{\partial z^2} \right| j, m_j = j \right\rangle = -\left\langle j, m_j = j \left| \frac{3z^2 - r^2}{r^5} \right| j, m_j = j \right\rangle \tag{5.89}$$

is the average gradient of the electric field produced by the electron at the nucleus.

The first-order energy shift due to the electric quadrupole interaction (5.87) is

$$\begin{aligned}\Delta E &= \langle jIFM_F | H'_{\mathrm{EQ}} | jIFM_F \rangle \\ &= \frac{B}{4} \frac{\frac{3}{2}K(K+1) - 2I(I+1)j(j+1)}{I(2I-1)j(2j-1)}\end{aligned} \tag{5.90}$$

where

$$K = F(F+1) - I(I+1) - j(j+1) \tag{5.91}$$

Since $\langle \partial^2 V_{\mathrm{e}}/\partial z^2 \rangle$ vanishes when the electron charge distribution is spherically symmetric, there is no quadrupole energy shift for s states. We recall that the nuclei having no spin ($I = 0$) or a spin $I = 1/2$ have no electric quadrupole moment, so that the energy shift (5.90) also vanishes in this case.

Adding the electric quadrupole correction (5.90) to the magnetic dipole energy shift (5.77) we find that the total hyperfine structure energy correction is given by

$$\Delta E = \frac{C}{2}K + \frac{B}{4} \frac{\frac{3}{2}K(K+1) - 2I(I+1)j(j+1)}{I(2I-1)j(2j-1)} \tag{5.92}$$

Because its dependence on the quantum number F is different from that of the magnetic dipole correction (5.77), we see that the electric quadrupole correction causes a *departure from the interval rule* (5.80).

---

[11] See for example Casimir (1963) or Ramsey (1953).

It is worth noting that the hyperfine energy levels obtained after the correction (5.92) has been applied are still independent of the quantum number M_F, and hence are $(2F+1)$-fold degenerate. This degeneracy can be removed by applying an external magnetic field. We shall return to this *Zeeman effect in hyperfine structure* in Chapter 9.

Isotope shifts

We now consider briefly the *isotope shifts*, which do not give rise to splittings of the energy levels. As we pointed out above, these isotope shifts are caused by two effects: the *mass effect* (due to the fact that the nuclear mass is finite) and the *volume effect* (arising from the distribution of the nuclear charge within a finite volume).

For one-electron atoms the mass effect is readily taken into account by the introduction of the *reduced mass* $\mu = mM/(m + M)$, as we saw in Chapters 1 and 3. For the case of atoms with more than one electron the finiteness of the nuclear mass gives rise to an additional energy shift called the *mass polarisation correction*, which will be examined in Chapter 7.

W. Pauli and R. Peierls first pointed out in 1931 that the difference in nuclear volume between isotopes can produce an isotope shift. Indeed, since the protons in the nucleus are distributed in a finite nuclear volume, the electrostatic potential inside the nucleus deviates from the $1/r$ law, and depends on the proton distribution within the nucleus. In order to obtain an estimate of this volume effect, let us consider a simple model of the nucleus, such that the nuclear charge is distributed in a uniform way within a sphere of radius

$$R = r_0 A^{1/3} \tag{5.93}$$

where A is the mass number of the nucleus, and r_0 is a constant whose value is given approximately by $r_0 \simeq 1.2 \times 10^{-15}$ m. In this model, the electrostatic potential $V(r)$ due to the nucleus is easily shown to be (Problem 5.8)

$$V(r) = \begin{cases} \dfrac{Ze^2}{(4\pi\varepsilon_0)2R}\left(\dfrac{r^2}{R^2} - 3\right) & r \leqslant R \\[2ex] -\dfrac{Ze^2}{(4\pi\varepsilon_0)r} & r \geqslant R \end{cases} \tag{5.94}$$

To simplify the problem further, we shall assume that the unperturbed Hamilionian H_0 is the hydrogenic Hamiltonian (5.1) and that the perturbation H' is just the difference between the interaction (5.94) and the Coulomb interaction $-Ze^2/(4\pi\varepsilon_0 r)$. Thus all other effects (such as the relativistic corrections) are neglected and we have

$$H' = \begin{cases} \dfrac{Ze^2}{(4\pi\varepsilon_0)2R}\left(\dfrac{r^2}{R^2} + \dfrac{2R}{r} - 3\right) & r \leqslant R \\[2ex] 0 & r \geqslant R \end{cases} \tag{5.95}$$

The first-order energy shift due to this perturbation is

$$\Delta E = \langle \psi_{nlm} | H' | \psi_{nlm} \rangle$$

$$= \frac{Ze^2}{(4\pi\varepsilon_0)2R} \int_0^R |R_{nl}(r)|^2 \left(\frac{r^2}{R^2} + \frac{2R}{r} - 3 \right) r^2 \, \mathrm{d}r \qquad \textbf{(5.96)}$$

where we have used (3.53) and the fact that the spherical harmonics are normalised on the unit sphere. Inside the small region $r \leqslant R$ we may write $R_{nl}(r) \simeq R_{nl}(0)$. Moreover, since $R_{nl}(0)$ vanishes except for s states ($l = 0$), we have, after a straightforward calculation (Problem 5.9),

$$\Delta E \simeq \frac{Ze^2}{4\pi\varepsilon_0} \frac{R^2}{10} |R_{n0}(0)|^2$$

$$\simeq \frac{Ze^2}{4\pi\varepsilon_0} \frac{2\pi}{5} R^2 |\psi_{n00}(0)|^2, \qquad l = 0 \qquad \textbf{(5.97)}$$

while $\Delta E \simeq 0$ for states with $l \neq 0$. Using (3.65) we have explicitly

$$\Delta E \simeq \frac{e^2}{4\pi\varepsilon_0} \frac{2}{5} R^2 \frac{Z^4}{a_\mu^3 n^3}, \qquad l = 0 \qquad \textbf{(5.98)}$$

The quantity which is measured experimentally is the difference δE of energy shifts between two isotopes, whose charge distributions have radii R and $R + \delta R$, respectively. We thus find to first order in δR

$$\delta E \simeq \frac{Ze^2}{4\pi\varepsilon_0} \frac{4\pi}{5} R^2 |\psi_{n00}(0)|^2 \frac{\delta R}{R}$$

$$\simeq \frac{e^2}{4\pi\varepsilon_0} \frac{4}{5} R^2 \frac{Z^4}{a_\mu^3 n^3} \frac{\delta R}{R} \qquad \textbf{(5.99)}$$

We note that the isotope with the larger radius has the higher energy value, and this is confirmed by experiment. We also see that δE increases when Z increases and n decreases, so that the most important volume effects occur for low-lying s states (and in particular the ground state) of hydrogenic atoms with larger Z.

So far we have only considered 'ordinary' hydrogenic atoms (ions) containing a nucleus and an electron. As we pointed out in Chapter 3, there exist also 'exotic atoms' such as muonic atoms, in which a muon μ^- forms a bound system with a nucleus. We also noticed in Chapter 3 that since the mass of the muon μ^- is about 207 times larger than the electron mass, the Bohr radius associated with muonic atoms is much smaller than for 'ordinary' (electronic) atoms (see Table 3.2). We therefore expect that *hyperfine effects will be much larger for muonic atoms than for the corresponding ordinary atoms*. In particular, using the fact that the quantity a_μ is roughly 200 times smaller for a muonic atom than for an ordinary atom, we deduce from the foregoing discussion that the volume effect will be considerably magnified for muonic atoms, as we pointed out in Section 3.6.

Problems

5.1 Starting from (5.14), obtain the result (5.15) for the energy shift ΔE_1.

5.2 Using (5.24) and (5.25), obtain the expressions (5.26) for the energy shift ΔE_2.

5.3 Obtain the total relativistic energy shift (5.28) by using (5.15), (5.26) and (5.27).

5.4 Show that the expression (5.29) agrees up to order $(Z\alpha)^2$ with the formula (5.30) obtained by solving the Dirac equation.

5.5 Show that the ratio of the probabilities of the transitions in atomic hydrogen $n\text{p}_{3/2} \to n'\text{s}_{1/2}$ and $n\text{p}_{1/2} \to n'\text{s}_{1/2}$ is 2:1.

5.6 Verify that the expression (5.47) follows from (5.45) and (5.46).

5.7 Prove the relations (5.52) and (5.68).

5.8 Consider an electron in the electrostatic field of a nucleus of charge Ze, and of mass number A. If the nuclear charge is distributed uniformly within a sphere of radius $R = r_0 A^{1/3}$ where $r_0 \simeq 1.2 \times 10^{-15}$ m,

(a) show that the potential is given by (5.94);
(b) verify that the first-order energy shift due to the perturbation (5.95) is given by (5.97).

6 Interaction of one-electron atoms with external electric and magnetic fields

In Chapter 4 we studied the interaction of hydrogenic atoms with electromagnetic radiation. In this chapter we shall study the effects of external static electric and magnetic fields on one-electron atoms. We shall start in Section 6.1 by considering the effect of a static electric field, called the *Stark effect*. In Section 6.2, we shall discuss the influence of a static magnetic field, known as the *Zeeman effect*.

6.1 The Stark effect

The effect of static electric fields on the spectrum of hydrogen and other atoms was studied by J. Stark and also by A. Lo Surdo in 1913. The splitting of spectral lines observed is known as the Stark effect. We shall assume that the external electric field is uniform over a region of atomic dimensions and we shall take it to be directed along the Z axis. We also suppose that the electric field strength $\mathscr{E}$ is large enough for fine structure effects to be unimportant [1]. Neglecting reduced mass effects, the Hamiltonian of the unperturbed hydrogenic atom is given by

$$H_0 = -\frac{\hbar^2}{2m}\nabla^2 - \frac{Ze^2}{(4\pi\varepsilon_0)r} \tag{6.1}$$

The perturbation due to the external electric field is

$$H' = e\mathscr{E}z \tag{6.2}$$

where we recall that $-e$ is the charge of the electron. Since H' does not depend on the electron spin we shall use for the zero-order wave functions the Schrödinger hydrogenic wave functions $\psi_{nlm}(\mathbf{r})$ given by (3.53).

Linear Stark effect

Since the ground state (100) is non-degenerate, we see from (2.308) and (6.2) that the first-order correction to its energy is given by

[1] This is a correct assumption for electric field strengths usually encountered, which are of the order of 10^7 V m^{-1}. On the other hand, the treatment given here must be modified for electric field strengths $\mathscr{E} < 10^5$ V m^{-1}, since in this case the Stark splittings are of the same order of magnitude as the fine structure splittings studied in Section 5.1.

$$E_{100}^{(1)} = e\mathscr{E}\langle\psi_{100}|z|\psi_{100}\rangle$$
$$= e\mathscr{E}\int |\psi_{100}(\mathbf{r})|^2 z \, d\mathbf{r} \tag{6.3}$$

In Section 3.3 we showed that the hydrogenic wave functions $\psi_{nlm}(\mathbf{r})$ have a *definite* parity (even when the orbital quantum number l is even, odd when l is odd). On the other hand the perturbation (6.2) is an *odd* operator under the parity operation since it changes sign when the coordinates are reflected through the origin. Thus we have

$$\langle\psi_{nlm}|z|\psi_{nlm}\rangle = 0 \tag{6.4}$$

since the matrix element $\langle\psi_{nlm}|z|\psi_{nlm}\rangle$ involves the product of the *even* function $|\psi_{nlm}(\mathbf{r})|^2$ times the odd function z under the parity operation. In particular, we see from (6.3) and (6.4) that $E_{100}^{(1)} = 0$, so that for the ground state there is no energy shift that is linear in the electric field $\mathscr{E}$. Remembering that a classical system having an electric dipole moment $\mathbf{D}$ will experience in an electric field $\mathscr{E}$ an energy shift of magnitude $-\mathbf{D}\cdot\mathscr{E}$, and noting that $-ez$ is the z component of the electric dipole moment operator in our case, we see that one-electron atoms in the ground state cannot possess a permanent electric dipole moment (energy change proportional to $\mathscr{E}$).

Let us now examine the Stark effect on the first excited level ($n = 2$) of one-electron atoms. Since we assume that $\mathscr{E}$ is large enough for fine structure effects to be neglected we may consider the unperturbed system in the $n = 2$ level to be fourfold degenerate, the four eigenfunctions

$$\psi_{200}, \qquad \psi_{210}, \qquad \psi_{211}, \qquad \psi_{21-1} \tag{6.5}$$

corresponding to the same unperturbed energy $E_{n=2} = -mc^2(Z\alpha)^2/8$ (see (3.30)). In principle we should therefore solve a homogeneous system (2.328) of four equations. However, we have already shown in our discussion of selection rules for electric dipole transitions (see Section 4.5) that matrix elements of the form $\langle nlm|z|n'l'm'\rangle$ vanish unless $m = m'$ and $l = l' \pm 1$. Thus the only non-vanishing matrix elements of the perturbation (6.2) are those connecting the 2s (200) and 2p_0 (210) states, and the linear homogeneous equations (2.328) reduce to a set of two equations which we write in matrix form as

$$\begin{pmatrix} -E^{(1)} & H'_{12} \\ H'_{21} & -E^{(1)} \end{pmatrix}\begin{pmatrix} c_1 \\ c_2 \end{pmatrix} = 0 \tag{6.6}$$

with

$$H'_{12} = H'_{21} = e\mathscr{E}\int \psi_{210}(\mathbf{r}) z \psi_{200}(\mathbf{r}) \, d\mathbf{r} \tag{6.7}$$

and we have used the fact that H'_{12} is real.

The reduction of the original homogeneous system of four equations to the two equations (6.6) may also be obtained easily by noting that (i) the operator H' commutes with L_z, the z component of the angular momentum, so that H' only connects states with the same value of the quantum number m, and (ii) H' is odd under the parity operation.

The matrix element H'_{12} can be evaluated by using the hydrogenic wave functions given in Table 3.1 of Chapter 3. Since $z = r\cos\theta$, we have

$$\begin{aligned}
H'_{12} &= e\mathscr{E}\frac{Z^3}{16\pi a_0^3}\int_0^\infty \mathrm{d}r r^3\left(\frac{Zr}{a_0}\right)\left(1-\frac{Zr}{2a_0}\right)\exp(-Zr/a_0)\int_0^\pi \mathrm{d}\theta \sin\theta\cos^2\theta\int_0^{2\pi}\mathrm{d}\phi \\
&= e\mathscr{E}\frac{Z^3}{8a_0^3}\frac{2}{3}\int_0^\infty \mathrm{d}r r^3\left(\frac{Zr}{a_0}\right)\left(1-\frac{Zr}{2a_0}\right)\exp(-Zr/a_0) \\
&= -3e\mathscr{E}a_0/Z
\end{aligned} \tag{6.8}$$

Thus the two roots of the determinantal equation

$$\begin{vmatrix} -E^{(1)} & H'_{12} \\ H'_{12} & -E^{(1)} \end{vmatrix} = 0 \tag{6.9}$$

are given by

$$E^{(1)} = \pm|H'_{12}| = \pm 3e\mathscr{E}\,a_0/Z \tag{6.10}$$

Upon returning to (6.6) we see that for the lower root $E_1^{(1)} = -3e\mathscr{E}a_0/Z$ one has $c_1 = c_2$. The corresponding normalised eigenstate ψ_1 is given by

$$\psi_1 = \frac{1}{\sqrt{2}}(\psi_{200} + \psi_{210}) \tag{6.11a}$$

The second root $E_2^{(1)} = +3e\mathscr{E}\,a_0/Z$ yields $c_1 = -c_2$ and a normalised eigenstate

$$\psi_2 = \frac{1}{\sqrt{2}}(\psi_{200} - \psi_{210}) \tag{6.11b}$$

It should be emphasised that the states (6.11) are neither eigenstates of the parity operator, nor of $\mathbf{L}^2$, so that neither parity nor l is a 'good' quantum number in this case. On the other hand, m is a good quantum number because H' commutes with L_z (that is, the system is invariant under rotation about the Z axis). We also remark that the wave number shifts $\delta\tilde{\nu}$ corresponding to the energy corrections $E^{(1)}$ are given by

$$\delta\tilde{\nu} = \pm\frac{3ea_0}{hc}\frac{\mathscr{E}}{Z} = \pm 12.8\left(\frac{\mathscr{E}}{Z}\right)10^{-7}\,\mathrm{cm}^{-1} \tag{6.12}$$

so that rather strong fields (of the order of 10^7 V m^{-1} in Stark's experiments) are required to demonstrate the effect.

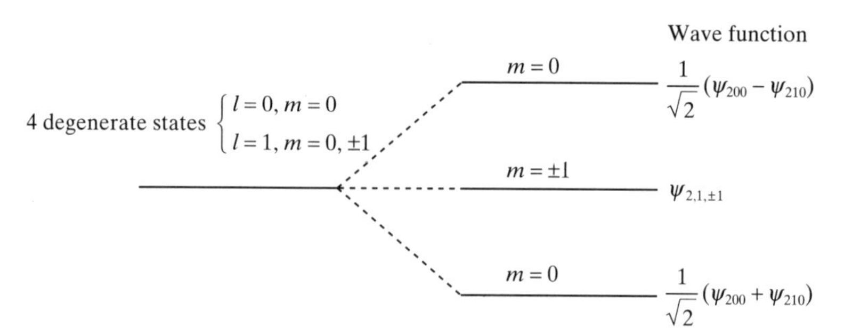

Figure 6.1 Splitting of the degenerate $n = 2$ level of a one-electron atom due to the linear Stark effect.

The splitting of the degenerate $n = 2$ level of a one-electron atom due to the linear Stark effect is illustrated in Fig. 6.1. The degeneracy is partly removed by the perturbation, the energies of the $2p_{\pm 1}$ (that is, 211 and 21–1) states remaining unaltered. Thus the level $n = 2$ splits in a symmetrical way into *three* sublevels, one of which (corresponding to $m = \pm 1$) is twofold degenerate.

At this point it is worth recalling that a classical system having an electric dipole moment $\mathbf{D}$ will experience in an electric field $\mathscr{E}$ an energy shift $-\mathbf{D}\cdot\mathscr{E}$). This suggests that the hydrogen atom in the *degenerate* unperturbed states $n = 2$ behaves as though it has a *permanent electric dipole moment* (independently of the value of $\mathscr{E}$), of magnitude $3ea_0$, which can be orientated in three different ways (that is, gives rise to spatial quantisation) in the presence of the electric field: one state (ψ_1) parallel to $\mathscr{E}$, one state (ψ_2) antiparallel to $\mathscr{E}$, and two states with no component along the electric field.

On the other hand, we found above that for the hydrogen ground state, which is *non-degenerate* and hence is an eigenstate of the *parity* operator, there is no energy shift linear in the electric field strength, and hence no permanent electric dipole moment. This conclusion may readily be generalised. Indeed, apart from tiny effects which we shall not consider here [2] all the systems studied in this book may be described by Hamiltonians which are unaffected by the parity operation (that is, the reflection of the coordinates of all the particles through the origin) and therefore any *non-degenerate* state of such systems has a definite parity (even or odd). Now for a system containing N particles of charges e_i ($i = 1, 2, \ldots, N$) and coordinates $\mathbf{r}_i$, the electric dipole moment operator

$$\mathbf{D} = \sum_{i=1}^{N} e_i \mathbf{r}_i \tag{6.13}$$

is odd under the parity operation, so that its expectation value in a state of given parity is zero. As a result, *systems in non-degenerate states cannot have permanent*

[2] Parity non-conserving effects occur in the so-called *weak interactions*, which are responsible for weak decay processes of (elementary) particles, such as those observed in beta decay.

electric dipole moments. Note, however, that if we have a positive ion A^+ located at the position $\mathbf{r}_1$ and a negative ion B^- at the position $\mathbf{r}_2$, the system ($A^+ B^-$) does possess an electric dipole moment. This does not contradict the previous argument since the configuration for which A^+ is at $\mathbf{r}_2$ and B^- at $\mathbf{r}_1$ has the same energy as the first arrangement and the system is necessarily degenerate. For the same reason, when atoms are bound together, the resulting molecules may possess permanent electric dipole moments [3].

Another remark concerns our use of degenerate perturbation theory for the treatment of the Stark effect on the $n = 2$ level. As we know from our discussion in Chapter 5 there are small effects (fine structure, Lamb shift) which remove some of the degeneracies of this level, so that the situation will then correspond to a *near-degenerate* case. We shall not treat this problem in detail [4] but consider instead the simple model problem in which two unperturbed states $\psi_1^{(0)}$ and $\psi_2^{(0)}$ do not correspond exactly to the same unperturbed energy $E^{(0)}$ ($E_{n=2}$ in our case) but to energies given respectively by $E_1^{(0)} = E^{(0)} - \varepsilon$ and $E_2^{(0)} = E^{(0)} + \varepsilon$, which differ by a small amount 2ε (with $\varepsilon > 0$). Instead of solving the equation (6.6), we must now solve the matrix equation

$$\begin{pmatrix} E^{(0)} - \varepsilon - E & H'_{12} \\ H'_{21} & E^{(0)} + \varepsilon - E \end{pmatrix} \begin{pmatrix} c_1 \\ c_2 \end{pmatrix} = 0 \qquad \textbf{(6.14)}$$

where $H'_{12} = H'_{21}$ is given by (6.8). Thus we have

$$E = E^{(0)} \pm [(H'_{12})^2 + \varepsilon^2]^{1/2} \qquad \textbf{(6.15)}$$

It is apparent from this result that for very weak fields (such that $|H'_{12}| \ll \varepsilon$ and the Stark splitting is small with respect to fine structure effects), there is no linear Stark effect. On the other hand, for strong field strengths $\mathscr{E}$ such that $|H'_{12}| \gg \varepsilon$ we retrieve the results found above by using degenerate perturbation theory (linear Stark effect). In what follows we shall continue to assume that the field strength $\mathscr{E}$ is large enough for the fine structure effects to be neglected.

The splitting of the $n = 3$ level due to the linear Stark effect may be treated in a way similar to the $n = 2$ case analysed above. It is found (Problem 6.1) that this level is split into *five* equally spaced levels, as shown in Fig. 6.2. Also displayed in Fig. 6.2 are the radiative transitions between the levels $n = 2$ and $n = 3$ (corresponding to the spectral line H_α) of atomic hydrogen in the presence of an electric field. The selection rules with respect to the magnetic quantum number m are the same as those we obtained in Chapter 4 without an external field. That is,

$$\Delta m = 0, \pm 1 \qquad \textbf{(6.16)}$$

---

[3] The kind of degeneracy to which we have referred is often removed in an 'exact' calculation of molecular ground states. However, because the splittings involved are very small, in experiments an average is taken over the ground and neighbouring states, of different parities, resulting in an effective permanent electric dipole moment.

[4] A comprehensive treatment may be found in Bethe and Salpeter (1957).

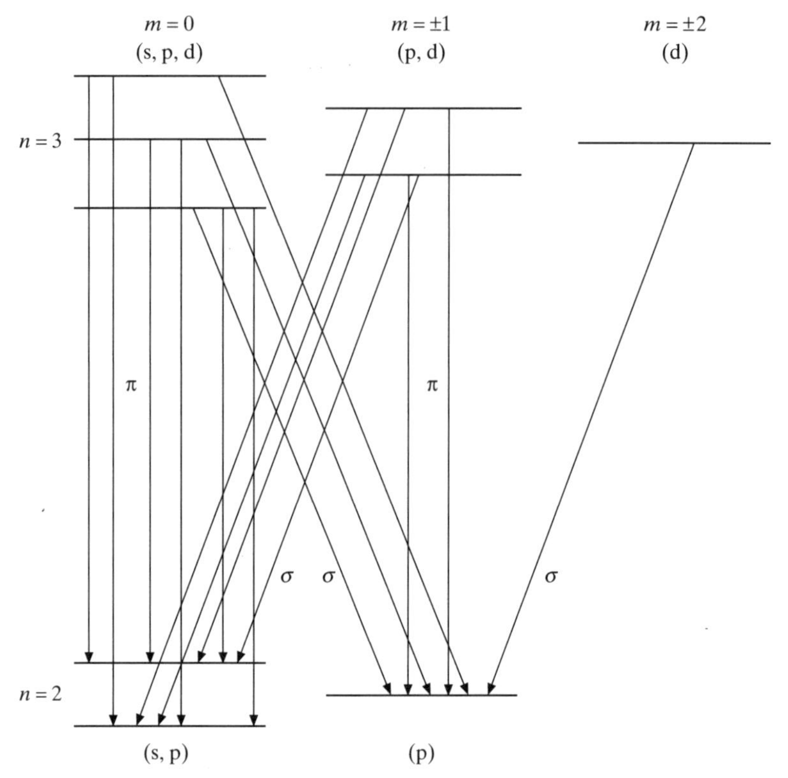

Figure 6.2 Splitting of the $n = 3$ and $n = 2$ levels of atomic hydrogen due to the linear Stark effect. The various possible transitions are shown; those with $\Delta m = 0$ correspond to π lines and those with $\Delta m = \pm 1$ to σ lines.

The $\Delta m = 0$ transitions are said to correspond to π components, and the $\Delta m = \pm 1$ to σ components.

On the other hand, since l is not a good quantum number in the presence of an external electric field, it is clear that the selection rules concerning l must be modified. In particular, because the operator $H' = e\mathcal{E}z$ has a non-vanishing matrix element between the 2s and $2p_0$ states, these two states are 'mixed' by the perturbation H' with the result that the metastable 2s state is 'contaminated' by the unstable 2p state. Thus a radiative transition from the 2s state to the 1s state can be induced by an external electric field [5], so that the lifetime of the 2s state is considerably shortened by comparison with its value (1/7 s) in the absence of an electric field.

In order to examine in more detail this process, which is called '*quenching* of the metastable 2s states', let us assume that at the initial time $t = 0$ the hydrogen atom is in the 2s (200) state. We then apply a constant electric field of strength

[5] It is worth noting that this external electric field need not be a static field, as in the case studied here, but can also be a time-dependent (oscillating) field.

$\mathscr{E}$ directed along the Z axis, and use the results (6.10) and (6.11) to write the time-dependent wave function of the atom at $t > 0$ as

$$\Psi(\mathbf{r}, t) = c_1\psi_1(\mathbf{r}) \exp\left[-\frac{i}{\hbar}(E_{n=2} - \Delta E)t\right] + c_2\psi_2(\mathbf{r}) \exp\left[-\frac{i}{\hbar}(E_{n=2} + \Delta E)t\right] \quad \textbf{(6.17)}$$

where $E_{n=2} = -mc^2\alpha^2/8$ and $\Delta E = |H'_{12}| = 3e\mathscr{E}a_0$ is the absolute value of the (first-order) energy shift. The coefficients c_1 and c_2 are easily found from the initial condition

$$\Psi(\mathbf{r}, t = 0) = \psi_{200}(r) \quad \textbf{(6.18)}$$

Using (6.11) and (6.18), we find that $c_1 = c_2 = 2^{-1/2}$, so that

$$\begin{aligned}\Psi(\mathbf{r}, t) &= \frac{1}{\sqrt{2}}\psi_1(\mathbf{r}) \exp\left[-\frac{i}{\hbar}(E_{n=2} - \Delta E)t\right] + \frac{1}{\sqrt{2}}\psi_2(\mathbf{r}) \exp\left[-\frac{i}{\hbar}(E_{n=2} + \Delta E)t\right] \\ &= \left[\psi_{200}(r) \cos\left(\frac{\Delta E}{\hbar}t\right) + i\psi_{210}(\mathbf{r}) \sin\left(\frac{\Delta E}{\hbar}t\right)\right] \exp\left(-\frac{i}{\hbar}E_{n=2}t\right)\end{aligned} \quad \textbf{(6.19)}$$

Thus the atom oscillates between the 200 (2s) and 210 ($2p_0$) states, with a period

$$T = \frac{\pi\hbar}{\Delta E} \quad \textbf{(6.20)}$$

For example, in the case of an electric field of strength $\mathscr{E} = 10^7$ V m^{-1}, we find from (6.20) that $T \simeq 1.3 \times 10^{-12}$ s which is much shorter than the time $\tau \simeq 1.6 \times 10^{-9}$ s corresponding to the radiative transition 2p–1s (that is, the lifetime of the 2p state in the absence of an external field). As a result, the average population of both states 2s and $2p_0$ is nearly equal during the entire decay time. This conclusion is easily seen to be true for initial conditions in which the atom is initially (at $t = 0$) in an arbitrary superposition of the 2s and $2p_0$ states (Problem 6.2). Thus, in the presence of a strong electric field the radiative transitions 2s–1s and 2p–1s have the same transition probability per unit time, which is equal to $1/(2\tau)$. It is apparent from this discussion that in general, an external electric field will be able to induce n's–ns radiative transitions.

The linear Stark effect in parabolic coordinates

It is also possible to evaluate the linear Stark effect of hydrogenic atoms by using parabolic coordinates, and this treatment is actually simpler than the one using spherical polar coordinates for principal quantum numbers $n \geqslant 3$. As we saw in Section 3.5, the Schrödinger equation for one-electron atoms in the absence of external fields can be separated in parabolic coordinates. This separability is maintained in the presence of an external electric field. Indeed, using (3.90), the interaction (6.2) can be expressed in parabolic coordinates as

$$H' = \frac{1}{2}\mathscr{E}(\xi - \eta) \tag{6.21}$$

where we are using atomic units. The atomic unit of electric field strength is the strength of the Coulomb field experienced by an electron in the first Bohr orbit of atomic hydrogen (with infinite nuclear mass). That is,

$$\mathscr{E}_a = \frac{e}{(4\pi\varepsilon_0)a_0^2} = 5.142\,21 \times 10^{11}\ \mathrm{V\ m^{-1}} \tag{6.22}$$

The electric field strengths considered in this section are such that $\mathscr{E} \ll 1$ (in a.u.).

The Schrödinger equation for a hydrogenic atom in the presence of a static electric field reads (in a.u.)

$$\left(-\frac{1}{2}\nabla^2 - \frac{Z}{r} + H'\right)\psi = E\psi \tag{6.23}$$

and has separable solutions of the form (3.95). The angular function $\Phi(\phi)$ is again given by (3.98), while the functions $f(\xi)$ and $g(\eta)$ satisfy the equations

$$\left[\frac{\mathrm{d}}{\mathrm{d}\xi}\left(\xi\frac{\mathrm{d}}{\mathrm{d}\xi}\right) + \left(\frac{1}{2}E\xi - \frac{m^2}{4\xi} - \frac{1}{4}\mathscr{E}\xi^2 + \nu_1\right)\right]f = 0 \tag{6.24a}$$

and

$$\left[\frac{\mathrm{d}}{\mathrm{d}\eta}\left(\eta\frac{\mathrm{d}}{\mathrm{d}\eta}\right) + \left(\frac{1}{2}E\eta - \frac{m^2}{4\eta} + \frac{1}{4}\mathscr{E}\eta^2 + \nu_2\right)\right]g = 0 \tag{6.24b}$$

where the separation constants ν_1 and ν_2 must fulfil the condition (see (3.99))

$$\nu_1 + \nu_2 = Z \tag{6.25}$$

Using the unperturbed hydrogenic wave functions in parabolic coordinates given in Section 3.5, and first-order perturbation theory, it is found [6] that the perturbed energies are (in a.u.)

$$E_{nn_1n_2} = -\frac{1}{2}\frac{Z^2}{n^2} + \frac{3}{2}\mathscr{E}\frac{n}{Z}(n_1 - n_2) \tag{6.26}$$

where n_1 and n_2 are the parabolic quantum numbers introduced in Section 3.5. The energies $E_{nn_1n_2}$ are seen to depend on the principal quantum number n and on the difference $n_1 - n_2$. Although there is no explicit dependence in (6.26) on the magnetic quantum number m, there is an implicit dependence because of the relation (3.110). For a given value of n, the Stark component having the highest energy is obtained by setting $n_1 = n - 1$ and $n_2 = 0$, which from (3.110) implies that

[6] See Bethe and Salpeter (1957).

$m = 0$. The lowest Stark component corresponds to $n_1 = 0$, $n_2 = n - 1$, which also implies that $m = 0$.

Quadratic Stark effect

We have shown above that for the ground state (100) of hydrogenic atoms there is no linear Stark effect. In order to investigate the effect of the perturbation (6.2) on that state we must therefore consider the second-order term of the perturbation series. Using (2.319) we see that in our case it reads

$$E_{100}^{(2)} = e^2 \mathscr{E}^2 \sum_{n \neq 1, l, m} \frac{|\langle \psi_{nlm} | z | \psi_{100} \rangle|^2}{E_1 - E_n} \tag{6.27}$$

where the sum implies a summation over the discrete set together with an integration over the continuous set of hydrogenic eigenfunctions. It is clear from (6.27) that the ground state energy will be *lowered* by the quadratic Stark effect, since the energy differences $E_1 - E_n$ $(n \geqslant 2)$ are always negative. In fact, we may readily obtain a lower limit for $E_{100}^{(2)}$ by replacing in (6.27) the energy differences $E_1 - E_n$ by $E_1 - E_2$. That is,

$$\begin{aligned} E_{100}^{(2)} &= -e^2 \mathscr{E}^2 \sum_{n \neq 1, l, m} \frac{|\langle \psi_{nlm} | z | \psi_{100} \rangle|^2}{E_n - E_1} \\ &> -e^2 \mathscr{E}^2 \frac{1}{E_2 - E_1} \sum_{n \neq 1, l, m} |\langle \psi_{nlm} | z | \psi_{100} \rangle|^2 \end{aligned} \tag{6.28}$$

The summation on the right of (6.28) may now be performed as follows. We first note that because $\langle \psi_{100} | z | \psi_{100} \rangle = 0$ we may write

$$\begin{aligned} \sum_{n \neq 1, l, m} |\langle \psi_{nlm} | z | \psi_{100} \rangle|^2 &= \sum_{n, l, m} |\langle \psi_{nlm} | z | \psi_{100} \rangle|^2 \\ &= \sum_{n, l, m} |\langle \psi_{100} | z | \psi_{nlm} \rangle \langle \psi_{nlm} | z | \psi_{100} \rangle \end{aligned} \tag{6.29}$$

Using the completeness of the hydrogenic states, namely

$$\sum_{nlm} |\psi_{nlm}\rangle \langle \psi_{nlm}| = I \tag{6.30}$$

where I is the unit operator, we have

$$\begin{aligned} \sum_{n, l, m} \langle \psi_{100} | z | \psi_{nlm} \rangle \langle \psi_{nlm} | z | \psi_{100} \rangle &= \langle \psi_{100} | z^2 | \psi_{100} \rangle \\ &= \langle z^2 \rangle_{100} \end{aligned} \tag{6.31}$$

But

$$\langle z^2 \rangle_{100} = \langle x^2 \rangle_{100} = \langle y^2 \rangle_{100} = \frac{1}{3} \langle r^2 \rangle_{100} = \frac{a_0^2}{Z^2} \tag{6.32}$$

so that from (6.28)–(6.32) and (3.29) we have

$$E^{(2)}_{100} > -\frac{8}{3}(4\pi\varepsilon_0)\frac{a_0^3}{Z^4}\mathscr{E}^2 \tag{6.33}$$

It is possible to obtain in a straightforward way another estimate for $E^{(2)}_{100}$ (Problem 6.3):

$$E^{(2)}_{100} \simeq -2(4\pi\varepsilon_0)\frac{a_0^3}{Z^4}\mathscr{E}^2 \tag{6.34}$$

The exact evaluation of the expression (6.27) can be carried out using parabolic coordinates. One finds (Bethe and Salpeter, 1957) that

$$\begin{aligned} E^{(2)}_{100} &= -2.25(4\pi\varepsilon_0)\frac{a_0^3}{Z^4}\mathscr{E}^2 \\ &= -3.71 \times 10^{-41}\left(\frac{\mathscr{E}^2}{Z^4}\right) \text{J} \end{aligned} \tag{6.35}$$

It is worth noting that about one-third of the result (6.35) arises from the contribution of the continuum in the summation (6.27). We also remark that the quadratic Stark effect given by (6.35) is generally very small, being approximately 0.02 cm^{-1} for atomic hydrogen in the case of a field strength $\mathscr{E} = 10^8$ V m^{-1}.

Upon differentiation of the expression (6.27) with respect to the electric field strength, we obtain for the magnitude of the dipole moment the result

$$D = -\frac{\partial E^{(2)}_{100}}{\partial \mathscr{E}} = \bar{\alpha}\mathscr{E} \tag{6.36}$$

where

$$\bar{\alpha} = 2e^2 \sum_{n\neq 1,l,m} \frac{|\langle \psi_{nlm}|z|\psi_{100}\rangle|^2}{E_n - E_1} \tag{6.37}$$

is the *static dipole polarisability* of the atom in the state (100). We note that because the matrix element $\langle \psi_{nlm}|z|\psi_{100}\rangle$ vanishes unless $m = 0$, the sum over m reduces to one term. We see from (6.36) that D is proportional to $\mathscr{E}$, so that we have an *induced dipole moment*. We also note from (6.27) and (6.37) that

$$E^{(2)}_{100} = -\frac{1}{2}\bar{\alpha}\mathscr{E}^2 \tag{6.38}$$

and the result (6.35) implies that

$$\begin{aligned} \bar{\alpha} &= 4.50(4\pi\varepsilon_0)\frac{a_0^3}{Z^4} \\ &= 7.42 \times 10^{-41}\, Z^{-4} \text{ F m}^2 \\ &= 4.50\, Z^{-4} \text{ a.u.} \end{aligned} \tag{6.39}$$

The quadratic Stark effect in parabolic coordinates

Starting from the equations (6.24), and using second-order perturbation theory, one obtains [6] for the perturbed energies (in a.u.)

$$E_{nn_1n_2m} = -\frac{1}{2}\frac{Z^2}{n^2} + \frac{3}{2}\mathscr{E}\frac{n}{Z}(n_1 - n_2) - \frac{1}{16}\mathscr{E}^2\left(\frac{n}{Z}\right)^4[17n^2 - 3(n_1 - n_2)^2 - 9m^2 + 19] \tag{6.40}$$

As seen from this result, the quadratic Stark effects always leads to a lowering of the levels. We also note that the second-order shift removes the m degeneracy, and remains unaltered under the interchange of n_1 and n_2.

The quadratic Stark effect is usually a small correction to the linear Stark effect. For example, in the case of the H_α line, where the separation of the outermost components is about 200 cm^{-1} for a field strength of 4×10^7 V m^{-1}, the corresponding (red) shift due to the quadratic Stark effect is only 1 cm^{-1}.

Ionisation by a static electric field

So far we have used perturbation theory to study the energy shifts and the spectral lines of hydrogenic atoms in the presence of a static electric field. We shall now consider another effect due to the presence of an external electric field, namely the removal of the electron from the atom.

To see how this comes about, we first note that the total potential energy V of the electron is obtained by adding the potential energy $e\mathscr{E}z$ arising from the external field (see (6.2)) to the Coulomb potential $-Ze^2/(4\pi\varepsilon_0 r)$ of the nucleus. Thus

$$V = -\frac{Ze^2}{(4\pi\varepsilon_0)r} + e\mathscr{E}z \tag{6.41}$$

A schematic drawing of V is shown in Fig. 6.3 as a function of z, for x and y fixed. It is apparent that the nucleus is not the only place at which V has a minimum, since V can become even more negative if z is negative enough, that is at large enough distances of the atom in the direction of the anode. Thus the potential V has two minima, one at the nucleus and the other at the anode, separated by a potential barrier. The electron, which is initially in a bound state of the atom, has therefore a finite probability of 'escaping' from the atom by means of the *tunnel effect*, and being accelerated towards the anode, so that ionisation will occur.

This possibility of ionisation by the electric field was first pointed out by J. Oppenheimer in 1928. Experimentally it can be observed when the external electric field is very strong and (or) for levels with high principal quantum number n. It is then seen that the spectral lines are *weakened* because of the competition between the radiative transitions and the ionisation process. Moreover, in the presence of an external electric field the lifetime of the discrete levels is decreased because of the 'tunnel effect', so that the width of the spectral lines is increased. This is known as *Stark broadening*. In particular, the ground state itself is no

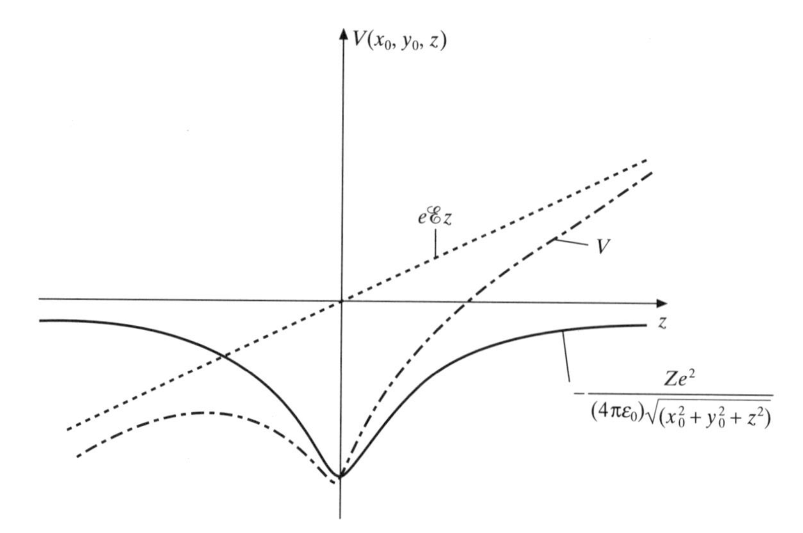

Figure 6.3 The potential V experienced by an electron interacting with a nucleus of charge Ze, in a uniform electric field of strength $\mathscr{E}$, as a function of z, for $x = x_0$ and $y = y_0$ fixed.

longer a stationary state, but becomes a metastable state when an external electric field is applied, and the perturbation series is found to be a semi-convergent (asymptotic) series. However, if the electric field is not too strong, the ground state is stable on a very large time scale, and the predictions of the first few terms of the perturbation series agree very well with experiment.

We shall now obtain an approximate expression for the rate of ionisation of a ground state hydrogen atom by a uniform electric field directed along the Z axis. It is convenient to use atomic units, and to perform the calculation in parabolic coordinates, starting from the equations (6.24). The field strengths that we shall consider are such that $\mathscr{E} \ll 1$.

By making in (6.24) the substitutions

$$f = \xi^{-1/2}F, \qquad g = \eta^{-1/2}G \tag{6.42}$$

we obtain for F and G the pair of equations

$$-\frac{1}{2}\frac{\mathrm{d}^2F}{\mathrm{d}\xi^2} + V_1(\xi)F = \frac{E}{4}F \tag{6.43a}$$

and

$$-\frac{1}{2}\frac{\mathrm{d}^2G}{\mathrm{d}\eta^2} + V_2(\eta)G = \frac{E}{4}G \tag{6.43b}$$

where V_1 and V_2 are defined by

$$V_1(\xi) = \frac{m^2-1}{8\xi^2} - \frac{\nu_1}{2\xi} + \frac{\mathscr{E}\xi}{8} \tag{6.44a}$$

and

$$V_2(\eta) = \frac{m^2 - 1}{8\eta^2} - \frac{\nu_2}{2\eta} - \frac{\mathscr{E}\eta}{8} \tag{6.44b}$$

It is seen that the equations (6.43) have the form of two one-dimensional Schrödinger equations in which $E/4$ plays the role of the energy, while V_1 and V_2 are potentials.

The unperturbed ground state wave function of atomic hydrogen, expressed in parabolic coordinates (see (3.119)) and in atomic units, namely

$$\psi_0 = \pi^{-1/2} \exp\left[-\frac{1}{2}(\xi + \eta)\right] \tag{6.45}$$

is a solution of the equations (6.43) when $\mathscr{E} = 0$. Since the magnetic quantum number for this state is $m = 0$, and because the perturbation (6.21) is independent of ϕ, the perturbed ground state wave function, which will be denoted by ψ_0^p, must also be such that $m = 0$. It must therefore satisfy the equations (6.43) with $\mathscr{E} > 0$, in which the potentials $V_1(\xi)$ and $V_2(\eta)$ are calculated from (6.44) by setting $m = 0$.

Let us write the unperturbed ground state wave function (6.45) in the form

$$\psi_0 = \pi^{-1/2}(\xi\eta)^{-1/2} F_0(\xi) G_0(\eta) \tag{6.46}$$

The functions

$$F_0(\xi) = \xi^{1/2} \exp(-\xi/2) \tag{6.47a}$$

and

$$G_0(\eta) = \eta^{1/2} \exp(-\eta/2) \tag{6.47b}$$

satisfy the equations (6.43) with $E = -1/2$, where in (6.44) $\mathscr{E} = 0$ and $\nu_1 = \nu_2 = 1/2$.

In Fig. 6.4, the potentials V_1 and V_2 are sketched for the cases $\mathscr{E} = 0$ and $\mathscr{E} > 0$. When $\mathscr{E} > 0$, it is seen that $V_1(\xi)$ has an infinite barrier as $\xi \to +\infty$ and this prevents the wave function $F(\xi)$ from penetrating into the region of large ξ. For $\mathscr{E} \ll 1$, the potential $V_1(\xi)$ does not differ appreciably over the extent of the unperturbed function $F_0(\xi)$ from its values when $\mathscr{E} = 0$. It follows that it is a good approximation to take

$$F(\xi) \simeq F_0(\xi) \tag{6.48}$$

for all ξ, where $F_0(\xi)$ is given by (6.47a). Since $F_0(\xi)$ only satisfies (6.47a) if $\nu_1 = 1/2$, we must also take $\nu_2 = 1/2$ because of the condition (6.25).

Referring to (6.44b), we see that if $\mathscr{E} > 0$, $V_2(\eta) \to -\infty$ as $\eta \to +\infty$ so that for all non-zero values of $\mathscr{E}$ the potential $V_2(\eta)$ presents a *finite barrier* through which tunnelling is possible to reach large values of η. The region in which ξ is small but η is large corresponds in Cartesian coordinates to finite values of x and y while $z \to -\infty$. This is the region in which we must calculate the flux of electrons which have been ionised by the electric field. To find an approximate solution of (6.43b),

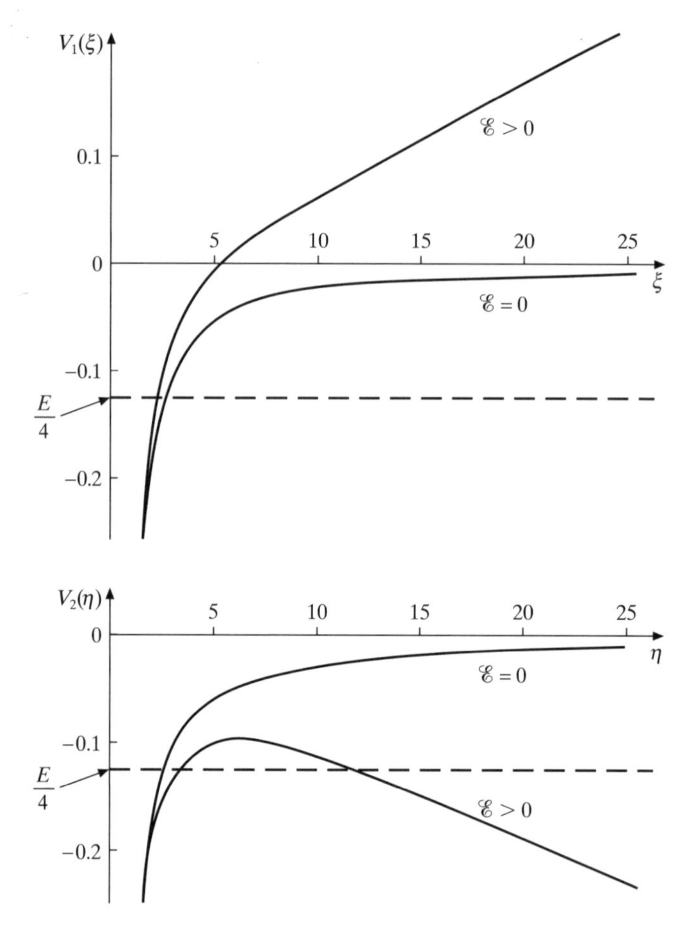

Figure 6.4 The form of the potentials $V_1(\xi)$ and $V_2(\eta)$ when $m = 0$ for $\mathscr{E} = 0$ and $\mathscr{E} > 0$.

we first choose a value η_0 of η, such that η_0 lies within the barrier presented by the potential $V_2(\eta)$ and satisfies the conditions

$$\eta_0 \gg 1 \tag{6.49}$$

and

$$\mathscr{E}\eta_0 \ll 1 \tag{6.50}$$

In view of the condition (6.50), the interaction $-\mathscr{E}\eta/8$ in (6.44b) can be neglected for $\eta \leqslant \eta_0$ so that

$$G(\eta) \simeq G_0(\eta), \qquad \eta \leqslant \eta_0 \tag{6.51}$$

In the region $\eta \geqslant \eta_0$, the WKB approximation can be employed since the potential $V_2(\eta)$ is smooth and slowly varying. Using (2.433), $G(\eta)$ can be approximated as

$$G(\eta) \simeq A|p(\eta)|^{-1/2} \exp\left[\mathrm{i}\int_{\eta_0}^{\eta} p(\eta')\,\mathrm{d}\eta'\right], \qquad \eta \geqslant \eta_0 \tag{6.52}$$

where $p(\eta)$ is defined by (see (2.421))

$$\begin{aligned} p(\eta) &= \left\{2\left[\frac{E}{4} - V_2(\eta)\right]\right\}^{1/2} \\ &= \left[-\frac{1}{4} + \frac{1}{4\eta^2} + \frac{1}{2\eta} + \frac{1}{4}\mathscr{E}\eta\right]^{1/2} \end{aligned} \tag{6.53}$$

since $E = -1/2$.

The constant A in (6.52) is obtained from the relation (6.51) and is given by

$$A = [\eta_0 |p(\eta_0)|]^{1/2} \exp(-\eta_0/2) \tag{6.54}$$

Neglecting in (6.53) small terms in η^{-2} and η^{-1}, $p^2(\eta)$ is seen to have a zero at $\eta = \eta_1$, where

$$\eta_1 \simeq \frac{1}{\mathscr{E}} \tag{6.55}$$

For $\eta > \eta_1$, $p^2(\eta) > 0$ and $p(\eta)$ is real, while for $\eta_0 \leqslant \eta \leqslant \eta_1$, $p^2(\eta) < 0$ and $p(\eta)$ is imaginary. Since $G(\eta)$ must decrease from η_0 to η_1, we must take $p(\eta) = +\mathrm{i}|p(\eta)|$ in this region.

To calculate the flux of outgoing electrons we require $|G(\eta)|^2$ in the limit $\eta \to +\infty$ and since $p(\eta)$ is real for $\eta > \eta_1$, we have

$$\begin{aligned} |G(\eta)|^2 &= A^2 \frac{1}{|p(\eta)|} \exp\left[-2\int_{\eta_0}^{\eta_1} |p(\eta')|\,\mathrm{d}\eta'\right] \\ &= \eta_0 \left|\frac{p(\eta_0)}{p(\eta)}\right| \exp\left[-2\int_{\eta_0}^{\eta_1} |p(\eta')|\,\mathrm{d}\eta' - \eta_0\right] \end{aligned} \tag{6.56}$$

In the factors multiplying the exponential, the approximations can be made that $|p(\eta_0)| \simeq 1/2$. Moreover, for large η (such that $\mathscr{E}\eta \gg 1$)

$$|p(\eta)| \simeq \frac{1}{2}(\mathscr{E}\eta - 1)^{1/2} \tag{6.57}$$

In the exponential, $|p(\eta')|$ can be approximated by retaining in (6.53) the small term $(2\eta')^{-1}$, but neglecting the term in $(4\eta'^2)^{-1}$, so that, for $\eta_0 \leqslant \eta' \leqslant \eta_1$,

$$\begin{aligned} |p(\eta')| &\simeq \frac{1}{2}\left(1 - \frac{2}{\eta'} - \mathscr{E}\eta'\right)^{1/2} \\ &\simeq \frac{1}{2}(1 - \mathscr{E}\eta')^{1/2} - \frac{1}{2\eta'}(1 - \mathscr{E}\eta')^{-1/2} \end{aligned} \tag{6.58}$$

Evaluating the integral in the exponent of (6.56), using the approximation (6.58), noting that $\mathscr{E}\eta_0 \ll 1$ and that $\eta_1 \simeq 1/\mathscr{E}$, it is found that

$$-2\int_{\eta_0}^{\eta_1} |p(\eta')|\,\mathrm{d}\eta' \simeq -\frac{2}{3\mathscr{E}}(1-\mathscr{E}\eta_0)^{3/2} + \log\left[\frac{1+(1-\mathscr{E}\eta_0)^{1/2}}{1-(1-\mathscr{E}\eta_0)^{1/2}}\right]$$

$$\simeq \eta_0 - \frac{2}{3\mathscr{E}} + \log\left(\frac{4}{\mathscr{E}\eta_0}\right) \tag{6.59}$$

Thus, from (6.56) and (6.59), we have approximately, for large η,

$$|G(\eta)|^2 \simeq \frac{4}{\mathscr{E}(\mathscr{E}\eta-1)^{1/2}}\exp\left(-\frac{2}{3\mathscr{E}}\right) \tag{6.60}$$

For small ξ and large η, the perturbed ground state wave function ψ_0^p is

$$\psi_0^p = \pi^{-1/2}(\xi\eta)^{-1/2}F_0(\xi)G(\eta) \tag{6.61}$$

Hence, using (6.47a) and (6.60), we find that for small ξ and large η

$$|\psi_0^p|^2 = \frac{1}{\pi}\frac{4}{\mathscr{E}\eta(\mathscr{E}\eta-1)^{1/2}}\exp\left(-\frac{2}{3\mathscr{E}}\right)\exp(-\xi) \tag{6.62}$$

The ionisation rate W_{ion} is equal to the number of electrons passing through a plane perpendicular to the Z axis per unit time as $z \to -\infty$, so that

$$W_{\text{ion}} = \int_0^{2\pi}\mathrm{d}\phi\int_0^\infty \mathrm{d}\rho\rho|\psi_0^p|^2 v_z \tag{6.63}$$

where v_z is the z component of the electron velocity, and

$$\rho = (x^2+y^2)^{1/2} = (\xi\eta)^{1/2} \tag{6.64}$$

A fixed value of $-z$, where $|z|$ is large, corresponds to a fixed value of η, since the significant values of ξ are small. Thus we may set

$$\mathrm{d}\rho = \frac{1}{2}\xi^{-1/2}\eta^{1/2}\,\mathrm{d}\xi \tag{6.65}$$

and since ψ_0^p is independent of ϕ, we have

$$W_{\text{ion}} = (2\pi)\frac{1}{2}\int_0^\infty \eta|\psi_0^p|^2 v_z\,\mathrm{d}\xi \tag{6.66}$$

In the large $|z|$ region the electron's motion is classical and using conservation of energy

$$E = \frac{1}{2}v_z^2 + \mathscr{E}z \tag{6.67}$$

so that with $E = -1/2$, and $-z \simeq \eta/2$,

$$v_z = (\mathscr{E}\eta - 1)^{1/2} \tag{6.68}$$

Making use of (6.62) and (6.68) in (6.66), it is found that

$$W_{\text{ion}} = \frac{4}{\mathscr{E}} \exp\left(-\frac{2}{3\mathscr{E}}\right) \tag{6.69}$$

and the lifetime of the hydrogen ground state in the electric field is (in a.u.)

$$\tau = \frac{1}{W_{\text{ion}}} = \frac{1}{4}\mathscr{E} \exp\left(\frac{2}{3\mathscr{E}}\right) \tag{6.70}$$

For small values of the electric field strength $\mathscr{E}$ the lifetime of the ground state is extremely long and this state is essentially stable. For example, if $\mathscr{E} = 10^{-2}$ a.u., then $\tau = 2 \times 10^{26}$ a.u. As $\mathscr{E}$ is increased, tunnelling through the barrier in the η coordinate becomes more probable and the lifetime diminishes exponentially. As can be seen from (6.44b) the height of the barrier presented by $V_2(\eta)$ decreases as $\mathscr{E}$ increases and a critical field strength can be defined at which the height of the barrier becomes less than the effective energy $E/4$. In this case ionisation can occur classically and the atom can no longer be considered to be in a quasi bound (resonant) state. For the ground state of hydrogen this critical field is about 6×10^{10} V m^{-1}.

The critical field above which ionisation occurs classically decreases like n^{-4} for excited states and, for a hydrogen atom in a Rydberg state with $n = 30$, the critical field strength is about 7×10^4 V m^{-1}. Experiments to detect the energies and properties of Rydberg atoms often make use of the sensitivity of these states to applied electric fields. For example, a beam of neutral atoms from an oven can be irradiated by a tunable laser and then passed through an electric field, the ions formed can be detected and the signal measured as a function of the laser frequency. When this frequency is the appropriate one to excite the atoms to a particular Rydberg level the ionic signal peaks, allowing the energy of the level to be determined.

A detailed account of the interaction of electric fields with Rydberg atoms can be found in the monograph by Gallagher (1994).

6.2 The Zeeman effect

In 1896, P. Zeeman observed that the spectral lines of atoms were split in the presence of an external magnetic field. In order to explain this effect, we shall discuss in this section the interaction of hydrogenic atoms with static magnetic fields, which can be taken to be uniform over atomic dimensions. The vector potential **A** can then be written as

$$\mathbf{A} = \frac{1}{2}(\mathcal{B} \times \mathbf{r}) \tag{6.71}$$

where $\mathcal{B}$ is a constant magnetic field, which satisfies the relation $\mathcal{B} = \nabla \times \mathbf{A}$.

The Hamiltonian of a hydrogenic atom in the presence of a constant magnetic field is given in the infinite nuclear mass approximation by (see Appendix 6)

$$\begin{aligned} H &= \frac{1}{2m}(\mathbf{p} + e\mathbf{A})^2 - \frac{Ze^2}{(4\pi\varepsilon_0)r} \\ &= -\frac{\hbar^2}{2m}\nabla^2 - \frac{Ze^2}{(4\pi\varepsilon_0)r} - \frac{i\hbar e}{m}\mathbf{A}\cdot\nabla + \frac{e^2}{2m}\mathbf{A}^2 \end{aligned} \tag{6.72}$$

where $\mathbf{A}$ is given by (6.71) and we have used the fact that $\nabla\cdot\mathbf{A} = 0$. The corresponding time-independent Schrödinger equation reads

$$\left[-\frac{\hbar^2}{2m}\nabla^2 - \frac{Ze^2}{(4\pi\varepsilon_0)r} - \frac{i\hbar e}{m}\mathbf{A}\cdot\nabla + \frac{e^2}{2m}\mathbf{A}^2\right]\psi(\mathbf{r}) = E\psi(\mathbf{r}) \tag{6.73}$$

The linear term in $\mathbf{A}$ becomes, in terms of $\mathcal{B}$,

$$\begin{aligned} -\frac{i\hbar e}{m}\mathbf{A}\cdot\nabla &= -\frac{i\hbar e}{2m}(\mathcal{B} \times \mathbf{r})\cdot\nabla \\ &= -\frac{i\hbar e}{2m}\mathcal{B}\cdot(\mathbf{r} \times \nabla) \\ &= \frac{e}{2m}\mathcal{B}\cdot\mathbf{L} \end{aligned} \tag{6.74}$$

where $\mathbf{L} = -i\hbar(\mathbf{r} \times \nabla)$ is the orbital angular momentum operator of the electron.

The quadratic term in $\mathbf{A}$ appearing in (6.73) can be written in the form

$$\begin{aligned} \frac{e^2}{2m}\mathbf{A}^2 &= \frac{e^2}{8m}(\mathcal{B} \times \mathbf{r})^2 \\ &= \frac{e^2}{8m}[\mathcal{B}^2 r^2 - (\mathcal{B}\cdot\mathbf{r})^2] \end{aligned} \tag{6.75}$$

The linear term (6.74) corresponds to the interaction energy of a magnetic field $\mathcal{B}$ with the orbital magnetic dipole moment

$$\mathcal{M}_L = -\frac{e}{2m}\mathbf{L} = -\mu_B\mathbf{L}/\hbar \tag{6.76}$$

where

$$\mu_B = \frac{e\hbar}{2m} \tag{6.77}$$

is the *Bohr magneton* (see (1.110)) and has the value $9.274\,01 \times 10^{-24}$ J T^{-1} or m^2 A. The interaction energy (6.74) then takes the form

$$H_1' = -\mathcal{M}_L \cdot \mathcal{B} \tag{6.78}$$

It is useful to express H_1' in various units. For example,

$$\begin{aligned} H_1' &= 2.13 \times 10^{-6}\, \mathcal{B} \cdot \mathbf{L} \text{ a.u.} \\ &= 0.4669\, \mathcal{B} \cdot \mathbf{L}/\hbar \text{ cm}^{-1} \end{aligned} \tag{6.79}$$

where in both cases the magnitude of $\mathcal{B}$ is to be given in tesla (T).

Until this point we have not taken into account the intrinsic magnetic moment of the electron, revealed by experiments of the Stern–Gerlach type (see Chapter 1). This intrinsic magnetic moment, due to the electron spin, is given by

$$\mathcal{M}_s = -g_s \frac{e}{2m} \mathbf{S} \tag{6.80}$$

or

$$\mathcal{M}_s = -g_s \mu_B \mathbf{S}/\hbar \tag{6.81}$$

where $\mathbf{S}$ is the spin operator of the electron and g_s its spin *gyromagnetic ratio*. Dirac's relativistic theory predicts for g_s the value $g_s = 2$ (see Appendix 7) which is in very good agreement with experiment [7]. The spin magnetic moment $\mathcal{M}_s$ gives rise to an additional interaction energy, linear in the magnetic field,

$$H_2' = -\mathcal{M}_s \cdot \mathcal{B} = g_s \mu_B \mathcal{B} \cdot \mathbf{S}/\hbar \tag{6.82}$$

The complete Schrödinger equation for a one-electron atom in a constant magnetic field, including the spin–orbit interaction, but neglecting the reduced mass effect, the relativistic kinetic energy correction and the Darwin term, is (with $g_s = 2$)

$$\left[-\frac{\hbar^2}{2m}\nabla^2 - \frac{Ze^2}{(4\pi\varepsilon_0)r} + \xi(r)\mathbf{L}\cdot\mathbf{S} + \frac{\mu_B}{\hbar}(\mathbf{L} + 2\mathbf{S})\cdot\mathcal{B} + \frac{e^2}{8m}(\mathcal{B} \times \mathbf{r})^2 \right] \psi(q) = E\psi(q) \tag{6.83}$$

where $\xi(r)$ is given by (5.18) and $\psi(q)$ is now a Pauli 'spin-orbital' with two components (see (5.9)).

In what follows it will be convenient to take the magnetic field $\mathcal{B}$ to be directed along the Z axis. In this case, the equation (6.83) becomes

$$\begin{aligned} &\left[-\frac{\hbar^2}{2m}\nabla^2 - \frac{Ze^2}{(4\pi\varepsilon_0)r} + \xi(r)\mathbf{L}\cdot\mathbf{S} + \frac{\mu_B}{\hbar}(L_z + 2S_z)\mathcal{B} + \frac{e^2}{8m}\mathcal{B}^2 r^2 \sin^2\theta \right] \psi(q) \\ &= E\psi(q) \end{aligned} \tag{6.84}$$

where θ is the angle between the vector $\mathbf{r}$ and the Z axis.

---

[7] The corrections to the Dirac result $g_s = 2$ come from quantum electrodynamics (see footnote [10] of Chapter 5).

The term linear in the magnetic field in equations (6.83) or (6.84), namely

$$\frac{\mu_B}{\hbar}(\mathbf{L}+2\mathbf{S})\cdot\mathcal{B} = \frac{\mu_B}{\hbar}(L_z+2S_z)\mathcal{B} \tag{6.85}$$

is called the *paramagnetic term* because it gives rise to the alignment of magnetic moments in an external magnetic field, which leads to paramagnetism. The term quadratic in the magnetic field,

$$\frac{e^2}{8m}(\mathcal{B}\times\mathbf{r})^2 = \frac{e^2}{8m}\mathcal{B}^2 r^2 \sin^2\theta \tag{6.86}$$

is called the *diamagnetic term*.

The relative magnitude of the linear (paramagnetic) and quadratic (diagmagnetic) terms can be estimated as follows. For states of low angular momentum (say about $\hbar$), we see by using (3.74) that the quadratic term is of the order $e^2a_0^2n^4\mathcal{B}^2/(8mZ^2)$. On the other hand, the order of magnitude of the linear term is $\mu_B\mathcal{B} = e\hbar\mathcal{B}/(2m)$. The ratio of the quadratic to the linear term is therefore given approximately by

$$\frac{ea_0^2n^4\mathcal{B}}{4\hbar Z^2} \simeq \frac{n^4}{Z^2}\mathcal{B}10^{-6} \tag{6.87}$$

where we recall that $\mathcal{B}$ is expressed in tesla (T). In the laboratory, the fields do not usually exceed a few tens of tesla, so that the quadratic term can be safely neglected unless n is large. However, in some astrophysical situations, very large magnetic fields are believed to exist. For example, in neutron stars, field strengths at the surface may exceed 10^8 T and in such a case the quadratic term cannot be neglected.

The atomic unit of magnetic field strength is

$$\mathcal{B}_a = \frac{\hbar}{ea_0^2} = 2.35052\times 10^5\ \text{T} \tag{6.88}$$

Using atomic units, the quantity (6.87) can be written as $n^4\mathcal{B}/(4Z^2)$, and the quadratic term can be ignored if

$$\frac{n^4\mathcal{B}}{Z^2} \ll 1 \tag{6.89}$$

The linear Zeeman effect

Let us first consider the case for which the quadratic (diamagnetic) term can be neglected, so that the equation (6.84) reduces to

$$\left[-\frac{\hbar^2}{2m}\nabla^2 - \frac{Ze^2}{(4\pi\varepsilon_0)r} + \xi(r)\mathbf{L}\cdot\mathbf{S} + \frac{\mu_B}{\hbar}(L_z+2S_z)\mathcal{B}\right]\psi(q) = E\psi(q) \tag{6.90}$$

The nature of the solution depends on whether the magnetic interaction is greater or less than the spin–orbit interaction. We shall first discuss the former case (strong magnetic fields) and then analyse the so-called 'anomalous Zeeman effect' which corresponds to weak fields.

Strong fields

The fine structure splitting of the $n = 2$ level of hydrogenic atoms is $(0.365Z^4)$ cm^{-1} and decreases for large n like n^{-3}. We see from (6.79) that the magnetic interaction energy will be greater than this for field strengths $\mathcal{B} > Z^4$ tesla. By laboratory standards, these are very strong fields even for hydrogenic atoms with small Z, but such fields can occur in certain astrophysical situations, such as in some stars. In the strong field limit, we first solve the Schrödinger equation without the spin–orbit coupling, which can be subsequently treated as a perturbation. Thus we have

$$\left(-\frac{\hbar^2}{2m}\nabla^2 - \frac{Ze^2}{(4\pi\varepsilon_0)r}\right)\psi(q) = \left[E - \frac{\mu_B\mathcal{B}}{\hbar}(L_z + 2S_z)\right]\psi(q) \tag{6.91}$$

The unperturbed hydrogenic spin-orbitals $\psi_{nlm_lm_s}$ defined by (5.9) are eigenfunctions of L_z and S_z and satisfy this equation if

$$E = E_n + \mu_B\mathcal{B}(m_l + 2m_s), \qquad m_s = \pm 1/2 \tag{6.92}$$

The introduction of the magnetic field does not remove the degeneracy in l, but by providing a preferred direction in space, it removes partially the degeneracy in m_l and m_s, splitting each level with a given n into equally spaced terms. This is illustrated in Fig. 6.5 for the case of a p level ($l = 1$). However, the energy of the states with $m_l = +1$ and $m_s = -1/2$ coincides with those with $m_l = -1$ and $m_s = +1/2$. In the strong-field limit we are considering here (no spin–orbit coupling) the orbital and spin angular momenta are constants of the motion and the eigenfunctions, written in Dirac notation, are of the form $|nlm_lsm_s\rangle$, with $s = 1/2$ and $m_s = \pm 1/2$.

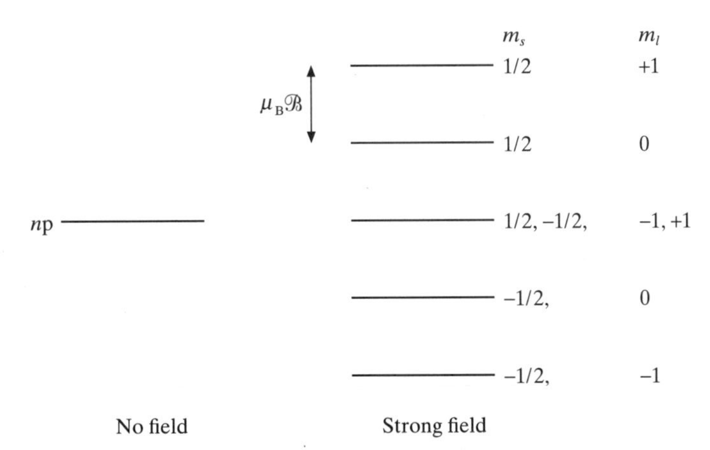

Figure 6.5 The splitting of a p level into five equally spaced levels by a strong magnetic field.

The selection rules for electric dipole transitions require that $\Delta m_s = 0$ and $\Delta m_l = 0, \pm 1$. Thus the spectral line corresponding to a transition $n \to n'$ is split into *three* components. The line corresponding to $\Delta m_l = 0$ has the original frequency $\nu_{n'n}$ and is called the π line, while the two lines with $\Delta m_l = \pm 1$ are called σ lines and correspond to frequencies

$$\nu^{\pm}_{n'n} = \nu_{n'n} \pm \nu_{\mathrm{L}} \tag{6.93}$$

where

$$\nu_{\mathrm{L}} = \frac{\mu_{\mathrm{B}} \mathcal{B}}{h} \tag{6.94}$$

is known as the *Larmor frequency*. This splitting is called the *normal Zeeman effect* and the three lines are said to form a *Lorentz triplet* (see Fig. 6.6). Apart from the

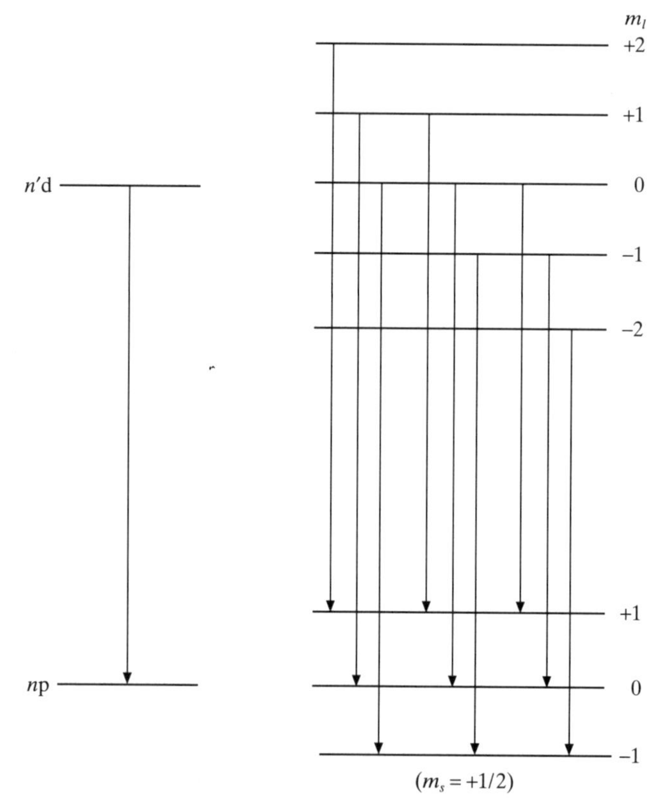

Figure 6.6 Illustration of the normal Zeeman effect for the transition $n'\mathrm{d} \to n\mathrm{p}$, with $m_s = +1/2$. In a strong magnetic field nine transitions are possible between the split levels consistent with $\Delta m_l = 0, \pm 1$ and $\Delta m_s = 0$. Of these, there are only three different frequencies corresponding to the π line ($\Delta m_l = 0$) and the two σ lines ($\Delta m_l = \pm 1$), which form a Lorentz triplet. The frequencies of the transitions associated with $m_s = -1/2$ are the same as those for $m_s = +1/2$.

case of very strong fields, Lorentz triplets can be observed in many-electron systems for which the total spin is zero, as in this case the spin–orbit coupling vanishes.

The polarisation of the radiation in each of the emission lines has interesting properties. The transition rate for spontaneous emission of radiation having a polarisation vector $\hat{\boldsymbol{\varepsilon}}$ is given in the dipole approximation by (4.79), namely

$$W_{ab}^{\mathrm{s,D}}\,\mathrm{d}\Omega = \frac{1}{2\pi\hbar c^3}\left(\frac{e^2}{4\pi\varepsilon_0}\right)\omega_{ba}^3\,|\hat{\boldsymbol{\varepsilon}}\cdot\mathbf{r}_{ba}|^2\,\mathrm{d}\Omega$$

$$= C(\omega_{ba})\,|\hat{\boldsymbol{\varepsilon}}\cdot\mathbf{r}_{ba}|^2\,\mathrm{d}\Omega \qquad \textbf{(6.95a)}$$

where we have set

$$C(\omega_{ba}) = \frac{1}{2\pi\hbar c^3}\left(\frac{e^2}{4\pi\varepsilon_0}\right)\omega_{ba}^3 \qquad \textbf{(6.95b)}$$

The vector $\hat{\boldsymbol{\varepsilon}}$ can be expressed in terms of two independent vectors $\hat{\mathbf{e}}_1$ and $\hat{\mathbf{e}}_2$ as in (4.123), where $\hat{\mathbf{e}}_1$, $\hat{\mathbf{e}}_2$ and $\hat{\mathbf{k}}$ form a right-handed system of axes (see (4.124)). If we take (as can always be done) $\hat{\mathbf{e}}_2$ to lie in the (X, Y) plane and if (Θ, Φ) are the polar angles of $\hat{\mathbf{k}}$ (see Fig. 6.7), we have

$$\begin{aligned} &(\hat{e}_1)_x = \cos\Theta\cos\Phi; \quad (\hat{e}_1)_y = \cos\Theta\sin\Phi; \quad (\hat{e}_1)_z = -\sin\Theta; \\ &(\hat{e}_2)_x = -\sin\Phi; \qquad\qquad (\hat{e}_2)_y = \cos\Phi; \qquad\qquad (\hat{e}_2)_z = 0 \end{aligned} \qquad \textbf{(6.96)}$$

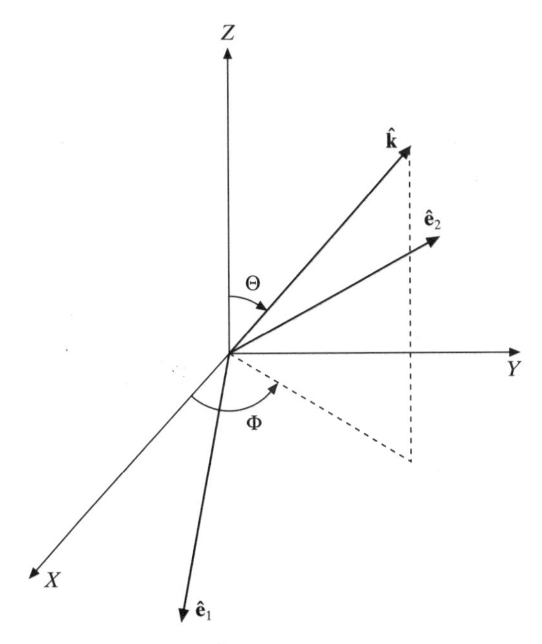

Figure 6.7 The unit vectors $\hat{\mathbf{e}}_1$, $\hat{\mathbf{e}}_2$ and $\hat{\mathbf{k}}$ form a right-handed set. The polar angles of $\hat{\mathbf{k}}$ are (Θ, Φ) and $\hat{\mathbf{e}}_2$ lies in the XY plane, $\hat{\mathbf{e}}_1$ pointing downwards.

Consider first the π line, with $\Delta m_l = 0$. From the discussion given in Section 4.5 we see that in this case $x_{ba} = y_{ba} = 0$ and we are only concerned with z_{ba}. The transition rate for emission in the solid angle dΩ of a photon with polarisation $\hat{\mathbf{e}}_1$ is then

$$W_{ab}^{s,D}\,d\Omega = C(\omega_{ba})\sin^2\Theta\,|z_{ba}|^2\,d\Omega \tag{6.97}$$

and the rate is zero for emission of a photon with polarisation $\hat{\mathbf{e}}_2$. When the light is viewed longitudinally, so that $\hat{\mathbf{k}}$ is in the direction of the magnetic field (which is parallel to the Z axis), $\Theta = 0$ and the π line is absent. In transverse observation ($\Theta = \pi/2$), in a direction at right angles to the magnetic field, the π radiation is plane-polarised with $\hat{\boldsymbol{\varepsilon}} = \hat{\mathbf{e}}_1$ in the direction of the negative Z axis.

Let us now consider the case in which $\Delta m_l = m_l' - m_l = -1$ which corresponds to the amplitude (see (4.110)–(4.112))

$$\varepsilon_{-1}^{*}\,I_{n'l'm_l-1;nlm_l}^{-1} = \frac{1}{\sqrt{2}}(\hat{\varepsilon}_x + i\hat{\varepsilon}_y)\frac{1}{\sqrt{2}}(x_{ba} - iy_{ba}) \tag{6.98}$$

The transition rate for emission of a photon with polarisation $\hat{\boldsymbol{\varepsilon}} = \hat{\mathbf{e}}_1$ is then

$$W_{ab}^{s,D}(1)\,d\Omega = C(\omega_{ba})\left|\frac{1}{\sqrt{2}}\cos\Theta\,e^{i\Phi}\right|^2\left|\frac{1}{\sqrt{2}}(x_{ba} - iy_{ba})\right|^2 d\Omega \tag{6.99}$$

and that for polarisation $\hat{\boldsymbol{\varepsilon}} = \hat{\mathbf{e}}_2$ is

$$W_{ab}^{s,D}(2)\,d\Omega = C(\omega_{ba})\left|\frac{i}{\sqrt{2}}e^{i\Phi}\right|^2\left|\frac{1}{\sqrt{2}}(x_{ba} - iy_{ba})\right|^2 d\Omega \tag{6.100}$$

Summing over both independent polarisation directions, the transition rate for the line corresponding to $\Delta m_l = m_l' - m_l = -1$, which is known as the σ^+ line, is

$$W_{ab}^{s,D}(\sigma^+)\,d\Omega = C(\omega_{ba})\frac{1}{2}(1 + \cos^2\Theta)\left|\frac{1}{\sqrt{2}}(x_{ba} - iy_{ba})\right|^2 d\Omega \tag{6.101}$$

In transverse observation $\Theta = \pi/2$ and the x and y components of $\hat{\mathbf{e}}_1$ vanish. In this case the σ^+ line is plane-polarised with $\hat{\boldsymbol{\varepsilon}} = \hat{\mathbf{e}}_2$, where $\hat{\mathbf{e}}_2$ lies in the (X, Y) plane. In contrast, in longitudinal observation along the direction of the magnetic field, we see from (4.131) that the radiation is left-hand circularly polarised, that is the emitted photon has helicity $+\hbar$.

In the same way, the transition rate for the σ^- line corresponding to $\Delta m_l = m_l' - m_l = +1$ is given by

$$W_{ab}^{s,D}(\sigma^-)\,d\Omega = C(\omega_{ba})\frac{1}{2}(1 + \cos^2\Theta)\left|-\frac{1}{\sqrt{2}}(x_{ba} + iy_{ba})\right|^2 d\Omega \tag{6.102}$$

The σ^- line is right-hand circularly polarised when viewed along the direction of the magnetic field and is plane-polarised in transverse observation.

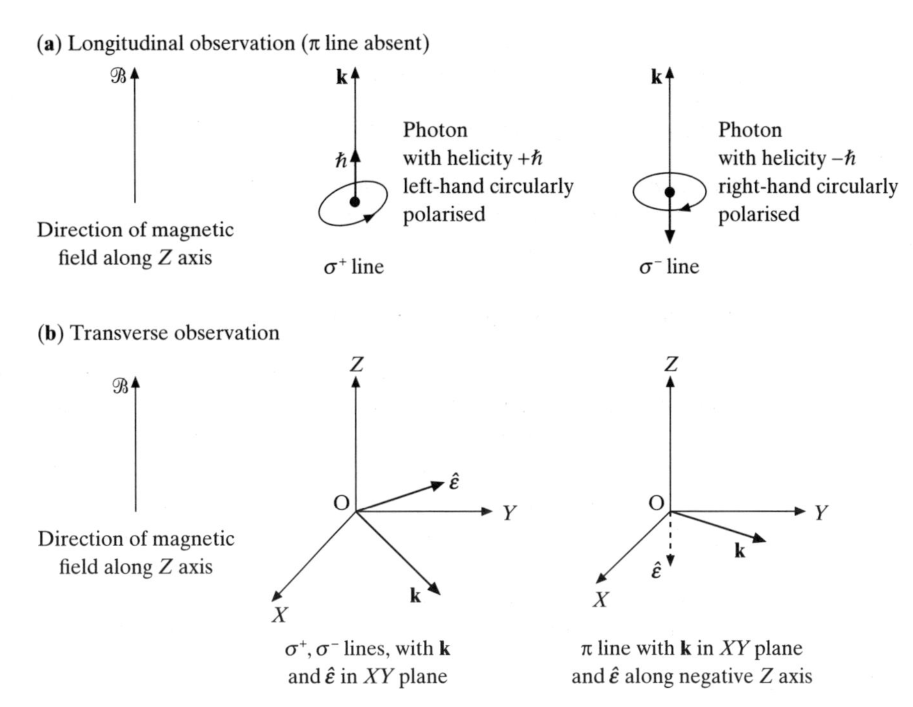

Figure 6.8 Polarisations of photons emitted in the direction of a magnetic field, or at right angles to a magnetic field.

The preceding discussion of the polarisations of the σ and π components is illustrated in Fig. 6.8. It is easily shown (Problem 6.4) that in the transverse direction the intensity of the π component is twice that of each σ component.

The Paschen–Back effect

At field strengths for which the spin–orbit interaction is appreciable, but still small compared with the term in $\mathcal{B}$ in (6.90), it can be treated in first-order perturbation theory. The perturbation is $\xi(r)\mathbf{L}\cdot\mathbf{S}$, and it can be verified (Problem 6.5) that it does not connect the degenerate states with $m_l = +1$, $m_s = -1/2$ and $m_l = -1$, $m_s = +1/2$. As a consequence, first-order non-degenerate perturbation theory can be used, and the contribution of the spin–orbit interaction $\xi(r)\mathbf{L}\cdot\mathbf{S}$ to the total energy is

$$\Delta E = \int_0^\infty \mathrm{d}r r^2 [R_{nl}(r)]^2 \xi(r) \left\langle l \frac{1}{2} m_l m_s \left| \mathbf{L}\cdot\mathbf{S} \right| l \frac{1}{2} m_l m_s \right\rangle$$

$$= \lambda_{nl} m_l m_s, \qquad l \neq 0 \tag{6.103}$$

while $\Delta E = 0$ for s states ($l = 0$). The quantity λ_{nl} is given by

$$\lambda_{nl} = \hbar^2 \int_0^\infty \mathrm{d}r r^2 [R_{nl}(r)]^2 \xi(r) = -\frac{\alpha^2 Z^2}{n} E_n \frac{1}{l(l+\frac{1}{2})(l+1)}, \qquad l \neq 0 \tag{6.104}$$

The degeneracy in l is removed, as we expect. The energy difference between levels (nlm_lm_s) and $(n'l'm_l'm_s')$ with $m_s = m_s'$ is

$$\delta E = E_n' - E_n + \mu_B \mathcal{B}(m_l' - m_l) + (\lambda_{n'l'} m_l' - \lambda_{nl} m_l) m_s \tag{6.105}$$

This expression gives the frequencies $\delta E/h$ of the observed lines, with $\Delta m_l = m_l' - m_l$ restricted to the values 0, ±1. The observed splitting in this case is known as the *Paschen–Back effect.*

Weak fields: the anomalous Zeeman effect

For historical reasons the case of a weak magnetic field is known as the *anomalous Zeeman effect*, although in fact this effect is the one most commonly encountered. In the early days of spectroscopy, before the electron spin was discovered, the normal Zeeman effect was predicted, on classical grounds, but the observations did not conform to the predictions and were said to be 'anomalous'. The explanation was finally given in terms of quantum mechanics and the electron spin.

When the interaction caused by the external magnetic field is small compared with the spin–orbit term, the unperturbed Hamiltonian can be taken to be

$$H_0 = -\frac{\hbar^2}{2m}\nabla^2 - \frac{Ze^2}{(4\pi\varepsilon_0)r} + \xi(r)\mathbf{L}\cdot\mathbf{S} \tag{6.106}$$

The unperturbed wave functions are eigenfunctions of $\mathbf{L}^2$, $\mathbf{S}^2$, $\mathbf{J}^2$ and J_z, but not of L_z and S_z. They are therefore products of radial functions times the 'generalised spherical harmonics' (see Appendix 4)

$$\mathcal{Y}_{ls}^{jm_j} = \sum_{m_l, m_s} \langle lsm_lm_s | jm_j \rangle Y_{lm_l}(\theta, \phi)\, \chi_{s,m_s} \tag{6.107}$$

where $\langle lsm_lm_s | jm_j \rangle$ are Clebsch–Gordan coefficients and $s = 1/2$. The perturbation is

$$\begin{aligned} H' &= \frac{\mu_B}{\hbar}(L_z + 2S_z)\mathcal{B} \\ &= \frac{\mu_B}{\hbar}(J_z + S_z)\mathcal{B} \end{aligned} \tag{6.108}$$

The additional energy due to this perturbation is

$$\Delta E = \mu_B \mathcal{B} m_j + \frac{\mu_B}{\hbar}\mathcal{B}\int d\Omega (\mathcal{Y}_{l,1/2}^{jm_j})^* S_z \mathcal{Y}_{l,1/2}^{jm_j} \tag{6.109}$$

where we have made use of the fact that $\mathcal{Y}_{l,1/2}^{jm_j}$ is a normalised eigenfunction of J_z belonging to the eigenvalue $m_j\hbar$.

Either of two methods can be used to evaluate the second term in (6.109).

1. The most straightforward procedure is to use the explicit expressions for the Clebsch–Gordan coefficients $\langle l1/2m_lm_s|jm_j\rangle$ given in Appendix 4. Setting $j = l \pm 1/2$, we have

$$\mathcal{Y}^{l+1/2,m_j}_{l,1/2} = \left(\frac{l+m_j+1/2}{2l+1}\right)^{1/2} Y_{l,m_j-1/2}(\theta,\phi)\chi_{1/2,1/2} + \left(\frac{l-m_j+1/2}{2l+1}\right)^{1/2} Y_{l,m_j+1/2}(\theta,\phi)\chi_{1/2,-1/2} \tag{6.110a}$$

and

$$\mathcal{Y}^{l-1/2,m_j}_{l,1/2} = -\left(\frac{l-m_j+1/2}{2l+1}\right)^{1/2} Y_{l,m_j-1/2}(\theta,\phi)\chi_{1/2,1/2} + \left(\frac{l+m_j+1/2}{2l+1}\right)^{1/2} Y_{l,m_j+1/2}(\theta,\phi)\chi_{1/2,-1/2} \tag{6.110b}$$

from which one readily obtains

$$\int d\Omega\,(\mathcal{Y}^{l+1/2,m_j}_{l,1/2})^* S_z \mathcal{Y}^{l+1/2,m_j}_{l,1/2} = \frac{m_j}{2l+1}\hbar \tag{6.111a}$$

and

$$\int d\Omega\,(\mathcal{Y}^{l-1/2,m_j}_{l,1/2})^* S_z \mathcal{Y}^{l-1/2,m_j}_{l,1/2} = -\frac{m_j}{2l+1}\hbar \tag{6.111b}$$

2. The same result can be obtained by operator methods. A *vector operator* **V** has three components (V_x, V_y, V_z) along three orthogonal axes, where V_x, V_y and V_z are operators which transform under rotations like the components of a vector (see Appendix 4). In particular, a vector operator **V** satisfies the identity (A4.47). The matrix element of the left-hand side of this identity with respect to states having the same value of j vanishes, so that

$$\langle lsjm_j|\mathbf{J}^2\mathbf{V} + \mathbf{V}\mathbf{J}^2|lsjm_j\rangle = 2\langle lsjm_j|(\mathbf{V}\cdot\mathbf{J})\mathbf{J}|lsjm_j\rangle \tag{6.112}$$

from which we have

$$j(j+1)\hbar^2\langle lsjm_j|\mathbf{V}|lsjm_j\rangle = \langle lsjm_j|(\mathbf{V}\cdot\mathbf{J})\mathbf{J}|lsjm_j\rangle \tag{6.113}$$

This relationship can also be obtained by using the Wigner–Eckart theorem discussed in Appendix 4.

Setting $\mathbf{V} = \mathbf{S}$ in (6.113) and taking the z component, we have

$$\begin{aligned} j(j+1)\hbar^2\langle lsjm_j|S_z|lsjm_j\rangle &= \langle lsjm_j|(\mathbf{S}\cdot\mathbf{J})J_z|lsjm_j\rangle \\ &= m_j\hbar\langle lsjm_j|\mathbf{S}\cdot\mathbf{J}|lsjm_j\rangle \end{aligned} \tag{6.114}$$

Since $\mathbf{S}\cdot\mathbf{J} = (\mathbf{J}^2 + \mathbf{S}^2 - \mathbf{L}^2)/2$, the matrix element of S_z is

$$\langle lsjm_j | S_z | lsjm_j \rangle = m_j\hbar \left[\frac{j(j+1) + s(s+1) - l(l+1)}{2j(j+1)} \right] \tag{6.115}$$

which agrees with (6.111) when $s = 1/2$ and $j = l \pm 1/2$.

The energy shift due to the magnetic field is seen from (6.109) and (6.111) (or (6.115)) to be proportional to $\mu_B \mathcal{B} m_j$ and may be written as

$$\Delta E_{m_j} = g\mu_B \mathcal{B} m_j \tag{6.116}$$

where g is called the *Landé g factor* and is given by

$$g = 1 + \frac{j(j+1) + s(s+1) - l(l+1)}{2j(j+1)} \tag{6.117}$$

The result (6.116) shows that, within a multiplet, the magnetic moment operator $\mathcal{M}$ of an atom is related to the total angular momentum operator $\mathbf{J}$ by

$$\mathcal{M} = -g\mu_B \mathbf{J}/\hbar \tag{6.118}$$

in conformity with (1.126).

Since in our case $s = 1/2$ we have

$$\Delta E_{m_j} = \frac{2l+2}{2l+1}\mu_B \mathcal{B} m_j, \qquad j = l + 1/2 \tag{6.119a}$$

$$= \frac{2l}{2l+1}\mu_B \mathcal{B} m_j, \qquad j = l - 1/2 \tag{6.119b}$$

The total energy of the level with quantum numbers n, j, m_j of a hydrogenic atom in a constant magnetic field is therefore

$$E_{n,j,m_j} = E_n + \Delta E_{n,j} + \Delta E_{m_j} \tag{6.120}$$

where E_n is the non-relativistic energy (3.29) (with $\mu = m$), $\Delta E_{n,j}$ is the fine structure correction (5.28) and ΔE_{m_j} is the correction due to the (weak) magnetic field.

The splitting of levels, δE, corresponding to the 'anomalous' Zeeman effect discussed above is illustrated in Fig. 6.9. We remark that since the splitting of the levels is not the same for each multiplet, there will be more lines in this case than the three lines (Lorentz triplet) corresponding to the normal Zeeman effect. This is shown in Fig. 6.10, where we display the allowed transitions (corresponding to $\Delta l = \pm 1$ and $\Delta m_j = 0, \pm 1$) between the $n = 2$ and $n = 1$ levels of atomic hydrogen occurring in the presence of a weak magnetic field.

As the magnitude of the magnetic field $\mathcal{B}$ increases from the weak field to the strong field limit, the energy changes smoothly. This is depicted in Fig. 6.11 for the 2p states of atomic hydrogen.

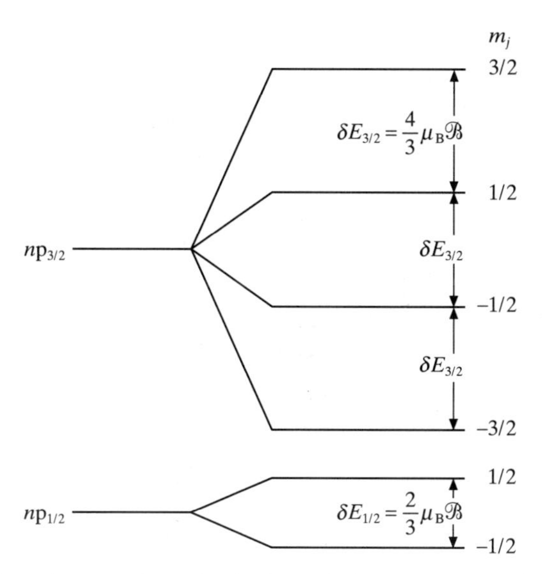

Figure 6.9 Splitting of $np_{3/2}$ and $np_{1/2}$ levels of atomic hydrogen in a weak magnetic field.

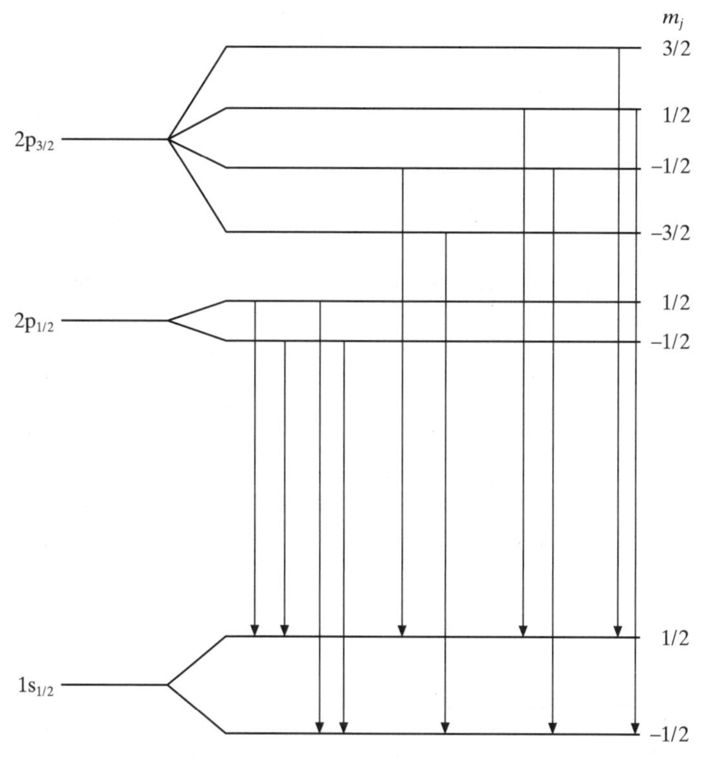

Figure 6.10 In electric dipole transitions between the $n = 2$ and $n = 1$ levels of atomic hydrogen, in a weak magnetic field, four lines result from the $2p_{1/2} \rightarrow 1s_{1/2}$ transitions and six lines from the $2p_{3/2} \rightarrow 1s_{1/2}$ transitions.

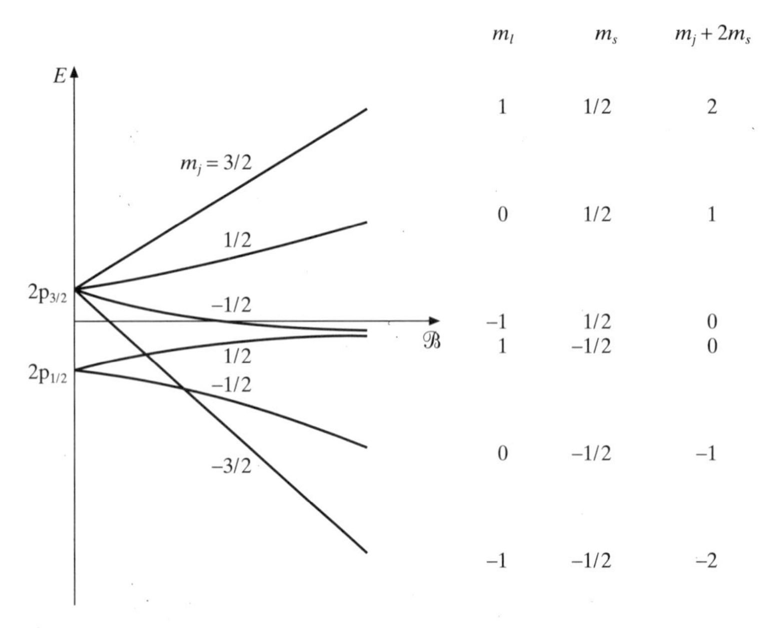

Figure 6.11 The energy of the levels of a hydrogen atom in a magnetic field are a smooth function of $\mathcal{B}$. For small $\mathcal{B}$, the splitting is uneven (the anomalous Zeeman effect), but for large $\mathcal{B}$, the splitting is even and only three lines are seen (the Lorentz triplet). A schematic diagram is shown for the 2p levels.

The quadratic Zeeman effect

As we have seen from the condition (6.89) the diamagnetic term proportional to $\mathcal{B}^2$, in the Schrödinger equation (6.84), can be neglected for the magnetic fields encountered in the laboratory, provided the principal quantum number n of the atom concerned is not too large. For Rydberg atoms for which n is large this is not the case and the diamagnetic term is significant for values of n greater than 10–20. The observed spectra of white dwarfs and neutron stars indicate that the surface magnetic fields can range from 10^4 to 10^8 T, or more, and for such fields the diamagnetic interaction cannot be ignored even for small principal quantum numbers.

For strong fields the spin–orbit interaction is small compared with the other terms in (6.84) and can be omitted. When this is done, the Schrödinger equation can be written as

$$\left[-\frac{\hbar^2}{2m}\nabla^2 - \frac{Ze^2}{(4\pi\varepsilon_0)r} + \frac{\mu_B}{\hbar}(L_z + 2S_z)\mathcal{B} + \frac{e^2}{8m}\mathcal{B}^2 r^2 \sin^2\theta\right]\psi(q) = E\psi(q) \tag{6.121}$$

Since L_z and S_z commute with H, ψ can be taken to be a simultaneous eigenfunction of L_z and S_z with eigenvalues $m_l\hbar$ and $m_s\hbar$ respectively, in which case (6.121) can be written as

$$\left[-\frac{\hbar^2}{2m}\nabla^2 - \frac{Ze^2}{(4\pi\varepsilon_0)r} + \mu_B(m_l + 2m_s)\mathcal{B} + \frac{e^2}{8m}\mathcal{B}^2 r^2 \sin^2\theta\right]\psi(q) = E\psi(q) \tag{6.122}$$

Three cases need to be considered. In the first $\mathcal{B}$ is so large that the Coulomb interaction can be neglected in comparison with the magnetic interaction, in the second the Coulomb and magnetic interactions are comparable, while in the third the diamagnetic term is sufficiently small to be treated as a perturbation.

Ultra-strong magnetic fields

If the magnetic interaction is so strong that the Coulomb potential can be neglected in comparison, the Schrödinger equation (6.122) can be expressed as

$$\left[-\frac{\hbar^2}{2m}\left(\frac{\partial^2}{\partial x^2} + \frac{\partial^2}{\partial y^2} + \frac{\partial^2}{\partial z^2}\right) + (m_l + 2m_s)\hbar\omega_L + \frac{1}{2}m\omega_L^2(x^2 + y^2)\right]\psi(q)$$
$$= E\psi(q) \tag{6.123}$$

where Cartesian coordinates have been used and $\omega_L = \mu_B\mathcal{B}/\hbar = 2\pi\nu_L$ is the Larmor angular frequency. Equation (6.123) admits solutions of the form

$$\psi = \Phi_q(x, y)\exp(ikz)\begin{cases}\alpha & \text{for } m_s = \frac{1}{2}\\ \beta & \text{for } m_s = -\frac{1}{2}\end{cases} \tag{6.124}$$

where Φ_q is the wave function for a two-dimensional harmonic oscillator. The eigenenergies of the Hamiltonian are correspondingly

$$E = \frac{\hbar^2 k^2}{2m} + \hbar\omega_L(m_l + 2m_s) + \hbar\omega_L(n_x + n_y + 1),$$
$$-\infty < k < \infty; \qquad m_l = 0, \pm 1, \pm 2, \ldots; \qquad m_s = \pm\frac{1}{2};$$
$$n_x = 0, 1, 2, \ldots; \qquad n_y = 0, 1, 2, \ldots \tag{6.125}$$

It is seen that the motion parallel to the Z axis is free, but the motion is bounded in the (X, Y) plane. Making use of the fact that the Hamiltonian can be written as a square (see Problem 6.6), it follows that the eigenvalues E are either positive or zero and that if $n = n_x + n_y$, then

$$n + m_l \geqslant 0 \tag{6.126}$$

The Hamiltonian for a two-dimensional harmonic oscillator is invariant under the parity operation $x \to -x$, $y \to -y$ so that the eigenfunctions $\Phi_q(x, y)$ are eigenstates of parity. The quantum number n is even for states of even parity and odd for those of odd parity. Using these facts it follows that $n + m_l$ is even,

$$n + m_l = 2r, \qquad r = 0, 1, 2, \ldots \tag{6.127}$$

Using (6.126) and (6.127) the eigenvalues E can be written as

$$E = \frac{\hbar^2 k^2}{2m} + \hbar\omega_{\rm L}(2r + 2m_s + 1)$$
$$-\infty < k < \infty; \qquad r = 0, 1, 2, \ldots; \qquad m_s = \pm\frac{1}{2} \tag{6.128}$$

For given values of k and m_s, the discrete energy levels labelled by the quantum number r are called *Landau levels*.

The intermediate region

We shall now consider the case in which both the Coulomb potential and the diamagnetic interaction are important. The ratio γ of the energy $\hbar\omega_{\rm L}$ to the ionisation potential of atomic hydrogen can be taken as a measure of the relative importance of the diamagnetic and Coulomb interactions. We have

$$\gamma = \frac{\hbar\omega_{\rm L}}{I_{\rm P}} = \frac{\mathcal{B}}{\mathcal{B}_a} \tag{6.129}$$

where $\mathcal{B}_a = 2.350\,52 \times 10^5$ T is the atomic unit of magnetic field strength (see (6.88)). For small values of the principal quantum number n the size of γ indicates the relative importance of the diamagnetic and Coulomb interactions. As we have already seen, in the case of magnetic fields encountered in the laboratory, for which γ is of the order of 10^{-5}, the influence of the diamagnetic interaction on low-lying atomic states is negligible. However, for Rydberg states with large n, the spacing of the energy levels is approximately $2I_{\rm P}/n^3$ and the diamagnetic effect will be significant for field strengths $\mathcal{B} \gg \mathcal{B}_a/n^3$. If γn^3 is much greater than unity the Coulomb interaction $-Ze^2/(4\pi\varepsilon_0 r)$ can be treated as a small perturbation of the Landau levels. The energy spectrum in the region in which the Coulomb and diamagnetic interactions are comparable ($\gamma n^3 \simeq 1$) is harder to calculate. One approach is to expand the wave function in a set of Landau states:

$$\psi(x, y, z) = \sum_{q=0}^{\infty} \Phi_q(x, y) F_q(z) \tag{6.130}$$

By inserting this expansion into the Schrödinger equation (6.121), and projecting successively with the functions Φ_q, a set of coupled equations for the expansion coefficients F_q is obtained, which can be solved numerically.

Interest was first aroused in the region in which the Coulomb and magnetic interactions are comparable by the discovery by W.R.S. Garton and T.S. Tomkins in 1969 that the absorption spectrum of Rydberg atoms in a strong magnetic field displays regular oscillations, now known as *quasi-Landau* resonances. Their measurements of the photoabsorption spectrum of barium in magnetic fields of different strengths are shown in Fig. 6.12. It is seen that the level spacing increases

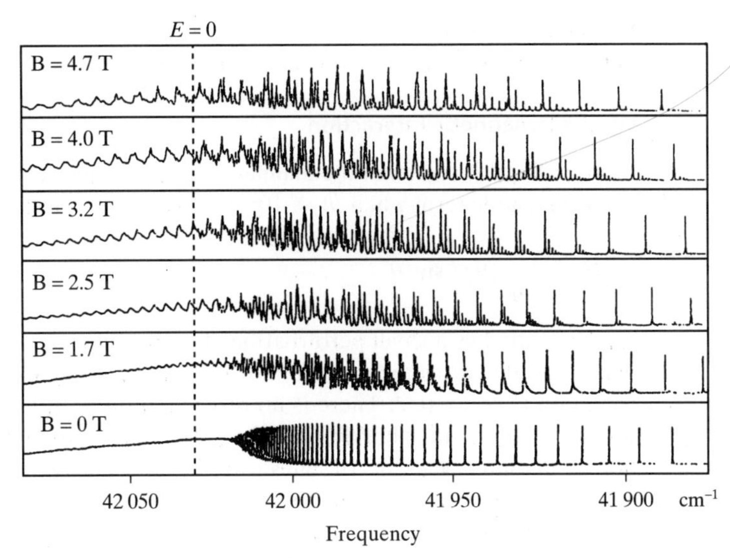

Figure 6.12 The photoabsorption spectrum of Rydberg Ba atoms for $n > 30$, for magnetic field strengths ranging from $\mathcal{B} = 0$ to 4.7 T, as observed by Garton and Tomkins.

with increasing field strengths $\mathcal{B}$, and it is found to be approximately $1.5\hbar\omega_L$ in the region below the ionisation threshold at $E = 0$. This spacing is 1.5 times as large as that of the Landau levels and was at first difficult to understand. However, A.R. Edmonds and A.F. Starace showed that the observed distribution of energy levels could be explained in terms of a simple semi-classical model. It was assumed that the energy levels near the ionisation threshold were associated with the motion in the (X, Y) plane and that the motion in the Z direction could be ignored. In the (X, Y) plane, with z set equal to zero, the potential energy can be expressed as

$$\begin{aligned} V(\rho) &= -\frac{Ze^2}{(4\pi\varepsilon_0)\rho} + \frac{e^2}{8m}\mathcal{B}^2\rho^2 \\ &= -\frac{Ze^2}{(4\pi\varepsilon_0)\rho} + \frac{1}{2}m\omega_L^2\rho^2 \end{aligned} \quad \textbf{(6.131)}$$

where $\rho^2 = x^2 + y^2$ and the linear term in $\mathcal{B}$ has been dropped. Moreover, since all neutral Rydberg atoms behave like Rydberg hydrogen atoms (see Section 3.6), one can set $Z = 1$ in (6.131). By using the WKB approximation to obtain the energy levels in the one-dimensional well $V(\rho)$, Edmonds and Starace found that the spacing was indeed $1.5\hbar\omega_L$, as experiment indicated. Further research has shown that the structure of the energy spectrum of the quasi-Landau resonances is extremely complicated when $\gamma n^3 \simeq 1$. This is of great interest since it is the region in which the corresponding classical motion becomes chaotic. The connection between the distribution of quantum energy in systems with large quantum numbers and the onset of classical chaos has been the subject of many studies [8].

Weak diamagnetic interaction

From (6.87) it is seen that the diamagnetic interaction is small compared with the linear magnetic term when $\mathcal{B} \ll 10^6 Z^2/n^4$ (with $\mathcal{B}$ in tesla)

$$H' = \frac{e^2}{8m}\mathcal{B}^2 r^2 \sin^2\theta \tag{6.132}$$

can be treated as a small perturbation. The matrix element of H' with respect to unperturbed hydrogenic states, $\langle n'l'm_l'|H'|nlm_l\rangle$, is diagonal in m_l, but satisfies the condition $|l' - l| = 0, 2$. There is no restriction on n. In hydrogenic atoms all the states with the same n but different l are degenerate, so that degenerate perturbation theory must be used. All states with the same parity but different l are coupled and l is no longer a good quantum number; the states are said to be *l-mixed.*

Alkali atoms can be treated as quasi one-electron atoms in which the valence electron moves in a Coulomb potential modified at short distances. In such atoms the degeneracy in l is removed and first-order non-degenerate perturbation theory can be employed. The shift in energy due to the magnetic interaction is given to first order by

$$\Delta E = (m_l + 2m_s)\hbar\omega_{\mathrm{L}} + \frac{1}{2}m\omega_{\mathrm{L}}^2\langle nlm_l|r^2 \sin^2\theta|nlm_l\rangle \tag{6.133}$$

Since the matrix element $\langle nlm_l|r^2|nlm_l\rangle$ behaves like n^4 for large n, the region in which first-order perturbation theory is accurate is extremely limited.

Problems

6.1 Show that in the linear Stark effect the $n = 3$ level of a hydrogen atom is split into five equally spaced components, and obtain an expression for the level separation (in cm^{-1}) as a function of the electric field strength (in $\mathrm{V\,m}^{-1}$).

6.2 Suppose that at time $t = 0$ a hydrogen atom is in an arbitrary superposition of the 2s and $2\mathrm{p}_0$ states. A constant electric field of strength $10^7\ \mathrm{V\,m}^{-1}$ is then applied along the Z axis. Show that during the lifetime of the $2\mathrm{p}_0$ state (due to radiative decay to the 1s state), the average population of the 2s level is nearly the same as that of the $2\mathrm{p}_0$ level.

6.3 Show that the second-order correction to the energy in perturbation theory can be written as

$$E_k^{(2)} = \frac{(H'^2)_{kk}}{E_k^{(0)}} - \frac{(H'_{kk})^2}{E_k^{(0)}} + \sum_{m\neq k}\frac{E_m^{(0)}|H'_{mk}|^2}{E_k^{(0)}(E_k^{(0)} - E_m^{(0)})}$$

[8] See for example Blümel and Reinhardt (1997) or Friedrich (1998).

By neglecting the sum of the right-hand side, obtain the approximation (6.34) for the quadratic Stark effect.

6.4 Show that in the limit of strong magnetic fields in transverse observation the intensity of the π Zeeman component is twice that of each σ component.

6.5 Verify that the perturbation $\xi(r)\mathbf{L}\cdot\mathbf{S}$ does not connect the degenerate states with $m_l = +1$, $m_s = -1/2$ and $m_l = -1$, $m_s = +1/2$.

(**Hint**: Use the raising and lowering operators $L_{\pm} = L_x \pm \mathrm{i}L_y$.)

6.6 (a) Show that the Hamiltonian of a free electron in a uniform time-independent magnetic field $\mathcal{B} = \mathcal{B}\hat{\mathbf{z}}$ is given by

$$H = H_{xy} + H_z$$

with

$$H_{xy} = \frac{1}{2m}(p_x^2 + p_y^2) + \frac{1}{2}m\omega_\mathrm{L}^2(x^2 + y^2)$$

and

$$H_z = \frac{1}{2m}p_z^2 + \omega_\mathrm{L}(L_z + 2S_z)$$

where $\omega_\mathrm{L} = (\mu_\mathrm{B}/\hbar)\mathcal{B} = 2\pi\nu_\mathrm{L}$ is the Larmor angular frequency.

(b) Using the fact that H can be written as a square, $H = (\mathbf{p} + e\mathbf{A})^2/(2m)$, and that the Hamiltonian H_{xy} of the harmonic motion in the XY plane is invariant under the reflection $x \to -x$, $y \to -y$, show that the energy eigenvalues are given by

$$E = \frac{\hbar^2 k^2}{2m} + \hbar\omega_\mathrm{L}(2r + 2m_s + 1)$$

where $-\infty < k < +\infty$, $r = 0, 1, 2, \ldots$ and $m_s = \pm 1/2$. For given k and m_s, the discrete energy levels labelled by the quantum number r are the *Landau levels* [see (6.128)].

(c) In neutron stars magnetic fields of the order of 10^8 T may occur. Find the energy separation between the adjacent Landau levels. What is the size of the region to which the motion in the XY plane is confined?

7 Two-electron atoms

In this chapter we begin our study of many-electron atoms by considering the simplest ones, namely atoms (or ions) consisting of a nucleus of charge Ze and *two* electrons. These include the negative hydrogen ion H^- ($Z = 1$), the helium atom ($Z = 2$), the singly ionised lithium atom Li^+ ($Z = 3$), and so on. These systems deserve particular attention for two reasons. It is for these systems that we shall first study in detail the implications of the *Pauli exclusion principle*, which plays a central role in atomic and molecular physics. Secondly, it is for two-electron atoms that various approximations used in atomic structure calculations can be explained most easily and tested accurately.

We shall limit our discussion to the non-relativistic theory of two-electron atoms. After analysing in succession the space and spin symmetries and the role of the Pauli exclusion principle, we will discuss the level scheme of two-electron atoms and introduce the independent particle model, which is of great importance in studying many-electron systems. We then study in some detail the ground state and the lowest excited states of two-electron atoms, using perturbation theory and the variational method, and conclude this chapter with a brief survey of autoionising states.

7.1 The Schrödinger equation for two-electron atoms. Para and ortho states

Let us consider an atom (or ion) consisting of a nucleus of charge Ze and mass M and two electrons of mass m. As in the case of one-electron atoms, we shall begin our treatment by neglecting all but the Coulomb interactions between the particles, and by writing down the Schrödinger equation for the spatial part of the wave function describing the relative motion. The separation of the centre of mass motion is somewhat more complicated than for the case of one-electron atoms, since we are now dealing with a three-body problem. This separation is performed in Appendix 8 for the general case of an atom (ion) having N electrons. Denoting by $\mathbf{r}_1$ and $\mathbf{r}_2$ the relative coordinates of the two electrons with respect to the nucleus (see Fig. 7.1) we see from the equations (A8.12) and (A8.16) of Appendix 8 that the Schrödinger equation for the spatial part $\psi(\mathbf{r}_1, \mathbf{r}_2)$ of the wave function describing the relative motion is for a two-electron atom

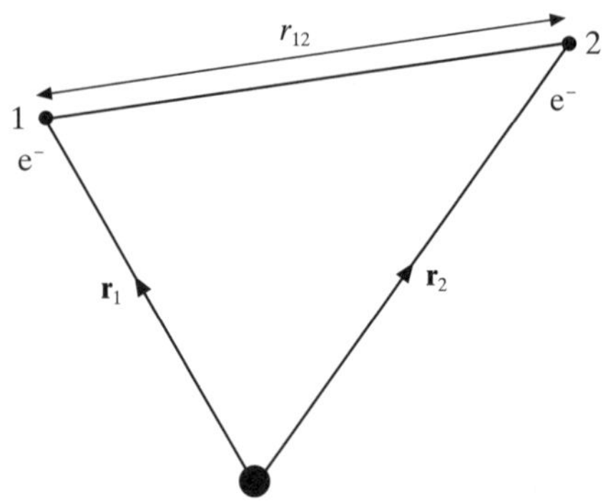

Figure 7.1 Coordinate system for two-electron atoms.

$$\left[-\frac{\hbar^2}{2\mu}\nabla^2_{\mathbf{r}_1} - \frac{\hbar^2}{2\mu}\nabla^2_{\mathbf{r}_2} - \frac{\hbar^2}{M}\nabla_{\mathbf{r}_1}\cdot\nabla_{\mathbf{r}_2} - \frac{Ze^2}{(4\pi\varepsilon_0)r_1} - \frac{Ze^2}{(4\pi\varepsilon_0)r_2} + \frac{e^2}{(4\pi\varepsilon_0)r_{12}}\right]\psi(\mathbf{r}_1, \mathbf{r}_2) = E\psi(\mathbf{r}_1, \mathbf{r}_2) \tag{7.1}$$

where $\mu = mM/(m + M)$ is the reduced mass of an electron with respect to the nucleus and $r_{12} = |\mathbf{r}_1 - \mathbf{r}_2|$.

We shall first consider the case of an 'infinitely heavy' nucleus ($M = \infty$) so that (i) $\mu = m$ and (ii) the 'mass polarisation' term $(-\hbar^2/M)\nabla_{\mathbf{r}_1}\cdot\nabla_{\mathbf{r}_2}$ can be omitted. We shall work in atomic units (a.u.), in which the Hamiltonian is

$$H = -\frac{1}{2}\nabla^2_{\mathbf{r}_1} - \frac{1}{2}\nabla^2_{\mathbf{r}_2} - \frac{Z}{r_1} - \frac{Z}{r_2} + \frac{1}{r_{12}} \tag{7.2}$$

and the Schrödinger equation for $\psi(\mathbf{r}_1, \mathbf{r}_2)$ becomes

$$\left[-\frac{1}{2}\nabla^2_{\mathbf{r}_1} - \frac{1}{2}\nabla^2_{\mathbf{r}_2} - \frac{Z}{r_1} - \frac{Z}{r_2} + \frac{1}{r_{12}}\right]\psi(\mathbf{r}_1, \mathbf{r}_2) = E\psi(\mathbf{r}_1, \mathbf{r}_2) \tag{7.3}$$

Because of the presence of the electron–electron interaction term $1/r_{12}$, this equation is not separable, so that an eigenfunction $\psi(\mathbf{r}_1, \mathbf{r}_2)$ of (7.3) cannot be written in the form of a single product of one-electron wave functions. Wave functions for systems of two or more particles which cannot be expressed as a single product of one-particle wave functions are said to be *entangled* (see Bransden and Joachain, 2000). The corresponding states are also said to be entangled: measurements cannot be made on one particle without affecting the others. It should be noted that wave functions for systems of mutually interacting particles can never be expressed (except approximately) as a single product of one-particle wave functions, and are therefore entangled.

We remark that the equation (7.3) is unchanged when the coordinates of the two electrons are interchanged. Thus, if we denote by P_{12} an interchange operator that permutes the spatial coordinates of the two electrons, the wave functions

$$\psi(\mathbf{r}_2, \mathbf{r}_1) = P_{12}\psi(\mathbf{r}_1, \mathbf{r}_2) \tag{7.4}$$

and $\psi(\mathbf{r}_1, \mathbf{r}_2)$ satisfy the same Schrödinger equation. Moreover, both functions $\psi(\mathbf{r}_1, \mathbf{r}_2)$ and $\psi(\mathbf{r}_2, \mathbf{r}_1)$ must be continuous, single-valued and bounded. If $\psi(\mathbf{r}_1, \mathbf{r}_2)$ corresponds to a non-degenerate eigenvalue, $\psi(\mathbf{r}_1, \mathbf{r}_2)$ and $\psi(\mathbf{r}_2, \mathbf{r}_1)$ can only differ by a multiplicative factor λ,

$$\psi(\mathbf{r}_2, \mathbf{r}_1) = P_{12}\psi(\mathbf{r}_1, \mathbf{r}_2) = \lambda\psi(\mathbf{r}_1, \mathbf{r}_2) \tag{7.5}$$

Applying the interchange operator P_{12} twice, we must obtain $\psi(\mathbf{r}_1, \mathbf{r}_2)$ again. Thus

$$\begin{aligned} P_{12}^2\psi(\mathbf{r}_1, \mathbf{r}_2) &= \lambda P_{12}\psi(\mathbf{r}_1, \mathbf{r}_2) \\ &= \lambda^2\psi(\mathbf{r}_1, \mathbf{r}_2) \\ &= \psi(\mathbf{r}_1, \mathbf{r}_2) \end{aligned} \tag{7.6}$$

so that $\lambda^2 = 1$, $\lambda = \pm 1$ and

$$\psi(\mathbf{r}_2, \mathbf{r}_1) = \pm\psi(\mathbf{r}_1, \mathbf{r}_2) \tag{7.7}$$

Wave functions which satisfy (7.7) with the plus sign (that is, whose spatial part remains unchanged upon permutating the spatial coordinates of the two electrons) are said to be *space-symmetric* and will be denoted by $\psi_+(\mathbf{r}_1, \mathbf{r}_2)$. On the other hand, wave functions satisfying (7.7) with the minus sign (that is, whose spatial part changes sign on interchanging the spatial coordinates of the two electrons) are said to be *space-antisymmetric* and will be written as $\psi_-(\mathbf{r}_1, \mathbf{r}_2)$. It is straightforward to show (Problem 7.1) that for degenerate eigenvalues the eigenfunctions of (7.3) can always be chosen so that (7.7) holds. Thus the eigenfunctions of a two-electron atom can be classified as being either space-symmetric or space-antisymmetric. The states described by space-symmetric wave functions are called *para* states; those corresponding to space-antisymmetric wave functions are known as *ortho* states.

7.2 Spin wave functions and the role of the Pauli exclusion principle

Until now we have not taken into account the spin of the two electrons. In the case of one-electron atoms we have seen in Chapter 5 that the electron spin only affects the fine and hyperfine structure of the spectrum. On the contrary, for two-electron atoms, we shall see that spin effects directly influence the spectrum because of the requirements of the *Pauli exclusion principle.*

Since we are dealing with the Hamiltonian (7.2) which is spin-independent, the atom can be completely described by specifying its spatial eigenfunction, together with the components of the electron spins in a given direction, which we choose as our Z axis. Thus the full eigenfunctions Ψ of the system must be products of the spatial eigenfunctions $\psi(\mathbf{r}_1, \mathbf{r}_2)$ satisfying the Schrödinger equation (7.3) times spin wave functions $\chi(1, 2)$ for the two-electron system. That is,

$$\Psi(q_1, q_2) = \psi(\mathbf{r}_1, \mathbf{r}_2)\chi(1, 2) \tag{7.8}$$

where q_i denotes collectively the space and spin coordinates of electron i.

Spin wave functions

The spin wave functions $\chi(1, 2)$ are easily constructed from our knowledge of the spin wave functions $\chi_{1/2,m_s}(1)$ and $\chi_{1/2,m_s}(2)$ of the individual electrons. Let us denote by $\mathbf{S}_1$ and $\mathbf{S}_2$ the spin operators of the two electrons, and by $(S_1)_z$ and $(S_2)_z$ the components of these operators along the Z direction. We also write the basic spin functions (see (2.209)) of the two electrons as $\alpha(1)$, $\beta(1)$ and $\alpha(2)$, $\beta(2)$, respectively. We emphasise that $\mathbf{S}_1$ only acts on $\alpha(1)$ and $\beta(1)$, while $\mathbf{S}_2$ operates only on $\alpha(2)$ and $\beta(2)$, so that the two operators $\mathbf{S}_1$ and $\mathbf{S}_2$ commute. The total spin angular momentum of the two electrons is represented by the operator

$$\mathbf{S} = \mathbf{S}_1 + \mathbf{S}_2 \tag{7.9}$$

whose z component is

$$S_z = (S_1)_z + (S_2)_z \tag{7.10}$$

Because $\mathbf{S}_1^2 = \mathbf{S}_2^2 = 3/4$ (in a.u.) we have

$$\begin{aligned}\mathbf{S}^2 &= \mathbf{S}_1^2 + \mathbf{S}_2^2 + 2\mathbf{S}_1\cdot\mathbf{S}_2 \\ &= \tfrac{3}{2} + 2\mathbf{S}_1\cdot\mathbf{S}_2\end{aligned} \tag{7.11}$$

Since there are no spin-dependent interactions each electron spin component in the Z direction can be directed either up ($\uparrow$) or down ($\downarrow$) independently of the other, and we have four independent spin states each of which can be represented as the product of two individual spin functions. That is,

$$\begin{aligned}\chi_1(1, 2) &= \alpha(1)\alpha(2)\uparrow\uparrow \\ \chi_2(1, 2) &= \alpha(1)\beta(2)\uparrow\downarrow \\ \chi_3(1, 2) &= \beta(1)\alpha(2)\downarrow\uparrow \\ \chi_4(1, 2) &= \beta(1)\beta(2)\downarrow\downarrow\end{aligned} \tag{7.12}$$

where the arrows illustrate the situation regarding the z components of the electron spins. Now

$$\begin{aligned}S_z\chi_1(1, 2) &= [(S_1)_z + (S_2)_z]\alpha(1)\alpha(2) \\ &= [(S_1)_z\alpha(1)]\alpha(2) + \alpha(1)[(S_2)_z\alpha(2)] \\ &= \tfrac{1}{2}\alpha(1)\alpha(2) + \tfrac{1}{2}\alpha(1)\alpha(2) \\ &= \chi_1(1, 2)\end{aligned} \tag{7.13}$$

Table 7.1 Values of $S_z\chi$ and $\mathbf{S}^2\chi$ (in a.u.) for the four two-electron spin functions (7.12).

Spin function χ	$S_z\chi$	$\mathbf{S}^2\chi$
$\chi_1 = \alpha(1)\alpha(2)$ ↑↑	χ_1	$2\chi_1$
$\chi_2 = \alpha(1)\beta(2)$ ↑↓	0	$\chi_2 + \chi_3$
$\chi_3 = \beta(1)\alpha(2)$ ↓↑	0	$\chi_2 + \chi_3$
$\chi_4 = \beta(1)\beta(2)$ ↓↓	$-\chi_4$	$2\chi_4$

Thus, if we write the eigenvalue equation for the operator S_z (in a.u.) as

$$S_z\chi = M_S\chi \tag{7.14}$$

we see that χ_1 is an eigenstate of S_z corresponding to the eigenvalue $M_S = +1$. Similarly, the spin states χ_2, χ_3 and χ_4 are easily shown to be eigenstates of S_z corresponding respectively to the values $M_S = 0$, 0 and -1 of the quantum number M_S [1]. Using (7.11) and the basic relations (2.206), the action of the operator $\mathbf{S}^2$ on the four spin functions (7.12) can also be studied (Problem 7.2). The results are given in Table 7.1. If we write the eigenvalue equation for $\mathbf{S}^2$ (in a.u.) as

$$\mathbf{S}^2\chi = S(S+1)\chi \tag{7.15}$$

we see that both χ_1 and χ_4 are eigenstates of $\mathbf{S}^2$ corresponding to the eigenvalue $S(S+1) = 2$, that is to the value $S = 1$ of the quantum number S associated with the magnitude of the total spin. On the other hand, χ_2 and χ_3 are *not* eigenstates of $\mathbf{S}^2$.

Looking back at the four spin functions (7.12), we also see that both χ_1 and χ_4 are symmetric in the exchange of the labels of the two electrons, while neither χ_2 nor χ_3 is symmetric or antisymmetric. As we shall see below, it is essential to deal with two-electron spin functions which are either symmetric or antisymmetric in the interchange of the electron labels. Fortunately, it is easy to form linear combinations of χ_2 and χ_3 which are respectively symmetric and antisymmetric in the exchange of the electron labels 1 and 2. That is,

$$\chi_+(1,2) = \frac{1}{\sqrt{2}}[\chi_2(1,2) + \chi_3(1,2)] \tag{7.16}$$

and

$$\chi_-(1,2) = \frac{1}{\sqrt{2}}[\chi_2(1,2) - \chi_3(1,2)] \tag{7.17}$$

where the subscripts + and – denote the symmetric and antisymmetric functions, respectively, and the factor $2^{-1/2}$ has been introduced so that both χ_+ and χ_- are normalised to unity. Using the results of Table 7.1, we see that the symmetric spin

[1] In what follows we shall use *capital* letters to denote the values of the quantum numbers for the *total* orbital angular momentum, total spin, etc.

Table 7.2 Values of the quantum numbers S and M_S for the antisymmetric spin function (7.18), and the three symmetric spin functions (7.19). Each of these spin functions is a simultaneous eigenstate of the operators $\mathbf{S}^2$ and S_z, with eigenvalues given respectively (in a.u.) by $S(S+1)$ and M_S. The antisymmetric spin function (7.18) corresponding to $S=0$ $M_S=0$ is a spin singlet, while the three symmetric spin states (7.19) corresponding to $S=1$ and $M_S=1, 0, -1$ are seen to form a spin triplet.

Spin funtion	S	M_S
$\frac{1}{\sqrt{2}}[\alpha(1)\beta(2)-\beta(1)\alpha(2)]$	0	0
$\alpha(1)\alpha(2)$	1	1
$\frac{1}{\sqrt{2}}[\alpha(1)\beta(2)+\beta(1)\alpha(2)]$	1	0
$\beta(1)\beta(2)$	1	−1

function χ_+ is an eigenstate of both operators $\mathbf{S}^2$ and S_z, with quantum numbers given by $S=1$ and $M_S=0$, respectively. The antisymmetric spin function χ_- is also an eigenstate of both $\mathbf{S}^2$ and S_z, with corresponding quantum numbers $S=0$ and $M_S=0$. In what follows, and by analogy with the one-electron case, we shall write the eigenstates common to both operators $\mathbf{S}^2$ and S_z as χ_{S,M_S}, so that in this new notation we have $\chi_{0,0}\equiv\chi_-$, $\chi_{1,1}\equiv\chi_1$, $\chi_{1,0}\equiv\chi_+$ and $\chi_{1,-1}\equiv\chi_4$.

The foregoing discussion shows that, starting with the four independent and normalised spin functions (7.12), we can construct four normalised and mutually orthogonal spin functions which are eigenstates of both operators $\mathbf{S}^2$ and S_z and possess a definite symmetry in the exchange of the two electrons. These are the *antisymmetric spin function*

$$\chi_{0,0}(1,2)=\frac{1}{\sqrt{2}}[\alpha(1)\beta(2)-\beta(1)\alpha(2)] \tag{7.18}$$

and the *three symmetric spin functions*

$$\begin{aligned}\chi_{1,1}(1,2)&=\alpha(1)\alpha(2)\\ \chi_{1,0}(1,2)&=\frac{1}{\sqrt{2}}[\alpha(1)\beta(2)+\beta(1)\alpha(2)]\\ \chi_{1,-1}(1,2)&=\beta(1)\beta(2)\end{aligned} \tag{7.19}$$

The antisymmetric spin function (7.18) corresponding to the quantum numbers $S=0$ and $M_S=0$ is called a (spin) *singlet*, while the three symmetric spin states (7.19) corresponding to the total spin $S=1$ and to the quantum numbers $M_S=1, 0, -1$, respectively, are said to form a (spin) *triplet* (see Table 7.2).

The role of the Pauli exclusion principle

Let us now return to the equation (7.8). At first sight, it would appear that by combining the four spin states (7.18) and (7.19) with the spatial eigenfunctions $\psi(\mathbf{r}_1,\mathbf{r}_2)$, we could obtain four times as many eigenstates $\Psi(q_1,q_2)$ for an atom (or ion) with two spin 1/2 electrons than if the electrons were spinless. However, this is not the case because, as we have seen in Chapter 2, *the Pauli exclusion principle requires that the total wave function* $\Psi(q_1, q_2, \ldots, q_N)$ *of a system of N electrons must be antisymmetric*. In other words $\Psi(q_1, q_2, \ldots, q_N)$ must change sign if all the coordinates (spatial and spin) of two electrons are interchanged. Hence, in our two-electron case, we see from (7.8) that in order to obtain total antisymmetric wave functions $\Psi(q_1, q_2)$ we must either multiply symmetric spatial (para) wave functions $\psi_+(\mathbf{r}_1, \mathbf{r}_2)$ by the antisymmetric (singlet) spin state (7.18),

$$\Psi(q_1, q_2) = \psi_+(\mathbf{r}_1, \mathbf{r}_2)\frac{1}{\sqrt{2}}[\alpha(1)\beta(2) - \beta(1)\alpha(2)] \quad \textbf{(7.20)}$$

or multiply antisymmetric spatial (ortho) wave functions $\psi_-(\mathbf{r}_1, \mathbf{r}_2)$ by one of the three symmetric spin functions (7.19) belonging to the spin triplet,

$$\Psi(q_1, q_2) = \psi_-(\mathbf{r}_1, \mathbf{r}_2) \times \begin{cases} \alpha(1)\alpha(2) \\ \dfrac{1}{\sqrt{2}}[\alpha(1)\beta(2) + \beta(1)\alpha(2)] \\ \beta(1)\beta(2) \end{cases} \quad \textbf{(7.21)}$$

Thus para states must always be spin singlets, while ortho states must be spin triplets, so that the Pauli exclusion principle introduces a *coupling* between the *space* and *spin* variables of the electrons.

7.3 Level scheme of two-electron atoms

We shall prove in Chapter 9 that radiative transitions between singlet and triplet spin states (known as *intercombination lines*) are *forbidden* in the electric dipole approximation, provided that spin–orbit interactions can be neglected. This is the case for atoms or ions with low enough Z, so that the energy spectrum of two-electron atoms (or ions) with $Z \lesssim 40$ consists of *two nearly independent systems of levels*, one made of para (singlet) states and the other of ortho (triplet) states. As an example, we show in Fig. 7.2 the first few (lowest) energy levels of helium, divided into singlets ($S = 0$) and triplets ($S = 1$). Because intercombination lines are absent in practice in the helium spectrum, spectroscopists spoke for a long time of two different species of helium, *parahelium* and *orthohelium*; this terminology is still used now.

Let $\mathbf{L} = \mathbf{L}_1 + \mathbf{L}_2$ be the sum of the two orbital angular momentum operators of the electrons. Using atomic units ($\hbar = 1$), we shall denote by $L(L+1)$ the eigenvalues of $\mathbf{L}^2$ and by M_L those of L_z, so that $M_L = -L, -L+1, \ldots, +L$. As seen from Fig. 7.2, the atomic energy levels, also called *terms* in spectroscopic language, are designated by symbols which generalise the ones we used for hydrogenic atoms. Thus each term is denoted as

$$^{2S+1}L$$

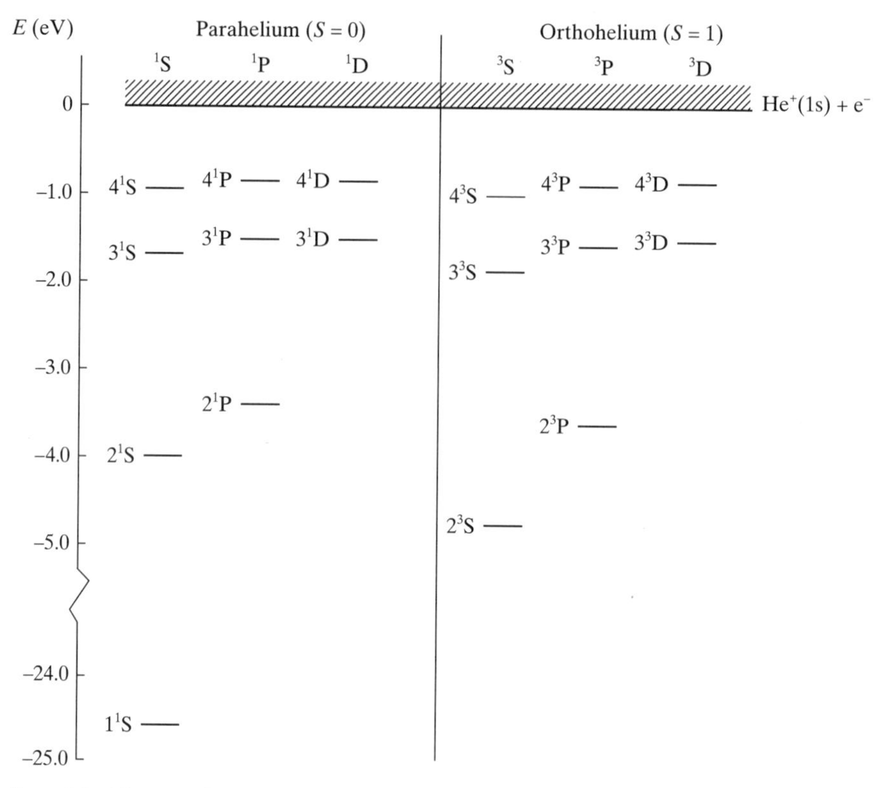

Figure 7.2 The experimental values of the lowest energy levels of helium. The energy scale is chosen so that $E = 0$ corresponds to the ionisation threshold. The configuration of each level is of the form 1s *nl*. The doubly excited states (for example, 2s *nl*) are at positive energies on this scale, within the $He^+(1s) + e^-$ continuum.

where a code letter is associated to the value of the total electronic orbital angular momentum quantum number L according to the correspondence

$$\begin{array}{cccccc} L=0 & 1 & 2 & 3 & 4 & 5 \\ \updownarrow & \updownarrow & \updownarrow & \updownarrow & \updownarrow & \updownarrow \\ \mathrm{S} & \mathrm{P} & \mathrm{D} & \mathrm{F} & \mathrm{G} & \mathrm{H} \end{array}$$

and so on. In addition, a superscript to the left gives the value of the quantity $2S + 1$, or *multiplicity*, which is equal to 1 for singlet ($S = 0$) states and 3 for triplet ($S = 1$) states.

We remark that Fig. 7.2. does not exhibit the fine structure splitting of the levels, due to the relativistic interaction between the spin and orbital angular momentum (spin–orbit effect) and to the magnetic interaction between the spins of the two electrons (spin–spin effect). Calling $\mathbf{J} = \mathbf{L} + \mathbf{S}$ the total electronic angular momentum and denoting by $J(J + 1)$ and M_J the eigenvalues (in a.u.) of the operators $\mathbf{J}^2$ and J_z, respectively, it may be shown that the spin–orbit and

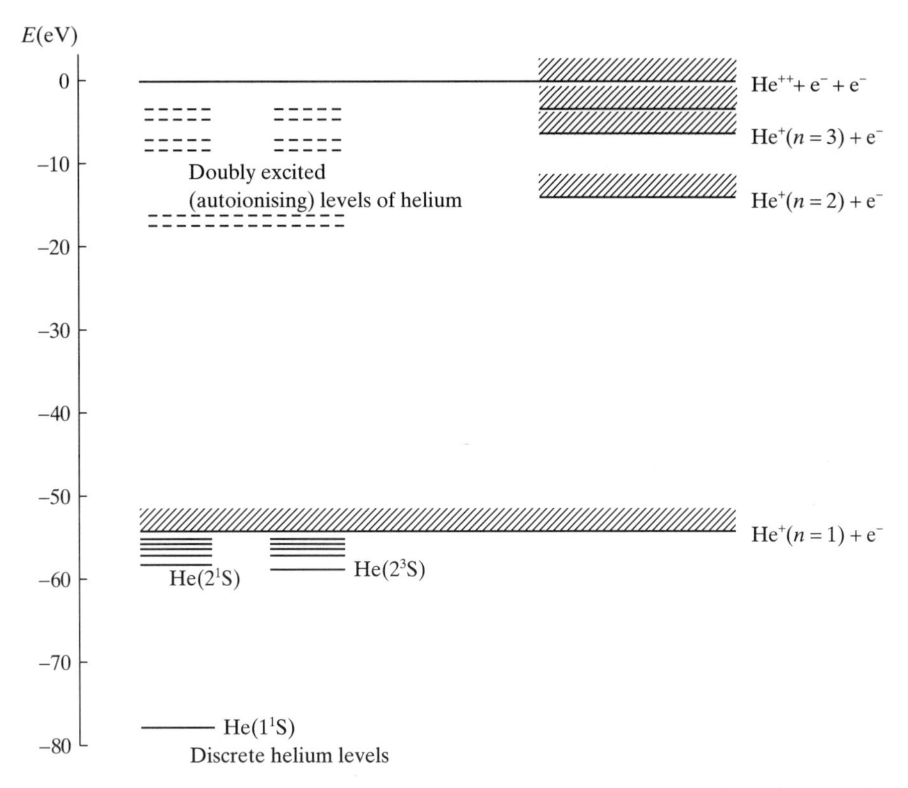

Figure 7.3 The 'complete' energy level spectrum of helium. The energy scale is relative to the threshold for the ionisation of both electrons and the zero of energy is 54.4 eV (the ionisation potential of He^+) above the zero energy of the scale of Fig. 7.2.

spin–spin interactions partially remove the degeneracy of the triplet states by splitting them (except the ^3S states) into three closely spaced levels corresponding to the three possible values $J = L + 1$, L or $L - 1$ of the total angular momentum quantum number J. We shall discuss this problem in Chapter 9.

It should be noted that Fig. 7.2 only represents the *discrete* part of the helium spectrum. A schematic diagram of the 'full' spectrum for the three-body system consisting of the He^{++} nucleus and two electrons is shown in Fig. 7.3. Choosing the origin of the energy scale in such a way that all three particles are unbound above $E = 0$, we see that the discrete levels of helium (displayed in more detail in Fig. 7.2) lie between the ground state value $E_0(\text{He}) \simeq -79.0$ eV and the value $E_0(\text{He}^+) \simeq -54.4$ eV of the ground state energy of the He^+ ion. Thus the ionisation potential numbers given in Fig. 7.2 correspond to the energy differences between the level $E_0(\text{He}^+)$ and a given energy level of the helium atom. For example, the ionisation potential corresponding to the helium ground state is

$$I_P = E_0(\text{He}^+) - E_0(\text{He}) \simeq 24.6 \text{ eV} \qquad \textbf{(7.22)}$$

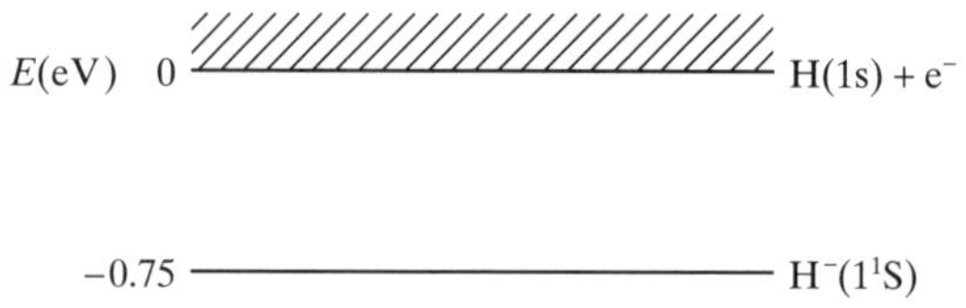

E(eV) 0 ——— H(1s) + e$^-$

−0.75 ——— H$^-$(1^{1}S)

Figure 7.4 The ground state of the hydrogen negative ion.

The spectrum of two-electron ions with $Z > 2$ is similar to that of helium which we have just discussed. On the other hand, the negative hydrogen ion H^-, for which $Z = 1$, constitutes an interesting special case. Indeed, as shown on Fig. 7.4, this ion has only *one* bound state. The corresponding ionisation potential is about 0.75 eV, so that the H^- ion is barely stable against detachment into a neutral hydrogen atom and a free electron. We shall return below to the H^- system, which is of great importance in astrophysics, and also provides a stringent test of the approximation methods used in the analysis of two-electron systems.

7.4 The independent particle model

Before discussing in detail the ground state and various excited states of two-electron atoms, we shall develop in this section a simple approach which yields a qualitative understanding of the main features of their spectrum, and which will also pave the way for our study of many-electron atoms in Chapter 8.

We begin by rewriting the basic Hamiltonian (7.2) as

$$H = H_0 + H' \tag{7.23}$$

where we choose our zero-order, 'unperturbed' Hamiltonian to be

$$H_0 = -\frac{1}{2}\nabla^2_{\mathbf{r}_1} - \frac{Z}{r_1} - \frac{1}{2}\nabla^2_{\mathbf{r}_2} - \frac{Z}{r_2} \tag{7.24}$$

while the 'perturbation'

$$H' = \frac{1}{r_{12}} \tag{7.25}$$

is the electron–electron interaction. We remark from (7.24) that H_0 is just the sum of two hydrogenic Hamiltonians, namely

$$H_0 = \hat{h}_1 + \hat{h}_2 \tag{7.26}$$

where

$$\hat{h}_i = -\frac{1}{2}\nabla^2_{\mathbf{r}_i} - \frac{Z}{r_i}, \qquad i = 1, 2 \tag{7.27}$$

In what follows we shall denote by E_{n_i} the energy eigenvalues and by $\psi_{n_il_im_i}(\mathbf{r}_i)$ the corresponding eigenfunctions of the hydrogenic Hamiltonian (7.27), normalised to unity. Thus

$$\hat{h}_i\psi_{n_il_im_i}(\mathbf{r}_i) = E_{n_i}\psi_{n_il_im_i}(\mathbf{r}_i) \tag{7.28}$$

with

$$E_{n_i} = -\frac{1}{2}\frac{Z^2}{n_i^2} \qquad \text{(in a.u.)} \tag{7.29}$$

Let us for the moment neglect the electron-electron repulsion term (7.25). The Schrödinger equation (7.3) for the spatial part of the two-electron wave function then reduces to the 'zero-order' equation

$$H_0\psi^{(0)}(\mathbf{r}_1, \mathbf{r}_2) = E^{(0)}\psi^{(0)}(\mathbf{r}_1, \mathbf{r}_2) \tag{7.30}$$

Using (7.26)–(7.29), we see that this equation is separable and that eigenfunctions of (7.30) can be written in the form of single products of hydrogenic wave functions. In particular, for discrete states, we have

$$\psi^{(0)}(\mathbf{r}_1, \mathbf{r}_2) = \psi_{n_1l_1m_1}(\mathbf{r}_1)\psi_{n_2l_2m_2}(\mathbf{r}_2) \tag{7.31}$$

the corresponding discrete energies being given (in a.u.) by

$$\begin{aligned} E^{(0)}_{n_1n_2} &= E_{n_1} + E_{n_2} \\ &= -\frac{Z^2}{2}\left(\frac{1}{n_1^2} + \frac{1}{n_2^2}\right) \end{aligned} \tag{7.32}$$

We note that the wave function

$$\psi^{(0)}(\mathbf{r}_2, \mathbf{r}_1) = \psi_{n_2l_2m_2}(\mathbf{r}_1)\psi_{n_1l_1m_1}(\mathbf{r}_2) \tag{7.33}$$

which differs from (7.31) only in an exchange of the electron labels, corresponds to the *same* energy $E^{(0)}_{n_1n_2}$. This particular case of degeneracy with respect to exchange of electron labels is called *exchange degeneracy*. According to the discussion of Section 7.2 the exact spatial wave functions of two-electron atoms must be either symmetric or antisymmetric with respect to the interchange of the coordinates $\mathbf{r}_1$ and $\mathbf{r}_2$ of the two electrons. The proper (zero-order) spatial wave functions of our simple independent-particle model must therefore be the symmetric (+) and antisymmetric (–) linear combinations

$$\psi^{(0)}_{\pm}(\mathbf{r}_1, \mathbf{r}_2) = \frac{1}{\sqrt{2}}[\psi_{n_1l_1m_1}(\mathbf{r}_1)\psi_{n_2l_2m_2}(\mathbf{r}_2) \pm \psi_{n_2l_2m_2}(\mathbf{r}_1)\psi_{n_1l_1m_1}(\mathbf{r}_2)] \tag{7.34}$$

where the factor $2^{-1/2}$ guarantees that the functions $\psi^{(0)}_{\pm}$ are normalised to unity. The functions $\psi^{(0)}_{+}$ are therefore approximations to the *para* wave functions, while the functions $\psi^{(0)}_{-}$ are approximations to the *ortho* wave functions. We see that the total orbital quantum number L can take the values $L = |l_1 - l_2|, \ldots, l_1 + l_2$, the possible values of the quantum number M_L being $M_L = -L, -L+1, \ldots, +L$. We remark that in our crude model the two states $\psi_\pm^{(0)}$ correspond to the *same* energy $E_{n_1 n_2}^{(0)}$. We shall see below that the electron–electron repulsion term $1/r_{12}$ removes this degeneracy.

It is worth stressing that the symmetric wave functions $\psi_+^{(0)}$ and the antisymmetric wave functions $\psi_-^{(0)}$ are not in the form of a single product of one-particle wave functions and hence represent *entangled states*.

An exception to the expression (7.34) occurs for the *ground state*, where both electrons are in the 1s state (that is, $n_1 = n_2 = 1$, $l_1 = l_2 = 0$, $m_1 = m_2 = 0$). The wave function $\psi_-^{(0)}$ for the ortho state is then seen to vanish, in agreement with the original formulation of the Pauli principle, according to which two electrons cannot be in the same state. Indeed, the spatial quantum numbers for both electrons having the same values $n = 1$, $l = 0$ and $m = 0$, the spin quantum numbers of the two electrons must be different, so that the two electrons must have antiparallel spin, and only the singlet (para) state is allowed. It is interesting to note that historically the argument was made the other way around, the experimental absence of the ground state triplet level of helium having provided key evidence that led W. Pauli to the discovery of the exclusion principle.

The normalised zero-order spatial wave function for the ground state of two-electron atoms is therefore given by the simple symmetric (para) wave function

$$\begin{aligned}\psi_0^{(0)}(r_1, r_2) &= \psi_{1s}(r_1)\psi_{1s}(r_2) \\ &= \frac{Z^3}{\pi} \exp[-Z(r_1 + r_2)] \qquad \textbf{(7.35)}\end{aligned}$$

where the subscript indicates that we are dealing with the ground state, and we have used the fact that the ground state wave function of the hydrogenic Hamiltonian (7.27) is

$$\psi_{100}(r_i) \equiv \psi_{1s}(r_i) = \left(\frac{Z^3}{\pi}\right)^{1/2} \exp(-Zr_i) \qquad \textbf{(7.36)}$$

The ground state energy corresponding to (7.35) is (see (7.32))

$$E_0^{(0)} = E_{n_1=1, n_2=1}^{(0)} = -Z^2 \quad \text{a.u.} \qquad \textbf{(7.37)}$$

Thus for helium ($Z = 2$) we find from (7.37) that $E_0^{(0)} = -4$ a.u. ($\simeq -108.8$ eV), which corresponds to an ionisation potential $I_\mathrm{P}^{(0)} = 2$ a.u. ($\simeq 54.4$ eV). The experimental values are $E_0^{\mathrm{exp}} = -2.90$ a.u. ($\simeq -79.0$ eV) and $I_\mathrm{P}^{\mathrm{exp}} = 0.90$ a.u. ($\simeq 24.6$ eV). As we should expect, our crude model gives an energy which is too low because we have neglected the repulsion term (7.25) between the two electrons, whose effect is clearly to raise the energy levels. It is also apparent that our simple independent particle approach should yield better results when Z is increased, since in that case the relative importance of the neglected term $1/r_{12}$ becomes smaller. For example, in the case of the C^{4+} ion, corresponding to $Z = 6$, the approximation (7.37) yields $E_0^{(0)} = -36$ a.u. ($\simeq -980$ eV) while the experimental value is $E_0^{\mathrm{exp}} = -32.4$ a.u.

($\simeq -882$ eV). On the other hand, for the negative hydrogen ion H^- ($Z = 1$), the value $E_0^{(0)} = -1$ a.u. ($\simeq -27.2$ eV) is in gross disagreement with the observed value $E_0^{\text{exp}} = -0.528$ a.u. ($\simeq -14.4$ eV).

Let us now examine the predictions of our simple model concerning excited states. The energy spectrum corresponding to (7.32) is illustrated in Fig. 7.5(a) for the case of helium ($Z = 2$). Also shown for comparison in Fig. 7.5(b) is the experimental spectrum. We first remark that the energy levels $E_{n_1,n_2}^{(0)}$ corresponding to states for which *both* electrons are excited (that is, $n_1 \geqslant 2$ *and* $n_2 \geqslant 2$) are higher than the ground state energy of the He^+ ion ($E_0(He^+) = -2$ a.u. $\simeq -54.4$ eV) plus a free electron. These doubly excited states therefore lie in the continuum of our simplified spectrum (see Fig 7.5(a)). Since the repulsion term $1/r_{12}$ can only raise the unperturbed energy levels $E_{n_1,n_2}^{(0)}$, the same property is also true for the actual He spectrum, and in fact it holds for all other He-like ions. We shall return at the

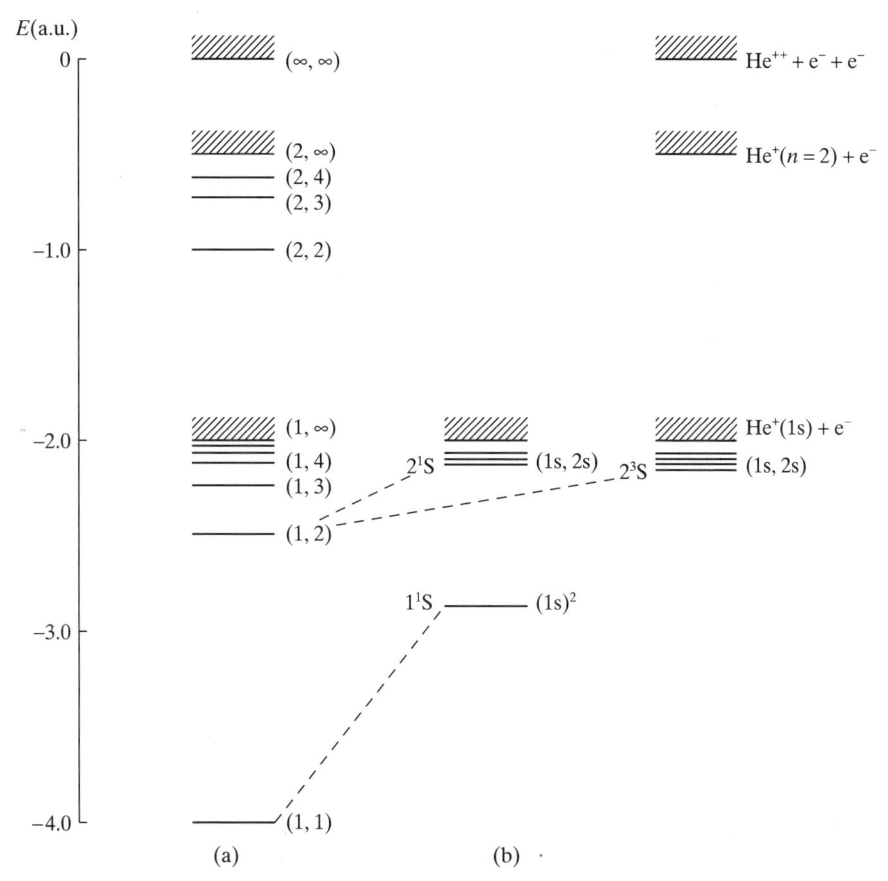

Figure 7.5 (a) The energy spectrum given by expression (7.32) with $Z = 2$. The levels are labelled by (n_1, n_2). (b) The energy spectrum of helium.

end of this chapter to these discrete states embedded in the continuum. For the moment, however, we focus our attention on the *genuinely discrete states* of two-electron atoms, for which one of the two electrons remains in the ground state. The properly symmetrised zero-order spatial wave functions for these states are given (see (7.34)) by

$$\psi_{\pm}^{(0)}(\mathbf{r}_1, \mathbf{r}_2) = \frac{1}{\sqrt{2}}[\psi_{100}(\mathbf{r}_1)\psi_{nlm}(\mathbf{r}_2) \pm \psi_{nlm}(\mathbf{r}_1)\psi_{100}(\mathbf{r}_2)], \qquad n \geqslant 2 \tag{7.38}$$

and are therefore characterised by three quantum numbers (n, l, m) as in the case of one-electron atoms. The total orbital angular momentum quantum number L is given by $L = l$ and the values of M_L $(= m)$ are $M_L = -l, -l+1, \dots, l$. The energy levels corresponding to the wave functions (7.38), namely (see (7.32))

$$E_{1,n}^{(0)} = -\frac{Z^2}{2}\left(1 + \frac{1}{n^2}\right) \quad \text{a.u.}, \qquad n \geqslant 2 \tag{7.39}$$

are degenerate in l and m. As pointed out above, they also exhibit the exchange degeneracy, according to which the para (+) and ortho (–) levels are degenerate in the 'zero-order' approximation (7.38). The electron–electron repulsion term $1/r_{12}$, which is ignored in the very simple approach leading to (7.39), will clearly raise these energy levels, as may be seen from Fig. 7.5. As we shall show at the end of this section, the term $1/r_{12}$ is responsible for removing the exchange degeneracy between the para and ortho states.

For the special case of the negative hydrogen ion H^-, corresponding to $Z = 1$, the repulsive term $1/r_{12}$ has an even more drastic effect on the spectrum, as Fig. 7.6 illustrates. Indeed, all the excited states (7.39) which are present in the 'unperturbed' spectrum shown in Fig. 7.6(a) are lifted into the continuum when the electron–electron repulsion $1/r_{12}$ is taken into account (see Fig. 7.6(b)). This spectacular effect of the 'perturbation' $1/r_{12}$ is obviously due to the small value $(Z = 1)$ of the nuclear charge. In what follows, when discussing the excited states of two-electron atoms, we shall assume implicitly that we are dealing with the case $Z \geqslant 2$.

So far we have used a very crude independent-particle approximation, in which the electron–electron repulsion term $1/r_{12}$ is completely omitted. While remaining within the convenient framework of the independent-particle model, we may improve our treatment by splitting the basic Hamiltonian (7.2) as

$$H = \tilde{H}_0 + \tilde{H}' \tag{7.40}$$

where

$$\tilde{H}_0 = -\tfrac{1}{2}\nabla_{\mathbf{r}_1}^2 + V(r_1) - \tfrac{1}{2}\nabla_{\mathbf{r}_2}^2 + V(r_2) \tag{7.41}$$

is the sum of the two individual Hamiltonians

$$h_i = -\tfrac{1}{2}\nabla_{\mathbf{r}_i}^2 + V(r_i) \tag{7.42}$$

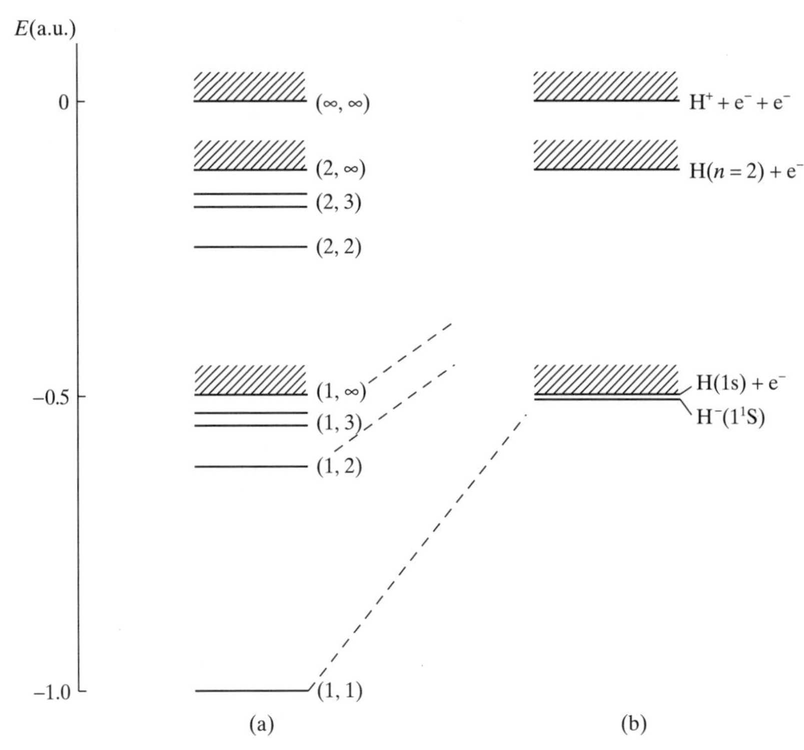

Figure 7.6 (a) The energy spectrum given by expression (7.32) with $Z = 1$. The levels are labelled by (n_1, n_2). (b) The energy spectrum of H^-.

and

$$\tilde{H}' = \frac{1}{r_{12}} - \frac{Z}{r_1} - V(r_1) - \frac{Z}{r_2} - V(r_2) \tag{7.43}$$

In the above formulae $V(r)$ is a central potential which should be chosen in such a way that the effect of the perturbation $\tilde{H}'$ is small. In Chapter 8 we shall study in detail this *central field approximation* for many-electron atoms, but for the moment we limit ourselves to simple qualitative considerations. Roughly speaking, the net effect of each electron on the motion of the other one is to screen somewhat the charge of the nucleus, so that a simple guess for $V(r)$ is

$$V(r) = -\frac{Z - S}{r} = -\frac{Z_e}{r} \tag{7.44}$$

where S is a 'screening constant' and the quantity $Z_e = Z - S$ may be considered as an 'effective charge'. Since the potential (7.44) is a Coulomb interaction, the corresponding individual electron energies are given (in a.u.) by (see (7.29))

$$E_{n_i} = -\frac{1}{2}\frac{(Z-S)^2}{n_i^2} = -\frac{1}{2}\frac{Z_e^2}{n_i^2} \tag{7.45}$$

and are independent of the quantum numbers l_i and m_i. Neglecting the perturbation $\tilde{H}'$, the total energy of the atom is just the sum of the individual electron energies (7.45). In particular, the ground state energy E_0 is then given approximately (in a.u.) by

$$E_0 \simeq -(Z-S)^2 = -Z_e^2 \tag{7.46}$$

the corresponding spatial part of the ground state wave function being (see (7.35))

$$\psi_0(r_1, r_2) = \frac{Z_e^3}{\pi}\exp[-Z_e(r_1 + r_2)] \tag{7.47}$$

In the next section we shall use the variational method to determine the 'optimum' value of the effective charge Z_e. Here we simply remark that the value $Z_e = 1.70$ would make the approximate expression (7.46) agree with the experimental value $E_0 = -2.903$ a.u. of the ground state energy of helium. Since $Z = 2$ in this case, the corresponding screening constant is $S = 0.30$. Thus, for the ground state of helium, we see that in this simple, 'average shielding' approximation, the screening effect of each electron on the other one is equivalent to about one-third of the electronic charge.

A better choice for the central potential $V(r)$ than the Coulomb form (7.44) is provided by an expression of the same type, but in which the 'screening constant' S varies with the distance r. Indeed, at small distances ($r \to 0$), the potential acting on an electron is essentially the Coulomb attraction $-Z/r$ of the nucleus, while for large r ($r \to \infty$), this potential is just the Coulomb field $-(Z-1)/r$ due to a net charge $(Z-1)$, namely the nuclear charge Z screened by the charge (-1) of the other electron. Thus we expect the quantity S in (7.44) to be in fact an increasing function of r, which takes on the values $S = 0$ at $r = 0$ and $S = 1$ at $r = \infty$. Since a potential of the form (7.44) where S is a function of r is no longer a Coulomb potential, the l degeneracy which is characteristic of the Coulomb field is removed. Thus the individual electron energies E_{nl} (where we have dropped the subscript i) are still degenerate with respect to the quantum number m, but now depend on both quantum numbers n and l. The principal quantum number n is defined as in the case of hydrogenic atoms, the number of nodes of the radial function being $n_r = n - l - 1$, with $n = 1, 2, \ldots$ and $l = 0, 1, \ldots, n-1$. Calling $u_{nlm}(\mathbf{r})$ an *individual electron orbital*, solution of the single-particle equation

$$\left[-\tfrac{1}{2}\nabla_{\mathbf{r}}^2 + V(r)\right]u_{nlm}(\mathbf{r}) = E_{nl}u_{nlm}(\mathbf{r}) \tag{7.48}$$

we see that $u_{nlm}(\mathbf{r})$ is just the product of a radial function and a spherical harmonic $Y_{lm}(\theta, \phi)$. It is important not to confuse the orbital $u_{nlm}(\mathbf{r})$, whose radial part depends on $V(r)$ and is likely to be a complicated function, with a hydrogenic function $\psi_{nlm}(\mathbf{r})$, whose radial part corresponds to the particular choice $V(r) = -Z/r$, and which has been obtained in Chapter 3.

Let us now return to the Hamiltonian (7.41). In terms of the individual electron orbitals $u_{nlm}(\mathbf{r})$, our new zero-order spatial wave functions, which are the properly symmetrised eigenstates of (7.41), are given by

$$\tilde{\psi}_0^{(0)}(r_1, r_2) = u_{100}(r_1)u_{100}(r_2) \qquad \textbf{(7.49)}$$

for the ground state and by

$$\tilde{\psi}_\pm^{(0)}(\mathbf{r}_1, \mathbf{r}_2) = \frac{1}{\sqrt{2}}[u_{100}(\mathbf{r}_1)u_{nlm}(\mathbf{r}_2) \pm u_{nlm}(\mathbf{r}_1)u_{100}(\mathbf{r}_2)], \qquad n \geqslant 2 \qquad \textbf{(7.50)}$$

for the genuinely discrete excited states. If we still neglect the perturbation $\tilde{H}'$, the total energy of the atom is just

$$\tilde{E}_{1s,nl}^{(0)} = E_{1s} + E_{nl} \qquad \textbf{(7.51)}$$

and the total orbital angular momentum quantum number is $L = l$. The para (+) and ortho (–) states (7.50) are still degenerate, but the degeneracy in l is removed. For a fixed value of n the value of E_{nl} is an increasing function of l. Indeed, electrons with a smaller value of l are more likely to penetrate at certain times the 'centrifugal barrier' (which is proportional to $l(l + 1)/r^2$) and hence to feel the fully unscreened attractive Coulomb potential $-Z/r$ of the nucleus. We therefore expect that the energy of the atom will be an increasing function of L (= l). That this is the case may be seen from Fig. 7.2. We note that in the central field approximation leading to (7.51) the energy of the atom is specified by the *electron configuration*, that is by the values of the quantum numbers n and l of the electrons. For the genuinely discrete states of two-electron atoms considered here, one electron remains in the ground state (that is, with $n = 1$ and $l = 0$) while the other, 'optically active' electron has the quantum numbers n and l. Following the convention used in spectroscopy the values of n and l are usually indicated by writing n as a number, and l as a letter, according to the code described in Chapters 2 and 3 (that is, s for $l = 0$, p for $l = 1$, d for $l = 2$, etc.). If there are k electrons having the same values of n and l, this is denoted as $(nl)^k$. For example, in this notation the ground state (7.49) is characterised by the configuration $(1s)^2$ [also written $1s^2$], the first excited states (7.50) by the configurations (1s)(2s) [or 1s2s], (1s)(2p) [or 1s2p], and so on. It is of course understood that when we write for example (1s)(3s) this does not mean that electron 1 (say) is in the state 1s and electron 2 in the state 3s, since we know that properly symmetrised spatial wave functions must be used in order to obtain two-electron wave functions which are fully antisymmetric in the space and spin coordinates of the two electrons.

From our discussion of the wave functions of one-electron atoms it is also clear that for states of the excited electron corresponding to large values of n and l the orbitals $u_{nlm}(\mathbf{r})$ are concentrated at much larger values of r than the ground state orbital $u_{100}(r)$. We may then speak of an 'inner' (1s) electron with spatial quantum numbers (1, 0, 0) which is moving in the unscreened Coulomb field $-Z/r$ of the nucleus, and an 'outer' electron, which moves in the fully screened potential

$-(Z-1)/r$. For such states of large n and l the 'zero-order' energy levels are then given approximately (in a.u.) by

$$E_n^{(0)} = -\frac{1}{2}Z^2 - \frac{1}{2}\frac{(Z-1)^2}{n^2} \tag{7.52}$$

Apart from the additive constant $-Z^2/2$, we see that these energy levels are identical to those of a hydrogenic atom of nuclear charge $Z-1$. This can be illustrated by drawing the energy levels of atomic hydrogen (shifted by the amount -2 a.u.) next to those of helium.

We shall not pursue further here the study of the central field approximation. From the above discussion it is clear that according to this approximation each electron moves independently of the other one in a net central potential $V(r)$ which represents the attraction of the nucleus plus some average central repulsive potential due to the other electron. This basic idea, first expressed by D.R. Hartree in 1928, will be fully developed in the next chapter. It is also apparent from the foregoing discussion that the averaged repulsive effect of the other electron depends on its dynamical state, so that a single potential $V(r)$ cannot, even approximately, account for the entire energy spectrum of the atom.

To conclude this section we shall now give a simple, qualitative argument showing how the exchange degeneracy is removed when the electron–electron repulsion term $1/r_{12}$ is taken into account. Returning to the symmetrised zero-order spatial wave functions (7.38), we first observe that the space-antisymmetric (ortho) wave functions $\psi_-^{(0)}$ given by (7.38) vanish for $\mathbf{r}_1 = \mathbf{r}_2$, so that in ortho (spin triplet) states the two electrons tend to 'keep away' from each other and hence, on the average, have a relatively small repulsion energy. On the other hand, the space-symmetric (para) wave functions $\psi_+^{(0)}$ do not vanish for $\mathbf{r}_1 = \mathbf{r}_2$, so that in para (spin singlet) states the two electrons may be very close at certain times and experience on the average a stronger repulsion than in the corresponding ortho state having the same values of the quantum numbers (nlm). Therefore the electron–electron repulsion term $1/r_{12}$ is more effective in raising the energy of the atom in the para (spin singlet) states, from which we conclude that an ortho (triplet) state must lie lower than the corresponding para (singlet) state having the same values of (nlm). That this is indeed the case may be seen for example in Fig. 7.2. Thus, as pointed out in Section 7.2, the Pauli exclusion principle introduces a coupling between the space and spin variables of the electrons, which act as if they were moving under the influence of a force whose sign depends on the relative orientation of their spins. Such a force, which has no classical analogue, is known as an *exchange force*, and its effects will be studied in detail in Section 7.6.

It is worth stressing at this point that exchange forces are negligible between two electrons which always remain far apart. Indeed, in that case the wave functions of the two electrons have a vanishingly small overlap, and the two electrons may be considered as distinguishable. An example of this situation is provided by the electrons of two hydrogen atoms which are located at a large distance from each other. Similarly, for excited states of two-electron atoms with high values of both n and l, the 'outer' orbital has a very small overlap with the 'inner' (1s) orbital. The exchange force is then very small and the para and ortho levels are nearly degenerate, as may be seen from Fig. 7.2. On the other hand, for small values of n and l (in particular for S states) the orbitals of the two electrons overlap significantly, and the energy difference between para (singlet) and ortho (triplet) states is appreciable, as shown in Fig. 7.2.

The results we have obtained in this section thus show that in addition to their symmetry property (para or ortho), the spatial wave functions for the genuinely discrete states of two-electron atoms may be characterised by the three quantum numbers n, l $(= L)$ and m $(= M_L)$. The ground state is a para state and is non-degenerate. Excluding the negative hydrogen ion H^- (which has no other bound state than the ground state), the energy levels of the excited states are degenerate with respect to M_L and depend on n, on l and on S (with $S = 0$ for para states, and $S = 1$ for ortho states). The energy levels or terms may therefore be labelled by the symbol

$$n^{2S+1}L$$

where the multiplicity $2S + 1$ takes on the values 1 or 3, and the code letters S, P, D, ... correspond to $L = 0, 1, 2, \ldots$ as we have seen in Section 7.3. Thus in this notation the ground state is denoted by 1^1S and the following energy levels (by order of increasing energy) are 2^3S, 2^1S, 2^3P, 2^1P, and so on. In the following two sections we shall study successively the ground state 1^1S and various excited states by using perturbation theory and the variational method developed in Chapter 2.

7.5 The ground state of two-electron atoms

We have seen in the previous sections that the ground state wave function of two-electron atoms, $\Psi_0(q_1, q_2)$, is a para (spin singlet) state whose general expression in the non-relativistic approximation is

$$\Psi_0(q_1, q_2) = \psi_0(\mathbf{r}_1, \mathbf{r}_2)\frac{1}{\sqrt{2}}[\alpha(1)\beta(2) - \beta(1)\alpha(2)] \tag{7.53}$$

where $\psi_0(\mathbf{r}_1, \mathbf{r}_2)$ is a space-symmetric function. We shall now focus our attention on this function and on the corresponding ground state energy E_0 of the Hamiltonian (7.2). The motion of the nucleus and other small corrections will be briefly discussed at the end of this section. It is worth noting that the quantum mechanical treatment of the ground state of helium has been of great historical importance since the 'old quantum theory' was unable to deal successfully with this problem.

Perturbation theory

We shall first use the time-independent perturbation theory of Section 2.8. As in the beginning of the previous section, we split the Hamiltonian (7.2) as $H = H_0 + H'$, where the unperturbed Hamiltonian H_0, given by (7.24), is the sum of two hydrogenic Hamiltonians (see (7.26)–(7.27)) and where $H' = 1/r_{12}$ is the perturbation. The 'zero-order' approximation to the wave function $\psi_0(\mathbf{r}_1, \mathbf{r}_2)$ is then given by the simple wave function $\psi_0^{(0)}(r_1, r_2)$ of equation (7.35) and the corresponding 'zero-order' ground state energy is $E_0^{(0)} = -Z^2$ a.u.

According to (2.308) the first-order correction to the ground state energy is

$$E_0^{(1)} = \langle \psi_0^{(0)} | H' | \psi_0^{(0)} \rangle \tag{7.54}$$

or, using (7.25) and (7.35),

$$E_0^{(1)} = \int |\psi_{1s}(r_1)|^2 \frac{1}{r_{12}} |\psi_{1s}(r_2)|^2 \, d\mathbf{r}_1 \, d\mathbf{r}_2 \tag{7.55}$$

We note that in SI units this quantity reads

$$E_0^{(1)} = \int |\psi_{1s}(r_1)|^2 \frac{e^2}{(4\pi\varepsilon_0) r_{12}} |\psi_{1s}(r_2)|^2 \, d\mathbf{r}_1 \, d\mathbf{r}_2 \tag{7.56}$$

and the integral on the right has a simple physical interpretation. Indeed, since $|\psi_{1s}(r_1)|^2$ is the probability density of finding the electron 1 at $\mathbf{r}_1$, we see that

$$\rho(r_1) = -e|\psi_{1s}(r_1)|^2 \tag{7.57}$$

may be interpreted as the charge density due to electron 1. A similar interpretation may be given of the quantity $\rho(r_2) = -e|\psi_{1s}(r_2)|^2$. Thus the integral in (7.56) is just the electrostatic interaction energy of two overlapping spherically symmetric distributions of electricity, of charge density $\rho(r_1)$ and $\rho(r_2)$ respectively.

Let us now return to (7.55), which we write explicitly (see (7.35)–(7.36)) as

$$E_0^{(1)} = \frac{Z^6}{\pi^2} \int \exp[-2Z(r_1 + r_2)] \frac{1}{r_{12}} \, d\mathbf{r}_1 \, d\mathbf{r}_2 \tag{7.58}$$

We shall calculate this integral by using a general procedure which is very useful in many atomic physics calculations. Using the generating function (2.168), we first expand $1/r_{12}$ in Legendre polynomials as

$$\begin{aligned} \frac{1}{r_{12}} &= \frac{1}{r_1} \sum_{l=0}^{\infty} \left(\frac{r_2}{r_1}\right)^l P_l(\cos\theta), \qquad r_1 > r_2 \\ &= \frac{1}{r_2} \sum_{l=0}^{\infty} \left(\frac{r_1}{r_2}\right)^l P_l(\cos\theta), \qquad r_1 < r_2 \end{aligned} \tag{7.59}$$

where θ is the angle between the vectors $\mathbf{r}_1$ and $\mathbf{r}_2$ (see Fig. 7.7), so that

$$\cos\theta = \cos\theta_1 \cos\theta_2 + \sin\theta_1 \sin\theta_2 \cos(\phi_1 - \phi_2) \tag{7.60}$$

Here (θ_1, ϕ_1) and (θ_2, ϕ_2) are the polar angles of the vectors $\mathbf{r}_1$ and $\mathbf{r}_2$, respectively. We can also write (7.59) in the more compact form

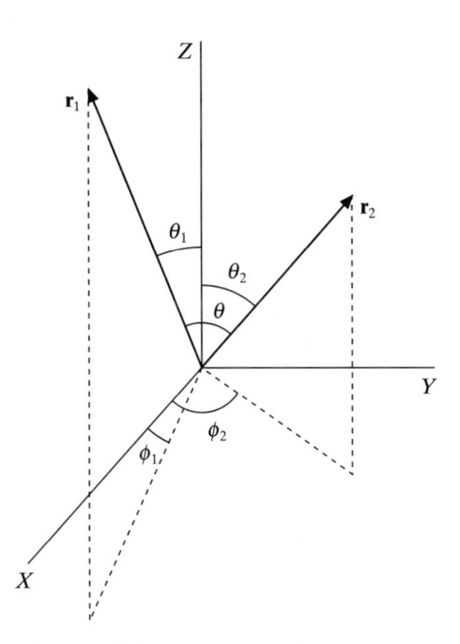

Figure 7.7 The polar coordinates of the vectors $\mathbf{r}_1$ and $\mathbf{r}_2$.

$$\frac{1}{r_{12}} = \sum_{l=0}^{\infty} \frac{(r_<)^l}{(r_>)^{l+1}} P_l(\cos\theta) \tag{7.61}$$

where $r_<$ is the smaller and $r_>$ the larger of r_1 and r_2. Using the 'addition theorem' of the spherical harmonics (see (A4.23)) we have

$$\frac{1}{r_{12}} = \sum_{l=0}^{\infty} \sum_{m=-l}^{+l} \frac{4\pi}{2l+1} \frac{(r_<)^l}{(r_>)^{l+1}} Y^*_{lm}(\theta_1, \phi_1) Y_{lm}(\theta_2, \phi_2) \tag{7.62}$$

Let us substitute this expansion in (7.58) and use the fact that the spherical harmonics are orthonormal on the unit sphere (see (2.182)). Since the function $\exp[-2Z(r_1 + r_2)]$ which appears under the integral sign in (7.58) is spherically symmetric, and because $Y_{00} = (4\pi)^{-1/2}$, we obtain at once by integrating over the polar angles (θ_1, ϕ_1) and (θ_2, ϕ_2)

$$\begin{aligned} E_0^{(1)} &= \frac{Z^6}{\pi^2} \sum_{l=0}^{\infty} \sum_{m=-l}^{+l} \frac{(4\pi)^2}{2l+1} \int_0^{\infty} \mathrm{d}r_1 r_1^2 \int_0^{\infty} \mathrm{d}r_2 r_2^2 \exp[-2Z(r_1 + r_2)] \frac{(r_<)^l}{(r_>)^{l+1}} \\ &\times \int \mathrm{d}\Omega_1 Y^*_{lm}(\theta_1, \phi_1) Y_{00} \int \mathrm{d}\Omega_2 Y_{00} Y_{lm}(\theta_2, \phi_2) \\ &= \frac{Z^6}{\pi^2} \sum_{l=0}^{\infty} \sum_{m=-l}^{+l} \frac{(4\pi)^2}{2l+1} \int_0^{\infty} \mathrm{d}r_1 r_1^2 \int_0^{\infty} \mathrm{d}r_2 r_2^2 \exp[-2Z(r_1 + r_2)] \\ &\times \frac{(r_<)^l}{(r_>)^{l+1}} \delta_{l,0}\, \delta_{m,0} \end{aligned} \tag{7.63}$$

Thus all the terms in this double sum vanish, except the first one, for which $l = m = 0$, and

$$\begin{aligned} E_0^{(1)} &= 16Z^6 \int_0^\infty \mathrm{d}r_1 r_1^2 \int_0^\infty \mathrm{d}r_2 r_2^2 \exp[-2Z(r_1 + r_2)] \frac{1}{r_>} \\ &= 16Z^6 \int_0^\infty \mathrm{d}r_1 r_1^2 \exp(-2Zr_1) \left[\frac{1}{r_1} \int_0^{r_1} \mathrm{d}r_2 r_2^2 \exp(-2Zr_2) + \int_{r_1}^\infty \mathrm{d}r_2 r_2 \exp(-2Zr_2) \right] \end{aligned} \quad \textbf{(7.64)}$$

The integrals are now straightforward, and yield the answer

$$E_0^{(1)} = \tfrac{5}{8} Z \quad \text{a.u.} \quad \textbf{(7.65)}$$

which is a positive contribution to the energy, as expected. We also remark that $E_0^{(1)}$ is linear in Z [2], while the unperturbed energy $E_0^{(0)} = -Z^2$ a.u. is quadratic in Z, so that the ratio $|E_0^{(1)}/E_0^{(0)}|$ decreases like Z^{-1} when Z increases. This is in line with our comments in the previous section concerning the decrease in relative importance of the electron–electron repulsion term $1/r_{12}$ with increasing Z.

Adding the first-order correction (7.65) to our zero-order result $E_0^{(0)} = -Z^2$ a.u., we find for the ground state energy E_0 the approximate value

$$E_0 \simeq E_0^{(0)} + E_0^{(1)} = -Z^2 + \tfrac{5}{8} Z \quad \text{a.u.} \quad \textbf{(7.66)}$$

Both this 'first-order' result and the unperturbed energy $E_0^{(0)}$ are given in Table 7.3 for various two-electron atoms (ions), from the negative hydrogen ion H^- ($Z = 1$) to four times ionised carbon C^{4+} ($Z = 6$). The 'exact' [3] values E_0^{ex} are also tabulated. Except for H^-, the simple first-order perturbation approach yields quite good results. If we define ΔE, where

$$\Delta E = E_0^{\text{ex}} - (E_0^{(0)} + E_0^{(1)}) \quad \textbf{(7.67)}$$

as the difference between the 'exact' value E_0^{ex} and the first-order result (7.66), we see that the ratio $|\Delta E/E_0^{\text{ex}}|$ varies from about 5 per cent for He to 0.4 per cent for C^{4+}. It is interesting to note that this first-order perturbation treatment, carried out by A. Unsöld in 1927, was the first quantum mechanical calculation of the helium ground state. In view of the large discrepancies shown by the old quantum theory

[2] The fact that $E_0^{(1)}$ is linear in Z may readily be understood by noting that each charge distribution in (7.56) contains a total charge $-e$ and extends over a region of space of linear dimension given approximately by $a = a_0/Z$. Their mutual interaction energy is therefore roughly given by $e^2/[(4\pi\varepsilon_0)a] = Ze^2/[(4\pi\varepsilon_0)a_0]$, which is indeed proportional to Z.

[3] The 'exact' results quoted in Table 7.3 are accurate values of the ground state energy E_0 of the Hamiltonian (7.2), obtained by using the Rayleigh–Ritz variational method with elaborate trial functions. Only a few of the presently available significant figures are given. Since these 'exact' values of E_0 must still be corrected for the motion of the nucleus, as well as relativistic and radiative corrections, they should not be confused with the experimental ground state energies E_0^{exp}.

Table 7.3 Values of the ground state energy E_0 of the Hamiltonian (7.2), for various two-electron atoms and ions (in atomic units).

	Ground state energy			
	Unperturbed $E_0^{(0)}$ (*equation* (7.37))	*First order* $E_0^{(0)} + E_0^{(1)}$ (*equation* (7.66))	*Simple variational* $\left(Z_e = Z - \frac{5}{16}\right)$ (*equation* (7.79))	*'Exact'*
H^-	−1	−0.375	−0.473	−0.528
He	−4	−2.750	−2.848	−2.904
Li^+	−9	−7.125	−7.222	−7.280
Be^{2+}	−16	−13.50	−13.60	−13.66
B^{3+}	−25	−21.88	−21.97	−22.03
C^{4+}	−36	−32.25	−32.35	−32.41

calculations, the relatively small difference of about 0.15 a.u. ($\simeq$ 4 eV) between his result and the experimental value was very promising.

The calculation of second- and higher order corrections, $E_0^{(n)}$ ($n \geqslant 2$), is a much more difficult problem, to which we shall return later, after having discussed the Rayleigh–Ritz variational approach.

Variational method

We have seen in Section 2.8 that if H denotes the Hamiltonian of a quantum system, and ϕ a physically admissible trial function, the functional

$$E[\phi] = \frac{\langle\phi|H|\phi\rangle}{\langle\phi|\phi\rangle} \tag{7.68}$$

provides a variational principle for the discrete eigenvalues of the Hamiltonian. Moreover, it also yields a minimum principle for the ground state energy. That is,

$$E_0 \leqslant E[\phi] \tag{7.69}$$

For the case of two-electron atoms the Hamiltonian H (neglecting the motion of the nucleus and all but the Coulomb interactions) is given by (7.2). Following the Rayleigh–Ritz variational method, we shall use trial functions ϕ depending on variational parameters, and carry out the variation $\delta E = 0$ with respect to these parameters.

It is apparent from the discussion of Section 7.4 that a basic defect of the 'zero-order' ground state wave function (7.35) (from which the first-order energy result (7.66) was obtained) is that each electron moves in the fully unscreened field of the nucleus. In order to take into account approximately the screening effect of each electron on the other one, we shall therefore choose a trial function of the form (7.47). That is (in a.u.),

$$\phi(r_1, r_2) = \frac{Z_e^3}{\pi} \exp[-Z_e(r_1 + r_2)] \tag{7.70}$$

or

$$\phi(r_1, r_2) = \psi_{1s}^{Z_e}(r_1)\psi_{1s}^{Z_e}(r_2) \tag{7.71a}$$

where

$$\psi_{1s}^{Z_e}(r) = \left(\frac{Z_e^3}{\pi}\right)^{1/2} \exp(-Z_e r) \tag{7.71b}$$

and the 'effective charge' Z_e is considered as a variational parameter.

The first step of the calculation consists in evaluating the expression $E[\phi]$. Since our trial function (7.70) is normalised ($\langle\phi|\phi\rangle = 1$) we have from (7.68) and (7.2)

$$E[\phi] = \left\langle \phi \left| T_1 + T_2 - \frac{Z}{r_1} - \frac{Z}{r_2} + \frac{1}{r_{12}} \right| \phi \right\rangle \tag{7.72}$$

where we have set $T_1 = -\nabla_{\mathbf{r}_1}^2/2$ and $T_2 = -\nabla_{\mathbf{r}_2}^2/2$. Using the equations (7.71) and the virial theorem (see (3.81)–(3.82)) we find that

$$\langle\phi|T_1|\phi\rangle = \langle\psi_{1s}^{Z_e}|T_1|\psi_{1s}^{Z_e}\rangle = \tfrac{1}{2}Z_e^2 \tag{7.73}$$

and $\langle\phi|T_2|\phi\rangle = \langle\phi|T_1|\phi\rangle$. We also have from (3.76) and (7.71)

$$\left\langle \phi \left| \frac{1}{r_1} \right| \phi \right\rangle = \left\langle \psi_{1s}^{Z_e} \left| \frac{1}{r_1} \right| \psi_{1s}^{Z_e} \right\rangle = Z_e \tag{7.74}$$

and $\langle\phi|1/r_2|\phi\rangle = \langle\phi|1/r_1|\phi\rangle$. The expression $\langle\phi|1/r_{12}|\phi\rangle$ has already been calculated when $Z_e = Z$, in which case it is identical to the first-order energy correction $E_0^{(1)}$ (see (7.54)–(7.58)). Using (7.65) we find that

$$\left\langle \phi \left| \frac{1}{r_{12}} \right| \phi \right\rangle = \frac{5}{8} Z_e \tag{7.75}$$

Putting together the above results, we have

$$E[\phi] \equiv E(Z_e) = Z_e^2 - 2ZZ_e + \tfrac{5}{8}Z_e \tag{7.76}$$

In the second step of the calculation, we shall now minimise $E(Z_e)$ with respect to the variational parameter Z_e. Hence we write

$$\frac{\partial E}{\partial Z_e} = 2Z_e - 2Z + \frac{5}{8} = 0 \tag{7.77}$$

so that

$$Z_e = Z - \frac{5}{16} \tag{7.78}$$

This 'optimum' value of Z_e corresponds to a 'screening constant' $S \simeq 0.31$. The lowest energy which can be obtained with a trial function of the form (7.70) is given by choosing $Z_e = Z - 5/16$ and substituting this value in (7.76), namely

$$E\left(Z_e = Z - \frac{5}{16}\right) = -Z^2 + \frac{5}{8}Z - \frac{25}{256} = -\left(Z - \frac{5}{16}\right)^2 \text{ a.u.} \tag{7.79}$$

We remark that with the choice $Z_e = Z$ the expression (7.76) reduces to the first-order perturbation theory value (7.66), which is therefore equivalent to a 'non-optimum' variation calculation. The variational result (7.79), corresponding to the 'optimum' choice $Z_e = Z - 5/16$, is lower, and hence more accurate, than the first-order perturbation theory result (7.66). This may be seen from Table 7.3, where the variational values obtained from (7.79) are given for various two-electron atoms (ions). It is clear from this table that the variational results are remarkably good, given the simplicity of the trial function we have used. If we denote by ΔE the difference between the 'exact' [3] value E_0^{ex} and the variational result (7.79), we see that the relative error $|\Delta E/E_0^{\text{ex}}|$ varies from about 2 per cent for He to less than 0.2 per cent for C^{4+}. This corresponds to a reduction by more than a factor of two with respect to the relative error made in the first-order perturbation calculation. For the delicate case of H^- the variational result is also a marked improvement over the first-order perturbation value, although the ground state energy of -0.473 a.u. given by our simple variational treatment still lies above the ground state energy of atomic hydrogen (-0.5 a.u.) and hence is not low enough to predict the existence of a stable bound state for H^-.

We now describe briefly how very accurate values of the non-relativistic ground state energy of two-electron atoms can be obtained by using the Rayleigh–Ritz variational method with elaborate trial functions. We first remark that for S states ($L = 0$) the spatial wave function does not depend on the Euler angles which specify the orientation of the triangle formed by the nucleus and the two electrons. Thus spatial wave functions for S states (and in particular for the ground state 1^1S) can only depend on the shape and size of this triangle, which is specified for example by the radial coordinates r_1 and r_2 of the two electrons and their relative distance $r_{12} = |\mathbf{r}_1 - \mathbf{r}_2|$. Another possible choice for the three variables describing this triangle is

$$\begin{aligned} s &= r_1 + r_2 & 0 &\leqslant s \leqslant \infty \\ t &= r_1 - r_2 & -\infty &\leqslant t \leqslant \infty \\ u &= r_{12} & 0 &\leqslant u \leqslant \infty \end{aligned} \tag{7.80}$$

The coordinates (7.80) have been used by E.A. Hylleraas to construct trial functions of the type

$$\phi(s, t, u) = e^{-ks} \sum_{l,m,n=0}^{N} c_{l,2m,n} s^l t^{2m} u^n \tag{7.81}$$

where the coefficients $c_{l,2m,n}$ are linear variational parameters and k is a 'non-linear' variational parameter similar to the 'effective charge' Z_e used in our simple trial function (7.70). We note that since the ground state is a para state (space-symmetric) the trial function (7.81) must be an even function of t. The number N which appears in (7.81) determines the maximum number of terms kept in the trial function. It is clear that for $N = 0$ the trial function (7.81) reduces to the simple form (7.70), so that $k = Z_e = Z - 5/16$ in that case.

The Hylleraas approach, which accounts explicitly for 'correlations' between the motion of the two electrons through the variable $u = r_{12}$, has been very successful in getting accurate values of the ground state energy E_0 of the Hamiltonian (7.2). For example, with a trial wave function of the type (7.81) containing six linear parameters, Hylleraas obtained in the case of helium the value $E_0 = -2.903\,24$ a.u., which differs from the 'exact' value $E_0^{\text{ex}} = -2.903\,72$ a.u. by only 0.000 48 a.u. ($\simeq$ 0.013 eV). By now, extensive variational calculations, using trial functions of a somewhat more general form [4] than (7.81), have been performed for a variety of two-electron atoms and ions. The corresponding 'exact' results are given in the last column of Table 7.3 for the systems H^- ($Z = 1$) through C^{4+} ($Z = 6$). It is worth stressing that the numbers quoted in that column do not contain all the significant figures which are presently available. For example, the value of E_0 for He is one of the most accurate theoretical numbers which has been calculated by quantum mechanical approximation methods. In particular, K. Frankowski and C.L. Pekeris obtained in 1966 the value $E_0 = -2.903\,724\,377\,032\,6$ a.u. Over the past years, the accuracy of the calculations has been significantly increased by 'doubling' the basis set so that each combination of powers (l, $2m$, n) in (7.81) is included twice with different exponential scale factors [5]. Optimization with respect to the non-linear variational parameters then leads to a 'natural' partition of the basis set into two distinct distance scales: one which is appropriate to describe the correlated motion of the electrons near the nucleus, and the other which is optimal to account for the long-range asymptotic behaviour of the wave function. Using such improved trial functions, the value of E_0 for helium has been calculated to be [5]

$$E_0 = -2.903\,724\,377\,034\,119\,598\,13 \text{ a.u.} \tag{7.82}$$

obtained by extrapolation from a 'doubled' basis set containing 2114 terms. The accuracy of the result (7.82) is about one part in 10^{19}.

The very accurate Hylleraas-type ground state wave functions determined by the Rayleigh–Ritz variational method have also been used to calculate expectation

[4] For example, T. Kinoshita has used trial functions which, in addition to the terms contained in (7.81), also include terms of the form $s^{h+1}(u/s)^i(t/u)^{2j}$. C. Schwartz has successfully tried functions of the type (7.81), but including also some half-integer powers. C.L. Pekeris has also obtained extremely accurate results by using trial functions expressed in terms of the three perimetric coordinates $u = \varepsilon(r_2 + r_{12} - r_1)$, $v = \varepsilon(r_1 + r_{12} - r_2)$ and $w = 2\varepsilon(r_1 + r_2 - r_{12})$, with $\varepsilon = (-E_0)^{1/2}$.

[5] See Drake (1996).

values of various operators, oscillator strengths, transition probabilities, and so on. In many cases of practical interest, however, it is convenient to use less accurate, but more tractable wave functions, which do not involve explicitly the interelectronic coordinate r_{12}, and we shall now briefly describe how such wave functions may be obtained.

As we pointed out above, the spatial wave functions for S states of two-electron atoms depend only on the shape and size of the triangle formed by the nucleus and the two electrons. Instead of using the three variables (r_1, r_2, r_{12}) or the Hylleraas coordinates (s, t, u), one can also choose the radial coordinates r_1 and r_2 of the two electrons, and the angle θ between the vectors $\mathbf{r}_1$ and $\mathbf{r}_2$. It is then natural to use for the ground state trial wave functions which are expansions in Legendre polynomials,

$$\phi(\mathbf{r}_1, \mathbf{r}_2) = \sum_{l=0}^{\lambda} F_l(r_1, r_2) P_l(\cos\theta) \qquad \textbf{(7.83)}$$

where the subscript l refers to *relative* partial waves. This approach is known as the '*configuration interaction*' (CI) method [6]. The (symmetric) functions $F_l(r_1, r_2)$ are expanded in some convenient basis set, for example

$$F_l(r_1, r_2) = \mathrm{e}^{-k(r_1+r_2)} \sum_{i \leqslant j} c_{ij}^{(l)} r_1^l r_2^l (r_1^i r_2^j + r_1^j r_2^i) \qquad \textbf{(7.84)}$$

where k is a non-linear variational parameter, and the coefficients $c_{ij}^{(l)}$ are linear variational parameters. If one sets $\lambda = 0$ in (7.83), so that the trial wave function is restricted to the pure relative s wave ($l = 0$), only *radial* correlations between the positions of the two electrons are introduced in the wave function. In this way a *radial limit* for the ground state energy is approached when an increasing number of parameters is included in the function $F_0(r_1, r_2)$. For example, in helium this radial limit is –2.879 a.u., and differs from the correct value of –2.904 a.u. by 0.025 a.u. (= 0.68 eV). This difference is due to *radial* and *angular* correlations distributed among the higher relative partial waves in the expansion (7.83).

A drawback of the CI approach is that the expansion (7.83) converges much less rapidly than the Hylleraas method, which makes explicit use of the interelectronic coordinate r_{12}. This is due to the fact that the expansion of r_{12} in Legendre polynomials $P_l(\cos\theta)$ converges very slowly. On the other hand, the calculation of matrix elements (such as those occurring for example in collision processes) is

[6] The name 'configuration interaction' (or 'configuration mixing') arises from the fact that a trial function of the form (7.83) may be written symbolically as

$$\phi = c_{10,10}(1\mathrm{s}, 1\mathrm{s}) + \sum_{nln'l'} c_{nl,n'l'}(nl, n'l')$$

Here (1s, 1s) is the $1\mathrm{s}^2$ configuration obtained for the ground state in the independent particle approximation, and $(nl, n'l')$ represent configurations 'mixed' into the ground state because the electron–electron interaction $1/r_{12}$ has non-zero matrix elements between the $1\mathrm{s}^2$ configuration and other ones.

considerably simpler when use is made of CI wave functions instead of Hylleraas-type wave functions.

Before closing our discussion of variational wave functions for the ground state of two-electron atoms, we give two explicit examples of simple wave functions which represent improvements over the one-parameter 'screened hydrogenlc' wave function (7.70), with $Z_e = Z - 5/16$. For helium, a useful wave function is that of F.W. Byron and C.J. Joachain,

$$\psi_0(r_1, r_2) = u_{1s}(r_1)u_{1s}(r_2) \quad \textbf{(7.85a)}$$

where the 1s orbital $u_{1s}(r)$ is given by

$$u_{1s}(r) = (4\pi)^{-1/2}(A\mathrm{e}^{-\alpha r} + B\mathrm{e}^{-\beta r}) \quad \textbf{(7.85b)}$$

with A = 2.605 05, B = 2.081 44, α = 1.41 and β = 2.61. This function, normalised to unity, is an analytical fit to the Hartree–Fock orbital, to be discussed in Chapter 8. The corresponding Hartree–Fock ground state energy is $E_0 = -2.861\,67$ a.u. For the negative hydrogen ion H^-, a convenient wave function, obtained by S. Chandrasekhar, is given by

$$\psi_0(r_1, r_2) = \frac{1}{4\pi}N(\mathrm{e}^{-\alpha r_1 - \beta r_2} + \mathrm{e}^{-\beta r_1 - \alpha r_2}) \quad \textbf{(7.86)}$$

with $N = 0.3948$, $\alpha = 1.039$ and $\beta = 0.283$. The wave function (7.86) is normalised to unity and yields an energy $E_0 = -0.514$ a.u.; it therefore correctly predicts a bound state for the H^- system. Although the ionisation potential $I_P = 0.014$ a.u. resulting from (7.86) is rather far from the correct value $I_P^{ex} = 0.028$ a.u., it is much better than the *negative* value $I_P = -0.027$ a.u. corresponding to the simple variational function (7.70). Wave functions of the type (7.86), with $\alpha \neq \beta$, are known as 'open shell' wave functions. We note that the values of α and β obtained in the present case imply that H^- can be viewed as a two-electron system in which one electron is weakly bound in the field of a polarized hydrogen atom.

Variation–perturbation method

Let us now return to the perturbation treatment which we started at the beginning of this section by calculating the first-order energy correction $E_0^{(1)}$. It is possible to obtain higher order corrections by using the variation–perturbation method discussed in Section 2.8. In order to display explicitly the Z dependence of the various orders of perturbation theory, it is convenient to choose $Za_0 = Z$ a.u. as our unit of length and $Z^2e^2/[(4\pi\varepsilon_0)a_0] = Z^2$ a.u. as our unit of energy. Setting

$$\boldsymbol{\rho}_1 = Z\mathbf{r}_1, \qquad \boldsymbol{\rho}_2 = Z\mathbf{r}_2, \qquad \rho_{12} = |\boldsymbol{\rho}_1 - \boldsymbol{\rho}_2| \quad \textbf{(7.87)}$$

the Schrödinger equation (7.3) then becomes

$$(H_0 + \lambda H')\psi(\boldsymbol{\rho}_1, \boldsymbol{\rho}_2) = \varepsilon\psi(\boldsymbol{\rho}_1, \boldsymbol{\rho}_2) \quad \textbf{(7.88)}$$

where

$$H_0 = -\frac{1}{2}\nabla^2_{\rho_1} - \frac{1}{2}\nabla^2_{\rho_2} - \frac{1}{\rho_1} - \frac{1}{\rho_2}$$
$$H' = \frac{1}{\rho_{12}}, \qquad \lambda = Z^{-1}, \qquad \varepsilon = E/Z^2 \tag{7.89}$$

With this choice of units, the ground state energy $\varepsilon_0 = E_0/Z^2$, expanded in powers of λ, reads

$$\varepsilon_0 = \sum_{n=0}^{\infty} \varepsilon_0^{(n)} Z^{-n} \tag{7.90}$$

Following the procedure outlined in Section 2.8, approximate values of $\varepsilon_0^{(n)}$ may be obtained by using trial functions $\phi^{(1)}, \phi^{(2)}, \ldots$ in the functionals $F_1[\phi^{(1)}]$, $F_2[\phi^{(2)}]$, and so on. Table 7.4 shows the first few values of $\varepsilon_0^{(n)}$ $(2 \leqslant n \leqslant 10)$ calculated in 1963 by C.W. Scherr and R.E. Knight from elaborate Hylleraas-type trial functions. The agreement between the values of $E_0 = Z^2\varepsilon_0$ obtained from (7.90) and the results of Table 7.4 and those calculated from the Rayleigh–Ritz method (quoted in the last column of Table 7.3) is excellent. It is interesting to note that for the case of H^- it is necessary to include the values of $\varepsilon_0^{(n)}$ up to order $n = 7$ to obtain agreement with the correct result ($E_0 = -0.528$ a.u.) within 0.001 a.u. ($\simeq 0.027$ eV).

Comparison with experiment

The extremely accurate result (7.82) is actually *lower* than the experimental value of the ground state of helium! This, however, is not a paradox since (7.82) is only an upper limit to the exact *non-relativistic* value of E_0 corresponding to an *infinitely heavy nucleus* (that is, the ground state of the Hamiltonian (7.2)). Before

Table 7.4 Values of $\varepsilon_0^{(n)}$ (in a.u.) for the first few orders of perturbation theory, as obtained by Scherr and Knight. We recall that $\varepsilon_0^{(0)} = -1$ and $\varepsilon_0^{(1)} = 5/8$. The corresponding values of $E_0^{(n)}$ are given (in a.u.) by $E_0^{(n)} = \varepsilon_0^{(n)} Z^{2-n}$.

Order n of perturbation theory	*Value of* $\varepsilon_0^{(n)}$
2	−0.157 66
3	0.008 69
4	−0.000 88
5	−0.001 03
6	−0.000 61
7	−0.000 37
8	−0.000 24
9	−0.000 16
10	−0.000 11

comparing theory and experiment several corrections must therefore be calculated, which we now discuss briefly.

We begin by considering the *motion of the nucleus*. From Appendix 8 and equation (7.1), we see that the Schrödinger equation in which this effect is ignored (that is, in which one sets $M = \infty$) is then modified in *two* ways. First, as in the case of one-electron atoms, the *reduced mass* $\mu = mM/(m + M)$ of the electron with respect to the nucleus replaces the mass m of the electron. Secondly, a 'mass polarisation' term $(-\hbar^2/M)\nabla_{\mathbf{r}_1} \cdot \nabla_{\mathbf{r}_2}$ appears in the Schrödinger equation.

The reduced mass correction is easily taken into account by using the method discussed in Chapters 1 and 3 for hydrogenic atoms. We introduce a new atomic unit of length $a_\mu = a_0 m/\mu$ and a new atomic unit of energy

$$\frac{e^2}{(4\pi\varepsilon_0)a_\mu} = \frac{e^2}{(4\pi\varepsilon_0)a_0} \frac{\mu}{m} \tag{7.91}$$

which differs from the usual atomic energy unit $e^2/[(4\pi\varepsilon_0)a_0]$ by the factor μ/m. After having calculated the energy E_∞ corresponding to infinite nuclear mass ($M = \infty$) we may therefore obtain the value E_μ appropriate to the reduced mass μ by writing

$$E_\mu = \frac{\mu}{m} E_\infty \tag{7.92}$$

Thus, as for one-electron atoms (see Chapters 1 and 3), the reduced mass effect modifies all the energy levels of the atom in the same manner, all the frequencies of the spectral lines being reduced by the factor $\mu/m \simeq 1 - m/M$. If we call ΔE_1 the reduced mass correction to be applied to the energy levels E_∞, we see that

$$\begin{aligned} \Delta E_1 &= E_\mu - E_\infty \\ &= E_\infty \left(\frac{\mu}{m} - 1 \right) \\ &\simeq -\frac{m}{M} E_\infty \end{aligned} \tag{7.93}$$

and we notice that ΔE_1 is positive. In particular, for the ground state of ^{4}He this correction reduces the ionisation potential I_P^∞ (corresponding to $M = \infty$) by an amount of about 27 cm^{-1}.

We now turn to the correction arising from the 'mass polarisation' term. Calling ΔE_2 that correction, and using first-order perturbation theory, we have (in a.u.)

$$\begin{aligned} \Delta E_2 &= -\frac{1}{M} \langle \psi_0 | \nabla_{\mathbf{r}_1} \cdot \nabla_{\mathbf{r}_2} | \psi_0 \rangle \\ &= \frac{1}{M} \langle \psi_0 | \mathbf{p}_1 \cdot \mathbf{p}_2 | \psi_0 \rangle \end{aligned} \tag{7.94}$$

where $\psi_0(\mathbf{r}_1, \mathbf{r}_2)$ is the spatial part of the ground state wave function, corresponding to the case $M = \infty$, and $\mathbf{p}_i$ is the momentum operator of electron i. We note that if ψ_0 is a simple product of individual orbitals $u_i(\mathbf{r}_i)$ $(i = 1, 2)$, then $\Delta E_2 = 0$, since the expectation value of the momentum operator, $\langle \mathbf{p} \rangle$, vanishes for a stationary bound (discrete) state. With an accurate Hylleraas-type variational wave function, it is found that for ^{4}He the correction ΔE_2 is about 4.8 cm^{-1}.

The discussion of the relativistic corrections and of the radiative corrections (Lamb shift) falls outside the scope of this book [7]. Calling ΔE_3 the former ones, and ΔE_4 the latter ones, we list in Table 7.5 the numerical values of these corrections for helium and several helium-like ions. Also given in Table 7.5 are the reduced mass correction ΔE_1 and the mass polarisation correction ΔE_2 discussed above. When subtracted from the non-relativistic ionisation potential I_P^∞ (corresponding to $M = \infty$), these four corrections yield the total theoretical ionisation potential I_P^{th}, which is seen from Table 7.5 to be in excellent agreement with experiment. Of particular interest is the comparison between the theoretical and experimental values of the ionisation potential for helium. From the work of Frankowski and Pekeris, and Schwartz, the theoretical value is

$$I_P^{\text{th}} = (198\,310.699 \pm 0.05) \text{ cm}^{-1} \tag{7.95}$$

while the experimental value obtained by G. Herzberg is

$$I_P^{\text{exp}} = (198\,310.82 \pm 0.15) \text{ cm}^{-1} \tag{7.96}$$

The remarkable agreement between (7.95) and (7.96) may be considered as one of the most striking triumphs of quantum mechanics.

7.6 Excited states of two-electron atoms

We now turn to the study of the excited states of helium-like atoms. Disregarding for the moment the 'doubly excited states' (which will be considered in the next section), we shall analyse various 'genuinely discrete' excited states of two-electron atoms with $Z \geqslant 2$. For the sake of simplicity we shall neglect all the small corrections (motion of the nucleus, relativistic and radiative corrections) mentioned at the end of the preceding section. Our starting point will therefore be the Schrödinger equation (7.3) and our discussion of Section 7.4 based on the independent particle model.

Perturbation theory

As in the case of the ground state, we shall first use perturbation theory. We again choose to split the Hamiltonian (7.2) as $H = H_0 + H'$, where the unperturbed Hamiltonian H_0 is given by (7.24) and $H' = 1/r_{12}$ is the perturbation. The properly symmetrised 'zero-order' spatial wave functions for the genuinely discrete excited

[7] A detailed account may be found in Bethe and Salpeter (1957).

Table 7.5 Values of the non-relativistic ionisation potential I_P^∞, the reduced mass correction ΔE_1, the mass polarisation correction ΔE_2, the relativistic correction ΔE_3, the radiative corrections ΔE_4, the theoretical ionisation potential $I_P^{th} = I_P^\infty - (\Delta E_1 + \Delta E_2 + \Delta E_3 + \Delta E_4)$ and the experimental ionisation potential I_P^{exp} (in cm^{-1}) for various two-electron atoms and ions.

	I_P^∞	ΔE_1	ΔE_2	ΔE_3	ΔE_4	I_P^{th}	I_P^{exp}
H^-	6090.644 289	3.315 791	3.928	0.304	0.0037	6083.092	6100 ± 100
He	198 344.580 143 48	27.192 711	4.785	0.562	1.341	198 310.699	198 310.82 ± 0.15
Li^+	610 120.488 2	47.768 9	4.960	−19.69	7.83	610 079.62	610 079 ± 25
Be^{2+}	1 241 253.351	75.681	5.619	−114.52	27.1	1 241 259.5	1 241 225 ± 100
B^{3+}	2 091 806.533	104.436	6.046	−372.88	65.7	2 092 003.2	2 091 960 ± 200
C^{4+}	3 161 805.752	144.864	6.878	−919.00	132	3 162 441	3 162 450 ± 300

states we are considering are therefore the wave functions $\psi_{\pm}^{(0)}(\mathbf{r}_1, \mathbf{r}_2)$ given by (7.38), the corresponding 'zero-order' energies being given by (7.39).

Since the energy levels (7.39) are degenerate with respect to l and m, and also exhibit the 'exchange degeneracy', we should in principle use degenerate perturbation theory. However, by using the expansion (7.62) of $1/r_{12}$ in spherical harmonics and the orthogonality property of the Y_{lm}, it is straightforward to prove that the matrix elements of H' between two degenerate zero-order wave functions $\psi_{\pm}^{(0)}$ corresponding to different l or m vanish. Moreover, since H' is invariant for a permutation of the spatial coordinates of the two electrons, it is clear that the matrix elements $\langle\psi_{\pm}^{(0)}|H'|\psi_{\mp}^{(0)}\rangle$ of H' between two degenerate zero-order wave functions corresponding respectively to a para and an ortho state vanish. Thus, from the point of view of degenerate perturbation theory, the wave functions (7.38) are already the correct 'zero-order' wave functions and we can use non-degenerate perturbation theory directly to calculate the first-order energy corrections. That is,

$$E_{\pm}^{(1)} = \langle\psi_{\pm}^{(0)}|H'|\psi_{\pm}^{(0)}\rangle \tag{7.97}$$

where the plus sign refers to para states and the minus sign to ortho states, as usual. Using the expression (7.38) of $\psi_{\pm}^{(0)}$ and the fact that $H' = 1/r_{12}$, we have

$$E_{+}^{(1)} = J + K, \qquad E_{-}^{(1)} = J - K \tag{7.98}$$

where

$$J = \int |\psi_{100}(\mathbf{r}_1)|^2 \frac{1}{r_{12}} |\psi_{nlm}(\mathbf{r}_2)|^2 \, \mathrm{d}\mathbf{r}_1 \, \mathrm{d}\mathbf{r}_2 \tag{7.99}$$

and

$$K = \int \psi_{100}^*(\mathbf{r}_1)\psi_{nlm}^*(\mathbf{r}_2) \frac{1}{r_{12}} \psi_{100}(\mathbf{r}_2)\psi_{nlm}(\mathbf{r}_1) \, \mathrm{d}\mathbf{r}_1 \, \mathrm{d}\mathbf{r}_2 \tag{7.100}$$

with $n \geqslant 2$.

The integral J is called the *Coulomb* (or *direct*) *integral*. By analogy with the discussion following (7.55), we see that it represents the Coulomb interaction between the charge distributions of the two electrons. The integral K is known as the *exchange integral*; it is the matrix element of $1/r_{12}$ between two states such that the electrons have exchanged their quantum numbers. Using the fact that the hydrogenic functions $\psi_{nlm}(\mathbf{r})$ are given by $\psi_{nlm}(\mathbf{r}) = R_{nl}(r)Y_{lm}(\theta, \phi)$ (see (3.53)), together with the expansion (7.62) and the orthonormality property of the spherical harmonics, we have

$$J_{nl} = \int_0^\infty \mathrm{d}r_2 r_2^2 R_{nl}^2(r_2) \int_0^\infty \mathrm{d}r_1 r_1^2 R_{10}^2(r_1) \frac{1}{r_>} \tag{7.101}$$

and

$$K_{nl} = \frac{1}{2l+1} \int_0^\infty \mathrm{d}r_2 r_2^2 R_{10}(r_2) R_{nl}(r_2) \int_0^\infty \mathrm{d}r_1 r_1^2 R_{10}(r_1) R_{nl}(r_1) \frac{(r_<)^l}{(r_>)^{l+1}} \tag{7.102}$$

where we recall that $r_<$ is the smaller and $r_>$ the larger of r_1 and r_2. In the above formulae we have explicitly indicated that the integrals J and K depend on n and l, but not on m. The first-order energy corrections (7.98) may therefore be written more explicitly as

$$E^{(1)}_{nl,\pm} = J_{nl} \pm K_{nl} \tag{7.103}$$

and we remark that the independence of these energy shifts with respect to m can also be deduced from the fact that the perturbation $H' = 1/r_{12}$ commutes with L_z.

Adding the corrections (7.103) to the zero-order energies (7.39), we thus obtain the 'first-order' energy values (in a.u.)

$$E_{nl,\pm} \simeq E^{(0)}_{1,n} + E^{(1)}_{nl,\pm} = -\frac{Z^2}{2}\left(1 + \frac{1}{n^2}\right) + J_{nl} \pm K_{nl} \tag{7.104}$$

The integrals J_{nl} and K_{nl} can be evaluated explicitly (see for example Problem 7.4 for the case $n = 2$, $l = 0, 1$), but we shall use here only a couple of their general properties. First of all, it is obvious from (7.101) that J_{nl} must be positive. Furthermore, for $l = n - 1$ the radial wave function $R_{n,n-1}$ has no nodes, so that we deduce directly from (7.102) that $K_{n,n-1} > 0$. It turns out that in the general case one has $K_{nl} > 0$, so that we see from (7.104) that *an ortho* (spin triplet) *state has a lower energy than the corresponding para* (spin singlet) *state* having the same value of n and l. This conclusion is in accordance with our discussion at the end of Section 7.4. In fact, the role of the spin-dependent *exchange force* introduced at that point can be made explicit in the following way. We first note from (7.11) and (7.15) that the operator

$$\mathbf{S}_1 \cdot \mathbf{S}_2 = \tfrac{1}{2}\mathbf{S}^2 - \tfrac{3}{4} \tag{7.105}$$

yields the value −3/4 when acting on singlet spin states ($S = 0$) and 1/4 when acting on triplet spin states ($S = 1$). The first-order energy correction (7.103) can therefore by symbolically written as

$$\begin{aligned} E^{(1)}_{nl,\pm} &= J_{nl} - \tfrac{1}{2}(1 + 4\mathbf{S}_1 \cdot \mathbf{S}_2)K_{nl} \\ &= J_{nl} - \tfrac{1}{2}(1 + \boldsymbol{\sigma}_1 \cdot \boldsymbol{\sigma}_2)K_{nl} \end{aligned} \tag{7.106}$$

where we have introduced the $\boldsymbol{\sigma}_i$ operators, which are related to the spin operators $\mathbf{S}_i$ by

$$\mathbf{S}_i = \tfrac{1}{2}\boldsymbol{\sigma}_i \qquad \text{(in a.u.)} \tag{7.107}$$

We see that in the sense specified by (7.106) the energy shifts $E^{(1)}_{nl,\pm}$ – and hence the first-order energy values (7.104) – depend explicitly upon the relative orientation of the electron spins, even though the Hamiltonian (7.2) does not contain spin-dependent terms. We also remark that the spin-dependent exchange force responsible for this effect is of the same order of magnitude as the electrostatic force. It is therefore much stronger than the spin-dependent forces arising from relativistic effects, such as the spin–orbit interaction which has been shown in

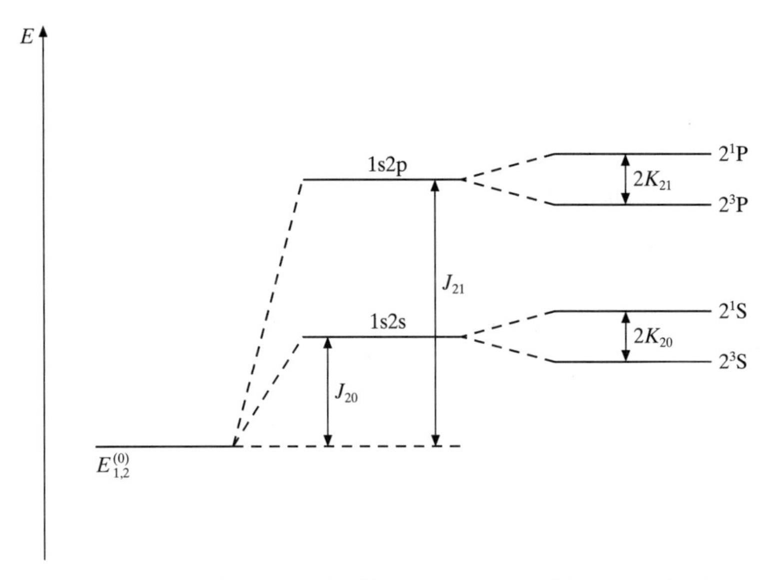

Figure 7.8 The splitting of the unperturbed helium level for $n = 2$ by the Coulomb integrals J_{2l} and the exchange integrals K_{2l}.

Chapter 5 to yield corrections of order $(Z\alpha)^2$ to the energy levels. In fact, as W. Heisenberg first observed, the exchange force is strong enough to keep the electron spins aligned in certain solids, and hence is responsible for the phenomenon of *ferromagnetism*.

The splitting of the unperturbed energy levels predicted by the first-order expression (7.104) is illustrated schematically in Fig. 7.8 for the case $n = 2$. The effects of the Coulomb integrals J_{2l} and of the exchange integrals K_{2l} are displayed separately. We remark that $J_{21} > J_{20}$, so that the energy of the configuration 1s2s is lower than that of the configuration 1s2p, as we expect from the discussion following (7.51). We also see from Fig. 7.8 that the triplet states lie below the singlet states corresponding to the same values of n and l, in agreement with the foregoing discussion.

Variational method

As we remarked in Section 2.8, the application of the Rayleigh–Ritz variational method to excited states of a quantum system is in general more difficult than for the ground state, because the trial wave function corresponding to an excited state must be orthogonal to the wave functions of all the eigenstates having a lower energy. However, we also noticed in Section 2.8 that an important simplification occurs when the Hamiltonian can be diagonalised simultaneously with a Hermitian operator A. Indeed, we saw that in that case a trial wave function entirely constructed from eigenfunctions of A corresponding to a given eigenvalue α_i is automatically orthogonal to all the eigenfunctions of A corresponding to eigenvalues α_j different from α_i; this trial wave function will therefore provide an upper bound for the lowest energy eigenvalue associated with the eigenvalue α_i.

We now apply these ideas to the case of two-electron atoms. Among the operators which can be diagonalised simultaneously with the Hamiltonian (7.2), we shall consider the interchange operator P_{12} (see (7.4)) as well as the operators $\mathbf{S}^2$ and $\mathbf{L}^2$. First of all, since para wave functions (which correspond to the eigenvalue +1 of P_{12} and to the value $S = 0$ of the total spin) are orthogonal to ortho wave functions (corresponding to the eigenvalue -1 of P_{12} and to the total spin $S = 1$), we can study separately the variational determination of the para and ortho energy levels, provided trial functions having the appropriate para or ortho symmetry are used in the variational principle. Moreover, because the eigenfunctions belonging to states with different values of L are orthogonal, we can consider a separate variational problem for each value of L by using trial functions which exhibit the angular dependence corresponding to a state of given L. In this way we see that a minimum principle for the lowest energy state belonging to a given L and S will always be obtained without imposing further orthogonality constraints on the trial function. Hence, if we adopt trial functions ϕ having the correct para or ortho symmetry and the appropriate angular dependence, we may use directly the Rayleigh–Ritz method – that is, substitute the trial function ϕ in the functional (7.68) – to obtain upper bounds for the energies of the states 2^3S, 2^1P, 2^3P, 3^1D, 3^3D, and so on. On the other hand, when calculating the energy of the 2^1S state by the variational method, one must explicitly constrain the trial function to be orthogonal to the wave function of the ground state 1^1S. Similarly, the trial wave function for the 3^1S state must be made orthogonal to the wave functions of both the 1^1S and 2^1S state, the trial wave function for the 3^1P state has to be made orthogonal to the wave function of the 2^1P state, and so on. It should be noted from our discussion at the end of Section 2.8 (see (2.402)) that even after the orthogonalisations have been performed we do not necessarily have upper bounds for the energies of these excited states, since the wave functions of the lower states are only known approximately.

As an illustration of the foregoing analysis we shall consider the lowest four excited states of helium. We begin by studying the 2^3S, 2^1P and 2^3P states, which can be treated by the 'standard' Rayleigh–Ritz variational procedure, without imposing additional orthogonality conditions; we then go on to discuss the 2^1S state, whose wave function must explicitly be orthogonalised to the wave function of the ground state.

The 2^3S state

This is the lowest ortho (space-antisymmetric) state, corresponding to the configuration 1s2s. On the basis of the independent-particle model discussed in Section 7.4 and of our first-order perturbation theory calculation, it is reasonable to adopt for the spatial part of the wave function a simple trial function which is the antisymmetrised product of an 'inner' (1s) orbital u_{1s} corresponding to the 'effective charge' Z_i and an 'outer' (2s) orbital v_{2s} corresponding to the effective charge Z_o. That is,

$$\phi_{2^3S}(r_1, r_2) = N[u_{1s}(r_1)v_{2s}(r_2) - v_{2s}(r_1)u_{1s}(r_2)] \tag{7.108a}$$

where

$$\begin{aligned} u_{1s}(r) &= \exp(-Z_i r) \\ v_{2s}(r) &= (1 - Z_o r/2)\exp(-Z_o r/2) \end{aligned} \tag{7.108b}$$

and N is a normalisation constant. Upon substituting (7.108) into the functional (7.68), and varying the variational parameters Z_i and Z_o to obtain the minimum energy, C. Eckart found the values $Z_i = 2.01$ and $Z_o = 1.53$, which yield the energy $E_{2^3S} = -2.167$ a.u. ($= -58.97$ eV).

Very accurate results for the energy of the 2^3S state may be obtained by using elaborate Hylleraas-type variational functions depending on the three variables (r_1, r_2, r_{12}) or the Hylleraas coordinates $s = r_1 + r_2$, $t = r_1 - r_2$ and $u = r_{12}$. Of course the trial wave function $\phi(s, t, u)$ must now be space-antisymmetric, and hence has to be an odd function of t. The 'exact' value obtained in this way by E.A. Hylleraas and B. Undheim and later by C.L. Pekeris *et al.* (using generalised Hylleraas-type wave functions) is $E^{ex}_{2^3S} = -2.175$ a.u. ($= -59.19$ eV). The agreement between the Eckart result and this 'exact' value is seen to be good, considering the simplicity of the trial function (7.108).

The 2^1P and 2^3P states

Since the 2^1P and 2^3P states are respectively the lowest para and ortho states corresponding to the configuration 1s2p, simple trial wave functions for these states may be written as

$$\phi_{2^{1,3}P}(\mathbf{r}_1, \mathbf{r}_2) = N_\pm[u_{1s}(r_1)v_{2pm}(\mathbf{r}_2) \pm v_{2pm}(\mathbf{r}_1)u_{1s}(r_2)] \tag{7.109a}$$

where the plus sign refers to the 2^1P state, the minus signs to the 2^3P state, $N_\pm$ are normalisation constants, and

$$\begin{aligned} u_{1s}(r) &= \exp(-Z_i r) \\ v_{2pm}(\mathbf{r}) &= r\exp(-Z_o r/2)\, Y_{1,m}(\hat{\mathbf{r}}), \qquad m = 1, 0, -1 \end{aligned} \tag{7.109b}$$

the symbol $\hat{\mathbf{r}}$ denoting the angular variables of $\mathbf{r}$. Substituting the trial functions (7.109) into the functional (7.68) and varying the parameters Z_i and Z_o to obtain a minimum for the energy, Eckart found for the 2^1P state the values $Z_i = 2.00$ and $Z_o = 0.97$, giving the energy $E_{2^1P} = -2.123$ a.u. ($= -57.77$ eV), which is in excellent agreement with the 'exact' result $E^{ex}_{2^1P} = -2.124$ a.u. ($= -57.80$ eV) of Pekeris. For the 2^3P state Eckart obtained for the variational parameters the optimum values $Z_i = 1.99$ and $Z_o = 1.09$, the corresponding energy being $E_{2^3P} = -2.131$ a.u. ($= -57.99$ eV), in very good agreement with the 'exact' result $E^{ex}_{2^3P} = -2.133$ a.u. ($= -58.04$ eV). It should be noted that for both the 2^1P and 2^3P states the value of

Z_o is much closer to $Z - 1 = 1$ (corresponding to complete screening) than for the 2^3S state which we have studied previously.

The 2^1S state

Since the 2^1S state is a para (space-symmetric) state corresponding to the configuration 1s2s, one could attempt to construct a simple trial wave function, similar to the 2^3S wave function (7.108), but which would be space-symmetric instead of space-antisymmetric. That is,

$$\phi_{2^1S}(r_1, r_2) = N[u_{1s}(r_1)v_{2s}(r_2) + v_{2s}(r_1)u_{1s}(r_2)] \tag{7.110a}$$

where u_{1s} is again an 'inner' (1s) orbital, v_{2s} an 'outer' (2s) orbital and N a normalisation constant. However, a trial wave function of the form (7.110a) is not necessarily orthogonal to the wave function of the ground state 1^1S. As we pointed out above, this orthogonality constraint must therefore be imposed explicitly. A simple variational wave function of the form (7.110a), which does satisfy the requirement of being orthogonal to the ground state wave function (7.85), is such that the orbitals $u_{1s}(r)$ and $v_{2s}(r)$ are given by

$$\begin{aligned} u_{1s}(r) &= \exp(-2r) \\ v_{2s}(r) &= \exp(-\tau_1 r) - Cr\exp(-\tau_2 r) \end{aligned} \tag{7.110b}$$

where the 'optimum' variational parameters, obtained by Byron and Joachain, are given by $\tau_1 = 0.865$, $\tau_2 = 0.522$ and $C = 0.432\,784$. The corresponding energy is $E_{2^1S} = -2.143$ a.u. $(= -58.31$ eV) which is close to the 'exact' result $E^{ex}_{2^1S} = -2.146$ a.u. $(= -58.40$ eV) obtained by Pekeris.

Another variational approach which may be applied to the study of the 2^1S state consists in using the Hylleraas–Undheim theorem (see Section 2.8). A trial function is first chosen, which has the symmetry of a ^{1}S state, and contains a number of variational parameters. According to the Hylleraas–Undheim theorem, the *second* root of the determinantal equation (2.409) then provides an upper bound for the energy of the 2^1S state. In this way very accurate results for the energy of the 2^1S state, and also for other excited states of helium, have been obtained.

7.7 Doubly excited states of two-electron atoms. Auger effect (autoionisation). Resonances

In the language of the independent particle model introduced in Section 7.4, the 'genuinely discrete' excited states which we have studied in the preceding section all correspond to *singly* excited states, for which one electron occupies the ground state (1s) orbital and one occupies an excited orbital. However, we noticed in Section 7.4 (see (7.34)) that there exist also *doubly excited states* of the Hamiltonian (7.24), in which both electrons occupy excited orbitals (for example,

$2s^2$, 2s2p, 3p4d, etc.). We remarked that all these states lie *above* the ionisation threshold and are therefore *discrete states embedded in the continuum.*

Let us now study the action of the perturbation $H' = 1/r_{12}$ on a given doubly excited state $|\alpha\rangle$, having an 'unperturbed' energy $E_\alpha^{(0)}$ and a corresponding zero-order wave function $\psi_\alpha^{(0)}$. Because of the presence of *continuum* energy levels in the neighbourhood of $E_\alpha^{(0)}$, an eigenfunction of the full Hamiltonian H in this energy region is

$$\Psi(E) \simeq a(E)\psi_\alpha^{(0)} + \int b(E, E')\psi(E')\,\mathrm{d}E' \tag{7.111}$$

where $\psi(E')$ is a properly symmetrised continuum wave function describing the system made of a *free* electron of kinetic energy E' and a one-electron atom (ion). It is clear that the perturbation H' will cause the discrete state $\psi_\alpha^{(0)}$ to interact with the nearby continuum states. Hence, in addition to an *energy shift* of the discrete state, a *radiationless transition* from the doubly excited state to a state of an *ionised* configuration will occur. Such a transition is known as the *Auger effect* or *autoionisation*, and the doubly excited states (which are unstable against ionisation) are called *autoionising states*. Direct application of the Fermi Golden Rule (see (2.362)) shows that, to first order of perturbation theory, the transition rate for autoionisation is given by

$$W = \frac{2\pi}{\hbar}|\langle\psi(E)|H'|\psi_\alpha^{(0)}\rangle|^2\rho_f(E) \tag{7.112}$$

where $\rho_f(E)$ is the density of final states corresponding to the continuum wave function $\psi(E)$, and the kinetic energy E of the free electron is specified by energy conservation. Not only helium-like atoms (ions) but all atoms or ions with two or more electrons possess such autoionising states.

As an example, let us consider the $(2s2p)^1P$ state of helium, in which the two electrons form a singlet spin state in the configuration 2s2p. This state can be excited from the ground state $(1s)^2\,{}^1S$ by the absorption of (ultra-violet) radiation, since the selection rules $\Delta L = \pm 1$ and $\Delta S = 0$ [8] are satisfied. Once excited, the $(2s2p)^1P$ state can decay via *radiative transitions* to bound states of He allowed by the selection rules (that is, ${}^1D$ and ${}^1S$ states, including of course the ground state). However, it can also decay via a *radiationless* (*autoionising* or *Auger*) *transition* into a free electron and a He^+ ion in the ground state, the kinetic energy of the free electron being determined by conservation of energy. Explicit calculations show that the autoionising transition is much more probable than a radiative transition to a He bound state, and this is also true for the other doubly excited states of helium. As a result, spectral lines of emission spectra corresponding to doubly excited states of helium are very weak.

---

[8] These selection rules will be proved in Chapter 9.

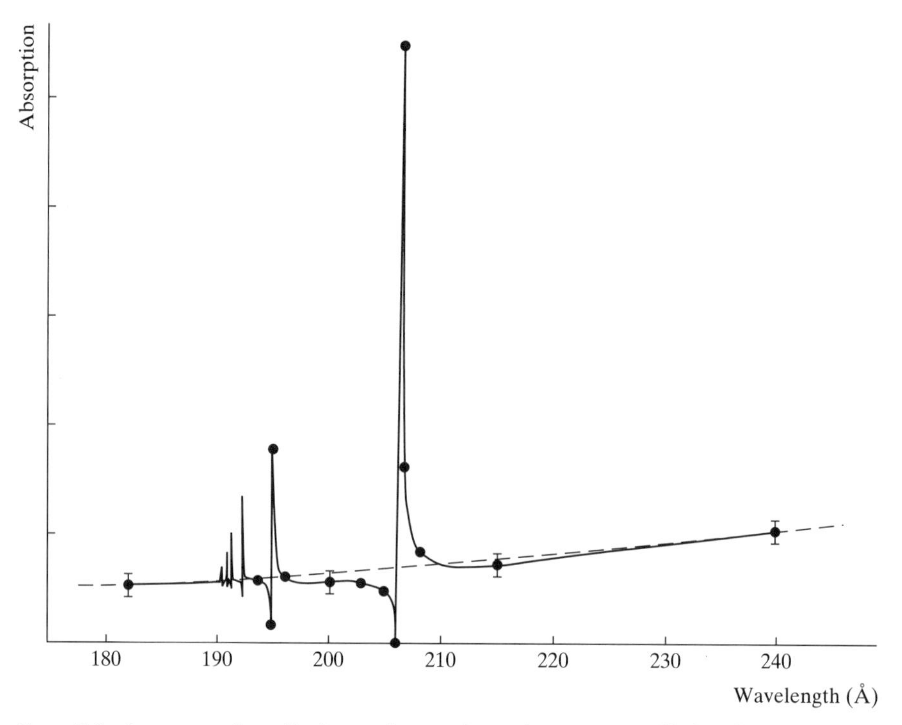

Figure 7.9 Resonances found in the continuous absorption spectrum of helium by R.B. Madden and K. Codling.

The presence of autoionising states in the continuum spectrum of two-electron atoms has several important consequences. First, these states will show up in the *absorption spectrum* of the system. For example, if one analyses the absorption spectrum of helium corresponding to the photoionisation process

$$h\nu + \text{He}(1^1\text{S}) \rightarrow \text{e}^- + \text{He}^+ \qquad \textbf{(7.113)}$$

one observes that sharp *peaks* or *resonances*, superimposed on the smooth continuum absorption, occur in the neighbourhood of the energy of autoionising states. This is illustrated in Fig. 7.9, which shows the resonances in the helium continuum absorption spectrum observed in 1963 by R.P. Madden and K. Codling, who used an electron synchrotron to obtain a continuous photon source in the wavelength region around 200 Å. For example, the large peak shown in Fig. 7.9 appears at the energy corresponding to the 2s2p level. In the vicinity of the energy of an autoionising state, which we denote by the symbol He** (doubly excited He state), the photoionisation reaction (7.113) may thus be viewed as proceeding mainly through the steps

$$h\nu + \text{He}(1^1\text{S}) \rightarrow \text{He}^{**} \rightarrow \text{e}^- + \text{He}^+ \qquad \textbf{(7.114)}$$

where the autoionising state He** corresponds to a temporarily bound or *resonant* state of the compound system ($e^- + He^+$). The shape of the resonance is determined by interference between the resonant process (7.114) and direct absorption.

A second way in which the autoionising states of two-electron atoms manifest themselves is in the *collisions of electrons* with the corresponding one-electron atom or ion. Consider for example the scattering of electrons by He^+ ions. If the incident electron has just the right kinetic energy, an autoionising state He** can be formed, and this compound state of the system ($e^- + He^+$) will subsequently decay into a free electron and a He^+ ion. Thus the scattering event may be analysed in terms of the two-step process

$$e^- + He^+ \rightarrow He^{**} \rightarrow e^- + He^+ \qquad \textbf{(7.115)}$$

in which a resonant state of the ($e^- + He^+$) system is temporarily obtained. As in the case of the absorption spectrum, a peak or *resonance* is then observed in the scattering cross-section. We shall return in Chapter 13 to the study of these resonances, which have become the subject of numerous investigations in recent years.

Problems

7.1 Prove that eigenfunctions $\psi(\mathbf{r}_1, \mathbf{r}_2)$ of the Schrödinger equation (7.3) corresponding to degenerate eigenvalues can always be chosen to be either space-symmetric $[\psi_+(\mathbf{r}_2, \mathbf{r}_1) = \psi_+(\mathbf{r}_1, \mathbf{r}_2)]$ or space-antisymmetric $[\psi_-(\mathbf{r}_2, \mathbf{r}_1) = -\psi_-(\mathbf{r}_1, \mathbf{r}_2)]$.

7.2 Study the action of the operator $\mathbf{S}^2 = 3/2 + 2\mathbf{S}_1 \cdot \mathbf{S}_2$ on the four spin functions (7.12) and obtain the results quoted in Table 7.1.

7.3 Calculate the average values of

(i) $r_1^2 + r_2^2$, (ii) $\delta(\mathbf{r}_1)$, (iii) $\delta(\mathbf{r}_{12})$

for the ground state of helium by using:

(a) the zero-order product of hydrogenic wave functions $\psi_0^{(0)}$ given by (7.35), with $Z = 2$;
(b) the simple 'screened' variational function given by (7.70), with $Z_e = 27/16$ (see (7.78));
(c) the Hartree–Fock wave function (7.85).

Compare your results with the 'exact' values $\langle r_1^2 + r_2^2 \rangle = 2.39$, $\langle \delta(\mathbf{r}_1) \rangle = 1.81$ and $\langle \delta(\mathbf{r}_{12}) \rangle = 0.106$ (in a.u.).

7.4 Evaluate explicitly the Coulomb integral J_{nl} and the exchange integral K_{nl}, given respectively by (7.101) and (7.102) for the cases $n = 2$, $l = 0, 1$. Using (7.104), obtain the corresponding 'first-order' energy values for the energies of the 2^1S, 2^1P, 2^3S and 2^3P states of helium.

7.5 A helium atom is excited from the ground state to the autoionising state 2s4p by absorption of ultra-violet light. Assuming that the 2s electron moves in the unscreened Coulomb field of the nucleus and the 4p electron in the screened Coulomb potential $-1/r$:

(a) Obtain the energy of this autoionising level and the corresponding wavelength of the ultra-violet radiation.
(b) Find the velocity of the electron emitted in the autoionising process in which the autoionising level 2s4p decays into a free electron and a He^+ ion in the ground state 1s.

8 Many-electron atoms

We have seen in the preceding chapter that Schrödinger's equation cannot be solved exactly for two-electron atoms or ions, so that approximation methods must be used. We also saw in Chapter 7 that very accurate results can be obtained for the energy levels and wave functions of helium-like atoms by performing variational calculations – such as those of the Rayleigh–Ritz type – in which elaborate trial functions containing a large number of variational parameters are used. Extended variational calculations of this kind have also been carried out for other light atoms – such as lithium, which contains three electrons [1] – but this approach becomes increasingly tedious when the number of atomic electrons increases. We shall therefore develop in this chapter some general methods which, at the expense of simplifying assumptions, can be applied to study the structure of many-electron atoms and ions.

The starting point of all calculations on many-electron atoms is the *central field approximation*, which we have already discussed in the previous chapter for the case of two-electron atoms. The basic idea of this approximation is that each of the atomic electrons moves in an effective spherically symmetric potential $V(r)$ created by the nucleus and all the other electrons. We shall first discuss some general properties of this central potential, and show that a number of qualitative features of many-electron atoms – including the periodicity in the properties of the chemical elements – can be understood without a detailed knowledge of the form of this central field. Next, we turn to the problem of the determination of the potential $V(r)$, for which we first discuss the simple semi-classical method of Thomas and Fermi, and then the more precise Hartree–Fock or self-consistent field approach. We conclude this chapter by considering the corrections which must be applied to the central field approximation.

8.1 The central field approximation

Let us consider an atom or ion containing a nucleus of charge Ze and N electrons. A detailed treatment of this system should take into account:

[1] See for example the review of King (1999).

1. The kinetic energy of the electrons and their potential energy in the electrostatic (Coulomb) attractive field of the nucleus (assumed to be point-like and infinitely massive).
2. The electrostatic (Coulomb) repulsion between the electrons.
3. The magnetic interactions of the electronic spins with their orbital motion (spin–orbit interactions).
4. Several small effects such as spin–spin interactions between the electrons, various relativistic effects, radiative corrections and nuclear corrections (due to the finite mass of the nucleus, its finite extension, nuclear magnetic dipole moments, etc.).

It is clear that such a detailed study of a many-electron atom is a very difficult task, in which approximations must be made. In what follows, we shall neglect all the 'small' effects mentioned in 4. The spin–orbit interactions (3 above) will be considered in Section 8.5 when we discuss the corrections to the central field approximation. Thus, keeping only the attractive Coulomb interactions between the electrons and the nucleus (which we assume to be infinitely heavy) and the Coulomb repulsions between the electrons, we write the Hamiltonian of the N-electron atom (ion) in the absence of external fields as

$$H = \sum_{i=1}^{N} \left(-\frac{\hbar^2}{2m} \nabla^2_{\mathbf{r}_i} - \frac{Ze^2}{(4\pi\varepsilon_0) r_i} \right) + \sum_{i<j=1}^{N} \frac{e^2}{(4\pi\varepsilon_0) r_{ij}} \tag{8.1}$$

where $\mathbf{r}_i$ denotes the relative coordinate of the electron i with respect to the nucleus, $r_{ij} = |\mathbf{r}_i - \mathbf{r}_j|$ and the last summation is over all pairs of electrons. As in the previous chapter, it is convenient to use atomic units, so that the Hamiltonian (8.1) becomes

$$H = \sum_{i=1}^{N} \left(-\frac{1}{2} \nabla^2_{\mathbf{r}_i} - \frac{Z}{r_i} \right) + \sum_{i<j=1}^{N} \frac{1}{r_{ij}} \tag{8.2}$$

and the Schrödinger equation for the N-electron atom wave function $\Psi(q_1, q_2, \dots, q_N)$ reads

$$\left[\sum_{i=1}^{N} \left(-\frac{1}{2} \nabla^2_{\mathbf{r}_i} - \frac{Z}{r_i} \right) + \sum_{i<j=1}^{N} \frac{1}{r_{ij}} \right] \Psi(q_1, q_2, \dots, q_N) = E\Psi(q_1, q_2, \dots, q_N) \tag{8.3}$$

where q_i denotes the ensemble of the (continuous) spatial coordinates $\mathbf{r}_i$ and (discrete) spin variable of electron i.

Since we are dealing with a system containing N indistinguishable particles, its Hamiltonian must be invariant under an interchange of the coordinates (spatial and spin) of any two particles. This is clearly the case for the Hamiltonian (8.2), which is independent of the spins of the electrons, and is symmetric in their spatial variables. Moreover, because electrons have spin 1/2 and hence are fermions, the *Pauli exclusion principle* requires that *the total wave function* $\Psi(q_1, q_2, \dots, q_N)$ *be completely antisymmetric*, that is it changes sign if the coordinates (spatial and spin) of any two-electrons are interchanged. This wave function can in general be constructed to be a sum of products of spatial wave functions with spin wave functions, as we shall see shortly.

Let us first ignore the spin of the electrons. The Schrödinger equation for the purely spatial wave function $\psi(\mathbf{r}_1, \mathbf{r}_2, \ldots, \mathbf{r}_N)$ then reads

$$\left[\sum_{i=1}^{N}\left(-\frac{1}{2}\nabla^2_{\mathbf{r}_i} - \frac{Z}{r_i}\right) + \sum_{i<j=1}^{N}\frac{1}{r_{ij}}\right]\psi(\mathbf{r}_1, \mathbf{r}_2, \ldots, \mathbf{r}_N) = E\psi(\mathbf{r}_1, \mathbf{r}_2, \ldots, \mathbf{r}_N) \qquad \textbf{(8.4)}$$

We see that this equation is a partial differential equation in $3N$ dimensions involving the coordinates $\mathbf{r}_1, \mathbf{r}_2, \ldots, \mathbf{r}_N$ of the electrons. Because of the presence of the terms $1/r_{ij}$, which express the mutual repulsion of the electrons, this equation is not separable. We have already encountered this situation in our study of helium-like atoms. However, in contrast to the case of two-electron atoms, where the term $1/r_{12}$ can be treated by perturbation theory, the term $\Sigma_{i<j} 1/r_{ij}$ which appears in (8.4) is in general too large to be treated as a perturbation. Indeed, even though for a fairly large value of Z any one of the terms $1/r_{ij}$ is small compared to Z/r_i, there are many $1/r_{ij}$ terms, and their total effect may become of the same size as that of the interaction between the electron i and the nucleus. Thus, if we want to apply perturbation theory in a meaningful way to the present problem, we must define a new 'unperturbed' Hamiltonian which is not just the sum of hydrogenic Hamiltonians $\Sigma_i(-(1/2)\nabla^2_{\mathbf{r}_i} - Z/r_i)$, but which includes, at least approximately, the mutual repulsion between the electrons. On the other hand this unperturbed Hamiltonian should be sufficiently simple so that the corresponding Schrödinger equation is tractable.

The answer to this problem, proposed by D.R. Hartree, V.A. Fock and J.C. Slater, is to use as our starting point the *central field approximation*. This approximation is based on an *independent-particle model*, in which each electron moves in an *effective potential* which represents the attraction of the nucleus and the *average* effect of the repulsive interactions between this electron and the $(N-1)$ other electrons. Moreover, since the overall effect of the $(N-1)$ other electrons is to *screen* the central Coulomb attraction between an electron and the nucleus, it is clear that the inter-electron repulsion term $\Sigma_{i<j} 1/r_{ij}$ contains a large spherically symmetric component, which we shall write as $\Sigma_i S(r_i)$. A good approximation to the effective potential energy of an electron is therefore provided by a spherically symmetric potential $V(r)$ such that

$$V(r) = -\frac{Z}{r} + S(r) \qquad \textbf{(8.5)}$$

We may readily obtain the form of $V(r)$ at large and small distances. Indeed, let us first consider an electron i whose distance r_i from the nucleus is large compared to the distances r_j associated with the $(N-1)$ other electrons. In this case we have $r_{ij} \simeq r_i$ and $1/r_{ij} \simeq 1/r_i$ so that the electron i moves in a potential given approximately by

$$-\frac{Z}{r_i} + \sum_{j=1}^{N-1} \frac{1}{r_i} = -\frac{Z - N + 1}{r_i} \tag{8.6}$$

which corresponds to the Coulomb field of the nucleus, screened by the $(N - 1)$ other electrons. As the distance r_i diminishes, this screening effect becomes less pronounced. In fact, when the electron i is near the nucleus, so that $r_{ij} \simeq r_j$, the potential felt by this electron is given approximately by

$$-\frac{Z}{r_i} + \left\langle \sum_{j=1}^{N-1} \frac{1}{r_j} \right\rangle = -\frac{Z}{r_i} + C \tag{8.7}$$

where the symbol $\langle\rangle$ denotes an average over the distances of the $(N - 1)$ other electrons, and C is a constant. Thus we see that in the limit $r_i \to 0$ the effective potential acting on the electron i is just the unscreened Coulomb potential $(-Z/r_i)$ due to the nucleus. We shall therefore require that the effective central potential $V(r)$ be such that

$$V(r) \to -\frac{Z}{r}, \qquad r \to 0 \tag{8.8a}$$

$$V(r) \to -\frac{Z - N + 1}{r}, \qquad r \to \infty \tag{8.8b}$$

In particular, for a neutral atom (such that $Z = N$), we have

$$V(r) \to -\frac{1}{r}, \qquad r \to \infty \tag{8.8c}$$

The determination of the effective potential at intermediate distances is a much more difficult problem, to which we shall return in Sections 8.3 and 8.4 where we discuss the Thomas–Fermi and Hartree–Fock methods, respectively. At this point, however, it is important to realise that for intermediate values of r the potential $V(r)$ – which represents the attraction of the nucleus plus the average repulsion of the other electrons – must depend on the details of the charge distribution of the electrons, or in other words on the dynamical state of the electrons. As a result, *the same effective potential $V(r)$ cannot account for the full spectrum of a complex atom (ion)*. However, if we restrict our attention to the ground state and the first excited states, it is reasonable to assume that a given central potential $V(r)$ can be used as a starting point. In fact, we shall see below that many important features concerning the structure of complex atoms or ions can be obtained by using only the information contained in equations (8.8), without a detailed knowledge of the central potential $V(r)$.

Let us now return to the Hamiltonian (8.2). From the foregoing discussion it is clear that a meaningful separation of H into an unperturbed part and a perturbation may be achieved by writing

$$H = H_c + H_1 \tag{8.9}$$

where

$$H_c = \sum_{i=1}^{N} \left(-\frac{1}{2}\nabla^2_{\mathbf{r}_i} + V(r_i) \right)$$

$$= \sum_{i=1}^{N} h_i, \qquad h_i = -\frac{1}{2}\nabla^2_{\mathbf{r}_i} + V(r_i) \tag{8.10}$$

is the Hamiltonian corresponding to the central field approximation, and

$$H_1 = \sum_{i<j=1}^{N} \frac{1}{r_{ij}} - \sum_i \left(\frac{Z}{r_i} + V(r_i) \right)$$

$$= \sum_{i<j=1}^{N} \frac{1}{r_{ij}} - \sum_i S(r_i) \tag{8.11}$$

is the remaining part of the full Hamiltonian (8.2) containing the residual spherical and all the non-spherical part of the inter-electron repulsion. All we have done, of course, is to add and subtract the expression $\Sigma_i V(r_i)$ in (8.2), but the perturbation H_1 defined by (8.11) is now much smaller than the term $\Sigma_{i<j} 1/r_{ij}$ representing the full mutual repulsion between the electrons.

We shall therefore begin by neglecting the perturbation H_1 and concentrate our attention on the central field Hamiltonian H_c which, as seen from (8.5) and (8.10), contains the kinetic energy, the potential energy in the field of the nucleus, and the average (spherical) inter-electron repulsion energy. The corresponding Schrödinger equation for the spatial part of the N-electron central field wave function $\psi_c(\mathbf{r}_1, \mathbf{r}_2, \dots, \mathbf{r}_N)$ then reads

$$H_c\psi_c = \sum_{i=1}^{N} \left[-\frac{1}{2}\nabla^2_{\mathbf{r}_i} + V(r_i) \right] \psi_c = E_c\psi_c \tag{8.12}$$

and is separable into N equations, one for each electron. A solution of (8.12) may therefore be written as

$$\psi_c = u_{a_1}(\mathbf{r}_1)u_{a_2}(\mathbf{r}_2) \dots, u_{a_N}(\mathbf{r}_N) \tag{8.13}$$

where the individual electron orbitals $u_{a_1}(\mathbf{r}_1)$, $u_{a_2}(\mathbf{r}_2)$, . . . , normalised to unity, are solutions of an equation having the form

$$[-\tfrac{1}{2}\nabla^2_{\mathbf{r}} + V(r)]u_{nlm_l}(\mathbf{r}) = E_{nl}u_{nlm_l}(\mathbf{r}) \tag{8.14}$$

and the symbol a_i in (8.13) refers to the three quantum numbers $(n_i l_i m_{l_i})$ of electron i.

With the substitution $m_l \to m$, the equation (8.14) is identical to (7.48), which we already discussed in our study of two-electron atoms. Since the potential $V(r)$ is central, the one-electron or *central field orbitals* $u_{nlm_l}(\mathbf{r})$ are products of radial functions times spherical harmonics,

$$u_{nlm_l}(\mathbf{r}) = R_{nl}(r)Y_{lm_l}(\theta, \phi) \tag{8.15}$$

where the radial functions satisfy the equation

$$-\frac{1}{2}\left(\frac{\mathrm{d}^2}{\mathrm{d}r^2} + \frac{2}{r}\frac{\mathrm{d}}{\mathrm{d}r} - \frac{l(l+1)}{r^2}\right)R_{nl}(r) + V(r)R_{nl}(r) = E_{nl}R_{nl}(r) \tag{8.16}$$

We recall that the principal quantum number n is defined to be $n = n_r + l + 1$, where n_r is the number of nodes of the radial function. The quantum numbers n, l and m_l can therefore take the values

$$\begin{aligned} n &= 1, 2, \ldots \\ l &= 0, 1, 2, \ldots, n-1 \\ m_l &= -l, -l+1, \ldots, +l \end{aligned} \tag{8.17}$$

As we already noted in connection with (7.48), the central field orbitals $u_{nlm}(\mathbf{r})$ should not be confused with the hydrogenic wave functions $\psi_{nlm}(\mathbf{r})$ of Chapter 3, since the radial functions $R_{nl}(r)$, solutions of (8.16), differ from the hydrogenic radial functions (3.58), which correspond to the particular choice $V(r) = -Z/r$.

Because the potential $V(r)$ in (8.14) is spherically symmetric, the energy eigenvalues E_{nl} do not depend on the quantum number m_l. However, in contrast to the hydrogenic case, the individual electron energies depend on both n and l. The total energy E_c in the central field approximation is of course the sum of the individual electron energies, namely

$$E_c = \sum_{i=1}^{N} E_{n_i l_i} \tag{8.18}$$

As in the case of two-electron atoms discussed in Chapter 7, we note that there is *exchange degeneracy*, since any spatial wave function obtained from (8.13) by a permutation of the electron coordinates is an equally good solution of (8.12) corresponding to the same energy (8.18).

Spin and the Pauli exclusion principle

Let us now take into account the fact that electrons are fermions of spin 1/2, so that the wave function $\Psi(q_1, q_2, \ldots, q_N)$ of equation (8.3) must be antisymmetric in the spatial and spin coordinates q_i ($i = 1, 2, \ldots, N$) of the electrons. In Section 7.2, we obtained the required antisymmetric wave functions $\Psi(q_1, q_2)$ for two-electron atoms (ions) by forming *products* of spatial wave functions $\psi_\pm(\mathbf{r}_1, \mathbf{r}_2)$ with spin functions χ_{S,M_S} having the appropriate symmetry (see (7.20) and (7.21)). For atoms (ions) with three or more electrons ($N \geqslant 3$), the wave function $\Psi(q_1, q_2, \ldots, q_N)$ is in general a *sum of products* of spatial wave functions and spin functions, as we shall now illustrate for the case $N = 3$.

Table 8.1 Spin functions $\chi_{S,M_S}(1, 2, 3)$ for three electrons.

Spin function	*S*	M_S
$\frac{1}{\sqrt{2}}[\alpha(1)\alpha(2)\beta(3) - \alpha(1)\beta(2)\alpha(3)]$	$\frac{1}{2}$	$\frac{1}{2}$
$\frac{1}{\sqrt{2}}[\beta(1)\beta(2)\alpha(3) - \beta(1)\alpha(2)\beta(3)]$	$\frac{1}{2}$	$-\frac{1}{2}$
$\frac{1}{\sqrt{6}}[\alpha(1)\alpha(2)\beta(3) + \alpha(1)\beta(2)\alpha(3) - 2\beta(1)\alpha(2)\alpha(3)]$	$\frac{1}{2}$	$\frac{1}{2}$
$\frac{1}{\sqrt{6}}[\beta(1)\beta(2)\alpha(3) + \beta(1)\alpha(2)\beta(3) - 2\alpha(1)\beta(2)\beta(3)]$	$\frac{1}{2}$	$-\frac{1}{2}$
$\alpha(1)\alpha(2)\alpha(3)$	$\frac{3}{2}$	$\frac{3}{2}$
$\frac{1}{\sqrt{3}}[\alpha(1)\alpha(2)\beta(3) + \alpha(1)\beta(2)\alpha(3) + \beta(1)\alpha(2)\alpha(3)]$	$\frac{3}{2}$	$\frac{1}{2}$
$\frac{1}{\sqrt{3}}[\beta(1)\beta(2)\alpha(3) + \beta(1)\alpha(2)\beta(3) + \alpha(1)\beta(2)\beta(3)]$	$\frac{3}{2}$	$-\frac{1}{2}$
$\beta(1)\beta(2)\beta(3)$	$\frac{3}{2}$	$-\frac{3}{2}$

In order to obtain the spin functions $\chi_{S,M_S}(1, 2, 3)$ for three electrons, we can first form the spin functions of two electrons, namely $\chi_{0,0}(1, 2)$, $\chi_{1,1}(1, 2)$, $\chi_{1,0}(1, 2)$ and $\chi_{1,-1}(1, 2)$ (see (7.18) and (7.19)). The three-electron system can then be regarded as a (1, 2) + 3 electron system, so that the three-electron spin functions $\chi_{S,M_S}(1, 2, 3)$ can be obtained by combining the spin functions $\chi_{1/2,\pm1/2}(3)$ (that is, $\alpha(3)$ or $\beta(3)$) with the two-electron spin functions, using the appropriate Clebsch–Gordan coefficients (see Appendix 4). The total spin quantum number can be either $S = 1/2$ or $S = 3/2$. In the case $S = 1/2$, the two possible values of M_S are +1/2 and –1/2, so that we have *doublets*, and there are two possibilities (see Table 8.1): there is a doublet spin function which is antisymmetric in the spin coordinates of 2 and 3, and a second doublet spin function which is symmetric in 2 and 3. For the case $S = 3/2$, there is a *quartet* of spin functions corresponding to the four possible values $M_S = 3/2, 1/2, -1/2, -3/2$, which are completely symmetric in the spin variables of the three electrons, as displayed in Table 8.1. It should be noted that the two doublet spin functions $\chi_{1/2,M_S}(1, 2, 3)$, $M_S = \pm1/2$, have no symmetry with respect to the interchange of electron 1 with electrons 2 and 3.

As an example, let us consider the ground state of the lithium atom, which has the electronic configuration $(1s)^2 2s$. This state is a spin doublet, the total spin quantum number being $S = 1/2$, and the two possible values of M_S being $M_S = \pm1/2$. Focusing our attention on the state with $S = 1/2$ and $M_S = 1/2$, we see from Table 8.1 that there are two distinct spin functions for which $S = 1/2$ and $M_S = 1/2$. The general form of the lithium ground state wave function corresponding to $S = 1/2$ and $M_S = 1/2$ is therefore

$$\Psi(q_1, q_2, q_3) = \psi_a(\mathbf{r}_1, \mathbf{r}_2, \mathbf{r}_3)\frac{1}{\sqrt{2}}[\alpha(1)\alpha(2)\beta(3) - \alpha(1)\beta(2)\alpha(3)]$$
$$+ \psi_b(\mathbf{r}_1, \mathbf{r}_2, \mathbf{r}_3)\frac{1}{\sqrt{6}}[\alpha(1)\alpha(2)\beta(3) + \alpha(1)\beta(2)\alpha(3) - 2\beta(1)\alpha(2)\alpha(3)] \quad \textbf{(8.19)}$$

where ψ_a and ψ_b must be such that Ψ is completely antisymmetric under the interchange of the coordinates (spatial and spin) of any two electrons. The lithium ground state wave function for $S = 1/2$ and $M_S = -1/2$ can be written down in a similar way. It should be noted that in the absence of external fields the energy of the atom does not depend on M_S.

Spin-orbitals and Slater determinants

Let us now assume that we are working within the framework of the central field approximation. The spin of the electron is then readily taken into account by multiplying each one-electron spatial orbital $u_{nlm_l}(\mathbf{r})$ by a spin-1/2 eigenfunction $\chi_{1/2,m_s}$, thus forming the (normalised) *spin-orbitals*

$$u_{nlm_lm_s}(q) = u_{nlm_l}(\mathbf{r})\chi_{1/2,m_s}$$
$$= R_{nl}(r)Y_{lm_l}(\theta, \phi)\chi_{1/2,m_s} \quad \textbf{(8.20)}$$

characterised by the *four* quantum numbers n, l, m_l and m_s. The three 'spatial' quantum numbers n, l and m_l can take the values given in (8.17), while the spin quantum number m_s is such that $m_s = \pm 1/2$. It is clear that the spin-orbitals (8.20) satisfy the equations

$$[-\tfrac{1}{2}\nabla_{\mathbf{r}}^2 + V(r)]u_{nlm_lm_s} = E_{nl}u_{nlm_lm_s} \quad \textbf{(8.21)}$$

Since the energy E_{nl} does not depend on the quantum numbers m_l and m_s, we see that each individual electron energy level is $2(2l + 1)$ times degenerate.

Slater determinants

Our next task is to build-up out of single-electron spin-orbitals a total N-electron (central field) wave function $\Psi_c(q_1, q_2, \ldots, q_N)$ which is antisymmetric in the (spatial and spin) coordinates of any two electrons, in order to satisfy the requirements of the *Pauli exclusion principle*. This may be accomplished in a simple way as follows. Let us designate the four quantum numbers (n, l, m_l, m_s) corresponding to given independent particle states in the atom by the single letters $\alpha, \beta, \ldots, \nu$. The total wave function Ψ_c describing an atom in which one electron is in state α, another in state β, and so on may then be written as an $N \times N$ determinant,

$$\Psi_c(q_1, q_2, \ldots, q_N) = \frac{1}{\sqrt{N!}}\begin{vmatrix} u_\alpha(q_1) & u_\beta(q_1)\ldots & u_\nu(q_1) \\ u_\alpha(q_2) & u_\beta(q_2)\ldots & u_\nu(q_2) \\ \vdots & & \\ u_\alpha(q_N) & u_\beta(q_N)\ldots & u_\nu(q_N) \end{vmatrix} \quad \textbf{(8.22)}$$

which is known as a *Slater determinant.* This wave function is obviously antisymmetric because if we interchange the (spatial and spin) coordinates of two electrons (say q_1 and q_2) this is equivalent to interchanging two rows, so that the determinant changes sign. The eigenvalue E_c of the central field Hamiltonian H_c corresponding to a given Slater determinant is just the sum (8.18) of the energies of the N individual states which are present in the determinant.

We note that since a determinant vanishes when two columns (or rows) are equal, the Slater determinant (8.22) will vanish if two electrons have the same values of the four quantum numbers n, l, m_l and m_s. Within the framework of the independent-particle model, we may therefore state the exclusion principle in the form originally discovered by W. Pauli in 1925, namely that *no two electrons in an atom can have the same set of four quantum numbers.*

The $(N!)^{-1/2}$ factor appearing in (8.22) is a normalisation factor, arising from the fact that there are $N!$ permutations of the electron coordinates $q_1, q_2, \ldots, q_N$. If we denote by P a permutation of the electron coordinates, we may rewrite the Slater determinant (8.22) as

$$\Psi_c(q_1, q_2, \ldots, q_N) = \frac{1}{\sqrt{N!}} \sum_P (-1)^P P u_\alpha(q_1) u_\beta(q_2) \ldots u_\nu(q_N) \tag{8.23}$$

where the symbol $(-1)^P$ is equal to +1 when P is an even permutation and to −1 when P is an odd permutation [2], and the sum is over all permutations P.

Let us now investigate the behaviour of the Slater determinant (8.22) under the reflection (or parity) transformation $\mathbf{r}_i \to -\mathbf{r}_i$. Since a spin orbital (8.20) has parity $(-1)^l$, the Slater determinant has the well-defined parity

$$(-1)^{l_1}(-1)^{l_2} \ldots (-1)^{l_N} = (-1)^{\Sigma_i l_i} \tag{8.24}$$

and will therefore be even or odd under the reflection transformation, depending on whether the sum of the orbital angular momentum quantum numbers of the electrons, $\Sigma_i l_i$, is even or odd.

As a simple example, let us consider the ground state $(1s)^2\ ^1S$ of helium. The spin-orbitals for the two electrons are then given by

$$u_{100,1/2} = u_{100}(r)\chi_{1/2,1/2} = u_{100}(r)\alpha \qquad \text{for} \quad m_s = +1/2 \tag{8.25a}$$

and

$$u_{100,-1/2} = u_{100}(r)\chi_{1/2,-1/2} = u_{100}(r)\beta \qquad \text{for} \quad m_s = -1/2 \tag{8.25b}$$

where α and β are the spin functions defined by (2.209), and the orbital $u_{100}(r)$ is that introduced in Section 7.4. According to (8.22), the two-electron wave function $\Psi_c(q_1, q_2)$ describing the helium ground state in the central field approximation is

[2] We recall that a permutation P is said to be even or odd depending on whether the number of interchanges leading to it is even or odd.

$$\Psi_c(q_1, q_2) = \frac{1}{\sqrt{2}} \begin{vmatrix} u_{100}(r_1)\alpha(1) & u_{100}(r_1)\beta(1) \\ u_{100}(r_2)\alpha(2) & u_{100}(r_2)\beta(2) \end{vmatrix}$$

$$= u_{100}(r_1)u_{100}(r_2)\frac{1}{\sqrt{2}}[\alpha(1)\beta(2) - \alpha(2)\beta(1)] \qquad \textbf{(8.26)}$$

in accordance with the results obtained in Chapter 7.

It is a simple matter to verify (Problem 8.1) that the central field Hamiltonian H_c commutes with both the total orbital angular momentum operator **L** and the total spin operator **S** of the electrons. That is,

$$[H_c, \mathbf{L}] = 0 \qquad \textbf{(8.27a)}$$

$$[H_c, \mathbf{S}] = 0 \qquad \textbf{(8.27b)}$$

where

$$\mathbf{L} = \sum_{i=1}^{N} \mathbf{L}_i \qquad \textbf{(8.28a)}$$

$$\mathbf{S} = \sum_{i=1}^{N} \mathbf{S}_i \qquad \textbf{(8.28b)}$$

Here $\mathbf{L}_i$ and $\mathbf{S}_i$ denote respectively the orbital angular momentum and the spin operator of the ith electron. As a consequence, it is possible to obtain eigenfunctions of H_c which are also eigenfunctions of the operators $\mathbf{L}^2$, $\mathbf{S}^2$, L_z and S_z, with eigenvalues given respectively by $L(L + 1)\hbar^2$, $S(S + 1)\hbar^2$, $M_L\hbar$ and $M_S\hbar$. Such eigenfunctions will be denoted by $|\gamma LSM_LM_S\rangle$, where γ is an index representing additional information (about the radial part of the wave function, the parity, etc.). It is precisely this 'coupled representation' L, S, M_L, M_S which we used in Chapter 7 to discuss the two-electron problem. Now, in obtaining the Slater determinant (8.22) we used spin-orbitals $u_{nlm_lm_s}$ expressed in the n, l, m_l, m_s representation. As a result, a Slater determinant (8.22) is an eigenfunction of the operators L_z and S_z, but not necessarily of $\mathbf{L}^2$ and $\mathbf{S}^2$, and we must in general construct *linear combinations of Slater determinants* to obtain an eigenfunction of $\mathbf{L}^2$, $\mathbf{S}^2$, L_z and S_z (see Problem 8.2). The helium ground state considered above is a simple case (with $L = S = 0$) for which a single Slater determinant is an eigenfunction of all four operators $\mathbf{L}^2$, $\mathbf{S}^2$, L_z and S_z.

Electron states in a central field. Configurations, shells and subshells

We have shown that within the framework of the central field approximation the energy levels E_c of an atom (ion) having N electrons are given by summing the individual electron energies E_{nl}, while the N-electron wave functions $\Psi_c(q_1, q_2, \ldots, q_N)$ are obtained by forming Slater determinants (or linear combinations of them) with the individual spin-orbitals $u_{nlm_lm_s}$. The problem of finding the eigenvalues and eigenfunctions of the central field Hamiltonian H_c is therefore reduced to, the determination of the individual energy levels E_{nl} and the spin-orbitals (8.20). Since the spherical harmonics $Y_{l,m_l}(\theta, \phi)$ and the spin eigenfunctions $\chi_{1/2,m_s}$ are known, this problem in turn comes down to solving the radial equation (8.16) for an attractive potential $V(r)$ satisfying the conditions (8.8).

The order of the energy levels E_{nl} does not depend in a crucial way on the detailed form of the potential $V(r)$. If $V(r)$ were simply the Coulomb field $-Z/r$ of the nucleus, all the levels $l = 0, 1, \ldots, n-1$ corresponding to a given value of n would coincide. The screening due to the other electrons results in a raising of the energy levels, this effect being more pronounced as n and l increase, since in this case the orbitals are concentrated at larger values of r. Thus, for a fixed value of l the energy E_{nl} is an increasing function of n and for a given value of n it is an increasing function of l (the orbitals with larger l values being 'forced' out by the centrifugal barrier). If we restrict our attention to the ground and lowest excited states, the order of succession of the individual energy levels E_{nl} (which can be inferred from spectroscopic evidence) is nearly the same for all atoms, and is given in Table 8.2. It is worth noting that this sequence is different from that in hydrogen, where the energy levels depend only on n, and $E_{n+1} > E_n$. For example,

Table 8.2 The ordering of the individual energy levels E_{nl}. The energy increases from bottom to top, the brackets enclosing levels which have so nearly the same energy that their order can vary from one atom to another. Also given is the spectroscopic notation for the subshell (nl) and the maximum number $2(2l+1)$ of electrons allowed in the subshell.

Quantum numbers n, l	*Spectroscopic notation for subshell* (nl)	*Maximum number of electrons allowed in the subshell* $= 2(2l+1)$
⎡ 6,2	⎡ 6d	10
⎢ 5,3	⎢ 5f	14
⎣ 7,0	⎣ 7s	2
6,1	6p	6
⎡ 5,2	⎡ 5d	10
⎢ 4,3	⎢ 4f	14
⎣ 6,0	⎣ 6s	2
5,1	5p	6
⎡ 4,2	⎡ 4d	10
⎣ 5,0	⎣ 5s	2
4,1	4p	6
⎡ 3,2	⎡ 3d	10
⎣ 4,0	⎣ 4s	2
3,1	3p	6
3,0	3s	2
2,1	2p	6
2,0	2s	2
1,0	1s	2

the 4s state, which has a higher energy than the 3d state in hydrogen, is depressed because of its low angular momentum, which causes this orbital to be large at small r, where it can feel the full nuclear attraction. We also see that as a rule the individual electron energy E_{nl} is an increasing function of the sum $n + l$.

In the central field approximation, the total energy (8.18) of the atom depends only on the number of electrons occupying each of the individual energy levels E_{nl}. Therefore, as we already noted in Section 7.4 for the particular case of two-electron atoms, this total energy is entirely determined by the *electron configuration*, that is by the distribution of the electrons with respect to the quantum numbers n and l. Thus the assignment of an electron configuration requires the enumeration of the values of n and l for all the electrons of the atom. We recall that in the usual spectroscopic notation the value of n is given as a number, that of l as a letter (s for $l = 0$, p for $l = 1$, and so on) and the number of electrons having given values of n and l as a numerical superscript (for example, $(2s)^2$ or $2s^2$, $(3p)^4$ or $3p^4$, etc.).

Electrons having the same value of n and l are said to belong to the same *subshell* [3]. According to our foregoing discussion, there are $2(2l + 1)$ states having the same value of n and l but different values of m_l and m_s. Such states are called equivalent. In spectroscopic jargon the electrons having the same value of n and l (that is, belonging to the same subshell) are known as *equivalent electrons*. Because of the Pauli exclusion principle there cannot be more than one electron in each individual state labelled by the quantum numbers (nlm_lm_s). Thus the maximum number of electrons in a subshell is $2(2l + 1)$, so that

For $l = 0$ (s electrons) this maximum number is 2
$l = 1$ (p electrons) this maximum number is 6
$l = 2$ (d electrons) this maximum number is 10
$l = 3$ (f electrons) this maximum number is 14
$l = 4$ (g electrons) this maximum number is 18
$l = 5$ (h electrons) this maximum number is 22

In addition to the ordering of the energy levels E_{nl}, Table 8.2 also gives the spectroscopic notation of the corresponding subshells, and the maximum number of electrons allowed in each subshell. An assembly of $2(2l + 1)$ equivalent electrons is called a *closed* (or filled) *subshell*.

Electrons which have the same value of the principal quantum number n are said to belong to the same *shell*. Following a notation introduced in Chapter 1 for the hydrogenic atoms, and commonly used in discussing X-ray spectra of complex atoms, the value of the principal quantum number n is sometimes specified by a capital letter according to the correspondence

[3] Some authors use the word 'shell' (instead of 'subshell') to characterise electrons having the same values of n and l. We shall reserve the word 'shell' for electrons having only the same value of n, the principal quantum number.

Value of n	1	2	3	4	5	6
	$\updownarrow$	$\updownarrow$	$\updownarrow$	$\updownarrow$	$\updownarrow$	$\updownarrow$
Code letter	K	L	M	N	O	P

The maximum number of electrons in a shell is $2n^2$, in which case we have a *closed* (or filled) *shell.*

Degeneracies

Since the assignment of an electron configuration requires only the enumeration of the values of n and l for all electrons, but not those of m_l and m_s, a given degeneracy g will be attached to a configuration. Let ν_i be the number of electrons occupying a given individual level $E_{n_i l_i}$ and $\delta_i = 2(2l_i + 1)$ be the degeneracy of that level. There are

$$d_i = \frac{\delta_i!}{\nu_i!(\delta_i - \nu_i)!} \tag{8.29}$$

ways of distributing the ν_i electrons among the δ_i individual states corresponding to the level $E_{n_i l_i}$, and we note that $d_i = 1$ for a closed subshell (such that $\nu_i = \delta_i$). The *total* degeneracy or statistical weight g of the configuration is then obtained by forming the product of d_i with the degeneracies corresponding to the electrons of the other subshells.

As an example, let us consider the ground state of the carbon atom, which has the configuration $1s^2 2s^2 p^2$. In this case:

for the two 1s electrons: $\nu = 2, \quad \delta = 2, \quad d = 1$

for the two 2s electrons: $\nu = 2, \quad \delta = 2, \quad d = 1$

for the two 2p electrons: $\nu = 2, \quad \delta = 6, \quad d = 15$

and the total degeneracy of the ground state configuration of carbon is $g = 15$.

8.2 The periodic system of the elements

We are now equipped with all the necessary information to discuss the electronic structure and the 'building up' (*aufbau*) of atoms. For the sake of simplicity we shall only consider the *ground state* of *neutral atoms*. The Z electrons of an atom of atomic number (nuclear charge) Z then occupy the lowest individual energy levels in accordance with the requirements of the Pauli exclusion principle discussed above; the ordering of the individual levels is that displayed in Table 8.2. The *ground state configuration* of an atom is therefore obtained by distributing the Z electrons among a certain number f of subshells, the first $(f - 1)$ subshells being filled and the last one – corresponding to the highest energy – generally not, except for particular values of Z (2, 4, 10, 12, 18, etc.). The least tightly bound electrons, which are in the subshell of highest energy, and in insufficient numbers to form another closed subshell, are called *valence* electrons. In going from one atom having atomic number Z to the next one, with atomic number $Z + 1$, the number of electrons increases by one, the $(Z + 1)$ electrons occupying the lowest energy levels allowed by the exclusion principle. In this way the subshells are progressively filled. This is illustrated in Table 8.3 which gives the electron

Table 8.3 Electron configuration, term value and ionisation potential of the atoms in their ground state.

Z	*Element*		*Electronic configuration*†	*Term*†	*Ionisation potential* (eV)
1	H	hydrogen	1s	$^2S_{1/2}$	13.60
2	He	helium	$1s^2$	$^1S_0$	24.59
3	Li	lithium	[He]2s	$^2S_{1/2}$	5.39
4	Be	beryllium	[He]$2s^2$	$^1S_0$	9.32
5	B	boron	[He]$2s^22p$	$^2P_{1/2}$	8.30
6	C	carbon	[He]$2s^22p^2$	$^3P_0$	11.26
7	N	nitrogen	[He]$2s^22p^3$	$^4S_{3/2}$	14.53
8	O	oxygen	[He]$2s^22p^4$	$^3P_2$	13.62
9	F	fluorine	[He]$2s^22p^5$	$^2P_{3/2}$	17.42
10	Ne	neon	[He]$2s^22p^6$	$^1S_0$	21.56
11	Na	sodium	[Ne]3s	$^2S_{1/2}$	5.14
12	Mg	magnesium	[Ne]$3s^2$	$^1S_0$	7.65
13	Al	aluminium	[Ne]$3s^23p$	$^2P_{1/2}$	5.99
14	Si	silicon	[Ne]$3s^23p^2$	$^3P_0$	8.15
15	P	phosphorus	[Ne]$3s^23p^3$	$^4S_{3/2}$	10.49
16	S	sulphur	[Ne]$3s^23p^4$	$^3P_2$	10.36
17	Cl	chlorine	[Ne]$3s^23p^5$	$^2P_{3/2}$	12.97
18	Ar	argon	[Ne]$3s^23p^6$	$^1S_0$	15.76
19	K	potassium	[Ar]4s	$^2S_{1/2}$	4.34
20	Ca	calcium	[Ar]$4s^2$	$^1S_0$	6.11
21	Sc	scandium	[Ar]$4s^23d$	$^2D_{3/2}$	6.54
22	Ti	titanium	[Ar]$4s^23d^2$	$^3F_2$	6.82
23	V	vanadium	[Ar]$4s^23d^3$	$^4F_{3/2}$	6.74
24	Cr	chromium	[Ar]$4s3d^5$	$^7S_3$	6.77
25	Mn	manganese	[Ar]$4s^23d^5$	$^6S_{5/2}$	7.44
26	Fe	iron	[Ar]$4s^23d^6$	$^5D_4$	7.87
27	Co	cobalt	[Ar]$4s^23d^7$	$^4F_{9/2}$	7.86
28	Ni	nickel	[Ar]$4s^23d^8$	$^3F_4$	7.64
29	Cu	copper	[Ar]$4s3d^{10}$	$^2S_{1/2}$	7.73
30	Zn	zinc	[Ar]$4s^23d^{10}$	$^1S_0$	9.39
31	Ga	gallium	[Ar]$4s^23d^{10}4p$	$^2P_{1/2}$	6.00
32	Ge	germanium	[Ar]$4s^23d^{10}4p^2$	$^3P_0$	7.90
33	As	arsenic	[Ar]$4s^23d^{10}4p^3$	$^4S_{3/2}$	9.81

Table 8.3 (*Cont.*)

Z	*Element*		*Electronic configuration*†	*Term*†	*Ionisation potential* (eV)
34	Se	selenium	$[Ar]4s^2 3d^{10} 4p^4$	$^3P_2$	9.75
35	Br	bromine	$[Ar]4s^2 3d^{10} 4p^5$	$^2P_{3/2}$	11.81
36	Kr	krypton	$[Ar]4s^2 3d^{10} 4p^6$	$^1S_0$	14.00
37	Rb	rubidium	$[Kr]5s$	$^2S_{1/2}$	4.18
38	Sr	strontium	$[Kr]5s^2$	$^1S_0$	5.70
39	Y	yttrium	$[Kr]5s^2 4d$	$^2D_{3/2}$	6.38
40	Zr	zirconium	$[Kr]5s^2 4d^2$	$^3F_2$	6.84
41	Nb	niobium	$[Kr]5s4d^4$	$^6D_{1/2}$	6.88
42	Mo	molybdenum	$[Kr]5s4d^5$	$^7S_3$	7.10
43	Tc	technetium	$[Kr]5s^2 4d^5$	$^6S_{5/2}$	7.28
44	Ru	ruthenium	$[Kr]5s4d^7$	$^5F_5$	7.37
45	Rh	rhodium	$[Kr]5s4d^8$	$^4F_{9/2}$	7.46
46	Pd	palladium	$[Kr]4d^{10}$	$^1S_0$	8.34
47	Ag	silver	$[Kr]5s4d^{10}$	$^2S_{1/2}$	7.58
48	Cd	cadmium	$[Kr]5s^2 4d^{10}$	$^1S_0$	8.99
49	In	indium	$[Kr]5s^2 4d^{10} 5p$	$^2P_{1/2}$	5.79
50	Sn	tin	$[Kr]5s^2 4d^{10} 5p^2$	$^3P_0$	7.34
51	Sb	antimony	$[Kr]5s^2 4d^{10} 5p^3$	$^4S_{3/2}$	8.64
52	Te	tellurium	$[Kr]5s^2 4d^{10} 5p^4$	$^3P_2$	9.01
53	I	iodine	$[Kr]5s^2 4d^{10} 5p^5$	$^2P_{3/2}$	10.45
54	Xe	xenon	$[Kr]5s^2 4d^{10} 5p^6$	$^1S_0$	12.13
55	Cs	caesium	$[Xe]6s$	$^2S_{1/2}$	3.89
56	Ba	barium	$[Xe]6s^2$	$^1S_0$	5.21
57	La	lanthanum	$[Xe]6s^2 5d$	$^2D_{3/2}$	5.58
58	Ce	cerium	$[Xe](6s^2 4f5d)$	$(^1G_4)$	5.47
59	Pr	praseodymium	$[Xe](6s^2 4f^3)$	$(^4I_{9/2})$	5.42
60	Nd	neodymium	$[Xe]6s^2 4f^4$	$^5I_4$	5.49
61	Pm	promethium	$[Xe](6s^2 4f^5)$	$(^6H_{5/2})$	5.55
62	Sm	samarium	$[Xe]6s^2 4f^6$	$^7F_0$	5.63
63	Eu	europium	$[Xe]6s^2 4f^7$	$^8S_{7/2}$	5.67
64	Gd	gadolinium	$[Xe]6s^2 4f^7 5d$	$^9D_2$	6.14
65	Tb	terbium	$[Xe](6s^2 4f^9)$	$^6H_{15/2}$	5.85
66	Dy	dysprosium	$[Xe](6s^2 4f^{10})$	$(^5I_8)$	5.93
67	Ho	holmium	$[Xe](6s^2 4f^{11})$	$(^4I_{15/2})$	6.02
68	Er	erbium	$[Xe](6s^2 4f^{12})$	$(^3H_6)$	6.10
69	Tm	thulium	$[Xe]6s^2 4f^{13}$	$^2F_{7/2}$	6.18
70	Yb	ytterbium	$[Xe]6s^2 4f^{14}$	$^1S_0$	6.25
71	Lu	lutetium	$[Xe]6s^2 4f^{14} 5d$	$^2D_{3/2}$	5.43
72	Hf	hafnium	$[Xe]6s^2 4f^{14} 5d^2$	$^3F_2$	7.0
73	Ta	tantalum	$[Xe]6s^2 4f^{14} 5d^3$	$^4F_{3/2}$	7.89
74	W	tungsten	$[Xe]6s^2 4f^{14} 5d^4$	$^5D_0$	7.98
75	Re	rhenium	$[Xe]6s^2 4f^{14} 5d^5$	$^6S_{5/2}$	7.88
76	Os	osmium	$[Xe]6s^2 4f^{14} 5d^6$	$^5D_4$	8.7
77	Ir	iridium	$[Xe]6s^2 4f^{14} 5d^7$	$(^4F_{9/2})$	9.1
78	Pt	platinum	$[Xe]6s4f^{14} 5d^9$	$^3D_3$	9.0

Table 8.3 (*Cont.*)

Z	*Element*		*Electronic configuration*†	*Term*†	*Ionisation potential* (eV)
79	Au	gold	[Xe]$6s4f^{14}5d^{10}$	$^2S_{1/2}$	9.23
80	Hg	mercury	[Xe]$6s^24f^{14}5d^{10}$	$^1S_0$	10.44
81	Tl	thallium	[Xe]$6s^24f^{14}5d^{10}6p$	$^2P_{1/2}$	6.11
82	Pb	lead	[Xe]$6s^24f^{14}5d^{10}6p^2$	$^3P_0$	7.42
83	Bi	bismuth	[Xe]$6s^24f^{14}5d^{10}6p^3$	$^4S_{3/2}$	7.29
84	Po	polonium	[Xe]$6s^24f^{14}5d^{10}6p^4$	$^3P_2$	8.42
85	At	astatine	[Xe]$(6s^24f^{14}5d^{10}6p^5)$	$^2P_{3/2}$	9.5
86	Rn	radon	[Xe]$6s^24f^{14}5d^{10}6p^6$	$^1S_0$	10.75
87	Fr	francium	[Rn]$7s$	$^2S_{1/2}$	4.0
88	Ra	radium	[Rn]$7s^2$	$^1S_0$	5.28
89	Ac	actinium	[Rn]$7s^26d$	$^2D_{3/2}$	6.9
90	Th	thorium	[Rn]$7s^26d^2$	$^3F_2$	
91	Pa	protactinium	[Rn]$(7s^25f^26d)$	$(^4K_{11/2})$	
92	U	uranium	[Rn]$7s^25f^36d$	$^5L_6$	4.0
93	Np	neptunium	[Rn]$7s^25f^46d$	$^6L_{11/2}$	
94	Pu	plutonium	[Rn]$7s^25f^6$	$^7F_0$	5.8
95	Am	americium	[Rn]$7s^25f^7$	$^8S_{7/2}$	6.0
96	Cm	curium	[Rn]$7s^25f^76d$	$^9D_2$	
97	Bk	berkelium	[Rn]$7s^25f^86d$	$^8H_{17/2}$	
98	Cf	californium	[Rn]$7s^25f^{10}$	$^5I_8$	
99	Es	einsteinium	[Rn]$7s^25f^{11}$	$^4I_{15/2}$	
100	Fm	fermium	[Rn]$(7s^25f^{12})$	$(^3H_6)$	
101	Md	mendelevium	[Rn]$(7s^25f^{13})$	$(^2F_{7/2})$	
102	No	nobelium	[Rn]$(7s^25f^{14})$	$(^1S_0)$	
103	Lw	lawrencium	[Rn]$(7s^25f^{14}6d)$	$(^2D_{3/2})$	

† Configurations and terms in parentheses are estimated.

configuration of the ground state of neutral atoms. Also given in Table 8.3 are the ionisation potential and the ground term, the latter being written according to the Russell–Saunders notation

$$^{2S+1}L_J$$

where J is the total angular momentum quantum number and the code letters S, P, D, . . . correspond to the values $L = 0, 1, 2, \ldots$ as described in Chapter 7. We shall return in detail in Section 8.5 to the assignment of the term values.

The table begins with hydrogen, which has the ground state configuration 1s. The ionisation potential, as we have learned in Chapter 3, is 13.6 eV and the ground term value is clearly $^2S_{1/2}$, since we have $L\ (= l) = 0$, $S\ (= s) = 1/2$ and $J\ (= j) = 1/2$.

The next element, helium, has the ground state configuration $1s^2$ and was studied in detail in Chapter 7. Since $L = 0$ and $S = 0$ in this case, the ground term value is $^1S_0$. We see by looking at Table 8.3 that helium has the largest ionisation potential (24.59 eV). We also note that the two electrons of helium fill the K shell ($n = 1$).

The third element, lithium, has the ground state configuration $1s^22s$ (abbreviated as [He]2s in the table) because the configuration $1s^3$ is forbidden by the exclusion principle. The ground term value of lithium is $^2S_{1/2}$, since we have one electron outside a closed shell. If the screening of the nuclear charge by the two inner 1s electrons were perfect, the outer electron would feel an effective charge $Z_e = 1$, the corresponding ionisation potential being then 13.6 eV/4 = 3.4 eV. In fact the screening is not perfect and, as a result, the ionisation potential is somewhat larger, being 5.39 eV.

With the next element, beryllium ($Z = 4$), the 2s subshell is filled and the ground state configuration is $1s^22s^2$. Since we have a closed subshell as in helium, the ground term is $^1S_0$. The ionisation potential (9.32 eV) is larger than for lithium, because of the increase of the nuclear charge.

The 2p subshell becomes progressively filled beginning with boron ($Z = 5$, ground state configuration $1s^22s^22p$) up to neon ($Z = 10$, ground state configuration $1s^22s^22p^6$). Since the 2p individual energy level is somewhat higher than the 2s level, the ionisation potential of boron (8.30 eV) is smaller than that of beryllium (9.32 eV). For neon the ionisation potential reaches the value of 21.56 eV, which is larger than any other one, except helium. We note that since the 2p subshell is closed for neon, its ground term value is $^1S_0$. In fact, neon has the maximum number of electrons allowed in the $n = 2$ (L) shell.

With $Z = 11$ (sodium) the eleventh electron must go into the 3s subshell. The ionisation potential of sodium (5.14 eV) is therefore much smaller than that of neon. From $Z = 11$ to $Z = 18$ the 3s and 3p subshells are progressively filled, the ground state configuration of argon ($Z = 18$) being $1s^22s^22p^63s^23p^6$.

The process of filling the $n = 3$ states is temporarily interrupted at $Z = 19$ (potassium). Indeed, looking at Table 8.2, we see that after the 3p subshell is filled the first departure from the ordering according to the lowest value of n occurs. The added electrons in potassium ($Z = 19$) and calcium ($Z = 20$) thus go into the 4s rather than the 3d subshell, the 3d level being energetically less favourable because of the screening by the argon core [Ar].

The filling of the 3d subshell is therefore deferred until scandium ($Z = 21$), which has the ground state configuration $1s^22s^22p^63s^23p^64s^23d$ (or $[Ar]4s^23d$ in abbreviated notation). It is the first element of the so-called *first transition* or *iron group*, extending from $Z = 21$ to $Z = 30$ (zinc). It is worth noting that because the 4s and 3d levels are very close in energy, a competition between these two levels develops, and the process of filling is not so regular as for the previous subshells. Thus from scandium ($Z = 21$) to vanadium ($Z = 23$) the added electrons occupy successively the levels $4s^23d$, $4s^23d^2$ and $4s^23d^3$, but chromium ($Z = 24$) has only one 4s electron, the state $4s3d^5$ being energetically more favourable than $4s^23d^4$. With manganese ($Z = 25$) the added electron goes into the 4s level left free in chromium, so that Mn has the ground state configuration $[Ar]4s^23d^5$. The 3d subshell then continues to be filled regularly with iron ($Z = 26$, ground state configuration [Ar]$4s^23d^6$), cobalt ($Z = 27$, ground state configuration [Ar]$4s^23d^7$) and nickel ($Z = 28$, ground state configuration [Ar]$4s^23d^8$), but this regularity is again broken with copper ($Z = 29$) which has only one 4s electron (as does chromium), its ground state configuration being [Ar]$4s3d^{10}$. The last element of the first transition or iron group, zinc ($Z = 30$), has the ground state configuration [Ar]$4s^23d^{10}$.

The elements of the first transition group have ground state configurations in which the outer electrons occupy levels of the type $(n + 1)s^2nd^x$ or $(n + 1)snd^{x+1}$ with $n = 3$. Two other sets of transition elements, the *second transition* or *palladium* group (from $Z = 39$ to $Z = 48$) and the *third transition* or *platinum* group (from $Z = 71$ to $Z = 80$), correspond to similar situations, but with $n = 4$ and $n = 5$, respectively. Here again a competition develops between the nd and the $(n + 1)$s levels, and irregularities in the filling of the subshells occur.

The *rare-earth* elements or *lanthanides* are the 14 elements, beginning with lanthanum ($Z = 57$), which correspond to the filling of the 4f subshell, the 4s subshell being already complete. We note that this filling of the 4f subshell is irregular, with a competition taking place between the 4f and 5d levels. Analogous to the rare-earth elements are the *actinides*, beginning with actinium ($Z = 89$) in which the 5s subshell is complete and a competition occurs between the 5f and the 6d levels.

The electron configurations of atoms with large values of Z are clearly difficult to explain on the basis of the simple qualitative arguments developed above. One reason, which we have already mentioned, is that various energy levels are very close in energy. Also, for large Z, relativisitic effects (such as spin–orbit coupling) become important and prevent the simple decoupling of the space and spin parts of the wave functions, which we have made here. We shall return to spin–orbit effects in Section 8.5.

Ionisation potentials

It is apparent from Table 8.3 and from Fig. 8.1 that in going from an element to the next one, the ionisation potential of the added electron does not vary in a monotonic way with Z. In particular, the ionisation potential is seen to reach maximum values for the *noble gases* (He, Ne, Ar, Kr, Xe, Rn) which have a full K shell or p subshell. It is smallest for the *alkalis* (Li, Na, K, Rb, Cs, Fr) whose electron configuration corresponds to that of a noble gas plus an s electron.

These features can be understood qualitatively by recalling that the electrons in the same subshell have equivalent spatial distributions, so that their screening of one another is rather small. For example, in the case of helium we saw in Section 7.4 that the screening constant corresponding to the shielding between the two 1s electrons is given approximately by $S \simeq 0.30$. As a result, the effective charge Z_e increases as Z increases during the filling of a subshell, and the ionisation potential is maximum for a closed subshell. On the other hand, when a subshell has been filled, the added electron must go into another state having a larger value of n or l, whose orbital is concentrated at larger r values, outside the closed subshells.

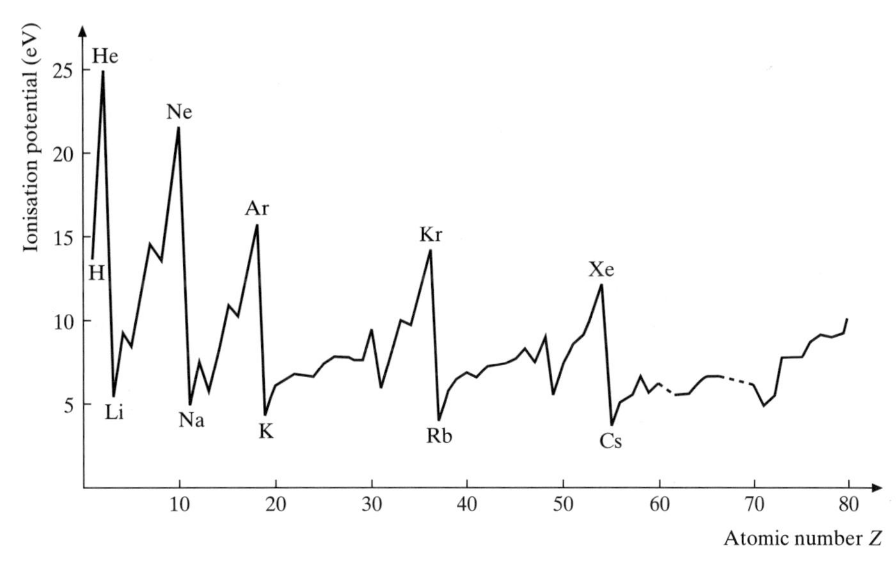

Figure 8.1 The ionisation potential as a function of the atomic number Z. The maxima occur at $Z = 2$ (He), $Z = 10$ (Ne), $Z = 18$ (Ar), $Z = 36$ (Kr), $Z = 54$ (Xe) and (not shown) $Z = 86$ (Rn).

Thus in this case the increase of the nuclear charge Z is more than compensated by the very effective screening of the inner electrons on the outer one, and the ionisation potential is significantly reduced.

Chemical properties and the Mendeleev classification of the elements

The chemical properties of an atom are related to the possible interactions of this atom with other ones, and in particular to its possibility of being bound with other atoms to form a molecule. At the low energies involved in chemical reactions, the interactions between atoms are mostly determined by the least tightly bound or *valence* electrons which, as we have seen, are in the 'outer' subshell. The key factors are the number of occupied electron states in this subshell, and the energy interval between this subshell and the next (empty) one. For example, an atom tends to be chemically *inert* if (i) its outer subshell is filled, and (ii) there is an appreciable energy difference between this subshell and the next higher one, so that it takes quite a lot of energy to perturb the atom. This is the case for the *noble gases* (He, Ne, Ar, Kr, Xe, Rn). On the other hand, the *alkalis* (Li, Na, K, Rb, Cs, Fr) which contain a single weakly bound s electron outside a 'noble gas' core are very active chemically, because they will frequently lose their valence electron in their interactions with other atoms. The *halogens* (F, Cl, Br, I) which have a p subshell lacking only one electron (that is, a *hole* in their outer p subshell) will also exhibit a large chemical reactivity, because of their high electron *affinity* – their tendency to capture an electron in order to reach the more stable arrangement corresponding to completely filled subshells. In particular, a halogen (such as F) will readily combine with an alkali (such as Li) to form an F^- ion and an Li^+ ion, which bond together. This is an example of ionic bonding, which will be discussed in Chapter 10.

The rare gases, alkalis and halogens provide examples of *recurrences* (as Z increases) *of similar chemical properties*, due to regularities in the structure of the outer electron shells. These recurrences led D.I. Mendeleev in 1869 to classify the elements in a table, the *periodic table*, such that the elements of a same column (group) have comparable chemical properties and those of a same row are said to form a *period*. The atomic number Z of an element was then obtained by ordering the spaces in the table. It is worth noting that when Mendeleev proposed his periodic table in 1869 neither electrons nor nuclei were known, and that he also had the foresight to leave some empty spaces in his table, to be filled by elements not yet discovered. Mendeleev subdivided the set of elements into *seven* periods; this subdivision is still kept today and includes elements discovered later. A modern version of the Mendeleev, or periodic, table of the elements is given in Table 8.4.

Table 8.4 The periodic table of the elements.

Period	IA	IIA	IIIB	IVB	VB	VIB	VIIB	VIIIB	VIIIB	VIIIB	IB	IIB	IIIA	IVA	VA	VIA	VIIA	VIIIA
1	1 H 1.008																	2 He 4.003
2	3 Li 6.939	4 Be 9.012											5 B 10.811	6 C 12.011	7 N 14.007	8 O 15.999	9 F 18.998	10 Ne 20.183
3	11 Na 22.990	12 Mg 24.312											13 Al 26.982	14 Si 28.086	15 P 30.974	16 S 32.064	17 Cl 35.453	18 Ar 39.948
4	19 K 39.102	20 Ca 40.08	21 Sc 44.956	22 Ti 47.90	23 V 50.942	24 Cr 51.996	25 Mn 54.938	26 Fe 55.847	27 Co 58.933	28 Ni 58.71	29 Cu 63.54	30 Zn 65.37	31 Ga 69.72	32 Ge 72.59	33 As 74.922	34 Se 78.96	35 Br 79.909	36 Kr 83.80
5	37 Rb 85.47	38 Sr 87.62	39 Y 88.905	40 Zr 91.22	41 Nb 92.906	42 Mo 95.94	43 Tc (99)	44 Ru 101.07	45 Rh 102.91	46 Pd 106.4	47 Ag 107.87	48 Cd 112.40	49 In 114.82	50 Sn 118.69	51 Sb 121.75	52 Te 127.60	53 I 126.90	54 Xe 131.30
6	55 Cs 132.91	56 Ba 137.34	57† La 138.91	72 Hf 178.49	73 Ta 180.95	74 W 183.85	75 Re 186.2	76 Os 190.2	77 Ir 192.2	78 Pt 195.09	79 Au 196.97	80 Hg 200.59	81 Tl 204.37	82 Pb 207.19	83 Bi 208.98	84 Po (210)	85 At (210)	86 Rn (222)
7	87 Fr (223)	88 Ra (226)	89‡ Ac (227)	104	105													

† Lanthanides	58 Ce 140.12	59 Pr 140.91	60 Nd 144.24	61 Pm (145)	62 Sm 150.35	63 Eu 151.96	64 Gd 157.25	65 Tb 158.92	66 Dy 162.50	67 Ho 164.93	68 Er 167.26	69 Tm 168.93	70 Yb 173.04	71 Lu 174.97
‡ Actinides	90 Th 232.04	91 Pa (231)	92 U 238.03	93 Np (237)	94 Pu (242)	95 Am (243)	96 Cm (247)	97 Bk (249)	98 Cf (251)	99 Es (254)	100 Fm (253)	101 Md (256)	102 No (253)	103 Lw (257)

We see that each of the periods begins with an alkali element and ends with a noble gas atom, except for the seventh period, which is incomplete. We remark, parenthetically, that nothing in atomic structure prevents atoms with $Z > 100$ existing. The reason that such atoms are not observed naturally is that their *nuclei* undergo spontaneous fission, and are unstable.

To conclude our discussion, we emphasise once more the importance of the Pauli exclusion principle. Indeed, the variety which we find in the periodic table is basically a consequence of the Pauli principle. If it were not obeyed, all the electrons of an atom in the ground state would be in the 1s subshell (which has the lowest energy), and all atoms would be more or less alike, with spherically symmetric charge distributions having very small radii.

8.3 The Thomas–Fermi model of the atom

We now turn to a basic problem in the central field approximation, namely the determination of the central potential $V(r)$. This problem will be analysed by using two approaches. The elaborate Hartree–Fock method will be studied in the next section, while the simpler Thomas–Fermi model, which is based on *statistical* and *semi-classical* considerations, is discussed here.

The Fermi electron gas

Before we analyse the theory developed by L.H. Thomas and E. Fermi for the ground state of multielectron atoms, it is convenient to consider the simpler problem of the Fermi electron gas, that is a system consisting of a large number N of free electrons confined to a certain region of space [4]. We shall suppose that the N free electrons of our system are confined to a large cube of side L. Each of the electrons is therefore moving independently in a potential which is constant (we may take this constant to be zero) inside the cube, and is assumed to be infinite at the boundary. Thus, the spatial part of the wave function describing the motion of an electron satisfies the free particle Schrödinger equation [5]

$$-\frac{\hbar^2}{2m}\left(\frac{\partial^2}{\partial x^2}+\frac{\partial^2}{\partial y^2}+\frac{\partial^2}{\partial z^2}\right)\psi(\mathbf{r}) = E\psi(\mathbf{r}) \qquad \textbf{(8.30)}$$

inside the cube, while $\psi = 0$ at the boundary.

Since the equation (8.30) is separable in Cartesian coordinates, we may use the results obtained in Section 2.4 for the one-dimensional infinite square well potential. Generalising these results to three dimensions, and moving the origin of our

[4] A good example of a Fermi electron gas is provided by the conduction electrons in a metal. A detailed discussion may be found in Kittel (1996).

[5] In this section we shall use SI units.

coordinate system from the centre of the box to one corner, we find that the eigenfunctions of (8.30) which vanish at the boundary (that is, the wave functions for a spinless particle in a cubical box of side L) are given by

$$\psi_{n_x n_y n_z}(\mathbf{r}) = C \sin\left(\frac{n_x \pi}{L} x\right) \sin\left(\frac{n_y \pi}{L} y\right) \sin\left(\frac{n_z \pi}{L} z\right) \tag{8.31}$$

where $C = (8/L^3)^{1/2}$ is a normalisation constant and n_x, n_y, n_z are positive integers [6]. The corresponding allowed values of the energy E of an electron are (see (2.131))

$$\begin{aligned} E &= \frac{\pi^2 \hbar^2}{2mL^2}(n_x^2 + n_y^2 + n_z^2) \\ &= \frac{\pi^2 \hbar^2}{2mL^2} n^2 \end{aligned} \tag{8.32}$$

where

$$n^2 = n_x^2 + n_y^2 + n_z^2 \tag{8.33}$$

We remark that each energy level (8.32) can in general be obtained from a number of different sets of values of (n_x, n_y, n_z), and is therefore usually degenerate.

Since electrons have spin 1/2, we must multiply the spatial part (8.31) of their wave function by the spin functions $\chi_{1/2,m_s}$, with $m_s = \pm 1/2$. The individual electron wave functions are therefore the spin-orbitals

$$\psi_{n_x n_y n_z m_s} = \psi_{n_x n_y n_z}(\mathbf{r}) \chi_{1/2,m_s} \tag{8.34}$$

and the quantum states of an electron are specified by the three spatial quantum numbers $(n_x n_y n_z)$ *and* the spin quantum number m_s. We note that for each energy level (8.32) labelled by the quantum numbers $(n_x n_y n_z)$, there are two spin-orbitals, one corresponding to spin up ($m_s = +1/2$) and one to spin down ($m_s = -1/2$), so that the degeneracy of the individual energy levels (8.32) is multiplied by two.

Because the energy spacings are very small for a macroscopic box of side L, it is a good approximation to consider that the energy levels are distributed nearly continuously. We may then introduce the *density of states* or *density of orbitals* $D(E)$, which is defined as the number of electron quantum states (that is, the number of spin-orbitals) per unit energy range. Thus $D(E)\,\mathrm{d}E$ is the number of electron states for which the energy of an electron lies between E and $E + \mathrm{d}E$.

In order to obtain the quantity $D(E)$, we consider the space formed by the axes n_x, n_y and n_z (see Fig. 8.2). Since n_x, n_y and n_z are positive integers, we are interested only in the octant for which $n_x > 0$, $n_y > 0$ and $n_z > 0$. As seen from Fig. 8.2 each set of spatial quantum numbers (n_x, n_y, n_z) corresponds to a point of a

[6] A zero value of n_x, n_y or n_z leads to the unacceptable trivial solution $\psi = 0$ everywhere, and negative values of n_x, n_y, n_z do not yield physically different wave functions from those given by (8.31).

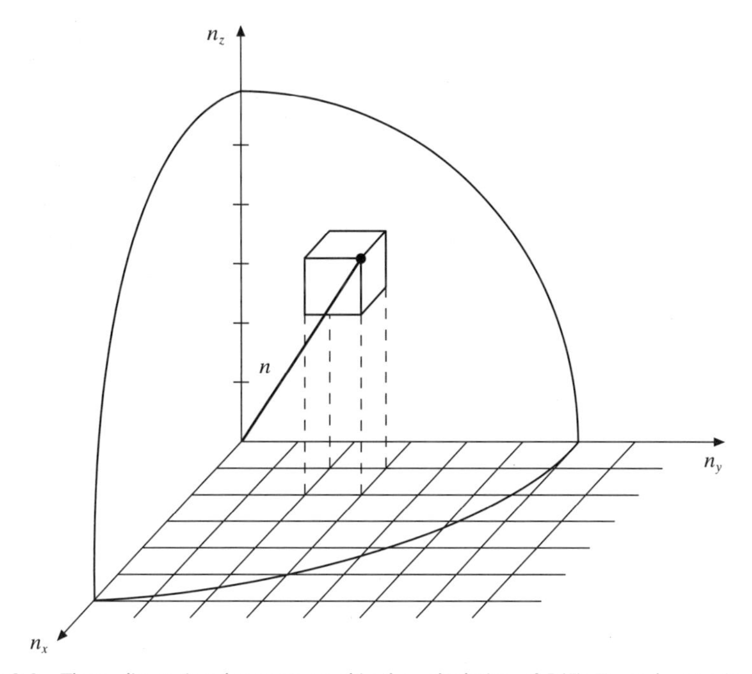

Figure 8.2 Three-dimensional n-space used in the calculation of $D(E)$. To each state (n_x, n_y, n_z) is associated a cube of unit volume. For fairly large values of (n_x, n_y, n_z) the total number of states within $n = (n_x^2 + n_y^2 + n_z^2)^{1/2}$ equals the volume of one octant of a sphere of radius n in n-space.

cubical lattice, and every elementary cube of the lattice has unit volume. Thus, for fairly large values of the quantum triplet (n_x, n_y, n_z), the total number of spatial orbitals for all energies up to a certain value E is closely equal to the volume of the octant of a sphere of radius $n = (n_x^2 + n_y^2 + n_z^2)^{1/2}$. The total number of individual electron states for energies up to E is therefore given approximately by

$$N_s = 2\frac{1}{8}\frac{4}{3}\pi n^3 = \frac{1}{3}\pi n^3 \tag{8.35}$$

where the factor 2 is due to the two spin states per spatial orbital. Using (8.32) and setting $V = L^3$, we may also write this result as

$$N_s = \frac{1}{3\pi^2}\left(\frac{2m}{\hbar^2}\right)^{3/2} VE^{3/2} \tag{8.36}$$

The number $D(E)\,\mathrm{d}E$ of electron states within the energy range $(E, E + \mathrm{d}E)$ is then obtained by differentiating (8.36), namely

$$\mathrm{d}N_s = D(E)\,\mathrm{d}E = \frac{1}{2\pi^2}\left(\frac{2m}{\hbar^2}\right)^{3/2} VE^{1/2}\,\mathrm{d}E \tag{8.37}$$

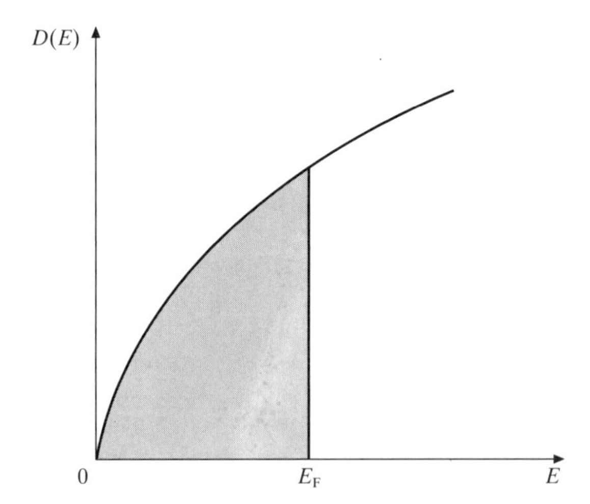

Figure 8.3 The density of states $D(E)$ as a function of the energy E. The occupied orbitals corresponding to the ground state of the Fermi electron gas are represented by the shaded area.

so that

$$D(E) = \frac{dN_s}{dE} = \frac{1}{2\pi^2}\left(\frac{2m}{\hbar^2}\right)^{3/2} VE^{1/2} \tag{8.38}$$

It may be shown that the above results remain valid for a volume of arbitrary shape, provided its minimum dimension is much larger than the average extension of the orbitals.

According to the Pauli exclusion principle, the total wave function describing the entire system of N electrons must be fully antisymmetric in the (spatial and spin) coordinates of the electrons, and will therefore be a Slater determinant constructed from the individual spin-orbitals (8.34). The corresponding total energy is the sum of the individual electron energies. Assuming that the system is in the *ground state* (that is, the Fermi electron gas is at an absolute temperature $T=0$), the lowest total energy is then obtained when the N electrons fill all the spin-orbitals up to an energy E_F, called the *Fermi energy*, the remaining spin-orbitals (with energies $E > E_F$) being vacant. This is illustrated in Fig. 8.3, which shows the density of states $D(E)$ as a function of E, the occupied spin-orbitals corresponding to the ground state of the Fermi electron gas being represented by the shaded area.

The Fermi energy may be evaluated by requiring that the total number N of electrons in the system should be equal to

$$N = \int_0^{E_F} D(E)\, dE \tag{8.39}$$

In writing this equation we have used the fact that the system contains many electrons, so that the integral (8.39) is a good approximation to the corresponding sum over discrete states. Moreover, since N is large, it does not matter whether the last level contains one or more electrons. Using the result (8.38), we have

$$N = \frac{1}{2\pi^2}\left(\frac{2m}{\hbar^2}\right)^{3/2} V \int_0^{E_F} E^{1/2}\,dE$$

$$= \frac{1}{3\pi^2}\left(\frac{2m}{\hbar^2}\right)^{3/2} V E_F^{3/2} \qquad \textbf{(8.40)}$$

so that

$$E_F = \frac{\hbar^2}{2m}(3\pi^2\rho)^{2/3} \qquad \textbf{(8.41)}$$

where

$$\rho = \frac{N}{V} \qquad \textbf{(8.42)}$$

is the number of electrons per unit volume, that is the density of electrons. We note that the total energy of a Fermi electron gas in the ground state (at absolute zero) is

$$E_{tot} = \int_0^{E_F} ED(E)\,dE$$

$$= \frac{1}{2\pi^2}\left(\frac{2m}{\hbar^2}\right)^{3/2} V \int_0^{E_F} E^{3/2}\,dE$$

$$= \frac{1}{5\pi^2}\left(\frac{2m}{\hbar^2}\right)^{3/2} V E_F^{5/2}$$

$$= \frac{3}{5} N E_F \qquad \textbf{(8.43)}$$

where we have used (8.38) and (8.40). The average electron energy at $T = 0$ K is therefore

$$\bar{E} = \frac{E_{tot}}{N} = \frac{3}{5} E_F \qquad \textbf{(8.44)}$$

It is also instructive to study the problem of the Fermi electron gas by imposing *periodic boundary conditions* on the (spatial) wave functions of the electrons, that is by requiring these wave functions to be periodic in x, y and z with period L. Instead of the standing waves (8.31) we then have *travelling wave* solutions of the Schrödinger equation (8.30), having the form

$$\psi_{\mathbf{k}}(\mathbf{r}) = \exp(i\mathbf{k}\cdot\mathbf{r}) \qquad \textbf{(8.45)}$$

As we already showed in Section 4.2, the allowed components of the wave vector **k** are then given by (see (4.57))

$$k_x = \frac{2\pi}{L} n_x, \qquad k_y = \frac{2\pi}{L} n_y, \qquad k_z = \frac{2\pi}{L} n_z \tag{8.46}$$

where n_x, n_y and n_z are positive or negative integers, or zero. The number of spatial orbitals in the range $\mathrm{d}\mathbf{k} = \mathrm{d}k_x\, \mathrm{d}k_y\, \mathrm{d}k_z$ is $(L/2\pi)^3\, \mathrm{d}k_x\, \mathrm{d}k_y\, \mathrm{d}k_z$ and this number must be multiplied by 2 to take into account the two possible spin states. A unit volume of **k**-space will therefore accommodate $V/(4\pi^3)$ electrons (with $V = L^3$). Thus, the individual electron states having energies up to $E = \hbar^2 k^2/(2m)$ will be contained within a sphere in **k**-space, of radius k, the number N_s of these states being given by

$$\begin{aligned} N_s &= \frac{V}{4\pi^3} \frac{4}{3} \pi k^3 = \frac{1}{3\pi^2} V k^3 \\ &= \frac{1}{3\pi^2} \left(\frac{2m}{\hbar^2} \right)^{3/2} V E^{3/2} \end{aligned} \tag{8.47}$$

in agreement with (8.36).

We have seen above that in the ground state of the Fermi electron gas the N electrons fill all the states up to the Fermi energy E_F. Thus in **k**-space all states up to a maximum value of k equal to k_F are then filled, while the states for which $k > k_F$ are empty. In other words all occupied spin-orbitals of a Fermi electron gas at 0 K fill a sphere in **k**-space having radius k_F. This sphere, which is called the *Fermi sphere*, obviously contains

$$\frac{1}{3\pi^2} V k_F^3 = N \tag{8.48}$$

spin-orbitals, so that

$$k_F = (3\pi^2 \rho)^{1/3} \tag{8.49}$$

At the surface of the Fermi sphere, known as the *Fermi surface*, the energy is the Fermi energy

$$E_F = \frac{\hbar^2}{2m} k_F^2 \tag{8.50}$$

and we note that the result (8.41) follows upon substitution of (8.49) in (8.50). It is also convenient to introduce the Fermi momentum $\mathbf{p}_F$, velocity $\mathbf{v}_F$ and temperature T_F such that

$$E_F = \frac{p_F^2}{2m} = \frac{1}{2} m v_F^2 = k_B T_F \tag{8.51}$$

where k_B is Boltzmann's constant.

The Thomas–Fermi theory for multielectron atoms and ions

The theory developed independently by L.H. Thomas and E. Fermi for the ground state of complex atoms (or ions) having a large number of electrons is based on *statistical* and *semi-classical* considerations. The N electrons of the system are treated as a Fermi electron gas in the ground state, confined to a region of space by a central potential $V(r)$ which vanishes at infinity. It is assumed that this potential is slowly varying over a distance which is large compared with the de Broglie wavelengths of the electrons, so that enough electrons are present in a volume where $V(r)$ is nearly constant, and therefore the statistical approach used in studying the Fermi electron gas can be applied. In addition, since the number of electrons is large, many of them have high principal quantum numbers, so that semi-classical methods should be useful.

The aim of the Thomas–Fermi model is to provide a method of calculating the potential $V(r)$ and the electron density $\rho(r)$. These two quantities can first be related by using the following arguments. The total energy of an electron is written as $p^2/(2m) + V(r)$, and this energy cannot be positive, otherwise the electron would escape to infinity. Since the maximum kinetic energy of an electron in a Fermi electron gas at 0 K is the Fermi energy E_F, we write for the total energy of the most energetic electrons of the system the classical equation

$$E_{max} = E_F + V(r) \tag{8.52}$$

It is clear that E_{max} must be independent of r, because if this were not the case electrons would migrate to that region of space where E_{max} is smallest, in order to lower the total energy of the system. Furthermore, we must have $E_{max} \leqslant 0$. We note from (8.50) and (8.52) that the quantity k_F is now a function of r. That is,

$$k_F^2(r) = \frac{2m}{\hbar^2}[E_{max} - V(r)] \tag{8.53}$$

Using (8.41) and (8.52) we then have

$$\rho(r) = \frac{1}{3\pi^2}\left(\frac{2m}{\hbar^2}\right)^{3/2}[E_{max} - V(r)]^{3/2} \tag{8.54}$$

and we see that ρ vanishes when $V = E_{max}$. In the classically forbidden region $V > E_{max}$ we must set $\rho = 0$, since otherwise (8.52) would yield a negative value of the maximum kinetic energy E_F. Let us denote by

$$\phi(r) = -\frac{1}{e}V(r) \tag{8.55}$$

the electrostatic potential and by $\phi_0 = -E_{max}/e$ a non-negative constant. Setting

$$\Phi(r) = \phi(r) - \phi_0 \tag{8.56}$$

we see that $\rho(r)$ and $\Phi(r)$ are related by

$$\rho(r) = \frac{1}{3\pi^2}\left(\frac{2m}{\hbar^2}\right)^{3/2}[e\Phi(r)]^{3/2}, \qquad \Phi \geqslant 0 \tag{8.57a}$$

$$= 0, \qquad \Phi < 0 \tag{8.57b}$$

The equation $\Phi = 0$ (that is, $\phi = \phi_0$ or $V = E_{max}$) may be thought of as determining the 'boundary' $r = r_0$ of the atom (ion) in this model. Now, for a neutral atom ($N = Z$) the electrostatic potential $\phi(r)$ vanishes at the boundary, so that we shall set $\phi_0 = 0$ in that case. On the other hand $\phi_0 > 0$ for an ion.

A second relation between $\rho(r)$ and $\Phi(r)$ may be obtained as follows. The sources of the electrostatic potential $\phi(r)$ are:

(i) the point charge Ze of the nucleus, located at the origin;
(ii) the distribution of electricity due to the N electrons.

Treating the charge density $-e\rho(r)$ of the electrons as continuous, we may use Poisson's equation of electrostatics to write

$$\nabla^2\Phi(r) = \frac{1}{r}\frac{\mathrm{d}^2}{\mathrm{d}r^2}[r\Phi(r)] = \frac{e}{\varepsilon_0}\rho(r) \tag{8.58}$$

The equations (8.57a) and (8.58) are two simultaneous equations for $\rho(r)$ and $\Phi(r)$. Eliminating $\rho(r)$ from these equations, we find that for $\Phi \geqslant 0$

$$\frac{1}{r}\frac{\mathrm{d}^2}{\mathrm{d}r^2}[r\Phi(r)] = \frac{e}{3\pi^2\varepsilon_0}\left(\frac{2m}{\hbar^2}\right)^{3/2}[e\Phi(r)]^{3/2}, \qquad \Phi \geqslant 0 \tag{8.59}$$

On the other hand, when $\Phi < 0$ we see from (8.57b) and (8.58) that

$$\frac{\mathrm{d}^2}{\mathrm{d}r^2}[r\Phi(r)] = 0, \qquad \Phi < 0 \tag{8.60}$$

For $r \to 0$ the leading term of the electrostatic potential must be due to the nucleus, so that the boundary condition at $r = 0$ reads

$$\lim_{r\to 0} r\Phi(r) = \frac{Ze}{4\pi\varepsilon_0} \tag{8.61}$$

On the other hand, since the N electrons of the system are assumed to be confined to a sphere of radius r_0, we must have the 'normalisation' condition

$$4\pi\int_0^{r_0}\rho(r)r^2\,\mathrm{d}r = N \tag{8.62}$$

In order to simplify the above equations, it is convenient to introduce the new dimensionless variable x and the function $\chi(x)$ such that

$$r = bx, \qquad r\Phi(r) = \frac{Ze}{4\pi\varepsilon_0}\chi(x) \tag{8.63}$$

where

$$b = \frac{(3\pi)^{2/3}}{2^{7/3}} a_0 Z^{-1/3} \simeq 0.8853\, a_0 Z^{-1/3} \tag{8.64}$$

and $a_0 = (4\pi\varepsilon_0)\hbar^2/(me^2)$ is the first Bohr radius. The relation (8.57) then becomes

$$\rho = \frac{Z}{4\pi b^3}\left(\frac{\chi}{x}\right)^{3/2}, \qquad \chi \geqslant 0 \tag{8.65a}$$

$$= 0, \qquad \chi < 0 \tag{8.65b}$$

and the important equation (8.59) may be written in dimensionless form as

$$\frac{d^2\chi}{dx^2} = x^{-1/2}\chi^{3/2}, \qquad \chi \geqslant 0 \tag{8.66}$$

This is known as the *Thomas–Fermi equation*. For negative χ we see from (8.60) and (8.63) that

$$\frac{d^2\chi}{dx^2} = 0, \qquad \chi < 0 \tag{8.67}$$

In addition, the boundary condition at $r = 0$, expressed by (8.61), now reads

$$\chi(0) = 1 \tag{8.68}$$

It is clear from (8.66) and (8.67) that $\chi(x)$ has at most one zero in the interval $(0, +\infty)$. Let x_0 be the position of this zero. From our above discussion we have $x_0 = r_0/b$, where r_0 is the 'boundary' of the system. We also note that $\chi > 0$ for $x < x_0$ and $\chi < 0$ for $x > x_0$. Moreover, the equation (8.67) has the solution $\chi = C(x - x_0)$, where C is a negative constant, which must be equal to $\chi'(x_0)$. As a result, the solution $\chi(x)$ is entirely determined if we know it for $\chi \geqslant 0$. We also remark that for any finite x_0 the quantity $\chi'(x_0)$ must be different from zero, since otherwise both χ and χ' would vanish at $x = x_0$, and the equation (8.66) would yield the unacceptable trivial solution $\chi = 0$.

The Thomas–Fermi equation (8.66) is a 'universal' equation, which does not depend on Z, nor on physical constants such as $\hbar$, m or e which have been 'scaled out' by performing the change of variables (8.63). We also note that it is a second-order, non-linear differential equation. Since the boundary condition at the origin (8.68) only specifies one constraint, there exist a whole family of solutions $\chi(x)$ satisfying the Thomas–Fermi equation (8.66) and the condition (8.68), which differ by their initial slope $\chi'(0)$. It is also clear from (8.66) that all these solutions must be concave upwards. As illustrated in Fig. 8.4, we can classify them into three categories:

1. a solution which is asymptotic to the x axis;
2. solutions which vanish for a finite value $x = x_0$;
3. solutions which never vanish and diverge for large x.

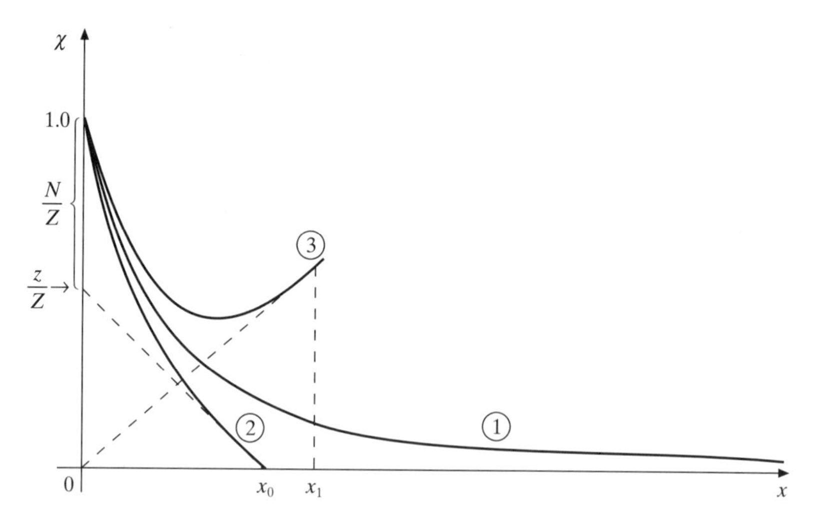

Figure 8.4 The three categories of solutions of the Thomas–Fermi equation: (1) neutral atom solution; (2) solution corresponding to a positive ion ($N < Z$); (3) solution corresponding to a neutral atom under pressure.

The physical meaning of the solutions belonging to the first two categories may be obtained by looking at the 'normalisation' condition (8.62). Taking into account (8.63), (8.65) and (8.66), this condition reads

$$\begin{aligned} N &= Z \int_0^{x_0} x^{1/2} \chi^{3/2} \, \mathrm{d}x \\ &= Z \int_0^{x_0} x \chi'' \, \mathrm{d}x \\ &= Z[x\chi' - \chi]_0^{x_0} \end{aligned} \tag{8.69}$$

Using the boundary condition (8.68) and the fact that $\chi(x_0) = 0$, we then have

$$x_0 \chi'(x_0) = \frac{N - Z}{Z} \tag{8.70}$$

Let us first consider *neutral atoms* for which $N = Z$. The condition (8.70) then requires that $\chi'(x_0) = 0$, so that χ' should vanish at the same point as χ. Since this condition cannot be satisfied for a finite value x_0 by non-trivial solutions, the point x_0 must be at infinity. As a consequence, the solution $\chi(x)$ corresponding to a neutral atom must be asymptotic to the x axis, namely

$$\chi(\infty) = 0 \tag{8.71}$$

and is therefore the (unique) solution classified above in the first category. We remark that since $\chi(x)$ vanishes only at infinity, there is no 'boundary' to the neutral atom in the Thomas–Fermi model.

Table 8.5 Values of the function $\chi(x)$ for neutral atoms.

x	$\chi(x)$	x	$\chi(x)$	x	$\chi(x)$	x	$\chi(x)$
0.00	1.000	0.9	0.453	3.4	0.135	9.0	0.029 5
0.02	0.972	1.0	0.425	3.6	0.125	9.5	0.026 8
0.04	0.947	1.2	0.375	3.8	0.116	10	0.024 4
0.06	0.924	1.4	0.333	4.0	0.108	11	0.020 4
0.08	0.902	1.6	0.298	4.5	0.0918	12	0.017 2
0.1	0.882	1.8	0.268	5.0	0.0787	13	0.014 7
0.2	0.793	2.0	0.242	5.5	0.0679	14	0.012 6
0.3	0.721	2.2	0.220	6.0	0.0592	15	0.010 9
0.4	0.660	2.4	0.201	6.5	0.0521	20	0.005 8
0.5	0.607	2.6	0.185	7.0	0.0461	25	0.003 5
0.6	0.561	2.8	0.171	7.5	0.0409	30	0.002 3
0.7	0.521	3.0	0.158	8.0	0.0365	40	0.001 1
0.8	0.485	3.2	0.146	8.5	0.0327	50	0.000 63

The Thomas–Fermi equation (8.66) and the boundary conditions (8.68) and (8.71) define a *universal function* $\chi(x)$ for all neutral atoms, Values of this function, obtained by numerical integration, are given in Table 8.5. We remark from this table that $\chi(x)$ is monotonically decreasing. It can be shown that the asymptotic form of $\chi(x)$ for large x is given by the function $144x^{-3}$. At $x = 0$ one has $\chi'(0) = -1.588$ so that in the vicinity of the origin

$$\chi(x) \simeq 1 - 1.588x + \dots \tag{8.72}$$

Knowing the universal function $\chi(x)$, we can obtain the function $\Phi(r)$, and hence the electrostatic potential $\phi(r)$, the potential energy $V(r)$ and the density $\rho(r)$. Using (8.55), (8.56), (8.63), and remembering that $\phi_0 = 0$ for a neutral atom, we see that in the Thomas–Fermi model the central potential $V(r)$ is given for neutral atoms by

$$V(r) = -\frac{Ze^2}{(4\pi\varepsilon_0)r}\chi \tag{8.73}$$

As $r \to 0$, we have $V(r) \to -Ze^2/(4\pi\varepsilon_0 r)$. More precisely, we deduce from (8.63), (8.72) and (8.73) that for small r

$$V(r) \simeq \frac{e^2}{4\pi\varepsilon_0}\left(-\frac{Z}{r} + 1.794\frac{Z^{4/3}}{a_0}\right) \tag{8.74}$$

or, using atomic units,

$$V(r) \simeq -\frac{Z}{r} + 1.794Z^{4/3} + \dots \tag{8.75}$$

The first term is the nuclear attraction while the second one, which is repulsive, arises from the contribution of the electrons. When $r \to \infty$, we see from (8.71) and (8.73) that $rV(r) \to 0$, so that the Thomas–Fermi potential (8.73) falls off more rapidly than $1/r$ for large r. This behaviour is at variance with the result (8.8c) which we obtained in our discussion of the central field approximation. The reason is that the potential V discussed in Section 8.1 is the one felt by an atomic electron, while the Thomas–Fermi potential (8.73) is that experienced by an infinitesimal negative test charge. The difference between the two potentials is due to the statistical and semi-classical approximations made in the Thomas–Fermi model, the Thomas–Fermi result becoming exact in the limit when $\hbar$ and e tend to zero, while the number N $(= Z)$ of electrons becomes infinite.

Turning now to the electron density $\rho(r)$, we see from (8.65a) that it is similar for all atoms, except for a different length scale, which is determined by the quantity b (see (8.64)) and is proportional to $Z^{-1/3}$. As a result, the radial scale of $\rho(r)$ contracts according to $Z^{-1/3}$ when Z increases. We remark that for fixed Z the Thomas–Fermi method is inaccurate at both small r $(r < a_0/Z)$ and large r $(r \gg a_0)$, where it overestimates the electron density. Indeed, the Thomas–Fermi electron density (8.65a) diverges at the origin (as $r^{-3/2}$) and falls off like r^{-6} as $r \to \infty$, while the correct electron density should remain finite at $r = 0$, and decrease exponentially for large r. Thus the application of the Thomas–Fermi method is limited to 'intermediate' distances r between a_0/Z and a few times a_0. It is worth noting, however, that in complex atoms most of the electrons are to be found precisely in this spatial region. Thus we expect the Thomas–Fermi method to be useful in calculating quantities which depend on the 'average electron', such as the total energy of the atom. On the other hand, quantities which rely on the properties of the 'outer' electrons (such as the ionisation potential) are poorly given in the Thomas–Fermi model.

We have shown above that a neutral atom has no 'boundary' in the Thomas–Fermi model. Nevertheless, it is possible to define in this case an atomic 'radius' $R(\alpha)$ as the radius of a sphere centred at the origin and containing a given fraction $(1 - \alpha)$ of the Z electrons. We then have (see (8.62))

$$4\pi \int_0^{R(\alpha)} \rho(r) r^2 \, \mathrm{d}r = (1 - \alpha) Z \tag{8.76}$$

Making the change of variable

$$R(\alpha) = bX(\alpha) \tag{8.77}$$

and taking into account (8.63), (8.65) and (8.66), we find for X the equation

$$\chi(X) - X\chi'(X) = \alpha \tag{8.78}$$

which must be solved numerically. If the same value of α is adopted for all atoms, (8.78) becomes a 'universal' equation and X is the same for all atoms. Using (8.64) and (8.77) we see that the atomic radius $R(\alpha)$ is then proportional to $Z^{-1/3}$. On the other hand, if we set $\alpha = Z^{-1}$, then $R(Z^{-1}) = bX(Z^{-1})$ is the radius of a sphere containing all the atomic electrons except one. The quantity $R(Z^{-1})$ is found to be a slowly increasing function of Z, such that $4a_0 < R(Z^{-1}) < 6a_0$. Thus in both cases the atomic radius is nearly independent of Z. Similarly, the energy of the 'outer' electrons – and hence the ionisation potential of the atom – is almost independent of Z. As a consequence, the Thomas–Fermi model cannot account for the periodic properties of atoms as a function of Z, discussed in Section 8.2.

Let us now briefly discuss the two other categories of solutions (see Fig. 8.4) mentioned in our discussion of the Thomas–Fermi equation (8.66). Returning to (8.69)–(8.70) we remark that solutions $\chi(x)$ which vanish at a finite value $x = x_0$ (that is, which belong to the second category) are such that $N \neq Z$, and hence correspond to *ions* of radius $r_0 = bx_0$. Moreover, since the slope of χ is negative at x_0 (see Fig. 8.4) the equation (8.70) implies that these ions must be *positive ions*, such that $Z > N$ [7]. Setting $z = Z - N$, so that ze is the net charge of the ion, we note from (8.70) that the quantity z/Z is readily obtained from the tangent to the curve χ at $x = x_0$, as shown in Fig. 8.4. Since $\chi(x_0) = 0$, the electron density $\rho(r)$ vanishes at $r = r_0 = bx_0$, as seen from (8.65). On the other hand, looking at (8.55), (8.56) and (8.63), and remembering that $\phi_0 > 0$ for an ion, we see that the potential $V(r)$ remains finite at $r = r_0$.

The solutions of the Thomas–Fermi equation belonging to the third category (that is, those which have no zero and diverge for large x) are more difficult to interpret. First of all, the electron density $\rho(r)$ does not vanish in this case, and one may consider that these solutions correspond to negative values of ϕ_0. As seen from Fig. 8.4, such solutions lie above the 'universal' curve of the neutral atom. Now the total charge inside a sphere of radius $r = bx$ is just

$$Ze - 4\pi e \int_0^r \rho(r')r'^2 \,\mathrm{d}r' = Ze[\chi(x) - x\chi'(x)] \tag{8.79}$$

Thus, at the point $r_1 = bx_1$, where

$$\chi(x_1) - x_1\chi'(x_1) = 0 \tag{8.80}$$

the total charge inside the sphere $r = r_1$ vanishes, and we note that the tangent to $\chi(x)$ at $x = x_1$ passes through the origin (see Fig. 8.4). For $x \leqslant x_1$, the curve $\chi(x)$ therefore corresponds to a neutral atom having a finite boundary at $r = r_1$, where the density $\rho(r)$ does not vanish. This may be interpreted as a representation of a *neutral atom under pressure* [8]. Further developments of the Thomas–Fermi method can be found in the monograph of Englert (1988).

---

[7] Negative ions cannot be handled by the Thomas–Fermi theory.

[8] We are not considering ions under pressure, since in dealing with an ensemble of such ions, difficulties due to the presence of the Coulomb forces between ions would arise.

8.4 The Hartree–Fock method and the self-consistent field

We shall now study a more elaborate approximation for complex atoms (ions), known as the *Hartree–Fock* or *self-consistent field* method. The starting point of this approach, formulated by D.R. Hartree in 1928, is the *independent-particle model*, discussed in Section 8.1, according to which each electron moves in an effective potential which takes into account the attraction of the nucleus and the average effect of the repulsive interactions due to the other electrons. Each electron in a multielectron system is then described by its own wave function. By using intuitive arguments, Hartree was able to write down equations for the individual electron wave functions. He also proposed an original iterative procedure, based on the requirement of *self-consistency*, to solve his equations. We shall return shortly to this self-consistent procedure, which is a key feature of the theory of many-electron atoms discussed in this section.

As we shall show at the end of this section, the Hartree total wave function for the atom (ion) is not antisymmetric in the electron coordinates. The generalisation of the Hartree method which takes into account this antisymmetry requirement – imposed by the Pauli exclusion principle – was carried out in 1930 by V.A. Fock and J.C. Slater. It is this generalisation of Hartree's theory, known as the *Hartree–Fock* method, which we now discuss.

The Hartree–Fock equations

In the Hartree–Fock approach, it is assumed, in accordance with the independent-particle approximation and the Pauli exclusion principle, that the N-electron wave function is a *Slater determinant* Φ, or in other words an antisymmetric product of individual electron spin-orbitals. The optimum Slater determinant is then obtained by using the *variational method* to determine the 'best' individual electron spin-orbitals. The Hartree–Fock method is therefore a particular case of the variational method, in which the trial function for the N-electron atom is a Slater determinant whose individual spin-orbitals are optimised. It should be noted that the N-electron atom wave function $\Psi(q_1, q_2, \ldots, q_N)$, solution of the Schrödinger equation (8.3), can only be represented by an infinite sum of Slater determinants, so that the Hartree–Fock method may be considered as a first step in the determination of atomic wave functions and energies. We also remark that the application of the Hartree–Fock method is not confined to atoms (ions), but can also be made to other systems such as the electrons in a molecule or a solid.

In what follows we shall limit our discussion to the *ground state* of an atom or ion having N electrons [9]. We start from the non-relativistic Hamiltonian (8.2), which we write (in a.u.) as

$$H = \hat{H}_1 + \hat{H}_2 \qquad \textbf{(8.81)}$$

[9] The derivation of the Hartree–Fock equations given here follows the treatment of Messiah (1968).

where

$$\hat{H}_1 = \sum_{i=1}^{N} \hat{h}_i, \tag{8.82}$$

$$\hat{h}_i = -\frac{1}{2}\nabla^2_{\mathbf{r}_i} - \frac{Z}{r_i} \tag{8.83}$$

and

$$\hat{H}_2 = \sum_{i<j=1}^{N} \frac{1}{r_{ij}}, \qquad r_{ij} = |\mathbf{r}_i - \mathbf{r}_j| \tag{8.84}$$

The first term of (8.81), $\hat{H}_1$, is the sum of the N identical *one-body* hydrogenic Hamiltonians $\hat{h}_i$, each individual Hamiltonian $\hat{h}_i$ containing the kinetic energy operator of an electron and its potential energy due to the attraction of the nucleus. The second term, $\hat{H}_2$, is the sum of $N(N-1)/2$ identical terms $1/r_{ij}$ which represent the *two-body* interactions between each pair of electrons.

Let us denote by E_0 the ground state energy of the system. According to the variational method (see Section 2.8) we have

$$E_0 \leqslant E[\Phi] = \langle\Phi|H|\Phi\rangle \tag{8.85}$$

where Φ is a trial function which we assume to be normalised to unity,

$$\langle\Phi|\Phi\rangle = 1 \tag{8.86}$$

In the Hartree–Fock method the trial function Φ is a Slater determinant, so that (see (8.22))

$$\Phi(q_1, q_2, \ldots, q_N) = \frac{1}{\sqrt{N!}} \begin{vmatrix} u_\alpha(q_1) & u_\beta(q_1) \ldots & u_\nu(q_1) \\ u_\alpha(q_2) & u_\beta(q_2) \ldots & u_\nu(q_2) \\ \vdots & & \\ u_\alpha(q_N) & u_\beta(q_N) \ldots & u_\nu(q_N) \end{vmatrix} \tag{8.87}$$

where we recall that each of the symbols $\alpha, \beta, \ldots, \nu$ represents a set of four quantum numbers (n, l, m_l, m_s). We require that all spin-orbitals be orthonormal, namely

$$\langle u_\mu|u_\lambda\rangle = \int u^*_\mu(q)u_\lambda(q)\,\mathrm{d}q = \delta_{\lambda\mu} \tag{8.88}$$

where the symbol $\int \mathrm{d}q$ implies an integration over the space coordinates and a summation over the spin coordinate. Since spin-orbitals corresponding to 'spin up' ($m_s = +1/2$) are automatically orthogonal to those corresponding to 'spin down' ($m_s = -1/2$), the requirement (8.88) reduces to the condition that the space orbitals corresponding to the same spin function be orthonormal. We note that (8.88) ensures that the Slater determinant (8.87) satisfies the normalisation condition (8.86).

It is convenient to rewrite the Slater determinant (8.87) in the more compact form (8.23). That is,

$$\Phi(q_1, q_2, \ldots, q_N) = \frac{1}{\sqrt{N!}} \sum_P (-1)^P P u_\alpha(q_1) u_\beta(q_2) \ldots u_\nu(q_N)$$
$$= \sqrt{N!}\, \mathscr{A} \Phi_{\mathrm{H}} \tag{8.89}$$

where Φ_{H} is the simple product of spin-orbitals

$$\Phi_{\mathrm{H}}(q_1, q_2, \ldots, q_N) = u_\alpha(q_1) u_\beta(q_2) \ldots u_\nu(q_N) \tag{8.90}$$

which will be referred to as a Hartree wave function. The operator which appears in (8.89) is the antisymmetrisation operator

$$\mathscr{A} = \frac{1}{N!} \sum_P (-1)^P P \tag{8.91}$$

It is a simple matter to show that the operator $\mathscr{A}$ is Hermitian and that it is also a projection operator, namely

$$\mathscr{A}^2 = \mathscr{A} \tag{8.92}$$

A further remark, which will be useful shortly, is that both operators $\hat{H}_1$ and $\hat{H}_2$ are invariant under permutations of the electron coordinates, and hence commute with $\mathscr{A}$,

$$[\hat{H}_1, \mathscr{A}] = [\hat{H}_2, \mathscr{A}] = 0 \tag{8.93}$$

Let us now calculate the functional $E[\Phi]$. Using (8.81) and (8.85), we have

$$E[\Phi] = \langle \Phi | \hat{H}_1 | \Phi \rangle + \langle \Phi | \hat{H}_2 | \Phi \rangle \tag{8.94}$$

The first expectation value $\langle \Phi | H_1 | \Phi \rangle$ is readily evaluated as follows. We first have

$$\langle \Phi | \hat{H}_1 | \Phi \rangle = N! \langle \Phi_{\mathrm{H}} | \mathscr{A} \hat{H}_1 \mathscr{A} | \Phi_{\mathrm{H}} \rangle$$
$$= N! \langle \Phi_{\mathrm{H}} | \hat{H}_1 \mathscr{A}^2 | \Phi_{\mathrm{H}} \rangle$$
$$= N! \langle \Phi_{\mathrm{H}} | \hat{H}_1 \mathscr{A} | \Phi_{\mathrm{H}} \rangle \tag{8.95}$$

where we have used (8.89), (8.92) and (8.93). Taking into account the fact that $\hat{H}_1$ is the sum of one-body operators (see (8.82)) together with (8.88), (8.90) and (8.91) we then find that

$$\langle \Phi | \hat{H}_1 | \Phi \rangle = \sum_{i=1}^{N} \sum_P (-1)^P \langle \Phi_{\mathrm{H}} | \hat{h}_i P | \Phi_{\mathrm{H}} \rangle$$
$$= \sum_{i=1}^{N} \langle \Phi_{\mathrm{H}} | \hat{h}_i | \Phi_{\mathrm{H}} \rangle$$
$$= \sum_\lambda \langle u_\lambda(q_i) | \hat{h}_i | u_\lambda(q_i) \rangle, \qquad \lambda = \alpha, \beta, \ldots, \nu \tag{8.96}$$

where the sum on λ runs over the N individual quantum states (that is, the N spin-orbitals) occupied by the electrons.

Defining

$$I_\lambda = \langle u_\lambda(q_i)|\hat{h}_i|u_\lambda(q_i)\rangle \tag{8.97}$$

to be the average value of the individual Hamiltonian $\hat{h}_i$ relative to the spin-orbital u_λ, we have

$$\langle\Phi|\hat{H}_1|\Phi\rangle = \sum_\lambda I_\lambda \tag{8.98}$$

The second expectation value $\langle\Phi|\hat{H}_2|\Phi\rangle$ can be calculated in a similar way. From (8.89), (8.92) and (8.93), we have

$$\begin{aligned}\langle\Phi|\hat{H}_2|\Phi\rangle &= N!\langle\Phi_\mathrm{H}|\mathcal{A}\hat{H}_2\mathcal{A}|\Phi_\mathrm{H}\rangle\\ &= N!\langle\Phi_\mathrm{H}|\hat{H}_2\mathcal{A}^2|\Phi_\mathrm{H}\rangle\\ &= N!\langle\Phi_\mathrm{H}|\hat{H}_2\mathcal{A}|\Phi_\mathrm{H}\rangle\end{aligned} \tag{8.99}$$

Using (8.91) and the fact that $\hat{H}_2$ is the sum (8.84) of two-body operators, we obtain

$$\begin{aligned}\langle\Phi|\hat{H}_2|\Phi\rangle &= \sum_{i<j}\sum_P(-1)^P\left\langle\Phi_\mathrm{H}\left|\frac{1}{r_{ij}}P\right|\Phi_\mathrm{H}\right\rangle\\ &= \sum_{i<j}\left\langle\Phi_\mathrm{H}\left|\frac{1}{r_{ij}}(1-P_{ij})\right|\Phi_\mathrm{H}\right\rangle\end{aligned} \tag{8.100}$$

where P_{ij} is an operator that interchanges the coordinates (spatial and spin) of the electrons i and j. Hence, taking into account (8.88) and (8.90), we find that

$$\begin{aligned}\langle\Phi|\hat{H}_2|\Phi\rangle = \sum_{\substack{\lambda,\mu\\ \text{(all pairs)}}}&\left[\left\langle u_\lambda(q_i)u_\mu(q_j)\left|\frac{1}{r_{ij}}\right|u_\lambda(q_i)u_\mu(q_j)\right\rangle\right.\\ &\left.-\left\langle u_\lambda(q_i)u_\mu(q_j)\left|\frac{1}{r_{ij}}\right|u_\mu(q_i)u_\lambda(q_j)\right\rangle\right]\end{aligned} \tag{8.101}$$

where the sum over λ and μ runs over the $N(N-1)/2$ pairs of orbitals. We may also write (8.101) as

$$\begin{aligned}\langle\Phi|\hat{H}_2|\Phi\rangle = \frac{1}{2}\sum_\lambda\sum_\mu&\left[\left\langle u_\lambda(q_i)u_\mu(q_j)\left|\frac{1}{r_{ij}}\right|u_\lambda(q_i)u_\mu(q_j)\right\rangle\right.\\ &\left.-\left\langle u_\lambda(q_i)u_\mu(q_j)\left|\frac{1}{r_{ij}}\right|u_\mu(q_i)u_\lambda(q_j)\right\rangle\right],\\ &\lambda,\mu=\alpha,\beta,\dots,\nu\end{aligned} \tag{8.102}$$

Let us define the *direct* term

$$J_{\lambda\mu} = \left\langle u_\lambda(q_i)u_\mu(q_j)\left|\frac{1}{r_{ij}}\right|u_\lambda(q_i)u_\mu(q_j)\right\rangle \tag{8.103}$$

which is the average value of the interaction $1/r_{ij}$ relative to the state $u_\lambda(q_i)u_\mu(q_j)$ such that electron i is in the spin-orbital u_λ and electron j in the spin-orbital u_μ. We also introduce the *exchange* term

$$K_{\lambda\mu} = \left\langle u_\lambda(q_i)u_\mu(q_j)\left|\frac{1}{r_{ij}}\right|u_\mu(q_i)u_\lambda(q_j)\right\rangle \tag{8.104}$$

which is the matrix element of the interaction $1/r_{ij}$ between the two states $u_\lambda(q_i)u_\mu(q_j)$ and $u_\mu(q_i)u_\lambda(q_j)$ obtained by interchanging the electrons i and j. We note that both $J_{\lambda\mu}$ and $K_{\lambda\mu}$ are real; they are also symmetric in λ and μ,

$$J_{\lambda\mu} = J_{\mu\lambda} \qquad K_{\lambda\mu} = K_{\mu\lambda} \tag{8.105}$$

In terms of $J_{\lambda\mu}$ and $K_{\lambda\mu}$, (8.102) reads

$$\langle\Phi|\hat{H}_2|\Phi\rangle = \frac{1}{2}\sum_\lambda\sum_\mu [J_{\lambda\mu} - K_{\lambda\mu}] \tag{8.106}$$

Using (8.94), (8.98) and (8.106), the total energy $E[\Phi]$ is seen to be given by

$$E[\Phi] = \sum_\lambda I_\lambda + \frac{1}{2}\sum_\lambda\sum_\mu [J_{\lambda\mu} - K_{\lambda\mu}] \tag{8.107}$$

Having obtained the functional $E[\Phi]$, we now proceed to the second step of the calculation, which consists in expressing that $E[\Phi]$ is stationary with respect to variations of the spin-orbitals u_λ ($\lambda = \alpha, \beta, \dots, \nu$), subject to the N^2 conditions (8.88) imposed by the orthonormality requirement on the u_λ's. To satisfy these conditions we introduce N^2 Lagrange multipliers which we denote by $\varepsilon_{\lambda\mu}$ ($\lambda, \mu = \alpha, \beta, \dots, \nu$). The variational equation then reads

$$\delta E - \sum_\lambda\sum_\mu \varepsilon_{\lambda\mu}\, \delta\langle u_\mu|u_\lambda\rangle = 0 \tag{8.108}$$

It is readily seen from (8.108) that $\varepsilon_{\lambda\mu} = \varepsilon^*_{\mu\lambda}$, so that the N^2 Lagrange multipliers $\varepsilon_{\lambda\mu}$ may be considered as the elements of a Hermitian matrix.

It is convenient at this point to make a unitary transformation on the spin-orbitals u_λ, namely

$$u'_\lambda = \sum_\mu U_{\mu\lambda}u_\mu \tag{8.109}$$

where $U_{\mu\lambda}$ are the elements of an $N \times N$ unitary matrix. The new Slater determinant Φ' formed with the spin-orbitals u'_λ differs from the previous one only by a phase factor, since

$$\Phi' = (\det U)\Phi \tag{8.110}$$

and $|\det U| = 1$ because U is unitary. Moreover, the functional $E[\Phi] = \langle\Phi|H|\Phi\rangle$ is clearly unaffected by this unitary transformation. Since any Hermitian matrix can be diagonalised by a unitary transformation, we may always choose U in such a way that the matrix $\varepsilon_{\lambda\mu}$ of Lagrange multipliers will become a *diagonal* matrix having elements $E_\lambda\delta_{\lambda\mu}$. In what follows we shall assume that this diagonalisation has already been made from the outset, so that the variational equation (8.108) then reads

$$\delta E - \sum_\lambda E_\lambda \, \delta\langle u_\lambda | u_\lambda\rangle = 0 \tag{8.111}$$

Let us now vary with respect to the spin-orbitals u_λ. Proceeding as in Section 2.8, and using the expression (8.107) of $E[\Phi]$ together with the relations (8.83), (8.97), (8.103) and (8.104), we find for the N spin-orbitals $u_\alpha, u_\beta, \ldots, u_\nu$ the system of integro-differential equations

$$\left[-\frac{1}{2}\nabla^2_{\mathbf{r}_i} - \frac{Z}{r_i}\right] u_\lambda(q_i) + \left[\sum_\mu \int u^*_\mu(q_j)\frac{1}{r_{ij}}u_\mu(q_j)\,\mathrm{d}q_j\right] u_\lambda(q_i)$$

$$-\sum_\mu \left[\int u^*_\mu(q_j)\frac{1}{r_{ij}}u_\lambda(q_j)\,\mathrm{d}q_j\right] u_\mu(q_i) = E_\lambda u_\lambda(q_i),$$

$$\lambda, \mu = \alpha, \beta, \ldots, \nu \tag{8.112}$$

where the summation over μ extends over the N occupied spin-orbitals. We recall that the symbol $\int \mathrm{d}q_j$ implies an integration over the spatial coordinates $\mathbf{r}_j$ and a summation over the spin coordinate of electron j. The equations (8.112) are known as the *Hartree–Fock equations*. Writing the spin-orbitals $u_\lambda(q_i)$ as

$$u_\lambda(q_i) = u_\lambda(\mathbf{r}_i)\chi_{1/2,m_s^\lambda} \tag{8.113}$$

and using the orthonormality property of the spin functions, namely

$$\langle \chi_{1/2,m_s^\lambda} | \chi_{1/2,m_s^\mu}\rangle = \delta_{m_s^\lambda, m_s^\mu} \tag{8.114}$$

we can also write the Hartree–Fock equations in a form that involves only the spatial part of the spin-orbitals. That is,

$$\left[-\frac{1}{2}\nabla^2_{\mathbf{r}_i} - \frac{Z}{r_i}\right] u_\lambda(\mathbf{r}_i) + \left[\sum_\mu \int u^*_\mu(\mathbf{r}_j)\frac{1}{r_{ij}}u_\mu(\mathbf{r}_j)\,\mathrm{d}\mathbf{r}_j\right] u_\lambda(\mathbf{r}_i)$$

$$-\sum_\mu \delta_{m_s^\lambda, m_s^\mu}\left[\int u^*_\mu(\mathbf{r}_j)\frac{1}{r_{ij}}u_\lambda(\mathbf{r}_j)\,\mathrm{d}\mathbf{r}_j\right] u_\mu(\mathbf{r}_i) = E_\lambda u_\lambda(\mathbf{r}_i),$$

$$\lambda, \mu = \alpha, \beta, \ldots, \nu \tag{8.115}$$

A more compact form of the Hartree–Fock equations may be written down in the following way. We define the *direct* operator

$$V_\mu^{\mathrm{d}}(q_i) = \int u_\mu^*(q_j)\frac{1}{r_{ij}}u_\mu(q_j)\,\mathrm{d}q_j$$

$$= \int u_\mu^*(\mathbf{r}_j)\frac{1}{r_{ij}}u_\mu(\mathbf{r}_j)\,\mathrm{d}\mathbf{r}_j \equiv V_\mu^{\mathrm{d}}(\mathbf{r}_i) \qquad \textbf{(8.116)}$$

which is just the electrostatic repulsion potential due to electron j, when the position of this electron is averaged over the orbital u_μ. We also define the *exchange* (non-local) operator $V_\mu^{\mathrm{ex}}(q_i)$ such that

$$V_\mu^{\mathrm{ex}}(q_i)f(q_i) = \left[\int u_\mu^*(q_j)\frac{1}{r_{ij}}f(q_j)\,\mathrm{d}q_j\right]u_\mu(q_i) \qquad \textbf{(8.117)}$$

where $f(q_i)$ is an arbitrary well-behaved function. In particular, when acting on a spin-orbital $u_\lambda(q_i)$, we see that the exchange operator V_μ^{ex} yields

$$V_\mu^{\mathrm{ex}}(q_i)\,u_\lambda(q_i) = \left[\int u_\mu^*(q_j)\frac{1}{r_{ij}}u_\lambda(q_j)\,\mathrm{d}q_j\right]u_\mu(q_i)$$

$$= \delta_{m_s^\lambda, m_s^\mu}\left[\int u_\mu^*(\mathbf{r}_j)\frac{1}{r_{ij}}u_\lambda(\mathbf{r}_j)\,\mathrm{d}\mathbf{r}_j\right]u_\mu(\mathbf{r}_i)\,\chi_{1/2,m_s^\mu}$$

$$= \delta_{m_s^\lambda, m_s^\mu}V_\mu^{\mathrm{ex}}(\mathbf{r}_i)u_\lambda(\mathbf{r}_i)\chi_{1/2,m_s^\mu} \qquad \textbf{(8.118)}$$

where the exchange operator $V_\mu^{\mathrm{ex}}(\mathbf{r}_i)$ only acts on the spatial coordinates, and is defined by

$$V_\mu^{\mathrm{ex}}(\mathbf{r}_i)f(\mathbf{r}_i) = \left[\int u_\mu^*(\mathbf{r}_j)\frac{1}{r_{ij}}f(\mathbf{r}_j)\,\mathrm{d}\mathbf{r}_j\right]u_\mu(\mathbf{r}_i) \qquad \textbf{(8.119)}$$

$f(\mathbf{r}_i)$ being an arbitrary well-behaved function. Using (8.116) and (8.117), the Hartree–Fock equations become

$$\left[-\frac{1}{2}\nabla_{\mathbf{r}_i}^2 - \frac{Z}{r_i} + \sum_\mu V_\mu^{\mathrm{d}}(\mathbf{r}_i) - \sum_\mu V_\mu^{\mathrm{ex}}(q_i)\right]u_\lambda(q_i) = E_\lambda u_\lambda(q_i) \qquad \textbf{(8.120)}$$

or

$$\left[-\frac{1}{2}\nabla_{\mathbf{r}_i}^2 - \frac{Z}{r_i} + \mathcal{V}^{\mathrm{d}}(\mathbf{r}_i) - \mathcal{V}^{\mathrm{ex}}(q_i)\right]u_\lambda(q_i) = E_\lambda u_\lambda(q_i) \qquad \textbf{(8.121)}$$

where we have introduced the *direct* and *exchange* potentials

$$\mathcal{V}^{\mathrm{d}}(\mathbf{r}_i) = \sum_\mu V_\mu^{\mathrm{d}}(\mathbf{r}_i) \qquad \textbf{(8.122a)}$$

and

$$\mathcal{V}^{\text{ex}}(q_i) = \sum_{\mu} V_{\mu}^{\text{ex}}(q_i) \tag{8.122b}$$

It is also interesting to write the Hartree–Fock equations in terms of the *density matrix* (see Bransden and Joachain, 2000)

$$\rho(q_i, q_j) = \sum_{\mu} u_{\mu}(q_i) u_{\mu}^{*}(q_j) \tag{8.123}$$

or the corresponding spinless density matrix

$$\rho(\mathbf{r}_i, \mathbf{r}_j) = \sum_{\mu} u_{\mu}(\mathbf{r}_i) u_{\mu}^{*}(\mathbf{r}_j) \tag{8.124}$$

The diagonal elements $\rho(q_i, q_i)$ and $\rho(\mathbf{r}_i, \mathbf{r}_i)$ of these density matrices will be denoted by $\rho(q_i)$ and $\rho(\mathbf{r}_i)$, respectively. We note that

$$\rho(\mathbf{r}) = \sum_{\mu} |u_{\mu}(\mathbf{r})|^2 \tag{8.125}$$

gives the probability density of finding an electron at the point $\mathbf{r}$ in one of the N occupied spin-orbitals. In terms of the density matrices, we have

$$\mathcal{V}^{\text{d}}(\mathbf{r}_i) = \int \rho(\mathbf{r}_j) \frac{1}{r_{ij}} \, \text{d}\mathbf{r}_j \tag{8.126a}$$

and

$$\begin{aligned} \mathcal{V}^{\text{ex}}(q_i) u_{\lambda}(q_i) &= \int \rho(q_i, q_j) \frac{1}{r_{ij}} u_{\lambda}(q_j) \, \text{d}q_j \\ &= \delta_{m_s^{\lambda}, m_s^{\mu}} \left[\int \rho(\mathbf{r}_i, \mathbf{r}_j) \frac{1}{r_{ij}} u_{\lambda}(\mathbf{r}_j) \, \text{d}\mathbf{r}_j \right] \chi_{1/2, m_s^{\mu}} \end{aligned} \tag{8.126b}$$

If we define the full Hartree–Fock potential as

$$\begin{aligned} \mathcal{V}(q_i) &= -\frac{Z}{r_i} + \sum_{\mu} V_{\mu}^{\text{d}}(\mathbf{r}_i) - \sum_{\mu} V_{\mu}^{\text{ex}}(q_i) \\ &= -\frac{Z}{r_i} + \mathcal{V}^{\text{d}}(\mathbf{r}_i) - \mathcal{V}^{\text{ex}}(q_i) \end{aligned} \tag{8.127}$$

the Hartree–Fock equations take the (deceptively) simple form

$$\left[-\frac{1}{2} \nabla_{\mathbf{r}_i}^2 + \mathcal{V}(q_i) \right] u_{\lambda}(q_i) = E_{\lambda} u_{\lambda}(q_i) \tag{8.128}$$

The Slater determinant constructed from the optimised spin-orbitals $u_{\lambda}(q_i)$ satisfying the Hartree–Fock equations (8.128) is called the *Hartree–Fock wave function* Ψ_{HF}. The corresponding Hartree–Fock energy is

$$E_{\text{HF}} = \langle \Psi_{\text{HF}} | H | \Psi_{\text{HF}} \rangle \tag{8.129}$$

Physical interpretation of the Hartree–Fock equations. Self-consistent field. Koopmans' theorem

A striking feature of the Hartree–Fock equations (8.128) is that they look similar to individual Schrödinger eigenvalue equations for each of the spin-orbitals u_λ. They are *not* genuine eigenvalue equations, however, since the Hartree–Fock potential $\mathcal{V}$ depends on the spin-orbitals themselves through the operators $V_\mu^{\rm d}$ and $V_\mu^{\rm ex}$. In fact, to solve the system of Hartree–Fock integro-differential equations, one proceeds by *iteration*. Starting from approximate individual spin-orbitals [10] $u_\alpha^{(1)}, u_\beta^{(1)}, \ldots, u_\nu^{(1)}$, one first calculates the corresponding approximate expression $\mathcal{V}^{(1)}$ of the Hartree–Fock potential. The Hartree–Fock equations are then solved with this potential $\mathcal{V}^{(1)}$ to obtain new spin-orbitals $u_\alpha^{(2)}, u_\beta^{(2)}, \ldots, u_\nu^{(2)}$, which in turn yield a new potential $\mathcal{V}^{(2)}$. This procedure is then repeated until the final spin-orbitals give a potential $\mathcal{V}^{(n)}$ which is identical (within the desired approximation) to the potential $\mathcal{V}^{(n-1)}$ obtained from the preceding cycle. The Hartree–Fock potential determined in this way is known as the *self-consistent field* (SCF) of the atom (ion).

Despite the fact that the Hartree–Fock equations (8.128) are not true eigenvalue equations, we shall now show that, when acting on the spin-orbital u_λ, the individual Hartree–Fock Hamiltonian

$$h_{\rm HF}(i) = -\tfrac{1}{2}\nabla_{\mathbf{r}_i}^2 + \mathcal{V}(q_i) \qquad \textbf{(8.130)}$$

may be interpreted as the energy operator of an electron in the state u_λ. To obtain this result, which is in accord with our basic independent-particle approximation, we first note from (8.116) and (8.117) that

$$V_\lambda^{\rm d}(q_i)u_\lambda(q_i) = V_\lambda^{\rm ex}(q_i)u_\lambda(q_i) \qquad \textbf{(8.131)}$$

Upon returning to the Hartree–Fock equations (8.120), we see that there is no self-energy ($\mu = \lambda$) contribution to the potential. It is therefore convenient to introduce the modified density matrix

$$\rho_\lambda(q_i, q_j) = \sum_{\mu\neq\lambda} u_\mu(q_i)u_\mu^*(q_j) \qquad \textbf{(8.132)}$$

and the corresponding modified spinless density matrix

$$\rho_\lambda(\mathbf{r}_i, \mathbf{r}_j) = \sum_{\mu\neq\lambda} u_\mu(\mathbf{r}_i)u_\mu^*(\mathbf{r}_j) \qquad \textbf{(8.133)}$$

whose diagonal elements will be denoted as $\rho_\lambda(q_i)$ and $\rho_\lambda(\mathbf{r}_i)$, respectively. We remark that

---

[10] Simple approximate spin-orbitals may be obtained by using screening arguments such as those discussed in Section 8.1. We shall also see below that the solutions of the Hartree equations can be used to provide a set of functions $u_\alpha^{(1)}, u_\beta^{(1)}, \ldots, u_\nu^{(1)}$.

$$\rho_\lambda(\mathbf{r}) = \sum_{\mu \neq \lambda} |u_\mu(\mathbf{r})|^2 \tag{8.134}$$

is the probability density of finding an electron at $\mathbf{r}$ in one of the $(N-1)$ occupied states other than u_λ. We also define the modified direct and exchange potentials

$$\begin{aligned} \mathcal{V}_\lambda^{\mathrm{d}}(\mathbf{r}_i) &= \sum_{\mu \neq \lambda} V_\mu^{\mathrm{d}}(\mathbf{r}_i) \\ &= \int \rho_\lambda(\mathbf{r}_j) \frac{1}{r_{ij}} \, \mathrm{d}\mathbf{r}_j \end{aligned} \tag{8.135}$$

and

$$\mathcal{V}_\lambda^{\mathrm{ex}}(q_i) = \sum_{\mu \neq \lambda} V_\mu^{\mathrm{ex}}(q_i) \tag{8.136}$$

with

$$\begin{aligned} \mathcal{V}_\lambda^{\mathrm{ex}}(q_i) u_\lambda(q_i) &= \int \rho_\lambda(q_i, q_j) \frac{1}{r_{ij}} u_\lambda(q_j) \, \mathrm{d}q_j \\ &= \delta_{m_s^\lambda, m_s^\mu} \left[\int \rho_\lambda(\mathbf{r}_i, \mathbf{r}_j) \frac{1}{r_{ij}} u_\lambda(\mathbf{r}_j) \, \mathrm{d}\mathbf{r}_j \right] \chi_{1/2, m_s^\mu} \end{aligned} \tag{8.137}$$

so that the Hartree–Fock equations (8.121) now read

$$h_{\mathrm{HF}}(i) u_\lambda(q_i) \equiv \left[-\frac{1}{2} \nabla_{\mathbf{r}_i}^2 - \frac{Z}{r_i} + \mathcal{V}_\lambda^{\mathrm{d}}(\mathbf{r}_i) - \mathcal{V}_\lambda^{\mathrm{ex}}(q_i) \right] u_\lambda(q_i) = E_\lambda u_\lambda(q_i) \tag{8.138}$$

In addition to the kinetic energy term $-\nabla_{\mathbf{r}_i}^2/2$ and the nuclear attraction term $-Z/r_i$, we see that the individual Hartree-Fock Hamiltonian h_{HF} contains a term $\mathcal{V}_\lambda^{\mathrm{d}}(\mathbf{r}_i)$ which represents the *average potential* due to the presence of the $(N-1)$ other electrons, and a (non-local) *exchange term* $\mathcal{V}_\lambda^{\mathrm{ex}}(q_i)$ which takes into account the exchange effects between the state u_λ and the $(N-1)$ other states occupied by the electrons. We may therefore interpret h_{HF} in (8.138) as the energy operator of the electron in the state u_λ.

According to this interpretation, the quantity E_λ has the meaning of a one-electron eigenvalue. To give a more precise significance to E_λ, we first remark that upon taking the scalar product of (8.112) with u_λ and using (8.97), (8.103) and (8.104) we find that

$$E_\lambda = I_\lambda + \sum_\mu [J_{\lambda\mu} - K_{\lambda\mu}] \tag{8.139}$$

Summing over λ, we then have

$$\begin{aligned} \sum_\lambda E_\lambda &= \sum_\lambda I_\lambda + \sum_\lambda \sum_\mu [J_{\lambda\mu} - K_{\lambda\mu}] \\ &= \langle \Phi | \hat{H}_1 | \Phi \rangle + 2 \langle \Phi | \hat{H}_2 | \Phi \rangle \end{aligned} \tag{8.140}$$

where we have used (8.98) and (8.106). We may also rewrite (8.140) as

$$E[\Phi] = \sum_{\lambda} E_\lambda - \langle \Phi | \hat{H}_2 | \Phi \rangle \tag{8.141}$$

and we see that the total energy is *not* the sum of the individual energies. This is because in summing the individual electron energies, each kinetic energy and each interaction energy with the nucleus is counted once, while the mutual interaction energy, which has the average value $\langle \Phi | \hat{H}_2 | \Phi \rangle$, is counted twice. The total energy must therefore be obtained by subtracting $\langle \Phi | \hat{H}_2 | \Phi \rangle$ from the sum of individual energies as expressed in (8.141).

Let us now imagine that the electron λ is removed from the N-electron system. For example, if the original system was a neutral atom, we now have a positive ion with $(N-1)$ electrons. If we assume that the spin-orbitals of the $(N-1)$-electron system are the same as those of the N-electron system, we see from (8.107) that the difference between the total energy of the two systems is

$$\begin{aligned} E_N - E_{N-1} &= I_\lambda + \sum_{\mu} [J_{\lambda\mu} - K_{\lambda\mu}] \\ &= E_\lambda \end{aligned} \tag{8.142}$$

Thus the quantity E_λ represents approximately the energy required to remove an electron from the spin-orbital u_λ, or in other words the *ionisation energy* of electron λ. This result is known as *Koopmans' theorem*. It is worth stressing that the identification of E_λ with the ionisation energy of the electron λ is not rigorous, since the readjustment of the spin-orbitals of the $(N-1)$-electron system, which occurs after the removal of an electron, has been ignored. We also remark that although E_N and E_{N-1} are upper bounds to the true ground state energy of the N-electron and the $(N-1)$-electron systems (see (8.85)), the quantity E_λ is not an upper bound for the ionisation energy, since we have taken the difference of two upper bounds.

Properties of the Hartree–Fock solutions

Let us return to the Hartree–Fock equations (8.128). We see that *for a given state* of the atom or ion (here the ground state), characterised by the occupied spin-orbitals, all the electrons move in the *same* Hartree–Fock potential. Moreover, for a given potential, two spin-orbitals u_λ and u_ν corresponding to different individual energies $E_\lambda \neq E_\nu$ are easily shown to be *orthogonal*. Indeed, if we take the scalar product of (8.112) with $u_\nu(q_i)$, we find that

$$\begin{aligned} &\left\langle u_\nu(q_i) \left| -\frac{1}{2}\nabla^2_{\mathbf{r}_i} - \frac{Z}{r_i} \right| u_\lambda(q_i) \right\rangle + \sum_{\mu} \left\langle u_\nu(q_i) u_\mu(q_j) \left| \frac{1}{r_{ij}} \right| u_\mu(q_j) u_\lambda(q_i) \right\rangle \\ &- \sum_{\mu} \left\langle u_\nu(q_i) u_\mu(q_j) \left| \frac{1}{r_{ij}} \right| u_\mu(q_i) u_\lambda(q_j) \right\rangle = E_\lambda \langle u_\nu(q_i) | u_\lambda(q_i) \rangle \end{aligned} \tag{8.143}$$

Similarly, from the Hartree–Fock equation satisfied by $u_\nu(q_i)$ and the fact that the Hamiltonian (8.2) is Hermitian, we deduce that

$$\left\langle u_\nu(q_i)\left|-\frac{1}{2}\nabla^2_{\mathbf{r}_i}-\frac{Z}{r_i}\right|u_\lambda(q_i)\right\rangle+\sum_\mu\left\langle u_\nu(q_i)u_\mu(q_j)\left|\frac{1}{r_{ij}}\right|u_\lambda(q_i)u_\mu(q_j)\right\rangle$$
$$-\sum_\mu\left\langle u_\nu(q_i)u_\mu(q_j)\left|\frac{1}{r_{ij}}\right|u_\mu(q_i)u_\lambda(q_j)\right\rangle=E_\nu\langle u_\nu(q_i)|u_\lambda(q_i)\rangle \tag{8.144}$$

Hence, upon subtracting (8.144) from (8.143), we have

$$(E_\lambda-E_\nu)\langle u_\nu(q_i)|u_\lambda(q_i)\rangle=0 \tag{8.145}$$

so that

$$\langle u_\nu|u_\lambda\rangle=0,\qquad E_\lambda\neq E_\nu \tag{8.146}$$

Brillouin's theorem

Let Ψ_{HF} be a Hartree–Fock wave function, that is an N-electron Slater determinant

$$\Psi_{\text{HF}}=\frac{1}{\sqrt{N!}}\sum_P(-1)^P Pu_\alpha(q_1)u_\beta(q_2)\dots u_\lambda(q_i)\dots u_\nu(q_N) \tag{8.147}$$

constructed of Hartree–Fock orbitals. Suppose that

$$\Psi'=\frac{1}{\sqrt{N!}}\sum_P(-1)^P Pu_\alpha(q_1)u_\beta(q_2)\dots u'_\lambda(q_i)\dots u_\nu(q_N) \tag{8.148}$$

is the N-electron wave function corresponding to the excitation of an electron from the spin-orbital u_λ to u'_λ. Using the fact that $\langle u'_\lambda|u_\lambda\rangle=0$, we have

$$\langle\Psi'|H|\Psi_{\text{HF}}\rangle=\langle u'_\lambda(q_i)|\hat{h}_i|u_\lambda(q_i)\rangle+\sum_{\mu\neq\lambda}\langle u'_\lambda(q_i)u_\mu(q_j)\left|\frac{1}{r_{ij}}\right|u_\lambda(q_i)u_\mu(q_j)\rangle$$
$$-\sum_{\mu\neq\lambda}\langle u'_\lambda(q_i)u_\mu(q_j)\left|\frac{1}{r_{ij}}\right|u_\mu(q_i)u_\lambda(q_j)\rangle \tag{8.149}$$

which can also be written as

$$\langle\Psi'|H|\Psi_{\text{HF}}\rangle=\left\langle u'_\lambda(q_i)\left|\left[-\frac{1}{2}\nabla^2_{\mathbf{r}_i}-\frac{Z}{r_i}+\mathcal{V}^{\text{d}}_\lambda(\mathbf{r}_i)-\mathcal{V}^{\text{ex}}_\lambda(q_i)\right]\right|u_\lambda(q_i)\right\rangle$$
$$=E_\lambda\left\langle u'_\lambda(q_i)\middle|u_\lambda(q_i)\right\rangle=0 \tag{8.150}$$

since the Hartree–Fock spin-orbitals u_λ and u'_λ are orthogonal. Hence, matrix elements of the Hamiltonian (8.2) between a state represented by an N-electron

Slater determinant constructed of Hartree–Fock spin-orbitals and a singly excited state vanish identically. This result is known as the *Brillouin theorem*.

Hartree–Fock potentials

We shall now show that for atoms or ions with *closed subshells* (such as He, Li^+, Be, B^+, C^{2+}, Ne, etc.) the Hartree–Fock potential is *spherically symmetric*, so that the electron spatial orbitals are solutions of a *central field* problem. To see this we first note that in the case of closed subshells the N spin-orbitals (with N even) are obtained from $N' = N/2$ spatial orbitals. Each of these N' spatial orbitals occurs twice in the Slater determinant (8.87), once with spin 'up' (α) and once with spin 'down' (β). The simplest example is that of helium, where there is only one spatial orbital $u_{100} \equiv u_{1s}$ (see (8.26)).

In order to prove the above statement, we shall assume that the N' spatial orbitals are of the form

$$u_{nlm}(\mathbf{r}) = r^{-1}P_{nl}(r)Y_{lm}(\theta, \phi) \tag{8.151}$$

where we have set

$$P_{nl}(r) = rR_{nl}(r) \tag{8.152}$$

and we shall then show that the resulting Hartree–Fock potential $\mathcal{V}$ is spherically symmetric. Remembering that $\mathcal{V}$ contains the (central) nuclear attraction term $-Z/r_i$, the 'direct' potential $\mathcal{V}^{\mathrm{d}}$ and the 'exchange' potential $\mathcal{V}^{\mathrm{ex}}$ (see (8.127)) we first look at the direct potential $\mathcal{V}^{\mathrm{d}}$. Using (8.116), (8.122a), (8.151) and (8.152), the part of $\mathcal{V}^{\mathrm{d}}$ coming from a closed subshell ($n'l'$) is seen to be

$$\begin{aligned}\mathcal{V}^{\mathrm{d}}_{n'l'} &= 2\sum_{m'=-l'}^{+l'} \int |u_{n'l'm'}(\mathbf{r}_j)|^2 \frac{1}{r_{ij}}\,\mathrm{d}\mathbf{r}_j \\ &= 2\int |P_{n'l'}(r_j)|^2 \frac{1}{r_{ij}} \sum_{m'=-l'}^{+l'} |Y_{l'm'}(\theta_j, \phi_j)|^2\,\mathrm{d}r_j\,\mathrm{d}\Omega_j\end{aligned} \tag{8.153}$$

where the factor of 2 arises from the two spin orientations, and $\mathrm{d}\Omega_j \equiv \sin\theta_j\,\mathrm{d}\theta_j\,\mathrm{d}\phi_j$. Using the addition theorem of the spherical harmonics (see equation (A4.23) of Appendix 4) we have

$$\sum_{m'=-l'}^{+l'} |Y_{l'm'}(\theta_j, \phi_j)|^2 = \frac{2l'+1}{4\pi} \tag{8.154}$$

so that

$$\mathcal{V}^{\mathrm{d}}_{n'l'} = \frac{2(2l'+1)}{4\pi}\int |P_{n'l'}(r_j)|^2 \frac{1}{r_{ij}}\,\mathrm{d}r_j\,\mathrm{d}\Omega_j \tag{8.155}$$

The integrals on the angular variables (θ_j, ϕ_j) can readily be performed by expanding the quantity $1/r_{ij}$ in spherical harmonics and using the fact that the function $P_{n'l'}(r_j)$ does not depend on the angles. Writing (see (A4.25))

$$\frac{1}{r_{ij}} = \sum_{l=0}^{\infty} \sum_{m=-l}^{+l} \frac{4\pi}{2l+1} \frac{(r_<)^l}{(r_>)^{l+1}} Y^*_{lm}(\theta_i, \phi_i) Y_{lm}(\theta_j, \phi_j) \qquad \textbf{(8.156)}$$

where $r_<$ is the smaller and $r_>$ the larger of r_i and r_j, and proceeding as in Section 7.5, we find that

$$\mathcal{V}^{\mathrm{d}}_{n'l'} = 2(2l'+1) \int_0^{\infty} |P_{n'l'}(r_j)|^2 \frac{1}{r_>} \,\mathrm{d}r_j \qquad \textbf{(8.157)}$$

This expression is clearly independent of the angles (θ_i, ϕ_i) of $\mathbf{r}_i$ so that $\mathcal{V}^{\mathrm{d}}_{n'l'}$ is *central*. For an atom or ion with closed subshells the direct potential $\mathcal{V}^{\mathrm{d}}$ is just the sum of spherically symmetric contributions of the form (8.157), coming from each subshell, so that $\mathcal{V}^{\mathrm{d}}$ itself is spherically symmetric.

Let us now turn to the exchange potential $\mathcal{V}^{\mathrm{ex}}$. Using (8.118) and (8.122b), we see that when acting upon a spin-orbital $u_\lambda(q_i)$ whose spatial orbital is of the form (8.151), the part of $\mathcal{V}^{\mathrm{ex}}$ coming from a closed subshell $(n'l')$ gives

$$\mathcal{V}^{\mathrm{ex}}_{n'l'}[r_i^{-1} P_{nl}(r_i) Y_{lm}(\theta_i, \phi_i)] = \sum_{m'=-l'}^{+l'} \left[\int P^*_{n'l'}(r_j) Y^*_{l'm'}(\theta_j, \phi_j) \frac{1}{r_{ij}} \right.$$
$$\left. \times P_{nl}(r_j) Y_{lm}(\theta_j, \phi_j) \,\mathrm{d}r_j \,\mathrm{d}\Omega_j \right] r_i^{-1} P_{n'l'}(r_i) Y_{l'm'}(\theta_i, \phi_i) \qquad \textbf{(8.158)}$$

The expression on the right-hand side can now be evaluated by expanding $1/r_{ij}$ in spherical harmonics (see (8.156)) and using the formula (A4.40) of Appendix 4, together with the orthonormality relations (A4.32) for the Clebsch–Gordan coefficients. The result is (Problem 8.8)

$$\mathcal{V}^{\mathrm{ex}}_{n'l'}[r_i^{-1} P_{nl}(r_i) Y_{lm}(\theta_i, \phi_i)] = (2l'+1) \left\{ \sum_{L=|l-l'|}^{l+l'} \frac{1}{2L+1} |\langle ll'00|L0\rangle|^2 \right.$$
$$\left. \times \int_0^{\infty} P^*_{n'l'}(r_j) \frac{(r_<)^L}{(r_>)^{L+1}} P_{nl}(r_j) \,\mathrm{d}r_j \right\} r_i^{-1} P_{n'l'}(r_i) Y_{lm}(\theta_i, \phi_i) \qquad \textbf{(8.159)}$$

Looking only at the angular dependence, we see that when acting on the spherical harmonic $Y_{lm}(\theta_i, \phi_i)$, the exchange operator $\mathcal{V}^{\mathrm{ex}}_{n'l'}$ gives back something proportional to $Y_{lm}(\theta_i, \phi_i)$, the proportionality factor being independent of the angles (θ_i, ϕ_i). The angular dependence of the exchange potential $\mathcal{V}^{\mathrm{ex}}$ (which is a sum of contributions of the form (8.159) arising from each subshell) is therefore such that it is equivalent to a central potential.

The above discussion shows that for atoms (ions) with closed subshells the central field approximation is exact within the framework of the Hartree–Fock method. The radial equations then read

$$\left[-\frac{1}{2}\frac{\mathrm{d}^2}{\mathrm{d}r_i^2} + \frac{l(l+1)}{2r_i^2} - \frac{Z}{r_i} + \mathcal{V}^{\mathrm{d}} - \mathcal{V}^{\mathrm{ex}} \right] P_{nl}(r_i) = E_{nl} P_{nl}(r_i) \qquad \textbf{(8.160)}$$

where

$$\mathcal{V}^{\mathrm{d}} = \sum_{n'l'} \mathcal{V}^{\mathrm{d}}_{n'l'}$$
$$= \sum_{n'l'} 2(2l'+1) \int_0^\infty |P_{n'l'}(r_j)|^2 \frac{1}{r_>} \mathrm{d}r_j \qquad \textbf{(8.161a)}$$

and

$$\mathcal{V}^{\mathrm{ex}} P_{nl}(r_i) = \sum_{n'l'} \mathcal{V}^{\mathrm{ex}}_{n'l'} P_{nl}(r_i)$$
$$= \sum_{n'l'} \sum_{L=|l-l'|}^{l+l'} \frac{2l'+1}{2L+1} |\langle ll'00|L0\rangle|^2 \left[\int_0^\infty P^*_{n'l'}(r_j) \frac{(r_<)^L}{(r_>)^{L+1}} P_{nl}(r_j)\, \mathrm{d}r_j \right] P_{n'l'}(r_i) \qquad \textbf{(8.161b)}$$

where

$$r_> = \max(r_i, r_j)$$
$$r_< = \min(r_i, r_j) \qquad \textbf{(8.161c)}$$
$$|l-l'| \leqslant L \leqslant l+l'$$

For atoms (ions) having incomplete subshells, the Hartree–Fock potential $\mathcal{V}$ is no longer spherically symmetric. However, this departure from spherical symmetry is often small, since in many cases (in particular for the ground state) it arises from only *one* incomplete subshell. An approximate Hartree–Fock central field $\bar{\mathcal{V}}$ is then obtained by averaging $\mathcal{V}$ over spin directions and angles.

An example: the Be ground state

We shall illustrate the Hartree–Fock method by considering the ground state $1s^2 2s^2\ {}^1S$ of beryllium. In this case the Slater determinant (8.87) reads

$$\Phi(q_1, q_2, q_3, q_4) = \frac{1}{\sqrt{4!}} \begin{vmatrix} u_{1s\uparrow}(q_1) & u_{1s\downarrow}(q_1) & u_{2s\uparrow}(q_1) & u_{2s\downarrow}(q_1) \\ u_{1s\uparrow}(q_2) & u_{1s\downarrow}(q_2) & u_{2s\uparrow}(q_2) & u_{2s\downarrow}(q_2) \\ u_{1s\uparrow}(q_3) & u_{1s\downarrow}(q_3) & u_{2s\uparrow}(q_3) & u_{2s\downarrow}(q_3) \\ u_{1s\uparrow}(q_4) & u_{1s\downarrow}(q_4) & u_{2s\uparrow}(q_4) & u_{2s\downarrow}(q_4) \end{vmatrix} \qquad \textbf{(8.162)}$$

and the Hartree–Fock potential is given by (see (8.127))

$$\mathcal{V} = -\frac{4}{r_i} + V^{\mathrm{d}}_{1s\uparrow} + V^{\mathrm{d}}_{1s\downarrow} + V^{\mathrm{d}}_{2s\uparrow} + V^{\mathrm{d}}_{2s\downarrow} - (V^{\mathrm{ex}}_{1s\uparrow} + V^{\mathrm{ex}}_{1s\downarrow} + V^{\mathrm{ex}}_{2s\uparrow} + V^{\mathrm{ex}}_{2s\downarrow}) \qquad \textbf{(8.163)}$$

where the notation is self-explanatory. Because we have closed (sub)shells, the spatial parts of $u_{1s\uparrow}$ and $u_{1s\downarrow}$ are the same, as are those of $u_{2s\uparrow}$ and $u_{2s\downarrow}$. We may thus write

$$u_{1s\uparrow}(q) = u_{1s}(r)\alpha; \quad u_{1s\downarrow}(q) = u_{1s}(r)\beta$$
$$u_{2s\uparrow}(q) = u_{2s}(r)\alpha; \quad u_{2s\downarrow}(q) = u_{2s}(r)\beta \qquad \mathbf{(8.164)}$$

From the foregoing discussion, the Hartree–Fock equations for the functions $u_{1s}(r)$ and $u_{2s}(r)$ are the two coupled integro-differential equations

$$\left[-\frac{1}{2}\nabla_{\mathbf{r}}^2 - \frac{4}{r} + V_{1s}^{\mathrm{d}}(r) + 2V_{2s}^{\mathrm{d}}(r) - V_{2s}^{\mathrm{ex}}(r)\right]u_{1s}(r) = E_{1s}u_{1s}(r) \qquad \mathbf{(8.165a)}$$

and

$$\left[-\frac{1}{2}\nabla_{\mathbf{r}}^2 - \frac{4}{r} + V_{2s}^{\mathrm{d}}(r) + 2V_{1s}^{\mathrm{d}}(r) - V_{1s}^{\mathrm{ex}}(r)\right]u_{2s}(r) = E_{2s}u_{2s}(r) \qquad \mathbf{(8.165b)}$$

where

$$V_{1s}^{\mathrm{d}}(r) = \int u_{1s}^*(r')\frac{1}{|\mathbf{r} - \mathbf{r}'|}u_{1s}(r')\,\mathrm{d}\mathbf{r}', \qquad \mathbf{(8.166a)}$$

and

$$V_{1s}^{\mathrm{ex}}(r)f(\mathbf{r}) = \left[\int u_{1s}^*(r')\frac{1}{|\mathbf{r} - \mathbf{r}'|}f(\mathbf{r}')\,\mathrm{d}\mathbf{r}'\right]u_{1s}(r) \qquad \mathbf{(8.166b)}$$

with similar definitions for $V_{2s}^{\mathrm{d}}(r)$ and $V_{2s}^{\mathrm{ex}}(r)$ (see (8.116) and (8.119)). The individual electron energies E_{1s} and E_{2s} introduced in (8.165) are such that

$$E_{1s} = E_{1s\uparrow} = E_{1s\downarrow} \quad E_{2s} = E_{2s\uparrow} = E_{2s\downarrow} \qquad \mathbf{(8.167)}$$

The spatial orbital $u_{1s}(r)$ is a solution corresponding to the lowest eigenvalue of (8.165a), whereas $u_{2s}(r)$ corresponds to the lowest eigenvalue of (8.165b), with the condition $\langle u_{1s}|u_{2s}\rangle = 0$. Writing the orbitals in the form (8.151), namely

$$u_{1s}(r) = r^{-1}P_{1s}(r)Y_{00}; \quad u_{2s}(r) = r^{-1}P_{2s}(r)Y_{00} \qquad \mathbf{(8.168)}$$

with $Y_{00} = (4\pi)^{-1/2}$, the two coupled Hartree–Fock equations (8.165) become

$$\left[-\frac{1}{2}\frac{\mathrm{d}^2}{\mathrm{d}r^2} + \frac{l(l+1)}{2r^2} - \frac{4}{r} + V_{1s}^{\mathrm{d}}(r) + 2V_{2s}^{\mathrm{d}}(r) - V_{2s}^{\mathrm{ex}}(r)\right]P_{1s}(r) = E_{1s}P_{1s}(r) \qquad \mathbf{(8.169a)}$$

and

$$\left[-\frac{1}{2}\frac{\mathrm{d}^2}{\mathrm{d}r^2} + \frac{l(l+1)}{2r^2} - \frac{4}{r} + V_{2s}^{\mathrm{d}}(r) + 2V_{1s}^{\mathrm{d}}(r) - V_{1s}^{\mathrm{ex}}(r)\right]P_{2s}(r) = E_{2s}P_{2s}(r) \qquad \mathbf{(8.169b)}$$

It is worth stressing that the Hartree–Fock potential corresponding to an excited state of the atom is different from the ground state potential (8.163), so that the Hartree–Fock equations for the orbitals of the excited atom will be different from (8.165) or (8.169). As a result, the excited state solutions of the coupled equations (8.165) or (8.169) do *not* represent orbitals corresponding to low-lying excited states of the atom.

Solution of the Hartree–Fock equations

We have already indicated above that the solution of the Hartree–Fock equations proceeds by iteration, starting from approximate individual spin-orbitals, and subject to the requirement of self-consistency. At each step of the iteration the coupled equations must be solved numerically, the final result of the computations being numerical values of the radial Hartree–Fock orbitals.

For practical purposes, however, analytical fits to the (numerical) solutions of the Hartree–Fock equations are very useful. For example, we already encountered in Chapter 7 a simple analytical fit to the Hartree–Fock ground state orbital for helium (see (7.85)). More generally, a convenient basis set for analytic fits to Hartree–Fock spatial orbitals is provided by *Slater orbitals*, the general form of which is

$$\chi_{nlm}(\mathbf{r}) = N r^{n-1} \mathrm{e}^{-\alpha r} Y_{lm}(\theta, \phi) \tag{8.170}$$

where n is a positive integer, α is the 'orbital exponent', and N is a normalisation constant, given by

$$N = \frac{(2\alpha)^{n+1/2}}{[(2n)!]^{1/2}} \tag{8.171}$$

Looking back at (3.53) and (3.58), and remembering that L_{n+l}^{2l+1} is a polynomial of degree $n - l - 1$, we see that the Slater orbitals (8.170) behave in the same way as hydrogenic wave functions for large r. However, in contrast to the hydrogenic radial functions (3.58), the Slater orbitals do not possess radial nodes. In terms of Slater orbitals $\chi_i(\mathbf{r})$, a Hartree–Fock spatial orbital $u(\mathbf{r})$ is then given by

$$u(\mathbf{r}) = \sum_{i=1}^{N} c_i \chi_i(\mathbf{r}) \tag{8.172}$$

where the quantities c_i are given coefficients. For example, the Hartree–Fock spatial orbitals of the neon ground state are [11]

$$\begin{aligned}
u_{1\mathrm{s}} &= r^{-1} P_{1\mathrm{s}}(r) Y_{00}(\theta, \phi) = 0.937\,17\chi_1 + 0.048\,99\chi_2 + 0.000\,58\chi_3 \\
&\quad - 0.000\,64\chi_4 + 0.005\,51\chi_5 + 0.019\,99\chi_6 \\
u_{2\mathrm{s}} &= r^{-1} P_{2\mathrm{s}}(r) Y_{00}(\theta, \phi) = -0.230\,93\chi_1 - 0.006\,35\chi_2 + 0.186\,20\chi_3 \\
&\quad + 0.668\,99\chi_4 + 0.309\,10\chi_5 - 0.138\,71\chi_6 \\
u_{2\mathrm{p}} &= r^{-1} P_{2\mathrm{p}}(r) Y_{10}(\theta, \phi) = 0.217\,99\chi_7 + 0.533\,38\chi_8 + 0.329\,33\chi_9 \\
&\quad + 0.018\,72\chi_{10}
\end{aligned} \tag{8.173}$$

[11] See Clementi and Roetti (1974).

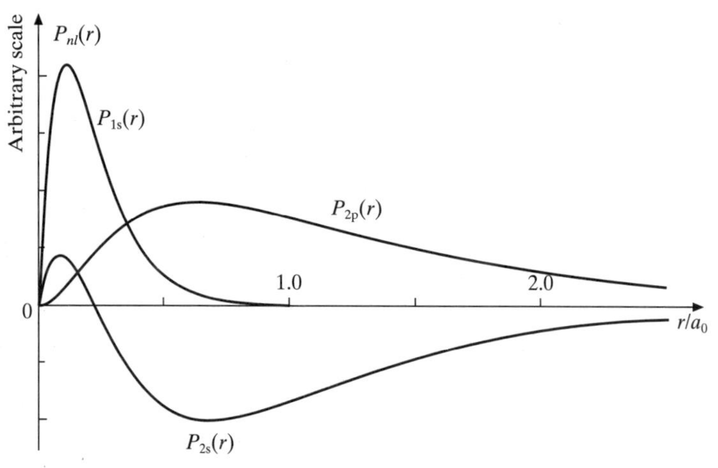

Figure 8.5 The radial functions $P_{1s}(r)$, $P_{2s}(r)$ and $P_{2p}(r)$ for neon.

where

$$
\begin{aligned}
\chi_1 &= N_1 \exp(-9.484\,86r) Y_{00}(\theta, \phi); \\
\chi_2 &= N_2 \exp(-15.565\,90r) Y_{00}(\theta, \phi) \\
\chi_3 &= N_3\, r \exp(-1.961\,84r) Y_{00}(\theta, \phi); \\
\chi_4 &= N_4\, r \exp(-2.864\,23r) Y_{00}(\theta, \phi) \\
\chi_5 &= N_5\, r \exp(-4.825\,30r) Y_{00}(\theta, \phi); \\
\chi_6 &= N_6\, r \exp(-7.792\,42r) Y_{00}(\theta, \phi) \\
\chi_7 &= N_7\, r \exp(-1.452\,08r) Y_{10}(\theta, \phi); \\
\chi_8 &= N_8\, r \exp(-2.381\,68r) Y_{10}(\theta, \phi) \\
\chi_9 &= N_9\, r \exp(-4.484\,89r) Y_{10}(\theta, \phi); \\
\chi_{10} &= N_{10}\, r \exp(-9.134\,64r) Y_{10}(\theta, \phi)
\end{aligned}
\tag{8.174}
$$

and the normalisation constants are given by (8.171).

The radial functions P_{1s}, P_{2s} and P_{2p} for neon are plotted in Fig. 8.5. The *radial density function* $D(r)$, defined as the probability per unit length of finding an electron at a distance r from the nucleus, is given by

$$
\begin{aligned}
D(r) &= r^2 \int \rho(\mathbf{r})\, \mathrm{d}\Omega \\
&= \sum_{nl} q_{nl} |P_{nl}(r)|^2
\end{aligned}
\tag{8.175}
$$

where q_{nl} is the number of equivalent electrons in a subshell (nl). Thus, for the case of neon considered here we have

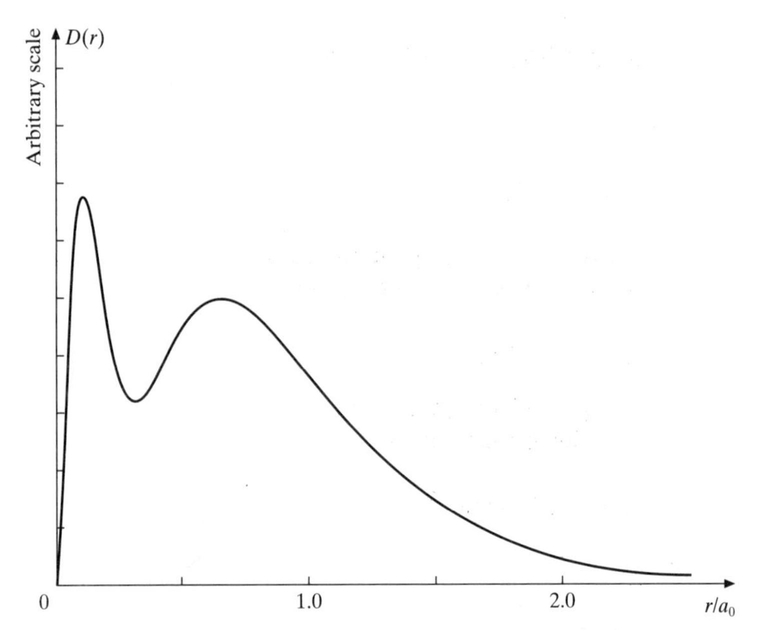

Figure 8.6 The radial density function $D(r)$ for neon.

$$D(r) = 2|P_{1s}(r)|^2 + 2|P_{2s}(r)|^2 + 6|P_{2p}(r)|^2 \tag{8.176}$$

This function is shown in Fig. 8.6.

The Hartree equations

Let us return to the Hartree–Fock equations (8.138). If the exchange potential $\mathcal{V}_\lambda^{\text{ex}}$ is neglected in (8.138), one obtains the system of equations

$$\left[-\frac{1}{2}\nabla_{\mathbf{r}_i}^2 - \frac{Z}{r_i} + \mathcal{V}_\lambda^{\text{d}}(\mathbf{r}_i)\right]u_\lambda(q_i) = E_\lambda u_\lambda(q_i),$$

$$\lambda = \alpha, \beta, \ldots, \nu \tag{8.177}$$

or, more explicitly,

$$\left[-\frac{1}{2}\nabla_{\mathbf{r}_i}^2 - \frac{Z}{r_i} + \sum_{\mu\neq\lambda}\int u_\mu^*(\mathbf{r}_j)\frac{1}{r_{ij}}u_\mu(\mathbf{r}_j)\,\mathrm{d}\mathbf{r}_j\right]u_\lambda(\mathbf{r}_i) = E_\lambda u_\lambda(\mathbf{r}_i),$$

$$\lambda = \alpha, \beta, \ldots, \nu \tag{8.178}$$

where we have used (8.113), (8.116) and (8.135). These equations were originally obtained by Hartree from the independent-particle approximation and the assumption that there is a charge density associated with each electron, which is equal to $-e$ times its position probability density. By making the central field approximation, which consists in replacing $\mathcal{V}_{\lambda}^{\mathrm{d}}(\mathbf{r}_i)$ by its average $\bar{\mathcal{V}}_{\lambda}^{\mathrm{d}}(r_i)$ taken over the angles of $\mathbf{r}_i$, one obtains for the radial functions $P_{nl}(r_i)$ the *radial Hartree equations*

$$\left[-\frac{1}{2}\frac{\mathrm{d}^2}{\mathrm{d}r_i^2} + \frac{l(l+1)}{2r_i^2} - \frac{Z}{r_i} + \bar{\mathcal{V}}_{\lambda}^{\mathrm{d}}(r_i)\right]P_{nl}(r_i) = E_{nl}P_{nl}(r_i) \tag{8.179}$$

From the derivation of the Hartree–Fock equations given above, it is apparent that the Hartree equations (8.177) can be obtained from the variational method by using as trial function the simple (*non*-antisymmetrised) product of spin-orbitals Φ_{H} given by (8.90) instead of the antisymmetrised product Φ given by the Slater determinant (8.87). Thus *the Hartree trial function* Φ_{H} *does not satisfy the requirement of antisymmetry* imposed by the Pauli exclusion principle, so that the Hartree equations (8.177) lack the exchange potential $\mathcal{V}_{\lambda}^{\mathrm{ex}}$. Nevertheless, the weaker condition of the exclusion principle, according to which 'no two electrons in an atom may have the same set of quantum numbers', can be satisfied in Hartree's method by requiring that only one electron populates each quantum state. It is also worth noting that in contrast to the Hartree–Fock equations, where all the electrons move in the same Hartree–Fock potential for a given state of the system, the effective potential in the Hartree equations depends on the particular orbital u_{λ} considered. It follows that in general the Hartree orbitals are not mutually orthogonal.

As in the case of the Hartree–Fock equations discussed above, the system of Hartree equations can be solved by iteration, subject to the self-consistency requirement. Because of the absence of exchange terms, the Hartree equations are in fact much simpler to solve than the corresponding Hartree–Fock equations, so that the Hartree spin-orbitals can be used as a first approximation to start the iterative solution of the Hartree–Fock equations.

8.5 Corrections to the central field approximation. Correlation effects. *L–S* coupling and *j–j* coupling

Correlation effects

We have seen above that the Hartree–Fock approach is a variational method in which the trial function for the N-electron atom (ion) is a Slater determinant, whose individual spin-orbitals are optimised. Let us call E_{HF} the Hartree–Fock energy of a given state, and Ψ_{HF} the corresponding Hartree–Fock wave function. It is clear that both E_{HF} and Ψ_{HF} are only approximations to the exact energy E_{exact} and the exact wave function Ψ_{ex} of the Hamiltonian (8.2). The difference

$$E_{\mathrm{corr}} = E_{\mathrm{exact}} - E_{\mathrm{HF}} \tag{8.180}$$

is known as the *correlation energy*. We emphasise that the quantity E_{exact} which appears in (8.180) is the exact energy of the *non-relativistic* Hamiltonian (8.2), and hence is not quite the same as the experimental energy. It should also be noted

Table 8.6 Values of E_{HF}, E_{exact}, E_{corr} (in a.u.) and of the ratio E_{corr}/E_{exact} for the ground state of several atomic systems.

Atomic system	E_{HF}	E_{exact}	E_{corr}	E_{corr}/E_{exact}
He	−2.862	−2.904	−0.042	0.014
Be	−14.573	−14.667	−0.094	0.006
B^+	−24.238	−24.349	−0.111	0.005
Ne	−128.55	−128.93	−0.38	0.003

that a certain amount of correlation is already included in the Hartree–Fock wave function because of the fact that it is totally antisymmetric. One usually denotes by *correlation effects* the remaining electron correlations, not included in the Hartree–Fock wave function, which are responsible for the correlation energy (8.180).

Since the Hartree–Fock method is a variational one, the energy E_{HF} must lie above E_{exact} for the ground state, so that E_{corr} is always negative in this case. As an example, we give in Table 8.6 the values of E_{HF}, E_{exact}, E_{corr} and the relative error E_{corr}/E_{exact} for the ground state of a few atomic systems. We see from this table that the relative error in the Hartree–Fock energy of an atom (ion) is only about 1 per cent. However, in particular regions of configuration space – which do not play an important role in the variational integral – the Hartree–Fock wave functions may be quite inaccurate. As a consequence, various matrix elements, such as those required to calculate transition rates or hyperfine structure effects, may sometimes be in serious error when evaluated with the Hartree–Fock wave functions.

Correlation effects can be studied by using *Hartree–Fock perturbation theory*, in which the unperturbed Hamiltonian H_0 is the total Hartree–Fock Hamiltonian

$$H_{HF} = \sum_{i=1}^{N} h_{HF}(i) \tag{8.181}$$

where $h_{HF}(i)$ are the individual Hartree–Fock Hamiltonians (8.130). The unperturbed wave function is therefore the Hartree–Fock wave function Ψ_{HF}. The perturbation is

$$H' = H - H_{HF} \tag{8.182}$$

where H is the total Hamiltonian (8.2).

Let us write E_{exact} in the form

$$E_{exact} = \sum_{n=0}^{\infty} E^{(n)} \tag{8.183}$$

where the index n refers to the order of the perturbation. Since the Hartree–Fock energy

$$\begin{aligned} E_{\rm HF} &= \langle \Psi_{\rm HF} | H | \Psi_{\rm HF} \rangle \\ &= \langle \Psi_{\rm HF} | H_{\rm HF} + H' | \Psi_{\rm HF} \rangle \\ &= E^{(0)} + E^{(1)} \end{aligned} \qquad \textbf{(8.184)}$$

already contains the contribution of the perturbation H' to first order, we see from (8.180) that

$$E_{\rm corr} = \sum_{n=2}^{\infty} E^{(n)} \qquad \textbf{(8.185)}$$

The quantities $E^{(n)}$ can be calculated by using the *variation–perturbation method* discussed in Section 2.8. For example, in 1967, F.W. Byron and C.J. Joachain calculated $E^{(n)}$ up to $n = 5$ for the ground state of Be and B^+. They found that $E_{\rm corr}(\text{Be}) = -0.0925$ a.u. and $E_{\rm corr}(\text{B}^+) = -0.1096$ a.u., in very good agreement with the exact non-relativistic values (obtained by using the experimental data) quoted in Table 8.6.

Correlation effects can also be calculated by using the variational method with a trial function Φ which is a linear combination of Slater determinants,

$$\Phi = \sum_i c_i \Phi_i \qquad \textbf{(8.186)}$$

where the coefficients c_i are variational parameters. The various Slater determinants Φ_i differ in the choice of the spin-orbitals occupied by the electrons, and correspond therefore to different electronic configurations. This approach is known as the *configuration–interaction method*. It is clear that the configuration–interaction approach reduces to the Hartree–Fock method when only one Slater determinant is used in (8.186).

Another approach which can be used to study correlation effects is the *density functional method*, which will be discussed in Section 8.6.

L–S coupling and *j–j* coupling

In the central field approximation, the non-relativistic Hamiltonian (8.2) of the N-electron atom (ion) is replaced by the Hamiltonian

$$H_c = \sum_{i=1}^{N} h_i \qquad \textbf{(8.187)}$$

where h_i is the individual Hamiltonian of electron i in the central field $V(r_i)$ (see (8.10)). For example, if we choose the Hartree–Fock potential $\bar{\mathcal{V}}$ (that is, the Hartree–Fock potential $\mathcal{V}$, averaged for the case of atoms or ions with incomplete subshells) as our central field, then

$$h_i = -\tfrac{1}{2}\nabla^2_{\mathbf{r}_i} + \bar{\mathcal{V}} \qquad \textbf{(8.188)}$$

Let us now consider the main corrections to the central field approximation. The first important correction to the central field Hamiltonian H_c is the term H_1, given by (8.11), which represents the *difference between the actual Coulomb repulsion between the electrons and the average electron repulsion* contained in the central field V. In particular, if V is a (central) Hartree–Fock potential, the term H_1 leads to the correlation effects discussed at the end of the previous section. In this section, however, we shall also consider the corrections due to the *spin–orbit interactions* of the electrons. Working within the framework of the independent-particle model, we write the spin–orbit correction term in a form which is an extension of (5.16)–(5.17). That is,

$$H_2 = \sum_i \xi(r_i)\mathbf{L}_i \cdot \mathbf{S}_i \tag{8.189}$$

where $\mathbf{L}_i = \mathbf{r}_i \times \mathbf{p}_i$ is the orbital angular momentum operator of electron i, $\mathbf{S}_i$ is its spin angular momentum, and the function $\xi(r_i)$ is given by

$$\xi(r_i) = \frac{1}{2m^2c^2}\frac{1}{r_i}\frac{\mathrm{d}V(r_i)}{\mathrm{d}r_i} \tag{8.190}$$

It can be shown that the contribution to the sum (8.189) coming from closed subshells vanishes, so that the summation in (8.189) is only over electrons outside closed subshells.

Adding the corrections H_1 and H_2 to the central field Hamiltonian, we obtain the total Hamiltonian

$$\mathcal{H} = H_c + H_1 + H_2 \tag{8.191}$$

Because this Hamiltonian describes an isolated atom, the *total parity* and the *total angular momentum* are constants of the motion. In the present case the total angular momentum of the electrons is given by

$$\mathbf{J} = \mathbf{L} + \mathbf{S} = \sum_{i=1}^{N} \mathbf{J}_i \tag{8.192}$$

where $\mathbf{L} = \Sigma_i \mathbf{L}_i$ is the total orbital angular momentum of the electrons, $\mathbf{S} = \Sigma_i \mathbf{S}_i$ is their total spin angular momentum, and

$$\mathbf{J}_i = \mathbf{L}_i + \mathbf{S}_i \tag{8.193}$$

In order to discuss the effects of the terms H_1 and H_2, we shall use *perturbation theory*, starting from the eigenfunctions and eigenenergies of H_c as our zero-order approximation. As we have seen in Section 8.1, these unperturbed eigenfunctions and eigenenergies correspond to given electron configurations. In what follows we shall assume that the matrix elements of the perturbation between two different electron configurations is small with respect to the energy intervals between unperturbed configuration energies. We can then study the effect of the perturbation on atomic states corresponding to a single configuration of the electrons.

The manner in which the perturbation calculation is to be carried out depends on the relative magnitude of the two perturbing terms H_1 and H_2. The case for which both perturbations H_1 and H_2 are of the same order of magnitude is difficult to handle because both terms must be treated on the same footing. This situation, which is known as *intermediate coupling*, will not be examined here [12]. Instead, we shall consider the two extreme situations for which $|H_1| \gg |H_2|$ and $|H_2| \gg |H_1|$. The first one is the most frequently encountered; it occurs for atoms (ions) with small and intermediate values of Z, and is called the L–S (or *Russell–Saunders*) *coupling case*. The second situation, which arises for atoms (ions) with large Z, is known as the *j–j coupling case*. We shall first discuss L–S coupling, and then turn to j–j coupling.

L–S coupling

Since in this case the electrostatic energy correction H_1 is large in comparison with the spin–orbit term H_2, the *first step* of the perturbation calculation consists of neglecting H_2 and obtaining approximate eigenfunctions and eigenenergies of the Hamiltonian

$$H = H_c + H_1 \tag{8.194}$$

In general, there are a number of degenerate states belonging to the same configuration, which differ by the values of the quantum numbers m_l and m_s of the individual electrons. The perturbation H_1 has the effect of removing – at least partly – this degeneracy. The energy levels of $H = H_c + H_1$ arising from a given unperturbed energy level E_c of H_c will thus be obtained by diagonalising the perturbation H_1 within the subspace of the degenerate states belonging to the eigenvalue E_c.

This diagonalisation is greatly simplified by taking into account the symmetry properties of the Hamiltonian $H = H_c + H_1$. Since this Hamiltonian does not contain spin–orbit energy terms, it commutes not only with the total angular momentum $\mathbf{J}$ of the electrons, but also separately with their total orbital angular momentum $\mathbf{L}$ (see Problem 8.9) and their total spin angular momentum $\mathbf{S}$ [13]. As a result, the eigenvalues of H can be characterised by the total electronic orbital angular momentum quantum number L and the total electronic spin quantum number S. Moreover, these energy eigenvalues are independent of the quantum numbers M_L and M_S, so that each energy level of H labelled by a pair (LS) is $(2L + 1)(2S + 1)$ times degenerate with respect to M_L and M_S. As we already saw in Section 7.3, energy levels corresponding to definite values of L and S are called *terms* and are denoted as ^{2S+1}L, with the capital letters S, P, D, F, . . . corresponding to the values $L = 0, 1, 2, 3, \ldots$. We also recall that the number $2S + 1$ is called the *multiplicity* of the term. If the multiplicity $2S + 1$ equals 1 (that is, if $S = 0$) the term is called a singlet. If $2S + 1 = 2, 3, 4, \ldots$ we have respectively a doublet, a triplet, a quartet, and so on. The parity of the term may be indicated by the addition of the

[12] The intermediate coupling case is treated for example in Condon and Shortley (1951) and Sobelman (1979) where a detailed account of the theory of atomic spectra can be found.

[13] It is worth noting that the use of antisymmetric wave functions (Slater determinants) couples the electron spins to the electrostatic energy (as we have seen in Section 7.6 in discussing the excited states of helium) despite the fact that $[H, \mathbf{S}] = 0$.

superscripts *e* (even) or *o* (odd), for example $^2P^o$, $^3D^e$, etc. Using the Dirac notation, the wave functions corresponding to a term will be denoted by $|\gamma LSM_L M_S\rangle$, where γ refers to the configuration of H_c to which the level belongs.

Determination of the possible terms of a multielectron configuration in *L–S* coupling

In order to determine all the possible terms corresponding to a given configuration (that is, the possible values of L and S), the rules for the *addition of angular momenta* must be used. However, in combining the individual electron orbital angular momenta $\mathbf{L}_i$ to obtain $\mathbf{L}$ and the individual electron spin angular momenta $\mathbf{S}_i$ to obtain $\mathbf{S}$, one must not forget to reject the values of L and S corresponding to states forbidden by the Pauli exclusion principle.

We first note that for a *closed subshell*, that is a configuration containing the maximum number $2(2l + 1)$ of equivalent electrons (having the same values of n and l), there is only one possible term, namely the ^1S term. This is a direct consequence of the fact that $M_L = \sum_i m_{l_i} = 0$ (since m_{l_i} runs over all the possible values $0, \pm 1, \pm 2, \dots, \pm l$) and similarly $M_S = \sum_i m_{s_i} = 0$. As a result, we have $L = S = 0$ for a filled subshell.

Let us now consider the case of atoms (ions) with *incomplete subshells*. Since $L = S = 0$ for closed subshells, we only need to consider the electrons outside closed subshells ('optically active' electrons) to determine the possible values of L and S. Three cases must be considered:

Electrons belonging to different subshells (non-equivalent electrons)

In this case no two 'optically active' electrons can have the same set of quantum numbers, so that the Pauli exclusion principle is automatically satisfied. The allowed values of the quantum numbers L and S are therefore obtained by adding the individual orbital angular momenta $\mathbf{L}_i$ of the 'optically active' electrons to form $\mathbf{L}$ and the spin angular momenta $\mathbf{S}_i$ of these electrons to form $\mathbf{S}$. We recall that on adding two orbital angular momenta $\mathbf{L}_1$ and $\mathbf{L}_2$, with corresponding orbital angular quantum numbers l_1 and l_2 (such that $l_1(l_1 + 1)\hbar^2$ is the eigenvalue of $\mathbf{L}_1^2$ and $l_2(l_2 + 1)\hbar^2$ is the eigenvalue of $\mathbf{L}_2^2$), the allowed values of the total orbital angular momentum quantum number L are

$$L = |l_1 - l_2|, |l_1 - l_2| + 1, \dots, l_1 + l_2 \tag{8.195}$$

Similarly, upon adding two spin angular momenta $\mathbf{S}_1$ and $\mathbf{S}_2$, with spin quantum numbers s_1 and s_2, the allowed values of the total spin quantum number S are

$$S = |s_1 - s_2|, |s_1 - s_2| + 1, \dots, s_1 + s_2 \tag{8.196}$$

Let us now consider two non-equivalent electrons, having orbital angular momentum quantum numbers l_1 and l_2, respectively. The allowed values of L are then given by (8.195). Moreover, since $s_1 = s_2 = 1/2$, we have $S = 0, 1$. We illustrate these considerations by two examples:

***Configuration* np *n*′p**

We have $l_1 = l_2 = 1$ and $s_1 = s_2 = 1/2$, so that $L = 0, 1, 2$ and $S = 0, 1$. The possible terms are therefore

$$^1S,\ ^1P,\ ^1D,\ ^3S,\ ^3P,\ ^3D \tag{8.197}$$

***Configuration* np *n*′d**

Here $l_1 = 1, l_2 = 2$ and $s_1 = s_2 = 1/2$. Thus $L = 1, 2, 3$ and $S = 0, 1$. The possible terms are thus given by

$$^1P,\ ^1D,\ ^1F,\ ^3P,\ ^3D,\ ^3F \tag{8.198}$$

If there are more than two electrons, the addition of orbital angular momenta and spin angular momenta is first performed for two electrons; the rules (8.195) and (8.196) are then used successively to add the orbital and the spin angular momentum of the third electron, the fourth electron, and so on. As an example, let us consider the configuration *n*p *n*′p *n*″d. Adding first the orbital and spin angular momenta of the two electrons *n*p and *n*′p, we find the terms listed in (8.197). Now the third electron *n*″d has $l = 2$ and $s = 1/2$. Using (8.195) and (8.196) we see that the addition of the electron *n*″d to the term ^1S of (8.197) gives a term with $L = 2$ and $S = 1/2$, namely ^2D. In the same way, adding the *n*″d electron

to the term ^1P yields the terms ^2P, ^2D, ^2F
to the term ^1D yields the terms ^2S, ^2P, ^2D, ^2F, ^2G
to the term ^3S yields the terms ^2D, ^4D
to the term ^3P yields the terms ^2P, ^2D, ^2F, ^4P, ^4D, ^4F
to the term ^3D yields the terms ^2S, ^2P, ^2D, ^2F, ^2G, ^4S, ^4P, ^4D, ^4F, ^4G

These results may be summarised by writing that the terms we have obtained are

$$\begin{matrix} ^2S, & ^2P, & ^2D, & ^2F, & ^2G, & ^4S, & ^4P, & ^4D, & ^4F, & ^4G \\ 2 & 4 & 6 & 4 & 2 & & 2 & 3 & 2 & \end{matrix} \tag{8.199}$$

where the number under the term symbol indicates the number of identical terms.

Electrons belonging to the same subshell (equivalent electrons)

The determination of the possible terms for equivalent ‘optically active’ electrons is more difficult. Indeed, in this case certain values of L and S are ruled out because of the Pauli exclusion principle.

The simplest case is that of two equivalent s electrons, corresponding to the configuration ns^2. As we have seen above in discussing closed subshells, we must have $M_L = M_S = 0$, which implies that the only possible term is ^1S. We note that if the Pauli principle were ignored, an additional term ^3S would appear.

Let us now consider the case of two equivalent p electrons, corresponding to the configuration np^2. We have seen in Section 8.1 that the degeneracy (or statistical weight) of a configuration np^2 is $g = 15$. The 15 possible states associated with the configuration np^2 can be inferred directly from Table 8.7, where the possible

Table 8.7 Possible quantum numbers for the configuration np^2.

Number	m_{l_1}	m_{s_1}	m_{l_2}	m_{s_2}	$M_L = m_{l_1} + m_{l_2}$	$M_S = m_{s_1} + m_{s_2}$
1	1	$\frac{1}{2}$	1	$-\frac{1}{2}$	2	0
2	1	$-\frac{1}{2}$	0	$\frac{1}{2}$	1	0
3	0	$\frac{1}{2}$	0	$-\frac{1}{2}$	0	0
4	0	$-\frac{1}{2}$	−1	$\frac{1}{2}$	−1	0
5	−1	$\frac{1}{2}$	−1	$-\frac{1}{2}$	−2	0
6	1	$\frac{1}{2}$	0	$\frac{1}{2}$	1	1
7	1	$-\frac{1}{2}$	0	$-\frac{1}{2}$	1	−1
8	0	$\frac{1}{2}$	−1	$\frac{1}{2}$	−1	1
9	0	$-\frac{1}{2}$	−1	$-\frac{1}{2}$	−1	−1
10	1	$\frac{1}{2}$	0	$-\frac{1}{2}$	1	0
11	1	$-\frac{1}{2}$	−1	$\frac{1}{2}$	0	0
12	0	$\frac{1}{2}$	−1	$-\frac{1}{2}$	−1	0
13	1	$\frac{1}{2}$	−1	$\frac{1}{2}$	0	1
14	1	$-\frac{1}{2}$	−1	$-\frac{1}{2}$	0	−1
15	1	$\frac{1}{2}$	−1	$-\frac{1}{2}$	0	0

quantum numbers $m_{l_1}, m_{s_1}, m_{l_2}, m_{s_2}$ of the two electrons are given. In obtaining this table, we note that the states for which the two electrons have the same values of m_l and m_s (for example, $m_{l_1} = m_{l_2} = 1$, $m_{s_1} = m_{s_2} = 1/2$) must be excluded because of the exclusion principle. Moreover, since the two electrons are indistinguishable, two pairs of values (m_{l_1}, m_{s_1}) and (m_{l_2}, m_{s_2}) which differ only by the electron label (1 or 2) only give one state. For example, $m_{l_1} = 1$, $m_{s_1} = 1/2$, $m_{l_2} = 0$, $m_{s_2} = -1/2$ yields the same state as $m_{l_1} = 0$, $m_{s_1} = -1/2$, $m_{l_2} = 1$, $m_{s_2} = 1/2$. We have also listed in Table 8.7 the values of the quantum numbers M_L and M_S, namely

$$M_L = m_{l_1} + m_{l_2}, \quad M_S = m_{s_1} + m_{s_2} \tag{8.200}$$

We shall now identify the possible terms corresponding to the 15 states listed in Table 8.7. We recall that in the case of the configuration np n'p (non-equivalent electrons) we had obtained the six possible terms given in (8.197). For the present $n\text{p}^2$ configuration, however, the pairs (M_L, M_S) appearing in the two last columns of Table 8.7 are only compatible with a restricted number of terms. Indeed, we first note that the pair ($M_L = 2$, $M_S = 1$) is missing so that the term ^{3}D is not present. On the other hand, the occurrence of the pairs ($M_L = 2$, $M_S = 0$) and ($M_L = -2$, $M_S = 0$) indicates that a term with $L = 2$ must exist. Since the term ^{3}D has just been ruled out, the term with $L = 2$ must be ^{1}D; it may be constructed with the states corresponding to the first five entries in Table 8.7. The states corresponding to the entries 6 to 14 are easily seen to correspond to a term ^{3}P. Finally, the last

Table 8.8 The possible terms for electron configurations $(nl)^k$, with $l = 0, 1, 2$.

Configuration								
	ns			^{2}S				
	ns^2		^{1}S					
np		np^5		^{2}P				
np^2		np^4	^{1}S, ^{1}D		^{3}P			
	np^3			^{2}P, ^{2}D		^{4}S		
	np^6		^{1}S					
nd		nd^9		^{2}D				
nd^2		nd^8	^{1}S, ^{1}D, ^{1}G		^{3}P, ^{3}F			
nd^3		nd^7		^{2}P, ^{2}D, ^{2}F, ^{2}G, ^{2}H		^{4}P, ^{4}F		
				2				
nd^4		nd^6	^{1}S, ^{1}D, ^{1}F, ^{1}G, ^{1}I		^{3}P, ^{3}D, ^{3}F, ^{3}G, ^{3}H		^{5}D	
			2 2 2		4 2			
	nd^5			^{2}S, ^{2}P, ^{2}D, ^{2}F, ^{2}G, ^{2}H, ^{2}I		^{4}P, ^{4}D, ^{4}F, ^{4}G		^{6}S
				3 2 2				
	nd^{10}		^{1}S					

entry in the table, with $M_L = M_S = 0$, corresponds to a term ^{1}S. Hence, for the configuration $n\mathrm{p}^2$, there are only three possible terms, namely

$$^1\mathrm{S},\ ^1\mathrm{D},\ ^3\mathrm{P} \tag{8.201}$$

instead of the six terms (8.197) we found for the configuration $n\mathrm{p}\ n'\mathrm{p}$.

A similar analysis can be carried out for other configurations containing equivalent electrons. In Table 8.8 are given the possible terms for configurations $(nl)^k$, with $l = 0, 1, 2$. We note in particular from this table that the configurations $(nl)^k$ and $(nl)^{2(2l+1)-k}$ lead to the same possible terms. In other words, the possible terms corresponding to a configuration in which there are k electrons in a subshell are the same as those in which k electrons are missing (that is, there are k 'holes') in this subshell.

Equivalent and non-equivalent electrons

If a configuration contains a group of equivalent electrons together with a number of non-equivalent electrons, the possible terms of the group of equivalent electrons must first be determined. The overall possible terms are then obtained by using the ordinary rules of addition of angular momenta to add the non-equivalent electrons. Similarly, if a configuration contains two or more groups of equivalent electrons, the possible terms of each group must first be obtained, and the overall possible terms are then determined by using the ordinary rules for the addition of angular momenta.

Hund's rules

Before leaving the subject of the effect of the electrostatic interaction H_1, we mention the so-called *Hund's rules*, which have been established empirically for the ground state configuration and for configurations containing equivalent electrons. According to these rules:

1. The term with the largest possible value of S for a given configuration has the lowest energy; the energy of the other terms increases with decreasing S.
2. For a given value of S, the term having the maximum possible value of L has the lowest energy.

Fine structure of terms in L–S coupling. Multiplet splitting and the Landé interval rule

Having obtained the energy levels of the Hamiltonian (8.194), we now proceed to the *second step* of the perturbation calculation, which consists in taking into account the spin–orbit term H_2 given by (8.189). We shall first examine how the additional perturbation H_2 further removes degeneracies. The total Hamiltonian $\mathcal{H} = H_c + H_1 + H_2$ does not commute with $\mathbf{L}$ and $\mathbf{S}$, but it does commute with $\mathbf{J} = \mathbf{L} + \mathbf{S}$. As a result, the $(2L + 1)(2S + 1)$ degeneracy associated with a term ^{2S+1}L corresponding to given values of L and S is partly removed by the perturbation H_2. The term ^{2S+1}L splits into a number of *fine structure components*, characterised by the value of J, the total electronic angular momentum quantum number, and written in the Russell–Saunders notation as $^{2S+1}L_J$. Because $|H_2| \ll |H_1|$ in L–S coupling, the energy separation between the fine structure components $^{2S+1}L_J$ of a term ^{2S+1}L is small with respect to the energy separation between terms. The various fine structure components are said to form a *fine structure multiplet*. Each fine structure term $^{2S+1}L_J$ is still $(2J + 1)$-fold degenerate with respect to M_J (where $M_J\hbar$ are the eigenvalues of J_z), the possible values of M_J for a given J being

$$M_J = -J, -J + 1, \ldots, J - 1, J \tag{8.202}$$

The degeneracy in M_J can only be removed if a preferred direction in space is introduced, for example if an external magnetic field is applied as in the Zeeman effect.

Using the rules of addition of angular momenta, the possible values of J corresponding to given values of L and S are

$$|L-S|, |L-S| + 1, \ldots, L + S \tag{8.203}$$

It is clear that the number of possible values of J is equal to the multiplicity $2S + 1$ if $L \geqslant S$, and to $2L + 1$ if $L < S$. It is worth noting that the word 'multiplicity' always denotes the number $2S + 1$, even in the case $L < S$, where the term ^{2S+1}L can only split into $2L + 1$ fine structure components. We also note that

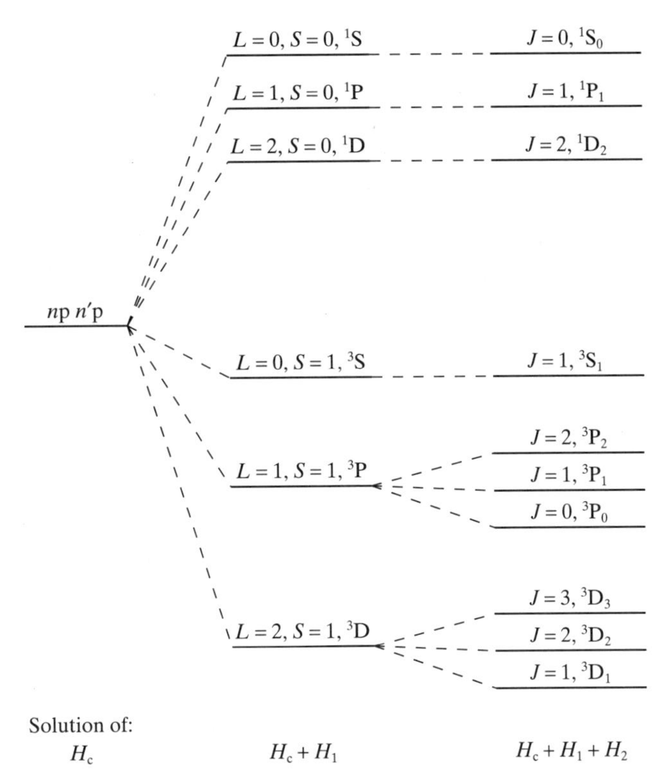

Figure 8.7 The splitting of the configuration $np\,n'\mathrm{p}$ by the electrostatic perturbation H_1 and the spin–orbit perturbation H_2.

$$\sum_{J=|L-S|}^{L+S} (2J + 1) = (2L + 1)(2S + 1) \tag{8.204}$$

so that by counting all the fine structure states (including their multiplicity) we retrieve the $(2L + 1)(2S + 1)$ degeneracy attached to the term ${}^{2S+1}L$.

As a first example, we show in Fig. 8.7 the partial removal of degeneracies, due successively to the perturbations H_1 and H_2, for a configuration containing two non-equivalent optically active electrons $np\,n'\mathrm{p}$. In this case the possible values of L, S, and J are given by

$$L = 0, 1, 2$$

$$S = 0, 1$$

$$J = 0, 1, 2, 3$$

In the absence of the spin–orbit perturbation H_2 the possible terms ${}^{2S+1}L$ are given by (8.197). When the effect of H_2 is included, the fine structure terms ${}^{2S+1}L_J$ are seen to be

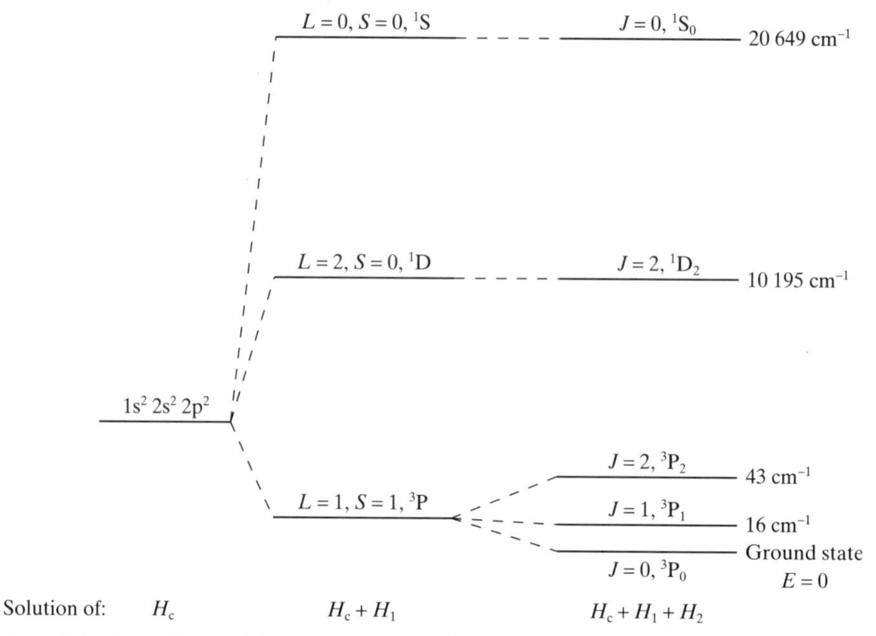

Figure 8.8 The splitting of the ground state configuration of carbon.

$$^1S_0,\ ^1P_1,\ ^1D_2,\ ^3S_1,\ ^3P_0,\ ^3P_1,\ ^3P_2,\ ^3D_1,\ ^3D_2,\ ^3D_3 \tag{8.205}$$

A second example is given in Fig. 8.8, where the case of a configuration containing two equivalent electrons np^2 is considered. This case is relevant in particular for the study of the ground state of the carbon atom, which has the configuration $1s^22s^22p^2$, the two 'optically active' electrons being the two 2p electrons. When the perturbation H_2 is taken into account the three terms obtained in (8.201) give rise to the five fine structure terms

$$^1S_0,\ ^1D_2,\ ^3P_0,\ ^3P_1,\ ^3P_2 \tag{8.206}$$

which are shown in Fig. 8.8. Both examples in Figs 8.7 and 8.8 also illustrate Hund's rules. Looking at Fig. 8.8 we also see that the 'true' ground state of the carbon atom is $1s^22s^22p^2\ ^3P_0$.

In order to obtain the fine structure energy shifts due to the spin–orbit interaction, the perturbation H_2 must be diagonalised in the subspace of the wave functions $|\gamma LSM_LM_S\rangle$ which correspond to a given term, that is to a given energy level (γLS). By using the Wigner–Eckart theorem (see Appendix 4) it may be shown that the matrix elements of H_2 in that subspace are the same as those of the operator $\bar{A}\mathbf{L}\cdot\mathbf{S}$, where $\bar{A}$ is a constant, characteristic of the unperturbed level (γLS). Thus

$$\langle\gamma LSM_LM_S|H_2|\gamma LSM_L'M_S'\rangle = \bar{A}\langle\gamma LSM_LM_S|\mathbf{L}\cdot\mathbf{S}|\gamma LSM_L'M_S'\rangle \tag{8.207}$$

The perturbation H_2 is not diagonal in the subspace of the wave functions $|\gamma LSM_LM_S\rangle$. However, by proceeding as in Section 5.1, we can form linear combinations of the functions $|\gamma LSM_LM_S\rangle$ to obtain new 'unperturbed' functions $|\gamma LSJM_J\rangle$ which are eigenstates of $\mathbf{L}^2$, $\mathbf{S}^2$, $\mathbf{J}^2$ and J_z in that subspace. Since $\mathbf{J}^2 = \mathbf{L}^2 + 2\mathbf{L}\cdot\mathbf{S} + \mathbf{S}^2$, we then have (with $A = \bar{A}\hbar^2$)

$$\begin{aligned}\langle \gamma LSJM_J|H_2|\gamma LSJM_J\rangle &= \tfrac{1}{2}\bar{A}\langle \gamma LSJM_J|\mathbf{J}^2 - \mathbf{L}^2 - \mathbf{S}^2|\gamma LSJM_J\rangle \\ &= \tfrac{1}{2}A[J(J+1) - L(L+1) - S(S+1)]\end{aligned} \quad \textbf{(8.208)}$$

so that the unperturbed level (γLS) splits into $2S+1$ (if $L \geqslant S$) or $2L+1$ (if $L < S$), fine structure levels labelled by J, as we have seen above. Moreover, we see from (8.208) that the energy separation between adjacent levels $E(J)$ and $E(J-1)$ of a given multiplet (L and S fixed) is

$$\begin{aligned}&E(J) - E(J-1) \\ &= \tfrac{1}{2}A[J(J+1) - L(L+1) - S(S+1)] - \tfrac{1}{2}A[(J-1)J - L(L+1) - S(S+1)] \\ &= AJ\end{aligned} \quad \textbf{(8.209)}$$

and is therefore proportional to J, the larger of the two J values characterising the pair of fine structure levels.

The result (8.209) is known as the *Landé interval rule*. It is well satisfied experimentally if the atom is well described by the L–S coupling case ($|H_2| \ll |H_1|$) and if in addition the perturbation H_2 only contains the spin–orbit interactions [14] as we have assumed here (see (8.189)). In that case the multiplets are called *regular multiplets*. When $A > 0$, we see from (8.208) that the multiplet component having the smallest possible value of J (that is, $J = |L - S|$) has the lowest energy value; these multiplets are called *normal*. On the other hand, if $A < 0$, the multiplet component having the largest possible value of J (namely $J = L + S$) has the lowest energy value; in this case one speaks of an *inverted* multiplet. The multiplets shown in Figs 8.7 and 8.8 are regular, normal multiplets. It has been established empirically that normal multiplets occur if there is a single open subshell that is *less than half-filled*, while inverted multiplets arise if that subshell is *more than half-filled*. When the subshell is just half-filled, there is no multiplet splitting.

j–*j* coupling

Let us now consider the case for which the spin–orbit energy H_2 is large with respect to the electrostatic energy correction H_1. In Section 5.1 we showed that the energy of the spin–orbit interaction is proportional to Z^4 (see (5.26)) while in Section 7.5 the first-order electrostatic correction $\langle 1/r_{12}\rangle$ was shown to be proportional to Z. We therefore expect that the importance of the spin–orbit term H_2 relative to the electrostatic correction term will increase as Z is increased, so that

[14] In some cases the fine structure of the levels may be very different because of the presence of other interactions, such as the spin–spin interactions between the electrons, mentioned at the beginning of Section 8.1.

j–j coupling could occur for atoms (ions) with large Z. In fact, j–j coupling is rarely found in pure form, although the atomic spectra of heavy atoms exhibit a structure which is close to that predicted by the j–j coupling scheme. The best examples of j–j coupling are provided by multiply charged ions the nuclei of which have high values of Z, since in these cases the relative importance of the electrostatic correction term H_1 is reduced because of the reduction in the number of electrons.

Since $|H_2| \gg |H_1|$ in the j–j coupling case, the *first step* of the perturbation calculation consists in neglecting H_1 and solving, approximately, the problem associated with the Hamiltonian

$$\tilde{H} = H_c + H_2 \tag{8.210}$$

Now, from (8.10) and (8.189), we see that $\tilde{H}$ is just the sum of individual Hamiltonians. That is,

$$\tilde{H} = \sum_{i=1}^{N} \tilde{h}_i \tag{8.211}$$

where

$$\begin{aligned}\tilde{h}_i &= h_i + \xi(r_i)\mathbf{L}_i\cdot\mathbf{S}_i \\ &= -\frac{1}{2}\nabla^2_{\mathbf{r}_i} + V(r_i) + \xi(r_i)\mathbf{L}_i\cdot\mathbf{S}_i\end{aligned} \tag{8.212}$$

The spin–orbit term $\xi(r_i)\mathbf{L}_i\cdot\mathbf{S}_i$ has the effect of partly removing the degeneracy of the individual electron energy levels E_{nl} (corresponding to the Hamiltonian h_i) by splitting each level E_{nl} with $l \neq 0$ into two components E_{nlj} having the total (individual) angular momentum quantum number $j = l \pm 1/2$. The corresponding individual wave functions are spin-orbitals u_{nljm_j}, labelled by the quantum numbers $(nljm_j)$ where m_j is the magnetic quantum number associated with the z component of the total angular momentum operator of an electron. Since m_j can take on the values

$$m_j = -j, -j+1, \dots, j \tag{8.213}$$

and the energy levels E_{nlj} do not depend on m_j, we see that each individual electron energy level E_{nlj} is $(2j+1)$-fold degenerate. When $j = l + 1/2$, this degeneracy is equal to $2l + 2$, and when $j = l - 1/2$, it is equal to $2l$. The energy levels $\tilde{E}$ associated with the Hamiltonian $\tilde{H}$ are obtained by summing the individual electron energies

$$\tilde{E} = \sum_{i=1}^{N} E_{n_i l_i j_i} \tag{8.214}$$

and the corresponding wave functions are antisymmetrised products (Slater determinants) formed from the spin-orbitals $u_{n_i l_i j_i m_{j_i}}$. Each configuration of H_c therefore yields a certain number of configurations of $\tilde{H}$, characterised by the values of the quantum numbers $(n_i l_i j_i)$ of the electrons.

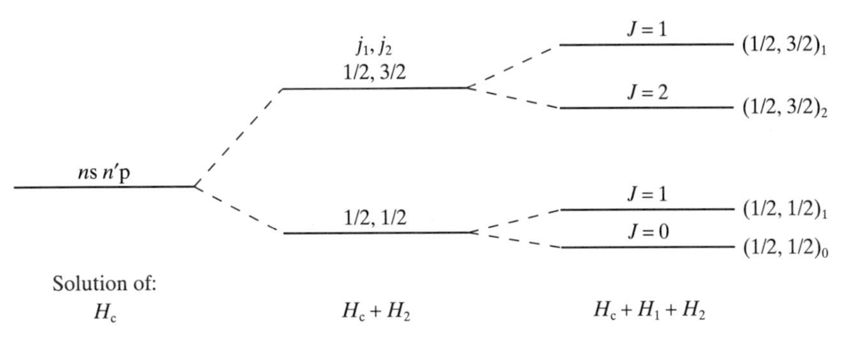

Figure 8.9 The splitting of levels in *j–j* coupling.

In the *second step* of the perturbation calculation, the effect of the electrostatic correction term H_1 is taken into account. Each of the levels $\tilde{E}$ of $\tilde{H}$, characterised by a set of quantum numbers $(n_i l_i j_i)$ of the electrons, will now be split by the additional perturbation H_1 into a certain number of levels labelled by the values of the total electronic angular momentum quantum number J of the system. Of course, as in the case of L–S coupling, each of these levels is still $(2J + 1)$-fold degenerate with respect to M_J. A typical example of the splitting of levels in j–j coupling is shown in Fig. 8.9.

In the case of j–j coupling the notation for the spectral terms must specify the quantum numbers $(n_i l_i j_i)$ of each electron and the total electronic angular momentum quantum number J. The values of the individual j_i's are usually written between parentheses, and J as a subscript. For example, one of the energy levels of the ground state configuration $6p^2$ of Pb, corresponding to the values $j_1 = 1/2$, $j_2 = 3/2$ and $J = 2$, is written as $6p^2\ (1/2, 3/2)_2$.

The possible values of J can be obtained by using methods similar to those we discussed above in order to obtain the possible terms in L–S coupling. In the case of non-equivalent electrons, it is sufficient to use the rule for addition of angular momenta. Let us consider for example the configuration $ns\,n'$p. For an s electron one has $j = 1/2$; while for a p electron $j = 1/2, 3/2$. Now, if $j_1 = 1/2$ and $j_2 = 1/2$, the possible values of J are $J = 0, 1$, while for $j_1 = 1/2$ and $j_2 = 3/2$ the allowed values of J are $J = 1, 2$. Thus the possible terms are

$$(1/2, 1/2)_0, \quad (1/2, 1/2)_1, \quad (1/2, 3/2)_1, \quad (1/2, 3/2)_2$$

In the case of equivalent electrons, it is necessary to take into account the effect of the Pauli exclusion principle, just as in the case of L–S coupling. The possible terms for a configuration $(j)^k$, with $j = 1/2$, $3/2$ and $5/2$, are given in Table 8.9. It is worth noting that the total number of levels having a given value of J for a given electron configuration must be the same in L–S and j–j coupling.

Finally, we recall that the two cases of L–S and j–j coupling which we have discussed here are *extreme* cases. Many atoms are not described accurately by

Table 8.9 The possible terms for electron configurations $(j)^k$, with $j = 1/2, 3/2, 5/2$.

Configuration			*J*
	$(1/2)^1$		1/2
	$(1/2)^2$		0
$(3/2)^1$		$(3/2)^3$	3/2
	$(3/2)^2$		0, 2
$(5/2)^1$		$(5/2)^5$	5/2
$(5/2)^2$		$(5/2)^4$	0, 2, 4
	$(5/2)^3$		3/2, 5/2, 9/2

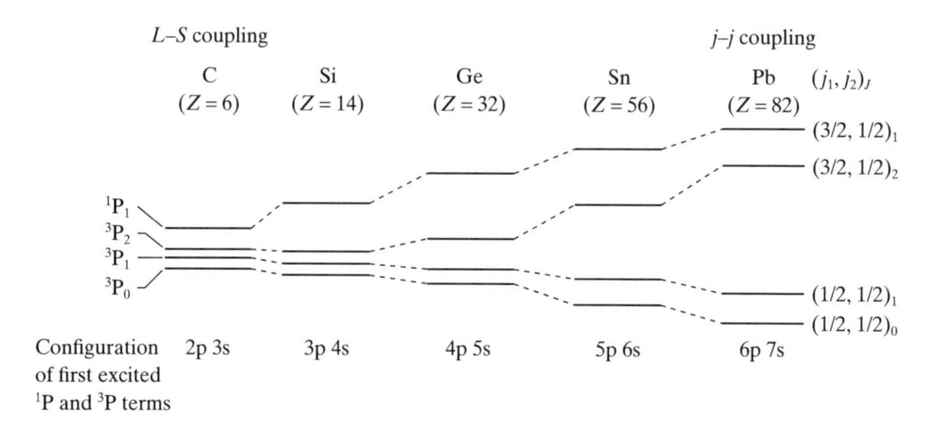

Figure 8.10 The splitting of levels in the first excited ^{1}P and ^{3}P terms of the carbon sequence.

either of these schemes, and some form of *intermediate coupling* scheme must be devised. This is illustrated in Fig. 8.10 where we see that the energy levels of the configurations 2p3s of C and 3p4s of Si are well described by *L–S* coupling, and the levels 6p7s of Pb by *j–j* coupling, while the 4p5s levels of Ge and 5p6s levels of Sn are intermediate in character.

8.6 Density functional theory

Both the variational–perturbation and configuration–interaction approaches to the physics of many-electron systems aim to include the correlation effects which are absent in the Hartree–Fock approximation. An alternative approach including correlation effects is the *density functional theory* in which the electron number density plays the central role rather than the many-electron wave function. We shall start from the Hamiltonian of an N-electron atom or ion with a nucleus of charge Z in the infinite nuclear mass approximation given by (8.2), which we shall write in the form

$$H = T + W + \hat{H}_2 \tag{8.215}$$

In this equation T is the total kinetic energy of the electrons,

$$T = -\frac{1}{2}\sum_{i=1}^{N} \nabla^2_{\mathbf{r}_i}, \tag{8.216}$$

the potential W is given by

$$W = \sum_{i=1}^{N} U(\mathbf{r}_i) \tag{8.217}$$

where

$$U(\mathbf{r}_i) = -\frac{Z}{r_i} \tag{8.218}$$

is the interaction energy between the ith electron and the nucleus, while $\hat{H}_2$ is the sum of the interactions between electron pairs given by (8.84). The spin variables will not be exhibited explicitly. If $\Psi_0(\mathbf{r}_1, \mathbf{r}_2, \ldots, \mathbf{r}_N)$ is the exact ground state wave function including all exchange and correlation effects, then the single particle number density $\rho(\mathbf{r})$ in the ground state is defined as

$$\rho(\mathbf{r}) = \int \mathrm{d}\mathbf{r}_2 \mathrm{d}\mathbf{r}_3 \ldots \mathrm{d}\mathbf{r}_N |\Psi_0(\mathbf{r}, \mathbf{r}_2, \mathbf{r}_3, \ldots, \mathbf{r}_N)|^2 \tag{8.219}$$

In 1960, P. Hohenberg and W. Kohn proved the basic theorems on which density functional theory is founded. These are:

(a) Every observable quantity of a stationary many-electron system is uniquely determined by the ground state number density $\rho(\mathbf{r})$.

(b) There exists a functional $I[\rho]$ so that for a given potential $U(\mathbf{r})$, both $\rho(\mathbf{r})$ and the ground state energy E_0 are obtained exactly if the functional

$$E_0[\rho] = I[\rho] + \int U(\mathbf{r})\rho(\mathbf{r})\, \mathrm{d}\mathbf{r} \tag{8.220}$$

is minimised by varying ρ, subject to the constraint

$$\int \rho(\mathbf{r})\, \mathrm{d}\mathbf{r} = N \tag{8.221}$$

Unfortunately, the Hohenberg–Kohn theorems provide no basis in themselves for practical calculations, since the functional form of $I[\rho]$ is not specified. Of course at the minimum the numerical value of I is equal to the expectation value of $T + \hat{H}_2$, as is apparent from (8.215).

The nature of $I[\rho]$ was clarified in 1979 by M. Levy who showed that $I[\rho]$ is obtained if the functional $\langle \Phi | T + \hat{H}_2 | \Phi \rangle$ is minimised by searching over all many-electron wave functions Φ which yield the given electron density $\rho(\mathbf{r})$. That is,

$$I[\rho] = {}_{\Phi \to \rho}\langle \Phi | T + \hat{H}_2 | \Phi \rangle \tag{8.222}$$

Attempts have been made to calculate $I[\rho]$ by using the Thomas–Fermi approximation for $\langle \Phi | T | \Phi \rangle$, however a better method has been developed by W. Kohn and L.J. Sham. It starts by introducing a set of N independent single electron orbitals $u_i(\mathbf{r})$, $i = 1, 2, \ldots, N$, which are the wave functions for motion in the ground state of the potential W. The single electron density of this system is

$$\rho(\mathbf{r}) = \sum_{i=1}^{N} |u_i(\mathbf{r})|^2 \tag{8.223}$$

and the corresponding expectation value of the kinetic energy is

$$T_s[\rho] = \sum_{i=1}^{N} \int \mathrm{d}\mathbf{r} \, \frac{1}{2} |\nabla_{\mathbf{r}} u_i(\mathbf{r})|^2 \tag{8.224}$$

The true kinetic energy of an interacting system cannot be written in this way, but when the orbitals are optimised, as will be described below, the resulting kinetic energy $T_s[\rho]$ turns out to be a good approximation.

Using (8.220) together with (8.222) and (8.224), the energy functional $E_0[\rho]$ can be written as

$$E_0[\rho] = T_s[\rho] + \frac{1}{2} \int \mathrm{d}\mathbf{r} \, \mathrm{d}\mathbf{r}' \frac{\rho(\mathbf{r})\rho(\mathbf{r}')}{|\mathbf{r} - \mathbf{r}'|} + \int \mathrm{d}\mathbf{r} \, U(\mathbf{r})\rho(\mathbf{r}) + E_{\mathrm{xc}}[\rho] \tag{8.225}$$

The second term on the right-hand side of (8.225) is the expectation value of the interaction $\hat{H}_2$ with respect to the set of independent orbitals and the last term $E_{\mathrm{xc}}[\rho]$ is just the difference between the exact ground state energy and the sum of the first three terms on the right-hand side of (8.225). The energy $E_{\mathrm{xc}}[\rho]$ is called the *exchange-correlation energy* and includes corrections to the expectation value both of the kinetic energy and of the interaction energy.

Minimising $E_0[\rho]$ given by (8.225) with respect to ρ (which is equivalent to minimising this functional with respect to the orbitals $u_i(\mathbf{r})$), it is found that the new, optimised, orbitals satisfy the *Kohn–Sham equation*

$$\left[-\frac{1}{2} \nabla_{\mathbf{r}}^2 + V_{\mathrm{eff}}(\mathbf{r}) \right] u_i(\mathbf{r}) = E_i u_i(\mathbf{r}) \tag{8.226}$$

where E_i is a Lagrange multiplier ensuring normalisation (see Section 2.8). The effective potential V_{eff} is given by

$$V_{\mathrm{eff}}(\mathbf{r}) = U(\mathbf{r}) + \int \frac{\rho(\mathbf{r}')}{|\mathbf{r} - \mathbf{r}'|} \mathrm{d}\mathbf{r}' + V_{\mathrm{xc}}(\mathbf{r}) \tag{8.227}$$

where

$$V_{xc}(\mathbf{r}) = \frac{\delta E_{xc}}{\delta \rho} \tag{8.228}$$

Given an approximation for $V_{xc}(\mathbf{r})$, equations (8.223), (8.226) and (8.227) can be solved self-consistently. This procedure is simpler than the self-consistent solution of the Hartree–Fock equations because the potential $V_{eff}(\mathbf{r})$ is local.

The success or failure of the density functional scheme depends on finding an adequate approximation for the exchange-correlation energy E_{xc}. Kohn and Sham introduced the *local density approximation* (LDA), in which

$$E_{xc} \simeq E_{xc}^{LDA} = \int \rho(\mathbf{r}) E_{xc}^{H} \, d\mathbf{r} \tag{8.229}$$

The energy E_{xc}^{H} is the exchange-correlation energy of an infinite homogeneous electron gas moving in a uniform positively charged background to neutralise the electronic charge. Even for this system the correlation energy cannot be found exactly, but good estimates can be obtained. The LDA has proved to be good for systems having a large number of electrons, such as large atoms or molecules, and particularly solids. Detailed accounts of the density functional theory and its further developments have been given by Dreizler and Gross (1990) and by Parr and Yang (1989).

Problems

8.1 Prove that the central field Hamiltonian H_c given by (8.10) commutes with the total orbital angular momentum $\mathbf{L}$ of the electrons.

8.2 Consider the wave functions of the 2^3S level of helium, which are given in the central field approximation by

$$\Psi_c(2^3S) = \phi_{-}(r_1, r_2) \begin{cases} \alpha(1)\alpha(2) & M_S = 1 \\ \dfrac{1}{\sqrt{2}}[\alpha(1)\beta(2) + \beta(1)\alpha(2)] & M_S = 0 \\ \beta(1)\beta(2) & M_S = -1 \end{cases}$$

with

$$\phi_{-}(r_1, r_2) = \frac{1}{\sqrt{2}}[u_{1s}(r_1)u_{2s}(r_2) - u_{2s}(r_1)u_{1s}(r_2)]$$

Write the three functions $\Psi_c(2^3S)$ in the form of Slater determinants (or a sum of Slater determinants) constructed with the spin-orbitals

$$u_{1s\uparrow} = u_{1s}(r)\alpha, \qquad u_{1s\downarrow} = u_{1s}(r)\beta$$

$$u_{2s\uparrow} = u_{2s}(r)\alpha, \qquad u_{2s\downarrow} = u_{2s}(r)\beta$$

8.3 (a) Show that the total energy of a neutral Thomas–Fermi atom can be written in terms of the electronic density $\rho(r)$ as

$$E = \kappa \int [\rho(r)]^{5/3} \, d\mathbf{r} - \frac{Ze^2}{4\pi\varepsilon_0} \int \frac{\rho(r)}{r} \, d\mathbf{r} + \frac{1}{2} \frac{e^2}{4\pi\varepsilon_0} \int \frac{\rho(r)\rho(r')}{|\mathbf{r} - \mathbf{r}'|} \, d\mathbf{r} \, d\mathbf{r}' \qquad [1]$$

with $\kappa = 3^{5/3}\pi^{4/3}\hbar^2/(10m)$. The first term on the right of [1] is the total kinetic energy of the electrons, the second term is the potential energy of their interaction with the nucleus, and the third term is the potential energy of their mutual interaction.

(b) Obtain the Thomas–Fermi equation (8.66) by minimising the expression [1], subject to the normalisation condition

$$4\pi \int_0^\infty \rho(r) r^2 \, dr = Z$$

8.4 Using the Thomas–Fermi model, obtain an estimate of the following quantities:

(a) average distance of an electron from the nucleus;
(b) average kinetic energy of an electron;
(c) total ionisation energy of the atom.

8.5 Obtain the two coupled equations (8.165) for the spatial orbitals $u_{1s}(r)$ and $u_{2s}(r)$ of the beryllium ground state.

8.6 Show that the Hartree–Fock and Hartree equations for the ground state orbital $u_{1s}(r)$ of helium are identical and given by

$$\left[-\frac{1}{2}\nabla_{\mathbf{r}}^2 - \frac{2}{r} + \int |u_{1s}(r')|^2 \frac{1}{|\mathbf{r} - \mathbf{r}'|} \, d\mathbf{r}' \right] u_{1s}(r) = E_{1s} u_{1s}(r)$$

Using the Hartree–Fock orbital (7.85b), evaluate the Hartree–Fock potential

$$\mathcal{V}(r) = -\frac{2}{r} + \int |u_{1s}(r')|^2 \frac{1}{|\mathbf{r} - \mathbf{r}'|} \, d\mathbf{r}'$$

Plot $\mathcal{V}(r)$ and interpret your results in terms of screening.

8.7 Obtain a pair of Hartree–Fock coupled equations for the spatial orbitals $u_{1s}(r)$ and $u_{2s}(r)$ corresponding to the 2^3S ($M = 1$) wave function of helium. Prove that the orbitals $u_{1s}(r)$ and $u_{2s}(r)$ are orthogonal.

8.8 Prove equation (8.159).

8.9 Prove that $[H, \mathbf{L}] = 0$, where H is the Hamiltonian (8.2) and $\mathbf{L} = \sum_i \mathbf{L}_i$ is the total orbital angular momentum of the electrons.

8.10 Assuming that L–S coupling holds:

(a) List the possible spectral terms ^{2S+1}L which result from the following electronic configurations
   (i) $ns\, n's$
   (ii) $ns\, n'p$
   (iii) $ns\, n'd$
   (iv) nd^2
   (v) $np\, n'p\, n''p$
   (vi) np^3.

(b) List the corresponding fine structure terms $^{2S+1}L_J$.

8.11 Assuming that j–j coupling holds, list the possible terms $(j_1, j_2)_J$ of the following electronic configurations

(a) $np\, nd$
(b) $(nl\ 3/2)^2$.

9 Interaction of many-electron atoms with electromagnetic radiation and with static electric and magnetic fields

In the previous two chapters, we studied the energy levels of atoms with more than one electron. In this chapter, we shall discuss the interaction of these atoms with electromagnetic radiation and analyse the main features of the spectra that result from radiative transitions between their energy levels. We shall also study the interaction of many-electron atoms with static electric and magnetic fields. For the most part, this is a straightforward generalisation of the material of Chapters 4 and 6, where the interaction of one-electron atoms and ions with electromagnetic radiation and with static electric and magnetic fields was discussed. After writing down in Section 9.1 the basic equations describing the interaction of many-electron atoms with electromagnetic radiation, we shall obtain in Section 9.2 the selection rules for electric dipole radiative transitions, and discuss oscillator and line strengths. We then consider in Section 9.3 magnetic dipole and electric quadrupole transitions, for which we also obtain the selection rules. In the next two sections, we analyse the simple spectra of the alkalis, followed by those of helium and the alkaline earths, which contain one and two optically active electrons respectively. We go on in Section 9.6 to describe some general features of multiplets, the group of spectral lines arising from transitions between two terms. This is followed in Section 9.7 by a discussion of X-ray spectra, including the Auger effect. Finally, the two last sections are devoted to the interaction of complex atoms with static electric and magnetic fields, respectively.

9.1 Many-electron atoms in an electromagnetic field

As in Chapter 4, we shall use a *semi-classical* theory in which the electromagnetic field is described classically, while the atomic system is treated by using quantum mechanics. Our starting point is therefore the time-dependent Schrödinger equation

$$i\hbar\frac{\partial}{\partial t}\Psi(X,t) = H(t)\Psi(X,t) \qquad \textbf{(9.1)}$$

where $X \equiv (q_1, q_2, \ldots, q_N)$ denotes the ensemble of the coordinates of the N electrons. The semi-classical Hamiltonian $H(t)$ which appears in this equation is readily obtained by generalising the arguments given in Section 4.1. Working in the Coulomb gauge, where $\nabla \cdot \mathbf{A} = 0$, and taking the scalar potential $\phi = 0$ in empty space, we have

$$H(t) = \frac{1}{2m} \sum_{i=1}^{N} [\mathbf{p}_i + e\mathbf{A}(\mathbf{r}_i, t)]^2 + V$$

$$= \frac{1}{2m} \sum_{i=1}^{N} \mathbf{p}_i^2 + V + \frac{e}{m} \sum_{i=1}^{N} \mathbf{A}(\mathbf{r}_i, t) \cdot \mathbf{p}_i + \frac{e^2}{2m} \sum_{i=1}^{N} \mathbf{A}^2(\mathbf{r}_i, t) \tag{9.2}$$

where $N = Z$ for a neutral atom and $N \neq Z$ for an ion. In this equation, $\mathbf{p}_i$ is the momentum operator of the electron i and V denotes the sum of all the interactions within the atom in the absence of the radiation field. For example, if all but the Coulomb interactions are neglected, we have (see (8.1))

$$V = -\sum_{i=1}^{N} \frac{Ze^2}{(4\pi\varepsilon_0)r_i} + \sum_{i<j=1}^{N} \frac{e^2}{(4\pi\varepsilon_0)r_{ij}} \tag{9.3}$$

Substituting (9.2) into (9.1), and using the fact that $\mathbf{p}_i = -i\hbar\nabla_{\mathbf{r}_i}$ in the position representation, we find that the time-dependent Schrödinger equation for an atom with N electrons in an electromagnetic field reads

$$i\hbar \frac{\partial}{\partial t} \Psi(X, t) = \left\{ -\frac{\hbar^2}{2m} \sum_{i=1}^{N} \nabla_{\mathbf{r}_i}^2 + V - i\hbar \frac{e}{m} \sum_{i=1}^{N} \mathbf{A}(\mathbf{r}_i, t) \cdot \nabla_{\mathbf{r}_i} \right.$$

$$\left. + \frac{e^2}{2m} \sum_{i=1}^{N} \mathbf{A}^2(\mathbf{r}_i, t) \right\} \Psi(X, t) \tag{9.4}$$

This equation may be rewritten in the form

$$i\hbar \frac{\partial}{\partial t} \Psi(X, t) = [H_0 + H_{\text{int}}(t)] \Psi(X, t) \tag{9.5}$$

where

$$H_0 = -\frac{\hbar^2}{2m} \sum_{i=1}^{N} \nabla_{\mathbf{r}_i}^2 + V = \frac{1}{2m} \sum_{i=1}^{N} \mathbf{p}_i^2 + V \tag{9.6}$$

is the time-independent Hamiltonian of the N-electron atom (ion) in the absence of an electromagnetic field, and

$$H_{\text{int}}(t) = -i\hbar \frac{e}{m} \sum_{i=1}^{N} \mathbf{A}(\mathbf{r}_i, t) \cdot \nabla_{\mathbf{r}_i} + \frac{e^2}{2m} \sum_{i=1}^{N} \mathbf{A}^2(\mathbf{r}_i, t)$$

$$= \frac{e}{m} \sum_{i=1}^{N} \mathbf{A}(\mathbf{r}_i, t) \cdot \mathbf{p}_i + \frac{e^2}{2m} \sum_{i=1}^{N} \mathbf{A}^2(\mathbf{r}_i, t) \tag{9.7}$$

is the Hamiltonian describing the interaction of the atom (ion) with the radiation field.

As in Chapter 4, we shall consider only the weak field case, such that the term in $\mathbf{A}^2$ can be neglected compared with the linear term in $\mathbf{A}$, which we shall consider as a small perturbation. Thus we write $H_{\text{int}}(t) \simeq H'(t)$, where

$$H'(t) = -\mathrm{i}\hbar\frac{e}{m}\sum_{i=1}^{N}\mathbf{A}(\mathbf{r}_i, t)\cdot\boldsymbol{\nabla}_{\mathbf{r}_i} = \frac{e}{m}\sum_{i=1}^{N}\mathbf{A}(\mathbf{r}_i, t)\cdot\mathbf{p}_i \tag{9.8}$$

will be treated to first order in perturbation theory. This means that we shall restrict our attention to radiative transitions in which a single photon is emitted or absorbed at a time [1]. The treatment of Chapter 4 can then be followed step by step. For example, let us consider the absorption of a photon in a transition $a \to b$ between the atomic states a and b, having energies E_a, E_b and wave functions ψ_a, ψ_b, respectively, with $E_b > E_a$. Assuming linear polarisation, the transition rate W_{ba} for this absorption process is given by the expression (4.45),

$$W_{ba} = \frac{4\pi^2}{m^2c}\left(\frac{e^2}{4\pi\varepsilon_0}\right)\frac{I(\omega_{ba})}{\omega_{ba}^2}|M_{ba}(\omega_{ba})|^2 \tag{9.9}$$

in which the matrix element M_{ba} is now a sum of N contributions of the type (4.40), one for each electron. That is,

$$M_{ba} = \sum_{i=1}^{N}\langle\psi_b|\exp(\mathrm{i}\mathbf{k}\cdot\mathbf{r}_i)\hat{\boldsymbol{\varepsilon}}\cdot\boldsymbol{\nabla}_{\mathbf{r}_i}|\psi_a\rangle \tag{9.10}$$

where $\mathbf{k}$ is the propagator vector and $\hat{\boldsymbol{\varepsilon}}$ is the unit polarisation vector. It is worth noting that the atomic wave functions ψ_a and ψ_b are only *approximate* eigenfunctions of the unperturbed N-electron Hamiltonian H_0, since exact eigenfunctions of H_0 can only be obtained for hydrogenic atoms. We also notice that since the electrons in the atom (ion) are indistinguishable,

$$M_{ba} = N\langle\psi_b|\exp(\mathrm{i}\mathbf{k}\cdot\mathbf{r}_1)\hat{\boldsymbol{\varepsilon}}\cdot\boldsymbol{\nabla}_{\mathbf{r}_1}|\psi_a\rangle \tag{9.11}$$

Proceeding in a similar way, we find that the transition rate $\bar{W}_{ab}$ for stimulated emission is given in the case of linear polarisation by (4.47), where the matrix element $\bar{M}_{ab}$ is now a sum of N contributions of the type (4.48). That is,

$$\bar{M}_{ab} = \sum_{i=1}^{N}\langle\psi_a|\exp(-\mathrm{i}\mathbf{k}\cdot\mathbf{r}_i)\hat{\boldsymbol{\varepsilon}}\cdot\boldsymbol{\nabla}_{\mathbf{r}_i}|\psi_b\rangle \tag{9.12}$$

As in the one-electron case, we have

$$\bar{M}_{ab} = -M_{ba}^{*} \tag{9.13}$$

[1] Multiphoton transitions occurring in intense laser fields are discussed in Chapter 15.

so that

$$\bar{W}_{ab} = W_{ba} \tag{9.14}$$

in agreement with the principle of detailed balancing.

Following again the treatment of Chapter 4, we find that the transition rate W^{s}_{ab} for spontaneous emission of a photon of linear polarisation $\hat{\boldsymbol{\varepsilon}}$ into the solid angle $\mathrm{d}\Omega$ in the direction (θ, ϕ) is given by (see (4.60))

$$W^{s}_{ab}(\theta, \phi)\, \mathrm{d}\Omega = \frac{\hbar}{2\pi m^2 c^3}\left(\frac{e^2}{4\pi\varepsilon_0}\right)\omega_{ba}\,|M_{ba}(\omega_{ba})|^2\, \mathrm{d}\Omega \tag{9.15}$$

where M_{ba} is the matrix element (9.10).

We also remark that the definitions (4.101) and (4.103) of the Einstein coefficients, as well as the relations (4.108) and (4.109) between them, are valid for an arbitrary atom.

Dipole approximation

Let us now assume that the wavelength λ of the radiation is large compared with the size a of the atom or ion, so that $ka \ll 1$. As explained in Section 4.3, we can then make the *dipole approximation*, which amounts to neglecting the spatial variation of the radiation field across the atom. In this approximation, the retardation factor $\exp(\mathrm{i}\mathbf{k}\cdot\mathbf{r}_i)$ is therefore set equal to unity in (9.10), and the matrix element M_{ba} is replaced by its dipole approximation

$$M^{\mathrm{D}}_{ba} = \sum_{i=1}^{N} \hat{\boldsymbol{\varepsilon}}\cdot\langle\psi_b|\nabla_{\mathbf{r}_i}|\psi_a\rangle \tag{9.16}$$

Proceeding as in Section 4.3, we find that

$$M^{\mathrm{D}}_{ba} = -\frac{m\omega_{ba}}{\hbar}\sum_{i=1}^{N} \hat{\boldsymbol{\varepsilon}}\cdot(\mathbf{r}_i)_{ba} \tag{9.17}$$

where

$$(\mathbf{r}_i)_{ba} = \langle\psi_b|\mathbf{r}_i|\psi_a\rangle \tag{9.18}$$

and we notice that since the electrons in the atom are indistinguishable

$$M^{\mathrm{D}}_{ba} = -N\frac{m\omega_{ba}}{\hbar}\hat{\boldsymbol{\varepsilon}}\cdot(\mathbf{r}_1)_{ba} \tag{9.19}$$

It is convenient to introduce the *electric dipole moment operator* of the atom

$$\mathbf{D} = \sum_{i=1}^{N}(-e)\mathbf{r}_i = -e\mathbf{R} \tag{9.20}$$

where

$$\mathbf{R} = \sum_{i=1}^{N}\mathbf{r}_i \tag{9.21}$$

is the sum of the coordinates of the N electrons. Let

$$\mathbf{D}_{ba} = \langle \psi_b | \mathbf{D} | \psi_a \rangle = -\sum_{i=1}^{N} e(\mathbf{r}_i)_{ba} = -e\mathbf{R}_{ba} \tag{9.22}$$

be the matrix element of the operator $\mathbf{D}$ between the two atomic states a and b, where

$$\mathbf{R}_{ba} = \langle \psi_b | \mathbf{R} | \psi_a \rangle \tag{9.23}$$

The expression (9.16) for M_{ba}^{D} can then be rewritten in terms of $\mathbf{D}_{ba}$ as

$$M_{ba}^{\mathrm{D}} = \frac{m\omega_{ba}}{\hbar e} \hat{\boldsymbol{\varepsilon}} \cdot \mathbf{D}_{ba} \tag{9.24}$$

The transition rates in the dipole approximation can be obtained by following a treatment analogous to that of Section 4.3, and the formulae written there in terms of the matrix element $\mathbf{D}_{ba}$ are directly applicable to the N-electron atom (ion) case, with $\mathbf{D}_{ba}$ given by (9.22). For example, the transition rate in the dipole approximation for absorption of a photon in the transition $a \to b$ (with $E_b > E_a$) is given for the case of linear polarisation by (see (4.75))

$$W_{ba}^{\mathrm{D}} = \frac{4\pi^2}{c\hbar^2} \left(\frac{1}{4\pi\varepsilon_0} \right) I(\omega_{ba}) |\hat{\boldsymbol{\varepsilon}} \cdot \mathbf{D}_{ba}|^2 \tag{9.25}$$

and the dipole approximation to the transition rate for spontaneous emission of a photon with linear polarisation $\hat{\boldsymbol{\varepsilon}}$ into a solid angle $\mathrm{d}\Omega$ in the direction (θ, ϕ) for the transition $b \to a$ is given by (see (4.79))

$$W_{ab}^{\mathrm{s,D}}(\theta, \phi)\, \mathrm{d}\Omega = \frac{1}{2\pi\hbar c^3} \left(\frac{1}{4\pi\varepsilon_0} \right) \omega_{ba}^3 |\hat{\boldsymbol{\varepsilon}} \cdot \mathbf{D}_{ba}|^2 \, \mathrm{d}\Omega \tag{9.26}$$

The velocity and acceleration forms of the dipole matrix elements

The matrix elements $\mathbf{D}_{ba}$ of the electric dipole operator have been written above in terms of the length matrix elements $\mathbf{R}_{ba}$ given by (9.23). As in Chapter 4, we can also write the velocity and acceleration forms of $\mathbf{D}_{ba}$. Let

$$\mathbf{P} = \sum_{i=1}^{N} \mathbf{p}_i \tag{9.27}$$

be the total momentum operator, and remember that V is the sum of all the interactions within the atom in the absence of the radiation field. The length, velocity and acceleration forms of $\mathbf{D}_{ba}$ are then given by straightforward generalisations of the results (4.87) obtained for hydrogenic atoms. That is,

$$\mathbf{D}_{ba}^{\mathrm{L}} = -e\mathbf{R}_{ba} \tag{9.28a}$$

$$\mathbf{D}_{ba}^{\mathrm{V}} = \frac{\mathrm{i}e}{m\omega_{ba}} \mathbf{P}_{ba} \tag{9.28b}$$

and

$$\mathbf{D}_{ba}^{\mathrm{A}} = \frac{e}{m\omega_{ba}^2}\left(\sum_{i=1}^{N} \nabla_{\mathbf{r}_i} V\right)_{ba} \tag{9.28c}$$

Since approximate wave functions must be used to describe the atomic states a and b of atoms (ions) with more than one electron, the three forms of $\mathbf{D}_{ba}$ yield different numerical values. However, the differences will be small if accurate atomic wave functions are used.

The Schrödinger equation in the velocity and length gauges

In Section 4.3, we remarked that, in the dipole approximation, one can perform gauge transformations to write the Schrödinger equation (4.25) for a hydrogenic atom interacting with an electromagnetic field in the velocity or length gauges. We shall now generalise this treatment to the case of an N-electron atom in an electromagnetic field.

Let us start from the time-dependent Schrödinger equation (9.1). In the dipole approximation the Hamiltonian $H(t)$ of equation (9.2) becomes

$$\begin{aligned} H(t) &= \frac{1}{2m}\sum_{i=1}^{N}[\mathbf{p}_i + e\mathbf{A}(t)]^2 + V \\ &= \frac{1}{2m}\sum_{i=1}^{N}\mathbf{p}_i^2 + V + \frac{e}{m}\mathbf{A}(t)\cdot\mathbf{P} + \frac{e^2 N}{2m}\mathbf{A}^2(t) \end{aligned} \tag{9.29}$$

where $\mathbf{P}$ is the total momentum operator (9.27). The interaction Hamiltonian (9.7) takes the simple form

$$H_{\mathrm{int}}(t) = \frac{e}{m}\mathbf{A}(t)\cdot\mathbf{P} + \frac{e^2 N}{2m}\mathbf{A}^2(t) \tag{9.30}$$

and the Schrödinger equation for an N-electron atom (ion) in an electromagnetic field can therefore be written as

$$\mathrm{i}\hbar\frac{\partial}{\partial t}\Psi(X,t) = \left[H_0 + \frac{e}{m}\mathbf{A}(t)\cdot\mathbf{P} + \frac{e^2 N}{2m}\mathbf{A}^2(t)\right]\Psi(X,t) \tag{9.31}$$

where the field-free Hamiltonian H_0 is given by (9.6).

The term in $\mathbf{A}^2$ can be eliminated from this equation by performing the gauge transformation

$$\Psi(X,t) = \exp\left[-\frac{\mathrm{i}}{\hbar}\frac{e^2 N}{2m}\int^{t}\mathbf{A}^2(t')\,\mathrm{d}t'\right]\Psi^{\mathrm{V}}(X,t) \tag{9.32}$$

which gives for the wave function $\Psi^{\mathrm{V}}(X, t)$ the Schrödinger equation in the $\mathbf{A}\cdot\mathbf{P}$ or *velocity gauge*

$$\mathrm{i}\hbar\frac{\partial}{\partial t}\Psi^{\mathrm{V}}(X,t)=\left[H_0+\frac{e}{m}\mathbf{A}(t)\cdot\mathbf{P}\right]\Psi^{\mathrm{V}}(X,t) \tag{9.33}$$

Thus, in the velocity gauge, the interaction Hamiltonian is

$$H^{\mathrm{V}}_{\mathrm{int}}(t)=\frac{e}{m}\mathbf{A}(t)\cdot\mathbf{P} \tag{9.34}$$

which is the generalisation of (4.91) to the N-electron atom case.

On the other hand, if we return to the Schrödinger equation (9.31) and perform the gauge transformation

$$\Psi(X,t)=\exp\left[-\frac{ie}{\hbar}\mathbf{A}(t)\cdot\mathbf{R}\right]\Psi^{\mathrm{L}}(X,t) \tag{9.35}$$

where $\mathbf{R}$ is defined by (9.21), we obtain the Schrödinger equation in the $\mathcal{E}\cdot\mathbf{R}$ or *length gauge*

$$\mathrm{i}\hbar\frac{\partial}{\partial t}\Psi^{\mathrm{L}}(X,t)=[H_0+e\,\mathcal{E}(t)\cdot\mathbf{R}]\,\Psi^{\mathrm{L}}(X,t) \tag{9.36}$$

where we recall that in the dipole approximation the electric field is given by $\mathcal{E}(t)=-\mathrm{d}\mathbf{A}(t)/\mathrm{d}t$. Looking at (9.36), we see that in the length gauge the interaction Hamiltonian is

$$H^{\mathrm{L}}_{\mathrm{int}}(t)=e\,\mathcal{E}(t)\cdot\mathbf{R}=-\mathcal{E}(t)\cdot\mathbf{D} \tag{9.37}$$

where $\mathbf{D}$ is the electric dipole moment operator (9.20). This generalises the result (4.94) obtained for hydrogenic atoms.

The 'velocity' or 'length' forms of the Schrödinger equation are particularly useful when the interaction between the atom and the radiation field must be treated beyond first order in perturbation theory. They will be used in Section 15.2 to study multiphoton processes in atoms.

9.2 Selection rules for electric dipole transitions. Oscillator and line strengths

The selection rules for electric dipole radiative transitions in one-electron atoms and ions were obtained in Chapter 4. These will now be generalised for atoms containing any number of electrons.

As in Section 4.5, we observe that in the dipole approximation, the transition rates for absorption, stimulated emission or spontaneous emission of radiation with a particular polarisation vector $\hat{\boldsymbol{\varepsilon}}$ depend on the key quantity $\hat{\boldsymbol{\varepsilon}}\cdot\mathbf{D}_{ba}$ (see (9.25) and (9.26)), which we shall therefore examine in detail. We first recall that since the total electronic [2] angular momentum operator $\mathbf{J}$ and the parity operator $\mathcal{P}$ commute

[2] We neglect here small hyperfine effects.

with H_0, each atomic state is an eigenstate of $\mathbf{J}^2$ and J_z (with quantum numbers J and M_J, respectively) and of $\mathcal{P}$. The states can be further specified by a non-angular momentum label γ. We shall therefore label the atomic states a and b as $a \equiv (\gamma J M_J)$ and $b \equiv (\gamma' J' M_J')$ and write the electric dipole matrix element $\mathbf{D}_{ba}$ explicitly as

$$\mathbf{D}_{ba} = \langle \gamma' J' M_J' | \mathbf{D} | \gamma J M_J \rangle \tag{9.38}$$

A similar notation can be used to denote the various transition rates. For example, the dipole approximation expression (9.26) of the transition rate for spontaneous emission of a photon of polarisation $\hat{\boldsymbol{\varepsilon}}$ into a solid angle $\mathrm{d}\Omega$ in the transition $b \to a$ (with $E_b > E_a$) will be written as

$$W^{\mathrm{s,D}}(\gamma J M_J, \gamma' J' M_J')\,\mathrm{d}\Omega = \frac{1}{2\pi\hbar c^3}\left(\frac{1}{4\pi\varepsilon_0}\right)\omega_{ba}^3 |\hat{\boldsymbol{\varepsilon}}\cdot\langle \gamma' J' M_J' | \mathbf{D} | \gamma J M_J \rangle|^2\,\mathrm{d}\Omega \tag{9.39}$$

The electric dipole moment is a vector operator (see Appendix 4) because it obeys the commutation relations (A4.45). The spherical components D_q $(q = 0, \pm 1)$ of $\mathbf{D}$ are given by

$$D_1 = -\frac{1}{\sqrt{2}}(D_x + \mathrm{i}D_y) = D\left(\frac{4\pi}{3}\right)^{1/2} Y_{1,1}(\alpha, \beta)$$

$$D_0 = D_z = D\left(\frac{4\pi}{3}\right)^{1/2} Y_{1,0}(\alpha, \beta) \tag{9.40}$$

$$D_{-1} = \frac{1}{\sqrt{2}}(D_x - \mathrm{i}D_y) = D\left(\frac{4\pi}{3}\right)^{1/2} Y_{1,-1}(\alpha, \beta)$$

where (α, β) are the polar angles of $\mathbf{D}$. The spherical components ε_q $(q = 0, \pm 1)$ of the polarisation vector $\hat{\boldsymbol{\varepsilon}}$ are given by (4.110), so that the scalar product $\hat{\boldsymbol{\varepsilon}}\cdot\mathbf{D}$ can be expressed in spherical components as

$$\hat{\boldsymbol{\varepsilon}}\cdot\mathbf{D} = \sum_{q=0,\pm 1} \varepsilon_q^* D_q \tag{9.41}$$

Now, the Wigner–Eckart theorem (see Appendix 4) states that the matrix elements of a vector operator with respect to eigenstates of $\mathbf{J}^2$ and J_z only depend on M_J, M_J' and q through the Clebsch–Gordan coefficients $\langle J 1 M_J q | J' M_J' \rangle$. We have

$$\langle \gamma' J' M_J' | D_q | \gamma J M_J \rangle = \frac{1}{\sqrt{2J' + 1}} \langle J 1 M_J q | J' M_J' \rangle \langle \gamma' J' \| \mathbf{D} \| \gamma J \rangle \tag{9.42}$$

where the reduced matrix element $\langle \gamma' J' \| \mathbf{D} \| \gamma J \rangle$ is independent of q, M_J, M_J'.

The Clebsch–Gordan coefficient $\langle J 1 M_J q | J' M_J' \rangle$ vanishes unless

$$\begin{aligned} &\text{(a)}\ M_J + q = M_J' \\ &\text{(b)}\ |J - 1| \leqslant J' \leqslant J + 1 \\ &\text{(c)}\ J + J' \geqslant 1 \end{aligned} \tag{9.43}$$

The selection rules for electric dipole (E1) transitions (also called allowed transitions) are therefore

(a) $\Delta M_J = 0, \pm 1$ **(9.44a)**

(b) $\Delta J = 0, \pm 1 \qquad (J = 0 \leftrightarrow J' = 0 \text{ forbidden})$ **(9.44b)**

In addition, since under the parity operation $\mathscr{P}\mathbf{r}_i = -\mathbf{r}_i$ $(i = 1, 2, \ldots, N)$ the operator $\mathbf{D}$ changes sign, and because the atomic states are eigenstates of parity, we see that

(c) The atomic states a and b must have opposite parity.
This is known as *Laporte's rule.* **(9.44c)**

L–S coupling

When the spin–orbit interaction is weak then, as we saw in Section 8.5, the Russell–Saunders, or L–S coupling, approximation is accurate, in which both the total electronic orbital angular momentum $\mathbf{L}$ and the total electronic spin angular momentum $\mathbf{S}$ are conserved. Since the operator $\mathbf{D}$ is independent of spin, we have

$$\langle \gamma' L' S' J' M_J' | \mathbf{D} | \gamma LSJM_J \rangle = \delta_{SS'} \langle \gamma' L' S' J' M_J' | \mathbf{D} | \gamma LSJM_J \rangle \qquad \textbf{(9.45)}$$

The selection rules for electric dipole transitions which hold in addition to those discussed above are then

(a) $\Delta M_L = 0, \pm 1$ **(9.46a)**

(b) $\Delta L = 0, \pm 1 \qquad (L = 0 \leftrightarrow L' = 0 \text{ forbidden})$ **(9.46b)**

(c) $\Delta S = 0$ **(9.46c)**

The most usual case is one in which the orbital of only one electron in the atom changes. If the orbital concerned has angular momentum quantum numbers (l_i, m_i), then our discussion of the one-electron atom case shows that

$$\Delta m_i = 0, \pm 1 \qquad \textbf{(9.47a)}$$

and

$$\Delta l_i = \pm 1 \qquad \textbf{(9.47b)}$$

The case $\Delta l_i = 0$ is not allowed because the parity of the states a and b must change. If two or more configurations are strongly mixed in a particular state, then more than one electron can make a transition simultaneously even though only one photon is emitted or absorbed, but this is a rather uncommon circumstance.

Electric dipole transitions between hyperfine structure components

Let us consider two levels (γJ) and $(\gamma' J')$ between which transitions are allowed. Electric dipole transitions between the components of the hyperfine structure of these two levels then obey the additional selection rules

$$\Delta F = 0, \pm 1 \qquad (F = 0 \leftrightarrow F' = 0 \text{ forbidden}) \qquad \textbf{(9.48a)}$$

and

$$\Delta M_F = 0, \pm 1 \tag{9.48b}$$

Electric dipole transitions between components of the hyperfine structure of the *same* level are forbidden by the parity selection rule. However, magnetic dipole and electric quadrupole transitions can occur in this case (see Section 9.3).

Oscillator and line strengths

In Section 4.6 it was noted that the relative intensities of radiative transitions from a given state a to various final states k, in the dipole approximation, could be discussed in terms of a dimensionless quantity called an *oscillator strength*, which was defined for a one-electron atom by (4.136). The oscillator strength for an atom containing N electrons can be defined similarly as

$$f_{ka} = \frac{2m\omega_{ka}}{3\hbar}\left|\sum_{i=1}^{N}(\mathbf{r}_i)_{ka}\right|^2 \tag{9.49}$$

where $\omega_{ka} = (E_k - E_a)/\hbar$. With this definition f_{ka} is positive for absorption ($E_k > E_a$), and negative for emission ($E_k < E_a$). By generalising the argument leading from (4.138) to (4.142), it is straightforward to show (Problem 9.1) that the Thomas–Reiche–Kuhn sum rule (4.137) becomes

$$\sum_k f_{ka} = N \tag{9.50}$$

where the sum runs over all states, including the continuum.

Now let us consider a transition between two states a and b where, as above, a and b have the quantum numbers $(\gamma J M_J)$ and $(\gamma' J' M'_J)$ respectively. Then the corresponding oscillator strength can be expressed as

$$f(\gamma' J' M'_J, \gamma J M_J) = \frac{2m\omega_{ba}}{3\hbar}\left|\sum_{i=1}^{N}\langle\gamma' J' M'_J|\mathbf{r}_i|\gamma J M_J\rangle\right|^2 \tag{9.51}$$

The transition rate for the spontaneous emission of a photon in the transition $a \to b$ (with $E_b < E_a$), integrated over all angles of emission and summed over all directions of polarisation, is given in the dipole approximation by

$$W^{\mathrm{s,D}}(\gamma' J' M'_J, \gamma J M_J) = \frac{2\hbar\alpha}{mc^2}\,\omega_{ba}^2\,|f(\gamma' J' M'_J, \gamma J M_J)| \tag{9.52}$$

where α is the fine structure constant. In the absence of any external fields the initial states $(\gamma J M_J)$ are degenerate with respect to the $(2J + 1)$ different values of M_J and similarly the final states $(\gamma' J' M'_J)$ are also degenerate with respect to the $(2J' + 1)$ different values of M'_J. There is an equal probability of finding the atom initially in any of the $(2J + 1)$ magnetic substates, so that the total rate for a spontaneous transition between the level (γJ) and the level $(\gamma' J')$ is given by

$$W^{s,D}(\gamma'J', \gamma J) = \frac{2\hbar\alpha}{mc^2}\omega_{ba}^2 |\bar{f}(\gamma'J', \gamma J)| \tag{9.53}$$

where $\bar{f}$ is the average oscillator strength (see (4.143))

$$\bar{f}(\gamma'J', \gamma J) = \frac{1}{2J+1}\sum_{M_J, M_J'} f(\gamma'J'M_J', \gamma JM_J) \tag{9.54}$$

Because of the degree of degeneracy factor $(2J+1)^{-1}$ the average oscillator strength is not symmetric between the levels $(\gamma'J')$ and (γJ). In fact we have

$$(2J+1)|\bar{f}(\gamma'J', \gamma J)| = (2J'+1)|\bar{f}(\gamma J, \gamma'J')| \tag{9.55}$$

For this reason a symmetrised quantity, called the *line strength*, is often introduced by the relations

$$\begin{aligned} \mathcal{S}(\gamma'J', \gamma J) &= \hbar c\alpha \sum_{M_J M_J'} \left| \sum_{i=1}^{N} \langle \gamma'J'M_J' | \mathbf{r}_i | \gamma JM_J \rangle \right|^2 \\ &= \frac{3\hbar^2 c\alpha}{2m}\frac{1}{|\omega_{ba}|}(2J+1)|\bar{f}(\gamma'J', \gamma J)| \end{aligned} \tag{9.56}$$

The transition rate for spontaneous emission is then given by

$$W^{s,D}(\gamma'J', \gamma J) = \frac{4|\omega_{ba}|^3}{3\hbar c^3}\frac{1}{2J+1}\mathcal{S}(\gamma'J', \gamma J) \tag{9.57}$$

The transition rates for absorption and for stimulated emission can also be written in terms of the line strength as

$$W^{D}(\gamma'J', \gamma J) = \frac{4\pi^2}{3c\hbar^2} I(|\omega_{ba}|)\frac{1}{2J+1}\mathcal{S}(\gamma'J', \gamma J) \tag{9.58}$$

where $I(|\omega_{ba}|)$ is the intensity of the radiation at the angular frequency $|\omega_{ba}|$. In both cases the initial level is (γJ) with degeneracy $(2J+1)$ and $(\gamma'J')$ is the final level, with degeneracy $(2J'+1)$.

A particular advantage of introducing the line strengths $\mathcal{S}(\gamma'J', \gamma J)$ is that the intensity of a spectral line due to spontaneous emission is proportional both to $W^{s,D}$ given by (9.57) and to the number of atoms which are present in the initial state. The latter number is proportional to the statistical weight of the level which is $(2J+1)$. The line intensity is thus proportional to $(2J+1)W^{s,D}$ which is in turn proportional to $\mathcal{S}$. Once the atomic wave functions have been calculated, for example by using the methods described in Chapter 8, the calculation of oscillator and line strengths is straightforward. Semi-empirical methods for the calculation of line strengths are also useful for some systems (see Sobelman, 1979) and the Thomas–Reiche–Kuhn sum rule provides an important test of the accuracy of such calculations.

In the L–S coupling approximation, we can express the transition rate for a spontaneous emission from a level (γLS) to a level $(\gamma'L'S')$ as

$$W^{s,D}(\gamma'L'S', \gamma LS) = \frac{4|\omega_{ba}|^3}{3\hbar c^3}\frac{1}{g}\mathcal{S}(\gamma'L'S', \gamma LS) \tag{9.59}$$

where g is the degree of degeneracy of the initial level,

$$g = (2L + 1)(2S + 1) \tag{9.60}$$

and $\mathcal{S}$ is the line strength

$$\mathcal{S}(\gamma'L'S', \gamma LS) = \hbar c\alpha \sum_{JJ'} \sum_{M_J M'_J} \left| \sum_{i=1}^{N} \langle \gamma'L'S'J'M'_J | \mathbf{r}_i | \gamma LSJM_J \rangle \right|^2 \tag{9.61}$$

where the sums over J and J' are over all the multiplet components. The line strengths have the useful additional property that

$$\mathcal{S}(\gamma'L'S', \gamma LS) = \sum_J \sum_{J'} \mathcal{S}(\gamma'L'S'J', \gamma LSJ) \tag{9.62}$$

where

$$\mathcal{S}(\gamma'L'S'J', \gamma LSJ) = \hbar c\alpha \sum_{M_J M'_J} \left| \sum_{i=1}^{N} \langle \gamma'L'S'J'M'_J | \mathbf{r}_i | \gamma LSJM_J \rangle \right|^2 \tag{9.63}$$

Dynamic and static dipole polarisabilities

The *dynamic* (or frequency-dependent) *dipole polarisability* of an atom in a state (γJM_J) is defined by the relation

$$\alpha(\gamma JM_J, \omega) = 2 \sum_{\gamma'J'}{}' \frac{(E_{\gamma'J'} - E_{\gamma J}) |\langle \gamma'J'M_J | D_z | \gamma JM_J \rangle|^2}{(E_{\gamma'J'} - E_{\gamma J})^2 - \hbar^2\omega^2} \tag{9.64}$$

In (9.64), the symbol Σ' means that the level (γJ) must be excluded from the sum.

Averaging over the M_J substates of the level (γJ), we obtain the quantity

$$\alpha(\gamma J, \omega) = \frac{1}{2J + 1} \sum_{M_J} \alpha(\gamma JM_J, \omega) \tag{9.65}$$

which can be related to the average oscillator strength $\bar{f}(\gamma'J', \gamma J)$ by

$$\alpha(\gamma J, \omega) = \frac{e^2\hbar^2}{m} \sum_{\gamma'J'} \frac{\bar{f}(\gamma'J', \gamma J)}{(E_{\gamma'J'} - E_{\gamma J})^2 - \hbar^2\omega^2} \tag{9.66}$$

In the limit $\omega \to 0$, the dynamic dipole polarisability $\alpha(\gamma JM_J, \omega)$ reduces to

$$\bar{\alpha}(\gamma JM_J) = 2 \sum_{\gamma'J'}{}' \frac{|\langle \gamma'J'M_J | D_z | \gamma JM_J \rangle|^2}{E_{\gamma'J'} - E_{\gamma J}} \tag{9.67}$$

which is called the *static dipole polarisability* of the atom in the state (γJM_J). The static dipole polarisability, averaged over all M_J substates of the level (γJ), is given by

Table 9.1 Static dipole polarisabilities $\bar{\alpha}$ (in atomic units) of various atoms and ions.

Atomic species	$\bar{\alpha}$	*Atomic species*	$\bar{\alpha}$
H(1s)	4.5	Ne(^{1}S)	2.68
H(2p)	176	Na(^{2}S)	162
H^-(1^{1}S)	206	Na^+(^{1}S)	12
He(1^{1}S)	1.383	Mg(^{1}S)	72
He(2^{3}S)	316	Ar (^{1}S)	11.08
He(2^{1}S)	803	K(^{2}S)	290
Li(^{2}S)	162	K^+(^{1}S)	8.1
Li^+(^{1}S)	0.1925	Ca(^{1}S)	170
Be(^{1}S)	38	Kr(^{1}S)	16.74
B(^{2}P)	20.5	Rb(^{2}S)	320
C(^{3}P)	11.8	Xe(^{1}S)	27.06
N(^{4}S)	7.5	Cs(^{2}S)	400
O(^{3}P)	5.41	Hg(^{1}S)	34
F(^{2}P)	3.76	Pb(^{3}P)	49

$$\bar{\alpha}(\gamma J) = \frac{1}{2J+1} \sum_{M_J} \bar{\alpha}(\gamma J M_J) \quad \textbf{(9.68)}$$

This is the generalisation of the expressions (4.236) or (6.37) which we obtained for one-electron atoms. The averaged static dipole polarisabilities of a few atoms and ions are given in Table 9.1.

9.3 Retardation effects. Magnetic dipole and electric quadrupole transitions

Let us return to the matrix element M_{ba} given by (9.10), which contains the retardation factor $\exp(i\mathbf{k}\cdot\mathbf{r}_i)$. As explained in Section 9.1, if the wavelength λ of the radiation is large compared to the size a of the atom (ion), so that $ka \ll 1$, one can make the dipole approximation, in which $\exp(i\mathbf{k}\cdot\mathbf{r}_i)$ is replaced by unity (neglect of retardation). The matrix element M_{ba} of equation (9.10) is then replaced by its dipole approximation M_{ba}^{D}, given by (9.16) or (9.24).

The neglect of retardation becomes inadequate when the condition $\lambda \gg a$ is not fulfilled. This is the case, for example, of X-ray radiation emitted or absorbed in 'bound–bound' transitions involving the inner electrons of atoms (ions) with large Z, or in the photoionisation of atoms by high-frequency photons whose wavelength λ is not large with respect to a.

In this section, we shall focus our attention on the discrete spectrum of atoms for which $Z\alpha \ll 1$. In this case $ka \ll 1$ and retardation effects are small. However, as we saw in Section 9.2, in the dipole approximation the transitions which do not obey the selection rules for 'allowed' (E1) transitions (that is, for which the electric dipole matrix element $\mathbf{D}_{ba}$ is zero) are 'forbidden'. In this case, higher terms in the expansion

$$\exp(\mathrm{i}\mathbf{k}\cdot\mathbf{r}_i) = 1 + \mathrm{i}\mathbf{k}\cdot\mathbf{r}_i + \frac{1}{2!}(\mathrm{i}\mathbf{k}\cdot\mathbf{r}_i)^2 + \ldots \tag{9.69}$$

must be taken into account. For some of the 'forbidden' transitions, these higher order terms will yield non-vanishing transition rates which are, however, much smaller than for the allowed (E1) transitions. As stated in Chapter 4, the transitions due to the higher order terms in (9.69) lead to radiation which is somewhat analogous to the types of radiation arising from a multipole expansion in classical radiation theory. In what follows we shall consider only the second term ($\mathrm{i}\mathbf{k}\cdot\mathbf{r}_i$) of the expansion (9.69), and show that it leads to *magnetic dipole* (M1) and *electric quadrupole* (E2) transitions.

Let us start, therefore, from the unapproximated expression (9.10) of the matrix element M_{ba}. Substituting (9.69) into (9.10), we have

$$M_{ba} = M_{ba}^{\mathrm{D}} + \tilde{M}_{ba} + \ldots \tag{9.70}$$

The first term, M_{ba}^{D}, arising from the first term (unity) of the expansion (9.69), is just the electric dipole approximation (9.16) (or (9.24)) to the matrix element M_{ba}. The second term, $\tilde{M}_{ba}$, generated by the next term ($\mathrm{i}\mathbf{k}\cdot\mathbf{r}_i$) on the right of (9.69), is given by

$$\tilde{M}_{ba} = \sum_{i=1}^{N} \langle \psi_b | (\mathrm{i}\mathbf{k}\cdot\mathbf{r}_i)(\hat{\boldsymbol{\varepsilon}}\cdot\boldsymbol{\nabla}_{\mathbf{r}_i}) | \psi_a \rangle \tag{9.71}$$

Let us take the direction of propagation, $\hat{\mathbf{k}}$, to be along the Z axis, and the polarisation vector $\hat{\boldsymbol{\varepsilon}}$ to be along the X axis. We then have

$$\tilde{M}_{ba} = \mathrm{i}k \sum_{i=1}^{N} \left\langle \psi_b \left| z_i \frac{\partial}{\partial x_i} \right| \psi_a \right\rangle = \mathrm{i}\frac{\omega_{ba}}{c} \sum_{i=1}^{N} \left\langle \psi_b \left| z_i \frac{\partial}{\partial x_i} \right| \psi_a \right\rangle \tag{9.72}$$

where we have used the fact that the wave number k of the absorbed or emitted photon is equal to ω_{ba}/c. It is convenient to rewrite the operator $\sum_{i=1}^{N} z_i \partial/\partial x_i$ appearing in (9.72) in the form

$$\sum_{i=1}^{N} z_i \frac{\partial}{\partial x_i} = \frac{1}{2}\sum_{i=1}^{N}\left(z_i \frac{\partial}{\partial x_i} - x_i \frac{\partial}{\partial z_i}\right) + \frac{1}{2}\sum_{i=1}^{N}\left(x_i \frac{\partial}{\partial z_i} + z_i \frac{\partial}{\partial x_i}\right) \tag{9.73}$$

The first (antisymmetric) operator on the right of (9.73) is given by

$$\frac{1}{2}\sum_{i=1}^{N}\left(z_i \frac{\partial}{\partial x_i} - x_i \frac{\partial}{\partial z_i}\right) = \frac{\mathrm{i}}{2\hbar} L_y \tag{9.74}$$

where L_y is the y component of the total electronic orbital angular momentum operator $\mathbf{L}$. The second (symmetric) operator on the right of (9.73) can be transformed by using the Heisenberg equation of motion

$$\dot{\mathbf{r}}_i = (\mathrm{i}\hbar)^{-1}[\mathbf{r}_i, H_0] \tag{9.75}$$

and noting that z_i and $\partial/\partial x_i$ commute. Thus we have

$$\begin{aligned}\frac{1}{2}\sum_{i=1}^{N}\left(x_i\frac{\partial}{\partial z_i} + z_i\frac{\partial}{\partial x_i}\right) &= \frac{1}{2}\sum_{i=1}^{N}\left(x_i\frac{\partial}{\partial z_i} + \frac{\partial}{\partial x_i}z_i\right)\\ &= \frac{m}{2\hbar^2}\sum_{i=1}^{N}(x_i[z_i, H_0] + [x_i, H_0]z_i)\\ &= \frac{m}{2\hbar^2}\sum_{i=1}^{N}(x_iz_iH_0 - H_0x_iz_i)\end{aligned} \tag{9.76}$$

Using (9.73)–(9.76) and the fact that ψ_a and ψ_b are eigenfunctions of the Hamiltonian H_0 with eigenvalues E_a and E_b, respectively, we may rewrite the matrix element $\tilde{M}_{ba}$ of equation (9.72) as

$$\tilde{M}_{ba} = \tilde{M}_{ba}^{(1)} + \tilde{M}_{ba}^{(2)} \tag{9.77}$$

where

$$\tilde{M}_{ba}^{(1)} = -\frac{\omega_{ba}}{2\hbar c}\langle\psi_b|L_y|\psi_a\rangle \tag{9.78}$$

and

$$\tilde{M}_{ba}^{(2)} = -\frac{\mathrm{i}m\omega_{ba}^2}{2\hbar c}\sum_{i=1}^{N}\langle\psi_b|x_iz_i|\psi_a\rangle \tag{9.79}$$

Magnetic dipole transitions

The matrix element $\tilde{M}_{ba}^{(1)}$ involves the total electronic orbital angular momentum **L**, which in turn is proportional to the *orbital magnetic dipole moment* operator $\mathcal{M}_L$ of the atom, since

$$\mathcal{M}_L = -\frac{e}{2m}\mathbf{L} = -\mu_B(\mathbf{L}/\hbar) \tag{9.80}$$

where μ_B is the Bohr magneton. Making the usual extension to include spin angular momentum, we replace **L** by **L** + 2**S**, where **S** denotes the total spin operator of the electrons. The matrix element $\tilde{M}_{ba}^{(1)}$ of equation (9.78) is then replaced by

$$\begin{aligned}M_{ba}^{\mathrm{M1}} &= -\frac{\omega_{ba}}{2\hbar c}\langle\psi_b|L_y + 2S_y|\psi_a\rangle\\ &= -\frac{\omega_{ba}}{2\hbar c}\langle\psi_b|J_y + S_y|\psi_a\rangle\end{aligned} \tag{9.81}$$

where J_y is the y component of the total angular momentum operator $\mathbf{J} = \mathbf{L} + \mathbf{S}$.

The matrix element (9.81) corresponds to a *magnetic dipole* (M1) transition. We note that the magnetic dipole matrix element M_{ba}^{M1} depends on the direction (along the Y axis) perpendicular to the plane containing the propagation direction $\hat{\mathbf{k}}$ (which we have chosen along the Z axis) and the polarisation direction $\hat{\boldsymbol{\varepsilon}}$ (chosen along the X axis) of the radiation. More generally, M_{ba}^{M1} is proportional to the component of $\mathbf{L} + 2\mathbf{S} = \mathbf{J} + \mathbf{S}$ in a direction $\hat{\mathbf{d}}$ perpendicular to the plane containing the propagation vector $\mathbf{k}$ and the polarisation vector $\hat{\boldsymbol{\varepsilon}}$.

The selection rules for magnetic dipole transitions can be obtained in the following way. First, from the Wigner–Eckart theorem, we have (Problem 9.2)

$$\text{(a)}\quad \Delta M_J = 0, \pm 1 \qquad \textbf{(9.82a)}$$

The case $\Delta M_J = 0$ occurs if the direction $\hat{\mathbf{d}}$ is parallel to the Z axis (the quantisation direction of the atom), while $\Delta M_J = \pm 1$ if the direction $\hat{\mathbf{d}}$ is perpendicular to the Z axis.

$$\text{(b)}\quad \Delta J = 0, \pm 1 \qquad (J = 0 \leftrightarrow J' = 0 \text{ forbidden}) \qquad \textbf{(9.82b)}$$

In addition, since the operator appearing in M_{ba}^{M1} is even under the parity operation, it follows that

(c) The atomic states a and b must have the same parity. **(9.82c)**

The selection rules (9.82) are independent of the angular momentum coupling scheme. If the Russell–Saunders (L–S coupling) approximation holds, the selection rules are much more restrictive. Indeed, in that approximation one forms linear superpositions of central field, product-type wave functions belonging to a given configuration, with given values of the quantum numbers (n_i, l_i) for each of the N electrons. For these Russell–Saunders wave functions L, S and J are good quantum numbers, as well as n_i and l_i. Since the operator $\mathbf{L} + 2\mathbf{S}$ commutes with $\mathbf{L}^2$, $\mathbf{S}^2$ and with $\mathbf{L}_i^2$ $(i = 1, 2, \ldots, N)$, it follows that a matrix element of $\mathbf{L} + 2\mathbf{S}$ must obey the selection rules

$$\text{(a)}\quad \Delta L = \Delta l_i = 0 \qquad (i = 1, 2, \ldots, N) \qquad \textbf{(9.83a)}$$

$$\text{(b)}\quad \Delta S = 0 \qquad \textbf{(9.83b)}$$

in addition to the general selection rules (9.82). Moreover, since the operator $\mathbf{L} + 2\mathbf{S}$ does not act on the radial part of the wave functions, and because two radial wave functions with the same l, but different n, are orthogonal, one has the additional selection rule

$$\text{(c)}\quad \Delta n_i = 0 \qquad (i = 1, 2, \ldots, N) \qquad \textbf{(9.83c)}$$

It follows that in the L–S coupling approximation, magnetic dipole transitions can only occur between two atomic states a and b belonging to the *same configuration* and having the *same* values of L and S. Two such states belong to the *same fine structure multiplet*, so that the value of ω_{ba} in (9.81) is very small and the transition is weak. Such transitions occur in the microwave or radio-frequency region. Because ω_{ba} is very small, it follows from the Einstein relation (4.108b) that spontaneous transitions of this type have a negligibly small probability compared to stimulated transitions. We note that if L–S coupling were exact (no spin–orbit interactions), the two states a and b would have exactly the same energy, so that ω_{ba} would vanish, and magnetic dipole transitions could not happen.

For an atom whose nucleus has a magnetic moment, magnetic dipole transitions can take place between two *hyperfine* structure components of the *same* fine structure level ($\Delta L = \Delta S = \Delta J = 0$), the selection rule on the hyperfine quantum number F being given by (9.48a). For example, the hyperfine transition between the states $F = 0$ and $F = 1$ of the atomic hydrogen ground state is an M1 transition.

Magnetic dipole transitions between *different* fine structure multiplets correspond to much higher frequencies (for example, optical frequencies). As we have seen above, such M1 transitions are forbidden in the L–S coupling approximation. They can be shown to have very small probabilities for atoms or ions with small Z, but can play an important role when Z is large. Of particular interest is the $2s_{1/2} \to 1s_{1/2}$ spontaneous transition in hydrogenic atoms or ions. As we learned in Chapter 4, this transition is forbidden in the electric dipole approximation. For small Z, the dominant mechanism for this transition is the simultaneous emission of two photons, the contribution of the M1 transition being negligible. However, the opposite happens for large Z. In that case the $2s_{1/2} \to 1s_{1/2}$ decay occurs mainly via an M1 transition, as illustrated in Fig. 5.9 for the U^{91+} hydrogenic ion.

Electric quadrupole transitions

Let us now consider the matrix element $\tilde{M}_{ba}^{(2)}$ given by (9.79). It is proportional to the matrix element of $\sum_{i=1}^{N} x_i z_i$, which is a component of the *electric quadrupole moment* operator of the atom. For this reason, $\tilde{M}_{ba}^{(2)}$ is said to describe an electric quadrupole (E2) transition; it will be denoted in what follows by M_{ba}^{E2}. We note that M_{ba}^{E2} depends on both the propagation direction $\hat{\mathbf{k}}$ (along the Z axis) and the polarisation direction $\hat{\boldsymbol{\varepsilon}}$ (along the X axis) of the radiation.

The selection rules for electric quadrupole transitions can be obtained as follows. First, using the Wigner–Eckart theorem, and noting that the operator appearing in the matrix element M_{ba}^{E2} $(= \tilde{M}_{ba}^{(2)})$ is an irreducible tensor of rank 2, one finds (Problem 9.3) that

(a) $\Delta M_J = 0, \pm 1, \pm 2$ **(9.84a)**

(b) $\Delta J = 0, \pm 1, \pm 2 \qquad (J + J' \geqslant 2)$ **(9.84b)**

In addition, because the operator in M_{ba}^{E2} is even under the parity operation, it follows that

(c) The atomic states a and b must have the same parity. **(9.84c)**

The selection rules (9.84) are independent of the coupling scheme used to couple the angular momenta. If the Russell–Saunders (L–S coupling) approximation is accurate,

$$\text{(a)}\ \Delta S = 0 \tag{9.85a}$$

$$\text{(b)}\ \Delta L = 0, \pm 1, \pm 2 \qquad (L + L' \geqslant 2) \tag{9.85b}$$

$$\text{(c)}\ \Delta M_L = 0, \pm 1, \pm 2 \tag{9.85c}$$

For the case in which a single electron orbital of angular quantum numbers (l_i, m_i) changes, one has

$$\Delta l_i = 0, \pm 2 \qquad (l = 0 \leftrightarrow l' = 0 \text{ forbidden}) \tag{9.86a}$$

and

$$\Delta m_i = 0, \pm 1, \pm 2 \tag{9.86b}$$

It is worth noting that transitions for which $\Delta J = \pm 2$, $\Delta L = \pm 2$ or $\Delta l_i = \pm 2$ (for a single electron transition) are purely of the electric quadrupole (E2) type. As an example, let us consider the oxygen atom. Its ground state electronic configuration is $1s^2\ 2s^2\ 2p^4$, leading to states ^3P, ^1D and ^1S which have the same (even) parity. The ground state is the ^3P state, the successive energy separations of the ^1D and ^1S states being approximately 1 eV. Electric dipole transitions between the three states are forbidden by Laporte's parity selection rule (9.44c). The $^1S \to {}^1D$ transition is a purely electric quadrupole transition (with $\Delta L = 2$), giving rise to the green line at a wavelength $\lambda = 5577$ Å which can be observed in the spectrum of the aurora.

Finally, we note that electric quadrupole transitions between components of the *hyperfine* structure of the *same* fine structure level can occur, and obey the selection rule

$$\Delta F = 0, \pm 1, \pm 2 \qquad (F + F' \geqslant 2) \tag{9.87}$$

It can be shown that all matrix elements for multipole transitions, either electric or magnetic, are proportional to the Clebsch–Gordan coefficient $\langle J\lambda M_J q | J'M'_J \rangle$, where $\lambda = 1$ for dipole transitions, $\lambda = 2$ for quadrupole transitions and so on. For electric multipole transitions, the parity changes if λ is odd and does not change if λ is even, while for magnetic multipoles the parity changes for even λ and does not change for odd λ. Combining this parity rule with the properties of the Clebsch–Gordan coefficient, the selection rules can be obtained (Problem 9.4) for any multipole.

9.4 The spectra of the alkalis

As we have seen in Chapter 8, the structure of the ground states of the alkali metals (Li, Na, K, Rb, Cs, Fr) is that of a single valence electron moving in an s orbital outside a *core* consisting of a *closed (sub)shell* system. In lithium, the configuration of the core is $(1s)^2$, in sodium $(1s)^2(2s)^2(2p)^6$, in potassium $(1s)^2(2s)^2(2p)^6(3s)^2(3p)^6$, and in each case the total electronic orbital angular momentum and spin angular momentum of the core are zero, designated as $^1S_0$.

The cores of rubidium, caesium and francium are also $^1S_0$. Because the core is spherically symmetric, the valence electron of a neutral alkali atom ($N = Z$) moves in an effective *central* potential $V(r)$, which at large distances approaches the Coulomb potential

$$V(r) \underset{r\to\infty}{\to} -\frac{e^2}{(4\pi\varepsilon_0)r} \quad \textbf{(9.88)}$$

since the nuclear charge Ze is screened by the core which contains $(Z - 1)$ electrons. The lowest state of the valence electron has zero orbital angular momentum ($l = 0$) and the corresponding orbital is designated (n_0s), where, to satisfy the Pauli exclusion principle, $n_0 = 2$ for Li, 3 for Na, 4 for K and so on. For small r the potential $V(r)$ is always more attractive than $-e^2/(4\pi\varepsilon_0 r)$, so that the binding energy of the orbital (n_0s) is always greater than that of the (n_0s) level of hydrogen. The excited states of the valence electron, moving with respect to the unexcited core, are designated (nl) with $n \geqslant n_0$. The binding energy of the orbital (nl) is again greater than that of the (nl) level of hydrogen. For highly excited states, for which the charge density of the orbital is almost entirely outside the region of the core, the orbitals approximate closely to true hydrogenic wave functions, and the energy levels are very close to those of hydrogen.

As $V(r)$ is not Coulombic at short distances, the energy levels are not degenerate in l, for a given n. Thus the (2s) level in lithium lies well below the (2p) level, and for sodium the (3s) level is below the (3p) level, which in turn is below the (3d) level (see Fig. 9.1). As the excitation increases and the levels become more hydrogenic in character, there is near degeneracy in l, for a given n. The valence electron is bound weakly compared with the core electrons. For each of the alkalis it requires about 5 eV to detach the valence electron, compared with over 20 eV for the least bound of the core electrons, and it follows that the optical spectrum is due to transitions involving the valence electron only, the core remaining inert. The term symbol of the ground state of an alkali can be written, omitting the core, as $(n_0\text{s})^2\text{S}_{1/2}$ where the superscript denotes the multiplicity of the spin state $(2s + 1)$, in this case with $s = 1/2$, and the subscript is the total angular momentum quantum number. Excited states are of the form $(n\text{s})^2\text{S}_{1/2}$, $(n\text{p})^2\text{P}_{1/2,3/2}$, $(n\text{d})^2\text{D}_{3/2,5/2}$ and so on.

The gross structure of the alkali spectra can be obtained by combining the selection rules for a one-electron atom (see Chapter 4) with a knowledge of the energies of the levels (nl) of the valence electron. The energy levels can be represented by the empirical formula

$$E_{nl} = -\frac{1}{2}\frac{1}{(n-\mu_{nl})^2} \text{ a.u.} \quad \textbf{(9.89)}$$

The quantities μ_{nl} are known as *quantum defects*. To a good approximation μ_{nl} is (for a particular alkali) a function of l only; $\mu_{nl} \simeq \alpha(l)$. Thus we can write the effective principal quantum number, in a form due to Rydberg, as

$$n^* = n - \alpha(l) \quad \textbf{(9.90)}$$

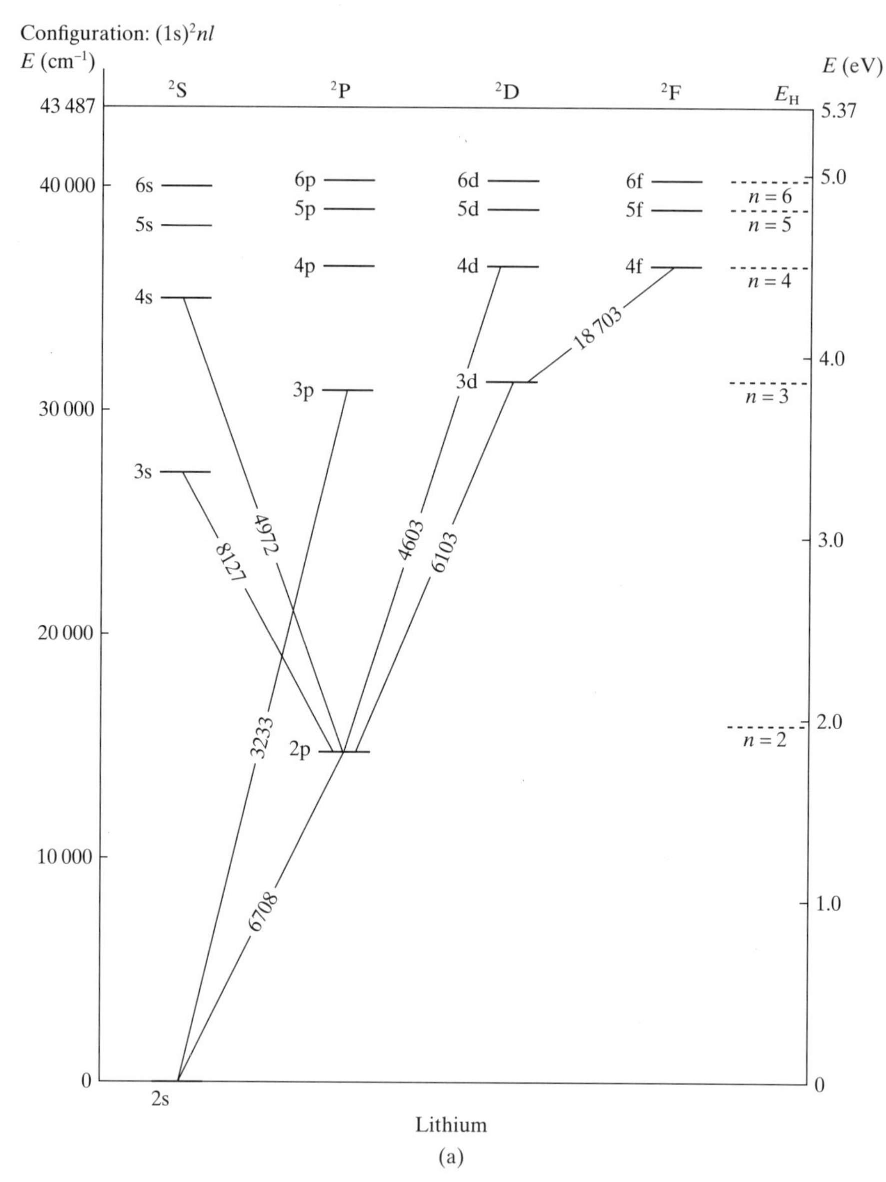

Figure 9.1 Grotrian diagrams of energy levels and transitions in (a) lithium and (b) sodium. Energies are shown relative to the ground state, with the horizontal line at the top of each diagram showing the ionisation potential of the ground state. The column headed E_H shows the corresponding positions of the levels of atomic hydrogen.

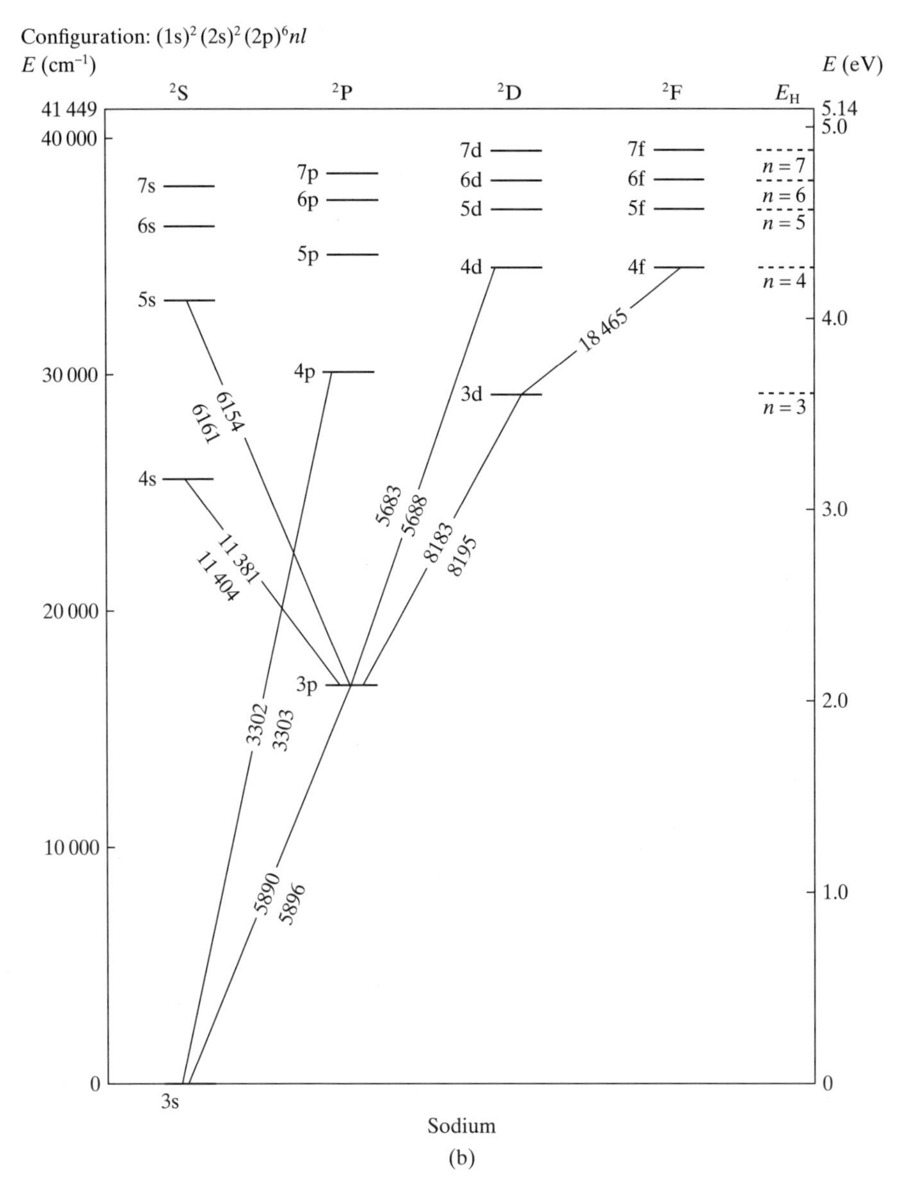

Figure 9.1 *(Cont.)*

for lithium, sodium and potassium. A small correction to this formula was made by Ritz, who wrote [3]

$$n^* = n - \alpha(l) - \beta(l)/n^2 \tag{9.91}$$

[3] Tables of the quantities n^* can be found in Kuhn (1970).

Table 9.2 Parameters of the energy levels of the alkalis.

(a) *quantum defects* $\alpha(l)$

Atom	$l =$ 0	1	2	3
Li	0.40	0.04	0.00	0.00
Na	1.35	0.85	0.01	0.00
K	2.19	1.71	0.25	0.00
Rb	3.13	2.66	1.34	0.01
Cs	4.06	3.59	2.46	0.02

(b) Effective principal quantum numbers n^* for the $(n_0\text{s})$ and $(n_0\text{p})$ levels

Level	Li ($n_0 = 2$)	Na ($n_0 = 3$)	K ($n_0 = 4$)	Rb ($n_0 = 5$)	Cs ($n_0 = 6$)
$(n_0\text{s})^2\text{S}_{1/2}$	1.588	1.626	1.771	1.805	1.869
$(n_0\text{p})^2\text{P}_{1/2}$	1.966	2.116	2.232	2.280	2.392
$(n_0\text{p})^2\text{P}_{3/2}$	1.966	2.117	2.235	2.293	2.362

(c) Ionisation potentials of the alkalis

	Li	Na	K	Rb	Cs
I_P (eV)	5.39	5.14	4.34	4.18	3.89

For example, in the case of the p levels of lithium one has $\alpha(1) = 0.040$ and $\beta(1) = 0.024$. Some parameters of the energy levels of the alkalis are given in Table 9.2.

Absorption spectra

The absorption spectra can be obtained by passing light through the vapour of the alkali metal. Except at very high temperatures, most atoms will be in the ground state, and the series of absorption lines correspond to transitions from the $(n_0\text{s})$ ground state to the $(n\text{p})$ levels, with frequencies

$$\nu = R\left[\left(\frac{1}{n^*_{n_0\text{s}}}\right)^2 - \left(\frac{1}{n^*_{n\text{p}}}\right)^2\right] \tag{9.92}$$

where R is the Rydberg constant. This series of lines is called the *principal series.*

Emission spectra

The *principal series* can be observed in emission as well as in absorption. By far the strongest line is that corresponding to the transition $n_0\text{p} \to n_0\text{s}$ and is called a *resonance line*. Because of spin–orbit coupling, which we shall discuss below, the resonance lines are doublets. The wavelengths of these lines are given in Table 9.3.

Table 9.3 The resonance lines of the alkali metals: Li, Na, K, Rb, Cs. The most prominent emission lines in the alkali spectra are the resonance lines corresponding to the transition $(n_0\mathrm{p}) \to (n_0\mathrm{s})$.

Atom	*Wave number* (cm^{-1})	*Wavelength* (Å)
Li	14 904	6708
Na†	16 956	5896
	16 973	5890
K	12 985	7699
	13 043	7665
Rb	12 582	7946
	12 817	7800
Cs	11 179	8943
	11 733	8521

† The yellow lines in this doublet are known as the Fraunhofer D-lines which occur in the spectrum of sunlight.

Several other series of emission lines can be observed (see Fig. 9.1), including the *sharp series*, corresponding to $n\mathrm{s} \to n_0\mathrm{p}$ transitions, the *diffuse series*, which corresponds to $n\mathrm{d} \to n_0\mathrm{p}$ and the *fundamental series*, which corresponds to $n\mathrm{f} \to n_0\mathrm{d}$.

The emission spectra of positive ions with one valence electron outside an inert core can also be observed in a spark discharge. For instance the sequence of ions iso-electronic with lithium consists of Li I, Be II, B III, C IV, . . . where I stands for the neutral atom, II for a singly ionised atom and so forth. The energy levels of such a sequence can be written as

$$E_{nl} = -\frac{1}{2}\frac{\tilde{Z}^2}{[n - \alpha(l)]^2}\ \text{a.u.} \tag{9.93}$$

where $\tilde{Z} = Z - N + 1$, Z being the nuclear charge and N the number of electrons. The quantity $\tilde{Z}$ is therefore the net charge of the nucleus and the core electrons, and is equal to the roman numeral in the symbols Be II, B III, C IV, and so on. The quantum defect $\alpha(l)$ decreases smoothly during such a sequence so that for large Z the spectrum becomes hydrogenic.

Fine structure

All the energy levels of the valence electron in an alkali (except for those with $l = 0$) are split into two, one level corresponding to a total angular momentum quantum number $j = l + 1/2$ and the other to $j = l - 1/2$. The interaction causing this splitting is the spin–orbit interaction, which was discussed in Section 5.1 in connection with one-electron atoms. The shift in energy due to this effect is given by (see (5.24))

Table 9.4 Spin–orbit splitting of the (n_0p) levels of the alkalis.

Atom	Li	Na	K	Rb	Cs
n_0	2	3	4	5	6
ΔE	0.337	17.2	57.7	238	554 cm^{-1}
	0.42	21	72	295	687×10^{-4} eV

$$\Delta E = \tfrac{1}{2}\lambda_{nl}[j(j+1) - l(l+1) - \tfrac{3}{4}] \qquad \textbf{(9.94)}$$

The constant λ_{nl} is proportional to the expectation value of $r^{-1}\,dV(r)/dr$, $V(r)$ being the effective central potential in which the valence electron moves:

$$\lambda_{nl} = \frac{\hbar^2}{2m^2c^2}\left\langle \frac{1}{r}\frac{dV(r)}{dr}\right\rangle = \hbar^2\langle\xi(r)\rangle \qquad \textbf{(9.95)}$$

The magnitude of λ_{nl} can be estimated by using $V(r)$ as calculated by the Hartree–Fock method. It turns out to be much larger than for atomic hydrogen (about 50 times larger for the 3p level of sodium than for the 2p level of hydrogen), and the other relativistic corrections which are important for hydrogen are negligible. The observed splitting of the (n_0p) levels of the neutral alkali atoms is shown in Table 9.4. For a given atom the splitting decreases with increasing n and l, while along an iso-electronic sequence of positive ions the splitting increases with the charge $\tilde{Z}e$ on the (nucleus + core), behaving like $\tilde{Z}^4$ for large $\tilde{Z}$.

Using the Hartree–Fock potential λ_{nl} is found to be positive, and the level with $j = 3/2$ has greater energy than that with $j = 1/2$. This is found to be true for all terms of lithium, and for the lower lying states of the other alkalis, although quantitative agreement with the observed magnitude of the splitting is not good. For the ^{2}D terms of sodium and potassium and for many other higher terms, the normal order is inverted, this effect being due to exchange interactions between the valence and core electrons and other small interactions with the core.

The one-electron selection rules (see Chapter 4) require $\Delta l = \pm 1$ and $\Delta j = 0, \pm 1$. Thus the principal and sharp series of lines corresponding to $^2P_{3/2,1/2} \leftrightarrow {}^2S_{1/2}$ transitions are doublets, while the diffuse series $^2D_{5/2,3/2} \to {}^2P_{3/2,1/2}$ and the fundamental series $^2F_{7/2,5/2} \to {}^2D_{5/2,3/2}$ are triplets. The wave numbers and wavelengths of the split resonance lines are given in Table 9.3.

9.5 Helium and the alkaline earths

The energy levels of helium and two-electron ions were discussed in Chapter 7. In the absence of spin–orbit and other spin-dependent interactions, which are in fact exceedingly small, the total spin S is a good quantum number, and the energy levels can be divided into singlet levels with $S = 0$ and triplet levels with $S = 1$.

The electric dipole operator cannot change the spin of an electron, and in the absence of spin-dependent forces, the selection rule $\Delta S = 0$ must hold. For helium and for helium-like ions with small nuclear charge Ze, the spin–orbit and spin–spin interactions are a small perturbation on the triplet states, but are not large enough to mix the triplet states with the singlet states to a detectable extent. Both the total electronic orbital angular momentum L and spin S remain conserved to a very good approximation, and the system is a good example of L–S or Russell–Saunders coupling. For helium-like ions with large Z, L–S coupling becomes a less accurate approximation and lines corresponding to singlet to triplet transitions can be observed. These are called *intercombination lines* and are usually weak. A term diagram for helium is shown in Fig. 9.2(a). The terms of the singlet and triplet states can be written $(1sns)^1S_0$, $(1snp)^1P_1$, $(1snd)^1D_2, \ldots$ and $(1sns)^3S_1$, $(1snp)^3P_J$, $(1snd)^3D_J, \ldots$. In the triplet states, the possible values of J for a given $L \neq 0$ are $J = L - 1, L, L + 1$ while $J = 1$ when $L = 0$. Spin–orbit coupling, which removes the degeneracy in J, splits each level with $L \neq 0$ into three components.

Helium can also be found in states in which both electrons are excited, for example $(2s2p)^1P_1$. Such states can decay by dipole radiation, for example

$$(2s2p)^1P_1 \rightarrow (2s1s)^1S_0$$

but a competing process is energetically possible in which an electron is ejected

$$He(2s2p) \rightarrow He^+(1s) + e^-$$

This process is called *autoionisation* and was discussed at the end of Chapter 7. It is generally a very rapid process, with the result that doubly excited levels have a very short lifetime and the corresponding spectral lines are very broad and weak.

The fine structure of helium and two-electron ions

The spin–orbit coupling is of the form $\bar{A}(\mathbf{L}\cdot\mathbf{S})$. The magnitude of the quantity $\bar{A}$ depends on the nuclear charge, and for helium it is so small that the magnetic interaction between the spins of the two electrons is comparable in magnitude. In this case the Landé interval rule (8.209) breaks down. The spin magnetic dipole moment of each electron is $-2\mu_B\mathbf{S}_i/\hbar$ $(i = 1, 2)$ and the interaction energy between them, when separated by a distance $r \neq 0$, is

$$V_S(r) = \frac{\mu_0}{4\pi}\frac{4\mu_B^2}{\hbar^2}\left[\frac{\mathbf{S}_1\cdot\mathbf{S}_2}{r^3} - 3\frac{(\mathbf{S}_1\cdot\mathbf{r})(\mathbf{S}_2\cdot\mathbf{r})}{r^5}\right] \qquad \textbf{(9.96)}$$

The energy shift due to this interaction can be calculated by first-order perturbation theory. It clearly vanishes for singlet states. There is one further interaction of significance in helium, namely the interaction between the spin of one electron and the orbital motion of another, called the spin–other orbit interaction. This interaction is of the form $A'(\mathbf{L}\cdot\mathbf{S})$ and so obeys the interval rule (for given L and S)

$$\frac{E(J) - E(J-1)}{J} = \text{constant} \qquad \textbf{(9.97)}$$

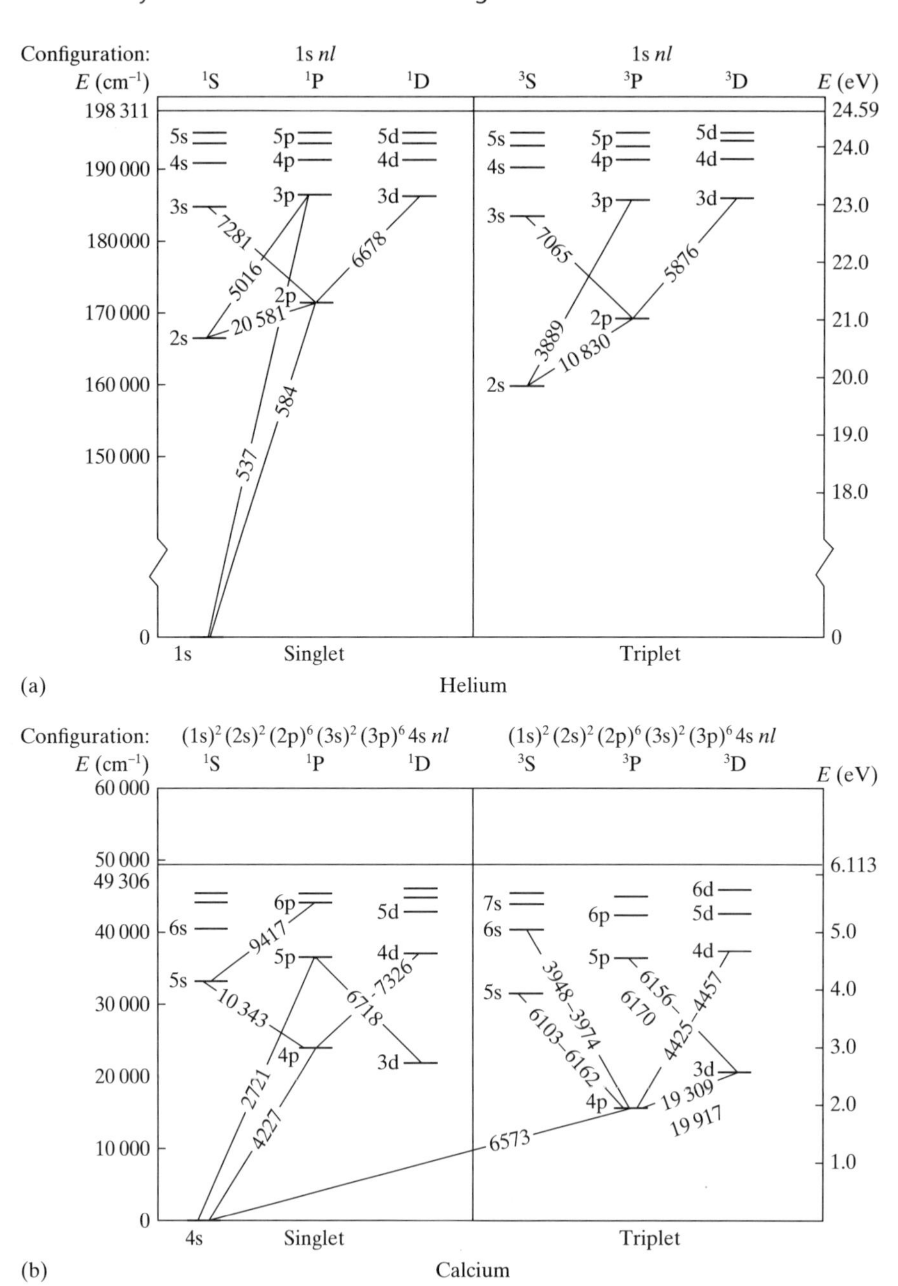

Figure 9.2 Simplified Grotrian diagrams showing transitions in (a) helium (b) calcium. Wavelengths in ångström units are shown against the lines representing the transitions. For multiplets with substantial splitting the extreme wavelengths are shown against the lines. Energies in cm^{-1} or eV are given relative to the ground state and the horizontal line indicates the ionisation potential. Notice the intercombination line shown for calcium. Only a few of the transitions which have been identified are shown, for clarity.

(although the sign of A' may be different from that of $\bar{A}$). The constant A' does not increase as fast as $\bar{A}$ does with Z and the spin–other orbit interaction is small compared with the spin–own orbit interaction in the heavier atoms.

To bring out some of these points it is interesting to compute the fine structure of helium and two-electron ions approximately, following the treatment of Bethe and Salpeter (1957). We are concerned with triplet states of the configuration (1snl) with $l \neq 0$. Since the charge distribution of the electron in the (nl) orbit (with $l > 0$) is at greater distances from the nucleus than that for the (1s) electron, exchange can be neglected and the (nl) electron (labelled 2) can be considered to be moving in a Coulomb potential V due to the nucleus (of charge Ze) fully screened by the inner electron:

$$V(r_2) = -\frac{Z-1}{r_2} \tag{9.98}$$

Here and in what follows we use atomic units. Since the average value of r_2 is much larger than the average value of r_1, we can replace the interelectron distance r in (9.96) by r_2. Then the energy due to the spin–spin interaction can be shown to be

$$\Delta E_{SS} = \langle nl|V_S(r_2)|nl\rangle = 2\alpha^2\mu_B^2\left\langle\left(\frac{1}{r_2}\right)^3\right\rangle X = \frac{\alpha^2}{2}\left\langle\left(\frac{1}{r_2}\right)^3\right\rangle X \tag{9.99a}$$

where

$$\begin{aligned} X &= l/(2l+3) && \text{for} \quad J = l+1 \\ X &= -1 && \text{for} \quad J = l \\ X &= (l+1)/(2l-1) && \text{for} \quad J = l-1 \end{aligned} \tag{9.99b}$$

The spin–orbit interaction is

$$\frac{1}{2}\alpha^2(Z-1)\left(\frac{1}{r_2}\right)^3 \mathbf{L}_2 \cdot \mathbf{S}_2 \tag{9.100}$$

and the spin–other orbit interaction is

$$-\frac{1}{2}\alpha^2 2\left(\frac{1}{r_2}\right)^3 \mathbf{L}_2 \cdot \mathbf{S}_1 \tag{9.101}$$

The two interactions (9.100) and (9.101) together provide an energy shift

$$\Delta E_{LS} = \left\langle\left(\frac{1}{r_2}\right)^3\right\rangle \frac{1}{8}\alpha^2(Z-3)[J(J+1) - l(l+1) - S(S+1)] \tag{9.102}$$

where $\langle \mathbf{S}_1 \rangle = \langle \mathbf{S}_2 \rangle = \langle \mathbf{S} \rangle/2$ and $S = 1$ for the triplet levels.

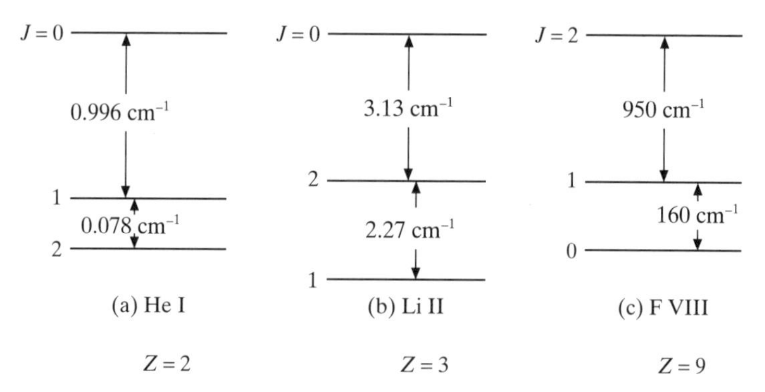

Figure 9.3 Diagram (not to scale) of the splitting of the 2 ^3P_J level in two-electron atoms. The experimental splitting is shown for He I, Li II and F VIII.

The level shift ΔE_{LS} obeys the Landé rule. If only the spin–orbit interaction had been considered $(Z - 3)$ would have been replaced by $(Z - 1)$. For neutral helium with $Z = 2$, the multiplet is inverted by the spin–other orbit interaction, but the addition of the spin–spin interaction completely alters the ratio of the level shifts (see Fig. 9.3). This is true for two-electron ions with small Z. We have in a ^{3}P multiplet, $E(J = 0) > E(J = 1) > E(J = 2)$ for helium; $E(J = 0) > E(J = 2) > E(J = 1)$ for Li^+; while for $Z \gtrsim 8$, $E(J = 2) > E(J = 1) > E(J = 0)$. For large Z the spin–orbit interaction dominates and the order of the levels is normal.

The alkaline earths (Be, Mg, Ca, Sr, Ba, Zn, Cd, Hg)

Just as the spectra of the alkalis have a close connection with those of one-electron atoms, the spectra of the alkaline earths can be compared with those of helium and the two-electron ions. The alkaline earths have two valence electrons moving outside an inert closed (sub)shell core, with $L = 0$ and $S = 0$. For example, the ground state of Be is $(1s)^2(2s)^2\ {}^1S_0$, and that of Mg is $(1s)^2(2s)^2(2p)^6(3s)^2\ {}^1S_0$. The energy required to ionise a valence electron is 9.32 eV for Be and 7.65 eV for Mg. As these electrons are more easily excited than core electrons, the optical spectrum is formed by transitions involving the valence electrons alone. As for helium, the energy levels are either singlet (total spin $S = 0$) or triplet (total spin $S = 1$) and the most prominent lines are the resonance lines between the lowest singlet states $(n_0 s n_0 p)^1P_1 \rightarrow (n_0 s)^2\ {}^1S_0$. The spin–orbit interaction splits the triplet series into three levels, except for S terms. A term diagram for Ca is shown in Fig. 9.2(b); the triplet splitting is too small to be indicated.

9.6 Atoms with several optically active electrons. Multiplet structure

The spectra of atoms with one and two valence electrons consist, as we have seen, of simple series and the same is largely true of the trivalent elements such as B and

Al. Such simple series can be readily identified and analysed, but in the complex spectra of more complicated systems, series are much more difficult, and often impossible, to identify. To identify the terms involved in a given region of the spectrum, the multiplets of lines which arise from transitions between the members of two terms must be analysed. In Chapter 8 we have seen how single configurations are split by the electrostatic energy into a number of terms ${}^{2S+1}L_J$, which are degenerate in the total electronic angular momentum J, for a given L and S. In L–S coupling the degeneracy in J is removed by the spin–orbit interaction, but L and S remain good quantum numbers. The relative spacing of the levels within a multiplet (L and S fixed) is given by the Landé interval rule (8.209)

$$E(J) - E(J-1) = AJ \tag{9.103}$$

Another aid to analysis of the observed spectra is the total splitting of the term, which is

$$\begin{aligned} E(J = L + S) - E(J = |L - S|) &= AS(2L+1), \qquad L \geqslant S \\ &= AL(2S+1), \qquad L < S \end{aligned} \tag{9.104}$$

For example, consider the triplet of lines which arise from the transition in Mg between the (3s4s) 4^3S_1 level and the (3s3p) $3^3P_{2,1,0}$ levels. The observed wave numbers are 19 286, 19 326 and 19 346 cm^{-1}. We see that $(1/2)\,[E(J=2) - E(J=1)] = 20$ cm^{-1} and $[E(J=1) - E(J=0)] = 20$ cm^{-1} which is consistent with the Landé rule (9.103) with $A = 20$ cm^{-1}. For heavy atoms on the other hand, L–S coupling breaks down and the Landé rule is not satisfied. An example is the corresponding transition in Hg, that is between the 7^3S_1 and $6^3P_{2,1,0}$ levels. The observed wave numbers are 18 307, 22 938, 24 705 cm^{-1}. From these data, $(1/2)\,[E(J=2) - E(J=1)] = 2315.5$ cm^{-1} and $[E(J=1) - E(J=0)] = 1767$ cm^{-1}. This time the two differences, which should be both equal to A if L–S coupling were satisfied accurately, are far from equal.

In general, the heavy elements have spectra which do not conform to the interval rule, because of the breakdown of $L - S$ coupling. As we have seen the interval rule can be violated in other circumstances, notably in helium, where the spin–orbit coupling is not large compared with the spin–spin interaction.

The number of spectral lines arising from transitions between two terms is determined by the selection rules. Consider the group of lines arising from transitions between members of two triplet terms, for example ${}^3D_{1,2,3}$ and ${}^3F_{2,3,4}$ terms. The selection rules $\Delta S = 0$, $\Delta L = \pm 1$ are satisfied, and the rule $\Delta J = 0, \pm 1$ allows the following six transitions:

$${}^3D_1 \leftrightarrow {}^3F_2; \quad {}^3D_2 \leftrightarrow {}^3F_2; \quad {}^3D_2 \leftrightarrow {}^3F_3; \quad {}^3D_3 \leftrightarrow {}^3F_2;$$

$${}^3D_3 \leftrightarrow {}^3F_3; \quad {}^3D_3 \leftrightarrow {}^3F_4$$

A particular example of a ${}^3D \rightarrow {}^3P$ transition in strontium is shown in Fig. 9.4 ($5^3D \rightarrow 5^3P$).

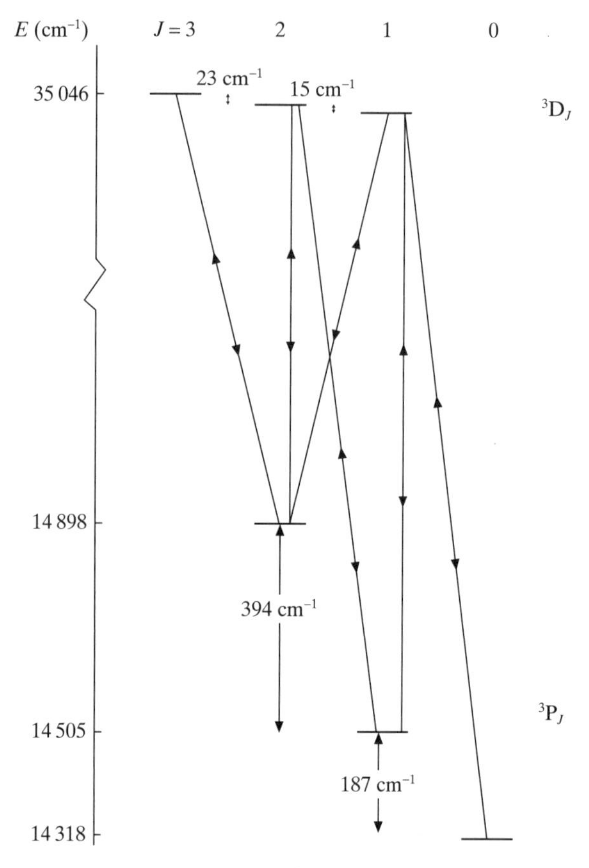

Figure 9.4 The transitions between the $(5s5d)^3D_J$ and the $(5s5p)\ ^3P_J$ terms of strontium. The left-hand side shows the energies above the $(5s)^2\ ^1S_0$ ground state in cm^{-1}.

Intensities

The radial integrals in the matrix element of the electric dipole moment (9.38) are the same for each transition between members of two terms, so the relative intensities of the lines forming the multiplet depend only on angular momentum factors. These have been tabulated and may be consulted in Kuhn (1970) or Condon and Shortley (1951). It is found that in each multiplet the strongest lines are those for which $\Delta J = \Delta L$, and if several lines satisfy this relation, the strength among these lines increases with J. Under normal conditions of excitation, the number of atoms in each level of a term is proportioned to the statistical weight $(2J + 1)$ of that level. This leads to the sum rule of Ornstein, Burger and Dorgelo, valid for L–S coupling, that the sum of the intensities of all the transitions from (or to) a given level of total angular momentum J in a multiplet is proportional to $(2J + 1)$. In some cases, this rule is sufficient to determine the relative strengths of the lines in a multiplet completely. For example, in a ^3P_J to $^3S_1$ multiplet, the intensities of the three lines are in the ratio 5:3:1. In other cases, where both terms are split, this is not so. An example is the $^3D_J \to {}^3F_J$ multiplet for which the sum rules provide only four equations, so the relative intensities between the six lines cannot be completely determined. On the other hand in a doublet system such as $^2D_{3/2,5/2} \leftrightarrow {}^2F_{5/2,7/2}$, the relative intensities for the three lines can be found as follows. Let a, b, c be the intensities of the lines $^2D_{3/2} \leftrightarrow {}^2F_{5/2}$, $^2D_{5/2} \leftrightarrow {}^2F_{5/2}$, $^2D_{5/2} \leftrightarrow {}^2F_{7/2}$. Applying the sum rule to each of the ^2D levels, we find

$$a = 4\lambda, \qquad b + c = 6\lambda \tag{9.105}$$

Similarly applying the sum rule to the ^2F levels we find

$$a + b = 6\mu, \qquad c = 8\mu \tag{9.106}$$

where λ and μ are constants of proportionality. Solving these equations, we find that a:b:c is 14:1:20.

Displaced terms

For the most part, we have considered situations in which only one electron in an atom is excited, the remaining electrons being in the ground state of the positive ion. In practice, lines can be observed which correspond to situations in which more than one electron is in an excited state, and we have already shown how some of these excited states are subject to autoionisation. Frequently, regular series of lines can be observed which are associated with a particular state of the excited ion. For example in Be, there is a doubly excited configuration $(1s)^2(2p)^2$, which gives rise to singlet and triplet series of lines corresponding to transitions $(1s)^2 2pns \to (1s)^2(2p)^2$. The series limit ($n \to \infty$) is at a higher energy than the series limit for the regular series $(1s)^2 2snp \to (1s)^2(2s)^2$, so the terms are said to be *displaced*. Transitions between the regular and displaced terms are possible. For example in Be, there is a line at 14 320 cm^{-1} (6983Å) corresponding to the transition

$$(1s)^2(2p)^2\ {}^1D_2 \to (1s)^2(2s2p)\ {}^1P_1$$

and another at 28 944 cm^{-1} (3455Å) corresponding to

$$(1s)^2(2p)^2\ {}^1S_0 \to (1s)^2(2s2p)\ {}^1P_1$$

Similarly, triplet transitions can be observed forming a multiplet of lines between the $(2p)^2\ {}^3D_J$ and the $(2s2p)\ {}^3P_J$ terms (see Fig. 9.5).

Hyperfine structure

We saw in Chapter 5, in the case of a one-electron atom, that the magnetic interaction between the nucleus and an electron splits an atomic term specified by the total electronic angular momentum **J** into a number of hyperfine levels characterised by the value of the quantum number of the total angular momentum of the whole atom, including the nucleus. The many-electron case is similar. The total

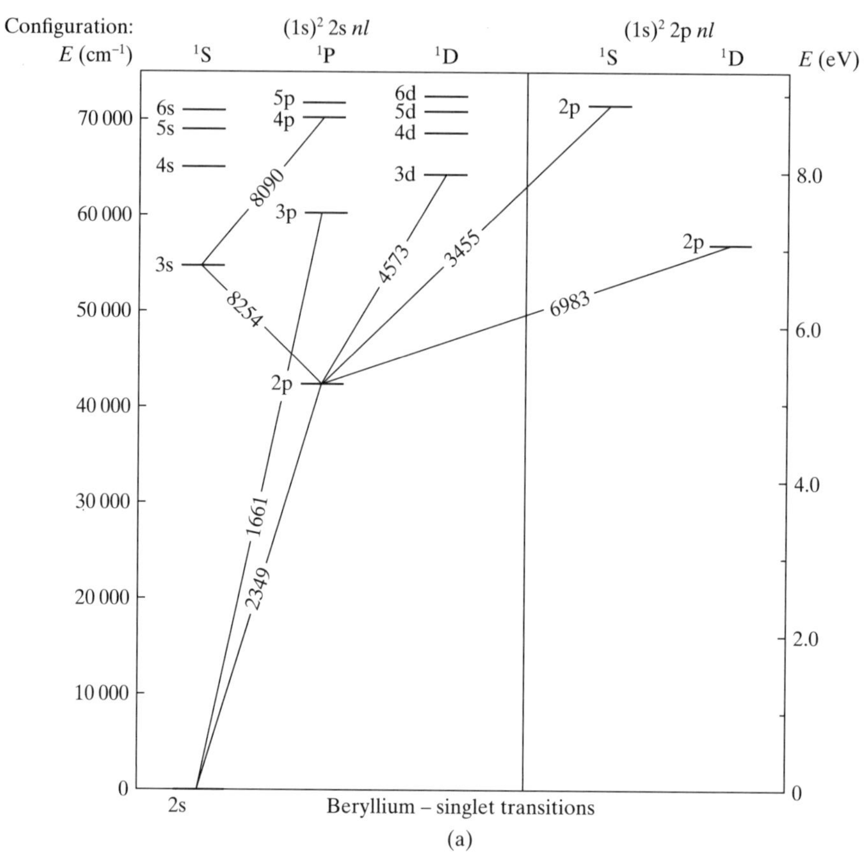

Figure 9.5 Grotrian diagram for transitions in beryllium (a) singlet (b) triplet; energies are shown relative to the $(1s)^2$ $(2s)^2$ ^1S ground state. Only a selection of transitions are shown, for clarity. The numbers against the lines show the wavelength in ångströms. The 'displaced terms' are shown on the right-hand side of each diagram.

angular momentum **F** of the atom is the sum of **J** and the angular momentum of the nucleus **I**,

$$\mathbf{F} = \mathbf{I} + \mathbf{J} \tag{9.107}$$

Each atomic level is an eigenstate of $\mathbf{F}^2$ and of F_z with quantum numbers F and M_F respectively. The eigenfunctions $|F, M_F\rangle$ satisfy

$$\begin{aligned} F^2|F, M_F\rangle &= F(F+1)\hbar^2|F, M_F\rangle \\ F_z|F, M_F\rangle &= M_F\hbar|F, M_F\rangle \end{aligned} \tag{9.108}$$

By recalling equation (5.62) and the discussion that follows it, we see that in the one-electron atom, the magnetic dipole interaction between the nucleus and the electron can be taken to be proportioned to $(\mathbf{I}\cdot\mathbf{J})$, where the constant of

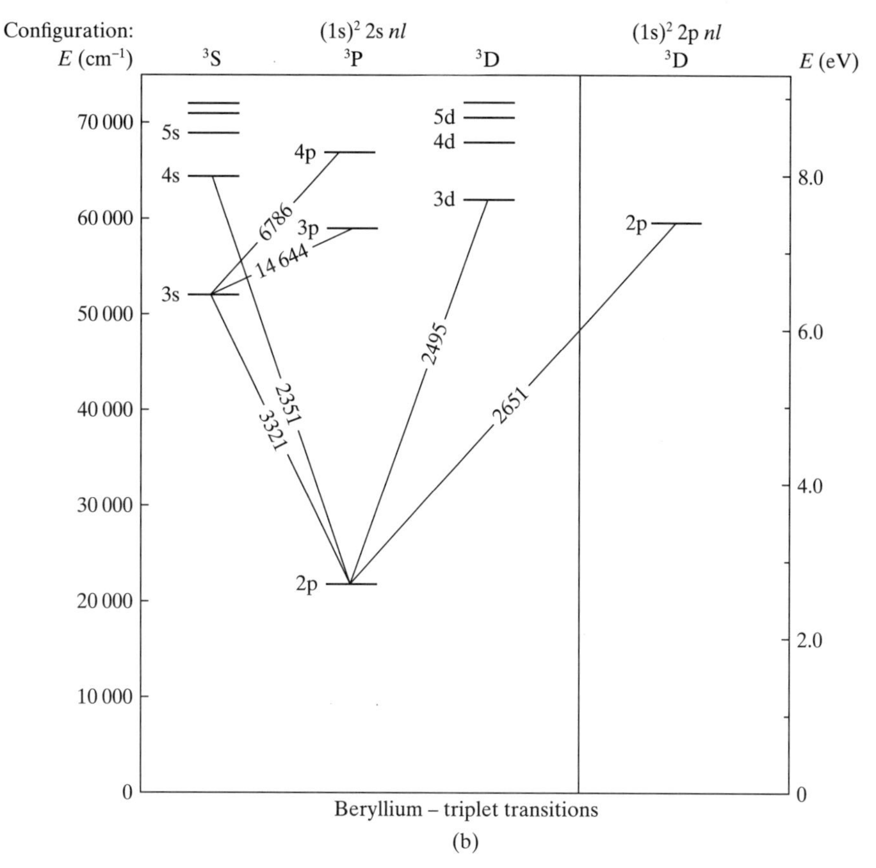

Figure 9.5 (Cont.)

proportionality depends only on the quantum numbers j, l and s (with $s = 1/2$). We shall assume this to be true for the many-electron case as well, so that the energy shift of each term will be determined by an interaction of the form

$$H' = \bar{C}(\mathbf{I}\cdot\mathbf{J}) \tag{9.109}$$

where $\bar{C}$ depends on J, L and S, and is known as the *hyperfine structure constant.* Since

$$\mathbf{I}\cdot\mathbf{J} = \tfrac{1}{2}(\mathbf{F}^2 - \mathbf{I}^2 - \mathbf{J}^2) \tag{9.110}$$

the energy shift of a particular term is (with $C = \bar{C}\hbar^2$)

$$\Delta E = \frac{C}{2}[F(F+1) - I(I+1) - J(J+1)] \tag{9.111}$$

The possible values of F are from $|I - J|$ to $(I + J)$, so the number of levels differing in F is $(2I + 1)$ if $I \leqslant J$, or $(2J + 1)$ if $J < I$. For example, if $J = 2$ and $I = 3/2$, the

Table 9.5 Hyperfine structure of selected levels.

Atom	*I*	*Term*	*F*	$[\Delta E(F+1) - \Delta E(F)]^{\dagger}$	$C^{\dagger}$
H	1/2	$(1s)^2S_{1/2}$	0	1420.405 751 766 7	1420.406
^{3}He	1/2	$(1s2s)^1S_{1/2}$	1/2	6739.70	4212.31
^{67}Zn	5/2	$(4s4p)^3P_2$	1/2	7810.865	521.24
^{133}Cs	7/2	$(6s)^2S_{1/2}$	3	9192.631 770	2298.2

† Energy splittings and C are given in MHz.

term will be split into four components with $F = 1/2, 3/2, 5/2, 7/2$. The energy difference between levels is given by the interval rule

$$\Delta E(F+1) - \Delta E(F) = C(F+1) \tag{9.112}$$

Some values of the hyperfine splitting for a number of levels are given in Table 9.5. Transitions between different hyperfine levels corresponding to the same term are forbidden in the dipole approximation by the parity selection rule. Nevertheless, such transitions may occur as magnetic dipole (M1), electric quadrupole (E2), . . . transitions (see Section 9.3). They can be measured by radio-frequency methods. The measurements can be made to great precision and for this reason can be used as time standards. In SI units the fundamental unit of time, the second, is defined in terms of the frequency of the transition between the levels with $F = 4$, $M_F = 0$ and $F = 3$, $M_F = 0$ of the ground state of ^{133}Cs. This frequency is assigned the value 9192.631 770 MHz.

The selection rules for electric dipole transitions between different terms must be supplemented by the rules $\Delta F = 0, \pm 1$ with the transition $F = 0 \leftrightarrow F' = 0$ forbidden and $\Delta M_F = 0, \pm 1$ (see (9.48)). The analysis of the hyperfine structure of particular transitions offers a method of determining the spin of the nucleus, and historically this has been of great importance.

Since the level shifts are small, of the order of 10^{-5} of the energy difference between the ground and first excited states of the atom, other effects may be of comparable magnitude. For example, the value of the constant C is different for different isotopes of the same element. For a given isotope, there is a further small interaction, which depends on F, and which is of experimental importance. This arises because the nucleus is not a point charge and so has a charge distribution which can be specified in terms of multipole moments (see Chapter 5). Being in a parity eigenstate, the nucleus can have no permanent electric dipole moment, but can possess an electric quadrupole moment. The interaction between this quadrupole moment and the electronic charge distribution leads to a further very small F-dependent shift, which can be detected by departures from the interval rule (9.112). This was discussed in Section 5.3 for one-electron atoms. For the generalisation to the many-electron atom case, reference may be made to Kuhn (1970).

9.7 X-ray spectra

The valence electrons in heavy neutral atoms move in an effective field, which at large distances approaches the Coulomb field due to a unit charge (e). The binding energies of these electrons are always of the same order of magnitude as the binding energy of atomic hydrogen – a few electron-volts. In contrast, the electrons in the inner shells, K, L, M, . . . with principal quantum numbers $n = 1, 2, 3, \ldots$, move in a potential field dominated by the charge of the nucleus (Ze). The nuclear charge is screened to some extent by the other electrons, and to a good approximation this can be represented by taking the effective potential in which the inner electrons move in a Coulomb potential due to a charge $(Z - \sigma_n)e$, where $\sigma_n \ll Z$. The energy levels in such a field are just those of a one-electron atom, and

$$E_n = -\frac{1}{2}\frac{(Z - \sigma_n)^2}{n^2} \text{ a.u.} \tag{9.113}$$

In contrast to the valence electrons, the binding energy of a K shell electron increases smoothly with increasing Z. To a fair approximation, $\sigma_1 = 1$. Thus for Fe with $Z = 26$, $|E_1| = 312$ a.u. $= 8.5$ keV, for Cs with $Z = 55$, $|E_1| = 1458$ a.u. $= 39.7$ keV and for Pb with $Z = 82$, $|E_1| = 3281$ a.u. $= 89.3$ keV. Similarly for the L shell, $n = 2$, an approximation for σ_2 is $\sigma_2 = 7.4$, and, for example, the binding energy of an L shell electron for Pb is 18.9 keV.

By bombarding a substance with fast electrons, some atoms become ionised through the ejection of an inner shell electron. Suppose a vacancy is created in the K shell in this way. Subsequently an electron from a higher shell, L, M, . . . , will make a radiative transition filling the vacancy. The energy of the photon emitted will be in the range from a few keV to a few hundred keV, and so lies in the X-ray region [4] which may be taken to be in the range of wavelengths from 0.1 to 10 Å (see Fig. 1.9). The emission spectrum which results is a line spectrum, forming a simple series. The lines originating from a K shell vacancy are called the K_α, K_β, K_γ, . . . lines and correspond to the $L \to K$, $M \to K$, $N \to K$, . . . transitions. The K_α line is the strongest. The early experiments of Moseley in 1913 and 1914 established that the square root of the frequency of the K_α line is a linear function of atomic number (see Fig. 1.20). Historically this was of considerable importance because it was consistent with the energy formula (9.113), which had been predicted by Bohr's model of a one-electron system, published in 1913. A similar series of lines, labelled L_α, L_β, L_γ, . . . in order of increasing frequency, arises from transitions which fill a vacancy in the L shell.

The spin–orbit interaction splits the energy levels of the shells (other than the K shell). In the L shell, the electrons with $l = 0$ have the greatest binding energy, those with $l = 1$, $j = 1/2$ come next, and then those with $l = 1$ and $j = 3/2$. The three levels are labelled L_I, L_{II} and L_{III} respectively. The energy splitting between the

[4] The wavelength λ, expressed in Å, is given by $\lambda \simeq 12.4/E$, where the energy E is in keV.

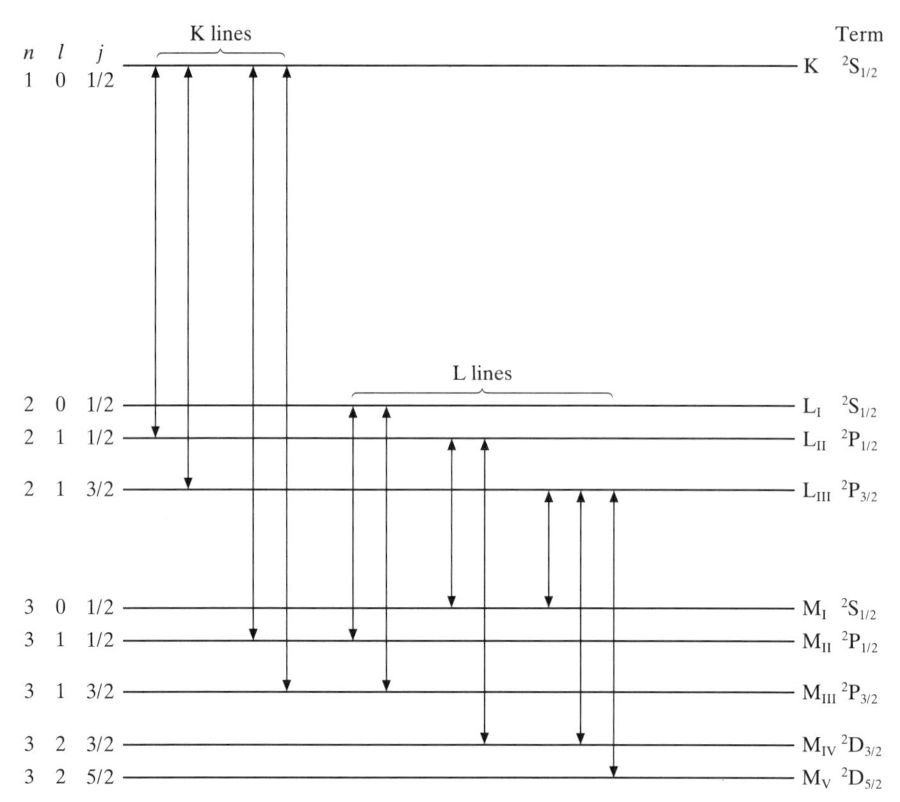

Figure 9.6 X-ray term diagram (not to scale). The allowed transitions are determined from the usual selection rules $\Delta l = \pm 1$, $\Delta j = 0, \pm 1$.

sublevels is large, because spin–orbit coupling is proportional to $(Z - \sigma_2)^4$ (see (5.26)) and the doublet $K_{\alpha 1}$ and $K_{\alpha 2}$, which corresponds to the transitions $L_{III} \rightarrow K$ and $L_{II} \rightarrow K$, can be resolved. For example, the wavelengths of the $K_{\alpha 1}$ and $K_{\alpha 2}$ lines in ^{26}Fe are 1.9360 Å and 1.9321 Å respectively. A typical X-ray term diagram (including the K, L and M shells) is shown in Fig. 9.6.

X-rays can be absorbed by ionising atoms through the photoelectric effect (see Section 4.8). The cross-section for this process decreases smoothly with increasing frequency, until sufficient energy is available to ionise an electron from the L_{III} shell. At this point, the cross-section increases sharply. Similar 'absorption edges' occur as successively sufficient energy is available to ionise electrons from the L_{II}, L_I and finally the K shell. This is illustrated in Fig. 9.7 for absorption in lead.

Auger transitions

When a vacancy is created in an inner shell it may be filled by an electron from a level with higher energy with emission of radiation. These radiative transitions compete with radiationless transitions in which one electron from a level with higher energy occupies the initial vacancy with the simultaneous ejection of a

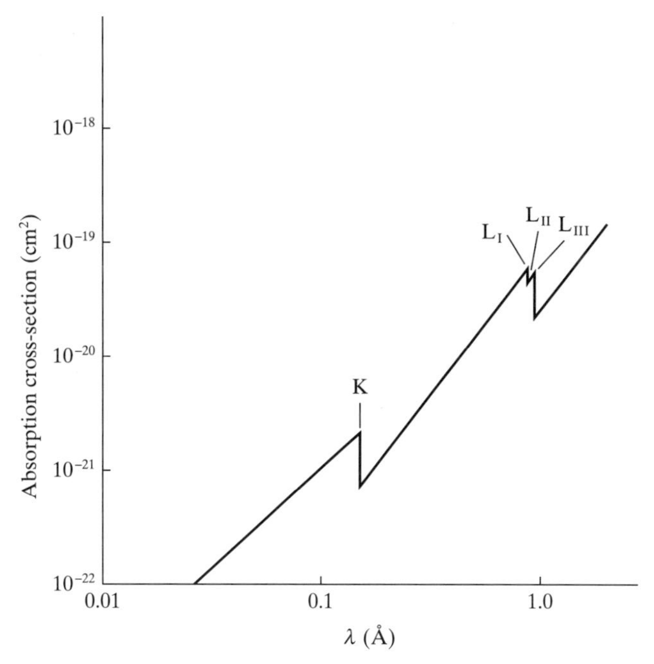

Figure 9.7 The cross-section for the absorption of X-rays by lead atoms shown as a function of the wavelength.

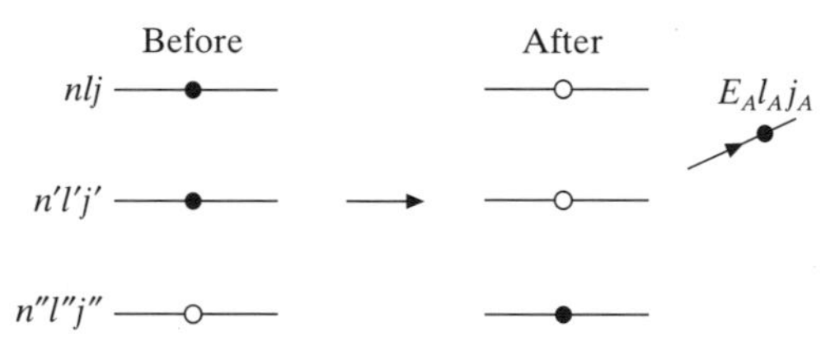

Figure 9.8 Schematic diagram of an Auger transition. Before the transition a vacancy (hole) with quantum numbers $(n''l''j'')$ is created by photoionisation or by particle impact; the upper levels (nlj) and $(n'l'j')$ are occupied. After the transition one electron occupies the original vacancy and the other is ejected with energy E_A and quantum numbers l_A and j_A, leaving two holes in the upper energy levels.

second electron. The transition can be denoted by $(n''l''j'') \to (nlj)(n'l'j')$ where $(n''l''j'')$ are the quantum numbers of the initial vacancy (hole) while (nlj) and $(n'l'j')$ are the quantum numbers of the two vacancies (holes) in the final state (see Fig. 9.8). As an alternative, the notation employed for X-ray spectra can be used, so for example, the $(1s) \to (2s)(2p_{1/2})$ and $(1s) \to (2s)(2p_{3/2})$ transitions are written as $K \to L_I\, L_{II}$ and $K \to L_I\, L_{III}$ respectively.

If the ejected electron has a kinetic energy E_A, then

$$E_A = E_{n''l''j''} - E_{nlj,n'l'j'} \qquad \textbf{(9.114)}$$

where $E_{n''l''j''}$ is the energy of the atom with the vacancy $(n''l''j'')$ relative to the energy of the neutral atom and $E_{nlj,n'l'j'}$ is the energy of the atom with two vacancies again relative to the energy of the neutral atom. These radiationless transitions are known as *Auger transitions* after their discovery by P. Auger in 1925. In light atoms Auger transitions are called autoionising transitions and these have been discussed in Section 7.7.

A simple non-relativistic theory of the Auger transition rate was given by G. Wentzel in 1928 taking the Coulomb interaction between the two electrons concerned as a small perturbation compared with the strong nuclear Coulomb field in which the inner shell electrons move. Since the two electrons are indistinguishable and obey the Pauli exclusion principle the initial unperturbed wave function must be antisymmetrical and can be written as

$$\Psi_a(q_1, q_2) = \frac{1}{\sqrt{2}}[\psi_{nlj}(q_1)\psi_{n'l'j'}(q_2) - \psi_{nlj}(q_2)\psi_{n'l'j'}(q_1)] \tag{9.115}$$

where $\psi_{nlj}(q_1)$ is a one-electron wave function in the central field model. Similarly the unperturbed wave function in the final state is written as

$$\Psi_b(q_1, q_2) = \frac{1}{\sqrt{2}}[\psi_{n''l''j''}(q_1)\psi_{E_A l_A j_A}(q_2) - \psi_{n''l''j''}(q_2)\psi_{E_A l_A j_A}(q_1)] \tag{9.116}$$

Using Fermi's Golden Rule (2.362), the Auger transition rate W_{ba}^{A} is given (in a.u.) by

$$W_{ba}^{\mathrm{A}} = 2\pi\left|\left\langle \Psi_b \left| \frac{1}{|\mathbf{r}_1 - \mathbf{r}_2|} \right| \Psi_a \right\rangle\right|^2 \rho_b(E_A) \tag{9.117}$$

where $\rho_b(E_A)$ is the density of final states for the ejected electron.

If the one-electron wave functions $\psi_{nlj}(q_i)$ are taken to be hydrogenic corresponding to an unscreened nuclear charge Z, the transitions rates W_{ba}^{A} are completely independent of Z. When more realistic wave functions are used, Z independence is found to be approximately true for large Z but the calculated transition rates are often very sensitive to the details of the wave function and these must be calculated accurately.

The number of Auger transitions which are available to fill a given vacancy increases sharply with Z, reaching nearly 3000 for a K shell vacancy in mercury ($Z = 80$). In contrast, the number of dipole allowed radiative spontaneous transitions available to fill a given vacancy is restricted by the selection rules, with the consequence that except for K shell vacancies with $Z < 30$, Auger transitions are far more probable than radiative transitions as a vacancy decay mechanism.

The total rate W_a at which a vacancy decays is given by

$$W_a = \sum_b W_{ba}^{\mathrm{A}} + \sum_c W_{ca}^{\mathrm{s}} \tag{9.118}$$

where the first term is the sum of all the Auger transition rates and the second is the sum of the allowed radiative spontaneous transition rates. The lifetime of the

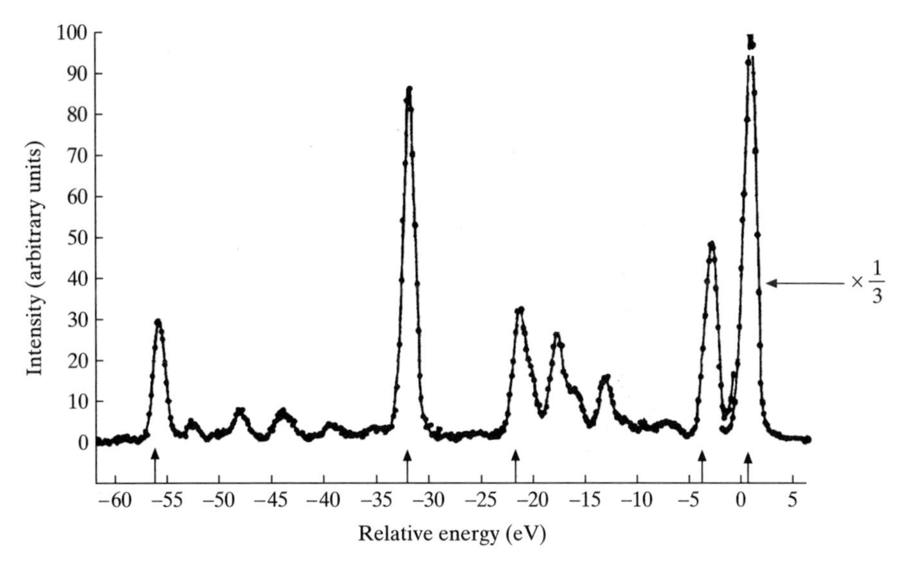

Figure 9.9 The K shell Auger spectrum of neon. K shell vacancies in neon caused by electron impact are filled by Auger transitions from the L shell (marked by the arrows). The subsidiary or satellite peaks are due to higher order processes, such as double Auger transitions involving three electrons. (After M.O. Krause, F.A. Stevie, L.J. Lewis, T.A. Carlson and W.E. Moddeman, *Physics Letters A* **31**, 81, 1970.)

vacancy is $\tau_a = W_a^{-1}$ and the width (in atomic units) is $\Gamma_a = W_a$. The K shell widths increase steadily with Z according to the approximate rule

$$\Gamma_K = 1.73Z^{3.93} \times 10^{-6}\ \text{eV} \tag{9.119}$$

reaching $\Gamma_K = 90$ eV for uranium.

The relative probability that a vacancy decays radiatively is given by a quantity called the *fluorescence yield*, defined as

$$Y = \frac{\Gamma^R}{\Gamma} \tag{9.120}$$

where

$$\Gamma^R = \sum_c W_{ca}^s \tag{9.121}$$

is the partial width due to radiative decay.

When vacancies are created in a sample, either by collision processes or by radiation, the kinetic energies and angular distributions of the resulting Auger electrons can be measured using *electron spectrometers*, and much information can be obtained in this way about the structure of the inner shells of atoms. Further information is given by measurements of the fluorescent yield.

In Fig. 9.9 the experimental $K \to LL$ spectrum of neon is shown in which the K shell vacancies were created by electron impact. The satellite lines shown arise from higher order processes, for example those in which three electrons take part.

Hollow atoms

Hollow atoms are short-lived multiply-excited neutral atoms in which the inner shells are empty. These atoms can be formed during the interaction of a slow, multiply-charged ion with a solid surface. Their lifetime is very short (of the order of 10^{-13} s) since the vacancies in the inner shells are rapidly filled by X-ray emission and Auger transitions. The experimental evidence for hollow atoms, and the theoretical aspects of their formation and decay, have been reviewed by H. Winter and F. Aumayr (1999).

9.8 The Stark effect

In the absence of external fields, there is no preferred direction in space, so that the energy of an isolated system, such as an atom, cannot depend on which direction we choose as the Z axis. As a consequence, the energy of an atom does not depend on M_J, and the atomic energy levels are $(2J + 1)$-fold degenerate. This degeneracy is removed, either partially or totally, by external electric or magnetic fields which destroy the isotropy of the situation.

In Chapter 6, we discussed the influence of a static and uniform electric field on atoms (the Stark effect) for the case of hydrogenic atoms. Since the excited energy levels of these atoms are degenerate in l, the perturbation mixes states of opposite parity and produces level shifts linear in the field strength.

Let us now consider an N-electron atom in a static, uniform electric field of strength $\mathscr{E}$ directed parallel to the Z axis. To the Hamiltonian H_0 of the unperturbed atom, given by (9.6), we must add the interaction energy

$$\begin{aligned} H' &= e\mathscr{E}\sum_{i=1}^{N} z_i \\ &= -\mathscr{E}D_z \end{aligned} \tag{9.122}$$

where $\mathbf{D}$ is the electric dipole moment operator defined by (9.20) and D_z its component along the Z axis.

The matrix elements of $\mathbf{D}$ with respect to atomic states of definite parity vanish, so that the first-order level shift also vanishes:

$$E^{(1)}_{\gamma JM_J} = -\mathscr{E}\langle \gamma JM_J | D_z | \gamma JM_J\rangle = 0 \tag{9.123}$$

where γ denotes the configuration.

The second-order perturbation theory gives a level shift

$$\begin{aligned} \Delta E = E^{(2)}_{\gamma JM_J} &= \mathscr{E}^2 \sum_{\gamma' J' M'_J}{}' \frac{|\langle \gamma' J' M'_J | D_z | \gamma JM_J \rangle|^2}{E_{\gamma J} - E_{\gamma' J'}} \\ &= \mathscr{E}^2 \sum_{\gamma' J'}{}' \frac{|\langle \gamma' J' M_J | D_z | \gamma JM_J \rangle|^2}{E_{\gamma J} - E_{\gamma' J'}} \end{aligned} \tag{9.124}$$

where we have used the fact that $\Delta M_J = 0$. The selection rules for the matrix element $\langle \gamma' J' M_J | D_z | \gamma J M_J \rangle$ require that the intermediate states $|\gamma' J' M_J\rangle$ must have opposite parity to the state $|\gamma J M_J\rangle$, and that $J' = J + 1, J$ or $J - 1$. The dependence of the matrix elements on M_J is then given by (9.42) with $q = 0$. It is entirely contained in the Clebsch–Gordan coefficient $\langle J 1 M_J 0 | J' M_J \rangle$. We have

$$\langle J1M_J0|J+1, M_J\rangle = \left[\frac{(J+1)^2 - M_J^2}{(2J+1)(J+1)}\right]^{1/2}$$

$$\langle J1M_J0|J, M_J\rangle = \left[\frac{M_J^2}{J(J+1)}\right]^{1/2} \tag{9.125}$$

$$\langle J1M_J0|J-1, M_J\rangle = -\left[\frac{J^2 - M_J^2}{J(2J+1)}\right]^{1/2}$$

From these expressions, we see that $E^{(2)}_{\gamma J M_J}$ is of the form

$$E^{(2)}_{\gamma J M_J} = \mathscr{E}^2(A + BM_J^2) \tag{9.126}$$

where A and B depend on γ and J. The degeneracy in M_J is not completely removed, because levels with $M_J = \pm|M_J|$ have the same energy. We also note that

$$E^{(2)}_{\gamma J M_J} = -\frac{1}{2}\bar{\alpha}(\gamma J M_J)\mathscr{E}^2 \tag{9.127}$$

where $\bar{\alpha}(\gamma J M_J)$ is the static dipole polarisability of the atom in the state $(\gamma J M_J)$, as defined by (9.67).

In the sum (9.124), the most important contribution is from the nearest level of opposite parity to the level (γJ). For example, in sodium, the shift in the ground state energy is very small because the nearest levels of opposite parity are the $2^2P_{1/2}$ and $2^2P_{3/2}$ levels and the corresponding energy differences $E_{\gamma' J'} - E_{\gamma J}$ are large. In fact the ground state energy is lowered and the energy shift in cm^{-1} is $7.8 \times 10^{-11}\mathscr{E}^2$, where $\mathscr{E}$ is given in kV m^{-1}. The energy shifts of the $2^2P_{1/2}$ and $2^2P_{3/2}$ levels are about twice as large, since these are perturbed both by the ground state and by the ^2D levels which lie closer. The two D lines are split into three by the field (see Fig. 9.10).

Apart from producing a level shift, the electric field has the effect of mixing states of opposite parity. A perturbed ^2P_J wave function, for example, now contains a small admixture of the ^2D_J wave function. As a consequence, lines are observed in the presence of an electric field that would normally be forbidden, for example the series $n\,^2S \to n'\,^2S$ and $n\,^2S \to n'\,^2D$.

Strong fields

The expression (9.124) for the quadratic Stark level shift can only be accurate if the matrix elements of the perturbation (9.122) are sufficiently small, which in

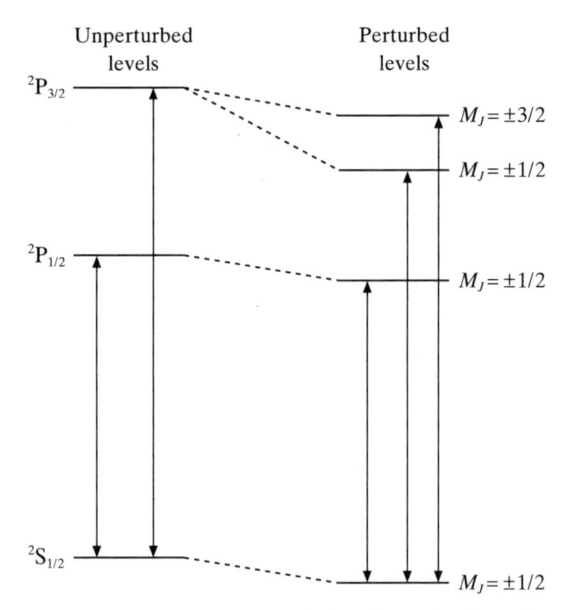

Figure 9.10 Schematic diagram of the splitting of the D lines of sodium due to an electric field (the Stark effect).

turn requires that the magnitude of the field strength $\mathscr{E}$ should not be too large. To quantify these statements and to see what happens when $\mathscr{E}$ becomes large, let us return to the case in which only one term makes a significant contribution to the sum on the right-hand side of (9.124). Labelling for convenience the state $(\gamma J M_J)$ by a and the state $(\gamma' J' M_J)$ by b (with $E_b > E_a$), we have in this approximation,

$$\Delta E_a = \mathscr{E}^2 \frac{|\langle b|D_z|a\rangle|^2}{E_a - E_b} \tag{9.128a}$$

and

$$\Delta E_b = \mathscr{E}^2 \frac{|\langle a|D_z|b\rangle|^2}{E_b - E_a} = -\Delta E_a \tag{9.128b}$$

Clearly a necessary requirement for these expressions to be accurate is that $|\Delta E_a| \ll |E_b - E_a|$. If $\mathscr{E}$ is increased so that this condition is violated a non-perturbative solution of the Schrödinger equation

$$(H_0 + H' - E)\psi = 0 \tag{9.129}$$

must be found, where H_0 is the atomic Hamiltonian in the absence of the electric field and H' is defined by (9.122). The exact wave function ψ can be expanded into the complete set of orthonormal atomic wave functions ϕ_i, where

$$(H_0 - E_i)\phi_i = 0 \tag{9.130}$$

but in view of our assumption that ϕ_a and ϕ_b are close in energy and strongly coupled compared with the coupling to any other state, to a good approximation we have that

$$\psi \simeq c_a\phi_a + c_b\phi_b \tag{9.131}$$

Substituting (9.131) into (9.129) and projecting in turn by ϕ_a and ϕ_b, we obtain coupled equations for the expansion coefficients c_a and c_b:

$$\begin{aligned} (E_a - E)c_a + H'_{ab}c_b &= 0 \\ H'_{ba}c_a + (E_b - E)c_b &= 0 \end{aligned} \tag{9.132}$$

where $H'_{ij} = \langle\phi_i|H'|\phi_j\rangle$, $i, j = a, b$, and we have used the fact that $H'_{aa} = H'_{bb} = 0$ since H' only connects states of opposite parity. The compatibility of equations (9.132) requires that

$$(E_a - E)(E_b - E) - |H'_{ba}|^2 = 0 \tag{9.133}$$

This is a quadratic equation for E with the solutions

$$\begin{aligned} E^{\pm} &= \frac{1}{2}(E_a + E_b) \pm \frac{1}{2}(E_a - E_b)\left[1 + \frac{4|H'_{ba}|^2}{(E_a - E_b)^2}\right]^{1/2} \\ &= \frac{1}{2}(E_a + E_b) \pm \frac{1}{2}(E_a - E_b)\left[1 + \frac{4\mathscr{E}^2|\langle b|D_z|a\rangle|^2}{(E_a - E_b)^2}\right]^{1/2} \end{aligned} \tag{9.134}$$

When $\mathscr{E} = 0$ we have $E^+ = E_a$ and $E^- = E_b$ as expected. Defining the level shifts as $\Delta E_a = E^+ - E_a$ and $\Delta E_b = E^- - E_b$ we find that

$$\Delta E_a = -\Delta E_b = \frac{1}{2}(E_b - E_a) + \frac{1}{2}(E_a - E_b)\left[1 + \frac{4\mathscr{E}^2|\langle b|D_z|a\rangle|^2}{(E_a - E_b)^2}\right]^{1/2} \tag{9.135}$$

When $\mathscr{E}$ is so small that

$$4\mathscr{E}^2|\langle b|D_z|a\rangle|^2 \ll (E_a - E_b)^2 \tag{9.136}$$

the expansion of the bracketed terms in (9.135) reproduces the expressions (9.128) which are quadratic in $\mathscr{E}$. On the other hand in the strong field case with

$$4\mathscr{E}^2|\langle b|D_z|a\rangle|^2 \gg (E_a - E_b)^2 \tag{9.137}$$

we obtain

$$\Delta E_a = -\Delta E_b = -\mathscr{E}\,|\langle b|D_z|a\rangle| \tag{9.138}$$

which is linear in the field strength. The transition from the quadratic Stark to the linear Stark effect is shown in Fig. 9.11.

The linear behaviour of the level shifts for strong fields is not quite the end of the story. At some point when $\mathscr{E}$ is increased further, levels additional to levels *a*

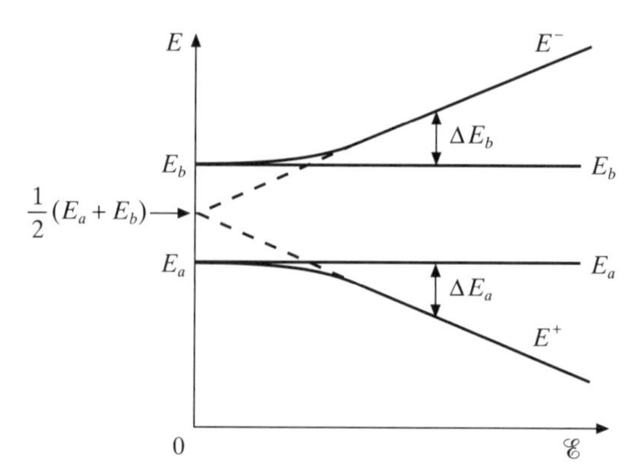

Figure 9.11 Schematic diagram showing the transition from the quadratic to the linear Stark effect as the electric field strength $\mathscr{E}$ is increased.

and b will become significant. In this case, the splittings will exhibit a complex non-linear variation with $\mathscr{E}$.

The AC Stark effect

So far we have discussed the level shifts when an atom is subjected to a static electric field. We shall now consider the level shifts due to an electric field that is varying in time, taking the particular case for which the variation is periodic with an angular frequency ω. The corresponding level shift is known as the 'alternating current' *(AC) Stark effect.*

Let us assume that the electric field is spatially homogeneous and is directed along the Z axis, so that we have

$$\boldsymbol{\mathscr{E}}(t) = \mathscr{E}(t)\hat{\mathbf{z}} \tag{9.139}$$

In addition, we shall assume that

$$\mathscr{E}(t) = \mathscr{E}_0 \cos(\omega t) \tag{9.140}$$

where $\mathscr{E}_0$ is the electric field strength. The interaction Hamiltonian is given by

$$H'(t) = -\mathscr{E}_0 D_z \cos(\omega t) \tag{9.141}$$

The wave function for the atom in the presence of the electric field, $\Psi(X, t)$, satisfies the time-dependent Schrödinger equation (9.5) with $H_{\text{int}} = H'(t)$ as given by (9.141). As in Chapter 4, we shall expand $\Psi(X, t)$ in the complete orthonormal set of energy eigenfunctions $\psi_k(X)$, where k represents a set of quantum numbers, so that

$$\Psi(X, t) = \sum_k c_k(t)\psi_k(X) \exp(-\mathrm{i}E_k t/\hbar) \tag{9.142}$$

Inserting (9.142) into the Schrödinger equation (9.5), a set of coupled equations for the amplitudes $c_k(t)$ can be obtained of the form (4.32).

Let us assume that the electric field is switched on at $t = 0$, and that the atom is in the state $a \equiv (\gamma J M_J)$ before the field is switched on. The initial conditions are therefore

$$c_k(t) = \delta_{ka}, \qquad t \leqslant 0 \tag{9.143}$$

The amplitude $c_a(t)$ can be written as

$$c_a(t) = |c_a(t)| \exp[-\mathrm{i}\eta(t)] \tag{9.144}$$

where $\eta(t)$ is a real phase and $\eta(0) = 0$. The probability of finding the atom in a state k at time t is given by $|c_k(t)|^2$. This quantity is very small for $k \neq a$ unless the resonance condition $\omega = |\omega_{ka}|$ is satisfied, when single photon transitions can occur (see Chapter 4). Here we shall study the case in which $\omega \neq |\omega_{ka}|$, so that $|c_k(t)| \ll 1$ for $k \neq a$ and $|c_a(t)| \simeq 1$. To understand the significance of the phase $\eta(t)$, we note that the term for which $k = a$ in the expansion (9.142) can be written in the form

$$c_a(t)\psi_a(X) \exp(-\mathrm{i}E_a t/\hbar) = |c_a(t)|\psi_a(X) \exp\left\{-\frac{\mathrm{i}}{\hbar}\int_0^t [E_a + \Delta E_a(t')]\,\mathrm{d}t'\right\} \tag{9.145}$$

with

$$\Delta E_a(t) = \hbar\dot{\eta}(t) \tag{9.146}$$

This suggests that $\Delta E_a(t)$ can be interpreted as the energy shift of the state a at time t due to the applied electric field.

To calculate $\Delta E_a(t)$ to the lowest (non-vanishing) order of perturbation theory (LOPT) the equations (4.32) can be solved to second order in the perturbation $H'(t)$, giving

$$c_a(t) \simeq 1 + (\mathrm{i}\hbar)^{-2} \sum_{k \neq a} \int_0^t H'_{ak}(t') \exp(\mathrm{i}\omega_{ak}t')\,\mathrm{d}t' \int_0^{t'} H'_{ka}(t'') \exp(\mathrm{i}\omega_{ka}t'')\,\mathrm{d}t'' \tag{9.147}$$

where the fact that $H'_{aa}(t) = 0$ has been employed. Noting that the unperturbed amplitude $c_a^{(0)} = 1$, we have

$$\begin{aligned} \dot{c}_a(t) &= \left(\frac{\mathrm{d}}{\mathrm{d}t}|c_a(t)|\right) \exp[-\mathrm{i}\eta(t)] - \mathrm{i}c_a(t)\dot{\eta}(t) \\ &\simeq -\mathrm{i}\dot{\eta}(t) \end{aligned} \tag{9.148}$$

and from (9.146)–(9.148)

$$\Delta E_a(t) = (\mathrm{i}\hbar)^{-1} \sum_{k \neq a} H'_{ak}(t) \exp(\mathrm{i}\omega_{ak}t) \int_0^t H'_{ka}(t') \exp(\mathrm{i}\omega_{ka}t')\,\mathrm{d}t' \tag{9.149}$$

The quantity of physical interest is the mean value of $\Delta E_a(t)$, denoted by $\overline{\Delta E_a}$, which is obtained by averaging $\Delta E_a(t)$ over time and is called the *AC Stark shift*. Thus, using (9.140) and (9.149), we have

$$\overline{\Delta E_a} = -\frac{\mathscr{E}_0^2}{4\hbar}\sum_{k\neq a}|\langle k|D_z|a\rangle|^2\left[\frac{1}{\omega_{ka}+\omega}+\frac{1}{\omega_{ka}-\omega}\right]$$

$$= -\frac{\mathscr{E}_0^2}{2\hbar}\sum_{k\neq a}|\langle k|D_z|a\rangle|^2\frac{\omega_{ka}}{\omega_{ka}^2-\omega^2}$$

$$= -\frac{\bar{\mathscr{E}}^2}{\hbar}\sum_{k\neq a}|\langle k|D_z|a\rangle|^2\frac{\omega_{ka}}{\omega_{ka}^2-\omega^2} \tag{9.150}$$

where $\bar{\mathscr{E}}^2$ is the average value of $\mathscr{E}^2(t)$ over a period $T = 2\pi/\omega$, namely

$$\bar{\mathscr{E}}^2 = \frac{1}{T}\int_0^T \mathscr{E}^2(t)\,\mathrm{d}t = \frac{1}{T}\mathscr{E}_0^2\int_0^T \cos^2(\omega t)\,\mathrm{d}t$$

$$= \frac{1}{2}\mathscr{E}_0^2 \tag{9.151}$$

By using (9.64) and (9.150), it is seen that the AC Stark shift $\overline{\Delta E_a}$ can be written in the form

$$\overline{\Delta E_a} \equiv \overline{\Delta E_{\gamma JM_J}} = -\frac{1}{2}\bar{\mathscr{E}}^2\alpha(\gamma JM_J,\,\omega) \tag{9.152}$$

where $\alpha(\gamma JM_J, \omega)$ is the dynamic dipole polarisability. We note that when $\omega \to 0$ the expression (9.150) reduces to the formula (9.124) for the quadratic Stark effect, since in that case $\bar{\mathscr{E}} = \mathscr{E}$.

Finally, we note that a time-dependent electric field $\mathscr{E}(t)$ can cause field ionisation by tunnelling or 'over the barrier ionisation', just as a static field can.

9.9 The Zeeman effect. The Hanle effect and level-crossing spectroscopy

In Chapter 6 we studied the interaction of a static and uniform magnetic field with atoms (the Zeeman effect) for hydrogenic atoms. We shall now extend the treatment to a multi-electron atom. The interaction energy between each electron and the magnetic field is of the same form as before (see (6.83)). Neglecting the term quadratic in the magnetic field, it is given, for the ith electron, by

$$H_i' = \frac{\mu_B}{\hbar}(\mathbf{L}_i + 2\mathbf{S}_i)\cdot\mathscr{B} \tag{9.153}$$

where $\mathscr{B}$ is a constant magnetic field, $\mathbf{L}_i$ and $\mathbf{S}_i$ are respectively the orbital and spin angular momentum operators of the electron i, and μ_B is the Bohr magneton. The total interaction energy between the magnetic field and an atom containing N electrons is obtained by summing over i. Thus

$$H' = \sum_{i=1}^{N} H_i' = \frac{\mu_B}{\hbar}(\mathbf{L} + 2\mathbf{S})\cdot\mathscr{B} \tag{9.154}$$

where $\mathbf{L}$ is the total electronic orbital angular momentum operator, $\mathbf{L} = \Sigma_i \mathbf{L}_i$, and $\mathbf{S}$ is the total electronic spin angular momentum operator, $\mathbf{S} = \Sigma_i \mathbf{S}_i$. This interaction is of exactly the same form as for one-electron atoms, and the matrix elements we worked out in Chapter 6 can be used here, provided the atom is described by the L–S (Russell–Saunders) coupling scheme.

As before, there are three cases, depending on whether the perturbation H' is much greater than the spin–orbit interaction, $\bar{A}(\mathbf{L}\cdot\mathbf{S})$, comparable with it, or (the most usual case) much less than it. In the strong field case (giving rise to the normal Zeeman effect), the spin–orbit energy can be ignored so that the atomic wave function becomes a simultaneous eigenfunction of $\mathbf{L}^2$, L_z, $\mathbf{S}^2$ and S_z. Taking the magnetic field to be in the Z direction, so that $\boldsymbol{\mathcal{B}} = \mathcal{B}\hat{\mathbf{z}}$, one has

$$\Delta E = \mu_B \mathcal{B}(M_L + 2M_S) \qquad \textbf{(9.155)}$$

In the intermediate case (the Paschen–Back effect), the spin–orbit interaction is now introduced as a perturbation, and we have (with $A = \bar{A}\hbar^2$)

$$\langle LSM_LM_S | \bar{A}\,\mathbf{L}\cdot\mathbf{S} | LSM_LM_S \rangle = AM_LM_S \qquad \textbf{(9.156)}$$

The level shift is the sum of (9.155) and (9.156), namely

$$\Delta E = \mu_B \mathcal{B}(M_L + 2M_S) + AM_LM_S \qquad \textbf{(9.157)}$$

Finally, in the weak field case (the anomalous Zeeman effect), we first take into account the spin–orbit interaction and subsequently treat the interaction with the external field as a perturbation. The degeneracy in J is removed by the interaction $\bar{A}(\mathbf{L}\cdot\mathbf{S})$ and the atom is in an eigenstate of $\mathbf{J}^2$, $\mathbf{L}^2$, $\mathbf{S}^2$and J_z. The magnetic interaction is treated as a small perturbation and

$$\Delta E = \frac{\mu_B}{\hbar}\mathcal{B}\,\langle JLSM_J | L_z + 2S_z | JLSM_J \rangle \qquad \textbf{(9.158)}$$

This matrix element was evaluated in Section 6.2 for one-electron atoms, with the result (6.116)–(6.117). In an identical way, one has in the present case

$$\Delta E = g\mu_B \mathcal{B} M_J \qquad \textbf{(9.159)}$$

where g is the Landé factor

$$g = 1 + \frac{J(J+1) + S(S+1) - L(L+1)}{2J(J+1)} \qquad \textbf{(9.160)}$$

The transition from the weak to the strong field limits is continuous, and an example was shown in Fig. 6.11 for the $2p_{1/2}$ and $2p_{3/2}$ levels of atomic hydrogen.

The Zeeman effect in hyperfine structure

The degeneracy of the hyperfine levels, corresponding to $(2F + 1)$ values of M_F, can be removed by the application of a magnetic field. If a nucleus possesses a magnetic moment $\mathcal{M}_N$ there is an additional interaction of the form $-(\mathcal{M}_N\cdot\boldsymbol{\mathcal{B}})$. We write (see (5.36))

$$\mathcal{M}_{\mathrm{N}} = g_I \mu_{\mathrm{N}} \mathbf{I}/\hbar \tag{9.161}$$

where $\mathbf{I}$ is the nuclear spin, μ_{N} is the nuclear magneton and g_I is the nuclear g factor. The total interaction of an atom with a magnetic field in the Z direction, of strength $\mathcal{B}$, is

$$H' = \left(\frac{\mu_{\mathrm{B}}}{\hbar}\right)(L_z + 2S_z)\mathcal{B} - g_I\left(\frac{\mu_{\mathrm{N}}}{\hbar}\right)I_z\mathcal{B} \tag{9.162}$$

We shall take the case of L–S coupling, and we can again identify a number of limiting cases. The first of these is the weak field case in which H' is smaller than the hyperfine interaction $\bar{C}(\mathbf{I}\cdot\mathbf{J})$. In this case, the atom is in an eigenstate of $\mathbf{F}^2$, $\mathbf{J}^2$, $\mathbf{I}^2$ and F_z and the energy shift is

$$\Delta E = \langle FJIM_F|H'|FJIM_F\rangle \tag{9.163}$$

By methods similar to those given in Chapter 6, it can be shown that

$$\Delta E = g_F \mu_{\mathrm{B}} \mathcal{B} M_F \tag{9.164}$$

where g_F is a Landé factor given by

$$g_F = g\frac{F(F+1) + J(J+1) - I(I+1)}{2F(F+1)} - g_I'\frac{F(F+1) + I(I+1) - J(J+1)}{2F(F+1)} \tag{9.165}$$

Here g is the Landé factor in the absence of a nuclear magnetic moment (9.160) and (see (5.40))

$$g_I' = \frac{\mu_{\mathrm{N}}}{\mu_{\mathrm{B}}} g_I \tag{9.166}$$

Since $\mu_{\mathrm{N}}/\mu_{\mathrm{B}} = m/M_{\mathrm{p}} \simeq 1/1836$, g_I' is small and can be neglected.

The second case arises when the magnetic interaction is larger than the hyperfine interaction $\bar{C}(\mathbf{I}\cdot\mathbf{J})$. This is the Back–Goudsmit effect and the theory can be developed in a similar way to that of the Paschen–Back effect. The interaction $\bar{C}(\mathbf{I}\cdot\mathbf{J})$ is first omitted. In this case the atom is in an eigenstate of $\mathbf{J}^2$, J_z and of $\mathbf{I}^2$, I_z which is $(2J+1) \times (2I+1)$-fold degenerate in M_J and M_I. The magnetic interaction causes a level shift

$$\Delta E = \mu_{\mathrm{B}}\mathcal{B}(M_J g - M_I g_I') \tag{9.167}$$

Introducing now the hyperfine interaction $\bar{C}(\mathbf{I}\cdot\mathbf{J})$ as a perturbation, the hyperfine interaction level shift becomes (with $C = \bar{C}\hbar^2$)

$$\Delta E = \mu_{\mathrm{B}}\mathcal{B}(M_J g - M_I g_I') + CM_I M_J \tag{9.168}$$

As in the Zeeman effect, the energy of the hyperfine levels is a smooth function of $\mathcal{B}$. A diagram of a particular case where $J = 1$ and $I = 1/2$ is shown in Fig. 9.12.

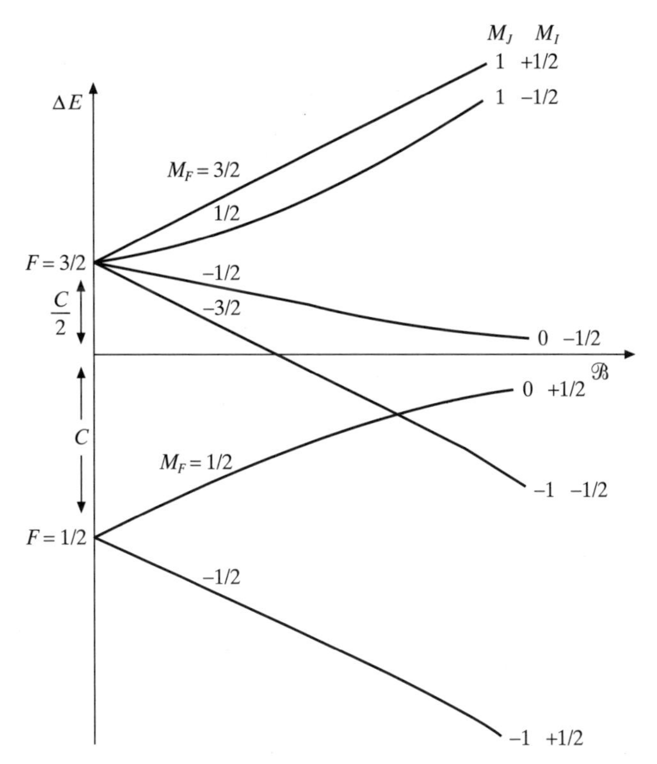

Figure 9.12 Behaviour of the hyperfine splitting as a function of $\mathcal{B}$ for $I = 1/2$, $J = 1$ (arbitrary scale). In the weak field case to the left of the diagram the states are classified by F, J, I, M_F and in the strong field case to the right by J, I, M_J and M_I (see equations (9.164) and (9.168), respectively). To connect the strong and weak field limits we must remember that $M_F = M_I + M_J$ and that levels with the same M_F or M_J never intersect each other. If the field strength increases even further, then the spin–orbit coupling breaks down and the states are classified by L, S, I, M_L, M_I, M_S (this is the extreme Paschen–Back effect).

The Hanle effect and level-crossing spectroscopy

Let us consider the elastic scattering of a beam of polarised light, with polarisation vector $\hat{\boldsymbol{\varepsilon}}$, by an atomic gas or vapour. It will be assumed that the intensity of the incident radiation is constant over the range of frequencies of interest (broad-band radiation). If the target atoms are in the state a and there exists an excited state n, then resonant scattering occurs at an angular frequency $\omega_{na} = (E_n - E_a)/\hbar$. If the states a and n are non-degenerate the cross-section for the beam to be scattered into the direction (θ, ϕ) with a polarisation vector $\hat{\boldsymbol{\varepsilon}}'$ is given by (4.226). At exact resonance, $\omega = \omega_{na}$, the differential cross-section is

$$\frac{d\sigma}{d\Omega}(\theta, \phi) = 4r_0^2\omega_{na}^4 \frac{m^2}{e^4\Gamma_n^2}|(\hat{\boldsymbol{\varepsilon}}'\cdot\mathbf{D}_{an})(\hat{\boldsymbol{\varepsilon}}\cdot\mathbf{D}_{na})|^2 \tag{9.169}$$

where Γ_n is the total width of the state n.

Now let us suppose that the state a is degenerate with substates defined by a magnetic quantum number M_J and that the intermediate state n is also degenerate with a magnetic quantum number $\bar{M}_J$. The excited intermediate states $(n, \bar{M}_J)$ are coherent, that is they all propagate in time like $\exp(-iE_n t/\hbar)$ and are in phase. It follows that to obtain the total scattering amplitude a sum must be made over the intermediate states $(n, \bar{M}_J)$. To achieve this we define the quantities

$$F_{\bar{M}_J,M_J} = \langle n, \bar{M}_J | \hat{\boldsymbol{\varepsilon}} \cdot \mathbf{D} | a, M_J \rangle \qquad \textbf{(9.170)}$$

and

$$G_{M'_J,\bar{M}_J} = \langle a, M'_J | \hat{\boldsymbol{\varepsilon}}' \cdot \mathbf{D} | n, \bar{M}_J \rangle \qquad \textbf{(9.171)}$$

The amplitude for scattering from an initial substate (a, M_J) to a final substate (a, M'_J) is obtained by replacing $(\hat{\boldsymbol{\varepsilon}}' \cdot \mathbf{D}_{an})(\hat{\boldsymbol{\varepsilon}} \cdot \mathbf{D}_{na})$ in (9.169) by $\sum_{\bar{M}_J} G_{M'_J,\bar{M}_J} F_{\bar{M}_J,M_J}$.

To find the differential cross-section, we assume that all the initial substates (a, M_J) are equally populated, so that summing over the final states (a, M'_J) and averaging over the initial states, the differential cross-section at resonance becomes

$$\frac{d\sigma}{d\Omega}(\theta, \phi) = 4r_0^2 \omega_{na}^4 \frac{m^2}{e^4 \Gamma_n^2} \mathcal{J} \qquad \textbf{(9.172)}$$

where

$$\mathcal{J} = \frac{1}{2J+1} \sum_{M_J,M'_J} \left| \sum_{\bar{M}_J} G_{M'_J,\bar{M}_J} F_{\bar{M}_J,M_J} \right|^2$$

$$= \frac{1}{2J+1} \sum_{M_J,M'_J} \sum_{\bar{M}_J,\bar{M}'_J} F^*_{\bar{M}'_J,M_J} F_{\bar{M}_J,M_J} G^*_{M'_J\bar{M}'_J} G_{M'_J\bar{M}_J} \qquad \textbf{(9.173)}$$

and $(2J+1)$ is the degree of degeneracy of the initial state.

If a weak magnetic field directed along the Z axis (the axis of quantisation) is applied to the atomic system, the magnetic substates will no longer be exactly degenerate. In particular, each intermediate state $(n, \bar{M}_J)$ will have a distinct energy $E_{\bar{M}_J}$ (see (9.159)), where

$$E_{\bar{M}_J} = E_n + g\mu_B \mathcal{B} \bar{M}_J \qquad \textbf{(9.174)}$$

If the differences between the energies $E_{\bar{M}_J}$ are large, then the scattered intensity will be the sum of the contributions from the different resonant frequencies, each contribution being proportional to $|G_{M'_J,\bar{M}_J} F_{\bar{M}_J,M_J}|^2$. In this case, there will be no interference between the different contributions which are *incoherent*. In contrast, if the differences between the energies $E_{\bar{M}_J}$ are small, then over the lifetime τ_n of the excited state n the magnetic substates $(n, \bar{M}_J)$ are approximately coherent. Each magnetic substate propagates in time like $\exp(-iE_{\bar{M}_J} t/\hbar)$, so that the phase differences between the substates will be small over the lifetime τ_n of the intermediate state n if

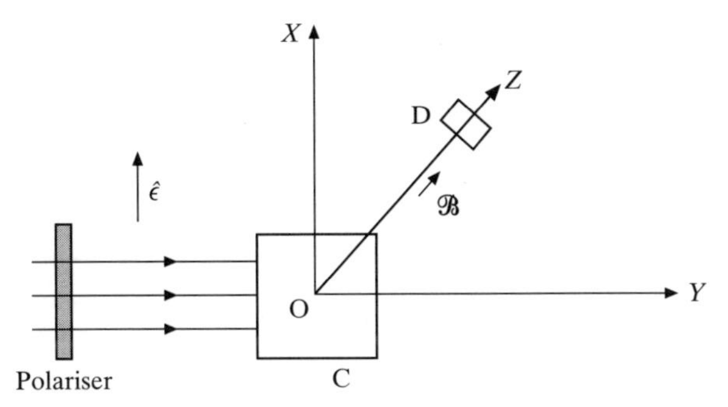

Figure 9.13 The Hanle effect. Light polarised parallel to the X axis propagates in a direction parallel to the Y axis and is scattered by the gas in the cell C. The scattered radiation emerging parallel to the Z axis is detected at D. The intensity and polarisation of the scattered light is measured as a function of a magnetic field directed parallel to the Z axis.

$$\frac{\left|E_{\bar{M}_J} - E_{\bar{M}'_J}\right|\tau_n}{\hbar} \ll 1 \tag{9.175}$$

for all the pairs of substates $\bar{M}_J$ and $\bar{M}'_J$. When this condition is satisfied, G. Breit in 1933 and P.A. Franken in 1961 showed that the intensity at resonance is proportional to the quantity

$$\begin{aligned}\bar{\mathcal{I}} &= \frac{1}{2J+1}\sum_{M_J,M'_J}\sum_{\bar{M}_J,\bar{M}'_J}\frac{F^*_{\bar{M}'_J,M_J}F_{\bar{M}_J,M_J}G^*_{M'_J,\bar{M}'_J}G_{M'_J,\bar{M}_J}}{1+\mathrm{i}(E_{\bar{M}_J}-E_{\bar{M}'_J})\tau_n/\hbar} \\ &= \frac{1}{2J+1}\sum_{M_J,M'_J}\sum_{\bar{M}_J,\bar{M}'_J}\frac{F^*_{\bar{M}'_J,M_J}F_{\bar{M}_J,M_J}G^*_{M'_J,\bar{M}'_J}G_{M'_J,\bar{M}_J}}{1+\mathrm{i}g\mu_{\mathrm{B}}\mathcal{B}(\bar{M}_J-\bar{M}'_J)\tau_n/\hbar}\end{aligned} \tag{9.176}$$

where in the last line the equation (9.175) has been employed.

The result (9.176) is known as the Breit–Franken relation and determines the intensity and polarisation of the radiation scattered in the direction (θ, ϕ) as a function of $\mathcal{B}$. It should be noted that the *total* scattered intensity given by the total cross-section is independent of $\mathcal{B}$.

The resonance radiation scattered from the 6 $^1S_0$ state of mercury at a wavelength $\lambda = 2537$ Å, corresponding to the 6 $^3P_1$ excited state, was studied by W. Hanle in 1924, by sweeping the external magnetic field from positive to negative values (see Fig. 9.13). He observed changes in polarisation and intensity which using (9.176) allowed the accurate determination of the natural lifetime τ_n, with no complication due to Doppler broadening. The changes in scattering parameters near zero field strength are known as the *Hanle effect.* It is interesting that Lord Rayleigh in 1919 had already noted that the polarisation of the 2537 Å mercury line changes significantly from one experiment to another. A few years later R.W. Wood and A. Ellett showed that this was related to the orientation of the apparatus with respect to the Earth's magnetic field. However, the first explanation of the effect was given by W. Hanle who employed a purely classical model.

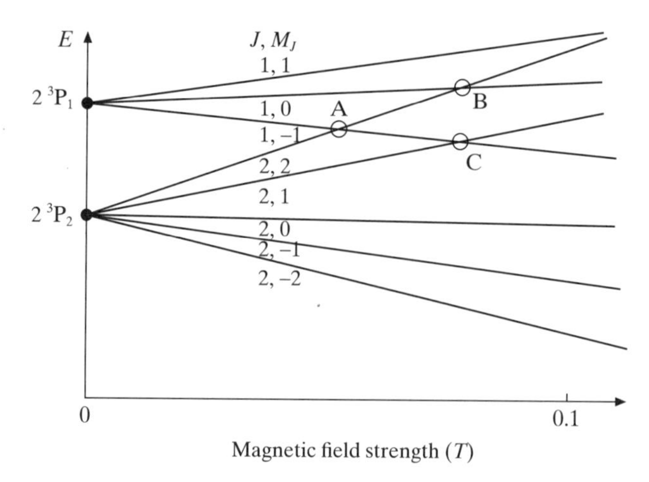

Figure 9.14 Schematic diagram of the 2 ^{3}P$_1$ and 2 ^{3}P$_2$ fine structure in helium as a function of the strength of an applied magnetic field directed parallel to the Z axis. Level crossings occur at A, B and C (see text).

Level-crossing spectroscopy

Magnetic substates which are non-degenerate in the absence of an applied magnetic field may become degenerate for certain values of an external field. This is known as *level crossing*. An example is shown in Fig. 9.12, where the hyperfine level with $F = 3/2$, $M_F = -3/2$ is seen to cross the level with $F = 1/2$ and $M_F = 1/2$. If the selection rules allow a resonant transition to be made from a lower state to the degenerate combination, then by sweeping the magnetic field from values below to values above the crossing point the changes in the scattering parameters can be studied in exactly the same way as in the Hanle effect. This has become an important method for determining the lifetimes of excited states. With stronger fields, using the Paschen–Back formula (9.157) or the Back–Goudsmit expression (9.167), the fine structure and hyperfine structure constants can also be measured [5].

The earliest level-crossing experiment was performed by F.D. Colegrove, P.A. Franken, R.R. Lewis and R.H. Sands in 1959 to measure the fine structure separation of the 2 ^{3}P$_1$ and 2 ^{3}P$_2$ levels of helium by resonant scattering from the 2 ^{3}S$_1$ level. The resonant 2 ^{3}S$_1$–2 ^{3}P$_2$ line is at 10 830.34 Å. The energy levels are shown schematically in Fig. 9.14 as functions of the applied field, with crossings at the points A, B and C. The dipole selection rules $\Delta M_J = 0, \pm 1$ show that there can be no coupling and hence no coherence between states with $\Delta M_J > 2$, so that the crossing at the point A does not produce a coherence effect. The study of the scattering from the crossings at B and C allowed Colegrove *et al.* to determine the separation of the 2 ^{3}P$_1$ and 2 ^{3}P$_2$ levels accurately.

[5] See for example Svanberg (2001).

Problems

9.1 Derive the Thomas–Reiche–Kuhn sum rule (9.50).

9.2 Using the Wigner–Eckart theorem, prove the selection rules (9.82) for magnetic dipole transitions.

9.3 Using the Wigner–Eckart theorem, prove the selection rules (9.84) for electric quadrupole transitions.

9.4 Write down the selection rules for

(a) electric multipole transitions of order $\lambda = 3, 4$.
(b) magnetic multipole transitions of order $\lambda = 2, 3, 4$.

9.5 Using Table 9.2, calculate the energies of the ten lowest levels of Li, Na, K and Cs, expressing the results both in eV and in cm^{-1}. Find the wave numbers of the resonance line in each case and compare the answer with the experimental values given in Table 9.3.

9.6 Use the expression (3.78) together with (9.102) to estimate the energy splitting of the 2^3P_J (1s2p), 3^3P_J (1s3p) and 3^3D_J (1s3d) levels of He I, C V, and Na X by the spin–orbit interactions, expressing the answer in cm^{-1}. Using the selection rules discuss the splitting of the spectral lines in the transitions from the 2^3P_J levels to the lowest triplet state and from the 3^3D_J levels to the 2^3P_J levels.
(**Hint**: The effective charge to be used in (3.78) is equal to the Roman numeral in the designation of the ions.)

9.7 Measurements made on the line spectrum of the neutral carbon atom show that a certain excited term consists of three fine structure levels having energies above the ground state given (in cm^{-1}) by 60 333, 60 353 and 60 393.

(a) With the help of the Landé interval rule, identify the values of L, S and J for these fine structure levels.
(b) The ground state term of carbon, $1s^22s^22p^2\ ^3P$, is split into three fine structure levels, with $J = 0, 1, 2$ (see Fig. 8.8). The levels with $J = 1$ and $J = 2$ are 16 cm^{-1} and 43 cm^{-1}, respectively, above the level $J = 0$. Discuss the multiplet structure of the lines arising from dipole transitions between the excited term considered in (a) (which has opposite parity to the ground state) and the ground state term. Write down the corresponding wavelengths.

9.8 (a) Find how many separate lines occur in the multiplet arising from the $^2P_{3/2} \to {}^2S_{1/2}$ and $^2P_{1/2} \to {}^2S_{1/2}$ transitions of an alkali placed in a weak magnetic field.
(b) Find the relative intensities of the lines in transverse observation.

10 Molecular structure

Bound systems of electrons and more than one nucleus are known as molecules. In this chapter, we shall discuss the structure of molecules, concentrating on the simplest diatomic systems which contain just two nuclei. We shall start by discussing the general nature of molecular structure, showing how the rotational, vibrational and electronic motions can essentially be treated independently. Next we shall study the electronic structure of diatomic molecules, and explain how chemical bonding (or binding) comes about. This is followed by an analysis of the vibrational and rotational motions. Finally we shall indicate how the more complicated polyatomic molecules can be treated.

10.1 General nature of molecular structure

The description of molecular structure is considerably more complicated than that of isolated atoms, but fortunately the problem is greatly simplified because *the mass of the electrons is much smaller than that of the nuclei*, while the forces to which the electrons and the nuclei are submitted are of comparable magnitude. As a result, the motion of the nuclei is much slower than that of the electrons, and the nuclei occupy nearly fixed positions within the molecule. This can be shown quite directly by observing the diffraction pattern in neutron scattering by molecules. The neutrons, having no charge, do not interact with the electrons, but interact strongly, through the nuclear force, with each of the nuclei within the molecule. For example, in the simplest neutral diatomic molecule H_2, formed by the association of two hydrogen atoms, the equilibrium spacing of the two protons is 0.74 Å, while in the molecule O_2 formed from two oxygen atoms, the internuclear spacing is 1.21 Å. In the methane molecule CH_4, composed of a single carbon atom and four hydrogen atoms, the carbon nucleus lies at the centre of a regular tetrahedron, with the protons at the vertices (see Fig. 10.1).

Evidence from X-ray diffraction and molecular spectra shows that when atoms associate to form molecules, the tightly bound inner shells of electrons are nearly undisturbed, and remain localised about each nucleus. The outer electrons, on the other hand, are distributed throughout the molecule, and it is the charge distribution of these *valence* electrons that provides the binding force. The order of magnitude of the separation of energy levels for the motion of the valence electrons can be estimated by using the following argument. Let a be a typical

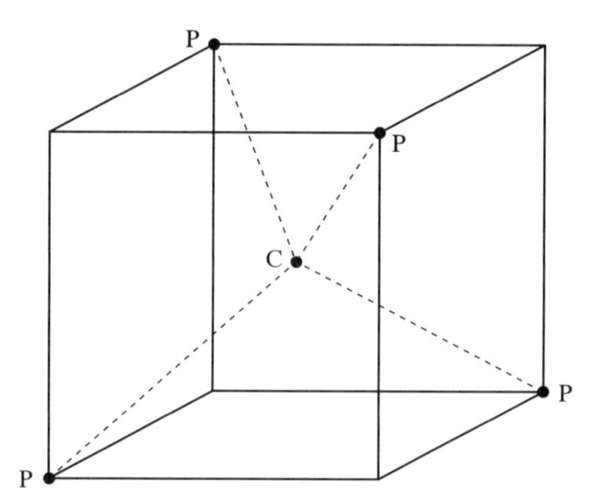

Figure 10.1 The methane molecule CH_4. The carbon nucleus lies at the centre of a regular tetrahedron inscribed in a cube with the protons at alternate vertices. The distance between the carbon nucleus and a proton (the bond length) is 2.067 a.u. (1.094 Å).

average distance of the nuclei in a molecule. From the uncertainty principle, the magnitude of the momentum of the valence electrons is of the order of $\hbar/a$, so that a rough estimate of their kinetic energy – and hence of the magnitude E_e of the electronic energies – is given by

$$E_e \simeq \frac{\hbar^2}{ma^2} \tag{10.1}$$

where m is the mass of the electron. Since $a \simeq 1$ Å we see that E_e is of the order of a few eV, which is similar to the binding energy of the outer electrons in isolated atoms. The result (10.1) clearly also gives an estimate of the energy separation between low-lying electronic energy levels of the molecule. The corresponding line spectra are observed in the ultra-violet and visible regions.

Let us now consider the nuclear motions. These can be classified into *translations* and *rotations* of the entire (quasi-rigid) equilibrium arrangement, and internal *vibrations* of the nuclei about their equilibrium positions. The translational motion can be separated by introducing the centre of mass, which moves as a free particle in the absence of external fields. In what follows we shall assume that the separation of the centre of mass has been performed, and we shall only consider the vibrational and rotational motions of the nuclei. To estimate the vibrational energy, let us assume that the vibrations are linear harmonic, with an angular frequency ω_N. If one of the nuclei is removed a distance a, the increase in potential energy of the molecule is $M\omega_N^2 a^2/2$, where M is of the order of a typical nuclear mass. Since the removal of a nucleus amounts to dissociating the molecule, this increase of energy must be of the order of the electronic energy E_e. Thus $M\omega_N^2 a^2 \simeq \hbar^2/(ma^2)$, and the energy associated with a low mode of vibration is given approximately by

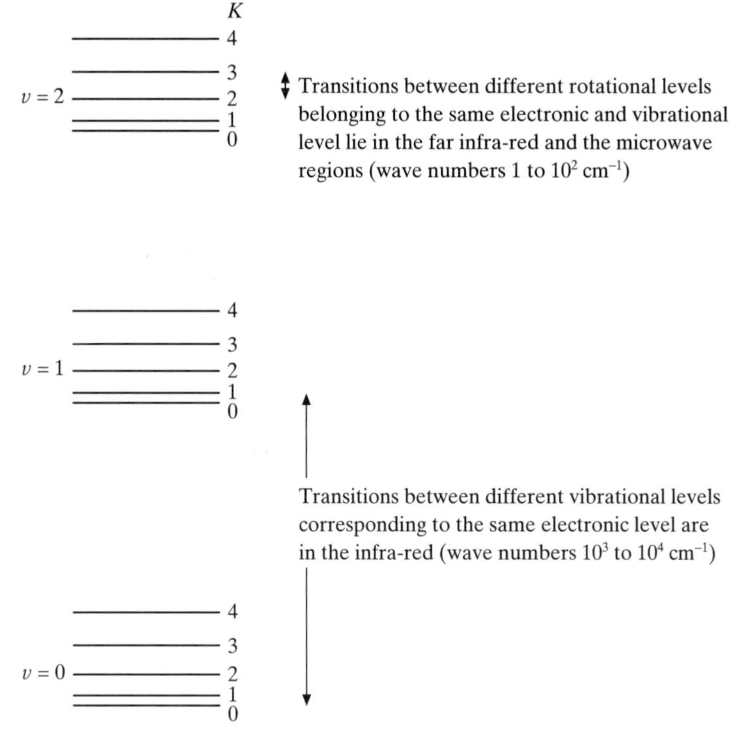

Figure 10.2 Schematic diagram of the energy levels of a diatomic molecule belonging to the same electronic level. The vibrational levels are labelled by the quantum number *v* and the rotational levels by the quantum number *K*.

$$E_{\mathrm{v}} \simeq \hbar\omega_{\mathrm{N}} \simeq \left(\frac{m}{M}\right)^{1/2} E_{\mathrm{e}} \tag{10.2}$$

The ratio m/M being in the range 10^{-3} to 10^{-5}, we see from (10.2) that E_{v} is roughly a hundred times smaller than E_{e}, so that typical vibrational transitions lie in the infra-red region (see Fig. 10.2). For example, the natural vibrational wave number of the molecule HCl is at approximately 3000 cm^{-1}. The vibrational motion of the nuclei also produces a 'first-order' splitting of the electronic lines.

To estimate the rotational energy E_{r}, we consider the simple case of a diatomic molecule, with each of the two nuclei having the same mass M and being a distance a apart, as shown in Fig. 10.3. The moment of inertia of the molecule is then $I = Ma^2/2$. Using the result (2.193) which we obtained in Chapter 2 for the rigid rotator, we see that the order of magnitude of the energy associated with a low mode of rotation is

$$E_{\mathrm{r}} \simeq \frac{\hbar^2}{Ma^2} \simeq \frac{m}{M} E_{\mathrm{e}} \tag{10.3}$$

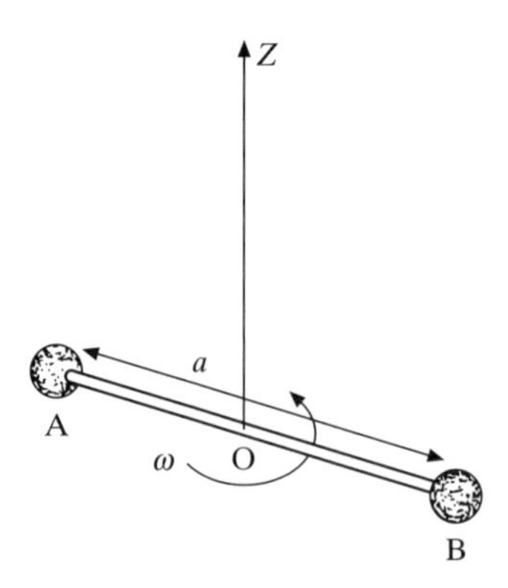

Figure 10.3 Rotation of a diatomic molecule composed of two nuclei A and B, each of the same mass M, a distance a apart. The rotation is of angular frequency ω about an axis OZ through the centre of mass. The classical energy of rotation is $E = \frac{1}{2}I\omega^2$, where the moment of inertia I is given by $I = \frac{1}{2}Ma^2$.

where we have used (10.1). Thus the rotational molecular energies are smaller than the electronic energies by a factor of the order of m/M, and smaller than the vibrational energies by a factor of the order of $(m/M)^{1/2}$. The rotational motion leads to a 'second-order' splitting of the electronic lines, the spacing being of the order of 0.001 eV, which is very small compared with the 'first-order' splitting of about 0.1 eV produced by the vibrational motion. Transitions between rotational levels, belonging to the same electronic and vibrational level, are observed in the far infra-red and the microwave regions at wave numbers of 1 to 10^2 cm^{-1}.

Because of the small ratio of the electronic mass to the nuclear mass ($m/M \simeq 10^{-3}$–10^{-5}), and since the period of a motion is of the order of h divided by its energy, we see from (10.1)–(10.3) that the nuclear periods are much longer than the electronic periods. Thus the electronic and nuclear motions can essentially be treated independently, and it is a good approximation to determine the electronic states at each value of the internuclear separation by treating the nuclei as fixed. The charge distribution of the electrons is then a function of the nuclear positions and determines the nuclear motion. In the next paragraphs we shall use these qualitative ideas to develop a mathematical formalism for diatomic molecules.

10.2 The Born–Oppenheimer separation for diatomic molecules

To see how the ideas of the previous paragraph can be implemented, we shall consider a diatomic molecule composed of nuclei A and B, of masses M_A and M_B together with a number N of electrons. The internuclear position vector will be denoted by $\mathbf{R}$ and the position vectors of the electrons with respect to O, the centre of mass of A and B, by $\mathbf{r}_1, \mathbf{r}_2, \ldots, \mathbf{r}_N$ (see Fig. 10.4). We shall write the position vectors of A and B with respect to O as $\mathbf{R}_A$ and $\mathbf{R}_B$ respectively, so that $\mathbf{R} = \mathbf{R}_A - \mathbf{R}_B$.

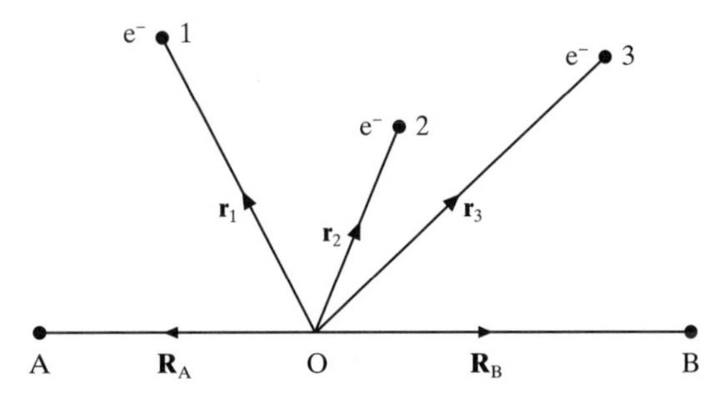

Figure 10.4 Coordinate system for a diatomic molecule. The nuclei are at A and B, with position vectors $\mathbf{R}_A$, $\mathbf{R}_B$ with respect to O the centre of mass. The position vectors of the electrons with respect to O are denoted by $\mathbf{r}_1$, $\mathbf{r}_2$, $\mathbf{r}_3$, ...

The Hamiltonian operator for the complete molecule can be expressed as

$$H = T_N + T_e + V \tag{10.4}$$

where T_N is the kinetic energy operator for the nuclei, T_e is the kinetic energy operator for the electrons and V is the total potential energy of the system. We shall take O to be the origin of a coordinate system fixed in space (the *laboratory or space-fixed* frame), and neglect the mass of an electron m with respect to the reduced mass μ of the nuclei A and B, where

$$\mu = \frac{M_A M_B}{M_A + M_B} \tag{10.5}$$

It follows that

$$T_N = -\frac{\hbar^2}{2\mu}\nabla^2_{\mathbf{R}} \tag{10.6}$$

and

$$T_e = \sum_{i=1}^{N}\left(-\frac{\hbar^2}{2m}\nabla^2_{\mathbf{r}_i}\right) \tag{10.7}$$

It will be convenient to express T_N in terms of the spherical polar coordinates (R, Θ, Φ) of $\mathbf{R}$, so that

$$T_N = -\frac{\hbar^2}{2\mu}\left[\frac{1}{R^2}\frac{\partial}{\partial R}\left(R^2\frac{\partial}{\partial R}\right) - \frac{\mathbf{N}^2}{\hbar^2 R^2}\right] \tag{10.8}$$

where $\mathbf{N}$ is the orbital angular momentum operator for the relative motion of the two nuclei A and B, and (see (2.159))

$$\mathbf{N}^2 = -\hbar^2\left[\frac{1}{\sin\Theta}\frac{\partial}{\partial\Theta}\left(\sin\Theta\frac{\partial}{\partial\Theta}\right) + \frac{1}{\sin^2\Theta}\frac{\partial^2}{\partial\Phi^2}\right] \tag{10.9}$$

If the small spin-dependent interactions are neglected, the potential energy consists of the sum of the Coulomb interactions between the electrons and the nuclei, between the electrons themselves, and between the two nuclei. Denoting the charges on the nuclei A and B by $Z_A e$ and $Z_B e$, we have

$$V(\mathbf{R}; \mathbf{r}_1, \mathbf{r}_2, \dots, \mathbf{r}_N) = -\sum_{i=1}^{N}\left[\frac{Z_A e^2}{(4\pi\varepsilon_0)|\mathbf{r}_i - \mathbf{R}_A|} + \frac{Z_B e^2}{(4\pi\varepsilon_0)|\mathbf{r}_i - \mathbf{R}_B|}\right] + \sum_{i<j=1}^{N}\frac{e^2}{(4\pi\varepsilon_0)|\mathbf{r}_i - \mathbf{r}_j|} + \frac{Z_A Z_B e^2}{(4\pi\varepsilon_0)R} \quad \textbf{(10.10)}$$

The time-independent Schrödinger equation for the molecule is

$$H\Psi(Q_A, Q_B; q_1, q_2, \dots, q_N) = E\Psi(Q_A, Q_B; q_1, q_2, \dots, q_N) \quad \textbf{(10.11)}$$

where Q_A and Q_B represent the space and spin coordinates of the nuclei collectively and $q_1, q_2, \dots, q_N$ the space and spin coordinates of the electrons collectively. The wave function Ψ must obey the Pauli exclusion principle and be antisymmetric under the interchange of the coordinates (both space and spin) of any pair of electrons. If the nuclei A and B are identical Ψ must also be either symmetric or antisymmetric under the interchange of A and B, for nuclei having integer (including zero) or half-odd integer spin, respectively. In what follows, to simplify the notation, the dependence of Ψ on the spin variables will not be displayed and we shall write $\Psi(\mathbf{R}; \mathbf{r}_1, \mathbf{r}_2, \dots, \mathbf{r}_N)$.

In order to construct a set of basis states into which the total wave function Ψ can be expanded, let us consider the time-independent Schrödinger equation for the N electrons moving in the Coulomb potential of the nuclei A and B, where $\mathbf{R}$ is held fixed, which is

$$H_e\Phi_q(\mathbf{R}; \mathbf{r}_1, \mathbf{r}_2, \dots, \mathbf{r}_N) = E_q(R)\Phi_q(\mathbf{R}; \mathbf{r}_1, \mathbf{r}_2, \dots, \mathbf{r}_N) \quad \textbf{(10.12)}$$

where H_e is the *electronic Hamiltonian*

$$H_e = T_e + V \quad \textbf{(10.13)}$$

Equation (10.12) is known as the *electronic wave equation*. The eigenvalues $E_q(R)$ and the wave functions $\Phi_q(\mathbf{R}; \mathbf{r}_1, \mathbf{r}_2, \dots, \mathbf{r}_N)$ for each electronic state q depend parametrically on the internuclear position vector $\mathbf{R}$, which is held fixed when E_q and Φ_q are calculated. Since the energies E_q cannot depend on the orientation of the internuclear line in the absence of any external fields, they depend parametrically only on the magnitude R of the vector $\mathbf{R}$. The wave functions Φ_q, on the other hand, depend on R and on the angles Θ and Φ as parameters.

For each (fixed) internuclear position vector $\mathbf{R}$, the electronic wave functions Φ_q form a complete set, which can be taken to be orthonormal

$$\langle\Phi_q|\Phi_p\rangle \equiv \int d\mathbf{r}_1\, d\mathbf{r}_2, \dots, d\mathbf{r}_N\, \Phi_q^*(\mathbf{R}; \mathbf{r}_1, \mathbf{r}_2, \dots, \mathbf{r}_N)\Phi_p(\mathbf{R}; \mathbf{r}_1, \mathbf{r}_2, \dots, \mathbf{r}_N) = \delta_{qp} \quad \textbf{(10.14)}$$

Since the set of functions Φ_q is complete, the exact wave function Ψ for the molecule can be expanded as

$$\Psi(\mathbf{R}; \mathbf{r}_1, \mathbf{r}_2, \ldots, \mathbf{r}_N) = \sum_q F_q(\mathbf{R})\Phi_q(\mathbf{R}; \mathbf{r}_1, \mathbf{r}_2, \ldots, \mathbf{r}_N) \tag{10.15}$$

where the expansion coefficients $F_q(\mathbf{R})$ are wave functions representing the nuclear motion (vibration and rotation) when the electronic system is in the state q.

To obtain equations satisfied by the wave functions $F_q(\mathbf{R})$, we first note that the Hamiltonian of the complete molecule can be written as

$$H = T_{\mathrm{N}} + H_{\mathrm{e}} \tag{10.16}$$

so that the Schrödinger equation (10.11) for the molecule becomes

$$(T_{\mathrm{N}} + H_{\mathrm{e}})\Psi(\mathbf{R}; \mathbf{r}_1, \mathbf{r}_2, \ldots, \mathbf{r}_N) = E\Psi(\mathbf{R}; \mathbf{r}_1, \mathbf{r}_2, \ldots, \mathbf{r}_N) \tag{10.17}$$

We now insert the expansion (10.15) into this equation, so that

$$(T_{\mathrm{N}} + H_{\mathrm{e}}) \sum_q F_q(\mathbf{R})\Phi_q(\mathbf{R}; \mathbf{r}_1, \mathbf{r}_2, \ldots, \mathbf{r}_N)$$

$$= E \sum_q F_q(\mathbf{R})\Phi_q(\mathbf{R}; \mathbf{r}_1, \mathbf{r}_2, \ldots, \mathbf{r}_N) \tag{10.18}$$

On projecting equation (10.18) successively with the functions Φ_s ($s = 0, 1, 2, \ldots$), we obtain the coupled equations

$$\sum_q \langle \Phi_s | T_{\mathrm{N}} + H_{\mathrm{e}} - E | \Phi_q \rangle F_q(\mathbf{R}) = 0, \qquad s = 0, 1, 2, \ldots \tag{10.19}$$

where

$$\langle \Phi_s | T_{\mathrm{N}} + H_{\mathrm{e}} - E | \Phi_q \rangle$$

$$\equiv \int \mathrm{d}\mathbf{r}_1 \mathrm{d}\mathbf{r}_2, \ldots, \mathrm{d}\mathbf{r}_N \, \Phi_s^*(\mathbf{R}; \mathbf{r}_1, \mathbf{r}_2, \ldots, \mathbf{r}_N)(T_{\mathrm{N}} + H_{\mathrm{e}} - E)\Phi_q(\mathbf{R}; \mathbf{r}_1, \mathbf{r}_2, \ldots, \mathbf{r}_N) \tag{10.20}$$

Using the electronic wave equation (10.12) together with (10.8) and (10.14), the coupled equations (10.19) can be rewritten in the form

$$\sum_q \left\langle \Phi_s \left| -\frac{\hbar^2}{2\mu} \frac{1}{R^2} \frac{\partial}{\partial R}\left(R^2 \frac{\partial}{\partial R} \right) + \frac{\mathbf{N}^2}{2\mu R^2} \right| \Phi_q \right\rangle F_q(\mathbf{R}) + [E_s(R) - E]F_s(\mathbf{R}) = 0,$$

$$s = 0, 1, 2, \ldots \tag{10.21}$$

In (10.21), since Φ_q depends on $\mathbf{R}$ as well as on $\mathbf{r}_1, \mathbf{r}_2, \ldots, \mathbf{r}_N$, the derivatives $\partial/\partial R$ and $\partial/\partial\Theta$, $\partial/\partial\Phi$ (contained in $\mathbf{N}$) operate on the product $\Phi_q F_q$.

The set of coupled equations (10.21) is exact and equivalent to the full Schrödinger equation (10.17). The *Born–Oppenheimer* or *adiabatic approximation* will now be introduced, which makes use of the fact that the nuclear motion is very slow compared with the electronic motion. As a consequence, the electronic wave functions Φ_q vary very slowly with respect to R, Θ and Φ. This fact allows $|\partial\Phi_q/\partial R|$ to be neglected with respect to $|\partial F_q/\partial R|$ and in addition it is sufficient to retain only the diagonal term $\langle\Phi_s|\mathbf{N}^2|\Phi_s\rangle$.

Making these approximations, the coupled equations (10.21) reduce to the set of uncoupled equations

$$\left[-\frac{\hbar^2}{2\mu}\frac{1}{R^2}\frac{\partial}{\partial R}\left(R^2\frac{\partial}{\partial R}\right)+\frac{\langle\Phi_s|\mathbf{N}^2|\Phi_s\rangle}{2\mu R^2}+E_s(R)-E\right]F_s(\mathbf{R})=0,$$

$$s = 0, 1, 2, \ldots \tag{10.22}$$

Each of these equations, which describes the vibrational and rotational motion of the molecule when the electronic system is in the state s, is called a *nuclear wave equation*. The uncoupled set of equations (10.22) constitutes the Born–Oppenheimer approximation in which the total wave function reduces to the single term for a given s,

$$\Psi_s = F_s(\mathbf{R})\Phi_s(\mathbf{R}; \mathbf{r}_1, \mathbf{r}_2, \ldots, \mathbf{r}_N) \tag{10.23}$$

In the Born–Oppenheimer approximation the electronic wave equation (10.12) is first solved at different values of R to obtain $E_s(R)$, and subsequently the nuclear wave equation (10.22) can be solved in which $E_s(R)$ plays the role of a potential. The general form of $E_s(R)$ for a bound state is shown in Fig. 10.5. As $R \to 0$, $E_s(R)$

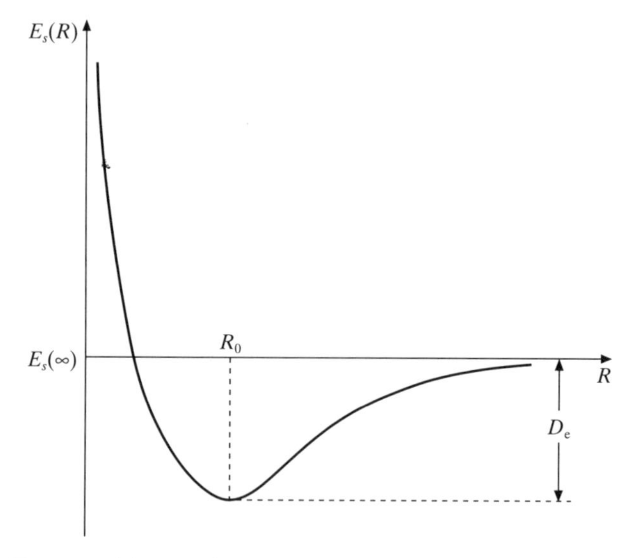

Figure 10.5 The general form of the electronic energy $E_s(R)$ for a bound state of a diatomic molecule. The quantity D_e is the electronic (or spectroscopic) dissociation energy of the molecule in the electronic state s.

is dominated by the Coulomb repulsion between the two nuclei A and B, while as $R \to \infty$, $E_s(R)$ tends to a constant energy $E_s(\infty)$ which is the sum of the energies of the two isolated atoms from which the molecule is composed. If there is a stable bound state, $E_s(R)$ exhibits a minimum at some internuclear separation $R = R_0$ and the nuclear wave function is confined to a limited region near $R = R_0$. The quantity

$$D_e = E_s(\infty) - E_s(R_0) \tag{10.24}$$

is often called the *electronic* (or *spectroscopic*) *dissociation energy* of the molecule in the electronic state s. As we shall see in Section 10.4, the quantity D_e differs slightly from the true (also called *chemical*) dissociation energy D_0 of the molecule, which takes into account the zero-point energy of the vibrational motion of the nuclei in the electronic state s under consideration (see (10.112)).

The accuracy of the Born–Oppenheimer approximation can be estimated in any particular case by treating the omitted terms as small perturbations and checking that their effect is small by using first-order perturbation theory. An important case in which the Born–Oppenheimer approximation must be improved is when the energy $E_s(R)$ is degenerate, because the coupling between degenerate states cannot be ignored. When the relative motion of the nuclei is not very slow, as in atom–atom collisions, the terms that are omitted in the Born–Oppenheimer approximation become important and are responsible for *inelastic transitions* between different electronic states. Finally, we note that for Rydberg molecules, in which an electron is in a highly excited electronic state (Rydberg state), the motion of this electron is slow, and the Born–Oppenheimer approximation fails.

10.3 Electronic structure of diatomic molecules

In this section we shall discuss the electronic wave functions of the simplest diatomic molecules. Our discussion will be based on the electronic wave equation (10.12), and we recall that all but the Coulomb interactions are neglected in the electronic Hamiltonian $H_e = T_e + V$. In solving the electronic wave equation (10.12) it is easiest to use a system of coordinates $\overline{OX}$, $\overline{OY}$, $\overline{OZ}$ which is fixed with respect to the molecule, taking the $\bar{Z}$ axis along the internuclear axis from B to A and choosing that the $\bar{Y}$ axis lies in the space-fixed XY plane. This frame of reference is called the *body-fixed* (or *molecular*) frame (see Appendix 9). The position vector $\mathbf{r}_i$ of an electron is the same in the space-fixed (or laboratory) and body-fixed frames, but the *components* of $\mathbf{r}_i$ are different in each frame. In the space-fixed frame the components of $\mathbf{r}_i$ along the X, Y and Z axes will be denoted by x_i, y_i, z_i and in the body-fixed frame along the $\bar{X}$, $\bar{Y}$, $\bar{Z}$ axes by $\bar{x}_i$, $\bar{y}_i$, $\bar{z}_i$.

As we have already noted the eigenvalues $E_s(R)$ are independent of the frame of reference, and can be calculated in the body-fixed frame for convenience. The electronic wave function is required to be expressed in terms of the space-fixed frame in order to evaluate the term $\langle \Phi_s | \mathbf{N}^2 | \Phi_s \rangle$ which occurs in the nuclear wave equation (10.22). Once having obtained the electronic wave function in the body-fixed frame coordinates a transformation to the space-fixed frame coordinates can be made without difficulty. This is discussed in Appendix 9.

We shall start our discussion of the electronic wave functions by considering some important symmetry properties.

Symmetry properties

Let us return to the Hamiltonian (8.2) which approximately describes an N-electron atom by taking into account all the Coulomb interactions, but neglecting spin–orbit and other corrections such as spin–spin and hyperfine structure effects. The three components of the total electronic orbital angular momentum $\mathbf{L}$ commute with the Hamiltonian (8.2) and so does $\mathbf{L}^2$. Taking the Z axis in an arbitrary direction, the atomic eigenfunctions of (8.2) can be constructed to be simultaneous eigenfunctions of H, $\mathbf{L}^2$ and L_z. Thus, both the total electronic orbital quantum number L and the quantum number M_L are conserved when the atom is described by the (approximate) Hamiltonian (8.2). As we have seen in Chapter 8, the fact that L is (approximately) a 'good quantum number' plays a central role in the classification of atomic energy levels (or terms).

In contrast, the internuclear axis of a diatomic molecule picks out a particular direction in space. If this direction is taken as the $\bar{Z}$ axis, then $L_{\bar{z}}$ commutes with the electronic molecular Hamiltonian $H_e = T_e + V$ of equation (10.13) (in which spin-dependent interactions have been neglected), but $L_{\bar{x}}$, $L_{\bar{y}}$ and $\mathbf{L}^2$ do not. This is due to the fact that the electronic Hamiltonian $H_e = T_e + V$ of a diatomic molecule is invariant under rotations about the internuclear line (the $\bar{Z}$ axis), but not under rotations about the $\bar{X}$ or $\bar{Y}$ axes. The same result can also be obtained by examining the commutators $[H_e, L_{\bar{x}}]$, $[H_e, L_{\bar{y}}]$ and $[H_e, L_{\bar{z}}]$ directly (Problem 10.1).

The electronic eigenfunctions Φ_s of a diatomic molecule can thus be constructed to be simultaneous eigenfunctions of H_e and $L_{\bar{z}}$. That is,

$$\begin{aligned} L_{\bar{z}}\Phi_s &= M_L\hbar\Phi_s, \qquad M_L = 0, \pm 1, \pm 2, \ldots \\ &= \pm\Lambda\hbar\Phi_s, \qquad \Lambda = 0, 1, 2, \ldots \end{aligned} \tag{10.25}$$

where $\Lambda = |M_L|$ is the absolute value (in a.u.) of the projection of the total electronic orbital angular momentum on the internuclear axis. The azimuthal part of the wave functions Φ_s is therefore of the form $(2\pi)^{-1/2}\exp(\pm i\Lambda\phi)$. By analogy with the spectroscopic notation S, P, D, F, . . . used for atoms, it is customary to associate code letters with the values of Λ according to the correspondence:

Value of Λ	0	1	2	3
	↕	↕	↕	↕
Code letter	Σ	Π	Δ	Φ, . . .

If we are dealing with individual electrons, we shall use the notation $\lambda = |m_l|$ and the correspondence:

Value of λ	0	1	2	3
	$\updownarrow$	$\updownarrow$	$\updownarrow$	$\updownarrow$
Code letter	σ	π	δ	$\phi, \ldots$

The electronic Hamiltonian H_e for a diatomic molecule is also invariant under reflections in all planes containing the internuclear line. The $(\bar{X}, \bar{Z})$ plane is such a plane, and reflection of the coordinates of the electrons in this plane corresponds to the operation $\bar{y}_i \to -\bar{y}_i$. If $A_{\bar{y}}$ is the operator that performs this reflection, then

$$[A_{\bar{y}}, H_e] = 0 \qquad \textbf{(10.26)}$$

In addition, we have

$$A_{\bar{y}} L_{\bar{z}} = -L_{\bar{z}} A_{\bar{y}} \qquad \textbf{(10.27)}$$

It follows immediately that if $\Lambda \neq 0$ the action of the operator $A_{\bar{y}}$ on a wave function corresponding to the eigenvalue $\Lambda\hbar$ of $L_{\bar{z}}$ converts this wave function into another one corresponding to the eigenvalue $-\Lambda\hbar$, and that both eigenfunctions have the same energy. The electronic terms such that $\Lambda \neq 0$ (that is, the Π, Δ, Φ, ... terms) are thus *doubly degenerate*, each value of the energy corresponding to two states which differ by the direction of the projection of the orbital angular momentum along the molecular axis. In fact, this twofold degeneracy is only approximate and it is possible to show that the interaction between the electronic and rotational motions leads to a splitting of the terms with $\Lambda \neq 0$ into two nearby levels, which is called *Λ-doubling* (see Section 10.5).

Let us now return to the relations (10.26)–(10.27) and consider the case $\Lambda = 0$ corresponding to the Σ states. These states are non-degenerate, so that the wave function of a Σ term can only be multiplied by a constant in a reflection through a plane containing the molecular axis. We also note that when $\Lambda = 0$ simultaneous eigenfunctions of H_e, $L_{\bar{z}}$ and $A_{\bar{y}}$ can be constructed. Since $A_{\bar{y}}^2 = 1$, the eigenfunctions of $A_{\bar{y}}$ have eigenvalues ± 1. To completely specify Σ states of diatomic molecules, one therefore distinguishes Σ^+ states, for which the wave function is left unchanged upon reflection in a plane containing the nuclei, from Σ^- states, for which it changes sign in performing that operation.

In the special case of a *homonuclear* diatomic molecule, namely a diatomic molecule containing identical nuclei (such as H_2, N_2, O_2, ...), there is an extra symmetry since in addition to the axis of symmetry provided by the internuclear axis, there is a *centre of symmetry* at the midpoint of the distance between the two nuclei [1]. Choosing this point as the origin of the coordinates, the Hamiltonian is invariant under a reflection of the coordinates of all electrons with respect to

[1] The symmetry discussed in this paragraph only depends on the two nuclear *charges* being the same. The two nuclei can therefore have different mass, that is they can be two isotopes of the same species such as the proton and the deuteron, or ^{16}O and ^{18}O, and so on.

that origin, namely in the operation $\mathbf{r}_i \to -\mathbf{r}_i$. Since the operator which effects this transformation also commutes with $L_{\bar{z}}$, we may then classify electronic terms having a given value of Λ according to their parity. The electronic wave functions therefore split into two sets: those that are even, that is remain unaltered by the operation $\mathbf{r}_i \to -\mathbf{r}_i$, and those that are odd, that is change sign in that operation. The former are denoted by a subscript g and are called *gerade* states, while the latter are denoted by a subscript u and are called *ungerade* states. The subscripts g or u are added to the term symbol, so that for homonuclear diatomic molecules we have Σ_g, Σ_u, Π_g, Π_u, . . . states. We note that a homonuclear diatomic molecule has four non-degenerate Σ states: Σ_g^+, Σ_u^+, Σ_g^-, Σ_u^-.

The two transformations we have considered ($\bar{y}_i \to -\bar{y}_i$ and $\mathbf{r}_i \to -\mathbf{r}_i$) are carried out keeping the internuclear coordinate $\mathbf{R}$ fixed. We now turn to the question of the behaviour of the electronic wave function of a homonuclear diatomic molecule if the two identical nuclei are interchanged, so that $\mathbf{R} \to -\mathbf{R}$ and

$$\Phi_s(\mathbf{R}; \mathbf{r}_1, \mathbf{r}_2, \ldots, \mathbf{r}_N) \to \Phi_s(-\mathbf{R}; \mathbf{r}_1, \mathbf{r}_2, \ldots, \mathbf{r}_N) \tag{10.28}$$

This operation can be achieved by first rotating the molecule as a whole through 180° about the $\bar{Y}$ axis, followed by a reflection in the $(\bar{X}\bar{Z})$ plane ($\bar{y}_i \to -\bar{y}_i$) and a reflection $\mathbf{r}_i \to -\mathbf{r}_i$ about the centre of symmetry. The rotation cannot alter the electronic part of the wave function because this is only a function of the *relative* position of the electrons and the nuclei. The net effect of the two reflections is to change the sign of the wave functions for the Σ_u^+ and Σ_g^- states, while leaving the sign of the wave functions for the states Σ_g^+ and Σ_u^- unaltered.

Spin

Let us denote by $\mathbf{S}$ the resultant of the individual electron spins, and as usual $S(S+1)\hbar^2$ are the eigenvalues of $\mathbf{S}^2$. As in the case of atoms, each electronic term of the molecule is also characterised by the value of S. Provided fine structure effects (spin–orbit coupling) are neglected, there is a degeneracy of order $2S+1$ associated with S. Just as for atoms, the quantity $2S+1$ is called the *multiplicity* of the term and is written as a (left) superscript, so that the term symbol is written $^{2S+1}\Lambda$, with the code letters for $\Lambda = 0, 1, 2, \ldots$ discussed above. For example, the symbol $^3\Delta$ denotes a term such that $\Lambda = 2$ and $S = 1$.

It is worth noting that the ground state (often labelled by the symbol X) of most diatomic molecules is such that $S = 0$ and exhibits maximum symmetry. Thus, in most cases it is a $^1\Sigma^+$ state (written as $X^1\Sigma^+$) for a heteronuclear molecule and a $^1\Sigma_g^+$ state (written as $X^1\Sigma_g^+$) for a homonuclear molecule [2]. Excited states of the same multiplicity as the ground state are usually, but not invariably, distinguished by the upper case letters A, B, C, . . . and those of different multiplicity by lower case letters a, b, c, ...

[2] Exceptions occur for the molecules O_2 and NO, for which the ground states are $X^3\Sigma_g^-$ and $X^2\Pi$, respectively.

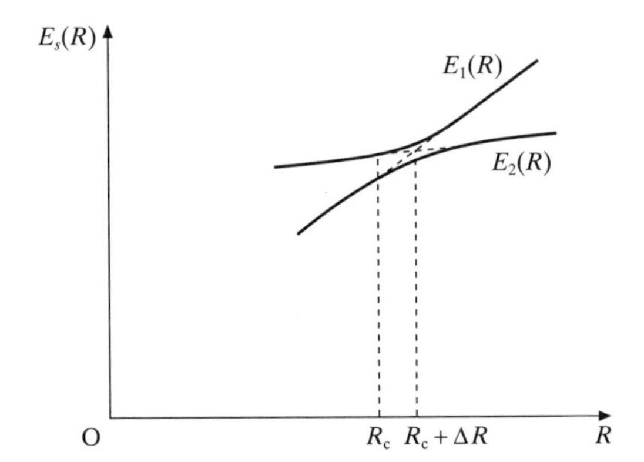

Figure 10.6 The non-crossing rule of von Neumann and Wigner. Two potential curves $E_1(R)$ and $E_2(R)$ cannot cross if the states 1 and 2 have the same symmetry.

So far we have not taken into account the spin of the nuclei. Hyperfine structure interactions due to the coupling between the nuclear spins and the orbital motion and spin of the electrons have very little effect on the molecular energies. However, *symmetry* effects related to the spin of the nuclei have an important influence on the structure of homonuclear molecules, as we shall see in detail in Section 11.5.

Intersection of potential curves and the von Neumann–Wigner non-crossing rule

The electronic terms or potential curves $E_s(R)$ of a diatomic molecule depend only on the internuclear distance R, and it is important to investigate the behaviour of these potential curves as R varies. We shall analyse below a few low-lying potential curves of simple molecular systems such as H_2^+ and H_2. Before we do this, however, we shall consider the important question of the intersection of two potential curves.

Let us denote by $E_1(R)$ and $E_2(R)$ two different electronic potential curves, and suppose that at the internuclear distance R_c the values $E_1(R_c)$ and $E_2(R_c)$ are *close*, but distinct (see Fig. 10.6). The energies $E_1^{(0)} = E_1(R_c)$ and $E_2^{(0)} = E_2(R_c)$ are eigenvalues of the Hamiltonian $H_0 \equiv H_e(R_c)$ describing the motion of the electrons in the field of the two nuclei located a distance R_c apart. The corresponding orthonormal electronic eigenfunctions will be denoted by $\Phi_1^{(0)}$ and $\Phi_2^{(0)}$ and are assumed to be real.

Let us see if the two potential curves $E_1(R)$ and $E_2(R)$ can be made to intersect by modifying the internuclear distance from R_c to $R_c + \Delta R$. The Hamiltonian now becomes $H_e(R_c + \Delta R) = H_0 + H'$, where $H' = \Delta R\, \partial H_0/\partial R_c$ is a small perturbation. Setting $H'_{ij} = \langle \Phi_i^{(0)} | H' | \Phi_j^{(0)} \rangle$, $i, j = 1, 2$, and referring to the equation (2.329) we deduce that in order for $E_1(R)$ and $E_2(R)$ to be equal at the point $R = R_c + \Delta R$ we must require that the two equations

$$E_1^{(0)} - E_2^{(0)} + H'_{11} - H'_{22} = 0 \qquad \textbf{(10.29a)}$$

and

$$H'_{12} = 0 \tag{10.29b}$$

be simultaneously satisfied. Remembering that we only have *one* parameter (ΔR) at our disposal, we must distinguish two cases:

1. The matrix element H'_{12} vanishes identically. It is then possible for the crossing to occur if, for a certain value of ΔR (that is, of R), equation (10.29a) is satisfied. Remembering that the operator H' or (H_e) commutes with the symmetry operators of the molecule, we see that this case will happen if the two functions $\Phi_1^{(0)}$ and $\Phi_2^{(0)}$ have *different symmetries* (for example, if they correspond to two electronic terms having different values of Λ, different parities g and u, different multiplicities, or to terms Σ^+ and Σ^-).
2. If $\Phi_1^{(0)}$ and $\Phi_2^{(0)}$ have the *same symmetry*, then H'_{12} will in general be non-zero. Except for accidental crossing which would occur if, by coincidence, the two equations (10.29) were satisfied at the same value of R, it is generally impossible to find a single value of ΔR (that is, of R) for which the two conditions are satisfied simultaneously. Thus we conclude that in general *two electronic curves belonging to the same symmetry species cannot cross*. This result, known as the non-crossing rule of von Neumann and Wigner, is illustrated in Fig. 10.6 [3].

The hydrogen molecular ion

The simplest of all molecules is the hydrogen molecular ion, H_2^+, which is composed of two protons and one electron. Using atomic units, the electronic Hamiltonian is given in this case by

$$H_e = -\frac{1}{2}\nabla_{\mathbf{r}}^2 - \frac{1}{r_A} - \frac{1}{r_B} + \frac{1}{R} \tag{10.30}$$

where R is the internuclear separation, and $\mathbf{r}_A$, $\mathbf{r}_B$ and $\mathbf{r}$ are the position vectors of the electron with respect to the protons A and B, and to the midpoint of the internuclear line, respectively (see Fig. 10.7). We note that the vectors $\mathbf{r}_A$, $\mathbf{r}_B$ and $\mathbf{r}$ are not independent since $\mathbf{r}_A = \mathbf{r} - \mathbf{R}/2$ and $\mathbf{r}_B = \mathbf{r} + \mathbf{R}/2$. The Schrödinger electronic wave equation reads

$$\left(-\frac{1}{2}\nabla_{\mathbf{r}}^2 - \frac{1}{r_A} - \frac{1}{r_B} + \frac{1}{R} - E_s\right)\Phi_s = 0 \tag{10.31}$$

This Schrödinger electronic equation for the hydrogen molecular ion is sufficiently simple so that it is possible to obtain numerical solutions to any desired degree of accuracy, as we shall see below. This is important because it provides

[3] It should be noted that the proof of the von Neumann–Wigner non-crossing rule assumes that the Born–Oppenheimer approximation can be made. If this is not the case the non-crossing rule may break down, the concept of potential energy curves becoming meaningless.

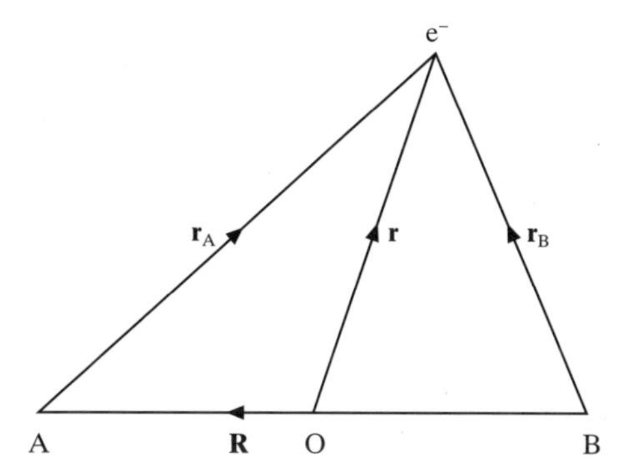

Figure 10.7 A coordinate system for the hydrogen molecular ion H_2^+.

a test of approximate methods, which have to be used for more complicated systems. However, we will first develop the approximate method of the linear combination of atomic orbitals (LCAO) because it gives considerable physical insight into the nature of the solution.

When the two protons A and B are far apart, the electron must be attached to one of them. If the system is in the ground state and the electron is attached to proton A, we might expect the spatial wave function to be (we drop the label s)

$$\Phi(\mathbf{R}; \mathbf{r}) = \psi_{1s}(r_A) \tag{10.32}$$

where $\psi_{1s}(r) = \pi^{-1/2} \exp(-r)$ is the normalised ground state wave function for atomic hydrogen. Clearly such a wave function does not have the required symmetry about the midpoint of the internuclear line, but we can construct the linear combinations

$$\Phi_g(\mathbf{R}; \mathbf{r}) = \frac{1}{\sqrt{2}} [\psi_{1s}(r_A) + \psi_{1s}(r_B)] \tag{10.33a}$$

and

$$\Phi_u(\mathbf{R}; \mathbf{r}) = \frac{1}{\sqrt{2}} [\psi_{1s}(r_A) - \psi_{1s}(r_B)] \tag{10.33b}$$

which are even and odd under reflection in the midpoint of AB, respectively. The first one is therefore a σ_g wave function and the second one is a σ_u wave function. Although these functions are only expected to be accurate in the asymptotic region of large R, we can use them as trial functions in the variational expression (2.383), namely

$$E_{g,u}(R) = \frac{\int \Phi^*_{g,u} H_e \Phi_{g,u} \, d\mathbf{r}}{\int |\Phi_{g,u}|^2 \, d\mathbf{r}} \tag{10.34}$$

Let us first work out the denominator of this expression:

$$D = \int |\Phi_{g,u}|^2 \, d\mathbf{r}$$

$$= \frac{1}{2}\int [|\psi_{1s}(r_A)|^2 + |\psi_{1s}(r_B)|^2 \pm 2\psi_{1s}(r_A)\psi_{1s}(r_B)] \, d\mathbf{r} \quad \textbf{(10.35)}$$

Since $\mathbf{R}$ is fixed, $\int d\mathbf{r} = \int d\mathbf{r}_A = \int d\mathbf{r}_B$ and, because $\psi_{1s}(r)$ is normalised to unity, we have

$$D = 1 \pm I(R) \quad \textbf{(10.36)}$$

where $I(R)$ is the overlap integral

$$I(R) = \int \psi_{1s}(r_A)\psi_{1s}(r_B) \, d\mathbf{r} \quad \textbf{(10.37)}$$

This expression can be evaluated by using the two-centre integrals given in Appendix 10, with the result

$$I(R) = \left(1 + R + \frac{1}{3}R^2\right) \exp(-R) \quad \textbf{(10.38)}$$

The numerator of (10.34) is equal to

$$N = H_{AA} \pm H_{AB} \quad \textbf{(10.39)}$$

where

$$H_{AA} = \int \psi_{1s}(r_A) H_e \psi_{1s}(r_A) \, d\mathbf{r}_A \quad \textbf{(10.40a)}$$

and

$$H_{AB} = \int \psi_{1s}(r_A) H_e \psi_{1s}(r_B) \, d\mathbf{r}_B \quad \textbf{(10.40b)}$$

Using the Schrödinger equation satisfied by $\psi_{1s}(r)$, which is

$$\left(-\frac{1}{2}\nabla_{\mathbf{r}}^2 - \frac{1}{r} - E_{1s}\right)\psi_{1s}(r) = 0 \quad \textbf{(10.41)}$$

where $E_{1s} = -0.5$ a.u. is the ground state energy of atomic hydrogen, and making use of the two-centre integrals of Appendix 10, we find that

$$H_{AA} = E_{1s} + \frac{1}{R}(1 + R) \exp(-2R) \quad \textbf{(10.42a)}$$

and

$$H_{AB} = \left(E_{1s} + \frac{1}{R}\right) I(R) - (1 + R) \exp(-R) \quad \textbf{(10.42b)}$$

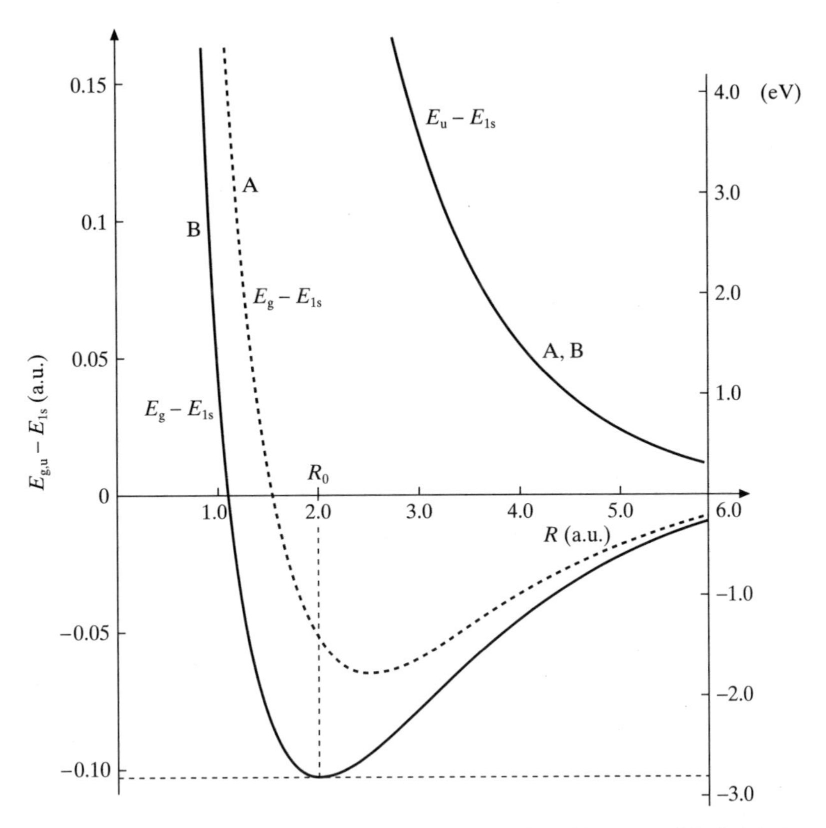

Figure 10.8 The lowest electronic potential energy curves of H_2^+. The curves labelled A show $E_g(R) - E_{1s}$ and $E_u(R) - E_{1s}$ calculated using the simple LCAO wave functions of (10.33) while the curves labelled B show the exact values of the same quantities calculated from equations (10.49) and (10.50).

Putting the various terms together, the expression (10.34) for the energies $E_g(R)$ and $E_u(R)$ becomes

$$E_{g,u}(R) = E_{1s} + \frac{1}{R}\frac{(1+R)\exp(-2R) \pm (1-2R^2/3)\exp(-R)}{1 \pm (1+R+R^2/3)\exp(-R)} \qquad \textbf{(10.43)}$$

The functions $E_g(R) - E_{1s}$ and $E_u(R) - E_{1s}$ are plotted in Fig. 10.8. The function $E_g(R)$ corresponding to the symmetrical (gerade) wave function exhibits a minimum at $R_0 = 2.49$ a.u. (1.32 Å) and it is found that the electronic dissociation energy $D_e = E_{1s} - E_g(R_0) = 0.065$ a.u. $= 1.77$ eV. As a result, this curve represents an attraction leading to the formation of a stable molecular ion. The corresponding molecular orbital Φ_g given by (10.33a) is said to be a *bonding* molecular orbital. In the present case it is an approximation to the ground state (that is, the lowest σ_g state) and is designated $\sigma_g 1s$.

In contrast, we see from Fig. 10.8 that the function $E_u(R)$ has no minimum and is repulsive at all distances; a molecular ion H_2^+ in this state will immediately dissociate into a proton and a hydrogen atom in the 1s state. The corresponding molecular orbital Φ_u (see (10.33b)) is called an *antibonding* orbital, and is denoted in the present case by σ_u^*1s, where the superscript (*) indicates that we are dealing with an antibonding orbital.

It is worth noting that the term H_{AB}, in which the matrix element of the Hamiltonian H_e between atomic orbitals centred on the protons A and B is evaluated, plays a vital role in obtaining bonding for the $\sigma_g 1s$ state. If this term and the overlap integral $I(R)$ were ignored, the two functions $E_g(R)$ and $E_u(R)$ would coincide, the resulting curve being repulsive at all distances.

The electron probability density in the states Φ_g and Φ_u is given by $|\Phi_g|^2$ and $|\Phi_u|^2$, respectively, so that the corresponding charge densities are $\rho_g = -e|\Phi_g|^2$ and $\rho_u = -e|\Phi_u|^2$ (or $\rho_g = -|\Phi_g|^2$ and $\rho_u = -|\Phi_u|^2$ in atomic units). If the charge density ρ_g is evaluated at points between A and B along the internuclear line, it is found to be greater (in absolute value) than the sum of the densities due to two isolated H atoms with their protons placed at A and B, normalised so that half an electron is associated with each proton. It is this *excess of negative charge* between the protons which causes the binding (or bonding). On the other hand, if the charge density ρ_u corresponding to the antibonding case is evaluated, a *deficiency of negative charge* is found between the protons. This is clearly seen in Fig. 10.9

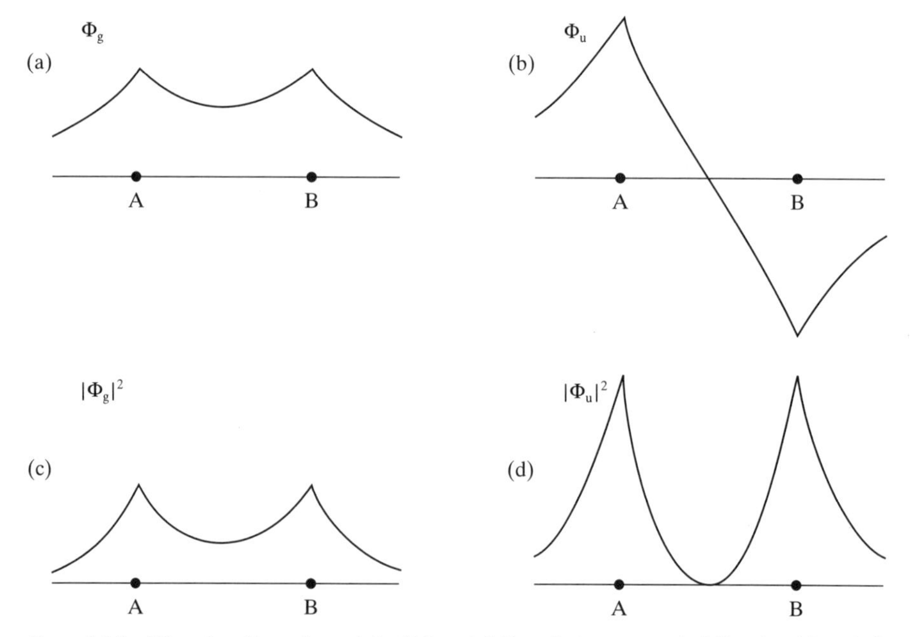

Figure 10.9 Wave functions Φ_g and Φ_u ((a) and (b)) and electron probability densities $|\Phi_g|^2$, $|\Phi_u|^2$ ((c) and (d)) for the hydrogen molecular ion H_2^+, plotted along the internuclear line to an arbitrary scale. The points A and B represent the positions of the two protons.

where the wave functions Φ_g and Φ_u as well as the absolute value (in a.u.) of the charge densities ρ_g and ρ_u are plotted along the internuclear line.

The *exact* binding energy of H_2^+ is a little greater than the result obtained above, with $D_e = E_{1s} - E_g(R_0) = 0.103$ a.u. $= 2.79$ eV, and the true equilibrium distance is $R_0 = 1.06$ Å. The principal failure of the approximate wave function $\Phi_g(\mathbf{R}; \mathbf{r})$ given by (10.33a) is that at small separations Φ_g should approach the wave function of $He^+(1s)$, the ground state of the positive helium ion with nuclear charge $Z = 2$, and in the approximation (10.33a) it does not. This defect can be remedied by using orbitals of variable charge, such as

$$u_{1s}(Z^*, r) = \left(\frac{Z^{*3}}{\pi}\right)^{1/2} \exp(-Z^* r) \tag{10.44}$$

where Z^* is a function of R, and by determining Z^* at each R with the help of the Rayleigh–Ritz variational method (see Chapter 2). We also note that at large R a dipole moment is induced in a hydrogen atom by the electrostatic field of a proton. This gives rise to an attractive *polarisation potential*, proportional to the static dipole polarisability of the atom, and falling off like R^{-4} for large R [4]. With the approximate wave functions discussed above this effect is not taken into account, and the energy $E_{g,u}(R) - E_{1s}$ decreases exponentially (see (10.43)). This feature is not important for bound-state calculations, but it plays an important role in elastic ion–atom scattering.

The exact solution

As we pointed out above, it is also possible to obtain accurate numerical solutions of the Schrödinger electronic equation (10.31) for the hydrogen molecular ion. The equation is first written in terms of confocal elliptic coordinates (ξ, η, ϕ), where

$$\begin{aligned} \xi &= \frac{1}{R}(r_A + r_B), \qquad 1 \leqslant \xi \leqslant \infty \\ \eta &= \frac{1}{R}(r_A - r_B), \qquad -1 \leqslant \eta \leqslant 1 \end{aligned} \tag{10.45}$$

and ϕ is the azimuthal angle with respect to the internuclear line, chosen as the Z axis. Using the fact that in these new coordinates the Laplacian operator ∇^2 is

$$\nabla^2 = \frac{4}{R^2(\xi^2 - \eta^2)} \left\{ \frac{\partial}{\partial \xi}\left[(\xi^2 - 1)\frac{\partial}{\partial \xi}\right] + \frac{\partial}{\partial \eta}\left[(1 - \eta^2)\frac{\partial}{\partial \eta}\right] + \frac{\xi^2 - \eta^2}{(\xi^2 - 1)(1 - \eta^2)}\frac{\partial^2}{\partial \phi^2} \right\} \tag{10.46}$$

[4] The polarisation potential will be discussed in Chapters 12, 13 and 14.

the Schrödinger electronic equation (10.31) becomes

$$\frac{\partial}{\partial\xi}\left[(\xi^2-1)\frac{\partial\Phi_s}{\partial\xi}\right]+\frac{\partial}{\partial\eta}\left[(1-\eta^2)\frac{\partial\Phi_s}{\partial\eta}\right]+\left(\frac{1}{\xi^2-1}+\frac{1}{1-\eta^2}\right)\frac{\partial^2\Phi_s}{\partial\phi^2}$$

$$+2R^2\left[\frac{1}{4}\left(E_s-\frac{1}{R}\right)(\xi^2-\eta^2)+\frac{1}{R}\xi\right]\Phi_s=0 \qquad \textbf{(10.47)}$$

An eigenfunction solution of this equation can be found in the form of the product

$$\Phi_s(\xi,\eta,\phi)=F(\xi)G(\eta)e^{im\phi}, \qquad m=0,\pm1,\pm2,\dots \qquad \textbf{(10.48)}$$

where the functions $F(\xi)$ and $G(\eta)$ are the normalisable solutions of the equations

$$\frac{d}{d\xi}\left[(\xi^2-1)\frac{dF}{d\xi}\right]+\left[\frac{R^2}{2}\left(E_s-\frac{1}{R}\right)\xi^2+2R\xi-\frac{m^2}{\xi^2-1}+\mu\right]F(\xi)=0 \qquad \textbf{(10.49)}$$

and

$$\frac{d}{d\eta}\left[(1-\eta^2)\frac{dG}{d\eta}\right]-\left[\frac{R^2}{2}\left(E_s-\frac{1}{R}\right)\eta^2+\frac{m^2}{1-\eta^2}+\mu\right]G(\eta)=0 \qquad \textbf{(10.50)}$$

μ being a separation constant. Each electronic term is therefore characterised by three quantum numbers, namely $\lambda=|m|$ and the quantum numbers n_ξ and n_η, giving respectively the number of zeros of the functions $F(\xi)$ and $G(\eta)$. As a result of this complete separation of variables, it is worth noting that two potential curves corresponding to the same λ but different values of the couple (n_ξ, n_η) are allowed to cross.

The equations (10.49) and (10.50) can be solved numerically for the ground state and excited states. This was done in particular in 1953 by D.R. Bates, K. Ledsham and A.L. Stewart. The accurate values of the quantities $E_g - E_{1s}$ and $E_u - E_{1s}$ obtained in this way for the states $\sigma_g 1s$ and $\sigma_u^* 1s$ are shown in Fig. 10.8. A similar treatment can be applied to any one-electron diatomic molecular ion (containing two nuclei A and B and one electron), whether homonuclear or heteronuclear.

Molecular hydrogen H_2

In much the same way as atomic wave functions are built in the Hartree–Fock method from one-electron atomic orbitals, electronic wave functions for molecular systems containing several electrons can be constructed from one-electron molecular orbitals. This approach is known as the Hund–Mulliken or *molecular orbital* (MO) method. It will be illustrated below for several diatomic molecules, starting with the neutral hydrogen molecule H_2, where the two-electron wave function will be built by using the one-electron molecular orbitals obtained for H_2^+. As in the case of the helium atom discussed in Chapter 7, the character of the two-electron wave function depends on the total electronic spin quantum

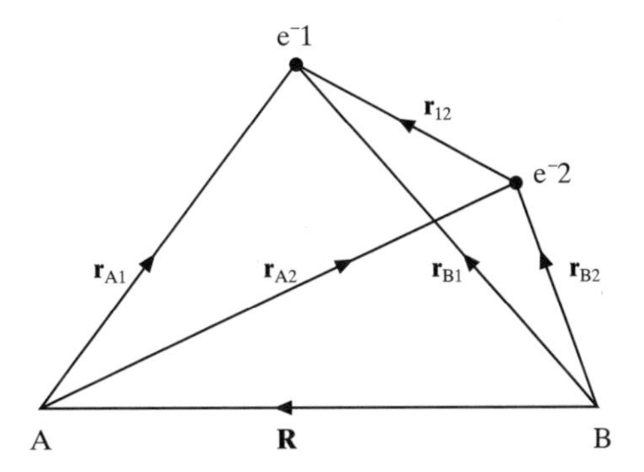

Figure 10.10 A coordinate system for the hydrogen molecule H_2.

number S. In fact, as we shall see shortly, the spin plays a key role as to whether bonding or antibonding takes place.

Molecular orbital treatment

Labelling the two electrons in H_2 as 1 and 2 and using the coordinate system of Fig. 10.10, we shall now build the lowest states of H_2 from the H_2^+ orbitals Φ_g and Φ_u discussed above. As shown in Chapter 7 (see (7.18)–(7.19)) the spin wave functions for the two-electron system must be either singlet ($S = 0$),

$$\chi_{0,0}(1, 2) = \frac{1}{\sqrt{2}}[\alpha(1)\beta(2) - \beta(1)\alpha(2)] \qquad \textbf{(10.51)}$$

or triplet ($S = 1$),

$$\begin{aligned} \chi_{1,1}(1, 2) &= \alpha(1)\alpha(2) \\ \chi_{1,0}(1, 2) &= \frac{1}{\sqrt{2}}[\alpha(1)\beta(2) + \beta(1)\alpha(2)] \\ \chi_{1,-1}(1, 2) &= \beta(1)\beta(2) \end{aligned} \qquad \textbf{(10.52)}$$

Since the total wave function must be antisymmetric in the coordinates (spatial and spin) of the two electrons, the spatial wave function corresponding to the singlet case must be symmetric, while the spatial wave function for the triplet case must be antisymmetric. Four combinations can be formed, namely

$$\Phi_A(1, 2) = \Phi_g(1)\Phi_g(2)\chi_{0,0}(1, 2) \qquad \textbf{(10.53a)}$$

$$\Phi_B(1, 2) = \Phi_u(1)\Phi_u(2)\chi_{0,0}(1, 2) \qquad \textbf{(10.53b)}$$

$$\Phi_C(1, 2) = \frac{1}{\sqrt{2}}[\Phi_g(1)\Phi_u(2) + \Phi_g(2)\Phi_u(1)]\chi_{0,0}(1, 2) \qquad \textbf{(10.53c)}$$

$$\Phi_D(1, 2) = \frac{1}{\sqrt{2}}[\Phi_g(1)\Phi_u(2) - \Phi_g(2)\Phi_u(1)]\chi_{1,M_S}(1, 2) \qquad M_S = 0, \pm 1 \qquad \textbf{(10.53d)}$$

Both Φ_A and Φ_B represent ${}^1\Sigma_g^+$ states, while Φ_C corresponds to a ${}^1\Sigma_u^+$ state and Φ_D to ${}^3\Sigma_u^+$ states.

The wave function Φ_A describes two electrons having opposite spins each occupying the bonding orbital Φ_g. By analogy with the He atom, we expect this wave function to yield the lowest energy of the four combinations, and to be therefore an approximation to the *ground state* wave function of H_2. In order to study this problem in more detail, we start from the electronic wave equation for the two electrons in the field of the two (fixed) protons, namely

$$(H_e - E_s)\Phi_s = 0 \tag{10.54}$$

where the electronic Hamiltonian H_e is given in atomic units by

$$H_e = -\frac{1}{2}\nabla_{\mathbf{r}_1}^2 - \frac{1}{2}\nabla_{\mathbf{r}_2}^2 - \frac{1}{r_{A1}} - \frac{1}{r_{A2}} - \frac{1}{r_{B1}} - \frac{1}{r_{B2}} + \frac{1}{r_{12}} + \frac{1}{R} \tag{10.55}$$

and may be written as

$$H_e = H_0(1) + H_0(2) + \frac{1}{r_{12}} + \frac{1}{R} \tag{10.56}$$

with

$$H_0(i) = -\frac{1}{2}\nabla_{\mathbf{r}_i}^2 - \frac{1}{r_{Ai}} - \frac{1}{r_{Bi}}, \qquad i = 1, 2 \tag{10.57}$$

The exact one-electron molecular orbitals satisfy the equation

$$H_0(i)\,\Phi_{g,u} = \left(E_{g,u} - \frac{1}{R}\right)\Phi_{g,u} \tag{10.58}$$

and we shall normalise them to unity (for all R), namely

$$\int |\Phi_{g,u}|^2\, d\mathbf{r} = 1 \tag{10.59}$$

The lowest energy, obtained by using the wave function Φ_A (given by (10.53a)) in the Rayleigh–Ritz variational expression (2.383), is then

$$\begin{aligned} E_A &= \int \Phi_A^* H_e \Phi_A \, d\mathbf{r}_1\, d\mathbf{r}_2 \\ &= 2E_g(R) - \frac{1}{R} + \int d\mathbf{r}_1\, d\mathbf{r}_2 \frac{|\Phi_g(1)\Phi_g(2)|^2}{r_{12}} \end{aligned} \tag{10.60}$$

The 'exact' molecular orbitals Φ_g (or elaborate variational approximations of them) can be used in this expression but it is instructive to use the simple LCAO form (10.33a). In this case the equilibrium distance for H_2 can be computed to be R_0 = 1.5 a.u. (0.8 Å), and the corresponding electronic dissociation energy

$D_e = 2E_{1s} - E_A(R_0)$ is found to be 0.098 a.u. (2.68 eV). The experimental values are $R_0 = 1.4$ a.u. (0.74 Å) and $D_e = 0.175$ a.u. (4.75 eV).

Using (10.33a) and writing out the approximate wave function (10.53a) in full, we have

$$\Phi_A = \frac{1}{2}[\psi_{1s}(r_{A1})\psi_{1s}(r_{B2}) + \psi_{1s}(r_{A2})\psi_{1s}(r_{B1}) + \psi_{1s}(r_{A1})\psi_{1s}(r_{A2}) + \psi_{1s}(r_{B1})\psi_{1s}(r_{B2})]\chi_{0,0}(1, 2) \quad \textbf{(10.61)}$$

or

$$\Phi_A = \Phi_A^{\text{cov}} + \Phi_A^{\text{ion}} \quad \textbf{(10.62)}$$

where

$$\Phi_A^{\text{cov}} = \frac{1}{2}[\psi_{1s}(r_{A1})\psi_{1s}(r_{B2}) + \psi_{1s}(r_{A2})\psi_{1s}(r_{B1})]\chi_{0,0}(1, 2) \quad \textbf{(10.63)}$$

and

$$\Phi_A^{\text{ion}} = \frac{1}{2}[\psi_{1s}(r_{A1})\psi_{1s}(r_{A2}) + \psi_{1s}(r_{B1})\psi_{1s}(r_{B2})]\chi_{0,0}(1, 2) \quad \textbf{(10.64)}$$

The wave function Φ_A^{cov} represents the situation in which one electron is associated with each nucleus. In the separated atom limit, this wave function yields two isolated neutral hydrogen atoms in the ground state: H(1s) + H(1s). The type of bonding associated with this wave function is called *covalent bonding* and Φ_A^{cov} is said to be the covalent part of Φ_A. On the other hand, the wave function Φ_A^{ion} corresponds to the situation in which both electrons are attached to one nucleus. In the separated atom limit, this wave function represents a proton and a negative hydrogen ion H^-. Wave functions like Φ_A^{ion} which represent an unequal division of charge between the nuclei are said to be of ionic character and the type of bonding corresponding to such wave functions is called *ionic bonding*. It is worth noting that each term of the wave function (10.64) yields a very poor representation of the bound state of H^- [5] and in any case the probability of this configuration of the system is not large. For this reason, the wave function Φ_A given by (10.61) is expected to be a poor approximation for large internuclear separations R.

A better approximation to the electronic ground state wave functions of H_2 can be obtained by using in the Rayleigh–Ritz variational principle a trial function Φ_T which is a mixture of Φ_A (see (10.53a)) and Φ_B (see (10.53b)), both of which have the correct symmetry $^1\Sigma_g^+$. That is,

$$\Phi_T = \Phi_A + \lambda\Phi_B \quad \textbf{(10.65)}$$

[5] See our discussion of H^- in Chapter 7, Section 7.5.

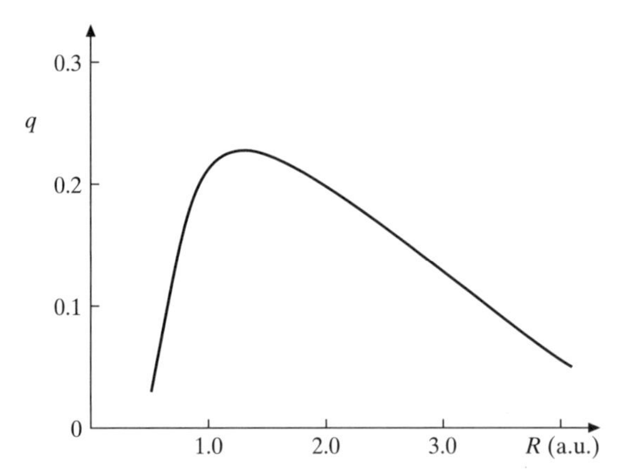

Figure 10.11 The ratio q of the ionic to the covalent part of the wave function (10.67) for the hydrogen molecule H_2, as a function of nuclear separation R.

The parameter λ can be determined by the variational method. We first obtain the energy as a function of λ (for a fixed value of R):

$$E(\lambda) = \frac{\int d\mathbf{r}_1\, d\mathbf{r}_2 \Phi_T^* H_e \Phi_T}{\int d\mathbf{r}_1\, d\mathbf{r}_2 \Phi_T^* \Phi_T} \quad \textbf{(10.66)}$$

and then require that $\partial E/\partial\lambda = 0$. The equilibrium distance is found to be $R_0 = 1.42$ a.u. (0.749 Å) and the electronic dissociation energy $D_e = 2E_{1s} - E(R_0)$ is given by 0.147 a.u. (4.00 eV), a considerable improvement over our former result. It is worth noting that the wave function Φ_T can be written in the form of a linear combination of covalent and ionic terms, namely

$$\Phi_T = (1 - \lambda)\Phi_A^{cov} + (1 + \lambda)\Phi_A^{ion} \quad \textbf{(10.67)}$$

The ratio of the ionic to the covalent part of the wave function is given by $q = (1 + \lambda)/(1 - \lambda)$, and is displayed in Fig. 10.11 as a function of R. We see from this figure that the maximum value of q is about 0.2 at $R_0 \simeq 1.5$ a.u. (0.8 Å) and that $q \to 0$ as $R \to \infty$.

We also note that the approximate H_2 wave functions discussed above do not take into account the existence of an attractive long-range van der Waals interaction between two neutral atoms, which falls off like R^{-6} [6].

The Heitler–London or valence bond method

An alternative approach, due to Heitler and London, is to approximate the wave function for H_2 by using orbitals based on the separated atom wave functions, just

[6] See Bransden and Joachain (2000).

as we did for H_2^+. In this approximation the trial wave function to be used in the Rayleigh–Ritz variational principle for the singlet case is Φ_A^{cov} (the covalent part of Φ_A) which is given by (10.63). The corresponding triplet wave function is

$$\Phi_D^{cov} = \frac{1}{2}[\psi_{1s}(r_{A1})\psi_{1s}(r_{B2}) - \psi_{1s}(r_{A2})\psi_{1s}(r_{B1})]\chi_{1,M_S}, \qquad M_S = 0, \pm 1 \tag{10.68}$$

Substituting Φ_A^{cov} (which corresponds to the *gerade* state $^1\Sigma_g^+$) and Φ_D^{cov} (which corresponds to the *ungerade* state $^3\Sigma_u^+$) in the Rayleigh–Ritz variational principle (2.383), we find that the corresponding gerade (g) and ungerade (u) expressions of the energy are given by

$$E_{g,u} = 2E_{1s} + \frac{J}{1 \pm I^2} \pm \frac{K}{1 \pm I^2} + \frac{1}{R} \tag{10.69}$$

where the upper signs correspond to g and the lower ones to u. The quantity I is the overlap integral

$$I = \int \psi_{1s}(r_{A1})\psi_{1s}(r_{B1})\, d\mathbf{r}_1 = \int \psi_{1s}(r_{A2})\psi_{1s}(r_{B2})\, d\mathbf{r}_2 \tag{10.70}$$

while J and K are defined as

$$J = \int d\mathbf{r}_1\, d\mathbf{r}_2 |\psi_{1s}(r_{A1})|^2 |\psi_{1s}(r_{B2})|^2 \left(\frac{1}{r_{12}} - \frac{1}{r_{A2}} - \frac{1}{r_{B1}}\right) \tag{10.71}$$

and

$$K = \int d\mathbf{r}_1\, d\mathbf{r}_2 \psi_{1s}(r_{A1})\psi_{1s}(r_{B2}) \left(\frac{1}{r_{12}} - \frac{1}{r_{A2}} - \frac{1}{r_{B1}}\right) \psi_{1s}(r_{A2})\psi_{1s}(r_{B1}) \tag{10.72}$$

The expression J is known as the *Coulomb integral* and represents the interaction between the charge densities $-|\psi_{1s}(r_{A1})|^2$ and $-|\psi_{1s}(r_{B2})|^2$, while K is called the *exchange integral*. It is found that K is negative, so that the $^1\Sigma_g^+$ state is lower than the state $^3\Sigma_u^+$. With the wave function Φ_A^{cov} used for the $^1\Sigma_g^+$ state, the equilibrium distance is given by $R_0 = 1.6$ a.u. (0.87 Å) and one has $D_e = 2E_{1s} - E(R_0) = 0.115$ a.u. (3.14 eV), which is better than the result 0.098 a.u. (2.68 eV) obtained above from the simple MO wave function Φ_A. In Fig. 10.12 we show the computed energy curves for the attractive ground state $X^1\Sigma_g^+$ and the repulsive state $^3\Sigma_u^+$.

If we add to the Heitler–London wave function (10.63) an extra term of the form $\mu\Phi_A^{ion}$ to represent ionic bonding (where the parameter μ is determined from the variational principle), the resulting energy is identical with that obtained from the improved MO wave function (10.65). When applied to more complicated systems, it is generally true that the MO method, using 'mixed' wave functions of the type (10.65), provides similar results to the valence-bond method, improved with additional variationally determined terms. When unimproved, the two

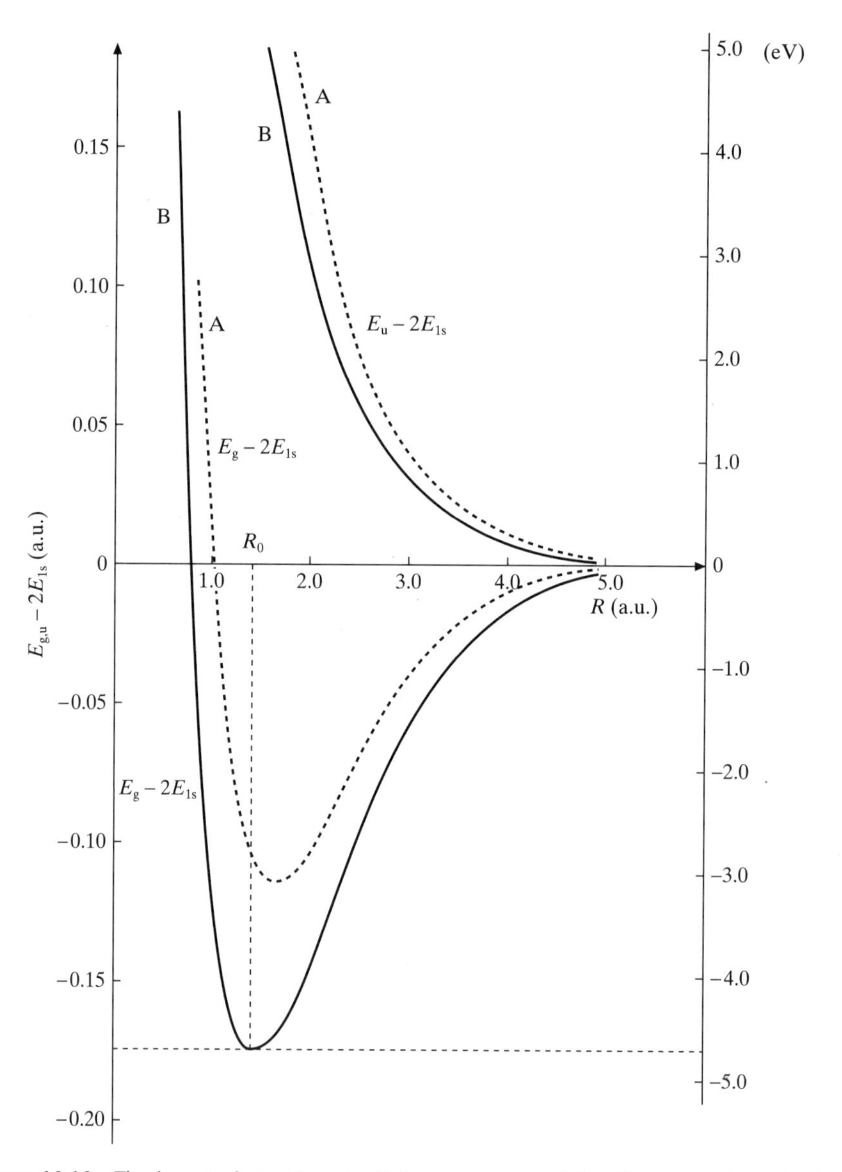

Figure 10.12 The lowest electronic potential energy curves of H_2. The curves A are in the Heitler and London approximation, and the curves B show the accurate values for the ground state X $^1\Sigma_g^+$ and the repulsive state $^3\Sigma_u^+$.

approximations are distinct, and the MO approximation is, in general, the better one. This, however, is not true for the H_2 molecule since in this case the nuclear charges ($Z_A = Z_B = 1$) are not large enough to dominate the electron–electron interaction.

The Hartree–Fock and configuration–interaction methods

Just as for atoms, the Hartree–Fock (HF) or self-consistent field (SCF) method can be applied to molecules. The Hartree–Fock electronic wave function Φ_{HF} is taken to be a single Slater determinant of electron spin-orbitals, which are determined so that the energy of the molecular state is a minimum. This approach, which neglects correlation effects between the electrons, gives for the H_2 ground state a dissociation energy $D_e = 3.6360$ eV at $R_0 = 1.4$ a.u., obtained in 1960 by W. Kolos and C.J.J. Roothaan.

The Hartree–Fock results can be improved by using configuration–interaction (CI) wave functions, which are linear combinations of Slater determinants. For example, A.D. McLean, A. Weiss and M. Yoshimine carried out in 1960 a CI calculation for the H_2 ground state with a wave function containing a linear combination of five Slater determinants. They obtained the value $D_e = 4.543\,06$ eV at $R_0 = 1.4013$ a.u., which accounts for about 90% of the correlation energy.

The HF (SCF) and CI methods have been extensively applied to a variety of molecules. In addition to these wave-function-based approximations, the density functional method offers an alternative approach to molecular structure calculations.

Improved variational treatments of the H_2 molecule

Elaborate variational calculations have been performed for the H_2 molecule, using trial wave functions which include explicitly the interelectronic coordinate r_{12}. In particular, H.M. James and A.S. Coolidge, employing confocal elliptic coordinates, found in 1933 the value $D_e = 4.72$ eV at $R_0 = 1.40$ a.u. With an extended basis set, W. Kolos and L. Wolniewicz obtained in 1968 the value $D_e = 4.747\,59$ eV at $R_0 = 1.4011$ a.u., in very good agreement with the experimental values.

To conclude our study of the electronic structure of H_2^+ and H_2, we illustrate in Fig. 10.13 the relationship between the energy levels of these two systems. From the examination of this figure it is clear that

$$I_P(H_2) + D_0(H_2^+) = I_P(H) + D_0(H_2) \qquad \textbf{(10.73)}$$

where I_P are ionisation potentials and D_0 dissociation energies.

Homonuclear diatomic molecules

We shall now discuss the electronic structure of more complicated diatomic molecules, and in particular their bonding properties. In general, to obtain a bound state of a molecule the negative charge density between the nuclei must be in excess of what would be expected if the charge distribution were due to two non-interacting atoms. The charge density depends on the overlap of the atomic wave functions centred on each nucleus, and only the wave functions of the electrons in the outer shells of the atoms will contribute to the overlap appreciably. The contributing electrons are known as *valence electrons*.

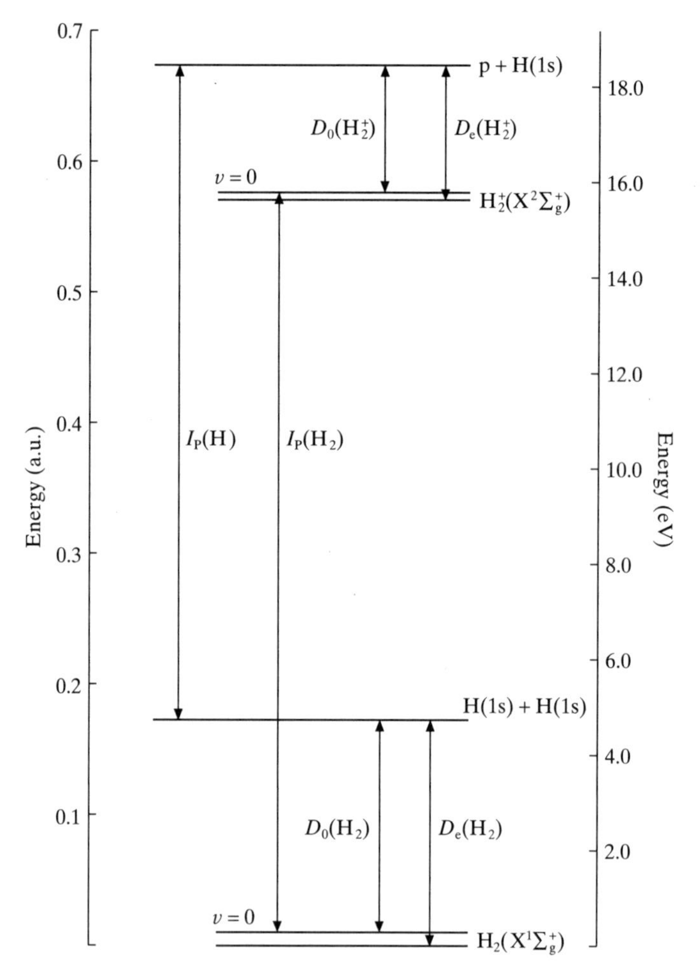

Figure 10.13 Relationships between the ground state energies of H_2, H_2^+ and H. The chemical dissociation energies and electronic dissociation energies denoted by D_0 and D_e, respectively, differ by the zero-point vibrational energy $\frac{1}{2}\hbar\omega_0$. The ionisation potentials of the hydrogen atom and hydrogen molecule are denoted by $I_P(H)$ and $I_P(H_2)$, respectively.

We begin by considering homonuclear diatomic molecules. Using the Hund–Mulliken molecular orbital approach, the full electronic wave function can be built from one-electron MOs. These in turn can be constructed in the LCAO approximation from atomic orbitals. For example, using one atomic orbital centred on each atom, the spatial MOs are given by

$$\Phi_{g,u}(i) = N_{g,u}[u_a(\mathbf{r}_{Ai}) \pm u_b(\mathbf{r}_{Bi})] \tag{10.74}$$

where $N_{g,u}$ is a normalisation factor and u_a, u_b are atomic orbitals.

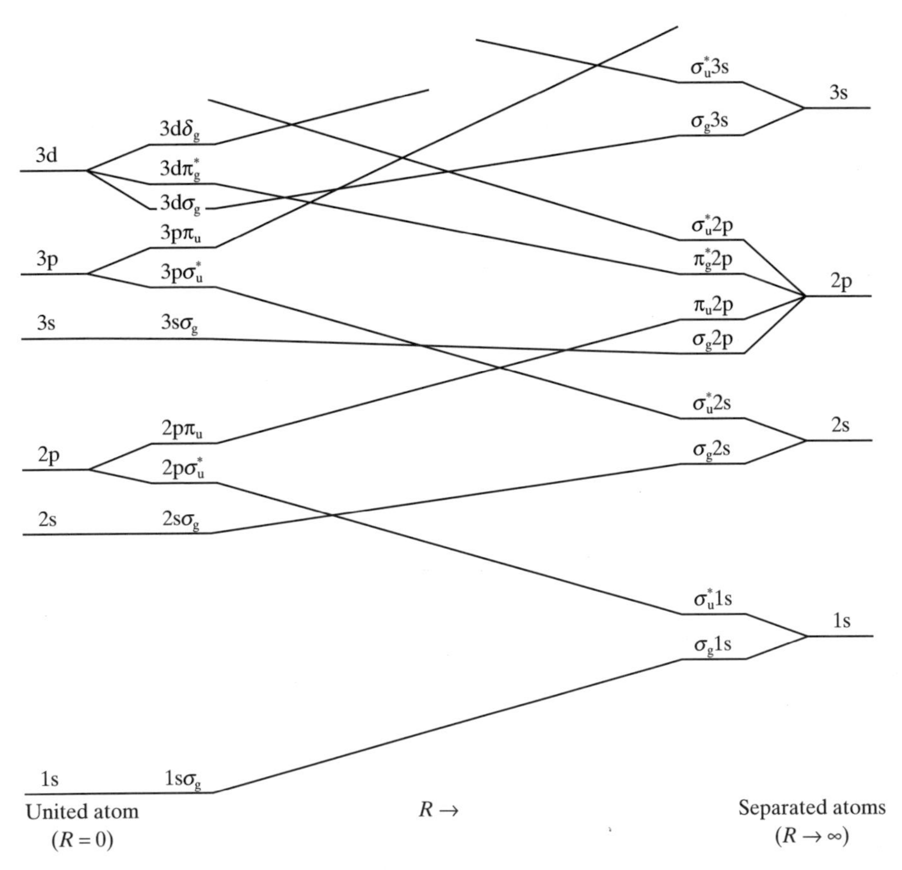

Figure 10.14 A diagram showing the correlation between united atom and separated atom states for homonuclear diatomic molecules. The diagram is not to scale and the actual energies of the united and separated atoms vary from molecule to molecule.

In order to analyse the nature of the MOs of a diatomic molecule, and in particular their behaviour when the internuclear distance varies, it is useful to draw a *correlation diagram*, which exhibits the qualitative features of the relative energies of the orbitals as a function of R. In establishing correlation diagrams it is convenient to subtract out the Coulomb repulsion between the nuclei. These diagrams therefore give both the *separated atom* limit ($R \to \infty$) and the *united atom* limit ($R \to 0$), with the intermediate region corresponding to $R \simeq R_0$, the equilibrium internuclear separation.

A simple example of a correlation diagram showing the lowest orbitals of a homonuclear diatomic molecule is displayed in Fig. 10.14. We see that this diagram represents a unique correspondence in going from the atomic orbitals of the united atom to the MOs of the molecule, and finally to the atomic orbitals of the separated atoms. The correspondence is based on the following features:

1. MOs having a given value of λ (the component of the orbital angular momentum along the internuclear axis) must connect with atomic orbitals having the same value of λ (that is, the same value of $|m|$).
2. The parity of the wave function (g or u) must be preserved as R varies from 0 to ∞.
3. The von Neumann–Wigner non-crossing rule must be obeyed, so that energy curves corresponding to orbitals having the same symmetry do not cross [7] as we let R vary from 0 to ∞.

A simple illustration of the two first conditions given above is provided by the H_2^+ problem we have already studied. We recall that in this case the exact ground state orbital Φ_g must approach for large R the orbital of atomic hydrogen in the 1s state. For small R, Φ_g must go over to the lowest state of He^+, also a 1s state. On the other hand, the exact lowest repulsive orbital Φ_u also correlates with H(1s) at large R, while as $R \to 0$ it must approach a wave function of He^+ with odd parity. The component of the angular momentum along the internuclear line (the Z axis) is $\lambda = 0$, and the lowest lying state of He^+ with odd parity and such that $\lambda = 0$ (that is, $m = 0$) is the $2p_0$ state. Thus Φ_u must go over to the $2p_0$ wave function in the united atom limit ($R \to 0$).

It should be noted that MOs can be labelled by either the separated or the united atom limits. For example, the lowest odd orbital Φ_u can be designated in the separated atom limit as σ_u^*1s (as was done above) or in the united atom limit as $2p\sigma_u^*$. We recall that the repulsive (or antibonding) orbitals are distinguished from the attractive (or bonding) orbitals by the addition of an asterisk.

From spatial MOs, $\Phi(i)$ for each electron, spin-orbitals $\Phi(i)\alpha(i)$ or $\Phi(i)\beta(i)$ can be formed, and a completely antisymmetric many-electron wave function is given by the Slater determinant of these spin-orbitals. For example, the $^1\Sigma_g^+$ wave function Φ_A for H_2, given in (10.53a), can be written in determinantal form as

$$\Phi_A = \frac{1}{\sqrt{2}} \begin{vmatrix} \Phi_g(1)\alpha(1) & \Phi_g(1)\beta(1) \\ \Phi_g(2)\alpha(2) & \Phi_g(2)\beta(2) \end{vmatrix} \qquad \textbf{(10.75)}$$

In this case, both electrons can be in the bonding orbital Φ_g which is σ1s in character. However, in the three-electron system He_2^+, the Pauli exclusion principle only allows two electrons to be in the σ1s orbital, and the third one must go into the next orbital σ*1s, so that the molecule has the configuration $(\sigma 1s)^2(\sigma^* 1s)$. The combination of one bonding and one antibonding orbital, with the same separated atom limit, leads to a small net repulsion. In the present case, with two bonding and one antibonding orbitals, the net effect is a weakly bound state.

The four-electron system He_2 must have the configuration $(\sigma 1s)^2(\sigma^* 1s)^2$ with two electrons bonding and two antibonding. The net effect is a repulsion and

[7] As we pointed out after the equation (10.50), there is an exception to this rule for one-electron systems (such as H_2^+) where a complete separation of variables can be made, so that quantum numbers other than λ are well defined for intermediate values of R.

there is no stable ground state in this approximation. However, a weakly bound ground state of the helium dimer $^4He_2$ exists, due to the long-range attraction of the van der Waals interaction, which falls off like R^{-6}. By diffracting a molecular beam consisting of small helium clusters from a 100 nm period transmission grating, J.P. Toennies and co-workers found in 2000 that the mean internuclear distance of $^4He_2$ is (52 ± 4) Å, and its binding energy is (1.1 + 0.3/– 0.2) mK ($\simeq 10^{-7}$ eV). It should be noted that excited states such as $(\sigma 1s)^2(\sigma^* 1s)(\sigma 2s)$ can exist as stable bound states.

Next in complexity is the lithium molecule Li_2. Atomic lithium has the configuration $1s^2 2s$. The two K shell electrons play a small role in the molecular structure, and the bonding is due to the 2s valence electron. The ground state has the configuration $(\sigma 2s)^2$. Continuing up the periodic table, the molecule Be_2 must be $(\sigma 2s)^2(\sigma^* 2s)^2$ and this state is not bound in this approximation. The boron atom has a configuration $1s^2 2s^2 2p$ and bonding results from a $(\sigma 2p)^2$ molecular configuration. The next case, carbon, with an atomic configuration $1s^2 2s^2 2p^2$ is interesting in that the molecule C_2 contains two separate bonding orbitals $\sigma_g 2p$ and $\pi_u 2p$.

Lastly we consider the case of the oxygen molecule O_2. Atomic oxygen has the configuration $1s^2 2s^2 2p^4$. Three of the p electrons in each atom can form the bonding orbitals $(\sigma_g 2p)^2(\pi 2p_x)^2(\pi 2p_y)^2$. The remaining pair of electrons must be associated with antibonding orbitals $\pi^* 2p$. It turns out that one electron goes into the $\pi^* 2p_x$ and one into the $\pi^* 2p_y$ orbital.

Pairing and valency

In order to form a bonding orbital, the electron from one atom must usually form a singlet spin state with the electron taken from the second atom, as in H_2. The triplet state, on the other hand, leads to a repulsion. Now consider what happens when a hydrogen atom is brought up to a helium atom. Both the electrons in the helium atom are in the 1s level, and are in a singlet state with $S = 0$, one electron with $m_s = +1/2$ and one with $m_s = -1/2$. The electron on the hydrogen atom cannot exchange with the electron of opposite spin state in the helium atom, for if this happened we would have two electrons in the same spin state in the 1s orbital, which would violate the Pauli exclusion principle. The electron in the hydrogen atom can only exchange with the electron of the same spin state in the helium atom. In this case, the corresponding spatial part of the wave function is antisymmetric and the orbital is antibonding. To see this in more detail, we can form a trial function, of the Heitler–London type, by taking two electrons to be in the 1s orbital of helium, v_{1s}, and one electron to be in the 1s orbital of atomic hydrogen, u_{1s}. The Slater determinant for the three-electron system is

$$\Phi = N \begin{vmatrix} v_{1s}(1)\alpha(1) & v_{1s}(1)\beta(1) & u_{1s}(1)\alpha(1) \\ v_{1s}(2)\alpha(2) & v_{1s}(2)\beta(2) & u_{1s}(2)\alpha(2) \\ v_{1s}(3)\alpha(3) & v_{1s}(3)\beta(3) & u_{1s}(3)\alpha(3) \end{vmatrix} \quad \textbf{(10.76)}$$

where N is a normalisation factor. Substituting (10.76) into the variational expression (2.383) of the energy, we find that

$$E(R) = J - K \tag{10.77}$$

where J is the direct integral

$$J = N^2 \int d\mathbf{r}_1\, d\mathbf{r}_2\, d\mathbf{r}_3\, v_{1s}(1)v_{1s}(2)u_{1s}(3)H_e v_{1s}(1)v_{1s}(2)u_{1s}(3) \tag{10.78}$$

and K is the exchange integral

$$K = N^2 \int d\mathbf{r}_1\, d\mathbf{r}_2\, d\mathbf{r}_3\, v_{1s}(1)v_{1s}(2)u_{1s}(3)H_e v_{1s}(3)v_{1s}(2)u_{1s}(1) \tag{10.79}$$

The only exchange is between electrons 1 and 3, which are in the same spin state, and the effect is to introduce a repulsion, so that a stable molecule of HHe does not exist. The two electrons of the helium atom are said to be *paired*. Only unpaired electrons contribute to chemical bonding, and the number of unpaired outer shell electrons is equal to the valency of the atom. As all the electrons in a closed subshell atom are paired, such atoms are chemically inert [8]. A chemical bond is formed from two unpaired electrons, one from each atom. The two bonding electrons are themselves in a singlet state, and therefore cannot form a bond with a third electron. Each bond uses up a different pair of electrons, and since each pair is in a singlet state, stable molecules generally are in states with overall spin $S = 0$, although exceptions to this rule occur, as in the case of O_2, where the two electrons in antibonding orbitals are in a relative triplet state and the total spin of O_2 is $S = 1$.

Heteronuclear diatomic molecules

The molecular orbital method can equally well be applied to heteronuclear molecules, formed from two dissimilar atoms. We now form spatial molecular orbitals by taking linear combinations of atomic orbitals, one from each atom

$$\Phi(i) = \lambda u_a(\mathbf{r}_{Ai}) + \mu v_b(\mathbf{r}_{Bi}) \tag{10.80}$$

Unlike the homonuclear case there is no reflection symmetry about the midpoint of the internuclear line, and the orbitals cannot be classified as g or u. Another difference is that the separated atom classification of the configurations is less convenient, and the orbitals can be either designated by the united atom configuration, for example $1s\sigma$, $1s\sigma^*$, $2p\pi$, etc., or just labelled 1σ, 2σ, 3σ, ... 1π, 2π, 3π, ... in order of increasing energy within the σ, π, ... levels. In general, the energies of the atomic levels must be close to each other, otherwise the overlap between the atomic wave functions is small and no bonding orbital can be formed.

[8] They can, however, form weakly bound dimers due to the attractive van der Waals interaction, as we have seen above for the case of $^4He_2$.

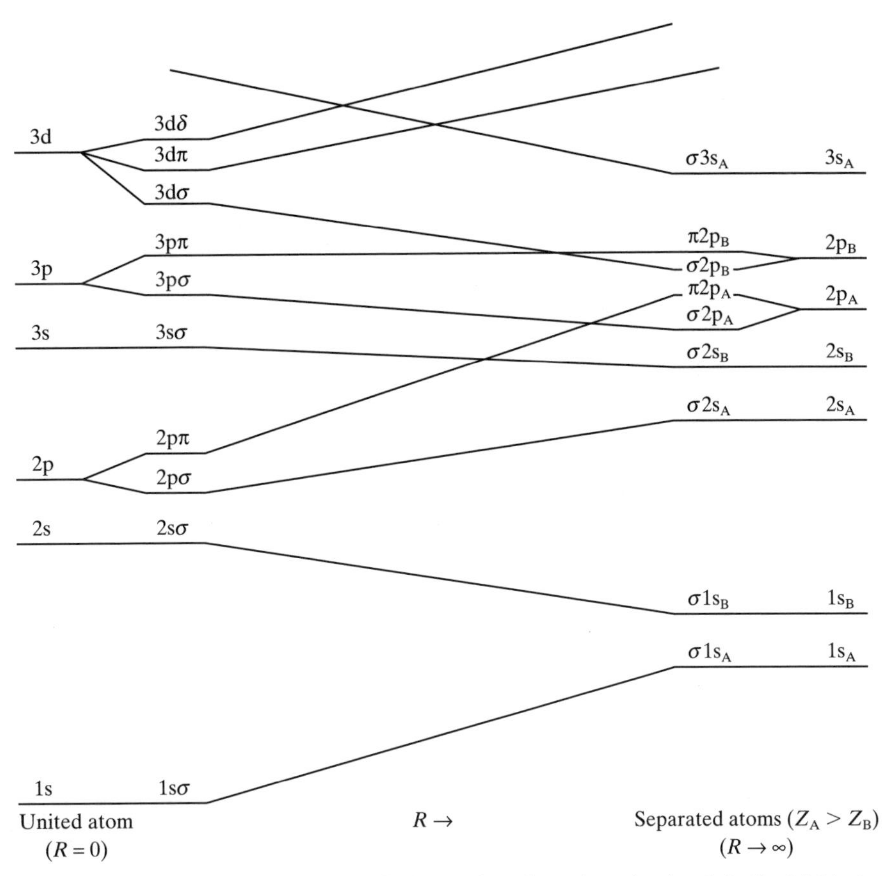

Figure 10.15 A correlation diagram for heteronuclear diatomic molecules. As in Fig. 10.14, the diagram is not to scale and the actual energies vary from molecule to molecule.

A correlation diagram for heteronuclear molecules such that the nuclear charges are not too different is shown in Fig. 10.15, while a diagram showing the energy splittings near the equilibrium distance is shown in Fig. 10.16. It should be emphasised that the very simple approximation to a molecular orbital in which only one orbital is taken from each atom (see (10.80)) cannot be expected to be very precise, and in general accurate approximations will require the combination of many atomic or other basis functions.

In order to illustrate some of these points, we shall now consider the examples of the LiH, HCl and NaCl molecules.

Lithium hydride LiH

In this system, which is one for which detailed calculations have been made, we have four electrons. An isolated lithium atom has the ground state configuration

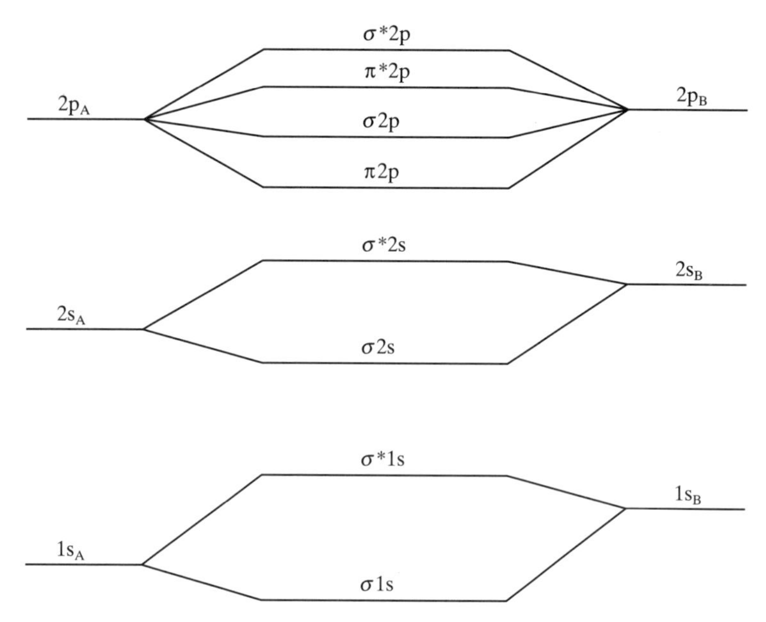

Figure 10.16 Energy levels of a diatomic molecule composed of two atoms A and B, where A is more electronegative than B. The energies of the molecular orbitals at equilibrium separation R_0 are shown in the middle with the energies of the atomic levels on either side. The diagram is not to scale.

$1s^2 2s$, and ground state atomic hydrogen is of course in the 1s state. The lowest lying molecular orbital is designated 1σ, and in accordance with the inert nature of the inner shells of atoms, it is practically identical with the 1s atomic orbital of the K shell of lithium.

It would be natural to suppose that the next highest molecular orbital would be a 2σ composed of the 2s atomic orbital of lithium and the 1s orbital of atomic hydrogen. In fact this is not quite correct, as it turns out that a lower energy is obtained from the variational method if a linear combination of the 2s and $2p_z$ atomic orbitals of lithium is used in place of just the 2s orbital. This phenomenon, which occurs when ns and np atomic orbitals are very close in energy, arises because the linear combination of a symmetrical s function $f_0(r)$ with a p function, such as $f_1(r) \cos \theta$, provides an asymmetric charge distribution (see Fig. 10.17) which leads to a large electron density in the region between the atoms. This combination of orbitals corresponding to different values of the angular momentum is called a *hybrid orbital* (an sp hybrid in the present case) and this phenomenon is known as *hybridisation*. In the separated atom limit the 2σ orbital approaches the 1s orbital of atomic hydrogen.

The ground state of the molecule, which has two electrons in each of the 1σ and 2σ orbitals, becomes $Li^+ + H^-$ in the separated atom limit, with two electrons in the 1s orbital of Li^+ and two in the 1s orbital of H^-. At the equilibrium distance,

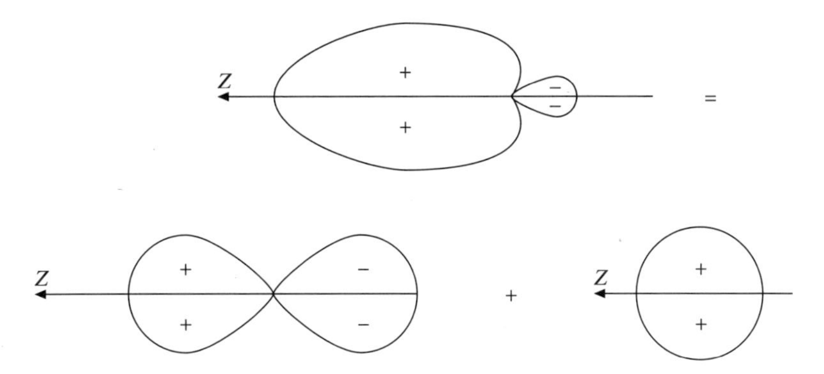

Figure 10.17 Charge distribution of a hybridised (s + p_z) orbital.

R_0 = 1.6 Å, excess *negative* charge is still associated with the proton. As a consequence, the molecule has a permanent electric dipole moment. It is interesting to remark that other hydrides such as BH, NH and HF also possess permanent electric dipole moments, but of opposite sign.

Hydrogen chloride HCl

In the chlorine atom the K and L shells are completely full and play no part in chemical activity. The valence shell with $n = 3$ has the configuration $3s^23p^5$. The 3s electrons do not mix significantly with the 1s orbital of atomic hydrogen, as the energies are not sufficiently commensurate. This leaves the $3p_x$, $3p_y$ and $3p_z$ orbitals to be considered. Of these only $3p_z$ can contribute to a σ orbital and the bonding orbital is thus a mixture of the $3p_z$ orbital of chlorine with the 1s orbital of hydrogen. This time the coefficient of the $3p_z$ atomic orbital of chlorine is larger than that of the 1s orbital of atomic hydrogen, so that the wave function represents $H^+ + Cl^-$ and the dipole moment has the polarity associated with the redistribution of charge. The bonding is *ionic* in character.

Sodium chloride NaCl

Particularly good examples of ionic bonding are given by the compounds of an alkali atom and a halogen. The alkalis (Li, Na, K, Rb, Cs, Fr) consist of a single valence electron moving outside closed (sub)shells. This electron is easily detached leaving a stable closed (sub)shell, singly charged positive ion. On the other hand, the halogens (F, Cl, Br, I) have a single vacancy, or hole, in an otherwise closed (sub)shell. These atoms are strongly electronegative, and readily combine with an electron to form a stable closed (sub)shell, singly charged negative ion. In the case of sodium, with the configuration $1s^22s^22p^63s$, and chlorine, with the configuration $1s^22s^22p^63s^23p^5$, it is necessary to supply 1.49 eV to convert (Na + Cl) to the ionic state ($Na^+ + Cl^-$) at infinite separation, but because of the Coulomb attraction between the ions, at separations less than about 18 a.u. the ionic system ($Na^+ + Cl^-$) has a lower energy than the atomic system (Na + Cl).

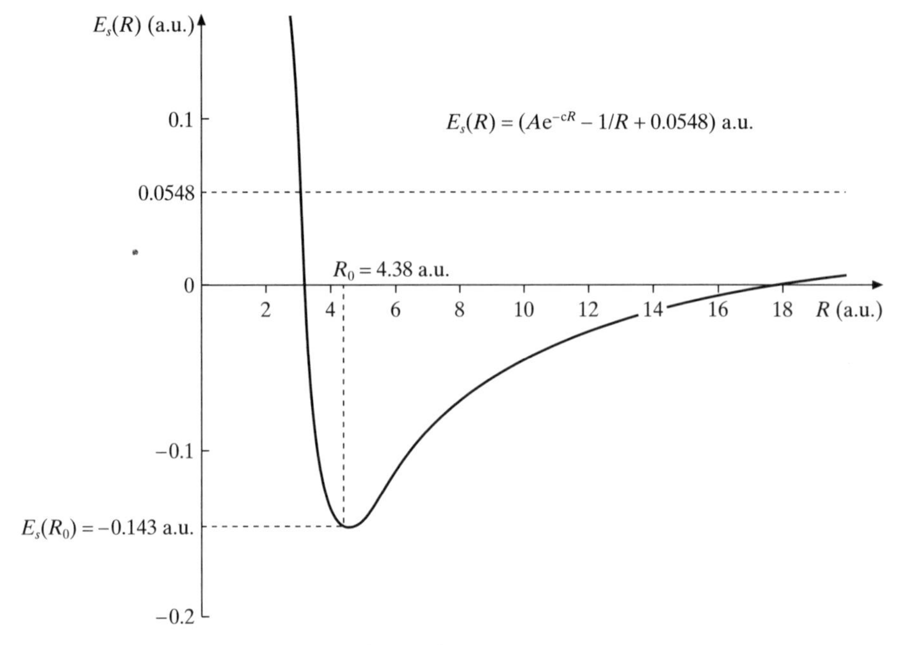

Figure 10.18 Potential energy curve for NaCl.

At small separations where the electron densities overlap, the interaction between the closed shell Na^+ and Cl^- systems is similar to that between the inert gases, He + He for example, and is strongly repulsive. It can be empirically represented by the potential $A \exp(-cR)$, where A and c are positive constants. The variation with the internuclear separation R of the energy of the ($Na^+ + Cl^-$) system relative to the sum of the energies of the isolated neutral atoms is given by

$$E_s(R) = E_s(\infty) - \frac{1}{R} + A \exp(-cR) \tag{10.81}$$

where atomic units have been used, and $E_s(\infty) = 0.0548$ a.u. (1.49 eV). The function $E_s(R)$ is displayed in Fig. 10.18. It has a minimum at an equilibrium distance R_0 determined by the constants A and c in the repulsive potential.

A simple ionic model of sodium chloride (NaCl) is obtained by viewing the molecule as a combination of Na^+ and Cl^- ions bound by the Coulomb attraction and with an energy $E_s(R_0)$, the energy at the minimum of (10.81). To establish the accuracy of this model, we need to find A and c, from which R_0, $E_s(R_0)$ and the dissociation energy can be computed. The chemically measured dissociation energy D_0 is 4.22 eV and this can be compared with the computed value of D_0. The rotational and vibrational spectra of the molecule can be used to find A and c, as we shall see in the next section.

The ionic model can be further confirmed by computing the electric dipole moment and comparing this with the experimental value. The model can be refined in various ways, the most important of which is to take into account the long-range attractive interaction between the Na^+ and Cl^- ions due to their polarisability (see Chapter 13). This very simple model of the ionic bond can be applied to the other alkali halides.

10.4 The rotation and vibration of diatomic molecules

Once the electronic Schrödinger equation (10.12) has been solved to obtain the wave functions Φ_q and the energy eigenvalues $E_q(R)$, the relative motion of the nuclei A and B can be determined. In the Born–Oppenheimer approximation the nuclear wave function satisfies equation (10.22) for the particular electronic state $q = s$. In this equation the first term represents the kinetic energy due to radial motion, the term $\langle\Phi_s|\mathbf{N}^2|\Phi_s\rangle/(2\mu R^2)$ the kinetic energy due to rotational motion and $E_s(R)$ acts as an effective potential. In order to simplify the rotational kinetic energy, the orbital angular momentum $\mathbf{N}$ of the relative motion of the nuclei A and B can be expressed as

$$\mathbf{N} = \mathbf{K} - \mathbf{L} \tag{10.82}$$

where $\mathbf{K}$ is the total orbital angular momentum of the whole molecule and $\mathbf{L}$ is the orbital angular momentum of the electrons. Since $\mathbf{N} = \mathbf{R} \times \mathbf{P}$, where $\mathbf{P}$ is the operator for the relative (linear) momentum of A and B and $\mathbf{R}$ is the internuclear position vector, it follows that (see Fig. 10.19)

$$\mathbf{R} \cdot \mathbf{N} = 0 \tag{10.83}$$

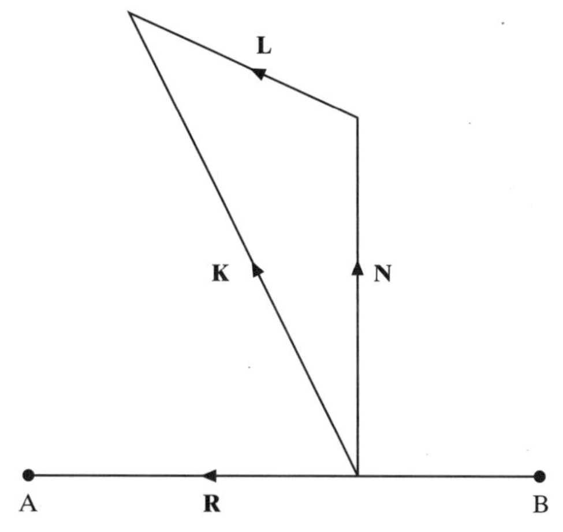

Figure 10.19 The angular momentum of a diatomic molecule with no coupling between the electronic and nuclear orbital motion and neglecting spin-dependent coupling. Since the angular momentum **N** of the nuclei A and B is at right angles to the internuclear line AB, the components of the electronic orbital angular momentum **L** and the total orbital angular momentum **K** along AB are equal, $L_{\hat{z}} = K_{\hat{z}}$.

As a consequence the projection of **K** along the internuclear line (the body-fixed $\overline{OZ}$ axis) must be equal to the projection of **L** along this line. That is,

$$K_{\bar{z}} \equiv \mathbf{K}\cdot\hat{\mathbf{R}} = L_{\bar{z}} \tag{10.84}$$

where $K_{\bar{z}}$ and $L_{\bar{z}}$ are the components of **K** and **L** respectively along the $\overline{OZ}$ axis.

The total spatial wave function in the absence of spin-dependent interactions is (see (10.23))

$$\Psi_s = F_s(\mathbf{R})\Phi_s(\mathbf{R}; \mathbf{r}_1, \mathbf{r}_2, \ldots, \mathbf{r}_N) \tag{10.85}$$

For an isolated molecule, Ψ_s must be a simultaneous eigenfunction of $\mathbf{K}^2$ and K_z, where K_z is the component of K along the OZ space-fixed axis. Thus we have that

$$\mathbf{K}^2\Psi_s = K(K+1)\hbar^2\Psi_s \tag{10.86}$$

where the *rotational quantum number* K is a positive integer, or zero, and

$$K_z\Psi_s = M_K\hbar\Psi_s \tag{10.87}$$

with $-K \leqslant M_K \leqslant K$.

In addition we know (see (10.25)) that Φ_s, the electronic wave function, is an eigenfunction of $L_{\bar{z}}$ belonging to the eigenvalue $M_L\hbar = \pm\Lambda\hbar$. Since $L_{\bar{z}}$ operates only on the electronic coordinates, Ψ_s must also be an eigenfunction of $L_{\bar{z}}$. We have

$$\begin{aligned} L_{\bar{z}}\Psi_s &= F_s(\mathbf{R})L_{\bar{z}}\Phi_s \\ &= \pm\hbar\Lambda\Psi_s \end{aligned} \tag{10.88}$$

In view of (10.84), Ψ_s must also be an eigenfunction of $K_{\bar{z}}$ belonging to the eigenvalue $\pm\Lambda\hbar$. The important result follows that as $|K| \geqslant K_{\bar{z}}$ we must have $K \geqslant \Lambda$ and the possible values of the quantum number K are

$$K = \Lambda, \Lambda+1, \Lambda+2, \ldots \tag{10.89}$$

We can now re-express the rotational kinetic energy term in equation (10.22) by using (10.82). We have

$$\begin{aligned} \frac{1}{2\mu R^2}\langle\Phi_s|\mathbf{N}^2|\Phi_s\rangle F_s(\mathbf{R}) &= \frac{1}{2\mu R^2}\langle\Phi_s|\mathbf{K}^2 + \mathbf{L}^2 - 2\mathbf{K}\cdot\mathbf{L}|\Phi_s\rangle F_s(\mathbf{R}) \\ &= \frac{\hbar^2}{2\mu R^2}[K(K+1) - \Lambda^2]F_s(\mathbf{R}) \\ &\quad + \frac{1}{2\mu R^2}\langle\Phi_s|L_{\bar{x}}^2 + L_{\bar{y}}^2|\Phi_s\rangle F_s(\mathbf{R}) \end{aligned} \tag{10.90}$$

where the eigenvalue equations (10.86) and (10.88) have been employed together with the fact that Φ_s is an eigenstate of $L_{\bar{z}}$. It has been assumed that Φ_s has been normalised to unity as in (10.14).

Using (10.87), the Schrödinger equation (10.22) for the relative notion of A and B can be expressed as

$$-\frac{\hbar^2}{2\mu}\left[\frac{1}{R^2}\frac{\partial}{\partial R}\left(R^2\frac{\partial}{\partial R}\right)-\frac{K(K+1)}{R^2}\right]F_s(\mathbf{R})+[E_s'(R)-E]F_s(\mathbf{R})=0 \qquad \textbf{(10.91)}$$

where

$$E_s'(R)=E_s(R)-\frac{\Lambda^2\hbar^2}{2\mu R^2}+\frac{1}{2\mu R^2}\langle\Phi_s|L_{\tilde{x}}^2+L_{\tilde{y}}^2|\Phi_s\rangle \qquad \textbf{(10.92)}$$

The energy $E_s'(R)$ acts as an effective potential in the one-body Schrödinger equation (10.91) for the wave function $F_s(\mathbf{R})$, and depends only on the electronic state s. The last two terms on the right of (10.92) are small compared with $E_s(R)$ since μ is much greater than the electronic mass, and in what follows we shall use the approximation $E_s'(R) \simeq E_s(R)$.

The wave function for the nuclear motion $F_s(\mathbf{R})$ can be expressed in terms of the product of a radial function $\mathcal{F}^s_{v,K}(R)$ and an angular function ${}^{\Lambda}\mathcal{H}_{KM_K}(\Theta, \Phi)$ by writing

$$F_s(\mathbf{R})=R^{-1}\mathcal{F}^s_{v,K}(R)\ {}^{\Lambda}\mathcal{H}_{KM_K}(\Theta,\Phi) \qquad \textbf{(10.93)}$$

where v, called the *vibrational quantum number*, has been added as a label to distinguish between the different solutions of (10.91). For a given electronic state s, the states labelled by the rotational quantum number K and the vibrational quantum number v are called *rovibronic states*. The angular functions ${}^{\Lambda}\mathcal{H}_{KM_K}(\Theta, \Phi)$ are eigenfunctions of $\mathbf{K}^2$ and K_z for a given value of Λ. The explicit form of ${}^{\Lambda}\mathcal{H}_{KM_K}(\Theta, \Phi)$ is obtained in Appendix 11. We note that for Σ states ($\Lambda = 0$) $\mathbf{K} = \mathbf{N}$. Since $\mathbf{N}^2$ is given by (10.9), the eigenfunctions of $\mathbf{K}^2$ and K_z in this case are the spherical harmonics $Y_{KM_K}(\Theta, \Phi)$. Inserting (10.93) into (10.91) and dividing throughout by ${}^{\Lambda}\mathcal{H}_{KM_K}(\Theta, \Phi)$, we obtain the radial equation

$$\left[-\frac{\hbar^2}{2\mu}\left(\frac{d^2}{dR^2}-\frac{K(K+1)}{R^2}\right)+E_s(R)-E_{s,v,K}\right]\mathcal{F}^s_{v,K}(R)=0 \qquad \textbf{(10.94)}$$

The eigenvalues E have been labelled by s, v and K and depend on the rotational quantum number K and vibrational quantum number v as well as on the electronic state s.

For a given effective potential $E_s(R)$ the equation (10.94) can be solved numerically with the boundary condition $\mathcal{F}^s_{v,K}(0) = 0$, but since the nuclear motion for a stable molecule is generally confined to a small interval of values of R close to the minimum in $E_s(R)$ at $R = R_0$ (see Fig. 10.5), an analytical approximation to $\mathcal{F}^s_{v,K}(R)$ can be found by expanding $E_s(R)$ about $R = R_0$. That is,

$$E_s(R)=E_s(R_0)+(R-R_0)\left.\frac{dE_s}{dR}\right|_{R=R_0}+\frac{1}{2}(R-R_0)^2\left.\frac{d^2E_s(R)}{dR^2}\right|_{R=R_0}+\ldots \qquad \textbf{(10.95)}$$

Since $E_s(R)$ has a minimum at $R = R_0$ the second term on the right-hand side of (10.95) vanishes. Neglecting terms of the third and higher orders in $(R - R_0)^n$, the potential well $E_s(R)$ can be represented by the parabolic approximation

$$E_s(R) \simeq E_s(R_0) + \frac{1}{2}k_s(R - R_0)^2 \tag{10.96}$$

where

$$k_s = \left.\frac{\mathrm{d}^2 E_s(R)}{\mathrm{d}R^2}\right|_{R=R_0}. \tag{10.97}$$

At the same time, we can approximate the term $\hbar^2 K(K + 1)/(2\mu R^2)$ in (10.94) by setting $R = R_0$. This term is now independent of R, and is called the *rotational energy* E_r. We have

$$E_\mathrm{r} = \frac{\hbar^2}{2\mu R_0^2}K(K + 1) = \frac{\hbar^2}{2I_0}K(K + 1) = BK(K + 1), \qquad K = \Lambda, \Lambda + 1, \Lambda + 2, \dots \tag{10.98}$$

where $I_0 = \mu R_0^2$ is the moment of inertia of the molecule for an equilibrium distance R_0 and reduced mass μ, and $B = \hbar^2/(2\mu R_0^2) = \hbar^2/(2I_0)$ is called the *rotational constant*. It should be noted that R_0, and hence I_0 and B, depends on the electronic state s.

Using equations (10.94)–(10.98), we see that the total energy $E_{s,v,K}$ is the sum of the electronic energy $E_s(R_0)$, the rotational energy E_r and a vibrational energy E_v,

$$E_{s,v,K} \equiv E_{s,v,\mathrm{r}} = E_s(R_0) + E_v + E_\mathrm{r} \tag{10.99}$$

where E_v is an eigenvalue of the equation

$$\left[-\frac{\hbar^2}{2\mu}\frac{\mathrm{d}^2}{\mathrm{d}R^2} + \frac{1}{2}k_s(R - R_0)^2 - E_v\right]\psi_v = 0 \tag{10.100}$$

corresponding to linear harmonic motion, with a force constant k_s. From Section 2.4, the eigenvalues E_v are given by

$$E_v = \hbar\omega_0\left(v + \frac{1}{2}\right) = h\nu_0\left(v + \frac{1}{2}\right), \qquad v = 0, 1, 2, \dots \tag{10.101}$$

with $\omega_0 = (k_s/\mu)^{1/2}$ and $\nu_0 = \omega_0/(2\pi)$. The corresponding eigenfunctions ψ_v are the linear harmonic oscillator wave functions which were studied in Section 2.4.

It should be noted that the energy of the molecule cannot depend on M_K, since in the absence of external fields there is no preferred direction in space. Thus taking into account the twofold degeneracy of the electronic energy for $\Lambda \neq 0$, the overall degeneracy of the energy levels $E_{s,v,K}$ is $2(2K+1)$ for states with $\Lambda \neq 0$ and $(2K+1)$ for Σ states.

The rotational energy can be discussed from a slightly different point of view. In the treatment given above, the approximation in which R is fixed to its equilibrium value R_0 was made at the last stage, but instead we could have started by considering the molecule to behave, as far as rotations are concerned, like a completely rigid body. In the principal axis system (Goldstein, 1980), the kinetic energy of a rotating body is

$$T = \frac{1}{2I_a}K_a^2 + \frac{1}{2I_b}K_b^2 + \frac{1}{2I_c}K_c^2 \qquad \textbf{(10.102)}$$

where I_a, I_b and I_c are the principal moments of inertia about axes a, b and c while K_a, K_b and K_c are the components of the angular momentum $\mathbf{K}$ about these axes. It should be noted that a, b and c form a system fixed in the rotating body, and the corresponding operators obey slightly different commutation relations from the components of $\mathbf{K}$ in a coordinate system fixed in space.

A diatomic molecule is axially symmetrical about the internuclear line, which can be taken as the axis c. Then $I_a = I_b$ and the system forms a *symmetrical top* with kinetic energy

$$T = \frac{1}{2I_a}(K_a^2 + K_b^2) + \frac{1}{2I_c}K_c^2 \qquad \textbf{(10.103)}$$

The component K_c is along the internuclear line $K_c = K_z$ and, as we have seen, the eigenvalues of K_z are $\pm\Lambda\hbar$. As usual the eigenvalues of $\mathbf{K}^2 = K_a^2 + K_b^2 + K_c^2$ are $K(K+1)\hbar^2$, so that the rotational energy levels are given by

$$E_r = \frac{\hbar^2}{2I_a}K(K+1) + \left(\frac{1}{2I_c} - \frac{1}{2I_a}\right)\Lambda^2\hbar^2, \qquad K = \Lambda, \Lambda+1, \Lambda+2, \ldots \qquad \textbf{(10.104)}$$

The moment of inertia of the molecule about the axis a, which is perpendicular to the internuclear line and goes through the centre of mass, is $I_a = \mu R_0^2 = I_0$, while I_c is the moment of inertia about the internuclear line, and depends on the electronic state. It is seen that the expression (10.104) is equivalent to the result (10.98), since the second term on the right of (10.104) can be included in the electronic energy.

In real systems, the function $E_s(R)$ is only represented accurately by the parabolic approximation (10.96) for small values of $|R - R_0|$, so that the vibrational energy E_v is only well represented by (10.101) for small values of the vibrational quantum number v. In general, for large values of v, the levels tend to become more closely spaced, as shown in Fig. 10.20. When the energy exceeds the depth of the well, the molecule dissociates, so that there are only a finite number of vibrational levels associated with each electronic level.

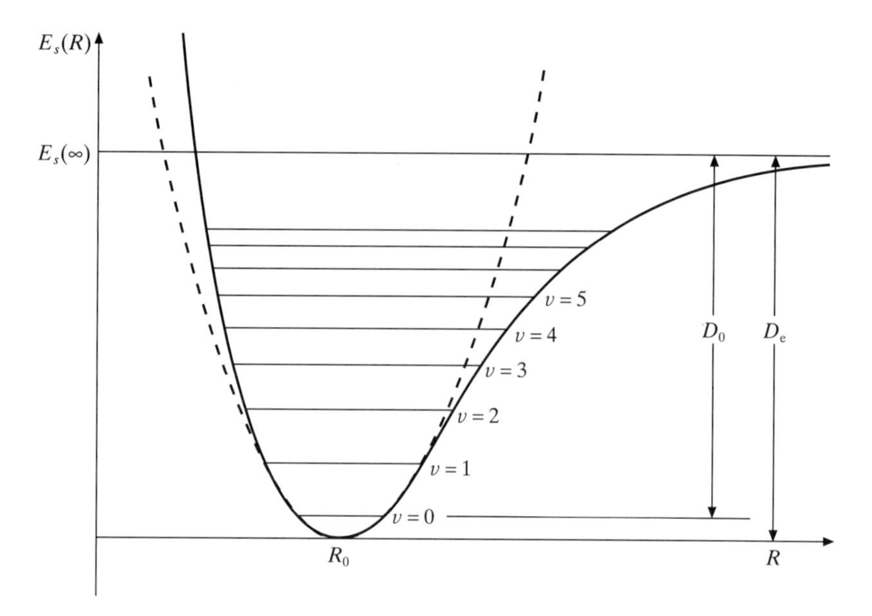

Figure 10.20 Vibrational energy levels of a diatomic molecule. The dashed line shows the parabolic approximation to the potential well, which is accurate in the region close to R_0, the equilibrium separation.

A better representation of the potential well $E_s(R)$ than the parabolic approximation (10.96) can be obtained as follows. Let us write $E_s(R)$ in the form

$$E_s(R) = E_s(\infty) + V(R) \tag{10.105}$$

where $V(R)$ is a potential to be determined. An empirical approximation to $V(R)$, due to P.M. Morse, is given by

$$V_M(R) = D_e\{\exp[-2\alpha(R - R_0)] - 2\exp[-\alpha(R - R_0)]\} \tag{10.106}$$

where R_0, D_e and α are positive constants for a given diatomic molecule. The values of R_0, D_e and α are given in Table 10.1 for a few molecules. We see from (10.106) that the Morse potential is attractive at large distances and has a minimum equal to $-D_e$ at the equilibrium distance R_0, so that

$$D_e = E_s(\infty) - E_s(R_0) \tag{10.107}$$

Table 10.1 Values of the constants R_0, D_e and α for a few typical molecules.

Molecule	R_0 (Å)	D_e (eV)	αR_0
H_2	0.742	4.75	1.44
I_2	2.66	1.56	4.95
HCl	1.27	4.62	2.38

is the electronic dissociation energy, in agreement with (10.24). At short distances, where the nuclei come close together, the Morse potential $V_M(R)$ exhibits a strong repulsion. A deficiency of the Morse potential is that it is finite at $R = 0$, in contrast to the true interaction $V(R)$ which is infinite at the origin. Another defect of the Morse potential is that it falls off exponentially at large R instead of exhibiting the R^{-6} fall-off due to the van der Waals interaction between two neutral atoms which are far apart. However, these inaccuracies of the Morse potential have little influence on the molecular energy levels.

The constant α appearing in the Morse potential can be related to the 'force constant' k_s introduced above (see (10.96)–(10.97)) by expanding $V_M(R)$ in powers of $(R - R_0)$. That is,

$$V_M(R) = D_e\,[-1 + \alpha^2(R - R_0)^2 + \dots] \tag{10.108}$$

so that, by comparing with (10.96) and using (10.107), we have

$$D_e\alpha^2 = \frac{1}{2}k_s \tag{10.109}$$

The lower energy levels in the potential $D_e + V_M(R)$, where $V_M(R)$ is the Morse potential (10.106), are given quite accurately by the expression

$$E_v = \hbar\omega_0\left[\left(v + \frac{1}{2}\right) - \beta\left(v + \frac{1}{2}\right)^2\right] \tag{10.110}$$

where β is a small number ($\beta \ll 1$). The quantity $\beta\omega_0$ is known as the *anharmonicity constant* and is found to be given by

$$\beta\omega_0 = \frac{\hbar\omega_0^2}{4D_e} \tag{10.111}$$

It is interesting to note that because of the zero-point energy ($v = 0$) of $\hbar\omega_0/2$, the true (chemical) dissociation energy of the molecule in the electronic state s is not exactly $E_s(\infty) - E_s(R_0)$, but is given by $E_s(\infty) - E_s(R_0) - \hbar\omega_0/2$ (see Fig. 10.20). In terms of D_e and ω_0, the true dissociation energy is therefore given by

$$D_0 = D_e - \hbar\omega_0/2 \tag{10.112}$$

In Table 10.2 we list the equilibrium distance R_0, the dissociation energy D_0, the fundamental vibrational frequency $\nu_0 = \omega_0/(2\pi)$ and the fundamental rotational constant $\hbar^2/(2\mu R_0^2)$ for a few typical diatomic molecules.

Centrifugal distortion

So far we have approximated the rotational energy E_r by the simple 'rigid rotator' expression (10.98) evaluated at the equilibrium distance R_0, so that there is no coupling between the rotational and vibrational motions. In a more accurate

Table 10.2 The equilibrium distance R_0, dissociation energy $D_0 = D_e - \hbar\omega_0/2$, fundamental vibrational frequency $\nu_0 = \omega_0/(2\pi)$, fundamental rotational constant $B = \hbar^2/(2\mu R_0^2) = \hbar^2/(2I_0)$ and magnitude $|D|$ of the electric dipole moment for some diatomic molecules.

Molecule	R_0 (Å)	D_0 (eV)	$\tilde{\nu}_0$ (cm^{-1})†	$\tilde{B}$ (cm^{-1})†	$10^{30}|D|$(C m)††
H_2^+	1.06	2.65	2297	29.8	–
H_2	0.742	4.48	4395	60.8	–
O_2	1.21	5.08	1580	1.45	–
Cl_2	1.99	2.48	565	0.244	–
N_2	1.09	9.75	2360	2.01	–
CO	1.13	9.60	2170	1.93	0.40
NO	1.15	5.3	1904	1.70	0.50
LiH	1.60	2.5	1406	7.51	19.4
HCl	1.28	4.43	2990	10.6	3.53
NaCl	2.36	4.22	365	0.190	28.1

† Note that for both quantities $\tilde{\nu}_0$ and $\tilde{B}$ the values have been converted in units of cm^{-1}; $\tilde{\nu}_0 = \nu_0/c$; $\tilde{B} = B/hc$.
†† Electric dipole moments are also given in units of debyes, where 1 debye = 3.36×10^{-30} C m (coulomb·metres).

treatment, we should solve the radial equation (10.94) directly. Adopting the Morse potential $V_M(R)$ for $E_s(R) - E_s(\infty)$, this equation reads

$$\left[-\frac{\hbar^2}{2\mu}\frac{d^2}{dR^2} + V_{\text{eff}}(R) - \bar{E}_{s,v,K}\right]\mathcal{F}^{s}_{v,K}(R) = 0 \qquad \textbf{(10.113)}$$

where $\bar{E}_{s,v,K} = E_{s,v,K} - E_s(\infty)$ and the effective interaction

$$V_{\text{eff}}(R) = V_M(R) + \frac{\hbar^2}{2\mu}\frac{K(K+1)}{R^2}, \qquad K = \Lambda, \Lambda+1, \Lambda+2, \ldots \qquad \textbf{(10.114)}$$

is the sum of the Morse potential (10.106) and the *centrifugal distortion* term $\hbar^2 K(K+1)/(2\mu R^2)$. It is clear that if we set $R = R_0$ in this centrifugal term we retrieve our earlier results. We also remark that for values of K which are not too large, the shape of $V_{\text{eff}}(R)$ is similar to that of $V(R)$. If we are interested in the lower energy levels we may expand $V_{\text{eff}}(R)$ about its minimum V_0 at $R = R_1$. Keeping terms up to order $(R - R_1)^4$, we have

$$V_{\text{eff}}(R) = V_0 + \frac{1}{2}\tilde{k}_s(R - R_1)^2 + c_1(R - R_1)^3 + c_2(R - R_1)^4 \qquad \textbf{(10.115)}$$

where the new force constant $\tilde{k}_s$ and the coefficients c_1 and c_2 can be expressed in terms of the rotational quantum number K and of the constants D_e, α and R_0 of the Morse potential (10.106).

It is important to notice that the value R_1 for which $V_{\text{eff}}(R)$ has its minimum only coincides with R_0 if $K = 0$. A simple approximation for R_1 may be obtained by setting $c_1 = c_2 = 0$ and $\tilde{k}_s = k_s$ in (10.115), and solving the equation $\mathrm{d}V_{\text{eff}}(R)/\mathrm{d}R = 0$ by iteration. Using (10.109), we then have (Problem 10.6)

$$R_1 \simeq R_0 + \frac{\hbar^2}{2\mu} \frac{K(K+1)}{\alpha^2 R_0^3 D_e} \tag{10.116}$$

which shows that the molecule 'stretches' because of the rotational motion.

Returning to the effective potential (10.115) and treating the c_1 and c_2 terms as perturbations it may be shown that the energy eigenvalues of (10.113) are given to second order in $(v + 1/2)$ and $(K(K + 1))$ by

$$\bar{E}_{s,v,K} = -D_e + \hbar\omega_0 \left[\left(v + \frac{1}{2} \right) - \beta \left(v + \frac{1}{2} \right)^2 \right] + \frac{\hbar^2}{2\mu R_0^2} K(K+1)$$

$$- a\left(v + \frac{1}{2} \right) K(K+1) - bK^2(K+1)^2 \tag{10.117a}$$

where

$$a = \frac{3\hbar^3\omega_0}{4\mu\alpha R_0^3 D_e} \left(1 - \frac{1}{\alpha R_0} \right) \tag{10.117b}$$

and

$$b = \frac{\hbar^4}{4\mu^2\alpha^2 R_0^6 D_e} \tag{10.117c}$$

The first three terms on the right of (10.117a) are identical to those discussed above in the absence of centrifugal distortion; they are respectively the potential depth ($-D_e$), the harmonic and anharmonic vibration terms, and the rigid rotator energy. The two additional terms are respectively a *rotation–vibration coupling* term (which is negative because for higher values of v the average internuclear distance is larger than R_0 due to the anharmonicity) and a *correction to the rigid rotator energy* (10.98), which is also negative since larger values of K lead to an increase in the average distance between the nuclei, as seen from (10.116).

Potential constants and the alkali halides

In Section 10.3 it was shown that a simple model of the ionic bond binding the $(Na^+ + Cl^-)$ system to form NaCl could be obtained by using the repulsive potential $A \exp(-cR)$ to represent the interaction together with the attractive Coulomb potential $-1/R$ (see (10.81)). To determine the constants A and c the observed rotational and vibrational spectrum of NaCl can be employed. First, from the rotational spectrum, the moment of inertia of the molecule can be obtained using

Table 10.3 Some properties of selected alkali halide molecules.

Molecule	R_0 (Å)	$\tilde{\nu}_0$ (cm^{-1})	$10^{29}\lvert D\rvert$(Cm)†	D_0 (eV)
LiF	1.564	910.3	2.11	5.99
LiCl	2.021	641.0	2.38	4.85
NaCl	2.361	364.6	3.00	4.22
NaBr	2.502	298.5	3.04	3.74
KCl	2.667	279.8	3.42	4.37
KBr	2.821	219.2	3.54	3.92
RbCl	2.787	223.3	3.51	4.31
CsCl	2.906	214.2	3.47	4.59

† Electric dipole moments are also given in units of debyes, where 1 debye = 3.36×10^{-30} C m (coulomb·metres).

(10.98) and this gives the equilibrium distance R_0, which is found to have the value 4.38 a.u. Then, from the vibrational spectrum, the fundamental wave number $\tilde{\nu}_0 = \omega_0/(2\pi c)$ can be determined (see (10.101)), which in turn yields the force constant k_s. The value of $\tilde{\nu}_0$ (see Table 10.3) is 364.6 cm^{-1}. Having obtained R_0 and $\tilde{\nu}_0$, the constants A and c can be calculated using the equations (10.96) and (10.97). The quantity $D_e = E_s(\infty) - E_s(R_0)$ and the chemical dissociation energy $D_0 = D_e - \hbar\omega_0/2$ can then be determined. Using the calculated values, the chemical dissociation energy D_0 is found to be 3.87 eV (see Problem 10.7), which is about 10 per cent less than the chemically measured value of 4.22 eV.

The simple model for the ionic bond which has been described for NaCl can be applied to other alkali halides (see Problem 10.8) and the properties of a number of these are shown in Table 10.3.

10.5 The electronic spin and Hund's cases

In molecules, as in atoms, magnetic interactions couple the spin and orbital angular momenta. The most important of these interactions is that between the electronic spin **S** and the electronic orbital angular momentum **L**. It can be expressed by an effective interaction of the form $\bar{A}\,\mathbf{L}\cdot\mathbf{S}$, where $\bar{A}$ is a constant within a given multiplet, called the spin–orbit coupling constant. The rotation of the molecule as a whole also generates a magnetic field which interacts with the magnetic moments of the electrons. This spin–rotation coupling can also be expressed as an effective interaction of the form $\bar{\Gamma}\,\mathbf{N}\cdot\mathbf{S}$, where $\bar{\Gamma}$ is a constant. In addition the nuclear spins also interact with **L**, **S** and **N**. However, as noted earlier, the interaction with the nuclear spins has a negligible effect on the energy levels, although, as we shall see in Chapter 11, the symmetry of the nuclear spin wave functions can be important in the spectra of homonuclear diatomic molecules.

If the nuclear spins are ignored, the total angular momentum operator **J** for a diatomic molecule is given by

$$\mathbf{J} = \mathbf{L} + \mathbf{N} + \mathbf{S} = \mathbf{K} + \mathbf{S} \qquad \textbf{(10.118)}$$

where **N**, the orbital angular momentum of the relative motion of the nuclei, is in a direction at right angles to the internuclear line (see Fig. 10.19). In the absence of external electric or magnetic fields, molecular wave functions must be eigenfunctions of $\mathbf{J}^2$ with eigenvalues $J(J+1)\hbar^2$ and of J_z with eigenvalues $M_J\hbar$ where J_z is the component of **J** in an arbitary direction in the space-fixed coordinate frame. An analysis of the couplings between the vectors **L**, **N** and **S** which add to form **J** and their effect on the energy levels was given by F. Hund who identified five limiting cases (a) to (e), which depend on the relative strength of the electrostatic, spin–orbit and spin–rotation energies. The electrostatic interaction between the electrons and the nuclei constrains the electronic wave function to rotate as the molecule rotates. A measure of the importance of this interaction is the magnitude of the difference in energy $|\Delta E|$ between two adjacent electronic levels with different values of Λ. The magnitude of the spin–orbit interaction is given by the absolute value of $A = \bar{A}\hbar^2$, while the importance of the spin–rotation coupling is governed by the rotational constant $B = \hbar^2/(2\mu R_0^2)$. Hund's cases correspond to different limiting values of the three energies, $|\Delta E|$, $|A|$ and B.

Hund's case (a), $|\Delta E| \gg |A| \gg B$

In this case, the electrostatic interaction is much larger than the spin–orbit interaction which in turn is much larger than the rotational energy. An example is given by the $A^2\Pi$ term of CO^+. The nearest level with a different value of Λ is the $X^2\Sigma^+$ ground state, from which it is found that $|\Delta E| = 20\,733$ cm^{-1} while the spin–orbit constant is $A = -117$ cm^{-1} and $B = 1.6$ cm^{-1}.

The electrostatic interaction has axial symmetry and, as we have seen, this causes **L** to precess about the internuclear axis, by which is meant that $L_{\bar{z}} = \pm\Lambda\hbar$ and $\langle L_{\bar{x}}\rangle = \langle L_{\bar{y}}\rangle = 0$. Since in this case the rotation of the molecule is slow, and the spin–orbit interaction is large compared wict the rotational energy, the spin angular momentum **S** will also tend to precess about the internuclear axis. This suggests that the spin should be quantised along the internuclear axis (the body-fixed $\overline{OZ}$ axis). Simultaneous eigenfunctions of $\mathbf{S}^2$ and $S_{\bar{z}}$ can be found with eigenvalues $S(S+1)\hbar^2$ and $\Sigma\hbar$ respectively [9]. Denoting these eigenfunctions by $|S, \Sigma\rangle$ we have that

$$\mathbf{S}^2|S, \Sigma\rangle = S(S+1)\hbar^2|S, \Sigma\rangle \qquad \textbf{(10.119a)}$$

and

$$S_{\bar{z}}|S, \Sigma\rangle = \Sigma\hbar|S, \Sigma\rangle \qquad \textbf{(10.119b)}$$

[9] The quantum number Σ used here should not be confused with the symbol Σ denoting the electronic term with $\Lambda = 0$.

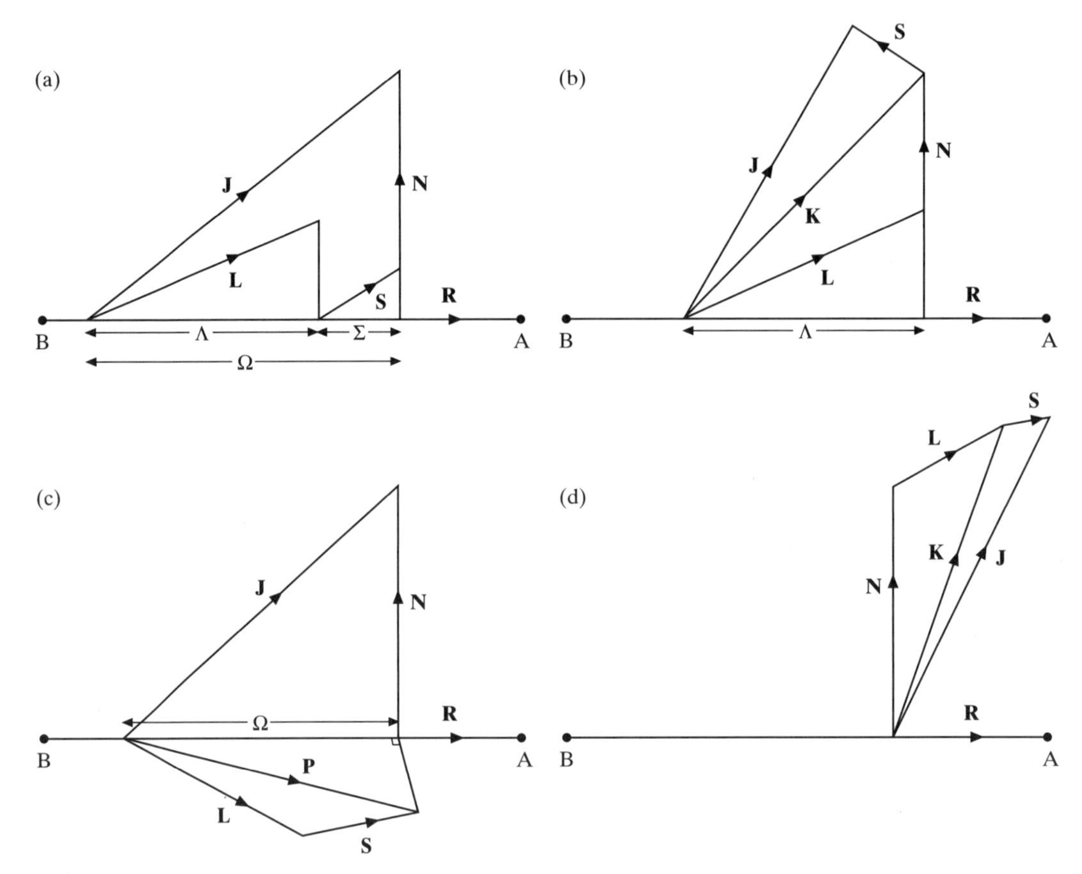

Figure 10.21 Hund's coupling cases (atomic units are used):
(a) **L** and **S** precess about **R** with well-defined components Λ and Σ, respectively, along **R**. **N** couples with $\Omega\hat{\mathbf{R}}$ to form **J**, where $\Omega = \Lambda + \Sigma$.
(b) **L** precesses about **R** with a well-defined component Λ along **R**. **N** couples with $\Lambda\hat{\mathbf{R}}$ to form **K** and **K** couples with **S** to form **J**.
(c) **L** and **S** couple to form **P** which precesses about **R**. The component of **P** along **R**, Ω, is well defined. **N** couples with $\Omega\hat{\mathbf{R}}$ to form **J**.
(d) **L** and **N** couple to form **K**. **J** results from coupling **K** to **S**.

where the quantum number Σ can take the $(2S + 1)$ values $-S, -S + 1, \ldots, S - 1, S$. Since $\mathbf{S}^2$ and S_z commute with the molecular Hamiltonian (10.4), simultaneous eigenfunctions of $\mathbf{S}^2$, S_z, L_z, $\mathbf{J}^2$ and J_z can be found specified by the quantum numbers S, Σ, $\pm\Lambda$, J and M_J. As **N** is perpendicular to the internuclear axis, $N_z = 0$ and hence (see Fig. 10.21(a))

$$J_z = L_z + S_z \qquad \textbf{(10.120)}$$

It follows that the molecular wave function is also an eigenfunction of J_z corresponding to the eigenvalues $\Omega\hbar$, where

$$\Omega = \pm\Lambda + \Sigma \tag{10.121}$$

It should be noted that $\mathbf{N}^2$ does not commute with $S_{\bar{z}}$, so that neither **N** nor **K** ($= \mathbf{N} + \mathbf{L}$) are conserved quantities. Writing the normalised eigenvectors of the molecular Hamiltonian as $|S, \Sigma, \pm\Lambda, J, \Omega, M_J\rangle$, the additional energy due to the spin–orbit coupling is E_{LS}, where

$$E_{LS} = \bar{A}\langle S, \Sigma, \pm\Lambda, J, \Omega, M_J|\mathbf{L}\cdot\mathbf{S}|S, \Sigma, \pm\Lambda, J, \Omega, M_J\rangle = A\Lambda\Sigma \tag{10.122}$$

To obtain this result, we recollect that as **L** and **S** are quantised along the body-fixed axis $\langle L_{\bar{x}}\rangle = \langle L_{\bar{y}}\rangle = \langle S_{\bar{x}}\rangle = \langle S_{\bar{y}}\rangle = 0$. The total electronic energy is then

$$E_s^T(R_0) = E_s(R_0) + A\Lambda\Sigma \tag{10.123}$$

For a given value of Λ the energy levels are split into $(2S + 1)$ multiplet components, which are equally spaced in energy. The corresponding electronic terms are labelled by a subscript giving the value of Ω. For example, in a $^3\Pi$ state with $\Lambda = 1$ and $S = 1$, the allowed values of Ω are $\Omega = 0, 1, 2$ and the corresponding terms are written as $^3\Pi_0$, $^3\Pi_1$ and $^3\Pi_2$. Each of these multiplet components gives rise to a series of rotational levels illustrated in Fig. 10.22.

To obtain the rotational energy spectrum we shall introduce the approximation in which R is set equal to the equilibrium separation R_0 of the nuclei. The rotational kinetic energy operator is then $\mathbf{N}^2/(2\mu R_0^2)$ and the rotational energy in Hund's case (a) is therefore E_r', where

$$E_r' = \frac{1}{2\mu R_0^2}\langle S, \Sigma, \pm\Lambda, J, \Omega, M_J|\mathbf{N}^2|S, \Sigma, \pm\Lambda, J, \Omega, M_J\rangle \tag{10.124}$$

Introducing a vector $\boldsymbol{\Omega}$ by setting

$$\boldsymbol{\Omega} = \Omega\hbar\,\hat{\mathbf{R}} \tag{10.125}$$

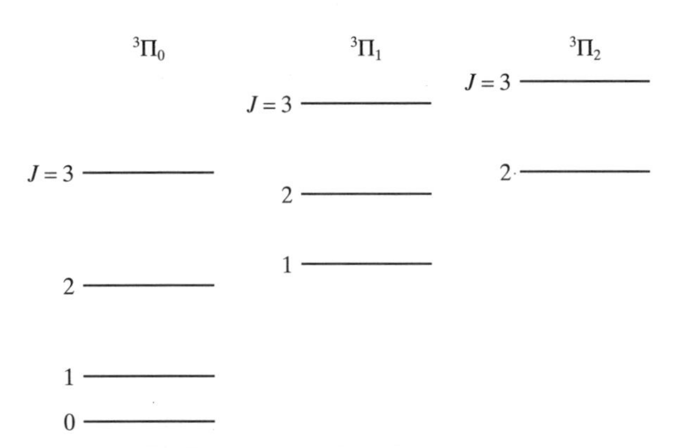

Figure 10.22 Spin–orbit splitting of a $^3\Pi_\Omega$ level. Note that $\Omega = 0$, 1 or 2 and $J \geqslant \Omega$. The electronic energy of a multiplet term is given by $A\,\Lambda\Sigma$ where A is a constant and Λ and Σ are the projections (in a.u.) of **L** and **S** on the internuclear line.

where $\hat{\mathbf{R}}$ is a unit vector along the body-fixed $\overline{OZ}$ axis (the internuclear line), we have (see Fig. 10.21(a))

$$\mathbf{N} = \mathbf{J} - \boldsymbol{\Omega} \quad \textbf{(10.126)}$$

From (10.124) and (10.126), we find immediately that

$$E'_{\mathrm{r}} = B[J(J+1) - \Omega^2] \quad \textbf{(10.127)}$$

Since $|\mathbf{J}| \geqslant J_{\bar{z}}$, the result (10.127) is subject to the important restriction that $J \geqslant \Omega$. The observation that spectral lines with lower values of J are missing provides an important method of determining the value of Ω when this is not known. The quantum numbers J can take integer values when the spin has integer values or half-odd integer values if the spin is a half-odd integer. The value of $-B\Omega^2$ just depends on the electronic state so, as we did in the spinless case (see (10.92)), this quantity can be combined with the energy $E_s^T(R_0)$, giving

$$E'_s(R_0) = E_s^T(R_0) - B\Omega^2 \quad \textbf{(10.128)}$$

in which case the rotational energy just becomes

$$E_{\mathrm{r}} = BJ(J+1), \qquad J = \Omega, \Omega + 1, \Omega + 2, \dots \quad \textbf{(10.129)}$$

In Hund's case (a) each level remains twofold degenerate with the two states with $L_{\bar{z}} = \pm\Lambda\hbar$ $(\Lambda \neq 0)$ having the same energy, and of course in the absence of external fields the $(2J + 1)$ substates with different values of M_J are always degenerate. The wave function can be determined by a procedure paralleling that given in Appendix 11 and is found to be ${}^{\Omega}\mathcal{H}_{JM_J}(\Theta, \Phi)$, which reduces to ${}^{\Lambda}\mathcal{H}_{KM_K}(\Theta, \Phi)$ in the case of zero spin.

Hund's case (b), $|\Delta E| \gg B \gg |A|$

In Hund's case (b), the spin–orbit interaction is either zero or is so small that it can be ignored. When this happens the most important spin interaction is that with the rotational motion of the nuclei. This can be represented by the effective interaction $\bar{\gamma}\mathbf{K}\cdot\mathbf{S}$, where $\mathbf{K} = \mathbf{L} + \mathbf{N}$. The energy shift is very small and can be calculated from first-order perturbation theory. The unperturbed Hamiltonian contains no spin interactions and the unperturbed energy levels are those given by (10.98), the rotational energies being given by

$$E_{\mathrm{r}} = BK(K+1), \qquad K = \Lambda, \Lambda + 1, \Lambda + 2, \dots \quad \textbf{(10.130)}$$

The uncoupled wave function for the unperturbed system is the product of the spin function $|S, M_S\rangle$ which is an eigenfunction of $\mathbf{S}^2$ and S_z and the spatial functions $|\pm\Lambda, K, M_K\rangle$ which are eigenfunctions of $L_{\bar{z}}$ $(= K_{\bar{z}})$, $\mathbf{K}^2$ and K_z. When the interaction $\bar{\gamma}\mathbf{K}\cdot\mathbf{S}$ is added to the Hamiltonian, the wave function is no longer an eigenfunction of S_z and K_z, but the total angular momentum $\mathbf{J} = \mathbf{K} + \mathbf{S}$ is always conserved (see Fig. 10.21(b)). For this reason a suitable basis for perturbative calculations is a simultaneous eigenfunction of $\mathbf{S}^2$, $\mathbf{K}^2$, $K_{\bar{z}}$, J^2 and J_z which we write as $|S, K, \pm\Lambda, J, M_J\rangle$. This function can be expressed as a linear combination of the uncoupled functions and

$$|S, K, \pm\Lambda, J, M_J\rangle = \sum_{M_K M_S} \langle KSM_K M_S|J, M_J\rangle |S, M_S\rangle |\pm\Lambda, K, M_K\rangle \qquad \textbf{(10.131)}$$

where $\langle KSM_K M_S|JM_J\rangle$ is a Clebsch–Gordan coefficient (see Appendix 4). The possible values of J for given values of K and S are

$$J = |K - S|, |K - S| + 1, \ldots, K + S \qquad \textbf{(10.132)}$$

with the restriction $J \geqslant 0$. The additional energy due to the spin–rotational coupling is then

$$\Delta E_r = \bar{\gamma}\langle S, K, \pm\Lambda, J, M_J|\mathbf{K}\cdot\mathbf{S}|S, K, \pm\Lambda, J, M_J\rangle \qquad \textbf{(10.133)}$$

where $\bar{\gamma}$ is called the *spin–rotation* constant. Since $\mathbf{K}\cdot\mathbf{S} = (\mathbf{J}^2 - \mathbf{K}^2 - \mathbf{S}^2)/2$, we have

$$\Delta E_r = \frac{\gamma}{2}[J(J+1) - K(K+1) - S(S+1)] \qquad \textbf{(10.134)}$$

where $\gamma = \bar{\gamma}\hbar^2$. The total rotational energy is

$$E_r = BK(K+1) + \frac{\gamma}{2}[J(J+1) - K(K+1) - S(S+1)]. \qquad \textbf{(10.135)}$$

For example, if the total spin is $S = 1/2$, there are two possible values of J, $J = K \pm 1/2$ $(K > 0)$. The degeneracy in J is removed and the rotational energies are

$$\begin{aligned} E_r &= BK(K+1) + \frac{\gamma}{2}K, & J &= K + 1/2 \\ &= BK(K+1) - \frac{\gamma}{2}(K+1), & J &= K - 1/2 \end{aligned} \qquad \textbf{(10.136)}$$

For diatomic molecules composed of light atoms, for example CH, OH and SiH, Hund's cases (a) and (b) are useful approximations; case (a) is a better approximation than (b) when the spin–orbit energy is greater than about twice the rotational energy. As the number of electrons increases, so does the spin–orbit energy and for diatomic molecules composed of heavy atoms Hund's case (c) is a more useful approximation.

Hund's case (c), $|A| \gg |\Delta E| \gg B$

In this case the spin–orbit coupling, and often the spin–spin coupling, is so strong that Λ and Σ are no longer good quantum numbers. To form a basis the vectors $\mathbf{L}$ and $\mathbf{S}$ can be combined to form a vector $\mathbf{P}$. Provided the rotation of the molecule as a whole is slow, $\mathbf{P}$ will precess about the internuclear axis and the basis states will be in eigenstates of $P_{\bar{z}}$, the component of **P** along the body-fixed $\overline{OZ}$ axis (see Fig. 10.21(c)). The eigenvalues of $P_{\bar{z}}$ can be denoted by $\Omega\hbar$ and the basis states must be simultaneous eigenstates of $P_{\bar{z}}$, $\mathbf{J}^2$ and $J_{\bar{z}}$. Since the total angular momentum is $\mathbf{J} = \mathbf{P} + \mathbf{N}$ and **N** is at right angles to the internuclear line, $J_{\bar{z}}$ has the same eigenvalues as $P_{\bar{z}}$ and the basis states can be denoted as $|\Omega, J, M_J\rangle$. An example of Hund's case (c) is the HgH molecule for which the spin–orbit coupling constant is nearly 4000 cm^{-1} and the $X^2\Sigma$ and $A^2\Pi$ states are strongly mixed. For such heavy molecules a relativistic treatment is required, which is beyond the scope of this book. A detailed account of the theory has been given by Mizushima (1975). Despite the difficulty of calculating the energy shifts, the rotational energy is easy to obtain, since a vector $\mathbf{\Omega} = \Omega\hbar\,\hat{\mathbf{R}}$ can be introduced, as in Hund's case (a). As $\mathbf{N} = \mathbf{J} - \mathbf{\Omega}$, we again find the result (10.127).

Hund's case (d), $B \gg |\Delta E| \gg |A|$

In this case the electrostatic interaction is so weak that the electronic wave function is not constrained to rotate with the molecule. This implies that **L** and **S** can be quantised along the space-fixed OZ axis. The electronic orbital angular momentum **L** can be coupled with **N** to form the orbital angular momentum **K**. In turn **K** can be coupled with **S** to form the angular momentum **J** (see Fig. 10.21(d)). The lower lying states of diatomic molecules do not conform to these conditions, but the Rydberg states may do so. In these states a valence electron is excited to an orbit which is so far from the centre of mass of the molecule that the remainder of the molecule acts like a centre of force on the valence electron and the axial symmetry is lost.

Hund's case (e), $|A| \gg B \gg |\Delta E|$

This case is similar to case (d), but as there are no examples among the bound states of diatomic molecules we shall not need to consider it.

Spin uncoupling. Λ-doubling

As we pointed out above, Hund's coupling cases are idealisations, to which many molecules approximately conform. However, deviations from these idealisations may occur, which represent a partial *uncoupling* or *decoupling* of some of the angular momenta. This uncoupling may increase as J increases since then the electrons may not follow the nuclear motion. Thus a molecule may fall approximately in one coupling case for low J, but in another case for high J, while for intermediate rotational states one has *intermediate coupling*.

A common example of intermediate coupling is provided by the transition from Hund's case (a) to (b). For low J, the spin **S** is coupled to the molecular axis, according to the coupling scheme (a). However, as J increases and the rotational frequency becomes larger, **S** uncouples from the molecular axis (spin uncoupling) and eventually couples with **K**, as in Hund's case (b).

Another type of uncoupling, produced by an interaction between the rotational and electronic motions, is that which decouples the electronic orbital angular momentum **L** from the molecular axis. This uncoupling has the effect of splitting the two otherwise degenerate levels corresponding to $\Lambda \neq 0$, and is known as *Λ-doubling*.

10.6 The structure of polyatomic molecules

Just as for diatomic molecules, the Born–Oppenheimer separation can be made for polyatomic molecules, and the motion divided into rotational, vibrational and electronic motions.

Rotational structure

The rotational motion may be treated approximately by supposing that the nuclei are fixed at their equilibrium positions, so that the molecule forms a rigid body. In this case, the kinetic energy operator can be expressed in terms of the principal axis system by equation (10.102). If the molecule possesses n-fold symmetry with $n \geqslant 3$, then two of the principal moments of inertia of the rigid body are equal and the molecular system forms a symmetrical top. We have already seen in Section 10.4 that diatomic molecules are examples of symmetrical top molecules. An example of a symmetrical top polyatomic molecule is the ammonia molecule NH_3 which has a threefold symmetry axis. The rotational kinetic energy is again given by (10.103), where now K_c is the component of the angular momentum along the *figure axis*, the axis of threefold symmetry. The eigenvalues of K_c^2 are $\Lambda^2\hbar^2$, so that the rotational energy levels are given by (10.104). If all three principal moments of inertia are equal, $I_a = I_b = I_c = I$, the molecule is called a *spherical top* and the expression (10.104) for the rotational energy levels reduces to

$$E_r = \frac{\hbar^2}{2I}K(K+1), \qquad K = \Lambda, \Lambda+1, \Lambda+2, \dots \tag{10.137}$$

All linear molecules are examples of spherical tops, for example CO_2. On the other hand, if all the three principal moments of inertia are different the molecule forms an *asymmetric top* for which an example is the water molecule H_2O. In this case no simple treatment of the rotational motion is possible and the energy levels and wave functions must be computed numerically.

Vibrational structure

The vibrational motion of a polyatomic molecule can be discussed by supposing that the nuclei execute small oscillations about their equilibrium positions. The *normal modes* of vibration can then be determined by the usual methods of classical mechanics, described for example by Goldstein (1980). Each normal mode is associated with a characteristic frequency ν_i and the corresponding quantised normal-mode energy is

$$E_{v_i} = h\nu_i\left(v_i + \frac{1}{2}\right) = \hbar\omega_i\left(v_i + \frac{1}{2}\right) \quad \textbf{(10.138)}$$

where $v_i = 0, 1, 2, \ldots$ and $\omega_i = 2\pi\nu_i$. The total vibrational energy is the sum of the individual vibrational energies associated with each normal mode, namely

$$E_{\rm v} = \sum_i \hbar\omega_i\left(v_i + \frac{1}{2}\right) \quad \textbf{(10.139)}$$

In general, the vibrational spectrum of a molecule with many degrees of freedom is extremely complex and the treatment of this problem is beyond the scope of this book [10]. In the next chapter, however, we shall consider in some detail a particularly interesting vibrational motion, which is responsible for the inversion spectrum of the ammonia molecule NH_3.

Electronic structure

The general principles of the molecular orbital approach can be applied to the electronic structure of polyatomic molecules. The molecular orbitals, as before, can be built from atomic orbitals corresponding to each centre (LCAO approximation). Except for very simple molecules containing a few centres and a few electrons, molecular orbital theory must be built up on a semi-empirical basis, but in simple cases elaborate calculations are possible, using modern computers, in which energy levels and wave functions are obtained *ab initio* [11].

The water molecule H_2O

As a simple example, let us consider the water molecule H_2O. From spectroscopic measurements, it is known that the oxygen nucleus O and the two protons p_1 and p_2 form an isosceles triangle, with the p_1Op_2 angle equal to 105° (see Fig. 10.23). The two protons are about twice as far apart (1.6 Å) as they are in the hydrogen molecule H_2 (0.7 Å), so that the important bonds are between the oxygen atom and each hydrogen atom. From our discussion of O_2, we expect that the bonding orbitals will be constructed from two of the 2p orbitals of atomic oxygen. If the plane of the molecule is the (XY) plane, we can, from linear combinations of the $2p_x$ and $2p_y$ orbitals of oxygen, construct orbitals v_1 and v_2 which are directed along the lines Op_1 and Op_2, respectively. If the hydrogenic 1s orbitals centred on the

[10] A comprehensive account of the rotational and vibrational spectra of polyatomic molecules may be found in Herzberg (1945).

[11] A detailed treatment of the electronic structure of polyatomic molecules is given by Pilar (1968).

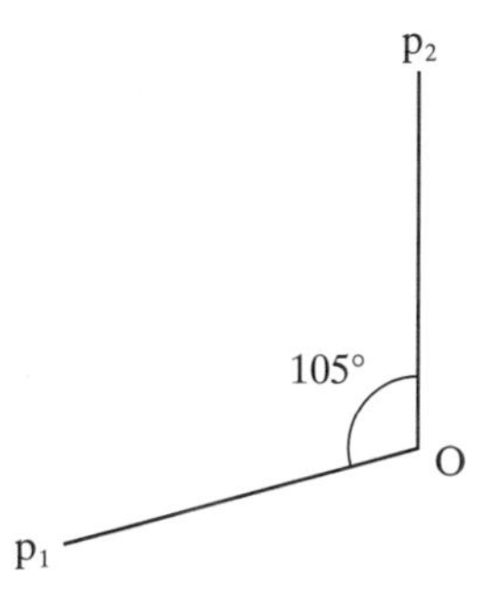

Figure 10.23 The water molecule. The bond angle with no hybridisation would be 90°.

protons p_1 and p_2 are denoted respectively by u_1 and u_2 then suitable LCAO molecular orbitals will be of the form

$$\begin{aligned} \Phi_A &= v_1 + \lambda u_1 \\ \Phi_B &= v_2 + \lambda u_2 \end{aligned} \qquad \textbf{(10.140)}$$

The full wave function for the two valence electrons coming from the oxygen atom and the two electrons from the hydrogen atoms can then be written as a Slater determinant, namely

$$\Phi = N \begin{vmatrix} \Phi_A(1)\alpha(1) & \Phi_A(1)\beta(1) & \Phi_B(1)\alpha(1) & \Phi_B(1)\beta(1) \\ \Phi_A(2)\alpha(2) & \Phi_A(2)\beta(2) & \Phi_B(2)\alpha(2) & \Phi_B(2)\beta(2) \\ \Phi_A(3)\alpha(3) & \Phi_A(3)\beta(3) & \Phi_B(3)\alpha(3) & \Phi_B(3)\beta(3) \\ \Phi_A(4)\alpha(4) & \Phi_A(4)\beta(4) & \Phi_B(4)\alpha(4) & \Phi_B(4)\beta(4) \end{vmatrix} \qquad \textbf{(10.141)}$$

where N is a normalisation constant.

Calculations based on wave functions of this kind show that the equilibrium angle p_1Op_2 is about 90°. The hybridisation of the 2s and 2p oxygen orbitals accounts for the slightly larger angle found experimentally. The orbitals Φ_A and Φ_B give rise to charge distributions which are localised along the lines Op_1 and Op_2. This localisation is the physical basis of the directional character of a chemical bond. If the hydrogen atom with nucleus p_1 were replaced by a different atom, then the wave function in the region between the oxygen atom and the proton p_2 would be hardly altered, so that we can speak of a characteristic O–H bond, associated with a molecular orbital of the type Φ_A or Φ_B.

The methane, ethylene and acetylene molecules

As a further example, we shall now look at the methane molecule, CH_4. The carbon atom in the ground state has the configuration $1s^22s^22p^2$, but there is an excited state with the configuration $1s^22s2p^3$ which is very close in energy and it is from this state that carbon bonds most readily. The orbitals are hybridised by forming linear combinations of the 2s, $2p_x$, $2p_y$ and $2p_z$ orbitals, which consist in this case of one 2s orbital and three 2p orbitals, and are therefore called sp^3 hybrids. Four combinations can be constructed:

$$\begin{aligned}
\Phi_1 &= v_{2s} + v_{2p_x} + v_{2p_y} + v_{2p_z} \\
\Phi_2 &= v_{2s} + v_{2p_x} - v_{2p_y} - v_{2p_z} \\
\Phi_3 &= v_{2s} - v_{2p_x} + v_{2p_y} - v_{2p_z} \\
\Phi_4 &= v_{2s} - v_{2p_x} - v_{2p_y} + v_{2p_z}
\end{aligned} \qquad \textbf{(10.142)}$$

Since the functions v_{2p_x}, v_{2p_y} and v_{2p_z} are proportional to x, y and z respectively, and v_{2s} is spherically symmetric and positive, the function Φ_1 has a maximum in the direction defined by a vector with Cartesian coordinates (1, 1, 1). Similarly, Φ_2, Φ_3 and Φ_4 have maxima in the directions (1, –1, –1), (–1, 1, –1) and (–1, –1, 1) respectively. The angles between each pair of directions are 109.6° and CH_4 forms a tetrahedral structure, as shown in Fig. 10.1.

Carbon is particularly rich in different structures, which arise because other types of directional bonds are possible. For example, sp^2 hybrid orbitals which are linear combinations of only v_{2s}, v_{2p_x} and v_{2p_y} orbitals can be formed, namely

$$\begin{aligned}
\Phi_1 &= v_{2s} + \sqrt{2}\, v_{2p_x} \\
\Phi_2 &= v_{2s} + \sqrt{3/2}\, v_{2p_y} - \sqrt{1/2}\, v_{2p_x} \\
\Phi_3 &= v_{2s} - \sqrt{3/2}\, v_{2p_y} - \sqrt{1/2}\, v_{2p_x}
\end{aligned} \qquad \textbf{(10.143)}$$

Here the bonds lie in a plane, with an angle of 120° between each bond. This structure is found in ethylene, C_2H_4 (see Fig. 10.24(a)). The four protons and the two carbon atoms lie in a plane, the four C–H bonds and one of the two C–C bonds being of the type considered in (10.143). For such bonds the component of the orbital angular momentum in the bond direction is zero, and by analogy with the terminology used in the theory of diatomic molecules these bonds are called σ bonds. The remaining pair of valence electrons are associated with the $2p_z$ orbitals in the carbon atom, the Z axis being perpendicular to the plane of the molecule. This pair is said to form a π bond, the component of the angular momentum in the bond direction being $\pm\hbar$.

Yet another possibility occurs in acetylene, C_2H_2, which is a linear molecule arranged as shown in Fig. 10.24(b). In this case one can form two hybrid orbitals consisting of the 2s orbital and only one 2p (for example, $2p_x$) carbon orbital. These sp hybrid orbitals are given by

$$\begin{aligned}
\Phi_1 &= v_{2s} + v_{2p_x} \\
\Phi_2 &= v_{2s} - v_{2p_x}
\end{aligned} \qquad \textbf{(10.144)}$$

This gives rise to σ bonds linking the hydrogen atoms with the carbon atom, and to a single σ bond linking the two carbon atoms. The angle between these bonds is 180°. The remaining four electrons form a pair of π bonds between the carbon atoms.

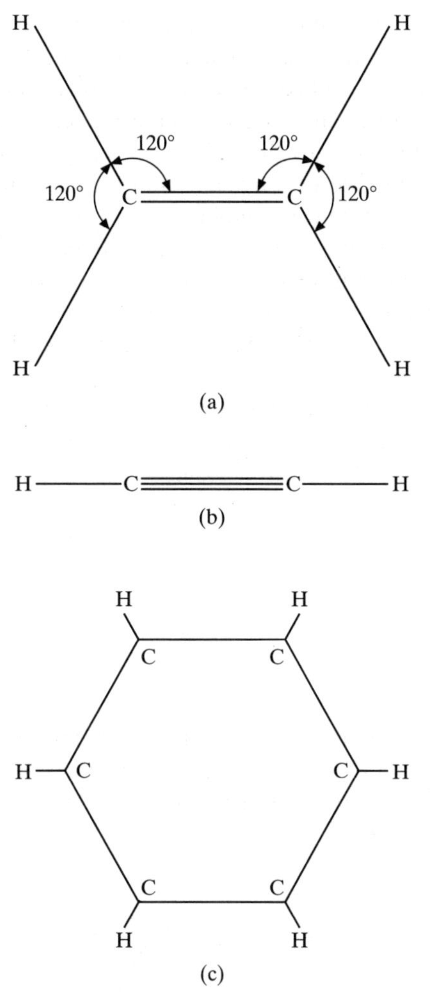

Figure 10.24 Three carbon–hydrogen bond structures: (a) Ethylene C_2H_4; (b) Acetylene C_2H_2; (c) Benzene C_6H_6.

The benzene molecule and non-localised orbitals

In the molecules we have been discussing, the bonds are localised and provide a directed interaction between a pair of atoms within a molecule. A single bond contains a pair of electrons in a singlet spin configuration. The most important bonds are designated σ, in which the magnitude of the component of the orbital angular momentum of each electron in the bond direction is zero, or π, in which this component has the magnitude $\hbar$. The σ bonds are usually stronger and more localised than π bonds, and we have also seen that hybridised bonds are stronger than simple bonds. Although this description is successful for many molecules, it must be extended and a new concept of *non-localised* bonds introduced for others. A typical example is benzene, C_6H_6. The six carbon atoms lie in a plane, which we shall take to be the (XY) plane, and form a regular hexagon, while the six hydrogen atoms also lie in this plane, as shown in Fig. 10.24(c). We can assume that three out of the four $n = 2$ electrons of carbon form sp^2 hybrid orbitals, as in the case of ethylene, the combinations of atomic orbitals being given by (10.143). A linear combination of one of these functions with the 1s orbital of atomic hydrogen forms a σ C–H bond, while the other two combine with the corresponding orbitals on the neighbouring carbon atoms to form C–C bonds which are also σ in character. The remaining atomic carbon orbitals are $2p_z$, and there is one of these per atom. Because of the symmetry of the structure, there is no unique way of associating these $2p_z$ orbitals in pairs, forming localised π bonds.

The way out of this difficulty is to form non-localised π-orbitals, each of which represents an electron moving in the potential due to all six carbon ions. This can be achieved by expressing a particular π-orbital $\Phi_\mu(\mathbf{r})$ by a linear combination of the atomic orbitals $u_{2p_z}(\mathbf{r} - \mathbf{R}_j)$ where $\mathbf{R}_j$ is the positron vector of the carbon nucleus j with respect to the centre of mass of the molecule as origin (see Fig. 10.25). Thus we write

$$\Phi_\mu(\mathbf{r}) = N_\mu \sum_{j=1}^{6} c_j^\mu u_{2p_z}(\mathbf{r} - \mathbf{R}_j) \qquad \textbf{(10.145)}$$

where N_μ is a normalisation factor and the c_j^μ are coefficients which can be found by the variational method. The π-orbitals Φ_μ extend over the whole molecule and we expect that six independent functions labelled by μ can be found from different combinations of the six atomic orbitals centred on the six carbon nuclei.

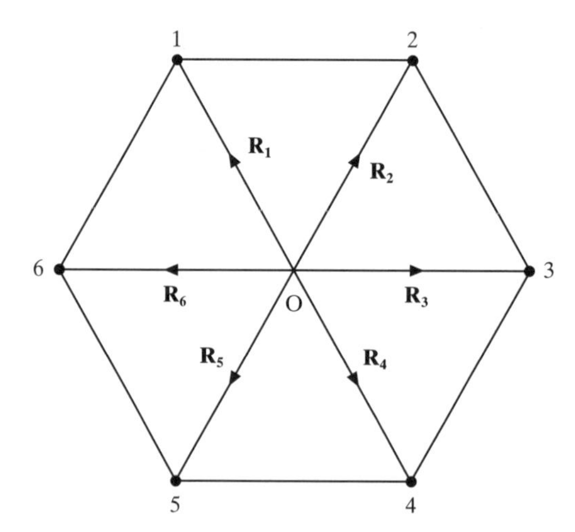

Figure 10.25 The six carbon nuclei in benzene are situated at the vertices of a regular hexagon, with position vectors $\mathbf{R}_1, \mathbf{R}_2, \ldots, \mathbf{R}_6$ with respect to the centre of mass O as origin.

As an alternative to the variational method, symmetry arguments can be employed to find the coefficients c_j^μ. We first note that the electronic Hamiltonian is clearly unaltered if the whole molecule is rotated about the Z axis through 60°, where the Z axis is perpendicular to the plane of the molecule. It follows that the rotated π-orbital denoted by $\mathcal{R}\Phi_\mu(\mathbf{r})$ must represent the same state as $\Phi_\mu(\mathbf{r})$ and therefore $\mathcal{R}\Phi_\mu(\mathbf{r})$ and $\Phi_\mu(\mathbf{r})$ can only differ by a phase factor of unit modulus. That is,

$$\mathcal{R}\Phi_\mu(\mathbf{r}) = \exp(\mathrm{i}\alpha_\mu)\Phi_\mu(\mathbf{r}) \tag{10.146a}$$

$$= \exp(\mathrm{i}\alpha_\mu)N_\mu \sum_{j=1}^{6} c_j^\mu u_{2\mathrm{p}_z}(\mathbf{r} - \mathbf{R}_j) \tag{10.146b}$$

where α_μ is a real number. If six rotations through 60° are performed successively, the original situation is regained so that

$$\exp(\mathrm{i}6\alpha_\mu) = 1 \tag{10.147}$$

from which we have that

$$6\alpha_\mu = 2\pi\mu \tag{10.148}$$

where μ is zero or an integer. All the distinct cases are covered by taking

$$\mu = 0, \pm 1, \pm 2, 3 \tag{10.149}$$

We also note that when a rotation through 60° is made, the carbon nucleus j which was at the position $\mathbf{R}_j$ moves to the position $\mathbf{R}_{j-1}$, so that we have

$$\mathcal{R}u_{2\mathrm{p}_z}(\mathbf{r} - \mathbf{R}_j) = u_{2\mathrm{p}_z}(\mathbf{r} - \mathbf{R}_{j-1}) \tag{10.150}$$

and from (10.145)

$$\mathcal{R}\Phi_\mu(\mathbf{r}) = N_\mu \sum_{j=1}^{6} c_j^\mu u_{2\mathrm{p}_z}(\mathbf{r} - \mathbf{R}_{j-1}) \tag{10.151}$$

where $\mathbf{R}_0 \equiv \mathbf{R}_6$. Since the atomic orbitals on different centres are independent of each other, the coefficients of the orbital $u_{2\mathrm{p}_z}(\mathbf{r} - \mathbf{R}_{j-1})$ can be equated in the expressions (10.146b) and (10.151), giving

$$\exp(\mathrm{i}\alpha_\mu)c_j^\mu = c_{j+1}^\mu, \qquad j = 1, 2, \ldots, 6 \tag{10.152}$$

with $c_7^\mu = c_1^\mu$. These equations are satisfied by taking

$$c_j^\mu = \exp(\mathrm{i}\alpha_\mu j) = \exp(\mathrm{i}\pi\mu j/3) \tag{10.153}$$

so that

$$\Phi_\mu(\mathbf{r}) = N_\mu \sum_{j=1}^{6} \exp(\mathrm{i}\pi\mu j/3)u_{2\mathrm{p}_z}(\mathbf{r} - \mathbf{R}_j) \tag{10.154}$$

The π-orbitals have been obtained using only symmetry arguments. They are analogous to the Bloch wave functions employed to describe the motion of an electron in a periodic crystal lattice [12].

To obtain the energy E_μ of each of the states μ, we use the variational expression (2.383) and evaluate

$$E_\mu = \langle \Phi_\mu | H_e | \Phi_\mu \rangle \tag{10.155}$$

where

$$\langle \Phi_\mu | \Phi_\mu \rangle = 1 \tag{10.156}$$

and H_e is the electronic Hamiltonian. This is not a simple task, but in 1931 E. Hückel introduced a semi-empirical method which although not very accurate has proved of great utility in understanding the structure of benzene and other more complicated organic molecules. The Hückel method consists of making three approximations:

(1) The overlap integrals $\langle u_{2p_z}(\mathbf{r} - \mathbf{R}_j) | u_{2p_z}(\mathbf{r} - \mathbf{R}_k) \rangle$ between an atomic orbital on one centre and an atomic orbital on another are ignored. The atomic orbitals are chosen to be normalised to unity, so that

$$\langle u_{2p_z}(\mathbf{r} - \mathbf{R}_j) | u_{2p_z}(\mathbf{r} - \mathbf{R}_k) \rangle = \delta_{jk} \tag{10.157}$$

It follows that

$$\langle \Phi_\mu | \Phi_\mu \rangle = 6N_\mu^2 \tag{10.158}$$

and hence, using (10.156), we have (apart from an arbitrary factor of modulus one)

$$N_\mu = \frac{1}{\sqrt{6}} \tag{10.159}$$

(2) The diagonal elements of H_e with respect to the atomic orbitals are all retained and are set equal to a constant α

$$\langle u_{2p_z}(\mathbf{r} - \mathbf{R}_j) | H_e | u_{2p_z}(\mathbf{r} - \mathbf{R}_j) \rangle = \alpha \tag{10.160}$$

(3) All the off-diagonal matrix elements of H_e are ignored, except those between nearest neighbours, which are set equal to a constant β. Thus for $j \neq k$

$$\begin{aligned} \langle u_{2p_z}(\mathbf{r} - \mathbf{R}_j) | H_e | u_{2p_z}(\mathbf{r} - \mathbf{R}_k) \rangle &= \beta \text{ for } j = k - 1 \text{ or } j = k + 1 \\ &= 0 \text{ otherwise} \end{aligned} \tag{10.161}$$

Using (10.154), together with (10.159)–(10.161), it follows that

$$\begin{aligned} E_\mu &= \langle \Phi_\mu | H_e | \Phi_\mu \rangle \\ &= \alpha + 2\beta \cos(\pi\mu/3) \end{aligned} \tag{10.162}$$

[12] See Bransden and Joachain (2000).

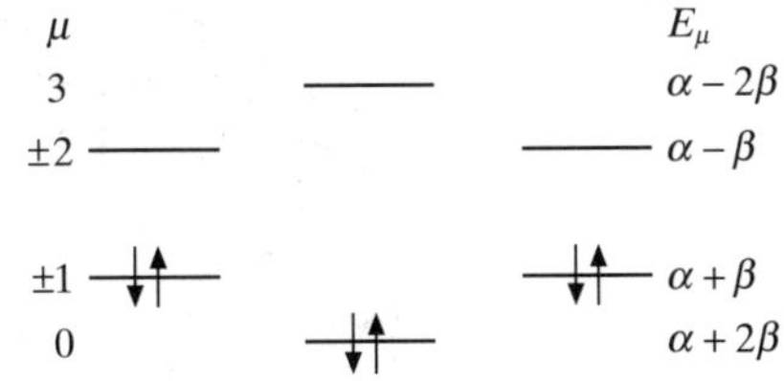

μ		E_μ
3	——	$\alpha - 2\beta$
± 2	—— ——	$\alpha - \beta$
± 1	↑↓ ↑↓	$\alpha + \beta$
0	↑↓	$\alpha + 2\beta$

Figure 10.26 Term diagram for the π electrons in benzene. The states with $\mu = 0$, -1 and $+1$ are each occupied by two bonding π-orbitals of opposite spin, while those with $\mu = -2$, $+2$ and 3 are anti-bonding and unoccupied. Note that β is a negative quantity.

Hence

$$\begin{aligned} E_0 &= \alpha + 2\beta \\ E_1 &= E_{-1} = \alpha + \beta \\ E_2 &= E_{-2} = \alpha - \beta \\ E_3 &= \alpha - 2\beta \end{aligned} \tag{10.163}$$

The constants α and β are both negative so that Φ_0 is the lowest energy state followed by the twofold degenerate states with $\mu = \pm 1$.

In benzene there are six π electrons to be distributed among the π-orbitals. Two electrons of opposite spin can occupy the ground state ($\mu = 0$), two more can be in the excited state with $\mu = +1$ and the last two in the state with $\mu = -1$ (see Fig. 10.26). Making the further assumptions that the contributions of the σ electrons and the electron–electron Coulomb interactions can be neglected, the energy of the π electrons in the ground state of benzene is

$$E = 6\alpha + 8\beta \tag{10.164}$$

The values of α and β can be estimated empirically and are given approximately by $\alpha = -7$ eV and $\beta = -24$ eV. They are of similar magnitude for all conjugated carbon systems, where conjugated means that the π electron wave function extends over the whole molecule.

Although the Hückel method and its extensions which allow for electron–electron interactions provide a useful insight into the structure of large organic molecules, the computing power which is now available makes it possible to obtain quantitative results from elaborate *ab initio* numerical calculations, based on the variational method, for the simpler organic molecules.

The fullerenes

In 1985, H.W. Kroto, J.R. Heath, S.C. O'Brien, R.F. Curl and R.E. Smalley discovered that the vaporisation of graphite by laser irradiation produced a particularly stable cluster of carbon, C_{60}. They proposed the structure for this C_{60} cluster to be a hollow truncated icosahedron composed of 32 faces of which 12

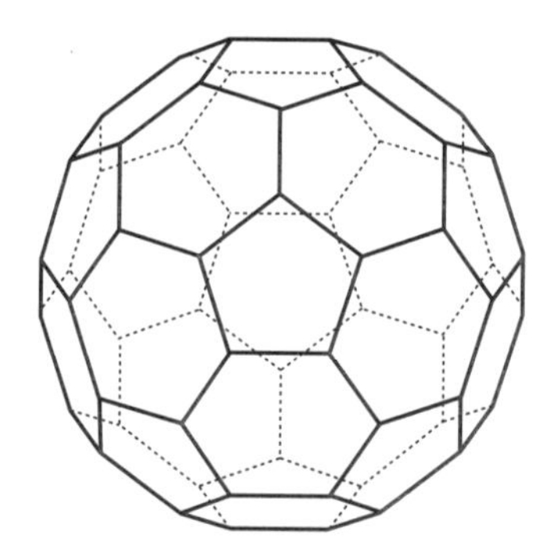

Figure 10.27 The hollow truncated icosahedron structure of C_{60}.

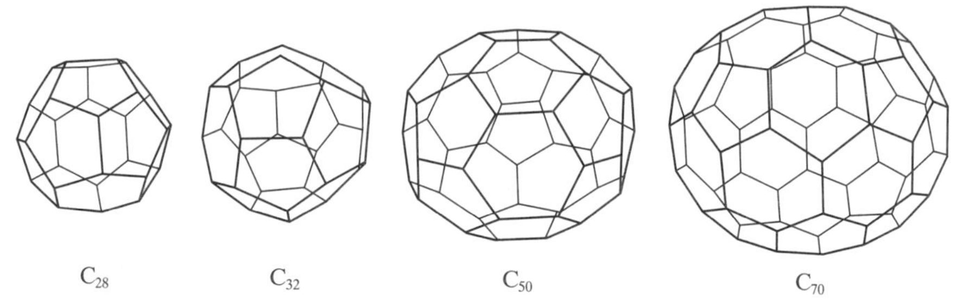

Figure 10.28 The fullerenes C_{28}, C_{32}, C_{50} and C_{70}.

are pentagonal and 20 are hexagonal (see Fig. 10.27), a structure geometrically similar to a soccerball and to the geodesic domes designed by Buckminster Fuller. For this reason, C_{60} is often referred to as 'buckminsterfullerene'. Because of its high degree of symmetry, C_{60} may be the most perfectly spherical molecule. Already in 1970, E. Osawa had suggested that such a molecule would be stable. In 1973, D.A. Bochvar and E.G. Gal'pern performed a Hückel calculation on C_{60} and in 1980 R.A. Davidson studied a number of highly symmetrical molecules including C_{60} by group thcory.

The C_{60} molecule is only one of an infinite number of closed nets (cages) composed of pentagons and hexagons. Starting at C_{20} (which has the form of a dodecahedron), and with the exception of C_{22}, any carbon cluster C_n with $n = 20 + 2k$ ($k = 0, 2, 3, \ldots$) can be formed into a closed net. These closed carbon cages are called *fullerenes*. As an example, the fullerenes C_{28}, C_{32}, C_{50} and C_{70} are shown in Fig. 10.28. In 1986, P.W. Fowler and J. Woolrich showed that fullerene cages C_n with $n + 60 + 6k$ ($k = 0, 2, 3, \ldots$) are electronically stable. Giant fullerenes with $n = 240, 560, 960, \ldots$ were studied in 1988 by H.W. Kroto and K.G. McKay, and shown to have quasi-icosahedral shapes.

The original detection of C_{60} was made on very small amounts of the species in graphite vaporisation experiments using a mass spectrometer as the detection device. In 1990, W. Krätschmer, K. Fostiropoulos and D.R. Huffmann, who for many years were attempting to produce laboratory analogues of interstellar dust, obtained spectroscopic evidence for C_{60} in soot from thermal evaporation of graphite rods. This was achieved by comparing the observed infra-red absorption spectrum with the theoretical predictions of the vibrational modes of C_{60}. Soon afterwards, W. Krätschmer, L.W. Lamb, K. Fostiropoulos and D.R. Huffmann reported the isolation of macroscopic quantities of pure C_{60} and C_{70}.

Since their discovery in 1985, the fullerenes have been the subject of intense investigations [13], stimulated by the prospect of exciting technological applications. For example, metal-doped fullerenes exhibit high-temperature conductivity and the large cousins of the fullerenes – the nanotubes – are about 50 times stronger than steel at one-sixth the weight, and can conduct electricity like copper.

Problems

10.1 If H_e is the electronic Hamiltonian of a diatomic molecule and **L** is the total orbital angular momentum of the electrons, show directly that $[H_e, L_{\bar{z}}] = 0$, $[H_e, L_{\bar{x}}] \neq 0$ and $[H_e, L_{\bar{y}}] \neq 0$, where the $\bar{Z}$ axis is along the internuclear line. The electronic Hamiltonian is given by $H_e = T_e + V$, where T_e and V are given by (10.7) and (10.10) respectively.

10.2 Obtain the function D and the matrix elements H_{AA} and H_{AB} for the hydrogen molecular ion from (10.35) and (10.40) by using the variable charge orbital (10.44) in place of the hydrogenic functions $\psi_{1s}(r_A)$ and $\psi_{1s}(r_B)$. Evaluate the integrals using the results of Appendix 10. Plot $E_g(R)$ as a function of the effective charge Z^* at the equilibrium distance $R_0 = 1.06$ Å.

10.3 The moment of inertia of $H^{79}Br$ is 3.30×10^{-47} kg m^2. Calculate the energies of the first five excited rotational levels of the molecule in eV and the corresponding wave numbers ($\tilde{\nu} = E/hc$) in units of cm^{-1}. Find the internuclear distance in atomic units and in ångströms.

10.4 The wave number $\tilde{\nu}_0$ of the fundamental vibrational motion of the molecule $H^{79}Br$ is 2650 cm^{-1}. Calculate (a) the energy of the lowest state and the first vibrational excited state in electron-volts; (b) the corresponding periodic times; (c) the force constant in SI units.

10.5 Find the dissociation energy D_0 of the deuterium molecule D_2, given that the dissociation energy of H_2 is 4.48 eV and that the energy of the lowest vibrational level of H_2 is 0.26 eV.

10.6 When a diatomic molecule rotates it stretches, so that the internuclear equilibrium distance is increased to R_1, where R_1 is given approximately by (10.116). Prove this result as indicated in the text.

[13] See Kroto, Fischer and Cox (1993).

10.7 Using $R_0 = 4.38$ a.u. and $\tilde{\nu}_0 = 364.6\ \text{cm}^{-1}$, determine the constants A and c in the potential (10.81) representing the energy of Na^{35}Cl. Calculate the depth of the well D_e and the chemical dissociation energy D_0.

10.8 Calculate the chemical dissociation energy D_0 for the molecules LiF, KF and KI using the following data:

$$\begin{aligned} &\text{LiF: } R_0 = 1.564\ \text{Å}, \qquad \tilde{\nu}_0 = 910.3\ \text{cm}^{-1} \\ &\text{KF: } R_0 = 2.172\ \text{Å}, \qquad \tilde{\nu}_0 = 426.0\ \text{cm}^{-1} \\ &\text{KI: } R_0 = 3.048\ \text{Å}, \qquad \tilde{\nu}_0 = 186.5\ \text{cm}^{-1} \end{aligned}$$

The ionisation potential of Li is 5.39 eV and that of K is 4.34 eV. The electron affinity of F is 3.45 eV and that of I is 3.08 eV.

(**Note**: $E_s(\infty)$ of (10.81) is the difference between the ionisation potential of the alkali and the electron affinity of the halogen.)

11 Molecular spectra

Transitions between the energy levels of a molecular system can take place with the emission or absorption of radiation. The molecular spectra are more complicated than those of atoms, and in this chapter we shall only analyse some of the simpler features of these spectra. For the most part we shall confine the discussion to diatomic molecules, pointing out in some cases how extensions to polyatomic systems can be made. We begin in Section 11.1 by analysing the rotational spectra of diatomic molecules, while their vibrational–rotational spectra and electronic spectra are discussed in Sections 11.2 and 11.3, respectively. In the next section, we examine how electric dipole transitions in diatomic molecules are modified when spin-dependent interactions are taken into account. The effect of the nuclear spin on the spectra of homonuclear diatomic molecules is studied in Section 11.5. Finally, in Section 11.6 we examine in some detail the inversion spectrum of the ammonia molecule NH_3, which as we shall see in Chapter 15 is of great interest in understanding the operation of the *ammonia maser*.

11.1 Rotational spectra of diatomic molecules

We have seen in the previous chapter that in the simplest approximation the couplings between the electronic, vibrational and rotational motions can be neglected. Disregarding small spin-dependent interactions and treating the vibrational motion of the two nuclei as that of a linear harmonic oscillator, we can write the complete molecular wave function Ψ_a of a diatomic molecule, corresponding to a given state a characterised by the quantum numbers (s, v, K, M_K, Λ), as

$$\Psi_a = \Phi_s(\mathbf{R}, \mathbf{r}_1, \mathbf{r}_2, \dots, \mathbf{r}_N)\, R^{-1}\psi_v(R - R_0)\, {}^{\Lambda}\mathcal{H}_{KM_K}(\Theta, \Phi), \qquad a \equiv (s, v, K, M_K, \Lambda) \tag{11.1}$$

where the spin variables of the electrons and nuclei have not been displayed explicitly. The energy of the state a is

$$E_a \equiv E_{s,v,K} \equiv E_{s,v,r} = E'_s(R_0) + E_v + E_r \tag{11.2}$$

where the electronic energy $E'_s(R_0)$ is given by (see (10.92))

$$E'_s(R_0) = E_s(R_0) - \frac{\Lambda^2\hbar^2}{2\mu R_0^2} + \frac{1}{2\mu R_0^2}\langle \Phi_s | L_{\bar{x}}^2 + L_{\bar{y}}^2 | \Phi_s \rangle \tag{11.3}$$

The vibrational energy E_v is (see (10.101))

$$E_v = \hbar\omega_0\left(v + \frac{1}{2}\right) = h\nu_0\left(v + \frac{1}{2}\right), \qquad v = 0, 1, 2, \ldots \tag{11.4}$$

with $\omega_0 = (k/\mu)^{1/2}$ and $\nu_0 = \omega_0/(2\pi)$. The rotational energy E_r is (see (10.98))

$$E_r = BK(K+1), \qquad K \geqslant \Lambda \tag{11.5}$$

where K is the quantum number of the total orbital angular momentum of the molecule, Λ is the modulus of the component of the electronic orbital angular momentum along the internuclear axis, and the rotational constant B is given by $B = \hbar^2/(2I_0)$, where $I_0 = \mu R_0^2$ is the moment of inertia of the molecule.

In the electric dipole approximation, which was discussed in Chapters 4 and 9, the transition amplitude for emission or absorption of radiation was shown to be proportional to the matrix element of the electric dipole operator $\mathbf{D}$. For a molecule this operator is given by

$$\mathbf{D} = e\left(\sum_i Z_i\mathbf{R}_i - \sum_j \mathbf{r}_j\right) \tag{11.6}$$

where the first sum is over the positions $\mathbf{R}_i$ and charges Z_ie of the nuclei and the second sum is over the positions $\mathbf{r}_j$ of the electrons. The diagonal matrix element of $\mathbf{D}$ in the state a is given by

$$\mathbf{D}_{aa} = \langle\Psi_a|\mathbf{D}|\Psi_a\rangle \tag{11.7}$$

and is the *permanent electric dipole moment* of the molecule in that state. This quantity always vanishes for non-degenerate levels of atoms because these are eigenstates of the parity operator. However, for heteronuclear molecules in which an excess of charge is associated with one of the nuclei, D_{aa} has a finite value (see Tables 10.2 and 10.3). In symmetrical homonuclear diatomic molecules, such as H_2, O_2, N_2, . . . , the permanent electric dipole moment vanishes. Since the rotational and vibrational motions preserve the symmetry of the molecule, the matrix elements of $\mathbf{D}$ between different rotational or vibrational states must vanish for symmetrical homonuclear molecules, unless the electronic state changes. As a result symmetrical molecules possess no rotational or vibrational spectrum without an electronic transition. In contrast, molecules which possess a permanent electric dipole moment such as HCl exhibit spectra corresponding to rotational and vibrational transitions without change in the electronic state.

The selection rules for a rotational transition are, if $\Lambda = 0$,

$$\Delta K = \pm 1 \tag{11.8a}$$

and

$$\Delta M_K = 0, \pm 1 \tag{11.8b}$$

If $\Lambda \neq 0$, the first selection rule (11.8a) must be replaced by $\Delta K = 0, \pm 1$. This is due to the fact that although the photon absorbed or emitted carries one unit of

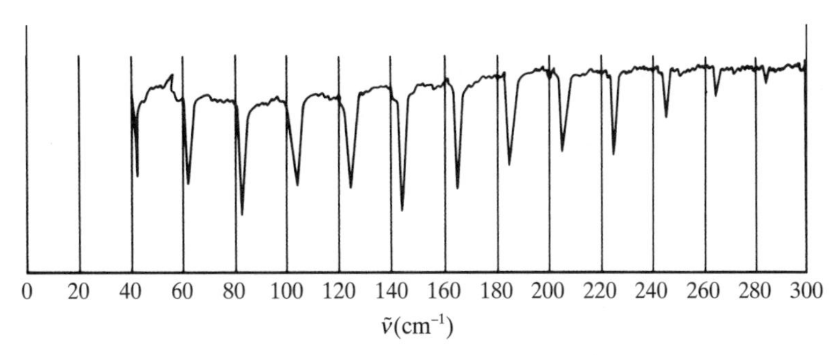

Figure 11.1 The rotational absorption spectrum of gaseous HCl. From (11.9) we see that the spacing in frequency between the absorption lines (corresponding to transitions from $K \to K+1$) is constant and equal to $2B/h$. For HCl $\tilde{B} = (B/hc) = 10.59$ cm^{-1} and the spacing in terms of the wave number $\tilde{\nu}$ is $2\tilde{B} = 21.18$ cm^{-1}.

angular momentum, the angular momentum due to the nuclear rotation can change with no change in K, provided that the electronic angular momentum makes an equal and opposite change. Symmetry considerations require that the electric dipole moment of a diatomic molecule is directed along the internuclear axis, and this leads to the additional selection rule $\Delta\Lambda = 0$. If no change in vibrational state occurs then we must have $\Delta K = \pm 1$.

The pure rotational spectrum of a diatomic molecule consists of lines in the far infra-red or the microwave region, the frequencies of these lines being given by

$$\begin{aligned} h\nu_{K+1,K} &= E_r(K+1) - E_r(K) \\ &= 2B(K+1) \end{aligned} \tag{11.9}$$

where $K \geqslant \Lambda$. Figure 11.1 illustrates the rotational spectrum of HCl. The constant B can be obtained with an accuracy of up to 1 part in 10^9 from the observed spectra, and this allows the equilibrium distance R_0 to be determined. The intensity of the line is proportional to the permanent electric dipole moment in the particular electronic state concerned.

11.2 Vibrational–rotational spectra of diatomic molecules

Vibrational transitions can occur, due to the interaction with the radiation field, if the matrix element

$$\mathbf{D}_{v'v} = \int_0^\infty \psi^*_{v'}(R - R_0)\mathbf{D}(R)\psi_v(R - R_0)\,\mathrm{d}R \tag{11.10}$$

does not vanish. In this expression $\mathbf{D}(R)$ is the matrix element of the electric dipole moment expressed as a function of the internuclear distance R, and ψ_v, $\psi_{v'}$ are linear harmonic oscillator eigenfunctions if the lowest approximation is used (see Section 10.4).

If $\mathbf{D}$ is independent of R, the integral (11.10) vanishes because of the orthogonality of ψ_v and $\psi_{v'}$. On the other hand, if $\mathbf{D}$ depends on R, we may expand it as

$$\mathbf{D}(R) = \mathbf{D}(R_0) + (R - R_0)\left.\frac{\mathrm{d}\mathbf{D}(R)}{\mathrm{d}R}\right|_{R=R_0} + \dots \qquad \textbf{(11.11)}$$

and we can neglect the higher order terms in first approximation since the amplitude of the vibrational motion is small compared with R_0. In this approximation the selection rule depends on the integral

$$I(v, v') = \int_0^\infty \psi^*_{v'}(R - R_0)[R - R_0]\psi_v(R - R_0)\,\mathrm{d}R \qquad \textbf{(11.12)}$$

We recall that the linear harmonic oscillator wave functions ψ_v are given by (see (2.148))

$$\psi_v(x) = N_v \mathrm{e}^{-\alpha^2x^2/2} H_v(\alpha x) \qquad \textbf{(11.13)}$$

where N_v is a normalisation constant, $x = R - R_0$ and $\alpha^2 = \mu\omega_0/\hbar$, μ being the reduced mass (10.5) of the two nuclei and ω_0 the angular frequency of the oscillator. Using the result (A3.15) of Appendix 3 we see that the integral (11.12) vanishes unless $v' = v \pm 1$. We note that this selection rule may also be derived immediately by using the recurrence relation for the Hermite polynomials. Using (2.147),

$$2(\alpha x)H_v(\alpha x) = 2vH_{v-1}(\alpha x) + H_{v+1}(\alpha x) \qquad \textbf{(11.14)}$$

it follows at once that the integral (11.12) vanishes unless

$$\Delta v = v' - v = \pm 1 \qquad \textbf{(11.15)}$$

Since the potential well is only approximately described by a linear harmonic oscillator, transitions with $\Delta v = v' - v = \pm 2, \pm 3, \dots$ can occur. However, these transitions are usually weaker, by at least an order of magnitude.

Because the absorption of a photon requires the molecule to take up one unit of angular momentum, vibrational transitions are accompanied by a change in the rotational state, which is subject to the same selection rules as for the pure rotational spectrum. For a molecule in a Σ state (with $\Lambda = 0$), the transitions between two rovibronic states (v, K) and (v', K'), with vibrational quantum numbers v and $v' = v + 1$, fall into two sets according to whether $\Delta K = +1$ or $\Delta K = -1$ (see Fig. 11.2). The first set is called the *R branch*. Using the linear harmonic oscillator expression (11.4) for the vibrational energy and the rigid rotator formula (11.5) with $\Lambda = 0$ for the rotational energy, we find that the corresponding frequencies ν^{R} are given by

$$\begin{aligned} h\nu^{\mathrm{R}} &= E(v + 1, K + 1) - E(v, K) \\ &= 2B(K + 1) + h\nu_0, \qquad K = 0, 1, 2, \dots \end{aligned} \qquad \textbf{(11.16)}$$

The second set is known as the *P branch*. In the same approximation (linear harmonic oscillator, rigid rotator), its frequencies ν^{P} are such that

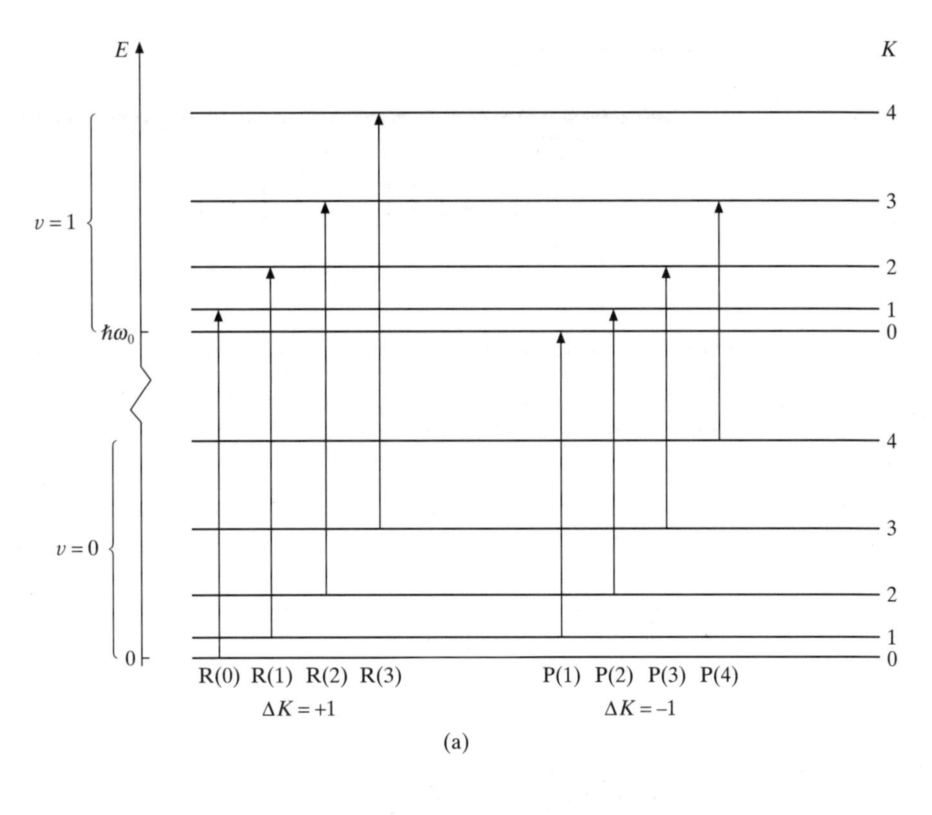

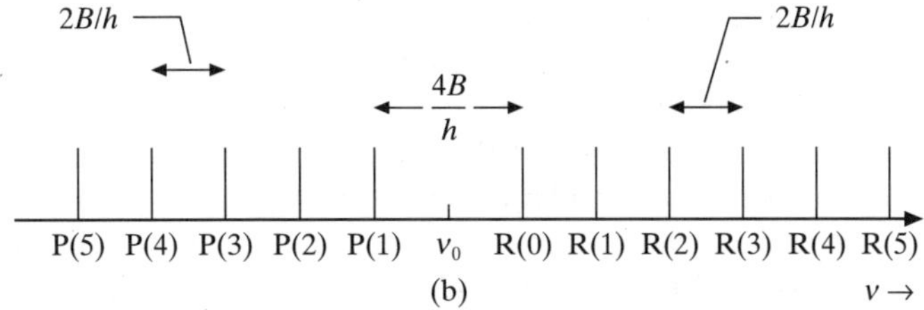

Figure 11.2 (a) Energy level diagram of the lowest vibrational–rotational levels in a diatomic molecule showing the absorptive transitions from the band with $v = 0$ to the band with $v = 1$. The band spectrum contains two branches: the R branch with $\Delta K = +1$ and the P branch with $\Delta K = -1$. The lines in each branch are labelled by the K value of the lower level. (b) A spectrogram showing the lines corresponding to the transitions shown in (a). It is assumed that the rotational constants of the $v = 0$ and $v = 1$ vibrational levels are the same so that the lines are equally spaced in frequency, except for a gap corresponding to the vibrational frequency ν_0.

$$\begin{aligned} h\nu^{\text{P}} &= E(v+1, K-1) - E(v, K) \\ &= -2BK + h\nu_0, \qquad K = 1, 2, 3, \ldots \end{aligned} \tag{11.17}$$

Both branches make up what is called a *vibrational–rotational band*. These bands are in the infra-red part of the spectrum. We note that according to (11.16)

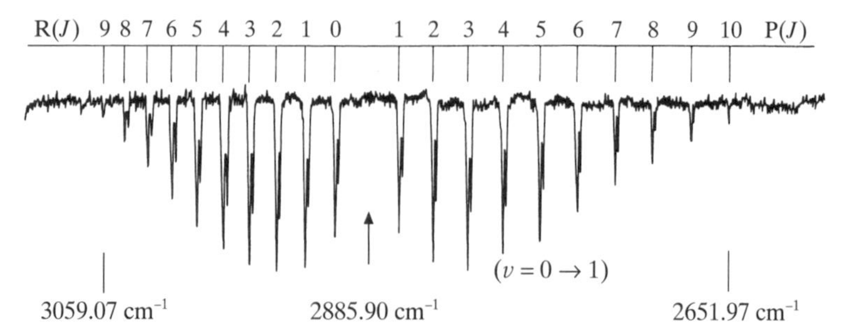

Figure 11.3 Absorption spectrum of HCl. The double peaks exist because naturally occurring chlorine is a mixture of the two isotopes ^{35}Cl (75.5 per cent) and ^{37}Cl (24.5 per cent). The position of the central gap in the spectrum determines an effective vibrational wave number (including the anharmonicity) $\nu_0 = 2885.9$ cm^{-1} ($\hbar\omega_0 = 0.369$ eV) and the spacing determines the rotational constant $\tilde{B} = B/(hc) = 10.59$ cm^{-1} ($B = 1.31 \times 10^{-3}$ eV). (By courtesy of R. Colin.)

and (11.17) such bands contain lines whose frequencies are equally spaced by $2B/h$, except that there is a gap of $4B/h$ at the vibrational frequency ν_0. This is illustrated in Fig. 11.2(b) and in Fig. 11.3, the latter showing the absorption spectrum of HCl. We see that a measurement of the position of the gap determines the vibrational frequency $\nu_0 = \omega_0/(2\pi)$, while a measurement of the spacing of the lines determines the rotational constant $B = \hbar^2/(2\mu R_0^2) = \hbar^2/(2I_0)$. Except for very heavy molecules, most molecules in a gas at room temperature are in the vibrational ground state with $v = 0$, so the most easily observed transitions are between the levels with $v = 0$ and $v = 1$. This band is called the fundamental band.

In actual spectra the difference in frequencies between adjacent lines is not quite constant, as in the simple model we have just described. This may be seen from Table 11.1, which displays the observed frequencies of the fundamental band in the spectrum of HCl. This departure from constancy is due to the fact that the rotational constant B is not quite the same in each of the vibrational states, so that we must consider two constants B_v and B_{v+1}. In this case the frequency of the lines is a quadratic function of K, and the second differences of the frequencies are constant. From the first and second differences both B_v and B_{v+1} can be calculated. We also note from Table 11.1 that the central line in the spectrum which should be at 2885.90 cm^{-1} is missing. The corresponding gap makes the identification of the lines of the P and R branches easy. In practice, the departure from linear harmonic motion in vibrational–rotational spectra can be observed and the anharmonicity constant (see (10.110)) can also be calculated.

If the molecule is not in a Σ state, so that $\Lambda \neq 0$, transitions with $\Delta K = 0$ are allowed. This gives rise to a further branch of the vibrational–rotational spectrum, called the *Q branch*. The frequencies ν^{Q} corresponding to the lines in this branch are given by a quadratic function of K if B_{v+1} and B_v are unequal. If $B_{v+1} = B_v$, they reduce to the single frequency ν_0, since

$$h\nu^{Q} = E(v+1, K) - E(v, K)$$
$$= h\nu_0 \qquad \textbf{(11.18)}$$

Table 11.1 Wave numbers of the central lines in the fundamental rotational–vibrational band ($v = 0 \leftrightarrow 1$) of HCl

	$\tilde{\upsilon}$ (cm^{-1})	A (cm^{-1})
R(5)	2997.78	
R(4)	2980.90	16.88
R(3)	2963.24	17.66
R(2)	2944.89	18.55
R(1)	2925.78	19.11
R(0)	2906.25	19.53
ν_0	missing at 2885.90	–
P(1)	2865.09	–
P(2)	2843.56	21.53
P(3)	2821.49	22.07
P(4)	2798.78	22.71
P(5)	2775.79	22.99
P(6)	2752.03	23.76

A = Wave number difference between neighbouring lines.

Rayleigh and Raman scattering

So far, the main experimental technique for obtaining information about the properties of atoms and molecules that we have considered is the observation of the emission or absorption of radiation. Important additional methods involve the *scattering* of electrons or of radiation by atomic or molecular systems. We shall study electron scattering in Chapter 13, while here we discuss briefly the scattering of radiation by molecules.

The scattering of radiation by an atomic system has been discussed in Section 4.9. The general expression for the cross-section corresponding to the emission of a photon in the direction (θ, ϕ) when the incident direction is along the Z axis is given, in the dipole approximation, by (4.223). In the case of molecular targets the dipole matrix elements $\mathbf{D}_{ba}$ which appear in (4.223) are calculated from the dipole operator (11.6) and the selection rule (4.225) which was given for a one-electron atom is replaced by

$$\Delta K = 0, \pm 2 \qquad \textbf{(11.19)}$$

As in the atomic case, energy conservation requires that

$$\omega' = \omega + (E_a - E_b)/\hbar \qquad \textbf{(11.20)}$$

where ω is the angular frequency of the incident photon, ω' is that of the emitted photon, while E_a and E_b are the energies of the initial and final states of the molecule. When the scattering is elastic ($\omega = \omega'$) it is called Rayleigh scattering and when it is inelastic ($\omega \neq \omega'$) one speaks of Raman scattering just as in the atomic case.

Raman scattering occurs with appreciable strength in molecular scattering because the rotational–vibrational levels have closely spaced energies E_n, which are also close to E_a and (or) E_b, so that the energy denominators in the expression for the cross-section (4.223) are small. When the initial state a of the molecule is the ground state, the final state b must have higher energy, so that $\omega' < \omega$. In this case the observed spectral line is called a *Stokes line*. If the initial state a is an excited state the final state b may have a higher or a lower energy. In the latter case $\omega' > \omega$ and the corresponding spectral line is called an *anti-Stokes line*. We note that the Raman effect does not require the existence of a permanent electric dipole moment, but rather that an electric dipole moment should be induced under the influence of the radiation field. For this reason, Raman lines are observed for symmetrical molecules like H_2, O_2, . . . which exhibit no rotational and vibrational spectra without a change in electronic state. A schematic energy level diagram for the rotational Raman spectrum is shown in Fig. 11.4, together with the corresponding spectrogram.

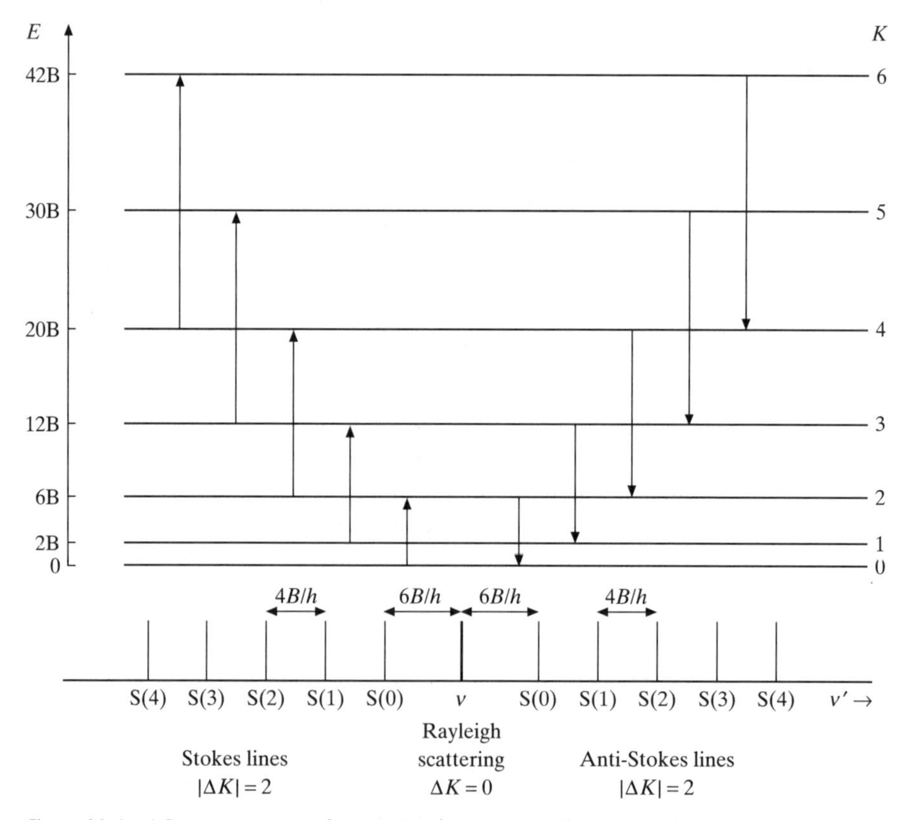

Figure 11.4 A Raman spectrum for rotational transitions. The scattered light contains a line at the incident frequency $\nu = \omega/(2\pi)$ due to Rayleigh scattering $\Delta K = 0$, and equally spaced lines on either side due to Raman scattering with $|\Delta K| = 2$. Transitions with $|\Delta K| = 2$ are often called the S branch and are numbered by the K value of the lower level.

11.3 Electronic spectra of diatomic molecules

Molecular spectra for which changes in the electronic as well as in the vibrational and rotational states of the molecule occur are called *electronic spectra*. The energy differences between electronic levels being much larger than those corresponding to transitions without a change in the electronic state, the lines associated with electronic spectra lie in the *visible* or the *ultra-violet* part of the spectrum. When observed with small dispersion, electronic spectra usually appear to consist of more or less broad *bands*, and one speaks of *electronic band spectra*. As an example we show in Fig. 11.5 various bands (photographed in emission) corresponding to the transition $A^2\Sigma^+ \rightarrow X^2\Sigma^+$ between the excited electronic state $A^2\Sigma^+$ and the ground state $X^2\Sigma^+$ of the AlO molecule. When spectrographs of larger resolving power are used, one finds that most of the bands exhibit a *fine structure* in the sense that they consist of many closely spaced individual lines. This is illustrated in Fig. 11.6, which shows the fine structure of a particular band of the electronic emission spectrum of the BeI molecule.

In order to understand the basic features of electronic spectra, we return to (11.1) which expresses the fact that the total energy $E_{s,v,r}$ of the molecule may be written (approximately) as the sum of the electronic energy E_s, the vibrational

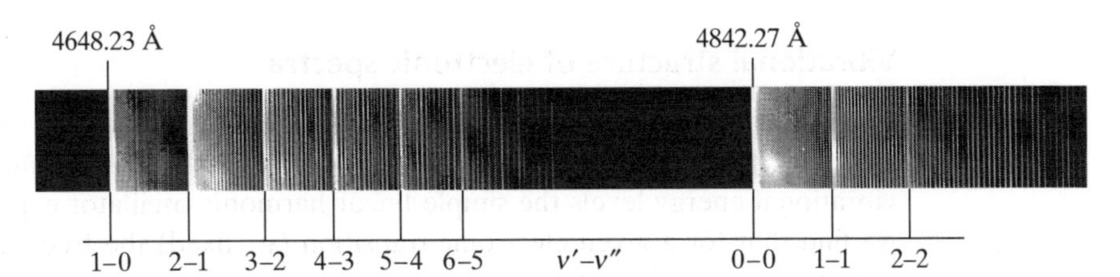

Figure 11.5 Electronic band spectra photographed in emission in transitions from the excited state $A^2\Sigma^+$ to the ground state $X^2\Sigma^+$ of the molecule AlO. (By courtesy of R. Colin.)

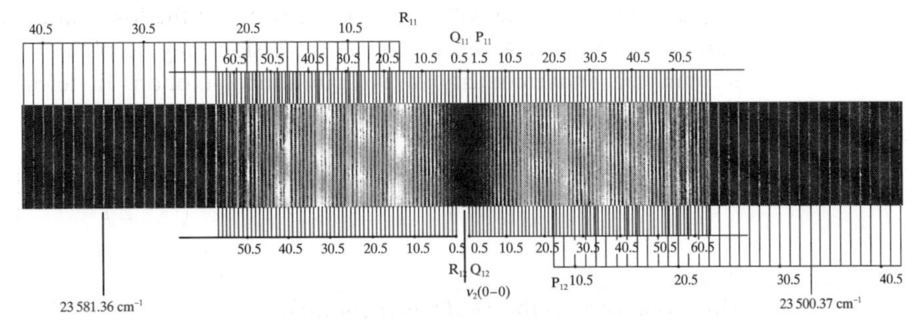

Figure 11.6 The fine structure of the 0–0 emission band of BeI, arising from the electronic transition $A^2\Pi_{1/2} - X^2\Sigma^+$. (By courtesy of R. Colin.)

energy E_v and the rotational energy E_r. Thus the emitted or absorbed frequencies of the various spectral lines corresponding to the transitions between two electronic states are given by

$$\nu = \frac{(E_{s'} + E_{v'} + E_{r'}) - (E_s + E_v + E_r)}{h} \quad \textbf{(11.21)}$$

where the primed letters refer to the upper electronic state. We may also write (11.21) as

$$\nu = \nu_{s's} + \nu_{v'v} + \nu_{r'r} \quad \textbf{(11.22)}$$

with

$$\nu_{s's} = \frac{E_{s'} - E_s}{h}, \qquad \nu_{v'v} = \frac{E_{v'} - E_v}{h}, \qquad \nu_{r'r} = \frac{E_{r'} - E_r}{h} \quad \textbf{(11.23)}$$

For a *given electronic transition* (s, s' fixed) the quantity $\nu_{s's}$ is a *constant*, which is positive since we have assumed that $E_{s'} > E_s$. On the other hand, $\nu_{v'v}$ and $\nu_{r'r}$ can vary and need not be positive. A given *band* corresponds to fixed values of both $\nu_{s's}$ and $\nu_{v'v}$ and all possible (allowed) values of $\nu_{r'r}$, while the *band system* of a given electronic transition ($\nu_{s's}$ fixed) is obtained by letting both $\nu_{v'v}$ and $\nu_{r'r}$ take all possible values. Thus $\nu_{v'v}$ characterises the *vibrational* or *band structure* of the spectrum, and $\nu_{r'r}$ its *rotational* structure, that is the *fine structure* of the band.

Vibrational structure of electronic spectra

Let us neglect for the moment the fine structure due to the rotation of the molecule. Keeping only the first two terms on the right of (11.22) and using for the vibrational energy levels the simple linear harmonic oscillator expression (11.4), we find that for a given electronic transition ($\nu_{s's}$ fixed) the frequencies ν of the transitions (s', $v' \leftrightarrow s$, v) are such that

$$h\nu = h\nu_{s's} + h\nu_0'\left(v' + \frac{1}{2}\right) - h\nu_0\left(v + \frac{1}{2}\right) \quad \textbf{(11.24)}$$

where ν_0 and ν_0' are the vibrational frequencies of the lower and upper electronic states, respectively. A more precise expression may be obtained by taking into account the anharmonicity of the vibrational motion. In terms of the anharmonicity constant introduced in (10.110), we have

$$h\nu = h\nu_{s's} + h\nu_0'\left(v' + \frac{1}{2}\right) - h\beta'\nu_0'\left(v' + \frac{1}{2}\right)^2 - h\nu_0\left(v + \frac{1}{2}\right) + h\beta\nu_0\left(v + \frac{1}{2}\right)^2 \quad \textbf{(11.25)}$$

which is known as the *Deslandres formula*.

It is customary to record the frequencies of the transitions (s', $v' \leftrightarrow s$, v) between pairs of vibrational levels v and v' (for a given pair of electronic states s and s')

in a *Deslandres table*. The rows of such a table are labelled by the vibrational quantum number v of the lower level and the columns by the vibrational quantum number v' of the upper level. A set of bands having the same v (or v') is called a v (or v') *progression*. At thermal energies (corresponding to room temperature) gases are primarily in the ground electronic state, with $v = 0$, and the absorption spectrum only contains the corresponding $v = 0$ progressions. Finally, a group of bands having the same value of $v' - v$ is called a band *sequence*; it occurs along the diagonal of the Deslandres table.

Rotational structure of electronic spectra

Let us consider a given electronic band, corresponding to fixed values of (s, v) and (s', v'). The rotational energies E_r and $E_{r'}$ are characterised by the quantum numbers K and K', respectively, with $K \geqslant \Lambda$ and $K' \geqslant \Lambda'$ (see (11.5)). If the electronic states are such that $\Lambda \neq 0$ and (or) $\Lambda' \neq 0$, the selection rules are $\Delta K = 0, \pm 1$ and $\Delta\Lambda = 0, \pm 1$. For transitions between Σ states ($\Lambda = \Lambda' = 0$) we have $\Delta K = \pm 1$, $\Sigma^+ \leftrightarrow \Sigma^+$ and $\Sigma^- \leftrightarrow \Sigma^-$, but $\Sigma^+ \leftrightarrow \Sigma^-$ is not allowed. For homonuclear molecules, we have the additional selection rule g $\leftrightarrow$ u which only allows transitions between gerade and ungerade states. There is no selection rule for the vibrational quantum number since the harmonic oscillator wave functions ψ_v and $\psi_{v'}$ which enter into (11.12) belong to different wells. In the case considered here, for which spin-dependent interactions are negligible, S is a 'good' quantum number. Since the electric dipole operator (11.6) is independent of the spin variables, it follows that the selection rule $\Delta S = 0$ must apply so that the multiplicity of the electronic state does not change in electronic dipole transitions.

As a result of the selection rule $\Delta K = 0, \pm 1$, the band divides into P, Q and R branches, the Q branch being absent for transitions between two Σ states, where $\Delta K = \pm 1$. The frequencies of the rotational lines are given by

$$\begin{aligned} h\nu^{\rm P} &= h\nu + B'K(K-1) - BK(K+1) \\ h\nu^{\rm Q} &= h\nu + B'K(K+1) - BK(K+1) \\ h\nu^{\rm R} &= h\nu + B'(K+1)(K+2) - BK(K+1), \; K \geqslant \Lambda \end{aligned} \tag{11.26}$$

where $h\nu$ is given either by the Deslandres formula (11.25) or – if the simple linear harmonic oscillator approximation is made – by (11.24). In these formulae, the lower level in each case has angular momentum K, while the upper level has angular momentum $K-1$, K or $K+1$ for the P, Q and R branches, respectively.

In contrast to the vibrational–rotational expressions (11.16)–(11.17), which are linear in K, or (11.18) which is independent of K, the formulae (11.26) for the rotational structure of electronic bands are *parabolic* in K. This is due to the fact that the rotational constants B and B' are in general different in the lower and upper levels. Plots of K against the frequencies of the rotational lines are called Fortrat diagrams. A typical example of a *Fortrat parabola* is shown in Fig. 11.7. We see that since $B \neq B'$ the lines are not equally spaced. In the example shown

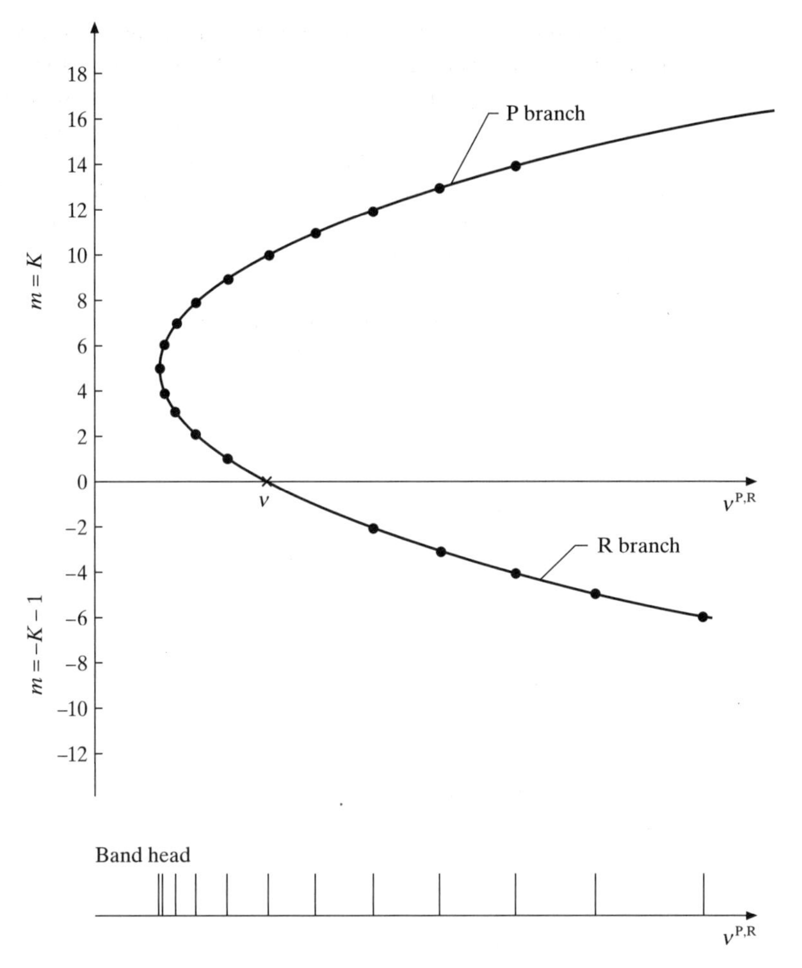

Figure 11.7 A Fortrat parabola for a $\Sigma \to \Sigma$ transition in a diatomic molecule (no Q branch). The spectrogram beneath shows the position of the band head.

the lines become closely spaced at the low-frequency end ('red end') of the band, and form what is known as a *band head* at the minimum frequency. As one goes to higher frequencies the intervals between the lines increase, and the intensity of the band falls off gradually. This is called *band degradation*. In other cases the band head can be at the highest frequency and the band degrades towards lower frequencies.

The Franck–Condon principle

We have seen in Chapter 10 that there is little interaction between the electronic and nuclear motions in a molecule, the nuclear periods being much longer than the electronic ones. Thus an electronic transition can be considered as taking place

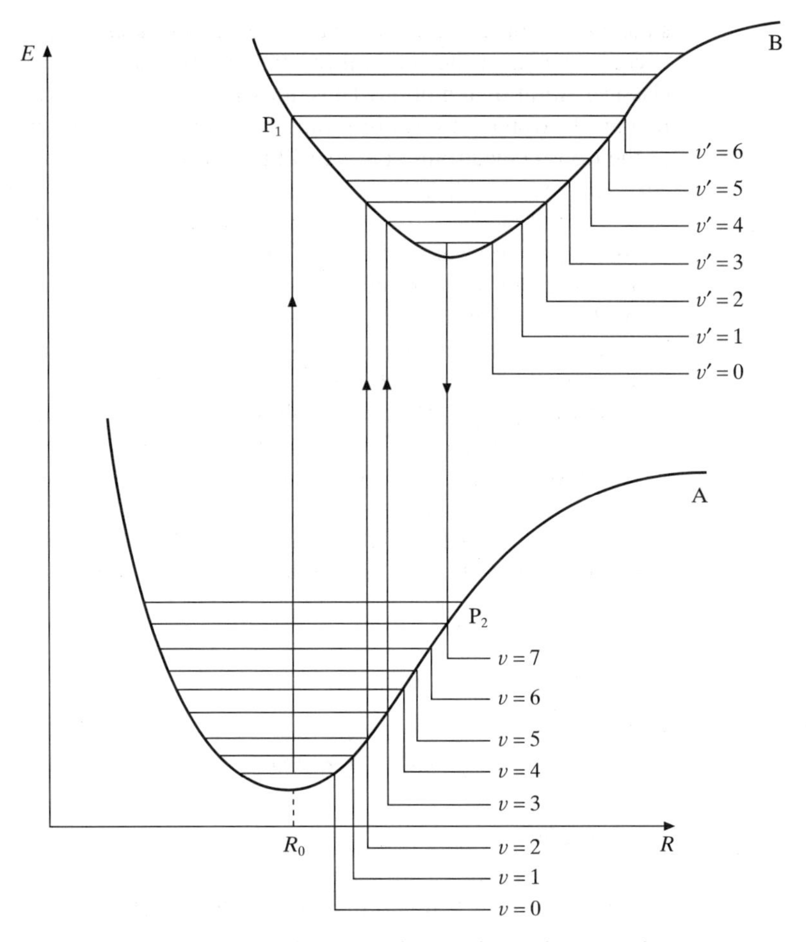

Figure 11.8 Electronic transitions illustrating the Franck–Condon principle.

almost 'instantaneously' on the time scale of vibrations, or in other words at a nearly *constant* value of the internuclear separation R. The heavy nuclei do not change their positions or momenta during the electronic transition, but only *after* it has occurred. Referring to Fig. 11.8, in which the potential energy curves of two electronic states A and B are shown with their associated vibrational levels, we see that an electronic transition between the states A and B is accurately represented by a *vertical line*, that is a line of constant R. This is known as the *Franck–Condon principle.*

The Franck–Condon principle can be combined with our discussion of the linear harmonic oscillator wave functions in Section 2.4 to understand the *intensity distribution* among the bands of a band system. Indeed, if the molecule is in the lowest vibrational state of a given electronic state, the probability distribution function for R is large only near the equilibrium separation of that electronic state. On the other hand, in excited vibrational states this probability distribution becomes larger at the classical turning points, that is at the extreme ends of the classical vibrational motion. Thus, looking again at Fig. 11.8, we see for example that an electronic (absorption) transition from the ground ($v = 0$) vibrational level of the (lower) electronic state A will lead to the point P_1 on the upper electronic curve B, corresponding to the vibrational level $v' = 6$. As a result, the band $v = 0 \to v' = 6$ will be the most intense member of the progression starting from the $v = 0$ vibrational level of the electronic state A. Similarly, we see from Fig. 11.8 that an (emission) electronic transition from the $v' = 0$ state of B will lead to the point P_2 on the lower electronic state, corresponding to $v = 7$. When both vibrational levels v and v' are excited states, the most favoured transitions will be those that can occur at constant R with the nuclei in the electronic states A and B near the classical end points of their vibrational motion. Examples in Fig. 11.8 are shown by the vertical lines from $v = 2$ to $v' = 2$ and $v = 3$ to $v' = 1$.

The above qualitative considerations can be expressed more precisely as follows. The total wave function Ψ_a of the molecule in a given state a (neglecting spin) is given by (11.1) and the transition amplitude is proportional to the matrix element of the electric dipole operator (11.6) between two wave functions Ψ_a and Ψ_b. That is,

$$\langle \Psi_b | \mathbf{D} | \Psi_a \rangle = e \int d\mathbf{R}\, R^{-2} \int d\mathbf{r}_1 d\mathbf{r}_2 \dots d\mathbf{r}_N\, \Phi_{s'}^* \psi_{v'}^* \, {}^{\Lambda'}\mathcal{H}_{K'M'_K}^* \times \left(\sum_{i=1}^{2} Z_i \mathbf{R}_i - \sum_{j=1}^{N} \mathbf{r}_j \right) \Phi_s \psi_v \, {}^{\Lambda}\mathcal{H}_{KM_K} \quad \textbf{(11.27)}$$

where $\mathbf{R} = \mathbf{R}_2 - \mathbf{R}_1$ and the primed quantities refer to the state b. Because of the orthogonality of the electronic wave functions Φ_s and $\Phi_{s'}$, the first term in (11.27) containing only nuclear coordinates vanishes, so that

$$\langle \Psi_b | \mathbf{D} | \Psi_a \rangle = \int \psi_{v'}^* \, {}^{\Lambda'}\mathcal{H}_{K'M'_K}^* \, \mathbf{D}_{\text{el}}(R) \psi_v \, {}^{\Lambda}\mathcal{H}_{KM_K} R^{-2} \, d\mathbf{R} \quad \textbf{(11.28)}$$

where

$$\mathbf{D}_{\text{el}}(R) = -e \int \Phi_{s'}^* \left(\sum_{j=1}^{N} \mathbf{r}_j \right) \Phi_s \, d\mathbf{r}_1 \, d\mathbf{r}_2 \dots d\mathbf{r}_N \quad \textbf{(11.29)}$$

is the electronic part of the matrix element of the electric dipole operator $\mathbf{D}$. The Franck–Condon principle amounts to assuming that $\mathbf{D}_{\text{el}}(R)$ is independent of R, so that the transition amplitude (11.28) is proportional to the *Franck–Condon factor*

$$f_{v'v} = \int_0^\infty \psi_{v'}^* \psi_v \, dR \quad \textbf{(11.30)}$$

This quantity is just the *overlap* integral between vibrational wave functions ψ_v and $\psi_{v'}$ in *different* electronic states. Thus we see that the most intense transitions will be those for which the overlap between ψ_v (determined from the electronic potential in the lower state) and $\psi_{v'}$ (obtained from the electronic potential for the upper state) is a maximum, in accordance with our foregoing qualitative discussion.

Dissociation and predissociation

In addition to the discrete molecular transitions which we have considered thus far, *continuous* molecular spectra are frequently observed, both in emission and absorption. These spectra result from transitions in which at least one of the states is a *dissociative state*. For example, we show in Fig. 11.9 two electronic absorption transitions leading to the *dissociation* of a molecule.

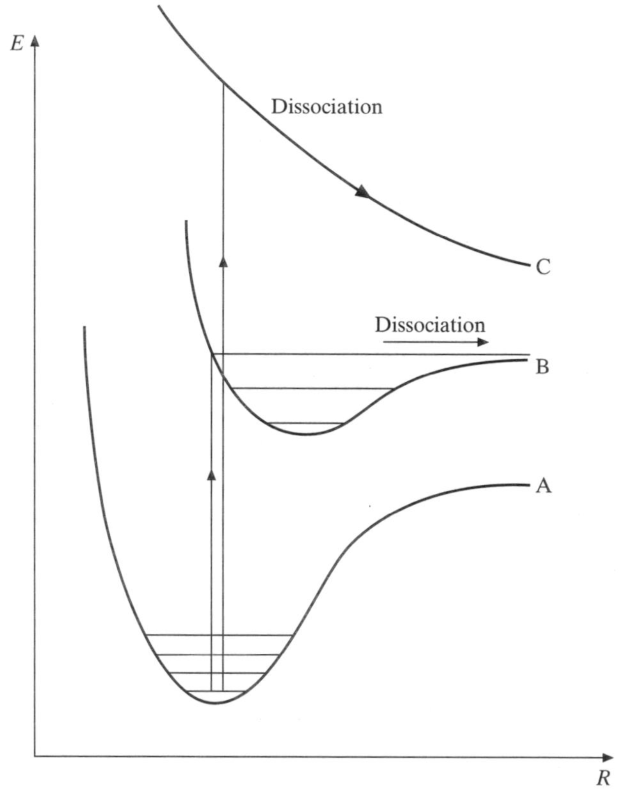

Figure 11.9 Electronic transitions leading to dissociation. A transition from level A to level C leads to dissociation because C is repulsive. The curve B exhibits a minimum, but the transition shown leads to a level with sufficient energy to surmount the barrier.

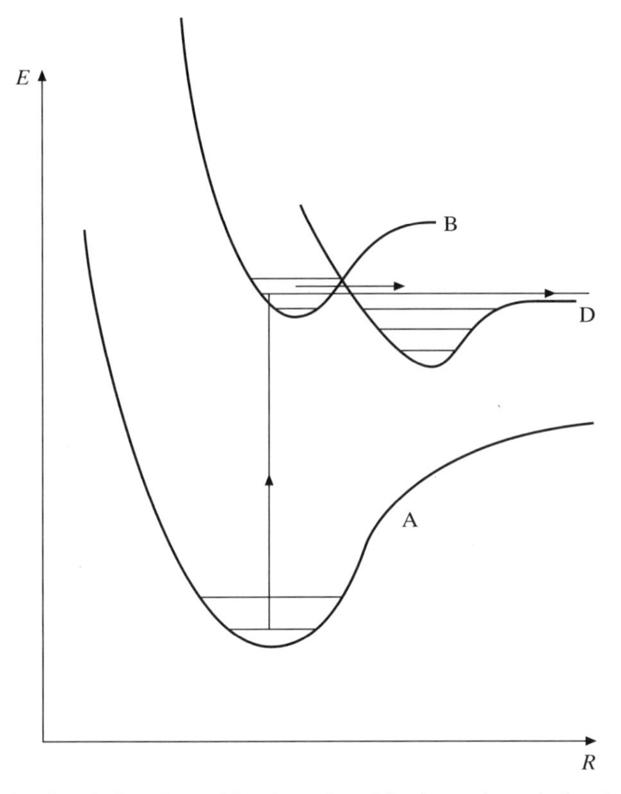

Figure 11.10 Predissociation. A transition from A to B leads to a bound vibrational level, but a further transition is made to a dissociating level of D.

It may also happen that an excited state B is coupled by internal perturbations (such as spin–orbit effects) or external ones (such as collisions) to a dissociative state D. In this case the excited state B can either decay to a lower electronic state A by a radiative transition (spontaneous emission of radiation) or be transferred to the dissociative state D by a *radiationless transition* due to the coupling between the states B and D. The latter process is called *predissociation* and is illustrated in Fig. 11.10. We remark that it is similar to autoionisation and to the Auger effect discussed in previous chapters for atoms. The effect of predissociation is to weaken and broaden the emission lines from the excited state.

Fluorescence and phosphorescence

Molecules which absorb radiation in the near ultra-violet and visible range may re-emit it at a longer wavelength. This phenomenon is known as *fluorescence* and is illustrated in Fig. 11.11(a). First, absorption transitions occur from a lower electronic state to some vibrational levels of an upper electronic state, these

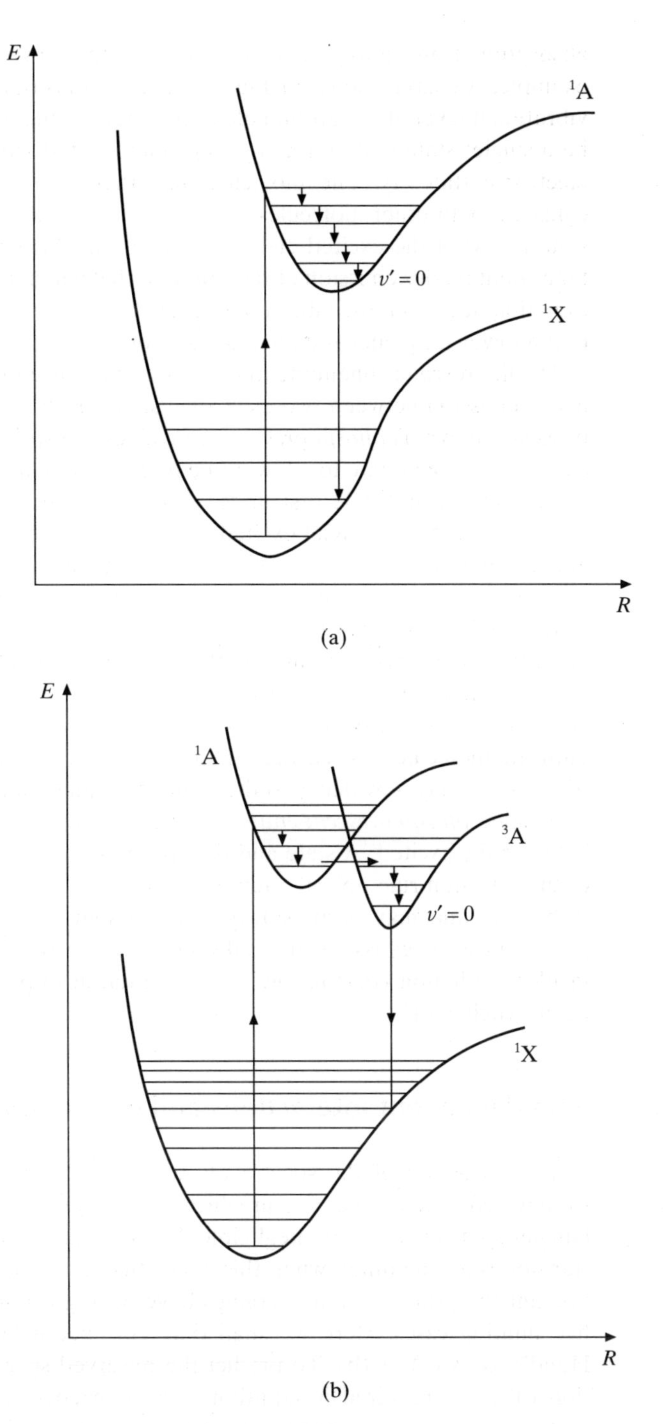

Figure 11.11 (a) Fluorescence. (b) Phosphorescence.

absorption transitions being governed by the Franck–Condon principle. For example, we have shown in Fig. 11.11(a) a transition leading from the lowest vibrational level of the ground electronic state of the molecule (which we take to be a singlet state and denote by ^{1}X) to an excited vibrational level of an upper electronic state ^{1}A. The molecule may then lose vibrational energy through collisions with other molecules (thermal decay) and reach the lowest vibrational state $v' = 0$ of the excited electronic level ^{1}A. This process is often more rapid than spontaneous emission of radiation, so that when the molecule finally re-emits radiation, it is from the vibrational level $v' = 0$. As a result, the emitted radiation is at a lower frequency than the incident one.

The fluorescence phenomenon which we have just discussed involves a spontaneous emission between two electronic states of the *same multiplicity*. A related process, known as *phosphorescence*, involves the spontaneous emission from an excited electronic state to a lower one of *different multiplicity*. For example, let us assume that a molecule first undergoes an absorption transition from a singlet ground state ^{1}X to a singlet excited state ^{1}A, as in the case of fluorescence studied above. Thermal degradation of the vibrational energy then occurs via collisions with other molecules. Suppose now that the state ^{1}A is coupled to a triplet state ^{3}A of similar energy, as shown in Fig. 11.11(b). Instead of going down the vibrational ladder of the state ^{1}A, the molecule may then undergo an *intersystem crossing* and be transferred by a radiationless transition into the triplet state ^{3}A. After the crossing has occurred, thermal decay will proceed within the ^{3}A well, until the molecule has reached the lowest vibrational level of the ^{3}A state. This triplet state may now decay to the ground (singlet) state ^{1}X by a radiative transition called *phosphorescent emission*. Indeed, although electric dipole transitions between the excited ^{3}A state and the ground state ^{1}X are spin-forbidden because of the selection rule $\Delta S = 0$, radiative transitions $^3\mathrm{A} \to {}^1\mathrm{X}$ may not be absolutely forbidden, and may occur slowly through spin–orbit interactions. As a result, phosphorescent emission may take place for seconds or even longer after the incident radiation (responsible for the initial absorption process $^1\mathrm{X} \to {}^1\mathrm{A}$) has been switched off.

11.4 Spin-dependent interactions and electric dipole transitions

In the discussion of the spectra of diatomic molecules given up to this point, we have considered the case in which spin-dependent interactions either vanish (singlet spin states) or are negligible. We shall now examine how electric dipole transitions are modified when this is not the case. The energy levels taking into account the principal spin couplings have been given in Section 10.5 for Hund's five limiting cases. Here we shall only consider light molecules which satisfy Hund's cases (a) or (b). To predict the observed spectra the appropriate selection rules are required in addition to a knowledge of the energy levels. Light molecules are in eigenstates of $\mathbf{S}^2$, corresponding to the eigenvalues $S(S+1)\hbar^2$, and since the electric dipole operator **D** is independent of the spin variables, the selection rules

$$\Delta S = \Delta\Sigma = \Delta M_S = 0 \tag{11.31}$$

are always satisfied, and transitions only occur between electronic states having the same multiplicity. It should be noted that as the molecular weight increases, the spin–orbit coupling becomes stronger, the spin is no longer conserved, and transitions violating the selection rule (11.31) begin to be seen. In fact for very heavy molecules transitions with $\Delta S = \pm 1$, called intersystem crossings, are strong.

Taking into account conservation of angular momentum and the fact that the electric dipole operator **D** only connects states of opposite parity, it can be shown, subject to some further restrictions, that

$$\Delta Q = 0, \pm 1 \tag{11.32}$$

where Q stands for any of the quantities J, M_J, K, Λ. The further restrictions are:

1. For Hund's case (a): $\Delta J = 0$ is not allowed if $\Omega' = \Omega = 0$.
2. For Hund's case (b): $\Delta K = 0$ is not allowed if $\Lambda' = \Lambda = 0$.

In addition, as we have seen earlier for transitions between Σ states ($\Lambda = \Lambda' = 0$), $\Sigma^+ \leftrightarrow \Sigma^+$ and $\Sigma^- \leftrightarrow \Sigma^-$ but $\Sigma^+ \leftrightarrow \Sigma^-$ is not allowed, while for homonuclear molecules only g $\leftrightarrow$ u transitions are allowed.

The spectra in Hund's case (a)

The main differences in Hund's case (a) between the previous analysis without spin–orbit coupling are (i) that the electronic energies $E_s'(R_0)$ are slighly shifted by the spin–orbit energies $A\Lambda\Sigma$ and contain Ω^2 in place of Λ^2 and (ii) the rotational energies are given by

$$E_r = BJ(J+1), \qquad J \geqslant \Omega \tag{11.33}$$

rather than by (11.5). It follows that the band structure can be analysed in terms of P, Q and R branches, just as before, and the frequencies of the rotational lines are given by (see (11.26))

$$\begin{aligned} h\nu^P &= h\nu + B'J(J-1) - BJ(J+1) \\ h\nu^Q &= h\nu + B'J(J+1) - BJ(J+1) \\ h\nu^R &= h\nu + B'(J+1)(J+2) - BJ(J+1), \qquad J \geqslant \Omega \end{aligned} \tag{11.34}$$

where $h\nu$ is given by (11.25) or (11.24). The lower level in each case has total angular momentum J, while the upper level has total angular momentum $J - 1$, J or $J + 1$ for the P, Q and R branches, respectively. The Q branch is absent for transitions between two states with $\Omega = 0$, for which $\Delta J = \pm 1$. It should be noted that when Ω is an integer (or zero), J takes on integer values (or zero), while if Ω is a half-odd integer, which occurs when $S = 1/2, 3/2, \ldots$, then J also takes on half-odd integer values. For singlet states $J = K$, $\Omega = \Lambda$ and the line frequencies are again given by (11.26).

The spectra in Hund's case (b)

The most important example of Hund's case (b) is afforded by Σ states ($\Lambda = 0$) for which the spin–orbit coupling is entirely absent and the main spin coupling is between the spin and the rotational motion of the molecule. We shall take the example of transitions between $^2\Sigma^+$ states, for which the rotational energy levels are given by (10.136). The selection rules for Σ states are $\Delta K = \pm 1$ and $\Delta J = 0, \pm 1$. It follows that there are six branches, P_1, P_2, Q_1, Q_2, R_1, R_2; the first two are for transitions with $J' = J - 1$, the second two with $J' = J$ and the third two with $J' = J + 1$. Under the assumption that the spin–rotation constant $\bar{\gamma}$ is the same for both the upper and lower levels, the frequencies of the rotational spectrum are given by

$$h\nu^{P_1} = h\nu + B'K(K-1) - BK(K+1) - \gamma/2 \begin{pmatrix} K' = K - 1 \\ J' = K - 1/2,\ J = K + 1/2 \end{pmatrix}$$

$$h\nu^{P_2} = h\nu + B'K(K-1) - BK(K+1) + \gamma/2 \begin{pmatrix} K' = K - 1 \\ J' = K - 3/2,\ J = K - 1/2 \end{pmatrix}$$

$$h\nu^{Q_1} = h\nu + B'K(K-1) - BK(K+1) + \gamma K \begin{pmatrix} K' = K - 1 \\ J' = K - 1/2 = J \end{pmatrix}$$

$$h\nu^{Q_2} = h\nu + B'(K+1)(K+2) - BK(K+1) - \gamma(K+1) \begin{pmatrix} K' = K + 1 \\ J' = K + 1/2 = J \end{pmatrix}$$

$$h\nu^{R_1} = h\nu + B'(K+1)(K+2) - BK(K+1) - \gamma/2 \begin{pmatrix} K' = K + 1 \\ J' = K + 3/2,\ J = K + 1/2 \end{pmatrix}$$

$$h\nu^{R_2} = h\nu + B'(K+1)(K+2) - BK(K+1) + \gamma/2 \begin{pmatrix} K' = K + 1 \\ J' = K + 1/2,\ J = K - 1/2 \end{pmatrix} \quad \mathbf{(11.35)}$$

As before, $h\nu$ is given by (11.25) or (11.24) while since the transition is between Σ states $K \geqslant 0$. Since $\gamma = \bar{\gamma}\hbar^2$ is always very small compared with B or B', the P_1, P_2 and Q_1 branches are very close to one another and in the limit $\gamma \to 0$ coalesce into a single (P) branch. In the same way, the R_1, R_2 and Q_2 branches are very close to each other and in the limit $\gamma \to 0$ coalesce into a single (R) branch.

In the examples given, it has been assumed that both the upper and lower energy levels belong either to Hund's case (a) or to Hund's case (b). This is not always the case. For example, in a transition from a Π to a Σ state, the Π state may fall into case (a) and the Σ state into case (b). This causes no difficulty in principle and the allowed transition frequencies can be determined from the appropriate energy levels and the selection rules.

11.5 The nuclear spin

The coupling between the spins of the nuclei and the magnetic fields, due to the orbital motion and the spin of the electrons makes no significant contribution to the molecular energies. Despite this fact, the nuclear spin has a large influence upon the spectra of molecules containing *identical* nuclei, because of symmetry considerations. A particularly interesting case is that of *homonuclear diatomic* molecules, which have two identical nuclei. Neglecting the very small coupling of the nuclear spins with the rest of the molecule, we can write the total wave function Ψ_{tot} of a homonuclear diatomic molecule as

$$\Psi_{\text{tot}} = \Psi(\mathbf{R}, q)\chi_N(1, 2) \tag{11.36}$$

where χ_N is the spin function of the two nuclei. The part of the wave function not depending on the nuclear spin, Ψ, is a function of the internuclear coordinate $\mathbf{R}$ and of the spatial and spin coordinates of the electrons, which we denote collectively by q. The total wave function Ψ_{tot} must of course be antisymmetric with respect to the interchange of any pair of electrons, in order to satisfy the Pauli exclusion principle. In addition, however, since the two nuclei 1 and 2 are *identical* in a homonuclear diatomic molecule, the total wave function Ψ_{tot} must be either *symmetric* or *antisymmetric* under the interchange $1 \leftrightarrow 2$ of the identical nuclei. According to the discussion in Section 2.7, the symmetrical case arises when the nuclei are *bosons* which have zero or integer spin (for example, $^{16}O_2$, $^{14}N_2$) and the antisymmetrical case occurs when the nuclei are *fermions*, having half-odd integer spin (for example, $^{1}H_2$, $^{19}F_2$, $^{127}I_2$).

On interchanging the nuclei 1 and 2, we see that $\Psi(\mathbf{R}, q) \to \Psi(-\mathbf{R}, q)$ while $\chi_N(1, 2) \to \chi_N(2, 1)$. Under these transformations, the factors Ψ and χ_N must each be either symmetric or antisymmetric, and we denote the corresponding functions by Ψ^{S}, Ψ^{A} and χ_N^{S}, χ_N^{A}, respectively. Thus, when the nuclei are bosons, the total wave function Ψ_{tot} must be of the form $\Psi^{\text{S}}\chi_N^{\text{S}}$ or $\Psi^{\text{A}}\chi_N^{\text{A}}$, while for fermions Ψ_{tot} must have the form $\Psi^{\text{S}}\chi_N^{\text{A}}$ or $\Psi^{\text{A}}\chi_N^{\text{S}}$.

Let us consider first the functions Ψ, which, neglecting spin–orbit coupling, can be written in the form of the product

$$\Psi = \Phi_s \chi_{SM_S} R^{-1} \psi_v \, {}^{\Lambda}\mathscr{H}_{KM_K} \tag{11.37}$$

where Φ_s is the spatial part of the wave function for the orbital motion of the electrons, χ_{SM_S} is the spin function of the electrons, and ψ_v and ${}^{\Lambda}\mathscr{H}_{KM_K}$ are the nuclear vibrational and rotational wave functions, respectively. The function Φ_s is even or odd under the operation $\mathbf{R} \to -\mathbf{R}$. Indeed, we saw in Section 10.4 that for Σ states ($\Lambda = 0$) the levels Σ_g^+ and Σ_u^- are even while Σ_g^- and Σ_u^+ are odd. Moreover, if the degeneracy of the levels with $\Lambda \neq 0$ ($\Pi, \Delta, \ldots$) is removed by magnetic interactions (Λ-doubling) then one has levels $\Pi_g^+, \Pi_u^-, \Delta_g^+, \Delta_u^-, \ldots$ which are even and levels $\Pi_g^-, \Pi_u^+, \Delta_g^-, \Delta_u^+, \ldots$ which are odd. The second factor, χ_{SM_S}, is clearly unaffected by the operation $\mathbf{R} \to -\mathbf{R}$, since it does not depend on the nuclear coordinates.

The vibrational wave function ψ_v depends only on the magnitude R of the vector **R**, and so is also unaltered when $\mathbf{R} \to -\mathbf{R}$. The rotational wave functions ${}^{\Lambda}\mathcal{H}_{KM_K}$, however, may be either even (if $K = 0, 2, 4, \ldots$) or odd (if $K = 1, 3, 5, \ldots$) under the transformation $\mathbf{R} \to -\mathbf{R}$. The overall symmetrical (S) or antisymmetrical (A) character of the function Ψ is therefore given as follows:

	g^+	g^-	u^+	u^-
K even	S	A	A	S
K odd	A	S	S	A

We now turn to the nuclear spin function χ_N. According to the rules of addition of angular momenta, the spins **I** of the two nuclei form a resultant **T**, which is the total nuclear spin of the molecule. The corresponding total nuclear spin quantum number T can therefore take on any one of the $2I + 1$ values $T = 0, 1, \ldots, 2I - 1, 2I$, where I is the spin quantum number of the individual nuclei.

We begin by looking at the simple case of spinless nuclei, for which $I = T = 0$ (for example, $^{16}O_2$). In this case χ_N is a constant and hence trivially symmetric in the interchange of the nuclei $1 \leftrightarrow 2$. Since the total wave function Ψ_{tot} has to be symmetric, it follows that only even rotational levels can occur if the electronic wave function is even, and only odd rotational levels are present if the electronic wave function is odd. For example, the ground electronic state of $^{16}O_2$ is a $^3\Sigma_g^-$ state; its (g^-) character requires that K should be odd. As a consequence, the ground state of the molecule is a $K = 1$ state. Moreover, in the rotational Raman spectrum of molecules like $^{16}O_2$, half of the expected levels are missing. Also, in transitions between electronic Σ states of homonuclear diatomic molecules with spin-zero nuclei (say $\Sigma_g^+ \to \Sigma_u^+$ or $\Sigma_g^- \to \Sigma_u^-$) alternate lines in the rotational fine structure are missing, since $\Delta K = \pm 1$ and transitions between symmetrical (S) and antisymmetrical (A) states are forbidden. If the nuclei are not identical, for example if one of the ^{16}O nuclei of the $^{16}O_2$ molecule is replaced by the isotope ^{17}O, then the missing transitions are restored. The appearance of these 'missing' lines is an important tool for determining the existence of isotopes like ^{17}O and ^{18}O.

Let us now analyse the case of the 'ordinary' hydrogen molecule $^1H_2$, which contains two protons. Since the spin of the proton is $I = \frac{1}{2}$, the total nuclear spin quantum number T can take on the values $T = 0$ or $T = 1$. When $T = 0$ the nuclear spin function is the singlet function

$$\chi_N^A(1, 2) = \frac{1}{\sqrt{2}}[\alpha(1)\beta(2) - \beta(1)\alpha(2)] \qquad \textbf{(11.38)}$$

which is antisymmetric (A) with respect to the interchange $1 \leftrightarrow 2$ of the two protons. For $T = 1$ we have the triplet of nuclear spin functions

$$\chi_N^S(1, 2) = \begin{cases} \alpha(1)\alpha(2) \\ \frac{1}{\sqrt{2}}[\alpha(1)\beta(2) + \beta(1)\alpha(2)] \\ \beta(1)\beta(2) \end{cases} \qquad \textbf{(11.39)}$$

which are symmetric (S) with respect to the interchange of the two protons. Transitions between the $T = 0$ and $T = 1$ states occur extremely rarely, because such transitions could result only from very small perturbations involving the nuclear spins. Thus one can consider molecular hydrogen as consisting of two distinct species, namely *para-hydrogen* for which $T = 0$ (antisymmetrical case) and *ortho-hydrogen* where $T = 1$ (symmetrical case). Remembering that the electronic ground state of the hydrogen molecule is a Σ_g^+ state, and that the total wave function Ψ_{tot} must be antisymmetric in the interchange of the two protons, we see that in this state para-hydrogen can only have rotational levels $K = 0, 2, 4, \ldots$ while ortho-hydrogen can only possess rotational levels with odd values $K = 1, 3, 5, \ldots$. In statistical equilibrium, at room temperature, three times as many hydrogen molecules will be in $T = 1$ (ortho) states as in $T = 0$ (para) states. As a consequence, the alternate lines in the rotational fine structure show a 3:1 ratio of intensities. On cooling to temperatures of the order of 20 K, at which hydrogen is liquid, the molecules of para-hydrogen are concentrated in their lowest allowed rotational state, $K = 0$, while those of ortho-hydrogen go over to their lowest allowed rotational state $K = 1$. Since the coupling between the $T = 0$ and $T = 1$ states is extremely small, the ortho-hydrogen molecules will remain in the $K = 1$ state for a long period of time. Eventually, after some months, the molecules will all be found in the $K = 0$ state of para-hydrogen. If now the hydrogen is allowed to warm to room temperature, hydrogen gas will be obtained in the pure para form.

Finally, let us consider the general case of two identical nuclei with spin quantum number $I \neq 0$. The possible values of the Z component of spin of each nucleus are $M_I = -I, -I + 1, \ldots, I - 1, I$; $(2I + 1)$ values in all. The total number of combinations of $M_I(1)$ and $M_I(2)$ is therefore $(2I + 1)^2$. Out of these, all the $2I + 1$ spin functions of the form $\chi_{M_I}(1)\chi_{M_I}(2)$ are symmetric. One-half of the remaining $(2I + 1)^2 - (2I + 1)$ functions can be combined in symmetrical states

$$\chi_N^S(1, 2) = \frac{1}{\sqrt{2}}[\chi_{M_I}(1)\chi_{M_I'}(2) + \chi_{M_I'}(1)\chi_{M_I}(2)], \qquad M_I \neq M_I' \tag{11.40}$$

and one-half in antisymmetrical states,

$$\chi_N^A(1, 2) = \frac{1}{\sqrt{2}}[\chi_{M_I}(1)\chi_{M_I'}(2) - \chi_{M_I'}(1)\chi_{M_I}(2)], \qquad M_I \neq M_I' \tag{11.41}$$

The total number of symmetrical states is then

$$2I + 1 + \frac{1}{2}[(2I + 1)^2 - (2I + 1)] = (2I + 1)(I + 1) \tag{11.42}$$

and the total number of antisymmetical states is

$$\frac{1}{2}[(2I + 1)^2 - (2I + 1)] = (2I + 1)I \tag{11.43}$$

As a result, the ratio of intensities in the lines of the rotational fine structure of homonuclear diatomic molecules with nuclei of spin quantum number I is $(I + 1)/I$; for bosons this represents the relative statistical weights of the states Ψ^S to the states Ψ^A, while for fermions it represents the relative statistical weights of the states Ψ^A to the states Ψ^S. The observation of this alternation of strong and weak lines provides an important method for the determination of nuclear spins.

11.6 The inversion spectrum of ammonia

The ammonia molecule NH_3 has the form of a pyramid, whose summit is occupied by the nitrogen atom, while the basis is an equilateral triangle formed by the three hydrogen atoms (see Fig. 11.12). At equilibrium, the distance NH is $d = 1.014$ Å, the distance of the nitrogen atom from the plane of the hydrogen atoms is $z_0 = 0.38$ Å and the angle α between an NH bond and the threefold axis of symmetry of the molecule is $\alpha = 67° \ 58'$.

There are many degrees of freedom in this system, involving electronic, vibrational and rotational motions, resulting in a variety of energy levels and various quantum numbers to specify them. In this section, however, we shall assume that the NH_3 molecule is in its lowest electronic state and analyse a particular vibrational motion which is associated with the *inversion* of the molecule. To see how this comes about, let us consider one of the possible vibrational motions of the NH_3 molecule, analogous to the movement of an umbrella which is being opened and closed, and during which the angle α oscillates around its equilibrium position. Neglecting all other degrees of freedom, the potential energy of the system is then a function $V(z)$ of the algebraic distance z between the plane of the hydrogen atoms (the XY plane) and the nitrogen atom. The curve $V(z)$ is sketched in Fig. 11.13(a). Because the system is symmetrical with respect to the plane $z = 0$ it is clear that the potential $V(z)$ must be an *even* function of z. The two minima

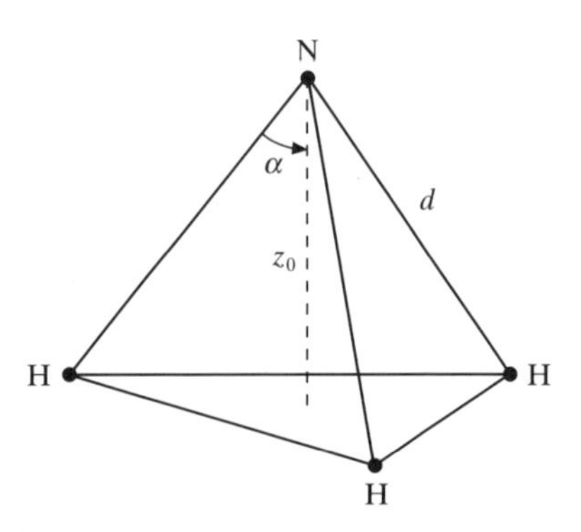

Figure 11.12 Schematic diagram of the ammonia molecule.

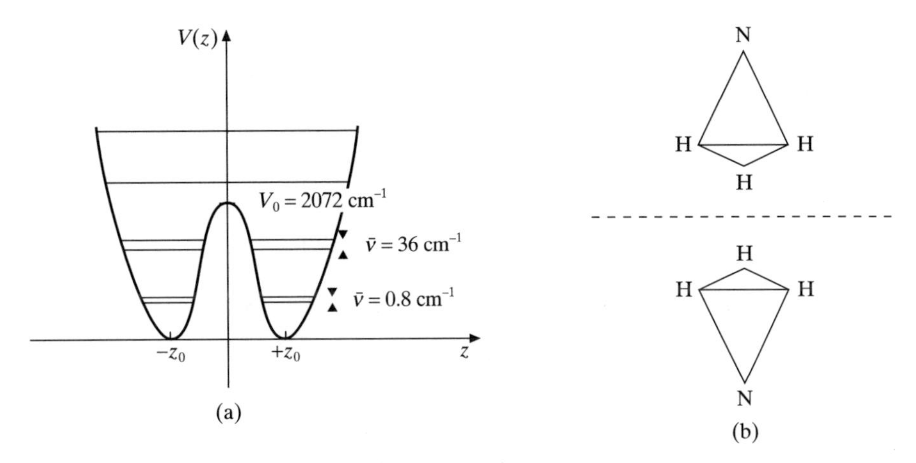

Figure 11.13 The potential well (a) for the inversion motion (b) of ammonia.

of $V(z)$ correspond to symmetrical configurations of the molecule such that the nitrogen atom is located respectively above and below the plane of the hydrogen atoms (see Fig. 11.13(b)) at the equilibrium positions $z = \pm z_0 = \pm 0.38$ Å. We shall refer to these two configurations as the 'up' and 'down' configurations, respectively. The molecule can vibrate in the manner indicated above in either of the two potential wells, with the nitrogen atom on one side of the plane of the hydrogen atoms. The wave number corresponding to this vibrational motion is $\tilde{\nu} = 950\ \text{cm}^{-1}$, which is in the infra-red region.

As seen from Fig. 11.13(a), the potential $V(z)$ forms a barrier about $z = 0$. This barrier is due to the Coulomb repulsion between the nitrogen nucleus and the three protons. If it were of infinite height, the nitrogen atom would never be able to penetrate the plane of the hydrogen atoms and be found on the other side of this plane. However, the barrier has a finite height $V_0 = 2072\ \text{cm}^{-1}$, so that there is a certain probability that the molecule will *invert* during the course of its vibrations, that is make transitions between the 'up' and 'down' configurations. It is important to emphasise that in the ground state ($v = 0$) as well as in the first excited state ($v = 1$) of the vibrational mode considered here, the energy of the molecule is lower than the potential height. As a result, the inversion of the molecule NH_3 in the vibrational states $v = 0$ and $v = 1$ is a classically forbidden (or hindered) motion which can only take place because of the quantum mechanical *tunnel effect.*

In order to understand the characteristics of this tunnelling motion let us write the one-dimensional Schrödinger equation for the motion along the Z axis, namely

$$-\frac{\hbar^2}{2m}\frac{\mathrm{d}^2\psi(z)}{\mathrm{d}z^2} + V(z)\psi(z) = E\psi(z) \qquad \textbf{(11.44)}$$

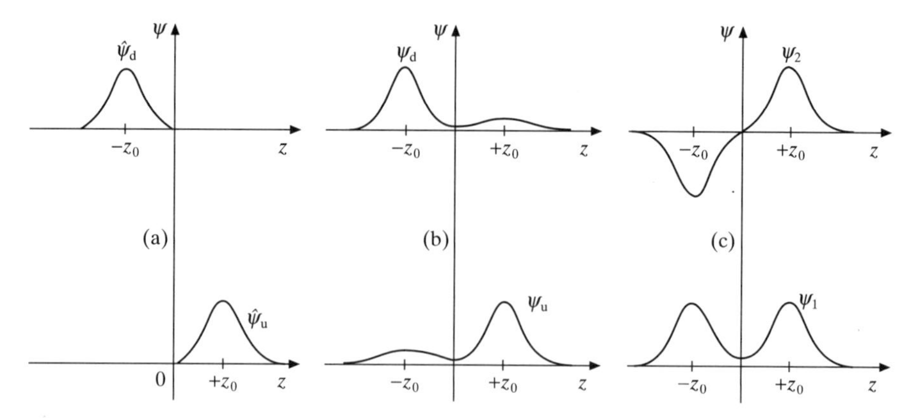

Figure 11.14 The wave functions (a) $\hat{\psi}_u$ and $\hat{\psi}_d$; (b) ψ_u and ψ_d; (c) ψ_1 and ψ_2.

where m is an effective mass [1]. If the potential barrier between the two wells were of infinite height, the two wells would be totally 'disconnected' and the energy spectrum would consist of the same set of energy eigenvalues in each well. Thus, each energy eigenvalue of the system would be doubly degenerate, and the eigenfunctions corresponding to a given energy would be linear combinations of the 'up' and 'down' wave functions $\hat{\psi}_u(z)$ and $\hat{\psi}_d(z)$ which vanish identically for $z \leqslant 0$ and $z \geqslant 0$, respectively, and which we take to be normalised to unity. A pair of wave functions $\hat{\psi}_u$ and $\hat{\psi}_d$ is sketched in Fig. 11.14(a) for the case of the lowest ($v = 0$) vibrational state.

In the real situation, with a finite barrier, there is a 'coupling' between the two wells which allows the inversion motion to occur. As a result, the degeneracy is removed, and the energy levels are split into *doublets* (see Problem 11.5). In the simple model considered here, the separation between the pair of energy levels forming a doublet depends only on the nature of the potential barrier and on the vibrational state of the molecule. As indicated in Fig. 11.13(a), the two energy levels forming the lowest ($v = 0$) doublet are separated by about 9.84×10^{-5} eV (0.8 cm^{-1}), while the next ($v = 1$) lowest pair of levels are about 4.4×10^{-3} eV (36 cm^{-1}) apart. We note that the inversion wave numbers $\tilde{\nu}(v = 0) = 0.8$ cm^{-1} and $\tilde{\nu}(v = 1) = 36$ cm^{-1} are much smaller than the wave number $\tilde{\nu} = 950$ cm^{-1} corresponding to the vibrational motion, since inversion is considerably inhibited by the presence of the potential barrier.

In what follows we shall focus our attention on the lowest doublet ($v = 0$) which we shall treat as a two-level system. It is clear from the above discussion that since the potential barrier is finite we do not have rigorous 'up' and 'down' eigenfunctions $\hat{\psi}_u(z)$ and $\hat{\psi}_d(z)$ vanishing identically for $z \leqslant 0$ and $z \geqslant 0$, respectively. Instead, we define the corresponding wave functions $\psi_u(z)$ and $\psi_d(z)$ – normalised

[1] Assuming that the distance between the hydrogen atoms remains constant during the motion, the effective mass m is then the reduced mass $m = 3M_H M_N/(3M_H + M_N)$, where M_H is the mass of the hydrogen atom and M_N the mass of the nitrogen atom.

to unity – to be those for which the nitrogen atom is *most probably* located above or below the plane of the hydrogen atoms. These wave functions are sketched in Fig. 11.14(b). It is important to realise that because the two wells are coupled, the functions ψ_u and ψ_d are *not* energy eigenfunctions, and are not orthogonal to each other. Indeed, the true energy eigenfunctions must be either symmetric or anti-symmetric with respect to the inversion operation $z \to -z$. In terms of ψ_u and ψ_d, the energy eigenfunctions (normalised to unity) are therefore given by

$$\psi_1 = \frac{1}{\sqrt{2}}(\psi_u + \psi_d) \tag{11.45a}$$

and

$$\psi_2 = \frac{1}{\sqrt{2}}(\psi_u - \psi_d) \tag{11.45b}$$

The symmetrical wave function ψ_1 corresponds to the lowest energy E_1 of the doublet, while the antisymmetrical wave function ψ_2 corresponds to the higher energy E_2. Both functions ψ_1 and ψ_2 are shown in Fig. 11.14(c).

Having obtained the energy eigenfunctions $\psi_1(z)$ and $\psi_2(z)$ we may write the general time-dependent wave function of our two-level problem as (see (2.85))

$$\Psi(z, t) = c_1\psi_1(z)\exp(-\mathrm{i}E_1t/\hbar) + c_2\psi_2(z)\exp(-\mathrm{i}E_2t/\hbar) \tag{11.46}$$

where

$$E_2 = E_1 + \Delta E \tag{11.47}$$

and $\Delta E = 9.84 \times 10^{-5}$ eV is the energy splitting of the doublet. Let us assume that at time $t = 0$ the wave function describing the system is ψ_u, so that the nitrogen atom is most probably to be found above the plane of the hydrogen atoms at that time. Using (11.45) and (11.46), we then have

$$\Psi(z, 0) = c_1\psi_1(z) + c_2\psi_2(z) = \psi_u(z) = \frac{1}{\sqrt{2}}[\psi_1(z) + \psi_2(z)] \tag{11.48}$$

so that

$$c_1 = c_2 = \frac{1}{\sqrt{2}} \tag{11.49}$$

Substituting (11.49) into (11.46) and using (11.47), we see that the wave function $\Psi(z, t)$ will evolve in time according to

$$\begin{aligned}\Psi(z, t) &= \frac{1}{\sqrt{2}}[\psi_1(z)\exp(-\mathrm{i}E_1t/\hbar) + \psi_2(z)\exp(-\mathrm{i}E_1t/\hbar)\exp(-\mathrm{i}(\Delta E)t/\hbar)] \\ &= \frac{1}{\sqrt{2}}[\psi_1(z) + \psi_2(z)\exp(-2\pi\mathrm{i}\nu t)]\exp(-\mathrm{i}E_1t/\hbar)\end{aligned} \tag{11.50}$$

where we have written $\Delta E = h\nu$. At the time $t = 1/(2\nu)$ the wave function (11.50) is given by

$$\Psi(z, t = 1/(2\nu)) = \frac{1}{\sqrt{2}}[\psi_1(z) - \psi_2(z)] \exp(-\mathrm{i}E_1 t/\hbar)$$
$$= \psi_{\mathrm{d}}(z) \exp(-\mathrm{i}E_1 t/\hbar) \qquad \mathbf{(11.51)}$$

so that

$$|\Psi(z, t = 1/(2\nu))|^2 = |\psi_{\mathrm{d}}(z)|^2 \qquad \mathbf{(11.52)}$$

and the nitrogen atom is most probably to be found under the plane of the hydrogen atoms at $t = 1/(2\nu)$. Since the energy difference $\Delta E = h\nu \simeq 9.84 \times 10^{-5}$ eV corresponds to a frequency $\nu \simeq 23\,870$ MHz, we see that the time required for the NH_3 molecule to invert is $t = 1/(2\nu) \simeq 2.1 \times 10^{-11}$ s.

The existence of the energy doublets of the ammonia molecule was first inferred from the analysis of its infra-red vibrational–rotational and pure rotational spectra. However, radiative transitions between the two states forming a doublet can also occur, the corresponding lines being in the microwave region. In 1934, the progress made in radio-frequency techniques allowed C.E. Cleeton and N.H. Williams to observe directly a peak in the absorption spectrum of the ammonia molecule at a wavelength $\lambda \simeq 1.25$ cm which corresponds to the inversion frequency $\nu \simeq 23\,870$ MHz of the lowest doublet. The experiment of Cleeton and Williams opened the new field of microwave spectroscopy [2], and eventually made possible the development in 1954 of the *ammonia maser* by C.H. Townes, J. Gordon and H. Zeiger (see Chapter 15). Of course, it should be realised that, in common with all molecular vibrational motions, the inversion spectrum of NH_3 contains a fine and hyperfine structure due respectively to the rotational motion of the molecule and to magnetic and quadrupole interactions involving the nuclei.

Problems

11.1 Show that the ratio of the number of molecules in the rotational level K to that in the lowest level, in a gas at temperature T, is $R(K) = (2K + 1) \exp[(E_0 - E_K)/(k_B T)]$ where $E_K = BK(K + 1)$. Make a graph of this expression as a function of K for $H^{35}Cl$ at room temperature. Show that $R(K)$ has a maximum and obtain the relative intensities of absorption lines in a pure rotational spectrum ($\tilde{B} = B/(hc) = 10.6$ cm^{-1} for HCl).

11.2 Find the ratio of the number of molecules in the first excited vibrational level to the number in the lowest vibrational level, in a gas at temperature T. For (a) $H^{35}Cl$ and (b) $D^{35}Cl$, calculate this ratio at 300 K, 1500 K and 3000 K ($\tilde{\nu}_0 = 2990$ cm^{-1} for $H^{35}Cl$).

[2] A detailed treatment of microwave spectroscopy may be found in Townes and Schawlow (1955).

11.3 In the fundamental band of $^{12}C^{16}O$ the spacing of the lines is found to be constant with the value 3.86 cm^{-1}. The band is centred on a missing line at 2170.21 cm^{-1}. Calculate the rotational constant $\tilde{B} = B/(hc)$, the internuclear separation R_0 and the force constant of the vibrational motion.

11.4 In the $^{14}N_2$ molecule the ground state $X^1\Sigma_g^+$ and the excited state $b'\ ^1\Sigma_u^+$ have the following constants:

	$\tilde{\nu}_0$	$\beta\tilde{\nu}_0$	$\tilde{B}$
X	2359.6	14.456	2.010
b′	751.7	4.82	1.145

where all values are given in units of cm^{-1} and $\tilde{\nu}_0 = \nu_0/c$, $\beta\tilde{\nu}_0 = \beta\nu_0/c$ and $\tilde{B} = B/(hc)$.

The wave number of the electronic transition is $\tilde{\nu}_{s's} = \nu_{s's}/c = 103\,678.9$ cm^{-1}.

(a) Construct a Deslandres table for the transitions between the vibrational states $v' = 0, 1, 2, 3, 4, 5$ and $v = 0, 1, 2, 3, 4, 5$.
(b) Calculate the wave numbers of the first few members of the R and P branches for the transition $v' = 0$ to $v = 0$.
(c) Draw the Fortrat parabola by using the variable $m = -K - 1$ for the R branch and $m = K$ for the P branch. Determine whether the band degrades towards the red or the violet and find the position of the band head.

11.5 Consider the one-dimensional Schrödinger equation (11.44) which describes the inversion motion of the ammonia molecule. Compare the solutions in two cases:

(a) An infinite barrier separating two wells, so that (with $0 < a < L$)

$$V(z) = \infty \qquad z < -L,\ -a < z < a,\ z > L$$

$$V(z) = 0 \qquad -L < z < -a,\ a < z < L$$

(b) A finite barrier separating the two wells:

$$V(z) = \infty \qquad z < -L \text{ and } z > L$$

$$V(z) = V_0 \qquad -a < z < a$$

$$V(z) = 0 \qquad -L < z < -a \text{ and } a < z < L$$

Show that the (degenerate) energy levels of case (a) are each split into two non-degenerate levels in case (b).

12 Atomic collisions: basic concepts and potential scattering

Atomic collisions phenomena are of fundamental importance in atomic and molecular physics and play also an important role in other fields such as astrophysics, chemistry, plasma physics and laser physics. Basically, these phenomena involve collisions between an 'elementary particle' (photon, electron, ...) and an atomic system (atom, ion, molecule) or between two atomic systems. We have already discussed in previous chapters various photon–atom (molecule) collision processes, such as the photoelectric effect, the Compton effect, Rayleigh and Raman scattering. In the next chapter, we shall study electron–atom collisions and atomic photoionisation, while Chapter 14 is devoted to atom (ion)–atom collisions. However, before analysing these complex collision processes, we shall introduce in this chapter various basic definitions, and discuss in some detail the simple problem of the quantum theory of scattering by a centre of force.

12.1 Types of collisions, channels, thresholds and cross-sections

Let us consider a typical atomic collision experiment [1] which is illustrated schematically in Fig. 12.1. A homogeneous, well-collimated beam of monoenergetic particles A is directed towards a target containing the scatterers B. We shall assume that the experimental conditions have been chosen in such a way that each target scatterer acts as if it were alone [2]. After the collision between a beam particle A and a target particle B, some or all the particles emerging from the interaction region are registered by detectors, which are placed outside the path of the incident beam, so that undeflected particles are not recorded. Several processes can occur:

[1] The type of experiment shown in Fig. 12.1 in which a beam of particles is scattered by a stationary target is not the only possible kind. It is sometimes more convenient to study the scattering of one beam of particles by another, for example, and more complicated experiments are concerned with two or more successive scattering processes.

[2] Single scattering conditions can be achieved by making the target sufficiently thin. Coherent scattering as in electron diffraction by crystals will not be considered here.

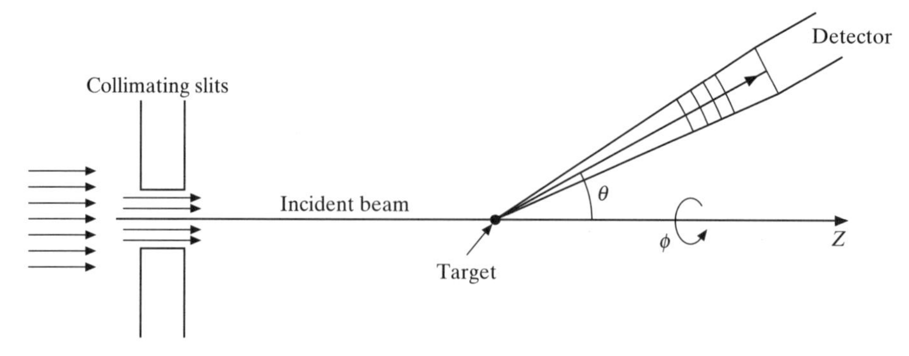

Figure 12.1 Schematic diagram of a scattering experiment.

1. The two particles A and B are scattered without change in their internal structure. That is,

$$\mathrm{A} + \mathrm{B} \rightarrow \mathrm{A} + \mathrm{B} \tag{12.1}$$

   This is known as *elastic scattering*.

2. The two particles A and (or) B undergo a change of their internal quantum state during the collision. For example,

$$\mathrm{A} + \mathrm{B} \rightarrow \mathrm{A} + \mathrm{B}^* \tag{12.2}$$

   where B* denotes an excited state of the particle B. Such processes are called *inelastic collisions*.

3. The composite system (A + B) splits into two particles C and D, different from A and B,

$$\mathrm{A} + \mathrm{B} \rightarrow \mathrm{C} + \mathrm{D} \tag{12.3}$$

   or into more than two particles:

$$\mathrm{A} + \mathrm{B} \rightarrow \mathrm{C}_1 + \mathrm{C}_2 + \cdots + \mathrm{C}_n \tag{12.4}$$

   These collision processes are known as *reactions*.

Channels

A channel is a possible mode of fragmentation of the composite system (A + B) during the collision. It is characterised by the number and the nature of the fragments into which the system (A + B) can be decomposed. In elastic collisions the two colliding particles A and B remain in the initial channel, while inelastic collisions or reactions are processes leading from a given initial channel to a different final channel. A channel is said to be *open* if the corresponding collision is allowed by known conservation laws (such as energy conservation); otherwise it is said to be closed.

Thresholds

Let us consider a general reaction of the type (12.4), and denote by E_A, E_B, E_1, E_2, $\ldots$, E_n the internal energies of the particles A, B, C_1, C_2, $\ldots$, C_n. Working in the *centre of mass* (CM) *system* (in which the centre of mass of the composite system (A + B) is at rest) we write the energy conservation law as

$$T_i + E_A + E_B = T_f + E_1 + E_2 \cdots + E_n \tag{12.5}$$

where T_i and T_f are the initial and final kinetic energies, respectively. This relation may also be written as

$$T_f = T_i + Q_{if} \tag{12.6}$$

where

$$Q_{if} = E_A + E_B - (E_1 + E_2 + \cdots + E_n) \tag{12.7}$$

is the change in internal energy which has occurred.

A necessary condition for the reaction to take place is that $T_f \geqslant 0$. As a result, if $Q_{if} \geqslant 0$ the reaction is always allowed from the point of view of energy conservation (it can be forbidden by other conservation laws) and is called *exothermic*. On the other hand, if $Q_{if} < 0$, the reaction is said to be *endothermic*. It can only occur if

$$T_i \geqslant T_i^t \tag{12.8}$$

where $T_i^t = -Q_{if}$ is known as the *threshold* of the reaction (in the CM system). In the *laboratory system*, where the target particle B is at rest before the collision, the corresponding threshold energy $(T_i^t)_L$ is given by (see (A2.9))

$$(T_i^t)_L = \frac{m_A + m_B}{m_B} T_i^t \tag{12.9}$$

where m_A and m_B are the masses of the particles A and B, respectively.

In order to illustrate the notions of channels and thresholds introduced above, let us consider the scattering of electrons by hydrogen atoms initially in the ground state 1s. We shall take the mass of the proton to be infinite with respect to the electron mass, so that the centre of mass and laboratory systems coincide. For incident electron energies E_i^e (= T_i) below the threshold energy for excitation of the $n = 2$ levels of atomic hydrogen (that is, for $E_i^e < 10.2$ eV) the only open final channel corresponds to the elastic scattering process:

$$e^- + H(1s) \rightarrow e^- + H(1s)$$

For incident electron energies E_i^e ranging between the $n = 2$ excitation threshold (10.2 eV) and the ionisation threshold at 13.6 eV, the inelastic channels $e^- + H^*$ – where H^* denotes an excited hydrogen atom – open successively, corresponding to the inelastic collisions

$$e^- + H(1s) \rightarrow e^- + H^*$$

Finally, if $E_i^e \geqslant 13.6$ eV, the ionisation channel ($H^+ + 2e^-$) becomes open, corresponding to the ionisation reaction

$$e^- + H(1s) \rightarrow H^+ + 2e^-$$

Cross-sections

The results of collision experiments are usually expressed in terms of quantities called *cross-sections.* Generalising the definition given in Appendix 1 for the simple case of scattering by a centre of force, we shall define *the cross-section for a certain type of event in a given collision as the ratio of the number of events of this type per unit time and per unit scatterer, to the flux of the incident particles with respect to the target.* Cross-sections are independent of the incident flux so that we may choose this number to be one, and cross-sections can also be defined as *transition probabilities per unit time, per unit target scatterer and per unit flux of the incident particles with respect to the target.*

Consider for example the reaction (12.3). Let N_A be the flux of incident particles relative to the target (defined as the number of particles A crossing per unit time a unit area placed at right angles to the direction of the incident beam and at rest with respect to the target) and let n_B be the number of particles B within the target interacting with the incident beam. Calling N_{tot}^C the total number of particles C which, together with the particles D, have been produced per unit time in the collision between the beam particles A and the target particles B, we have

$$N_{tot}^C = N_A n_B \sigma_{tot}^C \quad \textbf{(12.10)}$$

where σ_{tot}^C is the total cross-section for the reaction (12.3). Total cross-sections for elastic scattering, inelastic scattering and reactions of the type (12.4) may be defined in a similar way. Finally, if we denote by N_{tot} the total number of particles A which have interacted per unit time with target scatterers, we may define a total (complete) cross-section σ_{tot} by the relation

$$N_{tot} = N_A n_B \sigma_{tot} \quad \textbf{(12.11)}$$

If only elastic scattering between the particles A and B can occur, then σ_{tot} is just equal to the total elastic cross-section σ_{tot}^{el}. However, when channels other than the elastic one are open, then σ_{tot} is the sum of the total cross-sections corresponding to each of the open channels.

It is apparent from the foregoing discussion that total cross-sections have the dimensions of an area. Since in atomic units (a.u.) the unit of length is the first Bohr radius $a_0 = 5.29 \times 10^{-11}$ m, it is convenient to express atomic cross-sections in units of $a_0^2 = 2.80 \times 10^{-21}$ m^2. Units of $\pi a_0^2 = 8.80 \times 10^{-21}$ m^2 are also often used for atomic total cross-sections.

In order to define *differential* cross-sections, we must choose a coordinate system or reference framework. Two convenient choices are the *laboratory* system and *centre of mass* system which have been introduced above. Experiments are often, but not always [3], performed in the laboratory system, while calculations are usually carried out in the centre of mass system, where the three degrees of freedom attached to the centre of mass of the system (A + B) may be ignored. The relationship between the laboratory and centre of mass differential cross-sections is established in Appendix 2 for the simple case of elastic scattering. The results obtained in Appendix 2 can be generalised in a straightforward way (Problem 12.1) to inelastic collisions and reactions involving two particles (complex or not) in the initial or final states.

12.2 Potential scattering. General features

Let us start by considering the simplest collision problem, which is the *non-relativistic scattering of a spinless particle by a real potential* $V(\mathbf{r})$. The system is described by the Schrödinger equation

$$\mathrm{i}\hbar\frac{\partial}{\partial t}\Psi(\mathbf{r},t) = \left[-\frac{\hbar^2}{2m}\nabla^2 + V(\mathbf{r})\right]\Psi(\mathbf{r},t) \tag{12.12}$$

where m denotes the mass of the particle. From our discussion of Section 2.7 it is clear that this problem is equivalent to that of the elastic scattering between two spinless particles A and B, interacting through a potential $V(\mathbf{r})$ which depends only on their relative coordinate $\mathbf{r}$, provided that we work in the centre of mass system. It is then only necessary to replace m by the reduced mass $\mu = m_A m_B/(m_A + m_B)$ of the two particles.

The stationary state wave function for scattering

In the type of experiment schematically shown in Fig. 12.1, the incident beam of particles is switched on for times very long compared with the time a particle would take to cross the interaction region, so that steady-state conditions apply. Since the potential $V(\mathbf{r})$ does not depend on time, we can look for *stationary solutions* of (12.12) having the form

$$\Psi(\mathbf{r},t) = \psi(\mathbf{r})\exp(-\mathrm{i}Et/\hbar) \tag{12.13}$$

where $\psi(\mathbf{r})$ is a solution of the time-independent Schrödinger equation

$$\left[-\frac{\hbar^2}{2m}\nabla^2 + V(\mathbf{r})\right]\psi(\mathbf{r}) = E\psi(\mathbf{r}) \tag{12.14}$$

---

[3] Crossed beam experiments in which two beams, inclined at a certain angle, interact, are neither in the centre of mass nor the laboratory systems, but the measured quantities are easily transformed to one of the standard reference systems.

and E, the energy of the particle, has the well-defined value

$$E = \frac{p^2}{2m} = \frac{\hbar^2 k^2}{2m} = \frac{1}{2}mv^2 \tag{12.15}$$

Here

$$p = |\mathbf{p}_i| = |\mathbf{p}_f|, \qquad k = |\mathbf{k}_i| = |\mathbf{k}_f|, \qquad v = |\mathbf{v}_i| = |\mathbf{v}_f| \tag{12.16}$$

are respectively the magnitudes of the initial (or final) momentum, wave vector and velocity of the particle, with

$$\mathbf{p}_i = \hbar\mathbf{k}_i = m\mathbf{v}_i, \qquad \mathbf{p}_f = \hbar\mathbf{k}_f = m\mathbf{v}_f \tag{12.17}$$

It is convenient to introduce the reduced potential

$$U(\mathbf{r}) = \frac{2m}{\hbar^2}V(\mathbf{r}) \tag{12.18}$$

so that (12.14) may be rewritten as

$$[\nabla^2 + k^2 - U(\mathbf{r})]\psi(\mathbf{r}) = 0 \tag{12.19}$$

In what follows we shall assume that the potential tends to zero faster than $1/r$ as $r \to \infty$. At large r the scattering wave function must therefore satisfy the free-particle Schrödinger equation

$$(\nabla^2 + k^2)\psi(\mathbf{r}) = 0 \tag{12.20}$$

and in this region we can write

$$\psi(\mathbf{r}) \underset{r\to\infty}{\to} \psi_{\text{inc}}(\mathbf{r}) + \psi_{\text{sc}}(\mathbf{r}) \tag{12.21}$$

where ψ_{inc} represents the incident beam of particles and ψ_{sc} represents the scattered particles. Since the incident particles are monoenergetic and are travelling in the direction $\hat{\mathbf{k}}_i$, which we take to be parallel to the Z axis, the incident beam can be represented by the plane wave

$$\psi_{\text{inc}}(\mathbf{r}) = A\exp(\mathrm{i}\mathbf{k}_i\cdot\mathbf{r}) = A\exp(\mathrm{i}kz) \tag{12.22}$$

where A is an arbitrary normalisation constant. Since the number of particles per unit volume is $|\psi_{\text{inc}}|^2 = A^2$ and each particle has the velocity $v = \hbar k/m$, the incident flux F is

$$F = v|A|^2 \tag{12.23}$$

A plane wave is of infinite extent in a transverse direction, but in any real experiment the beam is collimated and has a finite transverse extension. However, the transverse dimensions of the beam, which may be of the order of 1 cm to 1 mm, are sufficiently large for the corresponding uncertainty in momentum to be negligible and for the wave function to be described accurately by a plane wave over the scattering region (which, for atoms, is of the order of 10^{-8} cm).

Far from the scatterer, the scattered wave function must represent an outward flow of particles from the scattering centre. It has the form of an *outgoing* spherical wave, the amplitude of which depends on the direction of $\mathbf{r}$ and on the energy E (that is, on k). We write

$$\psi_{sc}^{(+)}(\mathbf{r}) = Af(k, \theta, \phi)\frac{\exp(ikr)}{r} \tag{12.24}$$

where (θ, ϕ) are the polar angles of $\mathbf{r}$ with respect to the Z axis (the incident direction $\hat{\mathbf{k}}_i$) and f is known as the *scattering amplitude*. It is readily verified that the spherical wave $\psi_{sc}^{(+)}$ is an asymptotic solution (for large r) of the free particle Schrödinger equation (12.20). That this spherical wave is an outgoing wave can be checked by returning to the time-dependent wave function (12.13) and noting that the expression $r^{-1}\exp[i(kr - \omega t)]$ describes a spherical wave with positive (+) phase velocity $dr/dt = \omega/k$.

It follows from the above discussion that we may look for a particular solution of equation (12.14) (or equation (12.19)), which we shall call the *stationary state wave function for scattering* and denote by $\psi_{\mathbf{k}_i}^{(+)}(\mathbf{r})$, which satisfies the asymptotic boundary condition

$$\psi_{\mathbf{k}_i}^{(+)}(\mathbf{r}) \underset{r\to\infty}{\to} A\left[\exp(i\mathbf{k}_i\cdot\mathbf{r}) + f(k, \theta, \phi)\frac{\exp(ikr)}{r}\right] \tag{12.25}$$

Thus at large distances the stationary scattering wave function $\psi_{\mathbf{k}_i}^{(+)}(\mathbf{r})$ is the superposition of a plane wave of wave vector $\mathbf{k}_i$ and an outgoing spherical wave with positive phase velocity.

The cross-section

Using the general definition of cross-sections, given in Section 12.1, we introduce a *differential cross-section*, $d\sigma/d\Omega$. We imagine that the detector subtends at the scattering centre a small solid angle $d\Omega$ and is placed in the direction (θ, ϕ). Then $d\sigma/d\Omega$ is defined as the ratio of the outgoing flux of particles passing through the area $r^2\,d\Omega$ (with $r \to \infty$), per unit solid angle, to the incident flux. The detector is placed outside the incident beam, so that only scattered particles are recorded and the corresponding flux can be calculated from ψ_{sc} alone.

The probability current density $\mathbf{j}$ has been obtained in Section 2.2 (see (2.50)). For a stationary state (12.13), it is given by

$$\mathbf{j} = \frac{\hbar}{2mi}\left[\psi^*(\nabla\psi) - (\nabla\psi^*)\psi\right] \tag{12.26}$$

The gradient operator in spherical coordinates is

$$\nabla = \frac{\partial}{\partial r}\hat{\mathbf{r}} + \frac{1}{r}\frac{\partial}{\partial\theta}\hat{\boldsymbol{\theta}} + \frac{1}{r\sin\theta}\frac{\partial}{\partial\phi}\hat{\boldsymbol{\phi}} \tag{12.27}$$

so that the radial current is given by

$$\mathbf{j}\cdot\hat{\mathbf{r}} = \frac{\hbar}{2mi}\left(\psi^*\frac{\partial\psi}{\partial r} - \frac{\partial\psi^*}{\partial r}\psi\right) \tag{12.28}$$

If we substitute the expression (12.24) for $\psi_{sc}^{(+)}$ into (12.28), we find that at large r the corresponding scattered radial current, $\mathbf{j}_{sc}\cdot\hat{\mathbf{r}}$, is

$$\mathbf{j}_{sc}\cdot\hat{\mathbf{r}} = |A|^2 v |f(k, \theta, \phi)|^2/r^2 \tag{12.29}$$

where terms of higher order in $1/r$ have been neglected. The outgoing flux of particles passing through a spherical surface element $r^2\,\mathrm{d}\Omega$ for large r is given from (12.29) by $|A|^2 v\,|f(k, \theta, \phi)|^2\,\mathrm{d}\Omega$. Dividing this quantity by the incident flux (12.23), and by the solid angle $\mathrm{d}\Omega$, we obtain the *differential cross-section*

$$\frac{\mathrm{d}\sigma}{\mathrm{d}\Omega} = |f(k, \theta, \phi)|^2 \tag{12.30}$$

The *total cross-section* is obtained by integrating over all scattering angles, namely

$$\sigma_{\mathrm{tot}} = \int \frac{\mathrm{d}\sigma}{\mathrm{d}\Omega}\,\mathrm{d}\Omega = \int_0^{2\pi} \mathrm{d}\phi \int_0^{\pi} \mathrm{d}\theta \sin\theta\, |f(k, \theta, \phi)|^2 \tag{12.31}$$

A further cross-section, which arises for example in the study of the motion of electron swarms in gases, is the *momentum transfer cross-section*. It is defined by

$$\sigma_{\mathrm{M}} = \int (1 - \cos\theta)\frac{\mathrm{d}\sigma}{\mathrm{d}\Omega}\,\mathrm{d}\Omega = \int_0^{2\pi} \mathrm{d}\phi \int_0^{\pi} \mathrm{d}\theta \sin\theta\,(1 - \cos\theta)|f(k, \theta, \phi)|^2 \tag{12.32}$$

We remark that the cross-sections are determined by the modulus squared of the scattering amplitude f, and are independent of the normalisation coefficient A. The scattering amplitude f itself appears in the asymptotic behaviour (12.25) of the stationary scattering wave function; this corresponds to the experimental conditions because any detector will be placed at a very large distance from the scattering region, compared with atomic dimensions.

The optical theorem

Since the total number of particles entering the scattering region per unit time must be balanced by the number leaving it per unit time, we must have

$$r^2 \int (\mathbf{j}\cdot\hat{\mathbf{r}})\,\mathrm{d}\Omega = 0 \tag{12.33}$$

where the integration is over all angles and $\mathbf{j}\cdot\hat{\mathbf{r}}$ is the complete radial current, rather than the scattered radial current $\mathbf{j}_{sc}\cdot\hat{\mathbf{r}}$. The complete radial current is computed by substituting (12.21) into (12.28), which gives up to terms of order $(1/r^3)$

$$\mathbf{j}\cdot\hat{\mathbf{r}} = |A|^2 \frac{\hbar}{2mi}\left[\mathrm{e}^{-ikr\cos\theta} + f^*(k, \theta, \phi)\frac{\mathrm{e}^{-ikr}}{r}\right] \times \left[ik\cos\theta\,\mathrm{e}^{ikr\cos\theta} + f(k, \theta, \phi)\,ik\frac{\mathrm{e}^{ikr}}{r}\right] + \text{c.c.} \tag{12.34}$$

where c.c. denotes the complex conjugate and we have used the fact that $z = r\cos\theta$. By inserting this expression into (12.33), one finds (Problem 12.2)

$$\operatorname{Im} f(k, \theta = 0) = \frac{k}{4\pi} \int |f|^2 \, \mathrm{d}\Omega \tag{12.35}$$

or, using (12.30),

$$\sigma_{\text{tot}} = \frac{4\pi}{k} \operatorname{Im} f(k, \theta = 0) \tag{12.36}$$

This relation is known as the *optical theorem* and expresses *conservation of the probability flux.*

12.3 The method of partial waves

We now turn to the problem of calculating the scattering amplitude, from which the differential cross-section can be found from (12.30). We shall first consider the case of a *central* potential $V(r)$, which depends only on the magnitude r of the vector $\mathbf{r}$. As we have seen in Section 2.6 the Hamiltonian operator

$$H = -\frac{\hbar^2}{2m} \nabla^2 + V(r) \tag{12.37}$$

then commutes with the operators $\mathbf{L}^2$ and L_z, and the Schrödinger equation (12.14) is separable in spherical polar coordinates. In addition, the problem also possesses symmetry about the incident direction (which coincides with our Z axis), so that the wave function $\psi^{(+)}_{\mathbf{k}_i}$ – and hence the scattering amplitude and the differential cross-section – is independent of the azimuthal angle ϕ. The wave function $\psi^{(+)}_{\mathbf{k}_i}$ may therefore be expanded in a series of Legendre polynomials (which form a complete set in the interval $-1 \leqslant \cos\theta \leqslant +1$) as

$$\psi^{(+)}_{\mathbf{k}_i}(k, r, \theta) = \sum_{l=0}^{\infty} R_l(k, r) P_l(\cos\theta) \tag{12.38}$$

The radial equations

The equations satisfied by the radial functions $R_l(k, r)$ can be found as in Section 2.6. We have

$$\left[\frac{\mathrm{d}^2}{\mathrm{d}r^2} + \frac{2}{r}\frac{\mathrm{d}}{\mathrm{d}r} - \frac{l(l+1)}{r^2} - U(r) + k^2\right] R_l(k, r) = 0 \tag{12.39}$$

where $U(r) = 2mV(r)/\hbar^2$. For the special case of the Coulomb interaction, these are the same equations as we found in (3.6); but in Chapter 3, we were looking for solutions of negative energy. As shown in Section 2.6, the equations (12.39) can be simplified by introducing the new radial functions

$$u_l(k, r) = rR_l(k, r) \tag{12.40}$$

which satisfy the equations

$$\left[\frac{\mathrm{d}^2}{\mathrm{d}r^2} - \frac{l(l+1)}{r^2} - U(r) + k^2\right]u_l(k, r) = 0 \tag{12.41}$$

There is no loss of generality in assuming $u_l(k, r)$ to be real since both the real and imaginary parts of a complex $u_l(k, r)$ would separately satisfy the radial equations (12.41).

For potentials which are less singular than r^{-2} at the origin, the radial function $u_l(k, r)$ can be expanded in a power series,

$$u_l(k, r) = r^s \sum_{n=0}^{\infty} a_n r^n \tag{12.42}$$

and the examination of the indicial equation [4] for s shows that there are two solutions, one *regular* at the origin, which behaves like

$$u_l(k, r) \underset{r\to 0}{\sim} r^{l+1} \tag{12.43a}$$

and one *irregular*, such that

$$u_l(k, r) \underset{r\to 0}{\sim} r^{-l} \tag{12.43b}$$

In order to describe a physical scattering situation, the wave function $\psi_{\mathbf{k}_i}^{(+)}$ must be finite everywhere, so that we must choose for $u_l(k, r)$ the regular solution which behaves like r^{l+1} near the origin. It is clear from (12.40) that the corresponding regular radial function $R_l(k, r)$ behaves like r^l as $r \to 0$.

Let us now examine the behaviour of the functions $R_l(k, r)$ or $u_l(k, r)$ for large r. We shall assume that the potential may be neglected when r is larger than a given value a, in which case we shall say that it has a *range a*. In this *external region* $r > a$ the equation (12.39) then reduces to the free particle equation

$$\left[\frac{\mathrm{d}^2}{\mathrm{d}\rho^2} + \frac{2}{\rho}\frac{\mathrm{d}}{\mathrm{d}\rho} + \left(1 - \frac{l(l+1)}{\rho^2}\right)\right]R_l(\rho) = 0 \tag{12.44}$$

where we have set $\rho = kr$. This equation is known as the *spherical Bessel differential equation*. The general solution of this equation is a linear combination of the spherical Bessel function

$$j_l(\rho) = \left(\frac{\pi}{2\rho}\right)^{1/2} J_{l+1/2}(\rho) \tag{12.45}$$

[4] Series solutions of differential equations are discussed in Mathews and Walker (1973).

and the spherical Neumann function

$$n_l(\rho) = (-1)^{l+1}\left(\frac{\pi}{2\rho}\right)^{1/2} J_{-l-1/2}(\rho) \tag{12.46}$$

where $J_\nu(\rho)$ is a Bessel function of order ν. We have already encountered the spherical Bessel functions j_l in Section 2.6 when we discussed the expansion of a plane wave in Legendre polynomials, and we shall return to this point shortly. We recall here for convenience the expressions of the first three spherical Bessel functions, and also give the first three spherical Neumann functions, namely

$$\begin{aligned}
j_0(\rho) &= \frac{\sin\rho}{\rho}\\
n_0(\rho) &= -\frac{\cos\rho}{\rho}\\
j_1(\rho) &= \frac{\sin\rho}{\rho^2} - \frac{\cos\rho}{\rho}\\
n_1(\rho) &= -\frac{\cos\rho}{\rho^2} - \frac{\sin\rho}{\rho}\\
j_2(\rho) &= \left(\frac{3}{\rho^3} - \frac{1}{\rho}\right)\sin\rho - \frac{3}{\rho^2}\cos\rho\\
n_2(\rho) &= -\left(\frac{3}{\rho^3} - \frac{1}{\rho}\right)\cos\rho - \frac{3}{\rho^2}\sin\rho
\end{aligned} \tag{12.47}$$

The behaviour of the functions j_l and n_l for small ρ is given by

$$j_l(\rho) \underset{\rho\to 0}{\sim} \frac{\rho^l}{(2l+1)!!}\left[1 - \frac{\rho^2/2}{1!(2l+3)} + \frac{(\rho^2/2)^2}{2!(2l+3)(2l+5)} - \cdots\right] \tag{12.48a}$$

and

$$n_l(\rho) \underset{\rho\to 0}{\sim} -\frac{(2l-1)!!}{\rho^{l+1}}\left[1 - \frac{\rho^2/2}{1!(1-2l)} + \frac{(\rho^2/2)^2}{2!(1-2l)(3-2l)} - \cdots\right] \tag{12.48b}$$

where

$$(2l-1)!! = 1\cdot 3\cdot 5\cdot \ldots \cdot(2l-1); \qquad (2l-1)!! = 1 \quad \text{for} \quad l = 0$$

Thus the spherical Bessel function j_l, which is proportional to ρ^l as $\rho \to 0$, is a regular solution of (12.44). On the other hand, the function n_l, which has a pole of order $(l+1)$ at $\rho = 0$, is an irregular solution of (12.44).

For ρ somewhat larger than $l(l+1)/2$ one may use the asymptotic formulae

$$j_l(\rho) \underset{\rho\to\infty}{\rightarrow} \frac{1}{\rho}\sin(\rho - l\pi/2) \tag{12.49a}$$

and

$$n_l(\rho) \underset{\rho\to\infty}{\rightarrow} -\frac{1}{\rho}\cos(\rho - l\pi/2) \tag{12.49b}$$

Let us now return to the radial functions $R_l(k, r)$ and $u_l(k, r)$. We see that in the external region $r > a$, $R_l(k, r)$ must be a linear combination of $j_l(kr)$ and $n_l(kr)$:

$$R_l(k, r) = \frac{u_l(k, r)}{r} = B_l(k)j_l(kr) + C_l(k)n_l(kr) \tag{12.50}$$

where $B_l(k)$ and $C_l(k)$ are real 'constants of integration' which are independent of r. Using the asymptotic formulae (12.49), we see that

$$R_l(k, r) = \frac{u_l(k, r)}{r} \underset{r\to\infty}{\rightarrow} B_l(k)\frac{\sin(kr - l\pi/2)}{kr} - C_l(k)\frac{\cos(kr - l\pi/2)}{kr} \tag{12.51}$$

It is convenient to set

$$A_l(k) = [B_l^2(k) + C_l^2(k)]^{1/2} \tag{12.52a}$$

and

$$\tan\delta_l(k) = -C_l(k)/B_l(k) \tag{12.52b}$$

so that we obtain for $R_l(k, r)$ the simple asymptotic behaviour

$$R_l(k, r) = \frac{u_l(k, r)}{r} \underset{r\to\infty}{\rightarrow} \frac{A_l(k)}{kr}\sin[kr - l\pi/2 + \delta_l(k)] \tag{12.53}$$

It may be shown (Problem 12.3) that this expression is valid for all potentials which vanish faster than r^{-1} when $r \to \infty$. Using (12.50) and (12.52b), we see that for $r > a$ the radial function can be written as

$$R_l(k, r) = \frac{u_l(k, r)}{r} = B_l(k)[j_l(kr) - \tan\delta_l(k)n_l(kr)], \qquad r > a \tag{12.54}$$

The real quantities $\delta_l(k)$, which are called *phase shifts*, display the influence of the interaction. Indeed, in the complete absence of interaction the free particle equation (12.44) holds for all r and the boundary condition at $r = 0$ excludes the irregular solution n_l in (12.50), thus forcing us to set $C_l = 0$. As a result, the corresponding free particle radial functions R_l^0 are proportional to $j_l(kr)$, in agreement with the expansion (2.260) of a plane wave in Legendre polynomials,

$$A\exp(\mathrm{i}kz) = A\exp(\mathrm{i}kr\cos\theta) = A\sum_{l=0}^{\infty}(2l+1)\mathrm{i}^l j_l(kr)P_l(\cos\theta) \tag{12.55}$$

where A is an arbitrary normalisation coefficient as in (12.22). The asymptotic behaviour of R_l^0 is then

$$R_l^0(k, r) = r^{-1}u_l^0(k, r) \underset{r\to\infty}{\to} A_l(kr)^{-1} \sin(kr - l\pi/2) \tag{12.56}$$

To relate the scattering amplitude to the phase shifts, we use (12.55) to write the asymptotic form of $\psi_{\mathbf{k}_i}^{(+)}(\mathbf{r})$, given by (12.25), as

$$\begin{aligned}\psi_{\mathbf{k}_i}^{(+)}(\mathbf{r}) &\underset{r\to\infty}{\to} A\left[\sum_{l=0}^{\infty}(2l+1)\mathrm{i}^l(kr)^{-1}\sin(kr - l\pi/2)P_l(\cos\theta) + f(k,\theta)\frac{\mathrm{e}^{\mathrm{i}kr}}{r}\right] \\ &= A\left[\frac{\mathrm{e}^{\mathrm{i}kr}}{r}\left\{f(k,\theta) - \sum_{l=0}^{\infty}\left(\frac{\mathrm{i}}{2k}\right)(2l+1)P_l(\cos\theta)\right\}\right. \\ &\quad \left. + \frac{\mathrm{e}^{-\mathrm{i}kr}}{r}\left\{\sum_{l=0}^{\infty}(-1)^l\left(\frac{\mathrm{i}}{2k}\right)(2l+1)P_l(\cos\theta)\right\}\right]\end{aligned} \tag{12.57}$$

On the other hand, the asymptotic form of $\psi_{\mathbf{k}_i}^{(+)}(\mathbf{r})$ can be written from (12.53) and (12.38) as

$$\begin{aligned}\psi_{\mathbf{k}_i}^{(+)}(\mathbf{r}) &\underset{r\to\infty}{\to} \sum_{l=0}^{\infty}(kr)^{-1}A_l(k)P_l(\cos\theta)\frac{1}{2\mathrm{i}} \\ &\quad\times\{\exp[\mathrm{i}(kr - l\pi/2 + \delta_l)] - \exp[-\mathrm{i}(kr - l\pi/2 + \delta_l)]\}\end{aligned} \tag{12.58}$$

On equating the coefficients of $\exp(-\mathrm{i}kr)$ in (12.57) and (12.58) we find that

$$A_l(k) = A(2l+1)\mathrm{i}^l \exp[\mathrm{i}\delta_l(k)] \tag{12.59}$$

Next, by matching the coefficients of $\exp(\mathrm{i}kr)$ and using (12.59), we have

$$\begin{aligned}f(k,\theta) &= \frac{1}{2\mathrm{i}k}\sum_{l=0}^{\infty}(2l+1)\{\exp[2\mathrm{i}\delta_l(k)] - 1\}P_l(\cos\theta) \\ &= \frac{1}{k}\sum_{l=0}^{\infty}(2l+1)\exp[\mathrm{i}\delta_l(k)]\sin\delta_l(k)P_l(\cos\theta)\end{aligned} \tag{12.60}$$

which is the desired result. We remark that the scattering amplitude is independent of the choice of the normalisation coefficients $A_l(k)$. We may also rewrite (12.60) in the form

$$f(k,\theta) = \sum_{l=0}^{\infty} f_l(k)P_l(\cos\theta) \tag{12.61}$$

where the *partial wave amplitudes* $f_l(k)$ are such that

$$f_l(k) = \frac{2l+1}{2\mathrm{i}k}\{\exp[2\mathrm{i}\delta_l(k)] - 1\} = \frac{2l+1}{k}\exp[\mathrm{i}\delta_l(k)]\sin\delta_l(k) \tag{12.62}$$

Turning now to the differential cross-section, we have from (12.30) and (12.60)

$$\frac{\mathrm{d}\sigma}{\mathrm{d}\Omega} = |f(k, \theta)|^2$$

$$= \frac{1}{k^2}\sum_{l=0}^{\infty}\sum_{l'=0}^{\infty} (2l+1)(2l'+1)\exp\{\mathrm{i}[\delta_l(k) - \delta_{l'}(k)]\}$$

$$\times \sin\delta_l(k)\sin\delta_{l'}(k)P_l(\cos\theta)P_{l'}(\cos\theta) \qquad \textbf{(12.63)}$$

and the total cross-section is given by

$$\sigma_{\text{tot}}(k) = 2\pi\int_0^{\pi}\frac{\mathrm{d}\sigma}{\mathrm{d}\Omega}(k, \theta)\sin\theta\,\mathrm{d}\theta$$

$$= \frac{4\pi}{k^2}\sum_{l=0}^{\infty}(2l+1)\sin^2\delta_l(k) \qquad \textbf{(12.64)}$$

where we have used the result (2.171). We may also write (12.64) as

$$\sigma_{\text{tot}}(k) = \sum_{l=0}^{\infty}\sigma_l(k) \qquad \textbf{(12.65)}$$

where each partial wave cross-section $\sigma_l(k)$ is given by

$$\sigma_l(k) = \frac{4\pi}{k^2}(2l+1)\sin^2\delta_l(k) \qquad \textbf{(12.66)}$$

The maximum contribution of each partial wave to the total cross-section is

$$\sigma_l^{\text{max}}(k) = \frac{4\pi}{k^2}(2l+1) \qquad \textbf{(12.67)}$$

and occurs when $\delta_l(k) = (n + 1/2)\pi$, $n = 0, \pm1, \pm2, \ldots$. In contrast, when $\delta_l(k) = n\pi$ at a certain value of k there is no contribution to the scattering from the partial wave of order l at that value of k.

A partial wave expansion for the momentum transfer cross-section (12.32) can also be obtained. Substituting (12.60) into (12.32), one finds (Problem 12.4) that

$$\sigma_{\text{M}}(k) = \frac{4\pi}{k^2}\sum_{l=0}^{\infty}(l+1)\sin^2[\delta_{l+1}(k) - \delta_l(k)] \qquad \textbf{(12.68)}$$

The imaginary part of the scattering amplitude in the forward direction can be related to the total cross-section. Since $P_l(1) = 1$, we have from (12.60)

$$\text{Im}\, f(k, \theta = 0) = \frac{1}{k}\sum_{l=0}^{\infty}(2l+1)\sin^2\delta_l(k) \qquad \textbf{(12.69)}$$

and upon comparison of (12.64) with (12.69) we obtain immediately the optical theorem (12.36). More generally, one can prove the *unitarity relation* (Problem 12.5)

$$\text{Im}\, f(k, \theta) = \frac{k}{4\pi} \int f^*(k, \theta')f(k, \theta_0)\, \mathrm{d}\Omega' \tag{12.70}$$

where θ_0 is the angle between the directions $(\theta, 0)$ and (θ', ϕ'), and $\mathrm{d}\Omega' = \sin\theta'\, \mathrm{d}\theta'\, \mathrm{d}\phi'$. This 'unitarity relation', written here for the case of scattering by a central potential, is in fact a particular case of a general relation of quantum collision theory, which expresses the conservation of probability [5].

It is clear that the method of partial waves is most useful when only a small number of partial waves contribute to the scattering. This situation occurs at *low incident energies*. Indeed, the 'effective (reduced) potential' which occurs in the radial equations (12.39) or (12.41) is

$$U_{\text{eff}}(r) = U(r) + \frac{l(l+1)}{r^2} \tag{12.71}$$

Thus, as l increases, the centrifugal barrier term $l(l+1)/r^2$ becomes more important and the incident particle needs more energy to overcome this repulsion and to probe the interaction region where the potential acts. Hence at low energies we expect that only a few partial waves will be required in the partial wave expansion.

In order to obtain an estimate of the number of partial waves which are needed for a given value of the incident energy, we note that the first and most important maximum of the free radial wave function $j_l(kr)$ occurs approximately at $r_0 \simeq l/k$ while for small r, $j_l(kr)$ is small, and behaves like r^l as $r \to 0$. Thus, if the potential has a range a (that is, it acts in a region $r < a$), and if $a \ll l/k$, the function j_l will remain small in the scattering region, and the corresponding phase shift will be negligible. It is then reasonable to cut off the partial wave expansion at a value $l_{\text{max}} \simeq ka$. Hence, if ka is small only a small number of phase shifts must be calculated.

We can also come to the above conclusion by using a simple (non-rigorous) semi-classical argument. If a potential vanishes beyond a certain distance a, then according to classical mechanics an incident particle having an impact parameter b will be deflected or not according to whether $b < a$ or $b > a$ (see Fig. 12.2). Now the impact parameter is given by

$$b = \frac{L}{p} \tag{12.72}$$

where p and L are the magnitudes of the particle momentum and orbital angular momentum, respectively. Thus, according to classical mechanics, particles with orbital angular momentum

$$L > pa \tag{12.73}$$

are not scattered. Let us now assume that the quantity ka is somewhat larger than unity, so that the reduced de Broglie wavelength $\lambda\!\!\!^{-} = \lambda/(2\pi) = k^{-1}$ of the particle is small with respect to a and we are in a semi-classical situation. We may then write

[5] See Joachain (1983).

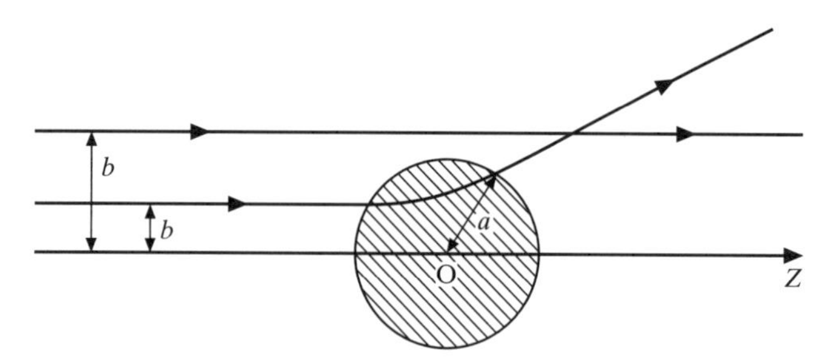

Figure 12.2 Classical scattering of a particle with impact parameter b, by a potential of finite range a. If $b > a$ no scattering occurs.

$L \simeq \hbar l$ and since $p = \hbar k$, we see that scattering is again expected to be small in angular momentum states for which

$$l > ka \tag{12.74}$$

The *S*-matrix, *K*-matrix and *T*-matrix elements

The radial equations (12.39) and (12.41) satisfied by the functions $R_l(k, r)$ and $u_l(k, r)$, respectively, are homogeneous equations. Hence these functions are only defined up to an arbitrary multiplicative constant (independent of r) and we can write

$$u_l(k, r) \underset{r\to\infty}{\to} N_l(k) \sin[kr - l\pi/2 + \delta_l(k)] \tag{12.75}$$

We note that the choice of $N_l(k)$ does not affect the expressions for the scattering amplitude and cross-sections. If $N_l(k)$ is chosen to be $-2\text{i} \exp[\text{i}\delta_l(k)]$, then

$$u_l(k, r) \underset{r\to\infty}{\to} \exp[-\text{i}(kr - l\pi/2)] - S_l(k) \exp[\text{i}(kr - l\pi/2)] \tag{12.76}$$

where the quantity

$$S_l(k) = \exp[2\text{i}\delta_l(k)] \tag{12.77}$$

is called an *S-matrix element*. Since the phase shifts $\delta_l(k)$ are real, we see that the corresponding S-matrix elements (12.77) are of unit modulus.

If the normalisation constant $N_l(k)$ in (12.75) is chosen to be $[\cos \delta_l(k)]^{-1}$, then

$$u_l(k, r) \underset{r\to\infty}{\to} \sin(kr - l\pi/2) + K_l(k) \cos(kr - l\pi/2) \tag{12.78}$$

where the quantity

$$K_l(k) = \tan \delta_l(k) \tag{12.79}$$

is called a *K-matrix element*. The S-matrix and K-matrix elements are related by

$$S_l(k) = \frac{1 + \text{i}K_l(k)}{1 - \text{i}K_l(k)} \tag{12.80}$$

It is also useful to define a *T-matrix element* by the relation

$$T_l(k) = S_l(k) - 1 = \exp[2i\delta_l(k)] - 1 \tag{12.81}$$

The phase shifts

Before we turn to the problem of the actual determination of the phase shifts, we shall discuss a few of their key properties. First of all, the relation between the phase shifts and the interaction potential may be analysed by considering the scattering first by a reduced potential $U(r)$ and then by a different reduced potential $\bar{U}(r)$. The radial equations are (see (12.41))

$$\left[\frac{d^2}{dr^2} - \frac{l(l+1)}{r^2} - U(r) + k^2\right]u_l(k, r) = 0 \tag{12.82a}$$

and

$$\left[\frac{d^2}{dr^2} - \frac{l(l+1)}{r^2} - \bar{U}(r) + k^2\right]\bar{u}_l(k, r) = 0 \tag{12.82b}$$

We shall adopt a normalisation such that in the external region $r > a$

$$\begin{aligned} R_l(k, r) &= r^{-1}u_l(k, r) = j_l(kr) - \tan\delta_l(k)n_l(kr) \\ &\underset{r\to\infty}{\to} \frac{1}{kr}[\sin(kr - l\pi/2) + \tan\delta_l(k)\cos(kr - l\pi/2)] \end{aligned} \tag{12.83}$$

with similar relations for $\bar{R}_l$ and $\bar{u}_l$, in which $\tan\delta_l$ is replaced by $\tan\bar{\delta}_l$.

The Wronskian of the two solutions u_l and $\bar{u}_l$ is defined as

$$W(u_l, \bar{u}_l) = u_l\bar{u}_l' - u_l'\bar{u}_l \tag{12.84}$$

where the prime denotes a derivative with respect to r. Premultiplying (12.82a) by $\bar{u}_l$, (12.82b) by u_l, subtracting term by term, and using (12.84), we have

$$\frac{d}{dr}W(u_l, \bar{u}_l) = -(U - \bar{U})u_l\bar{u}_l \tag{12.85}$$

Upon integration over the variable r from 0 to ∞, we obtain

$$\begin{aligned} \tan\delta_l(k) - \tan\bar{\delta}_l(k) &= -k\int_0^\infty \bar{u}_l(k, r)[U(r) - \bar{U}(r)]u_l(k, r)\,dr \\ &= -k\int_0^\infty \bar{R}_l(k, r)[U(r) - \bar{U}(r)]R_l(k, r)r^2\,dr \end{aligned} \tag{12.86}$$

where we have used the fact that $u_l(k, 0) = \bar{u}_l(k, 0) = 0$, together with (12.83).

We note from (12.86) that if the difference $(U - \bar{U})$ is small, so that $u_l\bar{u}_l \simeq (u_l)^2$, the difference $(\delta_l - \bar{\delta}_l)$ has the opposite sign to that of $(U - \bar{U})$. Moreover, by constructing a series of 'comparison potentials' between U and $\bar{U}$, this result remains true for all $(U - \bar{U})$. If we adopt an *absolute definition* of phase shifts by requiring that $\delta_l = 0$ if $U = 0$, we see from (12.86) that if the potential is everywhere repulsive (positive for all r) then $\delta_l < 0$. On the other hand, a potential which is everywhere attractive (negative for all r) yields phase shifts $\delta_l > 0$. By choosing $\bar{U} = 0$ we also deduce from (12.86) the important integral representation

$$\tan \delta_l = -k \int_0^\infty j_l(kr) U(r) R_l(k, r) r^2 \, \mathrm{d}r \tag{12.87}$$

which is valid provided the radial function, R_l, is normalised according to (12.83).

In order to compute the phase shifts, the radial equations (12.39) must be solved (in general by numerical methods) in the internal region ($r \leqslant a$) subject to the boundary conditions at the origin discussed above. We then require both R_l and $\mathrm{d}R_l/\mathrm{d}r$ to be continuous at $r = a$. An equivalent procedure consists in requiring that the logarithmic derivative $R_l^{-1}(\mathrm{d}R_l/\mathrm{d}r)$ be continuous at $r = a$. Let us call

$$\gamma_l(k) = [R_l^{-1}(\mathrm{d}R_l/\mathrm{d}r)]_{r=a} \tag{12.88}$$

the value of the logarithmic derivative of the internal solution of (12.39) at $r = a$. Since in the external region ($r \geqslant a$) the radial wave function $R_l(k, r)$ is given by (12.54), it follows that if the logarithmic derivative $[R_l^{-1}(\mathrm{d}R_l/\mathrm{d}r)]$ is continuous at $r = a$, then

$$\gamma_l(k) = \frac{k[j_l'(ka) - \tan \delta_l(k) n_l'(ka)]}{j_l(ka) - \tan \delta_l(k) n_l(ka)} \tag{12.89}$$

where

$$j_l'(ka) = \left[\frac{\mathrm{d}j_l(\rho)}{\mathrm{d}\rho}\right]_{\rho=ka}, \quad n_l'(ka) = \left[\frac{\mathrm{d}n_l(\rho)}{\mathrm{d}\rho}\right]_{\rho=ka} \tag{12.90}$$

Thus the K-matrix element (12.79) is given by

$$K_l(k) = \tan \delta_l(k) = \frac{kj_l'(ka) - \gamma_l(k) j_l(ka)}{kn_l'(ka) - \gamma_l(k) n_l(ka)} \tag{12.91}$$

A similar formula may clearly be obtained by using the radial functions $u_l(k, r) = rR_l(k, r)$ satisfying the radial equations (12.41). Let

$$\beta_l(k) = [u_l^{-1}(\mathrm{d}u_l/\mathrm{d}r)]_{r=a} \tag{12.92}$$

be the value of the logarithmic derivative of the internal solution of (12.41) at $r = a$. Introducing the functions

$$s_l(\rho) = \rho j_l(\rho) \tag{12.93a}$$

and

$$c_l(\rho) = -\rho n_l(\rho) \tag{12.93b}$$

such that

$$s_l(\rho) \underset{\rho\to\infty}{\to} \sin(\rho - l\pi/2) \tag{12.94a}$$

and

$$c_l(\rho) \underset{\rho\to\infty}{\to} \cos(\rho - l\pi/2) \tag{12.94b}$$

we see that in the external region the radial wave function $u_l(k, r)$ may be written as

$$u_l(k, r) = D_l(k)\,[s_l(kr) + \tan\delta_l(k)c_l(kr)], \qquad r > a \tag{12.95}$$

where $D_l(k)$ is a normalisation constant. Thus, if the logarithmic derivative $[u_l^{-1}(\mathrm{d}u_l/\mathrm{d}r)]$ is continuous at $r = a$, then

$$K_l(k) = \tan\delta_l(k) = \frac{ks_l'(ka) - \beta_l(k)s_l(ka)}{-kc_l'(ka) + \beta_l(k)c_l(ka)} \tag{12.96}$$

where

$$s_l'(ka) = \left[\frac{\mathrm{d}s_l(\rho)}{\mathrm{d}\rho}\right]_{\rho=ka}, \qquad c_l'(ka) = \left[\frac{\mathrm{d}c_l(\rho)}{\mathrm{d}\rho}\right]_{\rho=ka} \tag{12.97}$$

In terms of the spherical Bessel functions j_l and the spherical Neumann functions n_l we may also write (12.96) as

$$K_l(k) = \tan\delta_l(k) = \frac{ka\;j_l'(ka) + [1 - a\beta_l(k)]j_l(ka)}{ka\;n_l'(ka) + [1 - a\beta_l(k)]n_l(ka)} \tag{12.98}$$

The behaviour of the phase shifts at *low energies* is readily obtained by using the properties of the functions $j_l(\rho)$ and $n_l(\rho)$ for small values of ρ. For example, if we define the quantity

$$\hat{\gamma}_l = \lim_{k\to 0} \gamma_l(k) \tag{12.99}$$

it may be shown (Problem 12.9) that provided

$$a\hat{\gamma}_l \neq -(l + 1) \tag{12.100}$$

the quantity $\tan\delta_l(k)$ behaves like k^{2l+1} as $k \to 0$. Except for the s wave ($l = 0$) contribution which in general tends towards a non-zero constant, all partial cross-sections σ_l ($l \geqslant 1$) then vanish as k^{4l} when $k \to 0$. Thus the scattering is *isotropic* at very low energies and $\sigma_{\text{tot}} = \sigma_0$. Defining the *scattering length* α as

$$\alpha = -\lim_{k\to 0} \frac{\tan\delta_0(k)}{k} \tag{12.101}$$

the scattering amplitude is such that

$$f \underset{k\to 0}{\to} -\alpha \tag{12.102}$$

while the differential cross-section is given by

$$\frac{d\sigma}{d\Omega} \underset{k \to 0}{\longrightarrow} \alpha^2 \tag{12.103}$$

and the total cross-section becomes

$$\sigma_{\text{tot}} \underset{k \to 0}{\longrightarrow} 4\pi\alpha^2 \tag{12.104}$$

If by accident

$$a\hat{\gamma}_l = -(l+1) \tag{12.105}$$

one can prove (Problem 12.9) that $\tan \delta_l(k)$ now behaves like k^{2l-1} as $k \to 0$. In particular, if $a\hat{\gamma}_0 = -1$, so that the anomalous behaviour occurs in the s wave ($l = 0$), one has $\tan \delta_0 \sim k^{-1}$ as $k \to 0$, and the scattering length (12.101) diverges like k^{-2} when k tends to zero. The relations (12.102)–(12.104) break down, the scattering amplitude f behaves like i/k and the differential and total cross-sections diverge like k^{-2} when $k \to 0$. This singular case is often called a *zero-energy resonance.*

If $a\hat{\gamma}_1 = -2$, so that the 'accident' occurs in the p wave ($l = 1$), the scattering amplitude is given as $k \to 0$ by (Problem 12.9)

$$f \underset{k \to 0}{\longrightarrow} -\alpha + \beta \cos\theta \tag{12.106}$$

where α is the scattering length and β is a constant. Hence in this case the differential cross-section is not isotropic when $k \to 0$. On the contrary, if the equation (12.105) holds for $l \geqslant 2$, one finds (Problem 12.9) that the results (12.102)–(12.104) are unchanged.

Let us now examine the behaviour of the phase shifts at *high energies*. For fixed l and large k we expect that the importance of the potential will become vanishingly small, so that the radial wave function $R_l(k, r)$ will approach the corresponding free spherical wave $j_l(kr)$. Hence, using (12.87), we find that for fixed l and large k, we may write $\tan \delta_l \simeq (\tan \delta_l)^{\text{B1}}$, where

$$(\tan \delta_l)^{\text{B1}} = -k \int_0^\infty [j_l(kr)]^2 U(r) r^2 \, dr \tag{12.107}$$

is called the *first Born approximation* to $\tan \delta_l$.

Finally, let us investigate the behaviour of the phase shifts for large l such that $l \gg ka$, where a is the range of the potential. In this case, the centrifugal barrier term $l(l+1)/r^2$ is very important in the effective reduced potential (12.71), and it is a good approximation to replace the radial wave function $R_l(k, r)$ by $j_l(kr)$ in (12.87), so that for $l \gg ka$ we can again calculate the phase shifts by using the first Born approximation formula (12.107).

Examples

As a first example of the determination of the phase shifts, let us consider a *square well* potential

$$U(r) = \begin{cases} -U_0 \; (U_0 > 0), & r < a \\ 0, & r > a \end{cases} \qquad \textbf{(12.108)}$$

In the internal region $r < a$ the radial equation (12.39) becomes

$$\left[\frac{\mathrm{d}^2}{\mathrm{d}r^2} + \frac{2}{r}\frac{\mathrm{d}}{\mathrm{d}r} - \frac{l(l+1)}{r^2} + \kappa^2\right] R_l(k, r) = 0 \qquad \textbf{(12.109)}$$

where we have set $\kappa = (k^2 + U_0)^{1/2}$. Upon comparison with (12.44) we see that the regular solution of (12.109) is

$$R_l(k, r) = \tilde{C}_l(k) j_l(\kappa r) \qquad \textbf{(12.110)}$$

where $\tilde{C}_l(k)$ is a normalisation constant. In the external region $r > a$, the radial wave function R_l is given by (12.54), so we find that $\tan \delta_l(k)$ is given by (12.91), with

$$\gamma_l(k) = \frac{\kappa j_l'(\kappa a)}{j_l(\kappa a)} \qquad \textbf{(12.111)}$$

For example, in the case of s wave ($l = 0$) scattering, we have

$$\tan \delta_0 = \frac{k \tan(\kappa a) - \kappa \tan(ka)}{\kappa + k \tan(ka) \tan(\kappa a)} \qquad \textbf{(12.112)}$$

and

$$\delta_0 = -ka + \tan^{-1}\left[\frac{k}{\kappa} \tan(\kappa a)\right] \qquad \textbf{(12.113)}$$

where we have used the fact that $j_0(\rho) = (\sin \rho)/\rho$ and $n_0(\rho) = -(\cos \rho)/\rho$. We see from (12.101) and (12.112) that the scattering length is given by

$$\alpha = \left[1 - \frac{\tan(\lambda_0 a)}{\lambda_0 a}\right] a \qquad \textbf{(12.114)}$$

where $\lambda_0 = \sqrt{U_0}$. We note that for weak couplings, such that $\lambda_0 a < \pi/2$, there is no s wave bound state. In this case, the phase shift $\delta_0(k)$ tends to zero as $k \to 0$ and the scattering length α is negative. When $\lambda_0 a$ reaches the value $\pi/2$ such that the potential is nearly able to bind an s wave bound state, the phase shift $\delta_0(k)$ tends to $\pi/2$ as $k \to 0$. The scattering length then becomes infinite and the s wave cross-section diverges like k^{-2} at $k = 0$, thus providing an example of a 'zero-energy resonance'. If $\lambda_0 a$ is just above $\pi/2$, then $\delta_0(k)$ will tend to π as $k \to 0$. Repeating this argument, it is seen that if λ_0 is increased in such a way that the potential can support n_0 bound s states, then the s wave phase shift is such that

$$\lim_{k \to 0} \delta_0(k) = n_0 \pi \qquad \textbf{(12.115)}$$

Moreover, when $\lambda_0 a = (2n_0 + 1)\pi/2$, so that the potential is about to support its $(n_0 + 1)$th bound s state, we have

$$\lim_{k\to 0} \delta_0(k) = \left(n_0 + \frac{1}{2}\right)\pi \tag{12.116}$$

A similar study can also be carried out for higher angular momenta ($l > 0$). In particular, it may be shown that

$$\lim_{k\to 0} \delta_l(k) = n_l \pi \tag{12.117}$$

where n_l is the number of bound states of angular momentum $l > 0$ (including those of zero binding energy) which can be supported by the potential. The results (12.115)–(12.117) turn out to be true for more general interactions than the square well considered here; they are examples of *Levinson's theorem* (see Joachain, 1983).

Another simple, but interesting example is the '*hard sphere*' potential

$$U(r) = \begin{cases} +\infty, & r < a \\ 0, & r > a \end{cases} \tag{12.118}$$

Since in this case the particle cannot penetrate into the region $r < a$, the boundary condition is simply that the 'external' radial function (12.54) must vanish at $r = a$. Thus

$$\tan \delta_l = \frac{j_l(ka)}{n_l(ka)} \tag{12.119}$$

and γ_l is infinite. Using (12.48) we see that in the low-energy limit ($ka \ll 1$)

$$\tan \delta_l \simeq -\frac{(ka)^{2l+1}}{(2l+1)!!(2l-1)!!} \tag{12.120}$$

so that $|\tan \delta_l|$ quickly decreases as l increases. In fact the low-energy scattering is always dominated by the s wave ($l = 0$), the corresponding phase shift being

$$\delta_0 = -ka \tag{12.121}$$

As $k \to 0$, the differential cross-section is isotropic and given by $d\sigma/d\Omega = a^2$, and the total cross-section at zero energy becomes

$$\sigma_{\text{tot}} \underset{k\to 0}{\to} 4\pi a^2 \tag{12.122}$$

which is four times the classical value.

At high energies ($ka \gg 1$), we may use the asymptotic formulae (12.49) to obtain from (12.119) the approximate expressions of the phase shifts. That is,

$$\delta_l \simeq -ka + l\pi/2 \tag{12.123}$$

It is worth noting that because of the singular nature of the 'hard sphere' potential (12.118) the phase shifts (12.123) do not vanish as $k \to \infty$, but tend to $-\infty$ in that limit. Using (12.64) we write the total cross-section as

$$\sigma_{\text{tot}} \simeq \frac{4\pi}{k^2} \sum_{l=0}^{l_{\max}} (2l+1) \sin^2\left(\frac{l\pi}{2} - ka\right) \tag{12.124}$$

Taking $l_{\max} = ka$ in accordance with our general discussion of the partial wave method, and pairing successive terms in (12.124), we have

$$\begin{aligned}\sigma_{\text{tot}} &\simeq \frac{4\pi}{k^2}\left\{\left[\sin^2(ka) + \sin^2\left(\frac{\pi}{2} - ka\right)\right]\right.\\ &\left.\quad + 2\left[\sin^2\left(\frac{\pi}{2} - ka\right) + \sin^2(\pi - ka)\right] + \cdots\right\}\\ &\simeq \frac{4\pi}{k^2}\sum_{l=0}^{ka}(l) = 2\pi a^2\end{aligned} \tag{12.125}$$

This result is twice the classical value, which at first sight is surprising since we have $ka \gg 1$. However, because the 'hard sphere' potential has a sudden discontinuity at $r = a$ the scattering can never be treated classically. A detailed study of this problem shows that at high energies half of the total cross-section arises from 'diffraction' or 'shadow' scattering which is produced by interference between the incident wave and the outgoing scattered wave, and occurs within a narrow diffraction peak in the forward direction.

Resonances

In general the phase shifts – and therefore also the cross-sections – vary slowly as a function of the incident energy and of the strength of the potential. However, in certain cases it may happen that a phase shift δ_l will vary rapidly in a certain energy interval, causing a dramatic change in the corresponding partial cross-section σ_l in that energy range.

In order to study this problem, we first use (12.91) to write $\exp(2i\delta_l)$ in the form

$$\exp(2i\delta_l) = \exp(2i\xi_l)\frac{\gamma_l - r_l + is_l}{\gamma_l - r_l - is_l} \tag{12.126}$$

where γ_l is the logarithmic derivative of the internal solution of (12.39) at $r = a$ (see (12.88)) and the real quantities ξ_l, r_l and s_l are defined by

$$\exp(2i\xi_l) = -\frac{j_l(ka) - in_l(ka)}{j_l(ka) + in_l(ka)} \tag{12.127a}$$

and

$$r_l + \mathrm{i}s_l = k\frac{j_l'(ka) + \mathrm{i}n_l'(ka)}{j_l(ka) + \mathrm{i}n_l(ka)} \tag{12.127b}$$

Since the quantity $(\gamma_l - r_l + \mathrm{i}s_l)/(\gamma_l - r_l - \mathrm{i}s_l)$ is of unit modulus, it may be written as $\exp(2\mathrm{i}\rho_l)$, where

$$\begin{aligned}\rho_l &= \arg(\gamma_l - r_l + \mathrm{i}s_l)\\ &= \tan^{-1}\frac{s_l}{\gamma_l - r_l}\end{aligned} \tag{12.128}$$

and therefore we see from (12.126) that the phase shift δ_l can be decomposed as

$$\delta_l = \xi_l + \rho_l \tag{12.129}$$

The first term ξ_l on the right of this equation has an interesting significance. Indeed, by comparing (12.119) and (12.127a), we see that the quantity ξ_l corresponds to 'hard sphere' scattering by a potential of range a. The function ξ_l does not depend on the shape and depth of the potential. On the other hand, the term ρ_l depends on the details of the potential through the logarithmic derivative γ_l.

The quantities ξ_l, r_l and s_l vary, in general, slowly with the incident particle energy. On the other hand, the logarithmic derivative γ_l, and hence the contribution ρ_l to the phase shift, may in certain cases [6] vary rapidly in a small energy interval of width Γ about a given energy value E_r. In particular, if ρ_l increases rapidly through an odd multiple of $\pi/2$ as the energy increases through the value E_r, we may write $\rho_l \simeq \delta_l^\mathrm{r}$ in the energy interval $(E_\mathrm{r} - \Gamma/2, E_\mathrm{r} + \Gamma/2)$, with

$$\delta_l^\mathrm{r} = \tan^{-1}\frac{\Gamma/2}{E_\mathrm{r} - E} \tag{12.130}$$

This behaviour is called a *resonance*, with E_r being the *resonance energy*, Γ the *width* of the resonance and δ_l^r the *resonant phase shift*; it is illustrated in Fig. 12.3(a). In the small energy region $(E_\mathrm{r} - \Gamma/2, E_\mathrm{r} + \Gamma/2)$, the phase shift is therefore given approximately by

$$\delta_l = \xi_l + \delta_l^\mathrm{r} \tag{12.131}$$

and we see that it increases rapidly by π radians from a value ξ_l well below the resonance energy E_r to a value $\xi_l + \pi$ well above E_r. We also note that by using (12.131) we may rewrite (12.126) in the small energy interval $(E_\mathrm{r} - \Gamma/2, E_\mathrm{r} + \Gamma/2)$ as

$$\exp(2\mathrm{i}\delta_l) = \exp(2\mathrm{i}\xi_l)\frac{E - E_\mathrm{r} - \mathrm{i}\Gamma/2}{E - E_\mathrm{r} + \mathrm{i}\Gamma/2} \tag{12.132}$$

[6] A simple example is that of low-energy scattering by a strongly attractive square well. See for example Joachain (1983).

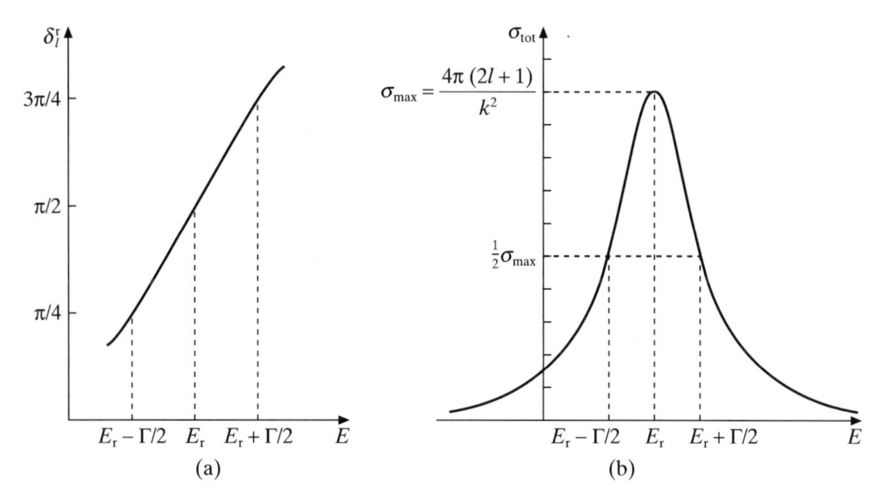

Figure 12.3 (a) Rapid increase of the resonant phase shift δ_l^r through $\pi/2$ at a narrow resonance. The energy scale is expanded to show only a narrow range of energies about E_r. (b) The 'pure resonance' total cross-section given by the Breit-Wigner formula (12.136).

Let us now return to the expression (12.62) of the partial wave amplitude $f_l(k)$. Using (12.131) and (12.132), we find that in the energy interval $(E_r - \Gamma/2, E_r + \Gamma/2)$,

$$f_l = \frac{2l+1}{k}\left[\exp(i\xi_l)\sin\xi_l + \exp(2i\xi_l)\frac{\Gamma/2}{E_r - E - i\Gamma/2}\right] \quad \textbf{(12.133)}$$

Thus near the resonance energy $E = E_r$, the behaviour of the quantity f_l will be governed by the second term on the right of (12.133). Let us assume for the moment that the 'hard sphere' (background) scattering in the lth partial wave (due to ξ_l) as well as the contribution of the other partial waves to the scattering amplitude (12.60) are negligible. In that idealised case, which corresponds to a *pure resonance*, the full scattering amplitude (12.60) can be written in the vicinity of $E = E_r$ as

$$f = \frac{2l+1}{k}\frac{\Gamma/2}{E_r - E - i\Gamma/2}P_l(\cos\theta) \quad \textbf{(12.134)}$$

and the corresponding differential cross-section is

$$\frac{d\sigma}{d\Omega} = \frac{(2l+1)^2}{k^2}\frac{\Gamma^2/4}{(E - E_r)^2 + \Gamma^2/4}P_l^2(\cos\theta) \quad \textbf{(12.135)}$$

We see that for any scattering angle θ the 'pure resonance' differential cross-section (12.135) exhibits a sharp peak of width Γ about the resonance energy $E = E_r$. It is also apparent that near $E = E_r$ the shape of the angular distribution does not depend on the energy, but only on the value of the angular momentum quantum number l. From (12.135), and using the result (2.171), we obtain the 'pure resonance' total cross-section

$$\sigma_{\text{tot}} \simeq \sigma_l = \frac{4\pi}{k^2}(2l+1)\frac{\Gamma^2/4}{(E-E_{\text{r}})^2+\Gamma^2/4} \tag{12.136}$$

which is called the (one-level) *Breit–Wigner formula* and is illustrated in Fig. 12.3(b). We note that at the energy $E = E_{\text{r}}$ the partial wave cross-section σ_l reaches its maximum value σ_l^{max} given by (12.67). However, as a function of E, the absolute maximum occurs at a lower energy because of the k^{-2} dependence occurring in (12.136). The shape of σ_l is the characteristic *Lorentzian shape* which we already encountered in Chapter 4.

If the contribution of the other partial waves ($\neq l$) is still neglected, but the 'hard sphere' (background) phase shift ξ_l is non-zero, then by using (12.133) we see that the total cross-section is given by

$$\sigma_{\text{tot}} \simeq \sigma_l = \frac{4\pi}{k^2}(2l+1)\left[\sin^2\xi_l + \frac{\Gamma^2/4}{(E-E_{\text{r}})^2+\Gamma^2/4}\right.$$
$$\left. + 2\text{Re}\left(\exp(\text{i}\xi_l)\sin\xi_l\frac{\Gamma/2}{E_{\text{r}}-E-\text{i}\Gamma/2}\right)\right] \tag{12.137}$$

The first term on the right-hand side accounts for pure hard sphere scattering, the second term represents the pure resonance contribution, and the third term is an interference term. Using (12.130) and (12.131), we may also write equation (12.137) as

$$\sigma_{\text{tot}} \simeq \sigma_l = \frac{4\pi}{k^2}(2l+1)\sin^2\delta_l = \frac{4\pi}{k^2}(2l+1)\sin^2\xi_l\frac{(q+\varepsilon)^2}{1+\varepsilon^2} \tag{12.138}$$

where ε is the *reduced energy*

$$\varepsilon = \frac{E-E_{\text{r}}}{\Gamma/2} = -\cot\delta_l^{\text{r}} \tag{12.139}$$

and q is the *resonance shape parameter* or *line profile index*

$$q = -\cot\xi_l \tag{12.140}$$

This quantity was introduced by U. Fano in the context of resonant atomic photoionisation processes. We see from (12.138) that the partial wave cross-section σ_l goes through a zero at $\varepsilon = -q$ and reaches its maximum value of $4\pi(2l+1)k^{-2}$ when $\varepsilon = q^{-1}$.

The behaviour of the cross-section (12.138) in the vicinity of an s-wave resonance at $E = E_{\text{r}}$ is illustrated in Fig. 12.4 for four different values of the resonance shape parameter, assuming a constant 'hard sphere' phase shift ξ_0 and no contribution to the scattering from the other partial waves. Case (a), with $q = \pm\infty$, corresponds to a pure Breit–Wigner resonance, where the 'hard sphere' (background) scattering is zero. Case (c) with $q = 0$ corresponds to the value $\xi_0 = \pi/2$ (modulo π), so that the background scattering cross-section has its maximum possible value of $4\pi(2l+1)k^{-2}$ and the cross-section σ_l given by (12.138) vanishes

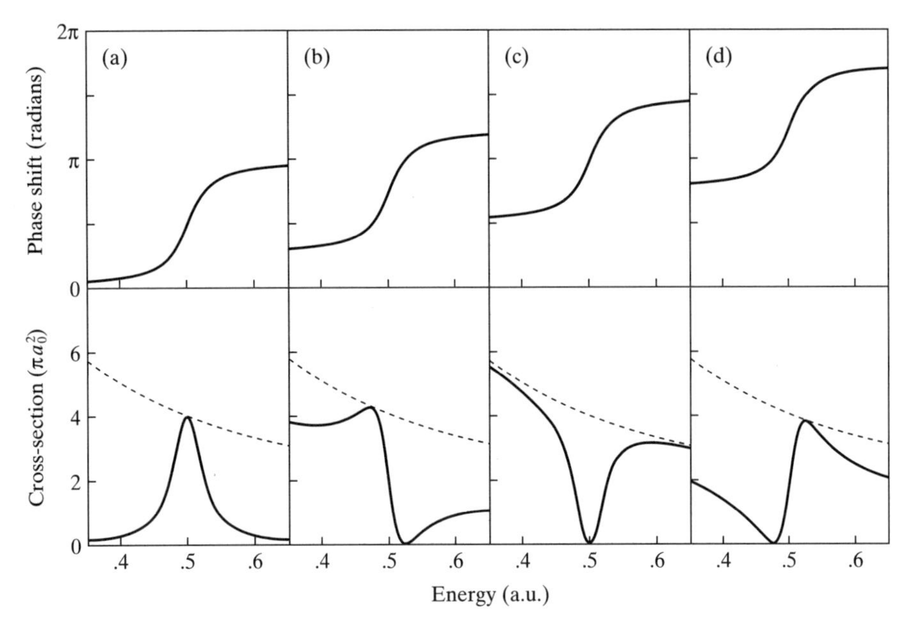

Figure 12.4 The phase shift and cross-section in the neighbourhood of an s-wave resonance with $E_r = 1$ a.u. Case (a), $\xi_0 = 0$, giving $q = \pm\infty$; case (b), $\xi_0 = \pi/4$ giving $q = -1$; case (c), $\xi_0 = \pi/2$, giving $q = 0$; case (d), $\xi_0 = 3\pi/4$, giving $q = 1$. The cross-section is given in units of πa_0^2 and the dashed line is its maximum value $4k^{-2}$.

at $E = E_r$; this is called a *window resonance*. Cases (b) and (d) correspond to intermediate cases where $q = -1$ and $q = 1$, respectively, so that the line shape is asymmetric.

The physical significance of a pure Breit–Wigner resonance can be inferred by examining the amplitude of the radial wave function inside the interaction region. Let us consider for example the case of a strongly attractive square well of depth U_0 (see (12.108)). If we normalise the radial wave function $u_l(k, r)$ outside the well to unit amplitude, namely

$$u_l(k, r) = \sin(kr - l\pi/2 + \delta_l), \qquad r \geqslant a \tag{12.141}$$

then near $E = E_r$ the corresponding 'internal' wave function, obtained by requiring that the function u_l and its first derivative be continuous at $r = a$, is given by

$$u_l(k, r) \simeq \frac{(2l + 1)!!}{(ka)^l} \frac{\Gamma/2}{[(E - E_r)^2 + \Gamma^2/4]^{1/2}} (\kappa r) j_l(\kappa r) \tag{12.142}$$

where $\kappa = (k^2 + U_0)^{1/2}$. We see from (12.142) that the probability of finding the scattered particle within the potential well is much larger near the resonance energy $E = E_r$, so that in this case the particle is *nearly bound* in the well.

In order to understand how this happens, we recall that for $l > 0$ the 'effective' reduced potential $U_{\text{eff}}(r) = U(r) + l(l + 1)/r^2$ appearing in the radial equations

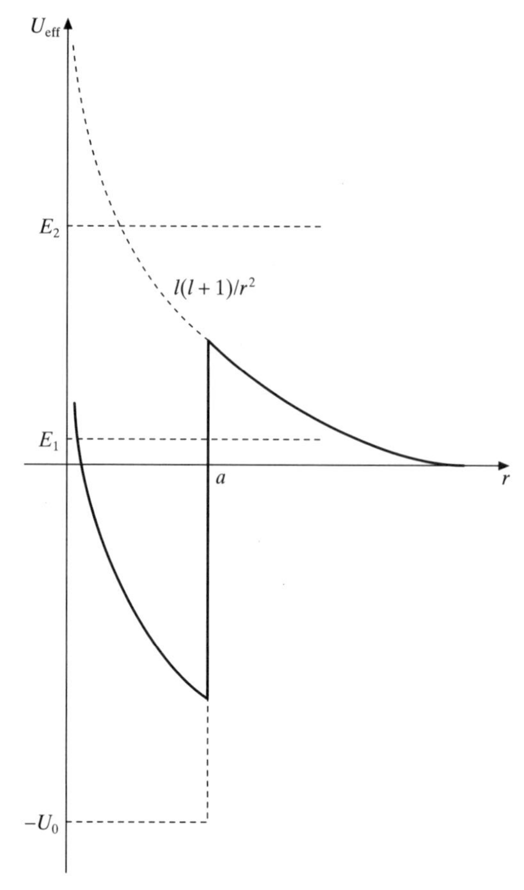

Figure 12.5 The effective reduced potential $U_{\text{eff}}(r) = U(r) + l(l+1)/r^2$ for the lth partial wave ($l \neq 0$) for an attractive square well.

(12.39) or (12.41) contains the centrifugal barrier term $l(l+1)/r^2$. This effective reduced potential is illustrated in Fig. 12.5 for the case of a deep attractive square well. If we imagine for a moment that the barrier at $r = a$ due to the repulsive centrifugal term $l(l+1)/r^2$ were infinitely high, then bound states with specific positive energies $E_1, E_2, \ldots$ (for a given value of U_0) would be allowed. In the real situation, where the barrier is finite, none of these positive energy bound states will remain genuine bound states. Those lying above the barrier, such as E_2, disappear completely. However, those lying below the barrier, such as E_1, will become *metastable* states (having a long lifetime τ). An incoming particle with an energy close to $E_r = E_1$, will therefore be trapped for a long time inside the well in a state that is nearly bound. It can only escape by tunnelling through the barrier, with a probability that is small at low energies. Thus a resonance may be considered as a metastable state whose lifetime, which is much longer than a typical collision time ($\sim a/v$), can be related to the resonance width Γ by using the uncertainty relation (2.23). Hence, with $\Delta t \simeq \tau$ and $\Delta E \simeq \Gamma$, we have

$$\tau \simeq \frac{\hbar}{\Gamma} \tag{12.143}$$

We also remark that if we increase the depth parameter U_0, the resonance energies will decrease and move in this way towards a region where the barrier is thicker (see Fig. 12.5). As a result, the lifetime of these resonances will increase, so that the corresponding widths Γ decrease and the resonances become narrower. Eventually, if U_0 becomes sufficiently large, a given resonance will reach the region $E < 0$ and become a bound state.

It is apparent from the above discussion that the repulsive centrifugal barrier term $l(l+1)/r^2$ is responsible for the occurrence of resonances in low-energy scattering by a strongly attractive square well. In fact, it can be shown [7] that no low-energy s-wave ($l = 0$) resonances occur for this potential. However, a potential $U(r)$ which contains its own repulsive tail is perfectly able to produce low-energy s-wave resonances. In all cases, it is the existence of a repulsive tail due to the *shape* of the effective potential $U_{\text{eff}}(r)$ which is important. For this reason, the resonances which we have discussed in this section, within the framework of potential scattering, are called *shape resonances*. In actual atomic collisions processes, where resonance phenomena are of great interest, the complex (many-body) structure of the particles involved in the collision must also be taken into account, and other types of resonances can occur. We shall return to this point in the next chapter, where the narrow resonances observed in the scattering of electrons by atoms will be discussed.

12.4 The integral equation of potential scattering

Let us return to the Schrödinger equation (12.19), which we rewrite as

$$(\nabla^2 + k^2)\psi(k, \mathbf{r}) = U(\mathbf{r})\psi(k, \mathbf{r}) \tag{12.144}$$

where we have indicated explicitly the k-dependence of the wave function. The general solution of this equation may be written as

$$\psi(k, \mathbf{r}) = \phi(k, \mathbf{r}) + \int G_0(k, \mathbf{r}, \mathbf{r}')U(\mathbf{r}')\psi(k, \mathbf{r}')\,\mathrm{d}\mathbf{r}' \tag{12.145}$$

where $\phi(k, \mathbf{r})$ is a solution of the homogeneous equation

$$[\nabla^2 + k^2]\phi(k, \mathbf{r}) = 0 \tag{12.146}$$

and $G_0(k, \mathbf{r}, \mathbf{r}')$ is a *Green's function* such that

$$[\nabla^2 + k^2]G_0(k, \mathbf{r}, \mathbf{r}') = \delta(\mathbf{r} - \mathbf{r}') \tag{12.147}$$

[7] See for example Joachain (1983).

The Green's function

In order to determine the Green's function $G_0(k, \mathbf{r}, \mathbf{r}')$, we first use the integral representation (2.31a) of the delta function, so that

$$\delta(\mathbf{r}-\mathbf{r}') = (2\pi)^{-3} \int \exp[\mathrm{i}\mathbf{k}'\cdot(\mathbf{r}-\mathbf{r}')]\, \mathrm{d}\mathbf{k}' \tag{12.148}$$

and we write

$$G_0(k, \mathbf{r}, \mathbf{r}') = (2\pi)^{-3} \int \exp(\mathrm{i}\mathbf{k}'\cdot\mathbf{r}) g_0(k, \mathbf{k}', \mathbf{r}')\, \mathrm{d}\mathbf{k}' \tag{12.149}$$

Substituting (12.148) and (12.149) in (12.147), we find that

$$g_0(k, \mathbf{k}', \mathbf{r}') = \frac{\exp(-\mathrm{i}\mathbf{k}'\cdot\mathbf{r}')}{k^2 - k'^2} \tag{12.150}$$

giving

$$G_0(k, \mathbf{r}, \mathbf{r}') = -(2\pi)^{-3} \int \frac{\exp[\mathrm{i}\mathbf{k}'\cdot(\mathbf{r}-\mathbf{r}')]}{k'^2 - k^2}\, \mathrm{d}\mathbf{k}' \tag{12.151}$$

The integrand in (12.151) has poles at $k' = \pm k$, so that a well-defined prescription is required to avoid these singularities. This may be done by using the boundary condition (12.25). Upon comparison of (12.25) and (12.145), we first note that the free wave $\phi(k, \mathbf{r})$ is just the incident plane wave $A \exp(\mathrm{i}\mathbf{k}_i\cdot\mathbf{r})$. In what follows it will be convenient to choose the normalisation constant to be $A = 1$ and write the corresponding incident plane wave as

$$\phi_{\mathbf{k}_i}(\mathbf{r}) = \exp(\mathrm{i}\mathbf{k}_i\cdot\mathbf{r}) \tag{12.152}$$

Comparison of (12.145) with (12.25) also shows that the Green's function $G_0(k, \mathbf{r}, \mathbf{r}')$ must be determined in such a way that it leads to an outgoing spherical wave for large r. This particular Green's function will be denoted by $G_0^{(+)}(k, \mathbf{r}, \mathbf{r}')$.

Setting $\mathbf{R} = \mathbf{r} - \mathbf{r}'$ and performing the angular integrations in (12.151) with $\mathbf{R}$ as the polar axis, we find that

$$G_0(k, R) = -\frac{1}{4\pi^2 R} \int_{-\infty}^{+\infty} \frac{k' \sin k'R}{k'^2 - k^2}\, \mathrm{d}k' \tag{12.153}$$

where we have used the fact that the integrand is an even function of k', so that the integral may be extended from $-\infty$ to $+\infty$. We may also write (12.153) as

$$G_0(k, R) = \frac{\mathrm{i}}{16\pi^2 R}(I_1 - I_2) \tag{12.154a}$$

where

$$I_1 = \int_{-\infty}^{+\infty} \exp(\mathrm{i}k'R) \left(\frac{1}{k'-k} + \frac{1}{k'+k} \right) \mathrm{d}k' \tag{12.154b}$$

and

$$I_2 = \int_{-\infty}^{+\infty} \exp(-\mathrm{i}k'R)\left(\frac{1}{k'-k} + \frac{1}{k'+k}\right) \mathrm{d}k' \tag{12.154c}$$

We can give a meaning to the integrals I_1 and I_2 by regarding them as contour integrals in the complex k' plane. Suppose, for example, that we avoid the poles at $k' = \pm k$ by choosing the path P shown in Fig. 12.6(a). The integral I_1 can then be evaluated by writing

$$I_1 = \oint_{\mathrm{C}} \exp(\mathrm{i}k'R)\left(\frac{1}{k'-k} + \frac{1}{k'+k}\right) \mathrm{d}k' \tag{12.155}$$

where the contour C consists of the path P plus an infinite semicircle C_1 in the upper-half k' plane (see Fig. 12.6(b)). Since $\exp(\mathrm{i}k'R)$ vanishes on C_1, the contribution to I_1 from the infinite semicircle C_1 is equal to zero, and the integral (12.155) is equal to its value along the path P. Using the Cauchy theorem [8], we then obtain for the integral I_1, evaluated along the path P, the value $I_1 = 2\pi\mathrm{i}\exp(\mathrm{i}kR)$.

The integral I_2 can be evaluated in a similar way by closing the contour with an infinite semicircle C_2 in the lower-half k' plane as shown in Fig. 12.6(c). Using again the Cauchy theorem, we find that I_2, evaluated along the path P, is given by $I_2 = -2\pi\mathrm{i}\exp(\mathrm{i}kR)$. Thus by choosing the path P to avoid the poles at $k' = \pm k$, we see that $I_1 - I_2 = 4\pi\mathrm{i}\exp(\mathrm{i}kR)$. Returning to (12.154) and remembering that $\mathbf{R} = \mathbf{r} - \mathbf{r}'$, we therefore obtain a Green's function

$$G_0^{(+)}(k, \mathbf{r}, \mathbf{r}') = -\frac{1}{4\pi}\frac{\exp(\mathrm{i}k|\mathbf{r}-\mathbf{r}'|)}{|\mathbf{r}-\mathbf{r}'|} \tag{12.156}$$

which exhibits the required purely outgoing wave behaviour (of the form $\exp(\mathrm{i}kr)/r$) when r is large. It is easy to verify that any other choice of integration contour which avoids the poles at $k' = \pm k$ in a way different from the path P shown in Fig. 12.6(a) leads to an *incoming* wave behaviour (of the form $\exp(-\mathrm{i}kr)/r$) in addition to or in place of the outgoing wave behaviour obtained above.

We remark that the choice of the path P is equivalent to keeping the integration path along the real axis, and shifting the two poles slightly so that the poles are now at

$$k' = \pm(k + \mathrm{i}\varepsilon'), \qquad \varepsilon' \to 0^+ \tag{12.157}$$

We can then write

$$G_0^{(+)}(k, \mathbf{r}, \mathbf{r}') = (2\pi)^{-3} \lim_{\varepsilon \to 0^+} \int \frac{\exp[\mathrm{i}\mathbf{k}'\cdot(\mathbf{r}-\mathbf{r}')]}{k^2 - k'^2 + \mathrm{i}\varepsilon} \mathrm{d}\mathbf{k}' \tag{12.158}$$

where we have set $\varepsilon = 2k\varepsilon'$ and neglected terms in ε'^2.

---

[8] The Cauchy theorem is discussed for example in Byron and Fuller (1969).

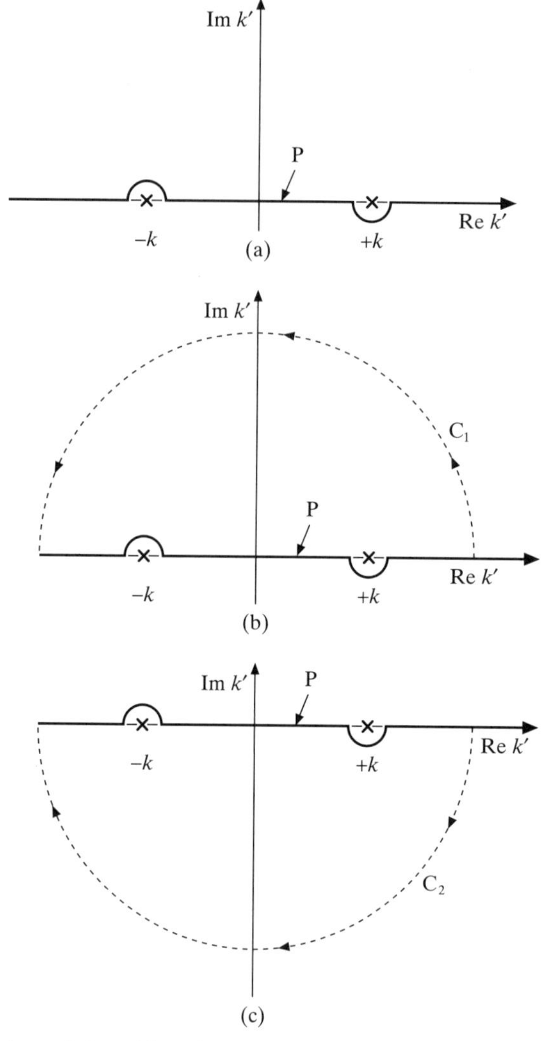

Figure 12.6 (a) The path P avoiding the poles at $k' = \pm k$. (b) Contour ($P + C_1$) for calculating the integral I_1. (c) Contour ($P + C_2$) for calculating the integral I_2.

The Lippmann–Schwinger equation

We now return to the integral equation (12.145). Using the Green's function $G_0^{(+)}(k, \mathbf{r}, \mathbf{r}')$ given by (12.156), and writing now $\psi(k, \mathbf{r}) \equiv \psi_{\mathbf{k}_i}^{(+)}(\mathbf{r})$, we have

$$\psi_{\mathbf{k}_i}^{(+)}(\mathbf{r}) = \exp(\mathrm{i}\mathbf{k}_i \cdot \mathbf{r}) + \int G_0^{(+)}(k, \mathbf{r}, \mathbf{r}')U(\mathbf{r}')\psi_{\mathbf{k}_i}^{(+)}(\mathbf{r}')\,\mathrm{d}\mathbf{r}' \qquad \textbf{(12.159a)}$$

or

$$\psi_{\mathbf{k}_i}^{(+)}(\mathbf{r}) = \exp(\mathrm{i}\mathbf{k}_i \cdot \mathbf{r}) - \frac{1}{4\pi} \int \frac{\exp(\mathrm{i}k|\mathbf{r} - \mathbf{r}'|)}{|\mathbf{r} - \mathbf{r}'|} U(\mathbf{r}') \psi_{\mathbf{k}_i}^{(+)}(\mathbf{r}') \, \mathrm{d}\mathbf{r}' \quad \textbf{(12.159b)}$$

This integral equation is known as the Lippmann–Schwinger equation of potential scattering. It is equivalent to the original Schrödinger equation (12.19), together with the boundary condition (12.25), which is taken care of through the Green's function $G_0^{(+)}(k, \mathbf{r}, \mathbf{r}')$.

Integral representations of the scattering amplitude

Let us examine the asymptotic behaviour of $\psi_{\mathbf{k}_i}^{(+)}(\mathbf{r})$. We note that for r large and $r' \ll r$,

$$|\mathbf{r} - \mathbf{r}'| \underset{r\to\infty}{\to} r - \hat{\mathbf{r}} \cdot \mathbf{r}' + \dots \quad \textbf{(12.160)}$$

so that

$$\frac{\exp(\mathrm{i}k|\mathbf{r} - \mathbf{r}'|)}{|\mathbf{r} - \mathbf{r}'|} \underset{r\to\infty}{\to} \frac{\exp(\mathrm{i}kr)}{r} \exp(-\mathrm{i}\mathbf{k}_f \cdot \mathbf{r}') + \dots \quad \textbf{(12.161)}$$

where terms of higher order in $1/r$ have been neglected, and we have introduced the final wave vector $\mathbf{k}_f = k\hat{\mathbf{r}}$, which points in the direction of the detector and has spherical polar coordinates (k, θ, ϕ). Thus, using (12.159b) and (12.161), we have

$$\psi_{\mathbf{k}_i}^{(+)}(\mathbf{r}) \underset{r\to\infty}{\to} \exp(\mathrm{i}\mathbf{k}_i \cdot \mathbf{r}) - \frac{1}{4\pi} \frac{\exp(\mathrm{i}kr)}{r} \int \exp(-\mathrm{i}\mathbf{k}_f \cdot \mathbf{r}') U(\mathbf{r}') \psi_{\mathbf{k}_i}^{(+)}(\mathbf{r}') \, \mathrm{d}\mathbf{r}' \quad \textbf{(12.162)}$$

Upon comparison with (12.25), we see that $\psi_{\mathbf{k}_i}^{(+)}$ exhibits the desired asymptotic behaviour (with $A = 1$) and we obtain for the scattering amplitude the integral representation

$$f = -\frac{1}{4\pi} \langle \phi_{\mathbf{k}_f} | U | \psi_{\mathbf{k}_i}^{(+)} \rangle = -\frac{1}{4\pi} \int \exp(-\mathrm{i}\mathbf{k}_f \cdot \mathbf{r}) U(\mathbf{r}) \psi_{\mathbf{k}_i}^{(+)}(\mathbf{r}) \, \mathrm{d}\mathbf{r} \quad \textbf{(12.163)}$$

where we have introduced the plane wave $\phi_{\mathbf{k}_f}$ corresponding to the final wave vector $\mathbf{k}_f$,

$$\phi_{\mathbf{k}_f}(\mathbf{r}) = \exp(\mathrm{i}\mathbf{k}_f \cdot \mathbf{r}) \quad \textbf{(12.164)}$$

In terms of the potential $V(\mathbf{r}) = [\hbar^2/(2m)]U(\mathbf{r})$, we may also write the scattering amplitude in the form

$$f = -\frac{m}{2\pi\hbar^2} \langle \phi_{\mathbf{k}_f} | V | \psi_{\mathbf{k}_i}^{(+)} \rangle \quad \textbf{(12.165)}$$

The transition matrix element T_{fi} is defined as

$$T_{fi} = \langle \phi_{\mathbf{k}_f} | V | \psi_{\mathbf{k}_i}^{(+)} \rangle \quad \textbf{(12.166)}$$

so that

$$f = -\frac{m}{2\pi\hbar^2} T_{fi} \tag{12.167}$$

and the differential cross-section is given by

$$\frac{d\sigma}{d\Omega} = |f|^2 = (2\pi)^{-2} m^2 \hbar^{-4} |T_{fi}|^2 \tag{12.168}$$

Let us return to the Lippmann–Schwinger equation (12.159), and write

$$\psi_{\mathbf{k}_i}^{(+)}(\mathbf{r}) = \phi_{\mathbf{k}_i}(\mathbf{r}) + \psi_{\text{sc}}^{(+)}(\mathbf{r}) \tag{12.169}$$

where the scattered wave $\psi_{\text{sc}}^{(+)}$ satisfies the inhomogeneous equation

$$[\nabla_{\mathbf{r}}^2 + k^2 - U(\mathbf{r})]\psi_{\text{sc}}^{(+)}(\mathbf{r}) = U(\mathbf{r})\phi_{\mathbf{k}_i}(\mathbf{r}) \tag{12.170}$$

Now, suppose that we know the *total* Green's function of the problem, $G^{(+)}(k, \mathbf{r}, \mathbf{r}')$, satisfying the equation

$$[\nabla_{\mathbf{r}}^2 + k^2 - U(\mathbf{r})]G^{(+)}(k, \mathbf{r}, \mathbf{r}') = \delta(\mathbf{r} - \mathbf{r}') \tag{12.171}$$

and such that

$$\psi_{\text{sc}}^{(+)}(\mathbf{r}) = \int G^{(+)}(k, \mathbf{r}, \mathbf{r}')U(\mathbf{r}')\phi_{\mathbf{k}_i}(\mathbf{r}')\, d\mathbf{r}' \tag{12.172}$$

corresponds to an outgoing spherical wave. Then we may write explicitly the solution of the Lippmann–Schwinger equation. Using (12.169) and (12.172) we have

$$\psi_{\mathbf{k}_i}^{(+)}(\mathbf{r}) = \phi_{\mathbf{k}_i}(\mathbf{r}) + \int G^{(+)}(k, \mathbf{r}, \mathbf{r}')U(\mathbf{r}')\phi_{\mathbf{k}_i}(\mathbf{r}')\, d\mathbf{r}' \tag{12.173}$$

In order to determine the total Green's function $G^{(+)}(k, \mathbf{r}, \mathbf{r}')$, let us rewrite (12.170) as

$$(\nabla_{\mathbf{r}}^2 + k^2)G^{(+)}(k, \mathbf{r}, \mathbf{r}') = \delta(\mathbf{r} - \mathbf{r}') + U(\mathbf{r})G^{(+)}(k, \mathbf{r}, \mathbf{r}') \tag{12.174}$$

With the help of (12.147) we therefore obtain for $G^{(+)}(k, \mathbf{r}, \mathbf{r}')$ the integral equation

$$G^{(+)}(k, \mathbf{r}, \mathbf{r}') = G_0^{(+)}(k, \mathbf{r}, \mathbf{r}') + \int G_0^{(+)}(k, \mathbf{r}, \mathbf{r}'')U(\mathbf{r}'')G^{(+)}(k, \mathbf{r}'', \mathbf{r}')\, d\mathbf{r}'' \tag{12.175}$$

so that the problem of solving the Lippmann–Schwinger equation (12.159) has been shifted to that of finding the solution of the integral equation (12.175) for the total Green's function $G^{(+)}(k, \mathbf{r}, \mathbf{r}')$.

Let us now return to the integral representation (12.163) of the scattering amplitude. With the help of (12.173) we can rewrite this expression as

$$f = -\frac{1}{4\pi}\langle\phi_{\mathbf{k}_f}|U + UG^{(+)}U|\phi_{\mathbf{k}_i}\rangle$$

$$= -\frac{1}{4\pi}\left[\int \mathrm{d}\mathbf{r}\, \exp(-\mathrm{i}\mathbf{k}_f\cdot\mathbf{r})U(\mathbf{r})\exp(\mathrm{i}\mathbf{k}_i\cdot\mathbf{r})\right.$$

$$\left. + \int \mathrm{d}\mathbf{r}\,\mathrm{d}\mathbf{r}'\, \exp(-\mathrm{i}\mathbf{k}_f\cdot\mathbf{r})U(\mathbf{r})G^{(+)}(k, \mathbf{r}, \mathbf{r}')U(\mathbf{r}')\exp(\mathrm{i}\mathbf{k}_i\cdot\mathbf{r}')\right] \quad \textbf{(12.176)}$$

Let us now consider the Lippmann–Schwinger equation

$$\psi^{(-)}_{\mathbf{k}_f}(\mathbf{r}) = \exp(\mathrm{i}\mathbf{k}_f\cdot\mathbf{r}) + \int G_0^{(-)}(k, \mathbf{r}, \mathbf{r}')U(\mathbf{r}')\psi^{(-)}_{\mathbf{k}_f}(\mathbf{r}')\,\mathrm{d}\mathbf{r}' \quad \textbf{(12.177)}$$

where

$$G_0^{(-)}(k, \mathbf{r}, \mathbf{r}') = -\frac{1}{4\pi}\frac{\exp(-\mathrm{i}k|\mathbf{r}-\mathbf{r}'|)}{|\mathbf{r}-\mathbf{r}'|} \quad \textbf{(12.178)}$$

or

$$G_0^{(-)}(k, \mathbf{r}, \mathbf{r}') = (2\pi)^{-3}\lim_{\varepsilon\to 0^+}\int \frac{\exp[\mathrm{i}\mathbf{k}'\cdot(\mathbf{r}-\mathbf{r}')]}{k^2 - k'^2 - \mathrm{i}\varepsilon}\,\mathrm{d}\mathbf{k}' \quad \textbf{(12.179)}$$

The wave function $\psi^{(-)}_{\mathbf{k}_f}(\mathbf{r})$ satisfies asymptotic boundary conditions corresponding to an incident plane wave of wave vector $\mathbf{k}_f$ plus an *incoming* (–) spherical wave. Introducing also the total Green's function $G^{(-)}(k, \mathbf{r}, \mathbf{r}')$ such that

$$G^{(-)}(k, \mathbf{r}, \mathbf{r}') = G_0^{(-)}(k, \mathbf{r}, \mathbf{r}') + \int G_0^{(-)}(k, \mathbf{r}', \mathbf{r}'')U(\mathbf{r}'')G^{(-)}(k, \mathbf{r}'', \mathbf{r}')\,\mathrm{d}\mathbf{r}'' \quad \textbf{(12.180)}$$

we can write explicitly the solution of the Lippmann–Schwinger equation (12.177) as

$$\psi^{(-)}_{\mathbf{k}_f}(\mathbf{r}) = \phi_{\mathbf{k}_f}(\mathbf{r}) + \int G^{(-)}(k, \mathbf{r}, \mathbf{r}')U(\mathbf{r}')\phi_{\mathbf{k}_f}(\mathbf{r}')\,\mathrm{d}\mathbf{r}' \quad \textbf{(12.181)}$$

We note from (12.156) and (12.178) that

$$G_0^{(+)}(k, \mathbf{r}, \mathbf{r}') = G_0^{(-)*}(k, \mathbf{r}, \mathbf{r}') \quad \textbf{(12.182)}$$

which, using (12.175) and (12.180), implies that

$$G^{(+)}(k, \mathbf{r}, \mathbf{r}') = G^{(-)*}(k, \mathbf{r}, \mathbf{r}') \quad \textbf{(12.183)}$$

Hence, from (12.176) and (12.183), we deduce another integral representation of the scattering amplitude, namely

$$f = -\frac{1}{4\pi}\langle\psi^{(-)}_{\mathbf{k}_f}|U|\phi_{\mathbf{k}_i}\rangle = -\frac{1}{4\pi}\int \psi^{(-)*}_{\mathbf{k}_f}(\mathbf{r})U(\mathbf{r})\exp(\mathrm{i}\mathbf{k}_i\cdot\mathbf{r})\,\mathrm{d}\mathbf{r} \quad \textbf{(12.184)}$$

We also remark that the transition matrix element (12.166), which is related to the scattering amplitude by (12.167), can also be written as

$$T_{fi} = \langle \psi^{(-)}_{\mathbf{k}_f} | V | \phi_{\mathbf{k}_i} \rangle \tag{12.185}$$

or

$$T_{fi} = \langle \phi_{\mathbf{k}_f} | V + VG^{(+)}V | \phi_{\mathbf{k}_i} \rangle \tag{12.186}$$

The integral equation for the radial functions

The integral equation (12.159) can also be analysed in partial waves. Assuming that we are dealing with a central potential, and expanding the scattering wave function $\psi^{(+)}_{\mathbf{k}_i}$ in Legendre polynomials as in (12.38), it can be shown (Problem 12.13) that each radial function $R_l(k, r)$, normalised according to (12.83), satisfies the radial integral equation

$$R_l(k, r) = j_l(kr) + \int_0^\infty G_l(k, r, r')U(r')R_l(k, r')r'^2 \, \mathrm{d}r' \tag{12.187}$$

where the radial Green's function $G_l(k, r, r')$ is given by

$$G_l(k, r, r') = kj_l(kr_<)n_l(kr_>) \tag{12.188}$$

Here $r_<$ and $r_>$ denote respectively the lesser and the greater of r and r'. By analysing the behaviour of the radial integral equation (12.187) as $r \to \infty$ and comparing with (12.83), the expression (12.87) for $\tan \delta_l$ can be re-derived.

12.5 The Coulomb potential

In the three previous sections we considered interaction potentials which fall off faster than $1/r$ as $r \to \infty$. The treatment given there must be modified when the potential behaves asymptotically as $1/r$. This is the case for the pure Coulomb potential

$$V_c(r) = \frac{q_A q_B}{(4\pi\varepsilon_0)r} \tag{12.189}$$

acting between two particles A and B having electric charges q_A and q_B, respectively. Clearly, it is also the case for any modified Coulomb potential such that a short-range potential is added to the pure Coulomb potential (12.189). The physical reason for the exceptional behaviour of the pure or modified Coulomb potentials is that these potentials have an infinite range, so that their influence is felt even for $r \to \infty$. As a result, the wave functions corresponding to scattering by these potentials do not satisfy the boundary condition (12.25).

In this section we shall consider the scattering by the pure Coulomb field (12.189), leaving aside the case of the modified Coulomb potential [9]. The Schrödinger equation in the centre of mass system is

$$[\nabla^2 + k^2 - U_c(r)]\psi_c(\mathbf{r}) = 0 \tag{12.190}$$

[9] This case is treated for example in Bransden (1983) or Joachain (1983).

where we have introduced the reduced Coulomb potential

$$U_c(r) = \frac{U_0}{r} \tag{12.191}$$

with

$$U_0 = \frac{2\mu}{\hbar^2}\frac{q_A q_B}{4\pi\varepsilon_0} \tag{12.192}$$

and μ denotes the reduced mass of the two particles A and B.

The Schrödinger equation (12.190) can be solved by introducing *parabolic coordinates* (ξ, η, ϕ) defined by the relations (3.90). Using the expression (3.92) of the operator ∇^2 in these coordinates, the Schrödinger equation becomes

$$\left\{\frac{4}{\xi+\eta}\left[\frac{\partial}{\partial\xi}\left(\xi\frac{\partial}{\partial\xi}\right) + \frac{\partial}{\partial\eta}\left(\eta\frac{\partial}{\partial\eta}\right)\right] + \frac{1}{\xi\eta}\frac{\partial^2}{\partial\phi^2} + k^2 - \frac{2U_0}{\xi+\eta}\right\}\psi_c = 0 \tag{12.193}$$

Since we are interested in a scattering solution ψ_c which exhibits azimuthal symmetry about the incident direction (the Z axis), we write

$$\psi_c = f(\xi)g(\eta) \tag{12.194}$$

and separate the Schrödinger equation (12.193) to obtain the two differential equations

$$\frac{d}{d\xi}\left(\xi\frac{df}{d\xi}\right) + \left(\frac{1}{4}k^2\xi - \tilde{\nu}_1\right)f = 0 \tag{12.195a}$$

and

$$\frac{d}{d\eta}\left(\eta\frac{dg}{d\eta}\right) + \left(\frac{1}{4}k^2\eta - \tilde{\nu}_2\right)g = 0 \tag{12.195b}$$

The two separation constants $\tilde{\nu}_1$ and $\tilde{\nu}_2$ are such that

$$\tilde{\nu}_1 + \tilde{\nu}_2 = \gamma k \tag{12.196}$$

where we have introduced the *Sommerfeld parameter*

$$\gamma = \frac{U_0}{2k} = \frac{\mu q_A q_B}{(4\pi\varepsilon_0)\hbar^2 k} = \frac{q_A q_B}{(4\pi\varepsilon_0)\hbar v} \tag{12.197}$$

and $v = \hbar k/\mu$ denotes the relative velocity of the two particles.

In order to determine the two functions $f(\xi)$ and $g(\eta)$, we first notice that

$$\exp(ikz) = \exp\left[\frac{1}{2}ik(\xi - \eta)\right] \tag{12.198}$$

while

$$\exp(\mathrm{i}kr) = \exp\left[\frac{1}{2}\mathrm{i}k(\xi + \eta)\right] \quad \textbf{(12.199)}$$

This suggests that we choose

$$f(\xi) = \exp\left(\frac{1}{2}\mathrm{i}k\xi\right) \quad \textbf{(12.200)}$$

which is indeed a solution of the equation (12.195a) provided that $\tilde{v}_1 = \mathrm{i}k/2$. Using (12.196) we may then rewrite the equation (12.195b) as

$$\frac{\mathrm{d}}{\mathrm{d}\eta}\left(\eta\frac{\mathrm{d}g}{\mathrm{d}\eta}\right) + \frac{1}{4}k^2\eta g - \left(\gamma k - \frac{1}{2}\mathrm{i}k\right)g = 0 \quad \textbf{(12.201)}$$

Let us introduce a new function $h(\eta)$ by the relation

$$g(\eta) = \exp\left(-\frac{1}{2}\mathrm{i}k\eta\right)h(\eta) \quad \textbf{(12.202)}$$

Substituting this expression for $g(\eta)$ in (12.201), we obtain for $h(\eta)$ the equation

$$\eta\frac{\mathrm{d}^2h}{\mathrm{d}\eta^2} + (1 - \mathrm{i}k\eta)\frac{\mathrm{d}h}{\mathrm{d}\eta} - \gamma kh = 0 \quad \textbf{(12.203)}$$

This last equation can be related to the Kummer–Laplace differential equation (3.35), whose solution regular at the origin is given, within a multiplicative constant, by the confluent hypergeometric function (3.36). Thus we find in the present case that the regular solution of (12.203) is

$$h(\eta) = C\,{}_1F_1(-\mathrm{i}\gamma, 1, \mathrm{i}k\eta) \quad \textbf{(12.204)}$$

where C is a constant. From (12.194), (12.200), (12.202) and (12.204), we see that the Schrödinger equation (12.190) admits the regular solution

$$\psi_{\mathrm{c}} = C\exp(\mathrm{i}kz)\,{}_1F_1(-\mathrm{i}\gamma, 1, \mathrm{i}k\eta) \quad \textbf{(12.205)}$$

The asymptotic form of the confluent hypergeometric function in (12.205) can be obtained by writing [10]

$${}_1F_1(a, c, z) = W_1(a, c, z) + W_2(a, c, z) \quad \textbf{(12.206)}$$

Here $W_1(a, c, z)$ and $W_2(a, c, z)$ are two linearly independent solutions of the Kummer–Laplace equation (3.35), irregular at the origin, and such that

$$W_1(a, c, z) \underset{|z|\to\infty}{\to} \frac{\Gamma(c)}{\Gamma(c - a)}(-z)^{-a}\,v(a, a - c + 1, -z), \qquad -\pi < \arg(-z) < \pi \quad \textbf{(12.207a)}$$

[10] See for example Whittaker and Watson (1935).

$$W_2(a, c, z) \underset{|z|\to\infty}{\to} \frac{\Gamma(c)}{\Gamma(a)} e^z z^{a-c} v(1-a, c-a, z), \qquad -\pi < \arg z < \pi \qquad \textbf{(12.207b)}$$

where Γ is Euler's gamma function, and v is the semi-convergent series

$$\begin{aligned} v(\alpha, \beta, z) &= 1 + \frac{\alpha\beta}{1!z} + \frac{\alpha(\alpha+1)\beta(\beta+1)}{2!z^2} + \cdots \\ &= \sum_{n=0}^{\infty} \frac{\Gamma(n+\alpha)\,\Gamma(n+\beta)}{\Gamma(\alpha)\,\Gamma(\beta)} \frac{(z)^{-n}}{n!} \end{aligned} \qquad \textbf{(12.208)}$$

Using these results, remembering that $\eta = r - z = r(1 - \cos\theta)$ and choosing

$$C = \exp(-\pi\gamma/2)\,\Gamma(1 + \mathrm{i}\gamma) \qquad \textbf{(12.209)}$$

so that

$$\psi_c = \exp(-\pi\gamma/2)\,\Gamma(1+\mathrm{i}\gamma)\exp(\mathrm{i}kz)\,{}_1F_1(-\mathrm{i}\gamma, 1, \mathrm{i}k(r-z)) \qquad \textbf{(12.210)}$$

we have

$$\begin{aligned} \psi_c \underset{|r-z|\to\infty}{\to} & \exp\{\mathrm{i}kz + \mathrm{i}\gamma\log[kr(1-\cos\theta)]\}\left(1 + \frac{\gamma^2}{\mathrm{i}kr(1-\cos\theta)} + \cdots\right) \\ & + f_c(k,\theta)\frac{\exp\{\mathrm{i}kr - \mathrm{i}\gamma\log(2kr)\}}{r}\left(1 + \frac{(1+\mathrm{i}\gamma)^2}{\mathrm{i}kr(1-\cos\theta)} + \cdots\right) \end{aligned} \qquad \textbf{(12.211)}$$

where

$$f_c(k,\theta) = -\frac{\gamma}{2k\sin^2(\theta/2)}\exp\{-\mathrm{i}\gamma\log[\sin^2(\theta/2)]\}\exp(2\mathrm{i}\sigma_0) \qquad \textbf{(12.212)}$$

is the *Coulomb scattering amplitude*. In the above equation

$$\exp(2\mathrm{i}\sigma_0) = \frac{\Gamma(1+\mathrm{i}\gamma)}{\Gamma(1-\mathrm{i}\gamma)} \qquad \textbf{(12.213)}$$

so that

$$\sigma_0 = \arg\Gamma(1+\mathrm{i}\gamma) \qquad \textbf{(12.214)}$$

It is clear that the asymptotic formula (12.211) does not hold for $\theta = 0$, since in this case $z = r$ and we cannot let $|r - z|$ tend to infinity. We also note from (12.211) that the choice of normalisation (12.209) corresponds to an incident wave of unit amplitude.

We see that the asymptotic form (12.211) of the Coulomb wave function ψ_c differs from the asymptotic expression (12.25) of the scattering wave function $\psi^{(+)}_{\mathbf{k}_i}$ corresponding to a potential which falls off faster than $1/r$ when $r \to \infty$. The infinite range of the Coulomb potential prevents the first term on the right of (12.211) from approaching a plane wave, while the second term does not exhibit the simple form $\exp(\mathrm{i}kr)/r$. However, for large $|r - z|$ the modifications due to the Coulomb interaction only affect the *phases* of these terms, so that the two terms on the right of (12.211) can again be interpreted respectively as an incident wave and an outgoing spherical scattered wave. For this reason, we shall denote the solution (12.210) as $\psi_c^{(+)}$. If we calculate the probability current density for $z < 0$ (with $|z|$ large), the first term of (12.211) yields a probability current density $\mathbf{j}_{\text{inc}}$ such that the incident flux through a unit area normal to the incident direction $\hat{\mathbf{k}}_i$ is given by

$$F = \mathbf{j}_{\text{inc}} \cdot \hat{\mathbf{k}}_i = v \tag{12.215}$$

and we remark that the factor $\exp\{i\gamma \log[kr(1 - \cos\theta)]\}$ gives a correction of order $1/r$ which can be neglected.

Similarly, the current density $\mathbf{j}_{\text{sc}}$ calculated for large r from the second term on the right of (12.211) is directed radially and is such that the outgoing flux through a unit area normal to $\hat{\mathbf{r}}$ is given by

$$\mathbf{j}_{\text{sc}} \cdot \hat{\mathbf{r}} = v|f_c(k, \theta)|^2/r^2 \tag{12.216}$$

the effect due to the presence of the factor $\exp[-i\gamma \log(2kr)]$ being negligible to lowest order in $1/r$.

The outgoing flux of particles passing through a spherical surface element $r^2\, d\Omega$ for large r is given from (12.216) by $v|f_c(k, \theta)|^2\, d\Omega$. Dividing this quantity by the incident flux (12.215) and by $d\Omega$, we find that the differential Coulomb scattering cross-section is given by

$$\frac{d\sigma_c}{d\Omega} = |f_c(k, \theta)|^2 \tag{12.217}$$

Substituting the expression (12.212) of the Coulomb scattering amplitude into the above formula, we find that

$$\frac{d\sigma_c}{d\Omega} = \frac{\gamma^2}{4k^2 \sin^4(\theta/2)} = \left(\frac{q_A q_B}{4\pi\varepsilon_0}\right)^2 \frac{1}{16E^2 \sin^4(\theta/2)} \tag{12.218}$$

where $E = \mu v^2/2$ is the kinetic energy of the colliding particles in the centre of mass system.

The result (12.218) is identical to the *Rutherford formula* (A1.33) derived in 1911 by E. Rutherford using classical mechanics, and discussed in Appendix 1. We simply recall here that the Coulomb differential cross-section (12.218) does not depend on the sign of the potential and is infinite in the forward direction ($\theta = 0$). Also, the total cross-section $\int (d\sigma_c/d\Omega)\, d\Omega$ is not defined for pure Coulomb scattering, since the integral diverges at $\theta = 0$. However, in actual physical situations the Coulomb field is screened at large distances because of the presence of other charges. As a consequence of this screening, the quantum mechanical differential cross-section becomes finite at $\theta = 0$, and the corresponding total cross-section is also finite. This will be illustrated in Section 12.7 for a screened Coulomb potential having the (Yukawa) form $U(r) = U_0\, r^{-1} \exp(-\alpha r)$.

It is interesting to mention at this point an important difference between classical and quantum scattering. As pointed out in Appendix 1, in classical mechanics both the differential cross-section in the forward direction and the total cross-section do not exist if the forces do not strictly vanish beyond a certain distance. On the contrary, in quantum mechanics, it can be shown (Problem 12.14) that the differential cross-section at $\theta = 0$ remains finite if the potential decreases faster than $1/r^3$ for large r, while the total cross-section is finite if the potential falls off faster than $1/r^2$.

We also note that the *phase* of the Coulomb scattering amplitude $f_c(k, \theta)$ has clearly no effect on the Coulomb differential cross-section (12.218). However, this phase plays an important role in processes for which the Coulomb amplitude $f_c(k, \theta)$ interferes with another amplitude. This is the case, for example, for the scattering by a modified Coulomb potential consisting of a pure Coulomb potential plus a short-range potential [9]. Another example is the Coulomb scattering of identical particles, where the phase of $f_c(k, \theta)$ leads to non-classical interference effects, as we shall see in Section 12.6.

Let us return to the Coulomb wave function (12.210). Remembering that ${}_1F_1(a, c, 0) = 1$, we see that the probability of finding the particle at $r = 0$ relative to that of finding it in the incident beam is given by

$$\frac{|\psi_c(0)|^2}{|\psi_c(\infty)|^2} = \exp(-\pi\gamma)|\Gamma(1 + i\gamma)|^2 = \frac{2\pi\gamma}{e^{2\pi\gamma} - 1} \tag{12.219}$$

If the Coulomb potential is repulsive ($\gamma > 0$) and strong compared to the kinetic energy of the colliding particles, so that $\gamma \gg 1$, we have

$$\frac{|\psi_c(0)|^2}{|\psi_c(\infty)|^2} \simeq 2\pi\gamma \exp(-2\pi\gamma) \tag{12.220}$$

The quantity

$$\exp(-2\pi\gamma) = \exp\left[-\frac{2\pi q_A q_B}{(4\pi\varepsilon_0)\hbar v}\right] \tag{12.221}$$

is called the *Gamow factor*. It plays an important role in the description of reactions between positive ions (in particular bare nuclei) at low relative velocities, as well as in the theory of alpha decay.

It is useful to write down the Coulomb wave function $\psi^{(+)}_{c,\mathbf{k}}$ corresponding asymptotically to an incident wave in the direction of an arbitrary vector $\mathbf{k}$, and exhibiting an outgoing spherical scattered wave behaviour. It is

$$\psi^{(+)}_{c,\mathbf{k}}(\mathbf{r}) = \exp(-\pi\gamma/2)\,\Gamma(1 + i\gamma)\,\exp(i\mathbf{k}\cdot\mathbf{r})\,{}_1F_1(-i\gamma, 1, i(kr - \mathbf{k}\cdot\mathbf{r})) \tag{12.222}$$

The corresponding Coulomb wave function $\psi^{(-)}_{c,\mathbf{k}}$ asymptotic to an incident wave in the direction of $\mathbf{k}$ and an incoming spherical scattered wave is given by

$$\psi^{(-)}_{c,\mathbf{k}}(\mathbf{r}) = \psi^{(+)*}_{c,-\mathbf{k}}(\mathbf{r}) = \exp(-\pi\gamma/2)\,\Gamma(1 - i\gamma)\,\exp(i\mathbf{k}\cdot\mathbf{r})\,{}_1F_1(i\gamma, 1, -i(kr + \mathbf{k}\cdot\mathbf{r})) \tag{12.223}$$

Both Coulomb wave functions (12.222) and (12.223) have been normalised so that the incident wave has unit amplitude.

12.6 Scattering of two identical particles

We have seen in Section 2.7 that, according to the symmetrisation postulate, systems of identical *bosons* are described by totally *symmetric* wave functions, while systems of identical *fermions* are described by totally *antisymmetric* wave functions. In this section, we shall examine the consequences of the symmetrisation postulate on the scattering of two identical particles. We shall consider first the elastic scattering of two identical bosons and then analyse elastic collisions between two identical fermions.

Scattering of two identical bosons

Let us consider the elastic scattering of two identical bosons of mass m, which interact through a potential $V(\mathbf{r})$ depending only on their relative coordinate $\mathbf{r}$. We begin by considering the case of *spinless* bosons. We work in the centre of mass system, so that the time-independent Schrödinger equation is

$$\left[-\frac{\hbar^2}{2\mu}\nabla^2 + V(\mathbf{r})\right]\psi(\mathbf{r}) = E\psi(\mathbf{r}) \tag{12.224}$$

where $\mu = m/2$ is the reduced mass and $\mathbf{r} = \mathbf{r}_1 - \mathbf{r}_2$ is the relative position vector of the two colliding particles 1 and 2. The situation in the centre of mass system is illustrated in Fig. 12.7. Two identical particles 1 and 2 approach one another, moving parallel to the Z axis in opposite directions. After an elastic collision the velocity of each particle is changed in direction but remains unchanged in magnitude. A detector counts the particles scattered into the direction characterised by the polar angles (θ, ϕ). Since the particles 1 and 2 are identical, there is no way of deciding whether a particle recorded by the detector results from a collision event in which particle 1 is scattered in the direction (θ, ϕ) (see Fig. 12.7(a)), or from a collision process in which particle 2 is scattered in that direction, so that particle 1 is scattered in the opposite direction $(\pi - \theta, \phi + \pi)$ (see Fig. 12.7(b)).

In classical mechanics the differential cross-section for scattering in the direction (θ, ϕ) would simply be the sum of the differential cross-sections for observation of particle 1 and particle 2 in that direction. If the same were to be true in quantum mechanics, we would obtain for the differential cross-section the 'classical' result

$$\frac{d\sigma_{cl}}{d\Omega} = |f(\theta, \phi)|^2 + |f(\pi - \theta, \phi + \pi)|^2 \tag{12.225}$$

where $f(\theta, \phi)$, the centre of mass amplitude for scattering in the direction (θ, ϕ), is related to the asymptotic behaviour of the solution $\psi^{(+)}_{\mathbf{k}_i}(\mathbf{r})$ of (12.224) satisfying the boundary condition (12.25)

$$\psi^{(+)}_{\mathbf{k}_i}(\mathbf{r}) \underset{r\to\infty}{\to} \exp(i\mathbf{k}_i \cdot \mathbf{r}) + f(\theta, \phi)\frac{\exp(ikr)}{r} \tag{12.226}$$

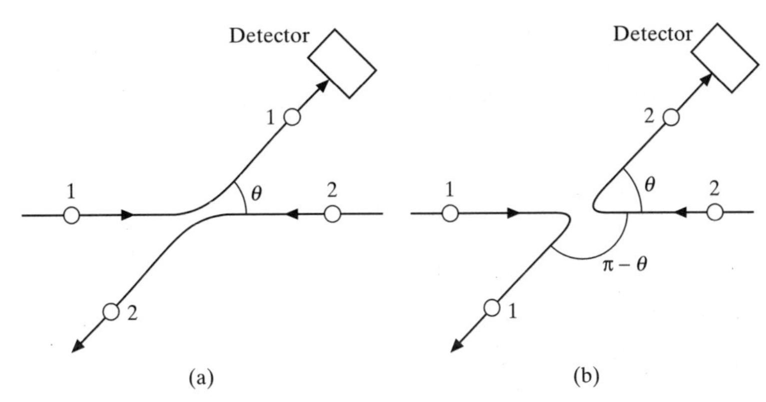

Figure 12.7 The scattering of two identical particles in the centre of mass system.

where we have taken the 'normalisation constant' A to be unity and we have omitted the k-dependence of the scattering amplitude for notational simplicity. However, we shall now show that the expression (12.225) for the differential cross-section is incorrect. Indeed, we have seen in Section 2.7 that a wave function describing a system of identical bosons must be completely symmetric. Thus, in the case of two identical spinless bosons, the wave function must be *symmetric* under the interchange of the spatial coordinates of the two particles. Now, the interchange $\mathbf{r}_1 \leftrightarrow \mathbf{r}_2$ corresponds to replacing the relative position vector $\mathbf{r}$ by $-\mathbf{r}$, which in spherical polar coordinates corresponds to (r, θ, ϕ) being replaced by $(r, \pi - \theta, \phi + \pi)$. The wave function $\psi^{(+)}_{\mathbf{k}_i}(\mathbf{r})$ satisfying the boundary condition (12.226) does not have the required symmetry, but the symmetric combination

$$\psi_+(\mathbf{r}) = \psi^{(+)}_{\mathbf{k}_i}(\mathbf{r}) + \psi^{(+)}_{\mathbf{k}_i}(-\mathbf{r}) \tag{12.227}$$

is also a solution of the Schrödinger equation (12.224) and does have the required symmetry: $\psi_+(-\mathbf{r}) = \psi_+(\mathbf{r})$. Using (12.226), the asymptotic form of $\psi_+(\mathbf{r})$ is seen to be

$$\psi_+(\mathbf{r}) \underset{r\to\infty}{\to} [\exp(\mathrm{i}\mathbf{k}_i\cdot\mathbf{r}) + \exp(-\mathrm{i}\mathbf{k}_i\cdot\mathbf{r})] + [f(\theta, \phi) + f(\pi - \theta, \phi + \pi)]\frac{\exp(\mathrm{i}kr)}{r} \tag{12.228}$$

The amplitude of the spherically outgoing wave is the symmetric amplitude

$$f_+(\theta, \phi) = f(\theta, \phi) + f(\pi - \theta, \phi + \pi) \tag{12.229}$$

so that the differential cross-section is

$$\frac{\mathrm{d}\sigma}{\mathrm{d}\Omega} = |f(\theta, \phi) + f(\pi - \theta, \phi + \pi)|^2 \tag{12.230}$$

a result which we can write in the form

$$\frac{\mathrm{d}\sigma}{\mathrm{d}\Omega} = |f(\theta, \phi)|^2 + |f(\pi - \theta, \phi + \pi)|^2 + 2\mathrm{Re}[f(\theta, \phi)f^*(\pi - \theta, \phi + \pi)] \tag{12.231}$$

It is important to note that this formula differs from the 'classical' result (12.225) by the presence of the third term on the right, which arises from the *interference* between the amplitudes $f(\theta, \phi)$ and $f(\pi - \theta, \phi + \pi)$. We also remark that the total cross-section

$$\sigma_{\text{tot}} = \int |f(\theta, \phi) + f(\pi - \theta, \phi + \pi)|^2 \, d\Omega \tag{12.232}$$

is equal to twice the number of particles removed from the incident beam per unit time and unit incident flux.

In the simple case for which the interaction potential is central, the scattering amplitude is independent of the azimuthal angle ϕ. The symmetrised scattering amplitude is then

$$f_+(\theta) = f(\theta) + f(\pi - \theta) \tag{12.233}$$

and the differential cross-section (12.230) becomes

$$\frac{d\sigma}{d\Omega} = |f(\theta) + f(\pi - \theta)|^2 \tag{12.234}$$

or

$$\frac{d\sigma}{d\Omega} = |f(\theta)|^2 + |f(\pi - \theta)|^2 + 2\text{Re}[f(\theta)f^*(\pi - \theta)] \tag{12.235}$$

so that the scattering is symmetric about the angle $\theta = \pi/2$ in the centre of mass system. Since

$$P_l[\cos(\pi - \theta)] = (-1)^l P_l(\cos\theta) \tag{12.236}$$

it is clear that the partial wave expansion of the symmetrised scattering amplitude $f_+(\theta)$ contains only *even* values of l. That is,

$$f_+(\theta) = \frac{2}{k} \sum_{l \text{ even}} (2l + 1) \exp(i\delta_l) \sin\delta_l P_l(\cos\theta) \tag{12.237}$$

Moreover, at $\theta = \pi/2$ we note that the quantum mechanical differential cross-section (12.234) is equal to

$$\frac{d\sigma(\theta = \pi/2)}{d\Omega} = 4|f(\theta = \pi/2)|^2 \tag{12.238}$$

and hence is four times as big as if the two colliding particles were distinguishable, and twice as big as the 'classical' result

$$\frac{d\sigma_{\text{cl}}(\theta = \pi/2)}{d\Omega} = 2|f(\theta = \pi/2)|^2 \tag{12.239}$$

following from (12.225). Furthermore, if there is only s-wave ($l = 0$) scattering, so that the scattering amplitude f is isotropic, we see from (12.234) that two colliding spinless bosons have a differential cross-section four times as big as if they were distinguishable particles, and twice as big as the 'classical' result (12.225).

The result (12.235) derived above for the scattering of two identical spinless bosons interacting by a central potential $V(r)$ can be generalised to the case of two identical bosons of any (integer) spin $s = 1, 2, \dots$. One finds (Problem 12.15) that

$$\frac{d\sigma}{d\Omega} = |f(\theta)|^2 + |f(\pi - \theta)|^2 + \frac{2}{2s+1}\mathrm{Re}[f(\theta)f^*(\pi - \theta)] \tag{12.240}$$

We note that the magnitude of the interference term decreases when the spin s increases.

Coulomb scattering of two identical spinless bosons

The non-classical effects due to the inferference term in (12.235) can be illustrated in a particularly simple and striking way for the case of *Coulomb scattering*. Let us consider two identical spinless bosons of charge Ze interacting only through the Coulomb force. This is the case for example in the scattering of two identical spinless nuclei (for example, the scattering of two alpha particles or two ^{12}C nuclei) at colliding energies which are low enough so that the nuclear forces between the two colliding particles can be neglected due to the presence of the Coulomb barrier. Equation (12.234) now reads

$$\frac{d\sigma}{d\Omega} = |f_c(\theta) + f_c(\pi - \theta)|^2 \tag{12.241}$$

where $f_c(\theta)$ is the Coulomb scattering amplitude (12.212). Therefore, we have

$$\frac{d\sigma}{d\Omega} = \left(\frac{\gamma}{2k}\right)^2 \left| \frac{\exp\{-2i\gamma \log[\sin(\theta/2)]\}}{\sin^2(\theta/2)} + \frac{\exp\{-2i\gamma \log[\cos(\theta/2)]\}}{\cos^2(\theta/2)} \right|^2 \tag{12.242}$$

or

$$\frac{d\sigma}{d\Omega} = \left(\frac{\gamma}{2k}\right)^2 \Big[\mathrm{cosec}^4(\theta/2) + \sec^4(\theta/2)$$

$$+ 2\mathrm{cosec}^2(\theta/2)\sec^2(\theta/2)\cos\{2\gamma \log[\tan(\theta/2)]\}\Big] \tag{12.243}$$

where (see (12.197))

$$\gamma = \frac{(Ze)^2}{(4\pi\varepsilon_0)\hbar v} \tag{12.244}$$

The result (12.243) is called the *Mott formula for the Coulomb scattering of two identical spinless bosons*. The corresponding 'classical' calculation would only yield the first two terms on the right of (12.243) while the Rutherford formula (12.218), which applies to the Coulomb scattering of two *distinct* particles, is obtained by retaining only the first term on the right of (12.243). We remark that the interference

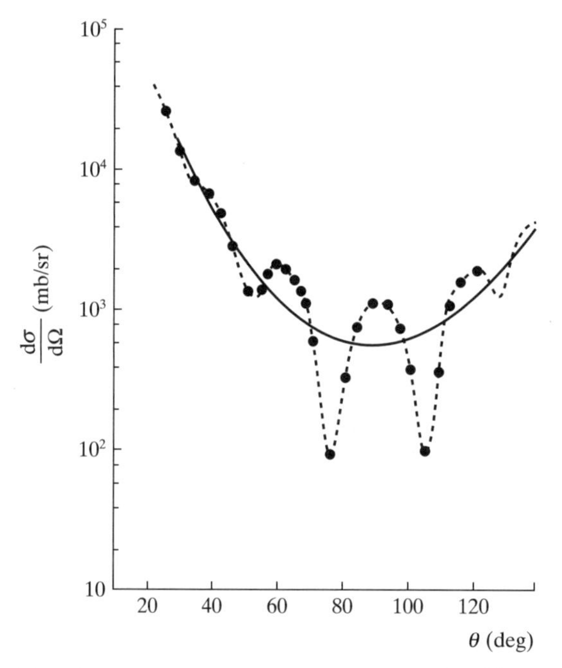

Figure 12.8 The differential cross-section (in mb/sr, where 1 mb ≡ 1 millibarn = 10^{-31} m^2), corresponding to the elastic scattering of two ^{12}C nuclei having a centre of mass energy of 5 MeV. The dashed curve represents the results obtained by using the Mott formula (12.243). The solid curve shows the corresponding 'classical' differential cross-section obtained by omitting the interference term in (12.243). The dots are the experimental data of D.A. Bromley, J.A. Kuehner and E. Almqvist.

term – the third term on the right of (12.243) – oscillates increasingly about zero as γ becomes larger and (or) one departs from the angle $\theta = \pi/2$. In the limit $\gamma \gg 1$, the differential cross-section (12.243) averaged over a small solid angle $\delta\Omega$ tends towards the 'classical' cross-section. We also verify that at $\theta = \pi/2$ the actual (quantum mechanical) cross-section (12.243) is twice as big as the 'classical' one.

As an example, let us consider the elastic scattering of two spinless ^{12}C nuclei having a centre of mass kinetic energy of 5 MeV. The corresponding relative velocity of the two colliding particles is $v \simeq 1.3 \times 10^7$ m s^{-1} and since $Z = 6$ for a carbon nucleus we have $\gamma \simeq 6$. As seen from Fig. 12.8, the experimental data are in excellent agreement with the Mott formula (12.243) and demonstrate very clearly the presence of the non-classical interference term.

Scattering of two identical fermions

The scattering of identical fermions is more difficult to analyse than that of spinless bosons because of the complications due to the spin. We begin by analysing the case of two identical spin-1/2 fermions interacting through *central* forces. Since the interaction is in general different in the singlet ($S = 0$) and triplet ($S = 1$) spin states of the two fermions, we shall start from two (unsymmetrised) scattering amplitudes ${}^1f(\theta)$ and ${}^3f(\theta)$ corresponding respectively to the singlet and triplet cases.

The full wave function describing a system of two identical spin-1/2 fermions must be *antisymmetric* in the interchange of the two particles, that is when all their coordinates (spatial and spin) are interchanged. Now, if the system is in a *singlet* spin state ($S = 0$), the spin part of the wave function is given by (7.18) and is antisymmetric. Hence the corresponding spatial part of the wave function must be symmetric in the interchange of the position vectors $\mathbf{r}_1$ and $\mathbf{r}_2$ of the two particles. As a result, the symmetrised singlet scattering amplitude is

$$ {}^1f_+(\theta) = {}^1f(\theta) + {}^1f(\pi - \theta) \qquad \textbf{(12.245)} $$

and the differential cross-section in the singlet spin state is

$$ \frac{\mathrm{d}^1\sigma}{\mathrm{d}\Omega} = |{}^1f(\theta) + {}^1f(\pi - \theta)|^2 \qquad \textbf{(12.246)} $$

If, on the other hand, the two spin-1/2 fermions are in a triplet spin state ($S = 1$) the corresponding three spin functions (7.19) are symmetric in the interchange of the spin coordinates of the two particles. The spatial part of the wave function must therefore be antisymmetric in the interchange of the position vectors $\mathbf{r}_1$ and $\mathbf{r}_2$, so that the symmetrised triplet scattering amplitude is given by

$$ {}^3f_-(\theta) = {}^3f(\theta) - {}^3f(\pi - \theta) \qquad \textbf{(12.247)} $$

and the differential cross-section in the triplet spin state is

$$ \frac{\mathrm{d}^3\sigma}{\mathrm{d}\Omega} = |{}^3f(\theta) - {}^3f(\pi - \theta)|^2 \qquad \textbf{(12.248)} $$

If the 'incident' and 'target' particles are unpolarised (that is, their spins are randomly orientated), the probability of obtaining triplet states is three times that of singlet states, so that the differential cross-section is given by

$$ \begin{aligned} \frac{\mathrm{d}\sigma}{\mathrm{d}\Omega} &= \frac{1}{4}\frac{\mathrm{d}^1\sigma}{\mathrm{d}\Omega} + \frac{3}{4}\frac{\mathrm{d}^3\sigma}{\mathrm{d}\Omega} \\ &= \frac{1}{4}|{}^1f(\theta) + {}^1f(\pi - \theta)|^2 + \frac{3}{4}|{}^3f(\theta) - {}^3f(\pi - \theta)|^2 \end{aligned} \qquad \textbf{(12.249)} $$

For the particular case of *spin-independent* central interactions, where

$$ {}^1f(\theta) = {}^3f(\theta) = f(\theta) \qquad \textbf{(12.250)} $$

we find from (12.249) that

$$ \frac{\mathrm{d}\sigma}{\mathrm{d}\Omega} = |f(\theta)|^2 + |f(\pi - \theta)|^2 - \mathrm{Re}[f(\theta)f^*(\pi - \theta)] \qquad \textbf{(12.251)} $$

We note that this formula differs from the 'classical' result by the presence of the third term on the right, which again is an *interference* term. We also remark that at $\theta = \pi/2$ the quantum mechanical differential cross-section (12.251) is given by

$$\frac{d\sigma(\theta = \pi/2)}{d\Omega} = |f(\theta = \pi/2)|^2 \tag{12.252}$$

and hence is equal to one-half of the 'classical' result $d\sigma_{cl}(\theta = \pi/2)/d\Omega = 2|f(\theta = \pi/2)|^2$. Furthermore, if there is only s-wave ($l = 0$) scattering, the differential cross-section (12.251) is four times smaller than the corresponding one (given by (12.235)) for the scattering of two identical spinless bosons.

The result (12.251) which we have obtained for the scattering of two identical spin-1/2 fermions interacting by a spin-independent central potential can be generalised to the case of two identical fermions of any half-odd integer spin s ($s = 1/2, 3/2, \dots$). It is found (Problem 12.15) that

$$\frac{d\sigma}{d\Omega} = |f(\theta)|^2 + |f(\pi - \theta)|^2 - \frac{2}{2s + 1}\text{Re}[f(\theta)f^*(\pi - \theta)] \tag{12.253}$$

a formula which differs from the boson–boson result (12.240) by the presence of the minus sign in front of the interference term. As in the case of boson–boson scattering, the magnitude of the interference term decreases when the spin s increases.

Coulomb scattering of two identical spin-1/2 fermions

As an illustration of equation (12.251), let us consider the case of *Coulomb scattering* of two identical spin-1/2 fermions of charge Ze, for example electron–electron scattering ($Z = -1$) or low-energy proton–proton scattering ($Z = +1$). Using the Coulomb scattering amplitude (12.212), we obtain from (12.251) the *Mott formula for the Coulomb scattering of two identical spin-1/2 fermions*:

$$\frac{d\sigma}{d\Omega} = \left(\frac{\gamma}{2k}\right)^2 \Big[\text{cosec}^4(\theta/2) + \sec^4(\theta/2)$$

$$- \text{cosec}^2(\theta/2)\sec^2(\theta/2)\cos\{2\gamma\log[\tan(\theta/2)]\}\Big] \tag{12.254}$$

where γ is given by (12.244). As in the case of equation (12.243), the corresponing 'classical' calculation would only yield the two first terms on the right of (12.254), and the interference term oscillates increasingly about zero as γ increases and (or) one departs from the scattering angle $\theta = \pi/2$. In contrast with the boson–boson case, however, there is a *minimum* in the differential cross-section (12.254) for spin-1/2 identical fermion–fermion scattering at $\theta = \pi/2$. At this value of the scattering angle the actual (quantum mechanical) differential cross-section (12.254) is indeed smaller than the 'classical' one by a factor of two, as we have seen above. The comparison between the classical and quantum mechanical differential

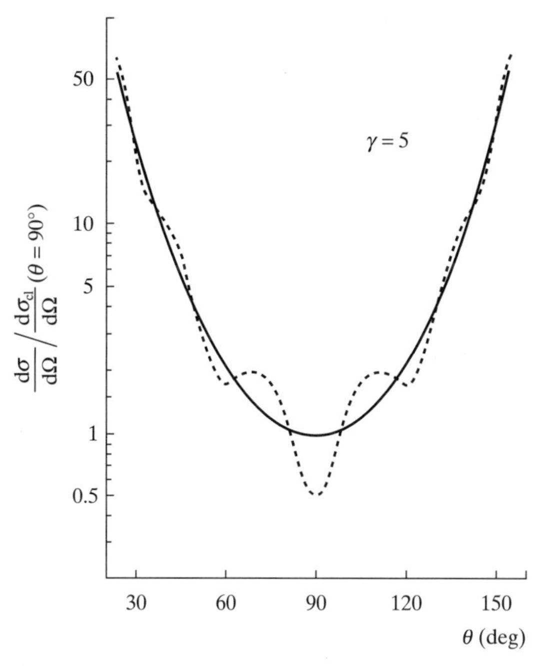

Figure 12.9 The differential cross-section $d\sigma/d\Omega$ corresponding to the scattering of two spin-$\frac{1}{2}$ fermions interacting through a Coulomb potential, divided by $d\sigma_{cl}(\theta = 90^0)/d\Omega$ for the value $\gamma = (Ze)^2/[(4\pi\varepsilon_0)\hbar v] = 5$. The dashed curve represents the results obtained by using the Mott formula (12.254). The solid curve shows the corresponding 'classical' results, obtained by omitting the interference term in (12.254).

cross-sections for the Coulomb scattering of two identical spin-1/2 fermions is illustrated in Fig. 12.9 for the case $\gamma = 5$.

12.7 Approximation methods

Approximation methods are clearly necessary in order to analyse complicated processes such as those occurring in atomic collisions, where exact solutions are not available. Whenever possible, it is useful to first discuss approximate treatments within the framework of potential scattering, where their interpretation is simpler and their accuracy can be checked easily. In this section, we shall outline some of the most useful approximation methods that have been devised. The first is the *Born series*, which is a perturbative expansion of the scattering wave function (or the scattering amplitude) in powers of the interaction potential. Next, we shall consider the *eikonal* and *WKB* approximations, which are *semi-classical* methods, and the *eikonal–Born series (EBS)* method. This will be followed by the *variational method* and the *R-matrix method*.

The Born series

Let us attempt to solve the Lippmann–Schwinger equation (12.159) by iteration, starting from the incident plane wave $\phi_{\mathbf{k}_i}(\mathbf{r})$ as our 'zero-order' approximation. We obtain in this way the sequence of functions

$$\psi_0(\mathbf{r}) = \phi_{\mathbf{k}_i}(\mathbf{r}) = \exp(\mathrm{i}\mathbf{k}_i\cdot\mathbf{r}),$$

$$\psi_1(\mathbf{r}) = \phi_{\mathbf{k}_i}(\mathbf{r}) + \int G_0^{(+)}(k, \mathbf{r}, \mathbf{r}')U(\mathbf{r}')\psi_0(\mathbf{r}')\,\mathrm{d}\mathbf{r}',$$

$$\vdots$$

$$\psi_n(\mathbf{r}) = \phi_{\mathbf{k}_i}(\mathbf{r}) + \int G_0^{(+)}(k, \mathbf{r}, \mathbf{r}')U(\mathbf{r}')\psi_{n-1}(\mathbf{r}')\,\mathrm{d}\mathbf{r}' \qquad \textbf{(12.255)}$$

where $G_0^{(+)}(k, \mathbf{r}, \mathbf{r}')$ is the Green's function given by (12.156). Assuming for the moment that the sequence (12.255) converges towards the exact stationary scattering wave function $\psi_{\mathbf{k}_i}^{(+)}(\mathbf{r})$, we may write for $\psi_{\mathbf{k}_i}^{(+)}(\mathbf{r})$ the *Born series*

$$\psi_{\mathbf{k}_i}^{(+)}(\mathbf{r}) = \sum_{n=0}^{\infty} \chi_n(\mathbf{r}) \qquad \textbf{(12.256)}$$

where $\chi_0 = \psi_0 = \phi_{\mathbf{k}_i}$ and

$$\chi_n(\mathbf{r}) = \int K_n(k, \mathbf{r}, \mathbf{r}')\chi_0(\mathbf{r}'), \qquad n \geqslant 1 \qquad \textbf{(12.257a)}$$

with

$$K_1(k, \mathbf{r}, \mathbf{r}') = G_0^{(+)}(k, \mathbf{r}, \mathbf{r}')U(\mathbf{r}') \qquad \textbf{(12.257b)}$$

and

$$K_n(k, \mathbf{r}, \mathbf{r}') = \int K_1(k, \mathbf{r}, \mathbf{r}'')K_{n-1}(k, \mathbf{r}'', \mathbf{r}')\,\mathrm{d}\mathbf{r}'', \qquad n \geqslant 2 \qquad \textbf{(12.257c)}$$

It is obvious from the above equations that the Born series (12.256) is a perturbative expansion in powers of the interaction potential. Substituting the series (12.256) into (12.163), we obtain the Born series for the scattering amplitude. That is,

$$f = \sum_{n=1}^{\infty} \bar{f}^{\mathrm{B}n} \qquad \textbf{(12.258)}$$

where

$$\bar{f}^{\mathrm{B1}} = -\frac{1}{4\pi}\langle\phi_{\mathbf{k}_f}|U|\phi_{\mathbf{k}_i}\rangle \qquad \textbf{(12.259)}$$

and the nth Born term $\bar{f}^{\mathrm{B}n}$ is given for $n \geqslant 2$ by

$$\bar{f}^{\mathrm{B}n} = -\frac{1}{4\pi}\langle\phi_{\mathbf{k}_f}|UG_0^{(+)}U\ldots|\phi_{\mathbf{k}_i}\rangle, \qquad n \geqslant 2 \qquad \textbf{(12.260)}$$

This last expression contains n times the reduced interaction potential U and $(n-1)$ times the Green's function $G_0^{(+)}$. We also define the jth-order Born approximation to the scattering amplitude as

$$f^{Bj} = \sum_{n=1}^{j} \bar{f}^{Bn} \quad . \tag{12.261}$$

For example, the first Born approximation is given by $f^{B1} = \bar{f}^{B1}$, while the second Born term $\bar{f}^{B2}$ reads

$$\bar{f}^{B2} = -\frac{1}{4\pi}\langle\phi_{\mathbf{k}_f}|UG_0^{(+)}U|\phi_{\mathbf{k}_i}\rangle \tag{12.262}$$

and the second Born approximation to the scattering amplitude, f^{B2}, is given by

$$f^{B2} = \bar{f}^{B1} + \bar{f}^{B2} \tag{12.263}$$

A Born series similar to that of the scattering amplitude f may be written down for the transition matrix element T_{fi} defined by (12.166).

In the particular case of scattering by a central potential, the integral equation (12.187) for the radial function $R_l(k, r)$ may be solved by perturbation theory in a similar way, starting from the 'unperturbed' radial wave function $R_l^{(0)}(k, r) = j_l(kr)$ as our zero-order approximation. Upon substitution in (12.87), we generate a Born series for tan δ_l. Its first term is the first Born approximation (12.107) for tan δ_l. By including the next term of the series, we obtain for tan δ_l the second Born approximation

$$(\tan\delta_l)^{B2} = -k\left[\int_0^\infty j_l^2(kr)U(r)r^2\,\mathrm{d}r + \int_0^\infty \mathrm{d}r\int_0^\infty \mathrm{d}r' j_l(kr)U(r)G_l(k, r, r')U(r')j_l(kr')r^2r'^2\right] \tag{12.264}$$

where $G_l(k, r, r')$ is the radial Green's function (12.188).

A simple physical picture of the Born series (12.258) may be obtained in the following way. Let us define in momentum space

$$\begin{aligned}\langle\mathbf{k}'|U|\mathbf{k}\rangle &= (2\pi)^{-3}\langle\phi_{\mathbf{k}'}|U|\phi_{\mathbf{k}}\rangle \\ &= (2\pi)^{-3}\int \exp[\mathrm{i}(\mathbf{k}-\mathbf{k}')\cdot\mathbf{r}]U(\mathbf{r})\,\mathrm{d}\mathbf{r}\end{aligned} \tag{12.265}$$

and use the integral representation (12.158) for the Green's function $G_0^{(+)}$. We then have

$$f^{B1} = \bar{f}^{B1} = -2\pi^2\langle\mathbf{k}_f|U|\mathbf{k}_i\rangle \tag{12.266}$$

and

$$\bar{f}^{\mathrm{B}n} = -2\pi^2 \int \mathrm{d}\mathbf{k}_1 \mathrm{d}\mathbf{k}_2 \dots \mathrm{d}\mathbf{k}_{n-1} \langle \mathbf{k}_f | U | \mathbf{k}_{n-1} \rangle \frac{1}{k^2 - k_{n-1}^2 + \mathrm{i}\varepsilon}$$

$$\times \langle \mathbf{k}_{n-1} | U | \mathbf{k}_{n-2} \rangle \dots \langle \mathbf{k}_2 | U | \mathbf{k}_1 \rangle \frac{1}{k^2 - k_1^2 + \mathrm{i}\varepsilon} \langle \mathbf{k}_1 | U | \mathbf{k}_i \rangle, \qquad n \geqslant 2 \quad \mathbf{(12.267)}$$

Thus the Green's function $G_0^{(+)}$ can be viewed as a propagator, while the vectors $\mathbf{k}_1, \mathbf{k}_2, \dots, \mathbf{k}_{n-1}$ may be interpreted as 'intermediate' wave vectors. The Born series (12.258) for the scattering amplitude can therefore be pictured as a *multiple scattering* series in which the projectile interacts repeatedly with the potential V and propagates freely between two interactions.

On the basis of this interpretation, we expect that the Born series for non-relativistic potential scattering should converge if the incident particle is fast enough so that it cannot interact too many times with the potential and (or) if the interaction is sufficiently weak. Detailed studies of the Born series, which lie outside the scope of this book, confirm these considerations. In particular, for a central potential $V(r)$ less singular than r^{-2} at the origin and decreasing faster than r^{-3} as $r \to \infty$, it may be proved that:

(1) the Born series converges at sufficiently high energies;
(2) the Born series converges for all energies if the potential $-|V(r)|$ cannot support any bound state.

A sufficient condition for convergence of the Born series at high energies may be obtained in a simple, non-rigorous way by requiring that the time τ_1 spent by the particle in the range a of the potential must be small with respect to a 'characteristic' time τ_2 necessary for the potential to have a significant effect. The time τ_1 is given approximately by $\tau_1 \simeq a/v$, where v is the velocity of the particle. On the other hand, if $|V_0|$ denotes a typical strength of the potential (which may be attractive or repulsive), and $|U_0| = 2m|V_0|/\hbar^2$ is the corresponding strength of the reduced potential, we may take $\tau_2 \simeq \hbar/|V_0| = 2m/(\hbar|U_0|)$. Thus, if we require that $\tau_1 \ll \tau_2$, we must have

$$\frac{|V_0|a}{\hbar v} = \frac{|U_0|a}{2k} \ll 1 \qquad \mathbf{(12.268)}$$

The first Born approximation

Let us study in more detail the first Born approximation (12.259). Using the explicit expression of $\phi_{\mathbf{k}_i}(\mathbf{r})$ and $\phi_{\mathbf{k}_f}(\mathbf{r})$ given respectively by (12.152) and (12.164), and remembering that $f^{\mathrm{B1}} = \bar{f}^{\mathrm{B1}}$, we have

$$f^{\mathrm{B1}} = -\frac{1}{4\pi} \int \exp[\mathrm{i}(\mathbf{k}_i - \mathbf{k}_f)\cdot\mathbf{r}] U(\mathbf{r})\, \mathrm{d}\mathbf{r} = -\frac{1}{4\pi} \int \exp(\mathrm{i}\boldsymbol{\Delta}\cdot\mathbf{r})\, U(\mathbf{r})\, \mathrm{d}\mathbf{r} \qquad \mathbf{(12.269)}$$

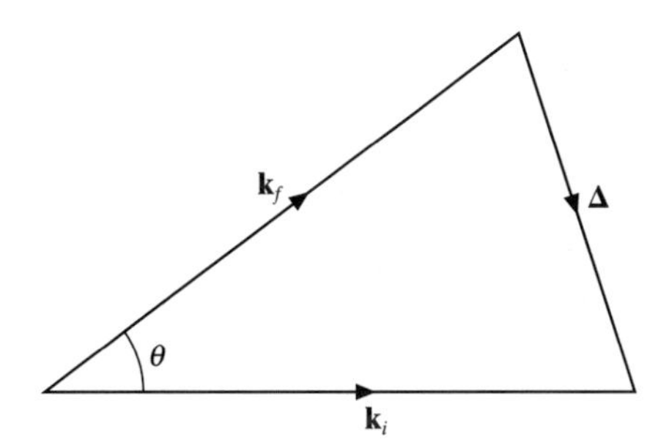

Figure 12.10 The vectors $\mathbf{k}_i$, $\mathbf{k}_f$ and Δ. The angle θ between $\mathbf{k}_i$ and $\mathbf{k}_f$ is the scattering angle.

where we have introduced the wave vector transfer

$$\boldsymbol{\Delta} = \mathbf{k}_i - \mathbf{k}_f \tag{12.270}$$

which is shown in Fig. 12.10, together with the vectors $\mathbf{k}_i$ and $\mathbf{k}_f$ and the scattering angle θ. The *momentum transfer* is given by $\hbar\boldsymbol{\Delta}$ and is of course identical to $\boldsymbol{\Delta}$ in systems of units (such as atomic units) where $\hbar = 1$. It is clear from Fig. 12.10 that since $k = |\mathbf{k}_i| = |\mathbf{k}_f|$ the magnitude of the vector $\boldsymbol{\Delta}$ is given by

$$\Delta = 2k \sin\frac{\theta}{2} \tag{12.271}$$

We note from (12.269) that the first Born scattering amplitude depends only on $\boldsymbol{\Delta}$ and is proportional to the Fourier transform of the potential corresponding to the wave vector transferred during the scattering. We also remark that the first Born differential cross-section

$$\frac{\mathrm{d}\sigma^{\mathrm{B1}}}{\mathrm{d}\Omega} = |f^{\mathrm{B1}}|^2 \tag{12.272}$$

remains unchanged when the sign of the interaction potential V is reversed.

Let us now examine in more detail the particular case of *central* potentials. We may then perform the angular integrations in (12.269) and obtain (Problem 12.18)

$$f^{\mathrm{B1}}(\Delta) = -\frac{1}{\Delta}\int_0^\infty r \sin(\Delta r) U(r)\, \mathrm{d}r \tag{12.273}$$

We see that this quantity is real and depends on k (that is, on the energy) and on the scattering angle θ only through the magnitude Δ of the wave vector transfer. The corresponding total cross-section in first Born approximation is given by

$$\begin{aligned}\sigma_{\mathrm{tot}}^{\mathrm{B1}} &= 2\pi \int_0^\pi |f^{\mathrm{B1}}(\Delta)|^2 \sin\theta\, \mathrm{d}\theta \\ &= \frac{2\pi}{k^2}\int_0^{2k} |f^{\mathrm{B1}}(\Delta)|^2\, \Delta\, \mathrm{d}\Delta \end{aligned} \tag{12.274}$$

where we have used the fact that $\sin\theta\, d\theta = \Delta\, d\Delta / k^2$. From (12.274) we deduce that

$$\lim_{k\to\infty} [k^2 \sigma_{tot}^{B1}(k)] = 2\pi \int_0^\infty |f^{B1}(\Delta)|^2\, \Delta\, d\Delta \quad \textbf{(12.275)}$$

Since the right-hand side of (12.275) is independent of k, we see that σ_{tot}^{B1} falls off like k^{-2} for large k. Because $E = \hbar^2 k^2/(2m)$, we may also write

$$\sigma_{tot}^{B1} \underset{E\to\infty}{\sim} AE^{-1} \quad \textbf{(12.276)}$$

where the quantity A is independent of the energy E. Thus σ_{tot}^{B1} tends to zero as E^{-1} at high energies.

As an illustration of the first Born approximation, we consider the Yukawa (or 'screened Coulomb') reduced potential

$$U(r) = U_0 \frac{e^{-\alpha r}}{r} \quad \textbf{(12.277)}$$

where $\alpha = a^{-1}$ is a screening parameter, a being the 'range' of the interaction and $|U_0|$ is a 'strength' parameter. The integral (12.273) is then straightforward, and yields

$$f^{B1} = -\frac{U_0}{\alpha^2 + \Delta^2} \quad \textbf{(12.278)}$$

The first Born differential cross-section

$$\frac{d\sigma^{B1}}{d\Omega} = \frac{U_0^2}{(\alpha^2 + \Delta^2)^2} \quad \textbf{(12.279)}$$

is plotted in Fig. 12.11 as a function of the scattering angle θ, for various values of ka. We see that for large ka this differential cross-section is essentially concentrated within a forward cone of angular aperture $\delta\theta \simeq (ka)^{-1}$. This behaviour is a direct consequence of (12.269) and of the fact that the Fourier transform of a function $U(r)$ which is negligible for $r \geqslant a$ is appreciable only for values of Δ such that $\Delta \leqslant a^{-1}$, that is (see (12.271)) for scattering angles θ given by

$$\theta \leqslant \frac{1}{ka} \quad \textbf{(12.280)}$$

We also note that the first Born differential cross-section (12.279) decreases like Δ^{-4} for large values of Δ.

By using (12.274) and (12.279), we also find that the total first Born cross-section is given by

$$\sigma_{tot}^{B1} = \frac{4\pi U_0^2}{\alpha^2(\alpha^2 + 4k^2)} \quad \textbf{(12.281)}$$

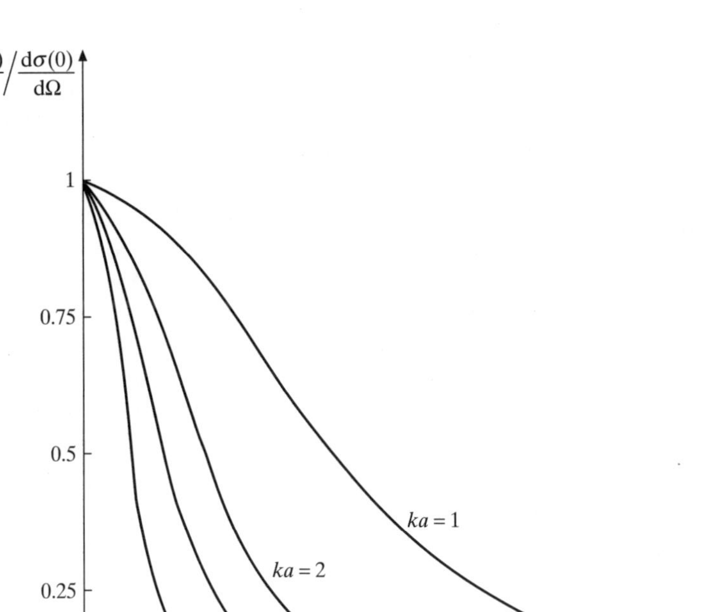

Figure 12.11 The angular distribution for scattering by a Yukawa potential in first Born approximation, for various values of *ka*.

Since $E = \hbar^2k^2/(2m)$, we see that for large E, $\sigma_{\text{tot}}^{\text{B1}} \simeq A/E$ with $A = \pi\hbar^2 U_0^2/(2m\alpha^2)$. Thus $\sigma_{\text{tot}}^{\text{B1}}$ falls off like E^{-1} for large E, in accordance with the statement following (12.276).

It is interesting to examine what happens in the above formulae when we let the 'screening constant' α tend to zero. The screened Coulomb reduced potential (12.277) then becomes the reduced Coulomb potential (see (12.191))

$$U_c(r) = \frac{U_0}{r} \tag{12.282}$$

The first Born scattering amplitude (12.278) now reads

$$f_c^{\text{B1}} = -\frac{U_0}{\Delta^2} \tag{12.283}$$

and the corresponding first Born differential cross-section is

$$\frac{d\sigma_c^{\text{B1}}}{d\Omega} = \frac{U_0^2}{\Delta^4} \tag{12.284}$$

Working in the centre of mass system, we recollect that for Coulomb scattering between two particles A and B, having charges q_A and q_B, respectively, one has (see (12.192)) $U_0 = 2\mu q_A q_B/(\hbar^2 4\pi\varepsilon_0)$, where μ is the reduced mass of the two particles. Remembering also that $\Delta = 2k \sin(\theta/2)$, and that in the centre of mass system the collision energy is given by $E = \hbar^2 k^2/(2\mu)$, we see that the first Born result (12.284) is identical to the Rutherford formula (12.218) obtained in Section 12.5. The first Born approximation therefore gives the correct differential cross-section for the Coulomb potential. However, the first Born amplitude for Coulomb scattering, f_c^{B1}, differs from the exact Coulomb scattering amplitude f_c of equation (12.212) by a phase factor. We also note that when the screening parameter α is non-zero, the corresponding screened Coulomb first Born differential cross-section (12.279), unlike the Rutherford formula, yields a finite result in the forward direction, and hence a finite total first Born cross-section.

Higher order terms of the Born series

Let us now analyse higher terms of the Born series. We begin by considering the Yukawa potential (12.277). Using (12.265) and (12.267) together with the fact that

$$\langle \mathbf{k}'|U|\mathbf{k}\rangle = (2\pi^2)^{-1}\frac{U_0}{\alpha^2 + (\mathbf{k}-\mathbf{k}')^2} \tag{12.285}$$

we see that the second Born term $\bar{f}^{B2}$ reads

$$\bar{f}^{B2} = \frac{U_0^2}{2\pi^2}\int d\mathbf{q}\frac{1}{q^2 - k^2 - i\varepsilon}\frac{1}{[\alpha^2 + (\mathbf{q}-\mathbf{k}_i)^2][\alpha^2 + (\mathbf{q}-\mathbf{k}_f)^2]}, \qquad \varepsilon \to 0^+ \tag{12.286}$$

where we have denoted by $\mathbf{q}$ the only intermediate momentum present in the second Born term.

The integral appearing on the right of (12.286) is a particular case of integrals studied by R.H. Dalitz in 1951, of the type

$$I_{m,n}(\alpha, \beta; \mathbf{k}_i, \mathbf{k}_f; \bar{k}) = \int d\mathbf{q}\frac{1}{q^2 - \bar{k}^2 - i\varepsilon}\frac{1}{[\alpha^2 + (\mathbf{q}-\mathbf{k}_i)^2]^m[\beta^2 + (\mathbf{q}-\mathbf{k}_f)^2]^n},$$
$$\varepsilon \to 0^+ \qquad (m, n = 1, 2, \dots) \tag{12.287}$$

These integrals can be evaluated by using a method introduced in 1949 by R.P. Feynman, and discussed in Appendix 12. For the case considered here, we have

$$\bar{f}^{B2} = \frac{U_0^2}{2\pi^2} I_{1,1}(\alpha, \alpha; \mathbf{k}_i, \mathbf{k}_f; k) \tag{12.288}$$

with $|\mathbf{k}_i| = |\mathbf{k}_f| = k$. The result is

$$\bar{f}^{B2}(k, \Delta) = \frac{U_0^2}{\Delta[\alpha^4 + 4k^2\alpha^2 + k^2\Delta^2]^{1/2}} \left\{ \tan^{-1} \frac{\alpha\Delta}{2[\alpha^4 + 4k^2\alpha^2 + k^2\Delta^2]^{1/2}} + \frac{1}{2} \mathrm{i} \log \left[\frac{(\alpha^4 + 4k^2\alpha^2 + k^2\Delta^2)^{1/2} + k\Delta}{(\alpha^4 + 4k^2\alpha^2 + k^2\Delta^2)^{1/2} - k\Delta} \right] \right\} \quad \mathbf{(12.289)}$$

We note from (12.289) that for large k we have

$$\bar{f}^{B2}(k, \Delta) = \mathrm{i} \left[\frac{A^{B2}(\Delta)}{k} + \mathrm{i} \frac{B^{B2}(\Delta)}{k^2} + O(k^{-3}) \right] \quad \mathbf{(12.290)}$$

where $A^{B2}(\Delta)$ and $B^{B2}(\Delta)$ are real functions of Δ. Thus, for large k the leading term of $\bar{f}^{B2}$ is purely imaginary and of order k^{-1}. We also see that in the limit of large Δ the quantity A^{B2} is proportional to $\Delta^{-2} \log \Delta$, while B^{B2} is proportional to Δ^{-2}.

Let us now consider briefly the general term $\bar{f}^{Bn}$ of the Born series. It can be shown [11] that for a Yukawa potential the result (12.290) can be generalized so that one has for large k

$$\bar{f}^{Bn}(k, \Delta) = \mathrm{i}^{n-1} \left[\frac{A^{Bn}(\Delta)}{k^{n-1}} + \mathrm{i} \frac{B^{Bn}(\Delta)}{k^n} \right] + O(k^{-n-1}), \qquad n \geqslant 2 \quad \mathbf{(12.291)}$$

where $A^{Bn}(\Delta)$ and $B^{Bn}(\Delta)$ are real functions of Δ. Moreover, this result can be extended to arbitrary superpositions of Yukawa potentials, or to potentials of the form

$$U(r) = \sum_{i=1}^{N} U_{0,i}\, r^{n_i} \exp(-\alpha_i r) \quad \mathbf{(12.292)}$$

where $n_1 = -1$, $n_i \geqslant -1$ for $i \geqslant 2$ and $\alpha_i > 0$.

It is apparent from equation (12.291) that for arbitrary superpositions of Yukawa potentials or potentials of the form (12.292) the Born series for the scattering amplitude converges towards its first term $\bar{f}^{B1}$ for large k and for all momentum transfers (that is, for all scattering angles). It is worth stressing that this property does not hold for all interaction potentials. For example, in the case of a Gaussian potential $U(r) = U_0 \exp(-\alpha^2 r^2)$, a Buckingham polarisation potential $U(r) = U_0/(r^2 + d^2)^2$ or a square well potential it is the second Born term $\bar{f}^{B2}$ which governs the scattering for large k when the momentum transfer Δ is large ($\Delta \geqslant k$); hence in these cases the Born series for the scattering amplitude converges in the large k limit towards the sum of its first two terms, $\bar{f}^{B1} + \bar{f}^{B2}$, when the entire range of momentum transfers ($0 \leqslant \Delta \leqslant 2k$) is considered.

[11] See Byron and Joachain (1977).

Semi-classical approximations

Semi-classical methods are useful when the de Broglie wavelength $\lambda = 2\pi/k$ of the projectile is sufficiently short with respect to the distance in which the potential varies by an appreciable amount. If the potential varies smoothly and has a finite range a, this short-wavelength condition is equivalent to the requirement that

$$ka \gg 1 \tag{12.293}$$

The eikonal approximation

We begin by considering the *eikonal approximation*, which was originally used in optics (the Greek word εικον means 'image'). Let us return to the Lippmann–Schwinger equation (12.159). In addition to the 'high wave number' condition (12.293), we also assume that the 'high-energy' condition

$$\frac{|V_0|}{E} = \frac{|U_0|}{k^2} \ll 1 \tag{12.294}$$

is satisfied, where $|V_0|$ is a typical strength of the potential $V(\mathbf{r})$ and $|U_0| = 2m|V_0|/\hbar^2$.

Since the potential varies slowly over the scale of the incident wavelength, it is natural to factor out the free incident plane wave from the wave function. Thus we write

$$\psi^{(+)}_{\mathbf{k}_i}(\mathbf{r}) = \exp(\mathrm{i}\mathbf{k}_i\cdot\mathbf{r})\varphi(\mathbf{r}) \tag{12.295}$$

where $\varphi(\mathbf{r})$ is a slowly varying function when $ka \gg 1$. Substituting (12.295) into the Lippmann–Schwinger equation (12.159), we find that the function $\varphi(\mathbf{r})$ satisfies the equation

$$\varphi(\mathbf{r}) = 1 + \int \exp\,[-\mathrm{i}\mathbf{k}_i\cdot(\mathbf{r}-\mathbf{r}')]G_0^{(+)}(k,\mathbf{r},\mathbf{r}')U(\mathbf{r}')\varphi(\mathbf{r}')\,\mathrm{d}\mathbf{r}' \tag{12.296}$$

where $G_0^{(+)}(k,\mathbf{r},\mathbf{r}')$ is the Green's function (12.156). Using its integral representation (12.158) and introducing the new variables $\mathbf{R} = \mathbf{r} - \mathbf{r}'$ and $\mathbf{p}' = \mathbf{k}' - \mathbf{k}_i$, we have

$$G_0^{(+)}(k,\mathbf{R}) = -(2\pi)^{-3}\exp(\mathrm{i}\mathbf{k}_i\cdot\mathbf{R})\int \frac{\exp(\mathrm{i}\mathbf{p}'\cdot\mathbf{R})}{2\mathbf{k}_i\cdot\mathbf{p}' + p'^2 - \mathrm{i}\varepsilon}\,\mathrm{d}\mathbf{p}' \tag{12.297}$$

where the limiting process $\varepsilon \to 0^+$ is implied. We may therefore write (12.296) as

$$\varphi(\mathbf{r}) = 1 - (2\pi)^{-3}\int \mathrm{d}\mathbf{R}\int \mathrm{d}\mathbf{p}'\frac{\exp(\mathrm{i}\mathbf{p}'\cdot\mathbf{R})}{2\mathbf{k}_i\cdot\mathbf{p}' + p'^2 - \mathrm{i}\varepsilon}U(\mathbf{r}-\mathbf{R})\varphi(\mathbf{r}-\mathbf{R}) \tag{12.298}$$

Since the product $U\varphi$ is slowly varying, the integral is dominated by small values of p'. It is therefore a good approximation to neglect the term p'^2 in the denominator of the integrand. This 'linearisation' of the denominator in (12.298) corresponds to the use of an approximate Green's function given by

$$\tilde{G}_0(k, \mathbf{R}) = -(2\pi)^{-3} \exp(\mathrm{i}\mathbf{k}_i \cdot \mathbf{R}) \int \frac{\exp(\mathrm{i}\mathbf{p}' \cdot \mathbf{R})}{2\mathbf{k}_i \cdot \mathbf{p}' - \mathrm{i}\varepsilon} \,\mathrm{d}\mathbf{p}' \quad \textbf{(12.299)}$$

The integral on the right of (12.299) may now be readily performed by contour integration. Returning to the original variables $\mathbf{r}$ and $\mathbf{r}'$, it is found that (Problem 12.21)

$$\tilde{G}_0(k, \mathbf{r}, \mathbf{r}') = -\frac{\mathrm{i}}{2k} \exp[\mathrm{i}k(z - z')]\delta^{(2)}(\mathbf{b} - \mathbf{b}')\theta(z - z') \quad \textbf{(12.300)}$$

where $\mathbf{r} = \mathbf{b} + z\hat{\mathbf{k}}_i$, $\mathbf{r}' = \mathbf{b}' + z'\hat{\mathbf{k}}_i$, and $\theta(x)$ is the step function such that $\theta(x) = 1$ for $x > 0$ and $\theta(x) = 0$ for $x < 0$. Substituting the approximate Green's function (12.300) into (12.296), we find that the resulting function $\tilde{\varphi}(\mathbf{r})$, which is an approximation to $\varphi(\mathbf{r})$, satisfies the integral equation

$$\tilde{\varphi}(\mathbf{r}) = 1 - \frac{\mathrm{i}}{2k} \int_{-\infty}^{z} U(\mathbf{b}, z')\tilde{\varphi}(\mathbf{b}, z') \,\mathrm{d}z' \quad \textbf{(12.301)}$$

This equation is readily solved to give

$$\tilde{\varphi}(\mathbf{r}) = \exp\left[-\frac{\mathrm{i}}{2k} \int_{-\infty}^{z} U(\mathbf{b}, z') \,\mathrm{d}z'\right] \quad \textbf{(12.302)}$$

Returning to (12.295), we see that the approximate scattering wave function which we have obtained, called the *eikonal wave function*, is given by

$$\begin{aligned} \psi^{\mathrm{E}}(\mathbf{r}) &= \exp(\mathrm{i}\mathbf{k}_i \cdot \mathbf{r})\tilde{\varphi}(\mathbf{r}) \\ &= \exp\left[\mathrm{i}\mathbf{k}_i \cdot \mathbf{r} - \frac{\mathrm{i}}{2k} \int_{-\infty}^{z} U(\mathbf{b}, z') \,\mathrm{d}z'\right] \end{aligned} \quad \textbf{(12.303)}$$

where the integration in the phase is along a straight line parallel to $\mathbf{k}_i$.

In order to obtain the scattering amplitude, we substitute (12.303) for $\psi^{(+)}_{\mathbf{k}_i}$ into the integral representation (12.163). This yields

$$f \simeq -\frac{1}{4\pi} \int \mathrm{d}\mathbf{r} \exp(\mathrm{i}\mathbf{\Delta} \cdot \mathbf{r})U(\mathbf{r}) \exp\left[-\frac{\mathrm{i}}{2k} \int_{-\infty}^{z} U(\mathbf{b}, z') \,\mathrm{d}z'\right] \quad \textbf{(12.304)}$$

where $\mathbf{\Delta} = \mathbf{k}_i - \mathbf{k}_f$ is the wave vector transfer.

In obtaining the eikonal wave function (12.303), we remarked that the integration in its phase should be performed along a straight line parallel to the incident momentum $\mathbf{k}_i$. In fact, since the phase of the corresponding semi-classical wave function must be evaluated along the classical trajectory, R.J. Glauber showed in 1959 that an improvement in the angular range of validity of the eikonal approximation can be achieved by carrying out the integration in the phase along a direction $\hat{\mathbf{n}}$ in the scattering plane (defined by the vectors $\mathbf{k}_i$ and $\mathbf{k}_f$) which is parallel to the bisector of the scattering angle. We note that this direction $\hat{\mathbf{n}}$ is perpendicular to the wave vector transfer $\mathbf{\Delta}$. Thus, if we work in a new coordinate system such that

$$\mathbf{r} = \mathbf{b} + z\hat{\mathbf{n}} \tag{12.305}$$

we have $\mathbf{\Delta}\cdot\mathbf{r} = \mathbf{\Delta}\cdot\mathbf{b}$. The integral over the variable z in (12.304) is now straightforward, and we obtain the *eikonal scattering amplitude*

$$f^{\mathrm{E}} = \frac{k}{2\pi \mathrm{i}}\int \mathrm{d}^2\mathbf{b}\, \exp(\mathrm{i}\mathbf{\Delta}\cdot\mathbf{b})\,\{\exp[\mathrm{i}\chi(k, \mathbf{b})] - 1\} \tag{12.306}$$

where

$$\chi(k, \mathbf{b}) = -\frac{1}{2k}\int_{-\infty}^{+\infty} U(\mathbf{b}, z)\,\mathrm{d}z \tag{12.307}$$

is the *eikonal phase shift function*. Even though the vector $\mathbf{b}$ appearing in (12.305) is now perpendicular to $\hat{\mathbf{n}}$ (and not to $\mathbf{k}_i$), we shall continue to call it an 'impact parameter vector'. We note that the two-dimensional integral in (12.306) must be performed over all such impact parameter vectors, which lie in a plane perpendicular to $\hat{\mathbf{n}}$. For further reference, it is useful to exhibit explicitly the k-dependence of the eikonal phase shift function, and to write

$$\chi(k, \mathbf{b}) = \frac{1}{k}\chi_0(\mathbf{b}) \tag{12.308a}$$

where

$$\chi_0(\mathbf{b}) = -\frac{1}{2}\int_{-\infty}^{+\infty} U(\mathbf{b}, z)\,\mathrm{d}z \tag{12.308b}$$

The differential cross-section is given in the eikonal approximation by

$$\frac{\mathrm{d}\sigma^{\mathrm{E}}}{\mathrm{d}\Omega} = |f^{\mathrm{E}}|^2 \tag{12.309}$$

and the total eikonal cross-section is obtained by integrating the eikonal differential cross-section (12.309) over all scattering angles.

For potentials which possess azimuthal symmetry, the expression (12.306) of the eikonal scattering amplitude simplifies to

$$f^{\mathrm{E}} = \frac{k}{\mathrm{i}}\int_0^{\infty} J_0(\Delta b)\{\exp[\mathrm{i}\chi(k, b)] - 1\}\, b\,\mathrm{d}b \tag{12.310}$$

where we have used the fact that the ordinary Bessel function J_0 has the integral representation

$$J_0(x) = (2\pi)^{-1}\int_0^{2\pi} \exp(\mathrm{i}x\cos\phi)\,\mathrm{d}\phi \tag{12.311}$$

For the case of pure Coulomb scattering, Glauber has shown that if a cut-off procedure is applied, the eikonal differential cross-section (12.309) agrees with the exact result (12.218).

The eikonal scattering amplitude (12.310) can also be related to the partial wave expansion of the scattering amplitude given by (12.60). To see how this can be done, we first write (12.60) in the form

$$f \simeq \frac{1}{\mathrm{i}k}\sum_{l=0}^{\infty}\left(l+\frac{1}{2}\right)J_0\left[\left(l+\frac{1}{2}\right)\theta\right]\left\{\exp\left[2\mathrm{i}\delta_l(k)\right]-1\right\} \tag{12.312}$$

where we have used the relation

$$P_l(\cos\theta) \simeq J_0\left[\left(l+\frac{1}{2}\right)\theta\right] \tag{12.313}$$

valid for $l \gg 1$ and $l\theta = \mathcal{O}(1)$, that is for small angles. Since the number of partial waves is large at high energies, we may replace the sum over l by an integral over the impact parameter b. Using the fact that

$$b \simeq [l(l+1)]^{1/2}/k \simeq \left(l+\frac{1}{2}\right)/k \tag{12.314}$$

we may then write

$$f \simeq \frac{k}{\mathrm{i}}\int_0^{\infty} J_0(\Delta b)\{\exp\left[2\mathrm{i}\delta(k,b)\right]-1\}\, b\, \mathrm{d}b \tag{12.315}$$

where

$$\delta(k,b) \equiv \delta_l(k), \qquad l+\frac{1}{2} \simeq kb \tag{12.316}$$

and we have replaced $kb\theta$ by $2kb\sin(\theta/2)$ in the argument of J_0. This replacement is justified since we are dealing here with small-angle scattering where the approximate formula (12.313) is valid. The expression (12.315) then reduces to (12.310), where

$$\delta(k,b) = \frac{1}{2}\chi(k,b) = -\frac{1}{4k}\int_{-\infty}^{+\infty} U(b,z)\,\mathrm{d}z \tag{12.317}$$

is called the *eikonal phase shift*.

By analogy with the Born series (12.258), we may define an *eikonal series* by expanding the eikonal amplitude (12.306) in powers of χ – and hence in powers of the potential strength. That is,

$$f^{\mathrm{E}} = \sum_{n=1}^{\infty} \bar{f}^{\mathrm{E}n} \tag{12.318}$$

where

$$\bar{f}^{\mathrm{E}n} = \frac{k}{2\pi \mathrm{i}} \frac{\mathrm{i}^n}{n!} \int \mathrm{d}^2\mathbf{b} \exp(\mathrm{i}\boldsymbol{\Delta}\cdot\mathbf{b})[\chi(k, \mathbf{b})]^n \tag{12.319}$$

The eikonal–Born series (EBS) method

Let us investigate the relationship between the terms of the Born series (12.258) and those of the eikonal series (12.318). First of all, we see from (12.307) and (12.319) that

$$\bar{f}^{\mathrm{E}1} = -\frac{1}{4\pi} \int \mathrm{d}^2\mathbf{b} \int_{-\infty}^{+\infty} \mathrm{d}z \exp(\mathrm{i}\boldsymbol{\Delta}\cdot\mathbf{b}) U(\mathbf{b}, z) \tag{12.320}$$

With the Glauber choice of coordinate system (12.305) such that the vector $\boldsymbol{\Delta}$ entirely lies in the plane of 'impact parameters' $\mathbf{b}$, we have $\boldsymbol{\Delta}\cdot\mathbf{b} = \boldsymbol{\Delta}\cdot\mathbf{r}$ and therefore

$$\begin{aligned} \bar{f}^{\mathrm{E}1} &= -\frac{1}{4\pi} \int \mathrm{d}\mathbf{r} \exp(\mathrm{i}\boldsymbol{\Delta}\cdot\mathbf{r}) U(\mathbf{r}) \\ &= \bar{f}^{\mathrm{B}1} \end{aligned} \tag{12.321}$$

Remarkable relationships also exist between the higher order terms of the eikonal and Born series. Let us focus our attention on central potentials. Expanding the eikonal amplitude (12.310) in powers of the potential strength, we find that

$$\bar{f}^{\mathrm{E}n} = \frac{k\, \mathrm{i}^{n-1}}{n!} \int_0^{\infty} J_0(\Delta b)[\chi(k, b)]^n b\, \mathrm{d}b \tag{12.322}$$

Moreover, using (12.308), we can rewrite the above expression in the form

$$\bar{f}^{\mathrm{E}n} = \left(\frac{\mathrm{i}}{k}\right)^{n-1} A^{\mathrm{E}n}(\Delta) \tag{12.323}$$

where

$$A^{\mathrm{E}n}(\Delta) = \frac{1}{n!} \int_0^{\infty} J_0(\Delta b)[\chi_0(b)]^n b\, \mathrm{d}b \tag{12.324}$$

and we note that for central potentials the terms $\bar{f}^{\mathrm{E}n}$ of the eikonal series are alternately purely real and purely imaginary. On the other hand, for superpositions of Yukawa potentials, or potentials of the form (12.292), it follows from (12.291) that to leading order in k^{-1} the term $\bar{f}^{\mathrm{B}n}$ is given by

$$\bar{f}^{\mathrm{B}n}(k, \Delta) = \left(\frac{\mathrm{i}}{k}\right)^{n-1} A^{\mathrm{B}n}(\Delta) + \mathcal{O}(k^{-n}) \tag{12.325}$$

and one finds (see Byron and Joachain, 1977) that

$$A^{\mathrm{B}n}(\Delta) = A^{\mathrm{E}n}(\Delta) \tag{12.326}$$

Table 12.1 Dependence of various terms of the Born and eikonal series for scattering by potentials of the form (12.292) as a function of (large) k and Δ. The dominant contributions are framed. The terms located above the dashed line contribute through order k^{-2} to the differential cross-section.

Order of perturbation theory	*Term*	*Small* Δ ($\Delta \leqslant 1$)	*Large* Δ ($\Delta \geqslant 1$)
First	$\bar{f}^{B1} = \bar{f}^{E1}$	$\boxed{1}$	$\boxed{\Delta^{-2}}$
Second	Re $\bar{f}^{B2}$	k^{-2}	$k^{-2}\Delta^{-2}$
	Re $\bar{f}^{E2}$	0	0
	Im $\bar{f}^{B2}$	k^{-1}	$k^{-1}\Delta^{-2} \log \Delta$
	Im $\bar{f}^{E2}$	k^{-1}	$k^{-1}\Delta^{-2} \log \Delta$
Third	Re $\bar{f}^{B3}$	k^{-2}	$k^{-2}\Delta^{-2} \log^2\Delta$
	Re $\bar{f}^{E3}$	k^{-2}	$k^{-2}\Delta^{-2} \log^2\Delta$
	Im $\bar{f}^{B3}$	k^{-3}	$k^{-3}\Delta^{-2} \log \Delta$
	Im $\bar{f}^{E3}$	0	0
nth ($n > 3$)	$\bar{f}^{Bn}$	$(ik)^{1-n}$	$(ik)^{1-n}\Delta^{-2} \log^{n-1}\Delta$
	$\bar{f}^{En}$	$(ik)^{1-n}$	$(ik)^{1-n}\Delta^{-2} \log^{n-1}\Delta$

for all values of Δ. This implies that the leading contribution to the Born term $\bar{f}^{Bn}$ at large k is given by the eikonal term $\bar{f}^{En}$. For other interaction potentials, the relation (12.326) only holds for large k and values of Δ of order $k^{1/2}$.

As an example, we show in Table 12.1 the dependence of the Born terms $\bar{f}^{Bn}$ (for large k) and of the eikonal terms $\bar{f}^{En}$ as a function of k and Δ, for potentials of the form (12.292). We see from Table 12.1 that for these potentials the dominant contribution at high energies is given at all wave vector transfers by the first Born term. We also remark that for large Δ the results quoted in Table 12.1 are given to leading order in k^{-1}, Δ^{-1} and $\log \Delta$. It is worth noting that because of the presence of powers of $\log \Delta$, the convergence of the Born and eikonal series is slower at large Δ than at small Δ. This is a general feature which also arises in the many-body electron– or positron–atom collision problem, as we shall see in Chapter 13.

It is also worth stressing that neither the eikonal amplitude f^E nor the second Born amplitude $f^{B2} = \bar{f}^{B1} + \bar{f}^{B2}$ are correct to order k^{-2}. Indeed, the amplitude f^E has no analogue to the second-order term Re$\bar{f}^{B2}$, while the amplitude f^{B2} lacks the third-order term Re$\bar{f}^{B3}$, both Re$\bar{f}^{B2}$ and Re$\bar{f}^{B3}$ being of order k^{-2} for large k.

In general, the calculation of the terms $\bar{f}^{Bn}$ of the Born series for $n \geqslant 3$ is a very difficult task. This discussion, however, shows that for large k the third Born term $\bar{f}^{B3}$ can be approximated by the third eikonal term $\bar{f}^{E3}$. The resulting eikonal–Born series (EBS) amplitude, proposed in 1973 by F.W. Byron and C.J. Joachain,

$$f^{EBS} = \bar{f}^{B1} + \bar{f}^{B2} + \bar{f}^{E3} \tag{12.327}$$

Table 12.2 Differential cross-sections (in a.u.) obtained by using the first Born approximation (B1), the second Born approximation (B2), the eikonal approximation (E), and the eikonal–Born series approximations (EBS and EBS′), compared with the exact results, for the static electron–atomic hydrogen potential (12.329). The wave number is k = 2.71 a.u., corresponding to an incident electron energy E = 100 eV. The numbers in parentheses indicate powers of 10.

θ (degrees)	B1	B2	E	EBS	EBS′	*Exact*
0	1.00	1.15	9.23(−1)	1.01	1.02	1.02
30	3.13(−1)	3.95(−1)	2.81(−1)	3.22(−1)	3.23(−1)	3.28(−1)
60	5.68(−2)	8.89(−2)	5.03(−2)	6.20(−2)	6.02(−2)	6.39(−2)
90	1.69(−2)	3.23(−2)	1.52(−2)	2.01(−2)	1.81(−2)	2.05(−2)
120	7.84(−3)	1.72(−2)	7.21(−3)	1.02(−2)	8.47(−3)	1.01(−2)
150	5.14(−3)	1.22(−2)	4.78(−3)	7.09(−3)	5.54(−3)	6.80(−3)
180	4.50(−3)	1.09(−2)	4.19(−3)	6.33(−3)	4.84(−3)	6.00(−3)

is seen to be correct through order k^{-2}. Similarly, their EBS′ amplitude

$$f^{\text{EBS}'} = f^{\text{E}} - \bar{f}^{\text{E2}} + \bar{f}^{\text{B2}} = \bar{f}^{\text{B1}} + \bar{f}^{\text{B2}} + \sum_{n=3}^{\infty} \bar{f}^{\text{E}n} \qquad \textbf{(12.328)}$$

is also correct through order k^{-2}.

As an illustration of the above discussion, we compare in Table 12.2 the differential cross-sections corresponding to various approximations with the exact values – obtained by using the partial wave method and solving numerically the radial Schrödinger equations (12.41) – for the reduced potential

$$U(r) = -2\left(1 + \frac{1}{r}\right)\exp(-2r) \qquad \textbf{(12.329)}$$

As we shall see in Chapter 13, the interaction (12.329) is the reduced static potential (in a.u.) felt by an electron in the field of a hydrogen atom in the ground state. For this potential, the second Born term can be calculated in closed form using the Feynman method discussed in Appendix 12, and the eikonal phase $\chi_0(b)$ is given by (Problem 12.22)

$$\chi_0(b) = 2K_0(2b) + 2bK_1(2b) \qquad \textbf{(12.330)}$$

where K_0 and K_1 are modified Bessel functions. The results of Table 12.2 have been obtained for a wave number $k = 2.71$ a.u., corresponding to an incident electron energy E = 100 eV. Since $U_0 = -2$ a.u. and $a = 0.5$ a.u. in this case, we have $|U_0|a/(2k) = 0.18$ which corresponds to a relatively weak coupling. Nevertheless, the second Born (B2) results are already quite inaccurate, especially at large angles. The EBS and EBS′ approximations provide improvements over the first Born, second Born and eikonal results, as expected from the above discussion. A detailed discussion of the eikonal–Born series method has been given by Byron and Joachain (1977). We shall see in Chapter 13 that the amplitudes f^{EBS} and $f^{\text{EBS}'}$, generalised to the multi-channel case, are very useful in analysing fast electron– and positron–atom collisions.

The WKB approximation

The basic features of the WKB approximation have been discussed in Section 2.8. Like the eikonal approximation, it is a semi-classical approach which is valid when the potential changes slowly over a de Broglie wavelength $\lambda = 2\pi/k$ of the incident particle. Since the WKB method is only simple when applied to one-dimensional problems, we shall use it here to solve approximately the radial equations for the scattering by a central potential. We shall obtain in this way the WKB phase shifts which may then be used in the partial wave series (12.60) or, when many partial waves contribute to the scattering amplitude, in an integral expression of the type (12.315).

We begin by rewriting the radial Schrödinger equation (12.41) in the form

$$\left[\frac{\mathrm{d}^2}{\mathrm{d}r^2} + W(r)\right]u_l(r) = 0 \qquad \textbf{(12.331)}$$

where

$$W(r) = k^2 - U(r) - \frac{l(l+1)}{r^2} \qquad \textbf{(12.332)}$$

and we have written $u_l(r) \equiv u_l(k, r)$ to simplify the notation. A general solution of (12.331) can be written as

$$u_l(r) = A(r)\exp[\mathrm{i}\alpha(r)] \qquad \textbf{(12.333)}$$

which on substitution into (12.331) gives

$$\frac{\mathrm{d}^2 A}{\mathrm{d}r^2} + WA - A\left(\frac{\mathrm{d}\alpha}{\mathrm{d}r}\right)^2 + \mathrm{i}\left(2\frac{\mathrm{d}A}{\mathrm{d}r}\frac{\mathrm{d}\alpha}{\mathrm{d}r} + A\frac{\mathrm{d}^2\alpha}{\mathrm{d}r^2}\right) = 0 \qquad \textbf{(12.334)}$$

Since $W(r)$ is real we can choose both $A(r)$ and $\alpha(r)$ to be real, so that the real and imaginary parts of equation (12.334) yield

$$\left(\frac{\mathrm{d}\alpha}{\mathrm{d}r}\right)^2 - W - A^{-1}\frac{\mathrm{d}^2 A}{\mathrm{d}r^2} = 0 \qquad \textbf{(12.335)}$$

and

$$\frac{\mathrm{d}^2\alpha}{\mathrm{d}r^2} + 2A^{-1}\frac{\mathrm{d}A}{\mathrm{d}r}\frac{\mathrm{d}\alpha}{\mathrm{d}r} = 0 \qquad \textbf{(12.336)}$$

A solution of (12.336) is given by the function $\zeta(r)$ such that

$$\zeta = A^{-2} = \frac{\mathrm{d}\alpha}{\mathrm{d}r} \qquad \textbf{(12.337)}$$

From (12.335) we then have

$$\zeta^2 = W + \zeta^{1/2}\frac{d^2}{dr^2}[\zeta^{-1/2}] \qquad \textbf{(12.338)}$$

and this equation provides an exact solution of the radial Schrödinger equation (12.41).

In the semi-classical regime the phase $\alpha(r)$ is slowly varying, so that the second term on the right-hand side of (12.338) is small. The WKB approximation is obtained by neglecting this term, so that $\zeta^2 = W$. In the classically allowed region, where $W(r) > 0$, the WKB radial wave functions have therefore the form

$$u_l(r) = [W(r)]^{-1/4} \exp\left[\pm i \int^r [W(r')]^{1/2} \, dr'\right] \qquad \textbf{(12.339)}$$

while in the classically forbidden region, where $W(r) < 0$, these wave functions are given by

$$u_l(r) = |W(r)|^{-1/4} \exp\left[\pm \int^r |W(r')|^{1/2} \, dr'\right] \qquad \textbf{(12.340)}$$

Appropriate linear combinations of these WKB solutions can be taken in order to satisfy the boundary conditions in each of these regions. It is worth noting that the WKB approximation becomes *exact* in the strong coupling limit. Indeed, the quantity $|W|$ increases linearly with the strength of the potential, while the correction term in (12.338), evaluated using the WKB choice $\zeta^2 = W$, is independent of the potential strength.

The solutions (12.339) and (12.340) cannot be used near a classical turning point r_0 such that $W(r_0) = 0$. It is therefore necessary to solve the original equation (12.331) – or a slightly modified version of it – in the vicinity of a turning point r_0. The derivation of formulae which connect the solutions (12.339) and (12.340) across this region is discussed for example in Bransden and Joachain (2000). Let us introduce the quantity

$$F(r) = k^2 - U(r) - \frac{(l + \frac{1}{2})^2}{r^2} \qquad \textbf{(12.341)}$$

and the corresponding turning point r_0 such that $F(r_0) = 0$. It is found that the solution which vanishes at the origin has the following form for $r > r_0$:

$$u_l(r) = \left[\frac{k^2}{F(r)}\right]^{1/4} \sin\left[\frac{\pi}{4} + \int_{r_0}^r [F(r')]^{1/2} \, dr'\right], \qquad r > r_0 \qquad \textbf{(12.342)}$$

and we have chosen the normalisation coefficient so that

$$u_l(r) \underset{r\to\infty}{\longrightarrow} \sin\left[\frac{\pi}{4} + \int_{r_0}^{\infty} ([F(r')]^{1/2} - k) \, dr' + k(r - r_0)\right] \qquad \textbf{(12.343)}$$

Comparing (12.343) and (12.75) in which we choose $N_l(k) = 1$, we find that the phase shifts are given in the WKB approximation by

$$\delta_l^{\text{WKB}} = \left(l + \frac{1}{2}\right)\frac{\pi}{2} - kr_0 + \int_{r_0}^{\infty} ([F(r)]^{1/2} - k)\,\mathrm{d}r \tag{12.344}$$

Integrating by parts and remembering that $F(r_0) = 0$, we find that

$$\delta_l^{\text{WKB}} = \left(l + \frac{1}{2}\right)\frac{\pi}{2} - \int_{r_0}^{\infty} r\frac{\mathrm{d}}{\mathrm{d}r}[F(r)]^{1/2}\,\mathrm{d}r \tag{12.345}$$

Let us analyse a few properties of the WKB phase shifts. First of all, we verify from (12.345) that $\delta_l^{\text{WKB}} = 0$ when $U(r) = 0$. Indeed, in that case, the function $F(r)$ reduces to

$$G(r) = k^2 - \frac{(l + \frac{1}{2})^2}{r^2} \tag{12.346}$$

and the corresponding turning point, which we call r_1, is given by $r_1 = (l + 1/2)/k$. The integral appearing on the right-hand side of (12.345) then reduces to

$$\begin{aligned}\int_{r_1}^{\infty} r\frac{\mathrm{d}}{\mathrm{d}r}[G(r)]^{1/2}\,\mathrm{d}r &= \left(l + \frac{1}{2}\right)\int_0^{\infty} \operatorname{sech} t\,\mathrm{d}t \\ &= \left(l + \frac{1}{2}\right)\frac{\pi}{2}\end{aligned} \tag{12.347}$$

where we have made the change of variables $r = r_1 \cosh t$ and used the fact that $\int \operatorname{sech} t\,\mathrm{d}t = 2\tan^{-1}[\exp(t)]$. The result (12.347) allows us to recast the WKB phase shift (12.345) in the form

$$\delta_l^{\text{WKB}} = \int_{r_1}^{\infty} r\frac{\mathrm{d}}{\mathrm{d}r}[G(r)]^{1/2}\,\mathrm{d}r - \int_{r_0}^{\infty} r\frac{\mathrm{d}}{\mathrm{d}r}[F(r)]^{1/2}\,\mathrm{d}r \tag{12.348}$$

It is worth remembering at this point that in applying the WKB method we have assumed that the short-wavelength condition $ka \gg 1$ is satisfied. Moreover, as we shall see below, quasi-classical scattering is dominated by large values of l. The magnitude L of the orbital angular momentum is then given by $\hbar[l(l+1)]^{1/2} \simeq \hbar(l + 1/2)$. In what follows, we shall write

$$L = \hbar\left(l + \frac{1}{2}\right) \tag{12.349}$$

so that the expression (12.344) for δ_l^{WKB} becomes

$$\delta_l^{\text{WKB}} = \frac{L\pi}{2\hbar} - kr_0 + \hbar^{-1}\int_{r_0}^{\infty}\left[\left\{2m[E - V(r)] - \frac{L^2}{r^2}\right\}^{1/2} - \hbar k\right]\mathrm{d}r \tag{12.350}$$

If we treat l as a continuous variable and differentiate the above equation with respect to it, we obtain with the help of (12.349)

$$\frac{\partial \delta_l^{\mathrm{WKB}}}{\partial l} = \hbar \frac{\partial \delta_l^{\mathrm{WKB}}}{\partial L} = \frac{\Theta}{2} \tag{12.351}$$

where

$$\begin{aligned}\Theta &= \pi + 2\int_{r_0}^{\infty} \frac{\partial}{\partial L}\left\{2m[E - V(r)] - \frac{L^2}{r^2}\right\}^{1/2} \mathrm{d}r \\ &= \pi - 2\int_{r_0}^{\infty} \frac{L}{mr^2}\left\{\frac{2}{m}[E - V(r)] - \frac{L^2}{m^2r^2}\right\}^{-1/2} \mathrm{d}r\end{aligned} \tag{12.352}$$

is the *classical deflection function* (see equations (A1.11) and (A1.12) of Appendix 1). Equation (12.351) is known as the *classical equivalence relation.*

Let us now return to the expression (12.348) of the WKB phase shift. For large l (and fixed k), the value of r_0 becomes large and the potential $U(r)$ remains small over the integration range (r_0, ∞). Expanding the quantity $[F(r)]^{1/2}$ as

$$[F(r)]^{1/2} = [G(r)]^{1/2}\left[1 - \frac{U(r)}{2G(r)} + \dots\right] \tag{12.353}$$

and using the fact that $r_0 \simeq r_1 = (l + 1/2)/k$, we may write the WKB phase shift (12.348) for large l as

$$\delta_l^{\mathrm{WKB}} \simeq -\frac{1}{2}\int_{r_1}^{\infty} \frac{U(r)}{[G(r)]^{1/2}} \mathrm{d}r \tag{12.354}$$

or

$$\delta_l^{\mathrm{WKB}} \simeq -\frac{1}{2k}\int_{b}^{\infty} \frac{rU(r)}{(r^2 - b^2)^{1/2}} \mathrm{d}r \tag{12.355}$$

where we have introduced the impact parameter b such that $l + 1/2 \simeq kb$ (see (12.314)). Changing the integration variable from r to $z = (r^2 - b^2)^{1/2}$, we obtain (for large l and $ka \gg 1$)

$$\delta_l^{\mathrm{WKB}} \simeq \delta(k, b) = -\frac{1}{4k}\int_{-\infty}^{+\infty} U(b, z)\,\mathrm{d}z, \qquad l + \frac{1}{2} \simeq kb \tag{12.356}$$

which is the *eikonal phase shift* (12.317).

As an example, we show in Table 12.3 the value of the phase shifts δ_l obtained from the eikonal approximation (12.317), those corresponding to the WKB formula (12.344) and the exact values, obtained from a numerical integration of the radial Schrödinger equations (12.41). This comparison is made for attractive Buckingham polarisation potentials of the form

Table 12.3 Comparison of the eikonal, WKB and exact phase shifts δ_l (in radians), for attractive 'polarisation' potentials of the form $U(r) = U_0/(r^2 + d^2)^2$, with $d = 1$ (unit of length) and various values of l. The wave number is $k = 5$. The numbers in parentheses indicate powers of 10.

	$U_0 = -1$			$U_0 = -250$		
l	*Eikonal*	*WKB*	*Exact*	*Eikonal*	*WKB*	*Exact*
0	7.74(−2)	7.69(−2)	7.80(−2)	1.93(1)	1.14(1)	1.14(1)
1	6.90(−2)	6.89(−2)	6.90(−2)	1.73(1)	1.12(1)	1.12(1)
2	5.62(−2)	5.63(−2)	5.59(−2)	1.40(1)	1.09(1)	1.09(1)
3	4.32(−2)	4.33(−2)	4.29(−2)	1.08(1)	1.04(1)	1.04(1)
4	3.23(−2)	3.24(−2)	3.21(−2)	8.06	9.78	9.78
5	2.39(−2)	2.40(−2)	2.38(−2)	5.98	8.95	8.96
10	6.24(−3)	6.26(−3)	6.26(−3)	1.56	2.32	2.31
15	2.27(−3)	2.28(−3)	2.28(−3)	5.68(−1)	6.17(−1)	6.19(−1)
20	1.04(−3)	1.05(−3)	1.05(−3)	2.61(−1)	2.69(−1)	2.69(−1)
25	5.60(−4)	5.60(−4)	5.60(−4)	1.40(−1)	1.42(−1)	1.42(−1)

$$U(r) = \frac{U_0}{(r^2 + d^2)^2} \tag{12.357}$$

with $d = 1$ (unit of length), $k = 5$ and two coupling strengths $U_0 = -1$ and $U_0 = -250$. We note that in the *weak coupling* situation ($U_0 = -1$) both the eikonal and the WKB phase shifts agree very well with the exact results for all values of l. For the *strong coupling* case ($U_0 = -250$) the WKB phase shifts are accurate for all values of l, while the eikonal phase shifts are only reliable for large values of l.

Having calculated the WKB phase shifts from (12.344), we may then insert them in the partial wave series (12.60), namely

$$f(k, \theta) = \frac{1}{2ik} \sum_{l=0}^{\infty} (2l + 1)\{\exp[2i\delta_l(k)] - 1\}P_l(\cos\theta) \tag{12.358}$$

It is clear from the above discussion that the resulting scattering amplitude should be accurate for smoothly varying potentials when the semi-classical condition (12.293) is satisfied. In fact, one may also elect to sum the partial wave series approximately by using the fact that we are in the semi-classical regime. Since $ka \gg 1$, large l values (of the order of ka) dominate the scattering. For *small scattering angles* such that $l\theta = \mathcal{O}(1)$ we may use the relation (12.313) and turn the sum over l into an integral on the impact parameter b (see (12.315)). Because of the result (12.356) we then obtain the eikonal scattering amplitude (12.310). We emphasize that this procedure is justified for scattering angles $\theta \leqslant l^{-1} \simeq (ka)^{-1}$ which correspond to diffraction scattering.

Let us now consider the angular region $\theta > (ka)^{-1}$. We first observe that the closure relation for the Legendre polynomials (see (2.172)) yields

$$\frac{1}{2} \sum_{l=0}^{\infty} (2l + 1)\, P_l(\cos\theta)P_l(1) = \delta(\cos\theta - 1) \tag{12.359}$$

so that the term in the partial wave series (12.358) which does not contain the phase shift δ_l only contributes in the forward direction. Since we are excluding the angular range $0 \leqslant \theta \leqslant (ka)^{-1}$, we may therefore omit this term and write

$$f(k, \theta) = \frac{1}{2ik} \sum_{l=0}^{\infty} (2l + 1) \exp[2i\delta_l(k)] P_l(\cos \theta), \qquad \theta \neq 0 \tag{12.360}$$

We shall study this expression in the *quasi-classical* limit, such that the phase shifts δ_l are large and the terms with large values of l dominate the sum. We may then use the asymptotic (large l) form of $P_l(\cos \theta)$, namely

$$P_l(\cos \theta) \approx \left(\frac{2}{\pi l \sin \theta}\right)^{1/2} \sin\left[\left(l + \frac{1}{2}\right)\theta + \frac{\pi}{4}\right] \tag{12.361}$$

to write

$$f(k, \theta) = -\frac{1}{k} \int_0^{\infty} \left(\frac{l}{2\pi \sin \theta}\right)^{1/2} \{\exp[i\phi_+(l, k)] - \exp[i\phi_-(l, k)]\}\, dl \tag{12.362}$$

where we have treated l as a continuous variable and set

$$\phi_{\pm}(l, k) = 2\delta_l(k) \pm \left(l + \frac{1}{2}\right)\theta \pm \frac{\pi}{4} \tag{12.363}$$

The exponential terms appearing in (12.362) are rapidly oscillating since the phases $\phi_{\pm}$ are large. Hence, the scattering amplitude (12.362) is effectively determined by the range of values of l near which the quantities $\phi_{\pm}$ have a stationary point. These values of l lie in the neighbourhood of the roots $l = l_i$ of the equation

$$\frac{\partial}{\partial l} \phi_{\pm}(l, k) = 0 \tag{12.364}$$

or

$$2\frac{\partial}{\partial l} \delta_l(k) \pm \theta = 0 \tag{12.365}$$

Let us assume that there is only *one* root of equation (12.364) at a large value $l = l_0$. We then have in the neighbourhood of this value

$$\phi_{\pm}(l, k) = \phi_{\pm}(l_0, k) + \beta_{\pm}(l_0, k)(l - l_0)^2 \tag{12.366}$$

with

$$\beta_{\pm}(l_0, k) = \frac{1}{2}\left[\frac{\partial^2 \phi_{\pm}(l, k)}{\partial l^2}\right]_{l=l_0} \tag{12.367}$$

and the scattering amplitude (12.362) becomes

$$f(k,\theta) = -\frac{1}{k}\left(\frac{l_0}{2\pi \sin\theta}\right)^{1/2} \exp[\mathrm{i}\phi_\pm(l_0,k)]\int_{l_0-\Delta l}^{l_0+\Delta l} \exp[\mathrm{i}\beta_\pm(l_0,k)(l-l_0)^2]\,\mathrm{d}l \quad \textbf{(12.368)}$$

where we have indicated explicitly that the main contribution to the integral arises from the interval $(l_0 - \Delta l, l_0 + \Delta l)$ centred about the value $l = l_0$. We may now change variables from l to $\xi = l - l_0$ and extend without appreciable error the integral from $-\infty$ to $+\infty$. This yields

$$f(k,\theta) = -\frac{1}{k}\left(\frac{l_0}{2\pi \sin\theta}\right)^{1/2}\left(\frac{\pi}{\mathrm{i}\beta_\pm(l_0,k)}\right)^{1/2} \exp[\mathrm{i}\phi_\pm(l_0,k)] \quad \textbf{(12.369)}$$

Before we analyse this expression in more detail, let us return to the relation (12.365). From the results (12.351) and (12.352), we see that if we use the WKB phase shifts δ_l^{WKB}, we obtain from (12.350) and (12.365) the *classical* result

$$\int_{r_0}^{\infty} \frac{\partial}{\partial L}\left\{2m[E - V(r)] - \frac{L^2}{r^2}\right\}^{1/2} \mathrm{d}r + \frac{\pi}{2} \pm \frac{\theta}{2} = 0 \quad \textbf{(12.370)}$$

or

$$\theta = \mp\Theta(l_0, k) \quad \textbf{(12.371)}$$

where the upper (−) sign corresponds to ϕ_+ and the lower (+) sign to ϕ_-, and the choice of sign is determined by the fact that the scattering angle θ is always positive. Calling $L_0 = (l_0 + 1/2)\hbar$ the root of equation (12.370), we have from (12.351), (12.363) and (12.367)

$$\beta_\pm(l_0,k) = \frac{1}{2}\left[\frac{\partial}{\partial l}\Theta(l,k)\right]_{l=l_0} = \left(\frac{\hbar}{2}\right)\left[\frac{\partial}{\partial L}\Theta(L,k)\right]_{L=L_0} \quad \textbf{(12.372)}$$

Let us now return to the scattering amplitude (12.369). In terms of L_0, we can write the corresponding differential cross-section as

$$\frac{\mathrm{d}\sigma}{\mathrm{d}\Omega} = |f(k,\theta)|^2 = \frac{L_0}{p^2 \sin\theta}\left|\frac{\mathrm{d}L_0}{\mathrm{d}\theta}\right| \quad \textbf{(12.373a)}$$

or

$$\frac{\mathrm{d}\sigma}{\mathrm{d}\Omega} = \frac{b_0}{\sin\theta}\left|\frac{\mathrm{d}b_0}{\mathrm{d}\theta}\right| \quad \textbf{(12.373b)}$$

where $p = \hbar k$ is the magnitude of the particle momentum and we have used the fact that $L_0 = b_0 p$ with $kb_0 \simeq l_0$. The differential cross-section given by equations (12.373) is simply the *classical* one (see (A1.19) and (A1.20)). We remark from the above derivation that the conditions for classical scattering through a given angle θ are that the value l_0 should be large and that δ_l should also be large for this value. We also assumed that there was only *one* stationary point at $l = l_0$, which by using (12.370) means that $\Theta(l, k)$ should be a monotonic function of l such that $0 \leqslant \Theta(l, k) \leqslant \pi$. If this is not the case, the angular distributions can differ considerably from the classical one, even though the semi-classical conditions are satisfied. An example of these more complicated situations, called 'rainbow scattering', will be discussed in Chapter 14.

In the derivation leading to the classical result (12.373) we assumed that the scattering angle θ should be such that $\theta > (ka)^{-1}$. It is easy to see why this condition must be satisfied. Indeed, a classical trajectory is only well defined if the uncertainty $\Delta\theta$ in the scattering angle θ is small compared to θ itself. Now, the uncertainty Δp in the transverse momentum of the scattered particle, as given by the Heisenberg uncertainty principle, is of the order of $\Delta p \simeq \hbar/a$. The corresponding uncertainty $\Delta\theta$ in the scattering angle is therefore given by

$$\Delta\theta \simeq \frac{\Delta p}{p} \simeq (ka)^{-1} \tag{12.374}$$

which implies that the scattering through small angles $\theta < (ka)^{-1}$ is never classical.

We may also express this idea in a more quantitative way by obtaining first an estimate of θ and then writing the condition $\Delta\theta \ll \theta$ for classical scattering. Since in the short-wavelength limit ($ka \gg 1$) which we are considering the bulk of the scattering occurs near the forward direction, we shall assume that θ is small (but larger than $(ka)^{-1}$). Now, the (small) deflection angle θ of a classical particle is roughly given by

$$\theta \simeq \frac{\delta p}{p} \tag{12.375}$$

where δp, the magnitude of the momentum transfer in the transverse direction, is given approximately by the force $|V_0|/a$ acting on the particle, multiplied by the 'collision time' a/v (where v is the magnitude of the particle velocity). Hence

$$\theta \simeq \frac{|V_0|}{pv} = \frac{|V_0|}{2E} \tag{12.376}$$

Using the relations (12.374) and (12.376), the condition $\Delta\theta \ll \theta$ for *classical scattering* yields

$$\frac{|V_0|a}{\hbar v} = \frac{|U_0|a}{2k} \gg 1 \tag{12.377}$$

We note that this condition corresponds to a *strong coupling* situation in which the coupling parameter $|V_0|a/(\hbar v)$ is large. This is in contrast to the condition of convergence (12.268) of the Born series, which corresponds to weak coupling.

Variational methods

In Chapter 2, the Rayleigh–Ritz variational method was introduced to obtain approximate energies of bound states. Variational methods of a different type can be developed, from which approximate phase shifts or scattering amplitudes can be found. As an illustration, we shall discuss the Hulthén–Kohn and Schwinger variational methods which are widely used in atomic collision theory. The former is based on the Schrödinger equation and hence requires trial functions satisfying the boundary conditions of the problem. The latter relies on the Lippman–Schwinger integral equation, so that the boundary conditions, being taken into account through the Green's function, need not be incorporated in the trial functions. A more detailed account of the subject may be found in the books by Joachain (1983) and Burke and Joachain (1995).

The Hulthén–Kohn variational method

We begin by considering the radial Schrödinger equation (12.41) describing scattering by a central potential, which we write in the form

$$L_l u_l(r) = 0 \tag{12.378}$$

where L_l is the operator

$$L_l \equiv \frac{\mathrm{d}^2}{\mathrm{d}r^2} - \frac{l(l+1)}{r^2} - U(r) + k^2 \tag{12.379}$$

The boundary condition at the origin is

$$u_l(0) = 0 \tag{12.380}$$

and we shall choose a normalisation such that

$$u_l(r) \underset{r\to\infty}{\to} \sin(kr - l\pi/2 + \gamma) + \tan(\delta_l - \gamma)\cos(kr - l\pi/2 + \gamma) \tag{12.381}$$

where γ is a fixed constant chosen so that $0 \leqslant \gamma < \pi$. For notational simplicity the k-dependence of the radial function u_l and phase shift δ_l has been omitted. We also introduce a *trial function* $u_l^t(r)$ satisfying the same boundary conditions as $u_l(r)$, namely

$$u_l^t(0) = 0 \tag{12.382}$$

and

$$u_l^t(r) \underset{r\to\infty}{\to} \sin(kr - l\pi/2 + \gamma) + \tan(\delta_l^t - \gamma)\cos(kr - l\pi/2 + \gamma) \tag{12.383}$$

where the quantity δ_l^t is a trial phase shift.

Let us now examine the functional I_l defined by

$$I_l[u_l^t] = \int_0^\infty u_l^t(r) L_l u_l^t(r)\,\mathrm{d}r \tag{12.384}$$

It is clear from (12.378) that

$$I_l[u_l] = 0 \tag{12.385}$$

where u_l is the exact radial function. Applying Green's theorem to the functions u_l and u_l^t gives

$$\int_0^\infty (u_l L_l u_l^t - u_l^t L_l u_l)\, dr = \left[u_l \frac{du_l^t}{dr} - u_l^t \frac{du_l}{dr} \right]_0^\infty \quad \textbf{(12.386)}$$

Hence, with the help of equations (12.378)–(12.385), we find that

$$I_l[u_l^t] - I_l[\Delta u_l] = k[\tan(\delta_l - \gamma) - \tan(\delta_l^t - \gamma)] \quad \textbf{(12.387)}$$

where we have defined

$$\Delta u_l(r) = u_l^t(r) - u_l(r) \quad \textbf{(12.388)}$$

The relation (12.387) is known as the *Kato identity*. If the trial function u_l^t is sufficiently close to the exact function u_l, then the functional $I_l[\Delta u_l]$, which is of second order of smallness, can be neglected. We may then write the variational principle

$$\delta[I_l + k \tan(\delta_l^t - \gamma)] = 0 \quad \textbf{(12.389)}$$

where

$$\delta I_l = I_l[u_l^t] - I_l[u_l] = I_l[u_l^t] \quad \textbf{(12.390)}$$

is the change in I_l under the infinitesimally small variation

$$\delta u_l(r) = u_l^t(r) - u_l(r) \quad \textbf{(12.391)}$$

and

$$\delta[\tan(\delta_l - \gamma)] = \tan(\delta_l^t - \gamma) - \tan(\delta_l - \gamma) \quad \textbf{(12.392)}$$

In practice, we start from a trial function $u_l^t(c_1, c_2, \ldots, c_N, \lambda_l^t, r)$ which depends on N parameters $c_1, c_2, \ldots, c_N$ and on the phase shift through the quantity

$$\lambda_l^t = \tan(\delta_l^t - \gamma) \quad \textbf{(12.393)}$$

Taking the variation in (12.389) with respect to these $N + 1$ parameters yields the equations

$$\frac{\partial I_l}{\partial c_i} = 0, \qquad i = 1, 2, \ldots, N \quad \textbf{(12.394a)}$$

and

$$\frac{\partial I_l}{\partial \lambda_l^t} = -k \quad \textbf{(12.394b)}$$

If the trial function u_l^t depends *linearly* on the parameters $c_1, c_2, \ldots, c_N$ and λ_l^t, the equations (12.394) will form a system of linear equations which can be solved to yield these parameters. We can then use the variational principle (12.389) to obtain a better estimate of $\lambda_l = \tan(\delta_l - \gamma)$ that is correct up to terms of second order. This estimate is given by

$$[\lambda_l] = \lambda_l^t + k^{-1} I_l[u_l^t] \qquad \textbf{(12.395)}$$

where the symbol $[\lambda_l]$ means that this quantity is a variational estimate of λ_l. The corresponding phase shift correct up to terms of second order is then obtained from the relation $[\lambda_l] = \tan(\delta_l - \gamma)$.

As a particular case of the above equations, let us choose $\gamma = 0$. The variational principle (12.389) then becomes

$$\delta[I_l + k \tan \delta_l^t] = 0 \qquad \textbf{(12.396)}$$

and from (12.383) we see that the trial function $u_l^t(r)$ satisfies the boundary condition

$$u_l^t(r) \underset{r\to\infty}{\to} \sin(kr - l\pi/2) + \tan \delta_l^t \cos(kr - l\pi/2) \qquad \textbf{(12.397)}$$

This choice was made by W. Kohn. It is convenient at this point to use the K-matrix element $K_l(k) = \tan \delta_l(k)$ and to define also the trial K-matrix element $K_l^t(k) = \tan \delta_l^t(k)$. If the trial function u_l^t depends on the $N + 1$ parameters c_1, c_2, . . . , c_N and K_l^t, then these parameters can be determined from the equations (12.394), with $\lambda_l^t = K_l^t$. Equation (12.395) now reads

$$[K_l] = K_l^t + k^{-1} I_l[u_l^t] \qquad \textbf{(12.398)}$$

which gives a variational estimate for K_l.

The choice of normalisation (12.397) made by Kohn is obviously not the only one possible. For example, if one takes $\gamma = \pi/2$, a variational principle for $\mu_l = \cot \delta_l = K_l^{-1}$ is obtained. This approach is known as the inverse Kohn method.

It is also possible to use the Kohn variational method to obtain a variational estimate of the S-matrix elements $S_l(k)$. To this end, we normalise the radial function u_l according to (12.76) and we introduce a trial function u_l^t satisfying the asymptotic boundary condition

$$u_l^t(r) \underset{r\to\infty}{\to} \exp[-\mathrm{i}(kr - l\pi/2)] - S_l^t(k) \exp[\mathrm{i}(kr - l\pi/2)] \qquad \textbf{(12.399)}$$

where $S_l^t(k)$ is the trial S-matrix element. As before, we consider the variation

$$\delta I_l = I_l[u_l^t] - I_l[u_l] = I_l[u_l^t] \qquad \textbf{(12.400)}$$

Using Green's theorem, remembering that both $u_l(r)$ and $u_l^t(r)$ vanish at the origin and satisfy the asymptotic boundary conditions (12.76) and (12.399), respectively, and neglecting terms of second order in $\Delta u_l = u_l^t - u_l$, we obtain the S-matrix form of the Kohn variational principle:

$$\delta[I_l + 2\mathrm{i}kS_l] = 0 \qquad \textbf{(12.401)}$$

where $\delta S_l = S_l^t - S_l$. If the trial function u_l^t depends on the $N + 1$ parameters c_1, c_2, . . . , c_N and S_l^t, then by taking the variation in (12.401) with respect to these parameters, one obtains the equations

$$\frac{\partial I_l}{\partial c_i} = 0, \qquad i = 1, 2, \ldots, N \tag{12.402a}$$

and

$$\frac{\partial I_l}{\partial S_l^t} = -2ik \tag{12.402b}$$

which can be solved to find the parameters. As before, one can then use the variational principle (12.401) to obtain an estimate of S_l which is correct up to terms of second order. That is,

$$[S_l] = S_l^t + (2ik)^{-1} I_l[u_l^t] \tag{12.403}$$

A similar method can be used to obtain a variational estimate of the T-matrix elements $T_l(k)$.

An alternative procedure was proposed by L. Hulthén. He noted that the exact solution satisfies the equation $I_l = 0$ and suggested that the approximate solution should also satisfy this equation. Instead of the equations (12.394), he therefore obtained the equations

$$\frac{\partial I_l}{\partial c_i} = 0, \qquad i = 1, 2, \ldots, N \tag{12.404a}$$

and

$$I_l = 0 \tag{12.404b}$$

This approach also guarantees that $\delta\lambda_l = 0$ and yields a value of λ_l which is correct up to terms of second order. However, unlike equations (12.394) and (12.402) of the Kohn variational method, the equations (12.404) are non-linear since I_l is a quadratic function of the $N + 1$ variational parameters $c_1, c_2, \ldots, c_N$ and λ_l. Hence two values of λ_l are obtained, while only one of them can be physical.

An example of the power of the variational method is given by the pioneering work of H.S.W. Massey and B.L. Moiseiwitsch performed in 1951. Using the Kohn variational principle (12.396), they calculated the s-wave phase shift δ_0 for the scattering of electrons by the static potential of atomic hydrogen in the ground state, the corresponding reduced potential being given (in a.u.) by (12.329). They used a simple trial function having the form

$$u_0^t(r) = \sin(kr) + (K_0^t + c_1\,e^{-r})(1 - e^{-r})\cos(kr) \tag{12.405}$$

which contains only two variational parameters: c_1 and K_0^t. Their variational values for the phase shift $\delta_0(k)$ are given in Table 12.4 for several values of k. They are seen to be in very good agreement with the exact phase shifts, obtained by numerical integration of the radial Schrödinger equation (12.41) for the reduced potential (12.329).

It is important to stress, however, that variational methods can lead to erroneous results if poor trial functions are used. Indeed, if the error $\Delta u_l = u_l^t - u_l$

Table 12.4 Comparison of the s-wave phase shift δ_0 (in radians) for electron scattering by the static potential of the hydrogen atom in the ground state using the Kohn variational method, with the exact results, for various values of the wave number k (in a.u.). The Kohn variational calculations are those performed by Massey and Moiseiwitsch, using the trial function (12.405).

k (a.u.)	*Variational*	*Exact*
0.1	0.721	0.722 22
0.2	0.972	0.972 52
0.3	1.045	1.045 55
0.4	1.057	1.057 50
0.5	1.044	1.044 66

cannot be made small, then the error in the variational estimate (which is of second order in Δu_l) can be large. In order to study this question, let us first examine how systematic procedures can be devised to perform Hulthén–Kohn variational calculations. Consider for example the Kohn variational principle (12.396) for $K_l = \tan \delta_l$. We may choose a trial function of the form

$$u_l^t(r) = \phi_l(r) + \sum_{i=1}^{N} c_i \chi_i(r) \tag{12.406}$$

where the function $\phi_l(r)$ yields the asymptotic part of $u_l^t(r)$, namely

$$\phi_l(r) \underset{r\to\infty}{\rightarrow} \sin(kr - l\pi/2) + K_l^t \cos(kr - l\pi/2) \tag{12.407}$$

The (linear) variational parameters are c_i ($i = 1, 2, \dots, N$) and $K_l^t = \tan \delta_l^t$, and the functions $\chi_i(r)$ are a set of basis functions, confined to the region where the potential acts. For example,

$$\chi_i(r) = \mathrm{e}^{-\kappa r} r^i, \qquad i = 1, 2, \dots, N \tag{12.408}$$

where the additional non-linear parameter κ is a range parameter which can be varied.

With the choice of basis functions (12.408) and if $\phi_l(r)$ satisfies the equation (12.407), the trial function (12.406) clearly obeys the asymptotic boundary condition (12.397). However, we must also have $u_l^t(0) = 0$. Since the basis functions (12.408) are such that $\chi_i(0) = 0$, we must also impose the condition $\phi_l(0) = 0$. For instance, in the case of s-wave scattering, we may choose

$$\phi_0(r) = \sin(kr) + K_0^t(1 - \mathrm{e}^{-\kappa r}) \cos(kr) \tag{12.409}$$

Using (12.406), (12.408) and (12.409), we see that the trial function u_0^t is then given explicitly by

$$u_0^t(r) = \sin(kr) + K_0^t(1 - \mathrm{e}^{-\kappa r}) \cos(kr) + \mathrm{e}^{-\kappa r} \sum_{i=1}^{N} c_i r^i \tag{12.410}$$

which is a generalisation of the trial function (12.405) used by Massey and Moiseiwitsch.

Let us now return to the Kohn variational principle (12.396), and use the trial function (12.406). For a given value of the non-linear variational parameter κ, the problem reduces to a simple one of matrix inversion. Indeed, defining

$$\begin{aligned} A_{ij} = A_{ji} &= k^{-1}\int_0^\infty \chi_i(r)L_l\chi_j(r)\,\mathrm{d}r \\ B_i &= k^{-1}\int_0^\infty \chi_i(r)L_l\phi_l(r)\,\mathrm{d}r \\ C &= k^{-1}\int_0^\infty \phi_l(r)L_l\phi_l(r)\,\mathrm{d}r \end{aligned} \tag{12.411}$$

and performing the variation with respect to the linear variational parameters c_i, we see that equation (12.394a) reduces to

$$\sum_{j=1}^{N} A_{ij}c_j = -B_i, \qquad i = 1, 2, \ldots, n \tag{12.412}$$

Hence, by inversion of the matrix (A_{ij}), one determines the variational parameters c_i. Varying also with respect to K_l^t, and substituting into (12.398), one obtains for K_l the variational estimate

$$[K_l] = K_l^t + \sum_{i=1}^{N}\sum_{j=1}^{N} c_i c_j A_{ij} + 2\sum_{i=1}^{N} c_i B_i + C \tag{12.413}$$

The calculation can then be repeated for other values of the non-linear parameter κ.

C. Schwartz has pointed out in 1961 that the finite matrix (A_{ij}) which represents the operator L_l in the space spanned by the N basis functions χ_i can occasionally have an eigenvalue very close to zero. In this case, *anomalous singularities* arise in $[K_l]$ – and more generally in $[\lambda_l]$ – which can invalidate the variational estimate. This is due to the fact that the variational principles for scattering discussed in this section, by contrast with those used in the study of bound states, are in general not extremum principles, but only *stationary* principles. It is therefore necessary to develop prescriptions for avoiding the singularities in order to obtain reliable results [12].

In particular, one can show that the S-matrix form (12.401) of the Kohn variational method does not contain the anomalous singularities associated with the Kohn and inverse Kohn methods. However, an additional constraint must be imposed to obtain real phase shifts.

Finally, we remark that the Hulthén–Kohn variational method can be generalised to the calculation of the full scattering amplitude f. The variational principle (12.389) now reads

---

[12] For a detailed discussion, see Burke and Joachain (1995) or Adhikari (1998).

$$\delta(I + 4\pi f) = 0 \tag{12.414}$$

where the functional I is given by

$$I\left[\psi_{\mathbf{k}_f}^{(-)}, \psi_{\mathbf{k}_i}^{(+)}\right] = \int \psi_{\mathbf{k}_f}^{(-)*}(\mathbf{r})[\nabla^2 + k^2 - U(\mathbf{r})]\psi_{\mathbf{k}_i}^{(+)}(\mathbf{r})\, \mathrm{d}\mathbf{r} \tag{12.415}$$

with

$$\psi_{\mathbf{k}_i}^{(+)}(\mathbf{r}) \underset{r\to\infty}{\to} \exp(\mathrm{i}\mathbf{k}_i\cdot\mathbf{r}) + f(\mathbf{k}_i\cdot\hat{\mathbf{r}})\frac{\exp(\mathrm{i}kr)}{r} \tag{12.416a}$$

and

$$\psi_{\mathbf{k}_f}^{(-)}(\mathbf{r}) \underset{r\to\infty}{\to} \exp(\mathrm{i}\mathbf{k}_f\cdot\mathbf{r}) + f^*(-\mathbf{k}_f\cdot\hat{\mathbf{r}})\frac{\exp(-\mathrm{i}kr)}{r} \tag{12.416b}$$

Since $I = 0$ for exact wave functions, we see from (12.414) that a variational estimate of the scattering amplitude, correct up to terms of second order, is given by

$$[f(\mathbf{k}_i, \mathbf{k}_f)] = f^t(\mathbf{k}_i, \mathbf{k}_f) + \frac{1}{4\pi} I[\psi_{\mathbf{k}_f}^t, \psi_{\mathbf{k}_i}^t] \tag{12.417}$$

where $\psi_{\mathbf{k}_i}^t$ and $\psi_{\mathbf{k}_f}^t$ are trial wave functions having the asymptotic behaviour given respectively by (12.416a) and (12.416b), and f^t is a trial scattering amplitude. We note that for the simple choice

$$\psi_{\mathbf{k}_i}^t(\mathbf{r}) = \exp(\mathrm{i}\mathbf{k}_i\cdot\mathbf{r}) \tag{12.418a}$$

and

$$\psi_{\mathbf{k}_f}^t(\mathbf{r}) = \exp(\mathrm{i}\mathbf{k}_f\cdot\mathbf{r}) \tag{12.418b}$$

we have $f^t = 0$ and

$$I = -\int \exp[\mathrm{i}(\mathbf{k}_i - \mathbf{k}_f)\cdot\mathbf{r}]U(\mathbf{r})\, \mathrm{d}\mathbf{r} \tag{12.419}$$

so that the scattering amplitude $[f]$ given by (12.417) reduces in this case to the first Born approximation (12.269).

The Schwinger variational method

We shall now derive a variational principle for the scattering amplitude, due to J. Schwinger, which is based on the Lippmann–Schwinger equation (12.159). We first recall two important integral representations of the scattering amplitude, given by (12.163) and (12.184), respectively. Moreover, from (12.184) and the Lippmann–Schwinger equation (12.159), we also find that

$$f = -\frac{1}{4\pi}\langle\psi_{\mathbf{k}_f}^{(-)}|U - UG_0^{(+)}U|\psi_{\mathbf{k}_i}^{(+)}\rangle \tag{12.420}$$

Using the three integral representations (12.163), (12.184) and (12.420), we can write down a stationary expression for f, namely

$$[f] = -\frac{1}{4\pi}[\langle\phi_{\mathbf{k}_f}|U|\psi^{(+)}_{\mathbf{k}_i}\rangle + \langle\psi^{(-)}_{\mathbf{k}_f}|U|\phi_{\mathbf{k}_i}\rangle - \langle\psi^{(-)}_{\mathbf{k}_f}|UG_0^{(+)}U|\psi^{(+)}_{\mathbf{k}_i}\rangle] \quad \textbf{(12.421)}$$

We first note that this expression yields the exact scattering amplitude when the correct wave functions $\psi^{(+)}_{\mathbf{k}_i}$ and $\psi^{(-)}_{\mathbf{k}_f}$ are used. Moreover, we have

$$\delta[f] = 0 \quad \textbf{(12.422)}$$

for arbitrary variations of the bra $\langle\psi^{(-)}_{\mathbf{k}_f}|$ and the ket $|\psi^{(+)}_{\mathbf{k}_i}\rangle$ about their correct values. Indeed, varying first the bra $\langle\psi^{(-)}_{\mathbf{k}_f}|$ by an amount $\langle\delta\psi^{(-)}_{\mathbf{k}_f}|$, we have

$$\delta[f] = -\frac{1}{4\pi}[\langle\delta\psi^{(-)}_{\mathbf{k}_f}|U|\phi_{\mathbf{k}_i}\rangle - \langle\delta\psi^{(-)}_{\mathbf{k}_f}|U - UG_0^{(+)}U|\psi^{(+)}_{\mathbf{k}_i}\rangle] \quad \textbf{(12.423)}$$

and we see that equation (12.422) is verified if $\psi^{(+)}_{\mathbf{k}_i}$ satisfies the Lippmann–Schwinger equation (12.159). Similarly, by varying the ket $|\psi^{(+)}_{\mathbf{k}_i}\rangle$ about its correct value, we find that equation (12.422) is verified if $\psi^{(-)}_{\mathbf{k}_f}$ satisfies the Lippmann–Schwinger equation (12.177). Conversely, if the variational equation (12.422) is satisfied, then $\psi^{(+)}_{\mathbf{k}_i}$ and $\psi^{(-)}_{\mathbf{k}_f}$ must satisfy the Lippmann–Schwinger equations (12.159) and (12.177), respectively.

The expression (12.421) is called the *bilinear form* of Schwinger's variational principle for the scattering amplitude, since it is bilinear in $|\psi^{(+)}_{\mathbf{k}_i}\rangle$ and $\langle\psi^{(-)}_{\mathbf{k}_f}|$. We note that with the choice of trial functions given by (12.418), the stationary expression (12.421) reduces to the second Born approximation (12.263). If the Born series converges, the Schwinger bilinear stationary expression (12.421) is therefore intrinsically superior to the Hulthén–Kohn variational principle (12.414). We also note that the Schwinger stationary expression (12.421) automatically incorporates the boundary conditions and uses trial functions which must only be specified in the region where the interaction potential U is non-zero.

A modified version of the Schwinger variational principle, called the *fractional form* of Schwinger's variational principle, may be obtained by starting from the bilinear form (12.421) and choosing trial functions such that

$$|\psi^{(+)}_{\mathbf{k}_i}\rangle = c_1|\tilde{\psi}^{(+)}_{\mathbf{k}_i}\rangle \quad \textbf{(12.424a)}$$

and

$$\langle\psi^{(-)}_{\mathbf{k}_f}| = c_2\langle\tilde{\psi}^{(-)}_{\mathbf{k}_f}| \quad \textbf{(12.424b)}$$

where c_1 and c_2 are variational parameters. Inserting the expressions (12.424) into the bilinear stationary quantity (12.421) and varying with respect to c_1 and c_2, we obtain the *fractional form* of Schwinger's variational principle for the scattering amplitude:

$$[f] = -\frac{1}{4\pi}\frac{\langle\tilde{\psi}^{(-)}_{\mathbf{k}_f}|U|\phi_{\mathbf{k}_i}\rangle\langle\phi_{\mathbf{k}_f}|U|\tilde{\psi}^{(+)}_{\mathbf{k}_i}\rangle}{\langle\tilde{\psi}^{(-)}_{\mathbf{k}_f}|U - UG_0^{(+)}U|\tilde{\psi}^{(+)}_{\mathbf{k}_i}\rangle} \quad \textbf{(12.425)}$$

As in the case of the bilinear form (12.421), the fractional form (12.425) of the Schwinger variational principle automatically incorporates the boundary conditions and uses the trial functions only in the region where the potential is non-vanishing. Furthermore, the fractional form (12.425) is clearly independent of the normalisation used for the trial functions.

As an example, let us use the simple trial functions given by (12.418). Upon substitution on the right-hand side of (12.425), we find that

$$[f] = \frac{f^{\mathrm{B1}}}{1 - \bar{f}^{\mathrm{B2}}/f^{\mathrm{B1}}} \tag{12.426}$$

Setting $x = \bar{f}^{\mathrm{B2}}/f^{\mathrm{B1}}$, we see that if $|x| < 1$ we may expand the function $(1 - x)^{-1}$ in powers of x, and write

$$[f] = f^{\mathrm{B1}} + \bar{f}^{\mathrm{B2}} + (\bar{f}^{\mathrm{B2}})^2/f^{\mathrm{B1}} + \dots \tag{12.427}$$

which agrees with the Born series through second order. In fact, the expression (12.426), based on the stationary expression (12.425), is often more accurate than the second Born approximation. However, this is not always the case because the variational principles for scattering are in general only stationary principles, not extremum principles.

We may also start from the stationary expressions (12.421) or (12.425) for the scattering amplitude to obtain stationary expressions for the phase shifts. Assuming that the interaction potential is central, we expand the functions $\psi^{(+)}_{\mathbf{k}_i}(\mathbf{r})$ and $\psi^{(-)}_{\mathbf{k}_f}(\mathbf{r})$ in Legendre polynomials and consider the coefficients of the expansion as variational parameters. We obtain in this way from the fractional form (12.425) of the Schwinger variational principle the stationary expression (Problem 12.23)

$$\left[\frac{\tan \delta_l}{k}\right] = \frac{\left[\int_0^\infty j_l(kr)U(r)R_l(r)r^2\, \mathrm{d}r\right]^2}{\int_0^\infty \mathrm{d}r \int_0^\infty \mathrm{d}r' R_l(r)U(r)G_l(k, r, r')U(r')R_l(r')r^2 r'^2 - \int_0^\infty R_l^2(r)U(r)r^2\, \mathrm{d}r} \tag{12.428}$$

where the radial Green's function $G_l(k, r, r')$ is given by (12.188). The stationary expression (12.428) can also be obtained (Problem 12.24) by using the integral equation (12.187) for the radial wave function $R_l(r)$, and the integral representation (12.87) for $\tan \delta_l$.

The Schwinger variational principle (12.428) for the phase shifts has the same advantages as those of the fractional form (12.425) for the scattering amplitude. In particular, the trial functions $R_l^t(r)$ used in (12.428) must only be specified in the region where the potential is non-zero, and hence do not have to satisfy the asymptotic boundary condition, in contrast to the trial functions employed in the Hulthén–Kohn method. However, double integrals involving the radial Green's function $G_l(k, r, r')$ occur in the Schwinger variational principle (12.428), while only single integrals appear in the Hulthén–Kohn approach.

We note that by choosing the simple trial function

$$R_l^t(r) = j_l(kr) \tag{12.429}$$

and substituting in the stationary expression (12.428), we obtain

$$\left[\frac{\tan\delta_l}{k}\right] = \frac{\left[\int_0^\infty j_l^2(kr)U(r)r^2\,\mathrm{d}r\right]^2}{\int_0^\infty \mathrm{d}r \int_0^\infty \mathrm{d}r' j_l(kr)U(r)G_l(k,r,r')U(r')j_l(kr')r^2r'^2 - \int_0^\infty j_l^2(kr)U(r)r^2\,\mathrm{d}r} \tag{12.430}$$

which agrees with the Born series up to second order (see (12.107) and (12.264)) and is often more accurate than the second Born approximation.

The *R*-matrix method

In many collision problems there exists an *internal region* where the interaction is complicated so that the solution cannot be found exactly, and an *external region* where the interaction is relatively simple and where often analytic solutions exist. For example, this is the case in electron collisions with atoms and ions, where at short distances the effective interaction potential contains complicated electron exchange and correlation terms, while at large distances it contains only the Coulomb potential and (or) other terms falling off as inverse powers of the distance between the electron and the atomic nucleus. The *R-matrix method* takes advantage of the fact that the dynamics of the projectile–target system depends on the distance r between the colliding particle and the target by dividing the complete range $0 \leqslant r < \infty$ of the radial coordinate r into an internal region $0 \leqslant r \leqslant a$ and an external region $a \leqslant r < \infty$. The two regions are linked by joining the internal and external wave functions at $r = a$ [13]. The R-matrix method was originally introduced by E. Wigner in 1946 in the theory of nuclear reactions. It was first applied to atomic collisions in 1971 by P.G. Burke, A. Hibbert and W. Robb.

We shall illustrate here the principle of the R-matrix method in the simple context of potential scattering, following the treatment of Burke and Joachain (1995).

Variational methods for the *R*-matrix

We begin by discussing variational methods in which the logarithmic derivative of the wave function on the boundary $r = a$ is determined in terms of a trial function defined only in the internal region $0 \leqslant r \leqslant a$. We first observe that the operator L_l defined by (12.379) is not Hermitian in the internal region for functions satisfying arbitrary boundary conditions at $r = a$ because of the presence of a non-vanishing surface term. Indeed, let us consider two functions $u_l^{(1)}(r)$ and $u_l^{(2)}(r)$ which are continuous and have continuous first derivatives in the internal region $0 \leqslant r \leqslant a$, and are such that $u_l^{(1)}(0) = u_l^{(2)}(0) = 0$. We then have

$$\int_0^a [u_l^{(1)}L_l u_l^{(2)} - u_l^{(2)}L_l u_l^{(1)}]\,\mathrm{d}r = \left[u_l^{(1)}\frac{\mathrm{d}u_l^{(2)}}{\mathrm{d}r} - u_l^{(2)}\frac{\mathrm{d}u_l^{(1)}}{\mathrm{d}r}\right]_{r=a} \tag{12.431}$$

[13] In contrast to the external region introduced in Section 12.3, it is not assumed here that the potential is negligible in the entire outer region $r \geqslant a$, although we shall suppose that $V(r)$ tends to zero faster than r^{-1} as $r \to \infty$.

C. Bloch has shown in 1957 how this difficulty can be overcome. Let us introduce the *Bloch operator*

$$\mathcal{L}(a, b) = \delta(r - a)\left(\frac{\mathrm{d}}{\mathrm{d}r} - \frac{b}{r}\right) \tag{12.432}$$

where b is an arbitrary constant. It is then a simple matter to show (Problem 12.26) that the operator $L_l - \mathcal{L}(a, b)$ is Hermitian, so that this operator has real eigenvalues and can be readily inverted. Since $L_l - \mathcal{L}(a, b)$ is Hermitian, we have

$$\int_0^a \{u_l^{(1)}[L_l - \mathcal{L}(a, b)]u_l^{(2)} - u_l^{(2)}[L_l - \mathcal{L}(a, b)u_l^{(1)}\}\,\mathrm{d}r = 0 \tag{12.433}$$

In this and subsequent equations the upper limit of the integral is defined by the limiting process $\lim_{\varepsilon\to 0^+}\int^{a+\varepsilon}$ so that the delta function $\delta(r - a)$ appearing in the Bloch operator (12.432) is included in the integral. The radial Schrödinger equation (12.41) can be rewritten in the form

$$[-L_l + \mathcal{L}(a, b)]u_l = \mathcal{L}(a, b)u_l \tag{12.434}$$

and has the formal solution

$$u_l = [-L_l + \mathcal{L}(a, b)]^{-1}\mathcal{L}(a, b)u_l \tag{12.435}$$

Let us now introduce a complete set of basis functions $\chi_i(r)$ in the internal region $0 \leqslant r \leqslant a$. These functions can be chosen without loss of generality to be real and such that they diagonalise the operator $-L_l + \mathcal{L}(a, b)$. That is,

$$\int_0^a \chi_i[-L_l + \mathcal{L}(a, b)]\chi_j\,\mathrm{d}r = (k_i^2 - k^2)\delta_{ij} \tag{12.436}$$

where the dependence of the functions χ_i on l has been omitted for notational simplicity. Equation (12.435) now becomes (Problem 12.27)

$$u_l(r) = \frac{1}{a}\sum_{i=1}^{\infty}\frac{\chi_i(r)\chi_i(a)}{k_i^2 - k^2}\left[a\frac{\mathrm{d}u_l}{\mathrm{d}r} - bu_l\right]_{r=a} \tag{12.437}$$

Setting $r = a$ in this equation, we find that

$$u_l(a) = R_l(k^2)\left[a\frac{\mathrm{d}u_l}{\mathrm{d}r} - bu_l\right]_{r=a} \tag{12.438}$$

where the R-matrix is defined by

$$R_l(k^2) = \frac{1}{a}\sum_{i=1}^{\infty}\frac{[\chi_i(a)]^2}{k_i^2 - k^2} \tag{12.439}$$

The two equations (12.438) and (12.439) are the basic equations of the R-matrix method. They enable the logarithmic derivative of the wave function on the boundary of the internal region to be calculated. We also note that in terms of the energies $E = \hbar^2k^2/(2m)$ and $E_i = \hbar^2k_i^2/(2m)$ the R-matrix can be written as

$$R_l(E) = \frac{\hbar^2}{2ma}\sum_{i=1}^{\infty}\frac{[\chi_i(a)]^2}{E_i - E} \qquad \textbf{(12.440)}$$

We note that the R-matrix is a meromorphic function of the energy (analytic in the complex plane, except for poles), with simple poles on the real energy axis.

A variational principle for the R-matrix was first obtained by W. Kohn in 1948. He defined the functional

$$F_l^1[u_l^t] = [u_l^t(a)]^{-2}\int_0^a u_l^t[-L_l + \mathcal{L}(a, b)]\,u_l^t\,\mathrm{d}r \qquad \textbf{(12.441)}$$

which has the value

$$F_l^1[u_l] = [aR_l(k^2)]^{-1} \qquad \textbf{(12.442)}$$

when $u_l(r)$ is an exact solution of the radial Schrödinger equation (12.41). The first-order variation of the functional (12.441) due to small variations $\delta u_l(r) = u_l^t(r) - u_l(r)$ about the exact solution, such that $\delta u_l(0) = 0$, is

$$\delta F_l^1 = [u_l(a)]^{-2}\,2\int_0^a \delta u_l[-L_l + \mathcal{L}(a, b)]u_l\,\mathrm{d}r - 2[u_l(a)]^{-1}F_l^1[u_l]\,\delta u_l(a) \qquad \textbf{(12.443)}$$

and by using (12.442) we see that $\delta F_l^1 = 0$. Thus F_l^1 is a variational functional which enables the R-matrix to be obtained from its stationary value.

Another functional for the R-matrix, which can be readily extended to multichannel scattering, was proposed in 1951 by J.L. Jackson. It is given by

$$F_l^2[u_l^t] = \int_0^a u_l^t[-L_l + \mathcal{L}(a, b)]u_l^t\,\mathrm{d}r - 2a^{-1}u_l^t(a) \qquad \textbf{(12.444)}$$

Let us consider the first-order variation of this functional due to small variations $\delta u_l(r) = u_l^t(r) - u_l(r)$, subject to the normalisation condition

$$\left[a\frac{\mathrm{d}u_l^t}{\mathrm{d}r} - bu_l^t\right]_{r=a} = 1 \qquad \textbf{(12.445)}$$

We have

$$\begin{aligned}\delta F_l^2 &= 2\int_0^a \delta u_l[-L_l + \mathcal{L}(a, b)]u_l\,\mathrm{d}r - 2a^{-1}\delta u_l(a)\\ &= 2a^{-1}\delta u_l(a)\left\{\left[a\frac{\mathrm{d}u_l}{\mathrm{d}r} - bu_l\right]_{r=a} - 1\right\}\\ &= 0\end{aligned} \qquad \textbf{(12.446)}$$

since the exact solution u_l satisfies the condition (12.445). Moreover, for an exact solution u_l, we have

$$\begin{aligned} F_l^2[u_l] &= \int_0^a u_l[-L_l + \mathcal{L}(a, b)]u_l \,\mathrm{d}r - 2a^{-1}u_l(a) \\ &= u_l(a)\left[\frac{\mathrm{d}u_l}{\mathrm{d}r} - \frac{b}{a}u_l\right]_{r=a} - 2a^{-1}u_l(a) \\ &= a^{-1}u_l(a) - 2a^{-1}u_l(a) \\ &= -a^{-1}R_l(k^2) \end{aligned} \qquad \textbf{(12.447)}$$

where we have used (12.438) and the normalisation condition (12.445) which imply that $R(k^2) = u_l(a)$ in the present case. Thus F_l^2 is also a variational functional allowing the R-matrix to be obtained from its stationary value.

A variational estimate for the R-matrix can be obtained by introducing the trial function

$$u_l^t(r) = \sum_{i=1}^{N} c_i \chi_i(r), \qquad 0 \leqslant r \leqslant a \qquad \textbf{(12.448)}$$

where the basis functions $\chi_i(r)$ vanish at the origin and are linearly independent over the internal region. As in equation (12.436), they can be chosen to diagonalise the operator $-L_l + \mathcal{L}(a, b)$. The corresponding expression of the functional F_l^2 is then

$$F_l^2[u_l^t] = \sum_{i=1}^{N} c_i^2(k_{N,i}^2 - k^2) - 2a^{-1}\sum_{i=1}^{N} c_i\chi_i(a) \qquad \textbf{(12.449)}$$

where the $k_{N,i}^2$ converge to the k_i^2 defined by (12.436) as the set of basis functions $\chi_i(r)$ in (12.448) is expanded to completeness. The quantity $F_l^2[u_l^t]$ is stationary with respect to variations of the c_i provided that

$$\frac{\partial F_l^2[u_l^t]}{\partial c_i} = 2c_i(k_{N,i}^2 - k^2) - 2a^{-1}\chi_i(a) = 0, \qquad i = 1, 2, \ldots, N \qquad \textbf{(12.450)}$$

so that

$$c_i = \frac{1}{a}\frac{\chi_i(a)}{k_{N,i}^2 - k^2}, \qquad i = 1, 2, \ldots, N \qquad \textbf{(12.451)}$$

By substituting this result in (12.449), we obtain the stationary value of F_l^2, which is given by

$$[F_l^2] = -\frac{1}{a^2}\sum_{i=1}^{N}\frac{[\chi_i(a)]^2}{k_{N,i}^2 - k^2} \qquad \textbf{(12.452)}$$

Hence, returning to (12.447), we find that the variational estimate of the R-matrix is

$$R_l^N(k^2) = \frac{1}{a}\sum_{i=1}^{N}\frac{[\chi_i(a)]^2}{k_{N,i}^2 - k^2} \qquad \textbf{(12.453)}$$

As the set of basis functions $\chi_i(r)$ is expanded to completeness, the variational estimate $R_l^N(k^2)$ converges to its exact value defined by the equation (12.439).

Calculation of the phase shifts

The phase shifts can be determined from the R-matrix if the solution of the radial equation (12.41) is known in the external region $r \geqslant a$. Let us assume that $\phi_{l,1}(r)$ and $\phi_{l,2}(r)$ are two linearly independent solutions of (12.41) in the external region satisfying the boundary conditions

$$\phi_{l,1}(r) \underset{r\to\infty}{\to} \sin(kr - l\pi/2) \qquad \textbf{(12.454a)}$$

and

$$\phi_{l,2}(r) \underset{r\to\infty}{\to} \cos(kr - l\pi/2) \qquad \textbf{(12.454b)}$$

We can then write

$$u_l(r) = \phi_{l,1}(r) + \phi_{l,2}(r)K_l, \qquad r \geqslant a \qquad \textbf{(12.455)}$$

where $K_l = \tan\delta_l$. Upon substitution in (12.438), we find that

$$K_l = \tan\delta_l = \frac{R_l(k^2)[a\phi'_{l,1}(a) - b\phi_{l,1}(a)] - \phi_{l,1}(a)}{R_l(k^2)[-a\phi'_{l,2}(a) + b\phi_{l,2}(a)] + \phi_{l,2}(a)} \qquad \textbf{(12.456)}$$

This result reduces to (12.96) if (i) one sets $b = 0$ so that equation (12.438) yields $R_l(k^2) = (a\beta_l)^{-1}$, where β_l is defined by (12.92), and (ii) the potential is negligible in the external region $r \geqslant a$, so that the functions $\phi_{l,1}(r)$ and $\phi_{l,2}(r)$ can be taken to be $s_l(\rho)$ and $c_l(\rho)$ respectively (see (12.93)).

The radial wave function defined by (12.448) and (12.455) can be made continuous at $r = a$ by a suitable choice of normalisation. However, its derivative at $r = a$ is in general discontinuous, and this leads to slow convergence in the expansion (12.453) for certain choices of basis functions. Several procedures can be used to correct for this defect. Extensions of the R-matrix method have also been developed for situations where a very large number of basis functions would be required to obtain convergence. In this case, the interaction region is divided into a number of subregions, and an independent expansion is used in each of them. The R-matrix is then propagated from one region to the next.

Other approximation methods

A number of other approximation methods have been proposed, which fall outside the scope of this book. We mention in particular the methods based on the *analytic properties* of the scattering amplitudes, considered as functions of complex energies or complex wave numbers. Of special interest are the *effective range expansions* for low-energy scattering, which are discussed in detail by Burke and Joachain (1995).

12.8 Absorption processes and scattering by a complex potential

Until now we have considered the scattering of spinless particles by a *real* potential $V(\mathbf{r})$, a problem which is equivalent to the elastic scattering between two structureless particles. However, we have seen in Section 12.1 that in general, when a particle collides with a target, *non-elastic* scattering may occur. This means that there may be one, several, or an infinite number of *open channels* in addition to the elastic one. For example, if electrons are incident on hydrogen atoms in the ground state, and if these electrons have a kinetic energy larger than the excitation energy of the $n = 2$ hydrogen states (10.2 eV), then electron impact excitation of the $n = 2$ states – which are inelastic collisions – will occur in addition to elastic scattering.

The fact that 'non-elastic' processes take place means that some particles have been removed from the incident (elastic) channel, or in other words have been 'absorbed'. This *absorption* of particles can be represented in a phenomenological way by introducing a *complex* or *optical* [14] potential having a negative imaginary part. Indeed, let

$$V(\mathbf{r}) = V_R(\mathbf{r}) + \mathrm{i}V_I(\mathbf{r}) \tag{12.457}$$

be such a potential. We still look for stationary solutions of the time-dependent Schrödinger equation (12.12) having the form (12.13), where E is the energy of the particle, given by (12.15). The wave function $\psi(\mathbf{r})$ then satisfies the time-independent Schrödinger equation

$$\left[-\frac{\hbar^2}{2m}\nabla^2 + V_R(\mathbf{r}) + \mathrm{i}V_I(\mathbf{r})\right]\psi(\mathbf{r}) = E\psi(\mathbf{r}) \tag{12.458}$$

and we note that $\psi^*(\mathbf{r})$ satisfies the complex conjugate equation

$$\left[-\frac{\hbar^2}{2m}\nabla^2 + V_R(\mathbf{r}) - \mathrm{i}V_I(\mathbf{r})\right]\psi^*(\mathbf{r}) = E\psi^*(\mathbf{r}) \tag{12.459}$$

Now, from (2.50) and (12.13), we have

$$\boldsymbol{\nabla}\cdot\mathbf{j} = \frac{\hbar}{2m\mathrm{i}}[\psi^*\nabla^2\psi - (\nabla^2\psi^*)\psi] \tag{12.460}$$

[14] The earliest application of the optical potential was indeed made in the field of optics, in particular to the analysis of the propagation of light through a refractive medium. In this case the use of a complex index of refraction (whose imaginary part accounts for the absorption of light in the medium) is equivalent to the introduction of a complex optical potential.

where $\mathbf{j}$ is the probability current density. Using (12.458) and (12.459), we find that

$$\nabla \cdot \mathbf{j} = \frac{2}{\hbar} V_I |\psi|^2 = \frac{2}{\hbar} V_I P \tag{12.461}$$

where $P = |\Psi|^2 = |\psi|^2$ is probability density. We see that for $V_I < 0$, so that the complex potential $V(\mathbf{r})$ has a negative imaginary part, we have a local absorption of the incident beam, the rate of absorption of incident particles per unit time per unit volume at the point $\mathbf{r}$ being $2V_I/\hbar$. The total number N_{abs} of particles which are absorbed per unit time within the volume V is therefore

$$N_{\text{abs}} = \frac{2}{\hbar} \int_V V_I |\psi|^2 \, \mathrm{d}\mathbf{r} \tag{12.462}$$

Thus an optical potential of the type defined by (12.457), with a negative imaginary part ($V_I < 0$), describes elastic scattering accompanied by absorption, the latter being understood as the sum of all non-elastic processes.

Let us now consider the particular case of a *central*, complex potential $V(r) = V_R(r) + \mathrm{i}V_I(r)$. The method of partial waves, discussed in Section 12.3 for real central potentials, can then be generalised as follows in order to deal with complex central potentials. We first write in analogy with equation (12.38)

$$\psi^{(+)}_{\mathbf{k}_i}(k, r, \theta) = \sum_{l=0}^{\infty} \tilde{R}_l(k, r) P_l(\cos\theta) \tag{12.463}$$

where $\tilde{R}_l(k, r)$ are radial functions satisfying the equation

$$\left[\frac{\mathrm{d}^2}{\mathrm{d}r^2} + \frac{2}{r}\frac{\mathrm{d}}{\mathrm{d}r} - \frac{l(l+1)}{r^2} - U_R(r) - \mathrm{i}U_I(r) + k^2\right] \tilde{R}_l(k, r) = 0 \tag{12.464}$$

and we have introduced the reduced complex potential

$$\begin{aligned} U(r) &= \frac{2m}{\hbar^2} V(r) \\ &= U_R(r) + \mathrm{i}U_I(r) \end{aligned} \tag{12.465}$$

As in Section 12.3, we define the new radial functions

$$\tilde{u}_l(k, r) = r\tilde{R}_l(k, r) \tag{12.466}$$

which satisfy the equation

$$\left[\frac{\mathrm{d}^2}{\mathrm{d}r^2} - \frac{l(l+1)}{r^2} - U_R(r) - \mathrm{i}U_I(r) + k^2\right] \tilde{u}_l(k, r) = 0 \tag{12.467}$$

Although this equation closely resembles equation (12.41), there is an important difference between the two. Indeed, the solutions $u_l(k, r)$ of equation (12.41) can always be taken to be real (apart from an irrelevant overall phase factor of modulus one), while the solutions $\tilde{u}_l(k, r)$ of equation (12.467) – and therefore also the radial functions $\tilde{R}_l(k, r)$ – are *complex*. To see this, we write $\tilde{u}_l(k, r)$ in the form

$$\tilde{u}_l(k, r) = \text{Re}\,\tilde{u}_l(k, r) + \text{i}\,\text{Im}\,\tilde{u}_l(k, r) \tag{12.468}$$

and we note that the real and imaginary parts of $\tilde{u}_l(k, r)$ satisfy the coupled equations

$$\left[\frac{\text{d}^2}{\text{d}r^2} - \frac{l(l+1)}{r^2} + k^2 - U_R(r)\right]\text{Re}\,\tilde{u}_l(k, r) = -U_I(r)\,\text{Im}\,\tilde{u}_l(k, r) \tag{12.469a}$$

and

$$\left[\frac{\text{d}^2}{\text{d}r^2} - \frac{l(l+1)}{r^2} + k^2 - U_R(r)\right]\text{Im}\,\tilde{u}_l(k, r) = U_I(r)\,\text{Re}\,\tilde{u}_l(k, r) \tag{12.469b}$$

Provided that $U_R(r)$ and $U_I(r)$ are less singular than r^{-2} at the origin, physical solutions of equations (12.469) can be found, which vanish at $r = 0$, as in the case of a real potential. Asymptotically, if $U_R(r)$ and $U_I(r)$ vanish faster than r^{-1}, then we can also write (see (12.53))

$$\tilde{R}_l(k, r) = \frac{\tilde{u}_l(k, r)}{r} \underset{r\to\infty}{\to} \frac{A_l(k)}{kr}\sin[kr - l\pi/2 + \delta_l(k)] \tag{12.470}$$

where the 'normalisation constants' $A_l(k)$ are independent of r. However, since $\tilde{u}_l(k, r)$ and $\tilde{R}_l(k, r)$ are now complex, we are no longer permitted to assume that the phase shifts $\delta_l(k)$ are real quantities as in equation (12.53). Instead, we must introduce the *complex phase shifts*

$$\delta_l(k) = \text{Re}\,\delta_l(k) + \text{i}\,\text{Im}\,\delta_l(k) \tag{12.471}$$

We remark that by choosing an appropriate 'normalisation', we can write in analogy with (12.76)

$$\tilde{u}_l(k, r) \underset{r\to\infty}{\to} \exp[-\text{i}(kr - l\pi/2)] - S_l(k)\exp[\text{i}(kr - l\pi/2)] \tag{12.472}$$

where

$$\begin{aligned} S_l(k) &= \exp[2\text{i}\delta_l(k)] \\ &= \exp[2\text{i}\,\text{Re}\,\delta_l(k)]\exp[-2\,\text{Im}\,\delta_l(k)] \end{aligned} \tag{12.473}$$

is the S-matrix element.

Since particles can now be removed from the incident (elastic) channel, the amplitude $S_l(k)$ of the outgoing wave in (12.472) will be either reduced in magnitude (if non-elastic processes occur, that is if $U_I < 0$) or left unchanged if there is only elastic scattering (corresponding to $U_I = 0$). Thus we have

$$|S_l(k)| \leqslant 1 \tag{12.474}$$

which implies that

$$\text{Im}\ \delta_l(k) \geqslant 0 \tag{12.475}$$

In both of the above equations the equality corresponds to pure elastic scattering.

It is convenient to introduce an 'absorption factor'

$$\eta_l(k) = \exp[-2\ \text{Im}\ \delta_l(k)] \tag{12.476}$$

so that the S-matrix element (12.473) can be rewritten as

$$S_l(k) = \eta_l(k) \exp[2\text{i}\ \text{Re}\ \delta_l(k)] \tag{12.477}$$

From (12.475) and (12.476) we have

$$0 \leqslant \eta_l(k) \leqslant 1 \tag{12.478}$$

the special case $\eta_l = 1$ corresponding to pure elastic scattering.

The calculation of the elastic scattering amplitude f_{el} proceeds as in Section 12.3. By requiring that in the elastic channel the wave function $\psi^{(+)}_{\mathbf{k}_i}(\mathbf{r})$ has the asymptotic form

$$\psi^{(+)}_{\mathbf{k}_i}(\mathbf{r}) \underset{r\to\infty}{\to} A\left[\exp(\text{i}\mathbf{k}_i\cdot\mathbf{r}) + f_{\text{el}}(k, \theta, \phi)\frac{\exp(\text{i}kr)}{r}\right] \tag{12.479}$$

it is found that f_{el} is independent of the azimuthal angle ϕ (as expected for a spherically symmetric potential), and is given by (Problem 12.28)

$$\begin{aligned} f_{\text{el}}(k, \theta) &= \frac{1}{2\text{i}k}\sum_{l=0}^{\infty}(2l+1)[S_l(k)-1]P_l(\cos\theta) \\ &= \frac{1}{2\text{i}k}\sum_{l=0}^{\infty}(2l+1)\{\eta_l(k)\exp[2\text{i}\ \text{Re}\ \delta_l(k)]-1\}P_l(\cos\theta) \end{aligned} \tag{12.480}$$

The elastic differential cross-section is

$$\frac{\text{d}\sigma_{\text{el}}}{\text{d}\Omega} = |f_{\text{el}}(k, \theta)|^2 \tag{12.481}$$

and the total elastic cross-section is given by

$$\begin{aligned} \sigma^{\text{el}}_{\text{tot}}(k) &= \frac{\pi}{k^2}\sum_{l=0}^{\infty}(2l+1)|\eta_l(k)\exp[2\text{i}\ \text{Re}\ \delta_l(k)]-1|^2 \\ &= \sum_{l=0}^{\infty}\sigma^{\text{el}}_l(k) \end{aligned} \tag{12.482}$$

where the partial elastic cross-sections σ^{el}_l are such that

$$\sigma^{\text{el}}_l(k) = \frac{\pi}{k^2}(2l+1)|\eta_l(k)\exp[2\text{i}\ \text{Re}\ \delta_l(k)]-1|^2 \tag{12.483}$$

The total number of particles which disappear from the incident channel per unit time is equal to minus the flux of the wave function $\psi_{\mathbf{k}_i}^{(+)}(\mathbf{r})$ through a sphere of large radius centred at the origin. Writing $\psi_{\mathbf{k}_i}^{(+)}(\mathbf{r})$ for large r in the form

$$\psi_{\mathbf{k}_i}^{(+)}(\mathbf{r}) \underset{r\to\infty}{\to} A\left[\frac{\exp(-ikr)}{r}C_{\text{in}}(k,\theta) + \frac{\exp(ikr)}{r}C_{\text{out}}(k,\theta)\right] \quad \textbf{(12.484)}$$

we find that

$$C_{\text{in}}(k,\theta) = -\frac{1}{2ik}\sum_{l=0}^{\infty}(-1)^l(2l+1)P_l(\cos\theta) \quad \textbf{(12.485a)}$$

and

$$C_{\text{out}}(k,\theta) = \frac{1}{2ik}\sum_{l=0}^{\infty}(2l+1)S_l(k)P_l(\cos\theta) \quad \textbf{(12.485b)}$$

Calling $\mathbf{j}$ the probability current density corresponding to the wave function (12.484), we have

$$\mathbf{j}\cdot\hat{\mathbf{r}} = |A|^2 v\frac{1}{r^2}\left[|C_{\text{out}}(k,\theta)|^2 - |C_{\text{in}}(k,\theta)|^2\right] \quad \textbf{(12.486)}$$

where we have neglected terms of higher order in r^{-1}. Since the incident flux through a unit area normal to $\mathbf{k}_i$ (the Z direction) is still given by $F = v|A|^2$ (see (12.23)), we see that the total absorption cross-section, corresponding to all non-elastic processes, is just

$$\begin{aligned}\sigma_{\text{tot}}^{\text{abs}} &= -\frac{1}{F}\int \mathbf{j}\cdot\hat{\mathbf{r}}\,r^2\,d\Omega \\ &= \int [|C_{\text{in}}(k,\theta)|^2 - |C_{\text{out}}(k,\theta)|^2]\,d\Omega \end{aligned} \quad \textbf{(12.487)}$$

Using equations (12.485) together with the orthogonality relation (2.171) of the Legendre polynomials, we then have

$$\begin{aligned}\sigma_{\text{tot}}^{\text{abs}}(k) &= \frac{\pi}{k^2}\sum_{l=0}(2l+1)[1-\eta_l^2(k)] \\ &= \sum_{l=0}^{\infty}\sigma_l^{\text{abs}}(k)\end{aligned} \quad \textbf{(12.488)}$$

where

$$\sigma_l^{\text{abs}}(k) = \frac{\pi}{k^2}(2l+1)[1-\eta_l^2(k)] \quad \textbf{(12.489)}$$

Using (12.482) and (12.488), we see that the total (complete) cross-section which includes both elastic and non-elastic (absorption) processes is given by

$$\begin{aligned}\sigma_{\text{tot}} &= \sigma_{\text{tot}}^{\text{el}} + \sigma_{\text{tot}}^{\text{abs}}\\ &= \frac{2\pi}{k^2}\sum_{l=0}^{\infty}(2l+1)\{1 - \text{Re}\exp[2\text{i}\delta_l(k)]\}\\ &= \frac{2\pi}{k^2}\sum_{l=0}^{\infty}(2l+1)\{1 - \eta_l(k)\cos[2\text{Re}\,\delta_l(k)]\}\\ &= \sum_{l=0}^{\infty}\sigma_l(k)\end{aligned} \tag{12.490}$$

where

$$\sigma_l(k) = \frac{2\pi}{k^2}(2l+1)\{1 - \eta_l(k)\cos[2\text{Re}\,\delta_l(k)]\} \tag{12.491}$$

From (12.480) and (12.490) we note that the *optical theorem* may be generalised to read

$$\sigma_{\text{tot}}(k) = \frac{4\pi}{k}\,\text{Im}\, f_{\text{el}}(k, \theta = 0) \tag{12.492}$$

where σ_{tot} is the *total* (complete) cross-section and Im $f_{\text{el}}(k, \theta = 0)$ is the imaginary part of the *elastic* scattering amplitude in the forward direction. This generalised version of the optical theorem plays an important role in the analysis of complex collision processes.

We also remark from (12.483) that the maximum value of the partial wave elastic cross-section σ_l^{el} is obtained when $\eta_l \exp(2\text{i}\ \text{Re}\ \delta_l) = -1$, namely in the absence of absorption ($\eta_l = 1$). Thus

$$\sigma_l^{\text{el}}(k) \leqslant \frac{4\pi}{k^2}(2l+1) \tag{12.493}$$

On the contrary, the partial wave absorption cross-section σ_l^{abs} given by (12.489) reaches its maximum when $\eta_l = 0$, and therefore

$$\sigma_l^{\text{abs}}(k) \leqslant \frac{\pi}{k^2}(2l+1) \tag{12.494}$$

Finally, we see from (12.491) that the partial wave total (complete) cross-section σ_l is such that

$$\sigma_l(k) \leqslant \frac{4\pi}{k^2}(2l+1) \tag{12.495}$$

Therefore pure elastic scattering in the lth partial wave is possible and occurs when $\eta_l = 1$, in which case $\sigma_l^{\text{abs}} = 0$. On the other hand, non-elastic processes (for which $0 \leqslant \eta_l < 1$) are always accompanied by elastic scattering, as we expect on physical grounds.

Problems

12.1 Consider a reaction of the type A + B → C + D. Let m_A, m_B, m_C, m_D be the masses of the particles A, B, C and D, and E_A, E_B, E_C, E_D be their internal energies. Using the methods of Appendix 2, show that the centre of mass and laboratory differential cross-sections for observation of the particle C in a given direction are related by

$$\frac{d\sigma_C}{d\Omega_L}(\theta_L, \phi_L) = \frac{(1 + \tau_C^2 + 2\tau_C \cos\theta)^{3/2}}{|1 + \tau_C \cos\theta|} \frac{d\sigma_C}{d\Omega}(\theta, \phi)$$

where the subscript L refers to laboratory quantities, and

$$\tau_C = \left[\frac{m_A m_C}{m_B m_D} \frac{T_i}{(T_i + Q_{if})}\right]^{1/2}$$

The quantity T_i is the initial centre of mass kinetic energy and Q_{if} is the change in internal energy which has occurred in the collision, namely

$$Q_{if} = E_A + E_B - (E_C + E_D)$$

12.2 Prove the optical theorem (12.36) by inserting the expression (12.34) into the equation (12.33), which expresses the conservation of the probability flux.

12.3 Prove that the asymptotic form (12.53) holds provided $r|V(r)| \to 0$ as $r \to \infty$. (**Hint**: Write solutions of the radial equation (12.41) for large r in the form

$$u_l(k, r) = F_l(k, r) \exp(\pm ikr)$$

where F_l is a slowly varying function of r. Obtain an equation for F_l and show that if $r|V(r)| \to 0$ as $r \to \infty$, then F_l is independent of r in the limit $r \to \infty$.)

12.4 Prove the equation (12.68) for the momentum transfer cross-section σ_M.

12.5 Prove the unitarity relation (12.70).

(**Hint**: Use the addition theorem (A4.23) and the orthonormality relations (A4.18) of the spherical harmonics to show that

$$\frac{k}{4\pi}\int f^*(k, \theta')f(k, \theta_0)\, d\Omega' = \frac{1}{k}\sum_{l=0}^{\infty}(2l + 1)\sin^2\delta_l(k)P_l(\cos\theta).)$$

12.6 Consider a repulsive (reduced) potential of the form $U(r) = A/r^2$ $(A > 0)$.

(a) Obtain the phase shifts δ_l and show that $\delta_l \simeq -\pi A/[2(2l + 1)]$ when l is large.
(b) Discuss the angular distribution. Is the differential cross-section finite in the forward direction? Is the total cross-section finite? (**Hint**: For given ν and large ρ one has $j_\nu(\rho) \simeq \sin(\rho - \nu\pi/2)/\rho$.)

12.7 Consider an attractive (reduced) potential of the form $U(r) = A/r^2$ $(A < 0)$. How must the treatment of Problem 12.6 be modified? Show that the radial equation (12.41) has physically acceptable solutions only if $A > -1/4$.

12.8 Suppose that in an elastic scattering experiment between two structureless particles the centre of mass differential cross-section may be represented by an expression of the type

$$\frac{d\sigma}{d\Omega} = A + BP_1(\cos\theta) + CP_2(\cos\theta) + \ldots$$

Express the coefficients A, B and C in terms of the phase shifts δ_l.
(**Hint**: Use the orthogonality relation (2.171) of the Legendre polynomials and the recurrence relation (2.170).)

12.9 Consider the scattering of a particle of mass m by a central potential $V(r)$ which is negligible for $r > a$.

(a) Starting from the expression (12.91) for $\tan\delta_l(k)$, and using the properties of the functions j_l and n_l, prove that

$$\tan\delta_l(k) \underset{k\to 0}{\to} \frac{(ka)^{2l+1}}{D_l}\frac{l-\hat{\gamma}_l a}{l+1+\hat{\gamma}_l a} \qquad \textbf{(1)}$$

where

$$D_l = (2l+1)!!(2l-1)!!, \qquad l > 0,$$

$$D_0 = 1 \qquad \textbf{(2)}$$

and

$$\hat{\gamma}_l = \lim_{k\to 0}\gamma_l(k) \qquad \textbf{(3)}$$

(b) Assuming first that $\hat{\gamma}_l a \neq -(l+1)$, prove that the partial wave amplitudes $f_l(k)$ exhibit the low-energy behaviour

$$f_l(k) \underset{k\to 0}{\sim} k^{2l} \qquad \textbf{(4)}$$

so that except for the s-wave ($l = 0$) contribution which in general tends to a non-zero constant, all partial cross-sections σ_l ($l \geqslant 1$) vanish as k^{4l} when $k \to 0$. The scattering is therefore isotropic at very low energies. Prove also that the scattering amplitude f is given as $k \to 0$ by

$$f \underset{k\to 0}{\to} -\alpha \qquad \textbf{(5)}$$

where α is the scattering length (12.101).

(c) Examine how the results obtained in (b) must be modified when by accident

$$\hat{\gamma}_l a = -(l+1) \qquad \textbf{(6)}$$

(**Hint**: Show first that the result (5) is unchanged if the 'accident' (6) occurs for $l \geqslant 2$. Then prove that if equation (6) holds for $l = 0$, one has $\tan\delta_0(k) \sim k^{-1}$ as $k \to 0$, so that

$$f_0(k) \underset{k\to 0}{\sim} \mathrm{i}/k \qquad \textbf{(7)}$$

and the scattering length α diverges like k^{-2} when $k \to 0$. Finally, if equation (6) holds for $l = 1$, prove that the scattering amplitude is given as $k \to 0$ by

$$f \underset{k \to 0}{\to} -\alpha + \beta \cos \theta \tag{8}$$

where α is the scattering length (12.101) and β is a constant.)

12.10 Consider the scattering of low-energy (0.7 eV) electrons by a rare gas atom. Suppose that the effective interaction acting on the projectile electron may be represented schematically by an attractive square well of strength V_0 and range $a \simeq 10^{-8}$ cm.

(a) Show that for a certain value of V_0 the s-wave phase shift may be such that $\delta_0 = \pi$, while the other phase shifts δ_l with $l \geqslant 1$ are very small (this is called the *Ramsauer–Townsend effect*).
(b) Using the first two partial waves, discuss the corresponding expression for the total cross-section.

12.11 Find the total cross-section for low-energy (s-wave) scattering by a potential barrier such that

$$U(r) = \begin{cases} U_0(>0), & r < a \\ 0, & r > a \end{cases}$$

Derive the 'hard sphere' zero-energy result (12.122) as a particular case.

12.12 Using the first Born approximation (12.107) for $\tan \delta_l$, find the $l = 0$ phase shift for scattering by

(a) the Yukawa (or 'screened Coulomb') potential $U(r) = U_0 \exp(-\alpha r)/r$
(b) the Buckingham 'polarisation' potential $U(r) = U_0/(r^2 + d^2)^2$, where d is a constant.

12.13 Obtain the radial integral equation (12.187) and the integral expression (12.87) for $\tan \delta_l$ by analysing the integral equation (12.159) in partial waves. (**Hint**: Use the expansions (A4.26) and (A4.27) and the addition theorem (A4.23) of the spherical harmonics.)

12.14 Consider the scattering by potentials which fall off like $1/r^n$ as $r \to \infty$. Using the fact that the first Born approximation formula (12.107) is accurate for large l, prove that

(a) the differential cross-section at $\theta = 0$ is finite if $n > 3$
(b) the total cross-section is finite if $n > 2$.

(**Hint**: Use the fact that for fixed n and large l

$$\int_0^{\infty} \frac{x^2[j_l(x)]^2}{x^n} \mathrm{d}x \simeq \frac{\pi}{2} \frac{\Gamma(n-1)}{[\Gamma(n/2)]^2} (2l)^{1-n}$$

where Γ is Euler's gamma function.)

12.15 Prove the results (12.240) and (12.253).

12.16 Two alpha particles having a centre of mass energy of 1 MeV scatter elastically.

(a) Use a simple classical argument to suggest that only Coulomb forces should be taken into account at that energy.
(b) Obtain and plot the Mott differential cross-section in the centre of mass system. Compare your results with those obtained (i) by ignoring the interference term (classical values) and (ii) by ignoring the fact that the two particles are identical.

12.17 Obtain the centre of mass differential cross-section or electron–electron scattering at a relative velocity $v = 10^9$ cm s^{-1}, assuming that the system is unpolarised. Compare your results with those obtained (i) by ignoring the interference term (classical values) and (ii) by ignoring the fact that the two particles are identical.

12.18 Prove that for the case of central potentials the expression (12.269) of the first Born amplitude can be reduced to (12.273).

12.19 Obtain in first Born approximation the scattering amplitude, the differential and the total cross-sections for scattering by the following (reduced) potentials:

(a) Exponential potential: $U(r) = U_0 \exp(-\alpha r)$
(b) Gaussian potential: $U(r) = U_0 \exp(-\alpha^2 r^2)$

(c) Square well: $U(r)\begin{cases} = U_0, & r < a \\ = 0, & r > a \end{cases}$

(d) Buckingham polarisation potential: $U(r) = U_0/(r^2 + d^2)^2$.

Discuss the angular distributions and compare your results with those obtained in the text for the Yukawa potential $U(r) = U_0 \exp(-\alpha r)/r$. Verify that $\sigma_{\text{tot}}^{\text{B1}} \sim AE^{-1}$ as $E \to \infty$, and find the coefficient A in each case.

12.20 Using the unitary relation (12.70), show that for a central potential

$$\text{Im}\, f^{\text{B1}}(\theta) = 0$$

$$\text{Im}\, \bar{f}^{\text{B2}}(\theta) = \frac{k}{4\pi}\int f^{\text{B1}}(\theta')f^{\text{B1}}(\theta_0)\, \text{d}\Omega'$$

$$\text{Im}\, \bar{f}^{\text{B3}}(\theta) = \frac{k}{2\pi}\int f^{\text{B1}}(\theta')\, \text{Re}\, \bar{f}^{\text{B2}}(\theta_0)\, \text{d}\Omega'$$

where the k-dependence of the amplitudes has been omitted.

12.21 (a) Perform the integral in (12.299) to obtain the result (12.300) for $\tilde{G}_0(k, \mathbf{r}, \mathbf{r}')$.

(b) Substituting this result into (12.296), prove that the resulting function $\tilde{\varphi}(\mathbf{r})$ satisfies the integral equation (12.302).

12.22 Prove that for the reduced potential (12.329) the eikonal phase $\chi_0(b)$ is given by (12.330).

(**Hint**: Use the fact that

$$K_0(\alpha b) = \int_0^\infty \frac{\exp[-\alpha b(u^2+1)^{1/2}]}{(u^2+1)^{1/2}} du$$

and

$$K_1(\alpha b) = \int_0^\infty \exp[-\alpha b(u^2+1)^{1/2}]\, du$$

where K_0 and K_1 are modified Bessel functions.)

12.23 Derive the stationary expression (12.428) for $k^{-1} \tan \delta_l$ from the fractional form (12.425) of the Schwinger variational principle for the scattering amplitude.

12.24 Obtain the stationary expression (12.428) for $k^{-1} \tan \delta_l$ by using the integral equation (12.187) for the radial wave function $R_l(r)$, and the integral representation (12.87) for $\tan \delta_l$.

12.25 Apply the Schwinger variational principle (12.428) to study zero-energy s-wave scattering by a square well (12.108) of depth U_0 and of range a. Using a simple trial function $u_0^t(k = 0, r) = r$, show that the variational estimate of the scattering length is

$$[\alpha] = -\frac{\beta^2/3}{1 - 2\beta^2/5} a$$

where $\beta = a\sqrt{U_0}$. Compare this result with the exact expression of the scattering length

$$\alpha = \left(1 - \frac{\tan \beta}{\beta}\right) a$$

and with the expressions $\alpha^{B1} = -(\beta^2/3)a$ and $\alpha^{B2} = -(\beta^2/3 + 2\beta^4/15)a$ obtained by using the first and second Born approximation, respectively.

12.26 Prove that the operator $L_l - \mathscr{L}(a, b)$ is Hermitian, where $\mathscr{L}(a, b)$ is the Bloch operator (12.432).

12.27 Prove the result (12.437).

12.28 Prove the equation (12.480).

12.29 Consider the scattering by the complex square well

$$U(r) = \begin{cases} -U_0, & r < a \\ 0, & r > a \end{cases}$$

with

$$U_0 = U_R + \mathrm{i}U_I$$

and $U_R a^2 = 1$, $U_I a^2 = 5$. The incident wave number is such that $ka = 0.1$.

(a) Find the complex phase shift δ_0 and the inelasticity factor η_0.

(b) Neglecting the contribution of other partial waves, obtain the total elastic cross-section $\sigma_{\text{tot}}^{\text{el}}$, the total absorption cross-section $\sigma_{\text{tot}}^{\text{abs}}$ and the total (complete) cross-section σ_{tot}.

13 Electron–atom collisions and atomic photoionisation

With the help of the results obtained in the last chapter for the scattering of a beam of particles by a potential, we are now ready to discuss electron collisions with atoms and ions. The knowledge of electron–atom collision cross-sections is required for many important applications in various areas such as astrophysics and plasma physics. In addition, electron scattering is also important in determining the structure of atoms or ions, since they can be excited by electron impact to levels that would be forbidden in excitation by photons. Moreover, global properties of atoms, such as the charge density, can be inferred from the cross-sections for high-energy electron scattering. After discussing in Section 13.1 the general features of electron–atom collisions, we shall develop in the next two sections the theory of elastic and inelastic electron–atom collisions, respectively at low and high electron impact energies. Electron impact ionisation of atoms will be considered in Section 13.4.

The photoionisation of atoms is a phenomenon which has attracted considerable interest. In Section 4.8 we discussed the photoionisation of one-electron atoms, and in Section 7.7 we showed how doubly excited states of two-electron atoms can be studied by performing photoionisation experiments using synchrotron radiation. In Section 13.5 we shall discuss the general theory of photoionisation. From the theoretical point of view, we shall see that photoionisation reactions (which are sometimes called 'half collisions') are closely linked to electron collisions.

13.1 Electron–atom collisions. General features

An electron scattering experiment can be carried out as shown schematically in Fig. 13.1. The apparatus consists of an electron gun with electrostatic focusing and collimating devices, a target containing the atoms to be studied and apparatus to detect and analyse the scattered electrons. The whole apparatus must be enclosed in a high vacuum, often at pressures of 10^{-6} to 10^{-9} Torr, requiring the most advanced vacuum technology. Since the energy range spanned by experiments is very large, from a few electron-volts to several MeV, the detailed design of the electron source may differ considerably from experiment to experiment. In

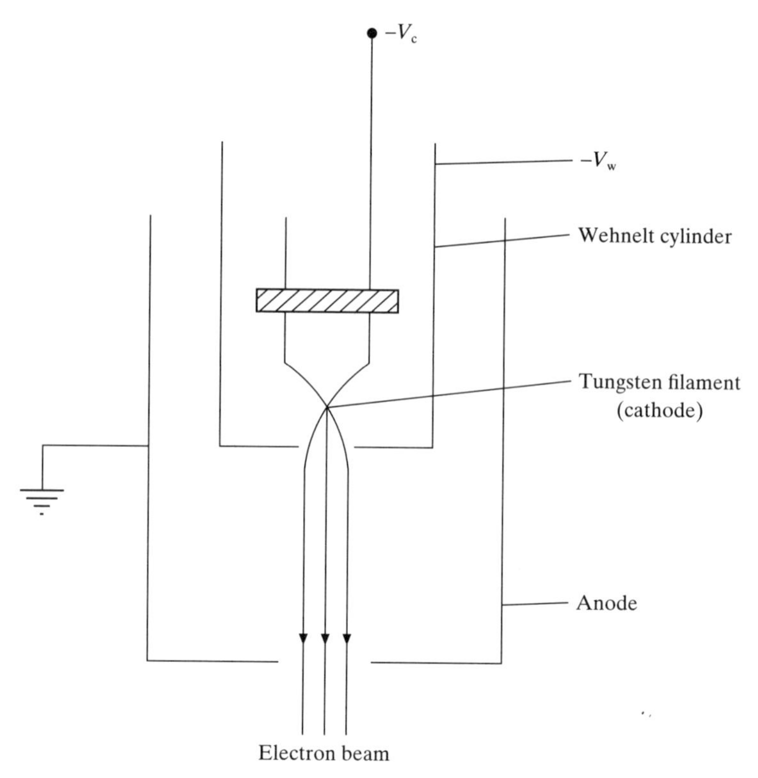

Figure 13.1 Schematic diagram of an electron gun. The potentials (with respect to the anode) on the cathode and the Wehnelt cylinder are denoted by $-V_c$ and $-V_w$, respectively.

general, the electron gun contains a heated filament (the cathode) which produces electrons by thermionic emission. The electrons are accelerated to an anode containing an aperture through which the beam is produced (see Fig. 13.1). The intensity of the beam is controlled by applying a voltage, negative with respect to the cathode, to a cylindrical electrode called a *Wehnelt cylinder* which surrounds the cathode except for a hole through which the beam passes. The Wehnelt cylinder acts as a converging 'lens' focusing the beam, and further *electrostatic lenses* can be added to produce a collimated beam in the region of interaction.

When studying scattering from the atoms of a monoatomic gas, such as helium, or argon, the target can consist of a cell containing a sample of the gas itself. This technique cannot be used when scattering by charged ions is to be investigated, or for neutral atoms which would normally occur bound in molecular form. In these cases a beam of target atoms, or ions, is produced and scattering takes place at the intersection of the atomic and electron beams. In the case of charged ions, the beam can be controlled and focused by passing it through suitable electrostatic fields, but for neutral atoms the procedure is different. Let us consider the example

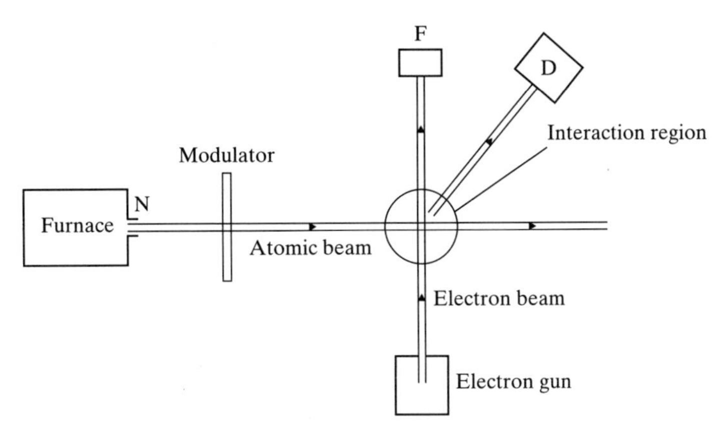

Figure 13.2 Diagram of an apparatus to measure electron scattering by atomic hydrogen. The atomic beam is produced by dissociating H_2 in a furnace at 3000 K. A jet of atomic hydrogen escapes through a nozzle N and passes through a modulator before reaching the interaction region, where it is intersected by an electron beam. The electron beam is monitored by collection in a Faraday cup F and the scattered electrons are detected and analysed at D.

of atomic hydrogen (see Fig. 13.2). By heating molecular hydrogen in a furnace, consisting of a tungsten atomic tube, to 3000 K, over 90 per cent of molecular hydrogen is dissociated to atomic hydrogen. If the hydrogen atoms are allowed to escape through a suitably shaped small nozzle, a jet of atomic hydrogen is produced, which can provide densities of up to 10^{14} atoms/m^3 in the interaction region. It should be noted that the velocities of the atoms are usually negligible compared with the electron beam velocities, so the collision can be treated as being between the moving electrons and a stationary target. To eliminate background effects, the atomic beam can be modulated by passing it through a slit in a rotating disc, thereby chopping the beam at a frequency of say 100 Hz. Only the signal in the detector varying with this frequency represents scattering by the atomic beam, and any other signal represents scattering by impurity atoms.

The electrons scattered from the interaction region at a certain angle must be analysed and measured. The velocity of the electrons can be measured, or selected, by deflecting their paths by electrostatic fields of known characteristics. The velocity of selected electrons can be detected using electron multipliers and the scattered current measured. The electron multiplier is a device containing multiple electrodes called dynodes, such that the secondary electrons ejected when the beam falls on the first electrode are accelerated towards the second electrode, producing further secondary electrons, which are in turn measured by collecting the electrons in a small metal container called a *Faraday cup* and amplifying the current produced electronically. Some results of an experiment performed in 1975 by J.F. Williams, using an experimental arrangement of the general kind described, are shown in Fig. 13.3.

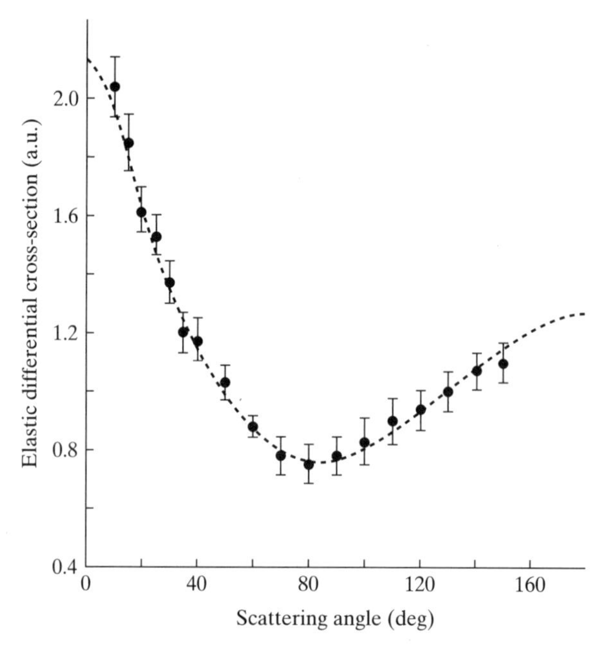

Figure 13.3 The differential cross-section for e$^-$–H(1s) elastic scattering at an incident electron energy of 4.89 eV. The theoretical results (dashed line) are obtained from variational calculations of the $l = 0$, 1 and 2 phase shifts, performed by C. Schwartz, R.L. Armstead, D. Register and R.T. Poe. The dots represent the absolute measurements of J.F. Williams.

Theoretical description

We shall now develop the theoretical description of non-relativistic collisions of electrons with target atoms or ions. Atomic units (a.u.) will be used. Our starting point is the Schrödinger equation describing the collision of an electron with a target atom or ion containing N electrons and having nuclear charge Z. That is,

$$H\Psi(q_0, q_1, \dots, q_N) = E\Psi(q_0, q_1, \dots, q_N) \tag{13.1}$$

where q_i denotes the ensemble of the spatial coordinates $\mathbf{r}_i$ and spin variable of electron i. In the above equation, E is the total energy of the electron–atom (ion) system, and H is the $(N+1)$-electron Hamiltonian of this system, which is given by

$$H = \sum_{i=0}^{N}\left(-\frac{1}{2}\nabla^2_{\mathbf{r}_i} - \frac{Z}{r_i}\right) + \sum_{i<j=0}^{N}\frac{1}{r_{ij}} \tag{13.2}$$

In writing this Hamiltonian, we have made the infinite nuclear mass approximation, so that the atomic nucleus can be considered to be at rest; it is taken to be the origin of coordinates. We also introduce the target eigenfunctions ψ_j such that

$$h\psi_j(q_1, q_2, \dots, q_N) = E_j\psi_j(q_1, q_2, \dots, q_N) \tag{13.3}$$

where h is given by (13.2), but with the summations starting at the index 1 instead of 0, and E_j are the target (internal) eigenenergies. The eigenvectors $|\psi_j\rangle$ of h will also be written as $|j\rangle$, the symbol j denoting collectively the quantum numbers which characterize this state.

In the following two sections we shall analyse the elastic and inelastic scattering of electrons by atoms and ions, first at low electron impact energies, and then at high (but non-relativistic) energies. In Section 13.4 we shall consider the ionisation of atoms and ions by electron impact, also called (e, 2e) reactions.

13.2 Elastic and inelastic electron–atom collisions at low energies

Let us consider a collision process of the type

$$e^- + A(i) \to e^- + A(j) \tag{13.4}$$

in which an electron of wave vector $\mathbf{k}_i$ (or momentum $\mathbf{k}_i$ since we are using atomic units) is incident upon a target atom or ion A in the eigenstate ψ_i, and in the final state a scattered electron of momentum $\mathbf{k}_j$ emerges and the target A is left in the eigenstate ψ_j. From energy conservation, we have

$$E = \frac{k_i^2}{2} + E_i = \frac{k_j^2}{2} + E_j \tag{13.5}$$

where $k_i^2/2 \equiv E_i^e$ is the kinetic energy of the incident electron and $k_j^2/2 \equiv E_j^e$ that of the scattered electron. We shall represent the initial conditions by the index $a \equiv (\mathbf{k}_i, i)$ and similarly we shall write $b \equiv (\mathbf{k}_j, j)$ to label the final state. It is convenient to consider separately the low-energy region such that the total energy $E < 0$, and the high-energy region for which $E > 0$.

We begin in this section by considering the low-energy region, so that $E_i^e < |E_i|$ and ionisation is forbidden. In that energy region only a *finite* number of channels are open, corresponding to elastic and inelastic (excitation or de-excitation) scattering processes. In the low-energy region the velocities of the incident and scattered electrons are less than the average velocities of the target electrons.

Electron collisions with atomic hydrogen

In order to introduce the basic concepts in a simple way, we shall first consider electron scattering by atomic hydrogen. The spatial part $\psi(\mathbf{r}_0, \mathbf{r}_1)$ of the wave function $\Psi(q_0, q_1)$ of the scattered and target electrons satisfies the Schrödinger equation

$$H\psi(\mathbf{r}_0, \mathbf{r}_1) = E\psi(\mathbf{r}_0, \mathbf{r}_1) \tag{13.6}$$

where

$$H = -\frac{1}{2}\nabla_{\mathbf{r}_0}^2 - \frac{1}{2}\nabla_{\mathbf{r}_1}^2 - \frac{1}{r_0} - \frac{1}{r_1} + \frac{1}{r_{01}} \tag{13.7}$$

Since we are dealing with an atomic hydrogen target, equation (13.3) simplifies to read

$$\left(-\frac{1}{2}\nabla_{\mathbf{r}}^2 - \frac{1}{r}\right)\psi_j(\mathbf{r}) = E_j\psi_j(\mathbf{r}) \tag{13.8}$$

The target eigenfunctions $\psi_j(\mathbf{r})$ are therefore the hydrogen atom eigenfunctions $\psi_{nlm}(\mathbf{r})$, the corresponding eigenenergies being given by $-1/(2n^2)$.

As explained in Section 7.1, the solutions of the Schrödinger equation (13.6) can always be chosen to be either symmetric or antisymmetric under the interchange of the spatial coordinates of the two electrons. The symmetric spatial wave function $\psi_+(\mathbf{r}_0, \mathbf{r}_1)$ corresponds to a singlet spin state ($S = 0$, $M_S = 0$) and the antisymmetric spatial wave function $\psi_-(\mathbf{r}_0, \mathbf{r}_1)$ to triplet spin states ($S = 1$, $M_S = 1, 0, -1$). For scattering in the singlet state we can therefore write the total wave function $\Psi^{S=0}(q_0, q_1)$ as the product

$$\Psi^{S=0}(q_0, q_1) = \psi_+(\mathbf{r}_0, \mathbf{r}_1)\chi_{0,0}(0, 1) \tag{13.9}$$

where $\chi_{0,0}(0, 1)$ is the antisymmetric spin function for a two-electron system (see (7.18)). For scattering in the triplet states we have instead

$$\Psi^{S=1}(q_0, q_1) = \psi_-(\mathbf{r}_0, \mathbf{r}_1)\chi_{1,M_S}(0, 1), \qquad M_S = 1, 0, -1 \tag{13.10}$$

where χ_{1,M_S} ($M_S = 1, 0, -1$) are the three symmetric spin functions for a two-electron system (see (7.19)).

It should be noted that since the energy E in (13.6) is a *degenerate* continuous eigenvalue, the wave functions $\psi(\mathbf{r}_0, \mathbf{r}_1)$ need not be symmetric or antisymmetric with respect to the interchange of the coordinates $\mathbf{r}_0$ and $\mathbf{r}_1$, although linear combinations of asymmetric wave functions can always be found, which are either space-symmetric or space-antisymmetric (see Problem 7.1). We shall return shortly to the asymmetric wave functions $\psi(\mathbf{r}_0, \mathbf{r}_1)$.

Boundary conditions, scattering amplitudes and differential cross-sections

Let us now consider the boundary conditions to be satisfied by the (anti)symmetrised wave functions $\psi_\pm(\mathbf{r}_0, \mathbf{r}_1)$. We begin by considering the particular case in which the target hydrogen atom is initially in the ground (1s) state described by the wave function $\psi_1(\mathbf{r}) \equiv \psi_{100}(\mathbf{r})$. The total energy of the system is given by (see (13.5))

$$E = \frac{k_i^2}{2} + E_1 \tag{13.11}$$

where $E_1 \equiv E_{1s} = -1/2$ a.u. ($\simeq -13.6$ eV). If the incident electron energy $k_i^2/2$ is insufficient to excite the hydrogen atom to the $n = 2$ level (that is, if $k_i^2/2 < 3/8$ a.u. $\simeq 10.2$ eV), the wave function, for $r_0 \gg r_1$, must represent electron 0 moving with respect to a ground state hydrogen atom containing electron 1, and we have

$$\psi_\pm(\mathbf{r}_0, \mathbf{r}_1) \underset{r_0\to\infty}{\to} F_1^\pm(\mathbf{r}_0)\psi_1(\mathbf{r}_1) \tag{13.12}$$

Because of the symmetry of the wave function, we must also have

$$\psi_{\pm}(\mathbf{r}_0, \mathbf{r}_1) \underset{r_1\to\infty}{\to} \pm F_1^{\pm}(\mathbf{r}_1)\psi_1(\mathbf{r}_0) \tag{13.13}$$

The wave function for the free electron, $F_1^{\pm}(\mathbf{r})$, must contain an incident plane wave and an outgoing spherical wave when $r \to \infty$, and hence satisfies the boundary condition (12.25). Taking as in Chapter 12 the direction of incidence along the Z axis and the normalisation constant A (see (12.25)) to be unity, we write

$$F_1^{\pm}(\mathbf{r}) \underset{r\to\infty}{\to} \exp(\mathrm{i}\mathbf{k}_i \cdot \mathbf{r}) + f_{11}^{\pm}(k_i, \theta, \phi)\frac{\exp(\mathrm{i}k_i r)}{r} \tag{13.14}$$

where $f_{11}^{\pm}$ are the scattering amplitudes for the *elastic* scattering process $1 \to 1$, the superscripts referring to scattering in the singlet (+) and triplet (–) states, respectively.

Proceeding as in Section 12.2, we find that the differential cross-sections for elastic scattering in the singlet and triplet spin states are determined by the scattering amplitudes f_{11}^{+} and f_{11}^{-}, respectively. If the electrons in the incident beam and in the target atom are *unpolarised* (so that the four initial spin states of the electron–hydrogen atom system are distributed at random), three times as many collisions occur in the triplet states as in the singlet state. If, in addition, no attempt is made to distinguish between the final spin states, the differential cross-section for the elastic scattering process $1 \to 1$ must be obtained by summing over the final spin states and averaging over the four possible initial spin states. Calling $(\mathrm{d}\sigma_{11}/\mathrm{d}\Omega)_{\mathrm{unp}}$ this differential cross-section, we have

$$\left(\frac{\mathrm{d}\sigma_{11}}{\mathrm{d}\Omega}\right)_{\mathrm{unp}} = \frac{1}{4}|f_{11}^{+}|^2 + \frac{3}{4}|f_{11}^{-}|^2 \tag{13.15}$$

Since the system is symmetrical about the Z axis, $f_{11}^{\pm}$ and $\mathrm{d}\sigma_{11}/\mathrm{d}\Omega$ do not depend on the azimuthal angle ϕ.

Let us now consider the boundary conditions at higher energies, such that it is possible to excite target states. Using (13.5), we see that the kinetic energy of an electron which emerges from the scattering region after exciting the target atom from the initial state of energy E_i (for example, the ground state) to an energy level E_j of higher energy is given by

$$\frac{k_j^2}{2} = E - E_j = \frac{k_i^2}{2} + (E_i - E_j) \tag{13.16}$$

For open channels we must have $k_j^2 \geqslant 0$, which implies that

$$\frac{k_i^2}{2} \geqslant E_j - E_i \tag{13.17}$$

In the asymptotic region $r_0 \gg r_1$, there must be terms in the wave function representing a free electron of energy $k_j^2/2$ moving with respect to the target in the state j, and in place of (13.12) the asymptotic form of $\psi_{\pm}(\mathbf{r}_0, \mathbf{r}_1)$ for $r_0 \to \infty$ is

$$\psi_\pm(\mathbf{r}_0, \mathbf{r}_1) \underset{r_0\to\infty}{\to} \sum_j F_j^\pm(\mathbf{r}_0)\psi_j(\mathbf{r}_1) \qquad (13.18)$$

where the sum runs over the open channels. The symmetry of the wave function also implies that

$$\psi_\pm(\mathbf{r}_0, \mathbf{r}_1) \underset{r_1\to\infty}{\to} \pm\sum_j F_j^\pm(\mathbf{r}_1)\psi_j(\mathbf{r}_0) \qquad (13.19)$$

In every open channel j (for which $k_j^2 \geqslant 0$), there must be spherical outgoing waves representing a current of electrons emerging from the scattering region, but only in the incident channel does the plane wave representing the incident electron of energy $k_i^2/2$ occur. The generalization of (13.14) is therefore

$$F_j^\pm(\mathbf{r}) \underset{r\to\infty}{\to} \exp(\mathrm{i}\mathbf{k}_i\cdot\mathbf{r})\delta_{ji} + f_{ji}^\pm(k_i, \theta, \phi)\frac{\exp(\mathrm{i}k_j r)}{r} \qquad (13.20)$$

where f_{ji}^+ and f_{ji}^- are called respectively the singlet (+) and triplet (−) scattering amplitudes for the transition $a \equiv (\mathbf{k}_i, i) \to b \equiv (\mathbf{k}_j, j)$. Using (13.18)–(13.20), we see that the asymptotic behaviour of the wave function $\psi_\pm^{(+)}(\mathbf{r}_0, \mathbf{r}_1)$ satisfying outgoing (+) spherical wave boundary conditions in all the open channels ($k_j^2 \geqslant 0$) is given for $r_0 \to \infty$ by

$$\psi_\pm^{(+)}(\mathbf{r}_0, \mathbf{r}_1) \underset{r_0\to\infty}{\to} \exp(\mathrm{i}\mathbf{k}_i\cdot\mathbf{r}_0)\psi_i(\mathbf{r}_1) + \sum_j f_{ji}^\pm(k_i, \theta_0, \phi_0)\frac{\exp(\mathrm{i}k_j r_0)}{r_0}\psi_j(\mathbf{r}_1) \qquad (13.21)$$

where (r_0, θ_0, ϕ_0) are the spherical polar coordinates of the vector $\mathbf{r}_0$. A similar asymptotic formula may be written down for $r_1 \to \infty$ by using (13.19) and (13.20).

The differential cross-section $\mathrm{d}\sigma_{ji}/\mathrm{d}\Omega$ for the transition $(\mathbf{k}_i, i) \to (\mathbf{k}_j, j)$ can be found by generalizing the reasoning given in Section 12.2. The outgoing flux of electrons of momentum $\mathbf{k}_j$ (that is, of energy $k_j^2/2$) passing through a spherical surface element $r^2\,\mathrm{d}\Omega$ for large r is given by $k_j|f_{ji}^\pm|^2\,\mathrm{d}\Omega$. Dividing by the incident flux (which is equal to k_i) and by $\mathrm{d}\Omega$, summing over the final spin states and averaging over the initial spin states, it is found that

$$\left(\frac{\mathrm{d}\sigma_{ji}}{\mathrm{d}\Omega}\right)_{\text{unp}} = \frac{k_j}{k_i}\left[\frac{1}{4}|f_{ji}^+|^2 + \frac{3}{4}|f_{ji}^-|^2\right] \qquad (13.22)$$

In obtaining this result we have assumed, as before, that the incident electron beam and the target atoms are unpolarised, and that no measurement is made to distinguish the final spin states.

It is also useful to examine the asymptotic form of the solution of equation (13.6) in a different way. Let us consider an *asymmetric* solution $\psi^{(+)}(\mathbf{r}_0, \mathbf{r}_1)$ satisfying outgoing (+) spherical wave boundary conditions in the open channels, where the incident electron is labelled 0 and the scattered electron is labelled either 0 (corresponding to *direct* scattering) or 1 (corresponding to *exchange* scattering, such that the incident electron 0 is captured by the atom and the initially bound atomic electron 1 is ejected). For large values of r_0 we shall therefore write

$$\psi^{(+)}(\mathbf{r}_0, \mathbf{r}_1) \underset{r_0\to\infty}{\to} \sum_j F_j(\mathbf{r}_0)\psi_j(\mathbf{r}_1) \tag{13.23}$$

where the sum runs over the open channels, and

$$F_j(\mathbf{r}_0) \underset{r_0\to\infty}{\to} \exp(\mathrm{i}\mathbf{k}_i\cdot\mathbf{r}_0)\,\delta_{ji} + f_{ji}(k_i, \theta_0, \phi_0)\frac{\exp(\mathrm{i}k_j r_0)}{r_0} \tag{13.24}$$

The amplitude f_{ji} is called a *direct scattering amplitude*. On the other hand, for large values of r_1, we have

$$\psi^{(+)}(\mathbf{r}_0, \mathbf{r}_1) \underset{r_1\to\infty}{\to} \sum_j G_j(\mathbf{r}_1)\psi_j(\mathbf{r}_0) \tag{13.25}$$

and

$$G_j(\mathbf{r}_1) \underset{r_1\to\infty}{\to} g_{ji}(k_i, \theta_1, \phi_1)\frac{\exp(\mathrm{i}k_j r_1)}{r_1} \tag{13.26}$$

where (r_1, θ_1, ϕ_1) are the spherical polar coordinates of the vector $\mathbf{r}_1$. The amplitude g_{ji} is called an *exchange scattering amplitude*. From (13.23)–(13.26), it follows that

$$\psi^{(+)}(\mathbf{r}_0, \mathbf{r}_1) \underset{r_0\to\infty}{\to} \exp(\mathrm{i}\mathbf{k}_i\mathbf{r}_0)\psi_i(\mathbf{r}_1) + \sum_j f_{ji}(k_i, \theta_0, \phi_0)\,\frac{\exp(\mathrm{i}k_j r_0)}{r_0}\,\psi_j(\mathbf{r}_1) \tag{13.27a}$$

and

$$\psi^{(+)}(\mathbf{r}_0, \mathbf{r}_1) \underset{r_1\to\infty}{\to} \sum_j g_{ji}(k_i, \theta_1, \phi_1)\,\frac{\exp(\mathrm{i}k_j r_1)}{r_1}\,\psi_j(\mathbf{r}_0) \tag{13.27b}$$

Upon comparison of the asymptotic forms (13.21) and (13.27), we see that

$$f^{\pm}_{ji} = f_{ji} \pm g_{ji} \tag{13.28}$$

and therefore the differential cross-section (13.22) can also be written in terms of direct and exchange amplitudes as

$$\left(\frac{\mathrm{d}\sigma_{ji}}{\mathrm{d}\Omega}\right)_{\text{unp}} = \frac{k_j}{k_i}\left[\frac{1}{4}|f_{ji} + g_{ji}|^2 + \frac{3}{4}|f_{ji} - g_{ji}|^2\right] \tag{13.29}$$

We remark that *two* scattering amplitudes (f_{ji} and g_{ji}, or f^+_{ji} and f^-_{ji}) are necessary to describe the electron–atomic hydrogen scattering process $(\mathbf{k}_i, i) \to (\mathbf{k}_j, j)$ occurring between two particular atomic states $|i\rangle$ and $|j\rangle$ [1]. This result remains valid for the more general case of electron scattering by spin-1/2 atoms, provided spin–orbit interactions are negligible. This is the case, for example, of electron scattering by the light alkalis.

[1] For transitions between two *energy levels*, one must of course take into account the degeneracy of these levels. For example, in the case of a transition between the 1s level and an np level, there are three direct amplitudes $f^1_{np,1s}, f^0_{np,1s}, f^{-1}_{np,1s}$ and three exchange amplitudes $g^1_{np,1s}, g^0_{np,1s}, g^{-1}_{np,1s}$ corresponding to the three values $m_l = 1, 0, -1$ of the magnetic quantum number of the np level.

Wave function expansions

We now turn to the problem of calculating the scattering amplitudes for electron–atomic hydrogen collisions. From the foregoing discussion, it is apparent that a natural way to proceed is to expand the asymmetric wave function $\psi^{(+)}(\mathbf{r}_0, \mathbf{r}_1)$ or the (anti)symmetrised wave functions $\psi_{\pm}^{(+)}(\mathbf{r}_0, \mathbf{r}_1)$ in terms of target eigenfunctions. We begin by considering asymmetric wave functions, writing

$$\psi^{(+)}(\mathbf{r}_0, \mathbf{r}_1) = \sum_j F_j(\mathbf{r}_0)\psi_j(\mathbf{r}_1) \tag{13.30a}$$

or

$$\psi^{(+)}(\mathbf{r}_0, \mathbf{r}_1) = \sum_j G_j(\mathbf{r}_1)\psi_j(\mathbf{r}_0) \tag{13.30b}$$

where the summation includes an integration over the continuum. L. Castillejo, I.C. Percival and M.J. Seaton showed in 1960 that the expansions (13.30) can only satisfy both boundary conditions (13.27) if suitable paths are chosen around singularities that occur in $F_j(\mathbf{r}_0)$ and $G_j(\mathbf{r}_1)$ in the continuum part of the expansion. In order to avoid the difficulties associated with these singularities, one can consider the alternative expansion

$$\psi^{(+)}(\mathbf{r}_0, \mathbf{r}_1) = \sum_j [F_j(\mathbf{r}_0)\psi_j(\mathbf{r}_1) + \psi_j(\mathbf{r}_0)G_j(\mathbf{r}_1)] \tag{13.31}$$

Clearly, the functions F_j and G_j are not uniquely defined in the above equation, since $\psi^{(+)}(\mathbf{r}_0, \mathbf{r}_1)$ is unchanged by the substitutions

$$F_j \to F_j + \sum_k a_{jk}\psi_k, \qquad G_j \to G_j - \sum_k a_{jk}\psi_k \tag{13.32}$$

where the constants a_{jk} are the elements of a symmetric matrix ($a_{jk} = a_{kj}$), but are otherwise arbitrary. This arbitrariness can be used to eliminate the singularities. Indeed, by choosing the constants a_{jk} so that the functions F_j and G_j are orthogonal to atomic states of lower energy,

$$\langle\psi_i|F_j\rangle = \langle\psi_i|G_j\rangle = 0, \qquad E_i < E_j \tag{13.33}$$

one can prove that for open channels ($k_j^2 \geqslant 0$) not involving ionisation, the asymptotic forms of the functions F_j and G_j are given by (13.24) and (13.26), respectively. In the closed channels ($k_j^2 < 0$), the functions F_j and G_j vanish asymptotically.

In order to achieve maximum simplicity in the form of the wave function expansion, we can use the (anti)symmetrised wave functions $\psi_{\pm}^{(+)}(\mathbf{r}_0, \mathbf{r}_1)$ which are either symmetric or antisymmetric under the interchange of the coordinates $\mathbf{r}_0$ and $\mathbf{r}_1$ of the two electrons, and write the wave function expansion as

$$\psi_{\pm}^{(+)}(\mathbf{r}_0, \mathbf{r}_1) = \sum_j [F_j^{\pm}(\mathbf{r}_0)\psi_j(\mathbf{r}_1) \pm F_j^{\pm}(\mathbf{r}_1)\psi_j(\mathbf{r}_0)] \tag{13.34}$$

where the functions $F_j^{\pm} = F_j \pm G_j$ do not contain singularities. For open channels ($k_j^2 \geqslant 0$) not involving ionisation, the functions $F_j^{\pm}$ satisfy the asymptotic boundary conditions (13.20), while for closed channels ($k_j^2 < 0$) they vanish asymptotically.

Since the functions $F_j^{\pm}$ contribute to the asymptotic form of $\psi_{\pm}^{(+)}$ only for open channels, it is possible to rewrite the expansion (13.34) in the alternative form

$$\psi_{\pm}^{(+)}(\mathbf{r}_0, \mathbf{r}_1) = \sum_j [F_j^{\pm}(\mathbf{r}_0)\psi_j(\mathbf{r}_1) \pm F_j^{\pm}(\mathbf{r}_1)\psi_j(\mathbf{r}_0)] + \sum_k c_k^{\pm}\chi_k^{\pm}(\mathbf{r}_0, \mathbf{r}_1) \tag{13.35}$$

where the first summation only runs over target eigenstates corresponding to *open* channels, and the $\chi_k^{\pm}(\mathbf{r}_0, \mathbf{r}_1)$ are square integrable *closed-channel* functions, also called *correlation* functions, which are either symmetric or antisymmetric with respect to the interchange of $\mathbf{r}_0$ and $\mathbf{r}_1$. Since for $r_0 \gg r_1$ or $r_1 \gg r_0$ there must be no electron flux associated with $\chi_k^{\pm}(\mathbf{r}_0, \mathbf{r}_1)$, we must have

$$r_0\chi_k^{(\pm)}(\mathbf{r}_0, \mathbf{r}_1) \underset{r_0\to\infty}{\longrightarrow} 0, \qquad r_1\chi_k^{(\pm)}(\mathbf{r}_0, \mathbf{r}_1) \underset{r_1\to\infty}{\longrightarrow} 0 \tag{13.36}$$

We remark that if the total energy E is such that $E < 0$, so that only a finite number of channels is open, then all the target eigenstates ψ_j which are energetically accessible can be included in the first expansion on the right of (13.35). If the closed-channel functions $\chi_k^{(\pm)}$ are chosen to span the space orthogonal to that corresponding to the first expansion, then the equation (13.35) provides an exact representation of the wave functions $\psi_{\pm}^{(+)}(\mathbf{r}_0, \mathbf{r}_1)$.

Another expansion of the wave functions $\psi_{\pm}^{(+)}(\mathbf{r}_0, \mathbf{r}_1)$ which has proved useful is the *pseudo-state* expansion proposed by P.G. Burke and H.M. Schey, and by R. Damburg and E. Karule. In this case the sum over energetically accessible target eigenstates is supplemented by a sum over pseudo-states, namely

$$\psi_{\pm}^{(+)}(\mathbf{r}_0, \mathbf{r}_1) = \sum_j [F_j^{\pm}(\mathbf{r}_0)\psi_j(\mathbf{r}_1) \pm F_j^{\pm}(\mathbf{r}_1)\psi_j(\mathbf{r}_0)] + \sum_l [H_l^{\pm}(\mathbf{r}_0)\bar{\psi}_l(\mathbf{r}_1) \pm H_l^{\pm}(\mathbf{r}_1)\bar{\psi}_l(\mathbf{r}_0)] \tag{13.37}$$

The pseudo-states $\bar{\psi}_l$ are not target eigenstates and there is no unique prescription for determining them. For example, they can be chosen to represent short-range *correlation effects* or long-range *polarisation effects*, the latter being characteristic of the interaction between a charged particle and a neutral polarisable system, as we shall see below. The pseudo-states can also be chosen to diagonalise the target Hamiltonian, namely

$$\left\langle \bar{\psi}_k \left| -\frac{1}{2}\nabla_{\mathbf{r}}^2 - \frac{1}{r} \right| \bar{\psi}_l \right\rangle = \bar{E}_k\delta_{kl} \tag{13.38}$$

and to be orthogonal to the target eigenstates ψ_j retained on the right of (13.37). If the pseudo-states $\bar{\psi}_l$ are expanded in the basis set of the target eigenstates,

$$\bar{\psi}_l = \sum_j b_{lj}\psi_j \tag{13.39}$$

this summation usually includes a contribution from the continuum. As a result, the continuum will then be partially represented by the expansion (13.37).

It is also possible to write down an expansion of the wave functions $\psi_{\pm}^{(+)}(\mathbf{r}_0, \mathbf{r}_1)$ in which target eigenstates ψ_j, pseudo-states $\bar{\psi}_l$ and square integrable closed-channel functions $\chi_k^{\pm}$ are all retained. Such an expansion may be written formally as in equation (13.35) provided that the first summation on the right of this equation is assumed to include both target eigenstates and pseudo-states.

Finally, we remark that since the functions F_j and G_j contribute to the asymptotic form of the asymmetric wave function $\psi^{(+)}(\mathbf{r}_0, \mathbf{r}_1)$ only for open channels, one can rewrite the expansion (13.31) for that wave function by including also square integrable closed-channel functions and (or) pseudo-states, as we have done above for the (anti)symmetrised wave functions $\psi_{\pm}^{(+)}(\mathbf{r}_0, \mathbf{r}_1)$.

Coupled equations

We shall now derive the equations satisfied by the unknown functions $F_j^{\pm}$. For simplicity, we shall assume that only target eigenstates are retained in the first expansion on the right of (13.35). We start from the Schrödinger equation (13.6) satisfied by the functions $\psi_{\pm}^{(+)}(\mathbf{r}_0, \mathbf{r}_1)$ and project this equation onto the target eigenstates ψ_j and correlation functions $\chi_k^{\pm}$. This gives

$$\int \psi_j^*(\mathbf{r}_1)[H - E]\psi_{\pm}^{(+)}(\mathbf{r}_0, \mathbf{r}_1)\, \mathrm{d}\mathbf{r}_1 = 0 \qquad \textbf{(13.40a)}$$

and

$$\int \chi_k^{\pm}(\mathbf{r}_0, \mathbf{r}_1)[H - E]\psi_{\pm}^{(+)}(\mathbf{r}_0, \mathbf{r}_1)\, \mathrm{d}\mathbf{r}_0\, \mathrm{d}\mathbf{r}_1 = 0 \qquad \textbf{(13.40b)}$$

We now substitute in these two equations the expansion (13.35) for $\psi_{\pm}^{(+)}(\mathbf{r}_0, \mathbf{r}_1)$. The first equation can be simplified by using equation (13.8) satisfied by the target eigenstates ψ_j. The second equation can be used to eliminate the coefficients $c_k^{\pm}$ in the expansion (13.35). Assuming that the square integrable functions $\chi_k^{\pm}$ are chosen to diagonalise the Hamiltonian H, so that

$$\langle \chi_k^{\pm} | H | \chi_l^{\pm} \rangle = \varepsilon_k^{\pm}\delta_{kl} \qquad \textbf{(13.41)}$$

one finds (Problem 3.1) that the functions $F_j^{\pm}(\mathbf{r})$ satisfy the coupled integro-differential equations

$$(\nabla^2 + k_j^2)F_j^{\pm}(\mathbf{r}) = 2\sum_k (V_{jk} + W_{jk}^{\pm} + X_{jk}^{\pm})F_k^{\pm}(\mathbf{r}) \qquad \textbf{(13.42)}$$

where V_{jk} is a direct potential matrix given by

$$\begin{aligned} V_{jk}(\mathbf{r}) &= \int \psi_j^*(\mathbf{r}')\left[-\frac{1}{r} + \frac{1}{|\mathbf{r} - \mathbf{r}'|}\right]\psi_k(\mathbf{r}')\, \mathrm{d}\mathbf{r}' \\ &= -\frac{1}{r}\delta_{jk} + \int \psi_j^*(\mathbf{r}')\frac{1}{|\mathbf{r} - \mathbf{r}'|}\psi_k(\mathbf{r}')\, \mathrm{d}\mathbf{r}' \end{aligned} \qquad \textbf{(13.43)}$$

and $W^{\pm}_{jk}$ are non-local exchange potential matrices such that

$$W^{\pm}_{jk}F^{\pm}_{k}(\mathbf{r}) = \pm \int \psi^{*}_{j}(\mathbf{r}')\left[\frac{1}{|\mathbf{r} - \mathbf{r}'|} + E_j + E_k - E\right]\psi_k(\mathbf{r})F^{\pm}_{k}(\mathbf{r}')\,\mathrm{d}\mathbf{r}' \qquad \textbf{(13.44)}$$

The non-local correlation potential matrices $X^{\pm}_{jk}$ appearing in (13.42) couple the closed channels with the open channels. They are found by using (13.41) to be such that

$$X^{\pm}_{jk}F^{\pm}_{k}(\mathbf{r}) = 2\sum_{m}\int Z^{\pm *}_{jm}(\mathbf{r})\frac{1}{E - \varepsilon^{\pm}_{m}}Z_{km}(\mathbf{r}')F^{\pm}_{k}(\mathbf{r}')\,\mathrm{d}\mathbf{r}' \qquad \textbf{(13.45)}$$

where

$$Z^{\pm *}_{jk}(\mathbf{r}) = \int \psi^{*}_{j}(\mathbf{r}')(H - E)\chi^{\pm}_{k}(\mathbf{r}, \mathbf{r}')\,\mathrm{d}\mathbf{r}' \qquad \textbf{(13.46)}$$

The coupled equations (13.42) are the basic equations describing low-energy electron scattering by atomic hydrogen. These equations can also be obtained from a variational principle (see Section 12.7) by considering the functional

$$I = \int \psi^{(+)*}_{\pm}(\mathbf{r}_0, \mathbf{r}_1)[H - E]\psi^{(+)}_{\pm}(\mathbf{r}_0, \mathbf{r}_1)\,\mathrm{d}\mathbf{r}_0\,\mathrm{d}\mathbf{r}_1 \qquad \textbf{(13.47)}$$

where $\psi^{(+)}_{\pm}$ is given by (13.35), and requiring that $\delta I = 0$ when $\psi^{(+)}_{\pm}$ is varied (Problem 13.2). We also note that the coupled equations (13.42) can equally well be formulated in momentum space, where they take the form of coupled integral Lippmann–Schwinger equations.

In order to solve the basic equations (13.42), subject to the boundary conditions discussed at the beginning of this section, a partial wave decomposition must be carried out. We shall return to this point below.

The static-exchange approximation

If only the initial eigenstate ψ_i of the target is retained in the first expansion on the right of (13.35), so that

$$\psi^{(+)}_{\pm}(\mathbf{r}_0, \mathbf{r}_1) = F^{\pm}_{i}(\mathbf{r}_0)\psi_i(\mathbf{r}_1) \pm F^{\pm}_{i}(\mathbf{r}_1)\psi_i(\mathbf{r}_0) \qquad \textbf{(13.48)}$$

the equations (13.42) reduce to the single equation

$$(\nabla^2 + k_i^2)F^{\pm}_{i}(\mathbf{r}) = 2(V_{ii} + W^{\pm}_{ii})F^{\pm}_{i}(\mathbf{r}) \qquad \textbf{(13.49)}$$

which is known as the *static-exchange approximation*. The direct potential V_{ii} is simply the *static potential* corresponding to the target eigenstate ψ_i, that is the interaction between the incident electron and the target hydrogen atom, averaged over this target eigenstate, namely

$$V_{ii}(\mathbf{r}) = \left\langle \psi_i \left| -\frac{1}{r} + \frac{1}{|\mathbf{r} - \mathbf{r}'|} \right| \psi_i \right\rangle = -\frac{1}{r} + \int \frac{|\psi_i(\mathbf{r}')|^2}{|\mathbf{r} - \mathbf{r}'|}\,\mathrm{d}\mathbf{r}' \qquad \textbf{(13.50)}$$

The first term is the interaction with the nucleus, and the second term is the interaction with the electronic charge density of the hydrogen atom in the eigenstate ψ_i. For example, in the case of the ground state (1s) of atomic hydrogen, one finds that the static potential acting on the incident electron is given by (Problem 13.3)

$$V_{11}(r) = -\left(1 + \frac{1}{r}\right)\exp(-2r) \quad \textbf{(13.51)}$$

The static-exchange approximation is clearly of limited validity. Firstly, it only describes elastic scattering, and even for this process it neglects the polarisation of the target by the unbound electron, which we shall consider below. Secondly, at electron impact energies higher than 3/8 a.u. ($\simeq 10{\cdot}2$ eV) where inelastic scattering is possible, the static-exchange approximation does not take into account the loss of probability flux into the inelastic channels.

The close-coupling approximation

When more than one target state is retained in the first expansion on the right of (13.35), one obtains the *close-coupling approximation* proposed in 1932 by H.S.W. Massey and C.B.O. Mohr. Thus, assuming that M target states are kept in this expansion, so that

$$\psi_{\pm}^{(+)}(\mathbf{r}_0, \mathbf{r}_1) = \sum_{j=1}^{M} [F_j^{\pm}(\mathbf{r}_0)\psi_j(\mathbf{r}_1) \pm F_j^{\pm}(\mathbf{r}_1)\psi_j(\mathbf{r}_0)] \quad \textbf{(13.52)}$$

the equations (13.42) reduce to the close-coupling equations

$$(\nabla^2 + k_j^2)F_j^{\pm}(\mathbf{r}) = 2\sum_k (V_{jk} + W_{jk}^{\pm})F_k^{\pm}(\mathbf{r}) \quad \textbf{(13.53)}$$

The close-coupling approximation is accurate for strong transitions between low-lying states whose energies are well separated from all other states. In other cases the convergence is often slow and the more general expansion (13.35) must be used, leading to the full coupled equations (13.42).

The optical potential

An alternative approach to elastic scattering, which goes beyond the static-exchange approximation, is to define effective or *optical* potentials $V_{\text{opt}}^{\pm}$ for scattering in the singlet (+) and triplet (–) states, respectively. These optical potentials are such that the exact channel functions $F_i^{\pm}(\mathbf{r})$ for elastic scattering from the hydrogen atom target in the state $|i\rangle$ satisfy the potential scattering equation

$$(\nabla^2 + k_i^2 - 2V_{\text{opt}}^{\pm})F_i^{\pm}(\mathbf{r}) = 0 \quad \textbf{(13.54)}$$

In the lowest (static-exchange) approximation, $V_{\text{opt}}^{\pm} \simeq V_{ii} + W_{ii}^{\pm}$ contain both the static potential V_{ii} and the non-local exchange potentials $W_{ii}^{\pm}$. The distortion of the target atom during the collision modifies the effective potential by the addition of interactions of long range. The potential of longest range is the most important and is due to the dipole induced in the target atom by the incident electron.

This effect is easiest to calculate at very low energies such that the kinetic energy of the incident electron can be neglected in first approximation. One then speaks of *adiabatic polarisation*. The Schrödinger equation for a hydrogen atom placed in the field of a *fixed* charge –1 (in a.u.) located at the position **r** is

$$[H_0 + V(\mathbf{r}, \mathbf{r}')]\Phi(\mathbf{r}, \mathbf{r}') = E(\mathbf{r})\Phi(\mathbf{r}, \mathbf{r}') \tag{13.55}$$

where

$$H_0 = -\frac{1}{2}\nabla^2_{\mathbf{r}'} - \frac{1}{r'} \tag{13.56}$$

is the Hamiltonian of the unperturbed hydrogen atom and

$$V(\mathbf{r}, \mathbf{r}') = -\frac{1}{r} + \frac{1}{|\mathbf{r} - \mathbf{r}'|} \tag{13.57}$$

is the interaction between the incident electron situated at **r** and the target atom. Both the eigenfunction Φ and the eigenvalue E depend on **r** parametrically. Let us consider the solution of (13.55) which, as $r \to \infty$, becomes equal to the unperturbed atomic hydrogen ground state wave function $\psi_{100}(r')$. Thus

$$\Phi(\mathbf{r}, \mathbf{r}') \underset{r\to\infty}{\to} \psi_{100}(r'), \qquad E(\mathbf{r}) \underset{r\to\infty}{\to} E_1 \tag{13.58}$$

where $E_1 = -1/2$ a.u.

Applying Rayleigh–Schrödinger perturbation theory (see Section 2.8) up to second order, we have

$$E(\mathbf{r}) = E_1 + E_1^{(1)}(\mathbf{r}) + E_1^{(2)}(\mathbf{r}) \tag{13.59}$$

The first-order correction $E_1^{(1)}$ is

$$\begin{aligned} E_1^{(1)}(\mathbf{r}) &= \langle \psi_{100} | V(\mathbf{r}, \mathbf{r}') | \psi_{100} \rangle \\ &= V_{11}(r) \end{aligned} \tag{13.60}$$

where V_{11} is the static potential between an electron and a ground state hydrogen atom, given by (13.51).

The second-order correction is

$$\begin{aligned} E_1^{(2)}(\mathbf{r}) &= \sum_{\substack{n\neq 1 \\ l,m}} \frac{|\langle \psi_{nlm} | V(\mathbf{r}, \mathbf{r}') | \psi_{100} \rangle|^2}{E_1 - E_n} \\ &= \sum_{j\neq 1} \frac{|V_{j1}(\mathbf{r})|^2}{E_1 - E_j} \end{aligned} \tag{13.61}$$

where in the second line we have used the notation of (13.43). The summation in (13.61) includes an integration over the continuum states of atomic hydrogen.

The direct part of the optical potentials $V^{\pm}_{\text{opt}}$ can then be identified at low energies with $E(\mathbf{r}) - E_1$. Adding the exchange potentials $W^{\pm}_{11}$, we find that the optical potentials for e^-–H(1s) elastic scattering are

$$V^{\pm}_{\text{opt}} = V_{11} + V_{\text{pol}} + W^{\pm}_{11} \tag{13.62}$$

where the *polarisation potential* V_{pol} is given approximately by

$$V_{\text{pol}}(\mathbf{r}) = E_1^{(2)}(\mathbf{r}) \tag{13.63}$$

With the optical potentials (13.62), the explicit form of the equation (13.54) for e^-–H(1s) scattering is

$$(\nabla^2 + k_1^2)F_1^{\pm}(\mathbf{r}) = 2[V_{11} + V_{\text{pol}} + W^{\pm}_{11}]F_1^{\pm}(\mathbf{r}) \tag{13.64}$$

We see from (13.61) and (13.63) that since $E_1 - E_n < 0$ for all $n \neq 1$, V_{pol} is negative, and is therefore an attractive potential.

For $r > r'$, we have, using the expansion (A4.24),

$$V(\mathbf{r}, \mathbf{r}') = \frac{1}{r}\sum_{l=1}^{\infty}\left(\frac{r'}{r}\right)^l P_l\left(\frac{\mathbf{r}\cdot\mathbf{r}'}{rr'}\right) \tag{13.65}$$

The term decreasing most slowly when r is large is that corresponding to $l = 1$. Neglecting higher order terms in r^{-1}, we have for large r,

$$V(\mathbf{r}, \mathbf{r}') \simeq \frac{1}{r^2}\hat{\mathbf{r}}\cdot\mathbf{r}' \tag{13.66}$$

Hence, using (13.61) and (13.63), the long-range behaviour of V_{pol} is given by

$$V_{\text{pol}} \underset{r\to\infty}{\to} -\frac{\bar{\alpha}}{2r^4} \tag{13.67}$$

where $\bar{\alpha}$ is the *static dipole polarisability* of the target hydrogen atom in the ground state, whose expression in atomic units is

$$\bar{\alpha} = 2\sum_{\substack{n\neq 1\\ l,m}} \frac{|\langle\psi_{nlm}|z|\psi_{100}\rangle|^2}{E_n - E_1} \tag{13.68}$$

and we recall that $\bar{\alpha} = 4.5$ a.u. (see Section 6.1). By studying the coupled equations (13.42) it can be shown that for large r the interaction of longest range is given by (13.67) not just at very low energies but at all energies.

The asymptotic form of V_{pol} given by (13.67) must be modified for values of r which are not large with respect to the 'radius' of the target atom. A simple phenomenological way of proceeding is to introduce a 'cut-off' parameter d, and to write

$$V_{\text{pol}}(r) = -\frac{\bar{\alpha}}{2(r^2 + d^2)^2} \tag{13.69}$$

This form of V_{pol} is called the *Buckingham polarization potential.* The cut-off parameter d has the dimension of a length, which is of the order of magnitude of the radius of the atom.

Above the first inelastic threshold, the optical potentials $V^{\pm}_{opt}$ acquire an imaginary part due to the *absorption* (loss of probability flux) from the initial channel (see Section 12.8). In this case the expressions (13.62) of the optical potentials for e^-–H(1s) elastic scattering become

$$V^{\pm}_{opt} = V_{11} + V_{pol} + W^{\pm}_{11} + iV_{abs} \tag{13.70}$$

where V_{abs} is called the *absorption potential.*

A powerful method, giving a general expression for the optical potential, is the *Feshbach projection operator formalism*, which we shall now describe for the case of electron–atomic hydrogen scattering. We assume that the space parts of the wave functions, $\psi^{(+)}_{\pm}(\mathbf{r}_0, \mathbf{r}_1)$, are expanded according to (13.35). Let us introduce a projection operator P, defined so that $P\psi^{(+)}_{\pm}$ contains the open-channel subspace represented by the sum over j in (13.35). Similarly, let Q be a projection operator such that $Q\psi^{(+)}_{\pm}$ contains the closed-channel subspace represented by the sum over k in (13.35), where the closed-channel functions $\chi^{\pm}_k$ have been taken to be orthogonal to the target functions ψ_j. Thus we have

$$P + Q = I \tag{13.71}$$

where I is the unit operator. Since P is a projection operator, we have

$$P^2 = P \tag{13.72}$$

and we verify that Q is also a projection operator, because

$$Q^2 = (I - P)^2 = I - 2P + P^2 = I - P = Q \tag{13.73}$$

From (13.71)–(13.73) we also deduce that the operators P and Q are orthogonal, so that

$$PQ = QP = 0 \tag{13.74}$$

The Schrödinger equation for $\psi^{(+)}_{\pm}$ can be rewritten as

$$(H - E)(P + Q)\psi^{(+)}_{\pm} = 0 \tag{13.75}$$

Projecting (13.75) with P and Q, and using (13.71)–(13.74), we obtain two coupled equations for $P\psi^{(+)}_{\pm}$ and $Q\psi^{(+)}_{\pm}$. That is,

$$(H_{PP} - E)P\psi^{(+)}_{\pm} + H_{PQ}Q\psi^{(+)}_{\pm} = 0 \tag{13.76a}$$

and

$$(H_{QQ} - E)Q\psi^{(+)}_{\pm} + H_{QP}P\psi^{(+)}_{\pm} = 0 \tag{13.76b}$$

where we have defined

$$H_{PP} = PHP, \; H_{QQ} = QHQ, \; H_{PQ} = PHQ, \; H_{QP} = QHP \tag{13.77}$$

Solving the equation (13.76b) for $Q\psi_{\pm}^{(+)}$, we have

$$Q\psi_{\pm}^{(+)} = (E - H_{QQ})^{-1}H_{QP}P\psi_{\pm}^{(+)} \tag{13.78}$$

and inserting this result into (13.76a), we obtain for $P\psi_{\pm}^{(+)}$ the equation

$$[H_{PP} + H_{PQ}(E - H_{QQ})^{-1}H_{QP} - E]P\psi_{\pm}^{(+)} = 0 \tag{13.79}$$

which is equivalent to the coupled equations (13.42). Introducing the *equivalent potential*

$$\mathcal{V} = H_{PQ}(E - H_{QQ})^{-1}H_{QP} \tag{13.80}$$

we can rewrite (13.79) in the form

$$(H_{PP} + \mathcal{V} - E)P\psi_{\pm}^{(+)} = 0 \tag{13.81}$$

The direct potentials V_{jk} and exchange potentials $W_{jk}^{\pm}$ are contained in H_{PP}, while the equivalent potential $\mathcal{V}$ contains the non-local correlation matrices $X_{jk}^{\pm}$ coupling the closed channels with the open channels. We shall see below that the equivalent potential $\mathcal{V}$ is closely related to the optical potential.

Since in the present context the Q subspace contains only closed channels at the energy E, the continuum spectrum of H_{QQ} must start at some energy $E' > E$. If follows that the operator $(E - H_{QQ})^{-1}$ is real and non-singular, except at possible discrete energies ε_r where H_{QQ} has eigenstates which are bound with respect to the continuum threshold at E'. At energies below the lowest eigenvalue of H_{QQ}, the operator $(E - H_{QQ})^{-1}$ is negative definite, so that $\mathcal{V}$ represents an attraction. If H_{QQ} has a discrete spectrum then, as we shall see later, this gives rise to resonances at energies E close to the eigenvalues ε_r.

Instead of including all the open channels in the projected wave function $P\psi_{\pm}^{(+)}$, we can include some of them in the projected wave function $Q\psi_{\pm}^{(+)}$. The wave function $P\psi_{\pm}^{(+)}$ will then include a subset of open channels usually defined to contain in particular the incident channel. In this case $Q\psi_{\pm}^{(+)}$ satisfies outgoing wave boundary conditions. The equivalent potential $\mathcal{V}$ is then given by

$$\mathcal{V} = \lim_{\varepsilon \to 0^+} H_{PQ}(E + i\varepsilon - H_{QQ})^{-1}H_{QP} \tag{13.82}$$

and is complex. We note that if the operator P is defined so that $P\psi_{\pm}^{(+)}$ includes M open channels corresponding to the expansion (13.52), then the equations

$$(H_{PP} - E)P\psi_{\pm}^{(+)} = 0 \tag{13.83}$$

obtained by neglecting the equivalent potential $\mathcal{V}$ in (13.81) are just the close-coupling equations (13.53). The operator $\mathcal{V}$ allows for the coupling with the other channels; it is complex if more than M channels are open.

As a particular case, let us include only the initial channel in $P\psi_{\pm}^{(+)}$. The equation (13.81) can then be reduced (Problem 13.5) to the form (13.54) describing elastic scattering from the hydrogen atom target in the state $|i\rangle$, with

$$V_{\text{opt}}^{\pm} = V_{ii} + W_{ii}^{\pm} + \mathcal{V}_i \tag{13.84}$$

where the equivalent potential $\mathcal{V}_i$ is a non-local operator which is complex when there are other open channels than the initial one. Thus, for elastic e$^-$–H(1s) scattering, $\mathcal{V}_i = \mathcal{V}_1$ becomes complex above the inelastic threshold at 10.2 eV. From (13.70) and (13.84) it follows that a good approximation to $\mathcal{V}_1$ is given by

$$\mathcal{V}_1 = V_{\text{pol}} + \text{i}V_{\text{abs}} \tag{13.85}$$

where we recall that V_{pol} is a polarisation potential falling off like r^{-4} for large r, and iV_{abs} accounts for absorption effects due to loss of probability flux from the initial channel.

Electron collisions with hydrogenic ions and quasi-one-electron atoms and ions

The basic theory of low-energy electron scattering by atomic hydrogen described above can be readily extended to light hydrogenic ions with small [2] nuclear charges Z, for example He$^+$, provided the asymptotic boundary conditions are modified to take into account the presence of an overall Coulomb interaction $-Z_R/r$ due to the 'residual charge' $Z_R = Z - 1$ of the ion. Thus, instead of (13.20), the functions $F_j^\pm$ must now satisfy the boundary conditions

$$F_j^\pm(\mathbf{r}) \underset{r\to\infty}{\to} \delta_{ji}\left[\exp\{\text{i}[k_i z + \gamma_i \log k_i(r-z)]\} + f_c(k_i,\theta)\frac{\exp\{\text{i}(k_i r - \gamma_i \log 2k_i r)\}}{r}\right] + \hat{f}_{ji}^\pm(k_i,\theta,\phi)\frac{\exp\{\text{i}(k_j r - \gamma_j \log 2k_j r)\}}{r} \tag{13.86}$$

where $f_c(k, \theta)$ is the Coulomb scattering amplitude (12.212) and

$$\gamma_j = -\frac{Z_R}{k_j} \tag{13.87}$$

is the Sommerfeld parameter (12.197), expressed in atomic units, which depends on the channel label j via k_j. The additional scattering amplitudes $\hat{f}_{ji}^\pm$ in (13.86) are due to the fact that the full interaction potential between the incident electron and the target ion is given by

$$V(\mathbf{r},\mathbf{r}') = -\frac{Z}{r} + \frac{1}{|\mathbf{r}-\mathbf{r}'|} \tag{13.88}$$

and hence deviates from the pure Coulomb potential $-(Z-1)/r$ when r is not large.

It follows from (13.86) that the singlet and triplet scattering amplitudes $f_{ii}^\pm$ for the elastic scattering process $(\mathbf{k}_i, i) \to (\mathbf{k}_j, i)$ with $k_i = k_j$ are given by

$$f_{ii}^\pm = f_c + \hat{f}_{ii}^\pm \tag{13.89}$$

[2] For heavy ions with larger nuclear charges, relativistic effects must be taken into account.

which is the sum of the pure Coulomb scattering amplitude f_c and the additional scattering amplitudes $\hat{f}^{\pm}_{ii}$. For an unpolarised electron–hydrogenic ion system, the corresponding elastic differential cross-section is then

$$\frac{d\sigma_{ii}}{d\Omega} = \frac{1}{4}|f_c + \hat{f}^{+}_{ii}|^2 + \frac{3}{4}|f_c + \hat{f}^{-}_{ii}|^2 \qquad (13.90)$$

We also deduce from (13.86) that the inelastic differential cross-sections for the process $(\mathbf{k}_i, i) \to (\mathbf{k}_j, j)$, with $j \neq i$, are given for an unpolarised electron–hydrogenic ion system by

$$\frac{d\sigma_{ji}}{d\Omega} = \frac{k_j}{k_i}\left[\frac{1}{4}|\hat{f}^{+}_{ji}|^2 + \frac{3}{4}|\hat{f}^{-}_{ji}|^2\right] \qquad (13.91)$$

The ground state of an alkali atom or alkali-like ions is composed of a single valence electron in an s state outside closed inner (sub)shells. Most features of electron scattering by the ground state of light alkali-like atoms or ions (such that relativistic effects are negligible) can be accounted for by considering them approximately as quasi-one-electron atoms or ions. In this model, the target atom (ion) is considered as a one-electron system, with the bound valence electron moving in an effective potential $V(r)$ due to the nucleus and the core of inner shell electrons. The theory of low-energy electron scattering by atomic hydrogen and hydrogenic ions can then be used, provided the Coulomb interaction binding the target electron to the nucleus is replaced by $V(r)$. However, there are qualitative differences between the hydrogenic and alkali systems. Notably, the energies of the ground and excited states of the alkali atoms are smaller in magnitude than those of atomic hydrogen. For instance, the ground state of lithium, with the valence electron in the 2s level, has the energy $E_{2s} = -5.39$ eV and the first excited state, with the valence electron in the 2p level, is at $E_{2p} = -3.54$ eV. As a consequence, the static dipole polarisability $\bar{\alpha}$ of lithium in the ground state is much larger than for atomic hydrogen (see Table 9.1) and the first excited p state provides 98% of the sum (9.67) which determines $\bar{\alpha}$. It follows that at low energies a two-state close-coupling approximation retaining the ground state and the first excited p state provides elastic scattering cross-sections of useful accuracy. Such two-state calculations were first carried out by E. Karule for Li, Na, K and Cs targets at incident electron energies below the first inelastic threshold. More accurate cross-sections have been obtained subsequently. For example, in the case of Li and Na targets, coupled channel calculations with larger basis sets have been performed by P.G. Burke and K.T. Taylor and by D.L. Moores and D.W. Norcross.

Electron collisions with complex atoms and ions

For electron–atom (ion) systems with three or more electrons it is not possible (except approximately for quasi-one-electron atoms or ions such as the alkalis) to write down the total wave function of the system in the simple forms (13.9) or (13.10). In order to satisfy the Pauli exclusion principle, we look for an expansion which is antisymmetric with respect to the interchange of the combined space and spin coordinates q_i of all the electrons. By analogy with the expansion (13.35) used for the two-electron problem, we shall write this expansion in the form

$$\Psi(q_0, q_1, \ldots, q_N) = \mathcal{A} \sum_j F_j(q_0)\psi_j(q_1, \ldots, q_N) + \sum_k c_k\chi_k(q_0, q_1, \ldots, q_N) \quad \textbf{(13.92)}$$

where $\mathcal{A}$ is the antisymmetrisation operator. The first summation goes over a limited number of target eigenstates and may also include some pseudo-states, while the second summation is over a set of square integrable functions χ_k, each of them being antisymmetric with respect to the interchange of the space and spin coordinates of the $N + 1$ electrons. In practice, the target eigenstates and pseudo-states cannot be obtained exactly, but instead are generally expanded in a configuration–interaction basis.

By substituting the expansion (13.92) into the Schrödinger equation (13.1) and projecting onto the target functions ψ_j and the square integrable functions χ_k, one obtains for the functions F_j a set of coupled integro-differential equations analogous to the equations (13.42) derived above for the two-electron problem.

Partial wave analysis

The partial wave decomposition of these coupled equations can be performed by remembering that the non-relativistic $(N + 1)$-electron Hamiltonian (13.2) commutes with the square of the total orbital angular momentum, $\mathbf{L}^2$, with the square of the total spin angular momentum operator, $\mathbf{S}^2$, and also with the operators L_z, S_z and the parity operator $\mathcal{P}$. Thus we can form solutions $\Psi^\Gamma(q_0, q_1, \ldots, q_N)$ of the Schrödinger equation (13.1) which are eigenfunctions of these operators belonging to the eigenvalues $\Gamma \equiv LSM_LM_S\pi$, where π is the parity quantum number. These eigenfunctions Ψ^Γ can be expanded in analogy with (13.92) as

$$\Psi^\Gamma(q_0, q_1, \ldots, q_N) = \mathcal{A} \sum_j F_j^\Gamma(r_0)\psi_j^\Gamma(q_1, \ldots, q_N; \hat{\mathbf{r}}_0, \sigma_0) + \sum_k c_k^\Gamma\chi_k^\Gamma(q_0, q_1, \ldots, q_N) \quad \textbf{(13.93)}$$

where r_0 is the radial coordinate of the scattered electron, $\hat{\mathbf{r}}_0$ denotes its angular coordinates (θ_0, ϕ_0) and σ_0 its (dichotomic) spin variable. The channel functions ψ_j^Γ are constructed by combining the target functions ψ_j with the angular and spin functions of the scattered electron to form eigenstates of the operators $\mathbf{L}^2$, $\mathbf{S}^2$, L_z, S_z and $\mathcal{P}$ corresponding to the eigenvalues $\Gamma \equiv LSM_LM_S\pi$.

By substituting (13.93) into the Schrödinger equation (13.1), projecting onto the channel functions ψ_j^Γ and onto the square integrable functions χ_k^Γ and after eliminating the coefficients c_k^Γ one obtains for the *reduced* radial functions

$$u_j^\Gamma(r) = rF_j^\Gamma(r) \quad \textbf{(13.94)}$$

the coupled integro-differential equations

$$\left(\frac{\mathrm{d}^2}{\mathrm{d}r^2} - \frac{l(l+1)}{r^2} + k_j^2\right)u_j^\Gamma(r) = 2\sum_k (V_{jk}^\Gamma + W_{jk}^\Gamma + X_{jk}^\Gamma)u_k^\Gamma(r) \quad \textbf{(13.95)}$$

where l is the orbital angular momentum of the scattered electron, and the potentials V^{Γ}_{jk}, W^{Γ}_{jk} and X^{Γ}_{jk} are partial wave decompositions of the direct, exchange and correlation potentials. The equations (13.95) were first derived in 1957 by I.C. Percival and M.J. Seaton for electron–atomic hydrogen scattering (in the absence of the square integrable functions). It should be noted that the potentials V^{Γ}_{jk}, W^{Γ}_{jk} and X_{jk} are independent of M_L and M_S, so that one must only solve the coupled equations (13.95) for the required $LS\pi$ symmetries.

In general, the expressions giving V^{Γ}_{jk}, W^{Γ}_{jk} and X^{Γ}_{jk} are very cumbersome, so that in practice these potentials are determined by computer programs. It is worth noting, however, that the direct potentials V^{Γ}_{jk} are local and have the asymptotic form

$$V^{\Gamma}_{jk}(r) = -\frac{(Z-N)}{r}\delta_{jk} + \sum_{\lambda=1}^{\lambda_{\max}} b^{\lambda}_{jk}\, r^{-\lambda-1}, \qquad r > a \tag{13.96}$$

where the quantities b^{λ}_{jk} are constants, and a is the range beyond which the bound orbitals in ψ^{Γ}_j and ψ^{Γ}_k are negligible. The dominant non-Coulombic long-range potential is the *dipole potential*, corresponding to $\lambda = 1$, which falls off like r^{-2} for $r > a$. This potential couples target states between which dipole transitions are allowed. It gives rise, in second order, to an equivalent diagonal ($j = k$) polarisation potential V_{pol} having the asymptotic form (13.67), where $\bar{\alpha}$ is now the static dipole polarisability of the target atom or ion in the state j.

The *S*-matrix, *K*-matrix and *T*-matrix

Let us suppose that for a given set of quantum numbers Γ there are M coupled channels, corresponding to retaining a finite number of target states and pseudostates in the expansion (13.93). Then both j and k in (13.95) will take the values $1, 2, \ldots, M$ and the M coupled equations (13.95) will in general have $2M$ linearly independent solutions. The requirement that the physically acceptable reduced radial functions $u^{\Gamma}_j(r)$ vanish at the origin reduces this number to M linearly independent solutions. We shall therefore introduce a second subscript on the functions $u^{\Gamma}_j(r)$ and denote these solutions by $u^{\Gamma}_{ji}(r)$, with $i, j = 1, 2, \ldots, M$. Thus

$$u^{\Gamma}_{ji}(0) = 0, \qquad i, j = 1, 2, \ldots, M \tag{13.97}$$

Assuming that there are M_a open channels (for which $k_j^2 \geqslant 0$) one has in these channels

$$u^{\Gamma}_{ji}(r) \underset{r\to\infty}{\to} k_j^{-1/2}(\delta_{ji}\mathrm{e}^{-\mathrm{i}\Theta_j} - S^{\Gamma}_{ji}\mathrm{e}^{\mathrm{i}\Theta_j}), \qquad j = 1, 2, \ldots, M_a \tag{13.98}$$

while in the remaining $M - M_a$ closed channels (for which $k_j^2 < 0$)

$$u^{\Gamma}_{ji}(r) \underset{r\to\infty}{\to} 0, \qquad j = M_a + 1, \ldots, M \tag{13.99}$$

with $i = 1, 2, \ldots, M$. The quantity Θ_j which appears in (13.98) is

$$\Theta_j = k_j r - l_j\pi/2 - \gamma_j \log(2k_j r) + \sigma_j \tag{13.100}$$

where γ_j is the Sommerfeld parameter given by using (13.87) with a 'residual charge' $Z_R = Z - N$, and

$$\sigma_j = \arg \Gamma(l_j + 1 + \mathrm{i}\gamma_j) \qquad \textbf{(13.101)}$$

are Coulomb phase shifts, obtained from a partial wave analysis of the Coulomb potential [3]. In (13.101), Γ is Euler's gamma function.

The quantities S^{Γ}_{ji} in (13.98) are the elements of the S-matrix. Since it is assumed that there are M_a open channels, we see that the S-matrix defined by (13.98) has dimensions $M_a \times M_a$. It can be shown from probability conservation that the S-matrix is unitary, while time-reversal invariance implies that it is symmetric.

Instead of the complex solutions satisfying the asymptotic boundary conditions (13.98), it is often convenient to introduce real solutions satisfying the boundary conditions

$$u^{\Gamma}_{ji}(r) \underset{r\to\infty}{\to} k_j^{-1/2}(\delta_{ji} \sin \Theta_j + K^{\Gamma}_{ji} \cos \Theta_j), \qquad j = 1, 2, \ldots, M_a \qquad \textbf{(13.102)}$$

in the M_a open channels ($k_j^2 \geqslant 0$) and

$$u^{\Gamma}_{ji}(r) \underset{r\to\infty}{\to} 0, \qquad j = M_a + 1, \ldots, M \qquad \textbf{(13.103)}$$

in the $M - M_a$ closed channels ($k_j^2 < 0$). The quantities K^{Γ}_{ji} are the elements of the K-matrix, which is related to the S-matrix by

$$\mathbf{S} = \frac{\mathbf{I} + \mathrm{i}\mathbf{K}}{\mathbf{I} - \mathrm{i}\mathbf{K}} \qquad \textbf{(13.104)}$$

Since the S-matrix is symmetric, it follows that the K-matrix is also symmetric. Moreover, equation (13.104) together with the unitary of the S-matrix and the symmetry of the K-matrix implies that the K-matrix is real.

One can also define solutions satisfying the outgoing wave boundary conditions

$$u^{\Gamma}_{ji}(r) \underset{r\to\infty}{\to} k_j^{-1/2}\left(\delta_{ji} \sin \Theta_j + \frac{1}{2\mathrm{i}} \mathrm{e}^{\mathrm{i}\Theta_j} T^{\Gamma}_{ji}\right), \qquad j = 1, 2, \ldots, M_a \qquad \textbf{(13.105)}$$

in the M_a open channels, while

$$u^{\Gamma}_{ji}(r) \underset{r\to\infty}{\to} 0, \qquad j = M_a + 1, \ldots, M \qquad \textbf{(13.106)}$$

in the $M - M_a$ closed channels. The quantities T^{Γ}_{ji} are the elements of the T-matrix, which is related to the S-matrix by

$$\mathbf{S} = \mathbf{I} + \mathbf{T} \qquad \textbf{(13.107)}$$

Scattering amplitudes, cross-sections and collision strengths

As in the case of potential scattering, which was studied in Chapter 12, the scattering amplitudes and hence the differential and total cross-sections can be

[3] See Joachain (1983).

related to the S-matrix (or the K-matrix or T-matrix). As an example, the total cross-section for a transition $i \to j$ is given by

$$\sigma_{\text{tot}}(i \to j) = \frac{\pi}{k_i^2} \sum_{LS\pi} \frac{(2L+1)(2S+1)}{2(2L_i+1)(2S_i+1)} \sum_{l_i l_j} |T_{ji}^{\Gamma}|^2 \quad \textbf{(13.108)}$$

where l_i and l_j are the orbital angular momentum quantum numbers of the incident and scattered electron, respectively.

A further quantity which is often used in applications is the *collision strength*. It is defined for a transition $i \to j$ by the relation

$$\Omega(i \to j) = g_i k_i^2 \sigma_{\text{tot}}(i \to j) \quad \textbf{(13.109)}$$

where $g_i = (2L_i + 1)(2S_i + 1)$ is the statistical weight of the initial state $|i\rangle$. We note that the collision strength is a dimensionless quantity and is symmetric, namely

$$\Omega(i \to j) = \Omega(j \to i) \quad \textbf{(13.110)}$$

Methods of solution and illustrative examples

Most of the results emerging from the coupled equations (13.95) have been obtained from one of the following methods. The *direct numerical solution* of these equations was first based on iterating the integral terms in the equations, but this gave rise to convergence problems. In the non-iterative approach proposed in 1973 by E.R. Smith and R.J.W. Henry, the equations (13.95) are transformed into a set of integral equations, while in the non-iterative method developed by M.J. Seaton in 1974 the differential and integral operators are represented by finite difference formulae following which a set of algebraic equations can be found for the values of the radial functions at a set of mesh points.

The *Hulthén–Kohn variational method* introduced in Section 12.7 within the framework of potential scattering has also been employed successfully to solve the coupled equations (13.95), in particular by C. Schwartz, J. Callaway and R.K. Nesbet. As a first illustration, we showed in Fig. 13.3 the differential cross-section for e^-–H(1s) elastic scattering at an incident electron energy of 4.89 eV. The theoretical results, obtained from accurate variational calculations of the $l = 0$, 1 and 2 phase shifts (performed respectively by C. Schwartz, R.L. Armstead and D. Register, and R.T. Poe), are seen to be in excellent agreement with the absolute experimental data of J.F. Williams. As a second example, the total and momentum transfer cross-sections for low-energy elastic electron–helium scattering are displayed in Fig. 13.4. The experimental data of D. Andrick and A. Bitsch, of W.E. Kauppila *et al.* and of R.W. Crompton *et al.* are seen to agree very well with the variational calculations of R.K. Nesbet and with the results of T.F. O'Malley, P.G. Burke and K.A. Berrington obtained by using the R-matrix method, to which we now turn our attention.

As we have seen in Section 12.7, the basic idea of the *R-matrix method* is that in most collision problems there exists an *internal region* where the interaction is complicated, and an *external region* where it is relatively simple, so that the dynamics of the projectile–target system depends on their relative distance. Thus,

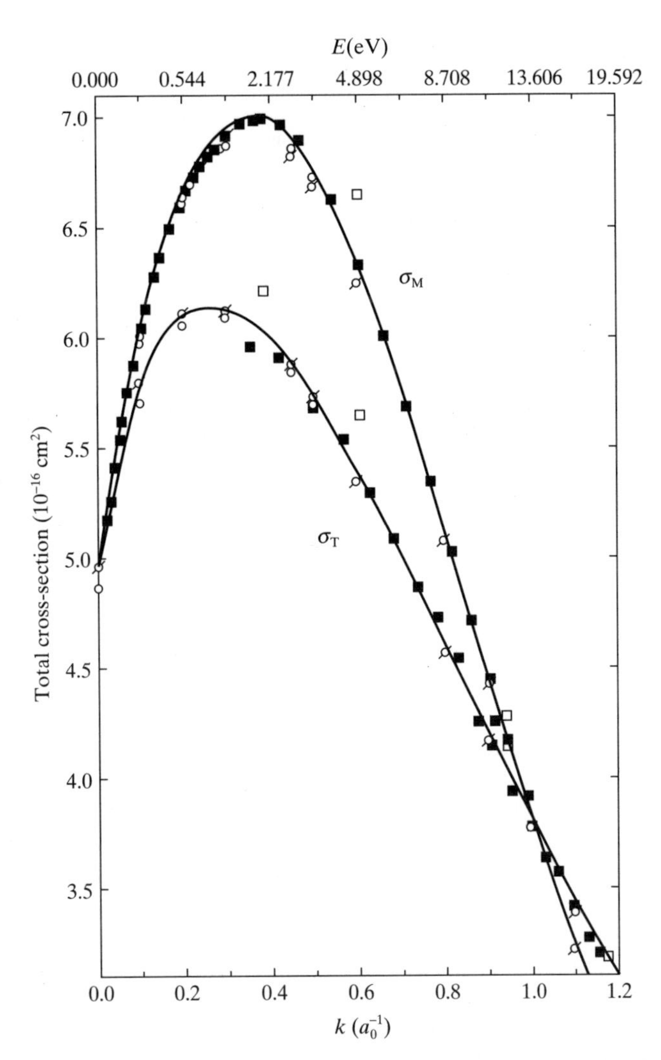

Figure 13.4 The total cross-section (σ_T) and momentum transfer cross-section (σ_M) for elastic electron–helium scattering.
The experimental data for σ_T are due to D. Andrick and A. Bitsch: □, and to W. Kauppila *et al.*: ■.
The experimental data for σ_M are due to D. Andrick and A. Bitsch: □, and to R.W. Crompton *et al.*: ■.
The theoretical cross-sections are the variational calcualtions of R.K. Nesbet: ——; and the *R*-matrix calculations of T.F. O'Malley *et al.*: ø and ○.

in the present case, if r denotes the relative distance of the colliding electron and the target atom (ion), it is useful to partition configuration space into two regions by a sphere of radius $r = a$ (see Fig. 13.5). The radius a is chosen so that the charge distribution of the target states of interest is contained within the sphere. In the internal region ($r < a$) the direct, exchange and correlation potentials in equations

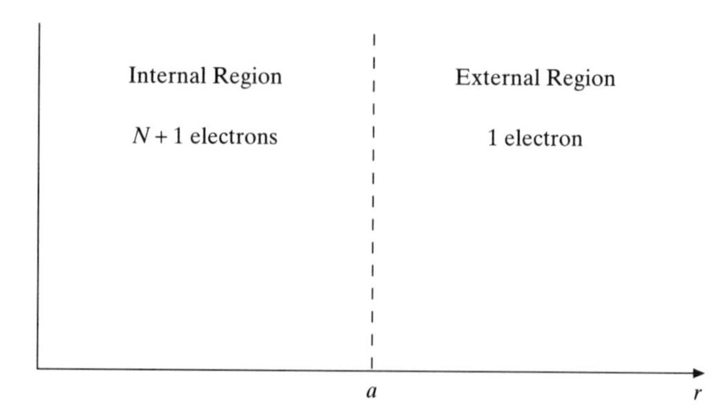

Figure 13.5 Partitioning of configuration space in the *R*-matrix method.

(13.95) are all non-zero and the $N+1$ electron system consisting of the projectile electron and the N-electron target has many of the properties of a bound state. By contrast, in the external region ($r > a$), the exchange and correlation effects between the unbound electron and the target can be neglected, so that the unbound electron moves in the direct potential V_{jk}^{Γ} which has reached its asymptotic form (13.96). The collision problem can then be solved in this region by using asymptotic methods.

As in the case of potential scattering discussed in Section 12.7, the link between the solutions in the internal and external regions is provided by the R-matrix, which enables the logarithmic derivative of the reduced radial functions $u_j^{\Gamma}(r)$ on the boundary $r = a$ of the internal region to be calculated. This is then matched at $r = a$ to the logarithmic derivative of the external region reduced radial functions, giving the K-matrix, the S-matrix or the T-matrix from which the scattering amplitudes and cross-sections can be obtained. It is important to note that, as in the case of potential scattering, the R-matrix is given *at all energies* by an expression of the type (12.439), so that the scattering quantities (phase shifts, scattering amplitudes and cross-sections) can be computed for a large number of energy values with little more numerical effort than for a few values. Comprehensive reviews of the R-matrix method for electron–atom (ion) collisions have been given by Burke and Robb (1975) and Burke and Berrington (1993).

As an example, we show in Fig. 13.6 the collision strength for electron–helium $1\,^1\text{S}$–$2\,^3\text{S}$ excitation, calculated using the R-matrix method by P.M.J. Sawey, K.A. Berrington, P.G. Burke and A.E. Kingston. The results obtained including the first 5, 11, 19 and 29 target eigenstates in the expansion (13.92) are shown. The various calculations agree close to threshold, but differ at higher energies. The 29-state calculation, which includes all the helium target eigenstates up to and including the 1s*nl* singlet and triplet states with $n = 5$, is accurate to about 10% up to and including the $n = 5$ singlet thresholds. This calculation exhibits *resonances* converging to each of the excited thresholds. We shall return below to the analysis of such resonances.

The R-matrix method has also been applied extensively to the calculation of electron–ion excitation cross-sections; a survey of these calculations has been given by Burke and Berrington (1993). We also remark that the fact that the R-matrix

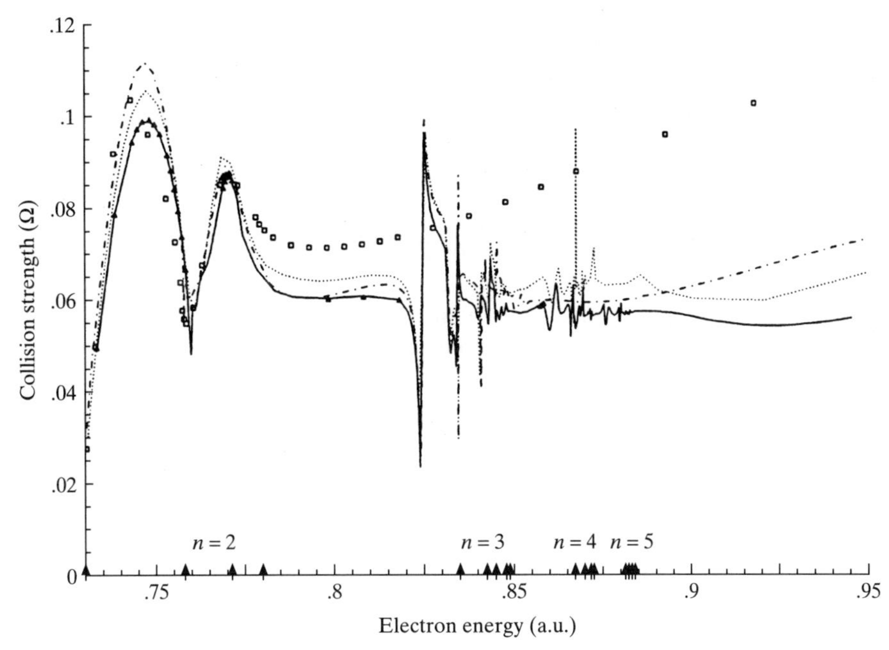

Figure 13.6 The collision strength for electron impact excitation of the ground state of helium to the 2^3S state. The theoretical calculations are due to P.M.J. Sawey *et al.*, who used the *R*-matrix method. Full curve: 29-state calculation; dotted curve: 19-state; chain curve: 11-state; and □: 5-state calculation. ▲: 11 states ($n \leqslant 3$) with $n \leqslant 5$ target orbitals. The arrows indicate the energies of the inelastic thresholds.

is a real *meromorphic* function of the energy, with simple poles on the real axis, provides the basis of the *quantum defect theory* developed by M.J. Seaton, M. Gailitis and U. Fano [4], in which the description of an electron in the field of a positive ion is made in terms of analytically known functions of the energy. In this way, electron–ion scattering can be analysed in terms of a few parameters.

The *J-matrix method* developed by E.J. Heller and H.A. Yamani, and by J.T. Broad and W.P. Reinhardt, uses a basis set of square integrable functions which span the target space. The functions which are usually chosen are non-orthogonal Laguerre functions obeying a three-term recurrence relation. Overlaps of the basis functions then form a tridiagonal or Jacobi (J) matrix. The full Hamiltonian is diagonalised in a finite subset of this basis, giving target eigenstates and pseudo-states. The channel functions themselves are also expanded in the same basis. The expansion coefficients are found to satisfy a recurrence relation from which the scattering quantities can be obtained. In common with the R-matrix method, the J-matrix approach has the advantage that the scattering quantities can be computed for many values of the energy with only slightly more numerical work than for a few values.

[4] A review of quantum defect theory has been given by Seaton (1983).

The *convergent close-coupling* (CCC) method is a generalised close-coupling approach developed by I. Bray and A.T. Stelbovics where, in addition to the discrete target states, the target continuum is treated with the help of square integrable pseudo-states. All of these states are obtained by diagonalising the target Hamiltonian in an orthogonal Laguerre basis. The resulting coupled integral Lippmann–Schwinger equations in momentum space are then partial wave analysed and solved numerically.

One of the earliest applications of the CCC method was to the Temkin–Poet model, a simplification of the full electron–atomic hydrogen scattering problem that treats only states with zero orbital angular momentum. This model problem was solved previously to high accuracy by A. Temkin for incident electron energies below the first inelastic threshold, and by R. Poet above the inelastic and ionisation thresholds. Comparison of the CCC results with the very accurate Temkin–Poet ones showed that the CCC method yielded convergent results as the basis size was increased, and that the convergence was to the correct results. The CCC method was then successfully applied to the full electron–atomic hydrogen scattering problem, as well as to electron scattering by other atomic targets. As an example, we show in Fig. 13.7 the total excitation cross-sections for the 2s

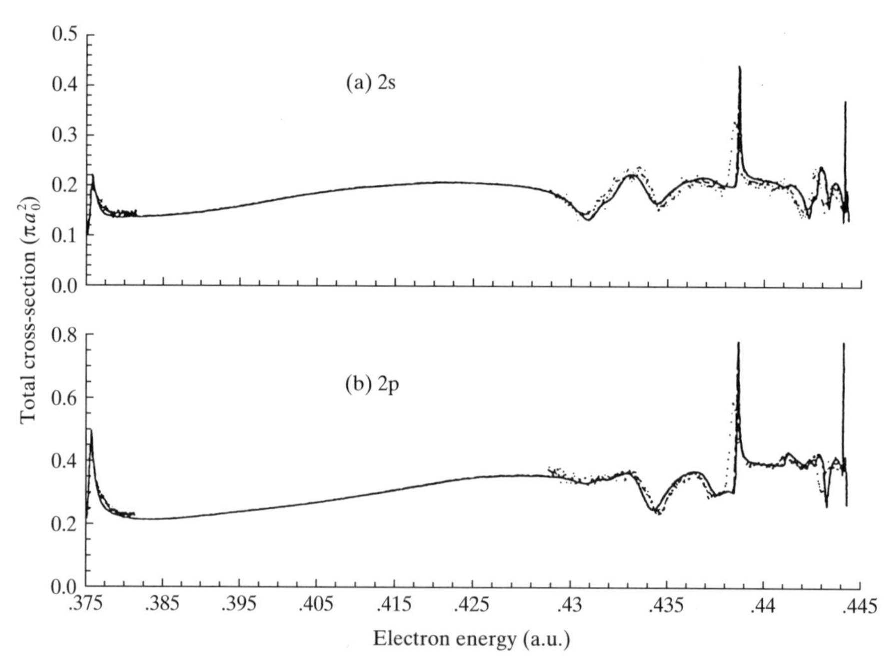

Figure 13.7 Total cross-sections for electron impact excitation of H(1s) to (a) the 2s level and (b) the 2p level, for collision energies between the $n = 2$ and $n = 3$ thresholds. The small dots represent the experimental data of J.F. Williams.
The theoretical calculations of K. Bartschat *et al.* using the *R*-matrix and CCC methods are represented by the solid line.

and 2p states of atomic hydrogen for electron impact energies between the $n = 2$ and $n = 3$ thresholds, calculated by K. Bartschat, I. Bray, P.G. Burke and M.P. Scott by using both the R-matrix and the CCC methods. The agreement between the two theoretical calculations is excellent, and so is the agreement with the experimental data of J.F. Williams. As in the case of low-energy electron–helium scattering discussed above (see Fig. 13.6), we see that the excitation cross-sections displayed in Fig. 13.7 exhibit a resonant behaviour. Resonances are ubiquitous in low-energy electron scattering by atoms and ions, and we shall now discuss them in some detail.

Resonances

We have seen in Chapter 12 within the context of potential scattering that if an effective potential well possesses metastable energy levels so that the incident particle is temporarily captured, then the cross-section will exhibit resonances at the corresponding energies.

In general, there are two kinds of collision processes in which resonant states can occur. The first one is a *formation* process such that the incident particle A combines with the target particle B to form a metastable complex $\tilde{C}$

$$A + B \rightarrow \tilde{C} \tag{13.111}$$

Subsequently, the metastable state $\tilde{C}$ may decay into the incident channel:

$$\tilde{C} \rightarrow A + B \tag{13.112}$$

or, if enough energy is available, other modes of decay may be possible. For example, $\tilde{C}$ may decay into excited states of A and (or) B:

$$\begin{aligned} \tilde{C} &\rightarrow A^* + B \\ &\rightarrow A + B^* \\ &\rightarrow A^* + B^* \end{aligned} \tag{13.113}$$

The resonances considered in Chapter 12 are particularly simple examples of formation processes in which only elastic scattering between the two particles A and B occurs, described by a central potential $V(r)$.

An example of resonance production in electron scattering involving the formation of a metastable complex is the process

$$e^- + He^+ \rightarrow He^{**} \rightarrow e^- + He^+ \tag{13.114}$$

considered in Section 7.7, in which a doubly excited, autoionising state He^{**} of helium is temporarily formed if the total energy E of the e^-–He^+ system coincides with the energy of the doubly excited state. If the energy of the metastable complex He^{**} is below the energy of He^+ ($n = 2$), then the resonance can only appear in the elastic scattering channel. On the other hand, if the energy of the metastable state He^{**} is higher than that of He^+ ($n = 2$), then resonances will appear in all the open channels.

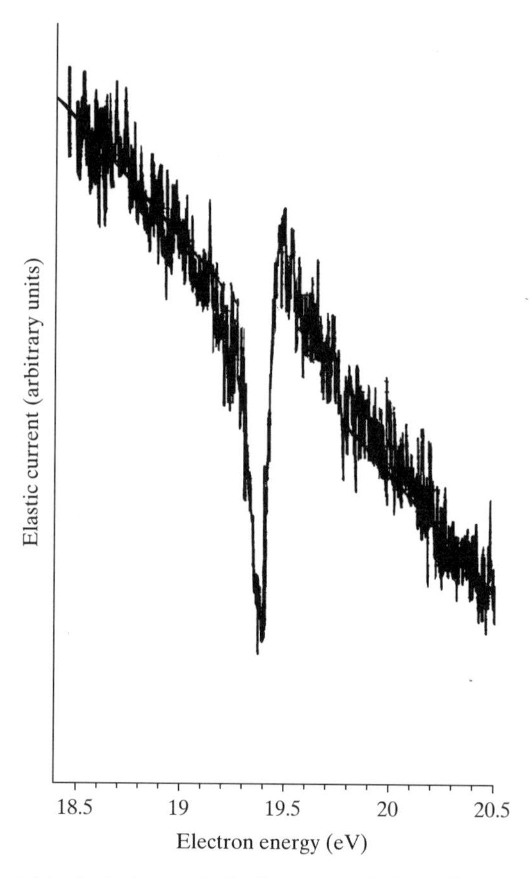

Figure 13.8 The yield of electrons elastically scattered through 72° by helium atoms, as observed by G.J. Schulz. A window resonance is seen to appear at an electron energy of 19.4 eV. (From G.J. Schulz, *Physical Review Letters* **10**, 104, 1963.)

Other examples of formation resonances are those occurring in electron scattering by neutral atoms, for instance in electron–helium collisions. In this case, the metastable complex is He^-. The simplest case occurs in the elastic channel

$$e^- + He(1\,^1S) \rightarrow He^- \rightarrow e^- + He(1\,^1S) \tag{13.115}$$

where a window resonance corresponding to a ^2S state of He^- appears at an electron energy of 19.4 eV, just below the energy of the first excited ($2\,^3S$) state of helium at 19.8 eV (see Fig. 13.8). This resonance was first observed in 1963 by G.J. Schulz. If enough energy is available, other metastable states of He^- can be formed, giving rise to resonances which can be observed in inelastic channels as well as in the elastic one. For instance, the prominent resonance appearing in Fig. 13.6 between the $2\,^3S$ threshold and the $2\,^1S$ threshold corresponds to a $^2P^0$ state of He^- having an energy of 20.3 eV.

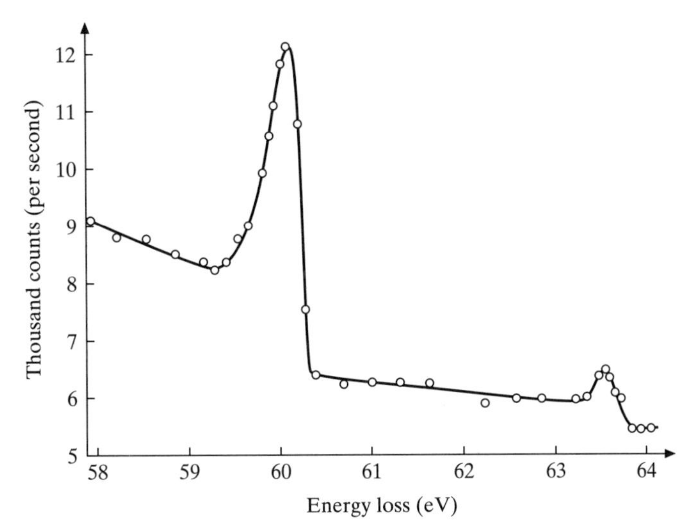

Figure 13.9 The energy loss spectrum of 500 eV electrons incident on helium, as observed by S.M. Silverman and E.N. Lassettre. Resonances are seen to occur at energies of 60.0 eV and 63.5 eV, respectively, above the helium ground state.

The second kind of collision process in which resonances can appear is a *production* process in which a metastable complex $\tilde{B}$ is produced, along with other particles. One simple case is when the target B is itself excited to a metastable state $\tilde{B}$,

$$A + B \to A + \tilde{B} \qquad \textbf{(13.116)}$$

and $\tilde{B}$ subsequently decays:

$$\tilde{B} \to C + D \qquad \textbf{(13.117)}$$

In this case, the resonance can be detected by measuring the energy loss of the scattered particle A, which is equal to the difference in energy of $\tilde{B}$ and B.

As an example of resonances occurring in a production process, we show in Fig. 13.9 the data of S.M. Silverman and E.N. Lassettre, who observed resonances corresponding to the process

$$e^- + \mathrm{He}(1\,^1S) \to e^- + \mathrm{He}^{**} \qquad \textbf{(13.118)}$$

As seen from Fig. 13.9, two resonances appear in the electron energy loss spectrum at energies of 60.0 eV and 63.5 eV, respectively, above the helium ground state.

There are several ways in which the theory of resonances can be developed. A convenient one is to use the Feshbach projection operator formalism introduced earlier. The wave function Ψ of the electron–atom system is divided into a part $P\Psi$ containing the open channels and another part $Q\Psi$ containing the closed channels. From (13.71)–(13.74), it follows that the Schrödinger equation (13.1) satisfied by Ψ is equivalent to the pair of coupled equations

$$(H_{PP} - E)P\Psi + H_{PQ}Q\Psi = 0 \quad \textbf{(13.119a)}$$

and

$$(H_{QQ} - E)Q\Psi + H_{QP}P\Psi = 0 \quad \textbf{(13.119b)}$$

where $H_{PP} = PHP$, $H_{QQ} = QHQ$, $H_{PQ} = PHQ$ and $H_{QP} = QHP$ (see (13.77)). Solving the equation (13.119b) for $Q\Psi$ and inserting this result into (13.119a), we obtain for $P\Psi$ the equation

$$(H_{PP} + \mathcal{V} - E)P\Psi = 0 \quad \textbf{(13.120)}$$

where the equivalent potential $\mathcal{V}$ is given by (13.80).

It is useful to distinguish two categories of resonances: those arising from the interaction contained in H_{PP}, and those coming from the equivalent potential $\mathcal{V}$. The former are called *shape* or *open-channel* resonances. If there is only one open channel, a shape resonance will occur if the effective interaction potential between the projectile electron and the target atom has the required strength and shape to have a metastable state, as discussed in Section 12.3.

The resonances arising from the equivalent potential $\mathcal{V}$ are called *Feshbach* or *closed-channel* resonances. We shall now show that they arise from the presence of zeros in the denominator of the expression (13.80) for $\mathcal{V}$. Indeed, let us assume that we have solved the eigenvalue problem

$$H_{QQ}\chi_{\mathrm{r}} = \varepsilon_{\mathrm{r}}\chi_{\mathrm{r}} \quad \textbf{(13.121)}$$

As we have seen previously, the eigenvalues ε_{r} are discrete below the excitation threshold. We also note that the wave functions χ_{r} are bound state solutions of an equation obtained from (13.119b) by omitting the coupling term $H_{QP}P\Psi$ which permits the decay of $Q\Psi$ into $P\Psi$. Now, a resonant state of the compound system (incident particle plus target) has a very long lifetime with respect to typical collision times, and its probability of re-entering an open channel is small. We may therefore, in first approximation, interpret a wave function χ_{r} as corresponding to a *resonant state*, while the resonance energy is given approximately by ε_{r}. When the coupling term $H_{QP}P\Psi$ is taken into account, and the equation (13.119a) is solved for the open-channel part $P\Psi$ of the wave function, resonances are found at energies $E_{\mathrm{r}} = \varepsilon_{\mathrm{r}} + \Delta_{\mathrm{r}}$ (where Δ_{r} are small *level shifts*) and with finite *widths* Γ_{r}.

As an example, the calculated positions and widths of some doubly excited states in helium giving rise to resonances in e^-–$He^+(1s)$ elastic scattering (below the $n = 2$ threshold) are given in Table 13.1. Also given are the corresponding configurations of the states. Similar sequences of levels give rise to resonances just below the $n = 3, n = 4, \ldots$ thresholds. Some of these resonances can be seen in Fig. 13.6.

Finally, we note that because of the strong Coulomb attraction, there are generally more resonances in electron scattering by positive ions than by neutral atoms.

Table 13.1 Resonances in e^-–$He^+(1s)$ elastic scattering below the $n = 2$ threshold.

^1S			^3S		
E_r (eV)	Γ (eV)	*Configuration*	E_r (eV)	Γ (eV)	*Configuration*
57.84	0.12	$2s^2$	62.62	2×10^{-4}	2s3s
62.13	0.007	$2p^2$	63.76	7×10^{-6}	2p3p
62.97	0.036	2s3s	63.95	9×10^{-5}	2s4s
^1P			^3P		
E_r (eV)	Γ (eV)	*Configuration*	E_r (eV)	Γ (eV)	*Configuration*
60.19	4.4×10^{-2}	2s2p	58.30	1×10^{-2}	2s2p
62.82	1.4×10^{-2}	2s3p – 2p3s	63.15	3×10^{-3}	2s3p + 2p3s
63.88	8.7×10^{-3}	2s3p + 2p3s	63.94	8×10^{-5}	2s3p – 2p3s

E_r = Energies of the doubly excited levels above the ground state of He.
Γ = Level widths.

Relativistic effects for heavy atoms and ions

As the nuclear charge Z of the target increases, relativistic effects become significant even for low-energy electron scattering. There are two ways in which relativistic effects manifest themselves. Firstly, the strong nuclear Coulomb potential gives rise to a relativistic distortion of the wave function describing the unbound electron, due to spin–orbit effects. Secondly, the structure of the target is modified by relativistic effects.

In order to take into account relativistic effects for high Z values, one must describe the system by using a relativistic Hamiltonian. For example, one can use the Breit–Pauli Hamiltonian [5]

$$H_{\mathrm{BP}} = H + H'_{\mathrm{rel}} \tag{13.122}$$

where H is the non-relativistic Hamiltonian (13.2) and H'_{rel} contains the spin–orbit, Darwin and mass correction terms discussed in Section 5.1. Alternatively, one can use a generalised Dirac Hamiltonian

$$H_{\mathrm{D}} = \sum_{i=0}^{N} \left(c\boldsymbol{\alpha} \cdot \mathbf{p}_i + \beta c^2 - \frac{Z}{r_i} \right) + \sum_{i<j=0}^{N} \frac{1}{r_{ij}} \tag{13.123}$$

where $\boldsymbol{\alpha}$ and β are the Dirac matrices defined by (A7.21).

[5] See Bethe and Salpeter (1957).

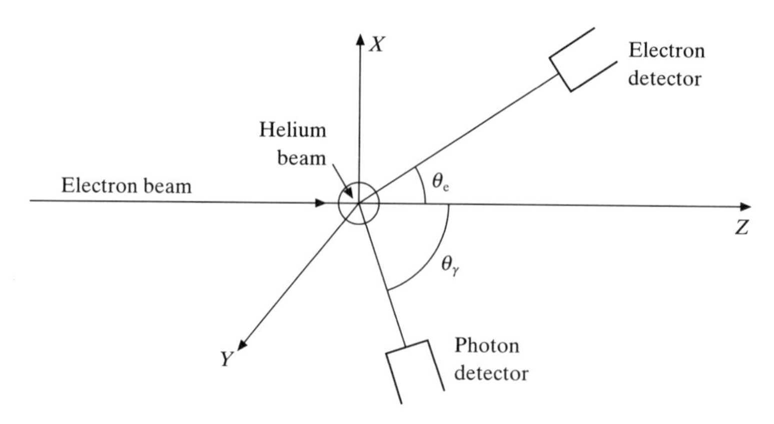

Figure 13.10 The geometry of the electron–photon coincidence experiment of M. Eminyan *et al.* The *Y* axis is perpendicular to the scattering (*XZ*) plane. The coordinate axes *XYZ* define the 'collision frame'.

Electron–photon coincidence and polarised electron experiments

One of the important developments which have taken place in the study of electron–atom collisions has been towards 'perfect scattering experiments', in which the complete scattering amplitudes are determined (apart from an unobservable phase factor). Since more than one amplitude is required for a complete description of the collision, several independent measurements must be performed. One class of experiments is that in which the target atoms are excited by electron impact, and the scattered electrons are observed in coincidence with the photons emitted by the excited atoms. A schematic diagram of the first experiment of this kind, performed in 1973 by M. Eminyan, K.B. MacAdam, J. Slevin and H. Kleinpoppen for the case of electron-helium excitation, is shown in Fig. 13.10. Such experiments have been discussed in detail by Slevin (1984) and Andersen (1997).

A second class of experiments is that in which polarised beams of electrons are used. As an example, let us focus our attention on the scattering of electrons by light spin-1/2 atoms for which the spin–orbit interaction can be neglected. We shall consider the scattering process $(\mathbf{k}_i, i) \rightarrow (\mathbf{k}_j, j)$ between two particular atomic states $|i\rangle$ and $|j\rangle$. As we have seen above, this process is described by two scattering amplitudes: f_{ji} and g_{ji}, or f_{ji}^+ and f_{ji}^-.

Let us first assume that the incident electrons and the target atoms are completely polarised. Denoting the spin directions by the symbols $\uparrow$ and $\downarrow$, the scattering of incident electrons with spin antiparallel or parallel to the spin of the target electron leads to the following possibilities:

Process	Differential cross-section	
$e^- \uparrow + A \downarrow \to e^- \uparrow + A \downarrow$	$\dfrac{k_j}{k_i} \vert f_{ji} \vert^2$	**(13.124a)**
$e^- \uparrow + A \downarrow \to e^- \downarrow + A \uparrow$	$\dfrac{k_j}{k_i} \vert g_{ji} \vert^2$	**(13.124b)**
$e^- \uparrow + A \uparrow \to e^- \uparrow + A \uparrow$	$\dfrac{k_j}{k_i} \vert f_{ji} - g_{ji} \vert^2$	**(13.124c)**

These results can readily be understood by noting that in the first two processes the incident and target electrons can be distinguished by their spin orientation. In the absence of spin–orbit interactions the process (13.124a) can only occur by direct scattering, while the spin-flip process (13.124b) can only happen by exchange scattering. On the other hand, in the process (13.124c) the incident and target electrons are truly indistinguishable, so that an antisymmetric spatial function must be used, leading to the scattering amplitude $f_{ji} - g_{ji}$.

In contrast with experiments performed with unpolarised particles, which generally result in the measurement of spin-averaged cross-sections (see (13.22) or (13.29)), we see from (13.124) that experiments using polarised beams and targets yield separate information about direct and exchange scattering. In fact, it is sufficient for the observation of the individual cross-sections listed in (13.124) to perform simpler experiments, in which either the electrons or the target atoms are initially polarised [6]. Furthermore, in practice one cannot use fully polarised electron beams or targets, but this does not introduce any basic difficulty. Indeed, by using the density matrix formalism [7], it can be shown that a partially polarised electron beam with a degree of polarisation P_e (such that $0 < P_e < 1$) can be considered to be made of a totally polarised fraction and an unpolarised fraction in the ratio $P_e/(1 - P_e)$.

Among the various experiments using polarised electron beams and (or) targets, let us consider one in which partially polarised electrons are scattered by partially polarised hydrogen atoms, and the scattered electron intensity is measured for the incident electron and atomic electron spins antiparallel ($\downarrow \uparrow$) and parallel ($\uparrow \uparrow$). The information provided by the experiment can be expressed in the form of the measured asymmetry

$$\Delta = P_e P_a A \tag{13.125}$$

where P_e and P_a are the degrees of polarisation of the incident electrons and of the atomic electrons, respectively, and A is the asymmetry parameter, defined for a transition from target state $|i\rangle$ to $|j\rangle$ by

[6] See Kessler (1985).

[7] See Bransden and Joachain (2000).

$$A = \frac{\dfrac{d\sigma_{ji}}{d\Omega}(\uparrow\downarrow) - \dfrac{d\sigma_{ji}}{d\Omega}(\uparrow\uparrow)}{\dfrac{d\sigma_{ji}}{d\Omega}(\uparrow\downarrow) + \dfrac{d\sigma_{ji}}{d\Omega}(\uparrow\uparrow)} \qquad \textbf{(13.126a)}$$

Here $d\sigma_{ji}(\uparrow\downarrow)/d\Omega$ and $d\sigma_{ji}(\uparrow\uparrow)/d\Omega$ are the differential cross-sections for the process $(\mathbf{k}_i, i) \to (\mathbf{k}_j, j)$ when the incident electron and target atom electron spins are antiparallel and parallel, respectively. We note that the asymmetry parameter (13.126a) may be rewritten in the forms

$$A = \frac{|f_{ji}|^2 + |g_{ji}|^2 - |f_{ji} - g_{ji}|^2}{|f_{ji}|^2 + |g_{ji}|^2 + |f_{ji} - g_{ji}|^2} \qquad \textbf{(13.126b)}$$

or

$$A = \frac{|f_{ji} + g_{ji}|^2 - |f_{ji} - g_{ji}|^2}{|f_{ji} + g_{ji}|^2 + 3|f_{ji} - g_{ji}|^2} = \frac{1 - R_{ji}}{1 + 3R_{ji}} \qquad \textbf{(13.126c)}$$

where $(d\sigma_{ji}/d\Omega)_{unp}$ is the unpolarised differential cross-section (13.29), and

$$R_{ji} = \frac{\dfrac{d\sigma_{ji}}{d\Omega}(S=1)}{\dfrac{d\sigma_{ji}}{d\Omega}(S=0)} \qquad \textbf{(13.126d)}$$

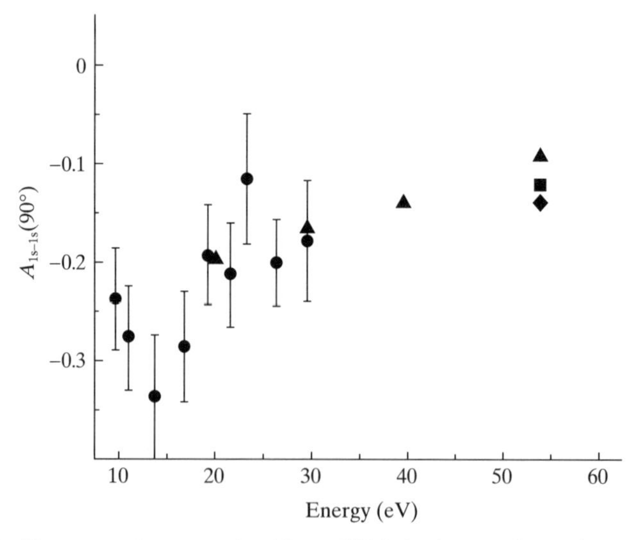

Figure 13.11 The asymmetry parameter A for e^-–H(1s) elastic scattering at the scattering angle $\theta = 90°$, as a function of the incident electron energy.
The experimental data are due to G.D. Fletcher *et al.*: ●.
The theoretical calculations of M.R.C. McDowell *et al.* are: static-exchange approximation: ▲; six-state close-coupling calculation: ■; UEBS calculation: ◆.

is the ratio of the triplet ($S = 1$) to the singlet ($S = 0$) differential cross-sections for the transition $(\mathbf{k}_i, i) \rightarrow (\mathbf{k}_j, j)$. From (13.126c) we see that the lower bound of A is equal to $-1/3$ and is reached for pure triplet scattering.

In Fig. 13.11, the experimental data of G.D. Fletcher, M.J. Alguard, T.J. Gay, V.W. Hughes, C.W. Tu, P.F. Wainwright and M.S. Lubell are compared with the theoretical calculations of M.R.C. McDowell, P.W. Edmunds, R.M. Potvliege, C.J. Joachain, R. Shingal and B.H. Bransden, for the differential asymmetry parameter A corresponding to e^-–H(1s) elastic scattering at a scattering angle $\theta = 90°$. The agreement between theory and experiment is seen to be good.

13.3 Elastic and inelastic electron–atom collisions at high energies

Let us now consider electron–atom (ion) elastic scattering and excitation at impact energies larger than the ionisation energy of the target. In this case an infinite number of channels are open and it is not possible to represent all these open channels explicitly in the total electron–atom wave function. In this section we shall examine various theoretical methods which have been proposed to deal with electron–atom(ion) collisions at high, but non-relativistic, energies.

The Born series

In the high-energy region it is reasonable to use an approach based on perturbation theory. It is convenient to calculate separately the direct amplitude f_{ji} and the exchange amplitude g_{ji}, since exchange effects become progressively unimportant as the incident electron energy increases.

For notational simplicity, we shall mainly discuss electron–atomic hydrogen collisions; the generalisation to complex target atoms will be indicated as necessary. For e^-–H collisions, the total Hamiltonian given by (13.7) can be split either as

$$H = H_d + V_d \qquad \textbf{(13.127a)}$$

or

$$H = H_e + V_e \qquad \textbf{(13.127b)}$$

where

$$H_d = T_0 + h(1) \qquad \textbf{(13.128a)}$$

is the direct channel Hamiltonian and

$$H_e = T_1 + h(0) \qquad \textbf{(13.128b)}$$

is the exchange channel Hamiltonian. Here $T_i = -\nabla^2_{\mathbf{r}_i}/2$ (with $i = 0, 1$) is the kinetic energy operator of electron i, and

$$h(i) = T_i - \frac{1}{r_i} \qquad \textbf{(13.129)}$$

is the Hamiltonian of the hydrogen atom for the electron i. In the direct channel, the interaction between the unbound electron and the hydrogen atom is

$$V_{\mathrm{d}} = -\frac{1}{r_0} + \frac{1}{r_{01}} \tag{13.130a}$$

and in the exchange channel it is

$$V_{\mathrm{e}} = -\frac{1}{r_1} + \frac{1}{r_{01}} \tag{13.130b}$$

The eigenfunctions of H_{d} are of the form

$$\Phi_{\mathbf{k}_j,j}(\mathbf{r}_0, \mathbf{r}_1) = (2\pi)^{-3/2} \exp(\mathrm{i}\mathbf{k}_j\cdot\mathbf{r}_0)\psi_j(\mathbf{r}_1), \tag{13.131a}$$

the corresponding energies being $k_j^2/2 + E_j$. Using the Dirac notation, we shall write the eigenvectors of H_{d} as

$$|\Phi_{\mathbf{k}_j,j}(0, 1)\rangle \equiv |\mathbf{k}_j, j\rangle \tag{13.131b}$$

Similarly, the eigenfunctions of H_{e} are given by

$$\tilde{\Phi}_{j,\mathbf{k}_j}(\mathbf{r}_0, \mathbf{r}_1) = \Phi_{\mathbf{k}_j,j}(\mathbf{r}_1, \mathbf{r}_0) = (2\pi)^{-3/2}\psi_j(\mathbf{r}_0) \exp(\mathrm{i}\mathbf{k}_j\cdot\mathbf{r}_1) \tag{13.132a}$$

and will be written in Dirac notation as

$$|\tilde{\Phi}_{j,\mathbf{k}_j}(0, 1)\rangle \equiv |j, \mathbf{k}_j\rangle \tag{13.132b}$$

The Born series for the direct scattering amplitude f_{ji} corresponding to the transition $a \equiv (\mathbf{k}_i, i) \rightarrow b \equiv (\mathbf{k}_j, j)$ is

$$f_{ji} = \sum_{n=1}^{\infty} \bar{f}_{ji}^{\mathrm{B}n} \tag{13.133}$$

where the first Born term is

$$\bar{f}_{ji}^{\mathrm{B1}} = -(2\pi)^2 \langle \mathbf{k}_j, j\,|V_{\mathrm{d}}|\,\mathbf{k}_i, i\rangle \tag{13.134}$$

and the nth Born term $\bar{f}_{ji}^{\mathrm{B}n}$ is given for $n \geqslant 2$ by

$$\bar{f}_{ji}^{\mathrm{B}n} = -(2\pi)^2 \langle \mathbf{k}_j, j\,|V_{\mathrm{d}}G_{\mathrm{d}}^{(+)}V_{\mathrm{d}} \ldots G_{\mathrm{d}}^{(+)}V_{\mathrm{d}}|\,\mathbf{k}_i, i\rangle, \qquad n \geqslant 2 \tag{13.135}$$

This expression contains n times the direct interaction V_{d} and $(n - 1)$ times the Green's operator $G_{\mathrm{d}}^{(+)}$, where

$$G_{\mathrm{d}}^{(+)} = \frac{1}{E + \mathrm{i}\varepsilon - H_{\mathrm{d}}}, \qquad \varepsilon \rightarrow 0^+ \tag{13.136}$$

We also define the sth-order Born approximation to the direct scattering amplitude as

$$f_{ji}^{\mathrm{B}s} = \sum_{n=1}^{s} \bar{f}_{ji}^{\mathrm{B}n} \tag{13.137}$$

In particular, the first Born approximation is given by $f_{ji}^{\mathrm{B1}} = \bar{f}_{ji}^{\mathrm{B1}}$. The second Born term is

$$\bar{f}_{ji}^{\mathrm{B2}} = -(2\pi)^2 \langle \mathbf{k}_j, j | V_{\mathrm{d}} G_{\mathrm{d}}^{(+)} V_{\mathrm{d}} | \mathbf{k}_i, i \rangle \tag{13.138}$$

and the second Born approximation to the direct scattering amplitude is given by

$$f_{ji}^{\mathrm{B2}} = \bar{f}_{ji}^{\mathrm{B1}} + \bar{f}_{ji}^{\mathrm{B2}} \tag{13.139}$$

Finally, we note that a Born series for the exchange amplitude g_{ji} can also be written down, in a similar manner to that for the direct amplitude f_{ji}.

The first Born approximation

We shall now study in more detail the first Born approximation for elastic and inelastic electron–atom collisions. We begin by analysing *direct* collisions, first for electron scattering by atomic hydrogen, and then for complex target atoms.

Atomic hydrogen

Using (13.134) and the expression (13.130a) for the direct potential V_{d}, we have

$$f_{ji}^{\mathrm{B1}} = -\frac{1}{2\pi} \int \mathrm{d}\mathbf{r}_0 \, \mathrm{d}\mathbf{r}_1 \exp(-\mathrm{i}\mathbf{k}_j \cdot \mathbf{r}_0)\, \psi_j^*(\mathbf{r}_1) \left(-\frac{1}{r_0} + \frac{1}{r_{01}} \right) \exp(\mathrm{i}\mathbf{k}_i \cdot \mathbf{r}_0)\, \psi_i(\mathbf{r}_1) \tag{13.140}$$

Introducing the momentum transfer

$$\boldsymbol{\Delta} = \mathbf{k}_i - \mathbf{k}_j \tag{13.141}$$

and using the definition (13.43) of the direct potential matrix, we can write (13.140) in the form

$$f_{ji}^{\mathrm{B1}} = -\frac{1}{2\pi} \int \exp(\mathrm{i}\boldsymbol{\Delta} \cdot \mathbf{r}_0) V_{ji}(\mathbf{r}_0) \, \mathrm{d}\mathbf{r}_0 \tag{13.142}$$

Elastic scattering

Let us first consider elastic direct scattering, so that $j = i$. In this case $k_i = k_j = k$, and the magnitude of the momentum transfer is given as in potential scattering (see (12.271)) by

$$\Delta = 2k \sin \frac{\theta}{2} \tag{13.143}$$

The corresponding first Born direct elastic scattering amplitude is

$$f_{ii}^{\mathrm{B1}} = -\frac{1}{2\pi} \int \exp(\mathrm{i}\boldsymbol{\Delta} \cdot \mathbf{r}_0) V_{ii}(\mathbf{r}_0) \, \mathrm{d}\mathbf{r}_0 \tag{13.144}$$

and we note that it is independent of k (that is, of the incident electron energy) in the forward direction ($\Delta = 0$). In particular, for elastic scattering from H(1s), the potential $V_{1s,1s} \equiv V_{11}$ is the static potential (13.51). Evaluating the integral in (13.144) one finds (Problem 13.6) that

$$f^{B1}_{1s,1s} \equiv f^{B1}_{11} = 2\frac{\Delta^2 + 8}{(\Delta^2 + 4)^2} \tag{13.145}$$

and the corresponding elastic first Born differential cross-section is given by

$$\frac{d\sigma^{B1}_{1s,1s}}{d\Omega} = |f^{B1}_{1s,1s}|^2 = 4\frac{(\Delta^2 + 8)^2}{(\Delta^2 + 4)^4} \tag{13.146}$$

We remark that in the forward direction this cross-section is equal to unity, and falls off like Δ^{-4} for large Δ. The differential cross-section (13.146) is illustrated in Fig. 13.12 for an incident electron energy of 500 eV. The total first Born cross-section for elastic scattering from H(1s) is given by

$$\begin{aligned}\sigma^{B1}_{tot}(1s \to 1s) &= 2\pi \int_0^\pi \frac{d\sigma^{B1}_{1s,1s}}{d\Omega} \sin\theta \, d\theta \\ &= \frac{2\pi}{k^2} \int_0^{2k} \frac{d\sigma^{B1}_{1s,1s}}{d\Omega} \Delta \, d\Delta\end{aligned} \tag{13.147}$$

where we have used the fact that $\sin\theta \, d\theta = k^{-2}\Delta \, d\Delta$. Performing the integral, we find that (Problem 13.6)

$$\sigma^{B1}_{tot}(1s \to 1s) = \pi\frac{7k^4 + 18k^2 + 12}{3(1 + k^2)^3} \tag{13.148}$$

and we see that at high energies (so that $k \gg 1$), the leading term of $\sigma^{B1}_{tot}(1s \to 1s)$ is given by

$$\sigma^{B1}_{tot}(1s \to 1s) \simeq \frac{7\pi}{3}k^{-2} \tag{13.149}$$

Since $E^e_i = k_i^2/2$ is the energy of the incident electron (with $k_i = k$), we see that $\sigma^{B1}_{tot}(1s \to 1s)$ is proportional to $(E^e_i)^{-1}$ when E^e_i is large, as in potential scattering (see Section 12.7).

Inelastic collisions

Let us now turn to inelastic electron scattering by atomic hydrogen. In that case the final target state ψ_j differs from the initial one ψ_i. Since target wave functions corresponding to different states are orthogonal, the term $(-1/r_0)$ arising from the projectile electron–nucleus interaction does not contribute to the integral on the right of (13.140). Thus, in first Born approximation, inelastic scattering can only occur because of the interaction between the incident electron and the bound electron, and the first Born direct amplitude for inelastic scattering reduces to

$$f^{B1}_{ji} = -\frac{1}{2\pi} \int d\mathbf{r}_0 \, d\mathbf{r}_1 \exp(i\mathbf{\Delta}\cdot\mathbf{r}_0)\, \psi^*_j(\mathbf{r}_1)\frac{1}{r_{01}}\psi_i(\mathbf{r}_1), \qquad j \neq i \tag{13.150}$$

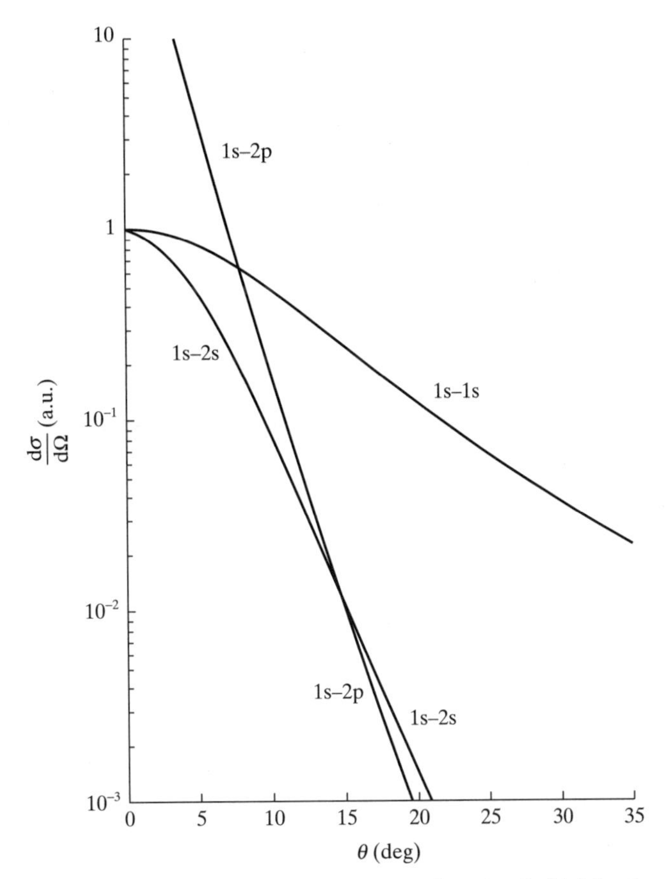

Figure 13.12 The first Born differential cross-sections (in units of a_0^2/sr) for the elastic scattering of electrons by H(1s), the excitation of H(1s) to the 2s level and the excitation of H(1s) to the 2p level by electron impact as a function of the scattering angle θ. The energy of the incident electrons is 500 eV.

We also note that for inelastic collisions the magnitude of the momentum transfer $\mathbf{\Delta} = \mathbf{k}_i - \mathbf{k}_j$ is given by

$$\Delta = (k_i^2 + k_j^2 - 2k_ik_j \cos\theta)^{1/2} \tag{13.151}$$

where θ is the scattering angle, and k_j is related to k_i by (13.5). In contrast to elastic collisions where $0 \leqslant \Delta \leqslant 2k$, in the inelastic scattering case Δ varies from a non-zero minimum value $\Delta_{\min} = |k_i - k_j|$ (reached for $\theta = 0$) to the maximum value $\Delta_{\max} = k_i + k_j$ (reached for $\theta = \pi$).

The integral over the variable $\mathbf{r}_0$ in (13.150) can be performed by using the *Bethe integral*

$$\int \frac{\exp(\mathrm{i}\boldsymbol{\Delta}\cdot\mathbf{r}_0)}{r_{01}}\,\mathrm{d}\mathbf{r}_0 = \exp(\mathrm{i}\boldsymbol{\Delta}\cdot\mathbf{r}_1)\int \frac{\exp(\mathrm{i}\boldsymbol{\Delta}\cdot\mathbf{r}_{01})}{r_{01}}\,\mathrm{d}\mathbf{r}_{01}$$
$$= \exp(\mathrm{i}\boldsymbol{\Delta}\cdot\mathbf{r}_1)\lim_{\alpha\to 0}\int \exp(\mathrm{i}\boldsymbol{\Delta}\cdot\mathbf{r}_{01})\frac{\exp(-\alpha r_{01})}{r_{01}}\,\mathrm{d}\mathbf{r}_{01}$$
$$= \frac{4\pi}{\Delta^2}\exp(\mathrm{i}\boldsymbol{\Delta}\cdot\mathbf{r}_1) \qquad \textbf{(13.152)}$$

so that (13.150) becomes

$$f_{ji}^{\mathrm{B1}} = -\frac{2}{\Delta^2}\int \psi_j^*(\mathbf{r}_1)\exp(\mathrm{i}\boldsymbol{\Delta}\cdot\mathbf{r}_1)\psi_i(\mathbf{r}_1)\,\mathrm{d}\mathbf{r}_1 \qquad \textbf{(13.153)}$$

As a first example, let us consider the excitation of the 2s state of atomic hydrogen from the ground state by fast electrons. In this case, the first Born direct scattering amplitude (13.153) becomes

$$f_{2\mathrm{s},1\mathrm{s}}^{\mathrm{B1}} = -\frac{2}{\Delta^2}\int \psi_{2\mathrm{s}}^*(r_1)\exp(\mathrm{i}\boldsymbol{\Delta}\cdot\mathbf{r}_1)\psi_{1\mathrm{s}}(r_1)\,\mathrm{d}\mathbf{r}_1 \qquad \textbf{(13.154)}$$

Performing the integral in (13.154), one obtains (Problem 13.7)

$$f_{2\mathrm{s},1\mathrm{s}}^{\mathrm{B1}} = -\frac{8\sqrt{2}}{(\Delta^2+\frac{9}{4})^3} \qquad \textbf{(13.155)}$$

The corresponding first Born differential cross-section is

$$\frac{\mathrm{d}\sigma_{2\mathrm{s},1\mathrm{s}}^{\mathrm{B1}}}{\mathrm{d}\Omega} = \frac{k_j}{k_i}\frac{128}{(\Delta^2+\frac{9}{4})^6} \qquad \textbf{(13.156)}$$

where

$$k_j = \left(k_i^2 - \frac{3}{4}\right)^{1/2} \qquad \textbf{(13.157)}$$

It is illustrated in Fig. 13.12 for an incident electron energy of 500 eV.

It is interesting to note that at high energies (where $k_j \simeq k_i$), the differential cross-section (13.156) is nearly constant as a function of k_i (that is, of the incident electron energy) in the forward direction. We also remark that for large values of Δ the expression (13.156) falls off like Δ^{-12}, much faster than the first Born elastic differential cross-section (13.146) which decreases like Δ^{-4} for large values of Δ. The much slower decrease of $\mathrm{d}\sigma_{1\mathrm{s},1\mathrm{s}}^{\mathrm{B1}}/\mathrm{d}\Omega$ at large scattering angles is due to the fact that the projectile–nucleus interaction ($-1/r_0$) contributes to the first Born amplitude (13.140) in the elastic scattering case ($i = j$), while it contributes nothing for inelastic collisions ($j \neq i$). We shall see below (see Table 13.3) that the leading contribution of the *second Born* term $\bar{f}_{2\mathrm{s},1\mathrm{s}}^{\mathrm{B2}}$ for the 1s → 2s transition behaves like $k_i^{-1}\Delta^{-2}$ when both k_i and Δ are large, so that the first Born differential cross-section (13.156) is useless for large momentum transfers, even at high energies.

The total first Born cross-section for the 1s → 2s transition is obtained by integrating the differential cross-section (13.156) over all scattering angles. Using (13.151), we have

$$\sigma_{\text{tot}}^{\text{B1}}(1\text{s} \rightarrow 2\text{s}) = 2\pi \frac{128}{k_i^2} \int_{\Delta_{\min}}^{\Delta_{\max}} \frac{\Delta}{(\Delta^2 + \frac{9}{4})^6} \, \text{d}\Delta \tag{13.158}$$

This integral can readily be evaluated analytically (Problem 13.7) but we shall only be interested here in obtaining the leading term of $\sigma_{\text{tot}}^{\text{B1}}(1\text{s} \rightarrow 2\text{s})$ at high energies ($k_i \gg 1$). To this end, we remark from (13.157) that for large values of k_i one has

$$k_j \simeq k_i \left(1 - \frac{3}{8k_i^2} + \cdots\right) \tag{13.159}$$

so that $\Delta_{\min} = k_i - k_j \simeq 3/(8k_i)$ and $\Delta_{\max} = k_i + k_j \simeq 2k_i$. Since most of the contribution to the integral in (13.158) comes from the region of small Δ, and because $\Delta_{\min}^2 \ll 9/4$, the leading term of $\sigma_{\text{tot}}^{\text{B1}}(1\text{s} \rightarrow 2\text{s})$ can be obtained by writing

$$\sigma_{\text{tot}}^{\text{B1}}(1\text{s} \rightarrow 2\text{s}) \simeq 2\pi \frac{128}{k_i^2} \int_0^{\infty} \frac{\Delta}{(\Delta^2 + \frac{9}{4})^6} \, \text{d}\Delta \tag{13.160}$$

Hence

$$\sigma_{\text{tot}}^{\text{B1}}(1\text{s} \rightarrow 2\text{s}) \simeq \frac{128\pi}{5} \left(\frac{2}{3}\right)^{10} k_i^{-2} \tag{13.161}$$

and we see that this cross-section decreases like $(E_i^{\text{e}})^{-1}$ for large incident electron energies, as in the elastic scattering case (see (13.149)).

As a second example of inelastic collisions, we shall consider the excitation of the 2p_m ($m = 0, \pm 1$) states of atomic hydrogen by fast electrons. In this case the first Born direct scattering amplitude is

$$f_{2\text{p}_m,1\text{s}}^{\text{B1}} = -\frac{2}{\Delta^2} \int \psi_{2\text{p}m}^*(\mathbf{r}_1) \exp(\text{i}\boldsymbol{\Delta}\cdot\mathbf{r}_1) \psi_{1\text{s}}(r_1) \, \text{d}\mathbf{r}_1 \tag{13.162}$$

Let us first choose the quantisation axis (the Z axis for the eigenfunctions) to be along the momentum transfer $\boldsymbol{\Delta}$. Upon performing the integration over the azimuthal angle ϕ_1, the factors $\exp(\pm\text{i}\phi_1)$ that appear in the wave functions $\psi_{2\text{p}_{\pm1}}$ for the magnetic substates $m = \pm 1$ prevent those substates from being excited. The only final substate which can be excited is the 2p_0 substate. The corresponding first Born direct scattering amplitude is (Problem 13.8)

$$f_{2\text{p}_0,1\text{s}}^{\text{B1}} = -\text{i} \frac{12\sqrt{2}}{\Delta(\Delta^2 + \frac{9}{4})^3} \tag{13.163}$$

The first Born differential cross-section for the transition $1\text{s} \rightarrow 2\text{p}_0$, obtained from (13.163), is

$$\frac{d\sigma^{B1}_{2p_0}}{d\Omega} = \frac{k_j}{k_i}\frac{288}{\Delta^2(\Delta^2 + \frac{9}{4})^6} \tag{13.164}$$

while the first Born differential cross-sections for exciting the $m = \pm 1$ substates are equal to zero. Because of the presence of the factor Δ^{-2}, the differential cross-section (13.164) exhibits at small momentum transfers a much stronger peak than the one corresponding to the elastic differential cross-section (13.146) or the 1s–2s differential cross-section (13.156) (see Fig. 13.12). On the other hand, we note that the first Born differential cross-section (13.164) falls off like Δ^{-14} for large Δ which is much faster than the Δ^{-4} decrease of the first Born elastic differential cross-section (13.146). As we pointed out above, the much slower decrease of $d\sigma^{B1}_{1s,1s}/d\Omega$ at large values of Δ is due to the non-vanishing contribution of the projectile–nucleus interaction $(-1/r_0)$ to the first Born scattering amplitude (13.140) for elastic collisions. It can be shown that the leading part of the second Born term for the 1s–2p transition behaves like $k_i^{-2}\Delta^{-1}$ when both k_i and Δ are large, so that the first Born differential cross-section (13.164) is useless for large momentum transfers, even at high energies.

It is also interesting to evaluate the first Born direct scattering amplitude (13.162) by choosing the quantisation axis along another direction. For example, if we choose this axis to lie along the incident direction $\hat{\mathbf{k}}_i$, we find that (Problem 13.8)

$$f^{B1}_{2p_m,1s} = -i\frac{24\sqrt{2\pi}}{\sqrt{3}}\frac{1}{\Delta(\Delta^2 + \frac{9}{4})^3}Y^*_{1m}(\alpha, \beta) \tag{13.165}$$

where (α, β) are the polar angles of $\mathbf{\Delta}$. The first Born differential cross-sections for the $1s \to 2p_m$ transitions, evaluated with the quantisation axis along $\hat{\mathbf{k}}_i$ are therefore given by

$$\frac{d\sigma^{B1}_{2p_0,1s}}{d\Omega} = \frac{k_j}{k_i}\frac{288}{\Delta^2(\Delta^2 + \frac{9}{4})^6}\cos^2\alpha \tag{13.166a}$$

and

$$\frac{d\sigma^{B1}_{2p_{\pm1},1s}}{d\Omega} = \frac{1}{2}\frac{k_j}{k_i}\frac{288}{\Delta^2(\Delta^2 + \frac{9}{4})^6}\sin^2\alpha \tag{13.166b}$$

The sum of these differential cross-sections for excitation of the three magnetic substates is of course equal to the corresponding sum calculated with any other choice of quantisation axis. In particular, it reduces to the result (13.164) if the quantisation axis is chosen to be along the momentum transfer $\mathbf{\Delta}$.

The total first Born cross-section for $1s \to 2p$ excitation is given by

$$\sigma^{B1}_{tot}(1s \to 2p) = 2\pi\frac{288}{k_i^2}\int_{\Delta_{min}}^{\Delta_{max}}\frac{d\Delta}{\Delta(\Delta^2 + \frac{9}{4})^6} \tag{13.167}$$

As in the case of the 1s–2s transition, we shall only be interested in obtaining the leading term of $\sigma_{\text{tot}}^{\text{B1}}(1\text{s} \to 2\text{p})$ at high energies ($k_i \gg 1$). Most of the contribution to the integral in (13.167) comes from the region of small values of Δ near $\Delta_{\min} = k_i - k_j \simeq 3/(8k_i)$, so that for large k_i one has (Problem 13.8)

$$\sigma_{\text{tot}}^{\text{B1}}(1\text{s} \to 2\text{p}) \simeq \frac{576\pi}{k_i^2}\left(\frac{2}{3}\right)^{12} \log k_i \tag{13.168a}$$

$$\simeq 144\pi \left(\frac{2}{3}\right)^{12} \frac{\log E_i^{\text{e}}}{E_i^{\text{e}}} \tag{13.168b}$$

We see that at high energies this cross-section behaves like $(E_i^{\text{e}})^{-1} \log E_i^{\text{e}}$. This is in contrast with the elastic and 1s–2s first Born cross-sections, which decrease more rapidly (like $(E_i^{\text{e}})^{-1}$) for large E_i^{e}. The presence of the logarithmic factor on the right of (13.168) is due to the fact that the differential cross-section (13.164) behaves like Δ^{-2} for small momentum transfers, so that the integrand on the right of (13.167) is proportional to Δ^{-1} for small Δ. It is interesting to note that this feature is in turn a consequence of the infinite range of the Coulomb interaction. As one might expect from the above discussion, the enhancement of the high-energy differential cross-section at small angles and the $(E_i^{\text{e}})^{-1} \log E_i^{\text{e}}$ fall-off of the total cross-section is exhibited not only in the 1s–2p transition, but in any 'dipole allowed' transition in which the change in the orbital angular momentum quantum number of the target hydrogen atom satisfies the selection rule $|\Delta l| = 1$.

Complex targets

Let us now consider the first Born approximation for electron scattering by complex atoms. For direct collisions, the interaction between the unbound electron, which we shall label 0, with coordinate $\mathbf{r}_0$, and the target atom containing the electrons $1, \dots, N$ is

$$V_{\text{d}} = -\frac{Z}{r_0} + \sum_{k=1}^{N} \frac{1}{r_{0k}} \tag{13.169}$$

where $r_{0k} = |\mathbf{r}_0 - \mathbf{r}_k|$. The first Born direct scattering amplitude is still given by the expression (13.142), where now

$$\begin{aligned} V_{ji} &= \langle j | V_{\text{d}} | i \rangle \\ &= -\frac{Z}{r_0}\delta_{ji} + \sum_{k=1}^{N} \left\langle j \left| \frac{1}{r_{0k}} \right| i \right\rangle \end{aligned} \tag{13.170}$$

As in the case of atomic hydrogen, the contribution of the projectile–nucleus term vanishes except for elastic scattering, to which we now turn our attention.

Elastic scattering

In that case, the first Born direct scattering amplitude is given by (13.144), where V_{ii} is the static potential

$$\begin{aligned} V_{ii} &= \langle i|V_{\mathrm{d}}|i\rangle \\ &= -\frac{Z}{r_0} + \int \frac{\rho_i(q)}{|\mathbf{r}_0 - \mathbf{r}|}\,\mathrm{d}q \end{aligned} \qquad \textbf{(13.171)}$$

As in Chapter 8, the symbol $\int \mathrm{d}q$ means an integration over the spatial coordinates $\mathbf{r}$ and a summation over the spin coordinate of the electron. In (13.171), $\rho_i(q)$ is the electron number density

$$\begin{aligned} \rho_i(q) &= \sum_{k=1}^{N} \int |\psi_i(q_1, q_2, \ldots, q_N)|^2 \delta(q - q_k)\,\mathrm{d}q_1 \ldots \mathrm{d}q_N \\ &= N \int |\psi_i(q_1, q_2, \ldots, q_N)|^2\,\mathrm{d}q_2 \ldots \mathrm{d}q_N \end{aligned} \qquad \textbf{(13.172)}$$

which is such that

$$\int \rho_i(q)\,\mathrm{d}q = N \qquad \textbf{(13.173)}$$

Substituting the expression (13.171) for V_{ii} into (13.144), and using the Bethe integral (13.152), it is found that

$$f_{ii}^{\mathrm{B1}} = \frac{2}{\Delta^2}\,[Z - \mathcal{S}_i(\Delta)] \qquad \textbf{(13.174)}$$

where

$$\mathcal{S}_i(\Delta) = \int \exp(\mathrm{i}\boldsymbol{\Delta}\cdot\mathbf{r})\rho_i(q)\,\mathrm{d}q \qquad \textbf{(13.175)}$$

is known as the *target elastic form factor*, also called the *X-ray scattering factor* [8]. Using (13.172), we have

$$\mathcal{S}_i(\Delta) = \sum_{k=1}^{N} \langle i|\exp(\mathrm{i}\boldsymbol{\Delta}\cdot\mathbf{r}_k)|i\rangle \qquad \textbf{(13.176)}$$

and we note that $\mathcal{S}_i(0) = N$. We also remark that if the quantisation axis for the eigenfunctions is chosen along the momentum transfer $\boldsymbol{\Delta}$, the target form factor may be written as

$$\mathcal{S}_i(\Delta) = \sum_{k=1}^{N} \langle i|\exp(\mathrm{i}\Delta z_k)|i\rangle \qquad \textbf{(13.177)}$$

[8] The term X-ray scattering factor is used because the function $\mathcal{S}_i(\Delta)$ was first introduced in the theory of X-ray scattering by atoms.

where z_k is the component of $\mathbf{r}_k$ along $\mathbf{\Delta}$. We note that $\mathcal{S}_i(\Delta)$ is an even function of Δ.

In terms of the target form factor $\mathcal{S}_i(\mathbf{\Delta})$, the first Born elastic differential cross-section is given by

$$\frac{\mathrm{d}\sigma_{ii}^{\mathrm{B1}}}{\mathrm{d}\Omega} = \frac{4}{\Delta^4} |Z - \mathcal{S}_i(\mathbf{\Delta})|^2 \tag{13.178}$$

The above derivation using the Bethe integral (13.152) fails in the forward direction ($\Delta = 0$). In this case, the value of $f_{ii}^{\mathrm{B1}}(\Delta = 0)$ may be obtained as follows. Choosing the Z axis along $\mathbf{\Delta}$ and expanding the form factor $\mathcal{S}_i(\Delta)$ for small Δ, we find from (13.177) that

$$\mathcal{S}_i(\Delta) = N - \frac{1}{2}\Delta^2 \sum_{k=1}^{N} \langle i | z_k^2 | i \rangle \tag{13.179}$$

where we have used the fact that $|\psi_i|^2$ has even parity. Thus, using (13.174), we have for a neutral atom ($N = Z$)

$$f_{ii}^{\mathrm{B1}}(\Delta = 0) = \sum_{k=1}^{N} \langle i | z_k^2 | i \rangle \tag{13.180}$$

which shows that in the forward direction the quantity $f_{ii}^{\mathrm{B1}}(\Delta = 0)$, and hence the elastic differential cross-section $\mathrm{d}\sigma_{ii}^{\mathrm{B1}}(\Delta = 0)/\mathrm{d}\Omega$, only depends on the matrix element $\langle i | \sum_{k=1}^{N} z_k^2 | i \rangle$ describing a purely structural feature of the target.

It is readily verified that for atomic hydrogen in the ground state the form factor is

$$\mathcal{S}_{1s}^{\mathrm{H}}(\Delta) = \left(1 + \frac{\Delta^2}{4}\right)^{-2} \tag{13.181}$$

Using (13.178), the elastic first Born differential cross-section (13.146) is regained.

For helium in the ground state, we find by using the approximate wave function (7.70) that the electron number density is given by

$$\rho_{1^1\mathrm{S}}(r) = 2\frac{Z_e^3}{\pi} \exp(-2Z_e r) \tag{13.182}$$

where $Z_e = 27/16$ (see (7.78)). Correspondingly,

$$\mathcal{S}_{1^1\mathrm{S}}^{\mathrm{He}}(\Delta) = 2\left(1 + \frac{\Delta^2}{4Z_e^2}\right)^{-2} \tag{13.183}$$

As we shall see below, the first Born approximation for direct elastic scattering becomes accurate at sufficiently high incident energies at all momentum transfers. Therefore, measurements of the differential cross-section at high energies determine the form factor $\mathcal{S}_i(\mathbf{\Delta})$. By inverting the Fourier transform (13.175), an 'experimental' electron number density $\rho_{\mathrm{exp}}(\mathbf{r})$ can be found. The 'experimental' values of $\rho_{\text{exp}}(\mathbf{r})$ obtained in this way may then be compared with theoretical calculations. In particular, theoretical estimates of $\rho_i(\mathbf{r})$ calculated by using Hartree–Fock wave functions can be compared with the 'experimental' values, and in this way the importance of electron correlation effects in the target can be established.

Inelastic scattering

Let us return to the first Born amplitude (13.142) for inelastic scattering, where V_{ji} is given by (13.170). Using the Bethe integral (13.152), we find that

$$f_{ji}^{\text{B1}}(\boldsymbol{\Delta}) = -\frac{2}{\Delta^2}\mathcal{S}_{ji}(\boldsymbol{\Delta}) \tag{13.184}$$

where

$$\mathcal{S}_{ji}(\boldsymbol{\Delta}) = \sum_{k=1}^{N}\langle j|\exp(\mathrm{i}\boldsymbol{\Delta}\cdot\mathbf{r}_k)|i\rangle, \qquad j \neq i \tag{13.185}$$

is called the *inelastic form factor*. If we choose the quantisation axis to be along the momentum transfer $\boldsymbol{\Delta}$, the equation (13.185) reduces to

$$\mathcal{S}_{ji}(\Delta) = \sum_{k=1}^{N}\langle j|\exp(\mathrm{i}\Delta z_k)|i\rangle \tag{13.186}$$

The *generalised oscillator strength* is defined as

$$\mathcal{F}_{ji}(\Delta) = \frac{2(E_j - E_i)}{\Delta^2}|\mathcal{S}_{ji}(\Delta)|^2, \qquad j \neq i \tag{13.187}$$

and the first Born differential cross-section for the inelastic transition $(\mathbf{k}_i, i) \rightarrow (\mathbf{k}_j, j)$ can be written as

$$\begin{aligned}\frac{\mathrm{d}\sigma_{ji}^{\text{B1}}}{\mathrm{d}\Omega} &= \frac{k_j}{k_i}|f_{ji}^{\text{B1}}(\Delta)|^2 \\ &= \frac{k_j}{k_i}\frac{2}{\Delta^2(E_j - E_i)}\mathcal{F}_{ji}(\Delta), \qquad j \neq i\end{aligned} \tag{13.188}$$

As in the case of atomic hydrogen, the first Born differential cross-section (13.188) is only accurate for large k_i and small values of the momentum transfer ($\Delta \lesssim 1$).

The corresponding total first Born cross-section is given by

$$\begin{aligned}\sigma_{\text{tot}}^{\text{B1}}(i \rightarrow j) &= \frac{2\pi}{k_i^2}\int_{\Delta_{\min}}^{\Delta_{\max}} |f_{ji}^{\text{B1}}(\Delta)|^2 \Delta\, \mathrm{d}\Delta \\ &= \frac{4\pi}{k_i^2}\frac{1}{E_j - E_i}\int_{\Delta_{\min}}^{\Delta_{\max}} \frac{\mathcal{F}_{ji}(\Delta)}{\Delta}\,\mathrm{d}\Delta\end{aligned} \tag{13.189}$$

Since at high incident energies the scattering is concentrated at small momentum transfers $\Delta \leqslant 1$, the dominant part of the total cross-section comes from this domain. Now, because for all direct inelastic transitions the first Born approximation is reliable at high energies for momentum transfers $\Delta \leqslant 1$, we can deduce that total first Born cross-sections, given by (13.189), are accurate at sufficiently high energies for direct inelastic collisions.

The generalised oscillator strength $\mathcal{F}_{ji}(\Delta)$ has some interesting properties which we shall now examine. First of all, we note that it is an *even* function of Δ. Secondly, if in (13.186) we expand the quantity $\exp(\mathrm{i}\Delta z_k)$ in powers of Δ, we find that for small Δ

$$\mathcal{F}_{ji}(\Delta) \simeq 2(E_j - E_i)\left| \mathrm{i} \sum_{k=1}^{N} \langle j|z_k|i\rangle - \frac{\Delta}{2} \sum_{k=1}^{N} \langle j|z_k^2|i\rangle + \ldots \right|^2 \qquad \textbf{(13.190)}$$

where we have used the fact that the initial and final target eigenfunctions are orthogonal. We see that in the limit $\Delta \to 0$

$$\mathcal{F}_{ji}(0) = 2(E_j - E_i)\left| \sum_{k=1}^{N} \langle j|z_k|i\rangle \right|^2 \qquad \textbf{(13.191)}$$

The right-hand side of this equation is just the optical oscillator strength (9.49), expressed in atomic units. Thus the equation (13.191) provides, within the framework of the first Born approximation, a connection between electron impact and photon impact phenomena.

It can be shown that the generalised oscillator strengths obey the *Bethe sum rule*

$$\sum_j \mathcal{F}_{ji}(\Delta) = N \qquad \textbf{(13.192)}$$

where the sum is over all target states, including the continuum. For the special case $\Delta = 0$, the Bethe sum rule reduces to the Thomas–Reiche–Kuhn sum rule (9.50) for optical oscillator strengths.

By using the measured values of the differential cross-section on the left-hand side of (13.188), and solving for $\mathcal{F}_{ji}(\Delta)$, we may obtain 'experimental' generalised oscillator strengths. If these values, obtained for different energies and momentum transfers, fall on the curve corresponding to the theoretical values of $\mathcal{F}_{ji}(\Delta)$, it is an indication that the first Born approximation is accurate at these energies and momentum transfers. The energy above which the first Born approximation becomes accurate (at small Δ) depends on the type of inelastic transition considered; it is generally lower for 'optically allowed' (dipole) transitions than for 'optically forbidden' transitions. As the incident energy increases, $\Delta_{\min}$ decreases, which allows the experimental generalised oscillator strengths to be extrapolated to $\Delta = 0$, thus giving an experimental estimate for the optical oscillator strength. As an example, the experimental generalised oscillator strength for the $1^1\mathrm{S} \to 2^1\mathrm{P}$ transition in helium is shown in Fig. 13.13, for incident electron energies ranging from 100 eV to 400 eV.

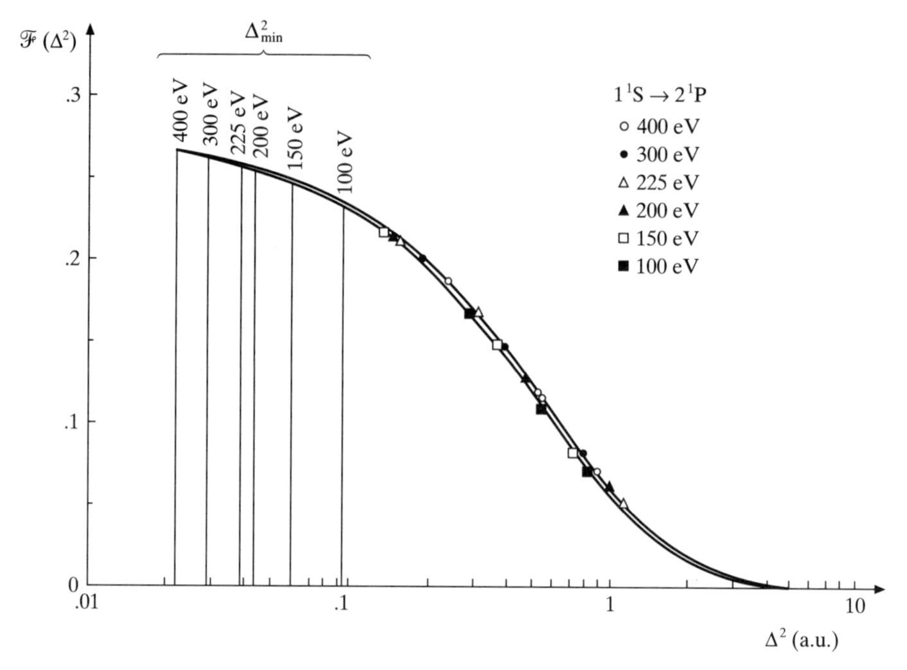

Figure 13.13 Generalised oscillator strength (in atomic units) for excitation of helium to the 2^1P level from the ground state, as a function of Δ^2.

Because the dominant contribution to the total cross-section comes from the small momentum transfer region, we expect that total cross-sections for 'optically allowed' (dipole) transitions will be enhanced with respect to those corresponding to 'optically forbidden' transitions at high energies. Indeed, let us denote by $\bar{\Delta}$ a fixed value of the momentum transfer Δ beyond which the first Born differential cross-section decreases rapidly. Using the fact that for 'optically allowed' transitions the dipole term is dominant at small Δ, so that we may write $\mathscr{F}_{ji}(\Delta) \simeq \mathscr{F}_{ji}(0)$ for $\Delta < \bar{\Delta}$, and noting that $\Delta_{\min} \simeq |E_j - E_i|/k_i$ for large k_i, we find from (13.189) and (13.191) that the first Born total cross-section for an 'allowed' transition is given approximately at high energies by

$$\begin{aligned}\sigma^{\mathrm{B1}}_{\mathrm{tot}}(i \to j) &\simeq \frac{8\pi}{k_i^2}\left|\sum_{k=1}^{N}\langle j|z_k|i\rangle\right|^2 \int_{\Delta_{\min}}^{\bar{\Delta}} \frac{\mathrm{d}\Delta}{\Delta} \\ &\simeq \frac{8\pi}{k_i^2}\left|\sum_{k=1}^{N}\langle j|z_k|i\rangle\right|^2 \log\frac{k_i\bar{\Delta}}{|E_j - E_i|} \\ &\simeq A\frac{\log E_i^{\mathrm{e}}}{E_i^{\mathrm{e}}}\end{aligned} \qquad \textbf{(13.193)}$$

where the quantity

$$A = 2\pi \left| \sum_{k=1}^{N} \langle j | z_k | i \rangle \right|^2 \tag{13.194}$$

is independent of the incident electron energy E_i^e and is proportional to the optical oscillator strength (13.191). It can be shown (Problem 13.10) that the correction to the leading term (13.193) of $\sigma_{tot}^{B1}(i \to j)$ at high energies is of the form $B(E_i^e)^{-1}$, where B is also energy-independent. Thus, for large E_i^e, the first Born total cross-section for an 'optically allowed' transition may be written as

$$\sigma_{tot}^{B1}(i \to j) \simeq A \frac{\log E_i^e}{E_i^e} + \frac{B}{E_i^e} \tag{13.195}$$

which is known as the *Bethe–Born approximation* to $\sigma_{tot}^{B1}(i \to j)$. The approximation (13.193) in which only the first term on the right of (13.195) is retained is often called the *Bethe approximation.*

If the transition is optically forbidden, so that the optical oscillator strength $\mathscr{F}_{ji}(0) = 0$, we can use (13.189), (13.190) and the fact that at high energies the scattering is concentrated at small momentum transfers $\Delta < \bar{\Delta}$ to write the first Born total cross-section for large E_i^e in the form

$$\begin{aligned} \sigma_{tot}^{B1}(i \to j) &\simeq \frac{2\pi}{k_i^2} \left| \sum_{k=1}^{N} \langle j | z_k^2 | i \rangle \right|^2 \int_{\Delta_{min}}^{\bar{\Delta}} \Delta \, d\Delta \\ &\simeq \frac{C}{E_i^e} \end{aligned} \tag{13.196}$$

where

$$C = \frac{\pi}{2} \left| \sum_{k=1}^{N} \langle j | z_k^2 | i \rangle \right|^2 \bar{\Delta}^2 \tag{13.197}$$

is an energy-independent quantity. The expression (13.196) is also due to H.A. Bethe.

The Bethe approximations (13.193) and (13.196) have been successfully used to analyse the *stopping power* of materials [9]. At high energies, the 'optically allowed' (dipole) transitions dominate (whether for discrete excitation or ionisation), and the energy loss per metre of an electron of non-relativistic velocity v passing through a material containing $\mathcal{N}$ atoms per cubic metre, each of atomic number Z, is given approximately (in SI units) by

$$\frac{dE}{dx} = -\frac{\mathcal{N} Z e^4}{4\pi\varepsilon_0^2 m v^2} \log\left(\frac{mv^2}{\bar{E}}\right) \tag{13.198}$$

where $\bar{E}$ is an empirical constant, which has the significance of a mean excitation energy.

[9] See for example Richtmyer, Kennard and Cooper (1969).

The first Born approximation for exchange scattering

We now turn to the first Born amplitude for exchange scattering. Our account will be brief for two reasons. Firstly, as we already noted, exchange scattering becomes increasingly less important compared to direct scattering as the incident electron energy increases. Secondly, as we shall see, the first Born approximation is inadequate to describe exchange scattering, except for elastic collisions at high energies.

The first Born approximation to the exchange amplitude, usually known as the *Born–Oppenheimer approximation*, is given for electron–atomic hydrogen collisions by

$$g_{ji}^{\mathrm{B1}} = -\frac{1}{2\pi}\int \mathrm{d}\mathbf{r}_0\,\mathrm{d}\mathbf{r}_1 \exp(-\mathrm{i}\mathbf{k}_j\cdot\mathbf{r}_1)\psi_j^*(\mathbf{r}_0)\left(-\frac{1}{r_1}+\frac{1}{r_{01}}\right)\exp(\mathrm{i}\mathbf{k}_i\cdot\mathbf{r}_0)\psi_i(\mathbf{r}_1) \quad \mathbf{(13.199)}$$

The Born–Oppenheimer approximation (13.199) suffers from the following defect. It can be shown [10] that the exact exchange amplitude g_{ji} must vanish if the electron–electron interaction $1/r_{01}$ is switched off. Since the expression (13.199) does not satisfy this requirement, we shall replace it by the modified first Born exchange expression

$$\tilde{g}_{ji}^{\mathrm{B1}} = -\frac{1}{2\pi}\int \mathrm{d}\mathbf{r}_0\,\mathrm{d}\mathbf{r}_1\, \exp(-\mathrm{i}\mathbf{k}_j\cdot\mathbf{r}_1)\psi_j^*(\mathbf{r}_0)\frac{1}{r_{01}}\exp(\mathrm{i}\mathbf{k}_i\cdot\mathbf{r}_0)\psi_i(\mathbf{r}_1) \quad \mathbf{(13.200)}$$

This procedure is indeed justified because it can be proved that the electron–nucleus term $(-1/r_1)$ is cancelled to leading order in k_i^{-1} by a contribution from the second Born exchange amplitude $\bar{g}_{ji}^{\mathrm{B2}}$.

In order to evaluate the integral on the right of (13.200), let us first set $k = 0$ in (12.156) and (12.158), so that we have

$$\frac{1}{r_{01}} = \frac{1}{2\pi^2}\int \frac{\exp[\mathrm{i}\mathbf{k}'\cdot(\mathbf{r}_0-\mathbf{r}_1)]}{k'^2}\,\mathrm{d}\mathbf{k}' \quad \mathbf{(13.201)}$$

Upon substitution in (13.200), we then have

$$\tilde{g}_{ji}^{\mathrm{B1}} = -2\int \frac{1}{k'^2}\phi_j^*(\mathbf{k}_i+\mathbf{k}')\phi_i(\mathbf{k}_j+\mathbf{k}')\,\mathrm{d}\mathbf{k}' \quad \mathbf{(13.202)}$$

where ϕ_i and ϕ_j are the Fourier transforms of the wave functions ψ_i and ψ_j, respectively.

The leading term of the expression (13.202) for large k_i can be obtained as follows. Setting $\mathbf{p} = \mathbf{k}_i + \mathbf{k}'$ in (13.202), we can write $\tilde{g}_{ji}^{\mathrm{B1}}$ in the form

$$\tilde{g}_{ji}^{\mathrm{B1}} = -2\int \frac{1}{(\mathbf{p}-\mathbf{k}_i)^2}\phi_j^*(\mathbf{p})\phi_i(\mathbf{p}-\Delta)\,\mathrm{d}\mathbf{p} \quad \mathbf{(13.203)}$$

[10] See Joachain (1983).

For large k_i, we may replace $(\mathbf{p} - \mathbf{k}_i)^2$ by k_i^2. Returning to configuration space, we find that for large k_i one has $\tilde{g}_{ji}^{\text{B1}} \simeq g_{ji}^{\text{Och}}$, where

$$g_{ji}^{\text{Och}} = -\frac{2}{k_i^2} \int \exp(\text{i}\boldsymbol{\Delta}\cdot\mathbf{r})\psi_j^*(\mathbf{r})\psi_i(\mathbf{r})\,\text{d}\mathbf{r} \tag{13.204}$$

is called the *Ochkur approximation to the exchange amplitude.*

For elastic exchange scattering from H(1s), it is readily found that (Problem 13.11)

$$g_{1\text{s},1\text{s}}^{\text{Och}} = -\frac{32}{k_i^2}\frac{1}{(\Delta^2 + 4)^2} \tag{13.205}$$

with $k_i = k_j = k$. We see that for momentum transfers $\Delta \leqslant 1$ this expression falls off like k_i^{-2} for large k_i. We shall see below that the Ochkur exchange amplitude must be included – together with higher order terms of the direct amplitude – to obtain a consistent approximation for the elastic differential cross-section to order k_i^{-2}.

For inelastic transitions in atomic hydrogen, we see by using (13.153) and (13.204) that

$$g_{ji}^{\text{Och}} = \frac{\Delta^2}{k_i^2} f_{ji}^{\text{B1}}, \qquad j \neq i \tag{13.206}$$

This amplitude is usually small compared with higher order contributions to the exchange amplitude.

Higher order terms of the Born series

Let us now consider higher order terms of the Born series. Since exchange effects are small at high energies, we shall consider only the terms $\bar{f}_{ji}^{\text{B}n}$ ($n \geqslant 2$) of the Born series for the direct amplitude. Using (13.136) and (13.138), one finds (Problem 13.12) that the second Born term $\bar{f}_{ji}^{\text{B2}}$ for a transition $(\mathbf{k}_i, i) \to (\mathbf{k}_j, j)$ in electron–atomic hydrogen collisions is given by

$$\bar{f}_{ji}^{\text{B2}} = 8\pi^2 \sum_n \int \text{d}\mathbf{q} \frac{\langle \mathbf{k}_j, j|V_\text{d}|\mathbf{q}, n\rangle\langle \mathbf{q}, n|V_\text{d}|\mathbf{k}_i, i\rangle}{q^2 - k_i^2 + 2(E_n - E_i) - \text{i}\varepsilon}, \qquad \varepsilon \to 0^+ \tag{13.207}$$

where we have introduced a complete set of states $|\mathbf{q}, n\rangle = |\Phi_{\mathbf{q},n}(0, 1)\rangle$ such that

$$\Phi_{\mathbf{q},n}(\mathbf{r}_0, \mathbf{r}_1) = (2\pi)^{-3/2} \exp(\text{i}\mathbf{q}\cdot\mathbf{r}_0)\psi_n(\mathbf{r}_1) \tag{13.208}$$

and the summation on n in (13.207) includes an integration over the continuum. The generalisation of (13.207) to complex atom targets is straightforward.

A few exact calculations of $\bar{f}_{ji}^{\text{B2}}$ have been performed for the simplest transitions in atomic hydrogen, but in general $\bar{f}_{ji}^{\text{B2}}$ must be evaluated approximately. At sufficiently high energies, a good approximation to $\bar{f}_{ji}^{\text{B2}}$ is obtained by replacing the energy differences $(E_n - E_i)$ of the target states by an average energy $\bar{E}$, so that the sum over the intermediate target states can be done by using the closure relation. The resulting integrals can then be performed by using the methods of

Table 13.2 Dependence of various terms of the Born and Glauber series for the direct scattering amplitude corresponding to elastic scattering by an atom in an S state, as a function of (large) k_i and of $\Delta = |\mathbf{k}_i - \mathbf{k}_j|$. The dominant contributions are framed. The terms located above the dashed line contribute through order k_i^{-2} to the differential cross-section.

Order of perturbation theory	*Term*	*Small* Δ ($\Delta < k_i^{-1}$)	*Intermediate* Δ ($k_i^{-1} < \Delta < 1$)	*Large* Δ $\Delta > k_i$
First	$\bar{f}_{ii}^{B1} = \bar{f}_{ii}^{G1}$	$\boxed{1}$	$\boxed{1}$	$\boxed{\Delta^{-2}}$
Second	Re $\bar{f}_{ii}^{B2}$	k_i^{-1}	k_i^{-2}	$k_i^{-2}\Delta^{-2}$
	Re $\bar{f}_{ii}^{G2}$	0	0	0
	Im $\bar{f}_{ii}^{B2}$	$k_i^{-1} \log k_i$	k_i^{-1}	$k_i^{-1}\Delta^{-2} \log \Delta$
	Im $\bar{f}_{ii}^{G2}$	$k_i^{-1} \log \Delta$	k_i^{-1}	$k_i^{-1}\Delta^{-2} \log \Delta$
Third	Re $\bar{f}_{ii}^{B3}$	k_i^{-2}	k_i^{-2}	$k_i^{-2}\Delta^{-2} \log^2\Delta$
	Re $\bar{f}_{ii}^{G3}$	k_i^{-2}	k_i^{-2}	$k_i^{-2}\Delta^{-2} \log^2\Delta$
	Im $\bar{f}_{ii}^{B3}$	k_i^{-3}	k_i^{-3}	$k_i^{-3}\Delta^{-2} \log \Delta$
	Im $\bar{f}_{ii}^{G3}$	0	0	0
nth ($n > 3$)	$\bar{f}_{ii}^{Bn}$	$(ik_i)^{1-n}$	$(ik_i)^{1-n}$	$(ik_i)^{1-n}\Delta^{-2} \log^{n-1}\Delta$
	$\bar{f}_{ii}^{Gn}$	$(ik_i)^{1-n}$	$(ik_i)^{1-n}$	$(ik_i)^{1-n}\Delta^{-2} \log^{n-1}\Delta$

Appendix 12. These methods can also be used to improve this procedure by evaluating exactly the first few terms in the sum on n in (13.207), while treating the remaining states by closure or by using pseudo-states.

The behaviour of the terms $\bar{f}_{ji}^{Bn}$ as a function of (large) k_i and of the magnitude Δ of the momentum transfer $\mathbf{\Delta} = \mathbf{k}_i - \mathbf{k}_j$ has been analysed by F.W. Byron and C.J. Joachain for a target atom whose initial state is an S state. Their results for *direct elastic* scattering are summarised in Table 13.2. This table also shows the dependence of the corresponding terms of the Glauber series, which will be considered shortly.

Looking at Table 13.2, we see that for *direct elastic* scattering the dominant contribution to the scattering amplitude at large k_i is given by the first Born term $\bar{f}_{ii}^{B1}$ at all momentum transfers. At small Δ, the real part of the second Born term, Re $\bar{f}_{ii}^{B2}$, is governed by polarisation effects and gives the dominant correction (of order k_i^{-1}) to the first Born differential cross-section $d\sigma_{ii}^{B1}/d\Omega$. The imaginary part of the second Born term, Im $\bar{f}_{ii}^{B2}$, gives at small Δ a correction to $d\sigma_{ii}^{B1}/d\Omega$ of order $(\log k_i)^2/k_i^2$ taking into account absorption effects due to the loss of probability flux into open channels other than the initial one. At large Δ, the terms $\bar{f}_{ii}^{Bn}$ ($n \geqslant 2$) of the direct elastic amplitude are dominated by processes in which the atom remains in its initial state $|i\rangle$ in all intermediate states; this reflects the fact that at large angles direct elastic scattering is governed by the static potential $V_{ii} = \langle i|V_d|i\rangle$.

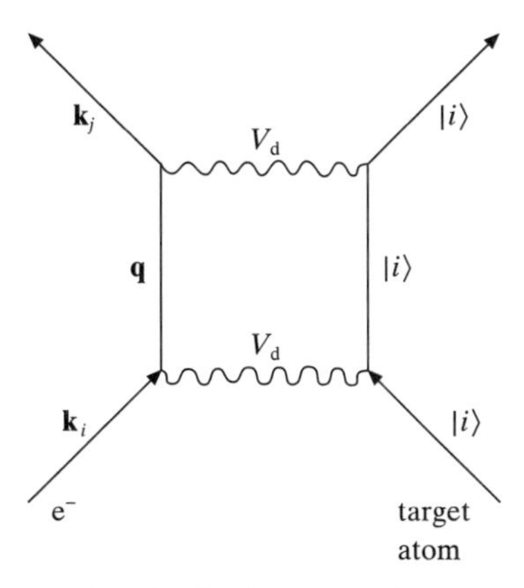

Figure 13.14 Diagram representing the dominant contribution to the second Born term $\bar{f}^{\text{B2}}_{ii}$ at large momentum transfers for an elastic scattering process $(\mathbf{k}_i, i) \to (\mathbf{k}_j, i)$. The target atom remains in the initial state $|i\rangle$ in the intermediate state. The vector $\mathbf{q}$ is the intermediate momentum of the projectile and V_d is the direct target–projectile interaction.

For example, we have represented in Fig. 13.14 the diagram corresponding to the dominant contribution to $\bar{f}^{\text{B2}}_{ii}$ at large momentum transfers. This dominant contribution is obtained by keeping in the sum on intermediate states on the right of (13.207) the only term $n = i$ for which the target atom remains in the initial state $|i\rangle$.

The situation is different for *direct inelastic* collisions $(j \neq i)$, where for large Δ the first Born term $\bar{f}^{\text{B1}}_{ji}$ falls off rapidly and the Born series is dominated at large k_i and large Δ by the second Born term $\bar{f}^{\text{B2}}_{ji}$. This is illustrated in Table 13.3 for inelastic S–S transitions but is valid for all inelastic transitions, and can be understood as follows. Large momentum transfer collisions can only take place if the projectile electron collides with the much heavier atomic nucleus. Now, as seen above, for inelastic scattering the orthogonality of the target initial and final states, $\langle j|i\rangle = 0$, removes the electron–nucleus interaction term $-Z/r_0$ from the first Born amplitude $\bar{f}^{\text{B1}}_{ji}$. As a result, the first Born differential cross-section for high-energy large-angle inelastic scattering is orders of magnitude too small. The fact that $\bar{f}^{\text{B2}}_{ji}$ falls off more slowly than $\bar{f}^{\text{B1}}_{ji}$ for inelastic collisions at large Δ is due to the possibility of elastic scattering in the intermediate states $|i\rangle$ and $|j\rangle$ (the initial and final target states of the inelastic transition), where the projectile electron can experience the Coulomb potential of the nucleus. The corresponding two contributions to $\bar{f}^{\text{B2}}_{ji}$ are obtained by keeping in the sum on intermediate states in (13.207) the terms $n = i$ and $n = j$. The two diagrams representing these contributions are shown in Fig. 13.15. We remark that since the small momentum transfer region $(\Delta < 1)$ where the first Born approximation is accurate for large k_i corresponds to scattering angles $\theta < k_i^{-1}$, the angular region in which the first Born approximation is valid shrinks with increasing energy. However, because the dominant contribution to total (integrated) cross-sections comes from the region $\Delta < 1$, the first Born values for total cross-sections are accurate at sufficiently high energies.

Table 13.3 Dependence of various terms of the Born and Glauber series for the direct scattering amplitude corresponding to inelastic S → S transitions as a function of (large) k_i and of $\Delta = |\mathbf{k}_i - \mathbf{k}_j|$. The dominant contributions are framed. The terms located above the dashed line contribute through order k_i^{-2} to the differential cross-section.

Order of perturbation theory	*Term*	*Small* Δ ($\Delta < k_i^{-1}$)	*Intermediate* Δ ($k_i^{-1} < \Delta < 1$)	*Large* Δ $\Delta > k_i$
First	$\bar{f}_{ji}^{B1} = \bar{f}_{ji}^{G1}$	$\boxed{1}$	$\boxed{1}$	Δ^{-6}
Second	Re $\bar{f}_{ji}^{B2}$	k_i^{-1}	k_i^{-2}	$k_i^{-2}\Delta^{-2}$
	Re $\bar{f}_{ji}^{G2}$	0	0	0
	Im $\bar{f}_{ji}^{B2}$	$k_i^{-1} \log k_i$	k_i^{-1}	$\boxed{k_i^{-1}\Delta^{-2}}$
	Im $\bar{f}_{ji}^{G2}$	$k_i^{-1} \log \Delta$	k_i^{-1}	$\boxed{k_i^{-1}\Delta^{-2}}$
Third	Re $\bar{f}_{ji}^{B3}$	k_i^{-2}	k_i^{-2}	$k_i^{-2}\Delta^{-2} \log \Delta$
	Re $\bar{f}_{ji}^{G3}$	k_i^{-2}	k_i^{-2}	$k_i^{-2}\Delta^{-2} \log \Delta$
	Im $\bar{f}_{ji}^{B3}$	k_i^{-3}	k_i^{-3}	$k_i^{-3}\Delta^{-2}$
	Im $\bar{f}_{ji}^{G3}$	0	0	0
nth ($n > 3$)	$\bar{f}_{ji}^{Bn}$	$(ik_i)^{1-n}$	$(ik_i)^{1-n}$	$(ik_i)^{1-n}\Delta^{-2} \log^{n-2}\Delta$
	$\bar{f}_{ji}^{Gn}$	$(ik_i)^{1-n}$	$(ik_i)^{1-n}$	$(ik_i)^{1-n}\Delta^{-2} \log^{n-2}\Delta$

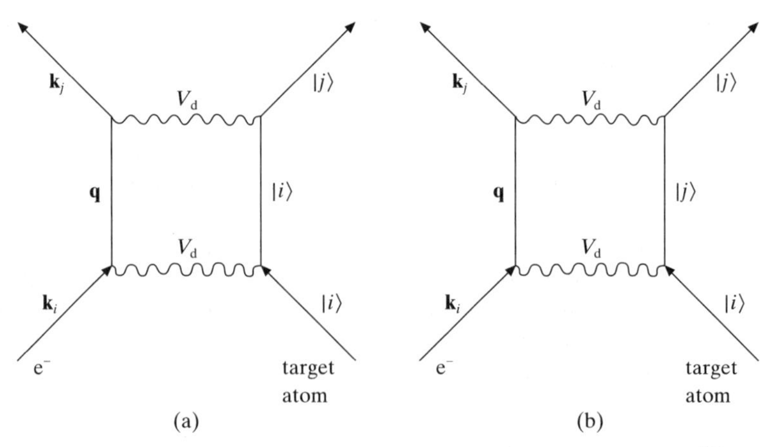

Figure 13.15 The two diagrams representing the dominant contributions to $\bar{f}_{ji}^{B2}$ at large momentum transfers for an inelastic transition $(\mathbf{k}_i, i) \rightarrow (\mathbf{k}_j, j)$. The diagram (a) corresponds to the contribution in which the target atom remains in the initial state $|i\rangle$ in the intermediate state. The diagram (b) represents the contribution in which the intermediate state of the target atom is the final state $|j\rangle$.

Finally, we note that as in the case of potential scattering analysed in Section 12.7, the convergence of the Born series for large k_i is slower at large Δ than at small Δ, for both elastic and inelastic collisions.

The Glauber approximation

The accurate evaluation of the higher Born terms $\bar{f}_{ji}^{\mathrm{B}n}$ ($n \geqslant 3$) is very difficult, but fortunately, useful information about these terms can be obtained by using the Glauber approximation, which is the simplest generalisation of the potential scattering eikonal approximation discussed in Section 12.7. For a direct transition $(\mathbf{k}_i, i) \to (\mathbf{k}_j, j)$ in electron scattering by atomic hydrogen, the Glauber scattering amplitude is given by

$$f_{ji}^{\mathrm{G}} = \frac{k_i}{2\pi \mathrm{i}} \int \mathrm{d}^2\mathbf{b}_0 \exp(\mathrm{i}\boldsymbol{\Delta}\cdot\mathbf{b}_0) \left\langle j \left| \exp\left[\frac{\mathrm{i}}{k_i}\chi_{\mathrm{G}}(\mathbf{b}_0, \mathbf{r}_1)\right] - 1 \right| i \right\rangle \qquad \textbf{(13.209)}$$

where we use a cylindrical coordinate system with $\mathbf{r}_0 = \mathbf{b}_0 + z_0\hat{\mathbf{z}}$. The Glauber phase χ_{G} is given in terms of the direct interaction V_{d} by

$$\chi_{\mathrm{G}}(\mathbf{b}_0, \mathbf{r}_1) = -\int_{-\infty}^{+\infty} V_{\mathrm{d}}(\mathbf{b}_0, z_0, \mathbf{r}_1)\, \mathrm{d}z_0 \qquad \textbf{(13.210)}$$

the integration being performed along a Z axis perpendicular to $\boldsymbol{\Delta}$ and lying in the scattering plane.

Insights into the properties of the Glauber method may be gained by expanding the Glauber amplitude (13.209) in powers of V_{d}, namely

$$f_{ji}^{\mathrm{G}} = \sum_{n=1}^{\infty} \bar{f}_{ji}^{\mathrm{G}n} \qquad \textbf{(13.211)}$$

where

$$\bar{f}_{ji}^{\mathrm{G}n} = \frac{k_i}{2\pi \mathrm{i}} \frac{1}{n!} \left(\frac{\mathrm{i}}{k_i}\right)^n \int \mathrm{d}^2\mathbf{b}_0 \exp(\mathrm{i}\boldsymbol{\Delta}\cdot\mathbf{b}_0) \langle j | \chi_{\mathrm{G}}^n(\mathbf{b}_0, \mathbf{r}_1) | i \rangle \qquad \textbf{(13.212)}$$

and comparing the nth Glauber term $\bar{f}_{ji}^{\mathrm{G}n}$ with the corresponding nth Born term $\bar{f}_{ji}^{\mathrm{B}n}$ (see Tables 13.2 and 13.3). We first note that $\bar{f}_{ji}^{\mathrm{B}1} = \bar{f}_{ji}^{\mathrm{G}1}$. For $n \geqslant 2$, the terms $\bar{f}_{ji}^{\mathrm{G}n}$ are alternately real or purely imaginary, while the corresponding Born terms are complex. This special feature of the Glauber amplitude leads to several defects such as the absence, for S–S transitions, of the important term $\mathrm{Re}\, \bar{f}_{ji}^{\mathrm{B}2}$ which, at small Δ, is of order k_i^{-1} and accounts for polarisation effects. Another deficiency of the Glauber amplitude is a logarithmic divergence for elastic scattering in the forward direction ($\Delta = 0$), which is due to the fact that $\mathrm{Im}\, \bar{f}_{ii}^{\mathrm{G}2}$ is proportional to $\log \Delta$ at small Δ (see Table 13.2). Moreover, the Glauber amplitude gives a poor description of inelastic transitions involving non-spherically symmetric states (for example, S–P transitions) at large Δ. Nevertheless, the Glauber amplitude (13.209) has the attractive property that it includes terms from all orders of perturbation theory.

The eikonal–Born series method

The eikonal–Born series (EBS) method, introduced in 1973 by F.W. Byron and C.J. Joachain, is based on an analysis of the terms of the Born series (13.133) and of the Glauber series (13.211), the aim being to obtain a consistent expansion of the scattering amplitude in powers of k_i^{-1}. The main results are summarised in Tables 13.2 and 13.3 for direct S–S elastic and inelastic transitions, respectively. We see that for these processes the Glauber term $\bar{f}_{ji}^{\mathrm{G}n}$ gives for large k_i the leading part of the corresponding Born term $\bar{f}_{ji}^{\mathrm{B}n}$ (for $j = i$ and $j \neq i$), at all momentum transfers, except in second order where $\mathrm{Im}\,\bar{f}_{ji}^{\mathrm{G2}}$ is proportional to $\log \Delta$ at small Δ.

It is clear from the examination of Tables 13.2 and 13.3 that the Glauber amplitude f_{ji}^{G} is not correct through order k_i^{-1} since it lacks the term $\mathrm{Re}\,\bar{f}_{ji}^{\mathrm{B2}}$. On the other hand, the second Born amplitude $f_{ji}^{\mathrm{B2}} = \bar{f}_{ji}^{\mathrm{B1}} + \bar{f}_{ji}^{\mathrm{B2}}$ is seen to be correct through order k_i^{-1}, but not through order k_i^{-2}. In fact, a consistent calculation of the direct scattering amplitude through order k_i^{-2} requires the terms $\bar{f}_{ji}^{\mathrm{B1}}$, $\bar{f}_{ji}^{\mathrm{B2}}$ and $\mathrm{Re}\,\bar{f}_{ji}^{\mathrm{B3}}$. Since $\mathrm{Re}\,\bar{f}_{ji}^{\mathrm{B3}}$ is very difficult to evaluate, and because $\bar{f}_{ji}^{\mathrm{G3}}$ is a good approximation to $\mathrm{Re}\,\bar{f}_{ji}^{\mathrm{B3}}$ for large enough k_i, one can use $\bar{f}_{ji}^{\mathrm{G3}}$ in place of $\mathrm{Re}\,\bar{f}_{ji}^{\mathrm{B3}}$. Thus one obtains in this way the *eikonal–Born series* (EBS) direct scattering amplitude

$$f_{ji}^{\mathrm{EBS}} = \bar{f}_{ji}^{\mathrm{B1}} + \bar{f}_{ji}^{\mathrm{B2}} + \bar{f}_{ji}^{\mathrm{G3}} \qquad \textbf{(13.213)}$$

which is correct through order k_i^{-2}. Similarly, the amplitude

$$f_{ji}^{\mathrm{EBS'}} = f_{ji}^{\mathrm{G}} - \bar{f}_{ji}^{\mathrm{G2}} + \bar{f}_{ji}^{\mathrm{B2}} = \bar{f}_{ji}^{\mathrm{B1}} + \bar{f}_{ji}^{\mathrm{B2}} + \sum_{n=3}^{\infty} \bar{f}_{ji}^{\mathrm{G}n} \qquad \textbf{(13.214)}$$

also gives a consistent approximation to the direct scattering amplitude through order k_i^{-2}. In addition, exchange effects are taken into account in the EBS theory by keeping the relevant terms in the Born series for the exchange amplitude. For example, in the case of elastic scattering (where $k_i = k_j = k$), the Ochkur amplitude, which is of order k_i^{-2} for large k_i and fixed Δ (see (13.205)), must be taken into account in order to perform a consistent calculation of the differential cross-section through order k_i^{-2}.

The EBS method is very successful when perturbation theory converges rapidly, namely at high incident energies, and for light atoms. As a first example, we show in Fig. 13.16(a) the differential cross-section for elastic e^-–H(1s) scattering at an incident electron energy of 50 eV, as obtained from the first Born approximation, the Glauber approximation and the EBS theory. It is seen that the EBS results are in excellent agreement with the experimental data. Also shown are the results of an 'exact' (partial-wave) treatment of the static potential $V_{1s,1s} \equiv V_{11}$ given by (13.51). These static results are seen to be accurate at large angles, as expected from the preceding discussion. As a second example of elastic scattering, we display in Fig. 13.16(b) the differential cross-section for the scattering of 500 eV electrons by helium atoms in the ground state. The experimental data are compared with the results of theoretical calculations using the first Born approximation, the Glauber approximation and the EBS method. The EBS theory is

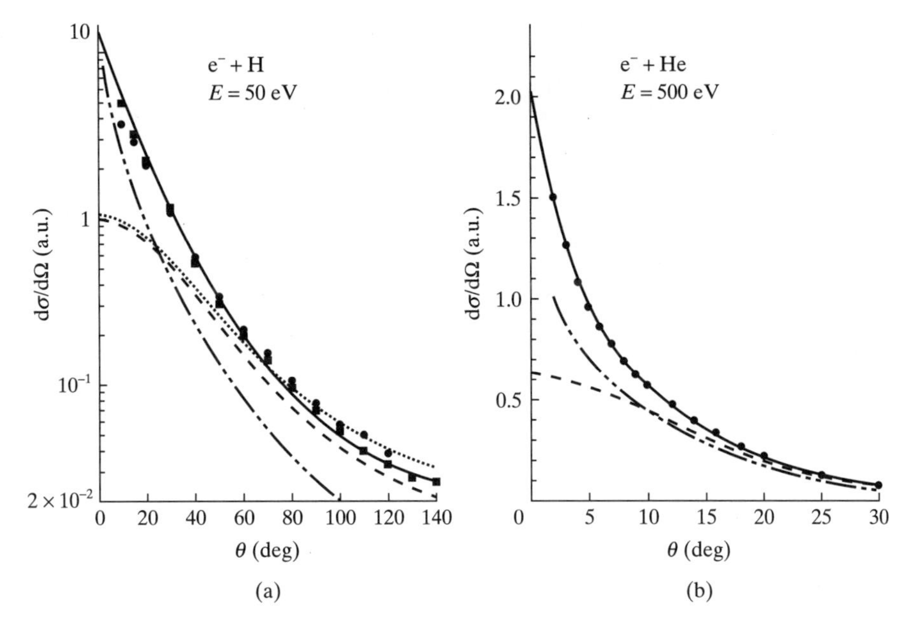

Figure 13.16 (a) Differential cross-section for the elastic scattering of 50 eV electrons by H(1s). The experimental data are due to C.R. Lloyd, P.J.O. Teubner, E. Weigold and B.R. Lewis: ●; and to J.F. Williams: ■.
The theoretical calculations by F.W. Byron and C.J. Joachain are: EBS theory: ——; first Born approximation: -----; Glauber approximation: — — — —; partial-wave treatment of the static potential (13.51): ·······.
(b) Differential cross-section for the elastic scattering of 500 eV electrons by He(1^1S).
The experimental data are due to J.P. Bromberg.
The theoretical calculations by F.W. Byron and C.J. Joachain are: EBS theory: ——; first Born approximation: -----; Glauber approximation: — — — —.

seen to agree extremely well with experiment. In Fig. 13.17, the differential cross-sections for the excitation of the 2^1S and 2^1P states of helium from the ground state are shown for an incident electron energy of 200 eV. As expected from the previous discussion, the first Born approximation fails at large angles. In contrast, the EBS results give a good account of the data.

When perturbation theory is more slowly convergent, improvements over the EBS method are necessary. These can be obtained by constructing methods which include terms from all orders of perturbation theory. The simplest method of this kind which is consistent to order k_i^{-2}, consists in using the EBS′ amplitude (13.214). A more elaborate way of extending the third-order perturbative EBS direct amplitude (13.213) to all orders is to use the unitarised eikonal–Born series (UEBS) method proposed in 1981 by F.W. Byron, C.J. Joachain and R.M. Potvliege [11].

[11] See Joachain (1989).

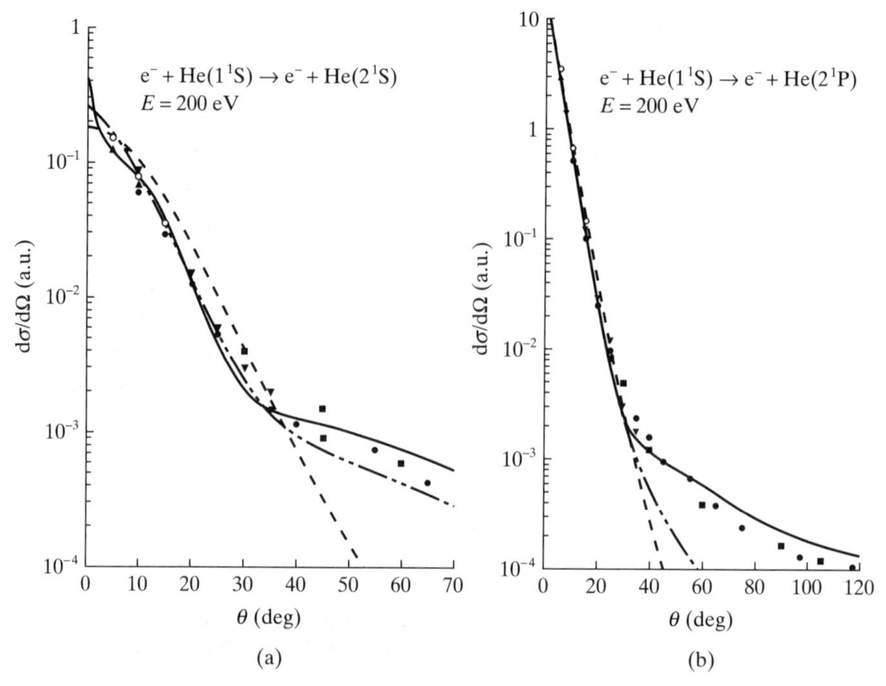

Figure 13.17 Differential cross-sections for the excitation of (a) the 2^1S and (b) the 2^1P levels of helium from the ground state by 200 eV electrons.
The experimental data are due to L. Vriens, J.A. Simpson and S.R. Mielczarek: o; G.E. Chamberlain, S.R. Mielczarek and C.E. Kuyatt: ▲; C.B. Opal and E.C. Beaty: ■; H. Suzuki and T. Takayanagi: ●; M.A. Dillon and E.N. Lassettre: ▼.
The theoretical calculations by F.W. Byron and C.J. Joachain are: EBS theory: ——; first Born approximation: -----; Glauber approximation: —-—-—.

Optical potentials

Another approach which is capable of including terms from all orders of perturbation theory is that based on the optical potential, introduced in Section 13.2. The central idea of the optical potential method is to analyse the elastic scattering of a particle from a complex target by replacing the complicated interactions between the projectile and the target by an effective potential in which the incident particle moves. Once the optical V_{opt} is determined, the original many-body elastic scattering problem reduces to a one-body situation. However, this reduction is in general a difficult task, and approximation methods are necessary.

Let us begin by considering direct elastic scattering, the corresponding direct part of the optical potential being denoted by $V^{\text{d}}_{\text{opt}}$. Formal expressions for $V^{\text{d}}_{\text{opt}}$ are readily obtained by using the Feshbach projection operator formalism. Let

$$P = |i\rangle\langle i| \tag{13.215}$$

be an operator projecting on the initial target state $|i\rangle$, and let $Q = I - P$. The total Hamiltonian of the electron–atom system can be decomposed as

$$H = H_\mathrm{d} + V_\mathrm{d} = T_0 + h + V_\mathrm{d} \tag{13.216}$$

where V_d is the direct interaction (13.169) between the unbound electron (labelled 0) and the target atom, T_0 is the kinetic energy operator of the unbound electron and h is the target Hamiltonian. Using the fact that the operators P and Q commute with $H_\mathrm{d} = T_0 + h$, we may write the Schrödinger equation for the system as

$$(T_0 + V^\mathrm{d}_\mathrm{opt} - E^\mathrm{e}_i)P\Psi = 0 \tag{13.217}$$

where $E^\mathrm{e}_i = E - E_i = k_i^2/2$ is the energy of the projectile electron, and we have used the fact that $h|i\rangle = E_i|i\rangle$. The direct optical potential $V^\mathrm{d}_\mathrm{opt}$ in (13.217) is given by

$$V^\mathrm{d}_\mathrm{opt} = PV_\mathrm{d}P + \mathcal{V} \tag{13.218}$$

where the equivalent potential $\mathcal{V}$ is given by (13.82).

At high energies it is appropriate to make a perturbative expansion of $V^\mathrm{d}_\mathrm{opt}$ in powers of V_d. That is,

$$V^\mathrm{d}_\mathrm{opt} = V^{(1)} + V^{(2)} + V^{(3)} + \ldots \tag{13.219}$$

where the first-order term is $V^{(1)} = PV_\mathrm{d}P$, the second-order term reads

$$V^{(2)} = PV_\mathrm{d}Q \frac{1}{E + \mathrm{i}\varepsilon - H_\mathrm{d}} QV_\mathrm{d}P, \qquad \varepsilon \to 0^+ \tag{13.220}$$

and so on. With the operator P given by (13.215), we see that when acting in P-space, $V^{(1)}$ is represented by the static potential $V_{ii} = \langle i|V_\mathrm{d}|i\rangle$, while $V^{(2)}$ is represented by the non-local, complex and energy-dependent potential

$$V^{(2)}_{ii} = \sum_{n \neq i} \frac{\langle i|V_\mathrm{d}|n\rangle\langle n|V_\mathrm{d}|i\rangle}{k_i^2/2 + \mathrm{i}\varepsilon - T_0 - (E_n - E_i)}, \qquad \varepsilon \to 0^+ \tag{13.221}$$

We recall that the static potential V_{ii} is real and of short range, and hence does not account for polarisation and absorption effects. However, for small values of the projectile coordinate r_0, the static potential correctly reduces to the Coulomb interaction $-Z/r_0$ acting between the projectile electron and the target nucleus, and hence gives a good account of large-angle direct elastic scattering.

Although the second- and higher order terms of the direct optical potential are in general complicated, non-local, complex operators, local approximations to them can be obtained at sufficiently high energies. For example, F.W. Byron and C.J. Joachain converted the lowest order terms of pertubation theory, calculated by using the EBS method for elastic scattering, into an *ab-initio* optical potential. The second-order part $V^{(2)}_{ii}$ of the direct optical potential may then be written approximately as

$$V^{(2)}_{ii} = V_\mathrm{pol} + \mathrm{i}V_\mathrm{abs} \tag{13.222}$$

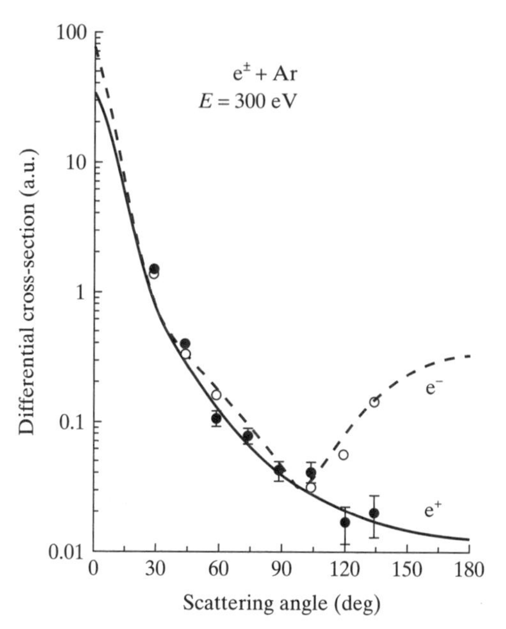

Figure 13.18 The differential cross-section for electron (-----) and positron (——) elastic scattering by argon at 300 eV, calculated by C.J. Joachain *et al.* from the *ab-initio* optical model theory.
The experimental data of G.M.A. Hyder *et al.* are: electron scattering: o; positron scattering: •.

where V_{pol} and V_{abs} are real and central but energy-dependent. The term V_{pol} (which falls off like r_0^{-4} at large r_0) accounts for dynamic polarisation effects and $\text{i}V_{\text{abs}}$ for absorption effects due to loss of flux from the incident channel.

Having obtained a local approximation for $V^{\text{d}}_{\text{opt}}$, exchange effects may be taken into account by using a local exchange pseudo-potential $V^{\text{ex}}_{\text{opt}}$. The full optical potential V_{opt}, containing the direct and exchange parts, is then treated in an essentially exact way by using the partial wave method.

As an illustration of the optical potential theory, we compare in Fig. 13.18 the theoretical predictions of C.J. Joachain *et al.* with the differential cross-sections measured by G.M.A. Hyder *et al.* for electron– and positron–argon elastic scattering at an incident energy of 300 eV. For incident positrons the direct interaction V_{d} has of course the opposite sign to that for electrons, and in addition there are no exchange effects.

The optical potential formalisn is readily generalised to the case in which one is interested in transitions between a certain number M of target states. Equations (13.217)–(13.220) are still valid provided the expression (13.215) of P is now replaced by

$$P = \sum_{n=1}^{M} |n\rangle\langle n| \qquad \textbf{(13.223)}$$

Of course, equation (13.217) now represents a set of coupled equations, and the optical potential is a potential matrix. In particular, the second-order potential (SOP) method of B.H. Bransden and J.P. Coleman is an approximation to these coupled equations, in which the optical potential matrix is treated to second order in V_d.

Distorted waves

Distorted wave treatments are characterised by the fact that the interaction is broken into two parts, one which is treated exactly and the other which is handled by perturbation theory. This separation is dictated by the physics of the problem, and consequently many kinds of distorted wave methods have been applied to electron–atom (ion) collisions. Distorted wave methods are discussed in Bransden (1983) and Joachain (1983).

13.4 Electron impact ionisation of atoms

In this section we shall discuss the process whereby electrons are ejected from the target atom, giving rise to ionisation. We shall limit our discussion to single ionisation or (e, 2e) reactions of the type

$$e^- + A(i) \rightarrow A^+(f) + 2e^- \quad \textbf{(13.224)}$$

where $A(i)$ is a neutral atom in state i and $A^+(f)$ is the corresponding ion in state f.

Kinematics

We begin our study by kinematical considerations. We consider an electron of momentum $\mathbf{k}_i$ and energy $E_i^e = k_i^2/2$ incident on a target atom A at rest in the eigenstate $|i\rangle$ with eigenfunction ψ_i and eigenenergy E_i. In the final state, two electrons emerge with momenta $\mathbf{k}_A$ and $\mathbf{k}_B$ and corresponding energies $E_A^e = k_A^2/2$ and $E_B^e = k_B^2/2$ (see Fig. 13.19), the remaining ion A^+ being left in the eigenstate $|f\rangle$, with eigenenergy ε_f. If $\mathbf{Q}$ denotes the recoil momentum of the ion, momentum conservation requires that

$$\mathbf{k}_i = \mathbf{k}_A + \mathbf{k}_B + \mathbf{Q} \quad \textbf{(13.225)}$$

The directions of the momenta $\mathbf{k}_A$ and $\mathbf{k}_B$ are specified respectively by the polar angles (θ_A, ϕ_A) and (θ_B, ϕ_B), with the incident direction $\hat{\mathbf{k}}_i$ as the polar or Z axis. The energy conservation condition gives

$$E = \frac{k_i^2}{2} + E_i = \frac{k_A^2}{2} + \frac{k_B^2}{2} + \varepsilon_f \quad \textbf{(13.226)}$$

In writing this equation we have neglected the recoil energy of the ion, $Q^2/(2M_{A^+})$, where M_{A^+} is the mass of the ion A^+. This is consistent with the infinite nuclear mass approximation made in this chapter. We shall denote by $\mathbf{\Delta} = \mathbf{k}_i - \mathbf{k}_A$ the momentum transfer, or more precisely the momentum lost by the faster (also

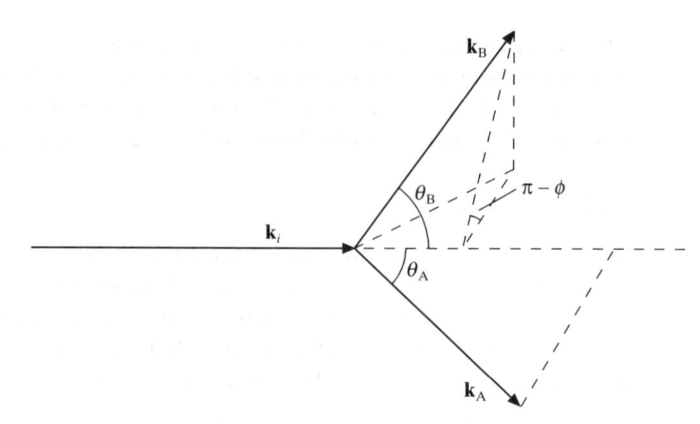

Figure 13.19 The kinematics of an (e, 2e) reaction. The incident electron momentum is $\mathbf{k}_i$ and the momenta of the outgoing electrons are $\mathbf{k}_A$ and $\mathbf{k}_B$, respectively. Also shown are the angles θ_A and θ_B with respect to the incident direction, and the angle $\pi - \phi = \pi - (\phi_A - \phi_B)$ measuring the deviation from the coplanar situation.

called 'scattered') electron A, with $E_A \geqslant E_B$. The slower electron B is usually called the 'ejected' electron.

The most detailed information about single ionisation reactions of the type (13.224) has been obtained by analysing the triple differential cross-section (TDCS) measured in (e, 2e) coincidence experiments. The TDCS is a measure of the probability that in an (e, 2e) reaction an incident electron of momentum $\mathbf{k}_i$ and energy E_i^e will produce on collision with the target two electrons having energies E_A^e and E_B^e and momenta $\mathbf{k}_A$ and $\mathbf{k}_B$, emitted respectively into the solid angles $d\Omega_A$ and $d\Omega_B$ centred about the directions (θ_A, ϕ_A) and (θ_B, ϕ_B). The TDCS is usually denoted by the symbol $d^3\sigma/(d\Omega_A\, d\Omega_B\, dE)$. For unpolarised incident electrons and targets, which is the case considered here, it is a function of the quantities E_i^e, E_A^e (or E_B^e), θ_A, θ_B and $\phi = \phi_A - \phi_B$. By integrating the TDCS over $d\Omega_A$, $d\Omega_B$ or dE one can form various double and single differential cross-sections. Finally, the total ionisation cross-section is obtained by integrating over all outgoing electron scattering angles and energies, taking into account the indistinguishability of the two electons. It depends only on E_i^e, the incident electron energy.

It is useful when studying (e, 2e) experiments to distinguish between several kinematical arrangements, since these have important implications on the theoretical analysis of the collision, as we shall see below. In coplanar geometries the momenta $\mathbf{k}_i$, $\mathbf{k}_A$ and $\mathbf{k}_B$ are in the same plane (so that $\phi_A = 0$ and $\phi_B = 0$ or π) while in non-coplanar geometries the momentum $\mathbf{k}_B$ is out of the $(\mathbf{k}_i, \mathbf{k}_A)$ reference plane. Another useful distinction can be made between asymmetric and symmetric geometries. In asymmetric geometries, a fast electron of energy E_i^e is incident on the target atom, and a fast ('scattered') electron A is detected in coincidence with a slow ('ejected') electron B. Experiments of this kind were first performed in 1969 by H. Ehrhardt and co-workers. Symmetric geometries are defined by the requirement that $\theta_A = \theta_B$ and $E_A^e = E_B^e$. The first (e, 2e) symmetric coincidence experiments were carried out by U. Amaldi and collaborators in 1969.

Basic theory

We shall now discuss the fundamental aspects of the theory of (e, 2e) reactions, following the work of R.K. Peterkop and of M.R.H. Rudge and M.J. Seaton. The key point which distinguishes ionisation from elastic scattering or excitation is the fact that in the case of ionisation the Coulomb interaction between the two outgoing electrons and between each outgoing electron and the residual ion is present even at large distances. In order to keep calculational details to a minimum, we shall concentrate on the ionisation of atomic hydrogen,

$$e^- + \mathrm{H}(i) \to \mathrm{H}^+ + 2e^- \tag{13.227}$$

where the symbol i denotes the initial state of the hydrogen atom target. Since the ion in the final state is just the bare nucleus H^+ of charge $Z = 1$, we may write the energy conservation equation (13.226) in the form

$$E = \frac{k_i^2}{2} + E_i = \frac{k_A^2}{2} + \frac{k_B^2}{2} = \frac{K^2}{2} \tag{13.228}$$

Asymptotic boundary conditions

Denoting by $\mathbf{r}_0$ the coordinates of the incident electron and by $\mathbf{r}_1$ those of the initially bound one, we look for solutions $\psi^{(+)}(\mathbf{r}_0, \mathbf{r}_1)$ of the Schrödinger equation corresponding to the initial state $|\mathbf{k}_i, i\rangle$ and to outgoing wave boundary conditions. When the target atom is neutral atomic hydrogen ($Z = 1$) these solutions satisfy the asymptotic boundary condition

$$\psi^{(+)}(\mathbf{r}_0, \mathbf{r}_1) \underset{r_0\to\infty}{\to} \exp(\mathrm{i}\mathbf{k}_i\cdot\mathbf{r}_0)\psi_i(\mathbf{r}_1) + \sum_j f_{ji}(k_i, \theta_0, \phi_0)\frac{\exp(\mathrm{i}k_j r_0)}{r_0}\psi_j(\mathbf{r}_1)$$

$$+ \int_{q_B\leqslant K} \mathrm{d}\mathbf{q}_B\, f_i(q_A\hat{\mathbf{r}}_0, \mathbf{q}_B)\frac{\exp\{\mathrm{i}[q_A r_0 + \rho(\mathbf{q}_B, \mathbf{r}_0)]\}}{r_0}\psi^{(-)}_{c,\mathbf{q}_B}(Z = 1, \mathbf{r}_1) \tag{13.229a}$$

which replaces (13.27a) and

$$\psi^{(+)}(\mathbf{r}_0, \mathbf{r}_1) \underset{r_1\to\infty}{\to} \sum_j g_{ji}(k_i, \theta_1, \phi_1)\frac{\exp(\mathrm{i}k_j r_1)}{r_1}\psi_j(\mathbf{r}_0)$$

$$+ \int_{q_B\leqslant K} \mathrm{d}\mathbf{q}_B\, g_i(q_A\hat{\mathbf{r}}_1, \mathbf{q}_B)\frac{\exp\{\mathrm{i}[q_A r_1 + \rho(\mathbf{q}_B, \mathbf{r}_1)]\}}{r_1}\psi^{(-)}_{c,\mathbf{q}_B}(Z = 1, \mathbf{r}_0) \tag{13.229b}$$

which replaces (13.27b). In these equations,

$$k_j = (k_i^2 + 2E_i - 2E_j)^{1/2}, \tag{13.230a}$$

$$q_A = (K^2 - q_B^2)^{1/2} \tag{13.230b}$$

and the phases $\rho(\mathbf{q}_B, \mathbf{r}_0)$ and $\rho(\mathbf{q}_B, \mathbf{r}_1)$ have been included since in ionisation the Coulomb interaction between the two outgoing electrons and between each electron and the nucleus is present even in the asymptotic region.

In equations (13.229), the wave functions ψ_j are negative energy solutions of the Schrödinger equation (13.8), that is they are eigenfunctions $\psi_{nlm}(\mathbf{r})$ of the target hydrogen atom corresponding to eigenenergies $-1/(2n^2)$. On the other hand, the wave functions $\psi^{(-)}_{\mathrm{c},\mathbf{q}_\mathrm{B}}(Z=1, \mathbf{r})$ are positive energy (Coulomb wave functions) solutions of the Schrödinger equation (13.8) defined as follows. Let us denote by $\psi^{(+)}_{\mathrm{c},\mathbf{k}}(Z, \mathbf{r})$ and $\psi^{(-)}_{\mathrm{c},\mathbf{k}}(Z, \mathbf{r})$ the Coulomb wave functions describing the motion of an electron of momentum $\mathbf{k}$ and (positive) energy $k^2/2$ (in a.u.) moving in a Coulomb field $-Z/r$ and exhibiting an outgoing (+) or incoming (−) spherical wave asymptotic behaviour. These Coulomb wave functions are normalised to a delta function in momentum space:

$$\langle \psi^{(\pm)}_{\mathrm{c},\mathbf{k}'}(Z, \mathbf{r}) | \psi^{(\pm)}_{\mathrm{c},\mathbf{k}}(Z, \mathbf{r}) \rangle = \delta(\mathbf{k} - \mathbf{k}') \tag{13.231}$$

and are therefore given by (see (12.222) and (12.223))

$$\psi^{(\pm)}_{\mathrm{c},\mathbf{k}}(Z, \mathbf{r}) = (2\pi)^{-3/2} \exp[\pi Z/(2k)]\Gamma(1 \mp \mathrm{i}Z/k) \times \exp(\mathrm{i}\mathbf{k}\cdot\mathbf{r})\, {}_1F_1(\pm \mathrm{i}Z/k, 1, \pm \mathrm{i}(kr \mp \mathbf{k}\cdot\mathbf{r})) \tag{13.232}$$

The Coulomb wave functions $\psi^{(-)}_{\mathrm{c},\mathbf{q}_\mathrm{B}}(Z = 1, \mathbf{r})$ appearing in (13.229) are therefore obtained by setting $\mathbf{k} = \mathbf{q}_\mathrm{B}$ and $Z = 1$ in (13.232), and by choosing the wave functions with incoming (−) spherical wave behaviour. It can be shown [12] that in order for f_i and g_i in equations (13.229) to have the meaning of direct and exchange ionisation amplitudes, respectively, one must use Coulomb wave functions with incoming (−) rather than outgoing (+) spherical wave asymptotic behaviour.

Finally, we note that if the target is a hydrogenic ion with a nucleus of charge Z one must modify the exponents in (13.229) by including the logarithmic phase factors due to the distortion by the Coulomb field of the target ion. Moreover, in that case the target wave functions ψ_j are hydrogenic ion wave functions $\psi_{nlm}(\mathbf{r})$ corresponding to eigenenergies $-Z^2/(2n^2)$, and one must use in (13.229) incoming spherical wave Coulomb wave functions $\psi^{(-)}_{\mathrm{c},\mathbf{q}_\mathrm{B}}(Z, \mathbf{r})$.

Cross-sections

Ionisation cross-sections are obtained by taking the ratio of the number of ionisation events per unit time and per unit target atom to the incident electron flux. For example, the triple differential cross-section (TDCS) for ionisation of a hydrogen atom in the initial state $|i\rangle$ is given for random electron spin orientations by

$$\frac{\mathrm{d}^3\sigma_i}{\mathrm{d}\Omega_\mathrm{A}\,\mathrm{d}\Omega_\mathrm{B}\,\mathrm{d}E} = \frac{k_\mathrm{A} k_\mathrm{B}}{k_i}\left[\frac{1}{4}|f_i + g_i|^2 + \frac{3}{4}|f_i - g_i|^2\right] \tag{13.233a}$$

$$= \frac{k_\mathrm{A} k_\mathrm{B}}{k_i}\left[\frac{1}{4}|f_i^+|^2 + \frac{3}{4}|f_i^-|^2\right] \tag{13.233b}$$

[12] See Peterkop (1977) or Byron and Joachain (1989).

where we have written $f_i \equiv f_i(\mathbf{k}_A, \mathbf{k}_B)$, $g_i \equiv g_i(\mathbf{k}_A, \mathbf{k}_B)$ and

$$f_i^{\pm} = f_i \pm g_i \tag{13.234}$$

As we mentioned above, various double and single differential cross-sections can be obtained by integrating the TDCS with respect to $d\Omega_A$, $d\Omega_B$ or dE. By integrating (13.233) over all outgoing electron scattering angles and energies, we obtain the total ionisation cross-section

$$\sigma_i = \frac{1}{k_i} \int_0^{E/2} dE_B^e \, k_A k_B \int d\Omega_A \int d\Omega_B \left[\frac{1}{4} |f_i^+|^2 + \frac{3}{4} |f_i^-|^2 \right] \tag{13.235}$$

where the upper limit of integration over the energy variable is $E/2$ because the two electrons are indistinguishable.

The first Born approximation

Let us consider an (e, 2e) reaction in which a fast (but non-relativistic) electron is incident on a neutral target atom, which we shall take to be atomic hydrogen in the eigenstate $|i\rangle$. By expanding the full wave function $\psi^{(+)}(\mathbf{r}_0, \mathbf{r}_1)$ in a Born series, one generates the corresponding Born series for the direct ionisation amplitude f_i and the exchange ionisation amplitude g_i. We begin by considering the direct ionisation amplitude f_i. The first Born approximation to f_i, which will be denoted by f_i^{B1} $(=\bar{f}_i^{B1})$, is given by

$$f_i^{B1} = -\frac{1}{2\pi} \int d\mathbf{r}_0 \, d\mathbf{r}_1 \exp(i\boldsymbol{\Delta}\cdot\mathbf{r}_0) \psi_{c,\mathbf{k}_B}^{(-)*}(Z=1, \mathbf{r}_1) \frac{1}{r_{01}} \psi_i(\mathbf{r}_1) \tag{13.236}$$

where $\boldsymbol{\Delta} = \mathbf{k}_i - \mathbf{k}_A$. This is the generalisation of the corresponding first Born direct amplitude (13.150) for inelastic e^-–H collisions studied in Section 13.3. Again we note that Coulomb wave functions with incoming (–) spherical wave asymptotic behaviour must be used. The corresponding first Born TDCS is obtained from (13.233) by replacing the amplitude f_i by its first Born approximation f_i^{B1} and neglecting the exchange amplitude g_i. It is given by

$$\frac{d^3\sigma_i^{B1}}{d\Omega_A \, d\Omega_B \, dE} = \frac{k_A k_B}{k_i} |f_i^{B1}|^2 \tag{13.237}$$

With the help of the Bethe integral (13.152), the integration over the variable $\mathbf{r}_0$ in (13.236) can be performed, so that

$$f_i^{B1} = -\frac{2}{\Delta^2} \mathcal{S}_i^{H}(\boldsymbol{\Delta}, \mathbf{k}_B) \tag{13.238}$$

where the atomic hydrogen ionisation form factor $\mathcal{S}_i^{H}(\boldsymbol{\Delta}, \mathbf{k}_B)$ is given by

$$\mathcal{S}_i^{H}(\boldsymbol{\Delta}, \mathbf{k}_B) = \int \psi_{c,\mathbf{k}_B}^{(-)*}(Z=1, \mathbf{r}_1) \exp(i\boldsymbol{\Delta}\cdot\mathbf{r}_1) \psi_i(\mathbf{r}_1) \, d\mathbf{r}_1 \tag{13.239}$$

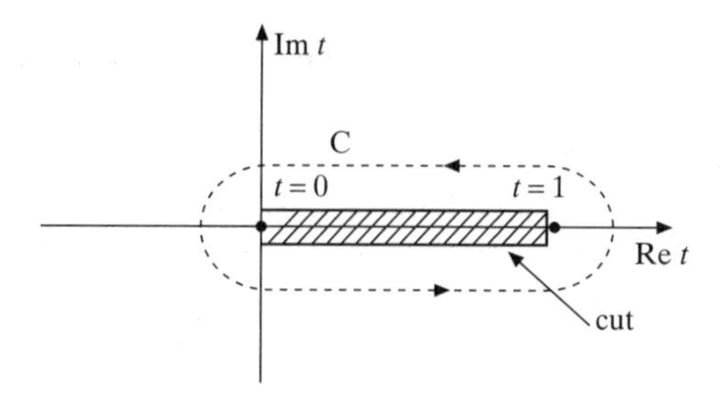

Figure 13.20 The integration contour C for the integral (13.243) in the complex t-plane.

As an example, let us consider the ionisation of atomic hydrogen from the ground state:

$$e^- + H(1s) \rightarrow H^+ + 2e^- \tag{13.240}$$

Using (13.232) with $Z = 1$, $\mathbf{r} = \mathbf{r}_1$ and $\mathbf{k} = \mathbf{k}_B$ and replacing $\psi_i(\mathbf{r}_1)$ by the ground state wave function $\psi_{1s}(r_1) = \pi^{-1/2}\exp(-r_1)$, we find that

$$\mathcal{S}^{H}_{1s}(\boldsymbol{\Delta}, \mathbf{k}_B) = -2^{-3/2}\pi^{-2}\exp[\pi/(2k_B)]\Gamma(1 - i/k_B)\left[\frac{\partial}{\partial\lambda}I(\mathbf{q}, \mathbf{k}_B, \lambda)\right]_{\lambda=1} \tag{13.241}$$

where

$$I(\mathbf{q}, \mathbf{k}_B, \lambda) = \int \exp(i\mathbf{q}\cdot\mathbf{r}_1)\frac{\exp(-\lambda r_1)}{r_1}\,{}_1F_1\left(\frac{i}{k_B}, 1, i(k_B r_1 + \mathbf{k}_B\cdot\mathbf{r}_1)\right)d\mathbf{r}_1 \tag{13.242}$$

with $\mathbf{q} = \boldsymbol{\Delta} - \mathbf{k}_B$. This integral can be evaluated by using parabolic coordinates. Alternatively, one may use the integral representation of the confluent hypergeometric function

$${}_1F_1(ia, 1, u) = \frac{1}{2\pi i}\oint_C t^{ia-1}(t-1)^{-ia}e^{ut}\,dt \tag{13.243}$$

where $a = 1/k_B$, $u = i(k_B r_1 + \mathbf{k}_B\cdot\mathbf{r}_1)$ and the integration contour C in the complex t-plane is one which encloses the points $t = 0$ and $t = 1$ where the integrand has branch points, the cut being chosen to join these points (see Fig. 13.20). The result is (Problem 13.13)

$$I(\mathbf{q}, \mathbf{k}_B, \lambda) = \frac{4\pi}{q^2 + \lambda^2}\exp\left[-\frac{i}{k_B}\log v(\mathbf{q}, \mathbf{k}_B, \lambda)\right] \tag{13.244a}$$

where

$$v(\mathbf{q}, \mathbf{k}_B, \lambda) = \frac{q^2 + \lambda^2 + 2\mathbf{q}\cdot\mathbf{k}_B - 2i\lambda k_B}{q^2 + \lambda^2} \tag{13.244b}$$

Using (13.241) and (13.244), we have

$$\mathcal{S}^{\mathrm{H}}_{1s}(\boldsymbol{\Delta}, \mathbf{k}_{\mathrm{B}}) = \frac{4\sqrt{2}}{\pi} \exp[\pi/(2k_{\mathrm{B}})]\Gamma(1 - \mathrm{i}/k_{\mathrm{B}}) \exp\left[-\frac{\mathrm{i}}{k_{\mathrm{B}}} \log v(\mathbf{q}, \mathbf{k}_{\mathrm{B}}, \lambda = 1)\right]$$

$$\times \frac{\boldsymbol{\Delta}\cdot[\boldsymbol{\Delta} - \mathbf{k}_{\mathrm{B}}(1 + \mathrm{i}/k_{\mathrm{B}})]}{[\Delta^2 - (k_{\mathrm{B}} + \mathrm{i})^2][(\boldsymbol{\Delta} - \mathbf{k}_{\mathrm{B}})^2 + 1]^2} \tag{13.245a}$$

with

$$v(\mathbf{q}, \mathbf{k}_{\mathrm{B}}, \lambda = 1) = \frac{\mathbf{q}\cdot(\mathbf{q} + 2\mathbf{k}_{\mathrm{B}}) + 1 - 2\mathrm{i}k_{\mathrm{B}}}{q^2 + 1} = \frac{\Delta^2 - k_{\mathrm{B}}^2 + 1 - 2\mathrm{i}k_{\mathrm{B}}}{(\boldsymbol{\Delta} - \mathbf{k}_{\mathrm{B}})^2 + 1} \tag{13.245b}$$

and according to (13.238) the first Born amplitude for electron impact ionisation from the ground state of atomic hydrogen is given by

$$f^{\mathrm{B1}}_{1s} = -\frac{2}{\Delta^2}\mathcal{S}^{\mathrm{H}}_{1s}(\boldsymbol{\Delta}, \mathbf{k}_{\mathrm{B}}) \tag{13.246}$$

It is interesting to note that since $\mathcal{S}^{\mathrm{H}}_{1s}(\boldsymbol{\Delta}, \mathbf{k}_{\mathrm{B}})$ is proportional to $\boldsymbol{\Delta}\cdot\mathbf{k}_{\mathrm{B}}$ at small momentum transfers, the first Born amplitude f^{B1}_{1s} behaves like Δ^{-1} in that region, reflecting the fact that at small Δ the ionisation is primarily into continuum p states. At momentum transfers large with respect to k_{B}, $\mathcal{S}^{\mathrm{H}}_{1s}(\boldsymbol{\Delta}, \mathbf{k}_{\mathrm{B}})$ is proportional to Δ^{-4}, so that f^{B1}_{1s} decreases like Δ^{-6}, a behaviour characteristic of ionisation into continuum s states.

Using (13.237), (13.245) and (13.246), we find that the first Born TDCS for electron impact ionisation from the ground state of atomic hydrogen is given by (Problem 13.14)

$$\begin{aligned}\frac{\mathrm{d}^3\sigma^{\mathrm{B1}}_{1s}}{\mathrm{d}\Omega_{\mathrm{A}}\,\mathrm{d}\Omega_{\mathrm{B}}\,\mathrm{d}E} &= \frac{k_{\mathrm{A}}k_{\mathrm{B}}}{k_i}|f^{\mathrm{B1}}_{1s}|^2 \\ &= \frac{2^8}{\pi}\frac{k_{\mathrm{A}}}{k_i}\frac{\Delta^2 - 2\Delta k_{\mathrm{B}}\cos\gamma + (k_{\mathrm{B}}^2 + 1)\cos^2\gamma}{\Delta^2[\Delta^2 + k_{\mathrm{B}}^2 - 2\Delta k_{\mathrm{B}}\cos\gamma + 1]^4[(\Delta + k_{\mathrm{B}})^2 + 1][(\Delta - k_{\mathrm{B}})^2 + 1]} \\ &\quad\times \frac{\exp\{-(2/k_{\mathrm{B}})\tan^{-1}[2k_{\mathrm{B}}/(\Delta^2 - k_{\mathrm{B}}^2 + 1)]\}}{1 - \exp(-2\pi/k_{\mathrm{B}})}\end{aligned} \tag{13.247}$$

where γ is the angle between the vectors $\boldsymbol{\Delta}$ and $\mathbf{k}_{\mathrm{B}}$. We shall return below to the analysis of this expression.

Integrated cross-sections

As explained above, a number of 'integrated' cross-sections can be obtained from the triple differential cross-section. In first Born approximation, and for the case of ionisation from the ground state of atomic hydrogen, one finds by integrating the triple differential cross-section (13.247) over the polar angles $(\theta_{\mathrm{B}}, \phi_{\mathrm{B}})$ of the ejected electron that the first Born double differential cross-section is given by (Problem 13.15)

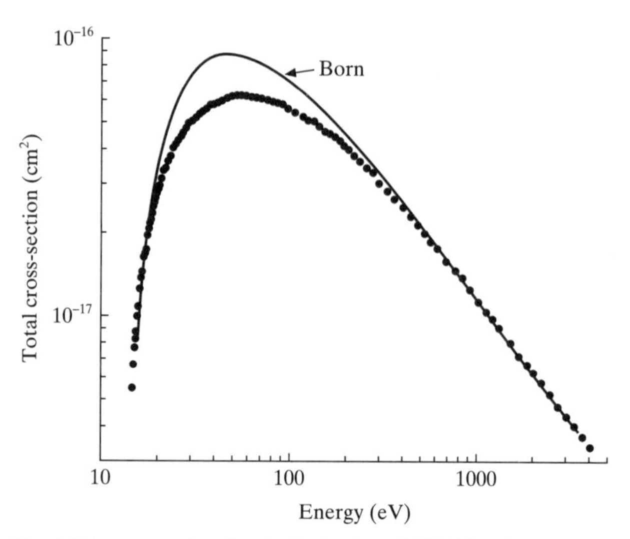

Figure 13.21 The total cross-section for the ionisation of H(1s) by electron impact. The experimental data are due to M.B. Shah, D.S. Elliott and H.B. Gilbody: ●. Calculated cross-section in the first Born approximation: ——.

$$\frac{d^2\sigma_{1s}^{B1}}{d\Omega_A\, dE} = \frac{2^{10}k_A}{k_i}\frac{[\Delta^2 + (1 + k_B^2)/3]\exp\{-(2/k_B)\tan^{-1}[2k_B/(\Delta^2 - k_B^2 + 1)]\}}{\Delta^2[(\Delta + k_B)^2 + 1]^3[(\Delta - k_B)^2 + 1]^3[1 - \exp(-2\pi/k_B)]} \tag{13.248}$$

By integrating (numerically) the double differential cross-section (13.248) over the angular variables (θ_A, ϕ_A) of the scattered electron, one obtains the first Born energy distribution of the ejected electron, $d\sigma_{1s}^{B1}/dE_B^e$.

Let us now consider the total ionisation cross-section. Replacing f_i by its first Born approximation f_i^{B1} and neglecting the exchange amplitude in (13.235), we obtain the first Born total ionisation cross-section

$$\sigma_{tot}^{B1}(i) = \frac{1}{k_i}\int_0^{E/2} dE_B^e\, k_A k_B \int d\Omega_A \int d\Omega_B\, |f_i^{B1}|^2 \tag{13.249}$$

where we recall that $E = k_i^2/2 + E_i$ is the total (positive) energy of the system.

The total first Born cross-section $\sigma_{tot}^{B1}(1s)$ for ionisation of atomic hydrogen from the ground state is compared in Fig. 13.21 with the experimental data of M.B. Shah, D.S. Elliott and H.B. Gilbody. We see that the first Born approximation agrees well with experiment at incident electron energies above 300 eV, but overestimates the total cross-section at lower energies, notably in the region where the cross-section has its maximum. Theoretical calculations performed by I. Bray and A.T. Stelbovics using the convergent close-coupling (CCC) approach discussed in Section 13.2 agree with the experimental data over the entire energy range.

As in the case of discrete inelastic transitions, we can derive a Bethe approximation to the first Born total ionisation cross-section (13.249). At high energies, most of the contribution to $\sigma_{\text{tot}}^{\text{B1}}(i)$ arises from small momentum transfer collisions. Thus, expanding $\exp(\text{i}\mathbf{\Delta}\cdot\mathbf{r}_1)$ in a power series in (13.239) and retaining only the first non-vanishing contribution due to the dipole term $(\text{i}\mathbf{\Delta}\cdot\mathbf{r}_1)$, we find that for large incident energies the first Born ionisation cross-section becomes

$$\sigma_{\text{tot}}^{\text{B1}}(i) \simeq \frac{8\pi}{k_i^2}\int_0^{E/2} \text{d}E_{\text{B}}^{\text{e}}\, k_{\text{B}} \int \text{d}\Omega_{\text{B}}\, |\langle \psi_{\text{c},\mathbf{k}_{\text{B}}}^{(-)}(Z=1,\mathbf{r}_1)|z_1|\psi_i(\mathbf{r}_1)\rangle|^2 \int_{\Delta_{\min}}^{\bar{\Delta}} \frac{\text{d}\Delta}{\Delta} \tag{13.250}$$

In writing this equation, we have taken the momentum transfer $\mathbf{\Delta}$ as the quantisation axis and changed the integration over θ_{A} to one over Δ, with

$$\Delta = (k_i^2 + k_{\text{A}}^2 - 2k_i k_{\text{A}} \cos\theta_{\text{A}})^{1/2} \tag{13.251}$$

Moreover, $\Delta_{\min} = k_i - k_{\text{A}}$ and $\bar{\Delta}$ is a value of the momentum transfer beyond which the contribution to $\sigma_{\text{tot}}^{\text{B1}}(i)$ decreases rapidly. Using the energy conservation equation (13.228) and the fact that at high energies and small momentum transfers $k_i + k_{\text{A}} \simeq 2k_i$, we have $\Delta_{\min} \simeq (E_{\text{B}}^{\text{e}} - E_i)/k_i$, with $E_{\text{B}}^{\text{e}} = k_{\text{B}}^2/2$. Hence, performing the integral over Δ in (13.250) and extending the upper limit of the integral over E_{B}^{e} to $+\infty$, we find that

$$\begin{aligned}\sigma_{\text{tot}}^{\text{B1}}(i) &\simeq \frac{8\pi}{k_i^2}\int_0^{\infty} \text{d}E_{\text{B}}^{\text{e}}\, k_{\text{B}} \int \text{d}\Omega_{\text{B}}\, |\langle \psi_{\text{c},\mathbf{k}_{\text{B}}}^{(-)}(Z=1,\mathbf{r}_1)|z_1|\psi_i(\mathbf{r}_1)\rangle|^2 \log\frac{k_i\bar{\Delta}}{E_{\text{B}}^{\text{e}} - E_i}\\ &\simeq A\frac{\log E_i^{\text{e}}}{E_i^{\text{e}}}\end{aligned} \tag{13.252}$$

where we recall that $E_i^{\text{e}} = k_i^2/2$ is the incident electron energy, and the quantity

$$\begin{aligned}A &= 2\pi\int_0^{\infty} \text{d}E_{\text{B}}^{\text{e}}\, k_{\text{B}} \int \text{d}\Omega_{\text{B}}\, |\langle \psi_{\text{c},\mathbf{k}_{\text{B}}}^{(-)}(Z=1,\mathbf{r}_1)|z_1|\psi_i(\mathbf{r}_1)\rangle|^2\\ &= 2\pi\int \text{d}\mathbf{k}_{\text{B}}|\langle \psi_{\text{c},\mathbf{k}_{\text{B}}}^{(-)}(Z=1,\mathbf{r}_1)|z_1|\psi_i(\mathbf{r}_1)\rangle|^2\end{aligned} \tag{13.253}$$

is independent of E_i^{e}. As in the case of discrete optically allowed transitions, it can be shown that the correction to the leading contribution $A(E_i^{\text{e}})^{-1}\log E_i^{\text{e}}$ to the total cross-section $\sigma_{\text{tot}}^{\text{B1}}(i)$ is of the form $B(E_i^{\text{e}})^{-1}$, where the quantity B is also energy-independent. Hence, at high incident energies, the first Born total ionisation cross-section can be written as

$$\sigma_{\text{tot}}^{\text{B1}}(i) \simeq A\frac{\log E_i^{\text{e}}}{E_i^{\text{e}}} + \frac{B}{E_i^{\text{e}}} \tag{13.254}$$

which is known as the *Bethe–Born approximation* to $\sigma_{\text{tot}}^{\text{B1}}(i)$. The approximation (13.252) in which only the leading term is retained is called the *Bethe approximation* to $\sigma_{\text{tot}}^{\text{B1}}(i)$.

By summing the Bethe expressions (13.193) and (13.252) corresponding respectively to optically allowed transitions from the initial state $|i\rangle$ to discrete and continuum final states, and remembering that optically allowed transitions dominate at high incident energies, we can obtain a simple high-energy expression for $\sigma_{\text{tot}}^{\text{B1,c}}(i)$, the complete (c) total first Born cross-section for electron scattering from atomic hydrogen in the state $|i\rangle$. Indeed, if $|k\rangle$ is a hydrogen atom state connected to the state $|i\rangle$ by a dipole-allowed transition, we have for large E_i^{e},

$$\sigma_{\text{tot}}^{\text{B1,c}}(i) \simeq \frac{8\pi}{k_i^2}(\log k_i)\sum_k \langle i|z_1|k\rangle\langle k|z_1|i\rangle \tag{13.255}$$

However, since $\langle k|z_1|i\rangle = 0$ for other states $|k\rangle$, we can extend the sum in (13.255) to all hydrogen atom states. Using the closure relation

$$\sum_k |k\rangle\langle k| = I \tag{13.256}$$

we then obtain at high energies

$$\sigma_{\text{tot}}^{\text{B1,c}}(i) \simeq \frac{8\pi}{k_i^2}(\log k_i)\langle i|z_1^2|i\rangle \tag{13.257}$$

In particular, if the initial state is the ground state, we have $\langle 1s|z_1^2|1s\rangle = 1$ and we find that for large values of E_i^{e}

$$\sigma_{\text{tot}}^{\text{B1,c}}(1\text{s}) \simeq \frac{8\pi}{k_i^2}\log k_i \tag{13.258}$$

Triple differential cross-sections

We now turn to a more detailed analysis of the triple differential cross-section (TDCS). We shall focus our attention on two kinematical arrangements known respectively as the Ehrhardt (coplanar) asymmetric geometry and the symmetric geometry.

Coplanar asymmetric (e, 2e) reactions

In the Ehrhardt (coplanar) asymmetric geometry, for a given energy E_i^{e} of the fast incident electron, a fast ('scattered') electron A is detected in coincidence with a slow ('ejected') electron B. The scattering angle θ_{A} of the fast electron is fixed and small, while the angle θ_{B} of the slow electron is varied. It is worth noting that in this geometry the magnitude $\Delta = |\mathbf{k}_i - \mathbf{k}_{\text{A}}|$ of the momentum transfer is small.

According to the first Born approximation (13.247), the TDCS exhibits maxima in the Ehrhardt asymmetric geometry when the direction of $\mathbf{k}_{\text{B}}$ coincides with $\hat{\boldsymbol{\Delta}}$ or $-\hat{\boldsymbol{\Delta}}$. In the first case, the maximum is referred to as the *forward peak*, while in the second case it is called the *recoil peak*. However, we see from Fig. 13.22 that for an incident electron energy $E_i^{\text{e}} = 250$ eV, an ejected electron energy $E_{\text{B}}^{\text{e}} = 5$ eV and a scattering angle $\theta_{\text{A}} = 3°$, the experimental forward peak is shifted towards larger

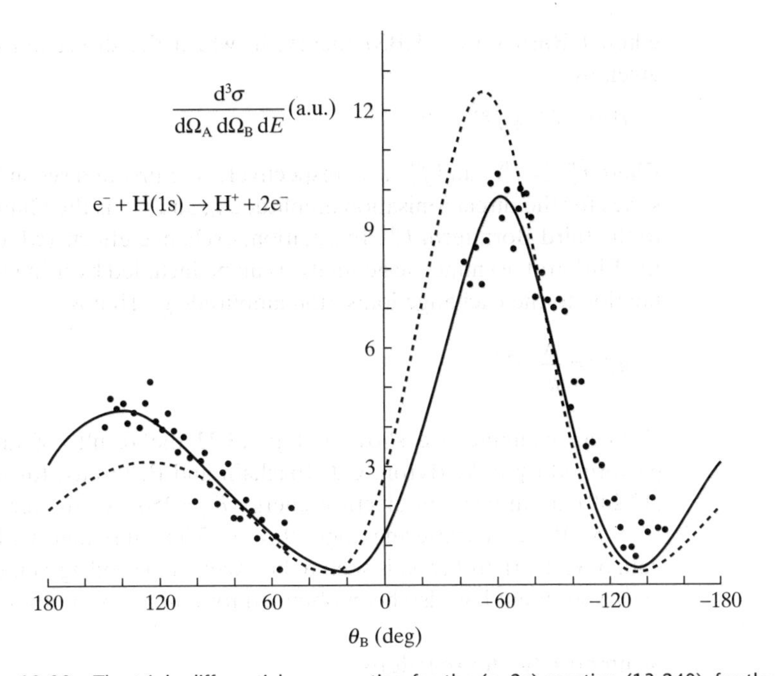

Figure 13.22 The triple differential cross-section for the (e, 2e) reaction (13.240), for the case $E_i^e = 250$ eV, $E_B^e = 5$ eV and $\theta_A = 3°$, as a function of θ_B.
The absolute experimental data are due to H. Ehrhardt *et al.*: ●.
The theoretical calculations of F.W. Byron, C.J. Joachain and B. Piraux are: EBS theory: ——; first Born approximation: -----.

angles $|\theta_B|$ with respect to the first Born prediction, and that the measured recoil peak is shifted by an even larger amount, also towards larger angles. Moreover, the ratio of the intensity of the forward peak to that of the backward peak is reduced with respect to the first Born prediction.

These features have been illustrated for the (e, 2e) reaction (13.240) in atomic hydrogen since in that case the wave function describing the target in the initial (1s) state is known exactly as is the Coulomb wave function corresponding to the final, unbound target state ($H^+ + e^-$). Moreover, the first Born TDCS (13.247) can be calculated in closed form, so that the comparison with absolute experimental data is unambiguous. In the case of (e, 2e) reactions in more complex atoms theoretical difficulties arise in describing accurately the target initial state and especially its final continuum state consisting of an ion and an unbound electron. These problems introduce additional complications in the interpretation of data on TDCS. Nevertheless, the measurements clearly show that at intermediate incident electron energies there are significant departures from the predictions of the first Born approximation.

In order to improve the agreement with experiment, more elaborate theories than the first Born approximation are required. For example, one can use the eikonal–Born series (EBS) theory, in which the direct ionisation amplitude is given by

$$f_i^{\text{EBS}} = \bar{f}_i^{\text{B1}} + \bar{f}_i^{\text{B2}} + \bar{f}_i^{\text{G3}} \tag{13.259}$$

where $\bar{f}_i^{\text{B1}}$ $(= f_i^{\text{B1}})$ and $\bar{f}_i^{\text{B2}}$ are, respectively, the first and second terms of the Born series for the direct ionisation amplitude f_i, and $\bar{f}_i^{\text{G3}}$ is the Glauber approximation to the third Born term $\bar{f}_i^{\text{B3}}$. In addition, exchange effects (which are very small in the Ehrhardt asymmetric geometry) can be included by using the Ochkur approximation to the exchange ionisation amplitude g_i. That is,

$$g_i^{\text{Och}} = \frac{\Delta^2}{k_i^2} f_i^{\text{B1}} \tag{13.260}$$

As an example, we show in Fig. 13.22 the results of an EBS calculation, performed by F.W. Byron, C.J. Joachain and B. Piraux, for the (e, 2e) reaction (13.240), at an incident electron energy $E_i^{\text{e}} = 250$ eV, an ejected electron energy $E_{\text{B}}^{\text{e}} = 5$ eV and a scattering angle $\theta_{\text{A}} = 3°$. The agreement with the experimental data of H. Ehrhardt *et al.* is seen to be excellent. Good agreement between theory and experiment has also been obtained for a number of other target atoms.

Symmetric (e, 2e) reactions

We shall now discuss fast (e, 2e) reactions in symmetric kinematical arrangements such that $\theta_{\text{A}} = \theta_{\text{B}} = \theta$ and $E_{\text{A}}^{\text{e}} = E_{\text{B}}^{\text{e}}$. We note that when the incident electron is fast the magnitude Δ of the momentum transfer is never small in a symmetric geometry. For example, in a coplanar symmetric geometry, we have

$$\Delta \simeq k_i \left(\frac{3}{2} - \sqrt{2} \cos \theta \right)^{1/2} \tag{13.261}$$

and we see that Δ is large when k_i is large, in contrast with the Ehrhardt asymmetric geometry. On the other hand, the magnitude of the recoil momentum of the ion, Q, can be either small or large. For example, in the coplanar symmetric geometry one has for large k_i

$$Q \simeq k_i |\sqrt{2} \cos \theta - 1| \tag{13.262}$$

As a result, the values of Q remain small or moderate for scattering angles $\theta \lesssim 70°$ (with $Q \simeq 0$ when $\theta = 45°$), while for $\theta \gtrsim 70°$ both Δ and Q become large.

Two angular regions can therefore be distinguished. Let us first consider the case $\theta \lesssim 70°$, for which Δ is relatively large, while Q remains small or moderate. In this region the momentum density distribution of the target atoms can be obtained. To see how this comes about let us consider the (e, 2e) reaction (13.227) in atomic hydrogen, in which the target is initially in the bound state $|i\rangle$, with eigenfunction $\psi_i(\mathbf{r}_1)$. Since the momentum $\mathbf{k}_{\text{B}}$ is large, we can in first approximation replace the Coulomb wave function $\psi_{\text{c},\mathbf{k}_{\text{B}}}^{(-)*}(Z = 1, \mathbf{r}_1)$ by the plane wave

$(2\pi)^{-3/2}\exp(-\mathrm{i}\mathbf{k}_\mathrm{B}\cdot\mathbf{r}_1)$ in the expression (13.236) of the first Born amplitude, which then reduces to the *plane wave Born* amplitude

$$f_i^{\mathrm{PWB}} = -(2\pi)^{-5/2}\int \mathrm{d}\mathbf{r}_0\,\mathrm{d}\mathbf{r}_1 \exp(\mathrm{i}\boldsymbol{\Delta}\cdot\mathbf{r}_0)\exp(-\mathrm{i}\mathbf{k}_\mathrm{B}\cdot\mathbf{r}_1)\frac{1}{r_{01}}\psi_i(\mathbf{r}_1) \qquad \textbf{(13.263)}$$

Using the Bethe integral (13.152), we find that

$$f_i^{\mathrm{PWB}} = -\frac{2}{\Delta^2}\phi_i(-\mathbf{Q}) \qquad \textbf{(13.264)}$$

where ϕ_i is the Fourier transform of the wave function ψ_i, and

$$\mathbf{Q} = \boldsymbol{\Delta} - \mathbf{k}_\mathrm{B} = \mathbf{k}_i - \mathbf{k}_\mathrm{A} - \mathbf{k}_\mathrm{B} \qquad \textbf{(13.265)}$$

is the recoil momentum of the ionised target (in the present case H^+). Looking at (12.283), we see that

$$f_\mathrm{c}^{\mathrm{B1}}(\Delta) = -\frac{2}{\Delta^2} \qquad \textbf{(13.266)}$$

is just the first Born amplitude corresponding to Coulomb scattering by the potential $V(r) = 1/r$. We may therefore recast (13.264) in the form

$$f_i^{\mathrm{PWB}} = f_\mathrm{c}^{\mathrm{B1}}(\Delta)\phi_i(\mathbf{p}) \qquad \textbf{(13.267)}$$

with $\mathbf{p} = -\mathbf{Q}$. From (13.265) we see that $\mathbf{p}$ can be interpreted as the momentum of the struck electron in the initial state.

The triple differential cross-section is given in the plane wave Born approximation by

$$\frac{\mathrm{d}^3\sigma_i^{\mathrm{PWB}}}{\mathrm{d}\Omega_\mathrm{A}\,\mathrm{d}\Omega_\mathrm{B}\,\mathrm{d}E} = \frac{k_\mathrm{A}k_\mathrm{B}}{k_i}|f_\mathrm{c}^{\mathrm{B1}}(\Delta)|^2\,|\phi_i(\mathbf{p})|^2 \qquad \textbf{(13.268)}$$

We see that it is proportional to a 'structure factor' which is the momentum probability density $|\phi_i(\mathbf{p})|^2$ of the target electron. This important property has given rise to the field of (e, 2e) *spectroscopy*.

As an example, we show in Fig. 13.23 the electron momentum probability density corresponding to the ground state of atomic hydrogen, extracted by B. Lohmann and E. Weigold from (e, 2e) experiments performed at various energies and compared with the calculated momentum probability density $|\phi_{1s}(p)|^2 = 8\pi^{-2}(1+p^2)^{-4}$.

Let us now consider the large-angle region ($\theta \gtrsim 70°$) for which both Δ and Q are large. In this region one expects on the basis of the calculations performed for inelastic (excitation) scattering that the second Born term should be very important, and that this term will be dominated by the contributions of the initial and final target states acting as intermediate states. This is confirmed by second Born calculations performed for large-angle symmetric (e, 2e) reactions in atomic hydrogen by F.W. Byron, C.J. Joachain and B. Piraux. Looking at Fig. 13.24(a),

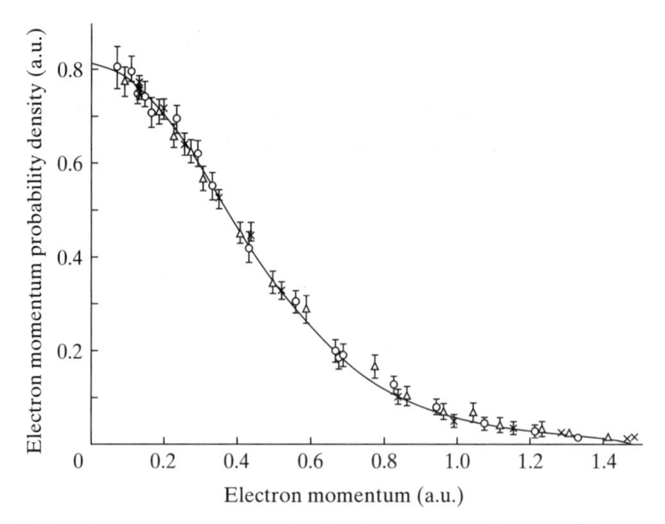

Figure 13.23 The electron momentum distribution corresponding to H(1s), as obtained by B. Lohmann and E. Weigold at 400 eV (△), 800 eV (○) and 1200 eV (×), compared with the calculated momentum probability density $|\phi_{1s}(p)|^2$. (Adapted from B. Lohmann and E. Weigold, *Physics Letters A* **86**, 139, 1981.)

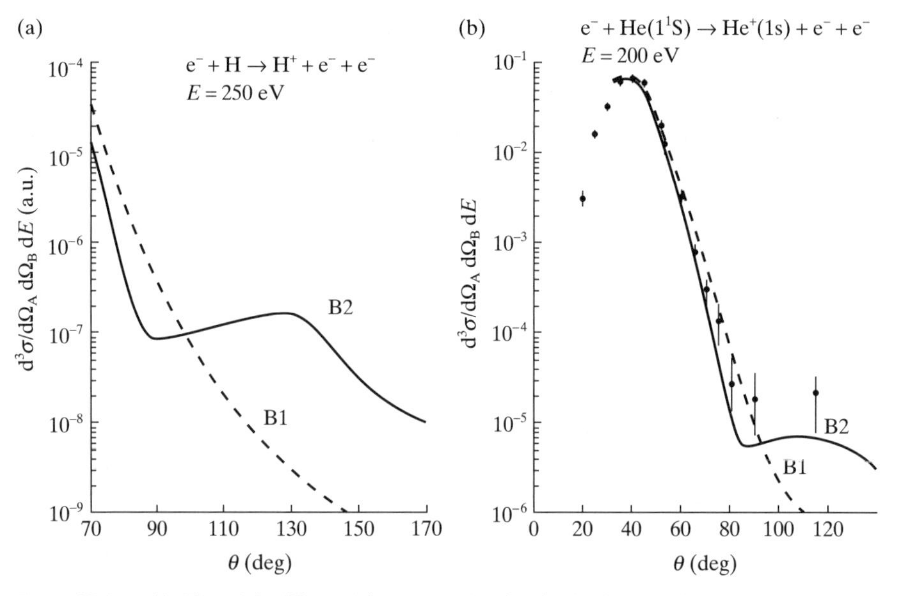

Figure 13.24 (a) The triple differential cross-section for the (e, 2e) reaction $e^- + H(1s) \to H^+ + 2e^-$, for a coplanar symmetric geometry, with $E_i^e = 250$ eV and $E_A^e = E_B^e = 118.2$ eV, as a function of the scattering angle $\theta = \theta_A = \theta_B$.
The theoretical calculations due to F.W. Byron, C.J. Joachain and B. Piraux are: second Born approximation (B2): ——; first Born approximation (B1): -----.
(b) The triple differential cross-section for the (e, 2e) reaction $e^- + He(1^1S) \to He^+(1s) + 2e^-$, for a coplanar symmetric geometry, with $E_i^e = 200$ eV and $E_A^e = E_B^e = 87.7$ eV, as a function of θ.
The experimental data are due to A. Pochat *et al.*: ●.
The theoretical first and second Born calculations are due to F.W. Byron, C.J. Joachain and B. Piraux: the notation is the same as for (a).

we see that the effect of the second Born term at large angles is very important. A similar striking behaviour of the calculated TDCS, due to second-order effects, arises for large-angle symmetric (e, 2e) reactions in helium, as seen from Fig. 13.24(b), and it is seen from this figure that the experimental data confirm the trend predicted by the second Born theoretical curve.

13.5 Atomic photoionisation

In this section, we shall study the photoionisation of atoms and ions. The basic process which we shall consider is the single photoionisation reaction

$$\hbar\omega + \mathrm{A}(i) \rightarrow \mathrm{A}^+(f) + \mathrm{e}^- \tag{13.269}$$

where $\mathrm{A}(i)$ is an atom in state i and $\mathrm{A}^+(f)$ is the residual ion in state f. The photoionisation reaction

$$\hbar\omega + \mathrm{A}^-(i) \rightarrow \mathrm{A}(f) + \mathrm{e}^- \tag{13.270}$$

involving a negative ion in the initial channel is known as *photodetachment*. The inverse processes of (13.269) and (13.270) are called *radiative recombination*. At the end of this section we shall also briefly consider double photoionisation reactions.

Basic theory

Let us consider the single photoionisation process (13.269). In the initial state a we have a photon of energy $\hbar\omega$ incident on a target atom A at rest in the eigenstate $|i\rangle$, with eigenfunction ψ_i and eigenenergy E_i. In the final state b, a photoelectron is emitted with momentum $\mathbf{k}_f$, the remaining ion A^+ being left in the eigenstate $|f\rangle$, with eigenfunction ψ_f and eigenenergy ε_f. From energy conservation we have

$$\hbar\omega + E_i = \frac{\hbar^2 k_f^2}{2m} + \varepsilon_f \tag{13.271}$$

where we have assumed that the photoelectron moves at a non-relativistic velocity, and we have neglected the recoil energy of the ion A^+. We remark that if the energy scale is chosen so that $E = 0$ corresponds to the ionisation threshold (which coincides with the ground state of A^+), and if the atom A is initially in the ground state, then $E_i = -I_\mathrm{P}$, where I_P is the ionisation potential of the atom.

In order to obtain the photoionisation cross-section for the process (13.269), we can generalise to an N-electron atom the method followed in Chapter 4 to study photoionisation by one-electron atoms. To first order in the coupling between the N-electron atom and the electromagnetic field (described classically by the vector potential $\mathbf{A}$ in the Coulomb gauge, for which $\boldsymbol{\nabla}\cdot\mathbf{A} = 0$), the interaction Hamiltonian is given by (see (9.8))

$$H_{\text{int}} = -i\hbar\frac{e}{m}\sum_{k=1}^{N}\mathbf{A}(\mathbf{r}_\mathbf{k}, t)\cdot\boldsymbol{\nabla}_{\mathbf{r}_k} \tag{13.272}$$

Using Fermi's Golden Rule, the differential or total transition rates for the 'bound-free' absorption transition $a \to b$ into a group of states at energy E_f can then be obtained, either in the dipole approximation or by keeping retardation terms. Dividing by the incident photon flux, one obtains the corresponding photoionisation cross-sections. In particular, the differential photoionisation cross-section for the process (13.269) is given in the dipole approximation and in atomic units by (Problem 13.16)

$$\frac{d\sigma}{d\Omega} = 4\pi^2 \alpha \omega k_f |\hat{\boldsymbol{\varepsilon}} \cdot \mathbf{D}_{ba}|^2 \tag{13.273}$$

where α is the fine structure constant, and it is assumed that the incident photons are linearly polarised, with polarisation vector $\hat{\boldsymbol{\varepsilon}}$.

The quantity $\mathbf{D}_{ba}$ appearing in (13.273) is the matrix element of the electric dipole operator $\mathbf{D}$. The above derivation of (13.273) makes use of its velocity form

$$\mathbf{D}_{ba}^{\mathrm{V}} = \frac{1}{\omega} \sum_{k=1} \langle \psi_{f\mathbf{k}_f}^{(-)} | \boldsymbol{\nabla}_{\mathbf{r}_k} | \psi_i \rangle \tag{13.274}$$

where ψ_i is the initial bound state wave function of the N-electron atom A, normalised to unity. The wave function $\psi_{f\mathbf{k}_f}^{(-)}$ appearing in (13.274) is the final (continuum) wave function describing the ion A^+ in state f and the ejected photoelectron; it satisfies incoming spherical wave boundary conditions and is normalised on the momentum scale according to

$$\langle \psi_{f\mathbf{k}_f}^{(-)} | \psi_{f'\mathbf{k}'_f}^{(-)} \rangle = \delta_{ff'} \delta(\mathbf{k}_f - \mathbf{k}'_f) \tag{13.275}$$

where we recall that the transition occurs into a group of states at energy $E_f = k_f^2/2$.

Substituting (13.274) into (13.273), we obtain for the differential photoionisation cross-section the *dipole velocity* expression

$$\frac{d\sigma^{\mathrm{V}}}{d\Omega} = \frac{4\pi^2 \alpha k_f}{\omega} \left| \sum_{k=1}^{N} \langle \psi_{f\mathbf{k}_f}^{(-)} | \hat{\boldsymbol{\varepsilon}} \cdot \boldsymbol{\nabla}_{\mathbf{r}_k} | \psi_i \rangle \right|^2 \tag{13.276}$$

Other formally equivalent expressions for this cross-section can be obtained by using in (13.273) the length or acceleration forms of the matrix element of the electric dipole operator $\mathbf{D}$. Thus, using the length form (see (9.28a))

$$\mathbf{D}_{ba}^{\mathrm{L}} = -\sum_{k=1}^{N} \langle \psi_{f\mathbf{k}_f}^{(-)} | \mathbf{r}_k | \psi_i \rangle \tag{13.277}$$

we obtain the *dipole length* expression for the differential photoionisation cross-section

$$\frac{d\sigma^{\mathrm{L}}}{d\Omega} = 4\pi^2 \alpha \omega k_f \left| \sum_{k=1}^{N} \langle \psi_{f\mathbf{k}_f}^{(-)} | \hat{\boldsymbol{\varepsilon}} \cdot \mathbf{r}_k | \psi_i \rangle \right|^2 \tag{13.278}$$

On the other hand, by using the acceleration form (see (9.28c))

$$\mathbf{D}^{\mathrm{A}}_{ba} = \frac{1}{\omega^2} \sum_{k=1}^{N} \langle \psi^{(-)}_{f\mathbf{k}_f} | (\nabla_{\mathbf{r}_k} V(\mathbf{r}_1, \mathbf{r}_2, \ldots, \mathbf{r}_N)) | \psi_i \rangle \tag{13.279}$$

where $V(\mathbf{r}_1, \mathbf{r}_2, \ldots, \mathbf{r}_N)$ is the atomic interaction potential, we find that

$$\frac{\mathrm{d}\sigma^{\mathrm{A}}}{\mathrm{d}\Omega} = \frac{4\pi^2 \alpha k_f}{\omega^3} \left| \sum_{k=1}^{N} \langle \psi^{(-)}_{f\mathbf{k}_f} | \hat{\boldsymbol{\varepsilon}} \cdot (\nabla_{\mathbf{r}_k} V(\mathbf{r}_1, \mathbf{r}_2, \ldots, \mathbf{r}_N)) | \psi_i \rangle \right|^2 \tag{13.280}$$

which is the *dipole acceleration* expression for the differential photoionisation cross-section.

The dipole length, velocity and acceleration forms of the photoionisation cross-section give identical results provided exact wave functions are used to describe the initial and final states of the transition. However, if approximate wave functions are used, different results are in general obtained, the magnitude of the difference being often an indication of the accuracy of the approximation.

The differential cross-section for photoionisation of an unpolarised atomic target by a linearly polarised photon beam, where the spin states of the photo-electron and the residual ion are not observed, is obtained from (13.276), (13.278) or (13.280) by summing over the final magnetic quantum numbers and averaging over the initial magnetic quantum numbers. This summation can be carried out by employing angular momentum algebra. It is found that

$$\frac{\mathrm{d}\sigma_{ba}}{\mathrm{d}\Omega} = \frac{\sigma_{\mathrm{tot}}(a \to b)}{4\pi} [1 + \beta P_2(\cos\gamma)] \tag{13.281}$$

where $\sigma_{\mathrm{tot}}(a \to b)$ is the total photoionisation cross-section for the transition $a \to b$, and γ is the angle of the direction $\hat{\mathbf{k}}_f$ of the ejected photoelectron relative to the polarisation direction $\hat{\boldsymbol{\varepsilon}}$ of the photon beam. The quantity β is often called the *asymmetry parameter*.

If the incident photons are unpolarised, one can obtain the angular distribution of the ejected photoelectrons from (13.281) by assuming that the incident photon beam is composed of an incoherent mixture of two beams linearly polarised at right angles, and averaging over the two polarisations. In this case it is found that

$$\frac{\mathrm{d}\sigma_{ba}}{\mathrm{d}\Omega} = \frac{\sigma_{\mathrm{tot}}(a \to b)}{4\pi} \left[1 - \frac{\beta}{2} P_2(\cos\theta)\right] \tag{13.282}$$

where θ is the angle of the direction $\hat{\mathbf{k}}_f$ of the ejected photoelectron relative to the incident photon propagation direction.

The fact that both differential cross-sections (13.281) and (13.282) must be non-negative at all angles implies that $-1 \leqslant \beta \leqslant 2$. Moreover, if the atom is initially in an S state one has $\beta = 2$ at all energies. As a result, the differential cross-section (13.282) then vanishes in the incident photon propagation direction.

The calculation of photoionisation cross-sections

We shall now examine how the cross-sections for single photoionisation can be calculated, first for hydrogenic atoms and ions, and then for complex atoms.

One-electron atoms and ions

It is only for hydrogenic atoms and ions that photoionisation cross-sections can be calculated analytically. In Section 4.8, we obtained the differential and total photoionisation cross-sections for one-electron atoms (ions), in the non-relativistic regime ($\hbar^2 k_f^2/(2m) \ll mc^2$), by using a simple plane wave (4.195) to describe the ejected photoelectron. This approximation, in which the final state Coulomb interaction between the ejected electron and the nucleus is neglected, is only satisfactory at high photoelectron energies such that $\hbar^2 k_f^2/(2m) \gg E_i$.

In fact, the correct wave function which takes into account the final state Coulomb attraction between the nucleus and the ejected electron (at non-relativistic energies) is a Coulomb wave function of the type studied in Section 12.5. Using the dipole velocity expression (13.276), the differential cross-section for photoionisation of a hydrogen atom or ion initially in the bound state i is then given for linearly polarised incident photons (in a.u.) by

$$\frac{\mathrm{d}\sigma^{\mathrm{V}}}{\mathrm{d}\Omega} = \frac{4\pi^2 \alpha k_f}{\omega} |\langle \psi^{(-)}_{\mathrm{c},\mathbf{k}_f}(Z, \mathbf{r}) | \hat{\varepsilon} \cdot \nabla | \psi_i(\mathbf{r}) \rangle|^2 \tag{13.283}$$

where $\psi^{(-)}_{\mathrm{c},\mathbf{k}_f}(Z, \mathbf{r})$ is a Coulomb wave function with incoming (–) spherical wave asymptotic behaviour, normalised on the momentum scale (see (13.231)–(13.232)), namely

$$\begin{aligned}\psi^{(-)}_{\mathrm{c},\mathbf{k}_f}(Z, \mathbf{r}) = {}& (2\pi)^{-3/2} \exp[\pi Z/(2k_f)] \Gamma(1 + \mathrm{i}Z/k_f) \\ & \times \exp(\mathrm{i}\mathbf{k}_f \cdot \mathbf{r})\, {}_1F_1(-\mathrm{i}Z/k_f, 1, -\mathrm{i}(k_f r + \mathbf{k}_f \cdot \mathbf{r}))\end{aligned} \tag{13.284}$$

If instead the dipole length expression (13.278) is used, we have

$$\frac{\mathrm{d}\sigma^{\mathrm{L}}}{\mathrm{d}\Omega} = 4\pi^2 \alpha \omega k_f |\langle \psi^{(-)}_{\mathrm{c},\mathbf{k}_f}(Z, \mathbf{r}) | \hat{\varepsilon} \cdot \mathbf{r} | \psi_i(\mathbf{r}) \rangle|^2 \tag{13.285}$$

and the two formulae (13.283) and (13.285) give identical results, since the initial hydrogenic bound state wave function $\psi_i(\mathbf{r})$ as well as the final Coulomb wave function $\psi^{(-)}_{\mathrm{c},\mathbf{k}_f}(Z, \mathbf{r})$ are known exactly.

As an example, let us consider the photoionisation of hydrogenic atoms and ions initially in the ground state 1s. First of all, we note that if one replaces in the dipole velocity expression (13.283) the Coulomb wave function $\psi^{(-)}_{\mathrm{c},\mathbf{k}_f}(Z, \mathbf{r})$ by the corresponding plane wave (see (4.195))

$$\psi_b(\mathbf{k}_f, \mathbf{r}) = (2\pi)^{-3/2} \exp(\mathrm{i}\mathbf{k}_f \cdot \mathbf{r}) \tag{13.286}$$

one retrieves the results obtained in Section 4.8 by neglecting the final state Coulomb interaction and making the dipole approximation. Denoting the corresponding differential cross-section by $\mathrm{d}\sigma^{(0)}/\mathrm{d}\Omega$, we have for an unpolarised photon beam (in a.u.)

$$\frac{\mathrm{d}\sigma^{(0)}}{\mathrm{d}\Omega} = 32\alpha Z^{-2}\left(\frac{|E_{1s}|}{\omega}\right)^{7/2} \sin^2\theta \tag{13.287}$$

in agreement with (4.221b). The corresponding total cross-section is

$$\sigma^{(0)}_{\text{tot}} = \frac{256\pi}{3}\alpha Z^{-2}\left(\frac{|E_{1s}|}{\omega}\right)^{7/2} \tag{13.288}$$

which agrees with (4.220b). It should be noted that different results are obtained if the Coulomb wave function $\psi^{(-)}_{c,\mathbf{k}_f}(Z, \mathbf{r})$ is replaced by the plane wave (13.286) in the dipole length form (13.285) of the cross-section.

We now turn to the calculation of the 'exact' (dipole) cross-sections for photoionisation of hydrogenic atoms (ions) from the ground state. Using the dipole length expression (13.285), the required matrix element $\langle\psi^{(-)}_{c,\mathbf{k}_f}(Z, \mathbf{r})|\hat{\varepsilon}\cdot\mathbf{r}|\psi_{1s}(r)\rangle$ can be evaluated by employing the expression (13.284) of $\psi^{(-)}_{c,\mathbf{k}_f}(Z, \mathbf{r})$ and the integral representation (13.243) of the confluent hypergeometric function. It is found that the corresponding differential photoionisation cross-section is given for an unpolarised photon beam by (Problem 13.17)

$$\frac{\mathrm{d}\sigma}{\mathrm{d}\Omega} = C\frac{\mathrm{d}\sigma^{(0)}}{\mathrm{d}\Omega} \tag{13.289a}$$

where $\mathrm{d}\sigma^{(0)}/\mathrm{d}\Omega$ is the dipole differential cross-section (13.287) without final state Coulomb interactions, and the factor C is given by

$$C = 2\pi\left(\frac{|E_{1s}|}{\omega}\right)^{1/2} \frac{\exp[-4(Z/k_f)\cot^{-1}(Z/k_f)]}{1-\exp(-2\pi Z/k_f)} \tag{13.289b}$$

It is readily checked (Problem 13.18) that

$$\lim_{k_f\to\infty} C = 1 \tag{13.290}$$

so that the final state Coulomb interaction is unimportant at high energies (within the dipole approximation). It should be noted, however, that because of the infinite range of the Coulomb interaction, the quantity C reaches its asymptotic value very slowly.

Let us now look at the behaviour of the differential cross-section (13.289) near threshold. In this case, one obtains (Problem 13.18)

$$C \simeq 2\pi\,\mathrm{e}^{-4}\left(1 + \frac{5}{6}\frac{\omega - |E_{1s}|}{|E_{1s}|} + \cdots\right) \tag{13.291}$$

Using (13.287), (13.289a) and (13.291), we then have

$$\frac{\mathrm{d}\sigma}{\mathrm{d}\Omega} = 64\pi\,\mathrm{e}^{-4}\alpha Z^{-2}\sin^2\theta\left(1 - \frac{8}{3}\frac{\omega - |E_{1s}|}{|E_{1s}|} + \cdots\right) \tag{13.292}$$

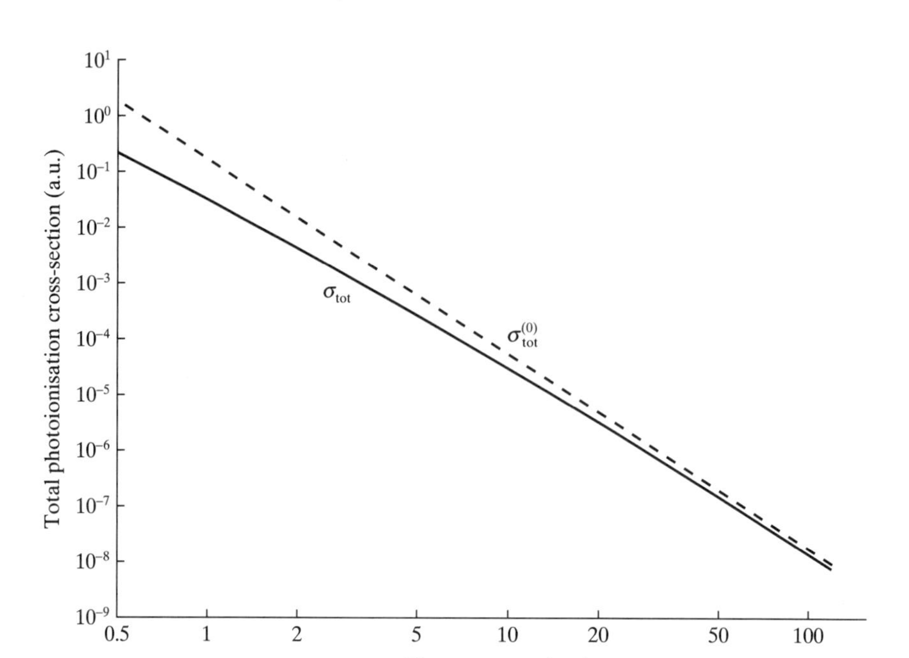

Figure 13.25 The total cross-section for photoionisation of atomic hydrogen (in a.u.), as a function of the incident photon energy. Dashed curve (-----): the quantity $\sigma_{\text{tot}}^{(0)}$ given by (13.288). Solid curve (——): the quantity σ_{tot}, given by (13.293), which includes the final state Coulomb interaction.

The total dipole cross-section, including the final state Coulomb interaction, is obtained by integrating the differential cross-section (13.289) over the angles. Thus we have

$$\sigma_{\text{tot}} = C\sigma_{\text{tot}}^{(0)} \tag{13.293}$$

where C is given by (13.289b) and $\sigma_{\text{tot}}^{(0)}$ by (13.288). Both cross-sections $\sigma_{\text{tot}}^{(0)}$ and σ_{tot} are displayed in Fig. 13.25 for the case of atomic hydrogen. At high (but non-relativistic) photoelectron energies it follows from (13.290) that σ_{tot} reduces to $\sigma_{\text{tot}}^{(0)}$, and hence falls off like $\omega^{-7/2}$ for high photon energies. We also note that σ_{tot} reaches its maximum at threshold, where it is given by

$$\sigma_{\text{tot}}^{\text{t}} = \frac{512\pi^2}{3}\text{e}^{-4}\alpha Z^{-2} \tag{13.294}$$

The fact that $\sigma_{\text{tot}}^{\text{t}}$ does not vanish at threshold is due to the Coulomb interaction between the two charged particles in the final state.

The foregoing calculations concerning the ground state can be extended to study the photoionisation of hydrogenic atoms (ions) from other initial bound states [13]. In particular, it is found that the total photoionisation cross-section of an hydrogenic atom (ion) from an initial state denoted by the principal quantum number n (averaged over the l substates) is given in atomic units by

$$\sigma_{\text{tot}} = \frac{8\pi}{3\sqrt{3}} \alpha \frac{Z^4}{n^5} \omega^{-3} g(\omega, n, Z) \quad \textbf{(13.295)}$$

where the quantity g is known as the *Kramers–Gaunt factor*. It is a slowly varying function of the photon energy ω, with a value almost always within 20% of unity near the absorption edge, while at high energies it falls off to zero as $\omega^{-1/2}$.

Complex atoms and ions

For atoms or ions having two or more electrons, approximations must be used to describe both the initial bound state wave function ψ_i and the final continuum state $\psi^{(-)}_{f\mathbf{k}_f}$.

The simplest approximation is based on the assumption that the photon is absorbed by an electron in the (nl) subshell of the atom, which is ejected into the continuum. The relaxation of the remaining electrons is neglected, so that the photoionisation process reduces to a one-electron problem. In particular, the motion of the active electron, both before and after photoionisation, can be represented by a model central potential $V(r)$. In this approach the length, velocity and acceleration expressions of the photoionisation cross-sections are essentially identical. This follows from the fact that a one-electron model atomic Hamiltonian $H_0 = -\nabla^2_{\mathbf{r}}/2 + V(r)$ is used, for which the initial bound state wave function ψ_i and final state continuum wave function $\psi^{(-)}_{f\mathbf{k}_f}$ can be obtained to arbitrary accuracy. The calculations based on this single-particle approximation, for example those performed in 1968 by S.T. Manson and J.W. Cooper, enable a general understanding of photoionisation cross-sections over a broad energy range to be obtained. However, the complete neglect of electron–electron correlation implies that the detailed structure of the cross-sections cannot be accurately predicted, in particular near the ionisation threshold.

In order to calculate precise values of photoionisation cross-sections for complex atoms and ions, electron correlation effects must be taken into account. A number of theoretical methods have been developed, which are capable of achieving this goal. We mention in particular the configuration–interaction approach, in which both the initial and final state wave functions in (13.276), (13.278) or (13.280) are expanded in terms of target states, pseudo-states and square integrable functions (see (13.92)). Another powerful method is the R-matrix theory.

The case of two-electron atoms and ions has been studied extensively, since it is possible in this instance to perform very accurate calculations. As an example, let us consider the photodetachment process

$$\hbar\omega + \mathrm{H}^-(1\,^1\mathrm{S}) \rightarrow \mathrm{H}(1s) + e^- \quad \textbf{(13.296)}$$

[13] See Bethe and Salpeter (1957), where non-dipole and relativistic corrections are also discussed.

It was pointed out in 1939 by R. Wildt that the photodetachment of H^- is of great importance in understanding the opacity of the atmospheres of certain stars including the Sun. The negative ion H^- is formed by radiative recombination of low-energy electrons and hydrogen atoms. Radiation from the surface of the star is absorbed by photodetachment. The continual formation and destruction of H^- conserves the total radiated energy but modifies the characteristics of the light emitted from the surface of the star. It should be noted that because of the weak binding energy of H^-, the photodetachment process (13.296) occurs throughout the visible region, down to the threshold value $\hbar\omega_t = 0.75$ eV, corresponding to a wavelength $\lambda_t = 16\,530$ Å lying in the infra-red part of the spectrum.

The work of Wildt aroused special theoretical interest in the photodetachment of H^-. Already in 1940, H.S.W. Massey and D.R. Bates had made a detailed calculation for the process (13.296) by using a Hylleraas-type wave function for the H^- bound state, and a final state accounting approximately for the effects of exchange and polarisation. In 1946, calculations were performed by S. Chandrasekhar and F.H. Breen, and increasingly accurate studies were made subsequently by using Hylleraas-type wave functions with many variational parameters for H^-, and final state wave functions of increasing precision. For example, in 1966 N.A. Doughty, P.A. Fraser and R.P. McEachran used a 70-term Hylleraas-type wave function for H^-, and a close-coupling expansion including the 1s, 2s, 2p, 3s, 3p and 3d states of H in the final state. A similar calculation, employing a polarised-orbital approximation for the final state, was performed in 1967 by K.L. Bell and A.E. Kingston. The results of these two calculations for the total cross-section, obtained by using both the dipole 'length' and 'velocity' forms of the cross-section, are shown in Fig. 13.26, together with the experimental data of S.J. Smith and D.S. Burch.

Looking at Fig. 13.26, we first note that the photodetachment cross-section vanishes at threshold. This is in contrast to the photoionisation cross-sections for neutral atoms and positive ions which (as we have seen above for the simple case of hydrogenic atoms and ions) are non-zero at threshold because of the infinite range of the Coulomb interaction between the two charged particles in the final state. We also see from Fig. 13.26 that the H^- photodetachment cross-section has a broad maximum extending from the infra-red to the ultra-violet. Not shown in Fig. 13.26 are *resonances* present in the ultra-violet part of the spectrum, which will be discussed below. It is also worth noting that the processes

$$\hbar\omega + H^-(1\,^1S) \to H(nl) + e^- \qquad \textbf{(13.297)}$$

must also be taken into account when the photon energy is large enough for the hydrogen atom to be excited into the $n = 2, 3, \ldots$ levels. In particular, J.H. Macek showed in 1967 that above the $n = 2$ threshold the $n = 2$ cross-section (hydrogen atom left in the $n = 2$ level) is comparable to the $n = 1$ cross-section (hydrogen atom left in the ground state). His treatment was extended in 1972 by H.A. Hyman, V.L. Jacobs and P.G. Burke, who found that photodetachment into the 2p state is much more probable than into the 2s state near threshold.

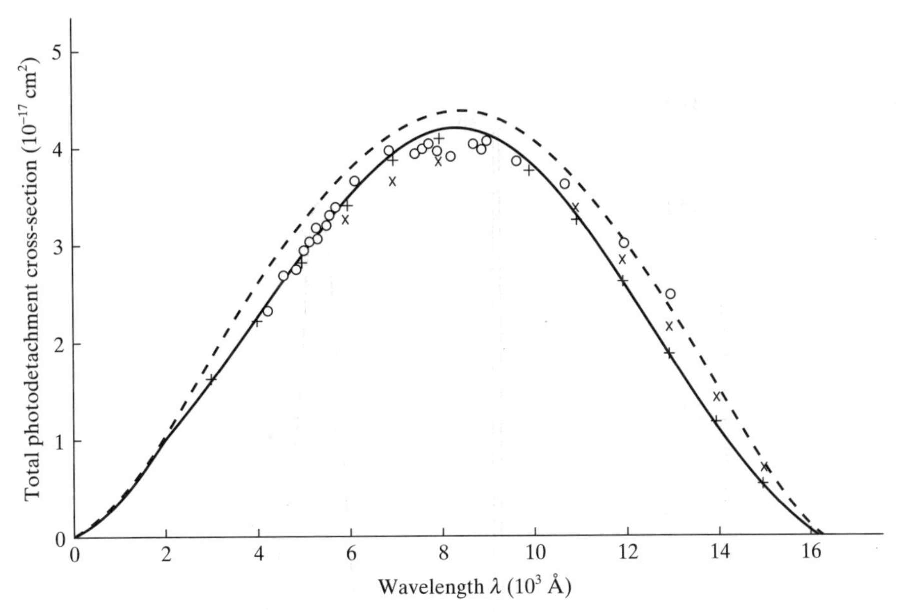

Figure 13.26 The total cross-section for photodetachment of H^-.
The experimental data (o) are due to S.J. Smith and D.S. Burch.
The theoretical calculations are due to N.A. Doughty *et al.* (length form: +; velocity form: ×) and to K.L. Bell and A.E. Kingston (length form: ——; velocity form: -----).

Resonances in photoionisation and photodetachment

Photoionisation or photodetachment cross-sections are dominated in certain energy ranges by resonances due to autoionising states. At energies close to resonance the photoionisation process can occur either directly, or through an intermediate resonant state $\tilde{\mathrm{A}}$ as follows:

$$\hbar\omega + \mathrm{A}(i) \rightarrow \tilde{\mathrm{A}} \rightarrow \mathrm{A}^+(f) + \mathrm{e}^- \tag{13.298}$$

The interference between these two mechanisms gives rise to the *absorption line profile*, which U. Fano and J.W. Cooper have parametrized by the formula

$$\sigma_{\mathrm{tot}} = \sigma_a \frac{(q+\varepsilon)^2}{1+\varepsilon^2} + \sigma_b \tag{13.299}$$

where

$$\varepsilon = \frac{E - E_{\mathrm{r}}}{\Gamma/2} \tag{13.300}$$

is the reduced energy introduced in Section 12.3, E_{r} being the resonance energy and Γ the resonance width. In (13.299) the quantity q is the line profile index which defines the line shape, and σ_a and σ_b are background cross-sections which vary slowly with energy.

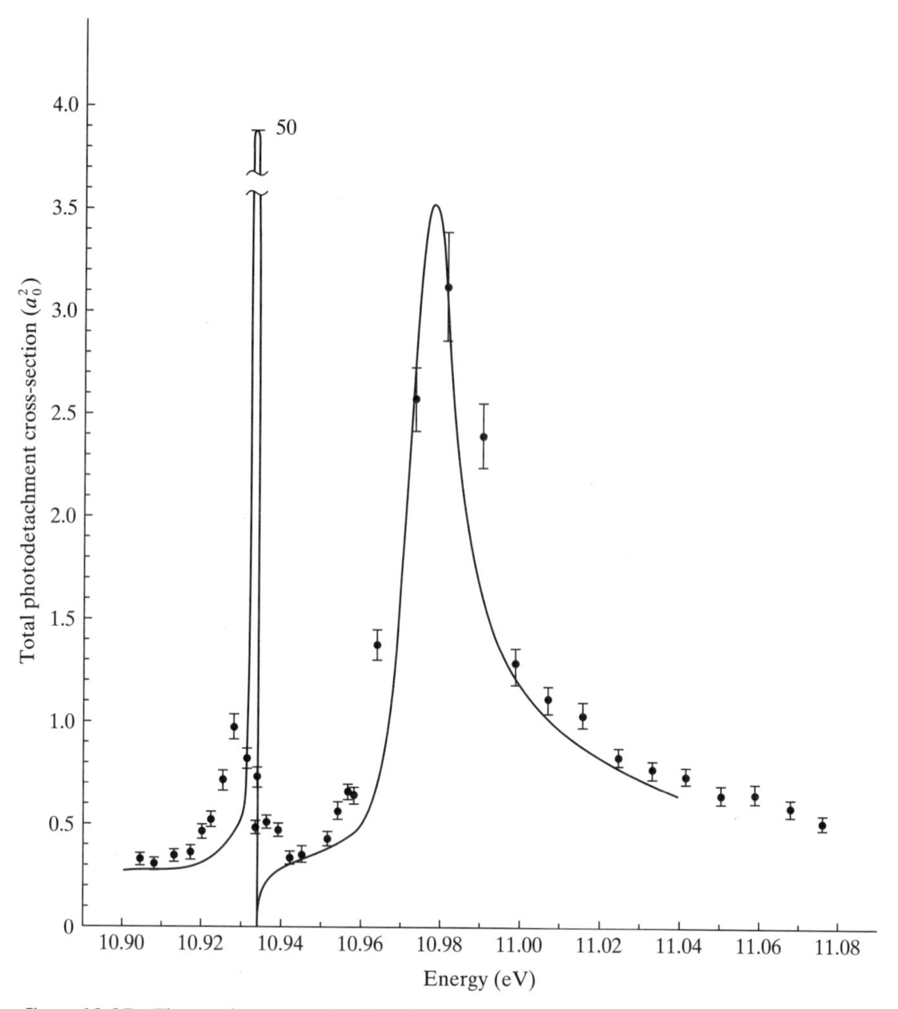

Figure 13.27 The total cross-section for photodetachment of H^-, in the energy range between 10.90 eV and 11.08 eV.
The experimental data are due to H.C. Bryant *et al.* They are normalised to the theoretical value at 10.90 eV.
The theoretical calculations are due to J.T. Broad and W.P. Reinhardt.

As discussed in Section 7.7, early examples of resonances were found in 1963 by R.P. Madden and K. Codling in the photoionisation of helium atoms. Using the continuous light from an electron synchrotron, they observed a number of ^{1}P Feshbach resonances below the $n = 2$ level of He^+ (see Table 13.1), in particular the 2s2p ^{1}P resonance and the Rydberg series $(2sn'p \pm 2pn's)$ ^{1}P with $n' = 3, 4, \ldots$ converging onto the threshold for exciting He^+ to the $n = 2$ level. Further faint

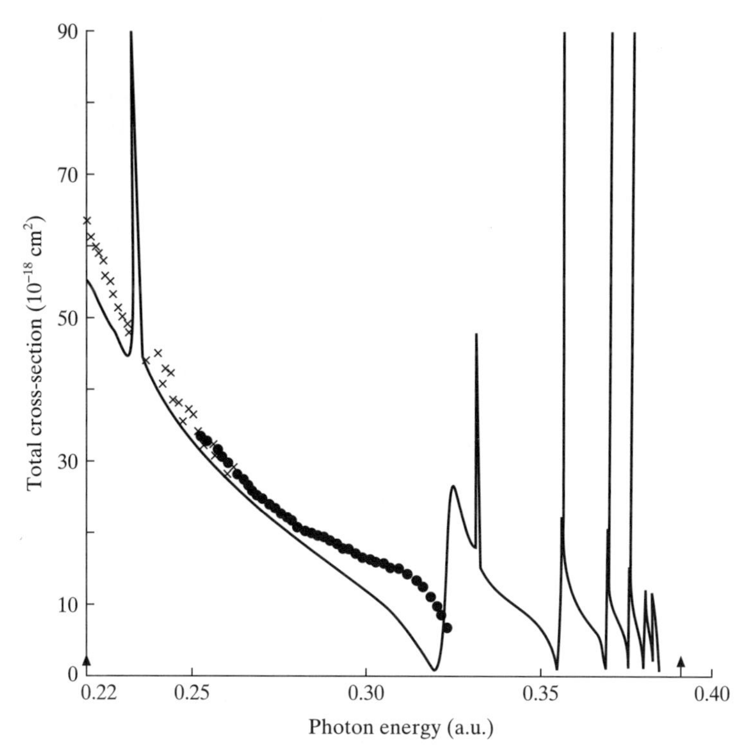

Figure 13.28 The total cross-section for photoionisation of the Al ground state between the $3s^2\ ^1S$ and $3s3p\ ^3P^o$ thresholds (indicated by the arrows) as a function of the incident photon energy.
The experimental data are due to J.L. Kohl and W.H. Parkinson: ×; and to R.A. Roig: •.
The solid curve corresponds to the *R*-matrix calculation of S.S. Tayal and P.G. Burke. (From S.S. Tayal and P.G. Burke, J. Phys. B *20*, 4715, 1987.)

series of ^{1}P Feshbach resonances can also be seen converging onto higher excited levels of He$^+$.

Let us now return to the H$^-$ photodetachment process (13.296). Series of ^{1}P Feshbach resonances, located below the threshold for exciting the H atom to the $n = 2, 3, \ldots$ levels, have been predicted to occur in the ultra-violet part of the photodetachment spectrum. In addition, there is a 2s2p ^{1}P shape resonance lying just above the $n = 2$ threshold of the H atom at a photon energy of 10.97 eV, corresponding to a wavelength of about 1129.5 Å. This shape resonance was predicted in 1967 by J.H. Macek, and then studied extensively, in particular by K.T. Taylor and P.G. Burke, by C.D. Lin and by J.T. Broad and W.P. Reinhardt. It was observed in electron scattering by atomic hydrogen, first by J.W. McGowan, J.F. Williams and F.K. Carley in 1969, and also by J.F. Williams and B.A. Willis in 1974. Three years later, H.C. Bryant *et al.* reported the observation of the 2s2p ^{1}P shape resonance and of a Feshbach resonance in the photodetachment cross-section of H$^-$, in the photon energy region near 11 eV (see Fig. 13.27).

They used a colliding beam method in which a laser beam is directed at variable angle across a beam of H^- ions having an energy of 800 MeV. By varying the angle of intersection between the beams, the Doppler-shifted energy seen by the H^- ions could be tuned continuously from 1.8 eV to 11.5 eV. The photodetached electrons were swept from the H^- beam by a magnetic field and brought to a detector. As seen from Fig. 13.27, the two measured resonances agree well with the theoretical predictions, although within the experimental resolution of 10 meV the Feshbach resonance could not be resolved.

As a final example, we show in Fig. 13.28 the total cross-section for photoionisation of the aluminium atom between the $3s^2\ {}^1S$ and $3s3p\ {}^3P^o$ threshold, calculated in 1987 by S.S. Tayal and P.G. Burke by using the R-matrix theory. This cross-section is seen to exhibit an important resonance structure. Also shown are the experimental data of J.L. Kohl and W.H. Parkinson, and of R.A. Roig, which are in fair agreement with the theoretical values.

Double photoionisation

Multiple ionisation processes are of special interest in atomic and molecular physics, since they are sensitive probes of electron–electron correlations. In what follows, we shall consider the double photoionisation of helium from its ground state:

$$\hbar\omega + \mathrm{He}(1\,{}^1\mathrm{S}) \rightarrow \mathrm{He}^{++} + 2\mathrm{e}^- \tag{13.301}$$

which is the most basic of these processes. We shall assume that the two outgoing electrons move at non-relativistic velocities, and use the dipole approximation. Atomic units will be employed.

Let us denote by $\mathbf{k}_A$ and $\mathbf{k}_B$ the momenta of the two outgoing electrons, and by $E^e_A = k^2_A/2$ and $E^e_B = k^2_B/2$ their corresponding energies. If $\mathbf{Q}$ is the recoil momentum of the He^{++} nucleus, we have from momentum conservation

$$\mathbf{Q} = -(\mathbf{k}_A + \mathbf{k}_B) = -\mathbf{K} \tag{13.302}$$

where $\mathbf{K} = \mathbf{k}_A + \mathbf{k}_B$ is the centre of mass momentum of the two-electron subsystem (called a 'dielectron') and we have set the photon momentum $\mathbf{k}$ equal to zero, in accordance with the dipole approximation. From energy conservation, we find that

$$E = \omega + E_i = E^e_A + E^e_B \tag{13.303}$$

where $E_i = -2.9037$ a.u. is the ground state energy of the helium atom. In writing the equation (13.303), we have neglected the recoil energy $Q^2/(2M_{\mathrm{He}^{++}})$ of the recoil nucleus.

The triple differential cross-section (TDCS) for the two electrons to emerge into the solid angles $d\Omega_A$ and $d\Omega_B$, centred respectively about the directions (θ_A, ϕ_A) and (θ_B, ϕ_B), is given for a linearly polarised photon beam by

$$\frac{d^3\sigma_{++}}{d\Omega_A\, d\Omega_B\, dE} = 4\pi^2\alpha\omega k_A k_B |\hat{\boldsymbol{\varepsilon}} \cdot \mathbf{D}_{ba}|^2 \tag{13.304}$$

where the prefactor depends on the normalisation of the final state wave function, which is chosen to be on the momentum scale, and $\mathbf{D}_{ba}$ is the dipole matrix element. Its velocity form is given by

$$\mathbf{D}_{ba} = \frac{1}{\omega}\langle \psi^{(-)}_{\mathbf{k}_A\mathbf{k}_B}(\mathbf{r}_1, \mathbf{r}_2)|\nabla_{\mathbf{r}_1} + \nabla_{\mathbf{r}_2}|\psi_i(\mathbf{r}_1, \mathbf{r}_2)\rangle \tag{13.305}$$

where $\psi_i(\mathbf{r}_1, \mathbf{r}_2)$ is the initial wave function for the ground state of helium, and $\psi^{(-)}_{\mathbf{k}_A\mathbf{k}_B}(\mathbf{r}_1, \mathbf{r}_2)$ is the final state wave function describing the two outgoing electrons, and having an incoming (−) spherical wave behaviour. By substituting (13.305) into (13.304), one obtains the dipole velocity expression of the TDCS,

$$\frac{d^3\sigma^V_{++}}{d\Omega_A\, d\Omega_B\, dE} = \frac{4\pi^2\alpha k_A k_B}{\omega}|\hat{\varepsilon}\cdot\langle \psi^{(-)}_{\mathbf{k}_A\mathbf{k}_B}(\mathbf{r}_1, \mathbf{r}_2)|\nabla_{\mathbf{r}_1} + \nabla_{\mathbf{r}_2}|\psi_i(\mathbf{r}_1, \mathbf{r}_2)\rangle|^2 \tag{13.306}$$

One can also write down the dipole lengths and acceleration expressions of the TDCS by using the corresponding forms of the dipole matrix element $\mathbf{D}_{ba}$. For example, by using the length form

$$\mathbf{D}^L_{ba} = -\langle \psi^{(-)}_{\mathbf{k}_A\mathbf{k}_B}(\mathbf{r}_1, \mathbf{r}_2)|\mathbf{r}_1 + \mathbf{r}_2|\psi_i(\mathbf{r}_1, \mathbf{r}_2)\rangle \tag{13.307}$$

one obtains for the TDCS the dipole length expression

$$\frac{d^3\sigma^L_{++}}{d\Omega_A\, d\Omega_B\, dE} = 4\pi^2\alpha\omega k_A k_B|\hat{\varepsilon}\cdot\langle \psi^{(-)}_{\mathbf{k}_A\mathbf{k}_B}(\mathbf{r}_1, \mathbf{r}_2)|\mathbf{r}_1 + \mathbf{r}_2|\psi_i(\mathbf{r}_1, \mathbf{r}_2)\rangle|^2 \tag{13.308}$$

As in the case of the (e, 2e) reactions discussed in the previous section, various double and single differential cross-sections can be formed by integrating the TDCS with respect to $d\Omega_A$, $d\Omega_B$ or dE. It should be noted that the single differential cross-section $d\sigma_{++}/dE$ gives a photoelectron energy distribution which is symmetric about the midpoint $E/2$ of the energy range. The total cross-section for double photoionisation, which we denote by σ^{++}_{tot}, is obtained by integrating $d\sigma_{++}/dE$ over a range of energies extending from zero to $E/2$, in order to take into account the fact that the two electrons are indistinguishable.

The first detailed theoretical analysis of the double ionisation process (13.301) was made in 1967 by F.W. Byron and C.J. Joachain, and in what follows we shall follow for the most part their treatment. First of all, we note that if the interaction between the two electrons is completely neglected (so that ψ_i is a product and $\psi^{(-)}_{\mathbf{k}_A\mathbf{k}_B}$ is a symmetrised product of hydrogenic wave functions with $Z = 2$), then the dipole matrix elements (13.305) or (13.307) vanish. As a result, the *double ionisation process* (13.301) *cannot occur in the absence of the electron–electron interaction*. This is clearly *not* the case for single ionisation, which would therefore be the only possible ionisation process if the electron–electron interaction was switched off.

It follows from this reasoning that we expect the double ionisation process (13.301) to be delicately dependent on electron correlation effects. Consequently, the accurate calculation of the dipole matrix elements (13.305) or (13.307) is a very difficult problem. While for ψ_i very accurate wave functions are available, the situation is much more complicated for $\psi^{(-)}_{\mathbf{k}_A\mathbf{k}_B}$, since one is dealing in this case with two electrons in the continuum. We will come back to this problem below, but at this point we first show that the problem simplifies considerably in the limit of large incident photon energies. Indeed, as $\omega \to \infty$, at least one of the ejected electrons must have a very large energy. The velocity form of the dipole matrix element for any ionisation process in which at least one electron is ejected can then be calculated by using an approximate final state wave function of the form

$$\psi_f(\mathbf{r}_1, \mathbf{r}_2) = (2\pi)^{-3/2} \frac{1}{\sqrt{2}} [\exp(i\mathbf{k}_A \cdot \mathbf{r}_1)\phi_n(\mathbf{r}_2) + \phi_n(\mathbf{r}_1) \exp(i\mathbf{k}_A \cdot \mathbf{r}_2)] \tag{13.309}$$

This is a symmetrised product of a plane wave times a single particle He^+ eigenfunction ϕ_n describing either a bound state (for discrete values of n) or a continuum state (in which case it is a positive energy Coulomb wave function). Under these circumstances, it can be shown (Problem 13.19) that the asymptotic cross-section $\sigma_{tot}(\omega \to \infty)$ for *any* ionisation process in helium in which at least one electron is ejected is given by

$$\sigma_{tot}(\omega \to \infty) = \frac{256\pi^2\alpha}{3\sqrt{2}} \omega^{-7/2} \sum_n \left| \int \phi_n^*(\mathbf{r}) \psi_i(0, \mathbf{r}) \, d\mathbf{r} \right|^2 \tag{13.310}$$

where the sum (integral) on final states is determined by the process of interest. For example, the asymptotic cross-section for *all* ionisation processes (both single and double) is given by

$$\sigma_{tot}^{+,++}(\omega \to \infty) = \frac{256\pi^2\alpha}{3\sqrt{2}} \omega^{-7/2} \int |\psi_i(0, \mathbf{r})|^2 \, d\mathbf{r} \tag{13.311}$$

where we have made use of the closure relation since the ϕ_n form a complete orthonormal set. Similarly, the asymptotic cross-section for single ionisation is just

$$\sigma_{tot}^{+}(\omega \to \infty) = \frac{256\pi^2\alpha}{3\sqrt{2}} \omega^{-7/2} \sum_n \left| \int \phi_n^*(\mathbf{r}) \psi_i(0, \mathbf{r}) d\mathbf{r} \right|^2 \tag{13.312}$$

where the sum runs over all bound states of the He^+ ion. Clearly,

$$\sigma_{tot}^{++}(\omega \to \infty) = \sigma_{tot}^{+,++}(\omega \to \infty) - \sigma_{tot}^{+}(\omega \to \infty) \tag{13.313}$$

The utility of the equation (13.313) lies in the fact that it allows an evaluation of $\sigma_{tot}^{++}(\omega \to \infty)$ even for very accurate initial state wave functions ψ_i of the Hylleraas type. Indeed, in computing the integrals appearing on the right-hand side of (13.311) and (13.312), the function ψ_i enters only through a simple function $\psi_i(0, \mathbf{r})$ of one variable, and final state electron correlations do not play any role.

Defining the ratio of the double to single ionisation total cross-sections

$$R(\omega) = \frac{\sigma_{tot}^{++}(\omega)}{\sigma_{tot}^{+}(\omega)} \tag{13.314}$$

Table 13.4 The asymptotic ratio $R(\omega \to \infty)$ calculated by F.W. Byron and C.J. Joachain using the equations (13.311)–(13.314), with various helium ground state wave functions ψ_i (see text).

Wave function ψ_i	$R(\omega \to \infty)$
Screened hydrogenic ($Z_e = 27/16$)	0.72%
Hartree–Fock	0.51%
Chandrasekhar–Herzberg	1.71%
Kinoshita	1.66%

we show in Table 13.4 the values of the asymptotic ratio $R(\omega \to \infty)$ calculated by F.W. Byron and C.J. Joachain using the equations (13.311)–(13.314), with various approximations to the helium ground state wave function ψ_i. Their best result, $R(\omega \to \infty) = 1.66\%$, was obtained with the very accurate 39-parameter Hylleraas wave function of T. Kinoshita, which gives for the ground state energy of helium the value $E_i = -2.903\,722$ a.u. The slightly less precise 18-parameter Hylleraas-type wave function of S. Chandrasekhar and G. Herzberg gives for $R(\omega \to \infty)$ a value of 1.71%, about 3% larger than the best result. The Hartree–Fock ground state wave function, on the other hand, gives a value of $R(\omega \to \infty)$ which is smaller than the best value by more than a factor of three. The screened hydrogenic (Z_e = 27/16) wave function also does very poorly. Thus the calculation of $R(\omega \to \infty)$ gives a striking illustration of the importance of initial state electron correlation effects in double ionisation phenomena. The theoretical value of 1.66% obtained for $R(\omega \to \infty)$ by Byron and Joachain has been confirmed by subsequent calculations (performed in 1992 by A. Dalgarno and H.R. Sadeghpour and in 1993 by L.R. Andersson and J. Burgdörfer) and by several experiments using synchrotron radiation. In particular, J.C. Levin *et al.* measured in 1991 the value $R(\omega) = (1.6 \pm 0.3)\%$ at a photon energy $\hbar\omega = 2.8$ keV, and L. Spielberger *et al.* found in 1995 that $R(\omega) = (1.72 \pm 0.12)\%$ at $\hbar\omega = 7$ keV.

We now return to the evaluation of the dipole matrix elements (13.305) or (13.307), which yield the various double photoionisation cross-sections of He at finite photon energies. The first calculations of these matrix elements by Byron and Joachain used an accurate configuration–interaction wave function for ψ_i (giving the value $E_i = -2.9020$ a.u. for the helium ground state) and a symmetrised product of two uncorrelated Coulomb wave functions (corresponding to a central charge Z = 2) for the final state $\psi^{(-)}_{\mathbf{k}_A\mathbf{k}_B}$. Good agreement was obtained with the data of T.A. Carlson, and the importance of electron correlation effects in the initial state at all photon energies was clearly established. However, the accurate determination of the final state wave function $\psi^{(-)}_{\mathbf{k}_A\mathbf{k}_B}$ could not be made until the development of modern computers. In 1993, F. Maulbetsch and J. Briggs successfully calculated angular distribution by approximating $\psi^{(-)}_{\mathbf{k}_A\mathbf{k}_B}$ as a product of three Coulomb wave functions which take final state correlations into account. A different approach, based on earlier work of M.R.H. Rudge and M.J. Seaton for electron impact ionisation, has been developed since 1993 by R. Shakeshaft, D. Proulx

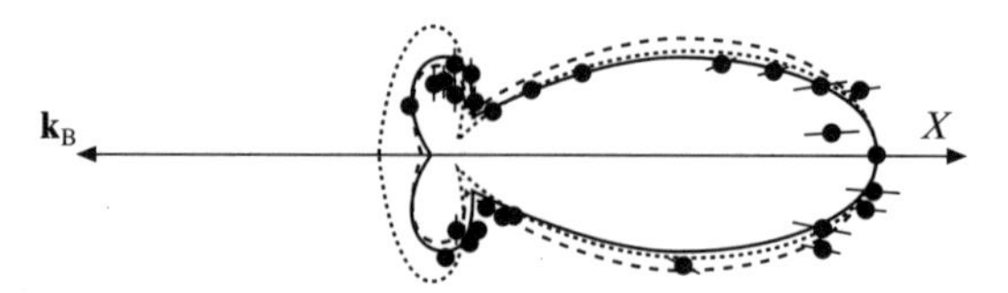

Figure 13.29 Polar plot of the relative triple differential cross-section for double photoionisation of He when electron A has an energy $E_A^e = 47.9$ eV and electron B emerges along the major polarisation axis of the light (the X axis) with an energy $E_B^e = 5$ eV.
The experimental data are due to O. Schwarzkopf *et al.*
The theoretical calculations are due to F. Maulbetsch and J. Briggs (length form: ·······; velocity form: -----) and to D. Proulx and R. Shakeshaft: ——.
All results have been normalised so that they agree at the maximum of the distribution.

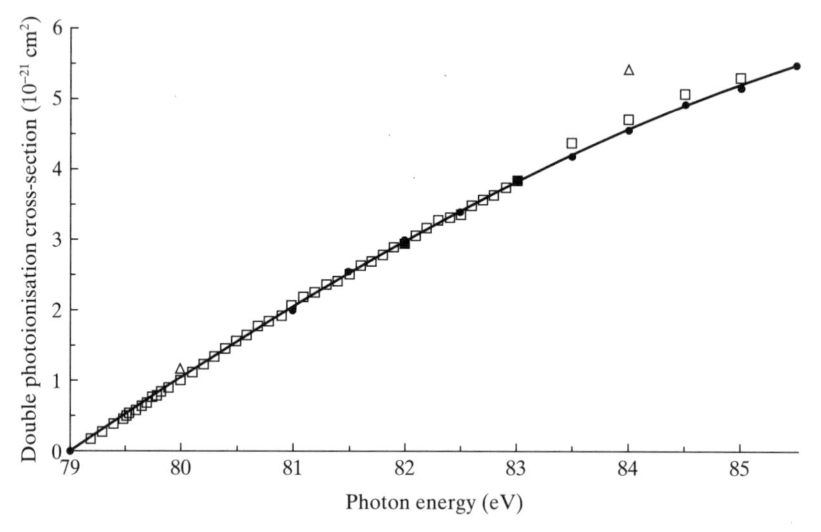

Figure 13.30 The total cross-section for double photoionisation of He as a function of the incident photon energy in the near-threshold region.
The experimental data are due to H. Kossmann *et al.*: □; and to J.M. Bizau and F. Wuilleumier: △.
The solid line is a fit through the calculated points (solid circles: ●) of M. Pont and R. Shakeshaft.

and M. Pont. It uses products of two screened Coulomb wave functions with effective charges, and has also given accurate results.

In Fig. 13.29 we show the angular distribution of electron A, for the case where this electron has an energy $E_A^e = 47.9$ eV, and electron B emerges along the major polarisation axis of the light with an energy $E_B^e = 5$ eV. The theoretical results of Maulbetsch and Briggs (using both the velocity and the length forms of the dipole matrix element) and those of Proulx and Shakeshaft are compared with the experimental data obtained in 1994 by O. Schwarzkopf *et al.* The overall agreement between theory and experiment is seen to be good.

The total cross-section for double photoionisation of He is shown in Fig. 13.30 in the near-threshold region. The theoretical calculations performed in 1995 by M. Pont and R. Shakeshaft are seen to be in excellent agreement with the absolute data from experiments using synchrotron radiation, performed in 1988 by H. Kossmann, V. Schmidt and T. Andersen, and in 1995 by J.M. Bizau and F.J. Wuilleumier.

Relation between photoionisation and electron impact ionisation

To conclude this section, we shall examine the link between the phenomena of photoionisation and electron impact ionisation. To this end, we return to the Bethe approximation (13.252) giving the leading term of the first Born total cross-section $\sigma_{\text{tot}}^{\text{B1}}(i)$ at high electron impact energies for the (e, 2e) process (13.227) in atomic hydrogen. This leading term has the form $A(E_i^{\text{e}})^{-1}\log E_i^{\text{e}}$, where E_i^{e} is the incident electron energy. Looking at the expression (13.253) for the coefficient A, and noting that the quantity $|\langle\psi_{\text{c},\mathbf{k}_\text{B}}^{(-)}(Z=1,\mathbf{r}_1)|z_1|\psi_i(\mathbf{r}_1)\rangle|^2$ integrated over the polar angles $(\theta_\text{B}, \phi_\text{B})$ of the ejected electron is proportional to the total photoionisation cross-section $\sigma_{\text{tot}}^i(\omega)$ from the initial state $|i\rangle$ (calculated in the dipole approximation) we find that

$$A = \frac{1}{2\pi\alpha}\int_{\omega_\text{t}}^{\infty}\frac{\sigma_{\text{tot}}^i(\omega)}{\omega}\,\text{d}\omega \qquad \textbf{(13.315)}$$

where $\omega_\text{t} = -E_i$ is the photon energy at threshold.

The relation (13.315) shows that the behaviour of the electron impact ionisation cross-section at high energies can be derived from the knowledge of the corresponding photoionisation cross-section over the entire range of photon energies. This relation can be extended to complex atoms, and to multiple ionisation processes. For example, in the case of double ionisation of helium, Byron and Joachain deduced from their double photoionisation calculations the value $A = 0.0033$. The measured value of A, obtained in 1969 by M.J. van der Wiel, Th.M. El-Sherbini and L. Vriens in experiments on the double ionisation of helium by fast electrons having energies in the range 2–16 keV, was found to be $A = 0.0028 \pm 0.0006$, in good agreement with the theoretical prediction.

Problems

13.1 Using the equations (13.40) and (13.41), derive the coupled integro-differential equations (13.42).

13.2 Derive the coupled integro-differential equations (13.42) by considering the functional (13.47), where $\psi_\pm^{(+)}$ is given by (13.35), and requiring that $\delta I = 0$ when $\psi_\pm^{(+)}$ is varied.

13.3 Using the definition (13.50), derive the expression (13.51) for the static potential acting on the incident electron when the target is atomic hydrogen in the ground state.
(**Hint**: Expand $|\mathbf{r}-\mathbf{r}'|^{-1}$ in spherical harmonics using equation (A4.25).)

13.4 Using the simple helium ground state wave function (7.70), namely

$$\phi(r_1, r_2) = \frac{Z_e^3}{\pi} \exp[-Z_e(r_1 + r_2)]$$

with $Z_e = 27/16$, calculate the static potential $V_{11}(r_0)$ for electron scattering by the ground state of helium, where

$$V_{11}(r_0) = \int d\mathbf{r}_1\, d\mathbf{r}_2 |\phi(r_1, r_2)|^2 \left(-\frac{2}{r_0} + \frac{1}{r_{01}} + \frac{1}{r_{02}}\right)$$

and

$$r_{0i} = |\mathbf{r}_0 - \mathbf{r}_i|, \qquad i = 1, 2$$

13.5 By including only the initial channel in $P\psi_{\pm}^{(+)}$, show that the equation (13.81) can be reduced to the form (13.54) with $V_{\text{opt}}^{\pm} = V_{ii} + W_{ii}^{\pm} + \mathcal{V}_i$, where $\mathcal{V}_i$ is the corresponding equivalent potential. Prove that $\mathcal{V}_i$ is complex when there are open channels other than the initial one.

13.6 Consider the elastic scattering of fast electrons by hydrogen atoms in the ground state.

(a) Evaluate the integral in (13.144) to obtain the first Born direct elastic scattering amplitude (13.145).
(b) Using the result (13.146), evaluate the integral (13.147) to obtain the total first Born cross-section $\sigma_{\text{tot}}^{\text{B1}}(1s \to 1s)$ given by (13.148).

13.7 Consider the excitation of the 2s state of atomic hydrogen from the ground state by fast electrons.

(a) Evaluate the integral in (13.154) to obtain the first Born direct scattering amplitude (13.155).
(b) Evaluate analytically the integral on the right of (13.158), and verify that the corresponding result for $\sigma_{\text{tot}}^{\text{B1}}(1s \to 2s)$ reduces to (13.161) for large k_i.

13.8 Consider the excitation of the $2p_m$ ($m = 0, \pm 1$) states of atomic hydrogen from the ground state by fast electrons.

(a) Evaluate the integral on the right of (13.162) by choosing the quantisation axis (the Z axis for the eigenfunctions) to be along the momentum transfer $\mathbf{\Delta}$ to obtain the result (13.163).
(b) Evaluate the integral on the right of (13.162) by choosing the quantisation axis along the incident direction $\hat{\mathbf{k}}_i$ to obtain the result (13.165).
(c) Using the fact that most of the contribution to the integral on the right of (13.167) comes from the region of small values of Δ near $\Delta_{\min} \simeq 3/(8k_i)$, obtain the expression (13.168a) of $\sigma_{\text{tot}}^{\text{B1}}(1s \to 2p)$ for large k_i.

13.9 Using the expression (13.144) and the static potential obtained in Problem 13.4, calculate the differential cross-section for the elastic scattering of electrons by the ground state of helium in the first Born approximation, at an incident electron energy of 500 eV.

13.10 Show that for an 'optically allowed' transition, the correction to the leading term (13.193) of $\sigma^{B1}_{tot}(i \to j)$ at high energies is of the form $B(E^e_i)^{-1}$, where B is energy-independent.

13.11 Using the equation (13.204), obtain the Ochkur approximation (13.205) for elastic exchange scattering from H(1s).

13.12 Using the equations (13.136) and (13.138), show that the second Born term $\bar{f}^{B2}_{ji}$ for a transition $(\mathbf{k}_i, i) \to (\mathbf{k}_j, j)$ in electron–atomic hydrogen collisions is given by the expression (13.207).

13.13 Perform the integral (13.242) to obtain the result (13.244).

13.14 Using the equations (13.237), (13.245) and (13.246), obtain the first Born TDCS (13.247) for electron impact ionisation of H(1s).

13.15 By integrating the first Born TDCS (13.247) over the polar angles (θ_B, ϕ_B) of the ejected electron, obtain the expression (13.248) for the first Born double differential cross-section for electron impact ionisation of H(1s).

13.16 Using first-order time-dependent perturbation theory, and making the dipole approximation, derive the expression (13.273) for the differential photoionisation cross-section.

13.17 Using the dipole length expression (13.285) of the differential photoionisation cross-section for hydrogenic atoms (ions), and employing the expression (13.284) of the Coulomb wave function $\psi^{(-)}_{c,\mathbf{k}_f}(Z, \mathbf{r})$ together with the integral representation (13.243) of the confluent hypergeometric function, obtain the result (13.289) for an unpolarised photon beam.

13.18 Starting from the expression (13.289b) for the quantity C:

(a) Prove the result (13.290) for $k_f \to \infty$.
(**Hint**: Show first that for large k_f, one has $C \simeq (|E_{1s}|/\omega)^{1/2}(k_f/Z)$. Then use the approximate relation (4.211) which gives $k_f/Z \simeq (\omega/|E_{1s}|)^{1/2}$ in a.u.)
(b) Prove the result (13.291) near threshold, and obtain the corresponding expression (13.292) of the differential photoionisation cross-section.

13.19 By using the approximate final state wave function (13.309) in the velocity form of the dipole matrix element, obtain the result (13.310) for the asymptotic cross-section $\sigma_{tot}(\omega \to \infty)$ for any ionisation process in helium in which at least one electron is ejected.

14 Atom–atom collisions

In Chapters 10 and 11, we discussed in some detail the properties of diatomic molecules, which are the bound states of two atoms or ions. We shall now turn to the case in which one atom is scattered by another. Such processes occur naturally in an assembly of atoms, such as in a gas, but can also be studied experimentally by allowing two beams of atoms to interact and studying the distribution of the reaction products.

Under all conditions *elastic scattering* occurs, in which atom A is scattered from atom B, without any change in internal energy of either A or B:

$$A + B \rightarrow A + B \tag{14.1}$$

Several other processes may also be possible. Indeed, provided enough energy is available, *inelastic collisions* can occur. For example, either or both A and B can be left in an *excited state* after the collision:

$$\begin{aligned} A + B &\rightarrow A^* + B \\ &\rightarrow A + B^* \\ &\rightarrow A^* + B^* \end{aligned} \tag{14.2}$$

Another process is one in which one, or more, electrons are transferred from one atom to another. For example,

$$\begin{aligned} A + B &\rightarrow A^- + B^+, \\ &\rightarrow A^{2-} + B^{2+} \end{aligned} \tag{14.3}$$

Such collisions are known as *charge exchange* reactions [1]. A further possibility is *ionisation*, in which one or more electrons are ejected. For example,

$$\begin{aligned} A + B &\rightarrow A^+ + B + e^- \\ &\rightarrow A + B^+ + e^- \\ &\rightarrow A^{2+} + B + 2e^- \end{aligned} \tag{14.4}$$

Other reactions of importance are *associative detachment*

$$A^- + B \rightarrow (AB) + e^- \tag{14.5}$$

and processes involving the emission of radiation.

[1] The terms *charge transfer*, *electron transfer* or *electron capture* are also used.

To illustrate various aspects of this large field of study, we shall choose a number of topics of particular interest, starting in Section 14.1 with elastic scattering between atoms in gases which have been cooled to very low temperatures of a few degrees K or less [2]. We shall then turn in Section 14.2 to elastic collisions between atoms in gases at higher temperatures, typically from 50 K to 1000 K. At these temperatures the kinetic energy of the colliding atoms is much less than 1 eV so that inelastic scattering is not energetically possible. The next topic to be studied in Sections 14.3 and 14.4 concerns collisions at energies of a few eV upwards. In this regime the full range of inelastic processes can take place. Finally collisions at high energies (> 100 keV) where perturbative methods can be employed will be discussed briefly in Section 14.5.

14.1 Collisions at very low energies

Modern experimental methods have made possible the study of atomic gases which have been cooled to very low temperatures [3] from a few kelvin down to 10^{-9} K. The viscosity and diffusion properties of such gases depend on the scattering parameters for elastic collisions between the gas atoms, and we shall discuss briefly how these can be calculated. Cold gases with temperatures of the order 1 K are also of interest in astrophysical contexts and for this temperature region some cross-section measurements have been made. Collisions occurring at temperatures below 1 μK are termed *ultra-cold*. They play an important role in the study of Bose–Einstein condensation, as we shall see in Section 15.5.

Let us consider the collisions of atoms A and B which are in ground states, which we shall assume are S states. Since the collision is very slow the Born–Oppenheimer approximation can be used to calculate the interaction potential $E_s(R)$ for the state of the molecule AB, which separates at large internuclear distances R into A + B. The relative motion of A and B is then described by the Schrödinger equation (10.22), which in the present case can be written in the simple form

$$\left(-\frac{\hbar^2}{2\mu}\nabla^2_{\mathbf{R}} + E_s(R) - E\right)F(\mathbf{R}) = 0 \qquad , s = 0 \tag{14.6}$$

where $\mu = M_A M_B/(M_A + M_B)$ is the reduced mass of the two nuclei. In Chapter 10, this Schrödinger equation was used to study the vibrational and rotational bound states of the molecule AB. Here we are concerned with the continuum states with energies

$$E = \frac{\hbar^2 k^2}{2\mu} + E_A + E_B \tag{14.7}$$

---

[2] A temperature of 1K corresponds to an energy of 8.62×10^{-5} eV (see Appendix 14).

[3] An extensive review of this subject has been given by Weiner *et al.* (1999).

where $\hbar^2k^2/(2\mu)$ is the kinetic energy of the relative motion, while E_A and E_B are the ground state energies of the atoms A and B, respectively. Defining

$$V(R) = E_s(R) - E_A - E_B \qquad , s = 0 \tag{14.8}$$

and

$$U(R) = \frac{2\mu}{\hbar^2} V(R) \tag{14.9}$$

the equation (14.6) can be written in the form

$$(\nabla^2_{\mathbf{R}} - U(R) + k^2)F(\mathbf{R}) = 0 \tag{14.10}$$

In Chapter 12 we have seen that this equation can be solved by using the partial wave method. The total elastic scattering cross-section is then given (see (12.65)) by

$$\sigma_{\text{tot}}(k) = \sum_{l=0}^{\infty} \sigma_l(k) \tag{14.11}$$

where the partial wave cross-sections $\sigma_l(k)$ are given by (see (12.66))

$$\sigma_l(k) = \frac{4\pi}{k^2}(2l + 1) \sin^2\delta_l(k) \tag{14.12}$$

For temperatures of a few kelvin, corresponding to energies of the order 10^{-4} eV, the series (14.11) converges rapidly. The $l = 0$ partial wave dominates and only a limited number of partial waves are significant. For ultra-cold collisions at temperatures of 1 μK and below, the total cross-section is given accurately by the $l = 0$ partial wave alone and in the limit $k \to 0$ one has in general

$$\lim_{k\to 0} \sigma_{\text{tot}} = 4\pi\alpha^2 \tag{14.13}$$

where α is the scattering length defined by (12.101).

At higher velocities, elastic scattering is dominated by high-order partial waves which are not very sensitive to the details of the potential for small values of R, and simple empirical potentials are often adequate. In contrast, the scattering length depends strongly on the details of the potential both at short and long distances and very precise Born–Oppenheimer potentials must be calculated. Indeed, to obtain accurate values of the cross-sections for cold and ultra-cold collisions it is often necessary to take into account non-adiabatic and relativistic corrections.

Very often cross-sections are required for collisions in a gas composed of identical atoms. In this case the wave function of a pair of colliding atoms must be either symmetric or antisymmetric under the interchange of the two atoms, and this requires some modification to the cross-section formulae. The atom–atom interaction potential $V(R)$ can be used to calculate phase shifts δ_l and a scattering amplitude

$$f(\theta) = \frac{1}{k}\sum_{l=0}^{\infty}(2l+1)\exp(\mathrm{i}\delta_l)\sin\delta_l P_l(\cos\theta) \tag{14.14}$$

where θ is the scattering angle in the centre of mass system, and we have omitted to indicate the k-dependence of δ_l and $f(\theta)$ for notational simplicity.

When the colliding atoms are identical two amplitudes $f_{\pm}(\theta)$ can be constructed which are either even or odd under the interchange $\mathbf{R} \to -\mathbf{R}$:

$$f_{\pm}(\theta) = f(\theta) \pm f(\pi - \theta) \tag{14.15}$$

Using (14.14), it is seen that in terms of the phase shifts

$$f_{+}(\theta) = \frac{2}{k}\sum_{l\text{ even}}(2l+1)\exp(\mathrm{i}\delta_l)\sin\delta_l P_l(\cos\theta) \tag{14.16a}$$

and

$$f_{-}(\theta) = \frac{2}{k}\sum_{l\text{ odd}}(2l+1)\exp(\mathrm{i}\delta_l)\sin\delta_l P_l(\cos\theta) \tag{14.16b}$$

where the relationship (12.236) has been used.

As an example let us consider the important case of the scattering of one ground state hydrogen atom by another. There are two molecular states of H_2 which separate to the configuration H(1s) + H(1s) at large R. These are the ${}^1\Sigma_g^+$ and ${}^3\Sigma_u^+$ states (see Chapter 10). To each of these correspond different Born–Oppenheimer potentials which will be labelled as ${}^1V(R)$ and ${}^3V(R)$, respectively.

For the singlet ($S = 0$) electronic spin state, the potential ${}^1V(R)$ gives rise to a set of phase shifts ${}^1\delta_l$ and two possible amplitudes ${}^1f_+(\theta)$ and ${}^1f_-(\theta)$. In the same way, for the triplet ($S = 1$) electronic spin state, the potential ${}^3V(R)$ gives rise to phase shifts ${}^3\delta_l$ and amplitudes ${}^3f_+(\theta)$ and ${}^3f_-(\theta)$.

We must now take into account the spin of the two protons for which the total spin can be either $T = 0$ or $T = 1$. Let us first consider scattering in the ${}^1\Sigma_g^+$ state. In Section 11.5, it was shown that in this state, if $T = 0$, only even rotational states are allowed for which the amplitude is ${}^1f_+(\theta)$, while if $T = 1$, only odd rotational states are allowed so that in this case the amplitude must be ${}^1f_-(\theta)$. If the nuclear spins are unpolarised the differential cross-section for scattering in the ${}^1\Sigma_g^+$ state must therefore be

$$\frac{\mathrm{d}{}^1\sigma}{\mathrm{d}\Omega} = \frac{1}{4}|{}^1f_+(\theta)|^2 + \frac{3}{4}|{}^1f_-(\theta)|^2 \tag{14.17}$$

Now, let us consider scattering in the ${}^3\Sigma_u^+$ state. By a similar analysis to that given for the ${}^1\Sigma_g^+$ state in Section 11.5, it is found that the rotational levels must be odd for $T = 0$ and even for $T = 1$. It follows that if the nuclear spins are unpolarised, the differential cross-section for scattering in the ${}^3\Sigma_u^+$ state must be

$$\frac{\mathrm{d}{}^3\sigma}{\mathrm{d}\Omega} = \frac{1}{4}|{}^3f_-(\theta)|^2 + \frac{3}{4}|{}^3f_+(\theta)|^2 \tag{14.18}$$

Finally, if the electronic spins are unpolarised, the overall differential scattering cross-section is

$$\frac{d\sigma}{d\Omega} = \frac{1}{4}\frac{d^1\sigma}{d\Omega} + \frac{3}{4}\frac{d^3\sigma}{d\Omega} \tag{14.19}$$

Integrating over all solid angles it is easily found that the partial cross-sections σ_l are given for even l by

$$\sigma_l = \frac{\pi}{k^2}(2l+1)[\sin^2(^1\delta_l) + 9\sin^2(^3\delta_l)], \qquad l \text{ even} \tag{14.20a}$$

while for odd l,

$$\sigma_l = \frac{3\pi}{k^2}(2l+1)[\sin^2(^1\delta_l) + \sin^2(^3\delta_l)], \qquad l \text{ odd} \tag{14.20b}$$

Defining the scattering lengths for the $^1\Sigma_g^+$ and $^3\Sigma_u^+$ potentials to be $^1\alpha$ and $^3\alpha$, respectively, we see that the total cross-section [4] for ultra-cold collisions is

$$\sigma_0 = \pi[|^1\alpha|^2 + 9|^3\alpha|^2] \tag{14.21}$$

Cross-sections in the temperature range 1–8 K based on very accurate potentials were calculated in 1999 by M.J. Jamieson, A. Dalgarno, B. Zygeman, P.S. Krstić and D.R. Schultz. They find for example at 1 K and 8 K total elastic cross-sections of 64.2×10^{-16} cm^2 and 110.6×10^{-16} cm^2, respectively.

14.2 Elastic collisions at low velocities

We shall now discuss atom–atom collisions at energies well below those at which inelastic scattering can take place, but for which a large number of partial waves are important. In collisions between the atoms of an atomic gas these conditions hold up to temperatures of a few thousand degrees K, and the elastic scattering cross-sections are important in understanding the viscosity and many other properties of gases. As in Section 14.1, we shall study collisions between atoms A and B which are in ground states, assumed to be S states. In the energy range concerned, the relative motion of the colliding atoms is much slower than that of the atomic electrons, so that the Born–Oppenheimer approximation can be made and the wave function is determined by the Schrödinger equation (14.6). In general there may be more than one Born–Oppenheimer potential acting between the colliding atoms, as in the case of the H(1s) + H(1s) system, but we shall concentrate on the case for which there is a single Born–Oppenheimer interaction $E_s(R)$ giving

[4] It should be pointed out that in the field of cold collisions a different normalisation of the cross-section for the scattering of identical particles than that given in Section 12.6 is sometimes adopted, leading to partial cross-sections which are one-half those given here.

rise to a potential $V(R)$ defined by (14.8). At short distances $V(R)$ is repulsive, both because of the Coulomb repulsion between the nuclei and because of the exchange interaction between the electronic distributions. For large R, if the atoms are both neutral the interaction is attractive; this is due to the electric dipole moment induced in each atom by the other [5]. This attractive long-range interaction between neutral atoms is called the *van der Waals interaction*. It is such that

$$V(R) \underset{R\to\infty}{\to} -\frac{C_W}{R^6} \tag{14.22}$$

where C_W is a positive constant known as the *van der Waals coefficient*. It can be calculated, in principle, by second-order perturbation theory [5]. If one atom, A, is neutral and the other, B, is charged the long-range interaction is the *polarisation potential*, such that

$$V(R) \underset{R\to\infty}{\to} -\frac{C_D}{R^4} \tag{14.23}$$

where C_D is a positive constant proportional to the static dipole polarisability of the neutral atom A.

The potential $V(R)$ is difficult to calculate for many-electron systems and a number of studies employ a semi-empirical potential depending on parameters which are determined from fits to the experimental scattering data. A widely used potential of this type is the *Lennard-Jones potential*

$$V(R) = \frac{m\varepsilon}{n-m}\left[\left(\frac{R_0}{R}\right)^n - \frac{n}{m}\left(\frac{R_0}{R}\right)^m\right] \tag{14.24}$$

If both atoms are neutral $m = 6$, but if one is charged $m = 4$. A frequently used value for n is $n = 12$. The parameters ε and R_0 are the depth of the well and the position of the minimum, respectively. The well depth ε varies widely for different pairs of atoms, ranging from 10^{-3} eV to 5 eV. The position of the minimum R_0 also varies, but less greatly, typical values being between 1 a.u. and 10 a.u. The Lennard-Jones potential is highly singular at $R = 0$. This, however, is unimportant for scattering at low energies since the atoms never penetrate into the region in which R is very small.

The semi-classical theory of atom–atom scattering

The potential $V(R)$ is smoothly varying and only changes slowly over distances of the order of 1 a.u. The de Broglie wavelength λ associated with the relative motion of the colliding atoms is much smaller than 1 a.u. over the energy range we are considering, where λ is given by

[5] See for example Bransden and Joachain (2000).

$$\lambda = \frac{2\pi}{\mu v} \tag{14.25}$$

Here μ is the reduced mass of the atoms A and B, and v is their relative velocity. Atomic units (a.u.) have been employed as will be the case throughout this chapter unless otherwise stated. The lightest atom is hydrogen, so that in all cases $\mu > 1836$. Thus $\lambda < 1$ if $v > v_c$ where $v_c = (2\pi/1836)$ a.u. $= 3.4 \times 10^{-3}$ a.u. which corresponds to 7.5×10^3 m s^{-1}. Hence for $v > v_c$ short-wavelength conditions hold which, as shown in Chapter 12, allow the semi-classical WKB approximation to be applied, provided the angle of scattering θ is well defined. This latter condition requires that $\theta > (\mu v b)^{-1}$, where b is the classical impact parameter. Remembering that $L = \mu v b$ is the magnitude of the classical angular momentum, we see that semi-classical conditions apply if $\theta > \theta_c$, where

$$\theta_c = \frac{1}{\mu v b} = \frac{1}{L} \tag{14.26}$$

For the energy range we are considering the important values of L are of the order 10^3 and above, so that semi-classical conditions apply at angles of scattering greater than a few milliradians.

The fraction of particles scattered through angles greater than θ_c is usually quite small, and almost all elastic scattering is confined to a narrow cone about the forward direction at all collision energies greater than a fraction of an electron volt. However, in most experiments the forward direction is excluded and semi-classical conditions apply. The small-angle scattering will not be discussed here, but reference may be made to Bransden and McDowell (1992).

When $v > v_c$ and $\theta > \theta_c$ the classical trajectories for motion in the potential $V(R)$ are well defined. If there is only one trajectory for each angle of scattering the semi-classical differential cross-section is given by (12.373) and is identical with the purely classical cross-section. This is not the case for an interaction with an inner repulsion and an outer attraction like the Lennard-Jones potential. To allow for this, the WKB theory can be extended. For such potentials, the classical deflection function Θ given by (12.352) has the form illustrated in Fig. 14.1. When L is small, Θ depends mainly on the inner, repulsive, part of the potential and is positive. For large L, corresponding to glancing collisions, Θ depends mainly on the outer part of the potential which is attractive and Θ is then negative. We recollect from (12.371) that the angle of scattering θ is equal to Θ if $\Theta > 0$ and equal to $-\Theta$ if $\Theta < 0$, provided $|\Theta| < \pi$. As a function of L, the classical deflection function Θ has a minimum at $L = L_m$ and the corresponding angle of scattering $\theta_m = -\Theta(L_m)$ is known as the *rainbow angle*. It should be noted that for some atom–atom interaction potentials Θ may become large corresponding to the scattered particle making several circuits around the scattering centre. Under certain circumstances Θ can become infinite. This phenomenon is called *orbiting*. Although a semi-classical theory can be developed in this situation it will not be considered here.

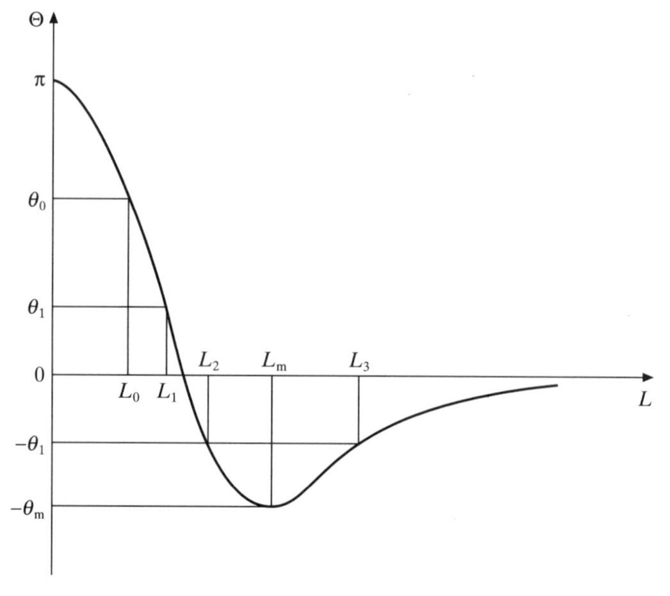

Figure 14.1 Schematic diagram of the deflection angle Θ as a function of angular momentum for a particle scattered by a Lennard-Jones potential. Note that the scattering angle θ is always positive ($0 \leqslant \theta \leqslant \pi$).

Referring to Fig. 14.1, it is seen that for $\theta > \theta_m$, for example at $\theta = \theta_0$, only one value of L, and hence only one classical trajectory, corresponds to each scattering angle. In contrast, for $\theta < \theta_m$, for example at $\theta = \theta_1$, three values of L (L_1, L_2 and L_3) correspond to the same angle of scattering. Before considering how the WKB method deals with this situation, let us look at the classical differential cross-section which we shall denote by $d\sigma_{cl}/d\Omega$. From (A1.22), it is the sum of the three terms that correspond to the angle θ, namely

$$\frac{d\sigma_{cl}}{d\Omega} = \sum_{i=1}^{3} I_i(\theta) \tag{14.27a}$$

where

$$I_i(\theta) = \frac{1}{\mu^2 v^2} \frac{L_i}{\sin\theta} \left| \frac{dL_i}{d\theta} \right| \tag{14.27b}$$

At the angle $\theta = \theta_m$ for which $\Theta(L)$ has a minimum, the classical cross-section is infinite. This is called the *rainbow singularity*. In addition, because of the factor $(\sin\theta)^{-1}$ the classical cross-section is infinite at $\theta = 0$ and $\theta = \pi$ unless $dL_i/d\theta$ vanishes. These singularities in the forward and backward directions are known as *glories*. None of these singularities occur in the quantum or semi-classical theory, but they are replaced by interesting interference effects which we shall examine in the case of rainbow scattering.

The WKB approximation and rainbow scattering

To apply the WKB approximation, the starting point is the expression (12.362) for the scattering amplitude

$$f(k, \theta) = -\frac{1}{k}\int_0^\infty \left(\frac{l}{2\pi \sin\theta}\right)^{1/2} \{\exp[i\phi_+(l, k)] - \exp[i\phi_-(l, k)]\}\, dl \tag{14.28}$$

where ϕ_+ and ϕ_- are given by (see (12.363))

$$\phi_\pm(l, k) = 2\delta_l(k) \pm \left(l + \frac{1}{2}\right)\theta \pm \frac{\pi}{4} \tag{14.29}$$

From (12.351) and (12.365), it is seen that the points of stationary phase in the integral of (14.28) occur in the ϕ_+ term when the deflection function Θ is negative $(\theta = -\Theta)$ and in the ϕ_- term when Θ is positive $(\theta = \Theta)$.

For a deflection function of the form shown in Fig. 14.1, it is seen that:

(a) If $\theta > \theta_m$, then ϕ_+ has no point of stationary phase and ϕ_- has one point of stationary phase at $L = L_0$, that is at $l = l_0 = L_0 - 1/2$.
(b) If $\theta < \theta_m$, then ϕ_+ has two points of stationary phase at $L = L_2$ and $L = L_3$ (corresponding to $l = l_2 = L_2 - 1/2$ and $l = l_3 = L_3 - 1/2$) while ϕ_- has one point of stationary phase at $L = L_1$, corresponding to $l = l_1 = L_1 - 1/2$).

Provided the points of stationary phase are well separated the integral (14.28) is the sum of the contributions from each point, each contribution being evaluated as in (12.366)–(12.369). For $\theta > \theta_m$, there is only one point of stationary phase and

$$f(k, \theta) = f_0(k, \theta) \tag{14.30a}$$

where

$$f_0(k, \theta) = [I_0(\theta)]^{1/2} \exp(i\gamma_0) \tag{14.30b}$$

In (14.30b), $I_0(\theta)$ is the classical cross-section arising from the stationary point at $l = l_0$ and the phase γ_0 is given by

$$\gamma_0 = \phi_-(l_0) - \frac{\pi}{4} = 2\delta_{l_0} - \left(l_0 + \frac{1}{2}\right)\theta - \frac{\pi}{2} \tag{14.31}$$

When $\theta < \theta_m$, the scattering amplitude is the sum of three terms

$$f(k, \theta) = \sum_{i=1}^{3} f_i(k, \theta) \tag{14.32a}$$

where

$$f_i(k, \theta) = [I_i(\theta)]^{1/2} \exp(i\gamma_i), \qquad i = 1, 2, 3 \tag{14.32b}$$

and the classical cross-section (see 14.27(b)) for $L = L_i$, that is $l = l_i = L_i - 1/2$, has been denoted by $I_i(\theta)$. Taking into account that $\mathrm{d}\Theta/\mathrm{d}L$ is negative at $L = L_1$ and $L = L_2$, but positive at $L = L_3$, and using (12.363), the phases γ_i are found to be

$$\gamma_1 = \phi_-(l_1) - \frac{\pi}{4} = 2\delta_{l_1} - \left(l_1 + \frac{1}{2}\right)\theta - \frac{\pi}{2}, \tag{14.33a}$$

$$\gamma_2 = \phi_+(l_2) - \frac{5\pi}{4} = 2\delta_{l_2} + \left(l_2 + \frac{1}{2}\right)\theta - \pi, \tag{14.33b}$$

$$\gamma_3 = \phi_+(l_3) - \frac{3\pi}{4} = 2\delta_{l_3} + \left(l_3 + \frac{1}{2}\right)\theta - \frac{\pi}{2} \tag{14.33c}$$

From (14.30) the differential cross-section when θ is greater than θ_m is the classical cross-section:

$$\frac{\mathrm{d}\sigma}{\mathrm{d}\Omega} = I_0(\theta) \tag{14.34}$$

while from (14.32), if $\theta < \theta_m$,

$$\begin{aligned}\frac{\mathrm{d}\sigma}{\mathrm{d}\Omega} &= \left|\sum_{i=1}^{3} f_i(k, \theta)\right|^2 \\ &= \sum_{i=1}^{3} I_i(\theta) + 2I_1(\theta)I_2(\theta)\cos(\gamma_1 - \gamma_2) \\ &\quad + 2I_1(\theta)I_3(\theta)\cos(\gamma_1 - \gamma_3) + 2I_2(\theta)I_3(\theta)\cos(\gamma_2 - \gamma_3)\end{aligned} \tag{14.35}$$

The first term on the right-hand side of (14.35) is just the classical differential cross-section (14.27a). The remaining terms are interference terms which result in oscillations in the differential cross-sections as a function of θ. The terms containing $\cos(\gamma_1 - \gamma_2)$ or $\cos(\gamma_1 - \gamma_3)$ oscillate with an angular separation between oscillations of $\Delta\theta = 2\pi(l_1 + l_2)^{-1}$ or $\Delta\theta = 2\pi(l_1 + l_3)^{-1}$ respectively, while the term in $\cos(\gamma_2 - \gamma_3)$ oscillates with an angular separation of $\Delta\theta = 2\pi(l_2 - l_3)^{-1}$. Thus there are two types of oscillation, a slow oscillation arising from $\cos(\gamma_2 - \gamma_3)$ and, superimposed on this, a rapid oscillation arising from the other two terms. These are illustrated in Fig. 14.2. It should be noted that experimentally unless the apparatus has a very high angular resolution only the slow oscillation is observed.

Near $\theta = \theta_m$ the condition that the points of stationary phase should be well separated is violated, and at $\theta = \theta_m$ the stationary phases at l_2 and l_3 coalesce. Consequently $f_2(k, \theta)$ and $f_3(k, \theta)$ cannot be calculated separately, but must be combined into an amplitude $f_m(k, \theta)$ for rainbow scattering. As we shall see $f_m(k, \theta)$ extends into the region $\theta > \theta_m$ and this accounts for the interference seen in Fig. 14.2 in the region $\theta > \theta_m$, but dying out for large θ.

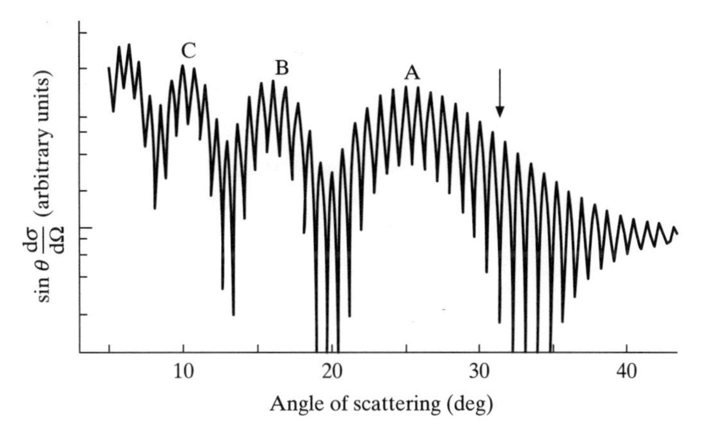

Figure 14.2 Illustrative diagram of a differential cross-section, weighted by $\sin\theta$, for a deflection function of the form shown in Fig. 14.1. The primary and secondary rainbow maxima are at A, B, C, . . . and the rainbow angle at $\theta = \theta_m$ is marked by an arrow.

Rainbow scattering

At $L = L_m$, the classical deflection function $\Theta(L)$ has a minimum and can be expanded in terms of l, where $l = L - 1/2$, as

$$\Theta(l) = \Theta_m + q(l - l_m)^2, \qquad q = \frac{1}{2}\left[\frac{d^2\Theta}{dl^2}\right]_{l=l_m} \tag{14.36}$$

From the classical equivalence relation (12.351) we have

$$\begin{aligned} 2\delta_l &= 2\delta_{l_m} + \int_{l_m}^{l} \Theta(l')\, dl' \\ &= 2\delta_{l_m} + \Theta_m(l - l_m) + \frac{1}{3}q(l - l_m)^3 \end{aligned} \tag{14.37}$$

where in the second line we have used (14.36).

Using (14.29) and (14.37), the function $\phi_+(l, k)$ can be expanded as

$$\phi_+(l, k) = \phi_+(l_m, k) + (\theta - \theta_m)(l - l_m) + \frac{1}{3}q(l - l_m)^3 + \ldots \tag{14.38}$$

where the relation $\theta_m = -\Theta_m$ has been used. From (14.28) and (14.38), the contribution from $\phi_+(l, k)$ to the scattering amplitude in the region around $l = l_m$ is

$$\begin{aligned} f_m(k, \theta) = -\frac{1}{k}\left[\frac{l_m + 1/2}{2\pi \sin\theta}\right]^{1/2} \exp[i\phi_+(l_m, k)] \int_{-\infty}^{\infty} \exp\Big[& i(\theta - \theta_m)(l - l_m) \\ & + \frac{i}{3}q(l - l_m)^3\Big] dl \end{aligned} \tag{14.39}$$

The integral in (14.39) can be expressed in terms of the Airy function $\mathrm{Ai}(x)$, where

$$\mathrm{Ai}(x) = \frac{1}{2\pi} \int_{-\infty}^{\infty} \exp[\mathrm{i}(xt + t^3/3)]\, \mathrm{d}t \tag{14.40}$$

by taking $x = q^{-1/3}(\theta - \theta_m)$. The Airy function decreases exponentially for $x > 0$ and oscillates for $x < 0$, with asymptotic forms

$$\begin{aligned}\mathrm{Ai}(x) &\underset{x\to\infty}{\rightarrow} \frac{1}{2}\pi^{-1/2}x^{-1/4}\exp[-2x^{3/2}/3] \\ &\underset{x\to-\infty}{\rightarrow} \pi^{-1/2}(-x)^{-1/4}\sin\left[\frac{2}{3}(-x)^{3/2} + \frac{\pi}{4}\right]\end{aligned} \tag{14.41}$$

From (14.39) and (14.40), the function $f_m(k, \theta)$ can be written as

$$f_m(k,\theta) = -\frac{1}{k}\left[\frac{2\pi(l_m + 1/2)}{\sin\theta}\right]^{1/2} \exp[\mathrm{i}\phi_+(l_m, k)]q^{-1/3}\mathrm{Ai}\left[\frac{\theta - \theta_m}{q^{1/3}}\right] \tag{14.42}$$

The differential cross-section for $\theta > \theta_m$ is

$$\frac{\mathrm{d}\sigma}{\mathrm{d}\Omega} = |f_0(k,\theta) + f_m(k,\theta)|^2 \tag{14.43}$$

and for $\theta < \theta_m$ it is

$$\frac{\mathrm{d}\sigma}{\mathrm{d}\Omega} = |f_1(k,\theta) + f_m(k,\theta)|^2 \tag{14.44}$$

The term in f_m produces the slow oscillations, called *supernumerary rainbows*, in the region $\theta < \theta_m$ and decreases exponentially for $\theta > \theta_m$. The interference between f_m and f_0 or f_1 produces the rapid oscillations we discussed earlier, because except near $\theta = \theta_m$, $f_m(k, \theta)$ is approximately equal to $f_2(k, \theta) + f_3(k, \theta)$.

Before leaving the semi-classical theory of elastic scattering, the case of collisions between two identical atoms should be noted. In Chapter 12, it was seen that if the colliding particles are identical the scattering amplitude $f(k, \theta)$ should be replaced by symmetrised scattering amplitudes $f(k, \theta) \pm f(k, \pi - \theta)$. This gives rise to additional interferences, but in the energy region under discussion these are only of importance near $\theta = \pi/2$, because it is the only region in which $f(k, \theta)$ is of comparable size to $f(k, \pi - \theta)$.

Experimental measurements

The principal objective of experiments in the region of thermal energies is to determine the parameters of semi-empirical atom–atom potentials, particularly the van der Waals coefficients. A knowledge of these potentials helps in finding the equations of state for bulk matter and in understanding the various phase transitions that can occur. Atomic beam experiments have been performed, which

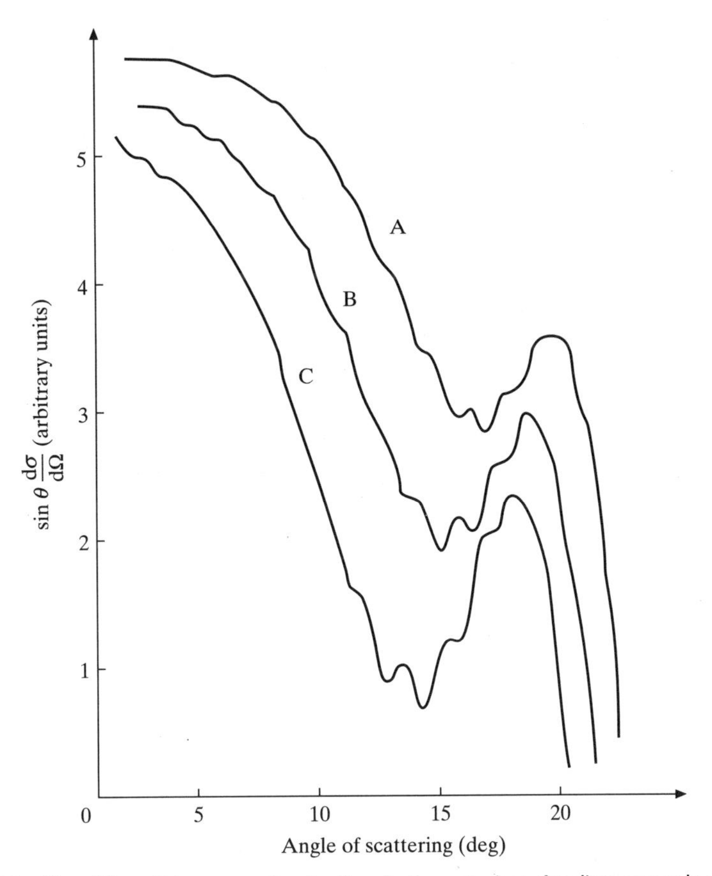

Figure 14.3 The differential cross-section for the elastic scattering of sodium atoms by krypton atoms measured by P. Barwig, U. Buck, E. Hundhausen and H. Pauly for relative velocities of (A) 767 ms^{-1}, (B) 817 ms^{-1} and (C) 870 ms^{-1}. Quantum interference effects are seen in the region of the rainbow angle.

allow measurements of the elastic differential cross-section with a high angular resolution. The parameters ε and R_0 of the Lennard-Jones potential (14.24) can be deduced by observing the position of the rainbow angle and the angles at which the supernumerary rainbows occur.

An example of the results of an experiment designed to determine the rainbow angle and the angle of the supernumerary rainbows is shown in Fig. 14.3. The principal rainbow and some of the supernumerary rainbows are clearly seen in the differential cross-section for the elastic scattering of sodium atoms by krypton atoms. The resolution in this experiment is not high enough to exhibit the rapid oscillations arising from the terms containing $\cos(\gamma_1 - \gamma_2)$ or $\cos(\gamma_1 - \gamma_3)$. However, observation of the rapid oscillations is not required for a determination of the potential constants.

14.3 Non-elastic collisions between atoms

As was noted at the beginning of this chapter, in an atom–atom collision a variety of non-elastic processes are possible above the various energy thresholds. Either atom can be excited or ionised and charge exchange (charge transfer or electron capture) can take place between one atom and the other. At very low energies charge exchange with the emission of radiation is important, and at very high energies the production of electron–positron pairs during the collision is also important, but neither of these processes will be discussed here. Experimental measurements of non-elastic cross-sections can be performed using colliding or merging beams of atoms, but it is convenient to express collision energies in terms of the laboratory kinetic energy $E_{\rm A}$ of an atom A incident on a target atom B which is initially at rest. In fact, because the velocity v of the incident atom A in the laboratory system is often a more useful collision parameter than its laboratory kinetic energy $E_{\rm A}$, it is convenient to define a *reduced laboratory energy* $E_{\rm L}$ by the relation

$$E_{\rm L} = \frac{E_{\rm A}}{M_{\rm A}} \tag{14.45}$$

where $E_{\rm A}$ is expressed in keV and $M_{\rm A}$, the mass of the atom A, is expressed in atomic mass units (a.m.u.). The reduced laboratory energy $E_{\rm L}$ depends only on the velocity v, which is also the *relative* velocity of the incident atom A and the target atom B. If v is expressed in atomic units (a.u.), then

$$E_{\rm L} = 25v^2 \text{ keV/a.m.u.} \tag{14.46}$$

For example, a proton having a laboratory energy of 25 keV moves with a velocity $v = 1$ a.u. relative to a target at rest.

It is useful to make distinctions between slow, intermediate and fast collisions according to whether $v \ll v_0$, $v \simeq v_0$ or $v \gg v_0$, where v_0 is a typical velocity of the active electron(s) in the collision between A and B. It is worth noting that even for slow collisions $E_{\rm L}$ can be quite large. For example, if we take $v_0 = 1$ a.u., we see that $E_{\rm L} - 1$ keV for a slow proton with $v = 0.2$ a.u.

It is instructive to examine first the measured total cross-sections for two simple systems, one of which, $\rm H^+ + H$, is homonuclear, and the other, $\rm H^+ + He$, is heteronuclear.

The system $\rm H^+ + H$

In Fig. 14.4, the total cross-sections for the collision processes

$$\rm H^+ + H(1s) \to H^+ + H(\mathit{n}) \qquad \mathit{n} = 2, 3 \tag{14.47a}$$

$$\rm \to H^+ + H^+ + e^- \tag{14.47b}$$

$$\rm \to H + H^+ \tag{14.47c}$$

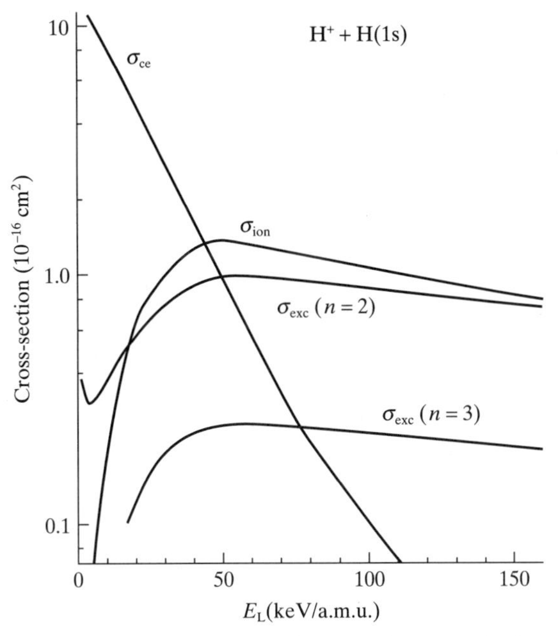

Figure 14.4 Total cross-sections for collisions of protons by atomic hydrogen in the ground state: $\sigma_{exc}(n)$ is the cross-section for excitation to the level n; σ_{ion} is the cross-section for ionisation; σ_{ce} is the cross-section for charge exchange into all bound states of atomic hydrogen.

are shown as a function of the reduced laboratory energy E_L of the incident proton. At low energies the charge exchange cross-section σ_{ce} is the largest, greatly exceeding the cross-sections for excitation (σ_{exc}) or ionisation (σ_{ion}). These latter cross-sections behave qualitatively like the corresponding cross-sections for electron impact at the same relative velocity, but the charge exchange cross-section is very different. For a homonuclear reaction σ_{ce} is proportional to $|C - \log v|^2$ as $v \to 0$, where C is a constant, but as v increases σ_{ce} decreases rapidly. For $v \gg 1$ a.u., σ_{ce} decreases at first like E_L^{-6} and then as $E_L^{-5.5}$, but the second region is in fact only reached at energies for which relativistic corrections become important ($E_L \geqslant 100$ MeV/a.m.u.). In contrast, the excitation and ionisation cross-sections, which are small at low energies, each increase to a maximum near 50 keV/a.m.u. and thereafter decrease, ultimately like $E_L^{-1} \log E_L$.

At low velocities, $v \leqslant 1$ a.u., electron capture is mainly into the ground state of atomic hydrogen:

$$H^+ + H(1s) \to H(1s) + H^+ \tag{14.48}$$

This is a case of *symmetrical resonance*, in which the initial and final systems are identical. It is only in this case that the capture cross-section increases as v decreases to zero. At first sight it would appear that no distinction can be made between capture and elastic scattering in homonuclear reactions of this kind.

However, this is not the case except at very low energies. This is due to the fact that the heavy particle motion is essentially classical ($v > v_c$) and for collision energies greater than a few electron volts the incident nucleus is only scattered through a very small angle of the order of milliradians. It follows that hydrogen atoms appearing close to the direction of incidence must be the product of charge exchange. The target proton recoils in a direction close to 90° to the direction of incidence, so that hydrogen atoms emerging near this direction are those corresponding to elastic scattering or excitation.

The system H^+ + He

As a second example, the cross-sections for charge exchange, excitation to the $2\,^1S$ and $2\,^1P$ levels and single ionisation of helium by proton impact are shown in Fig. 14.5 for the H^+ + He system. For this heteronuclear system, as with homonuclear systems, capture predominates at low energies, but the energy dependence is very different. The capture cross-section rises from zero at the threshold energy to a broad maximum near $E_L = 25$ keV/a.m.u., thereafter decreasing rapidly in the same way as in the case of homonuclear systems. Characteristically, charge exchange is the most important non-elastic process at low energies, ionisation and excitation dominating at higher energies. For this system the ionisation cross-section exceeds that for charge exchange at an energy E_L above 60 keV/a.m.u. The general shape of the charge exchange cross-section shown in Fig. 14.5 is typical of heteronuclear systems, although the magnitude and position of the maximum in the cross-section can vary widely. In general the greater the binding energy of the electron captured, the higher the energy at which the maximum of the capture cross-section occurs. For instance, the maximum of the cross-section for capture of the

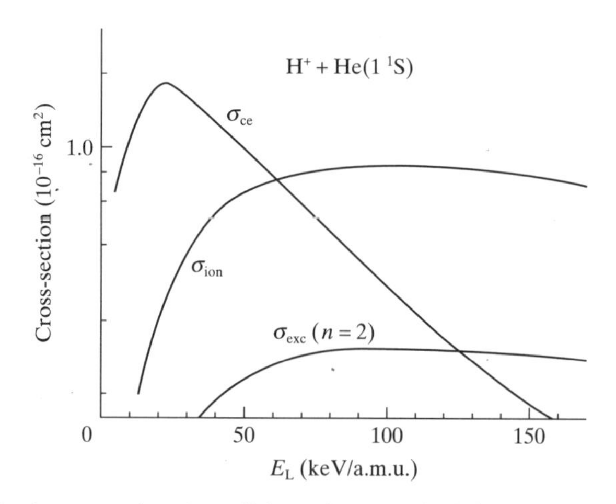

Figure 14.5 Total cross-sections for collisions of protons by helium in the ground state: σ_{exc} ($n = 2$) is the cross-section for excitation of the sum of the $2\,^1S$ and $2\,^1P$ levels; σ_{ion} is the cross-section for single ionisation; σ_{ce} is the cross-section for charge exchange into all bound states of atomic hydrogen.

2s valence electron from Li by He^+ ions occurs at $E_L = 2.5$ keV/a.m.u., but in the cross-section for capture of a 1s (K shell) electron it is at 125 keV/a.m.u.

State selectivity

In reactions like $H^+ + H(1s) \rightarrow H(nl) + H^+$, capture is mainly to the ground state of atomic hydrogen. However, ground state capture is not always predominant and usually the most important final state is the one for which the internal energy of the system changes the least. For example, in the reaction

$$He^{2+} + H(1s) \rightarrow He^+(nl) + H^+ \tag{14.49}$$

the most important channel is that leading to $He^+(n = 2)$, which is a *resonant reaction* with no change in internal energy. This property has an interesting application as a diagnostic mechanism in plasma research. Consider for example a plasma of protons and electrons which is contaminated by impurity ions such as C^{6+}. By injecting a beam of neutral hydrogen atoms, the reaction

$$H(1s) + C^{6+} \rightarrow H^+ + C^{5+}(nl) \tag{14.50}$$

takes place. Since the C^{5+} ion is formed selectively in the resonant level with $n = 4$, the subsequent line radiation can be measured and if the charge exchange cross-section is known the density of the impurity ions can be calculated.

Theoretical considerations

The theory of atom–atom inelastic collisions can be developed in a similar manner to the theory of electron–atom collisions. The wave function for the system can be expanded in a suitable set of basis functions and coupled channel equations found for the wave functions describing the relative motion of the two atoms. We have already obtained the set of coupled equations (10.21) by expanding the complete wave function for a diatomic system into the set of adiabatic electronic wave functions $\Phi_q(R; \mathbf{r}_1, \ldots, \mathbf{r}_N)$. These functions are asymptotic at large R to atomic wave functions centred on each nucleus (or combinations of atomic wave functions). Above threshold, for positive energies, the functions $F_q(\mathbf{R})$ satisfying scattering boundary conditions can be calculated to describe transitions from one channel q to another q'. In order to do this, a partial wave expansion can be made in terms of the functions ${}^{\Lambda}\mathcal{H}_{KM_K}(\Theta, \Phi)$ (see Chapter 10 and Appendix 11), leading to second-order differential equations for the radial functions which can be solved numerically. At low energies, transitions are only significant between channels q and q' if the corresponding potential energy curves $E_q(R)$ and $E_{q'}(R)$ cross or approach each other at a certain value of R and, in consequence, only a few terms in the coupled equations are required. This approach, originally developed in 1933 by H.S.W. Massey and R.A. Smith, is called the *perturbed stationary state* (PSS) approximation and is very successful for energies up to a few hundred eV/a.m.u.

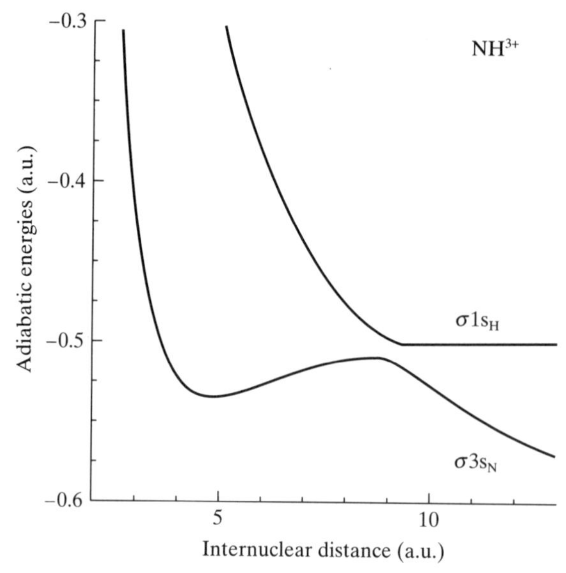

Figure 14.6 Potential energies for NH^{3+} showing an avoided crossing near $R = 9$ a.u.

As an example, let us consider the charge transfer reaction

$$N^{3+} + H(s) \rightarrow N^{2+}(3s) + H^{+} \tag{14.51}$$

The incident channel $N^{3+} + H(1s)$ is correlated to a σ molecular orbital designated as $\sigma 1s_H$, while the $N^{2+}(3s) + H^{+}$ channel is correlated to the $\sigma 3s_N$ molecular orbital. The corresponding energy curves approach each other near $R = 9$ a.u. giving rise to an *avoided crossing* (see Fig. 14.6). All the other energy curves are more distant, so that in this case a two-state approximation is adequate.

There are two difficulties with the PSS model which limit its applicability. The first is that when a partial wave expansion is used, the number of partial waves required increases very rapidly with increasing energy; this may be overcome by using a semi-classical approximation. The second difficulty is that the PSS method in its original form does not allow for the change in translational energy of an electron when it transfers from one atom to the other in a charge exchange reaction. This has the consequence that the boundary conditions cannot be satisfied exactly. At low velocities, however, the error made is rather small and the PSS method is usually sufficiently accurate up to $E_L \simeq 1$ keV/a.m.u.

14.4 The impact parameter method

For small values of the reduced laboratory energy E_L (of the order of the binding energy of the atomic electrons which are active in the collision), a full quantum mechanical treatment of the collision must be made, using the partial wave method. However, when E_L is much larger than the binding energy of these atomic electrons, the complexity of the partial wave expansion can be avoided, and one can use a semi-classical approach in which the motion of the nuclei is described by using classical trajectories, while the motion of the atomic electrons is treated by using quantum mechanics. At still higher reduced laboratory energies $E_L > 1$ keV/a.m.u., the motion of the nuclei can be taken to be along straight-line trajectories, using the *impact parameter method*. This is the case we shall consider in what follows. It is worth remembering that the condition $E_L > 1$ keV/a.m.u. does not imply that the collision is fast, as we noticed in the discussion following (14.46).

For simplicity, we shall consider a system containing one active electron, which is initially bound by an effective potential $V_B(r_B)$ to an inert ion core B. The incident ion A interacts with the active electron through an effective potential $V_A(r_A)$. The ion A will be taken to be inert. It will be assumed that there is no overall long-range Coulomb interaction either in the *direct channels* A + (B + e$^-$) or in the *rearranged channels* (A + e$^-$) + B. The two ion cores interact through a potential $V_{AB}(R)$. Examples of such systems are H^+ + H, H^+ + Li(2s), Li^+ + Na(3s), etc. where the closed inner shells of Li and Na are taken to be inert and to exert an effective potential on the valence electrons. It will be assumed that both A and the initial state of (B + e$^-$) have zero angular momentum, so that the system is axially symmetrical about the direction of incidence, taken as the Z axis. All these restrictions can be lifted straightforwardly, but at the expense of notational complexity.

A system of coordinates is now introduced, in which the mid-point O of the line AB is taken to be the origin and to be at rest, and the ion cores A and B follow classical undeflected straight-line paths, moving parallel to the Z axis, as illustrated in Fig. 14.7. If the relative velocity of A and B is $\mathbf{v}$, the internuclear position vector $\mathbf{R}(t)$ is given by

$$\mathbf{R}(t) = \mathbf{b} + \mathbf{v}t \tag{14.52}$$

where the impact parameter vector $\mathbf{b}$ is taken to be parallel to the X axis. The time $t = 0$ is chosen to be the time of closest approach, when $R = b$. To discuss the motion of the active electron, the coordinate system shown in Fig. 14.8 will be used, which is similar to the one employed in Chapter 10 to describe the hydrogen molecular ion H_2^+. The time-dependent Schrödinger equation for the electron is then

$$\mathrm{i}\frac{\partial}{\partial t}\Psi(\mathbf{r}, t) = H_e(t)\Psi(\mathbf{r}, t) \tag{14.53a}$$

where

$$H_e(t) = \left[-\frac{1}{2}\nabla_{\mathbf{r}}^2 + V_A(r_A) + V_B(r_B) + V_{AB}(R)\right] \tag{14.53b}$$

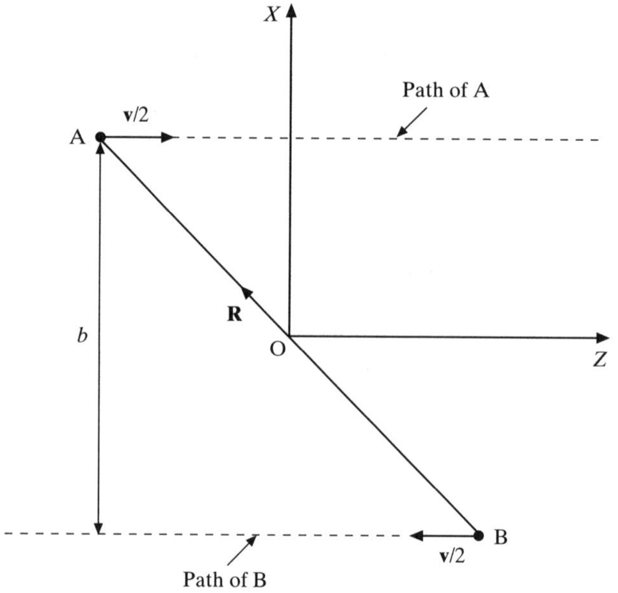

Figure 14.7 System of two ions A and B moving along rectilinear paths parallel to the Z axis with a relative velocity $\mathbf{v}$. The mid-point of AB is chosen as the origin O of the coordinate system. Relative to O, A has the velocity $\mathbf{v}/2$ and B has the velocity $-\mathbf{v}/2$. The ions A and B are taken to move in the XZ plane so that $\mathbf{R}$ rotates about the Y axis.

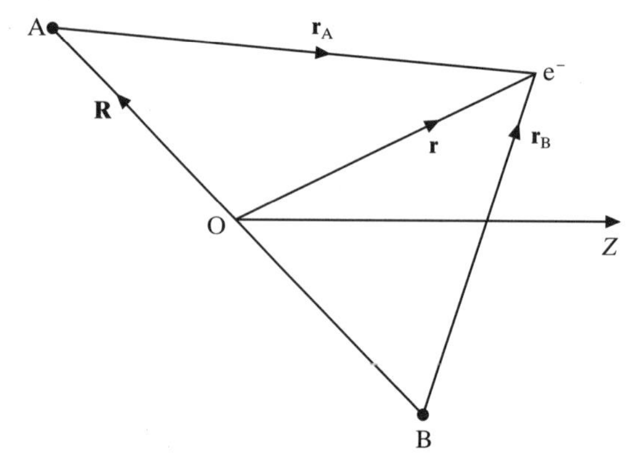

Figure 14.8 A coordinate system for the discussion of the motion of an electron in the field of two ions A and B.

In (14.53a) the derivative with respect to the time is taken with $\mathbf{r}$ held fixed in the laboratory frame of reference (see Fig. 14.8). As seen from Fig. 14.8, we have

$$\mathbf{r}_A = \mathbf{r} - \mathbf{R}/2 \tag{14.54a}$$

and

$$\mathbf{r}_B = \mathbf{r} + \mathbf{R}/2 \tag{14.54b}$$

It follows from (14.52) and (14.54) that the interaction potentials $V_A(r_A)$, $V_B(r_B)$ and $V_{AB}(R)$ are all time-dependent.

In the limits $t \to \pm\infty$ corresponding to $R \to \infty$, the active electron is bound either to A or to B. The asymptotic solutions of (14.53) can be expressed as linear combinations of the functions

$$X_k^A(\mathbf{r}, t) = \psi_k^A(\mathbf{r}_A) \exp(-iE_k^A t + i\mathbf{v}\cdot\mathbf{r}/2 - iv^2 t/8) \quad \textbf{(14.55a)}$$

for $R \to \infty$ and $r_A \ll r_B$, and of the functions

$$X_j^B(\mathbf{r}, t) = \psi_j^B(\mathbf{r}_B) \exp(-iE_j^B t - i\mathbf{v}\cdot\mathbf{r}/2 - iv^2 t/8) \quad \textbf{(14.55b)}$$

for $R \to \infty$ and $r_B \ll r_A$. In the above expressions $\psi_k^A(\mathbf{r}_A)$ and $\psi_j^B(\mathbf{r}_B)$ are the eigenfunctions (normalised to unity) of the atoms $(A + e^-)$ and $(B + e^-)$, respectively, and E_k^A, E_j^B are the corresponding eigenenergies. Thus

$$\left[-\frac{1}{2}\nabla_{\mathbf{r}_A}^2 + V_A(\mathbf{r}_A) - E_k^A\right]\psi_k^A(\mathbf{r}_A) = 0 \quad \textbf{(14.56a)}$$

and

$$\left[-\frac{1}{2}\nabla_{\mathbf{r}_B}^2 + V_B(\mathbf{r}_B) - E_j^B\right]\psi_k^B(\mathbf{r}_B) = 0 \quad \textbf{(14.56b)}$$

It is easily verified (Problem 14.2) that X_k^A and X_j^B satisfy the Schrödinger equation (14.53) when $R \to \infty$ and either $r_A \ll r_B$ or $r_B \ll r_A$. The phase factors $\exp(\pm i\mathbf{v}\cdot\mathbf{r}/2 - iv^2 t/8)$, called *translational factors*, express the fact that the electron shares the translational motion of the core, either A or B, to which it is bound. Since A has a velocity $\mathbf{v}/2$ parallel to the Z axis, the electron when bound to A is moving with a velocity $\mathbf{v}/2$ with respect to the origin O, and has a translational kinetic energy $v^2/8$. In the same way, when it is bound to B, the electron has a velocity $-\mathbf{v}/2$ with respect to the origin O, and a translational energy $v^2/8$. Translational factors were originally introduced in 1958 by D.R. Bates and R. McCarroll. If they are omitted, the boundary conditions cannot be satisfied exactly and the Galilean invariance of the theory is destroyed. However, if $\mathbf{v}$ is small enough, the translational factors can be replaced by unity to a good approximation, as in the PSS method.

To find solutions of the time-dependent Schrödinger equation (14.53), the wave function $\Psi(\mathbf{r}, t)$ can be approximated by $\Psi_a(\mathbf{r}, t)$, where

$$\Psi_a(\mathbf{r}, t) = \sum_{k=1}^{N} c_k^A(b, t) F_k^A(\mathbf{r}, t) + \sum_{j=1}^{M} c_j^B(b, t) F_j^B(\mathbf{r}, t) \quad \textbf{(14.57)}$$

where F_k^A and F_j^B are linearly independent basis functions which are required to have the asymptotic forms X_k^A and X_j^B, respectively, when $|t| \to \infty$. That is,

$$F_k^A(\mathbf{r}, t) \to X_k^A(\mathbf{r}, t), \qquad R \to \infty, r_A \ll r_B \quad \textbf{(14.58a)}$$

$$F_j^B(\mathbf{r}, t) \to X_j^B(\mathbf{r}, t), \qquad R \to \infty, r_B \ll r_A \quad \textbf{(14.58b)}$$

At low velocities, for a heteronuclear system, a frequent choice is to express the functions F_k^{A} and F_j^{B} in terms of molecular [6] orbitals $\Phi_q^{\mathrm{A,B}}(\mathbf{R}, \mathbf{r})$ satisfying the equation (see (10.12))

$$H_e\Phi_q^{\mathrm{A,B}}(\mathbf{R}, \mathbf{r}) = E_q^{\mathrm{A,B}}(R)\Phi_q^{\mathrm{A,B}}(\mathbf{R}, \mathbf{r}) \qquad \textbf{(14.59)}$$

The superscripts A and B have been added to $\Phi_q(\mathbf{R}, \mathbf{r})$ and $E_q(R)$ to indicate to which of the arrangements $(\mathrm{A} + \mathrm{e}^-)$ or $(\mathrm{B} + \mathrm{e}^-)$ the orbital q separates. Then $F_k^{\mathrm{A}}(\mathbf{r}, t)$ and $F_j^{\mathrm{B}}(\mathbf{r}, t)$ can be written as

$$F_k^{\mathrm{A}}(\mathbf{r}, t) = \Phi_k^{\mathrm{A}}(\mathbf{R}, \mathbf{r}) \exp(-\mathrm{i}E_k^{\mathrm{A}}t + \mathrm{i}\mathbf{v}\cdot\mathbf{r}/2 - \mathrm{i}v^2t/8) \qquad \textbf{(14.60a)}$$

and

$$F_j^{\mathrm{B}}(\mathbf{r}, t) = \Phi_j^{\mathrm{B}}(\mathbf{R}, \mathbf{r}) \exp(-\mathrm{i}E_j^{\mathrm{B}}t - \mathrm{i}\mathbf{v}\cdot\mathbf{r}/2 - \mathrm{i}v^2t/8) \qquad \textbf{(14.60b)}$$

where $\Phi_k^{\mathrm{A}}(\mathbf{R}, \mathbf{r}) \to \psi_k^{\mathrm{A}}(\mathbf{r}_{\mathrm{A}})$ and $\Phi_j^{\mathrm{B}}(\mathbf{R}, \mathbf{r}) \to \psi_j^{\mathrm{B}}(\mathbf{r}_{\mathrm{B}})$ as $R \to \infty$.

At higher velocities, F_k^{A} and F_j^{B} can be set equal to X_k^{A} and X_j^{B} respectively, thus forming a two-centre atomic orbital basis. As in the case of electron scattering, to obtain the correct long-range potentials, and to calculate ionisation cross-sections, the continuum must be represented in the two-centre atomic expansions. This can be achieved by including pseudo-states centred on A or B, or on both centres. Both for low and high energies other choices can be made for the basis functions; the book by Bransden and McDowell (1992) can be consulted for details of these. From (14.57), (14.54) and (14.55) it is seen that in the limit $|t| \to \infty$ the coefficients $c_k^{\mathrm{A}}(b, t)$ and $c_j^{\mathrm{B}}(b, t)$ are probability amplitudes for finding the system to be in the state k of the atom $(\mathrm{A} + \mathrm{e}^-)$ or in the state j of the atom $(\mathrm{B} + \mathrm{e}^-)$, respectively.

If in the initial channel the ion A is incident on the atom $(\mathrm{B} + \mathrm{e}^-)$ in the state i, the boundary conditions are

$$\lim_{t\to-\infty} c_k^{\mathrm{A}}(b, t) = 0, \qquad \lim_{t\to-\infty} c_j^{\mathrm{B}}(b, t) = \delta_{ij} \qquad \textbf{(14.61)}$$

After the collision, the probability that the atom $(\mathrm{B} + \mathrm{e}^-)$ has been excited to the state j is

$$P_j^{\mathrm{B}}(b) = \lim_{t\to\infty} |c_j^{\mathrm{B}}(b, t)|^2 \qquad \textbf{(14.62)}$$

and the probability that the electron has been captured into the state k of the atom $(\mathrm{A} + \mathrm{e}^-)$ is

$$P_k^{\mathrm{A}}(b) = \lim_{t\to\infty} |c_k^{\mathrm{A}}(b, t)|^2 \qquad \textbf{(14.63)}$$

The corresponding total cross-sections are found by integrating over all impact parameters. Keeping in mind that the system is axially symmetric, we have

$$\sigma_j^{\mathrm{B}}(v) = 2\pi \int_0^\infty P_j^{\mathrm{B}}(b)\, b\, \mathrm{d}b \qquad \textbf{(14.64)}$$

[6] For a homonuclear system a combination of molecular orbitals must be employed as will be seen later.

and

$$\sigma_k^{\mathrm{A}}(v) = 2\pi \int_0^{\infty} P_k^{\mathrm{A}}(b)\, b\, \mathrm{d}b \qquad \textbf{(14.65)}$$

The coupled equations

It remains to find equations from which to calculate the coefficients $c_k^{\mathrm{A}}(b, t)$ and $c_j^{\mathrm{B}}(b, t)$. To do this we require that

$$\int \mathrm{d}\mathbf{r}\, [F_k^{\mathrm{A}}(\mathbf{r}, t)]^* \left[\mathrm{i}\frac{\partial}{\partial t} - H_{\mathrm{e}}(t) \right] \Psi_{\mathrm{a}}(\mathbf{r}, t) = 0, \qquad k = 1, 2, \ldots, N \qquad \textbf{(14.66a)}$$

and

$$\int \mathrm{d}\mathbf{r} [F_j^{\mathrm{B}}(\mathbf{r}, t)]^* \left[\mathrm{i}\frac{\partial}{\partial t} - H_{\mathrm{e}}(t) \right] \Psi_{\mathrm{a}}(\mathbf{r}, t) = 0, \qquad j = 1, 2, \ldots, M \qquad \textbf{(14.66b)}$$

Inserting the wave function $\Psi_{\mathrm{a}}(\mathbf{r}, t)$ given by (14.57) into (14.66), a set of $(N + M)$ coupled first-order differential equations is obtained which can be written in the form

$$\begin{aligned} \mathrm{i}(\mathbf{S}_{\mathrm{AA}}\, \dot{\mathbf{c}}^{\mathrm{A}} + \mathbf{S}_{\mathrm{AB}}\, \dot{\mathbf{c}}^{\mathrm{B}}) &= \mathbf{M}_{\mathrm{AA}}\, \mathbf{c}^{\mathrm{A}} + \mathbf{M}_{\mathrm{AB}}\, \mathbf{c}^{\mathrm{B}} \\ \mathrm{i}(\mathbf{S}_{\mathrm{BA}}\, \dot{\mathbf{c}}^{\mathrm{A}} + \mathbf{S}_{\mathrm{BB}}\, \dot{\mathbf{c}}^{\mathrm{B}}) &= \mathbf{M}_{\mathrm{BA}}\, \mathbf{c}^{\mathrm{A}} + \mathbf{M}_{\mathrm{BB}}\, \mathbf{c}^{\mathrm{B}} \end{aligned} \qquad \textbf{(14.67)}$$

where $\mathbf{c}^{\mathrm{A}}$ is an N-dimensional column vector with elements $c_k^{\mathrm{A}}(b, t)$ and $\mathbf{c}^{\mathrm{B}}$ is an M-dimensional column vector with elements $c_j^{\mathrm{B}}(b, t)$. The matrices $\mathbf{S}_{\mathrm{AA}}$, $\mathbf{S}_{\mathrm{AB}}$, $\mathbf{S}_{\mathrm{BB}}$ and $\mathbf{S}_{\mathrm{BA}}$ are defined so that

$$\begin{aligned} (\mathbf{S}_{\mathrm{AA}})_{kk'} &= \langle F_k^{\mathrm{A}} | F_{k'}^{\mathrm{A}} \rangle, \qquad (\mathbf{S}_{\mathrm{AB}})_{kj} = \langle F_k^{\mathrm{A}} | F_j^{\mathrm{B}} \rangle \\ (\mathbf{S}_{\mathrm{BB}})_{jj'} &= \langle F_j^{\mathrm{B}} | F_{j'}^{\mathrm{B}} \rangle, \qquad (\mathbf{S}_{\mathrm{BA}})_{jk} = \langle F_j^{\mathrm{B}} | F_k^{\mathrm{A}} \rangle \end{aligned} \qquad \textbf{(14.68)}$$

where the symbol $\langle Y | Z \rangle$ denotes the integral

$$\langle Y | Z \rangle = \int Y^*(\mathbf{r}, t) Z(\mathbf{r}, t)\, \mathrm{d}\mathbf{r} \qquad \textbf{(14.69)}$$

With the basis functions (14.60) using molecular orbitals, and also with the two-centre atomic orbital basis functions, the matrices $\mathbf{S}_{\mathrm{AA}}$ and $\mathbf{S}_{\mathrm{BB}}$ are diagonal. However, because of the translational factors the matrices $\mathbf{S}_{\mathrm{AB}}$ and $\mathbf{S}_{\mathrm{BA}}$, known as *overlap matrices*, are not diagonal, but vanish as $|t| \to \infty$. The matrices $\mathbf{M}_{\mathrm{AA}}$, $\mathbf{M}_{\mathrm{AB}}$, $\mathbf{M}_{\mathrm{BB}}$, $\mathbf{M}_{\mathrm{BA}}$ are defined so that

$$\begin{aligned} (\mathbf{M}_{\mathrm{AA}})_{kk'} &= \left\langle F_k^{\mathrm{A}} \middle| H_{\mathrm{e}}(t) - \mathrm{i}\frac{\partial}{\partial t} \middle| F_{k'}^{\mathrm{A}} \right\rangle, \qquad (\mathbf{M}_{\mathrm{AB}})_{kj} = \left\langle F_k^{\mathrm{A}} \middle| H_{\mathrm{e}}(t) - \mathrm{i}\frac{\partial}{\partial t} \middle| F_j^{\mathrm{B}} \right\rangle \\ (\mathbf{M}_{\mathrm{BB}})_{jj'} &= \left\langle F_j^{\mathrm{B}} \middle| H_{\mathrm{e}}(t) - \mathrm{i}\frac{\partial}{\partial t} \middle| F_{j'}^{\mathrm{B}} \right\rangle, \qquad (\mathbf{M}_{\mathrm{BA}})_{jk} = \left\langle F_j^{\mathrm{B}} \middle| H_{\mathrm{e}}(t) - \mathrm{i}\frac{\partial}{\partial t} \middle| F_k^{\mathrm{A}} \right\rangle \end{aligned} \qquad \textbf{(14.70)}$$

Despite their somewhat complicated appearance all the matrices **S** and **M** can be calculated as functions of t. It should be noted that the matrix elements of $\mathbf{S}_{AB}$, $\mathbf{S}_{BA}$, $\mathbf{M}_{AB}$ and $\mathbf{M}_{BA}$ which govern charge exchange all contain the functions $\exp(\pm i\mathbf{v}\cdot\mathbf{r})$ arising from the translational factors. As a result, they decrease much more rapidly with increasing velocity than the matrix elements of $\mathbf{S}_{AA}$, $\mathbf{S}_{BB}$, $\mathbf{M}_{AA}$ and $\mathbf{M}_{BB}$ controlling elastic scattering, excitation or ionisation, where the translational factors are absent. This accounts for the predominance of elastic scattering, excitation and ionisation over charge exchange at high energies.

The coupled equations can be solved using a step by step algorithm starting from $t = -t_0$, where t_0 is positive and large enough for all the **M** matrices to be set equal to zero, and by setting $c_k^A(b, -t_0) = 0$, and $c_j^B(b, -t_0) = \delta_{ij}$. The integration stops at $t = t_0$ to provide the values of $c_k^A(b, t_0)$ and $c_j^B(b, t_0)$ from which the cross-sections can be evaluated using the equations (14.62)–(14.65). With the specified boundary conditions the wave function $\Psi_a(\mathbf{r}, t)$ is normalised to unity before the collision, as $t \to -\infty$. It is easy to show (see Problem 14.3) that as $t \to \infty$ the wave function remains normalised to unity so that the coupled channel method conserves probability and hence is unitary.

The general formulation of atom–atom collisions given here can be modified in various ways. For instance at very low energies ($E_L \leqslant 1$ keV/ a.m.u.), the ion cores A and B do not follow straight-line paths. This can be taken into account by using classical trajectories calculated from an effective atom–atom potential describing the relative motion of the atoms. Another modification arises from the expectation that at low energies the wave function should be molecular in character, so that at small separations the electron should be 'shared' equally by A and B. However, the functions $\exp(\pm i\mathbf{v}\cdot\mathbf{r}/2)$ associate the electron with one centre or the other. If the electron is associated with A the + sign is used, and if it is associated with B, the – sign is employed (see (14.60)). This may be overcome by using more general expressions $\exp[if(\mathbf{R}, \mathbf{r})]$ where $f(\mathbf{R}, \mathbf{r})$ is a *switching function* such that $f \to 0$ when $R \to 0$ and $f \to \mathbf{v}\cdot\mathbf{r}/2$ if $r_A \ll r_B$ and $R \to \infty$, while $f \to -\mathbf{v}\cdot\mathbf{r}/2$ if $r_B \ll r_A$ and $R \to \infty$. The switching functions, introduced in 1969 by S.M. Schneiderman and A. Russek, can be either chosen on the grounds of physical reasonableness or optimised by using the variational method. Many forms have been suggested, of which perhaps the simplest is

$$f(\mathbf{R}, \mathbf{r}) = \frac{R^2 \cos\theta}{R^2 + R_0^2}\frac{\mathbf{v}\cdot\mathbf{r}}{2} \tag{14.71}$$

where θ is the angle between **R** and **r**. When $r_A \ll r_B$ and $R \to \infty$, $\cos\theta = 1$ and $f(\mathbf{R}, \mathbf{r}) \to (\mathbf{v}\cdot\mathbf{r})/2$. On the other hand, when $r_B \ll r_A$ and $R \to \infty$, $\cos\theta = -1$ and $f(\mathbf{R}, \mathbf{r}) \to -(\mathbf{v}\cdot\mathbf{r})/2$ so that the function $\exp[if(\mathbf{R}, \mathbf{r})]$ has the correct form in the asymptotic region. The cut-off parameter R_0 is introduced so that $f(\mathbf{R}, \mathbf{r}) \to 0$ as $R \to 0$. The functions $F_k^A(\mathbf{r}, t)$ and $F_j^A(\mathbf{r}, t)$ are defined to have the same translational factor so that

$$F_k^A(\mathbf{r}, t) = \Phi_k^A(\mathbf{R}, \mathbf{r}) \exp[-iE_k^A t + if(\mathbf{R}, \mathbf{r}) - iv^2 t/8] \tag{14.72a}$$

and

$$F_j^{\mathrm{B}}(\mathbf{r}, t) = \Phi_j^{\mathrm{B}}(\mathbf{R}, \mathbf{r}) \exp[-\mathrm{i}E_j^{\mathrm{B}}t + \mathrm{i}f(\mathbf{R}, \mathbf{r}) - \mathrm{i}v^2t/8] \tag{14.72b}$$

which leads to considerable simplification. Unfortunately at higher energies ($E_{\mathrm{L}} \geqslant 25$ keV/a.m.u.) the use of a common translational factor is less successful because as the matrix elements of $\mathbf{S}_{\mathrm{AB}}$, $\mathbf{S}_{\mathrm{BA}}$, $\mathbf{M}_{\mathrm{AB}}$ and $\mathbf{M}_{\mathrm{BA}}$ do not then contain the factors $\exp(\pm\mathrm{i}\mathbf{v}\cdot\mathbf{r})$, the calculated charge exchange cross-section does not decrease rapidly at high velocities as it should.

The PSS method in the impact parameter approximation

To understand some of the properties of the expansion in terms of molecular orbitals, it is useful to look at the PSS method in which the translational factors are replaced by unity. This is a reasonable approximation at low velocities, for which $E_{\mathrm{L}} < 1$ keV/a.m.u. At such velocities the motion of the ion cores A and B is still classical, although to obtain accurate results the deviation from a straight-line trajectory should be taken into account. Omitting the superscripts A and B on Φ_j and $E_j(R)$, the wave function is expanded as

$$\Psi_{\mathrm{a}}(\mathbf{r}, t) = \sum_{j=1}^{N+M} c_j(b, t)\Phi_j(\mathbf{R}, \mathbf{r}) \tag{14.73}$$

where $c_j(b, t)$ is the probability amplitude for finding the system in the molecular state j. Then, corresponding to the coupled equations (14.67), we find that

$$\sum_{j=1}^{N+M} \int \mathrm{d}\mathbf{r}\, \Phi_k^*(\mathbf{R}, \mathbf{r}) \left[\mathrm{i}\frac{\partial}{\partial t} - H_{\mathrm{e}}(t)\right] \Phi_j(\mathbf{R}, \mathbf{r}) c_j(b, t) = 0 \tag{14.74}$$

so that

$$\mathrm{i}\dot{c}_k(b, t) = \sum_{j=1}^{N+M} \left\langle \Phi_k \left| -\mathrm{i}\frac{\partial}{\partial t} \right| \Phi_j \right\rangle c_j(b, t) + E_k(R)c_k(b, t) \tag{14.75}$$

where the equation (14.59) has been used together with the fact that the orbitals Φ_j form an orthonormal set, $\langle \Phi_k | \Phi_j \rangle = \delta_{kj}$.

By performing the phase transformation

$$c_k(b, t) = d_k(b, t) \exp\left[-\mathrm{i}\int_{-t_0}^{t} E_k(R)\, \mathrm{d}t\right] \tag{14.76}$$

we can rewrite the equations (14.75) in the form

$$\mathrm{i}\dot{d}_k(b, t) = \sum_{j=1}^{N+M} \left\langle \Phi_k \left| -\mathrm{i}\frac{\partial}{\partial t} \right| \Phi_j \right\rangle \exp\left[\mathrm{i}\int_{-t_0}^{t} [E_k(R) - E_j(R)]\, \mathrm{d}t\right] d_j(b, t) \tag{14.77}$$

If the system is initially in the molecular state i, these equations can be solved with the boundary conditions $d_k(b, -t_0) = \delta_{ik}$ to find the amplitude $d_k(b, t_0)$ and since $|d_k| = |c_k|$, the cross-sections can be found as before.

Looking at (14.77) it is clear that if E_j is not close to E_k over some interval of R, the exponential factor will oscillate rapidly over the range $-t_0 < t < t_0$ and the coupling between the states j and k will be weak. At low velocities only the coupling between states for which E_j and E_k cross for some value of R, or approach closely, is important, as we have seen before.

Another important feature of the equations (14.77) concerns the calculation of the coupling matrix element $\langle \Phi_k | \partial/\partial t | \Phi_j \rangle$. The molecular orbitals are usually expressed in terms of a body-fixed system, and are denoted by $\Phi_q(\mathbf{R}, \mathbf{r})$, as discussed in Chapter 10. However, the operation $\partial/\partial t$ must be performed with $\mathbf{r}$ held constant in the space-fixed system of coordinates. If Θ is the angle that the vector $\mathbf{R}$ makes with the space-fixed Z axis, the Cartesian coordinates (x, y, z) of $\mathbf{r}$ in the space-fixed frame are connected with those in the body-fixed frame $(\bar{x}, \bar{y}, \bar{z})$ by a rotation through the angle Θ about the Y axis, so that

$$\begin{aligned} \bar{x} &= x \cos \Theta - z \sin \Theta, \\ \bar{y} &= y, \\ \bar{z} &= x \sin \Theta + z \cos \Theta \end{aligned} \tag{14.78}$$

From (14.78), it is seen that

$$\begin{aligned} \frac{\partial}{\partial t} \Phi_j(\mathbf{R}; \bar{x}, \bar{y}, \bar{z}) &= \dot{R} \frac{\partial}{\partial R} \Phi_j(\mathbf{R}; \bar{x}, \bar{y}, \bar{z}) + \dot{\Theta} \left(\bar{x} \frac{\partial}{\partial \bar{z}} - \bar{z} \frac{\partial}{\partial \bar{x}} \right) \Phi_j(\mathbf{R}; \bar{x}, \bar{y}, \bar{z}) \\ &= \left(\dot{R} \frac{\partial}{\partial R} - \mathrm{i} \dot{\Theta} L_{\bar{y}} \right) \Phi_j(\mathbf{R}; \bar{x}, \bar{y}, \bar{z}) \end{aligned} \tag{14.79}$$

The coupling element contains two terms, namely

$$\left\langle \Phi_k \left| \frac{\partial}{\partial t} \right| \Phi_j \right\rangle = \dot{R} \left\langle \Phi_k \left| \frac{\partial}{\partial R} \right| \Phi_j \right\rangle - \mathrm{i} \dot{\Theta} \langle \Phi_k | L_{\bar{y}} | \Phi_j \rangle \tag{14.80}$$

where $\dot{R} = v^2 t/R$ and $\dot{\Theta} = -bv/R^2$ for a straight-line trajectory. The first term is called *radial coupling* and only connects states with the same value of λ, the projection of the electronic orbital angular momentum on the internuclear $\bar{Z}$ axis, while the second term is called *rotational* or *Coriolis coupling* and connects states for which $\lambda_j - \lambda_k = \pm 1$. It should be noted that by making a unitary transformation of the adiabatic basis functions Φ_j, a new basis of functions, Φ_k^{D}, can be formed in which the radial coupling is zero. This new basis is called a *diabatic basis*. It overcomes the difficulty of calculating the radial coupling near an avoided crossing or crossing, where the matrix elements of $\partial/\partial R$ vary rapidly.

Applications to homonuclear systems

So far we have assumed that each atomic wave function in the region of large R correlates to a single molecular orbital. For homonuclear systems, as we saw in Chapter 10, this is not correct and two orbitals of different symmetry correlate to each atomic wave function. Let us consider the simplest case which is the system $H^+ + H(1s)$. At very low energies only the elastic scattering channel

$$H_A^+ + H_B(1s) \rightarrow H_A^+ + H_B(1s) \tag{14.81}$$

and the *symmetrical resonance* channel

$$H_A^+ + H_B(1s) \rightarrow H_A(1s) + H_B^+ \tag{14.82}$$

are of importance. The subscripts A and B label the two protons, which can be treated as distinguishable in the impact parameter approximation, where the protons follow classical trajectories.

The two orbitals of lowest energy for the H_2^+ molecular ion are the $\sigma_g 1s$ and $\sigma_u^* 1s$, with wave functions and energies $\Phi_g(\mathbf{R}, \mathbf{r})$, $E_g(R)$ and $\Phi_u(\mathbf{R}, \mathbf{r})$, $E_u(R)$ respectively. At large R (see (10.33)) the combination $2^{-1/2}(\Phi_g + \Phi_u)$ is asymptotic to a situation in which the electron is attached to A and the combination $2^{-1/2}(\Phi_g - \Phi_u)$ to one in which the electron is attached to B. If the collision velocity is so low that the momentum of the electron relative to the origin can be ignored and the translational factors replaced by unity, the wave function $\Psi_a(\mathbf{r}, t)$ can be written in the form of the two-state expansion

$$\Psi_a(\mathbf{r}, t) = \frac{1}{\sqrt{2}}[\Phi_g(\mathbf{R}, \mathbf{r}) + \Phi_u(\mathbf{R}, \mathbf{r})]c^A(b, t) + \frac{1}{\sqrt{2}}[\Phi_g(\mathbf{R}, \mathbf{r}) - \Phi_u(\mathbf{R}, \mathbf{r})]c^B(b, t) \tag{14.83}$$

If the electron is initially attached to B, the boundary conditions are

$$\lim_{t\to-\infty} c^A(b, t) = 0, \qquad \lim_{t\to-\infty} c^B(b, t) = 1 \tag{14.84}$$

Let us rewrite the wave function (14.83) as

$$\Psi_a(\mathbf{r}, t) = A^+(b, t)\Phi_g(\mathbf{R}, \mathbf{r}) + A^-(b, t)\Phi_u(\mathbf{R}, \mathbf{r}) \tag{14.85}$$

where

$$A^\pm(b, t) = \frac{1}{\sqrt{2}}[c^A(b, t) \pm c^B(b, t)] \tag{14.86}$$

In order to satisfy the boundary conditions (14.84), we must have

$$\lim_{t\to-\infty} A^\pm(b, t) = \pm\frac{1}{\sqrt{2}} \tag{14.87}$$

The orbitals $\Phi_g(\mathbf{R}, \mathbf{r})$ and $\Phi_u(\mathbf{R}, \mathbf{r})$ are of opposite parity under the reflection $\mathbf{r} \to -\mathbf{r}$. This has the consequences that the coupling matrix element $\langle \Phi_u | \partial/\partial t | \Phi_g \rangle$ vanishes and that the coefficients $A^{\pm}(b, t)$ satisfy the uncoupled equations

$$\mathrm{i}\dot{A}^{+}(b, t) = E_g(R)\, A^{+}(b, t) \tag{14.88a}$$

and

$$\mathrm{i}\dot{A}^{-}(b, t) = E_u(R)\, A^{-}(b, t) \tag{14.88b}$$

The solutions of these equations satisfying the boundary conditions (14.87) are

$$A^{+}(b, t) = \frac{1}{\sqrt{2}} \exp\left[-\mathrm{i} \int_{-\infty}^{t} E_g(R)\, \mathrm{d}t' \right] \tag{14.89a}$$

and

$$A^{-}(b, t) = -\frac{1}{\sqrt{2}} \exp\left[-\mathrm{i} \int_{-\infty}^{t} E_u(R)\, \mathrm{d}t' \right] \tag{14.89b}$$

with $R(t') = (b^2 + v^2 t'^2)^{1/2}$ for a straight-line trajectory.

The probability of charge exchange $P_{ce}(b)$ is then given by

$$\begin{aligned} P_{ce}(b) &= \lim_{t \to \infty} |c^{A}(b, t)|^2 \\ &= \lim_{t \to \infty} \left| \frac{1}{\sqrt{2}} [A^{+}(b, t) + A^{-}(b, t)] \right|^2 \\ &= \sin^2 \left\{ \frac{1}{2} \int_{-\infty}^{\infty} [E_g(R) - E_u(R)]\, \mathrm{d}t \right\} \end{aligned} \tag{14.90}$$

Similarly the probability for elastic scattering is found to be

$$\begin{aligned} P_{el}(b) &= \lim_{t \to \infty} |c^{B}(b, t)|^2 \\ &= \cos^2 \left\{ \frac{1}{2} \int_{-\infty}^{\infty} [E_g(R) - E_u(R)]\, \mathrm{d}t \right\} \end{aligned} \tag{14.91}$$

It is important to notice that electron capture in homonuclear systems takes place as a result of the phase difference between scattering in the gerade and ungerade states even in the absence of coupling. When the translational factors are included an extra phase ξ which is a function of v and b is added to the integrals in (14.90) and (14.91). Thus, in the two-state approximation, the probabilities $P_{el}(b)$ and $P_{ce}(b)$ oscillate between 0 and 1 as b is varied for a given collision velocity or as v is varied for a fixed value of b.

To obtain the probability of charge exchange as a function of the scattering angle the assumption of a straight-line trajectory of the protons must be relaxed. For example, the average of the potentials $E_g(R) - E_{1s}$ and $E_u(R) - E_{1s}$ can be employed to calculate the classical trajectory which determines the impact parameter b corresponding to a given scattering angle θ. If $d\sigma/d\Omega$ is the (elastic) differential scattering cross-section for the assumed potential, the elastic scattering cross-section and charge exchange cross-sections for the collision $H^+ + H(1s)$ are

$$\frac{d\sigma_{el}}{d\Omega} = \frac{d\sigma}{d\Omega} P_{el}[b(\theta)] \tag{14.92}$$

and

$$\frac{d\sigma_{ce}}{d\Omega} = \frac{d\sigma}{d\Omega} P_{ce}[b(\theta)] \tag{14.93}$$

The experiments of G.T. Lockwood and E. Everhart, performed in 1962, and those of H.F. Helby and E. Everhart, carried out in 1965, demonstrate the oscillations predicted in the charge exchange probability over a wide energy range, from 130 eV/a.m.u. to 150 keV/a.m.u., and over a range of scattering angles. This is illustrated in Fig. 14.9 where the measured charge exchange probability $P_{ce}(\theta)$ for

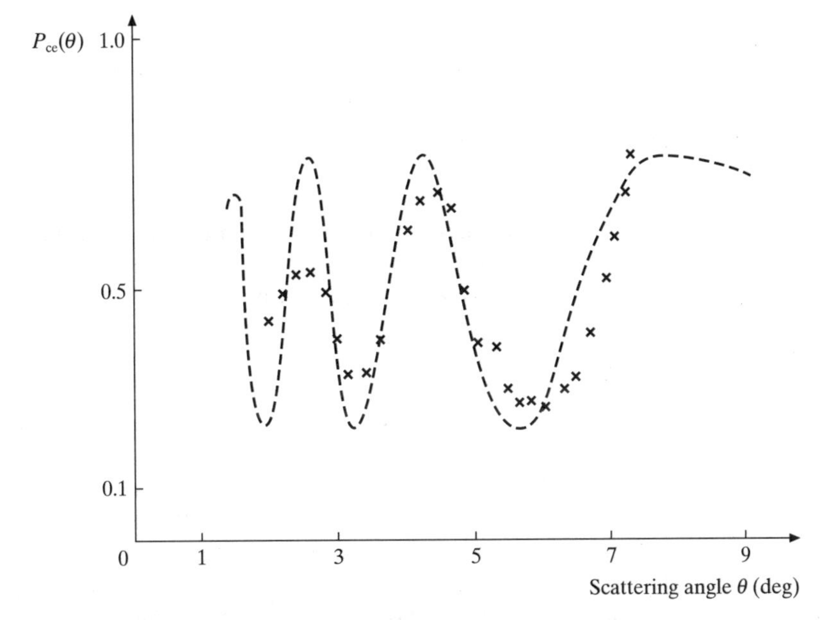

Figure 14.9 The probability of charge exchange in the collision of 250 eV protons with atomic hydrogen in the ground state, measured as a function of the scattering angle by G.T. Lockwood and E. Everhart. The dashed line shows the resuls of a coupled channel calculation in which the σ_g 1s, σ_u^* 1s and π_u 2p levels of the H_2^+ molecular ion are retained.

$E_L = 250$ eV/a.m.u. is shown for scattering angles between 1° and 9°. The maximum and minimum of the oscillations are near 0.8 and 0.2 rather than 1 and 0 as predicted by the two-state theory. This is not surprising since a glance at the correlation diagram (Fig. 10.14) shows that in the united atom limit ($R = 0$) the σ_u^*1s level is degenerate with the 2p level of He^+. In addition, the $\pi_u 2p$ level is degenerate with the 2p level of He^+ and in the small R region the rotational coupling between these two levels is very strong. At very low energies < 40 eV/a.m.u. the Coulomb repulsion between the protons prevents the system from entering the small R region and the two-state approximation is accurate. At higher energies the excitation of the $\pi_u 2p$ channel is significant. As the $\pi_u 2p$ orbital is asymptotic in the separated atom limit to 2p wave functions centred on both A and B, hydrogen atoms in the 2p level are produced by both excitation and charge exchange. Up to $E_L \simeq 1$ keV/a.m.u., only the $\pi_u 2p$ orbital needs to be included in the expansion of the wave function in addition to Φ_g and Φ_u, but as the energy increases further molecular orbitals with translational factors have to be included to obtain accurate differential and total cross-sections. Capture into the ground state remains the most important charge exchange channel up to at least 40 keV/a.m.u. Moreover, the total capture cross-section after integrating over all impact parameters is close to the predictions of the two-state approximation including translational factors up to approximately $E_L = 100$ keV/a.m.u.

As the impact energy is increased, excitation and ionisation become increasingly important. Ionisation can be treated in the molecular orbital model if orbitals asymptotic at large R to continuum states are included. Many studies of the $H^+ + H(1s)$ and other homonuclear systems have been made with molecular orbital basis sets. However, at the higher energies, it is more usual to employ the two-centre expansion into atomic states and pseudo-states. At energies above 75 keV/a.m.u. the capture cross-section decreases rapidly compared with those for excitation and ionisation which can be calculated with an expansion in hydrogen states and pseudo-states centred on the target proton B. At lower energies, to obtain rapid convergence with respect to the increase of the basis size, it is necessary to place states and pseudo-states on the projectile A when calculating excitation and ionisation in addition to those placed on B. Even if no states in the expansion are placed on the projectile, A.L. Ford, J.F. Reading and K.A. Hall have shown in 1993 that with a sufficiently large basis accurate excitation cross-sections can be obtained down to energies as low as 15 keV/a.m.u., which is a region where capture is important. In their work, up to 300 terms, both eigenstates and pseudo-states, were placed on the target, but convergence at the lower energies is slow. The flux into the positive energies' pseudo-states can be interpreted as giving the sum of the ionisation and capture cross-sections, but in the absence of states on the projectile these cannot be predicted separately. In Fig. 14.10, their calculated cross-section for the excitation process $H^+ + H(1s) \rightarrow H^+ + H(n = 2)$ is shown; it is seen to be in very good agreement with the experimental data.

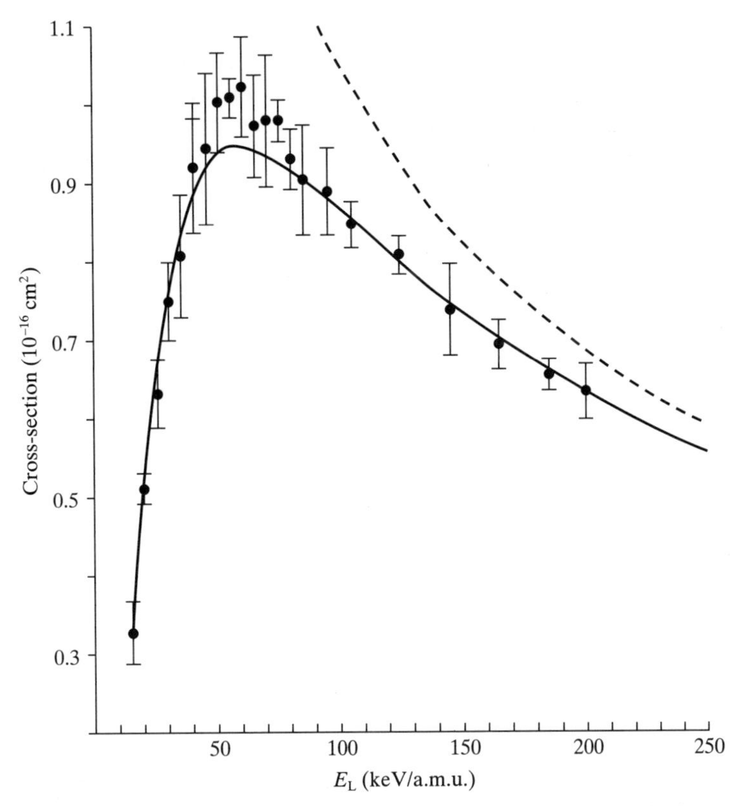

Figure 14.10 Cross-sections for excitation of the ground state of atomic hydrogen to the $n = 2$ level by proton impact:
——, calculated by A.L. Ford, J.F. Reading and K.A. Hall from a coupled channel method using a large single centred basis of eigenstates and pseudo-states on the target proton,
-----, calculated by using the first Born approximation,
●, measurements of J.T. Park, J.E. Alday, J.M. George and J.C. Peacher normalised to the coupled channel calculation of A.L. Ford *et al.* at 200 keV.

Heavy ion excitation

The theory of the $H^+ + H$ system has had an unexpected and surprising success in interpreting what at first glance would seem to be very complicated processes concerning the slow collisions of heavy ions, such as $Ne^+ + Ne$. In such collisions, it is found that excitation of an inner K shell electron, producing a vacancy, can be detected by the subsequent emission of X-rays when an outer shell electron makes a transition into the vacancy, or by the emission of 'Auger electrons'. The latter process is a radiationless mechanism in which an outer electron fills the K shell vacancy, giving up its energy to a second electron which is ejected. In 1965, U. Fano and W. Lichten realised that because of the large nuclear charge in a heavy ion, the K shell electrons could be considered as moving in the field of the two nuclei only, hardly influenced by the remaining electrons. The correlation diagram for the homonuclear system, composed of a K shell electron and the two nuclei, is the same as for H_2^+, and the system before the collision is composed of equal proportions of the $\sigma_g 1s$ and $\sigma_u^* 1s$ molecular orbitals. The $\sigma_u^* 1s$ orbital becomes degenerate with the $\pi_u 2p$ orbital in the united atom limit and at small separations the electron can be *promoted* easily from the $\sigma_u^* 1s$ to the $\pi_u 2p$ orbital. The cross-section for this excitation process can be calculated by scaling the corresponding cross-section for the proton–hydrogen atom system. This is possible because the Schrödinger equation for an electron moving in the field of two identical charges Z is

$$\mathrm{i}\frac{\partial}{\partial t}\Psi(\mathbf{r}, t) = \left[-\frac{1}{2}\nabla_{\mathbf{r}}^2 - \frac{Z}{|\mathbf{r} - \mathbf{R}/2|} - \frac{Z}{|\mathbf{r} + \mathbf{R}/2|}\right]\Psi(\mathbf{r}, t) \tag{14.94}$$

If we make the substitutions

$$\mathbf{r}' = Z\mathbf{r}, \qquad \mathbf{R}'(t') = Z\mathbf{R}(t), \qquad t' = Z^2 t \tag{14.95}$$

we regain the equation for the $H^+ + H$ system, namely

$$\mathrm{i}\frac{\partial}{\partial t'}\Psi(\mathbf{r}', t') = \left[-\frac{1}{2}\nabla_{\mathbf{r}'}^2 - \frac{1}{|\mathbf{r}' - \mathbf{R}'/2|} - \frac{1}{|\mathbf{r}' + \mathbf{R}'/2|}\right]\Psi(\mathbf{r}', t') \tag{14.96}$$

In fact all lengths scale from (14.94) as

$$l' \to Zl \tag{14.97a}$$

and the velocities scale as

$$v' \to v/Z \tag{14.97b}$$

The cross-section $\sigma(Z, v)$ for excitation of a particular state, for a system with nuclear charge Z, is related to that for excitation of the corresponding state in the $H^+ + H$ system by

$$\sigma(Z, v) = Z^{-2}\sigma(1, v/Z) \tag{14.98}$$

In fact, for the heavier systems, Z may be replaced by an effective charge to allow for the partial screening of the nuclear charge by the other electrons. In Fig. 14.11 it is shown how the measured X-ray emission probability as a function of the impact parameter is correlated with the computed $\sigma_u^* 1s \to \pi_u 2p$ transition probability, using the one-electron model.

Applications to heteronuclear systems

Many applications of the coupled channel impact parameter formulation to heteronuclear atom–atom collisions have been made. At low energy the PSS approximation has been applied successfully to many of the reactions important for understanding processes in stellar atmospheres, as well as processes in the

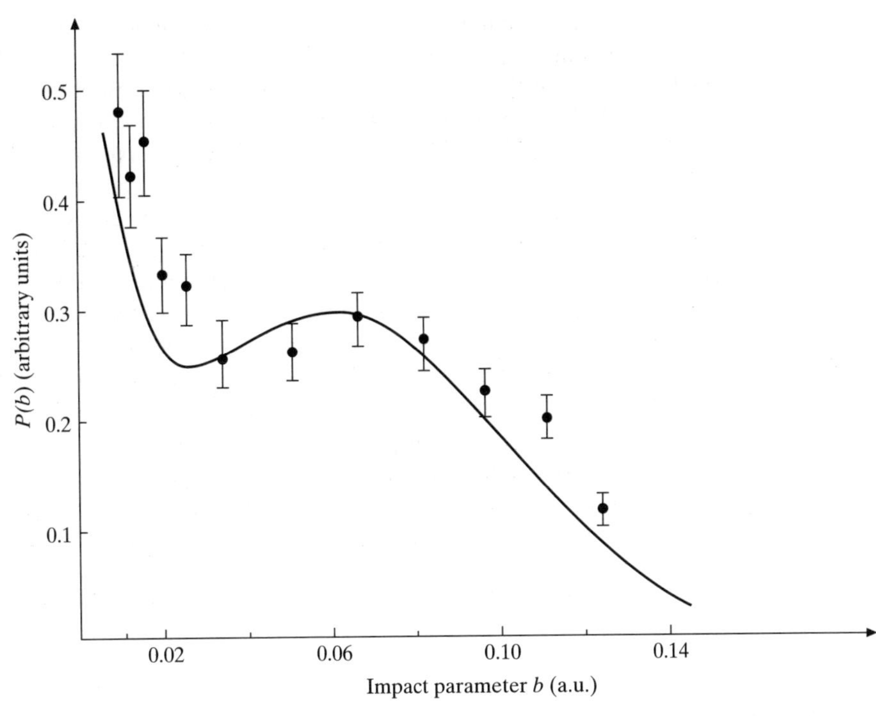

Figure 14.11 The probability of X-ray emission in an Ne^+–Ne collision at 363 keV, as a function of impact parameter.
The experimental data are due to S. Sackman, H.O. Lutz and J.S. Briggs, ●.
The solid curve shows the probability of the excitation of a K shell electron into a 2p level calculated as outlined in the text.

Earth's upper atmosphere, such as $N^{2+} + H$, $O^{4+} + H$ and so on. At high energies a very large variety of systems has been studied using basis sets, of both molecular orbital and two-centre atomic types, with translational factors of several different kinds. Other expansion sets have been employed, for example three-centre expansions in which additional pseudo-states are placed on the centre of charge with a view to representing the ionised states effectively. Direct numerical solution of the time-dependent Schrödinger equation (14.53a) has also been carried out.

In general, as we already noted, charge exchange is the main non-elastic channel at low impact energies $\leqslant 25$ keV/a.m.u., but there are significant differences between a reaction such as

$$H^+ + He^+(1s) \rightarrow H(1s) + He^{2+} \tag{14.99}$$

and one such as

$$He^{2+} + H(1s) \rightarrow He^+(n = 2) + H^+ \tag{14.100}$$

In the first case, the reaction is endothermic, the threshold energy being 1.5 a.u. The consequence of this large internal energy difference between the initial and final states is that the cross-section is small with a maximum of $\sigma_{ce} \simeq 3 \times 10^{-21}$ m^2 near $E_L = 50$ keV/a.m.u. The 1sσ orbital of (H He)$^{2+}$ is asymptotic to He$^+$(1s) and the 2pσ orbital to H(1s). The corresponding energy curves are well separated for all R, which explains the small cross-section and also accounts for the fact that a two-state approximation is not adequate even at low energies. Indeed, at least 30 states are required to obtain accurate cross-sections even at 2 keV/a.m.u. In contrast, in the He^{2+} + H(1s) reaction (14.100), the final state He$^+$(n = 2) + H$^+$ has the same internal energy as the initial state. This cross-section is much larger, rising to about 10^{-19} m^2 at $E_L = 10$ keV/a.m.u. As we noted earlier this is an example of state selectivity. It is a general characteristic of resonant or near resonant reactions, because the relevant molecular potentials approach each other in the large R region. The success of capture cross-section calculations with comparatively small basis sets is illustrated in Figs 14.12, 14.13 and 14.14 for the total cross-section of the reaction He^{2+} + H(1s) $\to$ He$^+$ + H$^+$, and the total cross-sections for electron capture into the 2s and 2p states of He$^+$. The convergence of the calculations as the size of the basis is increased has been studied by several authors using very large basis sets, for example in 1996 by K. Kuang and C.D. Lin who placed

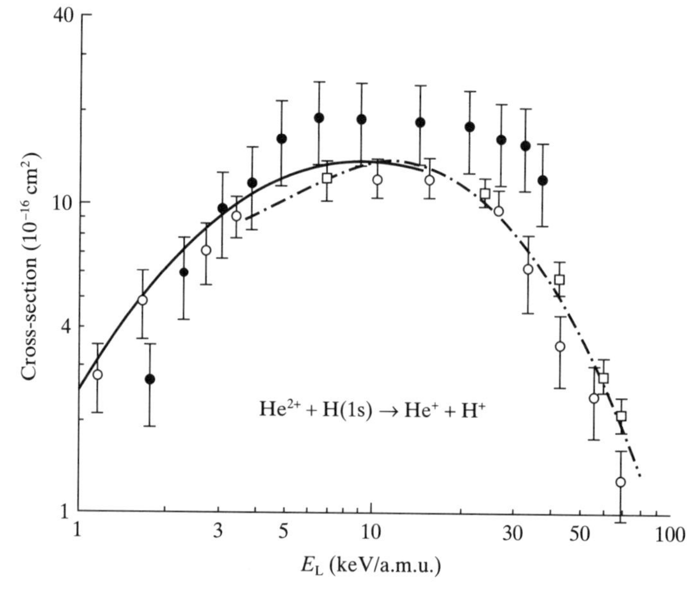

Figure 14.12 The total cross-section for electron capture by He^{2+} from H(1s) into all bound states of He$^+$.

The experimental data are due to M.B. Shah and H.B. Gilbody, o; J.E. Bayfield and G.A. Khayrallah, •; R.E. Olsen, A. Salop, R.A. Phaneuf and F.W. Meyer, □.

Calculated cross-sections from coupled channel impact parameter model: ——, ten-state molecular orbital basis due to T.G. Winter and G.J. Hatton; –·–·–, two-centre basis using atomic states with $n \leqslant 3$ on each centre due to B.H. Bransden, C.J. Noble and J. Chandler.

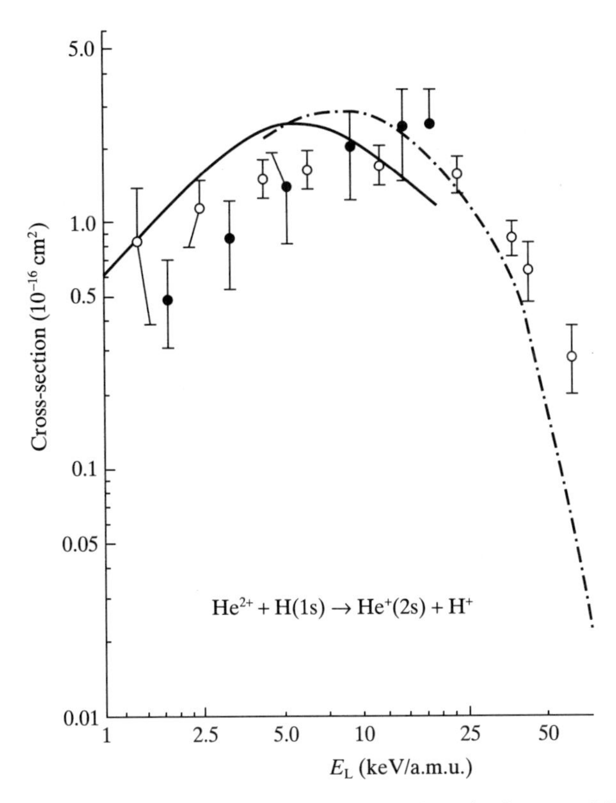

Figure 14.13 The total cross-section for electron capture into the 2s state of He^+ in collisions between He^{2+} ions and atomic hydrogen in the ground state. The notation is the same as in Fig. 14.12.

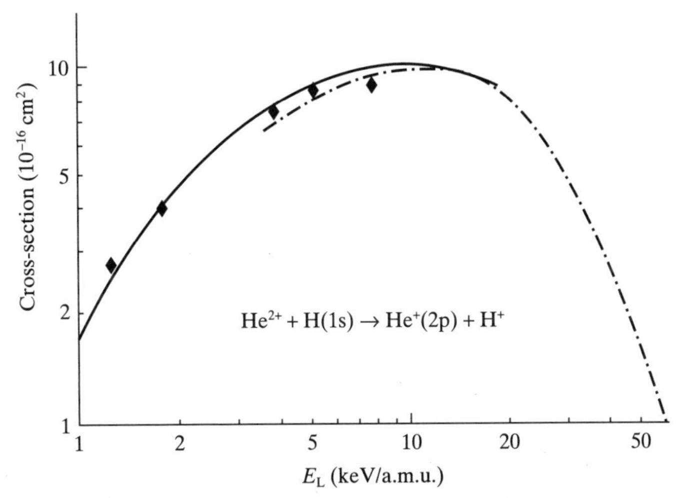

Figure 14.14 The total cross-section for electron capture into the 2p state of He^+ in collisions between He^{2+} ions and atomic hydrogen in the ground state.
The experimental data are due to D. Ciric, D. Dijkkamp, E. Vlieg and F.J. de Heer.
The other notation is the same as in Fig. 14.12.

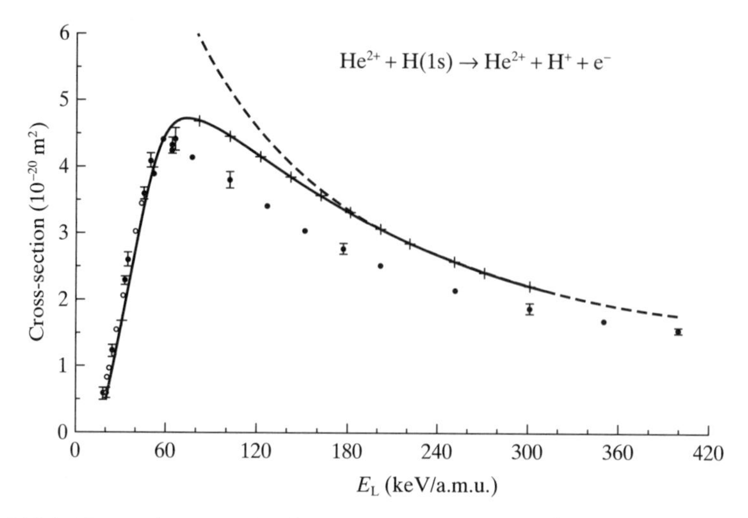

Figure 14.15 The total cross-section for ionisation of H(1s) by He^{2+} impact.
The experimental data are due to M.B. Shah *et al.*, ●, ○.
Calculated cross-sections: in the first Born approximation -----; in a coupled channel model using a large two-centre basis with 35 states and pseudo-states placed on the He^{2+} centre and 174 states and pseudo-states placed on the H^+ centre, by J. Kuang and C.D. Lin ——.

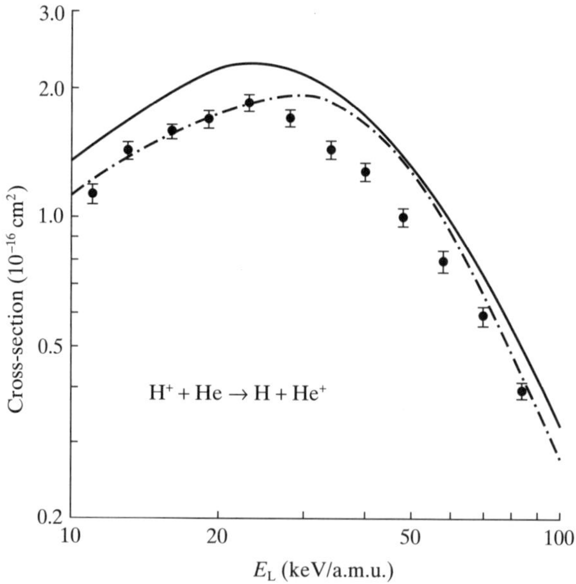

Figure 14.16 The total cross-section for electron capture by H^+ from He(1^1S).
The experimental are due to M.B. Shah, P. McCallion and H.B. Gilbody, ●.
Calculated cross-sections in the coupled channel impact parameter model:
–·–·–, a two-state atomic orbital basis representing the H^+ + He(1^1S) and H(1s) + He^+(1s) channels due to T.A. Green, H.J. Stanley and Y.C. Chang.
——, a two-centre basis containing 40 states and pseudo-states due to H. Slim, L. Heck, B.H. Bransden and D.R. Flower.

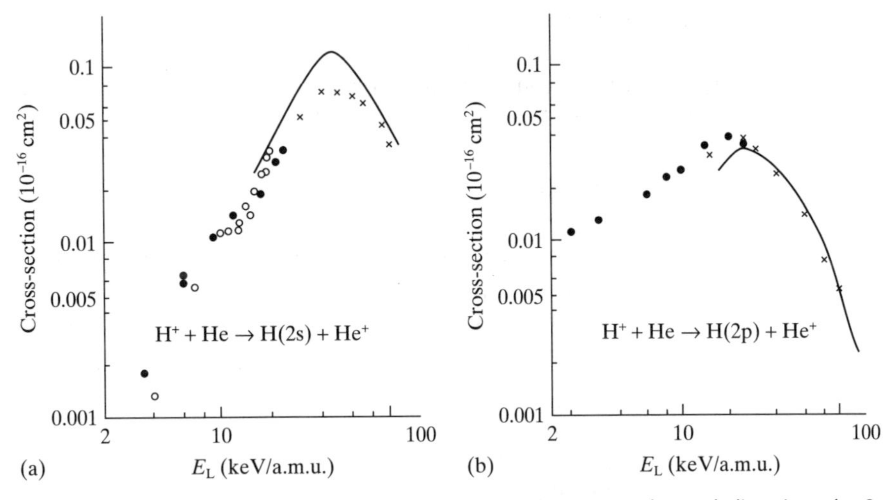

Figure 14.17 Total cross-sections for electron capture from ground state helium into the 2s and 2p states of atomic hydrogen.
The experimental data are due to D.W. Jaecks, B. van Zyl and R. Geballe, •; D.H. Crandall and D.W. Jaecks, o; R.H. Hughes, E.D. Stokes, S.S. Choe and T.J. King, ×.
Calculated cross-sections as in Fig. 14.16.

up to 35 states and pseudo-states on the He^{2+} centre and up to 175 on the H^+ centre. This convergence is slow for the ionisation cross-section (see Fig. 14.15) and there are difficulties in obtaining converged cross-sections for the small cross-sections.

Systems containing more than one electron are more demanding computationally unless treated as quasi-one-electron systems, but can be described by the same basic methods. For example, in Figs 14.16 and 14.17 some calculated total cross-sections for the $H^+ + He$ system are shown.

14.5 Atom–atom collisions at high velocities

Excitation and ionisation

When the velocity of the incident ion becomes greater than the orbital velocities of the target electrons, the cross-section for electron capture becomes much smaller than those for excitation or ionisation. It is then possible to ignore the electron capture channels when calculating cross-sections for excitation and ionisation. Total cross-sections for these processes can be calculated by using the first Born approximation at sufficiently high energies. The perturbative theory of charge exchange has some complications that will be discussed later, but the theory of excitation is straightforward. It can be formulated in much the same way as for excitation by electron impact using a full quantum mechanical treatment, but we shall continue to employ the impact parameter formulation. The resulting cross-sections from either formulation differ imperceptibly. To illustrate the impact parameter version of the first Born approximation, the simple case of the excitation of atomic hydrogen in the ground state by proton impact will be examined. The wave function for the electron satisfies the Schrödinger equation (14.53a) with the Hamiltonian $H_e(t)$ given by (14.53b). In the present case we have

$$H_e(t) = -\frac{1}{2}\nabla_{\mathbf{r}}^2 - \frac{1}{r_A} - \frac{1}{r_B} + \frac{1}{R} \tag{14.101}$$

Since we are going to omit the rearrangement channel, in which the electron is bound to the incident proton A, it is convenient to fix the origin of the coordinate system at the proton B, and to take $\mathbf{r}_B$ and t (see Fig. 14.8) as independent variables, with $\mathbf{r}_A$ given by

$$\mathbf{r}_A(t) = \mathbf{r}_B - \mathbf{R}(t) \tag{14.102}$$

where $\mathbf{R}(t)$ is again determined (in the straight-line approximation) by (14.52). The wave function $\Psi(\mathbf{r}_B, t)$ can be expanded in terms of a complete set of hydrogenic wave functions centred on the proton B, namely

$$\Psi(\mathbf{r}_B, t) = \sum_j a_j(b, t)\psi_j(\mathbf{r}_B)\exp(-\mathrm{i}E_j t) \tag{14.103}$$

Equations for the amplitudes a_j can be obtained by inserting the expansion (14.103) into the Schrödinger equation (14.53a). We find that

$$\mathrm{i}\sum_j \dot{a}_j(b, t)\psi_j(\mathbf{r}_B)\exp(-\mathrm{i}E_j t) = \sum_j \left(\frac{1}{R} - \frac{1}{r_A}\right) a_j(b, t)\psi_j(\mathbf{r}_B)\exp(-\mathrm{i}E_j t) \tag{14.104}$$

where we have made use of the Schrödinger equation satisfied by the hydrogen atom wave functions, namely

$$\left(-\frac{1}{2}\nabla_{\mathbf{r}_B}^2 - \frac{1}{r_B} - E_j\right)\psi_j(\mathbf{r}_B) = 0 \tag{14.105}$$

Premultiplying both sides of (14.104) by $\psi_k^*(\mathbf{r}_B)$, integrating over $\mathbf{r}_B$ and using the orthonormality property

$$\int \psi_k^*(\mathbf{r}_B)\psi_j(\mathbf{r}_B)\,\mathrm{d}\mathbf{r}_B = \delta_{kj} \tag{14.106}$$

we find the set of coupled equations

$$\mathrm{i}\dot{a}_k(b, t) = \sum_j V_{kj}(t)\exp[\mathrm{i}(E_k - E_j)t]a_j(b, t) \tag{14.107}$$

where

$$V_{kj}(t) = \int \psi_k^*(\mathbf{r}_B)\left(\frac{1}{R} - \frac{1}{r_A}\right)\psi_j(\mathbf{r}_B)\,\mathrm{d}\mathbf{r}_B \tag{14.108}$$

A coupled channel approximation can be obtained by retaining a number of terms on the right-hand side of (14.107) and solving the resulting equations numerically. A first-order perturbation (first Born) solution is found as follows. If the system was originally in the state labelled i, then

$$a_j(b, t) \to \delta_{ij}, \qquad t \to -\infty \tag{14.109}$$

If the perturbation is small, all the amplitudes $a_j(b, t)$ for $j \neq i$ will remain small and $a_i(b, t)$ will stay close to 1. The coupled equations (14.107) can then be approximated by

$$\mathrm{i}\dot{a}_k(b, t) = V_{ki}(t) \exp[\mathrm{i}(E_k - E_i)t] \tag{14.110}$$

and have the solution (for $k \neq i$)

$$a_k(b, t) = -\mathrm{i} \int_{-\infty}^{t} V_{ki}(t') \exp[\mathrm{i}(E_k - E_i)t'] \, \mathrm{d}t' \tag{14.111}$$

The probability amplitude of finding the system after the collision in the state k is given by

$$a_k(b, t = +\infty) = -\mathrm{i} \int_{-\infty}^{\infty} V_{ki}(t') \exp[\mathrm{i}(E_k - E_i)t'] \, \mathrm{d}t' \tag{14.112}$$

and the corresponding total cross-section is found by integrating over all impact parameters, namely

$$\sigma_k = 2\pi \int_0^{\infty} |a_k(b, t = +\infty)|^2 \, b \, \mathrm{d}b \tag{14.113}$$

As an example, let us consider the excitation of the 2s level of atomic hydrogen from the 1s level by proton impact. Then

$$V_{2\mathrm{s},1\mathrm{s}}(t) = \int \psi_{2\mathrm{s}}^*(\mathbf{r}_\mathrm{B}) \left(\frac{1}{R} - \frac{1}{|\mathbf{r}_\mathrm{B} - \mathbf{R}|} \right) \psi_{1\mathrm{s}}(r_\mathrm{B}) \, \mathrm{d}\mathbf{r}_\mathrm{B} \tag{14.114}$$

The contribution to the integral due to the term in $1/R$ vanishes because of the orthogonality of the wave functions $\psi_{1\mathrm{s}}$ and $\psi_{2\mathrm{s}}$, and the result of the integration is

$$V_{2\mathrm{s},1\mathrm{s}}(t) = -\frac{2^{3/2}}{27} [2 + 3R(t)] \exp[-3R(t)/2] \tag{14.115}$$

where $R(t) = (b^2 + v^2t^2)^{1/2}$. Since $V_{2\mathrm{s},1\mathrm{s}}(t)$ is even in t, we find that

$$a_{2\mathrm{s}}(b, t = +\infty) = \mathrm{i}\frac{2^{5/2}}{27} \int_0^{\infty} [2 + 3R(t')] \exp[-3R(t')/2] \cos(3t'/8) \, \mathrm{d}t' \tag{14.116}$$

The integral over t' can be done analytically in terms of Bessel functions, but the final integration over b to obtain the total cross-section (see (14.113)) must be done numerically. The total cross-section for the excitation of the sum of the 2s and 2p levels of atomic hydrogen has been measured and the experimental data are shown in Fig. 14.10, together with the results of the first Born approximation and those of the elaborate coupled channel calculation of A.L. Ford *et al.* discussed above. The experimental data are not absolute, but have been normalised to the coupled channel calculation at 200 keV. It is seen that the first Born approximation gives a fair representation of the data above 100 keV, and becomes quite accurate above 200 keV.

It is straightforward to generalise the above results for excitation in any atom(ion)–atom collision. Using the first Born approximation, the probability amplitude that the system is found in the state k after the collision is given again by

$$a_k(b, t = +\infty) = -\mathrm{i}\int_{-\infty}^{\infty} V_{ki}(t') \exp[\mathrm{i}(E_k - E_i)t]\, \mathrm{d}t' \qquad \textbf{(14.117)}$$

where E_i and E_k are the unperturbed atomic energies in the initial and final state, respectively, and

$$V_{ki}(t) = \int \psi_k^* V \psi_i \, \mathrm{d}\mathbf{r}_1 \, \mathrm{d}\mathbf{r}_2 \ldots \mathrm{d}\mathbf{r}_N \qquad \textbf{(14.118)}$$

Here V is the interaction potential between the two colliding systems, ψ_i and ψ_k are the unperturbed atomic wave functions and the integration is over the co-ordinates of all the N electrons of the system. Such calculations have been carried out for a variety of atom(ion)–atom excitation collisions.

Charge exchange at high velocities

Coupled channel calculations of charge exchange cross-sections have been performed successfully using the non-relativistic theory discussed earlier up to energies of a few MeV/a.m.u. for light atoms. The coupled channel approach has been extended for atoms with a large nuclear charge by replacing the non-relativistic Hamiltonian by the Dirac Hamiltonian. The radiative charge transfer reaction

$$\mathrm{A} + (\mathrm{B} + \mathrm{e}^-) \rightarrow (\mathrm{A} + \mathrm{e}^-) + \mathrm{B} + h\nu \qquad \textbf{(14.119)}$$

which is unimportant at low energies is ultimately the most important process, the corresponding cross-section decreasing like E^{-1} rather than the E^{-6} or $E^{-5.5}$ fall-off of the non-radiative mechanism. For collisions of highly charged ions the production of positron–electron pairs is also an important reaction channel.

Despite the possibility of carrying out coupled channel calculations over the whole of the non-relativistic region, there has always been much interest in developing simpler models based on perturbative or distorted wave expansions, which as we have already seen are very successful for the description of excitation. However, there are problems with the perturbation theory of rearrangement collisions. To understand these, let us consider a very simple one-electron system in which a fully stripped ion of charge Z_A interacts with a one-electron ion with nuclear charge Z_B, so that in the Hamiltonian $H_\mathrm{e}(t)$, given by (14.53b), we have

$$V_A(r_A) = -Z_A/r_A, \qquad V_B(r_B) = -Z_B/r_B, \qquad V_{AB}(R) = Z_A Z_B/R \tag{14.120}$$

The internuclear potential $V_{AB}(R)$ can be removed from the time-dependent Schrödinger equation (14.53a) by making the phase transformation

$$\Psi(\mathbf{r}, t) = \tilde{\Psi}(\mathbf{r}, t) \exp\left[-\mathrm{i}\int_{-t_0}^{t} V_{AB}(R)\, \mathrm{d}t\right] \tag{14.121}$$

where the wave function $\tilde{\Psi}(\mathbf{r}, t)$ satisfies the time-dependent Schrödinger equation

$$\mathrm{i}\frac{\partial}{\partial t}\tilde{\Psi}(\mathbf{r}, t) = \left[-\frac{1}{2}\nabla_{\mathbf{r}}^2 + V_A(r_A) + V_B(r_B)\right]\tilde{\Psi}(\mathbf{r}, t) \tag{14.122}$$

Since $\Psi(\mathbf{r}, t)$ and $\tilde{\Psi}(\mathbf{r}, t)$ only differ by a phase factor, inelastic cross-sections calculated from either wave function should be the same. This is consistent with the idea that as the relative motion of A and B is entirely predetermined by the use of a classical trajectory, the only role of the potential V_{AB} is in contributing to the effective potential from which $R(t)$ is found. If a straight-line trajectory is employed then the potential V_{AB} has not even this indirect effect and should play no role at all. Taking this into consideration and noting that in the arrangement $A + (B + e^-)$ the perturbation is $V_A(r_A)$ and in the arrangement $(A + e^-) + B$ the perturbation is $V_B(r_B)$, J.R. Oppenheimer in 1928 and H.C. Brinkman and H.A. Kramers in 1930 suggested that the first-order Born approximation amplitude for electron capture into the state k from the initial state i should be either

$$c_k^{\mathrm{OBK}}(b, t = +\infty) = -\mathrm{i}\int_{-\infty}^{\infty} \langle X_k^A(\mathbf{r}, t')|V_A(r_A)|X_i^B(\mathbf{r}, t')\rangle\, \mathrm{d}t' \tag{14.123}$$

called the *prior* form or

$$c_k^{\mathrm{OBK}}(b, t = +\infty) = -\mathrm{i}\int_{-\infty}^{\infty} \langle X_k^A(\mathbf{r}, t')|V_B(r_B)|X_i^B(\mathbf{r}, t')\rangle\, \mathrm{d}t' \tag{14.124}$$

called the *post* form, where X_k^A and X_i^B are defined by (14.55a) and (14.55b), respectively. These expressions are of the usual perturbative form analogous to (14.112) for excitation. In fact the post and prior forms of the amplitude give the same result (Problem 14.8). The resulting cross-sections in this approximation, called the Oppenheimer–Brinkman–Kramers (OBK) approximation, are very poor and do not improve with increasing energy.

The failure of the OBK approximation led D.R. Bates and A. Dalgarno in 1952 and J.D. Jackson and H. Schiff in 1953, who were investigating the $H^+ + H(1s)$ system, to suggest that the perturbation in the prior form should be taken as $V_A(r_A) + V_{AB}(R)$. They realised that $V_{AB}(R)$ should play no role in an exact theory, but suggested that the errors in the term containing $V_A = -1/r_A$ must be compensated by equal errors of opposite sign in the term containing $V_{AB} = 1/R$. In fact the corresponding cross-sections, in what is called the Jackson–Schiff or JS approximation, agree well with the experimental data for the $H^+ + H(1s)$ system.

However, attempts to apply the JS prescription to other systems resulted in failure. This interesting situation was resolved by I. Cheshire in 1964, who in the course of developing a distorted wave theory pointed out the importance of satisfying the proper boundary conditions of the problem.

Since, as we have seen above, the potential $V_{AB}(R)$ can be removed exactly from the time-dependent Schrödinger equation (14.53a), we can start from the resulting equation (14.122). In the limit $t \to -\infty$ and $r_A \gg r_B$, we have that $V_A(r_A) \to -Z_A/R$ and the equation (14.122) becomes

$$\mathrm{i}\frac{\partial}{\partial t}\tilde{\Psi}(\mathbf{r}, t) = \left[-\frac{1}{2}\nabla_{\mathbf{r}}^2 - \frac{Z_A}{R} + V_B(r_B)\right]\tilde{\Psi}(\mathbf{r}, t) \qquad \textbf{(14.125)}$$

In the OBK expression (14.123), it is assumed that the unperturbed wave function in the limit $t \to -\infty$ is $X_i^B(\mathbf{r}, t)$, but it is easy to show that X_i^B does not satisfy the equation (14.125). This is due to the fact that the term $V_A(r_A)$, which behaves like $-Z_A/R$ for large R, contains a long-range Coulomb potential which must be taken into account. By substitution it can be verified (Problem 14.9) that a solution of (14.125) is

$$\tilde{X}_i^{B(+)}(\mathbf{r}, t) = X_i^B(\mathbf{r}, t) \exp\{-\mathrm{i}\sigma(t)] \qquad \textbf{(14.126)}$$

where

$$\sigma(t) = \frac{Z_A}{v} \log[Rv - v^2 t] \qquad \textbf{(14.127)}$$

The exact solution of (14.122) which coincides with the unperturbed wave function $\tilde{X}_i^{B(+)}(\mathbf{r}, t)$ at $t \to -\infty$ is denoted by $\tilde{\Psi}_i^{(+)}(\mathbf{r}, t)$. That is,

$$\tilde{\Psi}_i^{(+)}(\mathbf{r}, t) \underset{t \to -\infty}{\to} \tilde{X}_i^{B(+)}(\mathbf{r}, t) \qquad \textbf{(14.128)}$$

In the limit $t \to \infty$ and $r_B \gg r_A$, the equation (14.122) reduces to

$$\mathrm{i}\frac{\partial}{\partial t}\tilde{\Psi}(\mathbf{r}, t) = \left[-\frac{1}{2}\nabla_{\mathbf{r}}^2 + V_A(r_A) - \frac{Z_B}{R}\right]\tilde{\Psi}(\mathbf{r}, t) \qquad \textbf{(14.129)}$$

This equation has a solution (Problem 14.9)

$$\tilde{X}_k^{A(-)}(\mathbf{r}, t) = X_k^A(\mathbf{r}, t) \exp[\mathrm{i}\eta(t)] \qquad \textbf{(14.130)}$$

where

$$\eta(t) = \frac{Z_B}{v} \log[Rv + v^2 t] \qquad \textbf{(14.131)}$$

A second solution of (14.122), denoted by $\tilde{\Psi}_k^{(-)}$, can be defined which is asymptotic to $\tilde{X}_k^{A(-)}$ as $t \to \infty$. That is,

$$\tilde{\Psi}_k^{(-)}(\mathbf{r}, t) \underset{t \to \infty}{\to} \tilde{X}_k^{A(-)}(\mathbf{r}, t) \qquad \textbf{(14.132)}$$

The probability amplitude for finding the system in the state k of the atom $(A + e^-)$ after the collision, as $t \to \infty$, is given by

$$
\begin{aligned}
c_k(b, t = +\infty) &= \lim_{t\to\infty} \langle \tilde{X}_k^{A(-)}(\mathbf{r}, t) | \tilde{\Psi}_i^{(+)}(\mathbf{r}, t)\rangle \\
&= \int_{-\infty}^{\infty} \frac{\partial}{\partial t'} \langle \tilde{X}_k^{A(-)}(\mathbf{r}, t') | \tilde{\Psi}_i^{(+)}(\mathbf{r}, t')\rangle \, \mathrm{d}t' \\
&\quad + \lim_{t\to-\infty} \langle \tilde{X}_k^{A(-)}(\mathbf{r}, t) | \tilde{\Psi}_i^{(+)}(\mathbf{r}, t)\rangle
\end{aligned}
\tag{14.133}
$$

The second term in (14.133) vanishes, since

$$
\lim_{t\to-\infty} \langle \tilde{X}_k^{A(-)}(\mathbf{r}, t) | \tilde{\Psi}_i^{(+)}(\mathbf{r}, t)\rangle = \lim_{t\to-\infty} \langle \tilde{X}_k^{A(-)}(\mathbf{r}, t) | \tilde{X}_i^{B(+)}(\mathbf{r}, t)\rangle \tag{14.134}
$$

and the wave functions $\psi_k^A(\mathbf{r}_A)$ and $\psi_i^B(\mathbf{r}_B)$ do not overlap as $R \to \infty$. Using (14.130), (14.131) and (14.122) the expression (14.133) for $c_k(b, t = +\infty)$ reduces to

$$
c_k(b, t = +\infty) = -\mathrm{i} \int_{-\infty}^{\infty} \left\langle \tilde{X}_k^{A(-)}(\mathbf{r}, t') \middle| V_B(r_B) + \frac{Z_B}{R} \middle| \tilde{\Psi}_i^{(+)}(\mathbf{r}, t') \right\rangle \mathrm{d}t' \tag{14.135}
$$

In a similar way, $c_k(b, t = +\infty)$ can be expressed in terms of the time-reversed wave function $\tilde{\Psi}_k^{(-)}(\mathbf{r}, t)$ as

$$
c_k(b, t = +\infty) = -\mathrm{i} \int_{-\infty}^{\infty} \left\langle \tilde{\Psi}_k^{(-)}(\mathbf{r}, t') \middle| V_A(r_A) + \frac{Z_A}{R} \middle| \tilde{X}_i^{B(+)}(\mathbf{r}, t') \right\rangle \mathrm{d}t' \tag{14.136}
$$

It is possible to develop a Born series starting from (14.135) in terms of the post perturbation $V_B(r_B) + Z_B/R$ or from (14.136) in terms of the prior perturbation $V_A(r_A) + Z_A/R$. Because the correct boundary conditions have been employed, both perturbations are of short range and vanish as $|t| \to \infty$. This ensures that the higher terms in the series are finite and that each term in the series, including the first, is meaningful. The first-order approximations to $c_k(b, t = +\infty)$, known as B1B approximations, are obtained by replacing $\tilde{\Psi}_i^{(+)}$ and $\tilde{\Psi}_k^{(-)}$ by the appropriate unperturbed wave functions. This gives the prior B1B form

$$
c_k^{\mathrm{B1B}}(b, t = +\infty) = -\mathrm{i} \int_{-\infty}^{\infty} \left\langle \tilde{X}_k^{A(-)}(\mathbf{r}, t') \middle| V_A(r_A) + \frac{Z_A}{R} \middle| \tilde{X}_i^{B(+)}(\mathbf{r}, t') \right\rangle \mathrm{d}t' \tag{14.137}
$$

and the post B1B form

$$
c_k^{\mathrm{B1B}}(b, t = +\infty) = -\mathrm{i} \int_{-\infty}^{\infty} \left\langle \tilde{X}_k^{A(-)}(\mathbf{r}, t') \middle| V_B(r_B) + \frac{Z_B}{R} \middle| \tilde{X}_i^{B(+)}(\mathbf{r}, t') \right\rangle \mathrm{d}t' \tag{14.138}
$$

Both forms are identical provided exact wave functions $\psi_k^A(r_A)$ and $\psi_i^B(r_B)$ are employed. These approximations were first written down in 1979 by Dz. Belkić, R. Gayet and A. Salin. Other distorted wave amplitudes obeying the proper boundary conditions have been successfully developed, in particular by I. Cheshire.

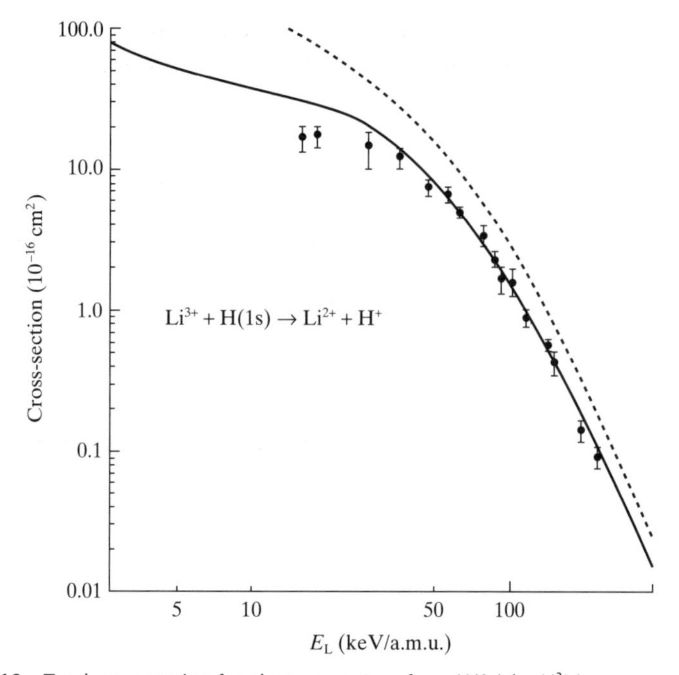

Figure 14.18 Total cross-section for electron capture from H(1s) by Li^{3+} ions. The experimental data are due to M.B. Shah, T.V. Goffe and H.B. Gilbody, ●. The calculated cross-sections are due to Dz. Belkić, S. Saini and H.S. Taylor: in the first Born approximation (B1B) —— and in the Jackson–Schiff model (JS) -----.

From (14.137) and (14.138), the success of the JS method using the perturbation $V_A(r_A) + V_{AB}(R)$ for the $H^+ + H$ system can be understood. When $Z_A = Z_B = 1$, $V_A(r_A) + V_{AB}(R) = V_A(r_A) + Z_A/R$, with the result that the JS amplitude employing the unperturbed functions $X_k^A(\mathbf{r}, t)$ and $X_i^B(\mathbf{r}, t)$ only differs from (14.137) by a time-independent phase factor. Hence the probabilities and cross-sections for charge exchange calculated by the JS and B1B models are the same. This is not the case for other values of Z_A and Z_B. As an illustration the calculated capture cross-sections in the B1B approximation for the $Li^{3+} + H(1s)$ and $C^{6+} + H(1s)$ systems are shown in Figs 14.18 and 14.19 compared with the JS approximation and with experimental data. For systems for which calculations have been made, it is found that the B1B approximation is accurate for total capture cross-sections over a wide energy range.

In 1955, R.M. Drisko using the JS perturbation series showed that the non-relativistic capture cross-section was given by the second term in the series as $E \to \infty$, which modifies the E^{-6} dependence of the first-order approximation to an $E^{-5.5}$ dependence [7]. However, the second- and higher order terms make little difference to the *total cross-section* until energies are reached which require a relativistic treatment. On the other hand, for the *angular distribution*, second-order theories provide a much more accurate cross-section at large angles (see Fig. 14.20).

[7] See the review by Shakeshaft and Spruch (1979).

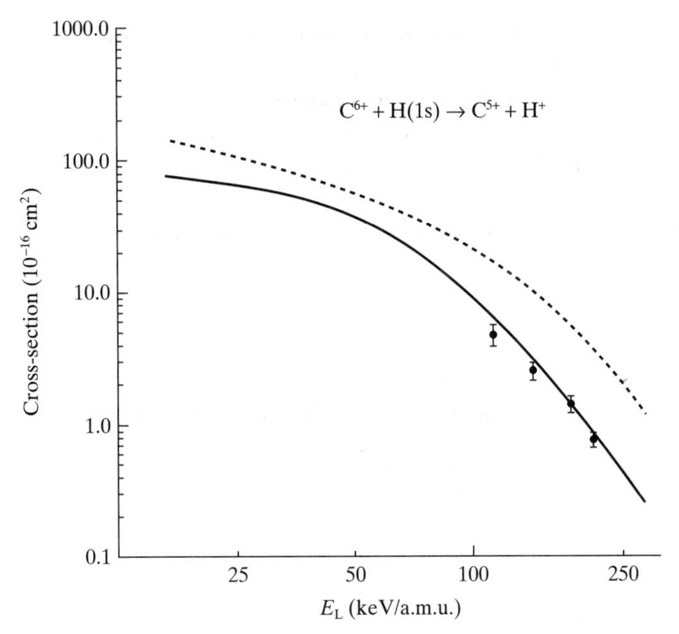

Figure 14.19 Total cross-section for electron capture from H(1s) by C^{6+} ions. The notation is the same as in Fig. 14.18.

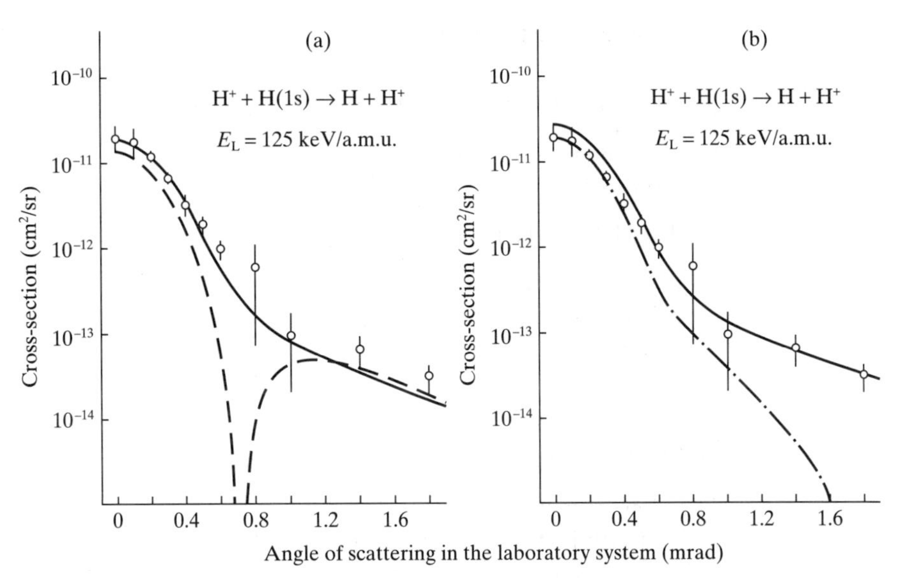

Figure 14.20 Differential cross-section for electron capture from H(1s) by protons at $E_L = 125$ keV/a.m.u.

The experimental data are due to P.J. Martin *et al.*, o.

The calculated cross-sections are due to F. Decker and J.K.M. Eichler: (a) -----, first Born approximation (B1B) showing an unphysical dip near 0.7 mrad.; ——, a first-order distorted wave approximation. (b) —·—·—, second Born approximation; ——, second-order distorted wave approximation.

Problems

14.1 The total spin F of a hydrogen atom can be either $F = 1$ (a triplet state) or $F = 0$ (a singlet state). In cold collisions between hydrogen and deuterium atoms in the ground state, the hydrogen atoms in the state with $F = 1$ can make transitions to the state $F = 0$. Show that the total cross-section for the transition is given by

$$\sigma_{\text{tot}}(F = 1 \to F = 0) = \frac{\pi}{2k^2} \sum_{l=0}^{\infty} (2l + 1) \sin^2({}^3\delta_l - {}^1\delta_l)$$

where ${}^3\delta_l$ and ${}^1\delta_l$ are the phase shifts for scattering in the electronic spin states with $S = 1$ and $S = 0$ respectively.

14.2 Verify that the functions X_k^{A} and X_j^{B} defined by (14.55a) and (14.55b) satisfy the time-dependent Schrödinger equation (14.53a) in the limit $|t| \to \infty$.

14.3 Show that if the wave function $\Psi_{\text{a}}(\mathbf{r}, t)$ satisfies the equations (14.66) and is normalised to unity when $t \to -\infty$, then it remains normalised to unity for all t, and hence that

$$\lim_{t\to\infty} \left[\sum_{k=1}^{N} |c_k^{\text{A}}(b, t)|^2 + \sum_{j=1}^{M} |c_j^{\text{B}}(b, t)|^2 \right] = 1$$

14.4 Show that in the coupled channel impact parameter approximation

$$\mathrm{i}\frac{\partial}{\partial t}(\mathbf{S}_{\text{AB}})_{kj} = (\mathbf{M}_{\text{BA}}^*)_{jk} - (\mathbf{M}_{\text{AB}})_{kj}$$

for all values of t, where $\mathbf{S}_{\text{AB}}$, $\mathbf{M}_{\text{BA}}$ and $\mathbf{M}_{\text{AB}}$ are defined by (14.68) and (14.70).

14.5 Obtain an expression for the probability of charge exchange $P_{\text{ce}}(b)$ for the reaction

$$\text{H}^+ + \text{H(1s)} \to \text{H(1s)} + \text{H}^+$$

using the two-state approximation for the wave function

$$\Psi(\mathbf{r}, t) = c^{\text{A}}(b, t)\psi_{1\text{s}}(r_{\text{A}}) \exp(-\mathrm{i}E_{1\text{s}}t + \mathrm{i}\mathbf{v}\cdot\mathbf{r}/2 - \mathrm{i}v^2t/8)$$
$$+ c^{\text{B}}(b, t)\psi_{1\text{s}}(r_{\text{B}}) \exp(-\mathrm{i}E_{1\text{s}}t - \mathrm{i}\mathbf{v}\cdot\mathbf{r}/2 - \mathrm{i}v^2t/8)$$

Show that $P_{\text{ce}}(b)$ oscillates between 0 and 1 as b is varied and v is held fixed.

14.6 Show that in the impact parameter method the integral

$$I = \int_{-\infty}^{\infty} \mathrm{d}t \int \mathrm{d}\mathbf{r}\, \Psi^*(\mathbf{r}, t)\left(H_{\text{e}} - \mathrm{i}\frac{\partial}{\partial t} \right)\Psi(\mathbf{r}, t)$$

is stationary under the independent variations

$$\Psi \to \Psi + \delta\Psi, \qquad \Psi^* \to \Psi^* + \delta\Psi^*$$

provided that $\delta\Psi \to 0$ and $\delta\Psi^* \to 0$ as $|t| \to \infty$.

14.7 In the 'one and a half' centre expansion of J.F. Reading, A.L. Ford and R.L. Becker the impact parameter wave function is expressed as

$$\Psi(\mathbf{r}, t) = \sum_{k=1}^{M} c_k^{\mathrm{A}}(b, \infty) f(t) X_k^{\mathrm{A}}(\mathbf{r}, t) + \sum_{j=1}^{N} c_j^{\mathrm{B}}(b, t) X_j^{\mathrm{B}}(\mathbf{r}, t)$$

where X_k^{A} and X_j^{B} are given by (14.55a) and (14.55b), respectively. The function $f(t)$ is predetermined to satisfy the conditions $f(t) \to 0$ as $t \to -\infty$ and $f(t) \to 1$ as $t \to \infty$. By varying the amplitudes $c_j^{\mathrm{B}}(b, t)$ and the time-independent coefficients $c_k^{\mathrm{A}}(b, \infty)$ in the stationary functional I of Problem 14.6, find the equations satisfied by $c_k^{\mathrm{A}}(b, \infty)$ and $c_j^{\mathrm{B}}(b, t)$.

14.8 Show that the prior form (14.123) and the post form (14.124) of the Oppenheimer-Brinkman-Kramers approximation give the same result.

14.9 Verify that the function $\tilde{X}_i^{\mathrm{B}(+)}$ defined by (14.126) is a solution of the equation (14.125) and that the function $\tilde{X}_k^{\mathrm{A}(-)}$ defined by (14.130) is a solution of the equation (14.129).

15 Masers, lasers and their interaction with atoms and molecules

The physical process of *stimulated emission*, first recognised by A. Einstein in 1916, is at the heart of maser or laser operation. Under normal conditions of thermal equilibrium in populations of atoms or molecules, the number of particles in a lower energy level always exceeds that in a higher level, so that radiation is absorbed rather than amplified. In Section 15.1, we shall examine various methods for obtaining a *population inversion* between two energy levels, which allows maser action (in the microwave region) or laser action (at higher frequencies). We shall see that, in contrast to other sources of light, the maser or laser radiation is *coherent*. Other characteristic properties of laser light will be examined, and the physical principles of some important masers and lasers will be discussed.

Masers and lasers are used nowadays in an enormous range of scientific and technological applications. For example, they play a central role in communications, and have revolutionised optics. In particular, studies of the statistical properties of maser or laser radiation inaugurated the field of *quantum optics*. In atomic and molecular physics, the advent of masers and lasers has opened the way to many developments. For example, the availability of tunable lasers has caused a revolution in spectroscopy. Intense monochromatic laser light has not only vastly increased spectroscopic resolution, but also made possible the observation of higher order radiation processes such as *multiphoton transitions* that are essentially unobservable with conventional light sources. Several methods of laser spectroscopy will be discussed in Section 15.2, while in Section 15.3 we shall consider laser interactions with atoms and molecules at *high intensities*.

In 1917, one year after his 1916 paper that introduced stimulated emission, Einstein studied the momentum interchange between atoms and radiation, and predicted the velocity change of an atom which has absorbed a photon. This process is at the basis of the manipulation of atoms by light, which is discussed in Section 15.4. In particular, we shall see that with the help of lasers, atoms can be *cooled* to temperatures in the microkelvin range, and *trapped*. Once in the trap, the atoms can be further cooled by evaporative cooling, which consists of selectively ejecting atoms having higher than average energy. In this way, the temperature of the atomic cloud can be reduced to the value required for the phenomenon of *Bose–Einstein condensation* (BEC) to occur in the lowest energy level of the trap potential. The BEC phenomenon is discussed in Section 15.5. It is the key ingredient of a new type of laser, called the *atom laser*, which emits coherent matter waves rather than photons, and promises to revolutionise atom optics. The atom laser will be discussed in Section 15.6.

15.1 Masers and lasers

In this section, we will show how the phenomenon of stimulated emission discussed in Chapter 4 can be used to construct amplifiers or oscillators for electromagnetic radiation. Suppose that 1 and 2 are two levels, with energies E_1, E_2 ($E_2 > E_1$), out of the many levels of a particular material. Consider a beam of electromagnetic radiation of intensity I and angular frequency $\omega = (E_2 - E_1)/\hbar$ passing through this material. The rate of change of the average energy density because of absorption from the beam is

$$\frac{\mathrm{d}\rho_a}{\mathrm{d}t} = -N_1\hbar\omega W_{21} \tag{15.1}$$

where N_1 is the number of atoms in the lower energy level per unit volume and W_{21} is the transition rate per atom for absorption. Similarly, the rate of change of the average energy density because of stimulated emission is

$$\frac{\mathrm{d}\rho_s}{\mathrm{d}t} = N_2\hbar\omega W_{12} \tag{15.2}$$

where N_2 is the number of atoms in the upper energy level per unit volume, and W_{12} is the transition rate per atom for stimulated emission. In Chapter 4, it was shown that W_{12} and W_{21} are equal and both are proportional to the intensity I of the incident radiation. The cross-section σ, defined as

$$\sigma = \hbar\omega W_{12}/I \tag{15.3}$$

is characteristic of the particular pair of levels, but is independent of the intensity of the beam radiation. In terms of σ, we can write the net rate of change of average energy per unit volume traversed by the beam as

$$\frac{\mathrm{d}\rho}{\mathrm{d}t} = \sigma I(N_2 - N_1) \tag{15.4}$$

If the beam is of cross-sectional area A, and is travelling parallel to the Z axis, then by using (4.13) we have

$$\frac{\mathrm{d}I}{\mathrm{d}z} = \sigma I(N_2 - N_1) \tag{15.5}$$

We see that if $N_1 > N_2$ the incident radiation is absorbed as it traverses the material, but if $N_2 > N_1$ the radiation is amplified. Spontaneous emission will also increase the number of transitions from the upper level 2 to the lower level 1, but the corresponding transition rate is independent of the intensity I, and provided that I is sufficiently large this contribution can be ignored.

Under thermal equilibrium, we know that for non-degenerate energy levels

$$\frac{N_2}{N_1} = \exp[-(E_2 - E_1)/(k_B T)] \qquad \textbf{(15.6)}$$

where k_B is Boltzmann's constant and T is the absolute temperature. Since $E_2 > E_1$ it follows that $N_2 < N_1$ and the material acts as an absorber. To achieve amplification, a *population inversion* must be obtained with $N_2 > N_1$, so that the substance cannot be in thermal equilibrium. A material in which this population inversion is produced is called an *active material* or *active medium*.

When amplification is achieved in the microwave region, one speaks of a *maser*, an acronym for 'microwave amplification by stimulated emission of radiation'. The first maser, constructed in 1954 by C.H. Townes, J. Gordon and H. Zeiger, will be described below. For radiation of higher frequencies one uses the term *laser*, an acronym for 'light amplication by stimulated emission of radiation'. The first laser was proposed in 1958 by A.L. Schawlow and C.H. Townes, who extended the maser concept into the optical frequency range. To specify the kind of radiation which is emitted by a laser, one often refers to infra-red, visible, ultraviolet or X-ray lasers.

To make an oscillator from an amplifier, one must introduce positive feedback so that the gain is increased. In the case of a maser this is done by placing the active material in a resonant cavity. In the case of a laser, the gain increase is often obtained by placing the active material between two mirrors. As an example, we show in Fig. 15.1 the laser scheme proposed in 1958 by Schawlow and Townes, who considered a plane-parallel (or Fabry–Perot) resonator such that an active material is placed in a cavity formed by two plane mirrors. An electromagnetic plane wave travelling in a direction perpendicular to the mirrors will then bounce back and forth between the two mirrors, and will be amplified on each passage through the active material. To extract an output beam from the cavity in a controlled way, an *output coupling* technique is needed. In the case of a laser, the output coupler is a partially transparent mirror. If the power generated within the material by stimulated emission is greater than the sum of the power output

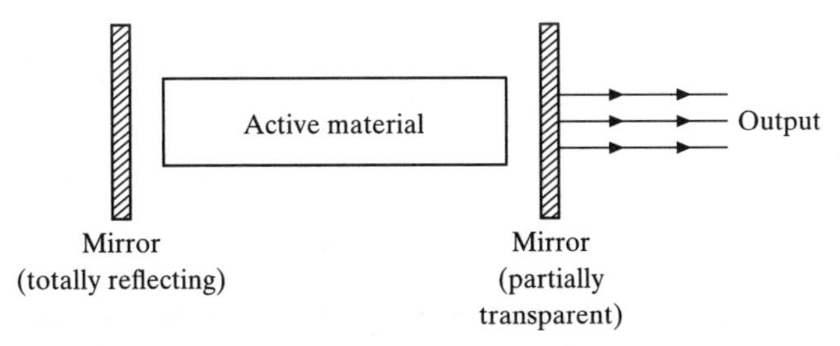

Figure 15.1 Schematic diagram of a laser in which the active material is confined between plane-parallel mirrors.

and the power losses, then the laser acts as an oscillator and the intensity of the radiation between the mirrors increases exponentially. This increase in intensity will be finally limited by the ability of the 'pumping' mechanism producing the population inversion to keep up the number of atoms or molecules in the upper level. The laser oscillations can be started by a single photon resulting from spontaneous emission from the upper to the lower level. In applications, masers and lasers are used either as amplifiers or as oscillators generating radiation.

The characteristic properties of a beam generated by a laser are (a) monochromaticity, (b) directionality, (c) brightness, (d) spatial coherence and (e) temporal coherence. The monochromaticity is a consequence of the fact that only light arising from a transition between a single pair of levels is amplified. The output of a laser is a parallel beam which emerges perpendicular to the plane of the mirrors in an arrangement such as that illustrated in Fig. 15.1. This is due to the fact that only electromagnetic waves propagating in this direction will be reflected back and forth between the mirrors. This directionality also accounts for the brightness (that is, the power emitted per unit area per unit solid angle) of a laser beam. The power output of other light sources is usually spread out into a large solid angle, but in a laser it is concentrated into a narrow unidirectional beam.

In stimulated emission, at each transition one photon is added to a *mode* of the resonator (that is, a stationary electromagnetic field configuration which satisfies both Maxwell's equations and the boundary conditions) containing N photons. The extra photon is completely in phase with the other photons and has the same polarisation. It follows that if laser (maser) action is initiated by a single photon, at each transition one extra photon will be produced and after N transitions, all $(N+1)$ photons will be in phase and contribute to the same mode of the electromagnetic radiation: the laser (maser) light is said to be *coherent*. This is in contrast to other sources of light (such as a lamp filament), where the dominant process is spontaneous emission, so that the phases and polarisations associated with each photon are different.

There are two independent concepts of coherence: *spatial coherence* and *temporal coherence*. If the phase difference between two points P_1 and P_2 lying on the same wave front (normal to the direction of propagation) is zero at all times, the wave is said to exhibit *perfect spatial coherence*. In practice, given any point P_1, the point P_2 must lie within some finite area around P_1, and the wave is said to exhibit *partial spatial coherence* over a *coherence area* $S_c(P_1)$. If the active material in a laser is homogeneous, the output beam exhibits spatial coherence over its whole cross-sectional area. It is in fact effectively a monochromatic plane wave with a single angular frequency ω and with all points on the wave front in phase.

In order to define temporal coherence, let us consider the phase difference of the electric field of an electromagnetic wave, at a given point P, between times t and $t+\Delta t$. If, for a given time delay Δt, this phase difference remains the same for any time t, one says that the wave exhibits temporal coherence over a time Δt. If this happens for any value of Δt, the wave is said to have *perfect temporal coherence*. If it occurs only for a time delay Δt such that $0 < \Delta t < \tau_c$, the wave is said to have *partial temporal coherence*, and τ_c is called the *coherence time*. For a perfectly coherent monochromatic plane wave, one clearly has $\tau_c = \infty$, while for a completely incoherent wave such as a black body source one has $\tau_c = 0$. A *coherence length* L_c can also be defined as $L_c = c\tau_c$. The stability of a laser determines the coherence time; it can be very long compared with the periodic time of the radiation. This allows the observation of interference effects between the beams of light originating from two different lasers. In contrast, using conventional light sources, interferences can only be observed by splitting and recombining the light from one source.

It is worth stressing that the two concepts of spatial and temporal coherence are independent of each other. Indeed, examples can be given of an electromagnetic wave having perfect spatial coherence but only limited temporal coherence, or vice versa [1].

Methods for obtaining a population inversion

We have seen that the key to maser or laser action is to achieve a *population inversion* between two levels, 1 and 2, of energy E_1 and E_2 with $E_1 < E_2$, so that more atoms (or molecules) are in the upper level 2 than in the lower level 1. Historically, the first method for obtaining a population inversion, employed in 1954 by C.H. Townes, J. Gordon and H. Zeiger in the original *ammonia maser*, was to use an inhomogeneous electric field which acts as a *quantum state selector*, removing most of the ammonia molecules in the lower of the two energy levels between which maser action is to occur. The principle of maser operation was recognised independently by N.G. Basov and A.M. Prokhorov, who shortly afterwards also achieved maser operation with ammonia. A related method, in which the quantum state selector is an inhomogeneous magnetic field, was used in 1960 by H.M. Goldenberg, D. Kleppner and N.F. Ramsey to construct the *hydrogen maser*. Both the ammonia maser and the hydrogen maser will be discussed in some detail below.

An alternative method is to use 'pumping' techniques. In the first of these, one looks for a three-level system (see Fig. 15.2) such that $E_1 < E_2 < E_3$ with a fast decay between levels 3 and 2 and a slow decay between 2 and 1. Incident radiation of angular frequency $\omega_{31} = (E_3 - E_1)/\hbar$ is used to raise as many atoms as possible from the level 1 to the level 3. This is known as *pumping* and can be realised for example by using a lamp of sufficient intensity or by producing an electrical discharge in the active medium. If level 3 decays rapidly to level 2, a population inversion can be obtained between levels 2 and 1 [2].

[1] See for example Svelto (1998).

[2] It should be noted that a population inversion cannot be obtained between levels 3 and 1, because when the number N_3 of atoms in level 3 equals the number N_1 of atoms in level 1, absorption is balanced by stimulated emission, and the material becomes transparent. This situation is called *two-level saturation*.

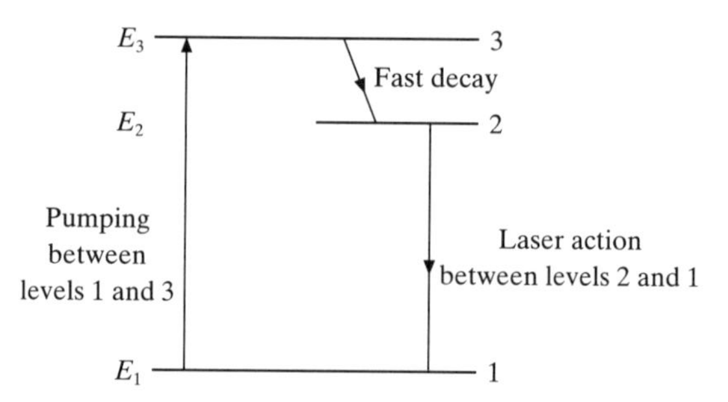

Figure 15.2 Principle of the three-level laser.

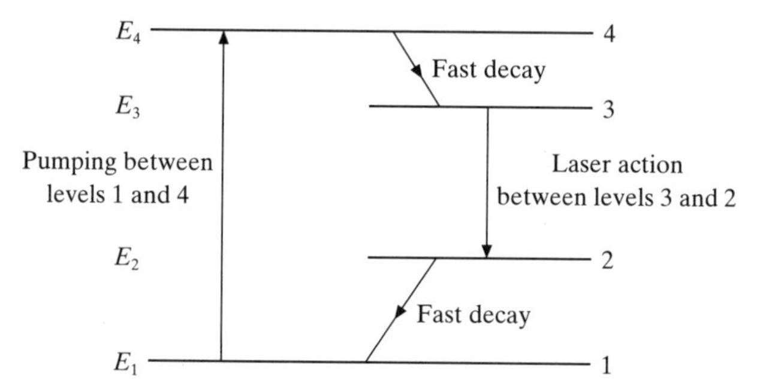

Figure 15.3 Principle of the four-level laser.

Except in special cases, it is difficult to produce a population inversion between an excited state and a ground state, because initially nearly all the atoms are likely to be in the ground state, and we have to get more than half the atoms in level 2 before a population inversion can be achieved. An easier approach is to use a four-level system (see Fig. 15.3) with $E_1 < E_2 < E_3 < E_4$, and attempt to create a population inversion between the two excited levels 3 and 2. We start with nearly all the atoms in the ground state 1. Level 4 is chosen so that it has a fast decay to level 3, and pumping between levels 1 and 4 then rapidly produces a population inversion between levels 3 and 2. As level 2 begins to fill up by stimulated emission at the angular frequency $\omega_{32} = (E_3 - E_2)/\hbar$, the population inversion will decrease. To minimise this, level 2 is chosen so that it has a fast decay to the ground state.

Many systems have been devised that exhibit maser or laser action in some region of the electromagnetic spectrum. In what follows we shall discuss the basic principles of a few important masers and lasers. A detailed account of the subject can be found in the books of Svelto (1998) and Svanberg (2001).

The ammonia maser

We begin by discussing the concepts behind the first maser to be constructed: the ammonia maser built by C. Townes *et al.* in 1954. In Chapter 11, we saw that the ammonia molecule NH_3 has two configurations, one in which the nitrogen atom is above the plane containing the three hydrogen atoms, and one in which it is on the other side. If the plane of the hydrogen atoms is the (XY) plane, we recall that the wave function for which the nitrogen atom is most probably above this plane (positive z coordinate) is denoted by ψ_u and the wave function for which the nitrogen atom is most probably below this plane is denoted by ψ_d. Because the potential barrier between the two configurations is finite, ψ_u and ψ_d are not energy eigenfunctions, but there are two closely spaced levels, with energies E_1, E_2 and corresponding eigenfunctions ψ_1 and ψ_2, such that (see (11.45))

$$\psi_1 = \frac{1}{\sqrt{2}}(\psi_u + \psi_d) \tag{15.7a}$$

and

$$\psi_2 = \frac{1}{\sqrt{2}}(\psi_u - \psi_d) \tag{15.7b}$$

The energy splitting is given by

$$\Delta E = E_2 - E_1 = 9.84 \times 10^{-5} \text{ eV} \tag{15.8}$$

corresponding to a frequency of 23 870 MHz which is in the microwave region. Of course the ammonia molecule has many other levels, but maser action is sought between just these two particular levels. The necessary coupling with the electromagnetic field takes place because the molecule possesses an electric dipole moment. In the configuration u (up) this moment is of magnitude D and is directed in the negative z direction, while in the configuration d (down), it is of the same magnitude but in the opposite direction.

Because the energy separation of the levels is so small a normal population in thermal equilibrium contains very nearly equal numbers of ammonia molecules in each of the energy eigenstates labelled E_1 and E_2. However, by passing a beam of ammonia molecules through an inhomogeneous electric field (see Fig. 15.4) a separation of the molecules in the two levels can be achieved, just as in a

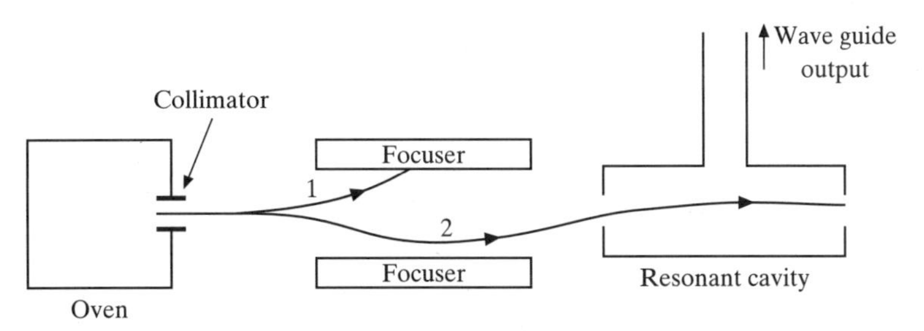

Figure 15.4 The ammonia maser. NH_3 molecules produced in an oven are collimated, selected in the upper level 2 by an inhomogeneous electric focusing field, and then passed through the maser cavity. The whole apparatus is placed in an evacuated vessel.

Stern–Gerlach magnet a separation is achieved between levels with different components of a magnetic dipole. To see this, consider a static electric field directed in the positive z direction of magnitude $\mathscr{E}$. The additional energy in configuration ψ_u is $+D\mathscr{E}$ and in configuration ψ_d is $-D\mathscr{E}$. This additional interaction alters the eigenenergies of the E_1 and E_2 levels slightly, which become

$$E_1' = E_1 - (D\mathscr{E})^2/\Delta E \tag{15.9a}$$

and

$$E_2' = E_2 + (D\mathscr{E})^2/\Delta E \tag{15.9b}$$

The force on the molecule in the z direction, F_z, depends on which of the states 1 or 2 is concerned, since

$$(F_z)_{1,2} = -\frac{\partial}{\partial z}(E_{1,2}') = \pm 2\left(\frac{D^2}{\Delta E}\right)\mathscr{E}\left(\frac{\partial \mathscr{E}}{\partial z}\right) \tag{15.10}$$

In a uniform field, the force vanishes but in an inhomogeneous field with $\partial\mathscr{E}/\partial z > 0$ the molecules in the 1 level are deflected in the positive z direction and those in the 2 level in the negative z direction.

Having obtained a population in the state of higher energy E_2, maser action is achieved by stimulated emission [3] of the transition from the level 2 to the level 1, which is reinforced by passing the beam through a cavity tuned to the required frequency. In the cavity, let us assume that the magnitude of the electric field varies in time like

$$\mathscr{E}(t) = \mathscr{E}_0 \cos(\omega t) \tag{15.11}$$

The wave function of an ammonia molecule in this field is a time-dependent mixture of the stationary states ψ_1 and ψ_2, and we can write

$$\Psi(t) = A_1(t)\psi_1 \exp(-iE_1t/\hbar) + A_2(t)\psi_2 \exp(-iE_2t/\hbar) \tag{15.12}$$

From the time-dependent Schrödinger equation, we find in a manner with which we are now familiar that the amplitudes $A_1(t)$ and $A_2(t)$ satisfy the equations

$$i\hbar\dot{A}_1(t) = \langle\psi_1|H'|\psi_2\rangle \exp(-i\omega_0 t)A_2(t) \tag{15.13a}$$

and

$$i\hbar\dot{A}_2(t) = \langle\psi_2|H'|\psi_1\rangle \exp(i\omega_0 t)A_1(t) \tag{15.13b}$$

where

$$\langle\psi_1|H'|\psi_2\rangle = \langle\psi_2|H'|\psi_1\rangle = -D\mathscr{E}(t) = -D\mathscr{E}_0\cos(\omega t) \tag{15.14}$$

and

$$\hbar\omega_0 = E_2 - E_1 = \Delta E \tag{15.15}$$

[3] We recall that the transition rate for spontaneous emission is proportional to ω^3, and is therefore negligible at microwave and radio frequencies.

The equations (15.13) can be solved approximately by using the *rotating wave approximation* discussed in Section 2.8. Assuming that at time $t = 0$ the molecules enter the cavity in the upper level 2, one finds that *at resonance* ($\omega = \omega_0$), the probabilities $P_{2\to1}(t)$ and $P_{2\to2}(t)$ for finding the system at time t in the levels 1 and 2, respectively, are given by

$$P_{2\to1}(t) = |A_1(t)|^2 = \sin^2(\bar{\omega}_0 t/4) \tag{15.16a}$$

and

$$P_{2\to2}(t) = |A_2(t)|^2 = \cos^2(\bar{\omega}_0 t/4) \tag{15.16b}$$

where

$$\hbar\bar{\omega}_0 = 2D\mathscr{E}_0 \tag{15.17}$$

We see from (15.16) that every ammonia molecule will make the transition from the upper to the lower level, giving up energy to the microwave field, if its transit time in the cavity, T, is such that

$$\frac{\bar{\omega}_0 T}{4} = \frac{\pi}{2} \tag{15.18}$$

Of course, not all the molecules entering the cavity have the same velocity, but if the most probable velocity in the direction of motion is v, then the length of the cavity, L, should be adjusted so that

$$L = vT = \frac{2\pi v}{\bar{\omega}_0} \tag{15.19}$$

More generally if $P_{2\to1}(T)$ is the probability that the transition $2 \to 1$ occurs while a molecule traverses the cavity, the power supplied to the microwave field is

$$P = N\Delta E P_{2\to1}(T) \tag{15.20}$$

where N is the number of molecules entering the cavity per second. The power lost from the microwave field in the cavity is

$$P_l = P_{\text{out}} + P_{\text{abs}} \tag{15.21}$$

where P_{out} is the power output and P_{abs} is the power absorbed by the walls:

$$P_{\text{abs}} = \frac{\omega_0 E}{Q} \tag{15.22}$$

Here E is the total energy of the field and Q is a constant known as the cavity quality factor (see Section 16.3). The power output can similarly be written as

$$P_{\text{out}} = \frac{\omega_0 E}{Q_1} \tag{15.23}$$

where Q_1 is a quality factor determined by the coupling between the cavity and the output wave guide.

The total energy in the field within the cavity is given by

$$E = \frac{1}{8}\varepsilon_0(\mathscr{E}_0)^2 V \tag{15.24}$$

where V is the cavity volume. If $P_{\text{out}} \ll P_{\text{abs}}$, the condition for the system to act as an oscillator is

$$P = P_l \simeq P_{\text{abs}} \tag{15.25}$$

from which we see that

$$N = \frac{\varepsilon_0 \omega_0 (\mathscr{E}_0)^2 V}{8Q\Delta E P_{2\to1}(T)} \tag{15.26}$$

If $(\bar{\omega}_0 T)$ is small, so that from (15.16) $P_{2\to1}(T) \simeq (\bar{\omega}_0 T/4)^2$, this condition becomes

$$N = \frac{\hbar v^2 \varepsilon_0 V}{2D^2 L^2 Q} \tag{15.27}$$

The mean velocity of molecules from an oven at temperature Θ is

$$v = \left(\frac{3k_{\text{B}}\Theta}{M}\right)^{1/2} \tag{15.28}$$

where M is the mass of the molecule and k_{B} is Boltzmann's constant. For an oven temperature of 700 K, $v \simeq 10^3$ m s^{-1} and with typical values of the other parameters, N is required to be about 10^{13} molecules per second, which can be achieved easily.

The actual magnitude $\mathscr{E}_0$ of the field in the cavity can be obtained from (15.26) if the Rabi flopping formula (2.381) is used to obtain $P_{2\to1}(T)$, rather than the small T approximation. In this case, one finds that

$$\frac{8ND^2}{\hbar\omega_{\text{R}}^2}\sin^2\left(\frac{\omega_{\text{R}}T}{2}\right) = \frac{\varepsilon_0 V}{Q} \tag{15.29}$$

where the Rabi flopping frequency ω_{R} is given by

$$\omega_{\text{R}} = \left[(\omega_0 - \omega)^2 + \frac{\bar{\omega}_0^2}{4}\right]^{1/2} = \left[(\omega_0 - \omega)^2 + \left(\frac{D\mathscr{E}_0}{\hbar}\right)^2\right]^{1/2} \tag{15.30}$$

This is an equation which can be solved numerically to find $\mathscr{E}_0$.

Power outputs of about 10^{-10} W can be obtained with a line width as small as 10^{-2} Hz (see Thorp, 1969), which made the ammonia maser the first of a new class of oscillators. However, as a primary frequency standard it proved to have fatal drawbacks because its frequency depended on various operating parameters (source pressure, operating voltage of the state-selector) which caused uncertainties and instabilities.

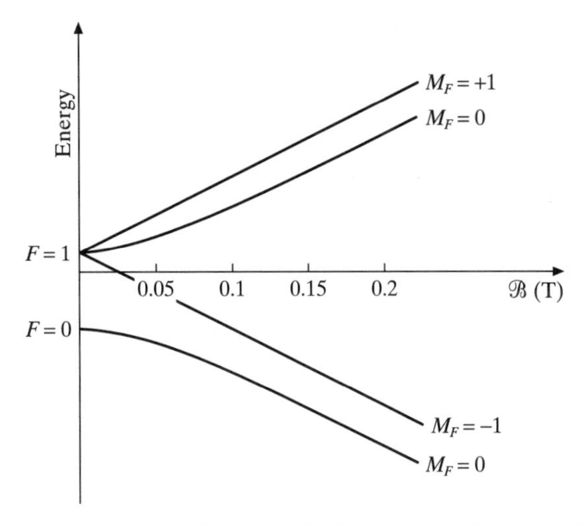

Figure 15.5 Hyperfine splitting of the atomic hydrogen ground state in the presence of a magnetic field.

The hydrogen maser

We have seen in Chapter 5 that because the proton spin is $I = 1/2$ and the ground state level $1s_{1/2}$ of 'ordinary' hydrogen (H) has a total electronic angular momentum quantum number $j = 1/2$, this level splits into two hyperfine components corresponding to the values $F = 0$ and $F = 1$ of the quantum number F, the state with $F = 0$ being the ground state (see Fig. 5.10). The degeneracy of the hyperfine levels, corresponding to the $(2F + 1)$ values of the quantum number $M_F = -F, -F + 1, \ldots, F$, can be removed by applying a magnetic field; this Zeeman effect in hyperfine structure was studied in Chapter 9 for a general atom. The behaviour of the hyperfine splitting of the ground state of atomic hydrogen as a function of the magnitude $\mathcal{B}$ of the magnetic field is shown in Fig. 15.5. It provides the conceptual basis of the hydrogen maser constructed by H.M. Goldenberg, D. Kleppner and N.F. Ramsey in 1960.

In analogy with the ammonia maser, the basic features of the hydrogen maser, shown schematically in Fig. 15.6, are a *source* of hydrogen atoms, collimated into a beam, a *quantum state selector* and a *resonant cavity*. An intense electrical discharge in the source converts molecular hydrogen (H_2) into atomic hydrogen (H). The atoms emerge from the source into a region evacuated to 10^{-6} Torr. They are collimated into a beam in a direction which we take as the Z axis. This beam penetrates into a *magnetic state selector*, which is a hexapole focusing magnet having three north poles alternating in a circle with three south poles. By symmetry, the magnetic field is zero along the Z axis, and increases in magnitude when the distance from this axis grows. Now, as seen from Fig. 15.5, the energy of the two states $F = 0$, $M_F = 0$ and $F = 1$, $M_F = -1$ is a decreasing function of the magnitude

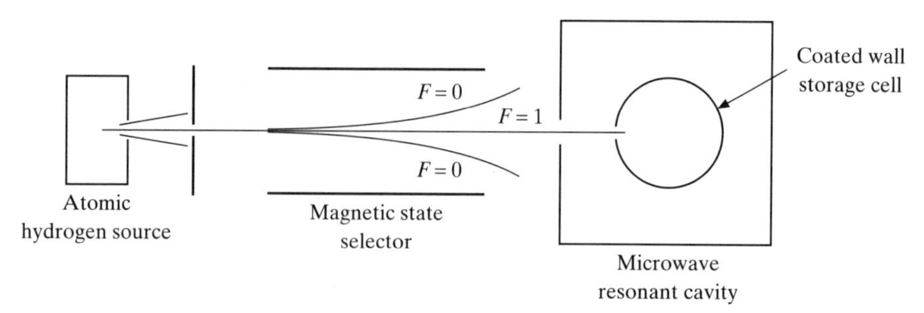

Figure 15.6 The hydrogen maser. H atoms produced by thermal dissociation of H_2 molecules are collimated, selected in the upper level $F = 1$ by the magnetic selector, and then sent into a storage bulb placed inside a resonant cavity.

$\mathcal{B}$ of the magnetic field. Since physical systems are accelerated towards regions of lower potential energy, the hydrogen atoms in these two states will be deflected away from the Z axis, towards regions of stronger magnetic field. Conversely, the hydrogen atoms in the states $F = 1$, $M_F = 0$ and $F = 1$, $M_F = 1$ will be focused onto the axis of the magnetic selector. Thus, at the exit of the selector, the atomic hydrogen beam along the Z axis will contain atoms predominantly in the upper level $F = 1$. The beam then penetrates by a small aperture into a *storage cell* placed inside a microwave resonant cavity. If the atomic hydrogen beam flows steadily, the storage cell will contain more atoms in the upper level $F = 1$ than in the lower level $F = 0$. Thus, if the resonant cavity is tuned at the frequency ν of the hyperfine transition between these two levels, maser action will take place by stimulated emission from the upper level $F = 1$ to the lower level $F = 0$. The energy released by the atoms makes the microwave radiation stronger, and an oscillation at the resonance frequency will increase in magnitude until an equilibrium value is reached. This oscillation will continue indefinitely, the energy to maintain it being supplied continuously by the hydrogen atoms in the $F = 1$ level. The device is then a free running maser oscillator at the frequency ν of the hyperfine transition. To minimise the loss of microwave power in the cavity through conversion to heat in the storage cell material, the latter is chosen to be a low-loss dielectric material such as quartz. Given the limited number of hydrogen atoms available, the power output of the hydrogen maser is very low, typically of the order of 10^{-12} W.

The atomic hydrogen maser allows the measurement of the hyperfine transition frequency to be made with extremely high precision, due to a combination of favourable features. The atoms stay in the storage cell for a comparatively long time (typically 10 s) so that the resonance line is very narrow. Indeed, by the Heisenberg time–energy uncertainty relation (2.23), the width of a resonance in an atomic beam experiment cannot be less than approximately the reciprocal of the time the atom is in the resonance region of the apparatus. Moreover, the atoms are stored at low pressure, so that they are relatively unperturbed when radiating. The first-order Doppler effect shift is removed since the atoms are exposed to a standing wave and their average velocity is very low. The main causes of inaccuracies are (a) the escape of atoms from the cavity, (b) collisions between atoms (which under ordinary operating conditions are spin-flip collisions in which the electrons of the two colliding H atoms exchange spin directions), (c) second-order Doppler effects, (d) residual magnetic effects and (e) collisions of the atoms with the wall coating of the storage cell, which cause a slight shift of the hyperfine frequency, called the wall shift. This is the major disadvantage of the hydrogen maser. However, the wall shift can be determined by measurements utilising storage bottles of different diameters, or with a deformable bulb whose surface to volume ratio can be changed. When all the required corrections are taken into account, and comparison is made with the frequency of the caesium clock defining the primary time standard (see Section 16.4), it is found that the hydrogen atom hyperfine frequency is given by $\nu = (1\,420\,405\,751.7667 \pm 0.0009)$ Hz, which is the value quoted in (5.84). A detailed discussion of the design and performance of the hydrogen maser may be found in the book by Major (1998).

Solid-state lasers

Lasers having ions introduced as 'active impurities' in an otherwise transparent dielectric–host material (in crystalline or glass form) are usually called solid-state lasers. The ions used as active impurities generally belong to one of the series of transition elements of the periodic table (see Section 8.2), in particular rare earth (lanthanides) or transition-metal ions. For host crystals, oxides or fluorides are most often employed.

The ruby laser

Constructed in 1960 by T.H. Maiman, the ruby laser was the first laser to operate. A ruby is a crystalline alumina (Al_2O_3) in which some of the Al^{3+} ions have been replaced by Cr^{3+} ions. The energy levels of interest for laser action are due to the three 3d electrons of the Cr^{3+} ion, under the influence of the field at the Al site in the Al_2O_3 lattice. These energy levels are displayed schematically in Fig. 15.7, which shows that the ruby laser operates as a (double) *three-level* laser. The Cr^{3+} ions can be excited from the ground state (level 1) to two main pump bands (that is, closely spaced levels) denoted by 3 and 3′, respectively. Transition peaks

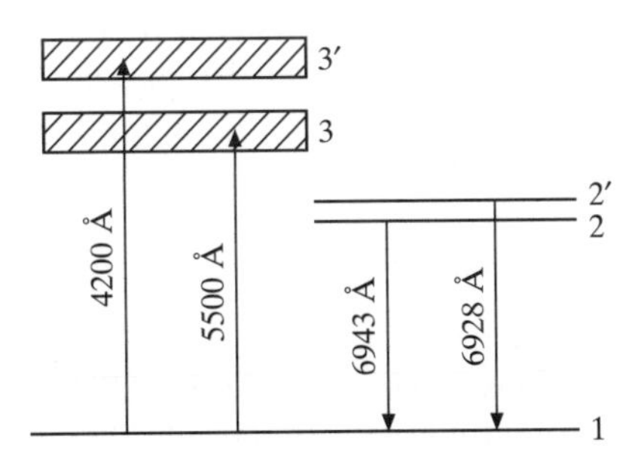

Figure 15.7 Relevant energy levels of ruby.

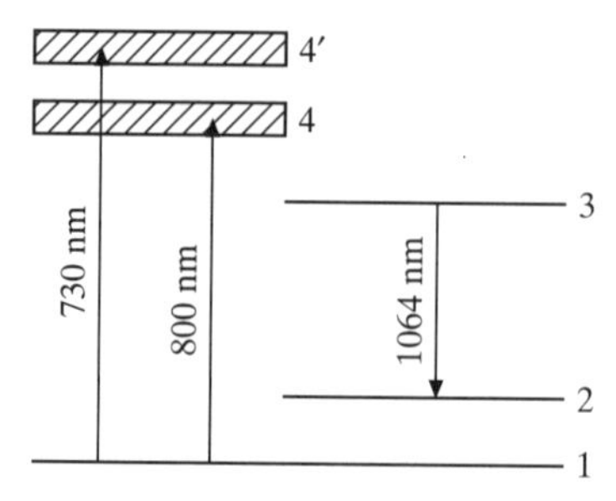

Figure 15.8 Relevant energy levels of Nd:YAG.

to these two bands occur at the wavelengths of 5500 Å (green) and 4200 Å (violet), respectively. Due to interaction with the crystal lattice, these two bands decay by a very fast (~10^{-12} s) non-radiative process to metastable levels 2 and 2′, whose lifetime is very long ($\tau \simeq 3 \times 10^{-3}$ s). Laser action usually occurs on the $2 \to 1$ transition at the wavelength $\lambda_1 = 6943$ Å (red), but also on the $2' \to 1$ transition at the wavelength $\lambda_2 = 6928$ Å (red).

Neodymium lasers

For these lasers, the host medium is often a crystal of $Y_3Al_5O_{12}$, called YAG (an acronym for 'yttrium aluminium garnet'), in which some of the Y^{3+} ions are replaced by Nd^{3+} ions. The relevant energy levels for laser action arise from the three inner-shell 4f electrons of the Nd^{3+} ion, which are screened by eight outer electrons ($5s^2$ and $5p^6$) and influenced by the crystal field of the YAG crystal. The Nd:YAG laser usually operates as a *four-level* laser (see Fig. 15.8), the main pump bands being at wavelengths of 730 nm and 800 nm, respectively. The strongest laser transition occurs at $\lambda = 1064$ nm (infra-red). Nd:YAG lasers are widely used, for example in material processing, medical applications, and high-intensity laser–matter interactions.

Other host media for the neodymium lasers include phosphate or silicate glasses. The energy levels of interest for the Nd:glass laser are approximately the same as those of the Nd:YAG laser, and the strongest laser transition occurs at nearly the same wavelength ($\lambda \simeq 1054$ nm) for phosphate glass. An important application of Nd:glass lasers is in very high-energy systems used in laser-driven, inertial confinement fusion (ICF) experiments. For example, the 'NOVA' Nd:glass laser built at the Lawrence Livermore Laboratory has delivered pulses with an energy of about 100 kJ and peak power of 100 TW (1 TW = 10^{12} W).

The titanium sapphire laser

The Ti:sapphire ($Ti:Al_2O_3$) laser is made by doping Ti_2O_3 into a crystal of Al_2O_3, so that Ti^{3+} ions replace some of the Al^{3+} ions in the crystal lattice. The two energy levels relevant for laser action result from the interaction of the 3d electron of Ti^{3+} with the crystal field, and are represented schematically in Fig. 15.9, as a function

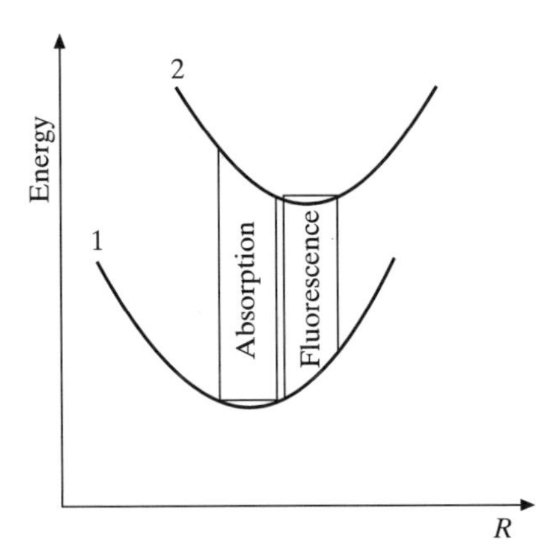

Figure 15.9 The two energy levels relevant for laser action in the Ti:sapphire laser, as a function of the Ti–O distance *R*.

of the Ti–O distance *R*. Since the upper state 2 has a much larger equilibrium distance than the lower state 1 it follows that the absorption and fluorescence bands are wide and well separated. Laser action takes place between the lowest vibrational level of the upper state 2 to some vibrational level of the ground state 1, giving rise to a laser transition with a large bandwidth. As a result, the Ti:sapphire laser is a *tunable* laser with a broad tuning range between 660 nm and 1100 nm. Ti:sapphire lasers are used in particular for the generation and amplification of femtosecond laser pulses.

Dye lasers

Laser action in organic dyes was discovered independently in 1966 by P.P. Sorokin and J.R. Lankard, and by F.P. Schäfer, W. Schmidt and J. Volze. The active material of a dye laser consists of a solution of an organic dye in a liquid solvent, such as methanol, ethanol, glycerol or water. Organic dyes are complex polyatomic molecules such as Rhodamine 6 G or Coumarin 2. The general structure of their electronic energy states is shown in Fig. 15.10 as a function of a coordinate *R* describing one of their many vibrational modes. We first note that in addition to a system of singlet states (^{1}X, ^{1}A, ^{1}B, . . .), there is a system of triplet states (^{3}A, ^{3}B, . . .). Also indicated in Fig. 15.10 are the vibrational sublevels of the electronic states. Due to the large number of vibrational and rotational levels involved, and because of the interaction with the liquid which broadens the lines, the vibrational–rotational structure of the electronic levels is unresolved at room temperature.

The transitions relevant for laser action are those occurring between the two lower electronic singlet states. In general, the molecules occupy the lowest vibrational sublevels of the ground state ^{1}X according to the Boltzmann distribution. In accordance with the selection rule $\Delta S = 0$, the molecules can be excited to the first excited state ^{1}A by absorbing light. As explained in Section 11.3, the molecules

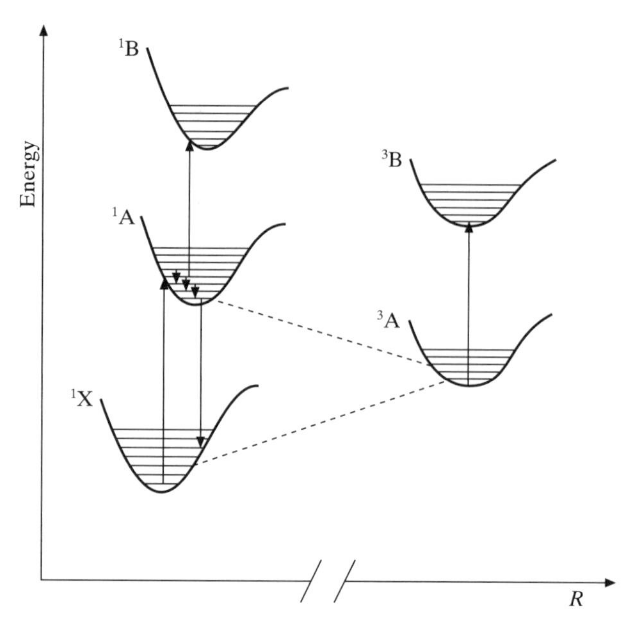

Figure 15.10 Simplified energy level diagram of an organic dye molecule.

can then undergo radiationless transitions and relax very quickly (in about 10^{-12} s) to populate the lowest vibrational level of the excited electronic state ^{1}A, which is the upper laser level. From there, they return to the vibrational sublevels of the ground electronic state ^{1}X (which is the lower laser level) by emitting fluorescent light, with a lifetime of about 10^{-9} s. Rapid radiationless decay within the ground state efficiently depopulates the lower laser level. Thus, dye lasers operate in a four-level scheme over the wavelength range of the fluorescence. Because of the relaxation in the ^{1}A state, the fluorescence curve is independent of the spectral distribution of the absorbed radiation.

The usefulness of an organic dye for the generation of laser light depends on the energies of higher singlet states (such as ^{1}B), and on the influence of the triplet states (^{3}A, ^{3}B, . . .). Strong absorption of fluorescent light can indeed occur if a higher singlet state (such as ^{1}B) has an inappropriate range of energies. Also, due to radiationless *intersystem crossing* transitions due to collisions, dye molecules in the state ^{1}A can be transferred to the ^{3}A state. In the same way, the transition $^3\mathrm{A} \to {}^1\mathrm{X}$, called *triplet quenching*, can occur by collisions with species within the solution. It should be noted that if enough molecules assemble in the lowest vibrational level of the ^{3}A state, then the fluorescent light can be strongly absorbed to cause transitions such as $^3\mathrm{A} \to {}^3\mathrm{B}$ in the triplet system, which are optically allowed ($\Delta S = 0$).

Because of their wavelength tunability and wide spectral range, organic dye lasers have been used in many areas of research. In particular, they provide a tunable source of laser radiation for high-resolution spectroscopy. Other areas of applications include laser photochemistry (in particular isotope separation) and medicine.

Semiconductor lasers

In 1962, it was discovered that laser action can be obtained in certain *semiconductor diode* arrangements. Although the basic requirement of a population inversion still exists, the physical principles at work in semiconductor laser operation differ from those considered so far, and fall outside the scope of this book. They are discussed in detail in Svelto (1998). For the active medium, many semiconductor lasers use a combination of elements of the third group of the periodic table (such as Al, Ga, In) and the fifth group (such as N, P, As, Sb). For example, a typical diode laser material is gallium arsenide (GaAs). Semiconductor lasers can be used in pulsed or continuum operation, and are tunable. They are employed in a variety of applications, for example in communications and as pumps for solid-state lasers.

Gas lasers

We shall now consider lasers using an active medium of low density, of which the gas lasers provide important examples. Because of the low pressure (about 1 Torr) often used, the broadening of the lines is mostly due to the Doppler effect. Gas lasers are usually pumped by passing a large current (such as an electrical discharge) through the gas. In what follows we shall briefly examine a few examples of gas lasers.

The helium–neon laser

The He–Ne laser, built in 1961 by A. Javan, W.R. Bennet and D.R. Herriott shortly after the introduction of the ruby laser, was the first gas laser. The active material is a mixture of helium and neon gases at low pressure (1 Torr). Laser action is obtained from transitions of the neon atom, while helium is added to facilitate the pumping process. The relevant energy levels are shown in Fig. 15.11. Using an electric discharge, the helium atoms are raised by electron impact excitation to the 2^1S and 2^3S levels, which are metastable. The ground state of neon has the configuration $1s^2 2s^2 2p^6$ and the lowest excited states have the configurations $1s^2 2s^2 2p^5 nl$. Each configuration gives rise to a number of sublevels, due to interaction with the remaining five electrons of the 2p orbitals. The nl = 4s and nl = 5s states coincide in energy with the 2^3S and 2^1S helium levels, respectively. As a result, in collisions between the excited helium atoms and ground-state neon atoms, there is a high probability that neon atoms will be excited to these levels, the helium atoms reverting to the ground state. The selection rules allow transitions to the lower lying neon 3p and 4p levels. Furthermore, the lifetimes of the 4s and 5s neon levels are of the order of 10^{-7} s, which is about ten times longer than the lifetimes of the 3p and 4p levels. The He–Ne mixture forms a 'double'

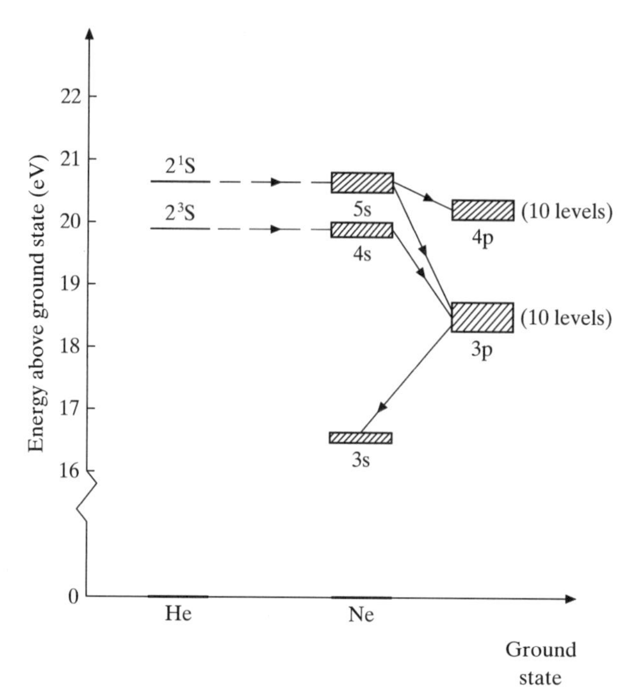

Figure 15.11 Energy level diagram for the He–Ne laser. The neon electronic configurations concerned are $1s^22s^22p^5nl$. Each configuration gives rise to a number of sublevels in the shaded regions.

laser system, which can show laser action between the 4s or 5s level and the 3p level, or between the 5s level and the 4p level. Each of the neon levels consists of several sublevels, and out of the various possible transitions the strongest are: (a) between the 5s and 4p levels at $\lambda = 33\,900$ Å; (b) between the 5s and 3p levels at $\lambda = 6330$ Å; and (c) between the 4s and 3p levels at $\lambda = 11\,500$ Å. The wavelength of the light generated in a He–Ne laser depends on the reflectivity of the mirrors between which the gas is placed. Oscillation occurs at the wavelength for which this reflectivity is a maximum. He–Ne lasers yield low output powers, typically a few mW.

Helium–neon lasers are used in many applications. In particular, those oscillating on the red transition at $\lambda = 6330$ Å or in the green part of the spectrum are employed when a low-power, visible beam is needed (e.g. for alignment or scanning purposes). He–Ne lasers are also frequently used in metrology and holography.

The argon and krypton lasers

The argon and krypton lasers are *gaseous ion lasers*, in which a population inversion between excited states of ionised argon or krypton is achieved. We begin by considering the argon-ion laser. The excited Ar^+ ions are produced in a two-step

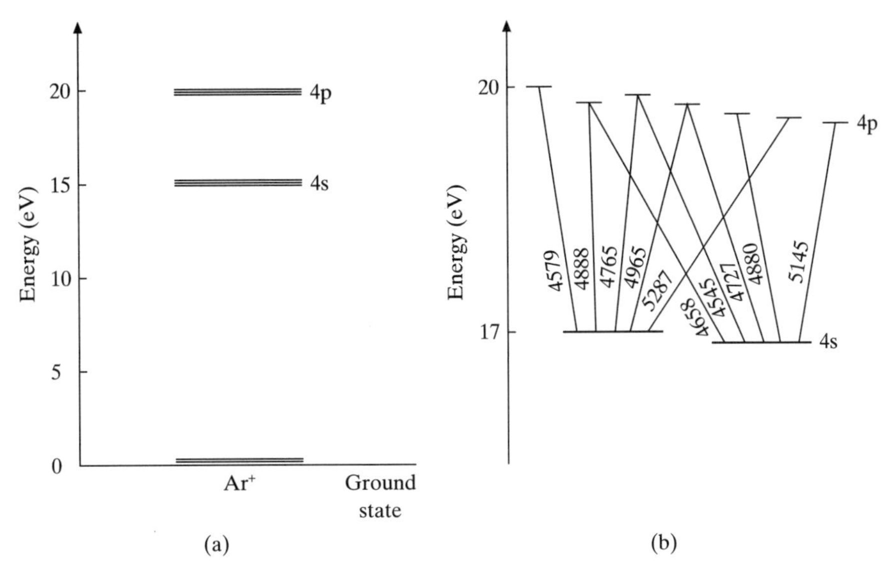

Figure 15.12 (a) Energy levels relevant for the Ar^+ laser. The two excited states denoted by 4s and 4p correspond to the electronic configurations $1s^2 2s^2 2p^6 3s^2 3p^4 4s$ and $1s^2 2s^2 2p^6 3s^2 3p^4 4p$, respectively. Each configuration gives rise to a number of sublevels. (b) Laser lines connecting the sublevels of the 4p and 4s states of Ar^+. The wavelengths of the lines are indicated in ångströms.

process involving electron impact collisions. In the first step, neutral Ar atoms are ionised by electrons having an energy larger than the ionisation potential (15.75 eV) of argon. In the second step, the ground state Ar^+ ion, whose electronic configuration is $1s^2 2s^2 2p^6 3s^2 3p^5$, is excited, again by electron impact. This pumping is accomplished by using a strong DC discharge in low-density gas.

The two excited states of Ar^+ relevant for laser action are the $1s^2 2s^2 2p^6 3s^2 3p^4 4s$ and the $1s^2 2s^2 2p^6 3s^2 3p^4 4p$ states, which we shall denote by 4s and 4p, respectively. As seen from Fig. 15.12(a), the 4p state is located approximately 20 eV above the Ar^+ ground state. Because the lifetime of the 4p state (about 10^{-8} s) is approximately an order of magnitude longer than the radiative lifetime of the transition from the 4s state to the Ar^+ ground state, the excited Ar^+ ions accumulate mostly in the 4p level, which can therefore be used as the upper level of the 4p $\rightarrow$ 4s laser transition. In fact, as shown in Fig. 15.12(b), due to interactions with the other $3p^4$ electrons, both the 4s and 4p states consist of various sublevels. In Fig. 15.12(b), the laser lines connecting the sublevels of the 4p and 4s states are indicated. Several blue and green lines are obtained, the strongest ones being the green line (λ = 5145 Å) corresponding to the 4p $^4D_{5/2} \rightarrow$ 4s $^2P_{3/2}$ transition and the blue line (λ = 4880 Å) corresponding to the 4p $^2D_{5/2} \rightarrow$ 4s $^2P_{3/2}$ transition.

Argon lasers are used in a number of applications, for example as a pump for solid-state lasers (in particular the Ti:sapphire laser) and dye lasers, but also in other areas such as medicine or industry.

The krypton-ion laser operates on the same principles as the argon-ion laser. In addition to blue and green lines, several red lines are also obtained, as well as three strong ultra-violet (UV) lines at 3564 Å, 3507 Å and 3375 Å.

The CO_2 laser

The active medium of the carbon dioxide laser is a mixture of CO_2, N_2 and He. Laser action takes place between vibrational levels of the electronic ground state of the CO_2 molecule. This linear molecule has three fundamental modes of vibration: symmetric stretching, bending and asymmetric stretching, with corresponding frequencies ν_1, ν_2 and ν_3 (see Fig. 15.13(a)). The vibrational state of the molecule is specified by a set of three quantum numbers (v_1, v_2, v_3), one for each vibrational mode. The energy levels relevant for the CO_2 laser are shown in Fig. 15.13(b). The level (001) corresponding to asymmetric stretching is the upper level of the laser transitions. This level can be efficiently populated by nearly resonant collisions with N_2 molecules in the electronic ground state, excited from the $v = 0$ to the $v = 1$ vibrational level; this excitation is accomplished by electron impact collisions, using an electric discharge. Laser action usually takes place on the (001) → (100) transition, corresponding to a wavelength $\lambda_1 = 10.6$ μm, but it can also occur on the (001) → (020) transition, corresponding to a wavelength $\lambda_2 = 9.6$ μm. In fact, both the upper and lower laser levels consist of many closely

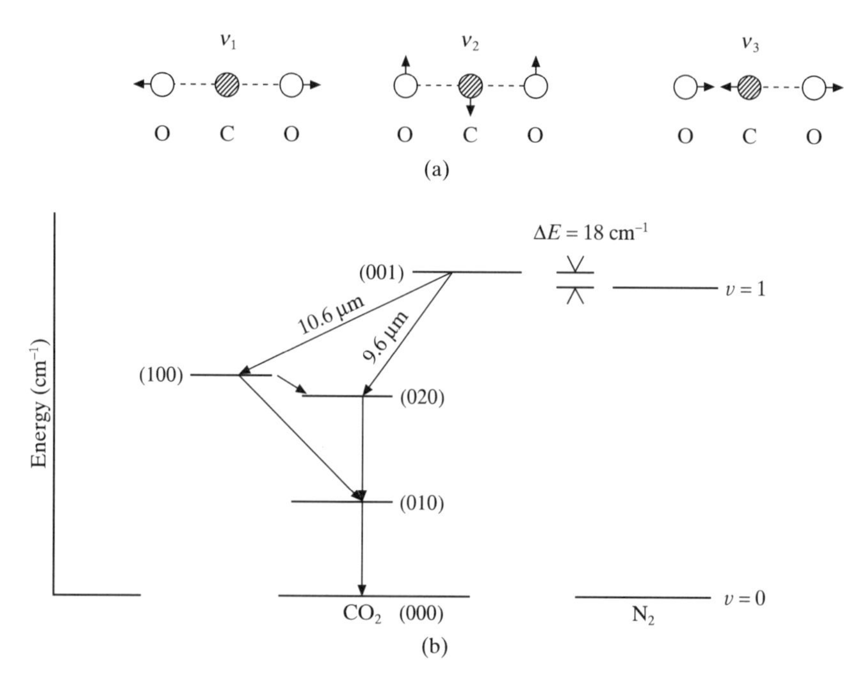

Figure 15.13 (a) The three fundamental modes of vibration of the CO_2 molecule. (b) Energy levels of the electronic ground state of the CO_2 and N_2 molecules relevant for the CO_2 laser.

spaced rotational levels, so that lasing can occur on several vibrational–rotational transitions belonging to either P or R branches, with the P branch having the strongest lines.

The relaxation of the lower levels (100) and (020) of the two lasing transitions occurs as follows. These two levels are strongly coupled, and they are also coupled to the (010) level by the nearly resonant collision processes

$$CO_2(100) + CO_2(000) \rightarrow CO_2(010) + CO_2(010) \tag{15.31a}$$

and

$$CO_2(020) + CO_2(000) \rightarrow CO_2(010) + CO_2(010) \tag{15.31b}$$

Having reached the (010) level, the CO_2 molecules can then relax to the ground state (000) by transferring their vibrational energy to translational energy in collisions with light atoms such as He.

In many applications, non-tunable CO_2 lasers are used. However, by using appropriate gratings, individual lines of the P and R branches of the lasing transitions can be selected, so that the CO_2 laser can be line-tuned. It can also be continuously tuned at high gas pressures. For example, at 10 atmospheres, the pressure broadening is about 2 cm^{-1}, so that the individual vibrational–rotational lines (which are typically separated by 1 to 3 cm^{-1}) merge, and continuous laser action is possible. It is also possible to obtain continuous laser action with some tunability by using *waveguide lasers* operating at a pressure of about one atmosphere. These waveguide lasers are such that the diameter of the laser tube is reduced to a few millimetres, so that laser radiation is guided by the inner walls of the tube.

The CO_2 laser has many applications in various areas of science and technology. In particular, powerful CO_2 lasers (with output powers up to 100 kW) have been constructed, which are used for industrial applications and for fusion research.

Excimer lasers

Excimer molecules are characterised by the absence of a stable ground state, while short-lived excited states exist in dimer form [4]. An excimer level diagram is shown in Fig. 15.14. If enough excimers are produced in the medium, laser action can occur on the transition between the upper (bound) state 2 and the lower (unbound) ground state 1. Once the molecule reaches the ground state, it quickly dissociates due to the repulsive potential of this state. Excimer molecules therefore constitute an ideal active material, with automatic population inversion when an excimer molecule has been created. Because the laser transition occurs between different electronic states, the corresponding wavelength is generally in the UV part of the electromagnetic spectrum. The observed transition is featureless (due to the absence of energy levels in the ground state) and rather broad ($\Delta\nu \simeq$ 20–100 cm^{-1}). The first excimer laser, operated in 1971 by N.G. Basov, V.A. Danilychev and Y.M. Popov, was the Ne_2^* laser.

[4] *Excimer* is a contraction of the words *excited dimer*.

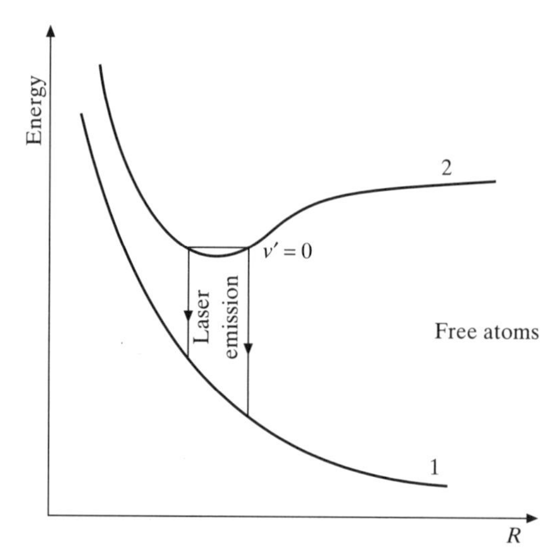

Figure 15.14 Energy levels of an excimer molecule, as a function of the internuclear distance *R*.

An important class of excimer lasers is that in which a rare gas atom (for example, Kr, Ar or Xe) is combined with a halogen atom (for example, Cl or F) to form a rare gas halide excimer. Examples are KrF, XeCl and ArF, which are formed in the excited state by passing a fast electrical discharge in a mixture of the inert gas and F_2 or HCl. Important laser emission lines are those of the KrF laser at $\lambda = 249$ nm, of the XeCl laser at 308 nm, and of the ArF laser at 193 nm. Excimer lasers can deliver average powers up to several kW and have found many important applications, for example in industry (lithography) and medicine.

Free electron lasers

In a free electron laser (FEL), an electron beam moving at a relativistic velocity goes through the magnetic field generated by a periodic structure called an *undulator* or *wiggler*. The alternating magnetic fields force the electrons to acquire a wiggly motion in the plane orthogonal to the magnetic field. This transverse acceleration causes spontaneous longitudinal emission of electromagnetic radiation of the synchrotron radiation type. This radiation interacts with the electron bunch and can stimulate further emission so that laser action can be obtained. It should be noted that this interaction is based on the *scattering* of light. Using suitable magnets, the electron beam can be injected into the laser cavity or deflected out of it.

The first FEL was demonstrated at a wavelength $\lambda = 34000$ Å, using the Stanford linear accelerator. Since then, several FELs have been operated at wavelengths ranging from the millimetre to the UV region, and others are planned, which will extend to the X-ray region. A picture of the FEL at the DESY Laboratory in Hamburg, producing UV light at a wavelength $\lambda = 930$ Å, is shown in Fig. 15.15. FELs have many attractive properties such as wide tunability and high laser

Figure 15.15 The free electron laser at the DESY Laboratory. (From Deutsches Elektronen-Synchrotron (DESY), www.desy.de)

power. However, they are large and expensive, since they involve using large electron beam accelerators.

High-intensity lasers

In recent years, intense laser fields have become available over a frequency range extending from the infra-red to the ultra-violet, in the form of short pulses of durations ranging from nanoseconds (10^{-9} s) to a few femtoseconds (10^{-15} s) yielding intensities up to 10^{20} W cm^{-2}.

To obtain high-intensity laser fields, one must concentrate large amounts of energy into short periods of time, and then focus the laser light onto small areas. In an intense laser system the oscillator produces a train of pulses of short duration. The amplifier then increases the energy of the pulses, which are subsequently focused. A successful method of amplification, called 'chirped pulse amplification' (CPA), was devised in 1985 by D. Strickland and G.A. Mourou. This method, which is illustrated in Fig. 15.16, consists of the following three steps. First, a short laser pulse to be amplified (produced by the oscillator) is stretched in time into its frequency components by a dispersive system such as a pair of diffraction gratings, so that a *chirped* pulse is generated with the red components of the pulse preceding the blue ones. This stretching in time of the pulse greatly reduces its peak intensity, so that in the second step the frequency components of the chirped pulse are sent in succession through the amplifier without distortions and damage. In the third step, the amplified chirped pulse is compressed in time by another pair of diffraction gratings

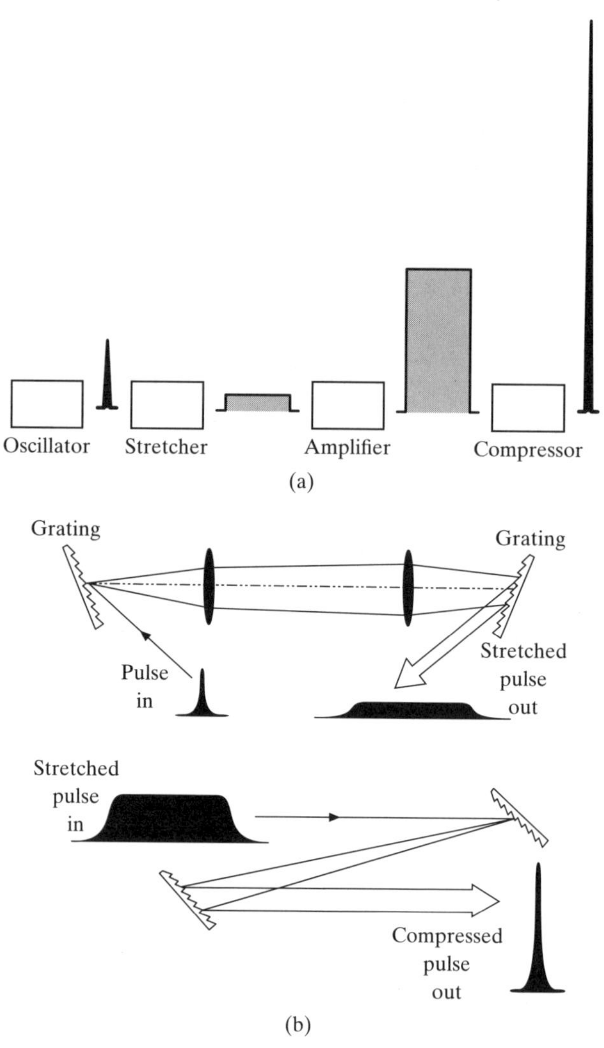

Figure 15.16 Illustration of the principle of the 'chirped pulse' amplification (CPA) method. (a) An oscillator produces a short pulse, which is then chirped: that is, stretched in time into its frequency components. In this way, the peak intensity of the pulse is lowered, so that amplification can take place without damage or distortions. The amplified chirped pulse is then compressed in time, resulting in a short pulse with a very high intensity. (b) The matched stretcher and compressor of the CPA method. The stretcher (top) consists of a telescope of magnification unity placed between two antiparallel gratings. In this configuration, the red components of the pulse have a shorter optical path than the blue ones. Conversely, the compressor (bottom) consists of a pair of parallel gratings, so that the optical path for the blue components of the pulse is shorter than for the red ones. (From G.A. Mourou, C.P.J. Barty and M.D. Perry, *Physics Today*, January 1998, p. 22.)

which recombine the dispersed frequencies, thus producing a short pulse with a very large peak intensity. Finally, the resulting amplified short pulse is focused on a small area. After focusing, the intensities currently available can exceed 10^{18} W cm^{-2}, and in a number of laboratories intensities up to 10^{20} W cm^{-2} have been obtained.

An important advantage of the CPA method is that it can yield very intense, short pulses by using a 'table-top' laser system. The first CPA high-intensity lasers to be constructed used Nd glass as the amplifying medium. As an example, in the CPA Nd:YAG laser system developed during the 1990s at Imperial College, London, a pulse from an Nd oscillator, of wavelength $\lambda = 1064$ nm, duration 1 ps (10^{-12} s) and energy 1 nJ was stretched by diffraction gratings to about 25 ps. It was then amplified by using Nd glass as the amplifying medium to an energy of about 1 J. This amplified chirped pulse was subsequently compressed to a duration close to its initial picosecond value by diffraction gratings, so that output powers of around 1 TW (10^{12} W) could be obtained. By focusing over an area having a diameter of 10 μm, intensities of the order of 10^{18} W cm^{-2} were reached. CPA Nd:YAG lasers of this kind have a repetition rate of about one shot per minute. More recently, CPA laser systems employing Ti:sapphire for the oscillator and the amplifying medium have been used extensively, because they can generate ultra-short pulses with high repetition rates. If only moderate intensities ($\sim 10^{14}$ W cm^{-2}) are required, such laser systems can produce pulses having a duration of about 30 fs, with a repetition rate of 300 kHz. CPA Ti:sapphire lasers can also yield very intense pulses. For example, at the ATLAS laser facility in the Max Planck-Institut für Quantenoptik in Garching, pulses of 100 fs in duration and 1 nJ in energy are stretched, amplified and compressed, giving output pulses of wavelength $\lambda = 790$ nm, duration 150 fs and energy 220 mJ at a repetition rate of 10 Hz. After focusing on a spot of 6 μm in diameter, the maximum intensity available from this laser is 4×10^{18} W cm^{-2}.

Recent technological advances in ultra-fast optics have also allowed the generation of very intense, *ultra-short* laser pulses comprising only a few oscillation cycles of the laser field. A detailed review of this subject has been given by Brabec and Krausz (2000). Ultra-short, very intense laser pulses play an important role in high-intensity laser–matter interactions, as we shall see in Section 15.3.

15.2 Lasers and spectroscopy

The decades since the discovery of the laser have witnessed a revolution in spectroscopy. Unlike other sources of light, laser light is *coherent* and very nearly *monochromatic*. The line width which can be achieved is often smaller than the line widths of the atomic or molecular system to be investigated, and this allows studies with a much higher resolution than could be obtained with a conventional spectrometer using a diffraction grating technique.

As a result of the advent of the tunable dye laser, spectroscopy has been transformed and methods have been devised which avoid the limitations of Doppler broadening and allow precise measurements of natural line widths and other quantities to be made. In what follows the principles of four of these methods will be discussed briefly. For experimental details and a fuller account than space will permit here the reader may consult the books by Demtröder (1996) and Svanberg (2001) or the collection of reprints of research papers made by Gupta (1993).

Atomic beam spectroscopy

Well-collimated beams of atoms can be produced for most elements. The simplest source is an oven containing an atomic gas or vapour which contains a small hole through which the gas effuses with a most probable velocity $v = (3k_B T/M)^{1/2}$. If the beam passes through a collimator and emerges within a small angular interval $\Delta\theta$, the atomic velocities transverse to the beam direction can be made very small, and consequently the Doppler broadening in the direction transverse to the beam is less than that parallel to the beam by a factor $\Delta\theta$. Values of $\Delta\theta$ of the order 10^{-2} radians can easily be produced so that the Doppler broadening is reduced to 10 MHz or less.

Resonance absorption can be studied by illuminating the beam in the transverse direction. With light from a conventional source too few atoms would be excited to be useful, but when a tunable dye laser is employed the high power density ensures that a sufficiently large number of atoms can be excited. The resonant absorption can be measured in several ways, the simplest of which is to observe the fluorescence light which is subsequently emitted (see Fig. 15.17). The intensity of the fluorescence is measured as the laser is tuned across the absorbing line. An early example of the power of the method is the resolution of the hyperfine structure of the sodium D_2 line, due to the $3^3S_{1/2} \rightarrow 3^3P_{3/2}$ transition, by W. Lange, J. Luther, B. Nottbeck and H.W. Schröder in 1973. The splitting of the ground state between the $F = 2$ and $F = 1$ levels is large (1772 MHz) and is easily observed.

The splitting of the excited $3^3P_{3/2}$ level is much smaller, the energy interval between the $F = 3$ and $F = 2$ levels being 62.5 MHz, that between the $F = 2$ and $F = 1$ levels being 36.1 MHz and that between the $F = 1$ and $F = 0$ levels being 16.4 MHz. Because of the high resolution of the experiment which is of the order of 10 MHz, these energy differences can be measured (see Fig. 15.18). In contrast these levels cannot be resolved by conventional spectroscopy for which the resolution is an order of magnitude smaller.

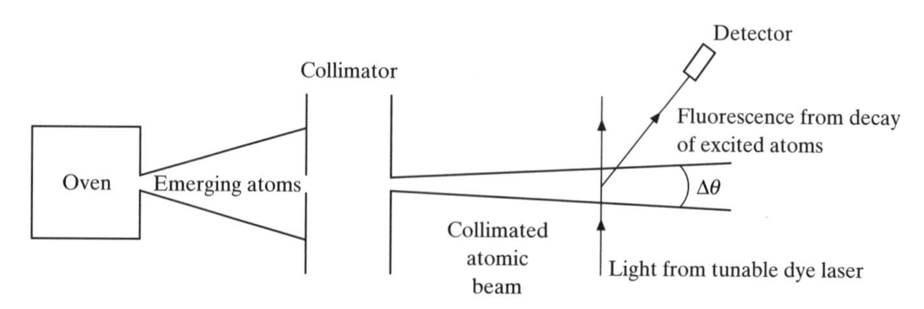

Figure 15.17 Schematic diagram of a laser–atomic beam spectroscopic experiment.

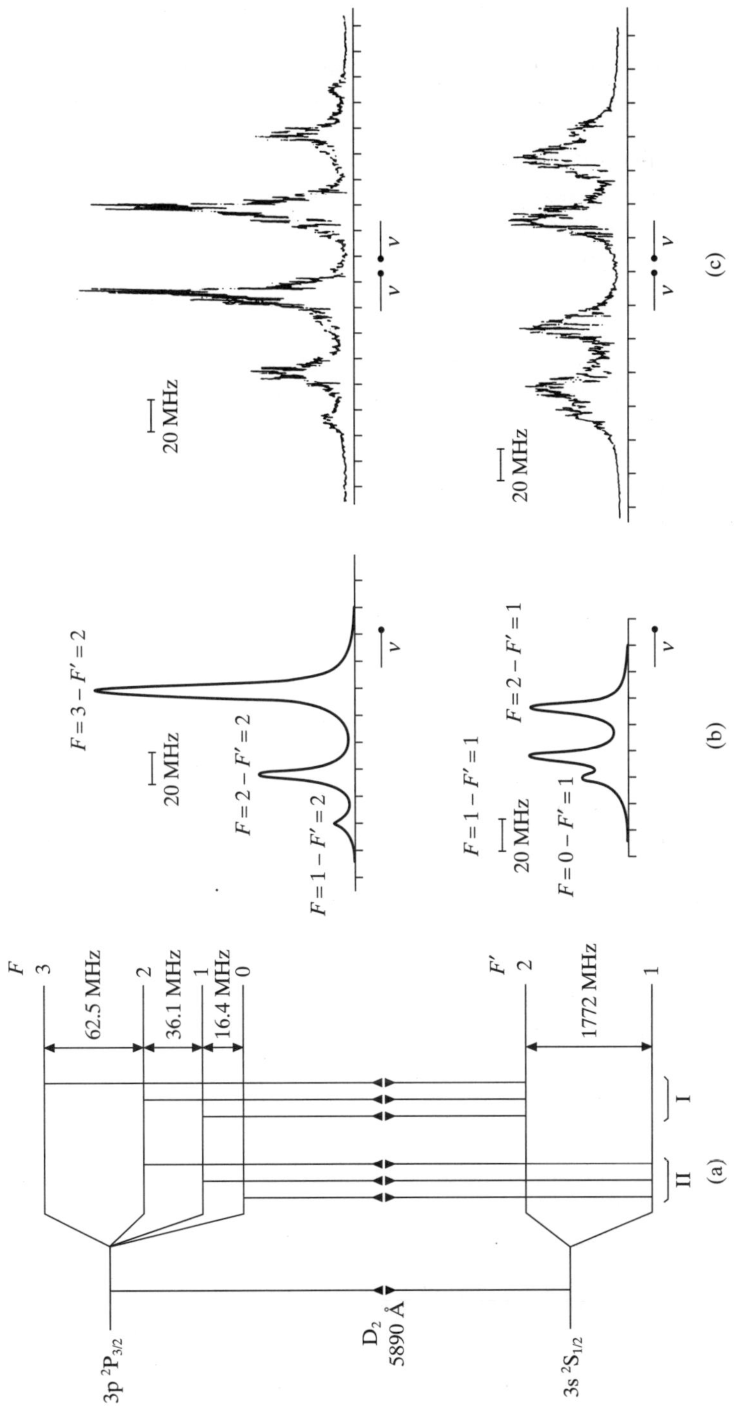

Figure 15.18 The hyperfine structure of the sodium D_2 line. (a) Level scheme. (b) Calculated line profiles. (c) Experimental results. (From W. Lange, J. Luther, B. Nottbeck and H.W. Schröder, *Optics Communications*, **8**, 157, 1973.)

Doppler-free two-photon spectroscopy

The second method of laser spectroscopy which we shall consider is Doppler-free two-photon spectroscopy. Because of the high intensity of laser radiation, transitions in which two photons are absorbed or emitted simultaneously by an atom can be observed, and this can be made the basis of several spectroscopic techniques. Let us first consider an atom which is irradiated by two laser beams of angular frequencies ω and ω' with unit polarisation vectors $\hat{\varepsilon}$ and $\hat{\varepsilon}'$, so that the vector potential is given by

$$\mathbf{A}(\mathbf{r}, t) = A_0\, \hat{\varepsilon} \cos(\omega t - \mathbf{k}\cdot\mathbf{r}) + A_0'\, \hat{\varepsilon}' \cos(\omega' t - \mathbf{k}'\cdot\mathbf{r}) \tag{15.32}$$

where $\mathbf{k} = (\omega/c)\hat{\mathbf{k}}$ and $\mathbf{k}' = (\omega'/c)\hat{\mathbf{k}}'$, $\hat{\mathbf{k}}$ and $\hat{\mathbf{k}}'$ being unit vectors in the direction of propagation of each beam. If the two beams are incident on a stationary atom situated at the origin, the transition rate for two-photon absorption can be calculated by solving the coupled equations (4.32) to second order, just as in the case of Raman and Rayleigh scattering. Two photons can be absorbed from one beam, or one photon can be absorbed from each beam. In the latter case, the transition rate for a transition from an initial atomic state a to a final state b, in the dipole approximation, is found to be

$$W_{ba} = \frac{4\pi^2}{c^2\varepsilon_0^2} I\, I' \left| \sum_n \left\{ \frac{(\hat{\varepsilon}'\cdot\mathbf{D}_{bn})(\hat{\varepsilon}\cdot\mathbf{D}_{na})}{\omega_{na} - \omega} + \frac{(\hat{\varepsilon}\cdot\mathbf{D}_{bn})(\hat{\varepsilon}'\cdot\mathbf{D}_{na})}{\omega_{na} - \omega'} \right\} \right|^2 \delta(\omega_{ba} - \omega - \omega') \tag{15.33}$$

where $\omega_{na} = (E_n - E_a)/\hbar$ and $\omega_{ba} = (E_b - E_a)/\hbar$. The factors I and I' are the intensities of the two beams given in terms of A_0 and A_0' (see (4.13)) by $I = \varepsilon_0\omega^2 cA_0^2/2$, $I' = \varepsilon_0\omega'^2 cA_0'^2/2$ and the factor $\delta(\omega_{ba} - \omega - \omega')$ expresses overall energy conservation. The matrix elements in (15.33) are similar to those for scattering (see (4.223)) and the sum runs over all states n of the atom with $n \neq a$, $n \neq b$. As in Raman scattering the selection rules are

$$\Delta J = 0, \pm 2 \tag{15.34}$$

and the parity of the atom does not change.

We shall limit the discussion to the case in which all the intermediate states are non-resonant so that $\hbar(\omega_{na} - \omega)$ and $\hbar(\omega_{na} - \omega')$ are large compared with the natural widths $\Gamma_n(E)$. As a consequence the finite lifetimes of the intermediate states need not be taken into account. In contrast, the final state b will usually decay by single photon transitions and to find the line profile of two-photon absorption this must be allowed for. If the excited state b is of width Γ_b and of lifetime $\tau_b = \hbar/\Gamma_b$, ω_{ba} must be replaced by $\omega_{ba} - \mathrm{i}\Gamma_b/(2\hbar)$. Using the relationship (4.170) we must make the replacement

$$\delta(\omega_{ba} - \omega' - \omega) \rightarrow \frac{1}{\pi} \frac{\Gamma_b/(2\hbar)}{(\omega_{ba} - \omega - \omega')^2 + (\Gamma_b/2\hbar)^2} \tag{15.35}$$

Since I, I' and the quantity inside the curly brackets in (15.33) vary slowly with ω and ω' (in the case of non-resonant intermediate states), the line profile for two-photon absorption is entirely determined by the factor $g(\omega, \omega')$ where

$$g(\omega, \omega') = \frac{\Gamma_b/(2\hbar)}{(\omega_{ba} - \omega - \omega')^2 + (\Gamma_b/2\hbar)^2} \tag{15.36}$$

This line profile is not realistic, because so far we have not allowed for the fact that the absorbing atoms are moving. If the velocity of an atom is $\mathbf{v}$ then we must set $\mathbf{r} = \mathbf{v}t$ in (15.32), so that the vector potential becomes

$$\mathbf{A}(\mathbf{r}, t) = A_0\,\hat{\boldsymbol{\varepsilon}}\cos\left[\omega\left(1 - \frac{\mathbf{v}\cdot\hat{\mathbf{k}}}{c}\right)t\right] + A_0'\,\hat{\boldsymbol{\varepsilon}}'\cos\left[\omega'\left(1 - \frac{\mathbf{v}\cdot\hat{\mathbf{k}}'}{c}\right)t\right] \tag{15.37}$$

Accordingly, the angular frequencies ω and ω' in the line profile (15.36) should be replaced by the Doppler-shifted frequencies $\omega(1 - \mathbf{v}\cdot\hat{\mathbf{k}}/c)$ and $\omega'(1 - \mathbf{v}\cdot\hat{\mathbf{k}}'/c)$, giving

$$g(\omega, \omega') = \frac{\Gamma_b/(2\hbar)}{[\omega_{ba} - \omega(1 - \mathbf{v}\cdot\hat{\mathbf{k}}/c) - \omega'(1 - \mathbf{v}\cdot\hat{\mathbf{k}}'/c)]^2 + [\Gamma_b/(2\hbar)]^2} \tag{15.38}$$

In general $g(\omega, \omega')$ must now be integrated over the velocity distribution of the atoms. This produces a profile of typical Doppler form. However, if the two beams propagate in opposite directions $\hat{\mathbf{k}} = -\hat{\mathbf{k}}'$ and both have the same angular frequency $\omega = \omega'$, the line profile becomes

$$g(\omega, \omega) = \frac{\Gamma_b/(2\hbar)}{(\omega_{ba} - 2\omega)^2 + [\Gamma_b/(2\hbar)]^2} \tag{15.39}$$

This expression is completely independent of the velocity so that the Doppler broadening has been eliminated. The maximum of this Lorentzian profile is obtained when the radiation is tuned to $\omega = \omega_{ba}/2$. This narrow peak of width $\Delta\omega = \Gamma_b/(2\hbar)$ is superimposed on a much broader Doppler distribution arising from the absorption of two photons from either one of the counter-propagating beams. This broad Doppler shape is of width

$$\Delta\omega^{\mathrm{D}} = \frac{\omega_{ba}}{c}\left(\frac{2k_{\mathrm{B}}T\log 2}{M}\right)^{1/2} \tag{15.40}$$

which is about two orders of magnitude greater that the width $\Gamma_b/(2\hbar)$. The maximum $(2\hbar/\Gamma_b)$ of the line profile (15.39) is much greater than the maximum of the Doppler distribution (see Fig. 15.19), so that the two contributions to the profile are easily distinguished. One method of using this property to make precision spectroscopic measurements is shown in the schematic diagram of Fig. 15.20. A beam from a tunable dye laser is directed into a cell containing the gas to be investigated. A counter-propagating beam is obtained by reflection in a mirror

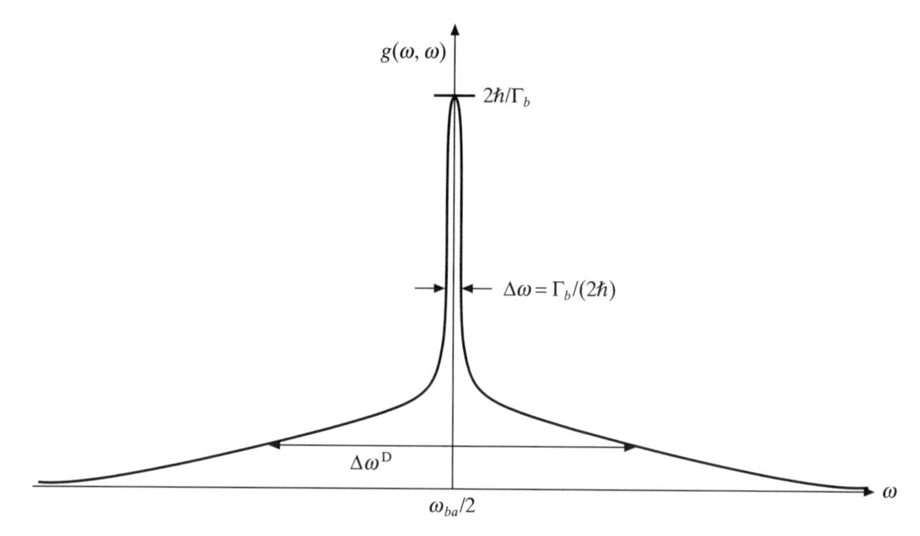

Figure 15.19 Two-photon absorption profile from counter-propagating laser beams of equal frequency: a narrow peak of width $\Delta\omega = \Gamma_b/(2\hbar)$ is superimposed on a background Doppler distribution of width $\Delta\omega^D$, with $\Delta\omega^D \gg \Delta\omega$.

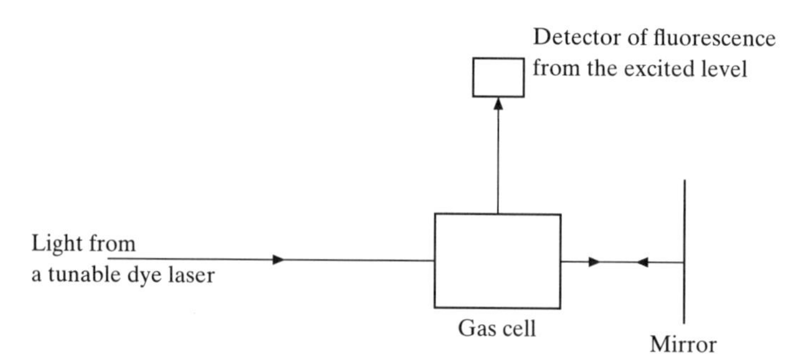

Figure 15.20 Schematic diagram of a two-photon absorption apparatus.

and the intensity of the fluorescence from the excited level b can be studied as the frequency of the laser is swept through the region centred on $\omega = \omega_{bn}/2$, allowing both the width and energy of the level b to be obtained with great precision. Many measurements of this type have been made on various gases and vapours. One of the most famous examples is a measurement of the excitation of the 2s level of atomic hydrogen from the ground state carried out in 1975 by T.W. Hänsch, S.A. Lee, C. Wallenstein and C.E. Wieman. They succeeded in detecting the Doppler-free two photon 1s–2s line and in resolving the hyperfine doublet for the $F = 1 \rightarrow 1$ and $F = 0 \rightarrow 0$ transitions. They also compared the frequencies of the 1s–2s transition in atomic hydrogen and deuterium to that of the Balmer

H_β line ($n = 4 \rightarrow n = 2$) at 4861 Å in order to obtain a value of the Lamb shift for the 1s state. Recently, T.W. Hänsch and co-workers have measured the frequency of the 1s–2s transition in atomic hydrogen, using Doppler-free two-photon spectroscopy of a cold beam of atomic hydrogen. They obtained the value $\nu_{1s,2s} = 246\,606\,141\,318\,710\,3(46)$ Hz, from which precise values of the Rydberg constant and the Lamb shift of the 1s ground state can be derived.

Saturation spectroscopy

A third example of spectroscopic methods relying on the strength and monochromaticity of a laser beam is based on the 'saturation' of a transition. This arises when a gas sample in a cell is illuminated by a laser beam tuned to the angular frequency of a transition to a level b from the ground state a. As the laser intensity is increased the number of atoms in the excited state increases and the number in the ground state decreases until the limiting condition is reached when one-half of the atoms are in level b and one-half in level a. Any further increase in laser intensity does not alter the ratio of the number of atoms in each level and the transition is said to be saturated.

If the angular frequency of a beam of monochromatic laser light is ω then, in the rest frame of an atom moving with a velocity v in a direction parallel to the propagation direction of the beam, the Doppler-shifted angular frequency is $\omega(1 - v/c)$. It follows that a transition from the ground state a to an excited state b can only take place if

$$\left|\omega\left(1 - \frac{v}{c}\right) - \omega_{ba}\right| < \Gamma_b/\hbar \tag{15.41}$$

where $\omega_{ba} = (E_b - E_a)/\hbar$ and Γ_b is the natural width of the level b. Before the laser light is switched on, the velocity distribution of the ground state atoms is Maxwellian and given by (4.189), but, when illuminated, some of the group of atoms with velocities v which satisfy (15.41) will be excited to the level b and the number in level a will be depleted. This causes a narrow dip to be formed in the velocity distribution of width of the order $\Delta v = c\Gamma_b/(\hbar\omega)$. The thermal velocity distribution is said to have a hole burnt into it, sometimes called a Bennet hole (see Fig. 15.21(a)).

Several techniques have been devised to make use of the formation of Bennet holes for Doppler-free spectroscopy. In one such method the incident laser beam is split into two beams which are arranged to counter-propagate through the cell containing the atomic gas to be investigated (see Fig. 15.22). The beam propagating in one direction, called the *pump beam*, is intense enough to saturate the transition and hence causes a large drop in the number of ground state atoms at the appropriate velocity. The counter-propagating beam, called the *probe beam*, is designed to be weak in intensity. It also causes a Bennet hole to be formed (see Fig. 15.21(b)), but because this beam is too weak to saturate the transition

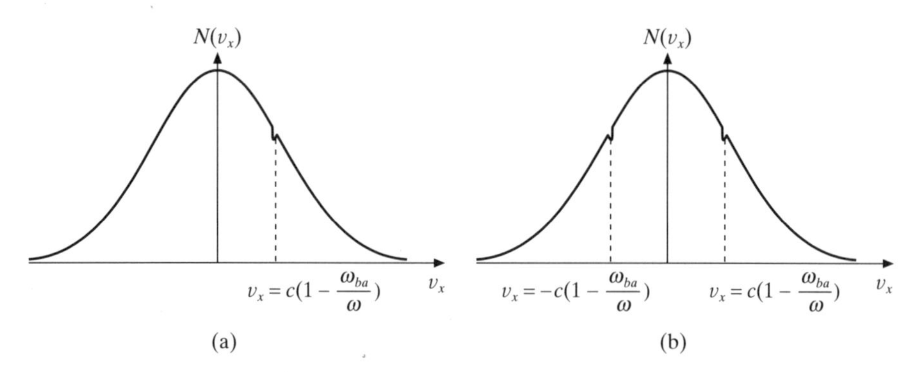

Figure 15.21 (a) A hole, called a Bennet hole, burnt into the velocity distribution of ground state atoms by interaction with a laser beam. (b) Bennet holes formed by counter-propagating laser beams.

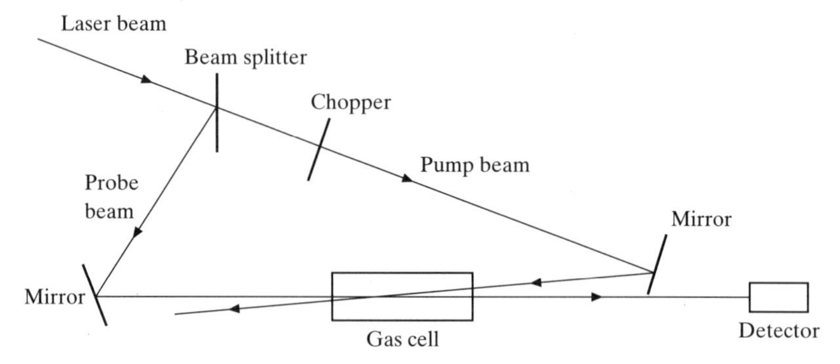

Figure 15.22 Schematic diagram of an apparatus for saturation spectroscopy.

the depth of the hole is small. If the laser is tuned so that $|\omega - \omega_{ba}| < \Gamma_b/\hbar$, both beams interact with the same group of atoms, namely those with $v = 0$. As a consequence, when $\omega \neq \omega_{ba}$ the probe beam is strongly absorbed since there are many atoms in the ground state, but at $\omega \simeq \omega_{ba}$, the pump beam will have saturated the transition so that the absorption of the probe beam will be much diminished and correspondingly the transmission of the probe beam is increased. By measuring the intensity of the transmitted probe beam as the angular frequency ω is changed, the frequency ω_{ba} can be determined to an accuracy limited by the natural width of the level b and not by Doppler broadening. To detect the dependence of the transmitted probe beam on the pump beam, the pump beam is switched rapidly on and off using a 'beam chopper', with which the probe detector is synchronised. This technique was used by T.W. Hänsch, I.S. Shahin and A.L. Schawlow in 1971 to resolve the D_1 line in sodium arising from the 3s $^2S_{1/2} \rightarrow$ 3p $^2P_{1/2}$ transition. For other methods of saturation spectroscopy reference may be made to Svanberg (2001).

Quantum beat spectroscopy

Let us consider an atom possessing a set of states closely spaced in energy. If these states can be excited coherently, the time dependence of the intensity of the radiation emitted subsequently is that of a decreasing exponential modulated by superimposed oscillations. These oscillations are called *quantum beats* and can be studied to provide information about the energy spacing of the atomic states. An example is given by the substates of an atomic level in the presence of a magnetic field (the Zeeman effect). These substates can be excited coherently in a variety of ways, such as passing an atomic beam through a thin foil, excitation by electron impact, or excitation by a short laser pulse. If the substates concerned consist of two energy eigenstates labelled $|1\rangle$ and $|2\rangle$ with energies $E_1 < E_2$ they may be excited coherently, provided the duration of the excitation mechanism Δt satisfies the condition

$$(E_2 - E_1)\Delta t \ll \hbar \qquad \textbf{(15.42)}$$

which ensures that the phase difference developed between the two states is small over the excitation period. When the two substates belong to the same Zeeman or hyperfine structure both substates have the same lifetime τ when decaying radiatively to a final state b. If the coherent mixture was created at time $t = 0$, the time-dependent wave function $|\Psi(t)\rangle$ is given for $t > 0$ by

$$|\Psi(t)\rangle = c_1|1\rangle \exp[-\mathrm{i}E_1 t/\hbar - t/(2\tau)] + c_2|2\rangle \exp[-\mathrm{i}E_2 t/\hbar - t/(2\tau)] \qquad \textbf{(15.43)}$$

where the coefficients c_1 and c_2 are time-independent. The intensity of the decay radiation $I(t)$ in the dipole approximation can be written as

$$I(t) = A|\langle b|\hat{\boldsymbol{\varepsilon}}\cdot\mathbf{D}|\Psi(t)\rangle|^2 \qquad \textbf{(15.44)}$$

where $\hat{\boldsymbol{\varepsilon}}$ is the unit polarisation vector of the emitted radiation, $\mathbf{D}$ is the electric dipole operator and A is a constant. Inserting (15.43) into (15.44) we obtain

$$\begin{aligned} I(t) = A\{&|c_1|^2|\hat{\boldsymbol{\varepsilon}}\cdot\mathbf{D}_{b1}|^2 + |c_2|^2|\hat{\boldsymbol{\varepsilon}}\cdot\mathbf{D}_{b2}|^2 \\ &+ 2|c_1|\,|c_2|\,|\hat{\boldsymbol{\varepsilon}}\cdot\mathbf{D}_{b1}|\,|\hat{\boldsymbol{\varepsilon}}\cdot\mathbf{D}_{b2}| \cos[(E_2 - E_1)t/\hbar + \phi]\} \exp(-t/\tau) \end{aligned} \qquad \textbf{(15.45)}$$

where ϕ is a constant phase. The third term on the right-hand side of (15.45) describes the interference between the two decaying substates and is responsible for modulating the overall exponential decay. Having excited the system at time $t = 0$, the decay intensity can be measured as a function of time by counting the decay photons with fast photomultipliers.

As an example, the modulation of the decay of the 5 $^3P_1$ level of Cd measured by J.N. Dodd, W.J. Sandle and D. Zissermann is shown in Fig. 15.23. The Cd atoms in a vapour were excited from the $J = 0$ ground state by a 200 ns pulse of laser radiation. A magnetic field of magnitude $\mathcal{B}$ produced a small energy separation between the substates with $M_J = 0$ and $M_J = \pm 1$ of the excited state. The observed decay curve resulted from the decay of the coherent mixture of

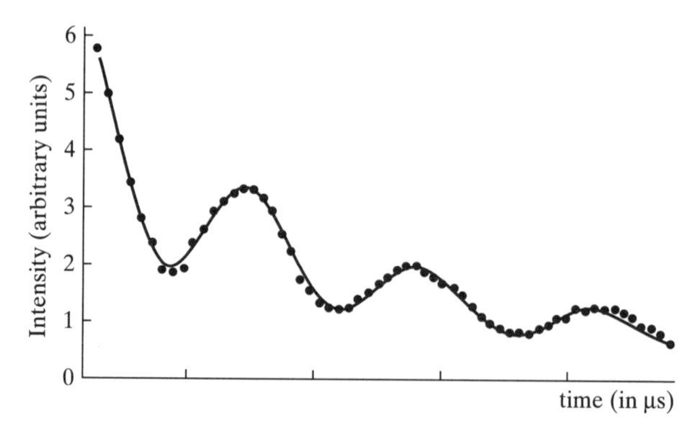

Figure 15.23 Quantum beats observed in the decay of the $M_J = \pm 1$ Zeeman-shifted substates of the 5 $^3P_1$ level of Cd. The dots correspond to the experimental data. (From J.N. Dodd, W.J. Sandle and D. Zissermann, *Proceedings of the Physical Society (London)* **92**, 497, 1967.)

the $M_J = \pm 1$ substates back to the 5 $^1S_0$ ground state. The angular frequency of the modulation is

$$\omega_{21} = (E_2 - E_1)/\hbar = g\mu_B \mathcal{B}/\hbar \tag{15.46}$$

so that the determination of ω_{21} enables the Landé factor g to be found. In addition, the lifetime of the excited state can be obtained from the overall exponential decay.

15.3 Atoms in intense laser fields

If laser fields of sufficient intensity interact with atoms or molecules, processes of higher order than the single photon absorption or emission can play a significant role, and even dominate. These higher order processes, called *multiphoton processses*, correspond to the net absorption or emission of more than one photon in an atomic or molecular transition. For example, an atom or a molecule can undergo a transition from a bound state (for example, the ground state) to another bound state of higher energy via the absorption of n photons ($n \geqslant 2$), a process known as *multiphoton excitation* (MPE). Also, an atom or a molecule in an excited state can emit n photons in a transition to a state of lower energy, either by spontaneous emission (which does not require the presence of an external radiation field) or by stimulated emission. Other examples include the *multiphoton ionisation* (MPI) of atoms and molecules, and the *multiphoton dissociation* (MPD) of molecules. An atomic or molecular system interacting with a strong laser field can also emit radiation at higher order multiples, or harmonics, of the frequency of the laser; this process is known as *harmonic generation* (HG). Finally, radiative collisions involving the exchange (absorption or emission) of n photons can occur in *laser-assisted atomic or molecular collisions* such as electron–atom, electron–molecule or atom–atom collisions in the presence of a laser field.

Except for spontaneous emission, which will not be considered here, the observation of multiphoton transitions requires relatively large radiation intensities that, for optical and nearby frequencies, have become available through the development of the laser. Typically, laser intensities of the order of 10^8 W cm^{-2} are required to observe multiphoton transitions in laser-assisted electron–atom collisions, while intensities of 10^{10} W cm^{-2} are the minimum necessary for the observation of multiphoton ionisation in atoms. In fact, such intensities are now considered to be very modest. Indeed, as we have seen at the end of Section 15.1, laser fields have become available in the form of short pulses having intensities of the order, or exceeding, the value

$$I_a = \frac{1}{2}\varepsilon_0 c \mathscr{E}_a^2 \simeq 3.5 \times 10^{16} \text{ W cm}^{-2} \tag{15.47}$$

corresponding to the atomic unit of electric field strength (see (6.22))

$$\mathscr{E}_a = \frac{e}{(4\pi\varepsilon_0)a_0^2} \simeq 5.1 \times 10^9 \text{ V cm}^{-1} \tag{15.48}$$

and we recall that $\mathscr{E}_a$ is the strength of the Coulomb field experienced by an electron in the first Bohr orbit of atomic hydrogen. Laser fields having intensities of the order, or larger than, I_a are strong enough to compete with the Coulomb forces in controlling the dynamics of atoms and molecules. Thus, while multiphoton processes involving laser fields with intensities $I \ll I_a$ can be studied by using perturbation theory, the effects of laser fields with intensities of the order, or exceeding, I_a must be analysed by using non-perturbative methods. The modified properties of atoms and molecules in intense laser fields generate new behaviour of bulk matter in intense laser fields, with wide ranging applications in plasma and condensed matter physics [5].

Theory of multiphoton processes in atoms

In what follows we shall give an introduction to the theory of multiphoton processes in atoms, and illustrate it subsequently on a few examples. We begin by observing that for all the laser fields which will be considered in this section, the number of photons in a laser mode is very large. For example, a laser field due to an Nd:YAG laser with photon energy $\hbar\omega = 1.17$ eV and modest intensity $I = 10^{12}$ W cm^{-2} in a coherence volume $V = \lambda^3$ (with $\lambda = 2\pi c/\omega = 1064$ nm) has a number of photons equal to

$$\mathcal{N} = \frac{IV}{c\hbar\omega} \simeq 2 \times 10^8 \tag{15.49}$$

---

[5] See for example Batani, Joachain, Martelluci and Chester (2001).

A classical description of the laser field is therefore adequate [6]. On the other hand, the atomic or molecular system will be treated by using quantum mechanics. This classical treatment of the radiation, combined with a quantum treatment of the atom or molecule, constitute the *semi-classical* theory used in previous parts of this book to study the interaction of atoms and molecules with radiation. In addition, since in the present case the number of photons per mode is very large, spontaneous emission can be neglected.

Except for ultra-strong laser fields, which will be considered briefly at the end of this section, we shall neglect relativistic as well as non-dipole effects in laser–atom interactions. The laser field will therefore be described in the dipole approximation by a spatially homogeneous vector potential $\mathbf{A}(t)$, or equivalently by a spatially homogeneous electric field $\mathscr{E}(t)$, with $\mathscr{E}(t) = -\mathrm{d}\mathbf{A}(t)/\mathrm{d}t$. We recall that the magnetic component $\mathscr{B}$ of the laser field, given by $\mathscr{B} = \nabla \times \mathbf{A}$, vanishes in the dipole approximation. For example, a linearly polarised laser field can be described, in the dipole approximation, by the electric field

$$\mathscr{E}(t) = \hat{\boldsymbol{\varepsilon}}\mathscr{E}_0 F(t) \sin(\omega t + \delta) \tag{15.50}$$

where $\hat{\boldsymbol{\varepsilon}}$ is the unit polarisation vector, $\mathscr{E}_0$ is the electric field strength, $F(t)$ is the pulse shape function and δ is a phase. We note that for a chirped pulse, either ω or δ is time-dependent. Laser fields in an arbitrary state of polarisation can be described in the dipole approximation by combining two independent linearly polarised electric fields of the form (15.50), with polarisation vectors $\hat{\boldsymbol{\varepsilon}}_\lambda$ ($\lambda = 1, 2$) perpendicular to the propagation vector $\mathbf{k}$ of the radiation. For the particular case of a linearly polarised *monochromatic* field, equation (15.50) reduces to

$$\mathscr{E}(t) = \hat{\boldsymbol{\varepsilon}}\mathscr{E}_0 \sin(\omega t + \delta) \tag{15.51}$$

the corresponding vector potential being $\mathbf{A}(t) = \hat{\boldsymbol{\varepsilon}} A_0 \cos(\omega t + \delta)$, with $A_0 = \mathscr{E}_0/\omega$.

Basic equations

Let us first write down the equations governing the dynamics of an atom or ion interacting with a laser field described by the electric field (15.50). Since we are neglecting relativistic effects, our starting point is the time-dependent Schrödinger equation

$$i\hbar \frac{\partial}{\partial t} \Psi(X, t) = H(t)\Psi(X, t) \tag{15.52}$$

where $X \equiv (q_1, q_2, \dots, q_N)$ denotes the ensemble of the coordinates of the N atomic electrons. In the above equation, $H(t)$ is the semi-classical Hamiltonian of

[6] A detailed justification of this statement can be given by using the fact that, to a good approximation, the radiation from a laser is in a *coherent state*, which is the quantum electrodynamic state approximating most closely the classical state of the field. For large values of the average number of photons in the coherent state, quantum corrections only cause small fluctuations about the classical vector potential $\mathbf{A}(\mathbf{r}, t)$. See for example Mittleman (1993).

the atom in the presence of the laser field. Following the treatment of Section 9.1, we can write $H(t)$ in the form

$$H(t) = H_0 + H_{\text{int}}(t) \tag{15.53}$$

where $H_0 = T + V$ is the time-independent field-free atomic Hamiltonian. Here T is the sum of the N-electron kinetic energy operators and V is the sum of the two-body Coulomb interactions. The laser–atom interaction is given in the dipole approximation by

$$H_{\text{int}}(t) = \frac{e}{m}\mathbf{A}(t)\cdot\mathbf{P} + \frac{e^2N}{2m}\mathbf{A}^2(t) \tag{15.54}$$

where $\mathbf{P}$ is the total momentum operator, given by (9.27). As in Section 9.1, we can perform the gauge transformation (9.32) and obtain the Schrödinger equation in the *velocity gauge*

$$\mathrm{i}\hbar\frac{\partial}{\partial t}\Psi^{\mathrm{V}}(X,t) = \left[H_0 + \frac{e}{m}\mathbf{A}(t)\cdot\mathbf{P}\right]\Psi^{\mathrm{V}}(X,t) \tag{15.55}$$

or we can perform the gauge transformation (9.35) and write the Schrödinger equation in the *length gauge* as

$$\mathrm{i}\hbar\frac{\partial}{\partial t}\Psi^{\mathrm{L}}(X,t) = [H_0 + e\boldsymbol{\mathscr{E}}(t)\cdot\mathbf{R}]\Psi^{\mathrm{L}}(X,t) \tag{15.56}$$

where $\mathbf{R}$ is given by (9.21).

As we shall see below, in the high-intensity and high-frequency regime it is useful to study the interaction of an atomic system with a laser field in an accelerated frame, called the *Kramers–Henneberger* (K–H) frame. Starting from the Schrödinger equation (15.55) in the velocity gauge, we perform the unitary transformation

$$\Psi^{\mathrm{V}}(X,t) = \exp\left[-\frac{\mathrm{i}}{\hbar}\boldsymbol{\alpha}(t)\cdot\mathbf{P}\right]\Psi^{\mathrm{A}}(X,t) \tag{15.57}$$

where

$$\boldsymbol{\alpha}(t) = \frac{e}{m}\int^t \mathbf{A}(t')\,\mathrm{d}t' \tag{15.58}$$

is a vector corresponding to the displacement of a 'classical' electron from its oscillation centre in the electric field $\boldsymbol{\mathscr{E}}(t)$. The K–H transformation (15.57) therefore corresponds to a spatial translation, characterised by the vector $\boldsymbol{\alpha}(t)$, to a new frame (the K–H frame) moving with respect to the laboratory frame in the same way as a 'classical' electron oscillating in the electric field $\boldsymbol{\mathscr{E}}(t)$. In this accelerated K–H frame, the new Schrödinger equation for the wave function $\Psi^{\mathrm{A}}(X,t)$ is

$$\mathrm{i}\hbar\frac{\partial}{\partial t}\Psi^{\mathrm{A}}(X,t) = \left[\frac{1}{2m}\sum_{i=1}^{N}\mathbf{p}_i^2 + V[\mathbf{r}_1 + \boldsymbol{\alpha}(t),\ldots,\mathbf{r}_N + \boldsymbol{\alpha}(t)]\right]\Psi^{\mathrm{A}}(X,t) \tag{15.59}$$

so that the interaction with the laser field is now incorporated via $\boldsymbol{\alpha}(t)$ into the potential V, which becomes time-dependent. We note that in the case of a linearly polarised monochromatic field (15.51), we have

$$\boldsymbol{\alpha}(t) = \hat{\boldsymbol{\varepsilon}}\alpha_0 \sin(\omega t + \delta) = \boldsymbol{\alpha}_0 \sin(\omega t + \delta) \tag{15.60}$$

where

$$\alpha_0 = \frac{e\mathscr{E}_0}{m\omega^2} \tag{15.61}$$

is called the 'excursion' amplitude of the electron in the laser field and we have defined $\boldsymbol{\alpha}_0 = \hat{\boldsymbol{\varepsilon}}\alpha_0$.

Perturbation theory

At low intensities (such that the electric field strength $\mathscr{E}_0$ is much smaller than the atomic fields relevant to the process considered), time-dependent perturbation theory can in general be used to study multiphoton processes. The simplest form of this approach, called *lowest order perturbation theory* (LOPT), consists in calculating the rate of an n-photon transition by using nth-order perturbation theory for the atom–field interaction. For example, in the case of n-photon ionisation from an initial field-free state ψ_i of energy E_i to final continuum states ψ_f with a density of states $\rho_f(E_f)$ with respect to the ejected electron energy $E_f = \hbar^2 k_f^2/(2m)$, the LOPT transition rate is

$$W_{fi}^{(n)} = \left(\frac{2\pi}{\hbar}\right)(2\pi\alpha\hbar)^n I^n |T_{fi}^{(n)}|^2 \rho_f(E_f) \tag{15.62}$$

where α is the fine structure constant and $T_{fi}^{(n)}$ is the LOPT transition matrix element for the absorption of n photons, given in the length gauge by (see for example Lambropoulos, 1976; Faisal, 1987)

$$T_{fi}^{(n)} = \sum_{\kappa_1}\sum_{\kappa_2}\cdots\sum_{\kappa_{n-1}} \frac{\langle\psi_f|\hat{\boldsymbol{\varepsilon}}\cdot\mathbf{R}|\psi_{\kappa_{n-1}}\rangle\ldots\langle\psi_{\kappa_2}|\hat{\boldsymbol{\varepsilon}}\cdot\mathbf{R}|\psi_{\kappa_1}\rangle\langle\psi_{\kappa_1}|\hat{\boldsymbol{\varepsilon}}\cdot\mathbf{R}|\psi_i\rangle}{(E_i + (n-1)\hbar\omega - E_{\kappa_{n-1}})\ldots(E_i + 2\hbar\omega - E_{\kappa_2})(E_i + \hbar\omega - E_{\kappa_1})} \tag{15.63}$$

In this expression, the summations over the intermediate states $\psi_{\kappa_1}, \psi_{\kappa_2}, \ldots, \psi_{\kappa_{n-1}}$ include integrations over the continuum for all sequences of one photon E1 transitions allowed by angular momentum and parity selection rules.

It is clear from the expression (15.63) that the n-photon atomic transition $i \to f$ is not limited by the single photon selection rules for the same atomic transition. As a result, the spectroscopy of multiphoton excitation is much richer than that of single photon excitation. This was illustrated in Section 15.2 for the case of the 1s–2s transition in atomic hydrogen, a 'forbidden' one-photon transition which can be induced by two-photon absorption.

The calculation of the LOPT transition matrix element $T_{fi}^{(n)}$ of equation (15.63) is in general a difficult task, particularly for high-order multiphoton processes and (or) for complex atoms. A relatively simple case is that of non-resonant multiphoton ionisation (MPI) in one-electron atoms, for which LOPT has been applied successfully for intensities $I < 10^{13}$ W cm^{-2}. Discrepancies from the perturbative I^n power law of equation (15.62) signal the breakdown of LOPT, as do other strong field phenomena which will be discussed below.

We also note that the LOPT expression (15.63) always fails for *resonant* multiphoton processes, such that one or several factors of the denominator vanishes, that is if

$$E_i + r\hbar\omega = E_{\kappa_r} \tag{15.64}$$

for a particular $r \in \{1, 2, \ldots, n-1\}$. In this case, modifications of the theory are required, in which the resonantly coupled states, also called *essential states*, are treated in a non-perturbative way, whereas the other states are treated by using perturbation theory (see Lambropoulos, 1976; Faisal, 1987). This approach, which belongs to the category of *semi-perturbative methods*, is conveniently formulated by using the Feshbach projection operator formalism (see Chapter 13), where the projection operator P now projects onto the space of the resonantly coupled essential states, while the projection operator $Q = I - P$ projects onto the complement space. Applications of the method of essential states to multiphoton phenomena are discussed in the review articles of Lambropoulos and Tang (1992) and Burnett, Reed and Knight (1993).

When resonances are present, multiphoton processes are usually dominated by the resonant contributions. This is the case, for example, for resonant enhanced multiphoton ionisation (REMPI). Measurements of the angular distribution of the photoelectrons can then be used to characterise the resonant intermediate state(s).

'Free' electron in a laser field. Volkov wave function. Ponderomotive energy

We now turn to fully non-perturbative methods. We begin by considering the simple case of a 'free' electron in the presence of a laser field described in the dipole approximation by the vector potential $\mathbf{A}(t)$. The corresponding time-dependent Schrödinger equation reads

$$\mathrm{i}\hbar\frac{\partial}{\partial t}X(\mathbf{r}, t) = \frac{1}{2m}[\mathbf{p} + e\mathbf{A}(t)]^2 X(\mathbf{r}, t) \tag{15.65}$$

It is convenient to perform the gauge transformation

$$X(\mathbf{r}, t) = \exp\left[-\frac{\mathrm{i}}{\hbar}\frac{e^2}{2m}\int^t \mathbf{A}^2(t')\,\mathrm{d}t'\right] X^{\mathrm{V}}(\mathbf{r}, t) \tag{15.66}$$

which gives for $X^{\mathrm{V}}(\mathbf{r}, t)$ the Schrödinger equation in the velocity gauge

$$\mathrm{i}\hbar \frac{\partial}{\partial t} X^{\mathrm{V}}(\mathbf{r}, t) = \left[\frac{\mathbf{p}^2}{2m} + \frac{e}{m} \mathbf{A}(t) \cdot \mathbf{p} \right] X^{\mathrm{V}}(\mathbf{r}, t) \qquad \mathbf{(15.67)}$$

Since $\exp(\mathrm{i}\mathbf{k}\cdot\mathbf{r})$ is an eigenfunction of the operator $\mathbf{p} = -\mathrm{i}\hbar\nabla$ corresponding to the eigenvalue $\hbar\mathbf{k}$ (where $\mathbf{k}$ is the wave vector of the electron), we can look for solutions of (15.67) having the form

$$X^{\mathrm{V}}_{\mathbf{k}}(\mathbf{r}, t) = \exp(\mathrm{i}\mathbf{k}\cdot\mathbf{r}) f(t) \qquad \mathbf{(15.68)}$$

Substituting (15.68) into (15.67), we obtain for the function $f(t)$ the first-order differential equation

$$\mathrm{i}\hbar \frac{\mathrm{d}}{\mathrm{d}t} f(t) = \left[\frac{\hbar^2 k^2}{2m} + \frac{e\hbar}{m} \mathbf{k}\cdot\mathbf{A}(t) \right] f(t) \qquad \mathbf{(15.69)}$$

which is readily solved to give

$$f(t) = C \exp[-\mathrm{i}\hbar k^2 t/(2m) - \mathrm{i}\mathbf{k}\cdot\boldsymbol{\alpha}(t)] \qquad \mathbf{(15.70)}$$

where $\boldsymbol{\alpha}(t)$ is given by (15.58) and C is a normalisation constant. Substituting (15.70) into (15.68), we find that the solution of (15.67) is the (non-relativistic) *Volkov wave function* [7]

$$X^{\mathrm{V}}_{\mathbf{k}}(\mathbf{r}, t) = (2\pi)^{-3/2} \exp\{\mathrm{i}\mathbf{k}\cdot[\mathbf{r} - \boldsymbol{\alpha}(t)] - \mathrm{i}Et/\hbar\} \qquad \mathbf{(15.71)}$$

where $E = \hbar^2 k^2/(2m)$ is the electron energy and we have chosen $C = (2\pi)^{-3/2}$ so that $X^{\mathrm{V}}_{\mathbf{k}}$ is normalised to a delta function. Using (15.66) and (15.71), the Volkov solution $X_{\mathbf{k}}(\mathbf{r}, t)$ of the Schrödinger equation (15.65) is

$$X_{\mathbf{k}}(\mathbf{r}, t) = (2\pi)^{-3/2} \exp\left\{ \mathrm{i}\mathbf{k}\cdot[\mathbf{r} - \boldsymbol{\alpha}(t)] - \mathrm{i}Et/\hbar - \frac{\mathrm{i}}{\hbar}\frac{e^2}{2m} \int^t \mathbf{A}^2(t')\,\mathrm{d}t' \right\} \qquad \mathbf{(15.72)}$$

We also note that in the K–H frame the Volkov wave function is simply the plane wave

$$X^{\mathrm{A}}_{\mathbf{k}}(\mathbf{r}, t) = (2\pi)^{-3/2} \exp[\mathrm{i}(\mathbf{k}\cdot\mathbf{r} - Et/\hbar)] \qquad \mathbf{(15.73)}$$

The Volkov wave functions have a number of applications in multiphoton physics. For example, in laser-assisted electron–atom collisions they play the role devoted to the ordinary plane waves in the corresponding field-free collision processes. Volkov wave functions have also been used to calculate approximate multiphoton ionisation rates when the Coulomb interaction between the ejected electron and the ionic core is neglected.

[7] Exact solutions of the Dirac equation for a charged particle in a plane electromagnetic wave were obtained by D.M. Volkov in 1935.

Before leaving the subject of a 'free' electron in a laser field, we note that the electron *ponderomotive energy* U_p is defined as the kinetic energy of a classical quivering electron in the laser field, averaged over a laser cycle. For non-relativistic electron velocities, and for a linearly polarised monochromatic field (15.51), one finds that

$$U_p = \frac{e^2\mathscr{E}_0^2}{4m\omega^2} \quad \textbf{(15.74)}$$

Since the intensity I is proportional to $\mathscr{E}_0^2$, we see that the ponderomotive energy is proportional to I/ω^2.

Floquet theory

We shall now consider a fully non-perturbative approach for studying laser–atom interactions, namely the Floquet theory [8], first used in this context by J.H. Shirley in 1965. The key feature of the Floquet theory is that it permits the time-dependent Schrödinger equation for an atomic system interacting with a periodical laser field to be reduced to an infinite set of *time-independent* coupled equations, in which the atom–field coupling is treated in a completely non-perturbative way.

Although the Floquet theory can be applied to polychromatic laser fields [9], we shall focus here on monochromatic laser fields of angular frequency ω and of arbitrary polarisation. The Hamiltonian $H(t)$ of the atom–laser field system is then periodic,

$$H(t+T) = H(t) \quad \textbf{(15.75)}$$

with a period $T = 2\pi/\omega$. The Floquet theory can therefore be used to write the wave function $\Psi(X, t)$ in the form [8]

$$\Psi(X, t) = \exp(-\mathrm{i}Et/\hbar)F(X, t) \quad \textbf{(15.76)}$$

where the time-independent quantity E is called the *quasi-energy*, and the function $F(X, t)$ is periodic in time, with period T, so that it can be expressed as the Fourier series

$$F(X, t) = \sum_{n=-\infty}^{+\infty} \mathrm{e}^{-\mathrm{i}n\omega t} F_n(X) \quad \textbf{(15.77)}$$

---

[8] This theory was originally developed in 1883 by G. Floquet to study differential equations with periodic coefficients. In quantum mechanics, the Floquet theory can be applied to the solution of the Schrödinger equation when the Hamiltonian is periodic in space or time. In the case of periodicity in space, for example in a crystal lattice, the Floquet approach leads directly to Bloch waves and conduction bands (see for example Bransden and Joachain, 2000). The case of interest to us here is the periodicity in time.

[9] See for example Chu (1985).

The functions $F_n(X)$ are called the *harmonic components* of $F(X, t)$. Using (15.76) and (15.77), we obtain for $\Psi(X, t)$ the Floquet–Fourier expansion

$$\Psi(X, t) = \exp(-\mathrm{i}Et/\hbar) \sum_{n=-\infty}^{+\infty} \mathrm{e}^{-\mathrm{i}n\omega t} F_n(X) \tag{15.78}$$

If we also make a Fourier analysis of the interaction Hamiltonian,

$$H_{\text{int}}(t) = \sum_{n=-\infty}^{+\infty} \mathrm{e}^{-\mathrm{i}n\omega t} (H_{\text{int}})_n \tag{15.79}$$

then substitute both (15.78) and (15.79) into the Schrödinger equation (15.52) and use (15.53), we obtain for the harmonic components $F_n(X)$ the *time-independent* infinite system of coupled equations

$$(E + n\hbar\omega - H_0)F_n(X) = \sum_{k=-\infty}^{+\infty} (H_{\text{int}})_{n-k} F_k(X) \tag{15.80}$$

with $n = 0, \pm 1, \pm 2, \ldots$. These equations, together with the boundary conditions, form an eigenvalue problem for the quasi-energies E, which we can write as

$$(\mathbf{H}_{\mathrm{F}} - E)\mathbf{F} = 0 \tag{15.81}$$

where the Floquet Hamiltonian matrix $\mathbf{H}_{\mathrm{F}}$ is an infinite matrix of operators. In general, this eigenvalue problem must be solved by keeping only a finite number of harmonic components, that is by truncating the sum on n in the equations (15.77) and (15.78).

In the case of multiphoton ionisation, the discrete states of the atom can ionise since they are coupled to the continuum states by the laser field. In other words, the discrete states become *resonant* states in the laser field. The quasi-energies are therefore complex and can be expressed as

$$E = E_i + \Delta - \mathrm{i}\frac{\Gamma}{2} \tag{15.82}$$

where E_i is the energy of the initial unperturbed (field-free) state and Δ is the AC Stark shift of the state. The physical meaning of Γ can be deduced by noting that the integral over a finite volume of the electron density, averaged over one cycle, decreases in time like $\exp(-\Gamma t/\hbar)$. Hence, the characteristic lifetime of an atom described by the Floquet state (15.76) is $\hbar/\Gamma$, which means that $\Gamma/\hbar$ is the total ionisation rate of that state. We also note that, in the velocity or length gauges, the interaction Hamiltonian can be written in the form

$$H_{\text{int}}(t) = H_+ \mathrm{e}^{-\mathrm{i}\omega t} + H_- \mathrm{e}^{\mathrm{i}\omega t} \tag{15.83}$$

where H_+ and H_- are time-independent operators. The coupled Floquet equations (15.80) then take the simpler form

$$(E + n\hbar\omega - H_0)F_n(X) = H_+ F_{n-1}(X) + H_- F_{n+1}(X) \tag{15.84}$$

and the Floquet Hamiltonian matrix $\mathbf{H}_{\mathrm{F}}$ is a tridiagonal matrix of operators.

The Sturmian–Floquet method

The Floquet theory has been used extensively to study multiphoton processes in atomic systems. In particular, detailed calculations have been performed for one-electron or quasi one-electron atoms. Denoting by $\mathbf{r}$ the electron position vector, and following the approach proposed in 1983 by A. Maquet, S.I. Chu and W.P. Reinhardt, the system of coupled equations (15.84) can be solved by expanding each harmonic component $F_n(\mathbf{r})$ on a *discrete* basis set:

$$F_n(\mathbf{r}) = \sum_{NLM} c_{NLM}^{(n)} r^{-1} S_{NL}^{\kappa}(r) Y_{LM}(\hat{\mathbf{r}}) \tag{15.85}$$

where the $Y_{LM}(\hat{\mathbf{r}})$ are spherical harmonics and the radial functions $S_{NL}^{\kappa}(r)$ are complex *Sturmian* functions defined as

$$S_{NL}^{\kappa}(r) = [(2L+1)!]^{-1} \left[\frac{(N+L)!}{(N-L-1)!}\right]^{1/2} (-2\mathrm{i}\kappa r)^{L+1} \times \exp(\mathrm{i}\kappa r)\, {}_1F_1(L+1-N, 2L+2, -2\mathrm{i}\kappa r) \tag{15.86}$$

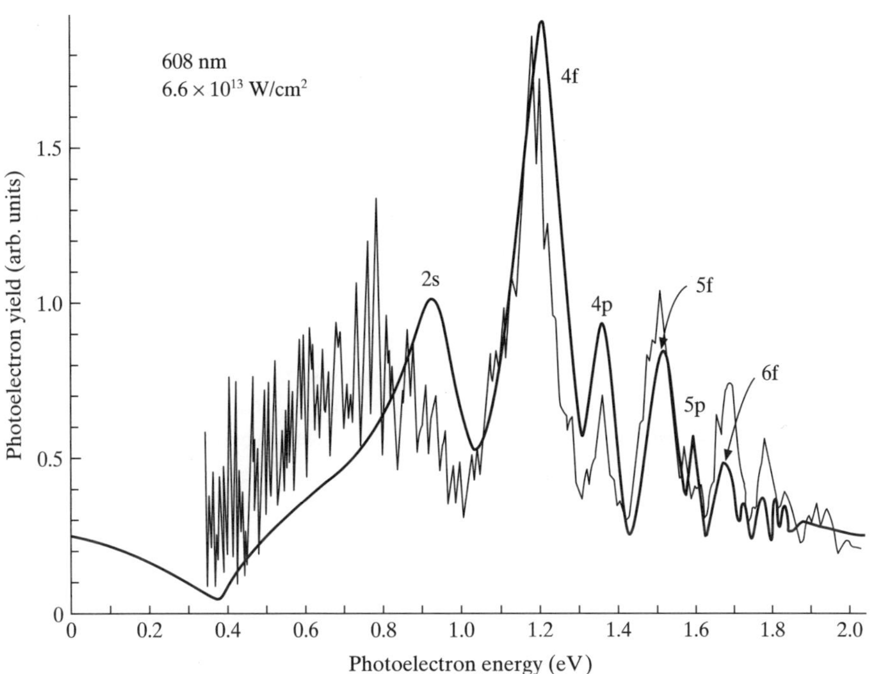

Figure 15.24 The yield of photoelectrons, into the lowest open channel, as a function of the photoelectron energy, for ionisation of H(1s) by a 608 nm laser pulse whose peak intensity is 6.6×10^{13} W cm^{-2} and whose duration is 0.5 ps. The thick line represents the theoretical results of M. Dörr *et al.* and the thin line corresponds to the experimental data of H. Rottke *et al.* (From R.M. Potvliege and R. Shakeshaft, in *Atoms in Intense Laser Fields*, edited by M. Gavrila, Academic Press, Boston (1992), p. 373.)

where the parameter κ is complex, allowing the computation of quasi-bound states (also called Siegert states). The Sturmian–Floquet method has been applied extensively to study multiphoton ionisation and harmonic generation for atomic hydrogen and other hydrogenic or quasi one-electron systems (such as alkali atoms), in monochromatic and bichromatic (two-colour) laser fields. A detailed review of the application of the Sturmian–Floquet method to multiphoton processes has been given by Potvliege and Shakeshaft (1992).

As an example, we show in Fig. 15.24 the low-energy part of the photoelectron spectrum for the multiphoton ionisation of H(1s) by a laser pulse of wavelength $\lambda = 608$ nm and peak intensity 6.6×10^{13} W cm^{-2}. The theoretical predictions of M. Dörr, R.M. Potvliege and R. Shakeshaft, using the Sturmian–Floquet method, are seen to agree well with the experimental data of H. Rottke, B. Wolff, M. Brickwedde, D. Feldmann and K.H. Welge. The peaks are due to *Stark-shift induced resonances*, a phenomenon which we shall analyse in detail subsequently.

R-matrix–Floquet theory

The *R*-matrix–Floquet (RMF) theory is a non-perturbative method proposed in 1990 by P.G. Burke, P. Francken and C.J. Joachain to analyse multiphoton processes in intense laser fields. The RMF theory treats multiphoton ionisation, harmonic generation and laser-assisted electron–atom collisions in a unified way. It is applicable to an arbitrary atomic system and allows an accurate description of electron correlation effects.

Let us return to the time-dependent Schrödinger equation (15.52) for an N-electron atom or ion interacting with a laser field. According to the R-matrix method (see Sections 12.7 and 13.2), configuration space is subdivided into two regions (see Fig. 15.25). As in Section 13.2, we shall define the internal region by the condition that the radial coordinates r_i of all the electrons of the system (N electrons in the present case) are such that $r_i \leq a$. The sphere of radius a is such that it envelops the charge distribution of the target atom states retained in the calculation. We recall that in the internal region exchange and correlation effects involving all N electrons are important. The external region is defined so that one

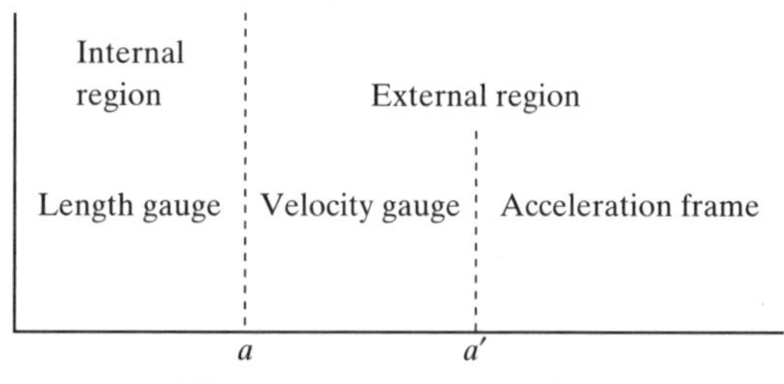

Figure 15.25 Partitioning of configuration space in the *R*-matrix–Floquet theory.

of the electrons (say electron N) has a radial coordinate $r_N \geqslant a$, and the remaining $N-1$ electrons are confined within the sphere of radius a. Hence in the external region exchange and correlation effects between the 'outer' electron and the remaining $(N-1)$ ones can be neglected. It is worth noting that by allowing only one electron to be present in the external region we restrict our analysis to processes involving at most one unbound electron, namely multiphoton single ionisation, harmonic generation and laser-assisted electron–atom (ion) elastic and inelastic collisions.

Having divided configuration space into an internal and an external region, we must solve the time-dependent Schrödinger equation (15.52) in these two regions separately. This is done by using the Floquet method which, as we have seen above, reduces the problem to solving an infinite set of coupled time-independent equations for the harmonic components $F_n(X)$ of the wave function $\Psi(X, t)$. The solutions in the internal and external regions are then matched on the boundary at $r = a$.

In the internal region it is convenient to use the length gauge, because in this gauge the laser–atom coupling tends to zero at the origin. We remark that in this region the Floquet Hamiltonian matrix $\mathbf{H}_\mathrm{F}$ is not Hermitian, due to surface terms at $r = a$ arising from the kinetic energy operator in H_0. These surface terms can be eliminated by introducing a Bloch operator $\mathbf{L}_\mathrm{B}$, so that $\mathbf{H}_\mathrm{F} + \mathbf{L}_\mathrm{B}$ is Hermitian in the internal region. A basis set is then constructed, in which the operator $\mathbf{H}_\mathrm{F} + \mathbf{L}_\mathrm{B}$ is diagonalised. One then obtains on the boundary a relation of the form (see (12.438))

$$\mathbf{u}(a) = {}^\mathrm{L}\mathbf{R}(E)\left[a\frac{\mathrm{d}\mathbf{u}}{\mathrm{d}r} - b\mathbf{u}\right]_{r=a} \tag{15.87}$$

where $\mathbf{u}(r)$ denotes the set of reduced radial wave functions (that is, radial wave functions multiplied by r), b is an arbitrary constant (which may be taken to be zero) and ${}^\mathrm{L}\mathbf{R}(E)$ is the R-matrix in the length gauge. The logarithmic derivatives of the reduced radial wave functions on the boundary $r = a$, which provide the boundary conditions for solving the problem in the external region, are then given by (15.87).

In the external region, one has only one electron $(r_N \geqslant a)$, whose dynamics are conveniently studied by using the velocity gauge. Here a simple close-coupling expansion neglecting exchange effects between this external electron and the remaining $N-1$ electrons can be used for the harmonic components. The resulting set of coupled differential equations is then solved, subject to boundary conditions at $r = a$ and $r \to \infty$. At $r = a$, the matching of the internal- and external-region solutions provides ${}^\mathrm{V}\mathbf{R}(E)$, the R-matrix in the velocity gauge. The coupled equations are then solved from $r = a$ to a large value $r = a'$ of the radial coordinate by propagating the R-matrix ${}^\mathrm{V}\mathbf{R}(E)$. The solutions obtained in this way are matched at $r = a'$ with the solutions satisfying given boundary conditions for $r \to \infty$, calculated by using asymptotic expansions.

The boundary conditions for $r \to \infty$ are formulated in the Kramers–Henneberger (K–H) frame, because it can be shown that in this frame the channels decouple asymptotically. These boundary conditions differ according to the process considered. For the case of multiphoton ionisation and harmonic generation, there are only outgoing waves. It is then found that solutions will only occur for certain complex values of the energy (see (15.82)). From the knowledge of the eigenvectors, one may obtain all the other physical quantities, such as the branching ratios into the channels and the angular distribution of the ejected electron. In the case of laser-assisted electron–atom (ion) collisions, one must impose S-matrix (or T- or K-matrix) asymptotic boundary conditions. The scattering amplitudes and cross-sections are then given in terms of the S-, T- or K-matrix elements.

The RMF theory has been used to analyse multiphoton processes in a number of atomic systems. In particular, it has provided multiphoton ionisation rates for complex atoms and ions including electron correlation effects. It has allowed the *ab-initio* study of a variety of resonance effects in multiphoton ionisation, harmonic generation and laser-assisted electron–atom collisions. It has also predicted the occurrence of *laser-induced degenerate states* (LIDS) involving autoionising states in complex atoms.

As an example of the application of the RMF theory to two-electron systems, we show in Fig. 15.26 the ionisation rate into the two-photon channel, as a function of the photoelectron energy, for He in a laser field of intensity 10^{12} W cm^{-2}. The Rydberg series of peaks visible below the $n = 1$ threshold correspond to one-photon resonances due to intermediate ^{1}P bound states, while the series of resonances below the $n = 2$ threshold are two-photon resonances due to ^{1}S and ^{1}D autoionising states.

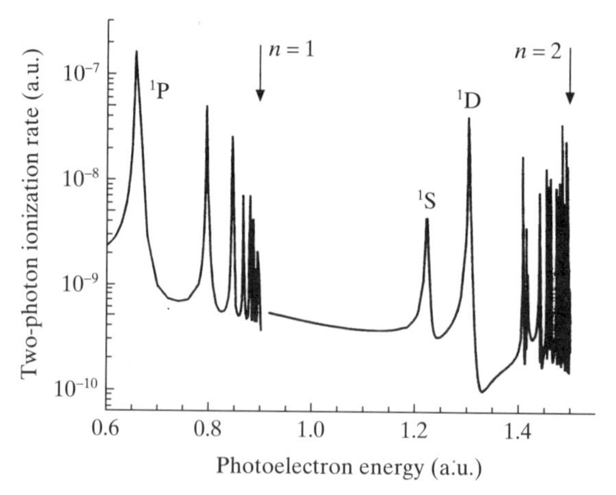

Figure 15.26 The ionisation rate of He as a function of the photoelectron energy, at an intensity of 10^{12} W cm^{-2}. The positions of the He$^+$(n = 1) and He$^+$(n = 2) thresholds are indicated by the arrows. (From J. Purvis, M. Dörr, M. Terao-Dunseath, C.J. Joachain, P.G. Burke and C.J. Noble, *Physical Review Letters* **71**, 3943, 1993.)

The RMF theory has been extended to describe two-colour multiphoton processes taking place when atoms or ions interact with two laser fields. An important application is the analysis of *laser-induced continuum structures* (LICS), a phenomenon in which a strong laser pulse (embedding laser) of angular frequency ω_E embeds a discrete state in the continuum. A weak laser pulse (probe laser) of angular frequency ω_P is then used to probe this enbedded state [10]. Another important application of two-colour processes is the study of multiply resonant multiphoton ionisation involving autoionising states of complex atoms. As an example, let us consider the two-colour doubly resonant multiphoton process in magnesium which is depicted in Fig. 15.27(a). A high-frequeny laser of angular frequency ω_H couples the $3s^2\,{}^1S$ ground state of Mg and the $3p^2\,{}^1S$ autoionising state by a two-photon process which is nearly resonant with the intermediate $3s3p\,{}^1P$ bound state. A second, low-frequency laser of angular frequency ω_L couples the $3p^2\,{}^1S$ and the $3p3d\,{}^1P$ autoionising states. The ionisation rates calculated by N.J. Kylstra, H.W. van der Hart, P.G. Burke and C.J. Joachain by using the RMF theory are shown in Fig. 15.27(b) as a function of ω_L, for three different values of ω_H, and for fixed intensities $I_L = 2 \times 10^9$ W cm^{-2} and $I_H = 5 \times 10^9$ W cm^{-2} of the two lasers. They are seen to be in good agreement with the experimental data of N.E. Karapanagioti *et al.* At higher intensities, higher order processes become

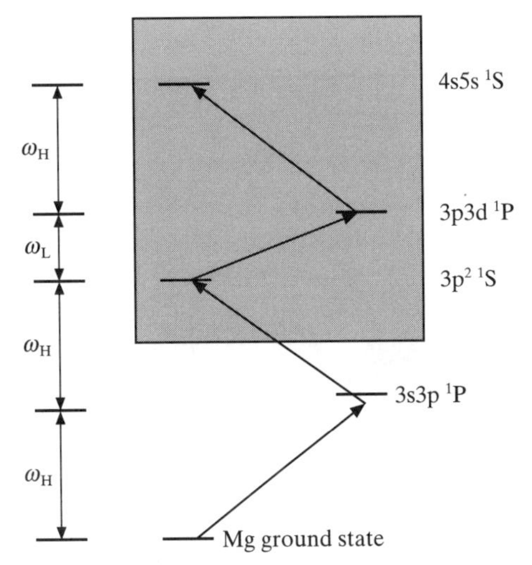

Figure 15.27(a) A high-frequency laser of angular frequency ω_H couples the ground state of magnesium and the $3p^2\,{}^1S$ autoionising state by a two-photon process which is nearly one-photon resonant with the intermediate $3s3p\,{}^1P$ state. The low-frequency laser of angular frequency ω_L couples the $3p^2\,{}^1S$ autoionising state and the $3p3d\,{}^1P$ autoionising state which in turn induces a resonant coupling between the $3p3d\,{}^1P$ and the $4s5s\,{}^1S$ autoionising states.

[10] A detailed discussion of laser-induced continuum structures can be found in the review article of Knight, Lauder and Dalton (1990).

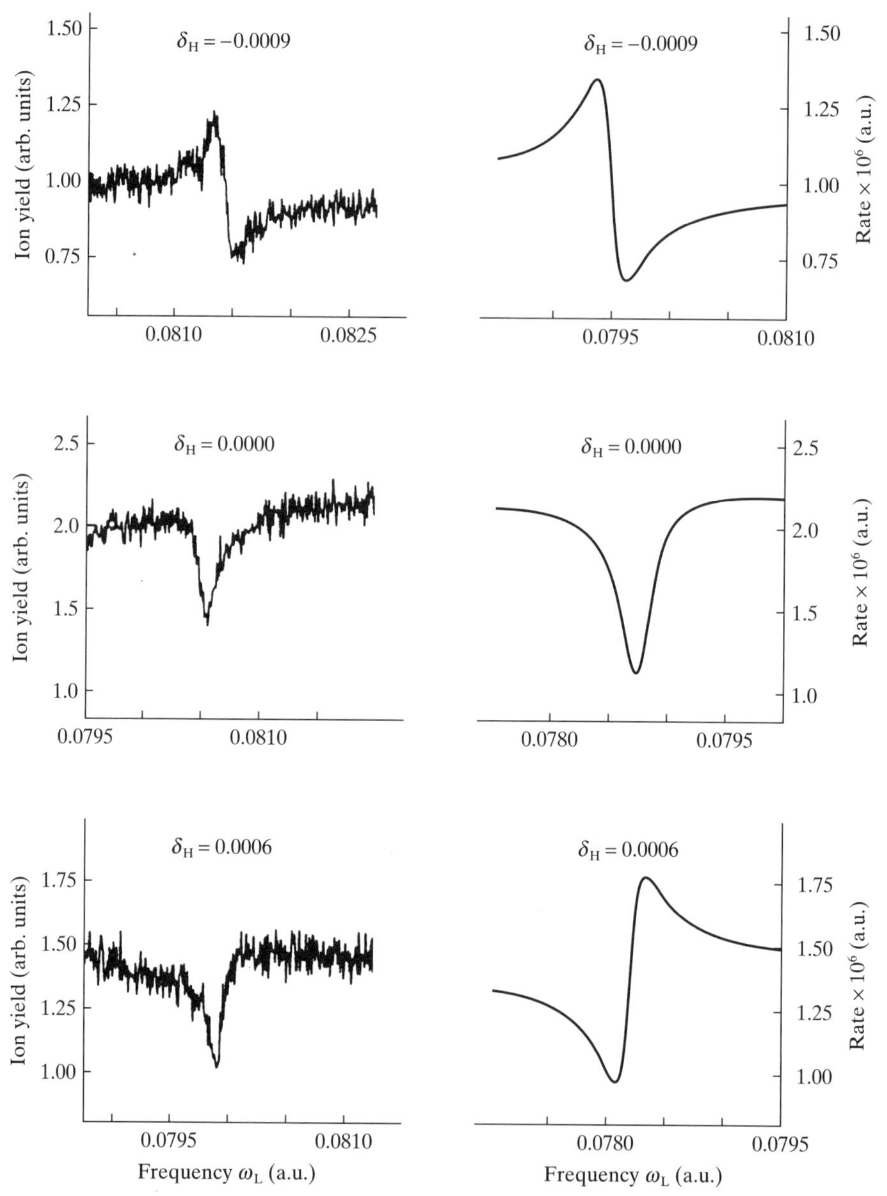

Figure 15.27(b) The experimental ionisation yields of N.E. Karapanagioti *et al.* (left) and the calculated values of N.J. Kylstra *et al.* (right) for Mg in the presence of two laser fields, as a function of ω_L for three different values of ω_H. The quantity $\delta_H = E_g - E_a + 2\omega_H$ denotes the detuning from resonance, where E_g and E_a refer to the field-free energies of the ground state and the $3p^2\ {}^1S$ autoionising state, respectively. (From N.J. Kylstra, H.W. van der Hart, P.G. Burke and C.J. Joachain, *Journal of Physics* B **31**, 3089, 1998.)

non-negligible. In particular, N.J. Kylstra *et al.* have performed non-perturbative RMF calculations including the effects of the 4s5s ^{1}S state which, as seen from Fig. 15.27(a), is resonantly coupled to the 3p3d ^{1}P state by the high-frequency laser. We also remark that within the context of these multiply resonant processes, *coherent control* of the ionisation can be exercised in the sense that by changing the laser parameters, the degree of interaction between the resonant processes – and hence the ionisation rate – can be varied.

Numerical solution of the time-dependent Schrödinger equation

The non-perturbative Floquet methods discussed above are based on the assumption that the Hamiltonian of the atomic system in the laser field is periodic (or multiperiodic for polychromatic laser fields) in time. Although this is not true for a realistic laser pulse, it is still possible to incorporate pulse shape effects into the Sturmian–Floquet or *R*-matrix–Floquet calculations for laser pulses which are very short. In particular, if the variation of the laser intensity is slow enough, the atom will remain in the Floquet eigenstate adiabatically connected to the initial state. Numerical studies indicate that this adiabaticity condition is robust for non-resonant multiphoton ionisation by short laser pulses.

For ultra-short laser pulses, typically in the femtosecond range, one must in general obtain information about the multiphoton processes by direct numerical integration of the time-dependent wave equations. In the non-relativistic regime, this amounts to solving the time-dependent Schrödinger equation (TDSE). This approach has the advantage that no restrictions need to be made about the type of laser pulse and that solutions can, in principle, be obtained for all regimes of intensity and frequency. It has the disadvantage that it is computationally intensive.

A straightforward way of reducing the computational load is to study one- or two-dimensional models. Since the one-dimensional models are relatively easy to solve numerically, it is possible to conduct 'numerical experiments' by investigating a large range of parameters. Two-dimensional models have also been used to study for example the dependence of multiphoton ionisation and harmonic generation on the ellipticity of the laser field.

Advances in computer technology have made possible the numerical integration of the TDSE for atoms or ions with one single active electron (SAE) interacting with laser fields. The first calculations of this kind were carried out by K.C. Kulander in 1987. These SAE calculations are 'exact' for hydrogenic systems. For atoms or ions with more than one electron, dynamic electron correlations are neglected in the SAE model.

As an illustration of the 'exact' TDSE calculations performed for atomic hydrogen, we show in Fig. 15.28 the estimated ionised fraction of atoms as a function of time, for H(1s) in high-frequency ($\omega = 2$ a.u.), linearly polarised, very intense laser fields which are turned on very rapidly (with a two-cycle ramp), as obtained by O. Latinne, C.J. Joachain and M. Dörr. The Floquet results at the corresponding

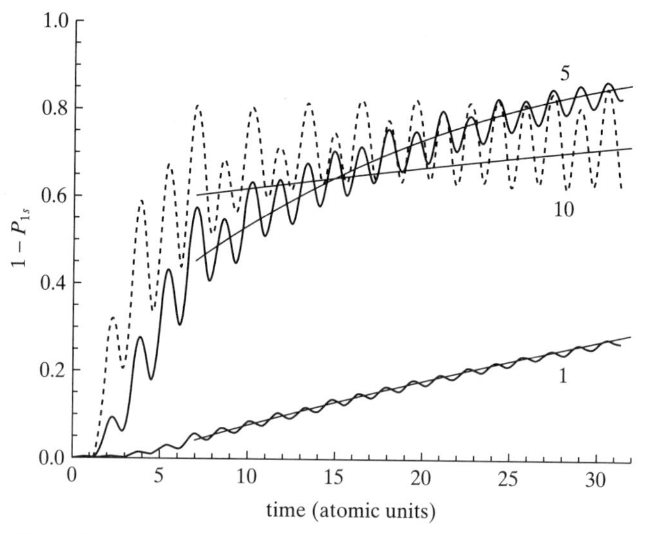

Figure 15.28 The estimated ionised fraction of atoms as a function of time for H(1s) in linearly polarised laser fields of angular frequency $\omega = 2$ a.u. with peak electric field strengths $\mathscr{E}_0 = 1$, 5 and 10 a.u. (indicated next to the curves) obtained by solving numerically the TDSE. The thin lines give the corresponding Floquet results. (From O. Latinne, C.J. Joachain and M. Dörr, *Europhysics Letters* **26**, 333, 1994.)

electric field strengths ($\mathscr{E}_0 = 1$ a.u., 5 a.u. and 10 a.u.) are also displayed, and are seen to be in good agreement with the values obtained by solving numerically the TDSE. The extra beat structure visible for the curve corresponding to $\mathscr{E}_0 = 10$ a.u. is due to interference between the 1s and higher lying Floquet states (mainly the 2s state) which are populated during the turn-on of the field. We note from Fig. 15.28 that both the Floquet and the time-dependent calculations show that the ionised fraction corresponding to $\mathscr{E}_0 = 10$ a.u. is inferior to that for $\mathscr{E}_0 = 5$ a.u. This decrease of the ionisation rate as a function of the intensity in the high-frequency, high-intensity regime is called *atomic stabilisation* and will be discussed in more detail below.

In the case of complex atoms, nearly all the TDSE calculations have been performed by using the SAE approximation. This model has worked well, being in agreement with most of the experimental data. An important test for theories going beyond the SAE model (that is, including electron correlation effects) is to calculate accurately double (and more generally multiple) ionisation yields for multielectron systems in intense laser fields. Double ionisation yields have been measured accurately over a wide range of laser intensities in several experiments involving different atoms. A striking feature of these experimental results, illustrated in Fig. 15.29 for the case of helium, is the existence of two distinct intensity regimes: one at high intensities where the double ionisation process proceeds predominantly sequentially (in accordance with the SAE approximation) and the other region at low intensities where it is mainly simultaneous. The large

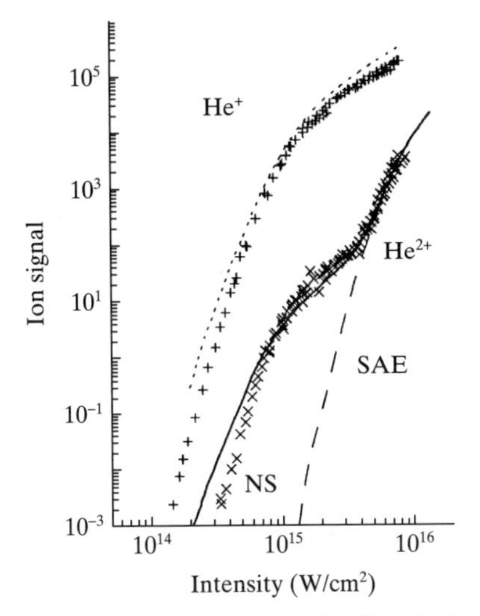

Figure 15.29 Comparison of the He^+ (dotted line) and He^{2+} (solid line) yields predicted by the theoretical calculations of J.B. Watson *et al.* with the experimental result of B. Walker *et al.* The dashed curve refers to the sequential (SAE) calculations. The symbol NS refers to the non-sequential mechanism. (From J.B. Watson, A. Sanpera, D.G. Lappas, P.L. Knight and K. Burnett, *Physical Review Letters*, **78**, 1884, 1997.)

enhancement due to the non-sequential (NS) mechanism at low intensities has been well reproduced by theoretical approaches, such as those of F.H.M. Faisal and A. Becker and of J.B. Watson, A. Sanpera, D.G. Lappas, P.L. Knight and K. Burnett, which incorporate electron correlation effects.

Finally, we note that using a massively parallel computer, K. Taylor and co-workers have studied multiphoton processes in helium beyond the SAE approximation by numerical integration of the TDSE.

Low-frequency methods

When the laser period is much longer than the typical 'orbital period' of the bound electron, the laser frequency can be characterised as being low. Thus, most experiments using short, intense laser pulses and noble gas atoms fall into this category, since the typical ground state binding energies are of the order of the atomic unit while the photon energy is typically an order of magnitude smaller.

We have seen while studying the Stark effect in Section 6.1 that ionisation by a static electric field can occur. Indeed, the initially bound electron can tunnel through the barrier in the total potential V, formed by adding to the atomic potential the potential energy due to the static electric field. In the same way, at sufficiently low frequencies, and for moderately high intensities, the initially bound electron can tunnel through an effective potential obtained by adding to the atomic potential the contribution due to the instantaneous laser field (see Fig. 15.30(a)). The key idea is that substantial ionisation occurs in a fraction

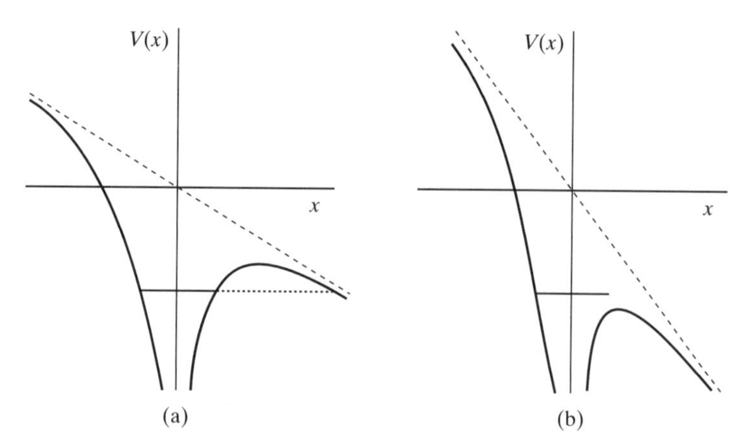

Figure 15.30 One-dimensional model showing (a) tunnelling ionisation and (b) over the barrier ionisation. The dashed line corresponds to the contribution to the potential energy due to the instantaneous laser electric field $\mathscr{E}(t) = \hat{\mathbf{x}}\mathscr{E}_0 \sin(\omega t + \delta)$. The solid line corresponds to the full effective potential energy. The position of a bound energy level (in the absence of the laser field) is indicated.

of an optical cycle, so that the electric field can be regarded as *quasi-static*. The method discussed in Section 6.1 to calculate the tunnelling ionisation rate for a static field can then be generalised to the quasi-static case corresponding to a low-frequency laser field. This approach was used in 1965 by L.V. Keldysh and extended by F.H.M. Faisal and H.R. Reiss; it is known as the KFR theory. A useful expression giving the instantaneous tunnelling ionisation rate for a general atom has been derived along these lines by M. Ammosov, N. Delone and V. Krainov. This result, called the ADK tunnelling formula, is given by

$$W = \omega_{\mathrm{P}} |C_{n^*l^*}|^2 G_{lm} \left(\frac{4\omega_{\mathrm{P}}}{\omega_{\mathrm{T}}} \right)^{2n^*-m-1} \exp[-4\omega_{\mathrm{P}}/(3\omega_{\mathrm{T}})] \qquad \textbf{(15.88a)}$$

where

$$\omega_{\mathrm{P}} = I_{\mathrm{P}}/\hbar, \qquad \omega_{\mathrm{T}} = e\mathscr{E}_0/(2mI_{\mathrm{P}})^{1/2}, \qquad n^* = (I_{\mathrm{P}}^{\mathrm{H}}/I_{\mathrm{P}})^{1/2},$$
$$|C_{n^*l^*}|^2 = 2^{2n^*}[n^*\Gamma(n^*+l^*+1)\Gamma(n^*-l^*)]^{-1}, \qquad \textbf{(15.88b)}$$
$$G_{lm} = (2l+1)(l+|m|)!(2^{-|m|})/[|m|)!(l-|m|)!]$$

In these equations, I_{P} is the ionisation potential of the atom of interest, $I_{\mathrm{P}}^{\mathrm{H}}$ is the ionisation potential of atomic hydrogen, l and m are the orbital angular momentum and magnetic quantum numbers, respectively, $\mathscr{E}_0$ is the electric field amplitude and Γ is Euler's gamma function. The effective quantum number l^* is given by $l^* = 0$ for $l \ll n$ and $l^* = n^* - 1$ otherwise. The probability of ionisation during the time interval $\mathrm{d}t$ is

$$P(t) = W[\mathscr{E}(t)]\, \mathrm{d}t \qquad \textbf{(15.89)}$$

where $\mathscr{E}(t)$ is the magnitude of the laser electric field. The tunnelling model therefore describes the formation of a sequence of outgoing electron wave packets, one near each peak of the laser electric field.

An important quantity in tunnelling ionisation studies is the *Keldysh parameter*, defined as the ratio of the laser and tunnelling frequencies, which is given by

$$\gamma = \left(\frac{I_P}{2U_p} \right)^{1/2} \tag{15.90}$$

where U_p is the electron ponderomotive energy. This parameter is an approximate indicator as to the applicability of the tunnelling model, in the sense that tunnelling ionisation will dominate when $\gamma < 1$.

As the laser intensity is increased, the barrier in the effective potential becomes thinner and lower until eventually, above a critical intensity I_c, the electron can classically 'flow over the top' of the barrier (see Fig. 15.30(b)). This is known as 'over the barrier ionisation' (OTBI). The tunnelling formulae then break down and the atom ionises in about one orbital period. The critical intensity I_c is obtained by equating the maximum induced by the electric field in the effective potential to the ionisation potential of the bound electron. It is found in this way that

$$I_c = 4 \times 10^9 (I_P[\text{eV}])^4 Z_R^2 \text{ W cm}^{-2} \tag{15.91}$$

where the ionisation potential I_P in expressed in eV and Z_R is the residual charge of the relevant atom or ion. For example, in the case of atomic hydrogen in the ground state, $I_P = 13.6$ eV and $Z_R = 1$, giving $I_c = 1.4 \times 10^{14}$ W cm^{-2}. For the case of neutral Xe in the ground state, one has $I_P = 12.1$ eV and $Z_R = 1$, so that $I_c = 8.6 \times 10^{13}$ W cm^{-2}.

In order to develop a successful model of strong field phenomena at low frequencies, it is necessary to go beyond tunnelling or over the barrier ionisation, and to take into account the possibility that the ionised electron will return to the vicinity of its parent ion or atom core. The semi-classical 'recollision model' developed by P.B. Corkum and by K.C. Kulander, K.J. Schafer and J.L. Krause is based on the idea that strong field ionisation or harmonic generation at low frequencies proceeds via several steps. In the first (bound–free) step, the active electron is detached from its parent ion (atom) core by tunnelling or over-the-barrier ionisation. In the second (free–free) step, the unbound electron interacts mainly with the laser field, so that its dynamics are essentially those of a free electron in the field, and can be treated to a good approximation by using classical mechanics. As the phase of the field reverses, the electron can be accelerated back towards the parent core. If the electron does not return to the core, single ionisation will occur. If it does return to the core, then a third step takes place in which scattering of the electron by the core leads to single or multiple ionisation, while radiative recombination leads to harmonic generation. As we shall see below, this semi-classical model has been very useful for explaining a number of experimental features of multiphoton ionisation and harmonic generation. A quantum mechanical formulation of the 'recollision' model, called the *low-frequency strong field approximation* (SFA), has been made by M. Lewenstein, P. Balcou, M. Yu Ivanov, A. L'Huillier and P.B. Corkum, and has been applied successfully to study high-order harmonic generation.

The high-frequency, high-intensity regime

The interaction of atoms with high-intensity monochromatic laser fields whose frequency is much larger than the threshold frequency for one-photon ionisation has been studied by M. Gavrila and co-workers, using a high-frequency version of the Floquet theory. Their approach is formulated in the Kramers–Henneberger (K–H) frame introduced above. For the case of an atom with one active electron, the Schrödinger equation (15.59) satisfied by the wave function $\Psi^{A}(\mathbf{r}, t)$ is

$$i\hbar\frac{\partial}{\partial t}\Psi^{A}(\mathbf{r}, t) = \left[\frac{\mathbf{p}^2}{2m} + V[\mathbf{r} + \boldsymbol{\alpha}(t)]\right]\Psi^{A}(\mathbf{r}, t) \tag{15.92}$$

where the displacement vector $\boldsymbol{\alpha}(t)$ is given by (15.60) for the case of a linearly polarised monochromatic field. Because the Hamiltonian (15.92) is periodic, one can seek solutions $\Psi^{A}(\mathbf{r}, t)$ having the Floquet–Fourier form (15.78), namely

$$\Psi^{A}(\mathbf{r}, t) = \exp(-iEt/\hbar) \sum_{n=-\infty}^{+\infty} e^{-in\omega t} F_n^{A}(\mathbf{r}) \tag{15.93}$$

Making also a Fourier analysis of the potential

$$V[\mathbf{r} + \boldsymbol{\alpha}(t)] = \sum_{n=-\infty}^{+\infty} e^{-in\omega t} V_n(\alpha_0, \mathbf{r}) \tag{15.94}$$

one obtains for the harmonic components $F_n^{A}(\mathbf{r})$ of $\Psi^{A}(\mathbf{r}, t)$ the system of coupled, time-independent equations

$$\left[E + n\hbar\omega - \left(\frac{\mathbf{p}^2}{2m} + V_0(\alpha_0, \mathbf{r})\right)\right] F_n^{A}(\mathbf{r}) = \sum_{\substack{k=-\infty \\ (k\neq n)}}^{+\infty} V_{n-k}(\alpha_0, \mathbf{r}) F_k^{A}(\mathbf{r}) \tag{15.95}$$

Gavrila and co-workers have shown that in the high-intensity, high-frequency limit, the atomic structure in the laser field is governed by the potential $V_0(\alpha_0, \mathbf{r})$, which is the static (time-averaged), 'dressed' potential associated with the interaction potential V in the K–H frame. Detailed calculations, performed for the hydrogen atom, show that as α_0 is increased the time-averaged potential $V_0(\alpha_0, \mathbf{r})$ becomes a double well, whose ground state eigenfunction exhibits the property of dichotomy, being increasingly localised around the end points $\pm\alpha_0\hat{\boldsymbol{\varepsilon}}$ of the 'classical' excursion amplitude. In addition, the ionisation potential of the ground state of $V_0(\alpha_0, \mathbf{r})$ decreases with increasing α_0.

An important prediction of the high-frequency Floquet theory (HFFT) is that, when the frequency of the laser field is larger than the threshold frequency for one-photon ionisation, the ionisation rate (for fixed ω) decreases as the intensity increases beyond some critical high value. This phenomenon is called *adiabatic stabilisation* (or quasi-stationary stabilisation). It can be understood qualitatively by noting that in the dipole approximation (where the photon momentum $\hbar\mathbf{k}$ is set equal to zero), momentum conservation requires the presence of the nucleus for photoionisation to occur. As α_0 increases, the wave functions become increasingly localised away from the nucleus, so that the ionisation probability decreases. 'Exact' Floquet calculations, performed for atomic hydrogen by using the Sturmian–Floquet or the R-matrix–Floquet methods, have confirmed the occurrence of adiabatic stabilisation for the (ideal) case of spatially homogeneous, monochromatic laser fields.

In reality, however, an atom is subjected to a laser pulse of finite duration. This implies that during the laser pulse turn-on, where the intensity is lower, substantial ionisation will occur. In other words, before the atoms experience a super-intense laser field where they can stabilise, they must pass through a 'death valley' where their lifetime is very short. Therefore, if adiabatic stabilisation is to be observed experimentally, the laser pulse rise-time must not be so slow that saturation (complete ionisation) occurs during the turn-on. However, it must be slow enough so that the atom adiabatically remains in the ground state in the K–H frame. The experimental verification of adiabatic stabilisation is therefore a very difficult task. Fortunately, the stabilisation condition can be met at lower intensities and frequencies if the atom is prepared in an initial state that is not the ground state, but a 'circular' Rydberg state with large n, l and $|m|$ quantum numbers, such that its lifetime in the 'death valley' is sufficiently large. Experiments of this kind have been performed by H.G. Muller and co-workers to study the stabilisation behaviour of the hydrogen-like, 'circular' 5g state of neon. A first pulse prepares the initial state. Comparison of the single-photon ionisation yield, due to a second laser pulse, for both short (0.1 ps) intense and longer (1 ps) less intense pulses, indicates a suppression of ionisation as the intensity increases, in accordance with the theoretical prediction of adiabatic stabilisation.

In the case of very short, intense laser pulses, stabilisation has been found by solving numerically the time-dependent Schrödinger equation (in the dipole approximation) in one dimension by Q. Su, J.H. Eberly and J. Javanainen, and for atomic hydrogen by K.C. Kulander, K.J. Schafer and J.L. Krause and by O. Latinne, C.J. Joachain and M. Dörr (see Fig. 15.28). Here stabilisation manifests itself as a decrease of the ionisation probability of an atomic electron at the end of the laser pulse beyond a certain critical high-intensity value, for a fixed high frequency. It is called *dynamic stabilisation*.

Finally, for super-intense laser fields, retardation (non-dipole) and relativistic effects must be taken into account in studying the stabilisation of atoms. We shall now briefly discuss these effects.

Non-dipole and relativistic effects

At sufficiently high intensities, an atom in a laser field can no longer be described using the Schrödinger equation in the dipole approximation. In particular, the magnetic field component of the laser field becomes important, spin effects must be considered, and eventually the dynamics of the electron become fully relativistic.

To estimate when non-dipole effects become important, one can use the classical Lorentz equations of motion for an electron in a laser pulse:

$$\frac{\mathrm{d}}{\mathrm{d}t}\mathbf{p} = -e(\mathscr{E} + \mathbf{v} \times \mathscr{B}) \tag{15.96}$$

where $\mathbf{v}$ is the velocity of the electron, and $\mathbf{p} = m\gamma\mathbf{v}$ is its momentum, with $\gamma = (1 - v^2/c^2)^{-1/2}$. These equations can then be solved to lowest order in $1/c$. Non-dipole effects due to the magnetic field component of the laser field are in general non-negligible when the electron drift per laser cycle in the propagation direction, which is given approximately by $\mathscr{E}_0^2/(c\omega^3)$, becomes comparable to the width of the electron wave packet in the propagation direction. Taking the latter to be one Bohr radius, it is found that for an angular frequency $\omega = 0.057$ a.u. (corresponding to a Ti:sapphire laser of wavelength $\lambda = 800$ nm), magnetic field effects can become important when the laser intensity I is larger than 10^{16} W cm^{-2}, while for $\omega = 1$ a.u. intensities in excess of 5×10^{18} W cm^{-2} are required.

Relativistic effects are expected to become important when the 'quiver' velocity of the electron approaches the velocity of light. By considering the non-relativistic classical equation of motion of an electron in a linearly polarised monochromatic field (15.51), it is readily seen that the magnitude of the quiver velocity is given by $v = e\mathscr{E}_0/(m\omega)$. A relativistic treatment therefore becomes necessary if the ratio $v/c = e\mathscr{E}_0/(m\omega c)$ approaches unity. Using the (non-relativistic) expression (15.74) of the electron ponderomotive energy U_p, we see that relativistic effects become important when U_p is of the order of the electron's rest mass energy mc^2, that is when the quantity

$$q = \frac{U_\mathrm{p}}{mc^2} = \frac{e^2\mathscr{E}_0^2}{4m^2\omega^2c^2} \tag{15.97}$$

is of the order of unity. If the electric field strength $\mathscr{E}_0$ and the angular frequency ω are expressed in atomic units, we have

$$q = 1.33 \times 10^{-5}\left(\frac{\mathscr{E}_0}{\omega}\right)^2 \tag{15.98}$$

Thus, for a Nd:YAG laser of angular frequency $\omega = 0.043$ a.u., one has $q = 1$ when $\mathscr{E}_0 = 11.8$ a.u., that is when the intensity $I = 4.9 \times 10^{18}$ W cm^{-2}. For a Ti:sapphire laser of wavelength $\lambda = 800$ nm, corresponding to an angular frequency $\omega = 0.057$ a.u., one has $q = 1$ when $\mathscr{E}_0 = 15.6$ a.u., so that $I = 8.5 \times 10^{18}$ W cm^{-2}. On the other hand, for a high angular frequency $\omega = 1$ a.u., we have $q = 1$ when $\mathscr{E}_0 = 274$ a.u., corresponding to the very large intensity $I = 2.6 \times 10^{21}$ W cm^{-2}.

In the low-frequency regime, as discussed above, a quasi-static model can be used to good approximation, in which the electron moves in an instantaneous effective potential obtained by adding to the atomic potential the instantaneous electric field. Therefore, at very high intensities, magnetic field and relativistic effects will essentially manifest themselves in the dynamics of the free, laser-driven electron wave packets. In this high-intensity, low-frequency regime, classical Monte-Carlo simulations have been carried out by C.H. Keitel, P.L. Knight and K. Burnett, and tunnelling calculations have been performed by V. Krainov.

The situation is different for the case of a high-frequency laser, that is in the stabilisation regime. The analysis of the stabilisation of atoms in intense, high-frequency laser fields discussed above was carried out in the non-relativistic limit, and in the dipole approximation. For sufficiently high intensities, the magnetic field component of the laser field modifies the stabilisation dynamics. Using a two-dimensional model, the breakdown of the dipole approximation has been demonstrated by N.J. Kylstra *et al.* Their quantum mechanical results confirmed relativistic, classical Monte-Carlo simulations of ionisation made by C.H. Keitel and P.L. Knight for atomic hydrogen. The magnetic component of the laser field induces a significant motion of the electron in the propagation direction of the laser field. As a result, a breakdown of stabilisation can occur.

One long-term goal is to study atoms interacting with ultra-intense laser pulses by numerically solving the time-dependent Dirac equation. This is a very difficult task since computationally the problem scales approximately as $\mathcal{E}_0^3/\omega^4$. For this reason, nearly all the quantum mechanical calculations of laser–atom interactions in the relativistic domain have been restricted until now to lower dimensional treatments and high frequencies. Magnetic field effects are clearly not included in one-dimensional model calculations. On the other hand, relativistic effects due to the dressing of the electron mass by the laser field can be investigated. We show in Fig. 15.31 the results of N.J. Kylstra, A.M. Ermolaev and C.J. Joachain for the Dirac and Schrödinger probability densities at the end of the ninth laser cycle,

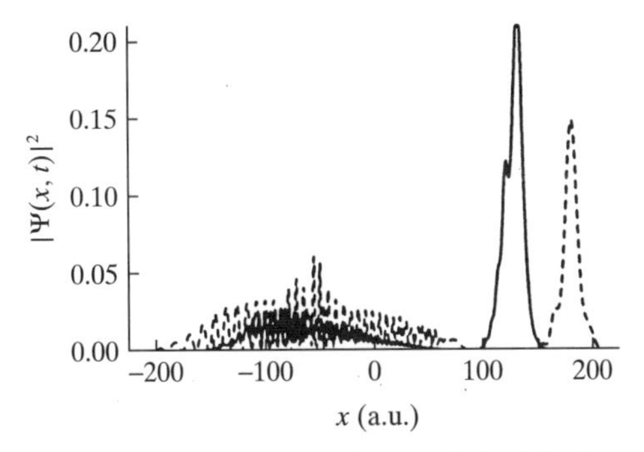

Figure 15.31 The Dirac (solid line) and Schrödinger (dashed line) probability densities, at the end of the ninth cycle, for a laser pulse with a four-cycle $\sin^2$ turn-on, an angular frequency $\omega = 1$ a.u. and a peak electric field strength $\mathcal{E}_0 = 175$ a.u. (From N.J. Kylstra, A.M. Ermolaev and C.J. Joachain, *Journal of Physics* B **30**, L 449, 1997.)

when the electric field is maximum. The peak in the Dirac probability density corresponds to the relativistic 'classical' excursion amplitude, $x_0 = 124$ a.u. In the same way, the peak in the Schrödinger probability density occurs at $x_0 = 175$ a.u., the non-relativistic classical excursion amplitude.

Multiphoton ionisation, above threshold ionisation, tunnelling and 'over the barrier' ionisation

We shall now apply the theory to analyse two important multiphoton processes in atoms and ions: multiphoton ionisation and harmonic generation. More detailed treatments can be found in Gavrila (1992), Joachain (1994), Protopapas, Keitel and Knight (1997) and Joachain, Dörr and Kylstra (2000).

We begin by considering the multiphoton single ionisation process

$$n\hbar\omega + \mathrm{A}^{q} \rightarrow \mathrm{A}^{q+1} + \mathrm{e}^{-} \tag{15.99}$$

where q is the charge of the target atomic system, expressed in atomic units.

This process was first observed in 1965 by G.S. Voronov and N.B. Delone, who used a ruby laser to ionise xenon via seven-photon absorption, and by J.L. Hall, E.J. Robinson and L.M. Branscomb, who measured the two-photon electron detachment from the negative ion I^{-}. In the following years, important results were obtained by several experimental groups, in particular at the atomic physics laboratory in Saclay, France, where the dependence of the ionisation rates on the laser intensity was studied. For the intensities $I \ll I_a$ available at that time, it was observed that the total n-photon ionisation rate was proportional to I^n, in agreement with the prediction of lowest order perturbation theory (LOPT). Subsequently, the resonant enhanced multiphoton ionisation (REMPI) phenomenon was also studied.

A crucial breakthrough was made when experiments detecting the energy-resolved photoelectrons were performed. In this way P. Agostini, F. Fabre, G. Mainfray, G. Petite and N. Rahman discovered in 1979 that at sufficiently high intensities ($I > 10^{11}$ W cm^{-2}), the ejected electron can absorb photons in excess of the minimum number required for ionisation to occur. This phenomenon is called 'above threshold ionisation' (ATI). The photoelectron spectra were seen to exhibit several peaks, separated by the photon energy $\hbar\omega$, and appearing at the energies

$$E_s = (n + s)\hbar\omega - I_{\mathrm{P}} \tag{15.100}$$

where n is the minimum number of photons needed to exceed the ionisation potential I_{P}, and $s = 0, 1, \ldots$ is the number of excess photons (or 'above threshold' photons) absorbed by the electron.

A typical example of ATI photoelectron energy spectra, measured in 1988 by G. Petite, P. Agostini and H.G. Muller, is shown in Fig. 15.32. Pulses of 130 ps duration obtained from an Nd:YAG laser of wavelength $\lambda = 1064$ nm were focused

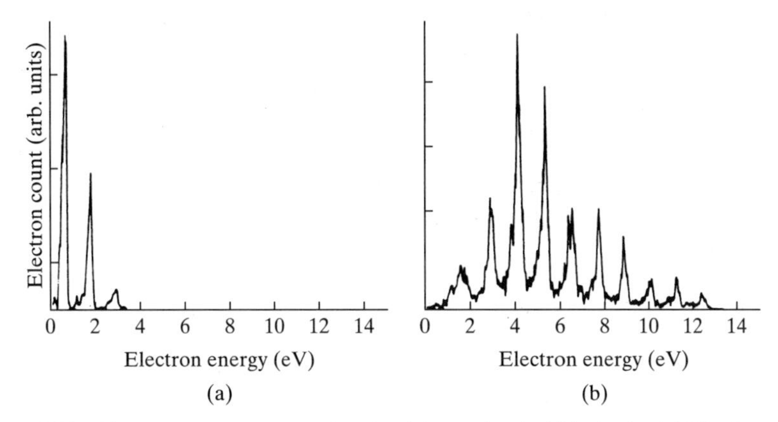

Figure 15.32 Electron energy spectra showing 'above threshold ionisation' (ATI) of xenon at a laser wavelength $\lambda = 1064$ nm. (a) $I = 2 \times 10^{12}$ W cm^{-2}, (b) $I = 10^{13}$ W cm^{-2}. (From G. Petite, P. Agostini and H.G. Muller, *Journal of Physics* B **21**, 4097, 1988.)

into a xenon vapour, and the electron energy spectrum was recorded using a time-of-flight spectrometer, with a 25 meV resolution. At relatively weak intensities, the intensity dependence of the peaks follows the LOPT prediction according to which the ionisation rate for an $(n + s)$ photon-process is proportional to I^{n+s} (see Fig. 15.32(a)). As the intensity increases, peaks at higher energies appear (see Fig. 15.32(b)) whose intensity dependence does not follow the I^{n+s} prediction of LOPT.

Another remarkable feature of the ATI spectra, also apparent in Fig. 15.32, is that as the intensity increases, the low-energy peaks are reduced in magnitude. The reason for this *peak suppression* is that the energies of the atomic states are *Stark-shifted* in the presence of a laser field. For low laser frequencies (for example, a Nd:YAG laser with $\hbar\omega = 1.17$ eV), the AC Stark shifts of the lowest bound states are small in magnitude. On the other hand, the induced Stark shifts of the Rydberg and continuum states are essentially given by the electron *ponderomotive energy* U_p (see (15.74)). We recall that the ponderomotive energy is proportional to I/ω^2, and we note that it can become quite large. For example, in the case of an Nd:YAG laser of wavelength $\lambda = 1064$ nm, the ponderomotive energy U_p given by (15.74) becomes equal to the photon energy $\hbar\omega = 1.17$ eV at an intensity $I \simeq 10^{13}$ W cm^{-2}. Since the energies of the Rydberg and continuum states are shifted upwards relative to the lower bound states by about U_p, there is a corresponding increase in the intensity-dependent ionisation potential $I_P(I)$ of the atom, so that $I_P(I) \simeq I_P + U_p$, where we recall that $I_P = -E_i$ denotes the ionisation potential of the field-free initial state of energy E_i. If this increase is such that $n\hbar\omega < I_P + U_p$, then ionisation by n photons is energetically forbidden (see Fig. 15.33). However, atoms interacting with smoothly varying pulses experience a range of intensities, so that the corresponding peak in the photoelectron spectrum will not completely disappear, as seen in Fig. 15.32(b).

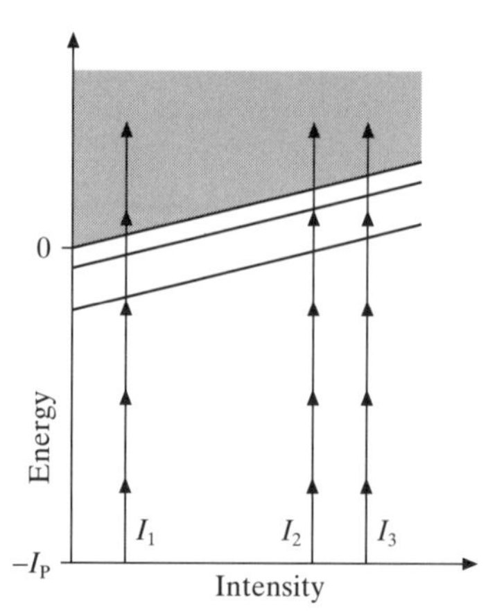

Figure 15.33 Illustration of the mechanism responsible for the suppression of low-energy peaks in ATI spectra. For low laser frequencies, the intensity-dependent ionisation potential of the atom, $I_P(I)$, is such that $I_P(I) \simeq I_P + U_p$, and hence increases linearly with the intensity I. Ionisation by 4 photons, which is possible at the intensity I_1 for which $4\hbar\omega \geqslant I_P + U_p$, is prohibited at the higher intensities I_2 and I_3, where 5 photons are needed to ionise the atom. Also illustrated is the mechanism responsible for the resonantly-induced structures appearing in ATI spectra for short laser pulses. At the intensity I_2, a Rydberg state has shifted into four-photon resonance with the ground state.

For relatively long pulses (in the picosecond range), the photoelectron escapes from the focal volume while the laser is still on, so that it experiences a force ∇U_p due to the laser inhomogeneity. The electron quiver motion is then converted into radial motion, increasing its kinetic energy by U_p, and hence exactly cancelling the decrease in energy caused by the (Stark-shifted) increase in the ionisation potential. As a result, the photoelectron energies are still given by equation (15.100). However, as noted above, the first ATI peak will nearly disappear if U_p exceeds the photon energy $\hbar\omega$. Similarly, the first two peaks will be weakened if U_p exceeds $2\hbar\omega$, etc. It should be noted that in this long-pulse limit, no photoelectron is observed with energy less than U_p.

For short (sub-picosecond) laser pulses, the laser field turns off before the photoelectron can escape from the focal volume. In this case, the quiver energy is returned to the field and the ATI spectrum becomes more complicated. The observed photoelectron energies are given by the values

$$\tilde{E}_s = (n+s)\hbar\omega - (I_P + U_p) \qquad \textbf{(15.101)}$$

relative to the *shifted* ionisation potential. Photoelectrons originating from different regions of the focal volume are thus emitted at different intensities with different energies. As a result, the ATI peaks exhibit a substructure which, as shown in 1987 by R.R. Freeman, P.H. Bucksbaum, H. Milchberg, S. Darack, D. Schumacher and M.E. Geusic, is due to the fact that the intensity-dependent Stark shifts bring different states of the atom into multiphoton resonance during the laser pulse (see Fig. 15.33). An example of substructure in the lowest ATI peak for multiphoton ionisation of H(1s) by a laser pulse of duration 0.5 ps is shown in Fig. 15.24. This substructure is not seen in long-pulse experiments because in that case, as explained above, the photoelectrons regain their ponderomotive energy deficit from the laser field as they escape from the focal volume.

For increasing laser field strengths approaching the Coulomb field binding the electron ($I > 10^{14}$ W cm^{-2}), and for low laser frequencies, the sharp ATI peaks of the photoelectron spectrum gradually blur into a continuous distribution, as observed in 1989 by S. August, D. Strickland, D.D. Meyerhofer, S.L. Chin and J.H. Eberly, and in 1993 by E. Mevel, P. Breger, R. Trainham, G. Petite, P. Agostini, A. Migus, J.P. Chambaret and A. Antonetti. As we have seen before, in this regime (which corresponds to a Keldysh parameter $\gamma < 1$), ionisation can be understood by using a quasi-static model in which the electron can be liberated from its parent atom by tunnelling ionisation (see Fig. 15.30(a)) or, at higher intensities, by 'over the barrier' ionisation (see Fig. 15.30(b)). It is interesting to note that early evidence of tunnelling was found in 1974 by J.E. Bayfield and P.M. Koch in the microwave ionisation of highly excited Rydberg atoms (see also Gallagher, 1994).

The semi-classical 'recollision model' discussed above has been very useful in explaining novel features observed in recent experiments performed with low-frequency lasers. In these experiments, the use of kilohertz-repetition rate high-intensity lasers has allowed precise measurements of photoelectron total yields and energy and angle-differential spectra over many orders of magnitude in yield.

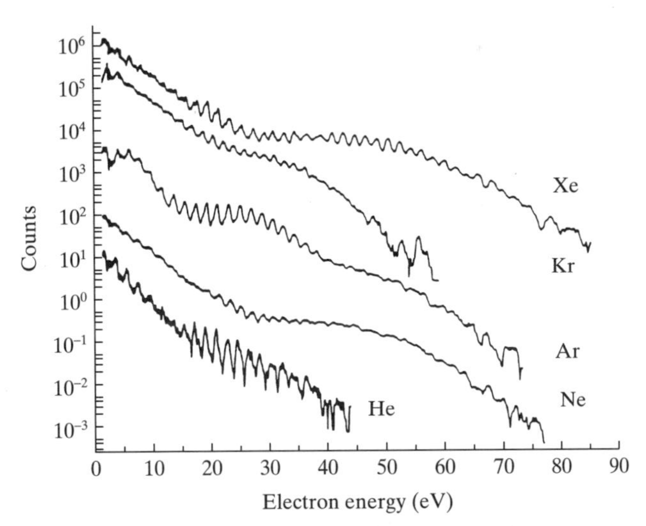

Figure 15.34 Photoelectron counts as a function of photoelectron energy, for various noble gases, at a laser wavelength $\lambda = 630$ nm and an intensity $I \simeq 2 \times 10^{14}$ W cm^{-2} (3×10^{14} W cm^{-2} for He). (From G.G. Paulus, W. Nicklich, H. Xu, P. Lambropoulos and H. Walther, *Physical Review Letters* **72**, 2851, 1994.)

Usually, the angular distributions of photoelectrons in ATI spectra produced by intense, linearly polarised laser fields are aligned along the axis of polarisation of the field. However, B. Yang, K.J. Schafer, B. Walker, K.C. Kulander, P. Agostini and L.F. DiMauro have shown in 1993 that a few high-energy peaks may be highly structured, and in some cases exhibit rings off the polarisation axis. Theoretical calculations have shown that these rings arise from rescattering of the electron wave packet from the parent ion. Other experiments, performed in 1994 by G.G. Paulus, W. Nicklich, H. Xu, P. Lambropoulos and H. Walther have revealed the existence of a 'plateau' in the ATI photoelectron energy spectra (see Fig. 15.34) which is also due to the rescattering of the detached electron by the parent ion. The observed dependence of the photoelectron yields in ATI spectra on the ellipticity of the laser field has also been explained in terms of quantum interference effects, within the framework of the low frequency strong field approximation of M. Lewenstein *et al.*

Harmonic generation

Atoms interacting with a strong laser field can emit radiation at higher order multiples, or *harmonics*, of the angular frequency ω of the 'pump' laser. Indeed, in response to the intense laser field, the electrons oscillate, and these oscillations cause the emission of radiation. Due to the inversion symmetry of the atom, the harmonic angular frequencies Ω are only emitted at *odd* multiples of the driving (also called 'fundamental') angular frequency ω. That is,

$$\Omega = q\omega, \qquad q = 3, 5, \ldots \tag{15.102}$$

The observation of the third harmonic in noble gases was recorded in 1967 by G.H.C. New and J.F. Ward. However, it was only in the late 1980s that the availability of intense, short laser pulses made it possible to observe high-order harmonics. Experiments performed in 1987 by C.K. Rhodes and co-workers at the University of Illinois showed the generation of up to the 17th harmonic of a 248 nm KrF laser in a neon vapour. The 33rd harmonic in argon was observed in 1988 at Saclay by C. Manus *et al.* using 30 ps pulses from a 1064 nm Nd:YAG laser at an intensity of 3×10^{13} W cm^{-2}. These results were subsequently extended to much higher harmonic values in a number of experiments. For example, in 1993 J.J. Macklin, J.D. Kmetec and C.L. Gordon reported the generation of up to the 109th harmonic in neon, using 125 fs pulses from an 806 nm Ti:sapphire laser at an intensity near 10^{15} W cm^{-2}.

Also in 1993, A. L'Huillier and P. Balcou detected high-order harmonic generation in rare gases, using 1 ps pulses from a 1053 nm Nd–glass laser at intensities between 10^{14} and 10^{15} W cm^{-2}. The experimental set-up of their interaction chamber is shown schematically in Fig. 15.35. The laser pulses, produced by using the chirped pulse amplification system, were focused below the nozzle of a pulsed gas jet. The light was analysed on axis by a monochromator consisting of a toroidal mirror and a grating. The mirror, placed at 1 m from the laser focal spot to

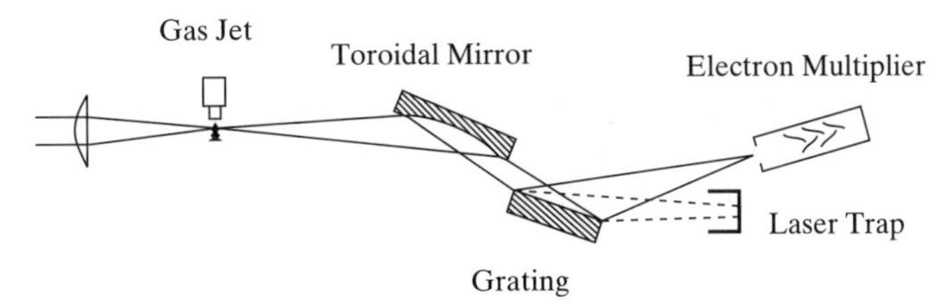

Figure 15.35 The experimental set-up used by A. L'Huillier and P. Balcou for the observation of harmonic generation in various noble gases. (From A. L'Huillier and P. Balcou, *Physical Review Letters* **70**, 774, 1993.)

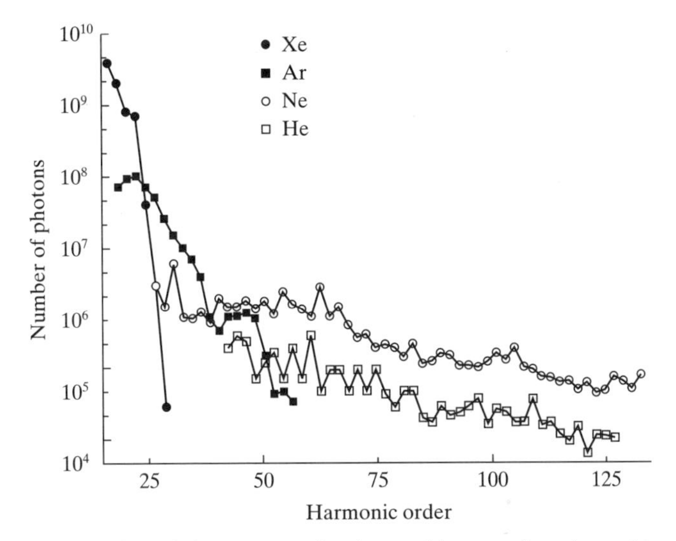

Figure 15.36 Harmonic emission spectra of various noble gases for a 'pump' laser of wavelength $\lambda = 1053$ nm and intensity $I \simeq 1.5 \times 10^{15}$ W cm^{-2}. (From A. L'Huillier and P. Balcou, *Physical Review Letters* **70**, 774, 1993.)

minimise the risk of damaging the optics with the incident laser light, refocused the radiation onto a 100 μm output slit. The laser light reflected by the grating was trapped before being focused too much in order to avoid plasma generation on the walls of the interaction chamber. The photons were detected by an electron multiplier. The number of photons obtained per laser pulse in xenon, argon, neon and helium at an intensity at best focus $I = 1.5 \times 10^{15}$ W cm^{-2} is shown in Fig. 15.36 as a function of the harmonic order. Of particular interest is the existence of *plateaux* of nearly constant conversion efficiency, which are seen to be particularly long for helium and neon. We shall see below that non-perturbative theories are required to explain the occurrence of such plateaux. L'Huillier and Balcou detected up to the 135th harmonic in neon at an energy of 160 eV, being then limited by their monochromator's resolution. The harmonic emission was observed to be directional and of short pulse duration (shorter than the pump pulse). The instantaneous power generated, at 20 eV (the 17th harmonic in Xe) for example reached about 30 kW, with a conversion efficiency of 10^{-6}. The instantaneous brightness was 10^{22} photons/(Å s), a number which is several orders of magnitude higher than that obtained with conventional light sources in this domain of the electromagnetic spectrum (but of course over a restricted time range).

The theoretical treatment of harmonic generation by an intense laser pulse focused into a gaseous medium has two main aspects. First, the *microscopic single atom* response to the laser field is analysed, and the spectra emitted by individual atoms are calculated for a range of pump laser intensities. Because different atoms interact with different parts of the focused laser beam, they are submitted to different peak field intensities and phases. The single atom responses must therefore be added to obtain the *macroscopic* harmonic signal generated from the coherent emission of all the atoms in the laser focus. This is done by solving Maxwell's equations to account for propagation and interference effects. The latter effects can suppress the harmonic signal completely unless a constructive *phase matching* takes place.

We shall only consider here the microscopic aspect of the problem. The emission of harmonics by the atom is due to the electron oscillations caused by the intense laser field. Let us introduce the *laser-induced dipole moment*

$$\mathbf{d}(t) = \langle\Psi(t)| - e\mathbf{R}|\Psi(t)\rangle \tag{15.103}$$

which is the expectation value of the electric dipole operator $\mathbf{D} = -e\mathbf{R}$ in the state $|\Psi(t)\rangle$, the corresponding atomic wave function in the presence of the laser field being $\Psi(X, t)$. Let

$$\mathbf{a}(t) = \frac{\mathrm{d}^2}{\mathrm{d}t^2}\mathbf{d}(t) \equiv \ddot{\mathbf{d}}(t) \tag{15.104}$$

be the acceleration of the induced dipole moment. The power spectrum of the emitted radiation is proportional to the quantity $|\mathbf{a}(\Omega)|^2$, where

$$\mathbf{a}(\Omega) = \frac{1}{2\tau}\int_0^\tau \exp(-\mathrm{i}\Omega t)\mathbf{a}(t)\,\mathrm{d}t \tag{15.105}$$

is the 'Fourier transform' of $\mathbf{a}(t)$ over the pulse duration τ.

If the atom is driven by a monochromatic laser field such as (15.51), the emitted harmonic radiation can be directly calculated from the induced dipole (15.103). Using the Floquet wave function (15.78), one finds that

$$\mathbf{d}(t) = \exp(-\Gamma t)\left\{\mathbf{d}_0 + 2\sum_{q=1}^{\infty}\mathrm{Re}\,[\mathbf{d}_q \exp(-\mathrm{i}q\omega t)]\right\} \tag{15.106}$$

where we have used the fact that $\mathbf{d}_q = \mathbf{d}^*_{-q}$ because $\mathbf{d}(t)$ is real. For an atomic state of given parity the quantities $\mathbf{d}_q$ vanish when q is even. If ionisation is neglected, so that the limit $\Gamma \to 0$ can be taken (see (15.82)), the rate of emission of photons of angular frequency $\Omega = q\omega$ (with q being an odd integer) is then found to be proportional to $|\mathbf{d}_q|^2$.

For weak laser fields, the harmonic emission rates can be calculated by using perturbation theory. It is then found that in general the harmonic intensity decreases from one order to the next. In contrast, we see from Fig. 15.36 that at high laser intensities, the harmonic intensity distribution exhibits a rapid decrease over the first few harmonics, followed by a plateau of approximately constant intensity, and then a cut-off corresponding to an abrupt decrease of harmonic intensity. The existence of the plateau can only be explained by using non-perturbative approaches such as the Sturmian–Floquet method, the R-matrix–Floquet method, the semi-classical or quantum mechanical versions of the 'recollision model' or the direct numerical integration of the time-dependent Schrödinger equation (TDSE). It was discovered in the framework of TDSE calculations that the cut-off angular frequency ω_c of the harmonic spectrum is given approximately by the relation

$$\omega_c \simeq (I_P + 3U_p)/\hbar \qquad \textbf{(15.107)}$$

Using the semi-classical 'recollision model', it is readily shown that the maximum returning kinetic energy of a classical electron colliding with the atomic core is $3.2U_p$, so that the highest energy that can be radiated is $I_P + 3.2U_p$, in good agreement with the TDSE calculations and with experiment.

An exciting new development is the possibility of using high-order harmonics to generate pulses of extremely short duration, in the range of hundreds of attoseconds (1 as = 10^{-18} s). Theoretical calculations performed by P. Antoine, P. Salières, A. L'Huillier and M. Lewenstein predict that high-order harmonics generated by an atom in a linearly polarised intense laser pulse form a train of ultra-short pulses, corresponding to different trajectories of electrons that tunnel out of the atom and recombine with the parent ion. Under appropriate geometrical conditions, the macroscopic propagation in an atomic jet allows the selection of one of these trajectories, leading to a train of attosecond pulses, with one pulse per half optical cycle. The observation of a train of attosecond pulses from high harmonic generation, based on theoretical work by V. Véniard, R. Taïeb and A. Maquet, was reported in 2001 by P. Agostini, H.G. Muller and co-workers.

The recent development of laser systems capable of delivering intense ultra-short pulses with peak intensities well above 10^{15} W cm^{-2} offers new possibilities for the generation of coherent, tunable, attosecond UV or X-ray pulses by frequency conversion in gases. Atoms or ions exposed to such pulses can experience much stronger laser fields before ionising than would be possible in longer pulses, and this in turn permits the generation of photons of much higher energies.

For example, in 1997 Z. Chang, A. Rundquist, H. Wang, M.M. Murnane and H.C. Kapteyn used a Ti:sapphire laser of wavelength 800 nm delivering pulses of 26 fs duration and peak intensity of about 6×10^{15} W cm^{-2} to generate coherent soft X-ray harmonics at wavelengths down to 2.7 nm (corresponding to an energy of 460 eV) in helium, and 5.2 nm (239 eV) in neon. In helium, they observed harmonic peaks up to order $q = 221$, and unresolved harmonic emission up to order $q = 297$. In 1998, T. Brabec, F. Krausz and co-workers reported the generation of coherent X-rays with wavelengths down to 2.5 nm (corresponding to an energy of 500 eV) in a helium gas irradiated by sub-10 fs pulses of peak intensity 4×10^{15} W cm^{-2} generated by a Ti:sapphire laser of wavelength 770 nm.

It is worth noting that when the driving pulse is only a few optical cycles long, the photons are emitted with a *continuous* distribution of frequencies, not at discrete harmonic frequencies. In addition, *multiple plateaux* [11] structures can appear in the photon emission spectra, which depend sensitively on the phase δ of the laser pulse.

Positive ions can survive higher laser intensities than neutral atoms because of their higher binding energies, and can therefore emit more energetic photons. As an example, we show in Fig. 15.37 some typical photon emission spectra of positive ions interacting with intense, ultra-short pulses. They have been calculated by N.J. Kylstra, R.M. Potvliege and C.J. Joachain, who generalised the low-frequency, strong field approximation (SFA) of M. Lewenstein *et al.* to treat the coupling of the ion with the incident pulse beyond the dipole approximation. In this way, the effect of the magnetic field component of the pulse on photon emission can be studied. The pulse, linearly polarised and travelling in the direction $\hat{\mathbf{k}}$ of the propagation vector $\mathbf{k}$, is described by a vector potential taking into account non-dipole effects. That is,

$$\mathbf{A}(\eta) = \hat{\boldsymbol{\varepsilon}}(\mathscr{E}_0/\omega)G(\eta)\sin(\eta + \delta) \qquad \mathbf{(15.108)}$$

where $\eta = \omega(t - \hat{\mathbf{k}}\cdot\mathbf{r}/c)$, $\mathscr{E}_0$ is the maximum electric field strength and $G(\eta)$ is a pulse shape function, which was taken to be

$$G(\eta) = \sin^2[\eta/(2N)] \qquad \mathbf{(15.109)}$$

where N is the number of optical cycles. The spectra displayed in Fig. 15.37 were calculated for two-cycle pulses ($N = 2$) with a wavelength $\lambda = 800$ nm (corresponding to a Ti:sapphire laser of angular frequency $\omega = 0.057$ a.u. and photon energy $\hbar\omega = 1.55$ eV), a phase $\delta = 0$ and peak intensities of 0.9, 1.8 and 3.6×10^{17} W cm^{-2}, incident on a Li^{2+} or a Be^{3+} ion. It should be noted that the 'relativistic parameter' $q = U_p/(mc^2)$ of equation (15.97) reaches the value $q = 1$ only at an intensity $I = 9 \times 10^{18}$ W cm^{-2} for the wavelength $\lambda = 800$ nm, so that the dynamics remains essentially non-relativistic for the peak intensities considered here.

Taking $\hat{\boldsymbol{\varepsilon}}$ and $\hat{\mathbf{k}}$ to coincide respectively with the unit vectors $\hat{\mathbf{x}}$ and $\hat{\mathbf{z}}$, the spectrum of the emitted photons was obtained by calculating $|\hat{\mathbf{x}}\cdot\mathbf{a}(\Omega)|^2$ for emission polarised parallel to the polarisation direction, and $|\hat{\mathbf{z}}\cdot\mathbf{a}(\Omega)|^2$ for emission polarised along the direction of propagation of the incident pulse. The quantity Ω is the angular frequency of the emitted photon and we recall that $\mathbf{a}(\Omega)$ is the 'Fourier transform' of $\mathbf{a}(t) = \ddot{\mathbf{d}}(t)$. The ratio Ω/ω is an effective 'harmonic order' since, as mentioned above, photons are emitted with a continuous distribution of frequencies for ultra-short pulses containing only a few optical cycles.

[11] Multiple plateaux also occur in several other cases, such as harmonic generation in an elliptically polarised laser field, in a two-colour field or in the presence of a static electric or magnetic field.

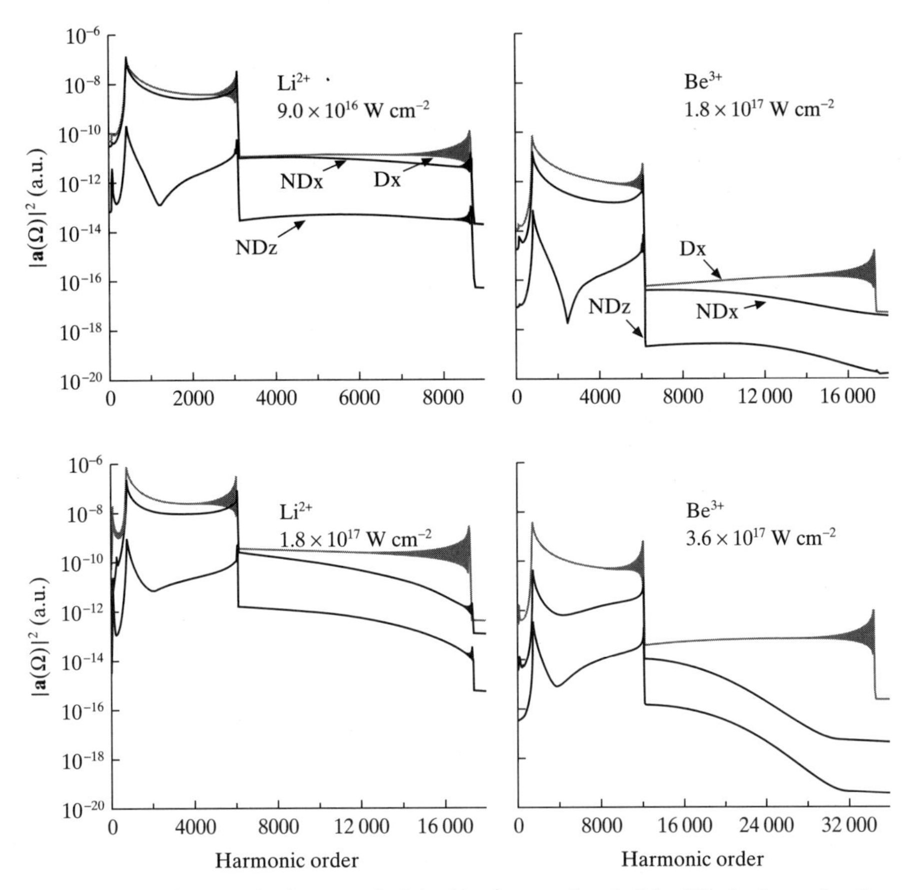

Figure 15.37 The magnitude squared of the 'Fourier transform' of the SFA dipole acceleration (in atomic units) of Li^{2+} (left) and Be^{3+} (right) as a function of the photon energy (in units of $\hbar\omega$). Dipole results (Dx) are displayed, as well as non-dipole results for photons polarised along $\hat{\mathbf{x}}$ (NDx) and $\hat{\mathbf{z}}$ (NDz). The incident laser pulse has a duration of two optical cycles, a carrier wavelength of 800 nm and the peak intensity indicated in each diagram. (From N.J. Kylstra, R.M. Potvliege and C.J. Joachain, *Laser Physics* **12**, 409, 2001.)

As seen from Fig. 15.37, emission polarised along the propagation ($\hat{\mathbf{z}}$) direction is weaker than emission polarised along the polarisation ($\hat{\mathbf{x}}$) direction; the difference is typically two orders of magnitude at the highest intensity considered, and is larger in weaker fields. It is worth noting that emission polarised in the z direction is forbidden by the dipole selection rules, and therefore would not occur if the magnetic field component of the pulse was neglected.

The influence of the magnetic field on photon emission polarised in the x direction can be seen by comparing the dipole and non-dipole results. Above a peak intensity of 10^{17} W cm^{-2} it significantly reduces the strength of photon emission, causes a 'bending over' of the plateaux, and suppresses some of the cut-offs which separate the plateaux in the dipole spectra. The decrease in the strength of photon emission should be expected as the displacement in the z direction due to the magnetic drift, 317 a.u. per half-cycle at 3.6×10^{17} W cm^{-2}, is large enough to reduce considerably the overlap of the returning wave packet with the nucleus. At peak intensities below 10^{17} W cm^{-2}, this displacement is not sufficiently large compared to the width of the returning electron wave packet to have an important effect. As a result, non-dipole effects in photon emission can be neglected for peak intensities below 10^{17} W cm^{-2} at the Ti:sapphire wavelength $\lambda = 800$ nm, but this is no longer the case for stronger laser fields.

The bending over of the plateaux has the same origin as the overall decrease in the strength of photon emission. The magnetic drift incurred by the detached electrons before they recombine depends on their trajectory in the continuum. Since at any given time fast electrons are accelerated further along the positive z direction than slow electrons, the emission of high-energy photons, which are produced by recombination of electrons that return to the nucleus with a high velocity, tends to be more suppressed than emission of low-energy photons. The suppression may not always increase monotonically with energy, though, since emission at a given frequency is normally due to the recombination of electrons detached at different times during the pulse, following different trajectories, and thus deflected differently by the magnetic field. In fact, the trajectories that contribute most to photon emission in the dipole approximation and are responsible for the plateau structure of the spectra are often more deflected from the nucleus than trajectories that contribute little, leaving the non-dipole spectra to be dominated by the latter; hence the changes in the number and positions of the cut-off frequencies.

Another interesting effect due to the magnetic field component of the pulse can be exhibited by examining the emission of photons of a given angular frequency Ω as a function of time. In accordance with the 'recollision model', this emission occurs in bursts precisely at the instants when detached electrons return to the nucleus with the velocity required for emission at this particular angular frequency. In the dipole approximation, photon emission at a given angular frequency Ω is usually dominated by several groups of trajectories. However, the magnetic drift, when taken into account, changes this picture dramatically. Indeed, in certain parts of the photon emission spectrum, all but one of the several returns that contribute most in the dipole approximation are suppressed. The only one remaining occurs towards the end of the pulse and dominates the spectrum. The corresponding trajectory is less deflected than the others, due to the decrease in the strength of the magnetic field in the trailing edge of the pulse. As a consequence, emission in this region essentially consists of a single burst of photons. For example, in studying the interaction of a Be^{3+} ion with a four-cycle Ti:sapphire pulse of 3.6×10^{17} W cm^{-2} peak intensity, N.J. Kylstra, R.M. Potvliege and C.J. Joachain found that in a narrow frequency window centred about $\Omega = 2500\omega$ (corresponding to an energy of about 3.9 keV), emission occurs in the form of a single burst of X-ray photons, the duration of the burst being about 20 attoseconds.

Multiphoton processes in molecules, clusters, solids and plasmas

The theory of multiphoton processes outlined above can be extended to analyse *molecular multiphoton processes*. For example, if we consider a diatomic molecule AB, the following multiphoton processes can occur:

1. Dissociation

$$n\hbar\omega + \mathrm{AB} \rightarrow \mathrm{A} + \mathrm{B} \tag{15.110}$$

2. Ionisation

$$n\hbar\omega + \mathrm{AB} \rightarrow \mathrm{AB}^{+} + \mathrm{e}^{-} \tag{15.111}$$

3. Dissociative ionisation

$$n\hbar\omega + \mathrm{AB} \rightarrow \mathrm{A} + \mathrm{B}^{+} + \mathrm{e}^{-} \tag{15.112}$$

4. Coulomb explosion

$$\begin{aligned} n\hbar\omega + \mathrm{AB} &\rightarrow \mathrm{AB}^{+} + \mathrm{e}^{-} \\ n\hbar\omega + \mathrm{AB}^{+} &\rightarrow \mathrm{A}^{+} + \mathrm{B}^{+} + \mathrm{e}^{-} \end{aligned} \tag{15.113}$$

5. Harmonic generation
6. Laser-assisted electron–molecule collisions.

All these multiphoton processes can also take place in polyatomic molecules. Reviews of high-intensity laser–molecule physics can be found in Bandrauk (1993), Codling and Frasinski (1994), and Giusti-Suzor, Mies, DiMauro, Charron and Yan (1995).

The interaction of clusters of atoms with strong laser fields constitutes a new area of multiphoton physics, where enhanced yields of high harmonics and the generation of very energetic ionisation fragments were observed in 1997 by T. Ditmire, J.W.G. Tisch, J. Marangos, M.H.R. Hutchinson and co-workers.

Solid targets and plasmas interacting with intense laser pulses also exhibit a wide range of interesting phenomena, which are relevant for applications such as particle acceleration, X-ray lasers, astrophysics and inertial confinement fusion [12].

15.4 Laser cooling and trapping of neutral atoms

The process which is at the basis of the manipulation of atoms by light is the *radiation pressure* exerted by light on material particles. In the 17th century, J. Kepler speculated that the repulsion of comet tails from the Sun was due to radiation pressure. Although the present explanation of this phenomenon is more complex, Kepler's hypothesis about the pressure exerted by light on material particles

[12] See Batani, Joachain, Martelluci and Chester (2001).

identified a significant effect. However, a proper theoretical basis for the concept was only established in 1873 when J.C. Maxwell formulated his electromagnetic theory of light. He showed that an electromagnetic field exerts a pressure $P = I/c = \rho$, where I is the light intensity and ρ its energy density.

Our present understanding of radiation pressure is based on the concept of light quanta or *photons* introduced by A. Einstein in 1905. As we saw in Sections 1.4 and 1.5, a photon of energy $E = h\nu$ carries a momentum $\mathbf{p}$ directed along the propagation direction of the light, whose magnitude is given by $p = E/c = h\nu/c = h/\lambda$. We also recall that $\mathbf{p} = \hbar\mathbf{k}$ where $\mathbf{k}$ is the wave vector of magnitude $k = 2\pi/\lambda$. At the microscopic level, the action of light on the motion of atoms can therefore be understood by remembering that when a photon is scattered by an atom, the recoil momentum $\mathbf{p}_R$ of the atom is given by the difference between the incident photon momentum $\mathbf{p}$ and the momentum $\mathbf{p}'$ of the scattered photon, with $p = h\nu/c$ and $p' = h\nu'/c$. Although each individual scattering event causes a very small change in the velocity of the atom, the accumulated momentum transfer per second can be large if the rate of photon–atom collisions is increased. This can be achieved by choosing the frequency of the incident light to be close to a *resonance* frequency $\nu_{ba} = (E_b - E_a)/h$ in the spectrum of the atom, so that the incident photons are *absorbed*.

If an atom of mass M is irradiated by resonant light, so that it absorbs a photon to make a transition from the state a to the state b, the photon energy is almost entirely converted into *internal energy*. The photon momentum $\mathbf{p} = \hbar\mathbf{k}$, however, causes the atom to recoil in the direction of the incident light and to change its velocity by an amount $\mathbf{v}_R = \hbar\mathbf{k}/M$. This velocity change per absorbed photon, called *recoil velocity*, was predicted by Einstein in 1917. It varies typically in magnitude from a few millimetres per second to a few centimetres per second, which is very small compared with atomic thermal velocities at room temperature ($\sim$300 m s^{-1}). The atom also recoils when it returns to the state a by spontaneously emitting a photon called a *fluorescence* photon. This time, the recoil momentum of the atom is opposite to that of the fluorescence photon, and is in a *random* direction because spontaneous emission is a random process. Thus, when an atom is irradiated with a resonant light beam, each photon absorption increases the atom's velocity in the direction of the beam, and each spontaneous emission leads to a recoil in a random direction. After averaging over many absorption–spontaneous emission *fluorescence cycles*, we see that an atom, on average, changes its velocity by an amount $\hbar\mathbf{k}/M$ per cycle. The net result, after many fluorescence cycles, is therefore an acceleration of the atom in the direction of the incident light beam. The rate of the fluorescence cycles is only limited by the lifetime of the excited atomic level b, and can reach 10^8 per second.

The first experimental demonstration of the action of light on the motion of atoms was made in 1933 by R. Frisch. He observed the deflection of a thermal beam of sodium atoms (with a mean velocity of 900 m s^{-1}) resonantly excited from the side with a sodium lamp. For the yellow Na resonance line at 5890 Å, the recoil corresponds to a velocity change of 3 cm s^{-1} per scattered photon. The results were consistent with an estimate that only one-third of the sodium atoms were excited. This low excitation rate was due to the fact that the intensity of the light produced by the lamp was insufficient. It prevented the atom from being manipulated in a useful way.

The situation changed completely with the advent, in the 1970s, of *narrow band tunable* lasers, which provide intense, highly directional, quasi-monochromatic and frequency adjustable sources of light. The high brightness of laser sources dramatically increases the rate of fluorescence cycles, resulting in a substantial radiation pressure force. In 1970, A. Ashkin showed that for a laser resonant with a strong optical transition (for example, the resonance transition of an alkali atom), this force could be used for isotope separation, velocity analysis and atom trapping. The magnitude F of the force is given by the magnitude of the momentum change $\Delta p = \hbar k$, divided by the time Δt that the atom is in the excited state b. If τ_b is the natural lifetime of the excited state and f is the fraction of time an atom spends in the excited state b, we have

$$F = \frac{\Delta p}{\Delta t} = \frac{\hbar k}{\tau_b / f} \qquad \textbf{(15.114)}$$

For example, in the case of the sodium resonance transition, $\tau_b = 16$ ns, and the maximum acceleration corresponding to $f = 1/2$ has a magnitude of 1.5×10^6 m s^{-2}, which is more than 10^5 times that of the gravitational acceleration on Earth. Following the advent of tunable dye lasers, H. Walther *et al.* in 1972 and P. Jacquinot *et al.* in 1973 were able to act efficiently on the motion of atoms.

Slowing of atomic beams

Let us now examine how radiation pressure produced by a quasi-resonant laser can slow an atomic beam. The average velocity of sodium atoms leaving an oven at 600 K is about 900 m s^{-1}. To stop an Na atom at this velocity requires the scattering of about 30 000 counter-propagating photons. In order to be resonant with an atom moving at 900 m s^{-1}, the light should be tuned about 1.5 GHz below the atomic resonance because of the Doppler effect. However, as the atom slows down, the Doppler shift decreases and resonance will no longer be achieved. To compensate the changing Doppler shift, V.I. Balykin, V.S. Letokhov and V.I. Mushin modified in 1979 the laser frequency by using a frequency chirping technique. To achieve the same goal, W.D. Phillips and H. Metcalf employed in 1982 the Zeeman effect to produce a spatial variation of the atomic resonance frequency.

Doppler cooling

In the methods described above, atoms travelling in a *given direction* are slowed. It is possible, however, to use laser light to cool a gas of atoms, that is to obtain a narrow velocity distribution around $v = 0$. In the proposal made in 1975 by T.W. Hänsch and A.L. Schawlow, atoms are cooled by using three orthogonal pairs of counter-propagating laser beams (see Fig. 15.38). The lasers are detuned below

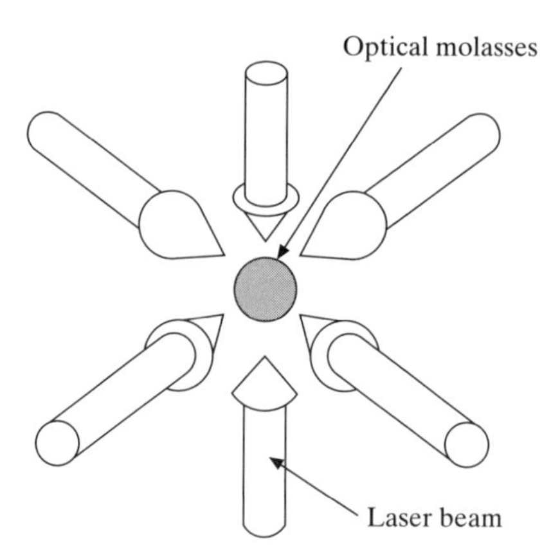

Figure 15.38 Schematic diagram of a laser arrangement producing optical molasses.

the resonance frequency ($\nu < \nu_{ba}$). As a result, an atom is more likely to absorb a photon when it is moving towards the laser source because only then is the frequency 'Doppler shifted' towards resonance. When the atom absorbs a photon, it recoils in the direction of the laser beam, and hence slows. The subsequent re-emission of the photon by spontaneous emission leads to a recoil in a random direction, so that the net effect is a deceleration of the atoms. The rate of these absorption–spontaneous emission cycles (fluorescence cycles) is only limited by the lifetime of the excited atomic level, and can reach 10^8 per second. The atoms then experience for small velocities a damping force

$$\mathbf{F} = -\alpha \mathbf{v} \tag{15.115}$$

where the friction coefficient α is positive for detuning to lower frequencies ('red detuning'). The atoms move as in a viscous medium, created by the laser light, called 'optical molasses'. On the other hand, the radiation pressure force has a fluctuating part, due to the discrete nature of the momentum exchanges between the atoms and the photons. This gives rise to a heating effect, which opposes the cooling mechanism. The equilibrium between heating and cooling occurs at a minimum temperature, called the *Doppler cooling limit*, which is of the order of

$$T_{\mathrm{D}} = \frac{\Gamma}{2k_{\mathrm{B}}} \tag{15.116}$$

where Γ is the width of the resonance line and k_{B} is Boltzmann's constant.

Using this Doppler-cooling method, S. Chu *et al.* were able in 1985 to cool a cloud of 10^5 sodium atoms to a temperature of about 240 μK. This temperature was estimated by using a *ballistic* method, called 'release and recapture', which consists in switching off the lasers for a given amount of time (a few milliseconds), and then applying the lasers again. The number of atoms having escaped from the optical molasses is evaluated by measuring the decrease of fluorescence after the switching. This number is related to the initial velocity distribution of the atoms in the molasses, which can therefore be deduced, thus yielding the temperature.

Sub-Doppler cooling

More precise temperature measurements in optical molasses showed that the temperatures attained were often much lower than the Doppler cooling limit (15.116) and that other cooling mechanisms were operating. The physical origin of these new mechanisms can be related to the *multilevel* structure of real atoms, in contrast to the model two-level atoms assumed in the simple theory used to obtain the result (15.116). In addition, one must allow for the existence of *light polarisation gradients* (that is, the spatial variation of the polarisation of the light) and *AC Stark shifts* also called *light shifts* in this context.

Sisyphus cooling

As an example of a sub-Doppler cooling mechanism, we shall describe *Sisyphus cooling*, proposed in 1989 by J. Dalibard and C. Cohen-Tannoudji. Let us consider alkali atoms interacting with laser light, the frequency of which being such that transitions can be made from the ground state g to an excited state e. Both the grounds state g and the excited state e consist of a number of Zeeman sublevels, which are degenerate in the absence of an external magnetic field. A particular case is illustrated in Fig. 15.39(a), in which the ground state has $J = 1/2$ and the excited state has $J = 3/2$. Which of the ground state sublevels with $M_J = \pm 1/2$ is going to be populated depends on the polarisation of the light. The excited state e will subsequently decay and make a transition to a different substate of the lower level. The relative transition probabilities are indicated on the diagram.

Let us take the case in which the atoms are illuminated by two laser beams of wavelength λ_L with orthogonal linear polarisations counter-propagating along the Z axis. The polarisation of the total radiation field, which varies from point to point, is shown in Fig. 15.39(b), changing from linear to circular over a distance of $\lambda_L/8$. As a consequence, the polarisation-dependent light shifts of the ground state sublevels, induced by the electric field of the light, will also vary from point to point. The light shift of the substate $g_{+1/2}$ is three times larger than that of the substate $g_{-1/2}$ at places where the polarisation is σ^+ while at places where the polarisation is σ^- the reverse is true (see Fig. 15.39(c)). Because transitions are most probable from the highest Zeeman sublevel of the excited state to the lowest sublevel of the ground state, at points where the polarisation is σ^+ the population of the substate $g_{+1/2}$ becomes larger than that of the $g_{-1/2}$ substate; the reverse is true at points where the polarisation is σ^-. The populations of the $g_{+1/2}$ and $g_{-1/2}$ substates are equal at points of linear polarisation. Thus, this combination of optical pumping and polarisation gradient produces a relative population of the $g_{+1/2}$ and $g_{-1/2}$ substates which varies from point to point. It is important to note that the optical pumping or fluorescence cycle of excitation and decay takes a finite time τ_p to produce the final population distribution.

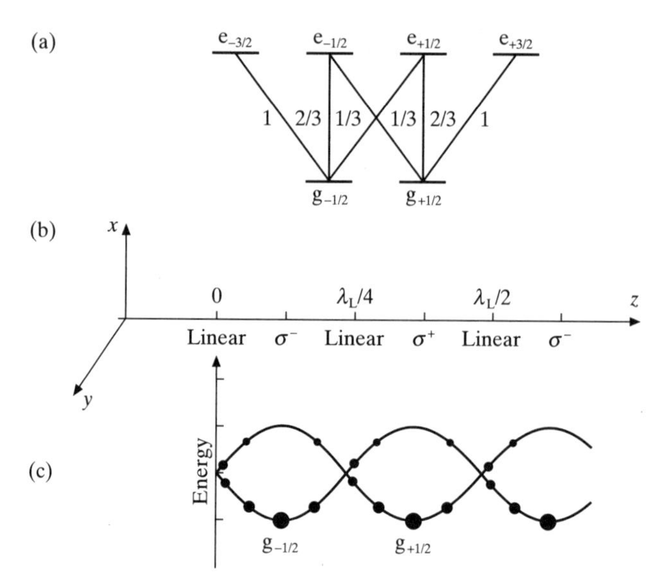

Figure 15.39 Light shifts (AC Stark shifts) in a polarisation gradient.
(a) The magnetic substates of the ground state g with $J = 1/2$ and of an excited state with $J = 3/2$ of an alkali atom. The relative transition probabilities are indicated next to the transitions.
(b) The polarisation of the total radiation field, produced by two laser beams with wavelength λ_L and orthogonal polarisations, counter-propagating along the Z axis.
(c) The light-shifted energies of the two ground state substates, as a function of the variable z. The size of the population of the two substates is proportional to the size of the solid circles. (From C. Cohen-Tannoudji and W.D. Phillips, *Physics Today*, October 1990, p. 33.)

Now, consider an atom moving to the right and initially at the bottom of a potential valley where the population for the $g_{+1/2}$ substate is less than that of the $g_{-1/2}$ substate (see Fig. 15.40). If the velocity of the atom is such that $v\tau_p \simeq \lambda_L/4$, the atom can move to the top of the potential well without making a transition (on the average) to a different substate. For this to occur, some of the kinetic energy of the atom must be transferred to potential energy. At the top of the potential well the atom has the largest probability of making a transition to the bottom of the next valley (see Fig. 15.40). This process will continue, with the atom continually climbing a potential hill and then transferring to the bottom of a valley by emitting a photon. We remark that the atom spends more time climbing hills. The net result is that kinetic energy is converted into potential energy, which is subsequently carried away by the fluorescence photons, so that the atoms are cooled. This cooling mechanism is called the *Sisyphus effect* after the unfortunate individual, in Greek mythology, who was forced for eternity to push a rock to the top of a hill, only to find it roll down again to the bottom.

In the discussion of Sisyphus cooling given above, we have taken the most favourable case, where $v\tau_p \simeq \lambda_L/4$. In fact, the cooling mechanism works over a range of velocities of the order of λ_L/τ_p. The optical pumping rate is proportional

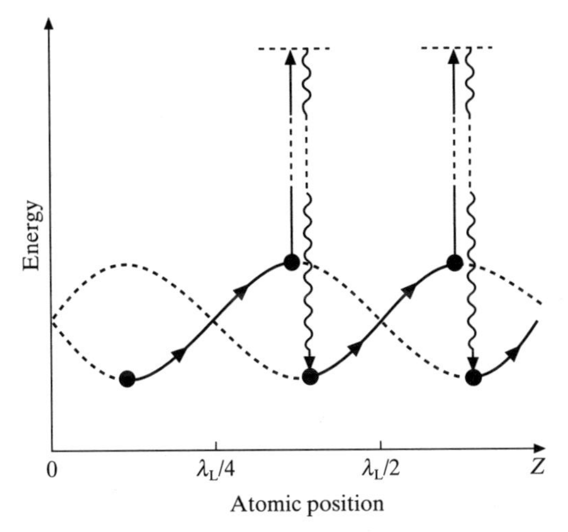

Figure 15.40 Illustration of the Sisyphus effect. An atom moving in the laser arrangement of Fig. 15.39 climbs out of a potential valley and reaches a potential hill before transferring to the bottom of the next valley by emitting a photon. This process is then repeated, the net effect being a cooling of the atom. (From C. Cohen-Tannoudji and W.D. Phillips, *Physics Today*, October 1990, p. 33.)

to the laser intensity I so that the range of velocities is also proportional to I and vanishes when the laser intensity is equal to zero. This is in contrast with Doppler cooling, for which the range of useful velocities is independent of I. However, in Doppler cooling, the friction coefficient is proportional to I, while in Sisyphus cooling it is independent of I, and large.

Sub-recoil cooling

In the Doppler and sub-Doppler cooling mechanisms, fluorescence cycles continue indefinitely. It would then seem impossible to avoid the random recoil of the atom due to spontaneously emitted photons and the corresponding *single photon recoil limit*, which corresponds to a temperature

$$T_R = \frac{E_R}{k_B} \tag{15.117}$$

where $E_R = \hbar^2 k^2/(2M)$ is the recoil energy of an atom having emitted one photon. However, this limitation can be overcome by using methods based on other physical mechanisms. One of these methods, used by A. Aspect *et al.* in 1988, is called 'velocity-selective coherent population trapping'. This method does not rely on a friction force. It is based on the fact that atoms may be optically pumped into a *non-coupled state* ψ_{nc} (also called a *dark state*) where they do not interact with the laser light. This can be accomplished by choosing the non-coupling state ψ_{nc} to be a coherent superposition of two ground state substates with absorption amplitudes interfering destructively. Once the atoms have been pumped into

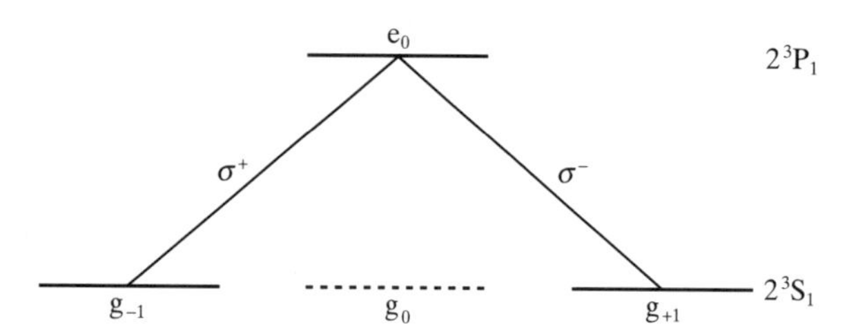

Figure 15.41 The three-level Λ configuration relevant for the discussion of the subrecoil cooling experiment of A. Aspect *et al.* in helium. The ground state g is the $2\,^3S_1$ state and the excited state e is the $2\,^3P_1$ state.

such a dark state, fluorescence stops. This phenomenon was already known from earlier experiments performed in 1976 by A. Gozzini and co-workers.

A perfect non-coupled state can only exist for atoms having zero velocity. Indeed, for velocities $v \neq 0$, complete destructive interference cannot occur, the absorption rate increasing with velocity. However, there is a finite probability that atoms originally in absorbing states with $v \neq 0$ will be pumped into the dark state with $v = 0$, where they are 'protected' from the 'bad' effects of spontaneous emission and where they accumulate.

In the experiment of A. Aspect *et al.*, one-dimensional subrecoil cooling was performed on a metastable helium atomic beam. The helium atoms crossed a light field resonant with the $2^3S_1 \rightarrow 2^3P_1$ $(J = 1 \rightarrow J' = 1)$ transition. The level scheme (see Fig. 15.41) involves a three-level Λ configuration, where two degenerate Zeeman sublevels $g_{\pm 1}(M_J = \pm 1)$ are coupled to an excited level $e_0(M_J = 0)$ by two counter-propagating circularly polarised σ^+ and σ^- laser beams with the same angular frequency ω and the same intensity. For an atom at rest, two-photon Raman processes yield a non-absorbing coherent superposition of g_{+1} and g_{-1}. If the atom is moving along the z direction, the Raman resonance condition is no longer satisfied, due to opposite Doppler shifts on the two counter-propagating laser beams. As explained above, the cooling scheme consists in accumulating atoms in the zero-velocity dark state, where they remain trapped.

It should be noted that this cooling scheme cannot be analysed by assuming that the centre of mass motion of the atoms can be described classically. Indeed, the de Broglie wavelength of the atoms, $\lambda = h/p$, becomes larger than the laser wavelength λ_L when p becomes smaller than $\hbar k$, as is the case for subrecoil cooling, and a fully quantum mechanical theory including both the external and internal degrees of freedom of the atoms is required. It predicts that atoms are trapped in non-coupled states having the form

$$|\psi_{\rm nc}\rangle = \frac{1}{\sqrt{2}}[|g_{-1}, -\hbar k\rangle + |g_{+1}, +\hbar k\rangle] \qquad \textbf{(15.118)}$$

where $|g_{\pm 1}, \pm\hbar k\rangle$ refer to g_{+1} (g_{-1}) Zeeman sublevels, together with a single photon with a momentum in the z direction given by $\pm\hbar k$. Since $|\psi_{\rm nc}\rangle$ is not an eigenstate

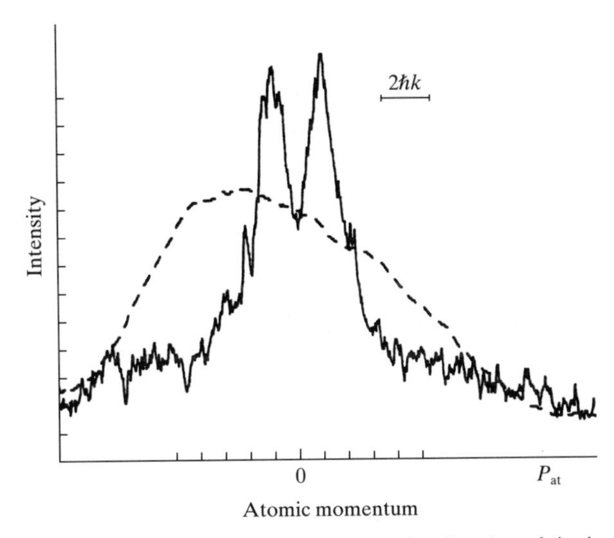

Figure 15.42 The momentum distribution of atoms in the direction of the laser beam at the end of the interaction region, with the laser on (solid line) and off (dashed line). A double peak structure is seen at about $\pm\hbar k$ and above the initial distribution. (From A. Aspect, E. Arimondo, R. Kaiser, N. Vansteenkiste and C. Cohen-Tannoudji, *Physical Review Letters* **61**, 826, 1988.)

of the atomic centre of mass momentum operator P_z, the atomic momentum distribution has two peaks at eigenvalues $\pm\hbar k$. This is confirmed by the experimental results, as seen from Fig. 15.42. A double peak structure is seen at about $\pm\hbar k$ and above the initial distribution. The width of each of the peaks is less than the magnitude of the photon momentum $\hbar k$, which shows that the width of the velocity distribution is less than the one-photon recoil limit. The width of the peaks in the corresponding velocity distribution is 6 cm s^{-1}, which corresponds to a temperature of 2 μK. This is lower than both the Doppler cooling limit of 23 μK and the single photon recoil limit of 4 μK.

Trapping of neutral atoms

Atom traps can be classified into three main categories depending on the physical mechanism employed to confine the atoms: (i) an induced atomic dipole moment; (ii) magnetic fields and (iii) a combination of radiation pressure and magnetic fields. It should be noted that cooling devices (such as optical molasses) do not trap atoms, since they are free to diffuse until they reach the boundary of the interaction regime, and escape.

Optical dipole traps

These traps rely on the fact that in a radiation field, a particle with a dynamic dipole polarisability, α, has an induced dipole moment of magnitude $\alpha\mathscr{E}$ and a potential energy $-\alpha\mathscr{E}^2$ where $\mathscr{E}$ is the electric field strength. Hence, if $\alpha > 0$, the particle is attracted towards regions of high intensity. The corresponding force is known as the dipole force. The possibility of trapping atoms with this force was considered in 1968 by V.S. Letokhov, who suggested that atoms could be confined in one dimension at the nodes or antinodes of a standing wave tuned far below or above the atomic transition frequency. For an atom in the ground state, the dynamic dipole polarisability α is positive for laser frequencies up to the first resonance (see (9.64)). Making use of this fact, A. Ashkin proposed in 1970 that atoms could be trapped at the focus of a strongly focused laser beam tuned below the atomic resonance (red detuned). A serious problem with this scheme is the strong heating of the atoms by the laser light. This difficulty was overcome in 1986 by S. Chu *et al.*, who demonstrated the first optical trap by periodically cooling the atoms using Doppler molasses to counterbalance the radiative heating. Further details about optical dipole draps can be found in the review by Grimm, Weidemüller and Ovchinnikov (2000).

Magnetic traps and evaporative cooling

The principle of a magnetic trap is to use the state-dependent force acting on a permanent dipole moment in an inhomogeneous magnetic field. Let us consider an atom having a magnetic dipole moment $\mathcal{M}$. In the presence of a magnetic field $\mathcal{B}$, it has a potential energy $-\mathcal{M}\cdot\mathcal{B}$ and experiences a force $\mathcal{M}\cdot\nabla\mathcal{B}$. For trapped states, $\mathcal{M}$ is antiparallel to $\mathcal{B}$, and the atom will experience a force directed towards a minimum of the magnetic field.

The simplest magnetic trap is formed by the magnetic quadrupole field created in the centre of two coils which carry currents in opposite directions. The magnetic field increases linearly with the distance from the centre, where the magnetic field vanishes. Magnetic traps are very well suited for confining atoms prepared by laser cooling.

A particularly attractive feature of magnetic traps is the possibility of ejecting atoms with a well-defined energy by driving a radio-frequency transition to the anti-trapped state, where $\mathcal{M}$ is parallel to $\mathcal{B}$. This idea was first suggested in 1986 by H.F. Hess as a technique to initiate *evaporative cooling* of magnetically trapped hydrogen atoms, and applied subsequently to magnetically trapped alkali atoms. The principle of evaporative cooling consists of selectively ejecting atoms having higher than average energy so that after thermalisation the temperature of the remaining atoms decreases. The high-energy tail of the thermal distribution from the trap must be constantly repopulated by collisions, so that thermal equilibrium is maintained and the cooling process is sustained. An essential condition for evaporative cooling is that the atomic sample must have a long lifetime compared to the collisional thermalisation time. It is important to note that trapped atom clouds are very dilute, so that collisional thermalisation can take seconds. Another requirement for the evaporative cooling process is a favourable ratio between the elastic collision rate (which is necessary to sustain the process) and the inelastic collision rate (which leads to trap loss and heating).

The magneto-optical trap

It can be shown that configurations with laser beams directed to a common centre and resonant with a two-level transition of the atom do not provide a stable trap. This is known as the *optical Earnshaw theorem* by analogy with Earnshaw's theorem of electrostatics, which states that it is impossible to arrange a set of charges in space to generate a point of stable equilibrium for a test charge. However, in 1986 J. Dalibard and D. Pritchard *et al.* suggested that the optical Earnshaw theorem could be bypassed by using the fact that real atoms may have a different behaviour from two-level atoms, if one uses the existence of several sublevels, and spatially varying electric or magnetic fields.

The most successful realisation of this concept is the *magneto-optical trap* (MOT), proposed in 1986 by J. Dalibard. The basic idea is to superimpose on the optical molasses an inhomogeneous static magnetic field such that the atoms which move away from the origin experience a restoring force. The principle of trapping in a MOT can most easily be understood in terms of a one-dimensional example (see Fig. 15.43), in which an atom with a $J = 0$ to $J' = 1$ transition is placed in the inhomogeneous magnetic field $\mathcal{B}(z) = bz$. As a result, the $M_J = +1$ and $M_J = -1$ sublevels of the excited state experience Zeeman shifts which are linear in z. The atom is acted upon by two counter-propagating laser beams tuned to the same frequency v below resonance in order to provide cooling. The two beams propagating in the $\pm z$ directions have $\sigma^{\pm}$ polarisations and induce $\Delta M_J = \pm 1$ transitions, respectively. An atom moving to the right will interact more with the σ^- than the σ^+ beam, becoming resonant with the σ^- beam at $z = z_r$. The net force on this atom pushes it back towards $z = 0$. Conversely, an atom moving to the left will interact more with the σ^+ beam than with the σ^- beam, becoming resonant at

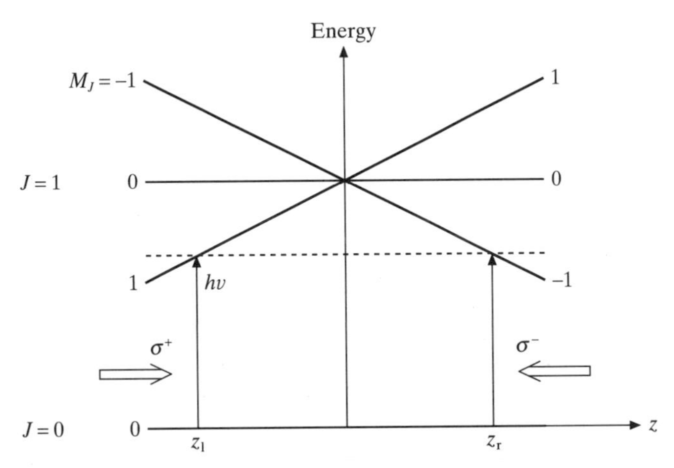

Figure 15.43 The principle of a magneto-optical trap (MOT). (From A. Aspect, in *Laser Interactions with Atoms, Solids and Plasmas*, edited by R.M. More, Plenum Press, New York (1994), p. 193.)

$z = z_1$, and will be pushed back to the centre of the trap. The net result in both cases is a restoring force towards $z = 0$ which leads to confinement of the atoms in the centre of the MOT.

This method was first demonstrated in 1987 by E.L. Raab *et al.* with sodium atoms. Since then, magneto-optical traps have proved to be extremely versatile, and many of the applications of atom trapping have been realised by using these traps. Temperatures of 10 μK can be achieved in a MOT.

Further details about laser cooling and trapping of neutral atoms can be found in the review articles by Sengstock and Ertmer (1995) and Adams and Riis (1997).

15.5 Bose–Einstein condensation

In 1924, S.N. Bose sent a paper to A. Einstein, in which he derived Planck's law for black body radiation by treating the photons as a gas of identical particles. Einstein saw to the paper's publication and generalised Bose's theory to an ideal gas of identical particles with mass, whose number is conserved. In a second paper, published in 1925, Einstein predicted that, as the temperature is lowered and (or) the particle density is increased, a large fraction of the particles would settle into the ground state of the system. This phenomenon is called *Bose–Einstein condensation* (BEC). It can only happen for bosons (particles having zero or integer spin $s = 0, 1, 2, \ldots$) which obey Bose–Einstein statistics [13]. In contrast, because of the Pauli exclusion principle, fermions (having half-odd integer spin $s = 1/2, 3/2, \ldots$) obey Fermi–Dirac statistics and are such that a particular quantum state cannot be occupied by more than one fermion.

Using quantum statistical mechanics [14], it can be shown that for an ideal gas of non-interacting bosons with mass, in the absence of any external potential, the fraction N_0/N of particles in the ground state of the system is given by

$$\frac{N_0}{N} = 1 - \left(\frac{T}{T_0}\right)^{3/2} \tag{15.119}$$

where

$$T_0(\rho) = \frac{\hbar^2}{2mk_B}\left(\frac{4\pi^2\rho}{2.315(2s+1)}\right)^{2/3} \tag{15.120}$$

is known as the *critical temperature*. In (15.120), $\rho = N/V$ is the particle density, where V is the volume of the container. Thus, when the temperature T and the particle density ρ are such that $T < T_0(\rho)$, we see from (15.119) that $N_0/N > 0$, so that a finite fraction of the particles occupies the ground state of the system. As T

[13] Bose–Einstein condensation does not happen in a photon gas, since photons can disappear when the energy of the system decreases.

[14] See Bransden and Joachain (2000).

is lowered and (or) ρ is raised, an increasing number of particles accumulate in the ground state, so that BEC occurs. It is worth stressing that the condensate behaves as a macroscopic quantum mechanical entity, described by a single wave function. Since the ground state has zero energy, the condensed system exerts no pressure, and is said to be 'degenerate'. It should be emphasised that the result (15.119) only holds for $T < T_0(\rho)$. If $T \geqslant T_0(\rho)$ the number of particles in the ground state is negligible.

Using (15.120), the condition $T < T_0(\rho)$ for BEC to occur can be recast in the form

$$\rho\lambda_{\mathrm{T}}^3 > 14.546(2s+1) \tag{15.121}$$

where we have introduced the *thermal de Broglie wavelength*

$$\lambda_{\mathrm{T}} = \frac{h}{(2mk_{\mathrm{B}}T)^{1/2}} \tag{15.122}$$

which is equal to the de Broglie wavelength of a particle of mass m and energy $k_{\mathrm{B}}T$.

The effect of interactions in BEC. The Gross–Pitaevskii equation

The Hamiltonian of a system of identical spin-zero particles of mass m in an external confining potential $V_{\mathrm{ext}}(\mathbf{r})$, which interact through a two-body interaction $W(\mathbf{r}_i - \mathbf{r}_j)$, is

$$H = \sum_{i=1}^{N}\left[-\frac{\hbar^2}{2m}\nabla_{\mathbf{r}_i}^2 + V_{\mathrm{ext}}(\mathbf{r}_i)\right] + \sum_{i<j=1}^{N} W(\mathbf{r}_i - \mathbf{r}_j) \tag{15.123}$$

In Section 8.4, we treated a system of N interacting fermions (electrons) by using the Hartree–Fock method, such that the N-electron wave function is approximated by a Slater determinant, that is an antisymmetric product of individual electron spin-orbitals. For a system of spinless bosons, the corresponding approximation is to write the N-particle wave function $\Psi(\mathbf{r}_1, \mathbf{r}_2, \ldots, \mathbf{r}_N)$ as a symmetric product of single-particle wave functions $\psi_i(\mathbf{r}_i)$. For the special case of a system of N spinless bosons in the ground state (at zero temperature), we assume that all particles occupy the same single particle state, $\psi_i(\mathbf{r}_i) = \psi(\mathbf{r}_i)$, $i = 1, 2, \ldots, N$, so that the N-particle wave function can be written as

$$\Psi(\mathbf{r}_1, \mathbf{r}_2, \ldots, \mathbf{r}_N) = \prod_{i=1}^{N}\psi(\mathbf{r}_i) \tag{15.124}$$

and we note that it is symmetrical since all the particles are in the same single particle state $\psi(\mathbf{r})$. By minimising the expectation value of the Hamiltonian (15.123) with respect to variations of the wave function $\psi(\mathbf{r})$, and requiring that $\psi(\mathbf{r})$ be normalised according to

$$\int |\psi(\mathbf{r})|^2\, \mathrm{d}\mathbf{r} = N \tag{15.125}$$

one finds for $\psi(\mathbf{r})$ the non-linear Schrödinger equation

$$\left[-\frac{\hbar^2}{2m}\nabla_{\mathbf{r}}^2 + V_{\text{ext}}(\mathbf{r})\right]\psi(\mathbf{r}) + \left[\int |\psi(\mathbf{r}')|^2 W(\mathbf{r}-\mathbf{r}')\,\mathrm{d}\mathbf{r}'\right]\psi(\mathbf{r}) = \mu\psi(\mathbf{r}) \qquad \textbf{(15.126)}$$

where μ is the energy of the single particle ground state (that is, the energy required to add one more atom to the condensate), which is also called the *chemical potential*. The equation (15.126) is known as the *time-independent Gross–Pitaevskii equation*. We see from (15.126) that if one neglects the two-body interactions between the atoms in the condensate, and takes only into account the external confining potential $V_{\text{ext}}(\mathbf{r})$, the wave function $\psi(\mathbf{r})$ would satisfy the one-body Schrödinger equation

$$\left[-\frac{\hbar^2}{2m}\nabla_{\mathbf{r}}^2 + V_{\text{ext}}(\mathbf{r})\right]\psi(\mathbf{r}) = \mu\psi(\mathbf{r}) \qquad \textbf{(15.127)}$$

It can be shown that in this case, for a non-interacting trapped gas in any geometry in which BEC occurs, the equation (15.119) is replaced by

$$\frac{N_0}{N} = 1 - \left(\frac{T}{T_0'}\right)^3 \qquad \textbf{(15.128)}$$

where $T_0' \simeq T_0$ and $T < T_0'$.

The term in (15.126) involving the two-body interaction $W(\mathbf{r} - \mathbf{r}')$ between atoms can be simplified as follows. We first note that atoms in the condensate interact by means of two-body *collisions*. Since the atoms are very cold, only s-wave collisions are important. Moreover, the de Broglie wavelength of the atoms is much larger than the range of the interatomic interactions. We can therefore model two-body collisions by using a contact interaction

$$W(\mathbf{r} - \mathbf{r}') = W_0\delta(\mathbf{r} - \mathbf{r}') \qquad \textbf{(15.129)}$$

where W_0 is given in terms of the two-body scattering length α by

$$W_0 = \frac{4\pi\hbar^2\alpha}{m} \qquad \textbf{(15.130)}$$

This interaction gives the exact low-energy scattering amplitude $-\alpha$ when used in the Born approximation. We note that when α is positive, the condensate atoms repel each other, while they attract each other if α is negative.

Using (15.129), the Gross–Pitaevskii equation becomes

$$\left[-\frac{\hbar^2}{2m}\nabla_{\mathbf{r}}^2 + V_{\text{ext}}(\mathbf{r}) + W_0|\psi(\mathbf{r})|^2\right]\psi(\mathbf{r}) = \mu\psi(\mathbf{r}) \qquad \textbf{(15.131)}$$

This equation can be solved numerically for given external potentials $V_{\text{ext}}(\mathbf{r})$ corresponding to the confining potential of the trap containing the condensate. We remark that if the kinetic energy term in (15.131) can be neglected, one finds that

$$|\psi(\mathbf{r})|^2 = \frac{\mu - V_{\text{ext}}(\mathbf{r})}{W_0} \tag{15.132}$$

provided the right-hand side is positive. This is called the *Thomas–Fermi limit.* The value of μ is determined by the normalisation condition (15.125).

Experimental verification of BEC in atomic gases

Although the phenomenon of Bose–Einstein condensation of an ideal gas was predicted in 1924, attempts to observe it met with great difficulties for many years. These difficulties arise from the fact that at a given temperature a sufficiently large density must be attained. However, if the density is too high, the interactions between the bosons dominate the quantum statistical effects responsible for the condensation phenomenon. These interactions result in phase transitions leading to liquids or solids and prevent BEC from occurring.

Recent experiments demonstrating BEC have therefore concentrated on using *dilute* atomic gases, in which the interparticle spacing is large compared with the range of interatomic interactions. The gas atoms must of course be bosons. As we saw in Section 2.7, a composite particle, such as an atom, is a boson if the total number of protons, neutrons and electrons it contains is even. The requirement that the atoms of the gas be bosonic is not too restrictive, since all stable elements with the exception of beryllium have at least one bosonic isotope. In addition, the atoms must be trapped and cooled to *very low* temperatures, so that their thermal de Broglie wavelength λ_T is large enough for several atoms to be contained in a volume λ_T^3. The choice of an atom for a BEC experiment is therefore determined mainly by the cooling and trapping methods.

The first observation of BEC was made in 1995 by M.H. Anderson, J.R. Ensher, M.R. Matthews, C.E. Wieman and E.A. Cornell, using a dilute gas of rubidium (^{87}Rb) atoms. First, atoms from a room temperature ($T \simeq 300$ K), 10^{-11} Torr rubidium vapour were Doppler-cooled and trapped in a magneto-optical trap (MOT), enabling the collection of about 10^7 atoms. The atom cloud was then compressed and cooled in the optical molasses to 20 μK by adjustment of the field gradient and laser frequency. Following this, a small magnetic field was applied, and a short pulse of circularly polarised light optically pumped the magnetic moments of all the atoms so that they were parallel with the magnetic field. All laser light was then removed and a magnetic trap was suddenly turned on. At that point, the number of atoms in the trap was about 4×10^6 and the temperature was roughly 90 μK. Next, the sample was evaporatively cooled for about 70 s, during which a radio-frequency oscillating magnetic field selectively drove the atoms having the higher energy (sampling the trap region with higher magnetic field) into an 'untrapped' spin state. In this way, the temperature of the atomic cloud of rubidium atoms was reduced to the value required for BEC to occur in the lowest energy level of the trap potential. It should be noted that for evaporative cooling to work, the rate of elastic collisions had to be about 100 times greater than the

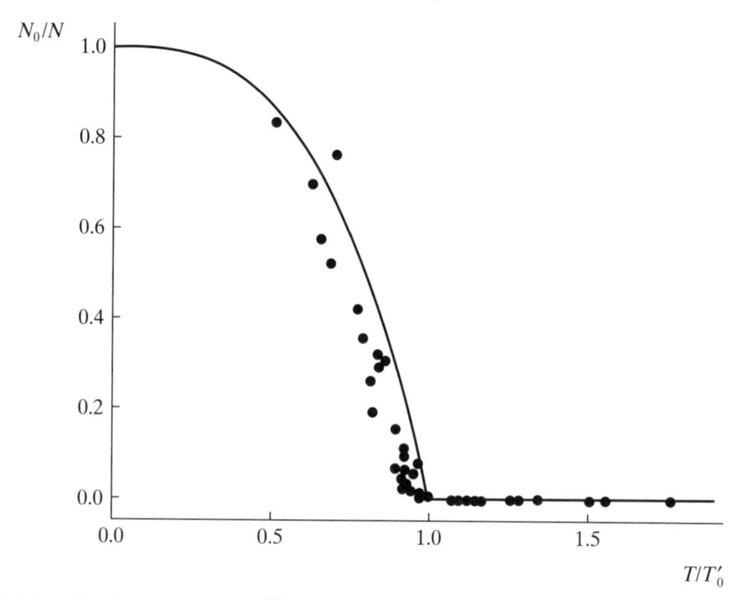

Figure 15.44 The fraction N_0/N of ^{87}Rb atoms in the ground state of the trap as a function of the scaled temperature T/T_0'. The solid curve corresponds to the theoretical prediction. The dots represent the experimental data of J.R. Ensher *et al.* At the transition, the atomic cloud contained 40 000 atoms at a temperature of 280 nK. (From J.R. Ensher, D.S. Jin, M.R. Matthews, C.E. Wieman and E.A. Cornell, *Physical Review Letters* **77**, 4984, 1996.)

rate of inelastic collisions, which, fortunately, turned out to be the case for the rubidium atoms used in the experiment. The condensate fraction was first observed near a temperature of 170 nK, and a number density of 2.5×10^{18} m^{-3}.

As an illustration, we show in Fig. 15.44 the data obtained by J.R. Ensher *et al.* in 1996 for a dilute gas of ^{87}Rb atoms. The onset of BEC is clearly apparent for temperatures $T < T_0'$, in agreement with the theoretical result (15.128). At temperatures $T \geqslant T_0'$, the fraction of the atoms in the ground state is negligible, as expected from the discussion following (15.120).

Following the experiment of Anderson *et al.*, BEC has been observed by a number of research groups. In particular, K.B. Davis, M.-O. Mewes, M.R. Andrews, M.J. Van Druten, D.S. Durfee, D.M. Kurn and W. Ketterle formed in 1995 a condensate of ^{23}Na atoms and in the same year C.C. Bradley, C.A. Sackett, J.J. Tollett and R.G. Hulet reported evidence of BEC of ^{7}Li atoms which, unlike ^{87}Rb and ^{23}Na, attract rather than repel each other. Since then, BEC has also been demonstrated in atomic hydrogen by T.J. Greytak, D. Kleppner *et al.*, in metastable helium by A. Aspect *et al.*, and in potassium by M. Inguscio *et al.* In 2001, T.W. Hänsch *et al.* reported the creation of a Bose–Einstein condensate of rubidium atoms in a microscopic magnetic trap built into a lithographically patterned chip. Such a device has potential applications in interferometry, holography and quantum information.

15.6 Atom lasers

As we have seen above, optical lasers have revolutionised the field of optics because of their high intensity, directionality, coherence and ability to emit light in a very narrow range of wavelengths. It is now possible to create laser-like sources capable of producing intense, highly directional, coherent beams of *atoms* having the same de Broglie wavelength. These sources of coherent matter waves are called *atom lasers*. They make use of the coherence property of Bose–Einstein condensates, which was demonstrated in 1997 by W. Ketterle and co-workers. Two condensates in a double-well potential were released from the trap and allowed to expand. They exhibited a well-defined interference pattern in their overlap region.

Research into atom lasers has relied on the analogy between those and optical lasers, the former emitting coherent matter waves while the latter emit coherent electromagnetic waves. As explained in Section 15.1, in an optical laser most of the photons occupy one or a few modes of a laser cavity, usually formed by two mirrors (see Fig. 15.1), and the active material is 'pumped' by an external source of energy to ensure that many photons are emitted into the selected mode(s). The 'output coupling' method to obtain a beam of coherent light consists in using a partially transparent mirror, so that photons can escape from the cavity. In the case of an atom laser, a Bose–Einstein condensate in a magnetic trap contains many atoms which occupy the ground state of the trap, and hence have the same energy and the same de Broglie wavelength. The active material consists of a thermal cloud of cold atoms. Atoms can be extracted from the trap by allowing them to tunnel through the potential barrier (see Fig. 15.45). The output coupling can be achieved by modifying an internal degree of freedom of the atoms so that they are transferred from states that are confined to others that are not. The output coupling scheme must of course be such that the coherence and monochromaticity of the atomic matter waves is maintained.

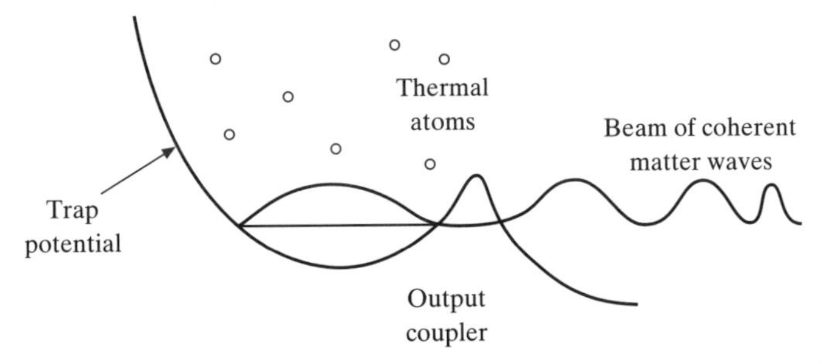

Figure 15.45 Principle of an atom laser. A Bose–Einstein condensate in a magnetic trap is a coherent source of matter waves. The active material is a thermal cloud of cold atoms. An output coupler allows atoms to be extracted from the trap in order to form a beam of coherent matter waves. (From K. Helmerson, D. Hutchinson, K. Burnett and W.D. Phillips, *Physics World*, August 1999, p. 31.)

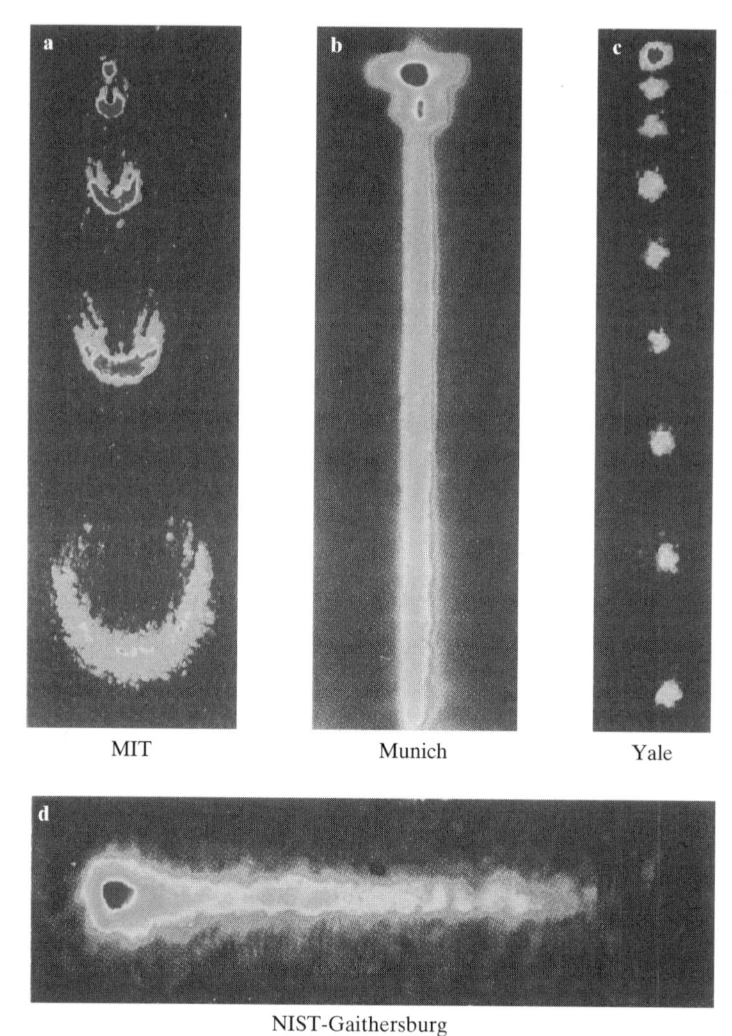

Figure 15.46 Output of four atom lasers using different output couplers (see text): (a) from the MIT group (b) from the Munich group (c) from the Yale group (d) from the NIST group. (From K. Helmerson, D. Hutchinson, K. Burnett and W.D. Phillips, *Physics World*, August 1999, p. 31.)

The first demonstration of output coupling from a Bose–Einstein condensate was made in 1997 by W. Ketterle and co-workers at MIT using sodium atoms in a magnetic trap. Short pulses of radio-frequency radiation were used to create Bose–Einstein condensates in a superposition of trapped and untrapped hyperfine states. The fraction of ejected atoms was adjusted by varying the amplitude of the radio-frequency radiation. The released atoms then accelerated away from the trap under the gravitational force. This atom laser produces output pulses of coherent matter waves expanding as they fall (see Fig. 15.46(a)).

In 1999, T.W. Hänsch and collaborators in Munich were able to demonstrate an atom laser with a continuous output. A Bose–Einstein condensate of rubidium atoms was produced in a magnetic trap which provided an extremely stable trapping potential. A weak radio-frequency field was used to continuously extract the atoms from the condensate over a period of up to 0.5 s. Again, the atoms were released to fall under the influence of gravity (see Fig. 15.46(b)).

Other coupling schemes have been devised to extract coherent beams of matter waves from a Bose–Einstein condensate. In an experiment performed by B. Anderson and M.A. Kasevich at Yale University in 1998, the condensate atoms were confined in a one-dimensional periodic optical trap, known as an *optical lattice*. This optical lattice was obtained by using two counter-propagating laser beams, which interacted to form a standing wave that trapped the atoms in the vertical direction. The potential was sufficiently shallow for the atoms to have a sizable probability of tunnelling downward and out of the wells due to gravity. Since the trap is aligned in the direction of gravity, the energy difference between adjacent potential wells is

$$\Delta E = Mg\frac{\lambda_L}{2} \tag{15.133}$$

where M is the mass of the atom, g is the magnitude of the acceleration due to gravity and λ_L is the laser wavelength. Hence, atoms leaving adjacent traps have an angular frequency difference $\Delta\omega = \Delta E/\hbar$. The various waves combine to give a wave function beating at the angular frequency

$$\bar{\omega} = \Delta\omega = \frac{Mg\lambda_L}{2\hbar} \tag{15.134}$$

When the periodic potential is loaded with atoms from a Bose–Einstein condensate, the initial relative phase of the atoms is identical and the beating is observed as a train of pulses of atoms (see Fig. 15.46(c)).

Another output-coupling scheme was demonstrated in 1999 by W.D. Phillips and co-workers at the US National Institute of Standards and Technology (NIST). The atoms from a Bose–Einstein condensate in a magnetic trap were extracted by using stimulated Raman scattering. The condensate atoms absorb photons from a laser at one wavelength and emit photons at a slightly different wavelength. The photon recoil expelled the atoms from the trap, producing a directional output coupler. Unlike the three other output-coupling schemes described above, the Raman technique does not rely on gravity to accelerate atoms out of the trap, so that the direction of the output beam can be selected. In Fig. 15.46(d), the clouds of atoms leaving the trap are seen to overlap in a quasi-continuous horizontal beam.

It is probable that atom lasers will find applications in many fields. These include atom optics (e.g. atom interferometry and atom holography), atom lithography, precision measurements of fundamental constants and atomic clocks [15].

[15] See for example K. Helmerson, D. Hutchinson, K. Burnett and W.D. Phillips (1999).

16 Further developments and applications of atomic and molecular physics

Atomic and molecular physics is a very active field of research, where many new developments are constantly occurring. Data from atomic and molecular physics are also necessary inputs for studying phenomena in other branches of physics such as condensed matter physics, astrophysics, plasma physics, chemical physics and geophysics. More generally, atomic and molecular physics has many applications to other areas of science and technology, for example in quantum electronics, space, environment and energy research, and in communication and medical technologies. Anything like a survey of all these subjects is not possible within the confines of a single chapter. Instead, we shall restrict our attention to the basic principles behind a few important developments and applications of atomic and molecular physics.

16.1 Magnetic resonance

The total electronic angular momentum **J** of an atom, arising from both the orbital and the spin angular momenta of the electrons, can either be zero, as for closed shell atoms, or non-zero, as for the ground state of atomic hydrogen ($j = s = 1/2$, $l = 0$). When the total electronic angular momentum **J** is non-zero, an atom possesses a permanent magnetic dipole moment $\mathcal{M}$ which can be expressed for weak fields in L–S coupling, consistently with (9.159), as

$$\mathcal{M} = -g\mu_B \mathbf{J}/\hbar \qquad \textbf{(16.1)}$$

where g is the Landé factor (9.160). In principle, the magnetic moment can be measured in a Stern–Gerlach experiment as described in Chapter 1. An important consequence of the existence of permanent magnetic dipole moments is the *paramagnetism* observed when the magnetic moments of the atoms in a material are partially aligned by a magnetic field.

If a single atom is placed in a uniform, constant, weak magnetic field directed along the Z axis, of magnitude $\mathcal{B}_z$, the system will possess stationary states, which are eigenstates of $\mathbf{J}^2$ and J_z. Let us denote by $J(J+1)\hbar^2$ the eigenvalues of $\mathbf{J}^2$ and by $M_J\hbar$ those of J_z, where M_J can take any one of the $(2J+1)$ values $-J, -J+1, \ldots, J$. As we saw before (see (9.159)), the energy due to the magnetic interaction $-\mathcal{M}\cdot\mathcal{B}$ is

$$E_{M_J} = g\mu_B \mathcal{B}_z M_J \tag{16.2}$$

The nuclear magnetic moments also contribute, in principle, to the total magnetic energy. However, since the nuclear magnetic moments are several orders of magnitude smaller than the atomic ones, this contribution is unimportant in the present context.

The time dependence of the wave function corresponding to a stationary state of energy E_{M_J} is

$$\Psi_{M_J}(t) = A\exp(-\mathrm{i}E_{M_J}t/\hbar) \tag{16.3}$$

which can be written as

$$\Psi_{M_J}(t) = A\exp(-\mathrm{i}\omega_0 M_J t) \tag{16.4}$$

where ω_0 is given by

$$\omega_0 = g\mu_B \mathcal{B}_z/\hbar = g\omega_L \tag{16.5}$$

ω_L being the Larmor angular frequency (1.117). Classically, a magnetic dipole placed in a constant magnetic field precesses about the direction of the field with the angular frequency ω_0.

We now consider the response of the system to an additional weak oscillating magnetic field, $\mathcal{B}_x \cos(\omega t)$, directed along the X axis. It will be shown that when the angular frequency ω of this second field is close to the angular frequency ω_0, the system is strongly disturbed and there is a large probability of a transition from the initial state. This is called *paramagnetic resonance* (or electron spin resonance, ESR) and detection of the resonant frequency affords an accurate method of measuring gyromagnetic ratios.

In the presence of both components of the magnetic field, the Schrödinger equation is

$$\mathrm{i}\hbar\frac{\partial}{\partial t}\Psi(t) = H\Psi(t) \tag{16.6}$$

where, using (16.1),

$$H = -\mathcal{M}\cdot\mathcal{B} = \frac{g\mu_B}{\hbar}[J_z\mathcal{B}_z + J_x\mathcal{B}_x\cos(\omega t)] \tag{16.7}$$

To simplify the discussion, consider a case for which $J = 1/2$, so that $\mathbf{J}$ can be written in terms of the Pauli spin matrices σ_x, σ_y, σ_z as

$$\mathbf{J} = \mathbf{S} = \frac{\hbar}{2}\boldsymbol{\sigma} \tag{16.8}$$

The wave function is now a two-component spinor:

$$\Psi(t) = \begin{pmatrix} a_+(t) \\ a_-(t) \end{pmatrix} \tag{16.9}$$

Assuming that this wave function is normalised to unity, the quantity $|a_+(t)|^2$ gives the probability of finding the system in the state with $M_J = +1/2$, and $|a_-(t)|^2$ yields the probability that the system is to be found in the state with $M_J = -1/2$. Hence we have, for all t,

$$|a_+(t)|^2 + |a_-(t)|^2 = 1 \qquad \textbf{(16.10)}$$

When the perturbation $\mathcal{B}_x \cos(\omega t)$ is absent, the unperturbed eigenfunctions are

$$\Psi_{1/2}(t) = \begin{pmatrix} \exp(-\mathrm{i}\omega_0 t/2) \\ 0 \end{pmatrix}, \qquad \Psi_{-1/2}(t) = \begin{pmatrix} 0 \\ \exp(\mathrm{i}\omega_0 t/2) \end{pmatrix} \qquad \textbf{(16.11)}$$

If the system was originally in the state $M_J = 1/2$ and the perturbation was switched on at time $t = 0$, we can calculate the probability of finding the system in the state $M_J = -1/2$ at some later time t. Since we shall be interested in large values of t, we cannot use perturbation theory. Instead, we shall proceed by using the explicit form of the Pauli matrices to write the Schrödinger equation as a pair of coupled equations for $a_\pm(t)$. That is,

$$\mathrm{i}\hbar \frac{\partial}{\partial t}\begin{pmatrix} a_+ \\ a_- \end{pmatrix} = \frac{g\mu_\mathrm{B}}{2}\left[\begin{pmatrix} 1 & 0 \\ 0 & -1 \end{pmatrix}\mathcal{B}_z + \begin{pmatrix} 0 & 1 \\ 1 & 0 \end{pmatrix}\mathcal{B}_x \cos(\omega t)\right]\begin{pmatrix} a_+ \\ a_- \end{pmatrix} \qquad \textbf{(16.12)}$$

In terms of $\omega_0 = g\mu_\mathrm{B}\mathcal{B}_z/\hbar$ and $\bar{\omega}_0 = g\mu_\mathrm{B}\mathcal{B}_x/\hbar$, we have

$$\begin{aligned} \mathrm{i}\dot{a}_+ &= \frac{1}{2}\omega_0 a_+ + \frac{1}{2}\bar{\omega}_0 \cos(\omega t) a_-, \\ \mathrm{i}\dot{a}_- &= -\frac{1}{2}\omega_0 a_- + \frac{1}{2}\bar{\omega}_0 \cos(\omega t) a_+ \end{aligned} \qquad \textbf{(16.13)}$$

A phase transformation

$$A_+ = a_+ \exp(\mathrm{i}\omega_0 t/2), \qquad A_- = a_- \exp(-\mathrm{i}\omega_0 t/2) \qquad \textbf{(16.14)}$$

removes the secular terms in $\pm\frac{1}{2}\omega_0$, giving

$$\begin{aligned} \mathrm{i}\dot{A}_+ &= \frac{1}{2}\bar{\omega}_0 \cos(\omega t) \exp(\mathrm{i}\omega_0 t) A_-, \\ \mathrm{i}\dot{A}_- &= \frac{1}{2}\bar{\omega}_0 \cos(\omega t) \exp(-\mathrm{i}\omega_0 t) A_+ \end{aligned} \qquad \textbf{(16.15)}$$

These equations cannot be solved exactly, but if ω is close to ω_0 an accurate approximation can be obtained by recognising that in the products $\cos(\omega t)\exp(\pm\mathrm{i}\omega_0 t)$, terms in $\exp[\pm\mathrm{i}(\omega - \omega_0)t]$ are much more important than those in $\exp[\pm\mathrm{i}(\omega + \omega_0)t]$. Indeed, the latter terms oscillate rapidly and on the average make little contribution to $\dot{A}_+$ or $\dot{A}_-$, so that we shall neglect them. This is the *rotating wave approximation* discussed in Section 2.8. We then find the approximate equations

$$\mathrm{i}\dot{A}_+ = \frac{1}{4}\bar{\omega}_0 \exp[\mathrm{i}(\omega_0 - \omega)t]A_-,$$
$$\mathrm{i}\dot{A}_- = \frac{1}{4}\bar{\omega}_0 \exp[-\mathrm{i}(\omega_0 - \omega)t]A_+ \tag{16.16}$$

Exact resonance

It is easy to verify that in the case of exact resonance $\omega = \omega_0$, the equations (16.16) have the general solution

$$A_+ = \lambda \cos(\bar{\omega}_0 t/4) + \mu \sin(\bar{\omega}_0 t/4),$$
$$A_- = \mathrm{i}\mu \cos(\bar{\omega}_0 t/4) - \mathrm{i}\lambda \sin(\bar{\omega}_0 t/4) \tag{16.17}$$

where λ and μ are constants, which are determined by the initial conditions.

Suppose that at $t = 0$ the system is in the state with $M_J = 1/2$. Then $A_+(0) = a_+(0) = 1$ and $A_-(0) = a_-(0) = 0$ so that $\lambda = 1$ and $\mu = 0$. Thus the probabilities $P(+ \to +)$ and $P(+ \to -)$ for finding the system in levels with $M_J = 1/2$ or $M_J = -1/2$ at time t are

$$P(+ \to +) = |A_+(t)|^2 = \cos^2(\bar{\omega}_0 t/4),$$
$$P(+ \to -) = |A_-(t)|^2 = \sin^2(\bar{\omega}_0 t/4) \tag{16.18}$$

Both $P(+ \to +)$ and $P(+ \to -)$ range between 0 and 1. Clearly one always has $P(+ \to +) + P(+ \to -) = 1$.

In the same way, if at $t = 0$ the system is the state with $M_J = -1/2$ so that $A_+(0) = a_+(0) = 0$ and $A_-(0) = a_-(0) = 1$, the probabilities $P(- \to +)$ and $P(- \to -)$ for finding the system in the levels with $M_J = 1/2$ and $M_J = -1/2$ at time t are

$$P(- \to +) = \sin^2(\bar{\omega}_0 t/4),$$
$$P(- \to -) = \cos^2(\bar{\omega}_0 t/4) \tag{16.19}$$

It should be noticed that, in conformity with time-reversal invariance,

$$P(- \to +) = P(+ \to -) \tag{16.20}$$

General solution

It is possible to solve the equations (16.16) exactly, even when $\omega \neq \omega_0$. The general solution is

$$A_+ = p \exp(\mathrm{i}\eta_+ t) + q \exp(\mathrm{i}\eta_- t),$$
$$A_- = -\frac{4}{\bar{\omega}_0}[p\eta_+ \exp(\mathrm{i}\eta_+ t) + q\eta_- \exp(\mathrm{i}\eta_- t)] \exp[\mathrm{i}(\omega - \omega_0)t] \tag{16.21}$$

where p and q are constants of integration (different from λ and μ) and $\eta_\pm$ are given by

$$\eta_{\pm} = \frac{1}{2}\{(\omega_0 - \omega) \pm [(\omega_0 - \omega)^2 + \bar{\omega}_0^2/4]^{1/2}\} \tag{16.22}$$

With the initial conditions at $t = 0$, $A_+(0) = 1$, $A_-(0) = 0$, we find that

$$p = \frac{\eta_-}{\eta_- - \eta_+}, \qquad q = \frac{-\eta_+}{\eta_- - \eta_+} \tag{16.23}$$

The probabilities $P(+ \to +)$ and $P(+ \to -)$ become

$$P(+ \to +) = \cos^2(\omega_R t/2) + \frac{(\omega_0 - \omega)^2}{(\omega_0 - \omega)^2 + \bar{\omega}_0^2/4} \sin^2(\omega_R t/2)$$
$$P(+ \to -) = \frac{\bar{\omega}_0^2/4}{(\omega_0 - \omega)^2 + \bar{\omega}_0^2/4} \sin^2(\omega_R t/2) \tag{16.24}$$

where ω_R is the Rabi 'flopping frequency' (see (2.379)) which is given by

$$\omega_R = \eta_+ - \eta_- = [(\omega_0 - \omega)^2 + \bar{\omega}_0^2/4]^{1/2} \tag{16.25}$$

Under the condition $\mathcal{B}_x \ll \mathcal{B}_z$ so that $\bar{\omega}_0 \ll \omega_0$, the probability that the system will be found in the second state with $M_J = -1/2$ remains small unless ω is close to ω_0. If the oscillating magnetic field $\mathcal{B}_x \cos(\omega t)$ is applied for a time T which is short ($\omega_R T \ll 1$), the equations (16.16) can be solved by using first-order perturbation theory.

Resonance occurs when the frequency of the oscillating field ω is such that $\hbar\omega$ is equal to the difference in energy of the two Zeeman levels of the system, ΔE. In our spin-1/2 case

$$\Delta E = \hbar\omega_0 = g\mu_B\mathcal{B}_z \tag{16.26}$$

In general, for atoms with non-zero angular momentum, resonance can be produced by matching the frequency of the applied field to the frequency of a transition between particular Zeeman sublevels, and the theory can be generalised to treat a system of $(2J + 1)$ equations rather than the pair of equations treated here.

Nuclear magnetic resonance

Although the magnetic moments of nuclei are smaller than those of atoms by a factor of the order of m/M_p (where m is the mass of the electron and M_p that of the proton) and nuclear paramagnetism is too small to be observed directly, nuclear magnetic resonance experiments are perfectly possible and are, in fact, of great importance. A substance is chosen for which the total angular momentum due to the electrons is zero, for otherwise the nuclear effect would be completely masked. For example, the resonant frequency for transitions between two states of a nucleus having a spin quantum number $I = 1/2$, with $M_I = \pm 1/2$, is

$$\omega_0 = |g_I|\mu_N\mathcal{B}_z/\hbar \tag{16.27}$$

where g_I is the nuclear Landé factor and μ_N is the nuclear magneton (5.37).

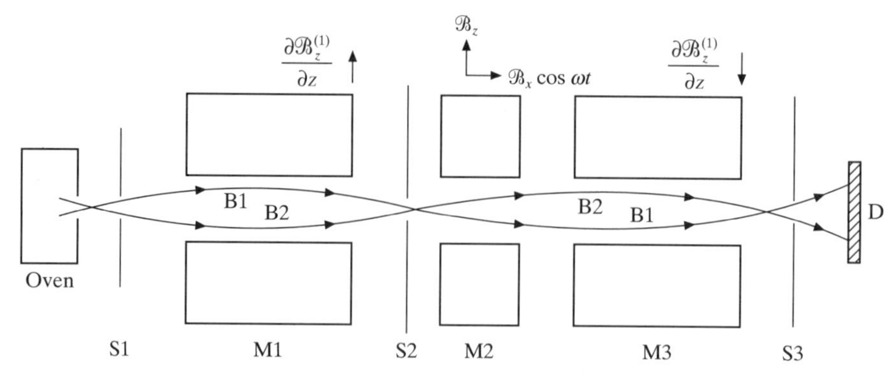

Figure 16.1 Schematic diagram of the Rabi molecular beam apparatus.

For a field of strength 0.5 T, the resonant frequency of a proton (having a Landé factor of 5.5883) is 21.3 MHz, which is in the radio-frequency region. The frequencies associated with paramagnetic resonance are about 2000 times higher and are in the microwave region.

The Rabi molecular beam apparatus

An atomic (molecular) beam experiment, which is much more accurate than the Stern–Gerlach experiment discussed in Chapter 1, was devised in 1938 by I.I. Rabi to measure magnetic moments of atoms based on paramagnetic resonance. A schematic diagram is shown in Fig. 16.1. A beam of atoms from an oven is passed through a system of three magnets M1, M2 and M3. We shall suppose that the atoms have total angular momentum one-half (examples of which are silver, the alkalis, or copper), although the principle of the experiment is the same for other non-zero values of the angular momentum. The magnets M1 and M3 produce inhomogeneous fields as in the Stern–Gerlach experiment, identical in magnitude but opposite in sign. If the field gradient in M1 is positive upwards, those atoms with $M_J = 1/2$ will be deflected downwards and those with $M_J = -1/2$ will be deflected upwards. If a slit S2 is placed, as shown, only two trajectories are possible from the source (slit S1) into the region to the right of S2. The trajectory B1 will contain the atoms with $M_J = 1/2$ and B2 will contain those with $M_J = -1/2$. Since the magnet M3 has an equal and opposite effect on the two trajectories, the atoms in B1 and B2 will be brought together at the slit S3 and detected at D.

Now let us see what happens if the magnet M2 is switched on, which produces a large uniform static field in the Z direction, $\mathcal{B}_z$, and a small microwave oscillating field, $\mathcal{B}_x \cos(\omega t)$, in the X direction. When ω is close to the resonance angular frequency ω_0, some of the atoms in the beam B1 will have the z component of their spin changed from $\hbar/2$ to $-\hbar/2$. These atoms will now be deflected downwards in the magnet M3 and will miss the slit S3. Similarly, atoms in the beam B2 for which the z component of the spin changes from $-\hbar/2$ to $\hbar/2$ will also miss the slit. The net effect is that as ω approaches the resonant angular frequency, the intensity of the beam entering the detector drops sharply. Under the condition $\omega_0 \gg \bar{\omega}_0$, the resonance region is very narrow and well defined, and since frequencies can be measured very accurately, this method can provide correspondingly accurate values for atomic or molecular magnetic moments.

Nuclear magnetic moments can also be measured in the Rabi apparatus by using the phenomenon of nuclear magnetic resonance. The experiment can be performed in the same way as for paramagnetic resonance, with an oscillating field $\mathcal{B}_x \cos(\omega t)$ which is now in the radio-frequency region.

Ramsey's method of separated oscillatory fields

The main difficulty with the Rabi method is its sensitivity to inhomogeneities in the static magnetic field, which is difficult to keep steady over the length of the path required for precision measurements. N.F. Ramsey thus proposed in 1949 the method of *separated oscillatory fields*, in which the single oscillatory magnetic field in the centre of the Rabi apparatus is replaced by *two* oscillatory fields, applied only over short distances, one at the entrance and the second at the exit of the space in which the properties of the atoms or molecules are studied [1]. The two oscillatory fields are driven *in phase* by the same oscillator. As illustrated in Fig. 16.2, one can consider that the two separated oscillatory fields act each during a time t_1, with a zero-amplitude oscillating field acting in the middle during a long time t_2. Because phase coherence is preserved between the two oscillating fields, it is as if the oscillation continued, but with zero amplitude during the time t_2. The second oscillatory field can therefore complete the spin-flip begun by the first only if the oscillator is exactly at the resonance frequency. The Ramsey method of separated oscillatory fields yields narrower resonance peaks than the Rabi method, and is not affected by inhomogeneities in the static magnetic field applied between the two oscillatory fields. Provided there is no unintended phase shift between the two oscillatory fields, first-order Doppler shifts and widths are eliminated. The earliest applications of the Ramsey method involved two oscillatory fields separated in *space*, but it was soon realised that the method could use successive oscillatory fields separated in *time* by using coherent pulses. Either way,

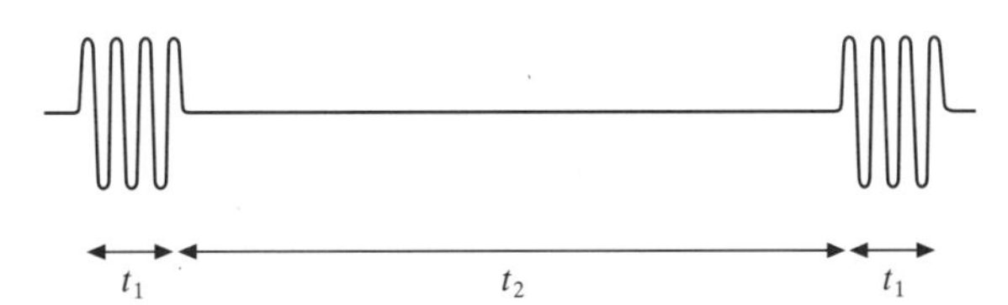

Figure 16.2 The principle of Ramsey's method of separated oscillatory fields. Two oscillatory fields each act during a time t_1, with a zero-amplitude oscillating field acting in between for a long time t_2. Phase coherence is preserved between the two oscillatory fields.

[1] For a detailed discussion of the method of separated oscillatory fields, see Ramsey (1985).

the longer the time t_2 between the oscillatory fields, the greater the accuracy of the measurement.

Ramsey's method has been extended to many experiments beyond those of magnetic resonance. For example, it has been used to synchronise a microwave oscillator to the internal oscillations of the caesium atom (see Section 16.4). For many purposes, the highest precision can be obtained with just two oscillatory fields separated by the maximum time, but in some instances it is better to utilise more than two separated oscillatory fields. In that case, the only requirement is that the oscillating fields be coherent. In particular, the separation of the pulses can even be at random, as in the case of the hydrogen maser discussed in Section 15.1. The atoms stimulated to emit move randomly in and out of the cavities with oscillatory fields and spend the intermediate time in the large container with no such fields.

Paramagnetic resonance in bulk samples

It is also possible to detect resonance phenomena in bulk samples of materials. Let us first consider the case of a material composed of atoms of total angular momentum $J = 1/2$. In the absence of a magnetic field, the two states of each atom with $M_J = \pm 1/2$ have the same energy, and in a bulk sample as many atoms have $M_J = 1/2$ as have $M_J = -1/2$. If the sample is placed in a uniform magnetic field $\mathcal{B}_z$, directed along the Z axis, then those atoms with $M_J = 1/2$ possess the energy (see (16.2))

$$E_+ = g\mu_B\mathcal{B}_z/2 \tag{16.28}$$

and those with $M_J = -1/2$ have the lower energy

$$E_- = -g\mu_B\mathcal{B}_z/2 \tag{16.29}$$

In thermal equilibrium, the ratio of the number N_+ of atoms per unit volume with $M_J = 1/2$ to the number N_- with $M_J = -1/2$ is

$$\frac{N_+}{N_-} = \frac{\exp[-E_+/(k_BT)]}{\exp[-E_-/(k_BT)]} \tag{16.30}$$

where T is the absolute temperature and k_B is Boltzmann's constant. Let $N = N_+ + N_-$ be the number of atoms per unit volume. Then

$$\begin{aligned} N_+ &= N\frac{\exp[-g\mu_B\mathcal{B}_z/(2k_BT)]}{\exp[g\mu_B\mathcal{B}_z/(2k_BT)] + \exp[-g\mu_B\mathcal{B}_z/(2k_BT)]}, \\ N_- &= N\frac{\exp[g\mu_B\mathcal{B}_z/(2k_BT)]}{\exp[g\mu_B\mathcal{B}_z/(2k_BT)] + \exp[-g\mu_B\mathcal{B}_z/(2k_BT)]} \end{aligned} \tag{16.31}$$

The atoms with $M_J = 1/2$ contribute a magnetic moment $(-g\mu_B/2)$ in the Z direction and those with $M_J = -1/2$ contribute a magnetic moment $(g\mu_B/2)$, so that the value of the net magnetic moment in the Z direction per unit volume is

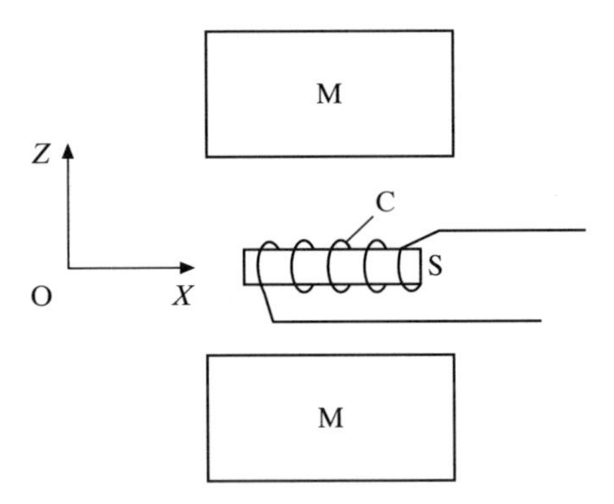

Figure 16.3 Schematic diagram of a paramagnetic resonance apparatus. The specimen S is placed between the poles M of a magnet producing a uniform static field in the *Z* direction. A small oscillating field in the *X* direction is produced by the coil C.

$$\begin{aligned}\mathcal{M}_z &= N_+(-g\mu_B/2) + N_-(g\mu_B/2) \\ &= \frac{g\mu_B}{2} N \frac{\exp[g\mu_B\mathcal{B}_z/(2k_BT)] - \exp[-g\mu_B\mathcal{B}_z/(2k_BT)]}{\exp[g\mu_B\mathcal{B}_z/(2k_BT)] + \exp[-g\mu_B\mathcal{B}_z/(2k_BT)]}\end{aligned} \quad \textbf{(16.32)}$$

Now, if a small field $\mathcal{B}_x \cos(\omega t)$ is applied in the X direction, when ω is close to the resonance frequency transitions are induced between the two states with $M_J = 1/2$ and $M_J = -1/2$. If these two levels were equally populated, these transitions could not be detected, but since the number of atoms in the state of lower energy is greater than the number in the state of higher energy, more transitions occur which absorb energy from the external field than transitions which feed energy into the external field. This net loss of energy at resonance is small, but can be detected in various ways. For example, the oscillating magnetic field may be produced by a coil which is placed in a bridge circuit (see Fig. 16.3). At resonance, the energy loss to the medium gives rise to an apparent change in the self-induction of the coil, which can be detected by the bridge.

The technique of paramagnetic resonance is important to research and industry in giving information about the constituents of a sample. On the other hand, if these are known, the apparatus can serve as a magnetometer to measure small fields $\mathcal{B}_x$.

As an example, an actual electron spin resonance spectrum is shown in Fig. 16.4. The material under investigation is Fe^{3+}/MgO containing 310 parts per million of Fe^{3+} ions, and it is these ions which give rise to the spectrum. The field $\mathcal{B}_x$ oscillates at a fixed frequency of 9.10 GHz and the energy absorption is measured as a function of the field $\mathcal{B}_z$. The ordinate on the trace is proportional to the derivative $(dI/d\mathcal{B}_z)$ where I is the energy loss, and resonance occurs at the points for which I is a maximum, that is for which $(dI/d\mathcal{B}_z) = 0$. The ion Fe^{3+} has angular momentum $J = 5/2$ so that various transitions can be observed corresponding to $M_J = 1/2 \leftrightarrow -1/2$, $\pm 3/2 \leftrightarrow \pm 5/2$ and $\pm 3/2 \leftrightarrow \pm 1/2$. For the transition $1/2 \leftrightarrow -1/2$, it is found (see Fig. 16.4) that $\mathcal{B}_z = 322.5$ mT. Using (16.5) and the fact that $\omega_0 = 9.10$ GHz, the value of g is found to be 2.016.

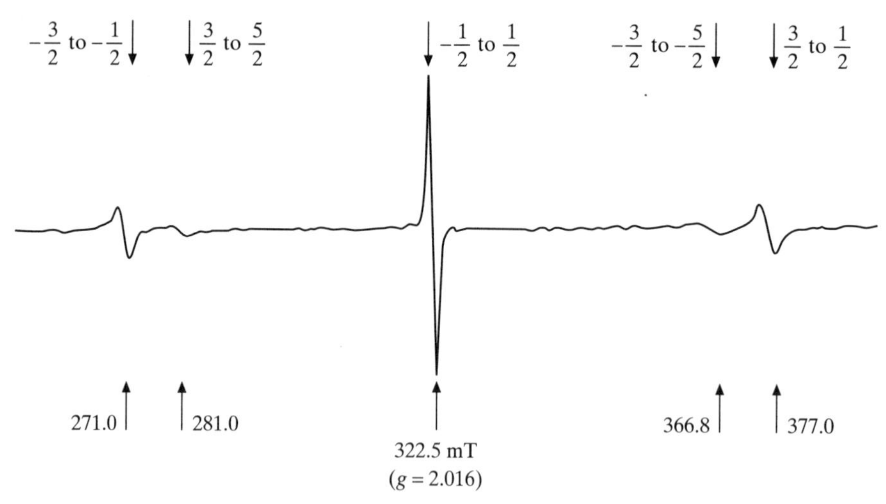

Figure 16.4 An electron spin resonance spectrum of Fe^{3+}/MgO. (By courtesy of J.S. Thorp.)

Nuclear magnetic resonance in bulk samples

Nuclear magnetic resonance (NMR) measurements in bulk samples are possible despite the small size of the nuclear moments. As mentioned above, the liquid or solid specimen must be composed of atoms for which the total electronic angular momentum is zero in order to avoid masking the nuclear effect. The measurements can then be performed in exactly the same way as for paramagnetic substances. The energy of a nucleus of spin **I** in a magnetic field directed along the Z axis is

$$E_{M_I} = -g_I \mu_N M_I \mathcal{B}_z \tag{16.33}$$

where M_I is the magnetic quantum number of the nucleus, which can take on the $(2I + 1)$ values from $-I$ to I. Measurements can be carried out on substances containing nuclei of any non-zero spin. Here we shall discuss the case of spin one-half which includes the proton (p), the common isotopes of fluorine (^{19}F) and phosphorus (^{31}P), and the less abundant isotope of carbon (^{13}C).

The Landé factor g_I is a constant for nuclei of a given kind and can be positive or negative. For the nuclei mentioned above g_I is positive (see Table 5.1), with the result that in the magnetic field those nuclei with $M_I = -1/2$ have the greater energy

$$E_{-1/2} = g_I \mu_N \mathcal{B}_z/2 \tag{16.34a}$$

while those with $M_I = 1/2$ have the smaller energy

$$E_{1/2} = -g_I \mu_N \mathcal{B}_z/2 \tag{16.34b}$$

The resonant angular frequency ω_0 for transitions between the two states with $M_I = \pm 1/2$ is given by (16.27). The magnitude of the net magnetic moment in the Z direction per unit volume in a bulk sample can be evaluated as in the case of paramagnetic resonance (see (16.32)). It is found to be $\mathcal{M}_z$, where if N is the number of nuclei per unit volume,

$$\mathcal{M}_z = \frac{|g_I|\mu_N}{2} N \frac{\exp[|g_I|\mu_N \mathcal{B}_z/(2k_B T)] - \exp[-|g_I|\mu_N \mathcal{B}_z/(2k_B T)]}{\exp[|g_I|\mu_N \mathcal{B}_z/(2k_B T)] + \exp[-|g_I|\mu_N \mathcal{B}_z/(2k_B T)]} \quad \textbf{(16.35)}$$

The magnetic field strength employed in NMR spectrometers does not usually exceed 10 T, so that at room temperature the factor $|g_I|\mu_N\mathcal{B}_z/(2k_B T)$ is very small. For example, in the case of protons and for a temperature $T = 290$ K, this factor is equal to $3.5 \times 10^{-6}\,\mathcal{B}_z$ (where $\mathcal{B}_z$ is expressed in tesla). It follows that to a good approximation

$$\mathcal{M}_z = \frac{|g_I|\mu_N}{4} N \frac{|g_I|\mu_N \mathcal{B}_z}{k_B T} = \frac{|g_I|\mu_N N}{4k_B T}\hbar\omega_0 \quad \textbf{(16.36)}$$

To obtain the greatest sensitivity, $\mathcal{M}_z$, and hence $\mathcal{B}_z$, must be as large as possible. As a consequence, the resonant frequencies encountered in NMR are much higher than those associated with paramagnetic resonance, being in the radio-frequency rather than the microwave region. To increase the magnitude of the signal the sample should be as large as possible. This, however, is limited by the need for the sample to be uniform.

The NMR spectrum of a sample containing nuclei of different species can be obtained either by varying the frequency ω of the applied oscillating magnetic field, or by varying the field strength $\mathcal{B}_z$ of the static magnetic field slowly over a given range. In either case the application of the oscillating field alters the population of the nuclear spin states with different values of M_I. This has the consequence that if the field is switched off thermal equilibrium is restored only gradually by interaction between the nuclei and the environment. This relaxation time is known as the *spin–lattice relaxation time* T_1 which can vary from a fraction of a second to several hundred seconds. During this relaxation period the NMR signal oscillates with a decreasing amplitude. To avoid these transient effects the field strength must be varied slowly on a time scale set by the time T_1. It also follows that the minimum width of an NMR spectral line is, by the Heisenberg uncertainty principle, $\Delta E = \hbar/T_1$. However, the observed line widths are often much greater than this, mainly because of magnetic dipole couplings between the nuclei under test. This can be taken into account by introducing a second relation time T_2, called the *spin–spin relaxation time*.

Because of the need to change the magnetic field strength adiabatically, NMR spectra obtained as described above (CW operation) can take an unacceptable time to obtain. As an alternative a pulsed mode of operation is now usually employed. In this case an RF field of frequency $\bar{\omega}$ is switched on and off to produce a short pulse of duration τ. Such a pulse can be Fourier analysed and is equivalent to applying a range of frequencies to the sample simultaneously, with

$$\bar{\omega} - \frac{2\pi}{\tau} < \omega < \bar{\omega} + \frac{2\pi}{\tau} \quad \textbf{(16.37)}$$

By Fourier analysing the NMR signal the whole spectrum over this frequency range can be obtained. In practice, to obtain a signal of sufficient strength, results are accumulated from a succession of short pulses.

Chemical shifts

Nuclear magnetic resonance has proved to be an important tool in the structural analysis of molecules and in the chemical analyses of both solid and liquid samples. This is because there is small dependence of the resonant frequency on the local environment of a nucleus of given type. The effect of the applied magnetic field on s electrons ($l = 0$) in the neighbourhood of a nucleus is to induce a current which creates a magnetic field which is (by Lenz's rule) antiparallel to the applied field. The effective magnetic field at the nucleus, $\mathcal{B}_z^E$, is then less than the applied field $\mathcal{B}_z$ and correspondingly the observed resonant frequency ω_0^E is less than ω_0. We can write

$$\omega_0^E = |g_I| \mu_N \mathcal{B}_z^E / \hbar = |g_I| \mu_N \mathcal{B}_z (1 - \sigma) / \hbar \tag{16.38}$$

where σ is known as the shielding factor. In addition to this *diamagnetic* shift, if there are p electrons ($l = 1$) in the neighbourhood of the nucleus a much larger magnetic field is induced which on the average increases the effective magnetic field at the nucleus and the corresponding *paramagnetic* shielding factor σ is negative.

For the very important case in which a hydrogen atom is part of a distinct group within a molecule, the diamagnetic effect is exhibited since the electron in the ground state of atomic hydrogen has zero orbital angular momentum. The effect is small, changing the resonant frequency in most cases by only a few parts per million (ppm), but the shielding factor is specific to the local environment. For example, the resonant frequency of a proton in an OH group is different from that of a proton in NH_2, which again is different from the signal observed from a proton in a CH_3 group. Thus the NMR proton spectrum shows the nature of the chemical groups containing protons within the sample. Still more information can be obtained by examining the relative intensity of the resonant lines and from fine structure effects which are due to interactions between neighbouring nuclei within a molecule. Similar information can be obtained from the spectra of many other nuclei. Since bare nuclei cannot be employed in an NMR spectrometer, the frequency shift due to the molecular environment is measured with respect to the observed resonant frequency of a nucleus of the same type in a standard substance. For protons this is often taken to be the organic solvent $(CH_3)_4Si$, called tetramethylsilane (TMS). The *chemical shift* δ is defined as

$$\delta = \frac{\omega_0^E - \omega_S}{\omega_S} 10^6 \text{ ppm} \tag{16.39}$$

where ω_0^E is the observed resonant angular frequency of a nucleus in the sample and ω_S the corresponding angular frequency in the standard substance.

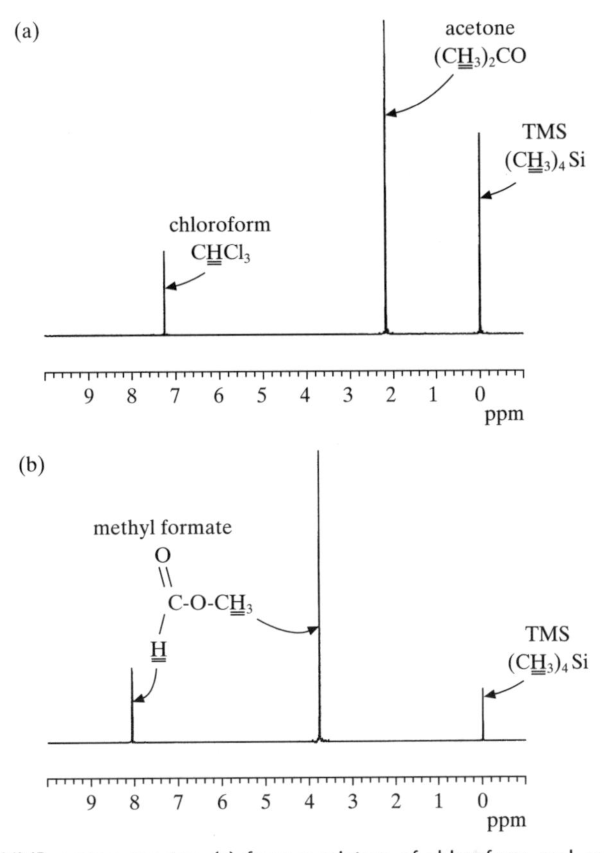

Figure 16.5 NMR proton spectra: (a) from a mixture of chloroform and acetone in TMS, (b) from a solution of methyl formate in TMS. (By courtesy of R.K. Harris and A.M. Kenwright.)

As an example, two NMR spectra are shown in Fig. 16.5, both from a spectrometer operating with a static magnetic field of 9.4 T, which corresponds to a Larmor frequency of 400 MHz for an isolated proton. In both spectra the strength of the signal is plotted against the frequency of the RF field, which is expressed in terms of the chemical shift δ, relative to the resonant frequency of the protons in the solvent TMS. In this example a shift of 1 ppm corresponds to a frequency interval of 400 Hz. The upper spectrum (a) is from a mixture of chloroform, $CHCl_3$, and acetone, $(CH_3)_2CO$, dissolved in TMS. The right-hand line arises from the protons in TMS, the middle line from those in acetone and the left-hand line from those in the chloroform molecules. The lower spectrum (b) is from a solution of methyl formate in TMS. The right-hand line again arises from the protons in TMS, but the other two lines are due to the protons in different groups in the methyl formate molecule (as indicated in the figure) rather than from groups in different molecules as in the upper spectrum.

16.2 Atom optics

Atom optics consists of acting on the trajectories and exploiting the wave properties of neutral atoms to obtain effects analogous to those exhibited by light: focusing, reflection, diffraction, interference. Its most salient feature is the small size of atomic de Broglie wavelengths compared to optical wavelengths. Soon after de Broglie's suggestion in 1924 that material particles possess wave-like properties, O. Stern and his colleagues performed in 1929 the first experiments in atom optics by demonstrating the diffraction of atoms from surfaces. Despite these early experiments, particle optics using electrons and neutrons developed earlier. Electrons had the advantage of a smaller mass (and hence a larger de Broglie wavelength), and their charge allowed control of their trajectory by the use of static electromagnetic fields. Neutrons had the advantage that Bragg diffraction in crystals could be used to realise efficient beam splitters.

In recent years, however, a rapid growth has occurred in atom optics. This is due to two main developments. First, advances in micro-fabrication technology have allowed the production of structures which are fine enough to diffract atomic beams. Secondly, as we saw in Section 15.4, important progress has been made in manipulating neutral atoms using laser light. In particular, 'cold' atoms are now available, which have a much larger de Broglie wavelength than thermal atoms.

A typical experiment in atom optics consists of a source, optical elements and a detector. The source must provide an intense, well-collimated and nearly monochromatic atomic beam. Atomic beams may be produced by effusion or supersonic expansion. These traditional sources offer the advantages of simplicity and high brightness (up to 10^{22} atoms/(sr cm^2 s)). Slow atomic beams can be obtained by launching laser-cooled atoms from a trap. The advantages of this method are larger de Broglie wavelengths and the possibility of long interaction times, leading to high sensitivity. Optical elements can be realised by either diffraction from micro-fabricated structures, or interaction with external electromagnetic fields.

Because of the short atomic de Broglie wavelengths, atom optics offers possibilities for accurate measurements and sub-nanometre fabrication. So far, the main applications of atom optics have been the construction and use of *atom interferometers* and the deposition of patterns of atoms on surfaces, called *atom lithography*.

In what follows we shall give a brief survey of the field of atom optics. Detailed accounts of the subject may be found in the review article by Adams, Sigel and Mlynek (1994) and the book by Meystre (2001).

Focusing of atomic beams. Atom lenses

The focusing of atomic beams can be accomplished by refracting the atomic matter waves from an effective parabolic potential, or by diffracting them. The diffraction-limited spot size is given approximately by

$$\mathscr{S} \simeq \frac{2f\lambda}{D} \tag{16.40}$$

where f is the focal length of the lens, D its diameter and λ is the de Broglie wavelength of the atom. In principle, since λ is very small, one could conceive of a lens producing a focus smaller than an atomic diameter. In practice, however, other factors limit the minimum spot size. In particular, for an atomic beam having a velocity spread $\Delta v/v$, the additional contribution to the spot size is approximately $D\Delta v/v$. In order for this undesirable contribution not to dominate, a velocity spread satisfying the condition

$$\frac{\Delta v}{v} \ll \frac{2f\lambda}{D^2} \tag{16.41}$$

is required.

Focusing using static electromagnetic fields

In 1951, H. Friedburg and W. Paul demonstrated that an atomic beam can be focused by using a *static hexapole magnetic field.* This field produces a Zeeman shift with a quadratic spatial dependence leading to a focusing effect for atoms in the appropriate magnetic sublevel. The same principle was used by N.F. Ramsey and co-workers to focus excited hydrogen atoms (in the upper hyperfine level $F = 1$) in the hydrogen maser, as explained in Section 15.1. In that section we also saw that C.H. Townes and co-workers used an inhomogeneous *static electric field* to focus excited ammonia molecules (in the upper level of energy E_2) in the ammonia maser.

The focusing of atoms with static electromagnetic fields suffers from two main disadvantages. First, the interaction is too weak to allow the realisation of compact lenses. Secondly, it is not possible to focus atoms in the ground state or lowest-lying magnetic sublevels. Indeed, the energy of these states decreases as the applied field strength increases, and there is no local maxima for static fields, in accordance with Earnshaw's theorem.

Focusing by using light forces

The interaction of near-resonant laser fields with atoms was discussed in Section 15.4. Such fields are well suited to focusing atomic beams, since the laser–atom interaction is relatively strong, and the atoms can experience an approximately parabolic effective potential for a variety of light field configurations. For example, close to the axis of a co-propagating Gaussian laser beam, atoms in an atomic beam experience such a potential U, due to the transverse dipole forces exerted by the intensity gradient of the superimposed laser beam (see Fig. 16.6). For red detuning ($\Delta < 0$) the net *dipole force* of resonance-radiation pressure acts towards the beam axis so that the atomic beam is focused. For blue detuning ($\Delta > 0$), the atomic beam is defocused. The first observation of focusing and

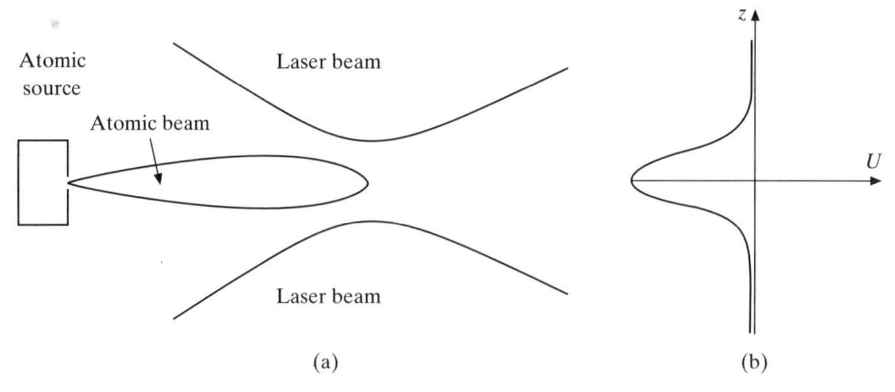

Figure 16.6 (a) Focusing of an atomic beam in a co-propagating Gaussian laser beam detuned to the red ($\Delta < 0$). (b) The effective potential U experienced by the atoms has the desired parabolic form close to the axis of the laser beam. (From C.S. Adams, M. Sigel and J. Mlynek, *Physics Reports*, **240**, 143, 1994.)

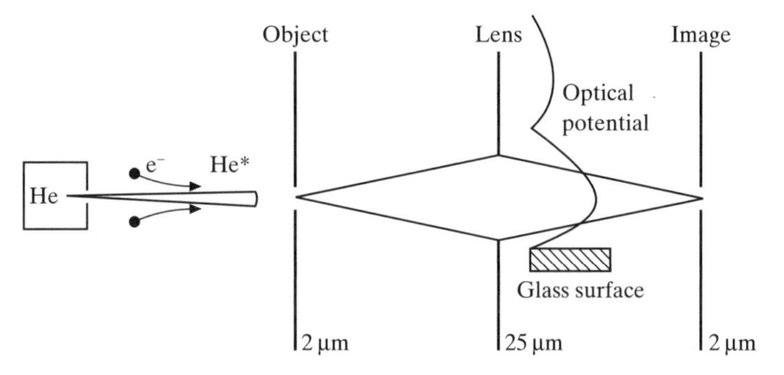

Figure 16.7 The experimental set-up of T. Sleator *et al.* for focusing a metastable helium beam using the antinode of a large-period optical standing wave produced by reflection of a laser beam from a glass surface under glancing incidence. The beam of metastable helium atoms is obtained by collinear electron impact excitation of an intense beam of ground state helium atoms, produced by supersonic gas expansion. (From T. Sleator, T. Pfau, V. Balykin and J. Mlynek, *Applied Physics* B **54**, 375, 1992.)

defocusing of neutral (sodium) atoms by the dipole force was made in 1978 by J.E. Bjorkholm, R.R. Freeman, A. Ashkin and D.B. Pearson.

An approximately parabolic potential can also be obtained close to the antinode of a transverse standing wave light field. In 1992, T. Sleator, T. Pfau, V. Balykin and J. Mlynek demonstrated the focusing of a metastable helium beam using the antinode of a large-period optical standing wave produced by reflection of a laser beam from a glass surface under glancing incidence (see Fig. 16.7). The focal length of this cylindrical lens could be changed by varying the intensity of the laser field. The minimum spot size obtained was 4 μm. Metastable helium atoms were used since they combine a number of advantages for atom optics experiments. The helium atom has a low mass and hence comparatively large de Broglie wavelength. In addition, helium is inert, so that delicate structures can be used and intense helium beams can be produced by using supersonic expansion techniques. Moreover, metastable helium atoms can be easily detected and have optical transitions in the near infra-red, so that laser light can be used to manipulate their degrees of freedom. The beam of metastable helium atoms in the states 2^1S_0 and 2^3S_1 was obtained by collinear electron impact excitation of an intense beam of ground state helium atoms (produced by supersonic expansion through a nozzle). After excitation, the beam of metastable helium atoms had a velocity spread $\Delta v/v \simeq 1/15$, where v denotes the mean velocity of the atoms and Δv is the full width at half maximum of the Gaussian velocity distribution.

The array of cylindrical lenses produced by a standing wave laser field can be used to deposit a series of regularly spaced atom lines on a substrate. For example, sodium and chromium line structures have been 'written' on a silicon substrate. Atom lithography based on this technique offers interesting possibilities in the nano-fabrication industry, for example the production of memory chips.

Focusing by using a Fresnel zone plate

A Fresnel zone plate consists of alternating transmitting and opaque concentric rings, the radius R_N of the Nth ring being given by

$$R_N = \sqrt{N} R_1 \tag{16.42}$$

It acts as a lens which images an object at a distance A onto images at distances B_n, where A and B_n obey the lens equation [2]

$$\frac{1}{A} + \frac{1}{B_n} = \frac{1}{f_n} = \frac{n\lambda}{R_1^2} \tag{16.43}$$

Here n is any integer, denoting the order of diffraction, and f_n is the focal length of the Fresnel zone plate for the nth diffraction order.

In 1991, O. Carnal, M. Sigel, T. Sleator, H. Takuma and J. Mlynek reported the focusing of metastable helium atoms using a Fresnel zone plate. The central ring had a radius $R_1 = 9.38$ μm and the total diameter d of the Fresnel zone plate was about 210 μm. The experimental set-up is shown schematically in Fig. 16.8. The beam of metastable helium atoms was obtained by electron impact excitation of a beam of ground state helium atoms, in the way described above. The atoms passed through an object structure located a few centimetres behind electron impact excitation. The objects were chosen to be either a single slit or a double slit with dimensions in the 10 μm range. They were imaged by using a Fresnel zone plate located at a distance $A = 0.96$ m. The intensity distributions were obtained by transversely scanning a single slit (10 μm in width and 3 mm in height) located at a distance $B = 0.84$ m behind the zone plate. Mounted on the same translation

[2] See for example Born and Wolf (1975).

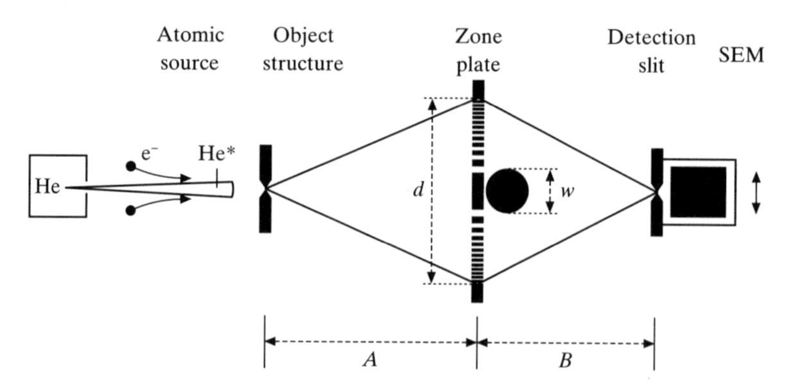

Figure 16.8 Schematic diagram of the experimental arrangement of O. Carnal *et al.* for focusing metastable helium atoms using a Fresnel zone plate. The dimensions are $A = 0.96$ m, $B = 0.84$ m and $d = 210$ μm. The wire diameter is $w = 50$ μm. (From O. Carnal, M. Sigel, T. Sleator, H. Takuma and J. Mlynek, *Physical Review Letters*, **67**, 3231, 1991.)

stage as the detection slit, a secondary electron multiplier (SEM) was used to detect the metastable helium atoms. The large number of helium atoms remaining in the ground state (approximately 10^6 times more than those in the two metastable states) have no effect on the experiments, since ground state atoms are not detected. In order for the first-order image distance B_1 to coincide with the distance B (for the fixed object distance A), the focal length of the lens had to be tuned to $f_1 = 0.45$ m, which by using (16.43) corresponds to a de Broglie wavelength $\lambda = 1.96$ Å. This in turn means that the nozzle through which the supersonic expansion of helium atoms occurs must be at a temperature $T = 20$ K. Using only the first order ($n = 1$) for imaging, all other orders (in particular the bright undiffracted zeroth order $n = 0$) contribute to an undesired background in the image plane which is blocked out by adjusting a 50 μm diameter metal wire in the centre of the zone plate (see Fig. 16.8). A disadvantage of a zone plate is that only about 10% of the atoms are diffracted into the first order. The images of a single and a double slit obtained by using the Fresnel zone plate are shown in Fig. 16.9. The resolution was mainly limited by the velocity spread $\Delta v/v \simeq 1/15$ of the metastable helium beam.

Atom mirrors

Atoms can be reflected from surfaces, or from effective potentials induced by static or optical electromagnetic fields.

Reflection from surfaces

The reflection of atoms by surfaces was first observed in 1929 by F. Knauer and O. Stern. It has now become an important technique in surface science. However, for atom optics purposes the interaction of an atom with a surface has

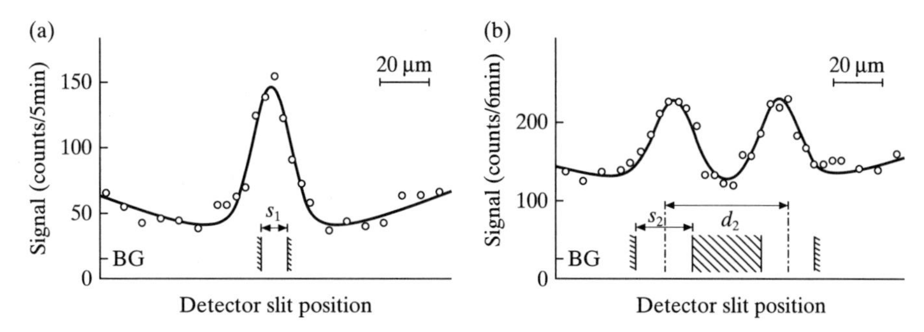

Figure 16.9 Images (a) of a single slit and (b) of a double slit obtained by using metastable helium atoms at a temperature of 20.5 K and a Fresnel zone plate. The circles represent the experimental data and the solid line is a fit through the data points. The slit dimensions are shown in the insets. The width of the single slit is s_1 = 10 μm. For the double slit, the slit distance is d_2 = 49 μm and the width of the individual slits is s_2 = 22 μm. BG denotes the detector background level. (From O. Carnal, M. Sigel, T. Sleator, H. Takuma and J. Mlynek, *Physical Review Letters*, **67**, 3231, 1991.)

the disadvantage of being a complicated phenomenon. First, the interaction potential contains an *attractive* long-range term due to the van der Waals interaction between atoms in addition to a strong *repulsive* term at short distances less than a few ångströms. In addition, the atom–surface interaction is complicated by surface impurities. Moreover, several processes may occur: elastic scattering, inelastic scattering and adsorption. For atom optics only elastic scattering yields a *coherent mirror*, such that the phase shift of the atomic wave function on reflection is constant. Unfortunately, for typical atomic kinetic energies, inelastic processes are the dominant ones.

Reflection from an evanescent light wave

An evanescent wave can be produced by total internal reflection of a laser beam at the interface between a dielectric and the vacuum (see Fig. 16.10). The evanescent wave intensity $I(z)$ decreases exponentially as a function of the distance z from the surface, namely

$$I(z) = I_0 \exp(-2z/\Lambda) \tag{16.44}$$

where the characteristic decay length

$$\Lambda = \frac{\lambda_L}{2\pi(n^2 \sin^2\theta - 1)^{1/2}} \tag{16.45}$$

depends on the wavelength λ_L of the laser radiation, the angle of incidence θ and the refractive index n of the dielectric. If the laser field is blue-detuned from the atomic resonance, the incident atomic beam will experience a repulsive effective potential with a similar exponential dependence.

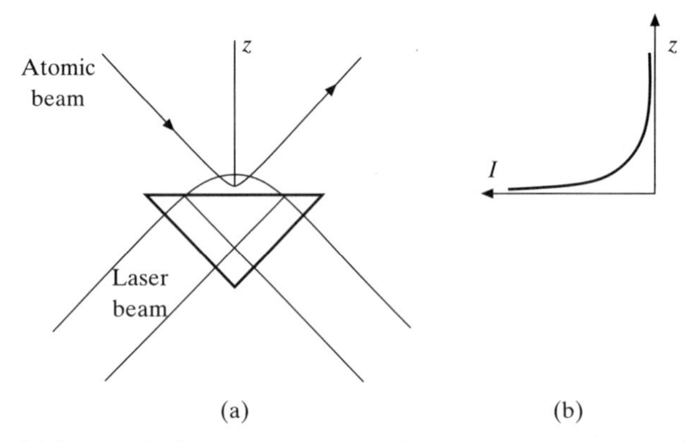

Figure 16.10 (a) Schematic diagram of an atom mirror using an evanescent wave produced by total-internal reflection of a laser beam at the interface between a dielectric medium and the vacuum. (b) The strong intensity gradient in the direction normal to the surface arises from the decrease of the light intensity within approximately one wavelength.

The use of an evanescent wave as a mirror for neutral atoms was suggested in 1982 by R.J. Cook and R.H. Hill. The first experimental demonstration of the reflection of atoms from an evanescent light wave was made in 1987 by V.I. Balykin, V.S. Letokhov, Yu.B. Ovchinnikov and A.I. Sidorov, who observed the grazing-incidence reflection of a thermal atomic beam. In 1990, M.A. Kasevich, D.S. Weiss and S. Chu observed the reflection of cold atoms at normal incidence. Since then, several experiments have been performed, in which bounces of the reflected atoms have been observed.

Interest in evanescent wave atom mirrors has been stimulated by their use in building *atom cavities*, as suggested in 1989 by V.I. Balykin and V.S. Letokhov. The simplest approach is to use a single mirror facing upwards and let gravity play the role of the second mirror. In such a *gravito-optical cavity* the atoms bounce as on a trampoline (see Fig. 16.11). A stable cavity can be realised by using a parabolic mirror to confine the transverse motion. Such gravito-optical cavities are well suited to trap a large number of cold atoms.

In 1993, C.G. Aminoff, A.M. Steane, P. Bouyer, P. Desbiolles, J. Dalibard and C. Cohen-Tannoudji observed the bouncing of atoms in a stable gravito-optical cavity. The evanescent wave atom mirror was formed by an 800 mW beam from a Ti:sapphire laser, reflected internally at the centre of the concave region of a glass prism. This laser beam was tuned between 1 and 10 GHz above the resonance transition between the ground state level $6S_{1/2}$, $F = 4$ and the excited level $6P_{3/2}$, $F = 5$ in caesium. About 10^7 cold caesium atoms were dropped onto the mirror from a magneto-optical trap located 3 mm above the prism. The bouncing atoms were detected by introducing a probe beam at the resonant atomic frequency and measuring the fluorescence it produces by using a photodiode. The experimental results shown in Fig. 16.12 exhibit up to 10 resolved bounces, after which the number of atoms dropped below the detection limit.

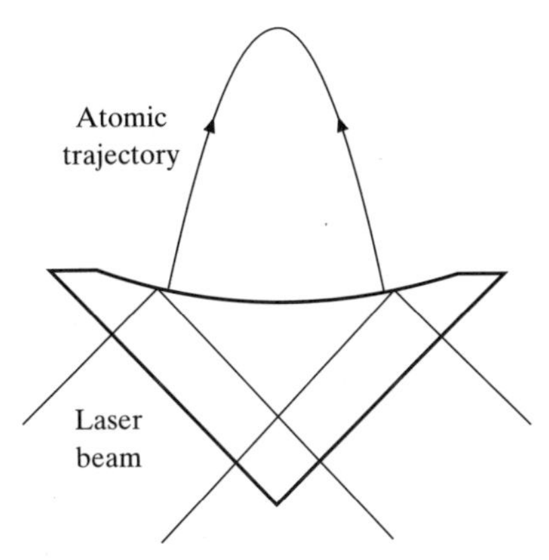

Figure 16.11 A gravito-optical cavity for atoms. The atoms are reflected by a parabolic evanescent wave mirror facing upwards. The gravitational field plays the role of the second mirror.

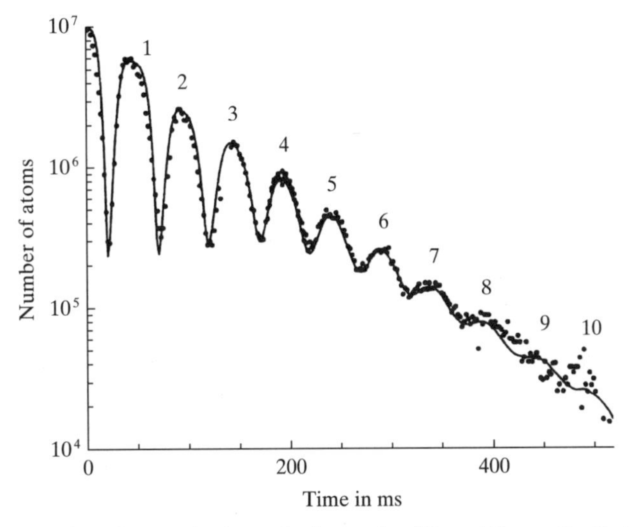

Figure 16.12 Number of atoms in the probe beam, for different times after their release. The dots represent the experimental data. The curve is a fit calculated by performing a Monte-Carlo simulation of the experiment. (From C.G. Aminoff, A.M. Steane, P. Bouyer, P. Desbiolles, J. Dalibard and C. Cohen-Tannoudji, *Physical Review Letters* **71**, 3083, 1993.)

Atomic beam splitters

In light optics, a beam splitter (also called beam divider or separatrix) can be obtained by using a system of slits (grating), a semi-reflecting plate (Mach–Zehnder or Michelson interferometers) or the interface between the vacuum and a crystal plate (polarisation interferometer). In atom optics the role of a beam splitter is to transform a single incoming state $|\Psi_{\text{in}}\rangle$ into an outgoing state $|\Psi_{\text{out}}\rangle$

which is a linear superposition of two or more states differing by at least one of the relevant degrees of freedom. An ideal beam splitter should produce two output states with a large splitting angle. However, in practice, it is often necessary to choose between two output beams with a small splitting, or several output beams among which two with a large enough splitting may be selected. The earliest demonstration of atomic beam splitting was the Stern–Gerlach experiment, performed in 1922 (see Section 1.8). For atoms with a ground state having a total angular momentum quantum number $J = S = 1/2$, a transverse magnetic field gradient deflects the two spin states $M_S = \pm 1/2$ in opposite directions. Since then, various techniques to realise atomic beam splitters have been developed. Some of them will now be briefly discussed.

Diffraction from micro-fabricated structures

An atomic beam can be diffracted by a periodic structure. The simplest technique uses the diffraction of atomic waves by slits or material gratings (see Fig. 16.13(a)). Since atomic de Broglie wavelengths are very small, the width of the slits and

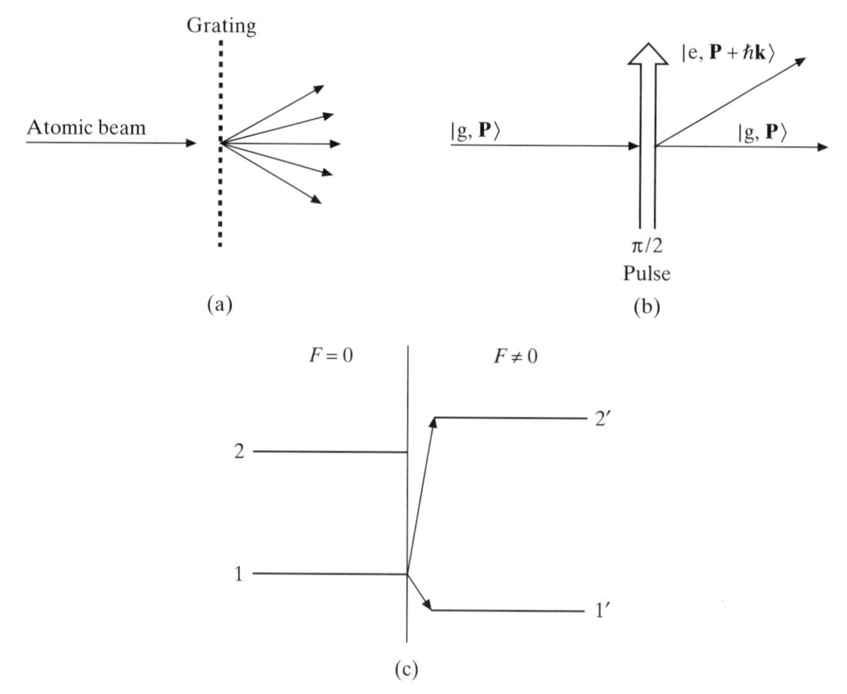

Figure 16.13 Atomic beam splitters. (a) Diffraction of an atomic beam by a material grating. (b) Photon recoil: a $\pi/2$ laser pulse excites the initial state $|g, \mathbf{P}\rangle$ into a linear superposition of the states $|g, \mathbf{P}\rangle$ and $|e, \mathbf{P} + \hbar\mathbf{k}\rangle$. As the excited state carries the photon momentum, the two components become spatially separated. (c) Sudden transitions: (1, 2) and (1′, 2′) are respectively eigenstates without and with an external field F. If this field is rapidly turned on, transitions are induced between the incoming free eigenstate 1 and the interaction eigenstates 1′ and 2′.

the period of the grating must be very small. For example, the diffraction of atoms by a transmission grating was first observed in 1988 by D.W. Keith, M.L. Schattenburg, H.I. Smith and D.E. Pritchard, using a grating made from a 0.5 μm thick old foil, and having a period of 0.2 μm. Micro-fabricated structures present several advantages: they are passive elements, require little maintenance and can be used with any atomic species. Unfortunately, their efficiency is low, the maximum intensity of the first diffraction order being about 10% of the incident intensity.

Diffraction from crystalline surfaces

The diffraction of atomic beams by crystalline surfaces was first observed in 1930 by I. Esterman and O. Stern. For thermal atomic beams the de Broglie wavelength is of the order of the lattice spacing, so that large diffraction angles can be obtained. Unfortunately, as discussed above for atom mirrors, the atom–surface interaction is complex, and is further complicated by surface impurities. For atom optics diffraction, high-quality surfaces are necessary, requiring ultra-high vacuum and elaborate surface preparation techniques.

Diffraction from an effective optical potential

Instead of material gratings, optical gratings can be used, which consist of a standing off-resonance optical wave behaving as a periodic dipole potential (of period $\lambda_{opt}/2$, where λ_{opt} is the wavelength of the optical wave) acting on the external atomic motion. Atomic waves can also be diffracted by an evanescent standing wave, produced by counter-propagating laser beams incident on a dielectric–vacuum interface.

Photon recoil beam splitter

A useful technique to split an atomic beam makes use of the recoil effect due to the interaction of an atom with resonant laser light (see Fig. 16.13(b)). Let us consider an atom initially in the ground state g and moving with momentum **P**. We shall denote this initial state by $|\Psi_{in}\rangle = |g, \mathbf{P}\rangle$. After crossing a laser beam, the frequency of which is tuned on a transition between the ground state and an excited state e, the outgoing atomic state $|\Psi_{out}\rangle$ has the required 'split' form

$$|\Psi_{out}\rangle = a_g|g, \mathbf{P}\rangle + a_e|e, \mathbf{P} + \hbar\mathbf{k}\rangle \tag{16.46}$$

where $\hbar\mathbf{k}$ is the photon momentum. The coefficients a_g and a_e depend on the laser power and the beam profile. The ideal superposition, for which $|a_g| = |a_e|$, is obtained with a so-called π/2-pulse. Since the excited state component in (16.46) carries the photon momentum, the excitation process behaves as a beam splitter.

In an experiment performed in 1991, M.A. Kasevich and S. Chu used stimulated Raman transitions to produce a coherent superposition of two ground state hyperfine levels $|1, \mathbf{P}\rangle$ and $|2, \mathbf{P} + 2\hbar\mathbf{k}\rangle$, with the advantage that the split beams contain atoms in the ground state and hence are not subjected to spontaneous emission.

Sudden transitions due to field gradients

As we have seen above, the role of a beam splitter is to induce transitions between states. These states are not necessarily free states, but can also be eigenstates of a Hamiltonian which contains an interaction with an external field F (see Fig. 16.13(c)). If this field is suddenly switched on, the incoming free eigenstate 1 is abruptly projected onto the interaction eigenstates $1'$ and $2'$. For example, sudden transitions from free states to Stark states, induced by the rapid passage of excited hydrogen atoms in the $2s_{1/2}$ state through a sharply increasing electric field, have been used since 1973 as an atomic beam splitter in the Sokolov interferometer, to be described below. In a similar way, sudden transitions between Zeeman states in a rapidly varying magnetic field have been employed since 1991 by J. Baudon *et al.* in their longitudinal Stern–Gerlach interferometer, which will also be considered shortly.

Atom interferometry

One of the most striking consequences of the wave properties of matter is the observation of interference effects with material particles. Although the *diffraction* of electrons and atoms was observed soon after L. de Broglie formulated in 1924 his hypothesis that material particles possess wave-like properties, it took some time before *interference* between matter waves was observed. Except for the pioneering work with neutrons carried out in 1938 by R. Frisch, H. von Halban and J. Koch, interferometry with matter waves began in 1956 with the electron interferometer of G. Möllenstedt and H. Dücker. Due to the considerable progress accomplished since that time in electron microscopy, electron interferometry has now reached a very high level of sophistication [3]. A neutron interferometer based on Bragg diffraction from crystals was built by H. Rauch, W. Treimar and U. Bonse in 1974. On the other hand, the realisation of an atom interferometer was hindered for a long time by the short de Broglie wavelength of atoms, and by the fact that atoms, unlike neutrons, do not pass through solid matter. The situation has changed considerably, due to the advances in microfabrication technology and in laser manipulation of atoms.

As in light optics, a matter-wave interferometer is a device in which a *source point* (in space and time) is 'joined' to a *detector point* by the propagation of the external motion wave through two or more different paths. This operation involves at least two degrees of freedom. In *separated arm* interferometers, the different paths are spatially distinct; their counterparts in light optics are for example Young's double slit interferometer, or the Mach–Zehnder interferometer. In other interferometers, the paths are identical in space but differ with

[3] See for example Tonomura (1993).

respect to another degree of freedom, such as the spin polarisation of atoms in a 'polarisation interferometer'; the analogue in light optics is a crystal plate interferometer. The ideal operation of an atom interferometer requires the use of a *pure* atomic state for the relevant degrees of freedom. This implies that the beam emitted by the source must be *prepared*, for instance by angular collimation or spin polarisation. Similarly, at the output of the interferometer, the beam must be *analysed* before the detection. Thus, in an ideal atom interferometer, a pure incoming atomic state is transformed into a linear superposition of two or more atomic states. To each of these quantum mechanical states corresponds a phase, and these phases will interfere with each other if the outgoing waves are brought back together. If the preparation and (or) analysis is not perfect, then an average must be made over a statistical ensemble [4], which in general causes a loss of contrast in the interference pattern.

In most atom interferometers, the *object* under consideration is an *ensemble* of atoms. Usually, an atomic beam is used; it is emitted by a source which can be a simple tube, a nozzle, an oven or a discharge. This beam is collimated, velocity selected or naturally monoenergetic as a nozzle beam. Another way to obtain an ensemble of atoms is to cool and trap them, for example in a magneto-optical trap. All these ensembles of atoms are incoherent: the individual atoms are mutually independent, their external wave functions being randomly phase-shifted with respect to each other. The interference pattern is then built up atom by atom, by the repetition of one-atom 'trials' each producing a signal proportional to the modulus squared of the wave function, averaged over all undefined experimental parameters (such as the direction and magnitude of the velocity). An alternative way to obtain an interference pattern is to use repeatedly the observation of a *single* system, for example to observe the photons emitted by the laser-induced fluorescence of a single trapped ion, as shown in 1999 by P.E. Toschek *et al.* Averaging over an ensemble of such individual measurements leads to the same result as obtained by performing one measurement on an ensemble of ions. A different situation occurs with Bose–Einstein condensates (BEC), since in this case all the condensed (bosonic) atoms occupy the ground state of the trap, and hence constitute a fully coherent source. In an experiment performed in 1997, W. Ketterle and co-workers observed interferences by merging two BEC initially separated in a double trap. This experiment provided evidence for the phase of a condensate rather than a tool for interferometry. However, as we have seen in Section 15.5, it is now possible to built atom lasers, which generate coherent atomic matter waves from BEC. The atom lasers are expected to greatly improve atom interferometry, just as the lasers did in light interferometry.

Atomic beam splitters, which have been described above, are key elements of atom interferometers. As we have seen, their role consists of transforming a single incoming state into a linear superposition of states differing by one (or more) of the relevant degrees of freedom. Two classes of interferometers can be defined

[4] See for example Bransden and Joachain (2000).

depending on the fact that one or two beam splitters are used. In *open* interferometers a *single* beam splitter produces a coherent superposition of waves which propagate along the different arms. The interference pattern resulting from the superposition of the outgoing waves is then recorded by moving in space (or time) an almost punctual detector. In *closed* interferometers, the outgoing waves are superposed again by a *second* beam splitter before being analysed and detected. In this case, the interference field does not exhibit any spatial (or temporal) dependence. The interference pattern is observed by scanning the phase difference accumulated between the two beam splitters.

In what follows we shall describe some of the atom interferometers which have been realised. Detailed accounts of the subject may be found in the review articles of Adams, Carnal and Mlynek (1994) and of Baudon, Mathevet and Robert (1999), as well as in the book edited by Berman (1997).

Two-slit interferometers

In 1991, O. Carnal and J. Mlynek demonstrated an atom interferometer based on an open double-slit arangement, which is the quantum mechanical analogue of Young's double-slit interferometer in classical optics. As we have seen in Chapter 1, such an interferometer had already been demonstrated for electrons by G. Möllenstedt and H. Dücker in 1956, and by C. Jönsson in 1961.

In the experiment of Carnal and Mlynek the slits were mechanical transmission structures with widths in the μm range. As atomic species, they used metastable helium atoms which, as we have seen previously, present several advantages for experiments in atom optics. The experimental set-up is shown schematically in Fig. 16.14. The atomic source is similar to that described in Figs 16.7 and 16.8. Collinear excitation of an intense beam of ground state helium atoms by electron impact yields metastable helium atoms in the states 2^1S_0 and 2^3S_1, with relative populations of 90% and 10%, respectively, and a velocity spread $\Delta v/v \simeq 1/15$. The mean velocity v of the atoms, and therefore their mean de Broglie wavelength, can be modified by changing the temperature of the gas reservoir and nozzle system. It is found that at $T = 295$ K the mean de Broglie wavelength of the He atoms is $\lambda = 0.56$ Å, while at $T = 83$ K it is $\lambda = 1.03$ Å. The metastable helium atoms are collimated by a slit s_0 of width 2 μm in plane A. After travelling a distance $L = 64$ cm, they illuminate the two slits s_1 and s_2 in plane B. The width of these two slits is 1 μm, and the distance d between them is 8 μm. The atomic matter waves emerging from the two slits s_1 and s_2 interfere in the region behind the plane B. The resulting interference fringes are recorded in the plane C located at a distance $L' = L$ behind the double slit. The detection system consists of a secondary electron multiplier (SEM) behind a gold foil with imprinted micro-fabricated slit and grating. As in the experiments on the focusing of metastable helium atoms described above, only metastable He atoms are detected. The interference pattern is monitored with a 2 μm slit s_3 identical to the entrance slit. To increase the count rate, the measurements were also performed using a grating moving in the

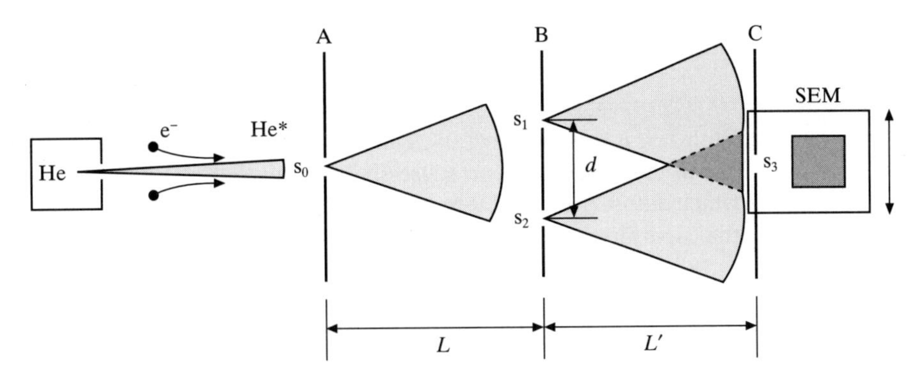

Figure 16.14 Schematic representation of the experimental set-up of O. Carnal and J. Mlynek for an atom interferometer based on Young's double-slit arrangement. A beam of helium atoms coming from a gas reservoir is produced by a supersonic gas expansion. Collinear excitation by electron impact yields a beam of metastable helium atoms, which is collimated by the entrance slit s_0 in plane A. It then reaches plane B containing the two slits s_1 and s_2. The resulting interference pattern is detected in plane C using a secondary electron multiplier (SEM) behind a gold foil with imprinted micro-fabricated slit and grating. (From O. Carnal and J. Mlynek, *Physical Review Letters* **66**, 2689, 1991.)

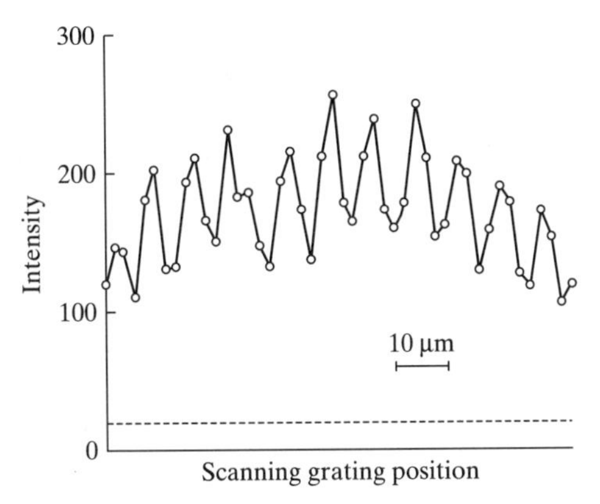

Figure 16.15 Intensity distribution observed in the two-slit experiment of O. Carnal and J. Mlynek, for a de Broglie wavelength $\lambda = 1.03$ Å. The interference pattern was monitored in the detector plane C with a grating of period 8 μm. The dashed line is the detector background. The line connecting the experimental points is a guide to the eye. (From O. Carnal and J. Mlynek, *Physical Review Letters* **66**, 2689, 1991.)

detector plane C. When the slits in the grating coincide with the interference maxima, the signal at the detector shows a maximum.

In Fig. 16.15 is shown a typical result of an experimental run performed for a de Broglie wavelength $\lambda = 1.03$ Å, with a grating of period 8 μm (matching the period of the interference pattern) in the detector plane C. A system of equidistant interference maxima was observed, separated by a periodic distance whose theoretical value is given to good approximation by

$$D = \frac{L'\lambda}{d} \tag{16.47}$$

With $L' = L = 64$ cm, $\lambda = 1.03$ Å and $d = 8$ μm, one finds that $D = 8.2$ μm. The average distance between two successive maxima was measured to be (7.7 ± 0.5) μm, which differs from the theoretical value by about 5% and is within the accuracy of the experiment. The *fringe visibility V* is defined by the relation

$$V = \frac{I_{max} - I_{min}}{I_{max} + I_{min}} \tag{16.48}$$

where I_{max} and I_{min} denote the maximum and minimum intensity of the interference pattern, respectively. In the experiment of Carnal and Mlynek, the fringe visibility was up to 60% and decreased off axis due to the velocity distribution of the atomic beam. Because of the low count rate, a phase shift of 1/3 rad required a measurement time of 10 min.

The period D of the interference pattern may be increased by decreasing the velocity of the atoms. In 1992, F. Shimizu, K. Shimizu and H. Takuma performed a Young's double-slit experiment with cold neon atoms trapped in a magneto-optical trap (MOT) and prepared in a particular metastable state Ne* by focusing a 598 nm laser beam into the MOT (see Fig. 16.16). The four laser beams used for the MOT had a wavelength of 640 nm. The temperature of the trapped atoms was approximately 2.5 mK. The atoms fell on a screen S placed 76 mm below the trap, having two slits, each of width 2 μm, and separated by a distance of 6 μm. The interference pattern was observed using a microchannel plate (MCP) detector located below the double slit. Since the atoms were accelerated by the force of gravity, the period D of the interference pattern depended on the vertical distance to the detection plane. Shimizu *et al.* measured periods in the 100 μm range. Because of the long de Broglie wavelength of the cold atoms, the accuracy requirement for the components of the interferometer is reduced by several orders of magnitude. Since atoms take a long time to move through a cold atom interferometer, the phase sensitivity is greatly increased.

A temporal version of the two-slit interferometer was realised in 1996 by P. Szriftgiser, D. Guéry-Odelin, M. Arndt and J. Dalibard (see Fig. 16.17). As we have seen previously, an evanescent optical wave whose frequency is blue detuned ($\Delta > 0$) with that of an atomic transition acts as a repulsive effective potential for the atomic motion, that is as a concave mirror if a hollow spherical interface is used. A cold atom cloud is released from a trap, and a first light pulse plays the double role of a velocity selector and of a collimating slit. Bouncing atoms return to the mirror and two successive light pulses make the temporal counterpart of the two Young's slits. They produce two coherent wave packets, the interference of which is recorded by means of a time-of-flight (TOF) measurement using a probe laser.

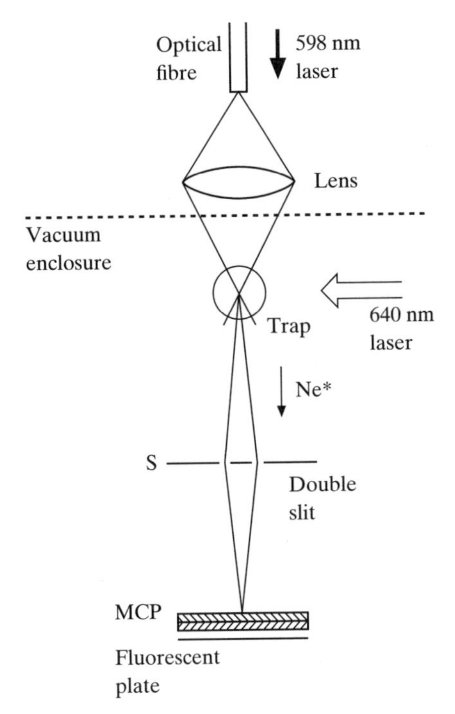

Figure 16.16 Schematic diagram of the experimental set-up of F. Shimizu *et al.* Cold atoms released from a magneto-optical trap fall under the force of gravity on a screen S having two slits. The impact pattern produced by the impact of the atoms is observed by using a microchannel plate detector placed below the double slit. (From F. Shimizu, K. Shimizu and H. Takuma, *Physical Review* A **46**, R 17, 1992.)

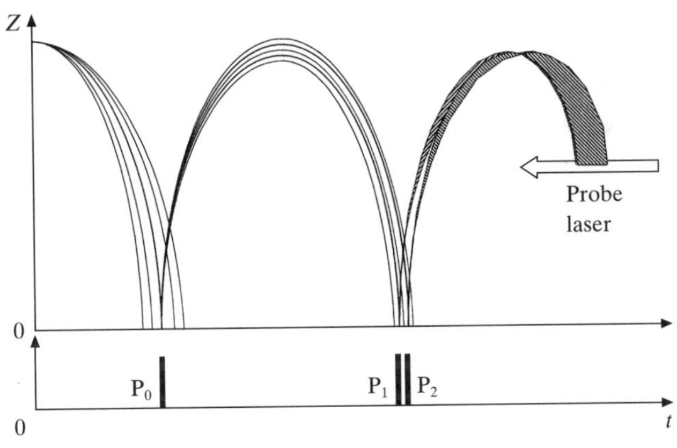

Figure 16.17 Illustration of the temporal interferometry experiment of P. Szriftgiser *et al.*; OZ is the vertical axis, t is the time. A cold atomic cloud is released above a mirror formed by an evanescent laser wave. The mirror intensity is chopped during three pulses. The first pulse P_0 plays the double role of a velocity selector and of a collimating and diffracting slit. The two following pulses P_1 and P_2 are the temporal equivalents of two Young's slits. The interference pattern is observed by performing a time-of-flight measurement of the fluorescence induced by a weak resonant probe laser beam.

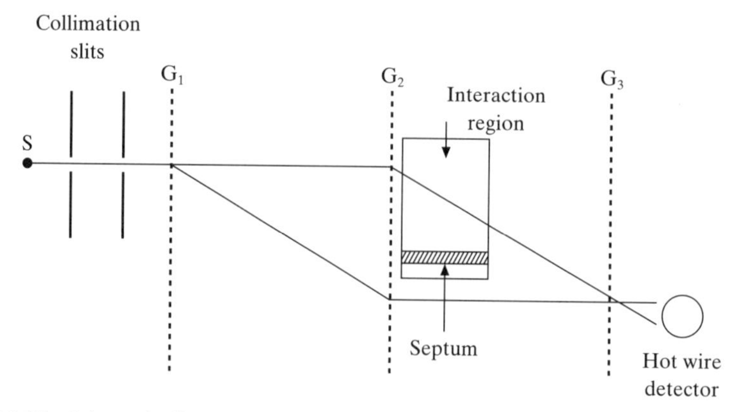

Figure 16.18 Schematic diagram of the three-grating atom interferometer of D.W. Keith *et al.* The first grating diffracts the atomic beam (the zero and +1 orders are shown). A second grating leads to spatial overlap and interference. The third grating is used to monitor the interference pattern, since in practice the detector is larger than the period of the interference pattern. The septum allows selective interaction with one path.

Three-grating interferometers

In 1991, D.W. Keith, C.R. Ekstrom, Q.A. Turchette and D.E. Pritchard demonstrated a three-grating atom interferometer based on microstructures (see Fig. 16.18). This interferometer is achromatic, giving 'white' fringes (that is, the position of the central fringe is independent of the de Broglie wavelength). This is convenient in atom optics, because of the broad velocity distributions of most of the atomic sources. In their experiment, Keith *et al.* used a beam of sodium atoms produced by supersonic expansion with an argon carrier gas. The atoms had a mean de Broglie wavelength $\lambda = 0.16$ Å and a velocity spread $\Delta v/v \simeq 1/9$. The incident beam was collimated with two slits, and had a diameter of about 20 μm. In a later version of the experiment, nanofabricated diffraction gratings with a period of 200 μm, stabilised by using optical interferometry, were placed at equal intervals of 0.6 m. Interference patterns were recorded as a function of the transverse positions of the second grating by using a 25 μm hot wire positioned behind the third grating to measure the count rate. The spatial separation of the zero and first diffraction order at the middle grating was 55 μm, more than twice the beam diameter, so that the edges of the two interfering beams did not overlap. The phase sensitivity was 15 mrad for a 1 min measurement time.

An interaction region consisting of a 10 μm metal foil (septum) positioned symmetrically between two side electrodes was inserted in the interferometer so that the atom waves in the two sides of the interferometer passed on opposite sides of the foil. By applying a static electric field on one side of the interaction region, the Stark shift of the atom wave on that side modifies the phase of the interference pattern. According to (9.127), the Stark shift is given by $\Delta E = -\bar{\alpha}\mathscr{E}^2/2$, where $\bar{\alpha}$ is the static dipole polarisability of the atom and $\mathscr{E}$ is the magnitude of the

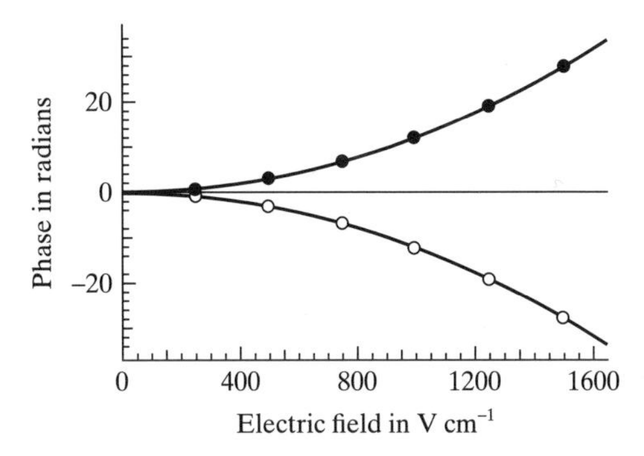

Figure 16.19 Stark phase shifts for static electric fields applied to the right (open circles) and the left (filled circles) of the interaction region. The phase shift per applied electric field squared (in (V cm^{-1})^2) is $1.220(7) \times 10^{-5}$ for the left side and $1.224(7) \times 10^{-5}$ for the right side. These measurements determine the static dipole polarisability of the ground state of sodium with a statistical uncertainty of 0.4%. (From D.E. Pritchard, in *Atomic Physics 13*, edited by H. Walther, T.W. Hänsch and B. Neizert, American Institute of Physics, New York, 1993.)

applied electric field. The Stark shift decreases slightly the potential energy, so that the wave number of the atom wave, and hence its spatial frequency, is increased compared to the wave that passes on the side of the septum with no field. The resulting differential phase shift is found to be

$$\delta\phi = \frac{|\Delta E|}{\hbar} t = \frac{\bar{\alpha} l \mathscr{E}^2}{2\hbar v} \qquad \textbf{(16.49)}$$

where t is the transit time and l is the length of the interaction region. The phase shift $\delta\phi$ measured in 1991 by D.E. Pritchard *et al.* is shown in Fig. 16.19 as a function of the applied electric field strength $\mathscr{E}$. In this way, the static dipole polarisability of sodium could be measured with an accuracy of about 1%. Measurements of the static dipole polarisability of atoms using the same principle were also performed in 1993 by V. Rieger, K. Sengstock, U. Sterr, J.H. Möller and W. Ertmer.

As explained previously, the periodic dipolar effective potential produced by an off-resonant standing optical wave of wavelength λ_{opt} acts as a grating of period $\lambda_{\text{opt}}/2$ on the atomic motion. In 1995, A. Zeilinger and co-workers built an atom interferometer using three identical and equidistant optical gratings (with λ_{opt} = 811 nm). It was operated with a beam of metastable argon atoms.

Ramsey–Bordé and stimulated Raman interferometers

These interferometers use photon recoil beam splitters, which have been described above. They are based on an extension of Ramsey's method of separated oscillatory fields (see Section 16.1). The original version of this method, used in the radio-frequency or microwave region, employs a two-zone configuration. In the optical region, the photon recoil is sufficiently large that it is necessary

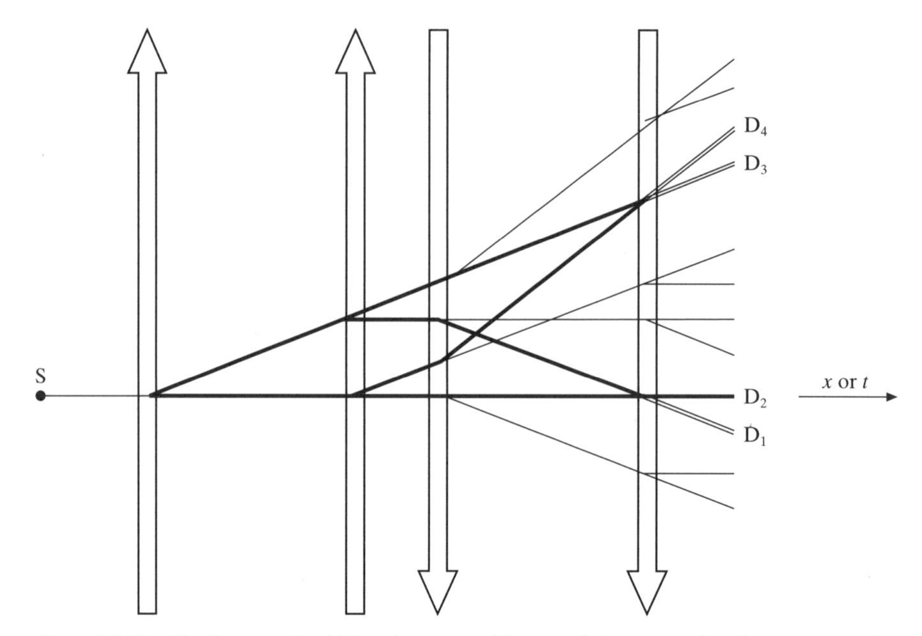

Figure 16.20 The Ramsey–Bordé interferometer. The atomic beam coming from a source S is successively diffracted by four coherent laser beams (open arrows). The heavy lines show the paths in the two closed interferometers. The four possible detection ports are indicated by D_1, D_2, D_3 and D_4. The horizontal axis is either the direction of propagation of atomic matter waves (x) in the spatial version, or the time (t) in the temporal version. In the latter case the atoms are irradiated with successive coherent light pulses. (From J. Baudon, R. Mathevet and J. Robert, *Journal of Physics* B **32**, R 173, 1999.)

to redirect the atomic trajectories in order to restore spatial overlap. This is achieved by repeating the usual Ramsey configuration with a second pair of laser beams propagating in the opposite direction (see Fig. 16.20). Optical Ramsey fringes were first observed in 1977 by J.C. Bergquist, S.A. Lee and J.L. Hall. In 1989, C.J. Bordé pointed out that the Ramsey fringes could be interpreted as due to atomic interference. Indeed, the structure formed by four coherent light beams perpendicular to the atomic beam, the first pair propagating in a direction opposite to the second one, is a Mach–Zehnder-type atom interferometer, in which the first and fourth zones are beam splitters, while the second and third zones act as mirrors. Sixteen atomic waves emerge from the fourth zone, of which eight of them issued from two closed loops form two times two (high- and low-frequency recoil components) independent closed interferometers, the final atomic state being either the ground state or the excited state. This Ramsey–Bordé interferometer has been operated with metastable magnesium atoms, metastable calcium atoms and iodine molecules.

Instead of sending the atoms through four successive laser beams, it is equivalent to irradiate them with successive coherent light pulses, as shown in 1991 by

M.A. Kasevich and S. Chu. In their experiments, a high sensitivity was achieved due to two improvements: the use of *stimulated Raman transitions* and of an *atomic fountain* of cold atoms. The advantage of using stimulated Raman transitions can be understood by remembering that the sensitivity of any measurement based on atom–light interactions is limited by the transition line width. To obtain a high sensitivity, narrow transitions are therefore required. However, the direct probing of narrow transitions requires ultra-stable lasers. To overcome this difficulty, Kasevich and Chu used velocity-selective stimulated Raman transitions, driven between two hyperfine levels of the ground state of alkali atoms by counter-propagating laser beams. Both lasers were detuned by the same amount from the optical transition frequency. As a result, only the *frequency difference* and not the *absolute* frequency of the laser beams must be stable, and this can be achieved by using a (radio-frequency, RF) oscillator. Since the initial and final levels are both ground states, the line width of the transition is extremely narrow, and, in practice, is limited only by the measurement time. Optical Ramsey excitation using stimulated Raman transitions can be obtained with a three-pulse sequence ($\pi/2$, π, $-\pi/2$). It can also be achieved by using the Ramsey–Bordé geometry with four $\pi/2$ pulses. It is worth noting that the momentum transfers are significantly increased by using stimulated Raman transitions. This increase can be magnified by adding a sequence of intermediate π-pulses.

The second improvement introduced by Kasevich and Chu was the use of an *atomic fountain* of cold atoms, first demonstrated in 1989 by M.A. Kasevich, E. Riis, S. Chu and R.G. De Voe. The idea was to cool and trap alkali atoms in a magneto-optical trap, and subsequently launch them on a vertical trajectory, so that they turn around and fall down due to gravity [5]. Combined with the very long lifetime of the Raman transition, the fountain geometry allows long interaction times, and hence increases considerably the sensitivity of the atom interferometer, compared with a conventional beam-type experiment. As we shall see below, high-precision measurements of gravity have been made with an atom interferometer using stimulated Raman transitions and an atomic fountain. This interferometer, operated in the Ramsey–Bordé configuration with additional π-pulses, was also used in 1993 by D.S. Weiss, B.C. Young and S. Chu to measure the frequency splitting between the high- and low-frequency recoil components for caesium. Their measurement provided a value of $\hbar/M$, where M is the mass of the caesium atom, accurate to 10^{-7}.

Atom interferometers using static electric or magnetic fields (polarisation interferometers)

In the atom interferometers considered so far, which use slits or light, the beam splitters operate by atom deflection, that is by acting on an extra *external* variable. By contrast, in polarisation interferometers, the second degree of freedom

---

[5] Atomic fountains will be discussed in more detail in Section 16.4, in connection with atomic clocks.

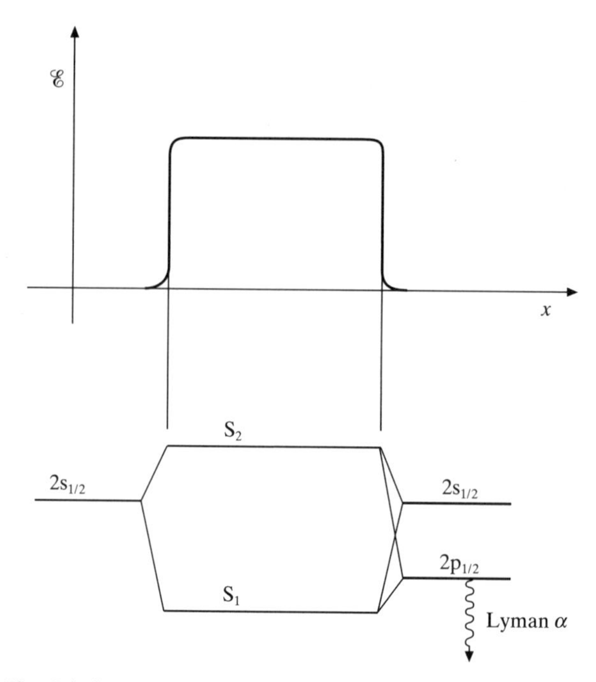

Figure 16.21 The Sokolov interferometer. Hydrogen atoms initially in the metastable $2s_{1/2}$ state pass through a quasi-rectangular profile of electrostatic field $\mathscr{E}(x)$. The incoming state $2s_{1/2}$ is suddenly projected onto the two interaction Stark eigenstates S_1 and S_2. At the output, when the electrostatic field is turned off, the two Stark states S_1 and S_2 are projected back onto the field-free $2s_{1/2}$ and $2p_{1/2}$ states. The outgoing atoms in the $2p_{1/2}$ state are detected at some distance through their spontaneous emission of Lyman α radiation.

involved in the beam splitters (the first one being an external variable x corresponding to the longitudinal motion) is an *internal* variable such as an induced electric dipole, or an atomic magnetic moment. As a result, the different paths in these closed interferometers are identical in space; they only differ by the internal atomic state, and hence by the velocity of the atoms along each arm.

The *Sokolov interferometer*, demonstrated in 1973 by Yu.L. Sokolov, uses the fact that interference between different internal states can be induced by the non-adiabatic passage of an atomic beam through a static electric field gradient (see Fig. 16.21). In its basic version, a beam of metastable hydrogen atoms $H^*(2s_{1/2})$, produced by resonant charge exchange of protons (having a kinetic energy of a few keV) in caesium vapour, passes through a quasi-rectangular profile of electrostatic field $\mathscr{E}(x)$. The two borders of this profile act as beam splitters. The sudden change in the basis sets of states adapted to the regions where $\mathscr{E} = 0$ ($2s_{1/2}$ and $2p_{1/2}$ states) and $\mathscr{E} \neq 0$ (Stark states S_1 and S_2) has the consequence that the incoming $2s_{1/2}$ state becomes suddenly a superposition of the two interaction Stark states S_1 and S_2. At the output of the profile, each of the states S_1 and S_2 becomes a superposition of the field-free $2s_{1/2}$ and $2p_{1/2}$ states. Interference fringes are obtained by

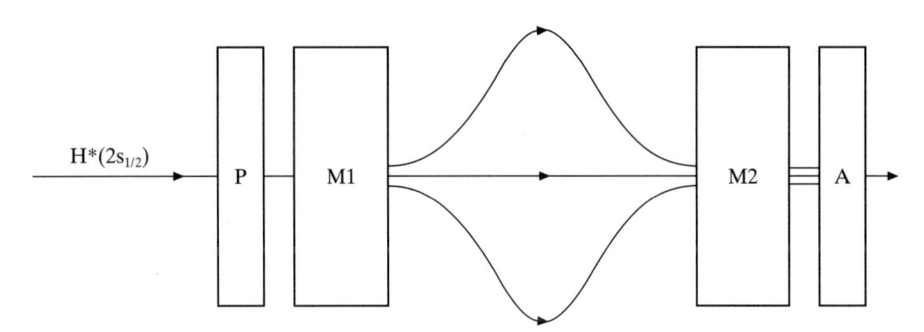

Figure 16.22 The longitudinal Stern–Gerlach interferometer of J. Baudon *et al.* A beam of metastable hydrogen atoms $H^*(2s_{1/2})$ is polarised by a transverse magnetic field gradient P. The initial state is projected onto the interaction Zeeman eigenstates by non-adiabatic passage through a magnetic field gradient M1. These eigenstates acquire different phase shifts in a central region where they experience different potentials due to a longitudinal magnetic field. They are projected back on the field-free eigenstates by passing through a second region of magnetic field gradient M2. Finally, the population of a particular state is monitored by an analyser A. (From C.S. Adams, O. Carnal and J. Mlynek, *Advances in Atomic, Molecular and Optical Physics* **34**, 1, 1994.)

detecting the Lyman α photons emitted due to the decay of the $2p_{1/2}$ state as a function of the electric field. The velocity distribution of the H* atoms is nearly identical to that of the protons, which is very narrow ($\Delta v/v < 10^{-3}$). This produces a very large number ($\simeq v/\Delta v$) of visible fringes, which allows accurate measurements, for example of the Lamb shift and of the hyperfine splitting in atomic hydrogen.

The *longitudinal Stern–Gerlach interferometer* demonstrated in 1991 by J. Baudon *et al.* uses a magnetic field gradient to induce sudden transitions among Zeeman states in a tiny and rapidly rotating magnetic field. A schematic diagram of their experiment is shown in Fig. 16.22. The interferometer operates with an incident beam of metastable hydrogen atoms $H^*(2s_{1/2})$ produced by electron bombardement of a thermal beam of H_2 molecules. This beam is polarised by a transverse magnetic field, which selects the $F = 1$, $M_F = 0$, 1 levels. A coherent superposition of Zeeman sublevels is prepared by non-adiabatic passage through a magnetic field orthogonal to the polarising field. In the central region, the Zeeman sublevels experience different potentials due to a longitudinal magnetic field and acquire different phase shifts. Finally, the sublevels are remixed and a particular state is selected by an analyser. Interference fringes are obtained by measuring the Lyman α photons emitted due to the decay of the $2p_{1/2}$ state as a function of the longitudinal magnetic field.

Applications of atom interferometry

Atom interferometers have been used successfully in three categories of applications: measurements of atomic and molecular properties, measurements of inertial and gravitational effects, and tests of quantum mechanics.

Let us first consider the applications to atomic and molecular properties. As we have seen above, the phase shifts induced by static electric fields have been used since 1993 by D.E. Pritchard *et al.* and by W. Ertmer and co-workers to measure the static dipole polarisability of atoms. Atom interferometry based on optical Ramsey excitation using stimulated Raman transitions was shown in 1993 by D.S. Weiss, B.C. Young and S. Chu to allow precise measurements of $\hbar/M$, where M is the atomic mass. Interferometry also provides additional phase information about collision processes.

Atom interferometers are sensitive to inertial effects because the atoms travel freely (provided the field gradients are small enough), whatever the acceleration of the apparatus. The difference in position between an interference pattern observed in an accelerating apparatus and an inertially stable apparatus can be measured accurately, giving therefore a precise measure of the non-inertial behaviour of the apparatus. As an example, let us consider a double-slit atom interferometer oriented vertically (see Fig. 16.14). Because of gravity, the fringes fall in the detector plane by an amount

$$l = ut - \frac{1}{2}gt^2 = -g\left(\frac{L}{v}\right)^2 \tag{16.50}$$

where g is the magnitude of the acceleration due to gravity, v is the magnitude of the particle longitudinal velocity and u is the magnitude of the initial vertical velocity required for the trajectories to pass through the same point in the double slit as with $g = 0$. The displacement of the fringes may be written as a phase shift

$$\delta\phi = \frac{2\pi|l|}{D} \tag{16.51}$$

where D is the fringe spacing. Using (16.47) and the fact that $L' = L$ for the experimental setting considered here, the gravitational phase shift is given by

$$\delta\phi = \frac{2\pi g A}{\lambda v^2} \tag{16.52}$$

where $A = Ld$ is the area of the interferometer. For interferometers based on diffraction by gratings or photon recoil, the area A is given approximately by $\lambda L^2/a$, where a is the grating period or the optical wavelength, respectively. In this case, the gravitational phase shift becomes

$$\delta\phi = \frac{2\pi g}{a}\left(\frac{L}{v}\right)^2 \tag{16.53}$$

and is seen to be independent of the mass of the test particle. The shift of the interference fringes due to gravity in a matter-wave interferometer was first observed in 1975 by R. Colella, A. Overhauser and S.A. Werner, who used a *neutron* interferometer; they achieved a sensitivity of 10^{-2} g.

In 1991, M.A. Kasevich and S. Chu showed that a higher sensitivity can be reached by measuring the rate of change of the Doppler shift of a falling *atom*. If T is the fall time, the Doppler shift is $k_L g T^2/2$, where k_L is the magnitude of the propagation vector of laser light tuned to an atomic resonance. The resolution of the measurement is limited by the line width of the transition. By using their atom interferometer based on stimulated Raman transitions and an atomic fountain of cold caesium atoms, S. Chu and co-workers achieved in 1999 an absolute uncertainty $\Delta g/g = 3 \times 10^{-9}$. They also compared their measurement with the value of g obtained at the same laboratory site using an optical Michelson interferometer gravimeter. They found that the macroscopic glass object used in this instrument falls with the same acceleration, to within 7 parts in 10^9, as the caesium atoms studied with the atom interferometer.

Let us now consider the case of *rotations*. In 1913, M.G. Sagnac showed that the phase in an optical ring interferometer is sensitive to rotation. For *light*, the *Sagnac effect* produces a phase shift

$$\delta\phi = \frac{8\pi A\Omega}{\lambda_{\text{opt}} c} \tag{16.54}$$

where A is the area of the interferometer, Ω is the magnitude of the rotation angular frequency and λ_{opt} is the optical wavelength. For *massive particles*, the Sagnac effect gives a phase shift

$$\delta\phi = \frac{8\pi A\Omega}{\lambda v} = \frac{4MA\Omega}{\hbar} \tag{16.55}$$

where M is the mass of the particle, λ its de Broglie wavelength and v is the magnitude of its velocity. Upon comparison of the expressions (16.54) and (16.55), we see that enhanced sensitivity to rotations can be achieved by using massive particles. Precise gyroscopes are needed for a wide range of applications, from inertial guidance systems for aircraft to tests of general relativity.

The Sagnac effect for matter waves was first observed in 1979 by S.A. Werner, J. Staudenmann and R. Colella, who used a neutron interferometer. In their experiment, $A \simeq 10^{-3}$ m^2, $\lambda \simeq 10^{-10}$ m and $v/c \simeq 10^{-5}$, giving a Sagnac phase shift about 10^9 larger than for an optical interferometer with the same area. However, for optical interferometers, the disadvantage of zero mass of the photon can be compensated in several ways. For example, an optical fibre ring interferometer may have an effective area of up to 10^8 m^2. In ring laser gyroscopes, high sensitivities are achieved, without using large areas, by introducing an active medium inside the interferometer. The phase is then determined very accurately by measuring the beat frequency between counter-propagating laser modes. In this way, sensitivities of the order of 10^{-8} rad s^{-1} can be achieved. It is interesting to note that the magnitude of the angular velocity of the Earth's rotation is $\Omega_{\text{Earth}} = 7.3 \times 10^{-5}$ rad s^{-1}.

The Sagnac effect for atoms was observed by C.J. Bordé and co-workers in 1991 using the optical Ramsey method and a beam of metastable Mg atoms. The apparatus was put on a rotatable platform. The sensitivity of the experiment was limited to about 10^{-2} rad s^{-1} by the stability of the laser, and the small enclosed area (about 6×10^{-6} m^2). This area can be increased by using cold atoms. For example, in 1997 T.L. Gustavson, P. Bouyer and M.A. Kasevich demonstrated a Sagnac-effect atom interferometer gyroscope using stimulated Raman transitions and cold caesium atoms. They achieved a short-term sensitivity to rotations of 2×10^{-8} rad s$^{-1}/\sqrt{\text{Hz}}$.

Let us now turn to tests of quantum mechanics. First of all, we note that matter-wave interferometry with electrons and neutrons has provided several interesting demonstrations of quantum mechanical effects. For instance, in 1983, G. Badurek, H. Rauch and J. Summhammer used a neutron interferometer to demonstrate the change of sign of a fermion's wave function after a rotation of angle 2π [6]. Another example is the demonstration of the role of potentials in quantum mechanics. In 1959, Y. Aharonov and D. Bohm showed that the fringe pattern in a two-slit interference experiment using charged particles should be shifted by altering the amount of magnetic flux passing between the two paths, even though the charged particle matter waves propagate only through regions where the magnetic field $\mathscr{B}$ vanishes (so that there is no classical force). However, the vector potential $\mathbf{A}$ is not zero, and is of direct physical relevance. In particular, an infinite solenoid piercing a charged particle interferometer (see Fig. 16.23) induces a phase shift proportional to the enclosed flux. This *vector Aharonov–Bohm*

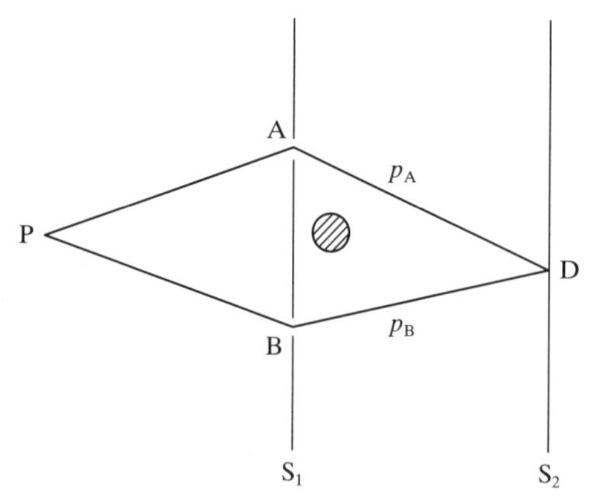

Figure 16.23 An Aharonov–Bohm experiment. Charged particles from a source P impinge on a screen S_1 with two slits A and B and are detected at D on the second screen S_2. The shaded area represents a thin solenoid which is perpendicular to the plane of the paper.

[6] The transformation properties of general spin functions under rotations are discussed in Bransden and Joachain (2000). When a rotation of angle 2π is made, the wave functions corresponding to particles of zero or integer spin (bosons) return to their original values, while those corresponding to half-odd integer spin $s = 1/2, 3/2, \ldots$ (fermions) change sign.

effect [7] was observed in an electron interferometer by R.G. Chambers in 1960, and by A. Tonomura *et al.* in 1982. An analogous effect, produced by a time-dependent scalar potential, is called the *scalar Aharonov–Bohm effect.*

The vector and scalar Aharonov–Bohm effects are particular examples of topological effects, due to the existence of topological phases, also called *Berry phases* [7]. Atom interferometers have been used to measure several topological phase shifts. For example, the scalar Aharonov–Bohm effect was investigated in 1994 by J. Baudon *et al.* using their longitudinal Stern–Gerlach interferometer, and in 1997 by W. Ertmer and co-workers using a Ramsey–Bordé interferometer. Ertmer *et al.* also studied the *Aharonov–Casher effect*, due to the atomic moment in an electric field.

Atom interferometers are ideally suited to study *entangled states*. An interesting example is the influence of spontaneous emission on the interference pattern [8]. On spontaneous decay, the centre of mass wave function describing the external atomic motion acquires a phase shift which depends on the direction of emission of the emitted photon. Assuming that every atom decays immediately after the double slit, and selecting atoms which emit a photon in a direction θ, then for a slit separation d, the wave function after spontaneous emission can be written in the form

$$\psi = \frac{1}{\sqrt{2}}\left[\exp\left(-\mathrm{i}\frac{k_{\mathrm{L}}d}{2}\cos\theta\right)\psi_1 + \exp\left(\mathrm{i}\frac{k_{\mathrm{L}}d}{2}\cos\theta\right)\psi_2\right] \qquad \textbf{(16.56)}$$

where ψ_1 and ψ_2 refer to the atomic centre of mass wave functions for the two paths, and k_{L} is the magnitude of the wave vector of the emitted photon. We see from (16.56) that the atomic centre of mass wave function and the state of the emitted photon form an entangled state. It can be shown [8] that the resulting interference pattern is displaced with respect to the interference pattern without spontaneous emission by $d \cos \theta/\lambda$ times the fringe separation.

Entangled states that persist after atom and field become spatially separated provide a way to study the *non-local* character of quantum theory. They can also be used to perform *delayed-choice* experiments [9]. The basic idea of such experiments [10] is to build an interference pattern particle by particle (with no more than one particle present at any time between the source and the detector) while the configuration of the interferometer, especially the beam splitters, is changed (chosen) just before the particle arrives in that part of the interferometer. According to the standard interpretation of quantum mechanics, the result should be identical to that obtained with an experimental setting already fixed in the final

[7] The vector Aharonov–Bohm effect and the Berry phase are discussed in Bransden and Joachain (2000).

[8] See Adams, Carnal and Mlynek (1994).

[9] See Wheeler (1978).

[10] See Baudon, Mathevet and Robert (1999).

state adopted in the delayed-choice experiment. The first investigations in this domain, made in 1987 by T. Hellmut, A.G. Zajong and H. Walther, used photons in a Mach–Zehnder interferometer whose second beam splitter was switched on and off just before the arrival of the photon. In this case, the delayed choice is either an interference when the beam splitter is switched on, or no interference when it is switched off. The experimental results were found to be in good agreement with the standard interpretation of quantum mechanics. A similar experiment was performed in 1996 by J. Baudon *et al.*, who used their longitudinal Stern–Gerlach atom interferometer; it led to the same conclusions.

16.3 Atoms in cavities and ions in traps

In this section, we shall discuss the behaviour of atoms or ions which are confined in small regions of space.

Atoms in cavities

The interaction of atoms with the electromagnetic field inside a cavity is the subject of *cavity quantum electrodynamics*. One of the most interesting physical effects, demonstrated in the 1970s, is that the spontaneous decay rate of an excited atom is modified in a cavity (see for example Hänsch and Walther, 1999). If W^{s} is the spontaneous decay rate of the atom in free space and W^{s}_{c} is the corresponding spontaneous decay rate in the cavity, it can be shown that

$$\frac{W^{s}_{c}}{W^{s}} = \frac{\lambda^3 Q}{4\pi^2 V_c} \tag{16.57}$$

where λ is the wavelength of the atomic transition, Q is the quality factor of the cavity [11] and V_c its volume. For low-order cavities in the microwave region, $V_c \simeq \lambda^3$ so that the spontaneous decay rate W^{s}_{c} is increased by about a factor of Q. On the other hand, if the cavity is detuned, the spontaneous decay rate W^{s}_{c} decreases, since the cavity can only accept resonant photons.

The one-atom maser

The simplest and most fundamental system in which the interaction of radiation with matter can be studied is a single two-level atom interacting with a

[11] The quality factor Q of a cavity is a measure of the sharpness of response of the cavity to external excitation. It is defined (see for example Jackson, 1998) to be 2π times the ratio of the time-averaged energy stored in the cavity to the energy loss per cycle. It is given by the dimensionless ratio

$$Q = \omega_0 \frac{\text{Energy stored}}{\text{Power loss}}$$

where ω_0 is the resonance angular frequency, assuming no losses.

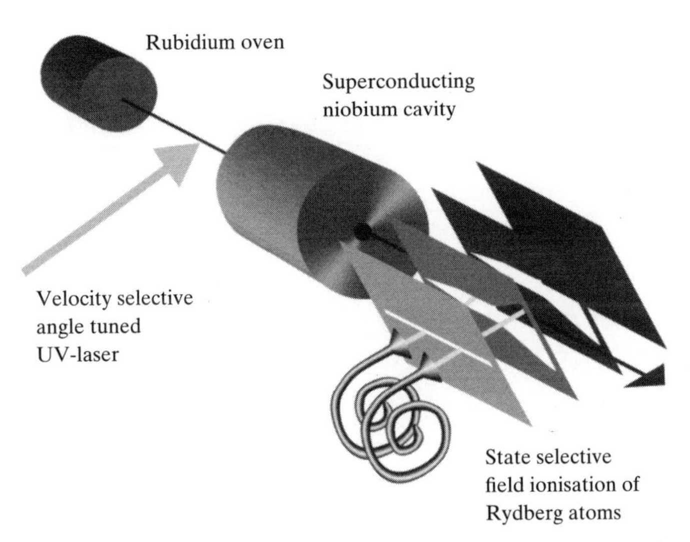

Figure 16.24 The one-atom maser experimental set-up. The velocity of the rubidium atoms is controlled by exciting a velocity subgroup of atoms in the atomic beam. The atoms in the uper and lower maser levels are selectively detected by field ionisation. (By courtesy of H. Walther.)

single-mode electromagnetic field. For many years, this system was not amenable to experimental study, since in the experiments it was necessary to have large numbers of atoms and photons. The situation changed drastically when the advent of tunable lasers allowed the excitation of highly excited Rydberg states of atoms. Such Rydberg atoms are very suitable for studying radiation–atom coupling for three reasons. First, these states are strongly coupled to the radiation field, the induced transition rates between neighbouring levels being proportional to n^4, where n denotes the principal quantum number. As a result, a single photon may be sufficient to saturate the transition. Second, the transitions to neighbouring levels are in the millimetre wave region, so that low-order mode cavities can be made large enough to ensure rather long interaction times. Third, highly excited Rydberg atoms have relatively long lifetimes with respect to spontaneous decay, so that only the coupling with the chosen cavity mode is important.

In 1985, D. Meschede, H. Walther and G. Müller demonstrated the first *one-atom maser* or *micromaser* by injecting Rydberg atoms into a superconducting cavity with a high quality factor Q. The injection rate was such that on the average there was less than one atom present in the cavity at any time. A recent version of the experimental arrangement is shown in Fig. 16.24. A highly collimated beam of rubidium atoms originating from an oven is velocity selected. The maser is pumped by ^{85}Rb atoms, the ^{85}Rb isotope being the most abundant (70%). Before entering the superconducting cavity, the atoms are excited from the 5 $^2S_{1/2}$ ground state to the upper maser level 63 $^2P_{3/2}$ by the UV light of a dye laser (see Fig. 16.25). This laser light is linearly polarised parallel to the electric field of the

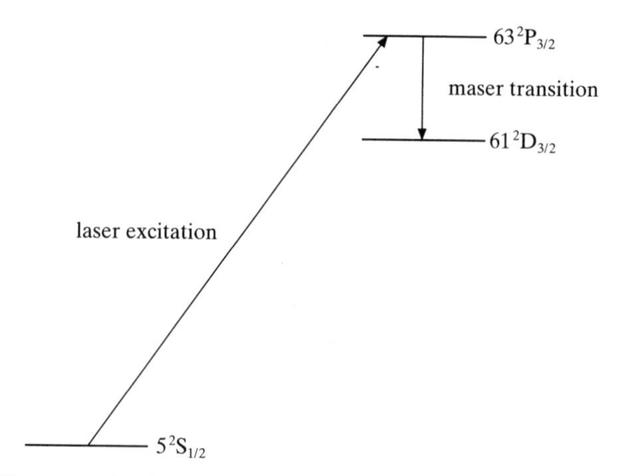

Figure 16.25 The energy levels relevant for the rubidium one-atom maser. The maser transition frequency is 21.50658 GHz.

cavity, so that only $\Delta M_J = 0$ transitions are excited by both the laser beam and the cavity microwave field. A superconducting niobium maser cavity resonant with the transition $63\,^2P_{3/2} \rightarrow 61\,^2D_{3/2}$ is used; the corresponding resonance frequency is 21.50658 GHz. The cavity is cooled to temperatures $T \simeq 0.5$ K where the number of thermal photons in the cavity is about 0.15 at the resonance frequency. The quality factor of the cavity is $Q = 3 \times 10^{10}$. The velocity of the Rydberg atoms and hence their interaction time with the cavity field are preselected by exciting a particular velocity subgroup with the laser. This is done by irradiating the atomic beam at an angle of approximately 82°. As a result, the UV laser light is blue-shifted by 50–200 MHz by the Doppler effect, depending on the velocity of the atoms. The Rydberg atoms in the upper and lower maser levels are detected after leaving the cavity by two separate field ionisation detectors. The field strength is adjusted so that in the first detector the atoms in the upper level are ionised, but not those in the lower level. To demonstrate maser operation, the cavity is tuned over the $63\,^2P_{3/2} \rightarrow 61\,^2D_{3/2}$ transition and the flux of atoms in the upper maser level is recorded simultaneously. Transitions from the initially prepared $63\,^2P_{3/2}$ state to the $61\,^2D_{3/2}$ state are detected by a reduction of the electron count rate.

As seen from Fig. 16.26, at a cavity temperature of 0.5 K, the reduction of the field ionisation signal from the $63\,^2P_{3/2}$ state at the resonance frequency is clearly seen for a range of atomic fluxes, down to values as small as 1750 atoms/s. An increase in flux causes broadening of the line and a small AC Stark shift, due mainly to the presence of neighbouring Rydberg states. Over the ranges of atomic fluxes, from 1750 to 28 000 atoms/s, the signal at resonance is independent of the atomic flux, showing that the transition is saturated. This effect, and the observed line broadening, demonstrate that there is a multiple exchange of photons between Rydberg atoms and the cavity field.

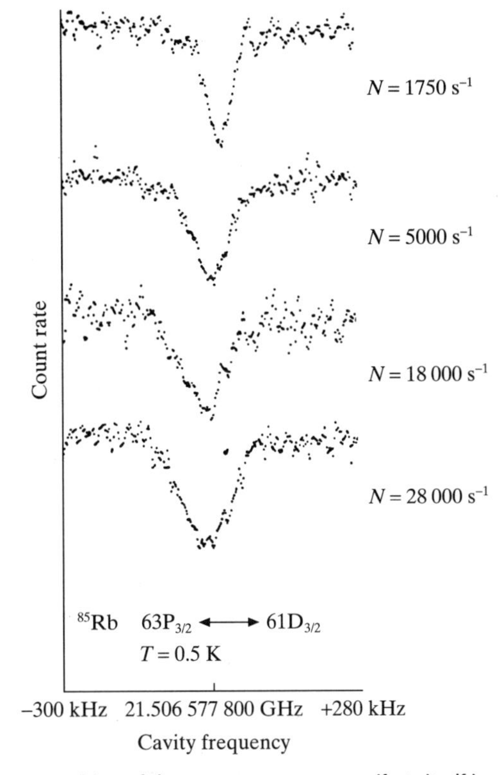

Figure 16.26 A maser transition of the one-atom maser manifests itself in a decrease of atoms in the excited state. The flux of excited atoms N governs the pump intensity. Power broadening of the resonance line demonstrates the multiple exchange of a photon between the cavity field and the atom passing through the resonator. (From G. Rempe, M.O. Scully and H. Walther, *Physica Scripta* T **34**, 5, 1991.)

The average transit time of the Rydberg atoms through the cavity is 50 μs, which implies that if the flux is 1750 atoms/s, then, on the average, about 0.09 Rydberg atoms are in the cavity at any time. It follows that more than 90% of the events are due to single atoms interacting with the microwave field. This clearly demonstrates that single atoms can sustain a continuous oscillation in the cavity.

Applications of the one-atom maser

The one-atom maser has been used by H. Walther and co-workers to make detailed studies of the dynamics of a single atom in a quantised electromagnetic field. The steady-state field of the micromaser has been shown to exhibit non-classical features. The one-atom maser has also been used to perform atomic interferometry experiments. A detailed account of the subject has been given by Walther (1994) and Scully and Zubairy (1997).

The one-atom laser

The concept of the one-atom maser can be extended to the optical region. In 1994, M.S. Feld and co-workers developed a one-atom laser. In their experiment, an atomic beam traversed a single-mode cavity (see Fig. 16.27(a)) formed by using two mirrors M1 and M2, each having a 10 cm radius of curvature, placed 1 mm apart. The mirrors used were highly reflecting, with reflectivities of 0.999 996, so that a single photon emitted into such a cavity was typically stored for about 800 ns before it escaped. The quality factor of the cavity was measured to be $Q = 8 \times 10^5$, corresponding to a line width of 190 kHz. The laser medium consisted

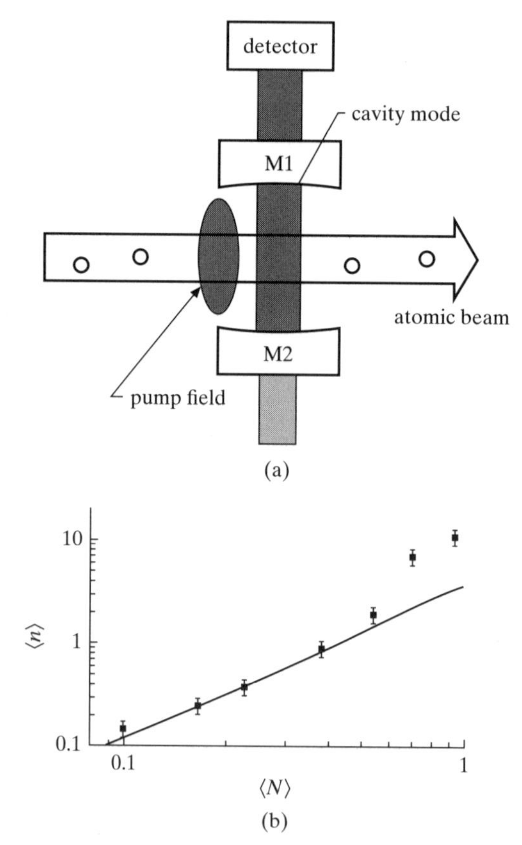

Figure 16.27 (a) Schematic diagram of the one-atom laser experiment of M.S. Feld *et al.* A beam of ^{138}Ba atoms traverses a single-mode cavity composed of two highly reflecting mirrors M1 and M2. The pump field is produced by a continuous-wave Ti:sapphire laser. The output of the microlaser is detected by using a photodiode. (b) The average number $\langle n \rangle$ of photons in the cavity mode, as a function of the average number $\langle N \rangle$ of atoms in the mode. The solid line corresponds to the results of a theoretical model developed by P. Filipowicz *et al.* (From K. An, J.J. Childs, R.R. Dasari and M.S. Feld, *Physical Review Letters* **73**, 3375, 1994.)

of ^{138}Ba atoms, produced in an atomic beam oven about 40 cm away from the cavity. Before they entered the cavity, the barium atoms were excited from the $^1S_0$ ground state to the $^3P_1$ excited state by means of a continuous-wave (cw) Ti:sapphire pump laser. The waist and intensity of the pump beam were chosen so that the velocity group of atoms with the most probable speed was subjected to a π pulse, thereby achieving maximum population in the excited $^3P_1$ state. Since their average velocity was about 320 m s^{-1}, the atoms were still excited when they entered the cavity.

Laser oscillation was obtained using the $^3P_1 \to {}^1S_0$ transition at the 'near visible' wavelength $\lambda = 791$ nm. Inside the cavity, the enhanced vacuum fluctuations between the mirrors triggered the atom to emit a photon of wavelength 791 nm into the resonator. Given the short transit time of about 250 ns and an initially empty cavity, the probability of photon emission is 0.25. Thus, four atoms were needed on the average to emit a first photon into the cavity. This first photon, stored in the cavity, could interact with the next atom entering the resonator, and stimulate the emission of a second photon with a probability of 0.5. This process went on until a steady state was established due to mirror transmission and loss, with an average number $\langle n \rangle$ of photons in the cavity mode ranging from 0.14 to 11 for an average number $\langle N \rangle$ of atoms in the cavity mode between 0.1 and 1.0.

The quantity $\langle n \rangle$ (which is proportional to the output power) is plotted in Fig. 16.27(b) as a function of the average number $\langle N \rangle$ of atoms in the mode (which is proportional to the pump power). The experimental data of M.S. Feld and co-workers are seen to be in good agreement with the results of a theoretical model developed in 1986 by P. Filipowicz, J. Javanainen and P. Meystre, for values of $\langle N \rangle$ less than 0.7. A striking property of the one-atom laser is that the plot of the output power versus pump power shows no distinctive threshold, unlike macroscopic lasers. The output power of the microlaser of M.S. Feld *et al.* is less than 3 pW.

One advantage of the one-atom laser compared with its microwave counterpart is that visible photons can be observed directly, while microwave photons are difficult to detect. However, it is easier to build and control a cavity at microwave wavelengths, which explains why the one-atom maser was developed first.

Ions in traps

In contrast to neutral atoms which we have been considering in this section, ions can easily be manipulated by electromagnetic fields because of their charge. The first ion trap was invented in 1958 by W. Paul, O. Osberghaus and E. Fischer. Since that time, several other kinds of traps have been developed [12] such as linear traps, ring traps and the electron beam ion trap (EBIT). Ion traps have been particularly useful in the fields of quantum optics, precision measurements and the physics of multiply charged ions.

[12] Detailed reviews have been given by Thompson (1993) and Walther (1994).

It is not possible to confine charged particles with an electrostatic field alone, because of Earnshaw's theorem. However, one can overcome this difficulty by using either a time-dependent electric field, as in the *Paul trap*, or a combination of static electric and magnetic fields, as in the *Penning trap*.

The Paul trap or radio-frequency trap

This trap uses three electrodes: two endcaps and one ring. It requires a combination of DC and AC voltages to be applied to the electrodes. The electrode arrangement shown in Fig. 16.28 gives rise to an electrostatic potential

$$V = A(r^2 - 2z^2) \tag{16.58}$$

which varies quadratically in every direction. In the above equation, A is proportional to the voltage applied to the ring electrode. The three electrodes are cut to follow the equipotentials of (16.58) and are conjugate hyperboloids of revolution about the Z axis.

Trapping can be obtained by applying an AC voltage between the ring and the endcaps, at a frequency Ω. As shown in 1959 by E. Fischer, this gives rise to a 'pseudopotential' which, unlike the electrostatic potential V, can have a three-dimensional minimum at the origin. The AC frequency Ω is typically in the radio-frequency region, between 100 kHz and 10 MHz, so that the Paul trap is also called a radio-frequency trap. An additional DC voltage can also be used to modify the relation of the potential depth along the symmetry axis (the Z axis) to that in a perpendicular direction. In a Paul trap, the radio-frequency micromotion of ions vanishes only for a single cooled ion at the trap centre.

Although originally Paul traps were made using electrodes having the shapes shown in Fig. 16.28, it turns out that for many applications this requirement is not essential. In particular, because of their small size, *miniature traps* (having an internal diameter of the order of 1 mm) have been shaped out of rounded wires

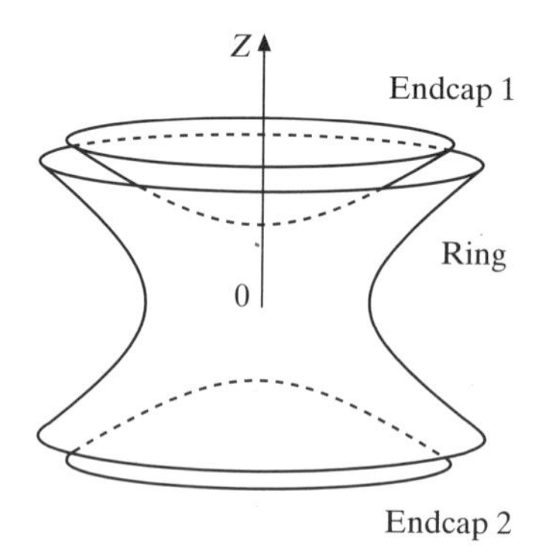

Figure 16.28 Electrode structure of the Paul trap.

for endcaps and a loop of wire for the ring. Such miniature traps, introduced in 1978 by W. Neuhauser, M. Hohenstatt, P.E. Toschek and H.G. Dehmelt, work very well, since they can be optimised to produce a potential that closely follows the desired quadratic dependence around the centre of the trap.

The Penning trap

This trap uses the same arrangement of electrodes as in the Paul trap, but a static axial magnetic field is applied in addition to a DC potential applied between the ring and cap electrodes. It was devised in 1967 by H.G. Dehmelt, who named it the Penning trap after F.M. Penning, who in 1936 had discovered the principles that could be used for trapping ions in this way. In contrast with the Paul trap, there is no radio-frequency heating effect, so that large clouds of ions (of the order of 10^5) can be cooled in a Penning trap using laser cooling.

Quantum jumps of a single ion

The use of ion traps, together with the developments of laser cooling and spectroscopy, have opened the possibility of studying the interaction of *single ions* with radiation in an unperturbed environment, over very long times. Indeed, it is now possible to isolate a single ion in an ultra-high vacuum, reduce its kinetic energy to the zero-point energy, and localise its position to better than 10^{-6} m. In this way, the internal states of the ion can be decoupled to a high degree of accuracy from perturbations due to collisions and Doppler shifts. Extremely high resolution, of the order of 10^{-15}, can be obtained if very weak transitions, for example transitions from the ground state to a long-lived (metastable) state, are studied in cooled ions by using narrow-band laser light. The main problem is then to detect such weak transitions, whose fluorescence yield is very low (for example, 1 photon per 10 seconds or even less). To solve this problem, H.G. Dehmelt proposed in 1975 to couple the weak transition via the ground state to a strong dipole-allowed transition whose fluorescence yield is much larger (for example, 10^8 photons per second). This strong transition is induced by a laser tuned to the transition resonance frequency. The key idea is that the fluorescence of the strong transition is interrupted as soon as the ion is excited to the metastable state, where it is 'shelved' on the average during the lifetime of that state, which therefore is called a *shelving state*. This can be accomplished by tuning a second laser to the very narrow resonance of the weak transition from the ground state to the metastable (shelving) state. The abrupt disappearance of the fluorescence signal indicates when the transition to the shelving state has taken place, that is when the ion has made a *quantum jump*. After a certain time, the ion will make another quantum jump to return to the ground state, and the fluorescence from the strong transition will resume. The method is illustrated in Fig. 16.29 for the case of the Hg^+ ion which was investigated in 1986 by D.J. Wineland and co-workers. The relevant energy levels are shown in Fig. 16.29(a). The strong transition at a wavelength of

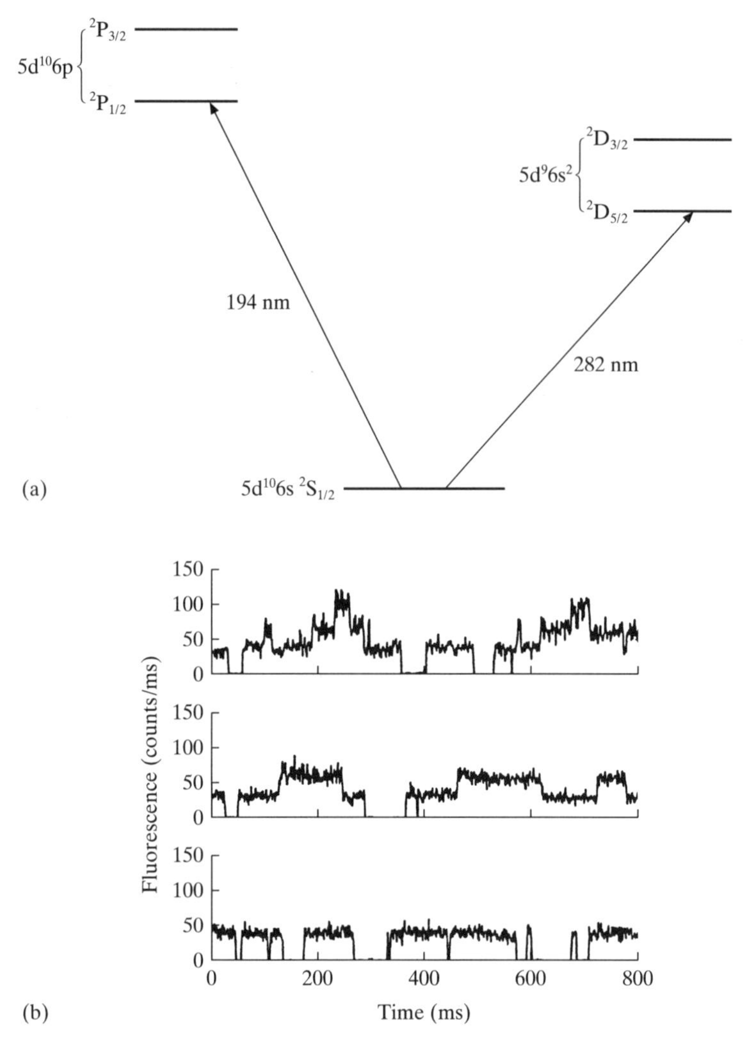

Figure 16.29 (a) The energy levels relevant for the discussion of quantum jumps in the Hg^+ ion. The strong transition at a wavelength of 194 nm and the weak transition at a wavelength of 282 nm are indicated. (b) Fluorescence light recorded at the wavelength of 194 nm, illustrating quantum jumps in three ions (top), two ions (middle) and one ion (bottom). (From J.C. Bergquist, F. Diedrich, W.M. Itano and D.J. Wineland, in *Laser Spectroscopy* IX, ed. M.S. Feld, J.E. Thomas and A. Mooradian, Academic Press, Boston, 1989.)

194 nm is between the $5d^{10}6s$ $^2S_{1/2}$ ground state and the $5d^{10}6p$ $^2P_{1/2}$ excited state, while the metastable (shelving) state is the $5d^96s^2$ $^2D_{5/2}$ state, the weak transition occurring at a wavelength of 282 nm. In Fig. 16.29(b) the fluorescence signal at 194 nm is shown as a function of time, illustrating quantum jumps in three ions, two ions and one ion.

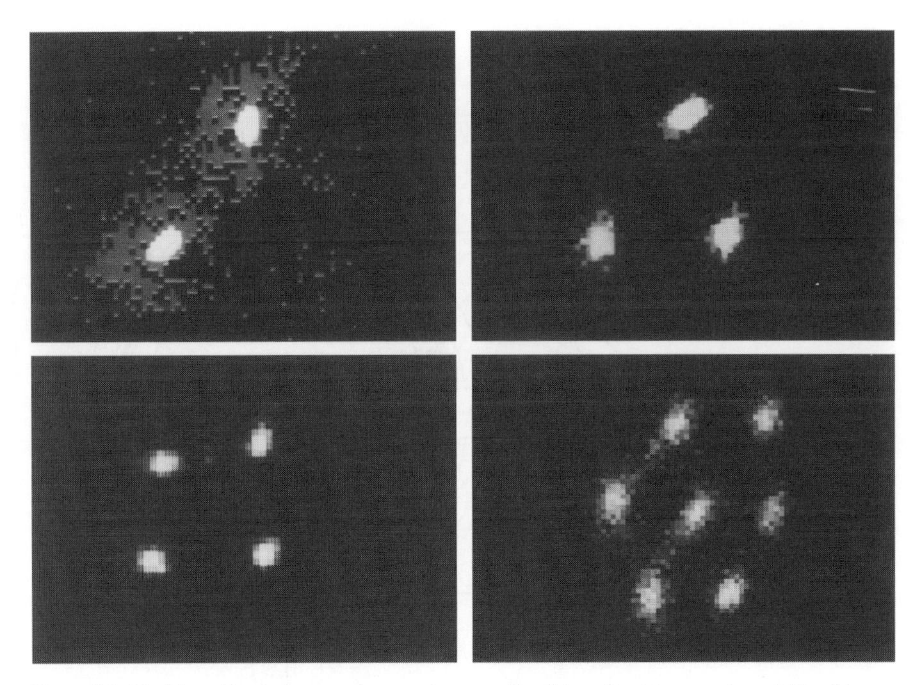

Figure 16.30 Two, three, four and seven ions confined in a Paul trap and crystallised to an ordered structure by using UV laser light tuned slightly below the resonance frequency. The average separation of the ions is 20 μm. (By courtesy of H. Walther.)

Crystallisation of laser-cooled ions in a trap. The ion storage ring

The first observation of the crystallisation of laser-cooled Mg^+ ions in a Paul trap (see Fig. 16.30) was reported in 1987 by F. Diedrich and H. Walther. The ions are laser-cooled to such a low velocity that the mutual Coulomb repulsion is compensated by the trap potential, and an ordered arrangement is obtained.

Phase transitions and ordered structures of laser-cooled Mg^+ ions stored in a miniature radio-frequency *quadrupole storage ring* were observed in 1992 by I. Waki, S. Kassner, G. Birkl and H. Walther. Their experimental set-up is shown in Fig. 16.31. The storage ring consisted of four circular electrodes and the diameter of the toroidal storage volume was 115 mm. The ions were confined in two dimensions by a quadrupole field and could move freely along the ring circumference. By cooling the ions with a laser beam, crystalline structures were observed along the centre line of the ring. Examples of these structures are shown in Fig. 16.32. At low ion densities, the ions form a string (see Fig. 16.32(a)). Increasing the ion density leads to a zigzag configuration (see Fig. 16.32(b)), while at still higher densities ordered helical structures are formed (see Fig. 16.32(c)). A detailed discussion of phase transitions of stored laser-cooled ions has been given by Walther (1993).

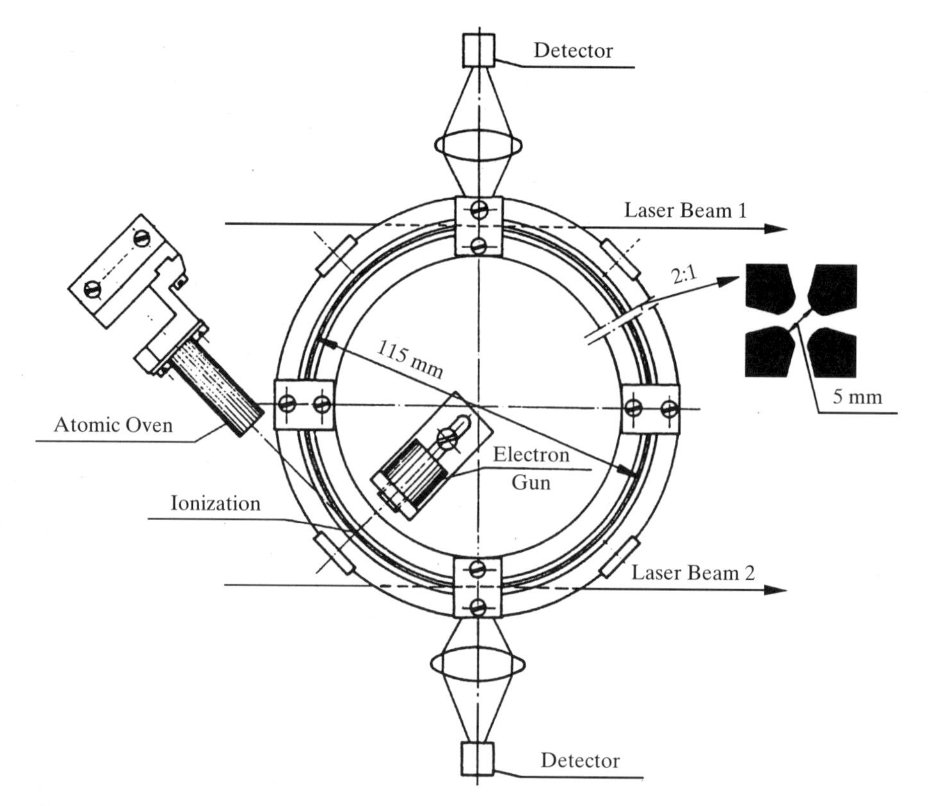

Figure 16.31 Experimental set-up of the miniature quadrupole storage ring of I. Waki *et al.*, including the atomic beam oven and electron gun. The storage ring consists of four circular electrodes. The diameter of the toroidal storage volume is 115 mm. The inset shows an enlarged cross-section with opposite electrodes having a separation of 5 mm. The laser beam enters the storage volume tangentially. Resonance fluorescence is detected with a photomultiplier tube or an imaging photon detector system. (From I. Waki, S. Kassner, G. Birkl and H. Walther, *Physical Review Letters* **68**, 2007, 1992.)

Cold trapped ions and quantum information

It was suggested in 1995 by J.I. Cirac and P. Zoller that a quantum computer can be implemented with cold ions confined in a linear trap and interacting with laser beams. In this scheme, quantum gates involving any pair, triplet or subset of ions are realised by coupling the ions through their collective quantised motion. As an illustration, a linear chain of ions is shown in Fig. 16.33. The position of the ions is marked by means of the fluorescence emitted when they are illuminated by a laser beam. Such an ion chain has been proposed as the register of a quantum computer.

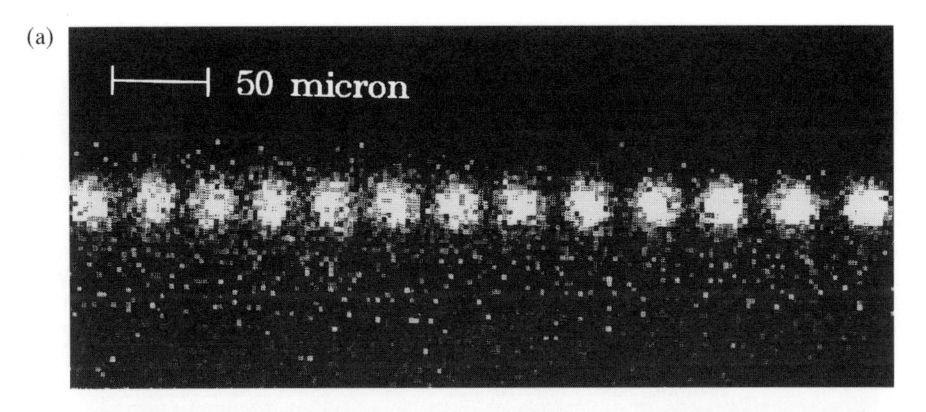

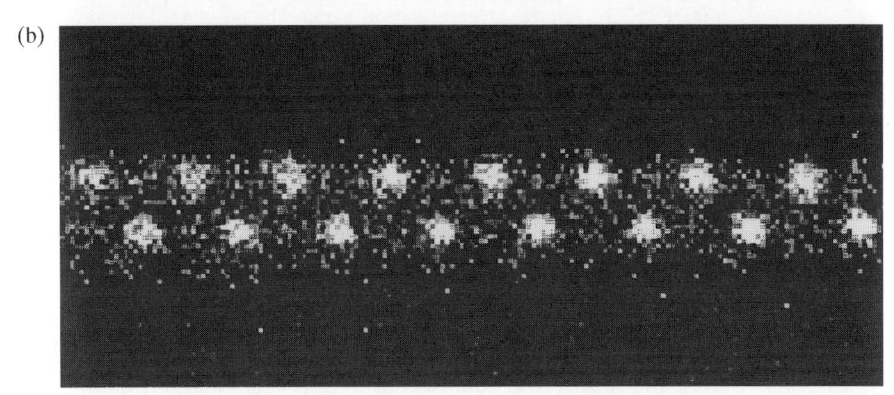

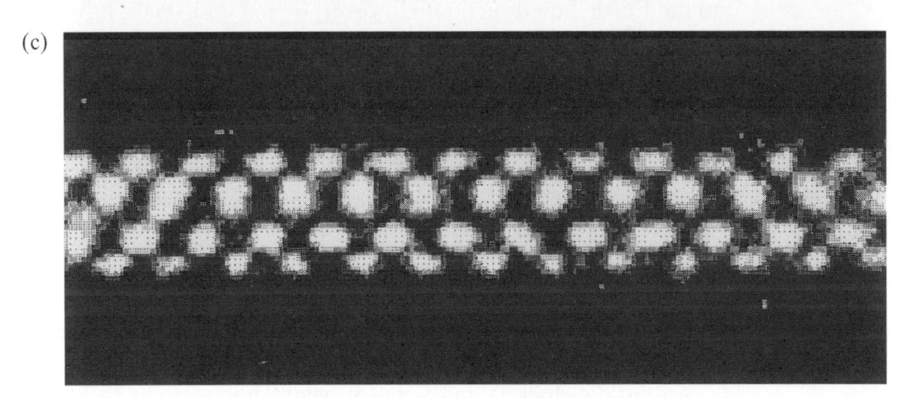

Figure 16.32 Images of crystalline structures of laser-cooled $^{24}Mg^{+}$ ions in the quadrupole storage ring. Individual ions could be resolved in these images. The ions arrange themselves in minimum energy configurations. (a) At a low ion density the ions form a string along the field axis. (b) Increasing the ion density changes the configuration to a zigzag. (c) At still higher ion densities, the ions form ordered helical structures on the surface of a cyclinder, for example three interwoven helices. (By courtesy of H. Walther.)

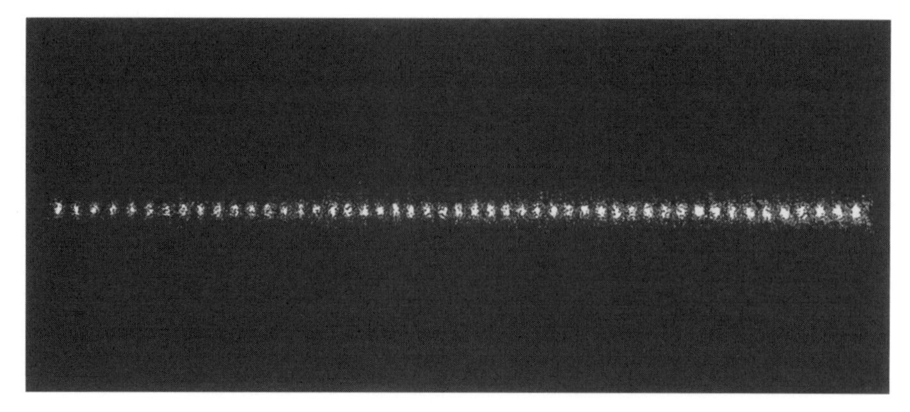

Figure 16.33 A linear chain of ions in a trap. The position of the ions is marked by the fluorescence radiation they emit when illuminated by laser light. (By courtesy of H. Walther.)

16.4 Atomic clocks

Any clock or frequency standard depends on some regular, periodic motion such as that of a pendulum. Astronomical observations were first used to define the international unit of time: the *second*. In particular, the definition of the 'ephemeris second', adopted in 1956, was based on the length of the 'tropical year', that is the length of time taken by the Earth to complete its orbit around the Sun and return to a point where its axis again makes the same angle with respect to the Earth–Sun direction. However, by the mid-1960s, it became clear that this definition was inadequate for accurate time keeping, and that atomic clocks [13] were far superior to all previous clocks. Indeed, any unperturbed transition is identical from atom to atom, so that two atomic clocks based on such a transition should generate the same time. Moreover, in contrast to macroscopic devices, atoms do not wear out. In 1967, the 13th General Conference on Weights and Measures adopted the following new definition of the international unit of time: 'The second is the duration of 9 192 631 770 periods of the radiation corresponding to the transition between the two hyperfine levels of the fundamental state of the atom of caesium-133'. The ten-digit number assigned in this new definition was chosen to agree with the then existing definition of the 'ephemeris second' to the extent that measurement allowed.

An atomic clock is an oscillator whose frequency is locked to the resonance frequency $\nu = (E_b - E_a)/h$ of an atomic transition between two states a and b having energies E_a and E_b, respectively, with $E_b > E_a$. Thus, in atomic clocks, the

[13] The first 'atomic clock' was in fact a molecular frequency standard, developed in 1949 by H. Lyons *et al.* at the US National Bureau of Standards. It was based on the inversion transition in ammonia, which, as we have seen in Sections 11.6 and 15.1, occurs at a frequency of about 23 870 MHz.

periodic motion is an internal one. The atoms interact with an electromagnetic wave generated by a macroscopic oscillator, and the transition rate between the two states a and b is measured, for example by using the deflection of the atoms in a magnetic field, or by laser-induced fluorescence. This transition rate is largest when the frequency ν of the electromagnetic wave is equal to the resonance frequency of an atomic transition. It can therefore be used to lock the frequency of the oscillator to the atomic resonance frequency. As we have seen above, for the caesium clock this resonance frequency is defined to be 9 192 631 770 Hz. The tuned signal is then used as a frequency reference, a time interval being measured by counting electronically the number of oscillations (which correspond to the 'ticks' of a conventional clock). For example, a time interval of one second corresponds to 9 192 631 770 oscillations of a signal whose frequency is locked to that of the caesium clock transition.

Stability and accuracy

The main characteristics on which the performance of a clock is judged are its *stability* and *accuracy*. The stability of a clock refers to the consistency of a set of measurements of the same time interval. Due to the clock's finite stability, all the measurements will be slightly different from each other. The frequency stability is determined by the width of the resonance and by the noise in the measurement of the transition rate. The accuracy of a clock is the precision with which each time interval can be determined. Because of the clock's finite accuracy, each time interval will be systematically too short or too long. It should be noted that accuracy and stability are not independent quantities. Indeed, a poor accuracy implies that some frequency shifts are not well controlled and may fluctuate, which limits the stability of the clock. Conversely, frequency shifts cannot be accurately measured if the clock is unstable.

Let us consider atoms which are illuminated with radiation from a tunable source operating near the resonance frequency ν of the clock transition between two atomic states. If a number of absorption measurements are performed for a range of frequencies, a resonance curve can be recorded, and the frequency corresponding to maximum absorption can be determined. In practice, however, this determination is limited by noise. With proper care, electronic and other sources of noise can be reduced, so that the measurement precision is ultimately limited by *quantum projection noise* [14]. This latter noise, which was studied in 1993 by W.M. Itano and co-workers, arises from the statistical nature of projecting a superposition of two quantum states into one state when a measurement is made. After interacting with the radiation, the atoms are in a coherent superposition of the two clock states. On detection, any single attempt to measure the state composition will yield only one of the states. It is impossible to predict with certainty the final result of the measurements due to *quantum fluctuations*. However, these

[14] See for example Bergquist, Jefferts and Wineland (2001).

fluctuations decrease proportionally to the square root of the number of detected atoms. As a result, the average of measurements, performed repeatedly over a single atom, or over many atoms in an ensemble, will yield a result with a precision which increases as the square root of the number M of measurements and the square root of the number N of atoms. The fractional frequency uncertainty of the measurement can then be written as [14]

$$\frac{\Delta\nu_{\mathrm{m}}}{\nu} = \frac{C}{2\pi\nu T\sqrt{NM}} = \frac{C}{2\pi\nu\sqrt{NT\tau}} \qquad \textbf{(16.59)}$$

where C is a constant of order unity, T is the interrogation time (that is, the time during which the clock radiation is applied before each measurement) and τ is the total measurement time (over many measurements). In deriving the result (16.59), it has been assumed that the time between interrogations is small compared to T. It follows from (16.59) that the stability of an atomic clock is increased when the frequency ν, the number of atoms N, the interrogation time T and the total measurement time τ are increased. The stability of a clock as a function of τ is usually specified in terms of a quantity $\sigma(\tau)$, known as the square root of the *Allan variance* [15], which is proportional to $\tau^{-1/2}$. Accuracy is maximised when the uncertainly $\Delta\nu_{\mathrm{a}}$ in determining the resonance frequency ν is minimised, or in other words when the fractional frequency inaccuracy $\Delta\nu_{\mathrm{a}}/\nu$ is minimised. Thus, in order to obtain stable and accurate atomic clocks, environmental perturbations affecting the measured transition frequency (such as atomic collisions, Doppler shifts or magnetic field effects) must be carefully controlled and minimised.

The atomic beam caesium clock

We shall now examine a few atomic clocks, starting with the caesium clock, which is used as a time and frequency standard. The only stable isotope of caesium is ^{133}Cs, which has a nuclear spin quantum number $I = 7/2$ (see Table 5.1). In the $6\,^2S_{1/2}$ ground state of the atom, the total electronic angular momentum is $J = 1/2$, so that the total angular momentum quantum number of the atom can take the values $F = 3$ and $F = 4$. The atomic ground level therefore splits into two hyperfine components, the state with $F = 3$ being the ground state. When a weak magnetic field is applied, each of the two hyperfine levels $F = 3$ and $F = 4$ is split into Zeeman sublevels.

The transition frequency used for the Cs atomic clock corresponds to the transition between the substates $F = 4$, $M_F = 0$ and $F = 3$, $M_F = 0$, extrapolated to zero magnetic field. The magnetic sublevels with $M_F = 0$ are chosen because, in first approximation, they are independent of magnetic field effects in the vicinity of zero field, and it is difficult in practice to shield out the perturbations due to a magnetic field. The transition frequency can be measured with a Ramsey separated oscillatory field magnetic resonance apparatus (see Section 16.1), the

[15] See for example Major (1998).

magnets being used to reject all Zeeman substates except for the $F = 4$, $M_F = 0$ and $F = 3$, $M_F = 0$ substates corresponding to the clock transition. The basic elements of an early caesium atomic beam clock using magnetic state selection are shown in Fig. 16.34(a). A beam of thermal caesium atoms coming from an oven travels in an evacuated tube where a high vacuum is maintained. This beam is passed through the strong magnetic field of a polarisation magnet A, where because of a steep transverse magnetic field gradient the atoms in the $F = 3$ state and those in the $F = 4$ state are deflected in opposite directions. By using suitable beam stops, the atoms in the $F = 3$ state are removed, leaving only those in the $F = 4$ state, among which are atoms in the desired $M_F = 0$ substate. After leaving the intense field of the state-selecting polarisation magnet A, the atoms in the $F = 4$ state pass through a transition region of gradually decreasing magnetic field (in order to avoid undesirable transitions between the magnetic substates), and enter

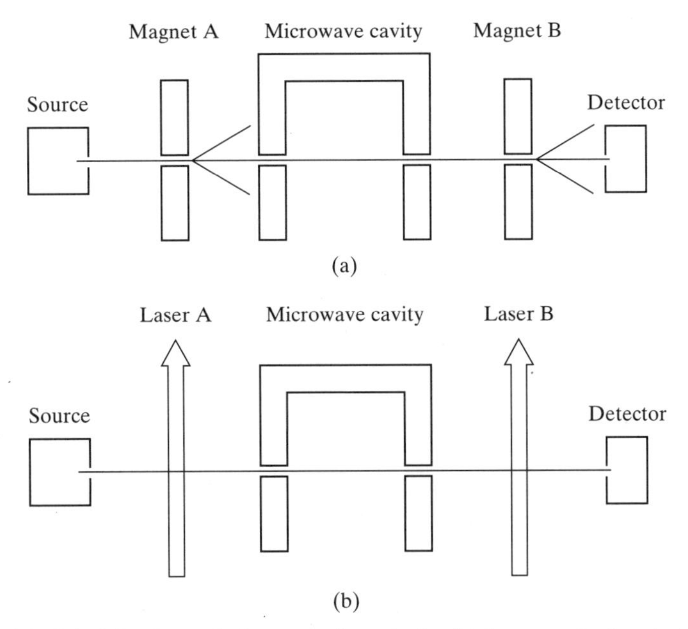

Figure 16.34 Principle of atomic beam caesium clocks. (a) In magnetically selected atomic beam caesium clocks, atoms from a thermal source are selected by the polarisation magnet A in one of the two clock states. After passing through a transition region of gradually decreasing magnetic field, they enter a microwave cavity. The microwave field oscillating at the resonance frequency drives some of the atoms into the second clock state. These atoms are deflected by the analysing magnet B towards the detector. (b) In optically pumped atomic beam caesium clocks, atoms from a thermal source are pumped into the first clock state $F = 3$ by passing through the beam of Laser A. They move through a transition region, passing into a microwave cavity where a microwave field oscillating at the resonance frequency drives some of the atoms into the second clock state $F = 4$. Only atoms having made the transition to the state $F = 4$ interact with the light from Laser B. The fluorescence radiation which they emit is detected by a photodetector.

a Ramsey separated oscillatory fields microwave cavity. If the microwave field applied there is off resonance with the atomic transition $F = 4$, $M_F = 0 \rightarrow F = 3$, $M_F = 0$, the atoms are deflected by the analysing magnet B away from the detector. On the other hand, if the microwave field is on resonance with the atomic transition, most of the atoms in the $F = 4$, $M_F = 0$ sublevel will make the transition to the $F = 3$, $M_F = 0$ sublevel by stimulated emission, and these atoms will be deflected by the analysing magnet B towards the detector. In this mode of design, called 'flop-in', only atoms that have made the transition are detected, the signal having a peak at the resonance frequency of 9.192 631 770 GHz, in the microwave region. Other designs are possible, such that the detector is exposed to the atoms that have not made the transition; in that case the detected signal exhibits a minimum when the microwave field is on resonance with the atomic transition.

The first commercial, magnetically selected, caesium beam atomic clock was developed in 1955 by J.R. Zacharias and co-workers. In the same year, L. Essen and V.L. Parry constructed the first apparatus of this kind that was used extensively as a frequency standard. A number of developments of the caesium beam atomic clock followed over the years. In particular, the stability and accuracy of magnetically selected atomic beam clocks were found to suffer in part from two defects. The first one is the relatively short interrogation time, limited by the time of flight of the thermal atoms in the apparatus, which is of the order of 10 ms for an atomic beam at room temperature. This leads to a resonance line width of the order of 100 Hz. We shall see below that this deficiency can be cured by using atomic fountains of cold atoms. The second defect of magnetically selected atomic beam clocks is inefficient state selection. It was remedied by using the *optical pumping* method (developed around 1950 by A. Kastler) for state preparation and detection of the ground state hyperfine levels. This method also avoids magnetic field inhomogeneities associated with magnetic state selection, and leads to a smaller value of the fractional frequency uncertainty $\Delta\nu_{\rm m}/\nu$. However, in the case of caesium, the optical pumping method became practical only after solid-state lasers had been developed at the end of the 1970s.

The first optically pumped caesium frequency standard constructed at the US National Bureau of Standards and Technology (NIST) became the US time and frequency standard in 1993. In this device, caesium atoms from a thermal source are pumped into the $F = 3$ hyperfine level by passing through a laser beam (see Fig. 16.34(b)). The atoms then transit through a 1 m long region, passing in a microwave cavity where a microwave field at the clock frequency is applied. Atoms which have made the clock transition to the $F = 4$ hyperfine level are detected by recording the fluorescence radiation they emit when interacting with another laser beam. The mean interrogation time in this apparatus is about 7 ms, which produces a resonance line width of approximately 70 Hz. This time standard yields values of $\Delta\nu_{\rm m}/\nu$ of about 5×10^{-15} for a total measurement time $\tau = 1$ day.

The rubidium clock

Another important atomic clock uses the ^{87}Rb isotope of rubidium. The nucleus of ^{87}Rb has a spin quantum number $I = 3/2$. In the 5 $^2S_{1/2}$ ground state of the atom, the total electronic angular momentum is $J = 1/2$, and therefore the total angular momentum quantum number of the atom can take the values $F = 1$ and $F = 2$. Thus the atomic ground level splits into two hyperfine components, the state with $F = 2$ being the ground state. If a weak magnetic field is applied, each of the two hyperfine levels $F = 1$ and $F = 2$ is split into Zeeman sublevels. The transition frequency used for the ^{87}Rb atomic clock corresponds to the transition between the substates $F = 2$, $M_F = 0$ and $F = 1$, $M_F = 0$. As in the caesium clock, the magnetic sublevels with $M_F = 0$ are chosen because, in first approximation, they are insensitive to magnetic field effects in the vicinity of zero field. The resonant frequency is 6.834 682 614 GHz, in the microwave region.

The rubidium clock uses rubidium vapour at low pressure, confined in a small absorption cell filled with a noble gas acting as a buffer gas which reduces the Doppler widths [16]. This method of confining the rubidium atoms has two advantages. The first one is that the interaction time of the atoms with the resonant field is increased. The second one is the compactness of the clock. In fact, miniature ^{87}Rb clocks have been developed for industrial applications. The disadvantages of size reduction are an increased rate of wall collisions or the need of a higher collision rate with the atoms of the buffer gas. In both cases, undesirable shifts in the measured transition frequency occur. As a result, the ^{87}Rb does not qualify as an absolute, primary time standard. Nevertheless, for many applications, it can be used as a secondary standard whose frequency must be set by reference to a primary standard.

The hydrogen clock

We have seen in Section 15.1 that the atomic hydrogen maser gives very accurate data for the frequency of the hyperfine splitting of the ground state of the hydrogen atom. The atomic hydrogen maser oscillator has also a very high stability (better than 10^{-15} over periods of several hours) which makes it one of the most stable atomic clocks. Its major disadvantage is that the hydrogen atoms collide with the walls at intervals, giving rise to a wall shift (see Section 15.1). This limits its absolute accuracy, so that the hydrogen clock is not used as a primary time standard.

Atomic fountain clocks

For a number of years, atomic clocks based on the principles described above have provided clocks of great accuracy and stability. However, as we shall see below, some new applications require even better clocks. Many developments are

---

[16] The narrowing of Doppler widths through collisions in an inert gas is known as the *Dicke effect*, after R.H. Dicke who made the first theoretical analysis of this phenomenon in 1953.

currently occurring in this area [17]. In what follows we shall briefly describe two routes to improved clocks.

The first one, which has led to a breakthrough in time metrology, is the advent of atomic clocks using fountains of laser-cooled atoms. Already in 1953, shortly after N.F. Ramsey had introduced the method of separated oscillatory fields, J.R. Zacharias proposed to make an atomic fountain in order to obtain a higher resolution. The basic idea is that the separated interactions can be achieved with a single microwave cavity. The atoms, launched vertically, pass once through the cavity on their way up and once again on their way down, following a ballistic trajectory. This allows a long interrogation time between the two passages, and hence would yield very narrow Ramsey resonances, particularly suitable for atomic clocks. Zacharias and co-workers attempted to build an atomic fountain 5 m high by directing upwards a thermal beam of caesium atoms coming from an oven. Unfortunately, they saw no signal and abandoned the experiment. It was later determined that the failure of their attempt was due to the lack of slow atoms. The faster atoms in the beam collided with the slower atoms and scattered them out of the beam.

The development of laser cooling of atoms revived the concept of atomic fountains, because all the atoms in the beam could be slowed. As we saw in Section 16.2, M.A. Kasevich *et al.* demonstrated the first atomic fountain of cold atoms. In their experiment, sodium atoms were captured in a magneto-optical trap, further cooled in optical molasses, and then pushed up by the radiation pressure of an upward-directed beam. They entered a waveguide, where the microwave transition corresponding to the hyperfine splitting of the sodium ground state was excited using Ramsey's method of separated oscillatory fields. After reaching their summit in the waveguide, the atoms fell because of gravity into a detection region. The relatively long measurement time of about 0.25 s allowed the hyperfine transition to be measured with a line width of only 2 Hz.

Two years later, an improved Zacharias atomic fountain was realised by A. Clairon, C. Salomon, S. Guellati and W.D. Phillips. Launched upwards from an optical molasses where they had been cooled to a temperature of 5.5 μK, caesium atoms were prepared in one of the two clock states, passed once through a microwave cavity where they interacted with the microwave field generated by an oscillator, reached the summit of their trajectory, and then fell through the same cavity, thus completing the Ramsey separated field interaction (see Fig. 16.35). The change of hyperfine state, due to the Ramsey resonance at the clock transition frequency, was detected by state-selective resonance fluorescence in the probe beam. In contrast to the radiation pressure launch performed in the experiment of M.A. Kasevich *et al.*, which has the disadvantage of heating the atoms, the launching of the atoms was done from a 'moving' molasses, created by detuning the two vertical molasses beams with opposite detunings while keeping the frequency ν_L of the horizontal laser beams unchanged. In the upward-travelling wave the laser

[17] See for example Lemonde (2001) and Bergquist, Jefferts and Wineland (2001).

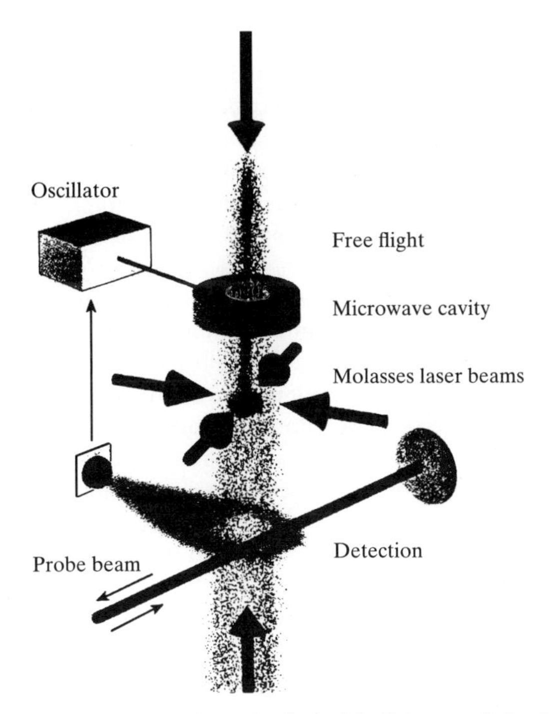

Figure 16.35 Principle of the atomic fountain clock of A. Clairon *et al.* Caesium atoms are first captured and cooled by three orthogonal pairs of counter-propagating laser beams producing optical molasses. They are then launched upwards from a 'moving' molasses obtained by shifting the frequency of the vertical laser beams while keeping the frequency ν_L of the horizontal beams unchanged. The light is then switched off and the atoms are prepared in one of the two clock states. They pass through a microwave cavity where they interact with the microwave field generated by an oscillator. They reach the summit of their trajectory, and then fall through the same microwave cavity. Finally, they are detected by state-selective resonance fluorescence, using a probe laser. The detection signal is used to lock the oscillator to the atomic frequency. (From P. Lemonde, *Physics World*, January 2001, p. 39.)

frequency is increased to $\nu_L + \Delta\nu_L$ and in the downward one it is decreased to $\nu_L - \Delta\nu_L$. This creates a standing wave moving upwards at a velocity $v = \lambda_L\Delta\nu_L$, where λ_L is the laser wavelength. Because of the Doppler shift, atoms moving up at the same velocity v 'see' all beams at the same frequency ν_L. In this way, a static molasses is created in the reference frame moving up with a velocity v. If, in addition, the detuning $\Delta\nu_L$ is slowly swept in time, starting from a zero value, the atoms, initially at thermal equilibrium in the laboratory frame at a given temperature, will remain at the same temperature in the moving frame. Another improvement of the atomic fountain of Clairon *et al.* is that the atoms pass completely out of the microwave cavity on their way up, and then fall back through it, so that errors associated with the motion of the atoms with respect to the residual microwave radiation in the cavity can be assessed.

In the 1991 experiment of A. Clairon *et al.*, the caesium atoms spent 0.25 s in free flight above the cavity, and Ramsey fringes with a width of 2 Hz were measured. About 2×10^5 atoms contributed to the Ramsey fringe signal. The frequency standard deviation was found to be 50 mHz, corresponding to a relative accuracy of 5.5×10^{-12}. The short-term stability of the clock, evaluated by the square root of the Allan variance, was found to be $\sigma(\tau) = 3 \times 10^{-12}\tau^{-1/2}$ (where τ is the total measurement time in seconds), corresponding to a stability of 10^{-15} over one day, a value at least as good as in thermal beam caesium clocks existing at that time.

The pioneering experiments of A. Clairon *et al.* clearly showed the potential of atomic clocks using fountains of cold atoms. After several technical improvements were made, Clairon and his colleagues at the Paris Observatory confirmed this potential when they built in 1993 a fountain clock using cold caesium atoms, which became the first cold-atom fountain used to define the international standard of time. The current accuracy of this clock is 10^{-15} and its stability is 10^{-16} over one day.

Although fountain clocks using cold caesium atoms are at the present time the most accurate clocks, there are several factors limiting their stability and accuracy. One of them is the frequency shift due to collisions between cold caesium atoms. Experiments performed in 1997 by J. Dalibard and co-workers showed that the rate of such collisions is high. In order to reduce this undesirable effect, B. Verhaar, D. Heinzen and K. Gibble independently suggested using cold rubidium atoms. Experiments performed by K. Gibble *et al.* and A. Clairon *et al.* have confirmed that the frequency shift in rubidium atomic fountains is about 50 times smaller than in caesium ones, and that cold rubidium fountains can reach an accuracy of 2×10^{-15}. Other proposals have been made to improve existing caesium fountain clocks. In particular, K. Gibble has suggested reducing the density of atoms in the fountain while preserving a high atomic flux, by using a 'juggling' fountain such that several clouds of atoms are launched during the same fountain cycle.

An ultimate limit for atomic fountains with cold atoms is gravity. For reasonable fountain heights of about 1 metre, the Ramsey interaction time cannot exceed 1 second. On the other hand, in microgravity, this interaction time can be considerably increased. For example, in a series of parabolic jet flights, each lasting about 20 seconds in reduced gravity, C. Salomon and co-workers showed in 1993 that laser-cooled atoms can be observed much longer than on Earth. In 1997, the group at the Paris Observatory operated a cold caesium space clock under the same conditions. Two cold caesium space clocks have been selected to fly in the microgravity environment of the future International Space Station: the European ACES (Atomic Clock Ensemble in Space) mission and the US PARCS (Primary Atomic Reference Clock in Space) mission.

Atomic clocks based on optical transitions

All the clocks described above are based on atomic transitions at relatively low frequencies, in the microwave or radio-frequency region of the spectrum. Another promising route to improved clocks is to use optical frequencies since, as seen from (16.59), the fractional frequency uncertainty $\Delta\nu_m/\nu$ is inversely proportional to the frequency ν. Since optical frequencies are larger than microwave frequencies by factors up to 10^5, they offer a very sizable potential gain.

Already in the 1960s, after the demonstration of the first laser, various optical atomic and molecular transitions that might be suitable for clocks had been investigated. However, difficult technical problems arise when using optical frequencies, since optical cycles are too fast to be counted directly by using electronic devices.

The solution to this problem has long been based on the use of an *optical frequency interval divider chain*. Given two lasers with frequencies ν_1 and ν_2, it is possible to force a third laser to oscillate at the precise midpoint $\nu_3 = (\nu_1 + \nu_2)/2$ by electronically phase-locking the second harmonic $2\nu_3$ to the sum frequency $\nu_1 + \nu_2$, using a low-frequency beat signal. With a chain of n cascaded divider stages, a given frequency interval can be divided by 2^n. For example, in 1997 T.W. Hänsch and collaborators measured the absolute frequency of the atomic hydrogen 1s–2s two-photon resonance with an accuracy of 3.4 parts in 10^{13} by using an optical frequency interval divider chain. In their experiment, the 1s–2s two-photon resonance frequency was compared with the 28th harmonic of a methane–stabilised 3.39 μm He–Ne laser. A frequency mismatch of 2.1 THz at the 7th harmonic was bridged with a phase-locked chain of five optical frequency interval dividers. It should be noted that a large number of frequency dividers are required to bridge the gap between a microwave frequency (or a radio frequency) and an optical frequency. Moreover, because of their technical complexity, only a few optical frequency interval divider chains have been built.

Recently, however, the problem of counting optical cycles has been solved by using a new method demonstrated in 2000 by T.W. Hänsch and co-workers at Garching, and by J.L. Hall, S.T. Cundiff *et al.* in Boulder. This method is based on the use of mode-locked pulsed femtosecond lasers spanning an octave from the infra-red to the ultra-violet with a grid of equidistant marker frequencies. As seen from Fig. 16.36, the spectrum of the pulse train from a mode-locked femtosecond laser consists of a comb of equidistant phase-locked frequency components whose spacing is given by the radio-frequency pulse repetition rate ν_r. In order to eliminate frequency fluctuations of the pulsed laser, one can, for example, lock one tooth of the comb to the frequency ν of a stable laser, and lock another tooth to the harmonic 2ν of the stable laser. Having spanned the frequency interval from ν to 2ν with an integer number n of phase-coherent frequencies equally spaced by ν_r, the frequency ν can be determined by measuring ν_r, because $\nu = 2\nu - \nu = n\nu_r$. If in addition the frequency ν_r is compared to a primary frequency standard (a caesium clock), the absolute frequency $\nu = n\nu_r$ of the optical standard can be obtained. Using this method, T.W. Hänsch and co-workers at Garching, working in collaboration with the Paris Observatory team of A. Clairon and C. Salomon, performed an absolute measurement of the frequency of the 1s–2s two-photon transition in an atomic hydrogen beam cooled to about 5 K. Their experimental result of

$$\nu_{1s,2s} = 246\,606\,141\,318\,710\,3(46)\ \text{Hz} \tag{16.60}$$

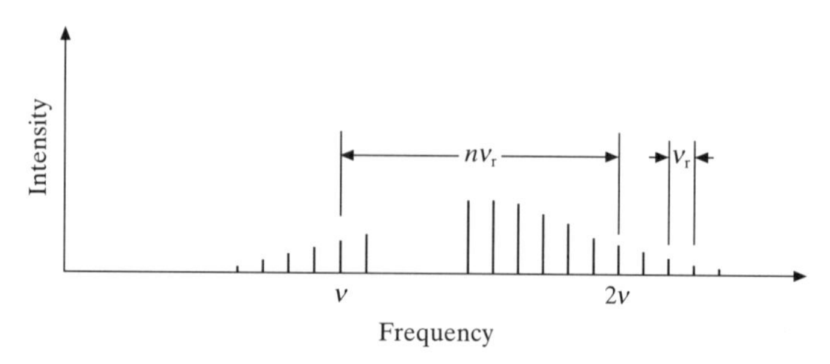

Figure 16.36 Principle of the optical frequency interval divider using a mode-locked, pulsed femtosecond laser. The spectrum of the pulse train from this laser corresponds to a comb of phase-locked frequency components equally spaced by ν_r. If an integer number n of phase-coherent frequencies, equally spaced by ν_r, can be made to exactly span the frequency interval from ν to 2ν, then the frequency ν can be obtained by measuring ν_r, because $\nu = 2\nu - \nu = n\nu_r$. The integer n can be determined from a relatively crude measurement of $\lambda = c/\nu$. (From J.C. Bergquist, S.R. Jefferts and D.J. Wineland, *Physics Today*, March 2001, p. 37.)

was obtained by phase-coherent comparison of the atomic hydrogen transition frequency with a transportable cold caesium fountain frequency standard. Both frequencies were locked with a comb of frequencies emitted by a mode-locked femtosecond laser. The result (16.60) corresponds to an accuracy of 2×10^{-14} and is the most accurate optical frequency determined so far. Ultimately, a source of colder hydrogen atoms (for example, an atomic fountain) could allow frequency measurements of the 1s–2s transition with a still better accuracy.

Optical atomic clocks based on laser-cooled single trapped ions

A single ion, stored in a radio-frequency trap and laser-cooled to a temperature in the microkelvin range, possesses unique properties for high-resolution spectroscopy. It is practically a motionless absorber in a nearly perturbation-free environment, not limited by finite interaction times, Doppler shifts or collisional broadening. As proposed in 1982 by H.G. Dehmelt, a narrow optical resonance of such an ion can therefore be used as a reference in an atomic clock of very high accuracy. For example, with a line width of 1 Hz at a frequency of 10^{15} Hz, and using a laser of sub-hertz line width, it seems possible to achieve a short-term frequency stability of 10^{-15} for 1 second and an accuracy of 10^{-18}. Such an accurate clock, combined with the new method described above for counting the optical cycles with a single mode-locked femtosecond laser, would be a very good candidate for an optical frequency standard. Dehmelt suggested using a very weak transition, for example a transition from the ground state to a long-lived metastable state, as the 'clock transition'. However, as we saw previously in discussing quantum jumps of a single ion, the main difficulty consists in the detection of such a weak transition whose fluorescence yield is very low. We also recollect that the solution proposed by Dehmelt was to couple the weak transition (in the present case the 'clock transition') via the ground state to a strong, dipole allowed transition whose fluorescence rate is much larger and can be easily detected.

A stringent condition in the choice of ions for atomic clocks is the availability of suitable transitions: a long-lived transition for the frequency standard (the clock transition) and a strong electric dipole transition for laser cooling. Moreover, both transitions have to be in the wavelength range accessible for continuous wave (cw) laser sources. A number of prospective clock transitions have been investigated using high-resolution laser spectroscopy. In particular, electric quadrupole (E2) transitions between S and D states have been studied in the alkali-like ions Hg^+, Ba^+, Sr^+ and Yb^+. As an example, let us consider the case of the ion $^{199}Hg^+$. Referring to Fig. 16.29(a), the clock transition is the weak transition from the ground state $5d^{10}\,6s\,^2S_{1/2}$, to the metastable 'shelving' state $5d^9\,6s^2\,^2D_{5/2}$, responsible for the quantum jumps of the ion. The wavelength of this transition is $\lambda = 282$ nm, corresponding to a frequency $\nu = 1.06 \times 10^{15}$ Hz. In an experiment performed in 2000, R.J. Rafac *et al.* measured this frequency with very high precision. Cooling, state preparation and state detection were done on the strong (allowed) transition $5d^{10}\,6s\,^2S_{1/2} \rightarrow 5d^{10}6p\,^2P_{1/2}$. A stabilised pulsed laser was used to measure the frequency of the clock transition with a fractional frequency uncertainty $\Delta\nu_m/\nu \simeq 10^{-14}$. The absorption spectrum measured by Rafac *et al.* is shown in Fig. 16.37. In subsequent work, a single tooth of the pulsed laser (having a repetition rate $\nu_r \simeq 1$ GHz) was locked to the Hg^+-stabilised laser to produce a time and frequency standard based on a single trapped $^{199}Hg^+$ ion.

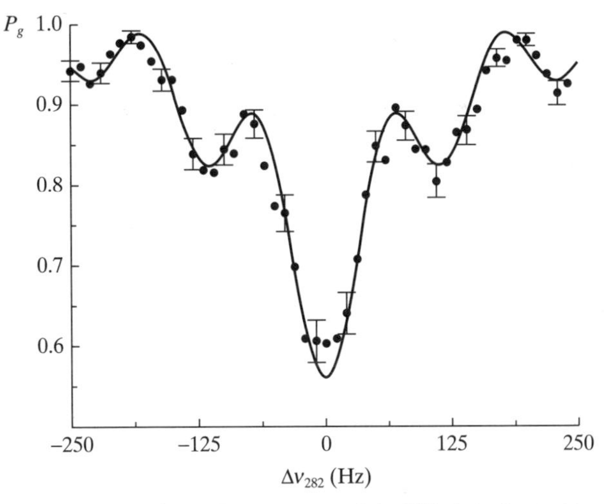

Figure 16.37 Quantum jump absorption spectrum of the $^{199}Hg^+$ clock transition $5d^{10}6s\,^2S_{1/2} \rightarrow 5d^9\,6s^2\,^2D_{5/2}$ at a frequency $\nu = 1.06 \times 10^{15}$ Hz, corresponding to a wavelength $\lambda = 282$ nm. $\Delta\nu_{282}$ is the frequency of the 282 nm probe laser detuning, and P_g is the probability of finding the ion in the ground state. The solid line is a least-squares fit to a Rabi line profile. The quantum projection noise-limited uncertainty is indicated by representative error bars plotted every third point. (From R.J. Rafac, B.C. Young, J.A. Beall, W.M. Itano, D.J. Wineland and J.C. Bergquist, *Physical Review Letters* **85**, 2462, 2000.)

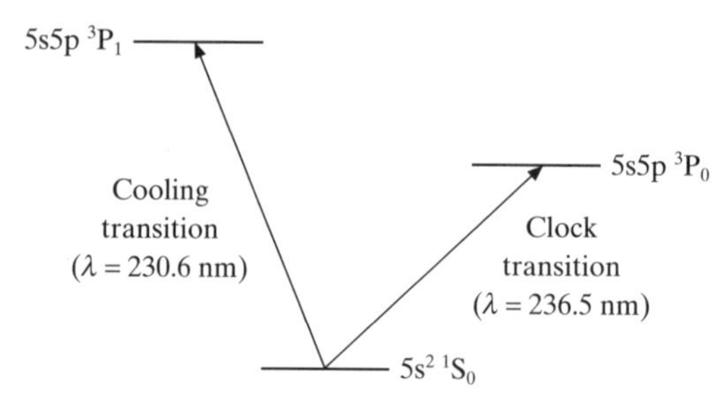

Figure 16.38 The relevant energy levels of the In^+ ion. The $5s^2\ ^1S_0 \to 5s5p\ ^3P_0$ transition at 236.5 nm is the clock transition. The $5s^2\ ^1S_0 \to 5s5p\ ^3P_1$ transition is the cooling transition, and is used to detect the ion by resonance fluorescence.

Electric quadrupole clock transitions between S and D states in alkali-like ions (such as Hg^+, Ba^+, Sr^+ and Yb^+) depend upon small external electromagnetic fields (such as the electric quadrupole field of a Paul trap), and this dependence may cause problems. To avoid these, H.G. Dehmelt suggested using a clock transition between two states with zero electronic angular momentum ($J = 0$), like the $6s^2\ ^1S_0 \to 6s6p\ ^3P_0$ transition between the two lowest levels of Tl^+ or the corresponding transitions of other alkaline-earth-like ions: B^+, Al^+, Ga^+ and In^+. The one-photon decay $^3P_0 \to {}^1S_0$ between two levels with $J = 0$ is strictly forbidden. However, if the spin of the nucleus is non-zero, the hyperfine interaction induces a weak electric dipole transition between these two states. This is due to the fact that the $^3P_0$ state is not a pure $J = 0$ state, but contains small hyperfine admixtures of the decaying states $^1P_1$ and $^3P_1$. These perturbations are responsible for the non-vanishing electric dipole moment between this state and the ground state.

As an example, let us consider the ion $^{115}In^+$ which has been studied recently by H. Walther and collaborators at Garching. Referring to Fig. 16.38, the lowest two levels $5s^2\ ^1S_0$ and $5s5p\ ^3P_0$ are connected by a hyperfine-induced electric dipole transition (the clock transition) at a wavelength of 236.5 nm. Absolute measurements of high precision were made in 2000 by J. von Zanthier *et al.* for the frequency of this transition, which was found to be 1 267 402 452 899.92 (0.23) kHz. Its natural line width is 1.1 Hz.

As a candidate for a primary optical frequency standard, $^{115}In^+$ has three advantages. Firstly, since both states of the clock transition $5s^2\ ^1S_0 \to 5s5p\ ^3P_0$ have zero electronic angular momentum ($J = 0$), the shifts of the transition frequency due to external electromagnetic fields are very small. In fact, as pointed out above, the $^3P_0$ is not a pure $J = 0$ state. It contains small hyperfine admixtures of the $5s5p\ ^1P_1$ and $5s5p\ ^3P_1$ states. These perturbations, which are responsible for the existence of the electric dipole moment between the two clock states, also lead to tiny Stark and Zeeman shifts of the clock transition (in the presence of electric and magnetic fields, respectively), which can be controlled in order to ensure a 10^{-18} reproducibility of the frequency standard. The influence of black body radiation on the clock frequency is also very small. At a trap temperature of 300 K, the control of

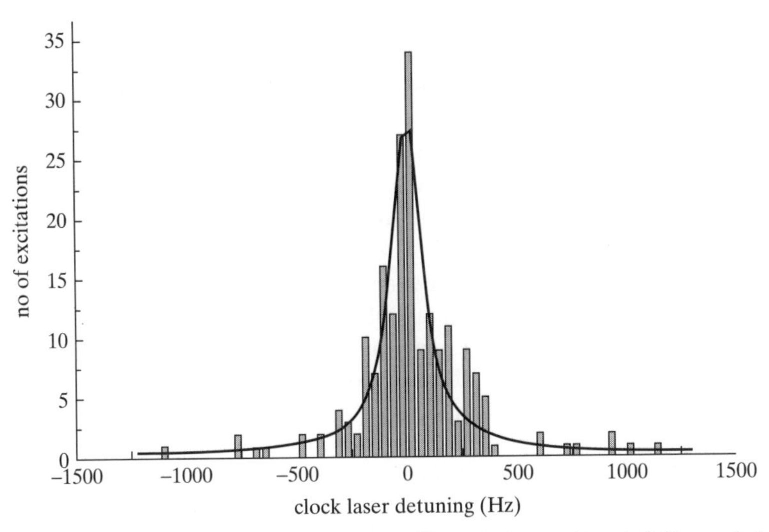

Figure 16.39 High-resolution spectrum of the $^{115}In^+$ clock transition $5s^2\ ^1S_0 \rightarrow 5s5p\ ^3P_0$, obtained with a single ion. The line width of the fitted Lorentzian curve is 170 Hz FWHM (full width at half maximum). The peak excitation probability is about 10% and the total measuring time is 30 minutes. (From Th. Becker, J. von Zanthier, A. Yu Nevsky, Ch. Schwedes, M.N. Skvortsov, H. Walther and E. Peik, *Physical Review* A **63**, R 051802, 2001.) (By Courtesy of H. Walther.)

the temperature to ±1 K is sufficient to reduce the uncertainty in the black body shift below 1 mHz. The electric field gradient of the trap has no influence since the two levels of the clock transition have no quadrupole moment. A second favourable feature of the $^{115}In^+$ clock transition is that its wavelength of 236.5 nm coincides with the fourth harmonic of the 946 nm line of the Nd:YAG laser. The availability of this reliable and frequency-stable laser source simplifies the experimental set-up. A third advantage of the $^{115}In^+$ ion is that laser cooling is performed by using the narrow intercombination line $5s^2\ ^1S_0 \rightarrow 5s5p\ ^3P_1$ (see Fig. 16.38). With a natural line width of 360 kHz, the photon scattering rate on this transition is high enough to detect a single ion via the resonance fluorescence. The ion can be cooled to the motional ground state of the trap, and temperatures below 100 μK can be reached. This reduces the fractional second-order Doppler shift to values below 10^{-19}.

In an experiment performed in 2001, Th. Becker *et al.* obtained for the clock transition at the frequency of 1267 THz a fractional resolution of 1.3×10^{-13}, corresponding to a line width of 170 Hz (see Fig. 16.39). This line width was limited by the frequency instability of the Nd:YAG laser system used for excitation of the clock transition.

Applications of atomic clocks

Atomic clocks are used for many different purposes. We shall consider a few applications where high-precision timing is required.

In fundamental physics, very accurate theoretical calculations can be made for simple physical systems, such as the hydrogen atom. The transition frequencies of atomic hydrogen can be calculated as a function of fundamental constants such as Planck's constant h, the velocity of light *in vacuo* c, the mass of the electron m and the absolute value of its charge e. Highly precise measurements of these transition frequencies, using atomic clocks as frequency standards, provide better values of the fundamental constants and very stringent tests of the theory.

It is also possible to test whether what we call 'fundamental constants' are indeed constants. For example, if the fine structure constant $\alpha = e^2/(4\pi\varepsilon_0\hbar c)$ were to vary in time, then the relative frequency of different atoms would also vary. By comparing over a year the frequencies of caesium and rubidium atoms using atomic fountains, the group of A. Clairon in Paris has shown that α did not change by more than the quantity 5×10^{-15} during that period of time.

Atomic clocks have also been used to test predictions of the special and general theories of relativity. For example, in an experiment performed in 1976 by R. Vessot and co-workers, a hydrogen maser was sent aboard a rocket and the small relativistic clock shift was measured within an accuracy of 7×10^{-5} at an altitude of 10 000 km. No deviation from Einstein's theory was found. Using space-based cold atom clocks, it is expected that the accuracy of this measurement can be improved by an order of magnitude.

In another test of relativity, highly stable atomic clocks have been used to measure the periods of pulsars, and the changes in these periods with time. Of special interest are millisecond pulsars, some of which have a remarkably stable period rivalling the stability of the best atomic clocks. In 1978, J.H. Taylor and co-workers found that the period of a binary-pulsar system is slowly varying by just the amount expected from the energy loss due to the radiation of gravity waves. This was the first experimental evidence for the existence of gravity waves.

Another important application of atomic clocks concerns radio-telescopes. We recall that the resolution (or resolving power) of a telescope depends on the ratio of the wavelength λ to the telescope aperture D. For a radio-telescope, D is the diameter of the effective antenna. Since ground-based radio-astronomy uses radiation having wavelengths in the range from about 2 cm to 30 m, even for the short-wavelength limit a resolution comparable to that of the human eye for visible radiation would require an antenna diameter D in excess of 100 m. For example, the radio telescope of Arecibo, in Puerto Rico, which has the largest fixed antenna ($D = 305$ m), is capable of detecting very faint radio sources, but its resolution is no better than the human eye at visible wavelengths. In order to overcome this practical limit on antenna size, let us consider *two* radio-telescopes separated by a distance D. If both telescopes are looking at the same radio-emitting object (for example, a star), and if the radio waves entering each of them are matched in time, these two telescopes become equivalent to a single radio-telescope whose aperture is their separation distance D. In particular, if the two radio-telescopes are on opposite sides of the Earth, the diameter D of the effective antenna has the size of the Earth, and the resolution of such a combination exceeds that of the largest single optical telescope. However, in order to ensure a precise matching in time, each of the two radio-telescopes needs a highly stable atomic clock.

Precise atomic clocks have also made possible the advent of a new radio navigation system, called the Global Positioning System (GPS). A number of satellites containing accurate atomic clocks transmit radio signals at specific times, so that any observer receiving and analysing the signals from four such satellites can determine his or her position to within 10 m and the correct time to better than 10^{-7} s. Two observers monitoring the same satellite can synchronise their clocks to within about 10^{-9} s.

A particular interesting navigation feat dependent on accurate atomic clocks is the long-distance tracking and data acquisition from interplanetary probes. For example, the success of the mission of the Voyager spacecraft to Neptune depended upon the ground controllers having a precise knowledge of Voyager's position. This was accomplished by using three radio-telescopes at different locations on Earth, each of them transmitting a coded signal to Voyager, which in turn transmitted the signals back to the telescopes. The distances from each of the telescopes to Voyager were deduced from the elapsed times so that Voyager's position could be determined. To achieve the required timing accuracy, two highly stable hydrogen masers were located at each of the radio-telescopes.

16.5 Astrophysics

Our knowledge of celestial objects such as the Sun, the comets, the stars and the galaxies is mainly based on the analysis of the electromagnetic radiation [18] which arrives to us from these objects. At most wavelengths, the electromagnetic radiation coming from outer space is absorbed by the Earth's atmosphere, except for three *windows*, or wavelength bands, through which ground-based astronomers can look at the universe.

The first window covers the optical or *visible* region (between 4000 and 7000 Å) and has been the source of most astronomical knowledge until recent times. The second window, in the *radio-frequency* region, extends from wavelengths of a few millimetres up to 100 m. Its existence was revealed in 1931 by K. Jansky, who discovered radio waves coming from the Milky Way. This discovery opened up the new field of radio-astronomy, where many remarkable discoveries about the universe – such as pulsars, quasars and the universal black body radiation at 3 K – have been made during the past few decades. The third window is located in the *infra-red* region, between the long-wavelength (red) end of the visible spectrum, at approximately 7000 Å, to about 1 mm. However, in contrast to the visible and the radio-frequency windows, the infra-red window is only partially transparent; it is opaque at many wavelengths, but contains a few transparent bands which

[18] The cosmic ray particles, neutrinos, meteorites, etc. entering the Earth's atmosphere also provide astronomical information.

have yielded interesting information about objects from which the radiation at visible wavelengths is absorbed by dust.

Astronomers began to be liberated from the difficulties of ground-based observation when instruments were first flown to high altitudes aboard balloons and aircraft. More recently, instruments aboard artificial satellites have been able to record the emissions of celestial objects in all regions of the electromagnetic spectrum. For example, an ultra-violet telescope satellite, called Copernicus, was launched in 1968 and provided the first observations of extra-terrestrial objects in the far ultra-violet region, while the 'Einstein X-ray Observatory', launched in 1978, has yielded exciting data about violent processes occurring in the universe, such as supernovae explosions.

We shall now discuss briefly a few topics closely connected with the phenomena studied in this book [19].

Stellar spectra

The first spectroscopic study of a star was made in 1802 by W.H. Wollaston, who observed that the Sun emitted a continuous spectrum interrupted by dark lines. In 1811, J. von Fraunhofer, using a diffraction grating, counted about six hundred of these lines, now called Fraunhofer lines, in the solar spectrum. The origin of the dark lines remained unexplained until G.R. Kirchhoff and R.G. Bunsen discovered that heated gas vapours emit spectra composed of bright lines characteristic of the elements from which the spectrum is emitted. Furthermore, Kirchhoff also showed that when continuous light is transmitted through a gas vapour, dark lines are observed at precisely the same wavelengths as the bright emission lines from the same gas (see Chapter 1). For example, sunlight passing through sodium vapour yields two dark lines at the same wavelengths at which heated sodium vapour emits two characteristic bright lines (the D lines). From these observations, Kirchhoff deduced that the gas vapour had *absorbed* its characteristic wavelengths, and in this way he could interpret the Fraunhofer lines of the solar spectrum as *absorption lines* of elements present in the atmosphere of the Sun. Since that time, thousands of dark absorption lines have been catalogued in the Sun's spectrum. By comparing them with the spectral lines emitted by chemical elements in laboratory experiments on the Earth, more than sixty elements have been identified in the Sun [20].

More generally, stellar spectra mainly consist of dark absorption lines superimposed on a continuous spectrum. In the interior of the star, where temperatures are very high, the atoms and ions undergo violent collisions, thus emitting many spectral lines which overlap because of collision broadening. The radiation emitted from the star's opaque surface, or *photosphere*, is therefore continuous. The

[19] A general introduction to astronomy can be found in the book by Jastrow and Thompson (1977).

[20] It is interesting to note that helium was discovered in 1868 through an unidentified dark line in the solar spectrum, before being discovered on the Earth in 1895.

atoms present in the cooler *atmosphere* of the star absorb their characteristic wavelengths, thus giving rise to the observed dark lines. The *chemical composition* of the star's atmosphere can therefore be deduced by analysing the dark lines in the star's spectrum.

Spectral classes. The Harvard classification system

When the absorption spectra of stars were first studied, it was soon realised that stars could be categorised into several different types, called *spectral classes*. At the beginning of the century, the 'Harvard classification system' was proposed, in which the stars were classified according to the strength of the hydrogen lines in their spectra. Letters of the alphabet were used to identify the classes, with class A corresponding to the stars having the strongest hydrogen lines, class B the next strongest, and so on. At that time it was thought that the amount of atomic hydrogen in the star decreased from class A to B, and so on. Today we know that the stars are nearly uniform in composition, being composed mainly of hydrogen and helium. As we shall see below the differences in their absorption spectra are due primarily to their *surface temperature* so that the spectral classes correspond in fact to different surface temperatures. However, the Harvard identification has been kept, and when the classes are arranged in order of decreasing temperature, the letters that designate each group form the sequence O B A F G K M. Astronomy students remember this sequence by using the mnemonic 'Oh Be A Fine Girl (Guy) Kiss Me'. Table 16.1 relates this classification to the range of temperatures involved, and also describes the main characteristics of the spectra. Each class is subdivided into 10 subclasses, for instance B is divided into B0, B1, . . . , B9 corresponding to smaller distinctions which do not merit the creation of a separate class.

The differences between the various stellar spectra can be understood in terms of ordinary atomic physics. The strength of a particular absorption line of a given

Table 16.1 Characteristics of the spectral classes of stars.

Spectral class (Harvard classification)	*Temperature range* (K)	*Main characteristics of the spectrum*
O	30 000–50 000	Lines of ionised helium
B	10 000–30 000	Lines of neutral helium
A	7500–10 000	Very strong hydrogen lines
F	6000–7500	Ionised calcium lines. Many metal lines (manganese, iron, titanium, strontium)
G	4500–6000	Very large number of metal lines. Strong ionised calcium lines, strong ionised and neutral iron lines.
K	3500–4500	Large number of neutral metal lines
M	2000–3500	Band spectra of molecules, particularly of the tightly bound titanium oxide molecule

Table 16.2 The temperature at which lines of certain elements have maximum strength.

Atomic species	*T*(K)
He^+	35 000
He	14 000
H	10 000
Fe	7 000
Ca^+	5 500
Na	4 000

element depends on the number of atoms of this element which are present in the required initial level a. This number will be depleted by inelastic and ionising collisions with electrons, and on the other hand will be enhanced by recombination between ions and electrons. Both of these processes depend on the temperature and on the electron and atom densities in the atmosphere of a star. Assuming that the stellar atmosphere is in a steady state, N.M. Saha obtained, in 1920, an equation relating the rate coefficients for ionisation with that for recombination, from which the line strengths can be deduced and expressed as a function of temperature and density. The temperatures and densities at which line strengths are at a maximum vary from element to element. Table 16.2 lists the temperature at which the absorption lines of various elements are most prominent. At high temperatures (O stars), most atoms are heavily ionised (that is, their outer electrons are missing) and only the lines of systems such as He^+, which is tightly bound, are seen. As the temperature decreases, the lines of metal atoms appear (F, G, K stars), and finally at still lower temperatures, molecular band spectra can be identified.

Stellar abundances

Making a reasonable initial guess about the surface temperature of a particular star, and with a knowledge of the relative intensities of absorption lines, the relative abundances of the elements can be deduced. From these values of the temperature and abundances, a theoretical spectrum can be calculated and compared with the observed spectrum. The initial values can then be adjusted until consistency is obtained. It is found that the relative abundances of the elements in most stars are nearly the same and the values that have been determined are shown in Table 16.3.

Colour and spectral type

Assuming that stars radiate like black bodies, the spectral distribution of the energy emitted is given by Planck's distribution law (1.30), the wavelength λ_{max} at which the distribution has a maximum being given by Wien's law (1.19). The hottest stars (O and B stars) radiate most energy in the ultra-violet region and only a small fraction in the visible region, mostly towards the blue end of the spectrum. As a result, these stars appear blue–white. The A-type stars radiate a bigger fraction

Table 16.3 Relative abundance of various elements in normal star atmospheres.

Element	Relative abundance (%) by number of atoms	Relative abundance (%) by mass
H	90.8	70
He	9	27
C, N, O, Ne	~0.1	~2
Na, Mg, Al, Si, Ca, Fe	~0.01	~1

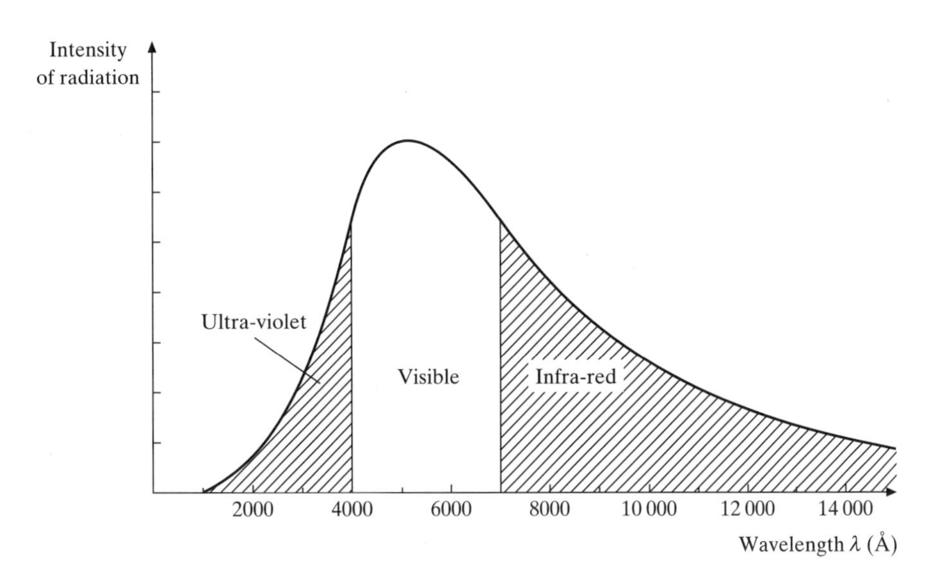

Figure 16.40 The spectral distribution of radiation from the Sun corresponding to a surface tempeature of about 6000 K as a function of the wavelength λ (in Å).

of energy in the visible part of the spectrum, and so appear white. Stars of lower temperature radiate more energy at lower frequencies. Thus F stars appear yellow–white, G stars appear yellow, K stars orange, and finally M stars have a reddish hue. Our Sun, being a G star having a surface temperature of about 6000 K, has a spectral distribution peaking at approximately $\lambda_{max} \simeq 4830$ Å, as shown in Fig. 16.40. The fraction of the Sun's energy radiated in the visible region is about 38%.

Doppler shift and radial velocities

From the analysis of the spectrum of a star, its *radial velocity*, that is its velocity along the line of sight, can be inferred. Indeed, by comparing the stellar spectral lines of a given element with those of a reference spectrum taken in the laboratory, the Doppler shift can be determined, which in turn provides the value of the radial velocity. If a star is moving towards the observer its lines are shifted towards the shorter wavelengths (blue shift) while if a star is moving away from the observer the wavelengths are longer (red shift).

Line shapes

Additional information about the state of a stellar atmosphere can be obtained by analysing the *shape* of the observed spectral lines. As we saw in Chapter 4, all spectral lines have finite widths. Apart from the natural line width, we discussed *Doppler broadening* and *collisional broadening* (also known as *pressure broadening*). Natural line widths are very small and are entirely outweighed in stellar spectra by collisional and Doppler broadening. In a stellar atmosphere, the Doppler broadening is caused by the random thermal motion of the atoms or ions, and in this context the effect is often called *thermal broadening*. While this effect depends entirely on temperature, collisional broadening also depends on the density. Each effect can be distinguished because, as we saw in Chapter 4, collisional broadening determines the wings of a spectral line, while Doppler broadening determines the central region. This allows separate information to be obtained about the density and the temperature.

Another source of Doppler broadening arises from the rotation of a star about its axis. This produces an effect called *rotational broadening*, which can be distinguished from other Doppler broadening mechanisms and can yield an estimate of the frequency of rotation.

Finally, in some stars the magnetic field is sufficiently large for the Zeeman effect to produce detectable changes in the appearance of the lines. Since the magnetic field varies from point to point on the stellar surface, the Zeeman components usually cannot be resolved, but produce a characteristic *magnetic broadening*. However, for some stars having a very large field (of the order of several tesla), the lines can be resolved and the magnetic field strengths determined accurately.

The 21 cm line of atomic hydrogen

In Chapter 5, we learned that the ground state of atomic hydrogen exhibits a hyperfine structure, being split into two levels with $F = 0$ and $F = 1$. The difference in energy between these levels corresponds to a frequency of about 1420 MHz (see (5.84)) or a wavelength of $\lambda \simeq 21$ cm. Radiation of this frequency can be detected by radio-frequency techniques and its observation can be used to map the concentration of atomic hydrogen in our galaxy. It is found that the spiral arms of the galaxy contain the greatest concentration of hydrogen atoms and the different arms can be identified by making use of the Doppler shift, since each arm possesses a different radial velocity with respect to the Earth. In contrast to atomic hydrogen, molecular hydrogen does not emit 21 cm radiation and the technique fails to reveal regions in which the concentration of hydrogen is so large that the formation of molecules is appreciable. Fortunately, in these regions of high density there exist concentrations of CO molecules which have a characteristic emission line at the shorter wavelength of 2.6 mm and this, too, can be detected and used in mapping the galaxy.

The interstellar medium and collision processes

In these brief paragraphs, we have explained how a whole range of spectroscopic measurements from the ultra-violet region to the radio-frequency region provides the basic information about the structure of both stars and the interstellar medium. New information is now coming from the X-ray and γ-ray parts of the spectrum as a result of high-altitude and satellite measurements. In order to interpret the results obtained, mathematical models must be built which describe the physical processes occurring in various layers of a star, or in the clouds of the interstellar medium. For this purpose, a new range of atomic and molecular physics is required, and in particular the collision physics discussed in previous chapters. For example, it has been discovered that the interstellar medium contains many molecules (in small concentrations) of surprising complexity. In dense interstellar clouds containing particle densities of about 10^{10} particles/m^3, typical molecular abundances are (relative to hydrogen) $CH^4 - 10^{-6}$; $CH - 10^{-8}$; $CN - 10^{-8}$; $OH - 10^{-6}$, $CO - 10^{-4}$, while of the more complicated systems the concentration of CH_3OH is 10^{-7}, that of CH_3HCO is 10^{-9}, and so on. These molecules are presumably formed in collision processes, some radiative, such as

$$C^+ + H \rightarrow CH^+ + h\nu \qquad \textbf{(16.61)}$$

and many of a radiationless character, for example

$$CH + O \rightarrow HCO^+ + e^- \qquad \textbf{(16.62)}$$

or

$$CH + O \rightarrow H + CO \qquad \textbf{(16.63)}$$

Dozens or even hundreds of possible reactions must be evaluated before a proper understanding of the physical conditions allowing the formation of large molecules can be achieved.

In general the solution of the problems posed by the wealth of spectral information from outer space gathered by modern instrumentation requires the astrophysicist to apply all the principles of atomic and molecular physics that we have studied in this book to systems with physical conditions far removed from those that can be created in terrestrial laboratories, and this can be done with great success.

Table 16.4 Stages in the history of the universe.

t	R	T	E	*Events*
10^{-43} s	10^{-5} m	10^{32} K	10^{19} GeV	Quantum gravity dominates
10^{-9} s	10^{12} m	10^{15} K	10^{2} GeV	Matter in the form of quarks and leptons
10^{-2} s	10^{15} m	10^{13} K	1 GeV	Nucleons formed and free quarks disappear
10^{2} s	10^{17} m	10^{9} K	0.1 MeV	Primordial nucleosynthesis: deuterium and helium formed
10^{3} s	10^{18} m	10^{8} K	10 keV	Primordial nucleosynthesis stops
3×10^{2} yrs	10^{21} m	1.6×10^{5} K	13.6 eV	Bound states of atomic hydrogen begin to be formed
3×10^{5} yrs	10^{22} m	10^{4} K	1 eV	Radiation decouples from matter
10^{10} yrs	10^{26} m	3 K	2.5×10^{-4} eV	Present day

Notes:
t: time from the big bang, R: scale factor specifying the size of the universe at time t,
T: temperature, E: energy per particle available for excitation, $E = k_B T$.

Atoms and molecules in the early universe

We shall conclude this section on astrophysics by examining how atoms and molecules first originated in the early universe. In the generally accepted *standard model* or *big bang* theory of modern cosmology, the universe is considered to have originated about 10^{10} years ago from a state of enormous density and temperature, that is an extremely hot fireball of particles and radiation occupying a small volume. This hot dense fireball then expanded and thereby cooled, the cooling and expansion continuing to the present day (see Table 16.4). The evidence for this picture is outside the scope of this book [21], but relies in part on the fact that since the velocity of light is finite, the observation of distant objects, such as galaxies and quasars, provides information about the past state of the universe, although as we shall see there is a time before which no information can be obtained without the use of theoretical models. In connection with the idea of an expanding universe, relativistic models require that all distances separating astronomical objects, such as galaxies, increase with time since space itself is expanding. This has the consequence that a light ray emitted by a distant galaxy with a wavelength λ is stretched by travelling through expanding space and is observed at the present time as having a longer wavelength λ'. If the size of the universe at the time of emission of a light ray is specified by a scale factor R and at the present time of observation by a scale factor R', then

$$\frac{\lambda'}{\lambda} = \frac{R'}{R} \tag{16.64}$$

[21] See for example Harrison (2000) and Bowers and Deeming (1984).

The spectral lines from distant objects are therefore shifted towards the red end of the spectrum. This cosmological red shift is defined as

$$z = \frac{\lambda' - \lambda}{\lambda} \tag{16.65}$$

so that, using (16.64), we have

$$z = \frac{R'}{R} - 1 \tag{16.66}$$

In its earliest highly compressed state the universe was extremely hot with most of the energy being in the form of thermal radiation obeying a black body distribution. Initially the intensity of the radiation was too strong for nucleons or nuclei to exist, let alone atoms or molecules. As the universe expanded and cooled, nucleons began to be formed when $k_B T$ fell below about 1 GeV, and at a later stage ($k_B T \simeq 0.1$ MeV) nucleons could combine through a series of reactions (nucleosynthesis) to produce significant quantities of ^{4}He (about 22% by weight) and small amounts of ^{2}D, ^{3}He and ^{7}Li. This primordial nucleosynthesis took place from about 10^2 s after the big bang until about 10^3 s, when the temperature fell below 10^8 K. The first atoms could not be formed until much later, since until $k_B T$ fell below the ionisation potential of atomic hydrogen (13.6 eV), any neutral hydrogen formed would be photoionised by energetic photons:

$$h\nu + \mathrm{H} \rightarrow \mathrm{H}^+ + \mathrm{e}^- \tag{16.67}$$

At this time, and earlier, the energy in the universe was mainly in the form of black body radiation with an isotropic distribution. The photons interacted with the electrons strongly by Thompson scattering and for this reason the universe was opaque and no observations of the early universe at this period can be made. After about 3×10^5 years, when the temperature dropped so that $k_B T$ became less than 13.6 eV, fewer and fewer photons would be sufficiently energetic to ionise any bound states of atomic hydrogen formed by radiative capture:

$$\mathrm{e}^- + \mathrm{H}^+ \rightarrow \mathrm{H} + h\nu \tag{16.68}$$

Cooling continued until after about 3×10^5 years, $k_B T$ fell below 1 eV so that no more photons could be absorbed by the atoms. At this stage the universe became transparent to radiation and the radiation and matter became *decoupled* and evolved separately. The photons continued to cool from the decoupling temperature of 10^4 K until the present time when the photon temperature is 2.75 K. Since during this time the radiation has not interacted with the matter in the universe, it has retained its black body spectrum and is isotropically distributed. This radiation is known as the *cosmic background radiation* and was detected in the microwave part of the spectrum by A.A. Penzias and R.W. Wilson in 1964. Since the universe has been transparent from the time of decoupling, observation of radiation emitted from distant objects after that time is, in principle, possible. It corresponds to red shifts in the range $z = 900$ to $z = 0$.

At the time that the first atomic hydrogen was formed, the universe contained protons as well as electrons, ^{2}D, ^{3}He, ^{4}He and ^{7}Li nuclei in small quantities. This had the consequence that further cooling allowed, in addition to the continued formation of atomic hydrogen, the successive formation of the positive ions Li^{2+}, He^+, Li^+ and the neutral atoms D and Li by radiative capture. The negative ions D^-, H^- and Li^- were also produced by the corresponding process of radiative recombination, for example

$$e^- + H \rightarrow H^- + h\nu \qquad \textbf{(16.69)}$$

At a still later stage the first molecules appeared through recombination reactions.

The uniformity of the cosmic background radiation shows that at the time of decoupling the universe contained a uniform distribution of both radiation and of matter, the latter being mainly in the form of neutral atomic hydrogen. Subsequently some fluctuations must have occurred allowing the formation of distinct clouds of gas which then started to condense under the influence of gravity, ultimately leading to the creation of the first galaxies. In the initial stages of condensation before stars are formed these gas clouds are called *protogalaxies*. The force of gravity which tends to compress the protogalaxies is opposed by the thermal pressure which depends on the temperature. If $k_B T$ is greater than the lowest excitation energy of the atoms composing the gas, the gas can cool by photon absorption. If the gas is entirely atomic, with excitation energies of the order of electron volts, the gas temperature cannot decrease by this mechanism much below 10^4 K. On the other hand, if any molecules were present, even in small concentrations, the temperature could fall to 10^2 K because the rotational excitation energies are of the order of 10^{-3} eV. Thus to understand the details of the early stages of galaxy formation it is important to consider what processes might lead to the existence of molecules. Extensive studies of processes leading to the formation of molecular hydrogen in the early universe were initiated by S. Lepp and M.J. Shull in 1980 and by A. Dalgarno and S. Lepp in 1987. More recently studies have also been made of the deuterium and lithium chemistry of the early universe.

Let us now review the reactions that can lead to the formation of molecular hydrogen in a gas composed of neutral atomic hydrogen. The first of these proceeds via the formation of negative hydrogen ions by the radiative process (16.69), followed by the associative detachment reaction

$$H^- + H \rightarrow H_2 + e^- \qquad \textbf{(16.70)}$$

Another process, involving those protons which have not yet been neutralised by radiative capture of electrons, is radiative capture

$$H + H^+ \rightarrow H_2^+ + h\nu \qquad \textbf{(16.71)}$$

followed by the charge exchange reaction

$$H + H_2^+ \rightarrow H^+ + H_2 \qquad \textbf{(16.72)}$$

To estimate the composition of a protogalaxy at different stages of gravitational compression, not only must the rate coefficients of the processes leading to the formation of H_2 be known, but also the corresponding rates for the destruction of H_2 or of H_2^+. For example, the population of H_2^+ is depleted by the reactions:

$$e^- + H_2^+ \rightarrow H + H \tag{16.73}$$

and

$$h\nu + H_2^+ \rightarrow H + H^+ \tag{16.74}$$

while H_2 can be dissociated in collisions with atomic hydrogen

$$H + H_2 \rightarrow H + H + H \tag{16.75}$$

It is clear that estimates of the composition of a protogalaxy having a particular age require an accurate knowledge of all the reaction rates mentioned above, together with those leading to molecules expected to occur in smaller concentrations such as HD, HeH^+ and LiH.

The general conclusion of Lepp and Shull, based on all the available theoretical or measured reaction rates, was that the fractional abundance of H_2 would rise from zero at the time of recombination to about 2×10^{-6} for $z = 100$ corresponding to a time of 3×10^7 years after the big bang. This is large enough to have a profound effect on galaxy formation. It is just one example of the many instances in which our understanding of the formation of astronomical objects depends on a detailed knowledge of atomic and molecular physics [22].

[22] Atomic and molecular processes in the early universe have been reviewed by Lepp, Stancil and Dalgarno (2002).

Appendices

1 Classical scattering by a central potential

In this appendix we show how particles are scattered from a central potential $V(r)$ using classical Newtonian mechanics, and we obtain the Rutherford scattering formula (1.58) for the scattering of a beam of particles by a repulsive Coulomb potential.

The path of a particle in the field of a central potential is confined to a plane (see Goldstein, 1980), which we can take to be the (XZ) plane. Let us introduce plane polar coordinates (r, ϕ), defined by (see Fig. A1.1)

$$\begin{aligned} r^2 &= x^2 + z^2 \\ r\cos\phi &= z \end{aligned} \qquad \textbf{(A1.1)}$$

In the scattering problem we are considering r is arbitrarily large before and after the collision event.

The kinetic energy of the particle which is being scattered is given by

$$T = \tfrac{1}{2}m(\dot{r}^2 + r^2\dot{\phi}^2) \qquad \textbf{(A1.2)}$$

where m is the mass of the particle and we have used the notation $\dot{r} \equiv \mathrm{d}r/\mathrm{d}t$ and $\dot{\phi} \equiv \mathrm{d}\phi/\mathrm{d}t$. When the particle is at any position, the component of its velocity perpendicular to $\mathbf{r}$ is $r|\dot{\phi}|$ and its angular momentum about the centre of force at O is therefore

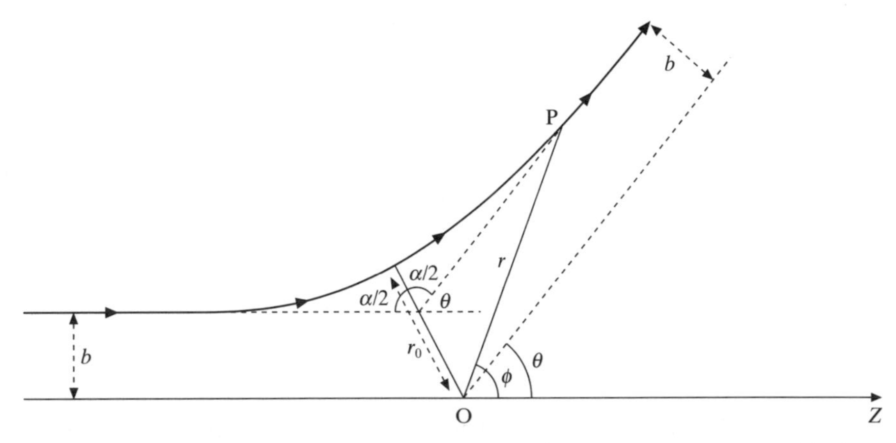

Figure A1.1 The scattering of a particle in the field of a repulsive central potential whose origin is at O.

$$L = mr^2|\dot{\phi}| \tag{A1.3}$$

Because the potential is central the angular momentum must remain constant along any trajectory. In particular, we have

$$L = mvb \tag{A1.4}$$

where b is the impact parameter of the particle (see Fig. A1.1) and v the magnitude of its initial velocity. The total energy E is also constant, and is equal to $mv^2/2$. Thus, since $E = T + V$, we have

$$\begin{aligned} E &= \tfrac{1}{2}m(\dot{r}^2 + r^2\dot{\phi}^2) + V(r) \\ &= \tfrac{1}{2}m\left(\dot{r}^2 + \frac{L^2}{m^2r^2}\right) + V(r) = \tfrac{1}{2}mv^2 \end{aligned} \tag{A1.5}$$

and we note that

$$\dot{r} = \pm\left\{\frac{2}{m}[E - V(r)] - \frac{L^2}{m^2r^2}\right\}^{1/2} \tag{A1.6}$$

To find an equation for the trajectory, or orbit, we write

$$\frac{\mathrm{d}\phi}{\mathrm{d}r} = \frac{\mathrm{d}\phi}{\mathrm{d}t}\frac{\mathrm{d}t}{\mathrm{d}r} = \pm\frac{L}{mr^2}\left\{\frac{2}{m}[E - V(r)] - \frac{L^2}{m^2r^2}\right\}^{-1/2} \tag{A1.7}$$

from which, upon integrating over r, we have

$$\phi - \phi_0 = \pm\int_{r_0}^{r}\frac{L}{mr^2}\left\{\frac{2}{m}[E - V(r)] - \frac{L^2}{m^2r^2}\right\}^{-1/2}\mathrm{d}r \tag{A1.8}$$

where we have chosen the constant of integration so that $\phi = \phi_0$ at $r = r_0$. The value of r_0 is conveniently taken to be the distance of closest approach, which is found from the equation

$$\dot{r} = 0 \tag{A1.9}$$

It is therefore the largest root of the equation

$$E - V(r) - \frac{L^2}{2mr^2} = 0 \tag{A1.10}$$

Having obtained r_0 from (A1.10), the orbit can be found by performing the integral in (A1.8).

The deflection function

Let us consider a particle incident upon a centre of force. We assume that the central force acting on it falls to zero at large distances. Thus, when the particle is very far from the centre of force, it will move along a straight-line trajectory. As it approaches the centre of force, it experiences an attractive or repulsive interaction and is therefore deflected from its original straight-line path. After the particle has passed the centre of force, the force acting on it will ultimately vanish, so that its orbit will again approach a straight line (see Fig. A1.1). In general the final direction of motion is different from the incident one, and the particle is said to have been *scattered* by the centre of force. Because of the symmetry of the orbit about the point ($r = r_0$, $\phi = \phi_0$), the angle between the asymptotes to the orbit (see Fig. A1.1) is given by

$$\begin{aligned}\alpha &= 2|\phi(r=\infty) - \phi(r=r_0)| \\ &= 2\int_{r_0}^{\infty} \frac{L}{mr^2}\left\{\frac{2}{m}[E - V(r)] - \frac{L^2}{m^2r^2}\right\}^{-1/2} \mathrm{d}r \qquad \textbf{(A1.11)}\end{aligned}$$

The *deflection function* Θ is then defined as

$$\Theta = \pi - \alpha \qquad \textbf{(A1.12)}$$

For central potentials, there is axial symmetry about the Z axis, and it is convenient to introduce an *angle of scattering* θ, which is defined so that $0 \leqslant \theta \leqslant \pi$. To obtain θ, we first form the quantity

$$\Phi = |\Theta| - 2\pi n \qquad \textbf{(A1.13)}$$

where n is 0 or an integer, chosen so that $0 \leqslant \Phi \leqslant 2\pi$. Then

$$\begin{aligned}\theta &= \Phi && \text{if} \quad \Phi \leqslant \pi \\ &= 2\pi - \Phi && \text{if} \quad \pi \leqslant \Phi \leqslant 2\pi \qquad \textbf{(A1.14)}\end{aligned}$$

Cross-sections

Let us now consider a uniform *beam* of non-interacting, monoenergetic particles incident upon a target made of n scattering centres. We shall assume that the target particles are far enough apart so that each collision process involves only one of them. We also suppose that the target is sufficiently thin so that multiple scattering by several scatterers may be neglected. The beam particles, of mass m, approach the target from infinity with an initial velocity of magnitude v and whose direction is parallel to the Z axis (see Fig. A1.2). Let N be the number of incident particles crossing per unit time a unit area perpendicular to the beam, so that N represents the flux of incident particles. Let $\mathrm{d}N'$ be the number of incident particles scattered per unit time in a small solid angle $\mathrm{d}\Omega$ centred about a direction $\Omega \equiv (\theta, \phi)$ having polar angles (θ, ϕ) with respect to the Z axis.

Under the experimental conditions described above the number $\mathrm{d}N'$ of incident particles emerging per unit time in the solid angle $\mathrm{d}\Omega$ is proportional to N, n and $\mathrm{d}\Omega$, so that one can write

$$\mathrm{d}N' = Nn\sigma(\theta, \phi)\,\mathrm{d}\Omega \qquad \textbf{(A1.15)}$$

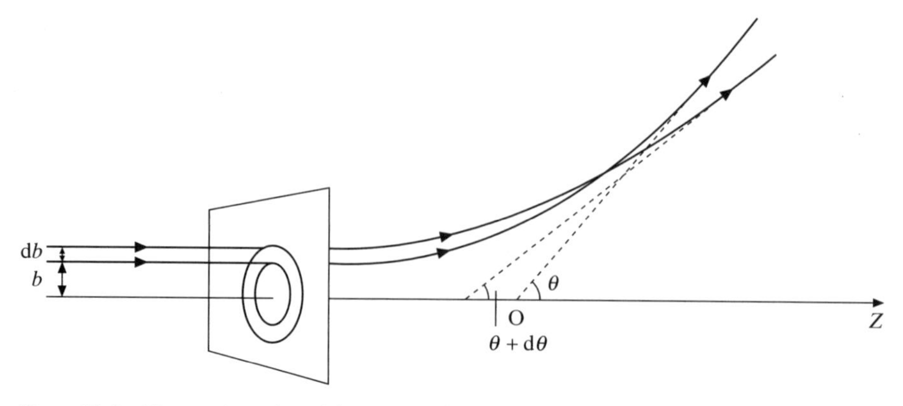

Figure A1.2 The number of particles scattered with impact parameters between b and $b + \mathrm{d}b$ is equal to $N2\pi b\,\mathrm{d}b$, where N is the incident flux. These particles are scattered with scattering angles between θ and $\theta + \mathrm{d}\theta$.

The proportionality factor $\sigma(\theta, \phi)$, which is also often written as

$$\sigma(\theta, \phi) \equiv \frac{\mathrm{d}\sigma}{\mathrm{d}\Omega}(\theta, \phi) \tag{A1.16}$$

is called the *differential scattering cross-section*. From this definition we see that $\mathrm{d}\sigma/\mathrm{d}\Omega$ is the ratio of the number $\mathrm{d}N'$ of particles scattered into the solid angle $\mathrm{d}\Omega$ per unit time per unit scatterer and per unit solid angle, to the flux of incident particles with respect to the target. It is also clear from (A1.15) that $\sigma(\theta, \phi)\,\mathrm{d}\Omega$ has the dimensions of an area.

The total scattering cross-section is obtained by integrating the differential cross-section over all scattering angles. That is,

$$\begin{aligned}\sigma_{\text{tot}} &= \int \frac{\mathrm{d}\sigma}{\mathrm{d}\Omega}(\theta, \phi)\,\mathrm{d}\Omega \\ &= \int_0^{2\pi} \mathrm{d}\phi \int_0^{\pi} \mathrm{d}\theta \sin\theta \frac{\mathrm{d}\sigma}{\mathrm{d}\Omega}(\theta, \phi)\end{aligned} \tag{A1.17}$$

It is worth noting that in defining the above cross-sections we have considered the simple case of *elastic* collisions of particles with scattering centres. A general definition of cross-sections, which applies to elastic collisions as well as other types of collisions, is given in Chapter 12.

Let us now consider the classical scattering of a beam of particles by a centre of force, assuming that the force is central and vanishes for large r. The position of each incident particle can then be specified by giving its cylindrical coordinates (b, ϕ, z), where b is the impact parameter and ϕ the azimuthal angle. Since the particles do not all have the same impact parameter b or angular momentum $L = mvb$, they will in general be scattered differently. Indeed, each value of L

(or b) defines a deflection function Θ and a corresponding angle of scattering θ. Thus the particles scattered at angles between θ and $\theta + \mathrm{d}\theta$, independently of ϕ, are those having an angular momentum between L and $L + \mathrm{d}L$ or an impact parameter between b and $b + \mathrm{d}b$. In other words they are the particles falling on the ring of area $2\pi b\,\mathrm{d}b$ shown on Fig. A1.2. Now, since N is the flux of incident particles, the number of particles passing per unit time through the ring is $N2\pi b\,\mathrm{d}b$, and the number of particles per unit time having angular momentum between L and $L + \mathrm{d}L$ is $N2\pi L\,\mathrm{d}L/(m^2v^2)$. This is the same as the number $\mathrm{d}N'$ of particles scattered per unit time within the solid angle $\mathrm{d}\Omega$, so that

$$\mathrm{d}N' = N2\pi b\,\mathrm{d}b = \frac{N2\pi L\,\mathrm{d}L}{m^2v^2} \tag{A1.18}$$

The differential scattering cross-section has been defined above (see (A1.15)) to be the ratio $\mathrm{d}N'/(Nn\,\mathrm{d}\Omega)$, where n is the number of scattering centres. Since $n = 1$ and $\mathrm{d}\Omega = 2\pi \sin\theta\,\mathrm{d}\theta$ in the present case, we have

$$\frac{\mathrm{d}\sigma}{\mathrm{d}\Omega} = \frac{b}{\sin\theta}\left|\frac{\mathrm{d}b}{\mathrm{d}\theta}\right| \tag{A1.19}$$

or

$$\frac{\mathrm{d}\sigma}{\mathrm{d}\Omega} = \frac{1}{m^2v^2}\frac{L}{\sin\theta}\left|\frac{\mathrm{d}L}{\mathrm{d}\theta}\right| \tag{A1.20}$$

If more than one value of b (or L) contributes to a given value of θ, the differential cross-section is the sum over all values of b (or L) that contribute, namely

$$\frac{\mathrm{d}\sigma}{\mathrm{d}\Omega} = \sum_i \frac{b_i}{\sin\theta}\left|\frac{\mathrm{d}b_i}{\mathrm{d}\theta}\right| \tag{A1.21}$$

or

$$\frac{\mathrm{d}\sigma}{\mathrm{d}\Omega} = \frac{1}{m^2v^2}\sum_i \frac{L_i}{\sin\theta}\left|\frac{\mathrm{d}L_i}{\mathrm{d}\theta}\right| \tag{A1.22}$$

In order to use the above formulae (A1.19)–(A1.22) one must first find b (or L) as a function of θ from (A1.8) and (A1.11)–(A1.14). There are only a few potentials for which this can be done analytically; in the other cases numerical methods must be employed.

Rutherford scattering

The most famous example of a potential for which an exact solution can be obtained is the *Coulomb potential*. As shown in Chapter 1, the solution for this case helped Rutherford to interpret his experiments, which established the existence of the nucleus and therefore provided the evidence for the nuclear model of the atom.

The Coulomb potential is that produced by a fixed charge q_B acting on incident particles having a charge q_A,

$$V_c(r) = \frac{q_A q_B}{(4\pi\varepsilon_0)r} \tag{A1.23}$$

and we shall treat the repulsive case for which $q_A q_B > 0$. According to (A1.10), the distance of closest approach r_0 is the largest root of the equation

$$E - \frac{q_A q_B}{(4\pi\varepsilon_0)r} - \frac{L^2}{2mr^2} = 0 \tag{A1.24}$$

Setting

$$A = \frac{q_A q_B}{(4\pi\varepsilon_0)2E}, \qquad B = \frac{1}{2mE} \tag{A1.25}$$

we find that

$$\begin{aligned} r_0 &= A + (A^2 + BL^2)^{1/2} \\ &= A + (A^2 + b^2)^{1/2} \end{aligned} \tag{A1.26}$$

From (A1.11) and (A1.12) the deflection function is given by

$$\Theta = \pi - 2\int_{r_0}^{\infty} \frac{L}{mr^2}\left\{\frac{2}{m}[E - V_c(r)] - \frac{L^2}{m^2r^2}\right\}^{-1/2} dr \tag{A1.27}$$

which we may also write as

$$\begin{aligned} \Theta &= \pi - 2\int_{r_0}^{\infty} \frac{b}{r^2}\left[1 - \frac{2A}{r} - \frac{b^2}{r^2}\right]^{-1/2} dr \\ &= \pi - 2\int_{r_0}^{\infty} \frac{b}{r}[r^2 - 2Ar - b^2]^{-1/2}\, dr \end{aligned} \tag{A1.28}$$

The integral on the right-hand side of this equation is a standard one (see for example Dwight, 1961) and we have

$$\begin{aligned} \Theta &= 2\cos^{-1}\left[\frac{1}{[1 + A^2/b^2]^{1/2}}\right] \\ &= 2\tan^{-1}\frac{A}{b} \end{aligned} \tag{A1.29}$$

This relation can be inverted to obtain b in terms of the deflection function Θ or, using (A1.13)–(A1.14), in terms of the scattering angle θ. That is,

$$b = A\cot\frac{\theta}{2} \tag{A1.30}$$

We note that the trajectory of the particle may be readily obtained by performing the required (indefinite) integral in (A1.8) by a method similar to the one we have used to calculate the definite integral in (A1.27). One finds that

$$\frac{1}{r} = \frac{A}{b^2}(\varepsilon \cos \phi - 1) \tag{A1.31}$$

where $\varepsilon = (1 + b^2/A^2)^{1/2}$. This is the equation of a hyperbola, with the centre of force being at the exterior focus.

The treatment of the attractive case ($q_A q_B < 0$) is similar to the one given above for the repulsive case. The quantity A is now negative, and one finds for the same value of $|A|$ a deflection angle Θ having the opposite sign of the one obtained above. The trajectory is then a hyperbola with the centre of force at the interior focus.

Let us now return to (A1.30). Since we know the relation between b and θ, we may directly obtain the differential cross-section $d\sigma/d\Omega$. Thus, using (A1.19) and (A1.30), we find that

$$\frac{d\sigma_c}{d\Omega} = \left(\frac{A}{2}\right)^2 \frac{1}{\sin^4(\theta/2)} \tag{A1.32}$$

or (see (A1.25))

$$\frac{d\sigma_c}{d\Omega} = \left(\frac{q_A q_B}{4\pi\varepsilon_0}\right)^2 \frac{1}{16E^2 \sin^4(\theta/2)} \tag{A1.33}$$

where the subscript c refers to the Coulomb potential (A1.23). This is the *Rutherford formula* for the differential cross-section corresponding to Coulomb scattering. The result (A1.33) is identical to that obtained from the quantum theory of Coulomb scattering. This coincidence is confined to the Coulomb potential and no other potential provides exactly the same cross-section in quantum theory and in classical theory.

The Rutherford formula (A1.33) for scattering by a Coulomb potential exhibits other remarkable features. Indeed, the differential cross-section (A1.33) does not depend on the *sign* of the potential. Moreover, since the energy E and the scattering angle θ enter into separate factors, $d\sigma_c/d\Omega$ is scaled at *all* angles by the factor $(q_A q_B)^2/(16\pi\varepsilon_0 E)^2$, so that the angular distribution is independent of the energy. We also note that at fixed θ the differential cross-section is proportional to E^{-2}. Finally, $d\sigma_c/d\Omega$ is infinite in the forward direction ($\theta = 0$), where it diverges like θ^{-4}. As a result, the total cross-section $\int(d\sigma_c/d\Omega)\, d\Omega$ is not defined for pure Coulomb scattering. When considering real scattering processes, we must remember that all Coulomb potentials will be modified at large distances because of the screening effect of the charges in other atoms and molecules. As a result of this screening, the *quantum mechanical* differential cross-section becomes finite in the forward direction, and the corresponding total cross-section is then defined. It is worth noting that in *classical mechanics* both the differential cross-section in the forward direction and the total cross-section do not exist for any potential that does not strictly vanish beyond a certain distance. This is due to the fact that two classical particles interacting through such potentials always suffer some deflection, even if they pass by each other at an arbitrary large distance. On the contrary, in quantum mechanics, it can be shown (see Problem 12.14) that the differential cross-section at $\theta = 0$ is finite if the potential tends to zero faster than $1/r^3$ when $r \to \infty$, and that the total cross-section is finite provided the potential falls off faster than $1/r^2$ when $r \to \infty$.

2 The laboratory and centre of mass systems

Let us consider a non-relativistic collision between a 'beam' particle A of mass m_A and a 'target' particle B of mass m_B. The *laboratory system* (L) is the framework in which the target particle B is at rest before the collision. In what follows we shall use the subscript L to denote quantities in the laboratory system. The *centre of mass* (CM) *system* is the coordinate system in which the centre of mass of the composite system (A + B) is at rest. Denoting by $\mathbf{v}_A$ and $\mathbf{v}_B$ the velocities of particles A and B in the CM system, and by $\mathbf{p}_A = m_A\mathbf{v}_A$ and $\mathbf{p}_B = m_B\mathbf{v}_B$ their CM momenta, we have

$$\mathbf{p}_A + \mathbf{p}_B = 0 \tag{A2.1}$$

Observations are often [1] made in the laboratory system, while *calculations* are frequently performed in the CM system, since the three degrees of freedom attached to the centre of mass of the system (A + B) may then be ignored. In this appendix we shall study the kinematical problem of passing from one frame of reference to the other.

Let us choose the incident direction as our Z axis in both laboratory and CM systems. Calling $\hat{\mathbf{z}}$ the unit vector along this axis, we write the velocity of the centre of mass in the laboratory system as

$$\mathbf{V}_L = V_L\hat{\mathbf{z}} \tag{A2.2}$$

We suppose that no external forces are present, so that the centre of mass keeps its uniform rectilinear motion and the laboratory and CM systems are in uniform translational motion of velocity $\mathbf{V}_L$ with respect to each other. Since the collision is non-relativistic, the velocity $\mathbf{v}_L$ of a particle in the laboratory system is therefore related to its velocity $\mathbf{v}$ in the CM system by

$$\mathbf{v}_L = \mathbf{v} + \mathbf{V}_L \tag{A2.3}$$

We shall choose the X and Y axes of the CM system to be parallel to the corresponding axes in the laboratory system. Then, if $\mathbf{v}$ points in the direction (θ, ϕ) in the CM system and $\mathbf{v}_L$ in the direction (θ_L, ϕ_L) in the laboratory system (see Fig. A2.1) we have

[1] Important exceptions are crossed beam, colliding beam or merged beam experiments, in which the 'target' consists of a beam of particles B.

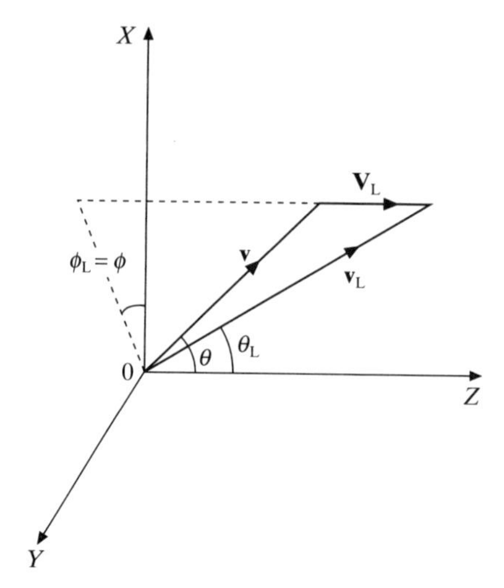

Figure A2.1 Illustration of the velocities $\mathbf{v}$, $\mathbf{v}_L$ and $\mathbf{V}_L$ and the angles (θ, ϕ) and (θ_L, ϕ_L).

$$\begin{aligned} v_L \cos \theta_L &= v \cos \theta + V_L \\ v_L \sin \theta_L &= v \sin \theta \\ \phi_L &= \phi \end{aligned} \tag{A2.4}$$

Eliminating v_L from the first two of these equations, we find that

$$\tan \theta_L = \frac{\sin \theta}{\cos \theta + \tau} \tag{A2.5}$$

where

$$\tau = \frac{V_L}{v} \tag{A2.6}$$

Since the particle B is at rest before the collision in the laboratory system, the total momentum $\mathbf{P}_L$ in that system is equal to the momentum $(\mathbf{p}_A)_L$ of the incident particle A,

$$\mathbf{P}_L = (\mathbf{p}_A)_L \tag{A2.7}$$

and the initial kinetic energy $(T_i)_L$ in the laboratory system is given by

$$(T_i)_L = \frac{(p_A)_L^2}{2m_A} = \frac{P_L^2}{2m_A} \tag{A2.8}$$

The corresponding initial kinetic energy T_i available in the CM system is obtained by subtracting from $(T_i)_L$ the kinetic energy of the centre of mass in the laboratory system. Using the fact that the centre of mass moves in the laboratory system as a free particle of mass $M = m_A + m_B$, momentum $\mathbf{P}_L$ and energy $P_L^2/2M$, we have

$$T_{\rm i} = (T_{\rm i})_{\rm L} - \frac{P_{\rm L}^2}{2M}$$

$$= \frac{m_{\rm B}}{m_{\rm A} + m_{\rm B}}(T_{\rm i})_{\rm L} \qquad \textbf{(A2.9)}$$

and we also note that the centre of mass velocity $V_{\rm L}$ in the laboratory system is given by

$$V_{\rm L} = \frac{(p_{\rm A})_{\rm L}}{m_{\rm A} + m_{\rm B}} = \frac{m_{\rm A}}{m_{\rm A} + m_{\rm B}}(v_{\rm A})_{\rm L}$$

$$= \left[\frac{2m_{\rm A}T_{\rm i}}{m_{\rm B}(m_{\rm A} + m_{\rm B})}\right]^{1/2} \qquad \textbf{(A2.10)}$$

It is also clear that before the collision the magnitudes of the CM velocities of the particles A and B are given by

$$v_{\rm A} = (v_{\rm A})_{\rm L} - V_{\rm L} = \frac{m_{\rm B}}{m_{\rm A} + m_{\rm B}}(v_{\rm A})_{\rm L} \qquad \textbf{(A2.11a)}$$

$$v_{\rm B} = V_{\rm L} = \frac{m_{\rm A}}{m_{\rm A} + m_{\rm B}}(v_{\rm A})_{\rm L} \qquad \textbf{(A2.11b)}$$

We remark that since the centre of mass always moves as a free particle, 'unaffected' by the collision process, it is the initial kinetic energy available in the CM system, $T_{\rm i}$, which physically characterises the collision. For example, in electron–atom scattering, where $m_{\rm A}$, the electron mass, is much smaller than $m_{\rm B}$, the mass of the atom, we deduce from (A2.9) and (A2.10) that $T_{\rm i} \simeq (T_{\rm i})_{\rm L}$ and $V_{\rm L} \simeq 0$. Thus in this case the laboratory and CM systems nearly coincide and almost all the initial kinetic energy in the laboratory system is available in the centre of mass system. On the contrary if we consider a collision between two atoms having the same mass ($m_{\rm A} = m_{\rm B}$) we see that $V_{\rm L} = (v_{\rm A})_{\rm L}/2$ and $T_{\rm i} = (T_{\rm i})_{\rm L}/2$, so that only one-half of the initial laboratory kinetic energy $(T_{\rm i})_{\rm L}$ is transformed into CM kinetic energy, the second half being 'dissipated' in the motion of the centre of mass.

A useful expression of $T_{\rm i}$ may be obtained in the following way. We first define the *relative momentum* $\mathbf{p}$ of two particles 1 and 2 having masses m_1 and m_2 and momenta $\mathbf{p}_1$ and $\mathbf{p}_2$ by the relation

$$\mathbf{p} = \frac{m_2\mathbf{p}_1 - m_1\mathbf{p}_2}{m_1 + m_2} \qquad \textbf{(A2.12)}$$

Using this definition, and evaluating the momenta of the colliding particles A and B successively in the centre of mass and in the laboratory system, we find that the *initial relative momentum* of the two particles A and B is given by

$$\mathbf{p}_{\rm i} = \mathbf{p}_{\rm A} = -\mathbf{p}_{\rm B} \qquad \textbf{(A2.13)}$$

or

$$\mathbf{p}_i = \frac{m_B}{m_A + m_B}(\mathbf{p}_A)_L \tag{A2.14}$$

As a result, the initial kinetic energy T_i available in the CM system may be written as

$$T_i = \frac{p_A^2}{2m_A} + \frac{p_B^2}{2m_B} = \frac{p_i^2}{2\mu} = \frac{1}{2}\mu v_i^2 \tag{A2.15}$$

where

$$\mu = \frac{m_A m_B}{m_A + m_B} \tag{A2.16}$$

is the reduced mass of the two particles A and B, $\mathbf{v}_i = (\mathbf{v}_A)_L$ is the initial relative velocity, of magnitude v_i, and we have used the fact that $\mathbf{p}_i = \mu(\mathbf{v}_A)_L = \mu\mathbf{v}_i$.

Elastic scattering

Thus far we have focused our attention on the initial situation, before the collision has occurred. The above analysis may also be repeated to obtain the relationships between the final centre of mass and laboratory quantities, that is *after* the collision has taken place. As an example, we shall consider the case of an *elastic scattering process*

$$A + B \to A + B \tag{A2.17}$$

in which the particles are simply scattered without any change in their internal structure [2]. This process is represented in Fig. A2.2(a) in the laboratory system, while in Fig. A2.2(b) it is described in the centre of mass system.

Let $\mathbf{v}'_A$ and $\mathbf{v}'_B$ be respectively the CM velocities of particles A and B after the collision, while $\mathbf{p}'_A = m_A\mathbf{v}'_A$ and $\mathbf{p}'_B = m_B\mathbf{v}'_B$ are the corresponding CM momenta. Similarly, we shall write the final laboratory velocities as $(\mathbf{v}'_A)_L$ and $(\mathbf{v}'_B)_L$, the corresponding momenta being $(\mathbf{p}'_A)_L = m_A(\mathbf{v}'_A)_L$ and $(\mathbf{p}'_B)_L = m_B(\mathbf{v}'_B)_L$. From momentum conservation we have in the laboratory system

$$\mathbf{P}_L = (\mathbf{p}_A)_L = (\mathbf{p}'_A)_L + (\mathbf{p}'_B)_L \tag{A2.18}$$

while in the centre of mass system

$$\mathbf{p}_A + \mathbf{p}_B = \mathbf{p}'_A + \mathbf{p}'_B = 0 \tag{A2.19}$$

We also note from (A2.12) that the final relative momentum is

$$\mathbf{p}_f = \mathbf{p}'_A = -\mathbf{p}'_B \tag{A2.20}$$

[2] The generalisation to inelastic collisions A + B → A′ + B′ or reactions A + B → C + D (see Problem 12.1) is straightforward; it may be found for example in Joachain (1983), where the case of relativistic collisions is also discussed.

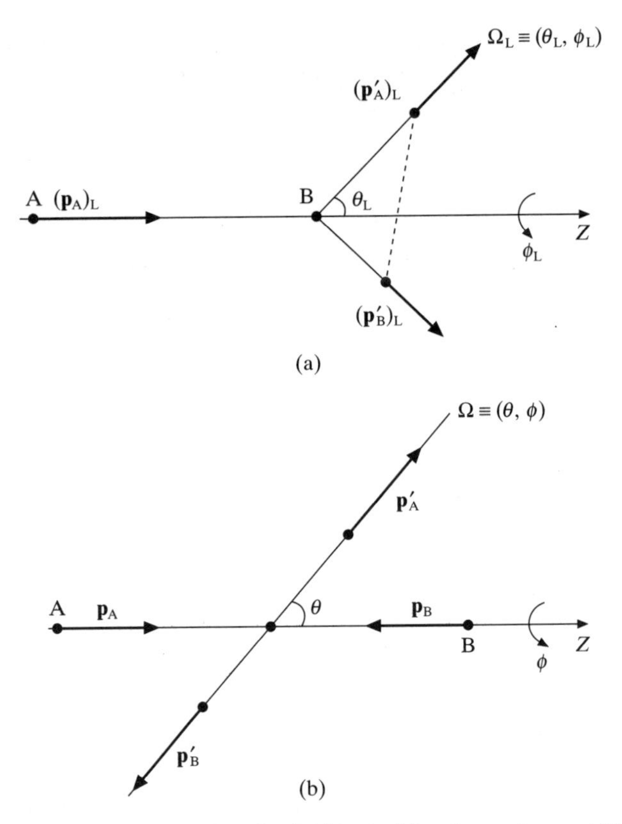

Figure A2.2 An elastic collision A + B → A + B (a) in the laboratory system and (b) in the centre of mass system.

or

$$\mathbf{p}_f = \frac{m_B(\mathbf{p}'_A)_L - m_A(\mathbf{p}'_B)_L}{m_A + m_B} \qquad \textbf{(A2.21)}$$

The final kinetic energy in the laboratory system is

$$(T_f)_L = \frac{(p'_A)^2_L}{2m_A} + \frac{(p'_B)^2_L}{2m_B} \qquad \textbf{(A2.22)}$$

and the corresponding quantity in the centre of mass system is

$$\begin{aligned} T_f &= (T_f)_L - \frac{P^2_L}{2M} \\ &= \frac{(p'_A)^2}{2m_A} + \frac{(p'_B)^2}{2m_B} = \frac{p^2_f}{2\mu} \end{aligned} \qquad \textbf{(A2.23)}$$

Table A2.1 The initial and final velocities of the particles A and B for an elastic collision $A + B \to A + B$. The spherical polar coordinates of a velocity $\mathbf{v}$ are denoted by (v, θ, ϕ).

	Initial velocities (before collision)		*Final velocities (after collision)*	
	A	B	A	B
In the laboratory system (L)	$(\mathbf{v}_A)_L = (v_A)_L \hat{\mathbf{z}}$	0	$(\mathbf{v}'_A)_L \equiv [(v'_A)_L, \theta_L, \phi_L]$	$(\mathbf{v}'_B)_L \equiv [(v'_B)_L, (\theta_B)_L, (\phi_B)_L]$
In the centre of mass system (CM)	$(\mathbf{v}_A)_L - \mathbf{V}_L = [(v_A)_L - V_L]\hat{\mathbf{z}} = v_A \hat{\mathbf{z}}$	$-\mathbf{V}_L = -V_L \hat{\mathbf{z}} = -v_B \hat{\mathbf{z}}$	$\mathbf{v}'_A \equiv (v_A, \theta, \phi)$	$\mathbf{v}'_B \equiv (v_B, \pi - \theta, \phi + \pi)$

where μ is the reduced mass (A2.16). Conservation of energy implies that for the elastic scattering event considered here we must have

$$T_i = T_f \quad \text{or} \quad p_i = p_f \tag{A2.24}$$

Let us now consider the velocities of the particles. We see from (A2.13), (A2.20) and (A2.24) that for an elastic collision

$$\begin{aligned} v'_A &= v_A \\ v'_B &= v_B \end{aligned} \tag{A2.25}$$

so that the magnitude of the CM velocities is unchanged. The initial and final velocities of the two particles are given in Table A2.1 in both the laboratory and centre of mass system.

From the definition (A1.15) of the differential cross-section, given in Appendix 1, we know that the same number dN'_A of particles A are emitted into the solid angle $d\Omega$ about the direction (θ, ϕ) in the CM system as are emitted into $d\Omega_L$ about (θ_L, ϕ_L) in the laboratory system. Thus the laboratory and CM differential cross-sections are related by

$$\frac{d\sigma}{d\Omega_L}(\theta_L, \phi_L)\, d\Omega_L = \frac{d\sigma}{d\Omega}(\theta, \phi)\, d\Omega$$

or

$$\frac{d\sigma}{d\Omega_L}(\theta_L, \phi_L) \sin\theta_L\, d\theta_L\, d\phi_L = \frac{d\sigma}{d\Omega}(\theta, \phi) \sin\theta\, d\theta\, d\phi \tag{A2.26}$$

Using the last of equations (A2.4), we have immediately $d\phi_L = d\phi$. Moreover, using (A2.5) and (A2.6) we have

$$\tan\theta_L = \frac{\sin\theta}{\cos\theta + \tau_A} \tag{A2.27}$$

where

$$\tau_{\mathrm{A}} = \frac{V_{\mathrm{L}}}{v_{\mathrm{A}}} = \frac{m_{\mathrm{A}}}{m_{\mathrm{B}}} \tag{A2.28}$$

for the elastic scattering case considered here. We may also write

$$\cos\theta_{\mathrm{L}} = \frac{\cos\theta + \tau_{\mathrm{A}}}{(1 + 2\tau_{\mathrm{A}}\cos\theta + \tau_{\mathrm{A}}^2)^{1/2}} \tag{A2.29}$$

so that the elastic laboratory and CM differential cross-sections are related by

$$\frac{\mathrm{d}\sigma}{\mathrm{d}\Omega_{\mathrm{L}}}(\theta_{\mathrm{L}}, \phi_{\mathrm{L}}) = \frac{(1 + \tau_{\mathrm{A}}^2 + 2\tau_{\mathrm{A}}\cos\theta)^{3/2}}{|1 + \tau_{\mathrm{A}}\cos\theta|}\frac{\mathrm{d}\sigma}{\mathrm{d}\Omega}(\theta, \phi) \tag{A2.30}$$

Finally, by integrating the differential cross-section over all scattering angles, we obtain the total cross-section σ_{tot}. Using (A2.26) we see that

$$\sigma_{\mathrm{tot}} = \int \frac{\mathrm{d}\sigma}{\mathrm{d}\Omega_{\mathrm{L}}}(\theta_{\mathrm{L}}, \phi_{\mathrm{L}})\,\mathrm{d}\Omega_{\mathrm{L}} = \int \frac{\mathrm{d}\sigma}{\mathrm{d}\Omega}(\theta, \phi)\,\mathrm{d}\Omega \tag{A2.31}$$

so that the total cross-section is independent of the reference frame.

In Appendix 1 we discussed the scattering of a beam of particles A by fixed scattering centres B, or in other words by infinitely massive target scatterers B. These results can easily be extended to the realistic case of target scatterers B which have a finite mass m_{B}, and hence can recoil. Indeed, it is shown in Section 2.7 that the problem of two particles interacting through a potential $V(\mathbf{r})$ which depends only on their relative coordinate is entirely equivalent, in the CM system, to the problem of a 'relative' particle moving in the potential $V(\mathbf{r})$, the mass of the 'relative' particle being the reduced mass μ of the two particles. This reduction of a two-body problem to an equivalent one-body problem in the CM system is valid in both classical and quantum mechanics. In particular, the results obtained in Appendix 1 for classical scattering of a beam of particles by a central potential can be used directly for the classical scattering of two particles A and B interacting via a central force, provided the mass m is replaced by the reduced mass $\mu = m_{\mathrm{A}}m_{\mathrm{B}}/(m_{\mathrm{A}} + m_{\mathrm{B}})$ and the scattering angle θ, energy E, velocities, etc. are understood to be the centre of mass quantities. For example, since the initial kinetic energy available in the CM system is given by $\mu v_{\mathrm{i}}^2/2$, the Rutherford differential cross-section (A1.33) becomes

$$\frac{\mathrm{d}\sigma_{\mathrm{c}}}{\mathrm{d}\Omega} = \left(\frac{q_{\mathrm{A}}q_{\mathrm{B}}}{4\pi\varepsilon_0}\right)^2 \frac{1}{4\mu^2 v_{\mathrm{i}}^4}\frac{1}{\sin^4(\theta/2)} \tag{A2.32}$$

where $v_{\mathrm{i}} = (v_{\mathrm{A}})_{\mathrm{L}}$ is the magnitude of the initial relative velocity of the two colliding particles. The result (A2.32) is the Rutherford differential cross-section for Coulomb scattering of a particle A of charge q_{A} by a particle B of charge q_{B}, written in the CM system. The corresponding laboratory differential cross-section may be obtained by using the relation (A2.30).

3 Evaluation of integrals by using generating functions

In this appendix we shall show how generating functions may be used to evaluate various integrals involving harmonic oscillator or hydrogenic wave functions.

Harmonic oscillator

We have seen in Section 2.4 that the Hermite polynomials $H_n(\xi)$ can be expressed in terms of a generating function $G(\xi, s)$ as

$$G(\xi, s) = \mathrm{e}^{-s^2+2s\xi} = \sum_{n=0}^{\infty} \frac{H_n(\xi)}{n!} s^n \tag{A3.1}$$

This generating function is useful in many calculations, and in particular for the evaluation of various integrals involving the harmonic oscillator wave functions

$$\psi_n(x) = N_n\, \mathrm{e}^{-\alpha^2 x^2/2} H_n(\alpha x) \tag{A3.2}$$

As a first example, we shall normalise the wave functions $\psi_n(x)$: that is, find the normalisation constant N_n such that

$$\int_{-\infty}^{+\infty} |\psi_n(x)|^2\, \mathrm{d}x = \frac{|N_n|^2}{\alpha} \int_{-\infty}^{+\infty} \mathrm{e}^{-\xi^2} H_n^2(\xi)\, \mathrm{d}\xi = 1 \tag{A3.3}$$

where we have set $\xi = \alpha x$. For this purpose, we consider the two generating functions $G(\xi, s)$ and

$$G(\xi, t) = \mathrm{e}^{-t^2+2t\xi} = \sum_{m=0}^{\infty} \frac{H_m(\xi)}{m!} t^m \tag{A3.4}$$

and use (A3.1) and (A3.4) to write

$$\int_{-\infty}^{+\infty} \mathrm{e}^{-\xi^2} G(\xi, s) G(\xi, t)\, \mathrm{d}\xi = \sum_{n=0}^{\infty} \sum_{m=0}^{\infty} \frac{s^n t^m}{n!m!} \int_{-\infty}^{+\infty} \mathrm{e}^{-\xi^2} H_n(\xi) H_m(\xi)\, \mathrm{d}\xi \tag{A3.5}$$

Using the fact that

$$\int_{-\infty}^{+\infty} \mathrm{e}^{-x^2}\, \mathrm{d}x = \sqrt{\pi} \tag{A3.6}$$

we see that the integral on the left-hand side of (A3.5) is simply

$$\int_{-\infty}^{+\infty} e^{-\xi^2} e^{-s^2+2s\xi} e^{-t^2+2t\xi} \, d\xi = \sqrt{\pi} \, e^{2st}$$

$$= \sqrt{\pi} \sum_{n=0}^{\infty} \frac{(2st)^n}{n!} \quad \textbf{(A3.7)}$$

Equating the coefficients of equal powers of s and t on the right-hand sides of (A3.5) and (A3.7), we find that

$$\int_{-\infty}^{+\infty} e^{-\xi^2} H_n^2(\xi) \, d\xi = \sqrt{\pi} \, 2^n n! \quad \textbf{(A3.8)}$$

and

$$\int_{-\infty}^{+\infty} e^{-\xi^2} H_n(\xi) H_m(\xi) \, d\xi = 0, \qquad n \neq m \quad \textbf{(A3.9)}$$

From (A3.3) and (A3.8) we see that apart from an arbitrary complex multiplication factor of modulus one the normalisation constant N_n is given by

$$N_n = \left(\frac{\alpha}{\sqrt{\pi} \, 2^n n!} \right)^{1/2} \quad \textbf{(A3.10)}$$

The second result (A3.9) implies that

$$\int_{-\infty}^{+\infty} \psi_m(x) \psi_n(x) \, dx = 0, \qquad n \neq m \quad \textbf{(A3.11)}$$

so that the (real) harmonic oscillator wave functions $\psi_n(x)$ and $\psi_m(x)$ are orthogonal if $n \neq m$, in agreement with the fact that they correspond to non-degenerate energy eigenvalues $E_n \neq E_m$.

As a second example, we calculate the integral

$$x_{nm} = \int_{-\infty}^{+\infty} \psi_n(x) x \psi_m(x) \, dx = \frac{N_n N_m}{\alpha^2} \int_{-\infty}^{+\infty} e^{-\xi^2} \xi H_n(\xi) H_m(\xi) \, d\xi \quad \textbf{(A3.12)}$$

Using again the two generating functions $G(\xi, s)$ and $G(\xi, t)$ given respectively by (A3.1) and (A3.4), we now look at the quantity

$$\int_{-\infty}^{+\infty} e^{-\xi^2} \xi G(\xi, s) G(\xi, t) \, d\xi = \sum_{n=0}^{\infty} \sum_{m=0}^{\infty} \frac{s^n t^m}{n! m!} \int_{-\infty}^{+\infty} e^{-\xi^2} \xi H_n(\xi) H_m(\xi) \, d\xi \quad \textbf{(A3.13)}$$

The integral on the left-hand side is simply

$$\int_{-\infty}^{+\infty} e^{-\xi^2} \xi \, e^{-s^2+2s\xi} e^{-t^2+2t\xi} \, d\xi = \sqrt{\pi} \, (s + t) e^{2st}$$

$$= \sqrt{\pi} \sum_{n=0}^{\infty} \frac{2^n}{n!} (s^{n+1} t^n + s^n t^{n+1}) \quad \textbf{(A3.14)}$$

Upon comparison of the coefficients of equal powers of s and t on the right-hand sides of (A3.13) and (A3.14), and using our previous result (A3.10), we find that

$$x_{nm} = \begin{cases} 0, & m \neq n \pm 1 \\ \dfrac{1}{\alpha}\left(\dfrac{n+1}{2}\right)^{1/2}, & m = n+1 \\ \dfrac{1}{\alpha}\left(\dfrac{n}{2}\right)^{1/2}, & m = n-1 \end{cases} \qquad \text{(A3.15)}$$

Hydrogenic atom

A similar method may be used to evaluate certain integrals involving the radial hydrogenic wave functions (see (3.52a))

$$R_{nl}(r) = N_{nl}\, \mathrm{e}^{-\rho/2} \rho^l L_{n+l}^{2l+1}(\rho), \qquad \rho = \frac{2Z}{na_\mu} r \qquad \text{(A3.16)}$$

Suppose, for example, that we want to evaluate integrals of the type

$$I^{\alpha}_{pq,p'q'} = \int_0^\infty \mathrm{e}^{-\rho} \rho^\alpha L_q^p(\rho) L_{q'}^{p'}(\rho)\, \mathrm{d}\rho \qquad \text{(A3.17)}$$

Using the generating function for the associated Laguerre polynomials

$$U_p(\rho, s) = \frac{(-s)^p\, \mathrm{e}^{-\rho s/(1-s)}}{(1-s)^{p+1}} = \sum_{q=p}^{\infty} \frac{L_q^p(\rho)}{q!} s^q \qquad \text{(A3.18)}$$

and a similar expression for $U_{p'}(\rho, t)$, we have

$$\int_0^\infty \mathrm{e}^{-\rho} \rho^\alpha U_p(\rho, s) U_{p'}(\rho, t)\, \mathrm{d}\rho = \sum_{q=p}^{\infty} \sum_{q'=p'}^{\infty} \frac{s^q t^{q'}}{q!(q')!} I^{\alpha}_{pq,p'q'} \qquad \text{(A3.19)}$$

The integral on the left-hand side is just

$$\begin{aligned} &\int_0^\infty \mathrm{e}^{-\rho} \rho^\alpha \frac{(-s)^p\, \mathrm{e}^{-\rho s/(1-s)}}{(1-s)^{p+1}} \frac{(-t)^{p'}\, \mathrm{e}^{-\rho t/(1-t)}}{(1-t)^{p'+1}}\, \mathrm{d}\rho \\ &= \frac{(-1)^{p+p'} s^p t^{p'}}{(1-s)^{p+1}(1-t)^{p'+1}} \int_0^\infty \mathrm{e}^{-\rho[1+s/(1-s)+t/(1-t)]} \rho^\alpha\, \mathrm{d}\rho \\ &= \frac{(-1)^{p+p'} s^p t^{p'}}{(1-s)^{p+1}(1-t)^{p'+1}} \frac{\Gamma(\alpha+1)}{\left(1 + \dfrac{s}{1-s} + \dfrac{t}{1-t}\right)^{\alpha+1}} \end{aligned} \qquad \text{(A3.20)}$$

where Γ is Euler's gamma function and we have used the fact that

$$\int_0^\infty \mathrm{e}^{-\beta x} x^\alpha\, \mathrm{d}x = \frac{\Gamma(\alpha+1)}{\beta^{\alpha+1}} \qquad \text{(A3.21)}$$

The required integral $I^{\alpha}_{pq,p'q'}$ may therefore be obtained by expanding the result (A3.20) as a power series in s and t and comparing the coefficients with those of the series on the right of (A3.19).

As a simple illustration of the method, we shall normalise the radial hydrogenic eigenfunctions. We see from (3.56) that for this purpose we need the integral

$$I^{p+1}_{pq,pq} = \int_0^{\infty} \mathrm{e}^{-\rho}\rho^{p+1}[L^p_q(\rho)]^2\,\mathrm{d}\rho \tag{A3.22}$$

with $p = 2l + 1$ and $q = n + l$. Using the generating functions $U_p(\rho, s)$ and $U_p(\rho, t)$ we therefore have in this case

$$\int_0^{\infty} \mathrm{e}^{-\rho}\rho^{p+1}U_p(\rho, s)U_p(\rho, t)\,\mathrm{d}\rho = \sum_{q=p}^{\infty}\sum_{q'=p}^{\infty}\frac{s^q t^{q'}}{q!(q')!}I^{p+1}_{pq,pq'} \tag{A3.23}$$

and also

$$\begin{aligned}
&\int_0^{\infty} \mathrm{e}^{-\rho}\rho^{p+1}U_p(\rho, s)U_p(\rho, t)\,\mathrm{d}\rho \\
&= \frac{(st)^p}{(1-s)^{p+1}(1-t)^{p+1}}\int_0^{\infty} \mathrm{e}^{-\rho[1+s/(1-s)+t/(1-t)]}\rho^{p+1}\,\mathrm{d}\rho \\
&= \frac{(p+1)!(st)^p(1-s)(1-t)}{(1-st)^{p+2}} \\
&= (p+1)!(1-s-t+st)\sum_{k=0}^{\infty}\frac{(p+k+1)!}{k!(p+1)!}(st)^{p+k}
\end{aligned} \tag{A3.24}$$

where we have used the binomial theorem to expand the quantity $(1 - st)^{-p-2}$. Upon comparison of the right-hand sides of (A3.23) and (A3.24), we see that the required integral $I^{p+1}_{pq,pq}$ is equal to $(q!)^2$ times the coefficient of $(st)^q$ in the series (A3.24). Thus

$$\begin{aligned}
I^{p+1}_{pq,pq} &= (q!)^2(p+1)!\left[\frac{(q+1)!}{(q-p)!(p+1)!} + \frac{q!}{(q-p-1)!(p+1)!}\right] \\
&= \frac{(q!)^3(2q-p+1)}{(q-p)!}
\end{aligned} \tag{A3.25}$$

and remembering that $p = 2l + 1$ and $q = n + l$, we finally have

$$\int_0^{\infty} \mathrm{e}^{-\rho}\rho^{2l}[L^{2l+1}_{n+l}(\rho)]^2\rho^2\,\mathrm{d}\rho = \frac{2n[(n+l)!]^3}{(n-l-1)!} \tag{A3.26}$$

which is the result quoted in (3.57).

4 Angular momentum – useful formulae and results

In this appendix, we collect useful relations and results concerning angular momentum in quantum mechanics. Elementary treatments of the subject can be found in the texts by Merzbacher (1998) and Bransden and Joachain (2000), while more advanced treatments are given in the monographs of Edmonds (1957) and Rose (1957).

Angular momentum operators

In Chapter 2, we discussed the representation of the Cartesian components of the orbital angular momentum $\mathbf{L}$ by differential operators, starting from the expression $\mathbf{L} = \mathbf{r} \times \mathbf{p}$. We also outlined the theory of particles of spin one-half, for which the Cartesian components of the spin angular momentum $\mathbf{S}$ are represented by two-by-two matrices. We shall now obtain a matrix representation for the Cartesian components J_x, J_y and J_z of a general angular momentum operator $\mathbf{J}$, which will include the particular cases such that the angular momentum is purely of orbital or of spin type.

The operators J_x, J_y and J_z are *defined* to be Hermitian operators satisfying the commutation relations

$$[J_x, J_y] = \mathrm{i}\hbar J_z, \qquad [J_y, J_z] = \mathrm{i}\hbar J_x, \qquad [J_z, J_x] = \mathrm{i}\hbar J_y \tag{A4.1}$$

Since $\mathbf{J}^2 = J_x^2 + J_y^2 + J_z^2$ commutes with each component of $\mathbf{J}$, simultaneous eigenfunctions ψ_{jm} of $\mathbf{J}^2$ and one component, say J_z, can be found, with

$$\mathbf{J}^2 \psi_{jm} = j(j+1)\hbar^2 \psi_{jm}; \tag{A4.2a}$$

$$J_z \psi_{jm} = m\hbar \psi_{jm} \tag{A4.2b}$$

We may normalise the ψ_{jm} to unity, in which case the orthonormality relations

$$\langle \psi_{j'm'} | \psi_{jm} \rangle = \delta_{jj'} \delta_{mm'} \tag{A4.3}$$

are satisfied, and J_z is represented by a diagonal matrix with elements

$$\langle \psi_{j'm'} | J_z | \psi_{jm} \rangle = \delta_{jj'} \delta_{mm'} m\hbar \tag{A4.4}$$

In this matrix representation, the eigenfunctions ψ_{jm} are, in fact, column vectors.

The eigenvalues, which we have written for later convenience in the form $j(j+1)\hbar^2$ and $m\hbar$, are real as $\mathbf{J}^2$ and J_z are Hermitian. They can be determined by the following argument, which we give in outline only.

Let us define the raising and lowering operators $J_\pm$ as

$$J_\pm = J_x \pm \mathrm{i}J_y \tag{A4.5}$$

with $J_+ = J_-^\dagger$; $J_- = J_+^\dagger$. We note the relations

$$[J_z, J_\pm] = \pm\hbar J_\pm \tag{A4.6}$$

$$[\mathbf{J}^2, J_\pm] = 0 \tag{A4.7}$$

$$J_+J_- = \mathbf{J}^2 - J_z^2 + \hbar J_z; \qquad J_-J_+ = \mathbf{J}^2 - J_z^2 - \hbar J_z \tag{A4.8}$$

From the commutation relation (A4.6), we have

$$\begin{aligned} J_z\{J_\pm\psi_{jm}\} &= J_\pm J_z\psi_{jm} \pm \hbar J_\pm\psi_{jm} \\ &= (m \pm 1)\hbar\{J_\pm\psi_{jm}\} \end{aligned} \tag{A4.9}$$

so that $(J_\pm\psi_{jm})$ are eigenfunctions of J_z belonging to the eigenvalues $(m \pm 1)$. Because of (A4.7), these functions are simultaneously eigenfunctions of $\mathbf{J}^2$, belonging to the eigenvalue $j(j+1)\hbar^2$.

For any wave function ϕ, $\langle\phi|\mathbf{J}^2|\phi\rangle \geqslant \langle\phi|J_z^2|\phi\rangle$, and setting $\phi = \psi_{jm}$, we find that

$$j(j+1) \geqslant m_j^2 \tag{A4.10}$$

By operating with J_+ or J_- repeatedly, sequences of eigenfunctions of J_z can be constructed, namely $(J_+)^n\psi_{jm}$, $(J_-)^{n'}\psi_{jm}$, with eigenvalues $(m+n)\hbar$ and $(m-n')\hbar$ respectively. In view of (A4.10), for each j there must be a maximum eigenvalue of J_z, say $\lambda\hbar$, and also a minimum eigenvalue, say $\lambda'\hbar$, such that $\lambda - \lambda'$ = an integer (or zero). If J_+ is applied to $\psi_{j\lambda}$, we must have $J_+\psi_{j\lambda} = 0$, for otherwise the sequence would not terminate and $\lambda\hbar$ would not be the maximum eigenvalue. Using (A4.8), we have

$$J_-J_+\psi_{j\lambda} = \{j(j+1) - \lambda^2 - \lambda\}\hbar^2\psi_{j\lambda} = 0 \tag{A4.11}$$

with the solution $\lambda = j$. In the same way $J_-\psi_{j\lambda'} = 0$, from which we find that $\lambda' = -j$. Since $(\lambda - \lambda')$ is an integer (or zero), $(2j)$ is an integer (or zero) and j must be one of the integers (or zero) or half-odd integers, $j = 0, 1/2, 1, 3/2, 2, \ldots$. For a given value of j, m can take the $(2j+1)$ values $-j, -j+1, \ldots, j-1, j$.

To find the matrix elements of J_x and J_y, or equivalently of J_+ and J_-, we note that

$$J_+\psi_{jm} = N\psi_{jm+1} \tag{A4.12}$$

where N is a constant. Since both ψ_{jm} and ψ_{jm+1} are normalised to unity, we have from (A4.3)

$$\begin{aligned} |N|^2 &= \langle J_+\psi_{jm}|J_+\psi_{jm}\rangle = \langle\psi_{jm}|J_-J_+|\psi_{jm}\rangle \\ &= \hbar^2(j(j+1) - m(m+1)) \end{aligned} \tag{A4.13}$$

where we have used (A4.8). Adopting the convention that N is real and positive, we then obtain

$$N = \hbar\sqrt{j(j+1) - m(m+1)} \qquad \textbf{(A4.14)}$$

From (A4.12), the matrix representing J_+ in the basis of eigenfunctions ψ_{jm} is

$$\langle\psi_{j'm'}J_+|\psi_{jm}\rangle = \sqrt{j(j+1) - m(m+1)}\,\hbar\delta_{jj'}\delta_{m'm+1} \qquad \textbf{(A4.15)}$$

In a similar way, we find

$$\langle\psi_{j'm'}J_-|\psi_{jm}\rangle = \sqrt{j(j+1) - m(m-1)}\,\hbar\delta_{jj'}\delta_{m'm-1} \qquad \textbf{(A4.16)}$$

As we saw in Chapter 2, if **J** is a pure orbital angular momentum (**L**), the wave function must be single valued as a function of position, and this excludes the half-integral values of j. For a spin angular momentum or when **J** is the sum of an orbital and a spin angular momentum, both the integral and half-integral values are allowed.

Spherical harmonics and Legendre polynomials

In Chapter 2 we introduced the spherical harmonics $Y_{lm}(\theta, \phi)$, which are the simultaneous eigenfunctions of the orbital angular momentum operators $\mathbf{L}^2$ and L_z,

$$\mathbf{L}^2 Y_{lm} = l(l+1)\hbar^2 Y_{lm}; \qquad \textbf{(A4.17a)}$$

$$L_z Y_{lm} = m\hbar Y_{lm} \qquad \textbf{(A4.17b)}$$

where $l = 0, 1, 2, \ldots$ and $m = -l, -l+1, \ldots, l-1, l$. They satisfy the orthonormality relations

$$\int Y^*_{l'm'}(\theta, \phi) Y_{lm}(\theta, \phi)\,\mathrm{d}\Omega = \delta_{ll'}\delta_{mm'} \qquad (\mathrm{d}\Omega = \sin\theta\,\mathrm{d}\theta\,\mathrm{d}\phi) \qquad \textbf{(A4.18)}$$

and the closure relation

$$\sum_{l=0}^{\infty}\sum_{m=-l}^{l} Y^*_{lm}(\theta, \phi) Y_{lm}(\theta', \phi') = \delta(\Omega - \Omega') \qquad \textbf{(A4.19)}$$

where

$$\delta(\Omega - \Omega') = \frac{1}{\sin\theta}\delta(\theta - \theta')\delta(\phi - \phi')$$

If the operators $L_\pm$ are defined as $L_\pm = L_x \pm \mathrm{i}L_y$ (see (A4.5)) we have

$$L_\pm = \hbar\,\mathrm{e}^{\pm\mathrm{i}\phi}\left[\pm\frac{\partial}{\partial\theta} + \mathrm{i}\cot\theta\frac{\partial}{\partial\phi}\right] \qquad \textbf{(A4.20)}$$

and

$$L_\pm Y_{lm} = \hbar\sqrt{l(l+1) - m(m\pm 1)}\,Y_{lm\pm 1} \qquad \textbf{(A4.21)}$$

in agreement with the general results of (A4.15) and (A4.16).

In the special case $m = 0$, the spherical harmonics are given by

$$Y_{l0} = \sqrt{\frac{2l + 1}{4\pi}} P_l(\cos\theta) \tag{A4.22}$$

where the functions $P_l(\cos\theta)$ are the Legendre polynomials defined in Section 2.5.

Let $\mathbf{r}_1$ and $\mathbf{r}_2$ be two vectors having polar angles (θ_1, ϕ_1) and (θ_2, ϕ_2) respectively, and let θ be the angle between them. It can be shown that

$$P_l(\cos\theta) = \frac{4\pi}{2l + 1} \sum_{m=-l}^{+l} Y^*_{lm}(\theta_1, \phi_1) Y_{lm}(\theta_2, \phi_2) \tag{A4.23}$$

which is known as the addition (or biaxial) theorem of the spherical harmonics. From the generating function (2.168) of the Legendre polynomials, we see that

$$\frac{1}{|\mathbf{r}_1 - \mathbf{r}_2|} = \sum_{l=0}^{\infty} \frac{(r_<)^l}{(r_>)^{l+1}} P_l(\cos\theta) \tag{A4.24}$$

which, using (A4.23), may also be written as

$$\frac{1}{|\mathbf{r}_1 - \mathbf{r}_2|} = \sum_{l=0}^{\infty} \sum_{m=-l}^{+l} \frac{4\pi}{2l + 1} \frac{(r_<)^l}{(r_>)^{l+1}} Y^*_{lm}(\theta_1, \phi_1) Y_{lm}(\theta_2, \phi_2) \tag{A4.25}$$

It can also be shown (Mathews and Walker, 1973) that

$$\frac{\exp[\mathrm{i}k|\mathbf{r}_1 - \mathbf{r}_2|]}{|\mathbf{r}_1 - \mathbf{r}_2|} = \mathrm{i}k \sum_{l=0}^{\infty} (2l + 1) j_l(kr_<)[j_l(kr_>) + \mathrm{i}n_l(kr_>)] P_l(\cos\theta) \tag{A4.26}$$

where j_l and n_l are spherical Bessel and Neumann functions respectively. Finally, we quote the formula giving the expansion of a plane wave in Legendre polynomials, namely

$$\exp(\mathrm{i}\mathbf{k}\cdot\mathbf{r}) = \sum_{l=0}^{\infty} (2l + 1)\mathrm{i}^l j_l(kr) P_l(\cos\theta) \tag{A4.27}$$

where θ is the angle between the vectors $\mathbf{k}$ and $\mathbf{r}$.

Addition of angular momenta. The Clebsch–Gordan coefficients

Consider a system described by two angular momenta $\mathbf{J}_1$ and $\mathbf{J}_2$, such that the components of $\mathbf{J}_1$ commute with the components of $\mathbf{J}_2$. For example, $\mathbf{J}_1$ and $\mathbf{J}_2$ could be the angular momenta of different particles, or the orbital and spin angular momenta of a single particle. The normalised simultaneous eigenfunctions of $\mathbf{J}_1^2$ and J_{1z} corresponding to eigenvalues $j_1(j_1 + 1)\hbar^2$ and $m_1\hbar$ will be denoted by $\psi_{j_1m_1}$ and similarly, the normalised simultaneous eigenfunctions of $\mathbf{J}_2^2$ and J_{2z} corresponding to eigenvalues $j_2(j_2 + 1)\hbar^2$ and $m_2\hbar$ will be denoted by $\psi_{j_2m_2}$. The simultaneous eigenfunctions of $\mathbf{J}_1^2$, J_{1z}, $\mathbf{J}_2^2$ and J_{2z} are then given by the product functions

$$\psi_{j_1m_1;j_2m_2} = \psi_{j_1m_1} \times \psi_{j_2m_2} \quad \textbf{(A4.28)}$$

and for a given j_1 and j_2, there are $(2j_1 + 1) \times (2j_2 + 1)$ of these functions.

Now consider the total angular momentum

$$\mathbf{J} = \mathbf{J}_1 + \mathbf{J}_2 \quad \textbf{(A4.29)}$$

Since $\mathbf{J}^2$, J_z, $\mathbf{J}_1^2$ and $\mathbf{J}_2^2$ all commute, these operators possess a set of simultaneous eigenfunctions, which we shall write as $\Phi^{jm}_{j_1j_2}$ where

$$\mathbf{J}^2\Phi^{jm}_{j_1j_2} = j(j+1)\hbar^2\Phi^{jm}_{j_1j_2}; \quad \textbf{(A4.30a)}$$

$$J_z\Phi^{jm}_{j_1j_2} = m\hbar\Phi^{jm}_{j_1j_2} \quad \textbf{(A4.30b)}$$

For a given j, there are $(2j + 1)$ values of m with $-j \leqslant m \leqslant j$ and j can take any of the values $|j_1 - j_2|, |j_1 - j_2| + 1, \dots, (j_1 + j_2)$. Again there are $(2j_1 + 1) \times (2j_2 + 1)$ of the functions $\Phi^{jm}_{j_1j_2}$, which can be related to the functions (A4.28) by a unitary transformation:

$$\Phi^{jm}_{j_1j_2} = \sum_{m_1m_2} \langle j_1j_2m_1m_2|jm\rangle \psi_{j_1m_1;j_2m_2} \quad \textbf{(A4.31)}$$

The coefficients $\langle j_1j_2m_1m_2|jm\rangle$ are called Clebsch–Gordan coefficients. These coefficients vanish unless $m = m_1 + m_2$ and $|j_1 - j_2| \leqslant j \leqslant j_1 + j_2$. The phase convention we use is such that they are *real*. They possess the following important properties: Orthonormality relations

$$\sum_{m_1m_2} \langle j_1j_2m_1m_2|jm\rangle\langle j_1j_2m_1m_2|j'm'\rangle = \delta_{jj'}\delta_{mm'}$$

$$\sum_{jm} \langle j_1j_2m_1m_2|jm\rangle\langle j_1j_2m_1'm_2'|jm\rangle = \delta_{m_1m_1'}\delta_{m_2m_2'} \quad \textbf{(A4.32)}$$

Symmetry properties

$$\begin{aligned}\langle j_1j_2m_1m_2|jm\rangle &= (-1)^{j_1+j_2-j}\langle j_2j_1m_2m_1|jm\rangle \\ &= (-1)^{j_1+j_2-j}\langle j_1j_2 - m_1 - m_2|j - m\rangle \\ &= (-1)^{j_1-m_1}\left(\frac{2j+1}{2j_2+1}\right)^{1/2}\langle j_1jm_1 - m|j_2 - m_2\rangle \quad \textbf{(A4.33)}\end{aligned}$$

In Table A4.1, the coefficients $\langle j_1j_2m_1m_2|jm\rangle$ are tabulated for the cases $j_2 = 1/2$ and $j_2 = 1$. By using the symmetry relations, all the coefficients with any one of j_1, j_2 or j equal to 1/2, or to 1, can be found.

Useful notations

When adding two orbital angular momenta $\mathbf{L}_1$ and $\mathbf{L}_2$, we shall write (A4.31) in the position representation as

$$\mathcal{Y}^{lm}_{l_1l_2}(\theta_1\phi_1; \theta_2\phi_2) = \sum_{m_1m_2} \langle l_1l_2m_1m_2|lm\rangle Y_{l_1m_1}(\theta_1, \phi_1)Y_{l_2m_2}(\theta_2, \phi_2) \quad \textbf{(A4.34)}$$

Table A4.1 Clebsch–Gordan coefficients for $j_2 = \frac{1}{2}$ and $j_2 = 1$.

	$\langle j_1 \frac{1}{2} m_1 m_2 \| jm \rangle$	
j	$m_2 = \frac{1}{2}$	$m_2 = -\frac{1}{2}$
$j_1 + \frac{1}{2}$	$\left(\dfrac{j_1 + m + \frac{1}{2}}{2j_1 + 1}\right)^{1/2}$	$\left(\dfrac{j_1 - m + \frac{1}{2}}{2j_1 + 1}\right)^{1/2}$
$j_1 - \frac{1}{2}$	$-\left(\dfrac{j_1 - m + \frac{1}{2}}{2j_1 + 1}\right)^{1/2}$	$\left(\dfrac{j_1 + m + \frac{1}{2}}{2j_1 + 1}\right)^{1/2}$

		$\langle j_1 1 m_1 m_2 \| jm \rangle$	
j	$m_2 = 1$	$m_2 = 0$	$m_2 = -1$
$j_1 + 1$	$\left[\dfrac{(j_1 + m)(j_1 + m + 1)}{(2j_1 + 1)(2j_1 + 2)}\right]^{1/2}$	$\left[\dfrac{(j_1 - m + 1)(j_1 + m + 1)}{(2j_1 + 1)(j_1 + 1)}\right]^{1/2}$	$\left[\dfrac{(j_1 - m)(j_1 - m + 1)}{(2j_1 + 1)(2j_1 + 2)}\right]^{1/2}$
j_1	$-\left[\dfrac{(j_1 + m)(j_1 - m + 1)}{2j_1(j_1 + 1)}\right]^{1/2}$	$\left[\dfrac{m^2}{j_1(j_1 + 1)}\right]^{1/2}$	$\left[\dfrac{(j_1 - m)(j_1 + m + 1)}{2j_1(j_1 + 1)}\right]^{1/2}$
$j_1 - 1$	$\left[\dfrac{(j_1 - m)(j_1 - m + 1)}{2j_1(2j_1 + 1)}\right]^{1/2}$	$-\left[\dfrac{(j_1 - m)(j_1 + m)}{j_1(2j_1 + 1)}\right]^{1/2}$	$\left[\dfrac{(j_1 + m + 1)(j_1 + m)}{2j_1(2j_1 + 1)}\right]^{1/2}$

where $\mathcal{Y}^{lm}_{l_1 l_2}$ is a simultaneous eigenfunction of $\mathbf{L}_1^2$, $\mathbf{L}_2^2$, $\mathbf{L}^2$ and L_z, with $\mathbf{L} = \mathbf{L}_1 + \mathbf{L}_2$. Similarly, when adding an orbital angular momentum $\mathbf{L}$ with a spin angular momentum $\mathbf{S}$, so that $\mathbf{J} = \mathbf{L} + \mathbf{S}$, we shall write

$$\mathcal{Y}^{jm}_{ls}(\theta, \phi) = \sum_{m_l m_s} \langle l s m_l m_s | jm \rangle Y_{lm_l}(\theta, \phi) \chi_{sm_s} \tag{A4.35}$$

where χ_{sm_s} is a spin wave function. The functions $\mathcal{Y}^{jm}_{ls}$ are called *generalised spherical harmonics*. They are simultaneous eigenfunctions of the operators $\mathbf{L}^2$, $\mathbf{S}^2$, $\mathbf{J}^2$ and J_z.

When taking matrix elements of operators with respect to the eigenfunctions $\Phi^{jm}_{j_1 j_2}$, $\mathcal{Y}^{jm}_{ls}, \ldots$, we shall frequently use the Dirac notation in which eigenvectors $|\Phi^{jm}_{j_1 j_2}\rangle$ are written in the form $|j_1 j_2 jm\rangle$. We then write

$$\langle \Phi^{jm}_{j_1 j_2} | A | \Phi^{j'm'}_{j'_1 j'_2} \rangle = \langle j_1 j_2 jm | A | j'_1 j'_2 j' m' \rangle \tag{A4.36}$$

and

$$\int d\Omega_1 \int d\Omega_2 \mathcal{Y}^{lm*}_{l_1 l_2}(\theta_1 \phi_1; \theta_2 \phi_2) A \mathcal{Y}^{l'm'}_{l'_1 l'_2}(\theta_1 \phi_1; \theta_2 \phi_2)$$

$$= \langle l_1 l_2 lm | A | l'_1 l'_2 l' m' \rangle \tag{A4.37}$$

where A is an operator. The defining relation of the Clebsch–Gordan coefficients (A4.31) can then be written as

$$|j_1 j_2 jm\rangle = \sum_{m_1 m_2} \langle j_1 j_2 m_1 m_2 | jm \rangle |j_1 m_1\rangle \times |j_2 m_2\rangle \tag{A4.38}$$

Integrals of products of spherical harmonics

It can be shown that the product $Y_{l_1m_1}(\theta, \phi)Y_{l_2m_2}(\theta, \phi)$ can be expressed as a series by

$$Y_{l_1m_1}(\theta, \phi)Y_{l_2m_2}(\theta, \phi) = \sum_{l=|l_1-l_2|}^{l_1+l_2} \sum_{m=-l}^{l} \left[\frac{(2l_1+1)(2l_2+1)}{4\pi(2l+1)}\right]^{1/2} \times \langle l_1 l_2 00 | l0\rangle \langle l_1 l_2 m_1 m_2 | lm\rangle Y_{lm}(\theta, \phi) \quad \textbf{(A4.39)}$$

This enables us to evaluate the integral of a product of three spherical harmonics. Using (2.181b) and the orthonormality property (A4.18) we have

$$\int Y_{l_1m_1}(\theta, \phi)Y_{l_2m_2}(\theta, \phi)Y_{l_3m_3}(\theta, \phi)\, d\Omega = (-1)^{m_3}\left[\frac{(2l_1+1)(2l_2+1)}{4\pi(2l_3+1)}\right]^{1/2} \langle l_1 l_2 00 | l_3 0\rangle \langle l_1 l_2 m_1 m_2 | l_3 - m_3\rangle \quad \textbf{(A4.40)}$$

Scalar and vector operators. Irreducible tensor operators

A *scalar* operator $\mathscr{S}$ is one for which the expectation values $\langle\phi|\mathscr{S}|\phi\rangle$ are unaltered by a rotation of the coordinate system. It can be shown (Merzbacher, 1998) that for an operator to be scalar it must commute with all the components of the total angular momentum **J**:

$$[\mathscr{S}, \mathbf{J}] = 0 \quad \textbf{(A4.41)}$$

from which it follows that if ψ_{jm} is a simultaneous eigenfunction of $\mathbf{J}^2$ and J_z belonging to the quantum numbers j and m, then $\langle\psi_{jm}|\mathscr{S}|\psi_{j'm'}\rangle$ vanishes unless $j = j'$ and $m = m'$, and

$$\mathscr{S}\psi_{jm} = \lambda\psi_{jm} \quad \textbf{(A4.42)}$$

where λ is an eigenvalue of $\mathscr{S}$. Since $J_+\psi_{jm} = N\psi_{jm+1}$ and as $[J_+, \mathscr{S}] = 0$, we must also have

$$\mathscr{S}\psi_{jm+1} = \lambda\psi_{jm+1} \quad \textbf{(A4.43)}$$

so that the eigenvalue λ is independent of m (but it does depend on j).

In general, wave functions depend on other quantum numbers in addition to angular momentum quantum numbers (for example, principal quantum numbers). Denoting these other quantum numbers collectively by α, we see that

$$\langle\phi_{\alpha jm}|\mathscr{S}|\phi_{\alpha' j'm'}\rangle \equiv \langle\alpha jm|\mathscr{S}|\alpha' j'm'\rangle = \lambda_{j\alpha\alpha'}\delta_{jj'}\delta_{mm'} \quad \textbf{(A4.44)}$$

This is the simplest example of the fact that matrix elements of operators having well-defined properties under rotations depend upon the magnetic quantum numbers through a 'geometrical' factor (equal in the present case to $\delta_{mm'}$) which is independent of the dynamics of the system.

In this book, we are mainly concerned with *vector operators*, whose components transform like the components of a vector under rotations. The condition for an operator **V** with Cartesian components V_x, V_y and V_z to be a vector operator is that it satisfies the commutation relations

$$\begin{aligned}
&[J_x, V_x] = 0, && [J_y, V_x] = -\mathrm{i}\hbar V_z, && [J_z, V_x] = \mathrm{i}\hbar V_y \\
&[J_x, V_y] = \mathrm{i}\hbar V_z, && [J_y, V_y] = 0, && [J_z, V_y] = -\mathrm{i}\hbar V_x \\
&[J_x, V_z] = -\mathrm{i}\hbar V_y, && [J_y, V_z] = \mathrm{i}\hbar V_x && [J_z, V_z] = 0
\end{aligned} \tag{A4.45}$$

where J_x, J_y and J_z are the Cartesian components of the total angular momentum **J**. The operators **L**, **S** and **J** are examples of vector operators. Using (A4.45) and the basic commutation relations (A4.1), it may be shown that a vector operator **V** satisfies the relations

$$\mathbf{J} \times \mathbf{V} + \mathbf{V} \times \mathbf{J} = 2\mathrm{i}\hbar\mathbf{V} \tag{A4.46}$$

and

$$[\mathbf{J}^2, [\mathbf{J}^2, \mathbf{V}]] = 2\hbar^2(\mathbf{J}^2\mathbf{V} + \mathbf{V}\mathbf{J}^2) - 4\hbar^2(\mathbf{V}\cdot\mathbf{J})\mathbf{J} \tag{A4.47}$$

The *spherical components* V_q ($q = 0, \pm 1$) of **V** are defined as

$$V_1 = -\frac{1}{\sqrt{2}}(V_x + \mathrm{i}V_y), \qquad V_0 = V_z, \qquad V_{-1} = \frac{1}{\sqrt{2}}(V_x - \mathrm{i}V_y) \tag{A4.48}$$

The set of three operators V_q satisfies the commutation relations (which follow from (A4.45))

$$\begin{aligned}
&[J_z, V_q] = q\hbar V_q \\
&[J_+, V_q] = [(1-q)(2+q)]^{1/2}\hbar V_{q+1} \\
&[J_-, V_q] = [(1+q)(2-q)]^{1/2}\hbar V_{q-1}
\end{aligned} \tag{A4.49}$$

where $V_q = 0$ if $q \neq 0, \pm 1$. The operators V_q are special cases of *irreducible tensor operators* T_q^k of rank k, which form a set of $(2k + 1)$ operators with q running over the values $-k, -k+1, \ldots, k-1, k$, and satisfying the commutation relations

$$\begin{aligned}
&[J_z, T_q^k] = q\hbar T_q^k \\
&[J_+, T_q^k] = [(k-q)(k+q+1)]^{1/2}\hbar T_{q+1}^k \\
&[J_-, T_q^k] = [(k+q)(k-q+1)]^{1/2}\hbar T_{q-1}^k
\end{aligned} \tag{A4.50}$$

For such operators, the matrix elements between eigenfunctions ψ_{jm} and $\psi_{j'm'}$ depend on m and m' only through a factor which can be shown to be equal to the Clebsch–Gordan coefficient $\langle jkmq | j'm'\rangle$. Thus

$$\langle \alpha'j'm' | T_q^k | \alpha jm\rangle = \frac{1}{\sqrt{2j'+1}}\langle jkmq | j'm'\rangle\langle \alpha'j' \| T^k \| \alpha j\rangle \tag{A4.51}$$

where the reduced matrix element $\langle \alpha' j' \| T^k \| \alpha j \rangle$ is a number depending on α, α', j and j' but not on m and m'. This result is called the *Wigner–Eckart theorem*. It should be noticed that the appearance of the factor $(2j' + 1)^{-1/2}$ on the right-hand side of (A4.51) is conventional; it could be absorbed into the reduced matrix element.

The application of the Wigner–Eckart theorem to a scalar operator $\mathcal{S} = T_0^0$ reproduces the result (A4.44). For vector operators we have $V_q \equiv T_q^1$. As an example of a vector operator we can take the total angular momentum $\mathbf{J}$. Then, definining J_q by using (A4.48) with J in place of V, we have

$$\langle \alpha' j' m' | J_q | \alpha j m \rangle = \frac{1}{\sqrt{2j' + 1}} \langle j 1 m q | j' m' \rangle \langle \alpha' j' \| J \| \alpha j \rangle \qquad \textbf{(A4.52)}$$

Setting $q = 0$ and noting that $J_0 \equiv J_z$, we find by using Table A4.1 that

$$\langle \alpha' j' \| J \| \alpha j \rangle = \sqrt{j(j+1)}\, \hbar \delta_{jj'} \delta_{\alpha\alpha'} \qquad \textbf{(A4.53)}$$

It follows from (A4.51) that if $\mathbf{V}$ is any vector operator, then

$$\langle \alpha' j' m' | \mathbf{V} | \alpha j m \rangle = C \, \langle \alpha' j' m' | \mathbf{J} | \alpha j m \rangle \qquad \textbf{(A4.54)}$$

where C is independent of m and m'. For the case $j' = j$, $\alpha' = \alpha$, C can be found by writing

$$\langle \alpha j m | \mathbf{V} \cdot \mathbf{J} | \alpha j m \rangle = \sum_{m'} \langle \alpha j m | \mathbf{V} | \alpha j m' \rangle \cdot \langle \alpha j m' | \mathbf{J} | \alpha j m \rangle \qquad \textbf{(A4.55)}$$

where we have used the closure relation

$$\sum_{m'} | \alpha j m' \rangle \langle \alpha j m' | = I \qquad \textbf{(A4.56)}$$

Thus

$$\begin{aligned} \langle \alpha j m | \mathbf{V} \cdot \mathbf{J} | \alpha j m \rangle &= C \sum_{m'} \langle \alpha j m | \mathbf{J} | \alpha j m' \rangle \cdot \langle \alpha j m' | \mathbf{J} | \alpha j m \rangle \\ &= C \, \langle \alpha j m | \mathbf{J}^2 | \alpha j m \rangle \\ &= C \, j(j+1)\hbar^2 \end{aligned} \qquad \textbf{(A4.57)}$$

Using (A4.54), we then obtain the useful equation

$$\begin{aligned} j(j+1)\hbar^2 \langle \alpha j m | \mathbf{V} | \alpha j m \rangle &= \langle \alpha j m | \mathbf{V} \cdot \mathbf{J} | \alpha j m \rangle \langle \alpha j m | \mathbf{J} | \alpha j m \rangle \\ &= \langle \alpha j m | (\mathbf{V} \cdot \mathbf{J}) \mathbf{J} | \alpha j m \rangle \end{aligned} \qquad \textbf{(A4.58)}$$

which relates the expectation values of the components of $\mathbf{V}$ to those of the components of $\mathbf{J}$.

5 Hydrogenic wave functions in momentum space

We have seen in Section 2.1 that the wave function in momentum space, $\phi(\mathbf{p})$, is defined as the Fourier transform of the ordinary wave function $\psi(\mathbf{r})$ in position space. That is,

$$\phi(\mathbf{p}) = (2\pi\hbar)^{-3/2} \int \exp(-\mathrm{i}\mathbf{p}\cdot\mathbf{r}/\hbar)\psi(\mathbf{r})\, \mathrm{d}\mathbf{r} \qquad \textbf{(A5.1)}$$

The wave function $\psi(\mathbf{r})$ may be deduced from $\phi(\mathbf{p})$ by the inverse Fourier transform

$$\psi(\mathbf{r}) = (2\pi\hbar)^{-3/2} \int \exp(\mathrm{i}\mathbf{p}\cdot\mathbf{r}/\hbar)\phi(\mathbf{p})\, \mathrm{d}\mathbf{p} \qquad \textbf{(A5.2)}$$

In what follows we shall use atomic units (see Appendix 14), the unit of length being a_0 and the unit of momentum being $p_0 = \hbar/a_0$. In these units the equations (A5.1) and (A5.2) become simply

$$\phi(\mathbf{p}) = (2\pi)^{-3/2} \int \exp(-\mathrm{i}\mathbf{p}\cdot\mathbf{r})\psi(\mathbf{r})\, \mathrm{d}\mathbf{r} \qquad \textbf{(A5.3)}$$

and

$$\psi(\mathbf{r}) = (2\pi)^{-3/2} \int \exp(\mathrm{i}\mathbf{p}\cdot\mathbf{r})\phi(\mathbf{p})\, \mathrm{d}\mathbf{p} \qquad \textbf{(A5.4)}$$

We note that if $\psi(\mathbf{r})$ is normalised to unity, that is

$$\int |\psi(\mathbf{r})|^2\, \mathrm{d}\mathbf{r} = 1 \qquad \textbf{(A5.5)}$$

we have from (A5.4)

$$\int \mathrm{d}\mathbf{r}\, \psi^*(\mathbf{r})\psi(\mathbf{r}) = (2\pi)^{-3} \int \mathrm{d}\mathbf{r} \int \mathrm{d}\mathbf{p} \int \mathrm{d}\mathbf{p}' \exp[\mathrm{i}(\mathbf{p}-\mathbf{p}')\cdot\mathbf{r}]\phi^*(\mathbf{p}')\phi(\mathbf{p}) = 1 \qquad \textbf{(A5.6)}$$

and since

$$(2\pi)^{-3} \int \mathrm{d}\mathbf{r} \exp[\mathrm{i}(\mathbf{p}-\mathbf{p}')\cdot\mathbf{r}] = \delta(\mathbf{p}-\mathbf{p}') \qquad \textbf{(A5.7)}$$

we see that

$$\int \mathrm{d}\mathbf{p} \int \mathrm{d}\mathbf{p}' \, \delta(\mathbf{p} - \mathbf{p}')\phi^*(\mathbf{p}')\phi(\mathbf{p}) = 1 \tag{A5.8}$$

Using the definition (2.28) of the Dirac δ-function, this last equation reduces to

$$\int |\phi(\mathbf{p})|^2 \, \mathrm{d}\mathbf{p} = 1 \tag{A5.9}$$

so that the wave function $\phi(\mathbf{p})$ is also normalised to unity.

Before using (A5.3) for a direct calculation of hydrogenic wave functions in momentum space, it is instructive to rewrite the Schrödinger equation itself in momentum space, namely as an equation involving $\phi(\mathbf{p})$ directly. Neglecting for the moment the reduced mass effect, we begin by considering an electron (of mass $m = 1$ in a.u.) in an arbitrary real potential $V(\mathbf{r})$. The corresponding time-independent Schrödinger equation in position space reads

$$\left[\frac{p^2}{2} + V(\mathbf{r})\right]\psi(\mathbf{r}) = E\psi(\mathbf{r}) \tag{A5.10}$$

We assume that $E < 0$, so that we are considering only the bound states. Pre-multiplying both sides by $(2\pi)^{-3/2}\exp(-\mathrm{i}\mathbf{p}\cdot\mathbf{r})$ and integrating over all space, we find by using (A5.3) that

$$\frac{p^2}{2}\phi(\mathbf{p}) + (2\pi)^{-3/2}\int \exp(-\mathrm{i}\mathbf{p}\cdot\mathbf{r})V(\mathbf{r})\psi(\mathbf{r})\,\mathrm{d}\mathbf{r} = E\phi(\mathbf{p}) \tag{A5.11}$$

Making use of (A5.4) and introducing the quantity

$$\tilde{V}(\mathbf{p} - \mathbf{p}') = (2\pi)^{-3}\int \exp[-\mathrm{i}(\mathbf{p} - \mathbf{p}')\cdot\mathbf{r}]V(\mathbf{r})\,\mathrm{d}\mathbf{r} \tag{A5.12}$$

we see that (A5.10) becomes

$$\left[\frac{p^2}{2} - E\right]\phi(\mathbf{p}) = -\int \tilde{V}(\mathbf{p} - \mathbf{p}')\phi(\mathbf{p}')\,\mathrm{d}\mathbf{p}' \tag{A5.13}$$

which is the required Schrödinger equation in momentum space (in a.u.). We note that it is an integral equation for $\phi(\mathbf{p})$.

If the potential $V(\mathbf{r})$ is central (that is, depends only on $r = |\mathbf{r}|$) the 'potential' $\tilde{V}(\mathbf{p} - \mathbf{p}')$ in momentum space may be simplified by setting $\mathbf{q} = \mathbf{p}' - \mathbf{p}$ and performing the integral in (A5.12) in spherical polar coordinates, with the polar axis in the direction of the vector $\mathbf{q}$. That is,

$$\tilde{V}(\mathbf{p}-\mathbf{p}') = (2\pi)^{-3}\int_0^\infty \mathrm{d}r r^2 \int_0^\pi \mathrm{d}\theta \sin\theta \int_0^{2\pi} \mathrm{d}\phi \exp(\mathrm{i}qr\cos\theta)V(r)$$

$$= (2\pi)^{-2}\int_0^\infty \mathrm{d}r r^2 V(r)\int_0^\pi \mathrm{d}\theta \sin\theta \exp(\mathrm{i}qr\cos\theta)$$

$$= (2\pi^2 q)^{-1}\int_0^\infty \mathrm{d}r r \sin(qr)V(r) \tag{A5.14}$$

Thus we see that when $V(\mathbf{r})$ is central the quantity $\tilde{V}$ is real and depends only on q, that is on $|\mathbf{p}-\mathbf{p}'|$. It is worth noting that the result (A5.14) may also be obtained by using the expansion of a plane wave in Legendre polynomials (see (2.260)), namely

$$\exp(\mathrm{i}\mathbf{q}\cdot\mathbf{r}) = \sum_{l=0}^{\infty}(2l+1)\mathrm{i}^l j_l(qr)P_l(\cos\theta) \tag{A5.15}$$

so that (A5.12) becomes

$$\tilde{V}(\mathbf{p}-\mathbf{p}') = (2\pi)^{-2}\sum_{l=0}^{\infty}(2l+1)\mathrm{i}^l\int_0^\infty \mathrm{d}r r^2 j_l(qr)V(r)\int_0^\pi \mathrm{d}\theta \sin\theta P_l(\cos\theta) \tag{A5.16}$$

Since

$$\int_0^\pi \mathrm{d}\theta \sin\theta P_l(\cos\theta)P_{l'}(\cos\theta) = \frac{2}{2l+1}\delta_{ll'} \tag{A5.17}$$

and $P_0 = 1$, we see that all the terms of the sum on the right of (A5.16) vanish, except the first one (for which $l = 0$). Hence $\tilde{V}(\mathbf{p}-\mathbf{p}')$ reduces to

$$\tilde{V}(|\mathbf{p}-\mathbf{p}'|) = (2\pi^2)^{-1}\int_0^\infty \mathrm{d}r r^2 j_0(qr)V(r) \tag{A5.18}$$

which agrees with (A5.14) since $j_0(qr) = \sin(qr)/(qr)$.

For a central potential $V(r)$ the Schrödinger equation (A5.10) admits solutions of the form $\psi(\mathbf{r}) = R_{E,l}(r)Y_{lm}(\theta, \phi)$. Similarly, if we denote by (p, θ_p, ϕ_p) the polar coordinates of the momentum $\mathbf{p}$, solutions of the Schrödinger equation in momentum space (A5.13) exist of the form

$$\phi(\mathbf{p}) = F_{E,l}(p)Y_{lm}(\theta_p, \phi_p) \tag{A5.19}$$

This may be checked directly by returning to the definition (A5.3) of $\phi(\mathbf{p})$, substituting in it $\psi(\mathbf{r}) = R_{E,l}(r)Y_{lm}(\theta, \phi)$ and using the expansion of a plane wave in spherical harmonics (see (2.261)). Explicitly, we have

$$\phi(\mathbf{p}) = (2\pi)^{-3/2}4\pi\sum_{l'=0}^{\infty}\sum_{m'=-l'}^{l'}(-\mathrm{i})^{l'}Y_{l'm'}(\theta_p, \phi_p)$$
$$\times\int_0^\infty \mathrm{d}r r^2 j_{l'}(pr)R_{E,l}(r)\int_0^\pi \mathrm{d}\theta \sin\theta\int_0^{2\pi}\mathrm{d}\phi\, Y^*_{l'm'}(\theta, \phi)Y_{lm}(\theta, \phi) \tag{A5.20}$$

Using the orthonormality of the spherical harmonics, we see that (A5.20) reduces to (A5.19), with

$$F_{E,l}(p) = N_l \int_0^\infty r^2 j_l(pr) R_{E,l}(r)\, \mathrm{d}r \qquad \textbf{(A5.21)}$$

N_l being a normalisation constant. We note that $F_{E,l}$ does not depend on the magnetic quantum number m. The quantity $|pF_{E,l}(p)|^2$ is called the *momentum distribution function*. The probability that the absolute value of the momentum lies between p and $p + \mathrm{d}p$ (independently of the direction) is given by $|pF_{E,l}(p)|^2\, \mathrm{d}p$ and we have

$$\int_0^\infty p^2 |F_{E,l}(p)|^2\, \mathrm{d}p = 1 \qquad \textbf{(A5.22)}$$

A one-dimensional integral equation for $F_{E,l}(p)$ may readily be obtained from the Schrödinger equation in momentum space (A5.13) by using (A5.19) and expanding $\tilde{V}(|\mathbf{p} - \mathbf{p}'|)$ in Legendre polynomials. That is,

$$\tilde{V}(|\mathbf{p} - \mathbf{p}'|) = \sum_{l=0}^{\infty} T_l(p, p') P_l(x) \qquad \textbf{(A5.23)}$$

with $x = \mathbf{p}\cdot\mathbf{p}'/(pp')$ and

$$T_l(p, p') = \frac{2l+1}{2} \int_{-1}^{+1} P_l(x) \tilde{V}(|\mathbf{p} - \mathbf{p}'|)\, \mathrm{d}x \qquad \textbf{(A5.24)}$$

The result is

$$\left[\frac{p^2}{2} - E\right] F_{E,l}(p) = -\int_0^\infty p'^2 K_l(p, p') F_{E,l}(p')\, \mathrm{d}p' \qquad \textbf{(A5.25)}$$

where

$$K_l(p, p') = 2\pi \int_{-1}^{+1} \tilde{V}([p^2 + p'^2 - 2pp'x]^{1/2}) P_l(x)\, \mathrm{d}x \qquad \textbf{(A5.26)}$$

is a symmetric kernel in p and p'.

Before we examine the particular case of the Coulomb potential, it is interesting to look at a screened Coulomb interaction which we write (in a.u.) as

$$V(r) = -\frac{Z}{r} \mathrm{e}^{-\alpha r}, \qquad \alpha > 0 \qquad \textbf{(A5.27)}$$

Using (A5.14), the corresponding potential $\tilde{V}$ in momentum space is easily obtained,

$$\tilde{V}(q) = -\frac{Z}{2\pi^2} \frac{1}{q^2 + \alpha^2} \qquad \textbf{(A5.28)}$$

where $q = |\mathbf{p} - \mathbf{p}'|$. The Coulomb potential $V(r) = -Z/r$ (in a.u.) may be considered as a limiting case of (A5.27) for which $\alpha \to 0$. Letting α tend to zero in (A5.28) we obtain

$$\tilde{V}(q) = -\frac{Z}{2\pi^2}\frac{1}{q^2} \tag{A5.29}$$

We note that this expression is singular at $q = 0$. By comparing (A5.28) and (A5.29) we see that this singularity may be traced to the 'infinite range' of the Coulomb potential, which in turn is due to the fact that the photon has zero mass.

Using the expression (A5.29) for $\tilde{V}(q)$, the kernel $K_l(p, p')$ given by (A5.26) becomes

$$\begin{aligned} K_l(p, p') &= -\frac{Z}{\pi}\frac{1}{2pp'}\int_{-1}^{+1} \frac{1}{(p^2 + p'^2)/(2pp') - x} P_l(x)\,\mathrm{d}x \\ &= -\frac{Z}{\pi}\frac{1}{pp'} Q_l\left(\frac{p^2 + p'^2}{2pp'}\right) \end{aligned} \tag{A5.30}$$

where Q_l is the Legendre function of the second kind, such that

$$Q_l(z) = \frac{1}{2}\int_{-1}^{+1} \frac{1}{z - x} P_l(x)\,\mathrm{d}x \tag{A5.31}$$

The first three functions $Q_l(z)$ are given by

$$\begin{aligned} Q_0(z) &= \frac{1}{2}\log\left(\frac{z+1}{z-1}\right) \\ Q_1(z) &= \frac{z}{2}\log\left(\frac{z+1}{z-1}\right) - 1 \\ Q_2(z) &= \frac{1}{4}(3z^2 - 1)\log\left(\frac{z+1}{z-1}\right) - \frac{3z}{2} \end{aligned} \tag{A5.32}$$

Substituting (A5.30) in (A5.25), we then obtain for the case of the Coulomb potential the one-dimensional integral equation

$$\left[\frac{p^2}{2} - E\right] F_{E,l}(p) = \frac{Z}{\pi p}\int_0^\infty p' Q_l\left(\frac{p^2 + p'^2}{2pp'}\right) F_{E,l}(p')\,\mathrm{d}p' \tag{A5.33}$$

This equation has been solved by V.A. Fock. For $E < 0$ it has discrete eigenvalues E_n which are of course identical to those which we obtained in Chapter 3 by solving the Schrödinger equation in position space. By analogy with the radial hydrogenic wave functions $R_{nl}(r)$, we shall denote the 'radial' hydrogenic momentum space wave functions by $F_{nl}(p)$ instead of $F_{E,l}(p)$.

The 'radial' functions $F_{nl}(p)$, corresponding to the hydrogen atom, obtained either by performing directly the Fourier transformation (see (A5.3) and (A5.21)) or by solving the radial momentum space equation (A5.33) and normalised according to (A5.22), are given (in a.u.) by

$$F_{nl}(p) = \left[\frac{2}{\pi}\frac{(n-l-1)!}{(n+l)!}\right]^{1/2} n^2 2^{2l+2} l! \frac{n^l p^l}{(n^2p^2+1)^{l+2}} C_{n-l-1}^{l+1}\left(\frac{n^2p^2-1}{n^2p^2+1}\right) \quad \textbf{(A5.34)}$$

where $C_N^{\alpha}(x)$ denotes the Gegenbauer polynomial, defined by the relation

$$(1-2xs+s^2)^{-\alpha} = \sum_{N=0}^{\infty} C_N^{\alpha}(x)s^N, \qquad |s|<1 \quad \textbf{(A5.35)}$$

For other hydrogenic atoms corresponding to a nucleus of charge Z the expressions for $F_{nl}(p)$ are identical, provided that p is expressed in units of Zp_0. The reduced mass effect is also easily taken into account by using 'reduced' units such that $p_0 = \hbar/a_0$ is replaced by $p_\mu = \hbar/a_\mu$, with $a_\mu = a_0(m/\mu)$ and μ being the reduced mass.

Let us write down the first few radial momentum space wave functions explicitly. Using the fact that

$$C_0^{\alpha}(x) = 1,$$
$$C_1^{\alpha}(x) = 2\alpha x, \quad \textbf{(A5.36)}$$
$$C_2^{\alpha}(x) = 2\alpha(\alpha+1)x^2 - \alpha$$

we find that

$$F_{10}(p) = \frac{2^{5/2}}{\sqrt{\pi}}\frac{1}{(p^2+1)^2}$$

$$F_{20}(p) = \frac{32}{\sqrt{\pi}}\frac{4p^2-1}{(4p^2+1)^3}$$

$$F_{21}(p) = \frac{128}{\sqrt{3\pi}}\frac{p}{(4p^2+1)^3}$$

$$F_{30}(p) = \frac{108\sqrt{2}}{\sqrt{3\pi}}\frac{81p^4-30p^2+1}{(9p^2+1)^4} \quad \textbf{(A5.37)}$$

$$F_{31}(p) = \frac{864}{\sqrt{3\pi}}\frac{p(9p^2-1)}{(9p^2+1)^4}$$

$$F_{32}(p) = \frac{5184}{\sqrt{15\pi}}\frac{p^2}{(9p^2+1)^4}$$

We note that in contrast with the position space hydrogenic radial functions $R_{nl}(r)$, which fall off *exponentially* at large r, the radial momentum space wave functions $F_{nl}(p)$ behave like *inverse powers* of p for large p. In particular, we see that for s states ($l = 0$) they decrease like p^{-4} at large p, for p states ($l = 1$) they fall off like p^{-5}, etc. More generally, since the argument of the Gegenbauer polynomial in (A5.34) is just + 1 when $p \to \infty$, we see that $F_{nl}(p)$ is proportional to $p^{-(l+4)}$ when p is large.

When p tends to zero the argument of the Gegenbauer polynomial in (A5.34) is –1 and we may use the fact that

$$\begin{aligned} C_N^{\alpha}(-1) &= (-1)^N C_N^{\alpha}(1) \\ &= (-1)^N \frac{(2\alpha + N - 1)!}{(2\alpha - 1)!N!} \end{aligned} \qquad \textbf{(A5.38)}$$

to deduce that for s states ($l = 0$)

$$F_{n0}(0) = (-1)^{n-1} 4 \left(\frac{2}{\pi} \right)^{1/2} n^{5/2} \qquad \textbf{(A5.39)}$$

For $l \neq 0$ we see from (A5.34) that F_{nl} vanishes in the limit $p \to 0$.

We recall that in all the expressions of $F_{nl}(p)$ written above the variable p is expressed in units of (Zp_0). If we want to use other units, we should simply make the substitution $p \to p/(Zp_0)$ in (A5.34) and recall that the normalisation condition (A5.22) introduces an extra factor of $(Zp_0)^{-3/2}$ in $F_{nl}(p)$. For example, the wave function $F_{10}(p)$ then becomes

$$F_{10}(p) = \frac{2^{5/2}}{\sqrt{\pi}} (Zp_0)^{5/2} \frac{1}{(p^2 + Z^2 p_0^2)^2} \qquad \textbf{(A5.40)}$$

Wave functions in momentum space are particularly useful to evaluate matrix elements involving various functions of p. For example, let us evaluate the average value of p^2 when the hydrogenic atom is in the ground state. We then have

$$\begin{aligned} \langle p^2 \rangle_{100} &= \int_0^{\infty} p^2 \, | p F_{10}(p) |^2 \, \mathrm{d}p \\ &= (Zp_0)^2 = (Z\hbar/a_0)^2 \end{aligned} \qquad \textbf{(A5.41)}$$

where we have used (A5.40). For a general hydrogenic state $\phi_{nlm}(\mathbf{p}) = F_{nl}(p) Y_{lm}(\theta_p, \phi_p)$ one finds from (A5.34) that

$$\langle p^2 \rangle_{nlm} = (Zp_0/n)^2 = [Z\hbar/(na_0)]^2 \qquad \textbf{(A5.42)}$$

In order to include the reduced mass effect we just have to replace a_0 by $a_\mu = a_0(m/\mu) = 4\pi\varepsilon_0 \hbar^2/(\mu e^2)$. The average value of the kinetic energy operator $T = p^2/(2\mu)$ is therefore given by

$$\langle T \rangle_{nlm} = \frac{1}{2\mu} [Z\hbar/(na_\mu)]^2 \text{ a.u.} = \frac{e^2}{(4\pi\varepsilon_0)a_\mu} \frac{Z^2}{2n^2} \qquad \textbf{(A5.43)}$$

in agreement with the result (3.81) of Chapter 3 and with the virial theorem proved in Section 3.4.

6 The Hamiltonian for a charged particle in an electromagnetic field

We start from Lagrange's equations of motion (Goldstein, 1980)

$$\frac{\mathrm{d}}{\mathrm{d}t}\left(\frac{\partial L}{\partial \dot{q}_i}\right) - \frac{\partial L}{\partial q_i} = 0, \qquad i = 1, 2, \ldots \tag{A6.1}$$

where q_i are generalised coordinates and L is the Lagrangian function. For a conservative system, $L = T - V$, where T is the kinetic energy and V is the potential energy. Since the electromagnetic field is not conservative, a generalisation is required. We have to find L such that (A6.1) provides the equation of motion of a particle of mass m, charge q and velocity $\mathbf{v}$ in an electromagnetic field specified by the electric field $\mathcal{E}(\mathbf{r}, t)$ and the magnetic field $\mathcal{B}(\mathbf{r}, t)$ and subject to the Lorentz force

$$\begin{aligned}\mathbf{F} &= q(\mathcal{E} + \mathbf{v} \times \mathcal{B}) \\ &= q\left[-\boldsymbol{\nabla}\phi - \frac{\partial \mathbf{A}}{\partial t} + \mathbf{v} \times (\boldsymbol{\nabla} \times \mathbf{A})\right]\end{aligned} \tag{A6.2}$$

where $\phi(\mathbf{r}, t)$ and $\mathbf{A}(\mathbf{r}, t)$ are respectively the scalar and vector electromagnetic potentials.

This can be achieved if we take L to be

$$L = \frac{1}{2}mv^2 - q\phi + q\mathbf{v}\cdot\mathbf{A} \tag{A6.3}$$

and work in a Cartesian system of coordinates, so that $q_1 = x$, $q_2 = y$ and $q_3 = z$. It is then easily shown that the equation (A6.1) reduces to the equation of motion

$$m\ddot{\mathbf{r}} = \mathbf{F} \tag{A6.4}$$

where the force $\mathbf{F}$ is given by (A6.2).

The generalised momenta are defined as

$$p_i = \frac{\partial L}{\partial \dot{q}_i} \tag{A6.5}$$

Using the Lagrangian L given by (A6.3), one finds that $p_1 = p_x$, $p_2 = p_y$ and $p_3 = p_z$, where p_x, p_y and p_z are the components of the vector $\mathbf{p}$, given by

$$\mathbf{p} = m\mathbf{v} + q\mathbf{A} \tag{A6.6}$$

The Hamiltonian H is defined by

$$H = \sum_{i=1}^{3} p_i \dot{q}_i - L \tag{A6.7}$$

and in terms of $\mathbf{p}$, $\mathbf{A}$ and ϕ we have

$$H = \frac{1}{2m}(\mathbf{p} - q\mathbf{A})^2 + q\phi \tag{A6.8}$$

7 The Dirac equation and relativistic corrections to the Schrödinger equation

In our discussion of atomic structure and of the interaction of atoms with external electromagnetic fields, we have introduced various interactions which are approximations to a complete relativistic theory and which could not be derived, without additional assumptions, from the non-relativistic Schrödinger equation for an atom interacting with an electromagnetic field. In this appendix, we shall show how the spin-dependent interactions as well as other terms in the Hamiltonian can be derived from Dirac's relativistic theory of the electron. Although the theory we shall discuss provides a highly accurate description of atomic structure, it is still an approximation to a fully self-consistent relativistic theory, because it does not allow for interactions in which the number of particles in a system changes. These interactions are characteristic of relativistic field theory, and are needed to provide a description of processes such as those in which photons turn into electron–positron pairs, or the reverse process, in which electron–positron pairs annihilate into photons.

The Schrödinger relativistic equation or Klein–Gordon equation

At the time when he was developing his non-relativistic wave equation, E. Schrödinger also proposed a relativistic generalisation of it, which is known as the *Schrödinger relativistic equation* or *Klein–Gordon equation*.

Free particle

We begin by considering the case of a free particle. The relativistic relationship between the energy E and momentum $\mathbf{p}$ of a free particle of rest mass m is

$$E^2 = m^2c^4 + \mathbf{p}^2c^2 \tag{A7.1}$$

Adopting for E and $\mathbf{p}$ the substitutions (see (2.11))

$$E \to i\hbar\frac{\partial}{\partial t}, \qquad \mathbf{p} \to -i\hbar\nabla \tag{A7.2}$$

and operating on a wave function $\Psi(\mathbf{r}, t)$, we obtain the Schrödinger relativistic equation (or Klein–Gordon equation) for a free particle, namely

$$-\hbar^2 \frac{\partial^2 \Psi}{\partial t^2} = m^2c^4\Psi - \hbar^2c^2\nabla^2\Psi \tag{A7.3}$$

It is worth noting that this is a second-order differential equation with respect to the time.

If the wave function Ψ is taken to have a single component, this equation describes a free particle with no internal degrees of freedom, and is a suitable equation for the description of a free particle of spin zero.

Charged particle in an electromagnetic field

If the spinless particle has an electric charge q, and it is placed in an external electromagnetic field described by a vector potential $\mathbf{A}$ and a scalar potential ϕ, we can make the replacements (see Appendix 6)

$$\mathbf{p} \to \mathbf{p} - q\mathbf{A}, \qquad E \to E - q\phi \tag{A7.4}$$

so that (A7.1) is replaced by

$$(E - q\phi)^2 = m^2c^4 + c^2(\mathbf{p} - q\mathbf{A})^2 \tag{A7.5}$$

Using the substitutions (A7.2), we then obtain the Klein–Gordon equation for a spinless particle of charge q in an electromagnetic field,

$$\left(i\hbar\frac{\partial}{\partial t} - q\phi\right)^2 \Psi = m^2c^4\Psi + c^2(-i\hbar\nabla - q\mathbf{A})^2\Psi \tag{A7.6}$$

In order to investigate the non-relativistic limit of this equation, we introduce the new function $X(\mathbf{r}, t)$ which is related to Ψ by

$$X(\mathbf{r}, t) = \Psi(\mathbf{r}, t)\exp(imc^2t/\hbar) \tag{A7.7}$$

and satisfies the equation

$$-\hbar^2\frac{\partial^2 X}{\partial t^2} + 2i\hbar(mc^2 - q\phi)\frac{\partial X}{\partial t} - \left[q\phi(2mc^2 - q\phi) + i\hbar q\frac{\partial \phi}{\partial t}\right]X$$

$$= c^2[-\hbar^2\nabla^2 + 2i\hbar q\mathbf{A}\cdot\nabla + i\hbar q(\nabla\cdot\mathbf{A}) + q^2\mathbf{A}^2]X \tag{A7.8}$$

In the limit in which $|q\phi| \ll mc^2$, $|[\hbar/(2mc^2)]\partial\phi/\partial t| \ll |\phi|$ and $|[\hbar^2/(2mc^2)]\partial^2 X/\partial t^2| \ll |\hbar\partial X/\partial t|$, this equation reduces to the non-relativistic Schrödinger equation for a spinless particle in an electromagnetic field, namely

$$i\hbar\frac{\partial X}{\partial t} = \left[\frac{1}{2m}(-i\hbar\nabla - q\mathbf{A})^2 + q\phi\right]X \tag{A7.9}$$

This is the equation we used in Chapter 4 (see (4.17)) with $q = -e$ and $q\phi = V(r) = -Ze^2/(4\pi\varepsilon_0 r)$.

The Dirac equation

To describe a particle of spin 1/2, we require a wave function having two components which allow for the two spin states, the z component S_z of the spin angular momentum taking on the values $m_s\hbar$, where $m_s = \pm 1/2$. However, since all spin-1/2 particles are associated with particles of the same mass and spin, but of opposite charge, known as antiparticles, we expect to need a four-component wave function. This was unknown when P.A.M. Dirac put forward his equation, and it was one of the great achievements of theoretical physics that Dirac was able to predict the existence of the positron, the antiparticle of the electron, from his theory.

Dirac started by looking for a wave equation of the form

$$i\hbar \frac{\partial}{\partial t} \Psi = H\Psi \tag{A7.10}$$

which like the (non-relativistic) Schrödinger equation (2.46) is *linear* in $\partial/\partial t$, and not quadratic like the Klein–Gordon equation. Since in a relativistic theory the spatial [1] coordinates (x_1, x_2, x_3) must enter on the same footing as $x_0 = ct$, the equation (A7.10) – and hence the Hamiltonian H – is expected to be linear in the space derivatives $\partial/\partial x_k$ $(k = 1, 2, 3)$. The wave function in (A7.10) is assumed to contain N components Ψ_i $(i = 1, 2, \ldots, N)$ and hence may be written in the form of a column matrix as

$$\Psi = \begin{pmatrix} \Psi_1 \\ \Psi_2 \\ \vdots \\ \Psi_N \end{pmatrix} \tag{A7.11}$$

Free particle

Let us first consider the case of a free particle. The Hamiltonian must then be independent of $\mathbf{r}$ and t (since there are no forces) and the simplest candidate, linear in the momentum and mass terms, may be written in the form

$$H = c\boldsymbol{\alpha}\cdot\mathbf{p} + \beta mc^2 \tag{A7.12}$$

where $\mathbf{p} = -i\hbar\nabla$ according to the correspondence rule (A7.2). The three components $(\alpha^1, \alpha^2, \alpha^3)$ of $\boldsymbol{\alpha}$ as well as the quantity β are independent of $\mathbf{r}$, t, $\mathbf{p}$ and E, but need not commute with each other.

Substituting (A7.12) into (A7.10) and remembering that $E = i\hbar\partial/\partial t$, we obtain the Dirac wave equation

$$(E - c\boldsymbol{\alpha}\cdot\mathbf{p} - \beta mc^2)\Psi = 0 \tag{A7.13}$$

[1] It is convenient to denote the Cartesian coordinates of a point by (x_1, x_2, x_3) in this appendix, rather than by (x, y, z) which we have used elsewhere.

or

$$i\hbar\frac{\partial}{\partial t}\Psi = -i\hbar c\boldsymbol{\alpha}\cdot\nabla\Psi + \beta mc^2\Psi \tag{A7.14}$$

More explicitly, we may write (A7.14) as

$$i\hbar\frac{\partial}{\partial t}\Psi_i = -i\hbar c\sum_{j=1}^{N}\sum_{k=1}^{3}\alpha_{ij}^k\frac{\partial}{\partial x_k}\Psi_j + \sum_{j=1}^{N}\beta_{ij}mc^2\Psi_j, \qquad i = 1, 2, \ldots, N \tag{A7.15}$$

where α_{ij}^1, α_{ij}^2, α_{ij}^3 and β_{ij} form $N \times N$ matrices. The equations (A7.15) are a set of N coupled equations for the N components Ψ_i of Ψ.

We require the Hamiltonian H to be Hermitian, $H = H^\dagger$, and for this reason the matrices α^1, α^2, α^3 and β must also be Hermitian:

$$\boldsymbol{\alpha} = \boldsymbol{\alpha}^\dagger, \qquad \beta = \beta^\dagger \tag{A7.16}$$

Further conditions to be satisfied by $\boldsymbol{\alpha}$ and β follow from the requirement that each component of Ψ must separately satisfy the Klein–Gordon equation (A7.3), which we rewrite as

$$[E^2 - \mathbf{p}^2c^2 - m^2c^4]\Psi = 0 \tag{A7.17}$$

Multiplying (A7.13) on the left by the operator $[E + c\boldsymbol{\alpha}\cdot\mathbf{p} + \beta mc^2]$, we obtain the second-order equation

$$\left\{E^2 - c^2\left[\sum_{k=1}^{3}(\alpha^k)^2p_k^2 + \sum_{\substack{k \\ (k<l)}}\sum_{l}(\alpha^k\alpha^l + \alpha^l\alpha^k)p_kp_l\right]\right.$$
$$\left. - mc^3\left[\sum_{k=1}^{3}(\alpha^k\beta + \beta\alpha^k)p_k\right] - m^2c^4\beta^2\right\}\Psi = 0 \tag{A7.18}$$

where p_k $(k = 1, 2, 3)$ denote the Cartesian components of $\mathbf{p}$. Comparing (A7.18) with (A7.17), we see that each component Ψ_i satisfies the Klein–Gordon equation provided that

$$\begin{gathered}(\alpha^1)^2 = (\alpha^2)^2 = (\alpha^3)^2 = \beta^2 = 1, \\ [\alpha^1, \alpha^2]_+ = [\alpha^2, \alpha^3]_+ = [\alpha^3, \alpha^1]_+ = 0, \\ [\alpha^1, \beta]_+ = [\alpha^2, \beta]_+ = [\alpha^3, \beta]_+ = 0\end{gathered} \tag{A7.19}$$

where $[\mathrm{A}, \mathrm{B}]_+$ denotes the anti-commutator

$$[\mathrm{A}, \mathrm{B}]_+ = \mathrm{AB} + \mathrm{BA} \tag{A7.20}$$

It can be shown that the minimum dimensions for the matrices α^1, α^2, α^3 and β required to satisfy the conditions (A7.16) and (A7.19) are 4×4. Correspondingly, the wave function Ψ must have at least four components $(N = 4)$. In view of our expectation that four components are required to accommodate a description of a particle and antiparticle each of spin 1/2, we shall assume that $N = 4$. The wave function Ψ is called a *four-component spinor*. The solution of equations (A7.16) and (A7.19) is not unique, but it can be shown that any set of matrices satisfying these conditions provides the same physical results. A representation of the matrices α^1, α^2, α^3 and β which is particularly useful for studying the non-relativistic limit of the Dirac equation is the *Dirac representation*, given by

$$\boldsymbol{\alpha} = \begin{pmatrix} 0 & \boldsymbol{\sigma} \\ \boldsymbol{\sigma} & 0 \end{pmatrix} \qquad \beta = \begin{pmatrix} I & 0 \\ 0 & -I \end{pmatrix} \tag{A7.21}$$

where I is the unit two-by-two matrix while σ_1, σ_2 and σ_3 are the three Pauli two-by-two spin matrices, namely (see (2.216))

$$\sigma_1 = \begin{pmatrix} 0 & 1 \\ 1 & 0 \end{pmatrix}, \qquad \sigma_2 = \begin{pmatrix} 0 & -\mathrm{i} \\ \mathrm{i} & 0 \end{pmatrix}, \qquad \sigma_3 = \begin{pmatrix} 1 & 0 \\ 0 & -1 \end{pmatrix} \tag{A7.22}$$

Using the properties (2.217) of $\boldsymbol{\sigma}$, the relations (A7.16) and (A7.19) are easily shown to be satisfied.

Charged particle in an electromagnetic field

To obtain the Dirac equation for a spin-1/2 particle of charge q in an electromagnetic field $(\mathbf{A}, \phi)$ we make the usual replacements (A7.4) in (A7.13). This gives

$$[(E - q\phi) - c\boldsymbol{\alpha}\cdot(\mathbf{p} - q\mathbf{A}) - \beta mc^2]\Psi = 0 \tag{A7.23}$$

or

$$i\hbar\frac{\partial}{\partial t}\Psi = [-i\hbar c\boldsymbol{\alpha}\cdot\boldsymbol{\nabla} - cq\boldsymbol{\alpha}\cdot\mathbf{A} + q\phi + \beta mc^2]\Psi \tag{A7.24}$$

Upon comparison with (A7.10), we see that the Dirac Hamiltonian in the presence of an external electromagnetic field $(\mathbf{A}, \phi)$ is given by

$$H = c\boldsymbol{\alpha}\cdot(\mathbf{p} - q\mathbf{A}) + q\phi + \beta mc^2 \tag{A7.25}$$

Adjoint equation. Continuity equation. Probability and current densities

We have seen above that the wave function Ψ may be taken to be a four-component spinor, that is a column matrix of the form (A7.11) with four components Ψ_i ($i = 1, \dots, 4$). We can define $\Psi^\dagger$ to be a row matrix with components Ψ_i^*, namely

$$\Psi^\dagger = (\Psi_1^* \quad \Psi_2^* \quad \Psi_3^* \quad \Psi_4^*) \tag{A7.26}$$

Using (A7.24), (A7.25) and the fact that $\boldsymbol{\alpha}$, β and $\mathbf{p}$ are Hermitian, we see that $\Psi^\dagger$ satisfies the *adjoint equation*

$$\begin{aligned} -i\hbar\frac{\partial}{\partial t}\Psi^\dagger &= \Psi^\dagger H \\ &= (i\hbar c\boldsymbol{\nabla} - cq\mathbf{A})\Psi^\dagger\cdot\boldsymbol{\alpha} + q\phi\Psi^\dagger + mc^2\Psi^\dagger\beta \end{aligned} \tag{A7.27}$$

The quantity

$$P(\mathbf{r}, t) = \Psi^\dagger\Psi = \sum_{i=1}^{4} |\Psi_i|^2 \tag{A7.28}$$

is clearly positive and can be interpreted as a *position probability density*, in the same way that $|\Psi|^2$ is the position probability density for the non-relativistic Schrödinger equation (2.46). By multiplying (A7.24) on the left by $\Psi^\dagger$ and (A7.27) on the right by Ψ, and taking the difference of the two results, it is found that

$$\frac{\partial P}{\partial t} + \nabla \cdot (\Psi^\dagger c\boldsymbol{\alpha}\Psi) = 0 \tag{A7.29}$$

If we interpret the vector

$$\mathbf{j}(\mathbf{r}, t) = \Psi^\dagger c\boldsymbol{\alpha}\Psi \tag{A7.30}$$

as a *probability current density*, the equation (A7.29) takes the form of a continuity equation,

$$\frac{\partial P}{\partial t} + \nabla \cdot \mathbf{j} = 0 \tag{A7.31}$$

and we see that $c\boldsymbol{\alpha}$ can be interpreted as a velocity operator.

Stationary solutions

Let us assume that $\mathbf{A}$ and ϕ are time-independent. We may then look for stationary solutions of the Dirac equation which we write as

$$\Psi(\mathbf{r}, t) = \chi(\mathbf{r}) \exp(-\mathrm{i}Et/\hbar) \tag{A7.32}$$

From (A7.24) and (A7.32) we obtain the time-independent equation

$$E\chi(\mathbf{r}) = [-\mathrm{i}\hbar c\boldsymbol{\alpha} \cdot \nabla - cq\boldsymbol{\alpha} \cdot \mathbf{A} + q\phi + \beta mc^2]\chi(\mathbf{r}) \tag{A7.33}$$

The four-component spinor $\chi(\mathbf{r})$ can be expressed in terms of two two-component spinors $\psi(\mathbf{r})$ and $\eta(\mathbf{r})$ by writing

$$\chi(\mathbf{r}) \equiv \begin{pmatrix} \psi(\mathbf{r}) \\ \eta(\mathbf{r}) \end{pmatrix} \tag{A7.34}$$

Using the Dirac representation (A7.21) of the matrices $\boldsymbol{\alpha}$ and β, the two-component spinors ψ and η are found to satisfy the two coupled equations

$$E\psi(\mathbf{r}) = c(-\mathrm{i}\hbar\nabla - q\mathbf{A}) \cdot \boldsymbol{\sigma}\eta(\mathbf{r}) + (q\phi + mc^2)\psi(\mathbf{r}) \tag{A7.35a}$$

$$E\eta(\mathbf{r}) = c(-\mathrm{i}\hbar\nabla - q\mathbf{A}) \cdot \boldsymbol{\sigma}\psi(\mathbf{r}) + (q\phi - mc^2)\eta(\mathbf{r}) \tag{A7.35b}$$

These equations will be used below to study the non-relativistic limit of the Dirac equation.

Central field. Spin angular momentum. Total angular momentum

Consider a spin-1/2 particle in a central field (such that $\mathbf{A} = 0$ and ϕ is spherically symmetric). Let $V(r) = q\phi(r)$ be the corresponding potential energy. The Dirac Hamiltonian (A7.25) is then

$$H = c\boldsymbol{\alpha}\cdot\mathbf{p} + \beta mc^2 + V(r) \tag{A7.36}$$

We saw in Chapter 2 that in the non-relativistic Schrödinger theory every component of the orbital angular momentum $\mathbf{L} = \mathbf{r} \times \mathbf{p}$, as well as $\mathbf{L}^2$, commutes with the (non-relativistic) Hamiltonian $H = p^2/(2m) + V(r)$ of a spinless particle in a central field. As a result, simultaneous eigenstates of the operators H, $\mathbf{L}^2$ and L_z exist in Schrödinger's theory, with eigenvalues given respectively by E, $l(l+1)\hbar^2$ and $m\hbar$. In Dirac's theory, however, neither the components of $\mathbf{L}$, nor $\mathbf{L}^2$, commute with the Dirac Hamiltonian (A7.36). Instead, one can readily show that

$$[H, \mathbf{L}] = -i\hbar c\boldsymbol{\alpha}\times\mathbf{p} \tag{A7.37}$$

Let us now consider the operator

$$\mathbf{S} = \frac{\hbar}{2}\boldsymbol{\Sigma} \tag{A7.38}$$

where $\boldsymbol{\Sigma} \equiv (\Sigma^1, \Sigma^2, \Sigma^3)$, the Σ^k being 4×4 matrices given by

$$\boldsymbol{\Sigma} = \begin{pmatrix} \boldsymbol{\sigma} & 0 \\ 0 & \boldsymbol{\sigma} \end{pmatrix} \tag{A7.39}$$

It follows from the properties (2.217) of the Pauli spin matrices that the three Cartesian components of $\mathbf{S}$ satisfy the commutation relations (2.194). Moreover, for any state Ψ,

$$\mathbf{S}^2\Psi = s(s+1)\hbar^2\Psi, \qquad s = \tfrac{1}{2} \tag{A7.40}$$

and the two possible eigenvalues of S_i $(i = 1, 2, 3)$ are $\pm\hbar/2$. We shall therefore refer to the operator (A7.38) as the *spin angular momentum* operator. We remark that any component of $\mathbf{S}$ commutes with any component of $\mathbf{L}$. Moreover, using (2.217) we have

$$[H, \mathbf{S}] = i\hbar c\boldsymbol{\alpha}\times\mathbf{p} \tag{A7.41}$$

where H is the Dirac Hamiltonian (A7.36)

We can now define a *total angular momentum* operator as the sum of the orbital and spin angular momentum operators,

$$\mathbf{J} = \mathbf{L} + \mathbf{S} \tag{A7.42}$$

We note that the three Cartesian components of $\mathbf{J}$ satisfy the commutation relations (2.223). In addition, using the results (A7.37) and (A7.41), we see that every component of $\mathbf{J}$ commutes with the Dirac Hamiltonian H given by (A7.36). Making use of (2.120c), we also remark that $\mathbf{J}^2$ commutes with H. We shall denote the eigenvalues of $\mathbf{J}^2$ by $j(j+1)\hbar^2$ and those of J_3 ($\equiv J_z$) by $m_j\hbar$. Thus, in Dirac's theory, simultaneous eigenstates of the Hamiltonian (A7.36) and of the operators $\mathbf{J}^2$ and J_z can be found, with eigenvalues given respectively by $E, j(j+1)\hbar^2$ and $m_j\hbar$.

The non-relativistic limit

Let us return to the two coupled equations (A7.35) for stationary states. In order to investigate its non-relativistic limit, we write

$$E = E' + mc^2 \tag{A7.43}$$

Substituting (A7.43) into (A7.35), we find that

$$E'\psi(\mathbf{r}) = c(-i\hbar\nabla - q\mathbf{A})\cdot\boldsymbol{\sigma}\eta(\mathbf{r}) + q\phi\psi(\mathbf{r}) \tag{A7.44a}$$

$$(E' + 2mc^2)\eta(\mathbf{r}) = c(-i\hbar\nabla - q\mathbf{A})\cdot\boldsymbol{\sigma}\psi(\mathbf{r}) + q\phi\eta(\mathbf{r}) \tag{A7.44b}$$

This pair of equations is still exact, but in the non-relativistic limit both $|E'|$ and $|q\phi|$ are small in comparison with mc^2. The equation (A7.44b) can then be solved approximately to give

$$\eta(\mathbf{r}) = \frac{1}{2mc}(-i\hbar\nabla - q\mathbf{A})\cdot\boldsymbol{\sigma}\psi(\mathbf{r}) \tag{A7.45}$$

and we see that η is smaller than ψ by a factor of order $p/(mc)$ (that is, v/c, where v is the magnitude of the velocity). The two-component spinors ψ and η are known in this case as the *large* and *small components*, respectively.

The Pauli equation

Substituting (A7.45) into (A7.44a), we find that

$$E'\psi(\mathbf{r}) = \frac{1}{2m}[(-i\hbar\nabla - q\mathbf{A})\cdot\boldsymbol{\sigma}]^2\psi(\mathbf{r}) + q\phi\psi(\mathbf{r}) \tag{A7.46}$$

The identity (2.218) satisfied by the Pauli spin matrices can be used to reduce the first term on the right-hand side of (A7.46). That is,

$$\frac{1}{2m}[(-i\hbar\nabla - q\mathbf{A})\cdot\boldsymbol{\sigma}]^2\psi = \frac{1}{2m}(-i\hbar\nabla - q\mathbf{A})^2\psi - \frac{q\hbar}{2m}\boldsymbol{\sigma}\cdot(\nabla\times\mathbf{A})\psi \tag{A7.47}$$

Now $\nabla\times\mathbf{A} = \mathcal{B}$, where $\mathcal{B}$ is the magnetic field, so that (A7.46) becomes

$$E'\psi(\mathbf{r}) = \left[\frac{1}{2m}(-i\hbar\nabla - q\mathbf{A})^2 - \frac{q\hbar}{2m}(\boldsymbol{\sigma}\cdot\mathcal{B}) + q\phi\right]\psi(\mathbf{r}) \tag{A7.48}$$

This equation is known as the *Pauli equation*. It differs from the non-relativistic form of the Klein–Gordon equation for spinless particles in predicting an interaction between the external magnetic field $\mathcal{B}$ and the spin $\mathbf{S} = \hbar\boldsymbol{\sigma}/2$ of the spin-1/2 particle [2]. We emphasize that the Pauli equation (A7.48) is an equation for a two-component spinor wave function ψ.

To apply the Pauli equation to an electron, we put $q = -e$, where e is the magnitude of the electronic charge. We then have

$$E'\psi(\mathbf{r}) = \left[\frac{1}{2m}(-i\hbar\nabla + e\mathbf{A})^2 + \frac{e\hbar}{2m}(\boldsymbol{\sigma}\cdot\mathcal{B}) - e\phi\right]\psi(\mathbf{r}) \qquad \textbf{(A7.49)}$$

The term $e\hbar(\boldsymbol{\sigma}\cdot\mathcal{B})/(2m)$ on the right of this equation corresponds to an interaction $-\mathcal{M}_s\cdot\mathcal{B}$ between the magnetic field $\mathcal{B}$ and an intrinsic magnetic moment $\mathcal{M}_s$ of the electron, due to its spin, with

$$\mathcal{M}_s = -\mu_B\boldsymbol{\sigma} = -g_s\mu_B\mathbf{S}/\hbar = -g_s\frac{e}{2m}\mathbf{S} \qquad \textbf{(A7.50)}$$

where

$$\mu_B = \frac{e\hbar}{2m} \qquad \textbf{(A7.51)}$$

is the Bohr magneton and the spin gyromagnetic ratio g_s has the value $g_s = 2$. We see that the Dirac theory not only predicts the existence of an intrinsic magnetic moment for the electron, but also gives essentially its correct value $\mathcal{M}_s = -(e/m)\mathbf{S}$ apart from very small corrections coming from quantum electrodynamics (see Section 5.3). The Pauli equation is used in Chapter 6 to analyse the Zeeman effect for one-electron atoms.

Higher order corrections for one-electron atoms and ions

We have shown above that to lowest order in v/c, the Dirac theory is equivalent to the two-component Pauli theory. We shall now investigate higher order corrections for the case of an electron in the Coulomb field of a nucleus, so that $\mathbf{A} = 0$ and $q\phi = -e\phi = V(r) = -Ze^2/(4\pi\varepsilon_0 r)$. The Dirac equation can be solved exactly in that case (see for example Bransden and Joachain, 2000) but here we shall only display the corrections of order $(v/c)^2$ which are needed in our discussion of fine structure effects in Chapter 5.

Let us return to the system of equations (A7.44). Solving (A7.44b) for $\eta(\mathbf{r})$, we obtain

$$\eta(\mathbf{r}) = \frac{1}{E' + 2mc^2 - V(r)}c(-i\hbar\boldsymbol{\sigma}\cdot\nabla)\psi(\mathbf{r}) \qquad \textbf{(A7.52)}$$

[2] We recall that the spin operator $\mathbf{S} = \hbar\boldsymbol{\Sigma}/2$ introduced in (A7.38) acts on four-component Dirac spinors. In order to simplify the notation we also denote by $\mathbf{S}$ the spin operator $\hbar\boldsymbol{\sigma}/2$ acting on two-component Pauli spinors.

Substituting in (A7.44a), we find that

$$E'\psi(\mathbf{r}) = c^2(\mathrm{i}\hbar\boldsymbol{\sigma}\cdot\boldsymbol{\nabla})\frac{1}{E' + 2mc^2 - V(r)}(\mathrm{i}\hbar\boldsymbol{\sigma}\cdot\boldsymbol{\nabla})\psi(\mathbf{r}) + V(r)\psi(\mathbf{r}) \tag{A7.53}$$

Expanding $[E' + 2mc^2 - V(r)]^{-1}$ in powers of $[E' - V(r)]/(2mc^2)$, we have to lowest order

$$[E' + 2mc^2 - V(r)]^{-1} \simeq \frac{1}{2mc^2}\left[1 - \frac{E' - V(r)}{2mc^2}\right] \tag{A7.54}$$

so that (A7.53) becomes

$$\begin{aligned} E'\psi(\mathbf{r}) = &-\frac{\hbar^2}{2m}\left[1 - \frac{E' - V(r)}{2mc^2}\right](\boldsymbol{\sigma}\cdot\boldsymbol{\nabla})^2\psi(\mathbf{r}) \\ &- \frac{\hbar^2}{4m^2c^2}[\boldsymbol{\sigma}\cdot\boldsymbol{\nabla}V(r)][\boldsymbol{\sigma}\cdot\boldsymbol{\nabla}\psi(\mathbf{r})] + V(r)\psi(\mathbf{r}) \end{aligned} \tag{A7.55}$$

Now, using the identity (2.218), we have $(\boldsymbol{\sigma}\cdot\boldsymbol{\nabla})^2 = \nabla^2$ and

$$(\boldsymbol{\sigma}\cdot\boldsymbol{\nabla}V)(\boldsymbol{\sigma}\cdot\boldsymbol{\nabla}\psi) = (\boldsymbol{\nabla}V)\cdot(\boldsymbol{\nabla}\psi) + \mathrm{i}\boldsymbol{\sigma}\cdot[(\boldsymbol{\nabla}V)\times(\boldsymbol{\nabla}\psi)] \tag{A7.56}$$

Moreover, since $V(r)$ is spherically symmetric,

$$\boldsymbol{\nabla}V(r) = \frac{\mathrm{d}V}{\mathrm{d}r}\hat{\mathbf{r}} \tag{A7.57a}$$

$$(\boldsymbol{\nabla}V)\cdot(\boldsymbol{\nabla}\psi) = \frac{\mathrm{d}V}{\mathrm{d}r}\frac{\partial\psi}{\partial r} \tag{A7.57b}$$

and

$$\mathrm{i}\boldsymbol{\sigma}\cdot[(\boldsymbol{\nabla}V)\times(\boldsymbol{\nabla}\psi)] = -\frac{2}{\hbar^2}\frac{1}{r}\frac{\mathrm{d}V}{\mathrm{d}r}\mathbf{L}\cdot\mathbf{S}\,\psi \tag{A7.57c}$$

where we have used the fact that $\mathbf{L} = \mathbf{r}\times\mathbf{p} = \mathbf{r}\times(-\mathrm{i}\hbar\boldsymbol{\nabla})$ and $\mathbf{S} = \hbar\boldsymbol{\sigma}/2$.

Collecting the various terms, we obtain

$$\begin{aligned} E'\psi(\mathbf{r}) = \Bigg[&-\frac{\hbar^2}{2m}\nabla^2 + V(r) + \frac{\hbar^2}{2m}\frac{E' - V(r)}{2mc^2}\nabla^2 + \frac{1}{2m^2c^2}\frac{1}{r}\frac{\mathrm{d}V}{\mathrm{d}r}\mathbf{L}\cdot \\ &- \frac{\hbar^2}{4m^2c^2}\frac{\mathrm{d}V}{\mathrm{d}r}\frac{\partial}{\partial r}\Bigg]\psi(\mathbf{r}) \end{aligned} \tag{A7.58}$$

Since $\mathbf{p} = -\mathrm{i}\hbar\boldsymbol{\nabla}$ and $E' - V(r) \simeq p^2/(2m)$, the third term on the right-hand side can be written as

$$\frac{\hbar^2}{2m}\frac{E' - V(r)}{2mc^2}\nabla^2 \simeq -\frac{p^4}{8m^3c^2} \tag{A7.59}$$

and is a *relativistic correction* (of order v^2/c^2) *to the kinetic energy* term $-\hbar^2\nabla^2/(2m) \equiv p^2/(2m)$. The fourth term is the *spin–orbit interaction* which is readily shown to be of order v^2/c^2 times the potential energy $V(r)$. The last term on the right-hand side of (A7.58) is a relativistic correction (or order v^2/c^2) to the potential energy which gives rise to some difficulty because it is non-Hermitian. The origin of the trouble is that if the original Dirac four-component spinor is normalised to unity, namely

$$\int (\psi^\dagger \psi + \eta^\dagger \eta)\, \mathrm{d}\mathbf{r} = 1 \tag{A7.60}$$

then the two-component spinor ψ only satisfies the normalisation condition approximately. C.G. Darwin has shown that the normalisation of ψ can be obtained correctly by replacing the last term in (A7.58) by the symmetrical combination (which is Hermitian)

$$\frac{1}{2}\left[\left(-\frac{\hbar^2}{4m^2c^2}\frac{\mathrm{d}V(r)}{\mathrm{d}r}\frac{\partial}{\partial r}\right) + \left(-\frac{\hbar^2}{4m^2c^2}\frac{\mathrm{d}V(r)}{\mathrm{d}r}\frac{\partial}{\partial r}\right)^\dagger\right] = \frac{\hbar^2}{8m^2c^2}\nabla^2 V(r) \tag{A7.61}$$

For a one-electron atom (ion) $V(r) = -Ze^2/(4\pi\varepsilon_0 r)$ and we have

$$\nabla^2 V(r) = 4\pi\left(\frac{Ze^2}{4\pi\varepsilon_0}\right)\delta(\mathbf{r}) \tag{A7.62}$$

so that the Darwin term (A7.61) is then given by

$$\frac{\hbar^2}{8m^2c^2}\nabla^2 V(r) = \frac{\pi\hbar^2}{2m^2c^2}\left(\frac{Ze^2}{4\pi\varepsilon_0}\right)\delta(\mathbf{r}) \tag{A7.63}$$

and we see that it only contributes to the energy of the states with $l = 0$.

Using the above results, we may write down the final form of the wave equation for one-electron atoms (ions), with relativistic corrections through order (v^2/c^2). That is,

$$H\psi(\mathbf{r}) = E'\psi(\mathbf{r}) \tag{A7.64}$$

where

$$H = \frac{p^2}{2m} + V(r) - \frac{p^4}{8m^3c^2} + \frac{1}{2m^2c^2}\frac{1}{r}\frac{\mathrm{d}V}{\mathrm{d}r}\mathbf{L}\cdot\mathbf{S} + \frac{\pi\hbar^2}{2m^2c^2}\left(\frac{Ze^2}{4\pi\varepsilon_0}\right)\delta(\mathbf{r}) \tag{A7.65}$$

and $V(r) = -Ze^2/(4\pi\varepsilon_0 r)$. It is the Hamiltonian (A7.65) which is the starting point of our discussion of the fine structure of hydrogenic atoms in Chapter 5.

8 Separation of the centre of mass coordinates for an N-electron atom

Let us consider an atom or ion containing a nucleus of mass M and charge Ze and N electrons of mass m and charge $(-e)$. We denote by $\mathbf{R}_0$ the coordinates of the nucleus with respect to a fixed origin O, and by $\mathbf{R}_1, \mathbf{R}_2, \ldots, \mathbf{R}_N$ those of the electrons. In the absence of an external field, and neglecting all but the Coulomb interactions, the non-relativistic Hamiltonian operator of this system is given by

$$H = T + V \tag{A8.1}$$

where the kinetic energy operator T reads

$$T = -\frac{\hbar^2}{2M}\nabla^2_{\mathbf{R}_0} + \sum_{i=1}^{N}\left(-\frac{\hbar^2}{2m}\nabla^2_{\mathbf{R}_i}\right) \tag{A8.2}$$

and the potential energy V is the sum of the Coulomb interactions between the $(N+1)$ particles of the system.

In order to separate the motion of the centre of mass, we change our coordinates from $(\mathbf{R}_0, \mathbf{R}_1, \ldots, \mathbf{R}_N)$ to $(\mathbf{R}, \mathbf{r}_1, \ldots, \mathbf{r}_N)$ where

$$\mathbf{R} = \frac{1}{M + Nm}(M\mathbf{R}_0 + m\mathbf{R}_1 + \cdots m\mathbf{R}_N) \tag{A8.3}$$

is the coordinate of the centre of mass and

$$\mathbf{r}_i = \mathbf{R}_i - \mathbf{R}_0, \qquad i = 1, 2, \ldots, N \tag{A8.4}$$

are the relative coordinates of the electrons with respect to the nucleus. It is apparent from (A8.3) and (A8.4) that

$$\boldsymbol{\nabla}_{\mathbf{R}_0} = \frac{M}{M + Nm}\boldsymbol{\nabla}_{\mathbf{R}} - \sum_{i=1}^{N}\boldsymbol{\nabla}_{\mathbf{r}_i} \tag{A8.5}$$

and

$$\boldsymbol{\nabla}_{\mathbf{R}_i} = \frac{m}{M + Nm}\boldsymbol{\nabla}_{\mathbf{R}} + \boldsymbol{\nabla}_{\mathbf{r}_i} \tag{A8.6}$$

Hence

$$\nabla_{\mathbf{R}_0}^2 = \left(\frac{M}{M+Nm}\right)^2 \nabla_{\mathbf{R}}^2 - \frac{2M}{M+Nm}\sum_{i=1}^{N} \nabla_{\mathbf{R}}\cdot\nabla_{\mathbf{r}_i} + \left(\sum_{i=1}^{N}\nabla_{\mathbf{r}_i}\right)^2 \quad \textbf{(A8.7)}$$

and

$$\nabla_{\mathbf{R}_i}^2 = \left(\frac{m}{M+Nm}\right)^2 \nabla_{\mathbf{R}}^2 + \frac{2m}{M+Nm}\nabla_{\mathbf{R}}\cdot\nabla_{\mathbf{r}_i} + \nabla_{\mathbf{r}_i}^2 \quad \textbf{(A8.8)}$$

Substituting the expressions (A8.7) and (A8.8) in (A8.2), we find that the kinetic energy operator becomes in the new coordinates

$$T = -\frac{\hbar^2}{2(M+Nm)}\nabla_{\mathbf{R}}^2 - \frac{\hbar^2}{2\mu}\sum_{i=1}^{N}\nabla_{\mathbf{r}_i}^2 - \frac{\hbar^2}{M}\sum_{i<j}\nabla_{\mathbf{r}_i}\cdot\nabla_{\mathbf{r}_j} \quad \textbf{(A8.9)}$$

where

$$\mu = \frac{mM}{m+M} \quad \textbf{(A8.10)}$$

is the reduced mass of the electron with respect to the nucleus. The Hamiltonian (A8.1) may therefore be written as

$$H = -\frac{\hbar^2}{2(M+Nm)}\nabla_{\mathbf{R}}^2 - \frac{\hbar^2}{2\mu}\sum_{i=1}^{N}\nabla_{\mathbf{r}_i}^2 - \frac{\hbar^2}{M}\sum_{i<j}\nabla_{\mathbf{r}_i}\cdot\nabla_{\mathbf{r}_j} + V(\mathbf{r}_1, \mathbf{r}_2, \dots, \mathbf{r}_N) \quad \textbf{(A8.11)}$$

The only term involving the coordinates **R** in (A8.11) is the first one, which represents the kinetic energy operator of the centre of mass. The second term represents the sum of the kinetic energy operators of the N electrons, each of them having their mass m replaced by the reduced mass μ because of the motion of the nucleus. The nuclear motion is also responsible for the existence of the third term, which is often called the *mass polarisation* term. We note that this term is only present if $N \geqslant 2$. Finally, the potential energy term V is readily expressed in terms of the relative coordinates as

$$V(\mathbf{r}_1, \mathbf{r}_2, \dots, \mathbf{r}_N) = -\sum_{i=1}^{N}\frac{Ze^2}{(4\pi\varepsilon_0)r_i} + \sum_{i<j}\frac{e^2}{(4\pi\varepsilon_0)r_{ij}} \quad \textbf{(A8.12)}$$

with $r_{ij} = |\mathbf{r}_i - \mathbf{r}_j|$.

The time-independent Schrödinger equation for the spatial part of the wave function corresponding to the total $((N+1)$-particle) system reads

$$H\psi_{\text{tot}}(\mathbf{R}, \mathbf{r}_1, \mathbf{r}_2, \dots, \mathbf{r}_N) = E_{\text{tot}}\psi_{\text{tot}}(\mathbf{R}, \mathbf{r}_1, \mathbf{r}_2, \dots, \mathbf{r}_N) \quad \textbf{(A8.13)}$$

where the Hamiltonian H is given by (A8.11) and E_{tot} is the total energy of the system. Since H is made up of a term involving only the centre of mass coordinates **R** and other terms which involve only the relative coordinates $(\mathbf{r}_1, \mathbf{r}_2, \dots, \mathbf{r}_N)$, the

Schrödinger equation (A8.13) possesses a complete set of eigensolutions of the form

$$\psi_{tot}(\mathbf{R}, \mathbf{r}_1, \mathbf{r}_2, \ldots, \mathbf{r}_N) = \Phi(\mathbf{R})\psi(\mathbf{r}_1, \mathbf{r}_2, \ldots, \mathbf{r}_N) \qquad \textbf{(A8.14)}$$

where the functions $\Phi(\mathbf{R})$ and $\psi(\mathbf{r}_1, \mathbf{r}_2, \ldots, \mathbf{r}_N)$ satisfy respectively the time-independent Schrödinger equations

$$-\frac{\hbar^2}{2(M + Nm)}\nabla^2_{\mathbf{R}}\Phi(\mathbf{R}) = E_{CM}\Phi(\mathbf{R}) \qquad \textbf{(A8.15)}$$

and

$$\left[-\frac{\hbar^2}{2\mu}\sum_{i=1}^{N}\nabla^2_{\mathbf{r}_i} - \frac{\hbar^2}{M}\sum_{i<j}\boldsymbol{\nabla}_{\mathbf{r}_i}\cdot\boldsymbol{\nabla}_{\mathbf{r}_j} + V(\mathbf{r}_1, \mathbf{r}_2, \ldots, \mathbf{r}_N)\right]\psi(\mathbf{r}_1, \mathbf{r}_2, \ldots, \mathbf{r}_N)$$

$$= E\psi(\mathbf{r}_1, \mathbf{r}_2, \ldots, \mathbf{r}_N) \qquad \textbf{(A8.16)}$$

with

$$E_{tot} = E_{CM} + E \qquad \textbf{(A8.17)}$$

Equation (A8.15) shows that the centre of mass moves as a *free* particle having a mass $(M + Nm)$ equal to the total mass of the system and a kinetic energy E_{CM}. On the other hand, equation (A8.16) describes the *relative* motion, which is the problem of physical interest in the study of atomic structure.

9 The space-fixed and body-fixed frames for a diatomic molecule

In Section 10.3, the calculation of the electronic wave functions of diatomic molecules was discussed using a frame of reference fixed with respect to the molecule with axes $\overline{OX}$, $\overline{OY}$ and $\overline{OZ}$, taking the $\overline{OZ}$ axis along the internuclear axis. In this frame, called the body-fixed or molecular frame, the Cartesian coordinates of the N electrons are denoted by $\bar{x}_i, \bar{y}_i, \bar{z}_i$ (with $i = 1, 2, \ldots, N$), and the electronic wave function in a state q is expressed as $\Phi_q(R; \bar{x}_i, \bar{y}_i, \bar{z}_i)$. The calculation of the electronic energies $E_q(R)$ is most conveniently carried out in the body-fixed (molecular) frame, but the original Schrödinger equation (10.11) and the expansion (10.15) were written for a space-fixed (laboratory) frame in which the axes OX, OY, OZ are in fixed directions in space. In the space-fixed frame the Cartesian coordinates of the N electrons are x_i, y_i, z_i, $i = 1, 2, \ldots, N$, and the wave functions $\Phi_q(\mathbf{R}; \mathbf{r}_1, \mathbf{r}_2, \ldots, \mathbf{r}_N)$ (see (10.15)) are expressed as

$$\Phi_q(\mathbf{R}; \mathbf{r}_1, \mathbf{r}_2, \ldots, \mathbf{r}_N) \equiv \Phi_q(\mathbf{R}; x_i, y_i, z_i) \qquad \textbf{(A9.1)}$$

We recall that $\mathbf{R}$, the relative position vector of the two nuclei, is in the direction specified by the polar coordinates (Θ, Φ), so that in the space-fixed system Φ_q depends on Θ and Φ as well as on R.

To obtain the body-fixed frame from the space-fixed frame a clockwise rotation through an angle Φ must be made about the OZ axis (see Fig. A9.1) producing new axes OX', OY', OZ'. This is followed by a rotation about OY' through an angle Θ which brings the axes into coincidence with $\overline{OX}$, $\overline{OY}$ and $\overline{OZ}$. The Cartesian components of the position vector $\mathbf{r}_i$ in the body-fixed frame $\bar{x}_i, \bar{y}_i, \bar{z}_i$ are then related to those in the space-fixed frame x_i, y_i, z_i by

$$\bar{x}_i = \cos\Theta\cos\Phi\, x_i + \cos\Theta\sin\Phi\, y_i - \sin\Theta\, z_i, \qquad \textbf{(A9.2a)}$$

$$\bar{y}_i = -\sin\Phi\, x_i + \cos\Phi\, y_i, \qquad \textbf{(A9.2b)}$$

$$\bar{z}_i = \sin\Theta\cos\Phi\, x_i + \sin\Theta\sin\Phi\, y_i + \cos\Theta\, z_i \qquad \textbf{(A9.2c)}$$

At a particular point in configuration space the numerical value of the electronic wave function Φ_q must be the same whichever coordinate system is employed, with the consequence that to find $\Phi_q(\mathbf{R}; x_i, y_i, z_i)$ expressed in space-fixed coordinates it is only necessary to substitute the expressions (A9.2) into $\Phi_q(R; \bar{x}_i, \bar{y}_i, \bar{z}_i)$ which has been calculated in the body-fixed frame. That is,

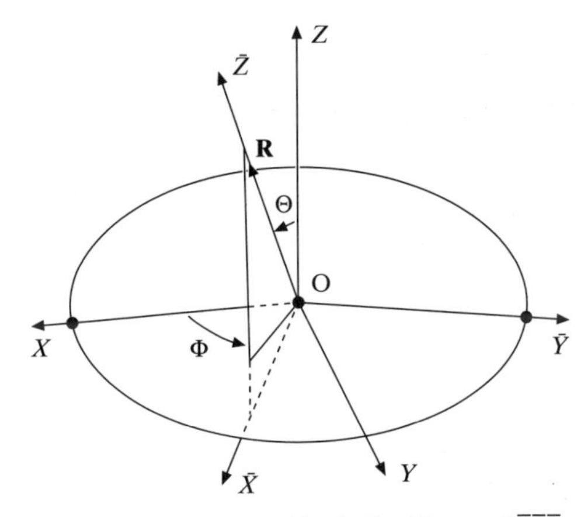

Figure A9.1 The space-fixed frame OXYZ and body-fixed frame O$\bar{X}\bar{Y}\bar{Z}$.

$$\Phi_q(\mathbf{R}; x_i, y_i, z_i) = \Phi_q(R; \bar{x}_i, \bar{y}_i, \bar{z}_i) \tag{A9.3}$$

where $\bar{x}_i$, $\bar{y}_i$ and $\bar{z}_i$ on the right-hand side are given by (A9.2). Usually it is not necessary to carry out this substitution explicitly. For example, in order to find the angular momentum eigenfunctions for the rotating molecule, the derivatives $\partial\Phi_q/\partial\Theta$ and $\partial\Phi_q/\partial\Phi$ are required. These can be calculated by noting that when applied to $\Phi_q(R; \bar{x}_i, \bar{y}_i, \bar{z}_i)$ we can write

$$\frac{\partial}{\partial\Theta} = \sum_{i=1}^{N}\left\{\frac{\partial\bar{x}_i}{\partial\Theta}\frac{\partial}{\partial\bar{x}_i} + \frac{\partial\bar{y}_i}{\partial\Theta}\frac{\partial}{\partial\bar{y}_i} + \frac{\partial\bar{z}_i}{\partial\Theta}\frac{\partial}{\partial\bar{z}_i}\right\} \tag{A9.4a}$$

and

$$\frac{\partial}{\partial\Phi} = \sum_{i=1}^{N}\left\{\frac{\partial\bar{x}_i}{\partial\Phi}\frac{\partial}{\partial\bar{x}_i} + \frac{\partial\bar{y}_i}{\partial\Phi}\frac{\partial}{\partial\bar{y}_i} + \frac{\partial\bar{z}_i}{\partial\Phi}\frac{\partial}{\partial\bar{z}_i}\right\} \tag{A9.4b}$$

From (A9.2) we find that

$$\frac{\partial\bar{x}_i}{\partial\Theta} = -\bar{z}_i, \qquad \frac{\partial\bar{y}_i}{\partial\Theta} = 0, \qquad \frac{\partial\bar{z}_i}{\partial\Theta} = \bar{x}_i \tag{A9.5a}$$

and

$$\frac{\partial\bar{x}_i}{\partial\Phi} = \bar{y}_i\cos\Theta, \qquad \frac{\partial\bar{y}_i}{\partial\Phi} = -(\bar{x}_i\cos\Theta + \bar{z}_i\sin\Theta), \qquad \frac{\partial\bar{z}_i}{\partial\Phi} = \bar{y}_i\sin\Theta \tag{A9.5b}$$

and combining (A9.4) and (A9.5), it is seen that

$$\frac{\partial\Phi_q}{\partial\Theta} = -\frac{\mathrm{i}}{\hbar}L_{\bar{y}}\Phi_q \tag{A9.6a}$$

and

$$\frac{\partial \Phi_q}{\partial \Phi} = \left[-\frac{i}{\hbar} \cos \Theta \, L_{\bar{z}} + \frac{i}{\hbar} \sin \Theta \, L_{\bar{x}} \right] \Phi_q \qquad \textbf{(A9.6b)}$$

where $L_{\bar{x}}$, $L_{\bar{y}}$ and $L_{\bar{z}}$ are the Cartesian components of the total orbital angular momentum operator $\mathbf{L}$ in the body-fixed frame,

$$L_{\bar{x}} = -i\hbar \sum_{i=1}^{N} \left(\bar{y}_i \frac{\partial}{\partial \bar{z}_i} - \bar{z}_i \frac{\partial}{\partial \bar{y}_i} \right),$$

$$L_{\bar{y}} = -i\hbar \sum_{i=1}^{N} \left(\bar{z}_i \frac{\partial}{\partial \bar{x}_i} - \bar{x}_i \frac{\partial}{\partial \bar{z}_i} \right), \qquad \textbf{(A9.7)}$$

$$L_{\bar{z}} = -i\hbar \sum_{i=1}^{N} \left(\bar{x}_i \frac{\partial}{\partial \bar{y}_i} - \bar{y}_i \frac{\partial}{\partial \bar{x}_i} \right)$$

In Appendix 11, the important result (A9.6) will be used to find the rotational eigenfunctions of a diatomic molecule.

10 Evaluation of two-centre integrals

The integrals required in the discussion of the hydrogen molecular ion can all be obtained from the basic integral

$$J = \int \frac{\exp(-pr_A)\exp(-qr_B)}{r_A r_B} \, d\mathbf{r} \tag{A10.1}$$

where $\mathbf{r}$, $\mathbf{r}_A$ and $\mathbf{r}_B$ are defined in Fig. 10.7.

This integral is most easily evaluated by introducing confocal elliptic coordinates defined by (see (10.45))

$$\begin{aligned} \xi &= \frac{1}{R}(r_A + r_B), \qquad 1 \leqslant \xi \leqslant \infty \\ \eta &= \frac{1}{R}(r_A - r_B), \qquad -1 \leqslant \eta \leqslant 1 \end{aligned} \tag{A10.2}$$

and ϕ, the azimuthal angle about the Z axis. We recall that the quantity R which appears in (A10.2) is the internuclear separation. The volume element $d\mathbf{r}$, expressed in terms of the confocal elliptic coordinates (ξ, η, ϕ), is given by

$$d\mathbf{r} = \frac{R^3}{8}(\xi^2 - \eta^2) \, d\xi \, d\eta \, d\phi \tag{A10.3}$$

so that

$$J = \frac{R}{2}\int_1^\infty d\xi \int_{-1}^{+1} d\eta \int_0^{2\pi} d\phi \, \exp(-a\xi - b\eta) \tag{A10.4}$$

where

$$\begin{aligned} a &= \frac{R}{2}(p + q) \\ b &= \frac{R}{2}(p - q) \end{aligned} \tag{A10.5}$$

The integral (A10.4) is now elementary and is given by

$$J = \frac{\pi R}{ab} e^{-a}(e^{b} - e^{-b}) = \frac{4\pi}{R}\frac{1}{p^2 - q^2}(e^{-qR} - e^{-pR}) \tag{A10.6}$$

The other relevant integrals are obtained by differentiating the above result with respect to p and q. That is,

$$K = \int \frac{\exp(-pr_{\mathrm{A}})\exp(-qr_{\mathrm{B}})}{r_{\mathrm{A}}}\,\mathrm{d}\mathbf{r} = -\frac{\partial}{\partial q}J$$

$$= \frac{4\pi}{R}\left[\frac{R}{p^2-q^2}\mathrm{e}^{-qR} + \frac{2q}{(p^2-q^2)^2}(\mathrm{e}^{-pR}-\mathrm{e}^{-qR})\right] \quad \textbf{(A10.7)}$$

and

$$L = \int \exp(-pr_{\mathrm{A}})\exp(-qr_{\mathrm{B}})\,\mathrm{d}\mathbf{r} = -\frac{\partial}{\partial p}K$$

$$= \frac{8\pi}{R(p^2-q^2)^2}\left[R(p\,\mathrm{e}^{-qR} + q\,\mathrm{e}^{-pR}) + \frac{4pq}{p^2-q^2}(\mathrm{e}^{-pR}-\mathrm{e}^{-qR})\right] \quad \textbf{(A10.8)}$$

In the particular case $p = q$, we have

$$J = \frac{2\pi}{p}\mathrm{e}^{-pR} \quad \textbf{(A10.9)}$$

$$K = \frac{\pi}{p^2}(1 + pR)\,\mathrm{e}^{-pR} \quad \textbf{(A10.10)}$$

and

$$L = \frac{\pi}{p^3}\left(1 + pR + \frac{1}{3}p^2R^2\right)\mathrm{e}^{-pR} \quad \textbf{(A10.11)}$$

11 The rotational wave functions of a diatomic molecule

In this appendix, the rotational wave functions of a diatomic molecule will be studied in the case that spin-dependent interactions can be neglected. In the Born–Oppenheimer approximation the spatial wave function can be written as (see (10.23) and (10.93))

$$\Psi_s = R^{-1}\mathcal{F}^s_{v,K}(R)\,{}^{\Lambda}\mathcal{H}_{KM_K}(\Theta, \Phi)\Phi_s(R; \bar{x}_i, \bar{y}_i, \bar{z}_i) \qquad \textbf{(A11.1)}$$

where $\mathcal{F}^s_{v,K}(R)$ is a radial wave function, ${}^{\Lambda}\mathcal{H}_{KM_K}(\Theta, \Phi)$ is a rotational wave function and Φ_s is the electronic wave function. We have chosen to write the electronic wave function in (A11.1) in terms of the body-fixed coordinates of the electrons $\bar{x}_i$, $\bar{y}_i$, $\bar{z}_i$ (with $i = 1, 2, \ldots, N$) so that Φ_s has an implicit dependence on (Θ, Φ) through the equations (A9.2).

For an isolated molecule, and in the absence of spin-dependent interactions, the complete wave function Ψ is a simultaneous eigenfunction of $\mathbf{K}^2$, K_z and $L_{\bar{z}}$ where $\mathbf{K}$ is the total orbital angular momentum of the molecule and $L_{\bar{z}}$ is the component of the total orbital angular momentum $\mathbf{L}$ of the electrons along the internuclear axis, with eigenvalues $K(K+1)\hbar^2$, $M_K\hbar$ and $\Lambda\hbar$ respectively. These eigenvalue equations will be used to obtain a differential equation from which ${}^{\Lambda}\mathcal{H}_{KM_K}$ can be calculated. First we note that

$$\mathbf{K} = \mathbf{N} + \mathbf{L} \qquad \textbf{(A11.2)}$$

where $\mathbf{N}$ is the angular momentum operator for the nuclear motion, with

$$N_{\pm} = N_x \pm \mathrm{i}N_y = \hbar\, \mathrm{e}^{\pm \mathrm{i}\Phi}\left(\pm\frac{\partial}{\partial\Theta} + \mathrm{i}\cot\Theta\,\frac{\partial}{\partial\Phi}\right) \qquad \textbf{(A11.3a)}$$

and

$$N_z = -\mathrm{i}\hbar\frac{\partial}{\partial\Phi} \qquad \textbf{(A11.3b)}$$

The components of $\mathbf{L}$ in the space-fixed frame L_x, L_y, L_z are related to those in the body-fixed frame $L_{\bar{x}}$, $L_{\bar{y}}$, $L_{\bar{z}}$ by the inverse transformation of (A9.2), which is

$$L_x = \cos\Theta\cos\Phi\, L_{\bar{x}} - \sin\Phi\, L_{\bar{y}} + \sin\Theta\cos\Phi\, L_{\bar{z}}, \quad \textbf{(A11.4a)}$$

$$L_y = \cos\Theta\sin\Phi\, L_{\bar{x}} + \cos\Phi\, L_{\bar{y}} + \sin\Theta\sin\Phi\, L_{\bar{z}}, \quad \textbf{(A11.4b)}$$

$$L_z = -\sin\Theta\, L_{\bar{x}} + \cos\Theta\, L_{\bar{z}} \quad \textbf{(A11.4c)}$$

From (A9.6) it is seen that

$$\frac{\partial}{\partial\Theta}[{}^{\Lambda}\mathcal{H}_{KM_K}(\Theta, \Phi)\Phi_s(R; \bar{x}_i, \bar{y}_i, \bar{z}_i)]$$

$$= \left[\frac{\partial}{\partial\Theta}{}^{\Lambda}\mathcal{H}_{KM_K}(\Theta, \Phi)\right]\Phi_s(R; \bar{x}_i, \bar{y}_i, \bar{z}_i) + {}^{\Lambda}\mathcal{H}_{KM_K}(\Theta, \Phi)\left(-\frac{\mathrm{i}}{\hbar}\right)L_{\bar{y}}\Phi_s(R; \bar{x}_i, \bar{y}_i, \bar{z}_i)$$

(A11.5a)

and

$$\frac{\partial}{\partial\Phi}[{}^{\Lambda}\mathcal{H}_{KM_K}(\Theta, \Phi)\Phi_s(R; \bar{x}_i, \bar{y}_i, \bar{z}_i)] = \left[\frac{\partial}{\partial\Phi}{}^{\Lambda}\mathcal{H}_{KM_K}(\Theta, \Phi)\right]\Phi_s(R; \bar{x}_i, \bar{y}_i, \bar{z}_i)$$

$$+ {}^{\Lambda}\mathcal{H}_{KM_K}(\Theta, \Phi)\left[-\frac{\mathrm{i}}{\hbar}\cos\Theta\, L_{\bar{z}} + \frac{\mathrm{i}}{\hbar}\sin\Theta\, L_{\bar{x}}\right]\Phi_s(R; \bar{x}_i, \bar{y}_i, \bar{z}_i) \quad \textbf{(A11.5b)}$$

Using (A11.2) with (A11.3b), (A11.4c) and (A11.5b) we see that

$$K_z[{}^{\Lambda}\mathcal{H}_{KM_K}(\Theta, \Phi)\Phi_s(R; \bar{x}_i, \bar{y}_i, \bar{z}_i)] = \Phi_s(R; \bar{x}_i, \bar{y}_i, \bar{z}_i)(-\mathrm{i}\hbar)\frac{\partial}{\partial\Phi}{}^{\Lambda}\mathcal{H}_{KM_K}(\Theta; \Phi)$$

(A11. 6)

Since $[{}^{\Lambda}\mathcal{H}_{KM_K}(\Theta, \Phi)\Phi_s(R; \bar{x}_i, \bar{y}_i, \bar{z}_i)]$ must be an eigenfunction of K_z belonging to the eigenvalue $M_K\hbar$, ${}^{\Lambda}\mathcal{H}_{KM_K}(\Theta, \Phi)$ must satisfy the equation

$$-\mathrm{i}\frac{\partial}{\partial\Phi}[{}^{\Lambda}\mathcal{H}_{KM_K}(\Theta, \Phi)] = M_K\, {}^{\Lambda}\mathcal{H}_{KM_K}(\Theta, \Phi) \quad \textbf{(A11.7)}$$

so that ${}^{\Lambda}\mathcal{H}_{KM_K}(\Theta, \Phi)$ must be of the form

$${}^{\Lambda}\mathcal{H}_{KM_K}(\Theta, \Phi) = {}^{\Lambda}h_{KM_K}(\Theta)\exp(\mathrm{i}M_K\Phi) \quad \textbf{(A11.8)}$$

where ${}^{\Lambda}h_{KM_K}$ is a function of Θ only. Setting

$$\mathbf{K}^2 = \frac{1}{2}(K_+K_- + K_-K_+) + K_z^2 \quad \textbf{(A11.9)}$$

where

$$K_\pm = K_x \pm \mathrm{i}K_y = N_x \pm \mathrm{i}N_y + L_x \pm \mathrm{i}L_y \quad \textbf{(A11.10)}$$

and again making use of the relations (A11.3), (A11.4) and (A11.5) it is found that

$$\mathbf{K}^2[{}^{\Lambda}\mathcal{H}_{KM_K}(\Theta; \Phi)\Phi_s(R; \bar{x}_i, \bar{y}_i, \bar{z}_i)] = -\hbar^2\Phi_s(R; \bar{x}_i, \bar{y}_i, \bar{z}_i)\left[\frac{\partial^2}{\partial\Theta^2} + \cot\Theta\frac{\partial}{\partial\Theta}\right.$$

$$\left. - \frac{1}{\sin^2\Theta}(M_K - \Lambda\cos\Theta)^2 - \Lambda^2\right]{}^{\Lambda}\mathcal{H}_{KM_K}(\Theta, \Phi) \qquad \textbf{(A11.11)}$$

where the fact that $L_{\bar{z}}\Phi_s = \pm\Lambda\hbar\Phi_s$ has been employed. Making use of the eigenvalue equation (10.86) for $\mathbf{K}^2$, ${}^{\Lambda}h_{KM_K}(\Theta)$ is found to satisfy the differential equation

$$-\left[\frac{\mathrm{d}^2}{\mathrm{d}\Theta^2} + \cot\Theta\frac{\mathrm{d}}{\mathrm{d}\Theta} - \frac{1}{\sin^2\Theta}(M_K - \Lambda\cos\Theta)^2 - \Lambda^2\right]{}^{\Lambda}h_{KM_K}(\Theta)$$

$$= K(K+1){}^{\Lambda}h_{KM_K}(\Theta) \qquad \textbf{(A11.12)}$$

The solutions of this equation have been given by Rose (1957) and Edmonds (1957) in the form, with $\mu = \cos\Theta$,

$${}^{\Lambda}h_{KM_K}(\Theta) = N(1-\mu)^{(\Lambda-M_K)/2}(1+\mu)^{-(\Lambda+M_K)/2}$$

$$\times\left[\left(\frac{\partial}{\partial\mu}\right)^{K-M_K}\left\{(1-\mu)^{K-\Lambda}(1+\mu)^{K+\Lambda}\right\}\right] \qquad \textbf{(A11.13)}$$

where N is a normalisation factor. If N is taken to be

$$N = (-1)^K(2\pi)^{-1/2}\left[\frac{(2K+1)(K+M_K)!}{2^{2K+1}(K+\Lambda)!(K-\Lambda)!(K-M_K)!}\right]^{1/2} \qquad \textbf{(A11.14)}$$

the functions ${}^{\Lambda}\mathcal{H}_{KM_K}$ satisfy the orthonormality relations

$$\int [{}^{\Lambda}\mathcal{H}_{K'M'_K}(\Theta, \Phi)]^* \, {}^{\Lambda}\mathcal{H}_{KM_K}(\Theta, \Phi)\,\mathrm{d}\Omega = \delta_{KK'}\delta_{M_KM'_K} \qquad \textbf{(A11.15)}$$

where $\mathrm{d}\Omega = \sin\Theta\,\mathrm{d}\Theta\,\mathrm{d}\Phi$. In the case $\Lambda = 0$, the rotational eigenfunctions reduce to spherical harmonics:

$${}^{0}\mathcal{H}_{KM_K}(\Theta, \Phi) = Y_{KM_K}(\Theta, \Phi) \qquad \textbf{(A11.16)}$$

A further important property of the rotational wave functions is that under the parity transformation ($\mathbf{R} \to -\mathbf{R}$) so that $\Theta \to \pi - \Theta$, $\Phi \to \Phi + \pi$, we have

$${}^{\Lambda}\mathcal{H}_{KM_K}(\pi - \Theta, \Phi + \pi) = (-1)^{K+\Lambda}\,{}^{\Lambda}\mathcal{H}_{KM_K}(\Theta, \Phi) \qquad \textbf{(A11.17)}$$

12 Dalitz integrals

In this appendix we shall study Dalitz integrals of the type

$$I_{m,n}(\alpha, \beta; \mathbf{k}_i, \mathbf{k}_f; \bar{k}) = \int d\mathbf{q} \frac{1}{q^2 - \bar{k}^2 - i\varepsilon} \frac{1}{[\alpha^2 + (\mathbf{q} - \mathbf{k}_i)^2]^m} \frac{1}{[\beta^2 + (\mathbf{q} - \mathbf{k}_f)^2]^n} \quad \textbf{(A12.1)}$$

with $\varepsilon \to 0^+$. Using a method introduced by R.P. Feynman in 1949, we set

$$\begin{aligned} a &= \alpha^2 + (\mathbf{q} - \mathbf{k}_i)^2, \\ b &= \beta^2 + (\mathbf{q} - \mathbf{k}_f)^2 \end{aligned} \quad \textbf{(A12.2)}$$

and use the integral representation

$$\frac{1}{ab} = \int_0^1 \frac{dt}{[at + b(1-t)]^2} \quad \textbf{(A12.3a)}$$

By differentiating both sides of (A12.3a) with respect to a or (and) b, we also have

$$\frac{1}{a^2 b} = 2\int_0^1 \frac{t}{[at + b(1-t)]^3}\, dt, \quad \textbf{(A12.3b)}$$

$$\frac{1}{ab^2} = 2\int_0^1 \frac{1-t}{[at + b(1-t)]^3}\, dt, \quad \textbf{(A12.3c)}$$

$$\vdots$$

$$\frac{1}{a^m b^n} = \frac{(m+n-1)!}{(m-1)!(n-1)!} \int_0^1 \frac{t^{m-1}(1-t)^{n-1}}{[at + b(1-t)]^{m+n}}\, dt \quad \textbf{(A12.3d)}$$

so that we may rewrite (A12.1) as

$$I_{m,n}(\alpha, \beta; \mathbf{k}_i, \mathbf{k}_f; \bar{k}) = \frac{(m+n-1)!}{(m-1)!(n-1)!} \int_0^1 dt\, t^{m-1}(1-t)^{n-1}$$

$$\times \int d\mathbf{q} \frac{1}{q^2 - \bar{k}^2 - i\varepsilon} \frac{1}{[\alpha^2 t + (\mathbf{q} - \mathbf{k}_i)^2 t + \beta^2(1-t) + (\mathbf{q} - \mathbf{k}_f)^2(1-t)]^{m+n}}$$

(A12.4)

We now observe that

$$\alpha^2 t + (\mathbf{q} - \mathbf{k}_i)^2 t + \beta^2(1 - t) + (\mathbf{q} - \mathbf{k}_f)^2(1 - t) = \Gamma^2 + (\mathbf{q} - \boldsymbol{\Lambda})^2 \quad \textbf{(A12.5)}$$

where

$$\boldsymbol{\Lambda} = t\mathbf{k}_i + (1 - t)\mathbf{k}_f \quad \textbf{(A12.6)}$$

and

$$\begin{aligned}\Gamma^2 &= \alpha^2 t + \beta^2(1 - t) + t(1 - t)(\mathbf{k}_i - \mathbf{k}_f)^2 \\ &= \alpha^2 t + \beta^2(1 - t) + t(1 - t)\Delta^2\end{aligned} \quad \textbf{(A12.7)}$$

and we recall that $\boldsymbol{\Delta} = \mathbf{k}_i - \mathbf{k}_f$ is the momentum transfer.

Apart from a one-dimensional integral on the t variable, the calculation of $I_{m,n}(\alpha, \beta; \mathbf{k}_i, \mathbf{k}_f; \bar{k})$ therefore reduces to the evaluation of integrals of the type

$$L_S = \int \mathrm{d}\mathbf{q} \frac{1}{q^2 - \bar{k}^2 - \mathrm{i}\varepsilon} \frac{1}{[\Gamma^2 + (\mathbf{q} - \boldsymbol{\Lambda})^2]^S} \quad \textbf{(A12.8)}$$

Let us begin by considering the case $S = 1$. Using spherical coordinates (q, θ_q, ϕ_q) in $\mathbf{q}$-space, taking the Z axis along the vector $\boldsymbol{\Lambda}$ and performing the integration over the azimuthal angle ϕ_q, we find that

$$L_1 = 2\pi \int_0^{\pi} \mathrm{d}\theta_q \sin\theta_q \int_0^{\infty} \mathrm{d}q q^2 \frac{1}{q^2 - \bar{k}^2 - \mathrm{i}\varepsilon} \frac{1}{\Gamma^2 + q^2 + \Lambda^2 - 2q\Lambda\cos\theta_q} \quad \textbf{(A12.9)}$$

Upon changing the integration variables to $q' = -q$ and $\theta'_q = \pi - \theta_q$ we can also write

$$L_1 = 2\pi \int_0^{\pi} \mathrm{d}\theta'_q \sin\theta'_q \int_{-\infty}^{0} \mathrm{d}q' q'^2 \frac{1}{q'^2 - \bar{k}^2 - \mathrm{i}\varepsilon} \frac{1}{\Gamma^2 + q'^2 + \Lambda^2 - 2q'\Lambda\cos\theta'_q} \quad \textbf{(A12.10)}$$

so that, by comparing (A12.9) and (A12.10), we have

$$L_1 = \pi \int_0^{\pi} \mathrm{d}\theta_q \sin\theta_q \int_{-\infty}^{+\infty} \mathrm{d}q q^2 \frac{1}{q^2 - \bar{k}^2 - \mathrm{i}\varepsilon} \frac{1}{\Gamma^2 + q^2 + \Lambda^2 - 2q\Lambda\cos\theta_q} \quad \textbf{(A12.11)}$$

The integral on the q variable may be performed by considering q as a complex variable and closing the contour with a semi-circle of infinite radius in the upper-half complex q-plane. The poles of the denominator in this upper-half q-plane are located at q_1 and q_2, with

$$q_1 = \bar{k} + \mathrm{i}\varepsilon, \qquad q_2 = \Lambda\cos\theta_q + \mathrm{i}(\Gamma^2 + \Lambda^2\sin^2\theta_q)^{1/2} \quad \textbf{(A12.12)}$$

Using the residue theorem, we therefore have

$$L_1 = \pi^2 \mathrm{i}\bar{k} \int_{-1}^{+1} \frac{\mathrm{d}w}{\Gamma^2 + \bar{k}^2 + \Lambda^2 - 2\bar{k}\Lambda w} + \frac{\pi^2\mathrm{i}}{\Lambda} \int_{\mathrm{i}\Gamma-\Lambda}^{\mathrm{i}\Gamma+\Lambda} \frac{q_2}{q_2^2 - \bar{k}^2 - \mathrm{i}\varepsilon} \mathrm{d}q_2 \quad \textbf{(A12.13)}$$

where we have set $w = \cos\theta_q$ in the first integral. Performing the integrals in (A12.13), we obtain

$$L_1(\bar{k}, \Gamma, \Lambda) = \frac{\pi^2 \mathrm{i}}{\Lambda} \log\left(\frac{\bar{k} + \Lambda + \mathrm{i}\Gamma}{\bar{k} - \Lambda + \mathrm{i}\Gamma}\right) \tag{A12.14}$$

The integrals L_S for $S = 2, 3, \ldots$ may be readily obtained from L_1 by successive differentiations with respect to Γ. Thus we have

$$L_2(\bar{k}, \Gamma, \Lambda) = -\frac{1}{2\Gamma}\frac{\partial}{\partial\Gamma} L_1(\bar{k}, \Gamma, \Lambda)$$

$$= -\frac{\pi^2}{\Gamma(\bar{k}^2 - \Gamma^2 - \Lambda^2 + 2\mathrm{i}\bar{k}\Gamma)} \tag{A12.15a}$$

$$\vdots$$

$$L_S(\bar{k}, \Gamma, \Lambda) = -\frac{1}{2(S-1)\Gamma}\frac{\partial}{\partial\Gamma} L_{S-1}(\bar{k}, \Gamma, \Lambda) \tag{A12.15b}$$

Let us now return to the expression for $I_{m,n}(\alpha, \beta; \mathbf{k}_i, \mathbf{k}_f; \bar{k})$ given by (A12.4). In certain cases simple closed form expressions may be obtained for the integration on the variable t. For example, when $m = n = 1$ one has

$$I_{1,1}(\alpha, \beta; \mathbf{k}_i, \mathbf{k}_f; \bar{k}) = \pi^2(A^2 - B)^{-1/2} \log\left[\frac{A + (A^2 - B)^{1/2}}{A - (A^2 - B)^{1/2}}\right] \tag{A12.16a}$$

where

$$A = -\mathrm{i}\bar{k}[\Delta^2 + (\alpha + \beta)^2] + \alpha(k_f^2 + \beta^2 - \bar{k}^2) + \beta(k_i^2 + \alpha^2 - \bar{k}^2) \tag{A12.16b}$$

and

$$B = [\Delta^2 + (\alpha + \beta)^2][k_i^2 + (\alpha - \mathrm{i}\bar{k})^2][k_f^2 + (\beta - \mathrm{i}\bar{k})^2] \tag{A12.16c}$$

It should be noted that the function on the right of (A12.16a) is single valued, even when we cross a branch cut of $(A^2 - B)^{1/2}$, that is either square root can be chosen. This function is therefore analytic, the only problem being the specification of the branch of the logarithm; examination shows that one must take the arguments of numerator and denominator from $-\pi$ to $+\pi$.

Let us look in more detail at the particular case for which $\alpha = \beta \neq 0$ and $k_i = k_f = \bar{k} = k$. Using equations (A12.16), we find that

$$I_{1,1}(\alpha, \alpha; \mathbf{k}_i, \mathbf{k}_f; k) = \frac{2\pi^2}{\Delta[\alpha^4 + 4k^2\alpha^2 + k^2\Delta^2]^{1/2}} \left\{ \tan^{-1}\frac{\alpha\Delta}{2[\alpha^4 + 4k^2\alpha^2 + k^2\Delta^2]^{1/2}} \right.$$

$$\left. + \frac{1}{2}\mathrm{i}\log\left[\frac{(\alpha^4 + 4k^2\alpha^2 + k^2\Delta^2)^{1/2} + k\Delta}{(\alpha^4 + 4k^2\alpha^2 + k^2\Delta^2)^{1/2} - k\Delta}\right]\right\} \tag{A12.17}$$

We remark that this result may also be obtained by using the equations (A12.4), (A12.15a) and the fact that we have here $\Gamma^2 + \Lambda^2 = k^2 + \alpha^2$. Thus we may write

$$I_{1,1}(\alpha, \alpha; \mathbf{k}_i, \mathbf{k}_f; k) = \pi^2 \int_0^1 \frac{\mathrm{d}t}{\Gamma(\alpha^2 - 2ik\Gamma)} \qquad \textbf{(A12.18)}$$

with $\Gamma = (\alpha^2 + t(1 - t)\Delta^2)^{1/2}$. The integral (A12.18) is then readily performed in closed form to yield the expression given by (A12.17). Substitution of this result in (12.288) yields the second Born term (12.289) corresponding to scattering by the Yukawa potential (12.277).

13 Solutions of selected problems

Chapter 1

1.1 (a) $v = 4.55 \times 10^7$ m s^{-1}
(b) $e/m = 1.65 \times 10^{11}$ C kg^{-1}.

1.3 The value of λ_{max} is obtained by solving the equation $\mathrm{d}\rho(\lambda)/\mathrm{d}\lambda = 0$, where $\rho(\lambda)$ is given by (1.30). Setting $x = hc/(\lambda k_B T)$, it is found that the maximum of $\rho(\lambda)$ occurs at a value x_0 of x such that $x_0 = 5(1 - \mathrm{e}^{-x_0})$. This equation is conveniently solved by writing $x_0 = 5 - \varepsilon$, so that $\varepsilon \simeq (\mathrm{e}^5/5 - 1)^{-1} = 0.034\,86$ and $x_0 = 4.965$. Thus $\lambda_{max} = hc/(4.965 k_B)$, and from the values of h, c and k_B given in Appendix 14 one finds $b = 2.898 \times 10^{-3}$ mK.

1.4
$$\rho_{tot} = \int_0^\infty \rho(\lambda)\,\mathrm{d}\lambda = 8\pi hc \int_0^\infty \lambda^{-5}\left\{\exp[hc/(\lambda k_B T)] - 1\right\}^{-1} \mathrm{d}\lambda$$
$$= 8\pi hc[k_B T/(hc)]^4 \int_0^\infty x^3(\mathrm{e}^x - 1)^{-1}\,\mathrm{d}x$$

where we have written $x = hc/(\lambda k_B T)$. Thus, using the result

$$\int_0^\infty x^3(\mathrm{e}^x - 1)^{-1}\,\mathrm{d}x = \pi^4/15$$

we obtain

$$\rho_{tot} = aT^4, \qquad a = \frac{8\pi^5}{15}\frac{k_B^4}{h^3c^3}$$

1.5 (a) $\lambda_t = 5391$ Å
(b) Maximum kinetic energy $\frac{1}{2}mv_{max}^2 = 3.90$ eV; stopping voltage $V_0 = 3.90$ V.

1.7 (a) $\Delta\lambda = \lambda_c(1 - \cos 30°) = 0.003\,25$ Å
(b) $\phi = 15°$
(c) $T_2 = 80$ eV.

1.9 $E_1 = 7.6$ GeV.

1.10 (a) The maximum momentum transferred to the atomic electron (of mass m) is $\Delta p = m(2v)$ where $v = 2.2 \times 10^7$ m s^{-1} is the velocity of the α particle. Thus $\Delta p = 4.0 \times 10^{-23}$ kg m s^{-1}.

(b) $\theta_{\text{max}} = \Delta p/p = m(2v)/(M_\alpha v) = 2.7 \times 10^{-4}$ radians.

1.11 (a) From (1.59) we have $r_0 = 2.3 \times 10^{-14}$ m.

(b) From (A1.25) and (A1.30) we have $\theta = 2\cot^{-1}\left(\dfrac{2b}{r_0}\right) = 16.5°$.

1.12 Using (1.59), departures from Rutherford scattering occur if the CM energy of the α particle is larger than 16.7 MeV, that is if its laboratory energy $E > 17.8$ MeV.

1.15 $\Delta\lambda = 1.8$ Å.

1.18 $P_2P_3 = 2P_1P_2 = 2|\mathcal{M}_z|\dfrac{\partial\mathcal{B}_z}{\partial z}\dfrac{L(L/2+l)}{Mv^2}$

$$= 2\mu_{\text{B}}\frac{\partial\mathcal{B}_z}{\partial z}\frac{L(L/2+l)}{3k_{\text{B}}T} = 0.078 \text{ m}$$

Chapter 2

2.1 From (2.13)

$$\psi(x) = (2\pi\hbar)^{-1/2}C\int_{p_0-\gamma}^{p_0+\gamma}\exp(\text{i}p_x x/\hbar)\,\text{d}p_x$$

$$= 2\left(\frac{\hbar}{2\pi}\right)^{1/2}C\frac{\sin(\gamma x/\hbar)}{x}\exp(\text{i}p_0 x/\hbar)$$

Using the result

$$\int_{-\infty}^{\infty}\frac{\sin^2 y}{y^2}\,\text{d}y = \pi$$

we find $C = 1/\sqrt{2\gamma}$. The function $|\psi(x)|^2$ has a maximum at $x = 0$ and falls to zero at $x = \pm\pi\hbar/\gamma$. Taking $\Delta x = \pi\hbar/\gamma$, and $\Delta p_x = \gamma$, we find $\Delta x\,\Delta p_x = \pi\hbar$ so that $\Delta x\,\Delta p_x > \hbar$.

2.2 From the uncertainty principle $\Delta p \gtrsim \hbar/\Delta r$. Making the assumptions $\Delta p \simeq p$, $\Delta r \simeq r$ and writing $rp = \hbar$, we obtain

$$E = \frac{\hbar^2}{2m}\frac{1}{r^2} - \left(\frac{e^2}{4\pi\varepsilon_0}\frac{1}{r}\right)$$

The radius at which E is a minimum is given by the condition $\mathrm{d}E/\mathrm{d}r = 0$, from which we find that

$$r = \frac{(4\pi\varepsilon_0)\hbar^2}{me^2} = a_0$$

The corresponding lowest value of the energy is

$$E_0 = -\frac{e^4 m}{(4\pi\varepsilon_0)^2 2\hbar^2} = -13.6 \text{ eV}$$

It is interesting, but not significant, that the particular choice $\Delta r = r$ and $\Delta p = p$, together with the assumption $\Delta r\, \Delta p = \hbar$, gives the *exact* value of E_0. The correct order of magnitude would have been obtained with any reasonable choice for Δr and Δp, with $\Delta r\, \Delta p \gtrsim \hbar$.

2.3 (a) Given the momentum eigenfunction

$$\psi_{p_x}(x) = C \exp(\mathrm{i}p_x x/\hbar)$$

we have

$$\int \psi^*_{p'_x}(x)\psi_{p_x}(x)\,\mathrm{d}x = |C|^2 \int \exp\left[\frac{\mathrm{i}}{\hbar}(p_x - p'_x)x\right]\mathrm{d}x$$

$$= |C|^2\, 2\pi\hbar\,\delta(p_x - p'_x)$$

where we have used the representation (2.31a) of the Dirac delta function and the property (2.33d). In order to normalise $\psi_{p_x}(x)$ according to the p-normalisation condition

$$\int \psi^*_{p'_x}(x)\psi_{p_x}(x)\,\mathrm{d}x = \delta(p_x - p'_x)$$

we must therefore choose the normalisation constant C so that

$$|C| = (2\pi\hbar)^{-1/2}$$

(b) From the relations $E = p_x^2/(2m)$ and $E' = p_x'^2/(2m)$ and the representation (2.31a) of the Dirac delta function, we have

$$\int \psi^*_{p'_x}(x)\psi_{p_x}(x)\,\mathrm{d}x = |C|^2\, 2\pi\delta\left[\left(\frac{2mE}{\hbar^2}\right)^{1/2} - \left(\frac{2mE'}{\hbar^2}\right)^{1/2}\right]$$

Using the property (2.33g) with

$$g(E) = \left(\frac{2mE}{\hbar^2}\right)^{1/2} - \left(\frac{2mE'}{\hbar^2}\right)^{1/2}$$

we have

$$\delta\left[\left(\frac{2mE}{\hbar^2}\right)^{1/2} - \left(\frac{2mE'}{\hbar^2}\right)^{1/2}\right] = \left(\frac{\hbar^2}{2m}\right)^{1/2} 2\sqrt{E}\,\delta(E - E')$$

Thus, in order to normalise $\psi_{p_x}(x)$ on the energy scale according to

$$\int \psi^*_{p'_x}(x)\psi_{p_x}(x)\,\mathrm{d}x = \delta(E - E')$$

we must choose the normalisation constant C so that

$$|C| = (2\pi\hbar)^{-1/2}\left(\frac{m}{2E}\right)^{1/4}$$

2.6 From (2.56) $\mathrm{i}\hbar(\mathrm{d}/\mathrm{d}t)\langle\mathbf{r}\rangle = \langle[\mathbf{r}, H]\rangle$. We take each of the Cartesian coordinates in turn, thus

$$\mathrm{i}\hbar\frac{\mathrm{d}}{\mathrm{d}t}\langle x\rangle = \langle[x, H]\rangle$$

Now,

$$[x, H] = x\left\{-\frac{\hbar^2}{2m}\left(\frac{\partial^2}{\partial x^2} + \frac{\partial^2}{\partial y^2} + \frac{\partial^2}{\partial z^2}\right) + V(x, y, z)\right\}$$

$$- \left\{-\frac{\hbar^2}{2m}\left(\frac{\partial^2}{\partial x^2} + \frac{\partial^2}{\partial y^2} + \frac{\partial^2}{\partial z^2}\right) + V(x, y, z)\right\}x$$

$$= -\frac{\hbar^2}{2m}\left\{x\frac{\partial^2}{\partial x^2} - \frac{\partial^2}{\partial x^2}x\right\}$$

$$= \frac{\hbar^2}{m}\frac{\partial}{\partial x}$$

Since $p_x = -\mathrm{i}\hbar(\partial/\partial x)$, we have $\mathrm{d}\langle x\rangle/\mathrm{d}t = (1/m)\langle p_x\rangle$, and similarly with y and z, which proves the result. The result $\mathrm{d}\langle\mathbf{p}\rangle/\mathrm{d}t = -\langle\boldsymbol{\nabla}\cdot V\rangle$ can be proved in a similar way.

2.7 (a) From (2.65)

$$\int \Phi^*(AB)^\dagger\Psi\,\mathrm{d}\mathbf{r} = \int (AB\Phi)^*\Psi\,\mathrm{d}\mathbf{r} = \int (B\Phi)^*A^\dagger\Psi\,\mathrm{d}\mathbf{r}$$

$$= \int \Phi^*B^\dagger(A^\dagger\Psi)\,\mathrm{d}\mathbf{r}. \quad \text{Hence} \quad (AB)^\dagger = B^\dagger A^\dagger.$$

(b) Using the result (a), we find that (i) and (ii) are not Hermitian while (iii), (iv) and (v) are Hermitian.

2.8 (a) $$\Psi(x, t) = \frac{1}{\sqrt{2}}\exp(\mathrm{i}\alpha_1)\psi_1(x)\exp(-\mathrm{i}E_1t/\hbar)$$

$$+\frac{1}{\sqrt{3}}\exp(\mathrm{i}\alpha_2)\psi_2(x)\exp(-\mathrm{i}E_2t/\hbar) + \frac{1}{\sqrt{6}}\exp(\mathrm{i}\alpha_3)\psi_3(x)\exp(-\mathrm{i}E_3t/\hbar)$$

(b) Probability = 1/3

(c) $\langle x\rangle$ and $\langle p_x\rangle$ are time-dependent; $E = \langle H\rangle$ is time-independent.

2.11 (a) $$P_n = \frac{960}{\pi^6 n^6},\ n \text{ odd}$$

$$= 0,\ n \text{ even}$$

(b) $E = \langle H\rangle = 5\hbar^2/(4ma^2)$.

2.17 Since **L** and **S** operate in different spaces all the components of **L** commute with all the components of **S** and hence $[\mathbf{L}^2, \mathbf{S}^2] = 0$. Now $\mathbf{J}^2 = \mathbf{L}^2 + \mathbf{S}^2 + 2\mathbf{L}\cdot\mathbf{S}$ so that since $\mathbf{L}^2$ commutes with all the components of **L** and $\mathbf{S}^2$ commutes with all the components of **S**, one has $[\mathbf{L}^2, \mathbf{J}^2] = [\mathbf{S}^2, \mathbf{J}^2] = 0$. Since $J_z = L_z + S_z$, we have $[\mathbf{L}^2, J_z] = [\mathbf{S}^2, J_z] = 0$. Finally since **J** satisfies the commutation relations (2.223), it follows that $\mathbf{J}^2$ commutes with all its components and $[\mathbf{J}^2, J_z] = 0$, from which $[(\mathbf{L}\cdot\mathbf{S}), J_z] = 0$.

2.18 Using the orthogonality properties of $P_l(\cos\theta)$ we have

$$\left(\frac{2}{2l+1}\right)c_l j_l(kr) = \int_{-1}^{+1} \mathrm{d}(\cos\theta)\ \mathrm{e}^{ikr\cos\theta}P_l(\cos\theta)$$

Integrating by parts we find

$$\left(\frac{2}{2l+1}\right)c_l j_l(kr) = \left[\frac{\mathrm{e}^{ikr\cos\theta}}{ikr}P_l(\cos\theta)\right]_{\cos\theta=-1}^{\cos\theta=1} - \int_{-1}^{+1}\mathrm{d}(\cos\theta)\frac{\mathrm{e}^{ikr\cos\theta}}{ikr}\frac{\mathrm{d}P_l(\cos\theta)}{\mathrm{d}(\cos\theta)}$$

For large r, the second term on the right-hand side is of order r^{-2} (which can be seen by integrating by parts a second time), so that

$$\left(\frac{2}{2l+1}\right)c_l\frac{1}{kr}\sin(kr - \tfrac{1}{2}l\pi) = \frac{1}{ikr}(\mathrm{e}^{ikr} - (-1)^l\,\mathrm{e}^{-ikr})$$

Hence we find $c_l = (2l+1)\mathrm{i}^l$.

2.21 (b) The first excited level is twofold degenerate with $n_x = 1$, $n_y = 0$ and $n_x = 0$, $n_y = 1$. The determinantal equation (2.329) becomes

$$\begin{vmatrix} \langle 1|\lambda xy|1\rangle - E^{(1)} & \langle 1|\lambda xy|2\rangle \\ \langle 2|\lambda xy|1\rangle & \langle 2|\lambda xy|2\rangle - E^{(1)} \end{vmatrix} = 0$$

Using equation (A3.15), $\langle 1|\lambda xy|1\rangle = \langle 2|\lambda xy|2\rangle = 0$ and $\langle 1|\lambda xy|2\rangle = \langle 2|\lambda xy|1\rangle = \lambda(2\alpha^2)^{-1}$, where $\alpha^2 = (mk/\hbar^2)^{1/2}$, and we find $E^{(1)} = \pm\lambda(2\alpha^2)^{-1}$. Hence to first order in λ the first excited level splits into two levels with energies $E = 2\hbar(k/m)^{1/2} \pm \lambda(2\alpha^2)^{-1}$.

2.22 Since $H'_{ba}(t) = -q\mathscr{E}(t)\langle\psi_b|x|\psi_a\rangle$, it follows from (A3.15) that the only non-vanishing matrix element is H'_{10}. From (2.343) and (2.344) we find that

$$P_{10}(t = +\infty) = \frac{q^2\mathscr{E}_0^2}{2m\hbar\omega}\left|\int_0^\infty e^{i\omega t}\, e^{-t/\tau}\, dt\right|^2$$

where $\omega = (k/m)^{1/2}$. Thus

$$P_{10}(t = +\infty) = \frac{q^2\mathscr{E}_0^2}{2m\hbar\omega}\frac{\tau^2}{(\tau\omega)^2 + 1}$$

Chapter 3

3.2 (a) $r \geqslant 2a_0$
(b) Probability is $13e^{-4} = 0.238$.

3.3 (a) Yes, since ψ_{100}, ψ_{200} and ψ_{322} have even parity.
(b) $P_{100} = 2/7$, $P_{200} = 9/14$, $P_{322} = 1/14$. For other states the probability is zero.
(c) Using atomic units,

$$\langle H\rangle = (2/7)E_{n=1} + (9/14)E_{n=2} + (1/14)E_{n=3} = -0.227$$

$$\langle \mathbf{L}^2\rangle = 6(1/14) = 0.43$$

$$\langle L_z\rangle = 2(1/14) = 0.14$$

3.4 Let us assume that the decay occurs at $t = 0$. At times $t \leqslant 0$, the wave function of the system is (in a.u.)

$$\Psi(\mathbf{r}, t) = \psi_{1s}^{Z=1}(r)\exp(-iE_{1s}t)$$

where

$$\psi_{1s}^{Z=1}(r) = \frac{1}{\sqrt{\pi}}\exp(-r)$$

is the ground state wave function for the tritium atom ($Z = 1$). At times $t \geqslant 0$, we have

$$\Psi(\mathbf{r}, t) = \sum_k c_k\psi_k^{Z=2}(\mathbf{r})\exp(-iE_k t)$$

where $\psi_k^{Z=2}$ are hydrogenic wave functions for $Z = 2$ (^{3}He). Thus the probability of finding the ^{3}He ion in the state $\psi_k^{Z=2}$ at $t \geqslant 0$ is $P_k = |c_k|^2$, where

$$c_k = \langle\psi_k^{Z=2}|\psi_{1s}^{Z=1}\rangle$$

As a result, we have that

(a) $$c_{1s} = \int \psi_{1s}^{Z=2}(r)\psi_{1s}^{Z=1}(r)\,d\mathbf{r} = \frac{16\sqrt{2}}{27} \text{ and}$$

$$P_{1s} = |c_{1s}|^2 = \frac{512}{729} = 0.702$$

(b) Total probability for excitation and ionisation $= 1 - P_{1s} = 0.298$.

(c) $$P_{2s} = |c_{2s}|^2 = \int \psi_{2s}^{Z=2}(r)\psi_{1s}^{Z=1}(r)\,d\mathbf{r} = 0.25$$

(d) For states with $l \neq 0$ the coefficients c_k vanish since $\psi_{1s}^{Z=1}$ is spherically symmetric, and $P_k = 0$.

3.5 The perturbation H' is the potential energy of the electron in the gravitational field of the proton. That is,

$$H' = -G\frac{mM_p}{r}$$

where G is the gravitational constant, m is the electron mass and M_p the proton mass. The energy shift ΔE due to H' is given to first order in perturbation theory by

$$\Delta E = \langle \psi_{1s}|H'|\psi_{1s}\rangle = \frac{1}{\pi a_\mu^3}\int \exp(-2r/a_\mu)\left(-\frac{GmM_p}{r}\right)d\mathbf{r}$$

$$= -\frac{GmM_p}{a_\mu}, \qquad a_\mu = \frac{4\pi\varepsilon_0\hbar^2}{\mu e^2}$$

On the other hand, since $E_{1s} = -[e^2/(4\pi\varepsilon_0)]/(2a_\mu)$, we have

$$\frac{\Delta E}{E_{1s}} = \frac{8\pi\varepsilon_0 GmM_p}{e^2} \simeq 8.8 \times 10^{-40}$$

Chapter 4

4.1 The total number of photons radiated per second is:

(a) 5×10^{25} (b) 5×10^{23} (c) 3×10^{18} (d) 5×10^{14};

The corresponding fluxes and densities are:

(a) 4×10^{22} photons/(m^2 s); 1.3×10^{14} photons/m^3
(b) 4×10^{20} photons/(m^2 s); 1.3×10^{12} photons/m^3
(c) 2.4×10^{15} photons/(m^2 s); 0.8×10^{7} photons/m^3
(d) 4×10^{11} photons/(m^2 s); 1.3×10^{3} photons/m^3.

4.2 (a) To obtain (4.78), we assume the radiation to be randomly orientated, with a uniform distribution of polarisation vectors. The average required is

$$A = \frac{1}{4\pi}\int \cos^2\theta \, d\Omega_\varepsilon$$

where $d\Omega_\varepsilon$ is an element of solid angle about the direction of $\hat{\boldsymbol{\varepsilon}}$.

Taking $\mathbf{r}_{ba}$ to be fixed in space as axis,

$$A = \frac{1}{4\pi}\int_{-1}^{+1} \cos^2\theta \, d(\cos\theta) \int_0^{2\pi} d\phi = \frac{1}{3}$$

(b) In considering the transition rate for spontaneous emission (4.79), let us choose the Z axis to be the direction of emission of the photon. The two independernt polarisation vectors can then be taken to be $\hat{\boldsymbol{\varepsilon}}_1 = \hat{\mathbf{x}}$ and $\hat{\boldsymbol{\varepsilon}}_2 = \hat{\mathbf{y}}$, unit vectors along the X and Y axes. The sum over polarisation directions is proportional to

$$|\hat{\boldsymbol{\varepsilon}}_1 \cdot \mathbf{r}_{ba}|^2 + |\hat{\boldsymbol{\varepsilon}}_2 \cdot \mathbf{r}_{ba}|^2 = |\mathbf{r}_{ba}|^2(\cos^2\theta_1 + \cos^2\theta_2)$$

where θ_1 is the angle between $\mathbf{r}_{ba}$ and the X axis and θ_2 is the angle between $\mathbf{r}_{ba}$ and the Y axis. If the polar angles of $\mathbf{r}_{ba}$ are θ and ϕ, $\cos\theta_1 = \sin\theta\cos\phi$, $\cos\theta_2 = \sin\theta\sin\phi$ and $\cos^2\theta_1 + \cos^2\theta_2 = \sin^2\theta$. We now have to integrate over all directions of emission, which is equivalent to keeping the direction of emission fixed and integrating over all orientations of $\mathbf{r}_{ba}$. We have

$$\int d\Omega \sin^2\theta = \int_{-1}^{+1} d(\cos\theta)(1 - \cos^2\theta) \int_0^{2\pi} d\phi = 8\pi/3$$

from which we find the result (4.80).

4.5 We assume that each of the degenerate levels a are populated with an equal probability, in which case the transition rate per atom for absorption is obtained by averaging over the g_a initial states and summing over the g_b final states, so that

$$B_{ba} = \frac{1}{g_a}\sum_a \sum_b W_{ba}/\rho$$

Similarly for stimulated emission, since $\bar{W}_{ab} = W_{ba}$

$$B_{ab} = \frac{1}{g_b}\sum_b \sum_a W_{ba}/\rho$$

from which $g_a B_{ba} = g_b B_{ab}$ (the principle of detailed balance). Hence equation (4.104) becomes

$$\frac{g_b}{g_a} B_{ab}\rho N_a = (A_{ab} + B_{ab}\rho)N_b$$

Combining this with Boltzmann's relation, which in the case of degenerate levels is

$$\frac{N_a}{N_b} = \frac{g_a}{g_b} \exp[\hbar\omega_{ba}/(k_B T)]$$

and using the Planck distribution law (4.107) for ρ, the expression (4.108b) for A_{ab} is found.

4.6 The explicit expressions for $\mathscr{A}$ are

$$\mathscr{A}(l, m; l+1, m; 0) = \left(\frac{3}{4\pi}\right)^{1/2} \left[\frac{(l+1)^2 - m^2}{(2l+1)(2l+3)}\right]^{1/2}$$

$$\mathscr{A}(l, m; l-1, m; 0) = \left(\frac{3}{4\pi}\right)^{1/2} \left[\frac{l^2 - m^2}{(2l-1)(2l+1)}\right]^{1/2}$$

$$\mathscr{A}(l, m; l+1, m \pm 1; \pm 1) = \left(\frac{3}{4\pi}\right)^{1/2} \left[\frac{(l \pm m + 1)(l \pm m + 2)}{2(2l+1)(2l+3)}\right]^{1/2}$$

$$\mathscr{A}(l, m; l-1, m \pm 1; \pm 1) = -\left(\frac{3}{4\pi}\right)^{1/2} \left[\frac{(l \mp m)(l \mp m - 1)}{2(2l-1)(2l+1)}\right]^{1/2}$$

4.7 The sums $\sum_{m'} |\mathscr{A}(l, m; l', m'; m' - m)|^2$ can be performed using the explicit forms of $\mathscr{A}$ found in Problem 4.6. The absorption rate from a level (nlm) to the $(2l' + 1)$ degenerate levels $(n'l'm')$ is given by (4.71) and (4.112), as

$$W_{ba}^{D} = \frac{4\pi}{c\hbar^2}\left(\frac{e^2}{4\pi\varepsilon_0}\right) \sum_{m'} |\varepsilon^*_{m'-m} I^{m'-m}_{n'l'm';nlm}|^2$$

$$= \frac{4\pi}{c\hbar^2}\left(\frac{e^2}{4\pi\varepsilon_0}\right) \sum_{m'} |\varepsilon^*_{m'-m}|^2 |I^{m'-m}_{n'l'm';nlm}|^2$$

Now $|\varepsilon^*_q|^2 = \left|\sqrt{\frac{4\pi}{3}}\, Y_{1,q}(\hat{\varepsilon})\right|^2$

and for unpolarised isotropic light, the average of $|\varepsilon^*_q|^2$ is

$$\text{Av}|\varepsilon^*_q|^2 = \frac{1}{4\pi}\int d\Omega |\varepsilon^*_q|^2 = \frac{1}{3}$$

We have

$$W_{ba}^{\mathrm{D}} = \frac{4\pi}{c\hbar^2}\left(\frac{e^2}{4\pi\varepsilon_0}\right)\frac{1}{3}\sum_{m'}|I_{n'l'm';nlm}^{m'-m}|^2$$

$$= \frac{4\pi}{3c\hbar^2}\left(\frac{e^2}{4\pi\varepsilon_0}\right)\left[\int_0^\infty R_{n'l'}(r)R_{nl}(r)r^3\,\mathrm{d}r\right]^2 \frac{1}{2l+1}\begin{cases} l+1, & \text{if } l' = l+1 \\ l, & \text{if } l' = l-1 \end{cases}$$

which is independent of m. In the same way, it follows that the lifetime of a level nlm is independent of m.

4.10 In the electric dipole approximation, the left-hand side of the required relation is

$$\text{left-hand side} = \frac{4\pi^2}{c\hbar}\left(\frac{e^2}{4\pi\varepsilon_0}\right)\left[\sum_b \omega_{ba}|\hat{\boldsymbol{\varepsilon}}\cdot\mathbf{r}_{ba}|^2 - \sum_b{}' \omega_{ab}|\hat{\boldsymbol{\varepsilon}}\cdot\mathbf{r}_{ba}|^2\right]$$

where $\hbar\omega_{ba} = (E_b - E_a)$. Averaging over all directions of polarisation gives a factor of 1/3, and using the definition of the oscillator strength (4.136) we find that

$$\text{left-hand side} = \frac{4\pi^2}{c\hbar}\left(\frac{e^2}{4\pi\varepsilon_0}\right)\frac{1}{3}\frac{3\hbar}{2m}\left(\sum_b + \sum_b{}'\right) f_{ba}$$

Since

$$\left(\sum_b + \sum_b{}'\right) f_{ba} = \sum_k f_{ka} = 1$$

and $r_0 = e^2/(4\pi\varepsilon_0\, mc^2)$, the result follows.

Chapter 5

5.5 The wave functions for the $np_{3/2}$, $np_{1/2}$ and $n's_{1/2}$ levels of atomic hydrogen are

$$\psi(np_{3/2}) = R_{n1}(r)\sum_{m_s}\left\langle 1\frac{1}{2}m_l m_s \middle| \frac{3}{2}m_j\right\rangle Y_{1,m_l}\chi_{1/2m_s}$$

$$\psi(np_{1/2}) = R_{n1}(r)\sum_{m_s}\left\langle 1\frac{1}{2}m_l m_s \middle| \frac{1}{2}m_j\right\rangle Y_{1,m_l}\chi_{1/2m_s}$$

$$\psi(n's_{1/2}) = R_{n'0}(r)\chi_{1/2m_s'}$$

The selection rules require that $m_s' = m_s$. Taking the case $m_s = 1/2$ the possible transitions from the states of the $np_{3/2}$ level to the $n's_{1/2}$ level and the corresponding transition rates are given (apart from an overall constant) in the following table

Initial states				Transition rate
q	m_l	m_s	m_j	
1	1	1/2	3/2	$A^2\left\vert\left\langle 1\frac{1}{2}1\frac{1}{2}\middle\vert\frac{3}{2}\frac{3}{2}\right\rangle\right\vert^2 = A^2$
0	0	1/2	1/2	$A^2\left\vert\left\langle 1\frac{1}{2}0\frac{1}{2}\middle\vert\frac{3}{2}\frac{1}{2}\right\rangle\right\vert^2 = \frac{2}{3}A^2$
−1	−1	1/2	−1/2	$A^2\left\vert\left\langle 1\frac{1}{2}-1\frac{1}{2}\middle\vert\frac{3}{2}-\frac{1}{2}\right\rangle\right\vert^2 = \frac{1}{3}A^2$

where $A = CI^q_{n'00,n1m_l}$, C being a constant and $I^q_{n'00,n1m_l}$ is given by (4.113) and has the same value in each case. The transition rates for the case $m_s = -1/2$ are the same as for $m_s = +1/2$, so that the total rate is $4A^2$. For transitions from the states of the $n\text{p}_{1/2}$ level the possible transition rates are, for $m_s = m_s' = 1/2$,

Initial states				Transition rate
q	m_l	m_s	m_j	
0	0	1/2	1/2	$A^2\left\vert\left\langle 1\frac{1}{2}0\frac{1}{2}\middle\vert\frac{1}{2}\frac{1}{2}\right\rangle\right\vert^2 = \frac{A^2}{3}$
−1	−1	1/2	−1/2	$A^2\left\vert\left\langle 1\frac{1}{2}-1\frac{1}{2}\middle\vert\frac{1}{2}-\frac{1}{2}\right\rangle\right\vert^2 = \frac{2A^2}{3}$

where A has the same value as before. The total rate from both the $m_s = 1/2$ and $m_s = -1/2$ states is thus $2A^2$ and the ratio of the transition rate $n\text{p}_{3/2} \to n'\text{s}_{1/2}$ to the transition rate $n\text{p}_{1/2} \to n'\text{s}_{1/2}$ is 2:1.

Chapter 6

6.4 From (6.97), (6.101) and (6.102) the transition rates for the π and $\sigma^\pm$ lines are, using (4.113),

$$W^{\text{s,D}}_{ab}(\pi) = C(\omega_{ba})\sin^2\Theta\langle l1m0|l'm'\rangle^2 B$$

$$W^{\text{s,D}}_{ab}(\sigma^\pm) = C(\omega_{ba})\tfrac{1}{2}(1+\cos^2\Theta)\langle l1m\mp 1|l'm'\rangle^2 B$$

where $B = \dfrac{2l+1}{2l'+1}\langle l100|l'0\rangle^2\left[\displaystyle\int_0^\infty \text{d}r r^3 R_{n'l'}(r)R_{nl}(r)\right]^2$

and $l - l' = \pm 1$.

Each of the $(2l' + 1)$ states of the initial level will be, in general, populated equally, so that the intensities of the π and $\sigma^{\pm}$ lines are

$$I(\pi) = \frac{\hbar\omega_{ba}}{2l' + 1} \sum_{m'} W_{ab}^{s,D}(\pi)$$

$$I(\sigma^{\pm}) = \frac{\hbar\omega_{ba}}{2l' + 1} \sum_{m'} W_{ab}^{s,D}(\sigma^{\pm})$$

Using the orthogonality relation for the Clebsch–Gordan coefficients (A4.32), together with (A4.33), we see that

$$\sum_{m'} \langle l1m0|l'm'\rangle^2 = \sum_{m'} \langle l1m \mp 1|l'm'\rangle^2 = \tfrac{1}{3}(2l' + 1)$$

It follows that with $\Theta = \pi/2$

$$I(\pi):I(\sigma^+):I(\sigma^-) = 2:1:1$$

A similar argument shows that in the anomalous Zeeman effect the average intensity of the π lines is twice that of the σ^+ (or σ^-) lines (see Problem 9.8).

Chapter 7

7.3 The following table summarises the results (in a.u.)

	'Zero-order' wave function $\psi_0^{(0)}$ (7.35)	Screened variat. function (7.70) with $Z_e = 27/16$	Hartree–Fock wave function (7.85)	'Exact' (variat.) wave function
Energy	−2.750	−2.848	−2.862	−2.904
$\langle r_1^2 + r_2^2\rangle$	1.50	2.11	2.37	2.39
$\langle\delta(\mathbf{r}_1)\rangle$	2.55	1.53	1.80	1.81
$\langle\delta(\mathbf{r}_{12})\rangle$	0.318	0.191	0.188	0.106

Note the improvement in going from the zero-order wave function $\psi_0^{(0)}$ to the Hartree–Fock wave function. Nevertheless, the Hartree–Fock wave function provides a poor value (too large by almost a factor of two) of $\langle\delta(\mathbf{r}_{12})\rangle$. This is to be expected since the Hartree–Fock wave function is an independent particle wave function, and $\delta(\mathbf{r}_{12})$ probes the region $\mathbf{r}_1 = \mathbf{r}_2$ where the electrostatic repulsion term $1/r_{12}$ is most important.

7.5 (a) $E_{2s4p} = \left(-\frac{1}{2} - \frac{1}{32}\right)$ a.u. $= -0.531\,25$ a.u. $\simeq -14.5$ eV; $\lambda = 192$ Å;

(b) $v = 1.71$ a.u. $= 3.75 \times 10^6$ m s^{-1}.

Chapter 8

8.1 $H_c = \sum_{i=1}^{N} h_i, \qquad h_i = -\frac{1}{2}\nabla^2_{\mathbf{r}_i} + V(r_i)$

Since the angular part of $\nabla^2_{\mathbf{r}_i}$ is proportional to $\mathbf{L}_i^2$, we have $[\nabla^2_{\mathbf{r}_i}, \mathbf{L}_i] = 0$. Moreover, we know from Problem 2.13 that $\mathbf{L}_i$ commutes with any function of r_i. Therefore $\mathbf{L}_i$ commutes with h_i, and hence $\mathbf{L} = \sum_i \mathbf{L}_i$ commutes with H_c.

8.2 $$\Phi_{2^3\mathrm{S}}(M_S = 1) = \frac{1}{\sqrt{2}}\begin{vmatrix} u_{1s\uparrow}(1) & u_{2s\uparrow}(1) \\ u_{1s\uparrow}(2) & u_{2s\uparrow}(2) \end{vmatrix}$$

$$\Phi_{2^3\mathrm{S}}(M_S = 0) = \frac{1}{\sqrt{2}}\left\{\frac{1}{\sqrt{2}}\begin{vmatrix} u_{1s\uparrow}(1) & u_{2s\downarrow}(1) \\ u_{1s\uparrow}(2) & u_{2s\downarrow}(2) \end{vmatrix} + \frac{1}{\sqrt{2}}\begin{vmatrix} u_{1s\downarrow}(1) & u_{2s\uparrow}(1) \\ u_{1s\downarrow}(2) & u_{2s\uparrow}(2) \end{vmatrix}\right\}$$

$$\Phi_{2^3\mathrm{S}}(M_S = -1) = \frac{1}{\sqrt{2}}\begin{vmatrix} u_{1s\downarrow}(1) & u_{2s\downarrow}(1) \\ u_{1s\downarrow}(2) & u_{2s\downarrow}(2) \end{vmatrix}$$

8.4 (a) $Z^{-1/3}$ a.u.
(b) $Z^{4/3}$ a.u.
(c) $Z^{7/3}$ a.u.

8.7 From Problem 8.2, we have

$$\Phi_{2^3\mathrm{S}}(M_S = 1) = \frac{1}{\sqrt{2}}\begin{vmatrix} u_{1s\uparrow}(1) & u_{2s\uparrow}(1) \\ u_{1s\uparrow}(2) & u_{2s\uparrow}(2) \end{vmatrix}$$

with $u_{1s\uparrow} = u_{1s}(r)\alpha, \qquad u_{2s\uparrow} = u_{2s}(r)\alpha$

The coupled Hartree–Fock equations for $u_{1s}(r)$ and $u_{2s}(r)$ are

$$\left[-\frac{1}{2}\nabla^2 - \frac{2}{r} + V^{\mathrm{d}}_{2s} - V^{\mathrm{ex}}_{2s}\right]u_{1s}(r) = E_{1s}u_{1s}(r) \qquad \textbf{(1)}$$

and

$$\left[-\frac{1}{2}\nabla^2 - \frac{2}{r} + V^{\mathrm{d}}_{1s} - V^{\mathrm{ex}}_{1s}\right]u_{2s}(r) = E_{2s}u_{2s}(r) \qquad \textbf{(2)}$$

with

$$V^{\mathrm{d}}_{1s}(r) = \int u_{1s}(r')\frac{1}{|\mathbf{r} - \mathbf{r}'|}u_{1s}(r')\,\mathrm{d}\mathbf{r}'$$

$$V^{\mathrm{ex}}_{1s}(r)f(\mathbf{r}) = \left[\int u_{1s}(r')\frac{1}{|\mathbf{r} - \mathbf{r}'|}f(\mathbf{r}')\,\mathrm{d}\mathbf{r}'\right]u_{1s}(r)$$

and similar relations for $V^{\mathrm{d}}_{2\mathrm{s}}$ and $V^{\mathrm{ex}}_{2\mathrm{s}}$. Taking the scalar product of (1) with $u_{2\mathrm{s}}$, we have

$$\left\langle u_{2\mathrm{s}} \left| -\frac{1}{2}\nabla^2 - \frac{2}{r} + V^{\mathrm{d}}_{2\mathrm{s}} - V^{\mathrm{ex}}_{2\mathrm{s}} \right| u_{1\mathrm{s}} \right\rangle = E_{1\mathrm{s}} \langle u_{2\mathrm{s}} | u_{1\mathrm{s}} \rangle \qquad \textbf{(3)}$$

Similarly, taking the scalar product of (2) with $u_{1\mathrm{s}}$ and using the fact that the operators are Hermitian, we find that

$$\left\langle u_{2\mathrm{s}} \left| -\frac{1}{2}\nabla^2 - \frac{2}{r} + V^{\mathrm{d}}_{2\mathrm{s}} - V^{\mathrm{ex}}_{2\mathrm{s}} \right| u_{1\mathrm{s}} \right\rangle = E_{1\mathrm{s}} \langle u_{2\mathrm{s}} | u_{1\mathrm{s}} \rangle \qquad \textbf{(4)}$$

Subtracting (4) from (3) and remembering that

$$(V^{\mathrm{d}}_{1\mathrm{s}} - V^{\mathrm{ex}}_{1\mathrm{s}})u_{1\mathrm{s}} = 0, \qquad (V^{\mathrm{d}}_{2\mathrm{s}} - V^{\mathrm{ex}}_{2\mathrm{s}})u_{2\mathrm{s}} = 0$$

we see that

$$\langle u_{2\mathrm{s}} | u_{1\mathrm{s}} \rangle = 0$$

8.10 (a) (i) ^{1}S; ^{3}S

(ii) ^{1}P; ^{3}P

(iii) ^{1}D; ^{3}D

(iv) ^{1}S; ^{1}D; ^{1}G; ^{3}P; ^{3}F

(v) ^{2}S; ^{2}P; ^{2}D; ^{2}F; ^{4}S; ^{4}P; ^{4}D; ^{4}F

2 6 4 2 3 2

(vi) ^{2}P; ^{2}D; ^{4}S

(b) (i) $^1S_0$; $^3S_1$

(ii) $^1P_1$; $^3P_0$; $^3P_1$; $^3P_2$

(iii) $^1D_2$; $^3D_1$; $^3D_2$; $^3D_3$

(iv) $^1S_0$; $^1D_2$; $^1G_4$; $^3P_0$, $^3P_1$, $^3P_2$; $^3F_2$, $^3F_3$, $^3F_4$

(v) $^2S_{1/2}$; $^2P_{1/2}$, $^2P_{3/2}$; $^2D_{3/2}$, $^2D_{5/2}$; $^2F_{5/2}$, $^2F_{7/2}$; $^4S_{3/2}$; $^4P_{1/2}$, $^4P_{3/2}$, $^4P_{5/2}$; $^4D_{1/2}$, $^4D_{3/2}$, $^4D_{5/2}$, $^4D_{7/2}$; $^4F_{3/2}$, $^4F_{5/2}$, $^4F_{7/2}$, $^4F_{9/2}$

(vi) $^2P_{1/2}$, $^2P_{3/2}$; $^2D_{3/2}$, $^2D_{5/2}$; $^4S_{3/2}$.

8.11 (a) We have $j_1 = 1/2, 3/2$ for a p electron and $j_2 = 3/2, 5/2$ for a d electron. For the configuration np nd the possible values of J and terms $(j_1 j_2)_J$ are then given by

j_1	j_2	J	Terms $(j_1 j_2)_J$
1/2	3/2	1, 2	$(1/2\ 3/2)_1, (1/2\ 3/2)_2$
1/2	5/2	2, 3	$(1/2\ 5/2)_2, (1/2\ 5/2)_3$
3/2	3/2	0, 1, 2, 3	$(3/2\ 3/2)_0, (3/2\ 3/2)_1, (3/2\ 3/2)_2, (3/2\ 3/2)_3$
3/2	5/2	1, 2, 3, 4	$(3/2\ 5/2)_1, (3/2\ 5/2)_2, (3/2\ 5/2)_3, (3/2\ 5/2)_4$

(b) For the configuration $(nl\ 3/2)^2$ the only allowed values of J are $J = 0, 2$ and the terms are

$$(3/2\ 3/2)_0, (3/2\ 3/2)_2$$

Chapter 9

9.4 (a) $\Delta M_J = \pm 1, 0$ for all λ and
$|\Delta J| = \lambda, \lambda - 1, \ldots, 0$ with $J + J' > \lambda$
For odd λ the parity changes while for even λ the parity does not change.

(b) The rules for M_J and J are the same as for (a) but for odd λ the parity does not change, while for even λ the parity changes.

9.7 (a) The intervals between adjacent levels are 20 cm^{-1} and 40 cm^{-1}. Thus, if J is the total angular momentum quantum number of the highest level of the multiplet, the Landé interval rule gives $40 = AJ$ and $20 = A(J-1)$. Hence $J = 2$, and the two other levels have $J = 1$ and $J = 0$. Since $J = 0$ is possible, $|L - S| = 0$ from which $L = S$. From (9.104) we then see that $L = S = 1$. The three fine structure levels of the multiplet are therefore $^3P_0$, $^3P_1$, $^3P_2$.

(b) Since both the ground state term and the excited term are ^3P, we have $\Delta L = \Delta S = 0$. The allowed transitions are given by $\Delta J = 0, \pm 1$, with $J = 0 \rightarrow J' = 0$ forbidden. Thus there are six lines

J		J'	λ
$J = 2$	to	$J' = 2$	$\lambda = 1657.0$ Å
1		1	$= 1657.4$ Å
2		1	$= 1658.1$ Å
1		0	$= 1657.9$ Å
1		2	$= 1656.3$ Å
0		1	$= 1656.9$ Å

9.8 (a) From (9.160), the Landé factors are

$$^2P_{3/2},\ g = 4/3; \quad ^2P_{1/2},\ g = 2/3; \quad ^2S_{1/2},\ g = 2$$

The possible angular frequencies are $\omega = \omega_0 + \Delta\omega$, where ω_0 is the angular frequency in the absence of magnetic field and

$$\Delta\omega = \frac{\mu_B \mathscr{B}}{\hbar}(g'M'_J - gM_J)$$

where the prime denotes the upper level b.

For $^2P_{1/2} \rightarrow {}^2S_{1/2}$ we find separate lines in order of increasing frequency:

M'_J	M_J	$g'M'_J - gM_J$	Type of line	q
$-1/2$	$1/2$	$-4/3$	σ^+	-1
$1/2$	$1/2$	$-2/3$	π	0
$-1/2$	$-1/2$	$2/3$	π	0
$1/2$	$-1/2$	$4/3$	σ^-	$+1$

For $^2P_{3/2} \to {}^2S_{1/2}$ we find that

M'_J	M_J	$g'M'_J - gM_J$	Type of line	q
−1/2	1/2	−5/3	σ^+	−1
−3/2	−1/2	−1	σ^+	−1
1/2	1/2	−1/3	π	0
−1/2	−1/2	1/3	π	0
3/2	1/2	1	σ^-	+1
1/2	−1/2	5/3	σ^-	+1

(b) The transition rates for π and σ lines are from (9.26) (compare with (6.97), (6.101) and (6.102))

$$W_{ab}^{s,D}(\pi) = \tilde{C}(\omega_{ba}) \sin^2\Theta \, |\langle \gamma' J' M'_J | D_0 | \gamma J M_J \rangle|^2$$

$$W_{ab}^{s,D}(\sigma^\pm) = \tilde{C}(\omega_{ba}) \tfrac{1}{2}(1 + \cos^2\Theta) \, |\langle \gamma' J' M'_J | D_{\mp 1} | \gamma J M_J \rangle|^2$$

where $\tilde{C}(\omega_{ba}) = e^{-2}C(\omega_{ba})$ and $\Theta = \pi/2$ for transverse observation. Assuming that the upper levels are equally populated, the relative intensities of the lines are given by using (9.42), through the factors $\sin^2\Theta \, |\langle J1M_J0|J'M'_J\rangle|^2$ for the π lines and $\frac{1}{2}(1 + \cos^2\Theta) \, |\langle J1M_J \mp 1|J'M'_J\rangle|^2$ for the $\sigma^\pm$ lines. With $\Theta = \pi/2$, we find by using Table A4.1 that

(*i*) the intensities of all the lines in the $^2P_{1/2} \to {}^2S_{1/2}$ transition are equal and

(*ii*) the intensities of the lines in the $^2P_{3/2} \to {}^2S_{1/2}$ transition are in the ratios 1:3:4:4:3:1.

Note that (see Problem 6.4) the average intensity of the π lines is twice the average intensity of the σ^+ or of the σ^- lines.

Chapter 10

10.3 The energies of these five levels are: 2.1×10^{-3}; 6.3×10^{-3}; 12.6×10^{-3}; 21×10^{-3}; 31.5×10^{-3} eV and the corresponding wave numbers are: 16.9; 50.8; 101.5; 169.2; 253.8 cm^{-1}. The internuclear distance is 1.43 Å or 2.71 a.u.

10.4 (a) 0.164 eV and 0.493 eV (b) 2.5×10^{-14} s and 0.84×10^{-14} s. (c) 4.12×10^2 N m^{-1}.

10.5 The depth of the potential well in H_2 is $D_e = D_0 + \hbar\omega_0/2 = (4.48 + 0.26)$ eV and this is the same for the D_2 molecule. The force constant for the vibrational motion k is the same for H_2 and D_2, hence the zero-point energy for $D_2 = 0.26 \times (M_p/M_D)^{1/2} = 0.18$ eV (see (10.101)). Hence for D_2, $D_0 = (4.48 + 0.26 - 0.18)$ eV $= 4.56$ eV.

10.6 From (10.115), with $\tilde{k}_s \simeq k_s$ and $c_1 = c_2 = 0$, we have

$$\frac{\mathrm{d}V_{\mathrm{eff}}}{\mathrm{d}R} \simeq k_s (R - R_1) \qquad \textbf{(1)}$$

and from (10.114)

$$\frac{\mathrm{d}V_{\mathrm{eff}}}{\mathrm{d}R} = \frac{\mathrm{d}V_{\mathrm{M}}(R)}{\mathrm{d}R} - \frac{\hbar^2}{\mu}\frac{K(K+1)}{R^3} \qquad \textbf{(2)}$$

Equating (1) and (2) at the point $R = R_0$ for which $\mathrm{d}V_{\mathrm{M}}(R)/\mathrm{d}R = 0$, we find that

$$R_1 = R_0 + \frac{\hbar^2}{k_s \mu}\frac{K(K+1)}{R_0^3}$$

from which (10.116) follows on substituting for k_s using (10.109).

10.7 Using the reduced mass $\mu = 2.3042 \times 10^{-26}$ kg and $\tilde{\nu}_0$ we calculate the force constant k_s. Converting to atomic units this is $k_s = 0.0698$ a.u. From (10.81), at $R = R_0$

$$\frac{\partial E_s}{\partial R} = 0 \quad \text{or} \quad \left(\frac{1}{R_0^2} - cA\exp(-cR_0)\right) = 0 \qquad \textbf{(1)}$$

which gives $A = \exp(cR_0)/(cR_0^2)$. Now

$$k_s = \left.\frac{\mathrm{d}^2 E_s}{\mathrm{d}R^2}\right|_{R=R_0} = -\frac{2}{R_0^3} + Ac^2\exp(-cR_0) \qquad \textbf{(2)}$$

From (1) and (2) and using the computed value of k_s, we find $c = 1.837$ a.u. (or 3.47 Å^{-1}). Finally from (1) $A = 98.9$ a.u. (or 2692 eV) and $E_s(R_0) = -0.142$ a.u. (or -3.87 eV), giving $D_0 = (3.87 - 0.02)$ eV $= 3.85$ eV.

Chapter 11

11.1 The ratio is given by Boltzmann's distribution $R(K) = (g_K/g_0)\exp[(E_0 - E_K)/(k_B T)]$ where g_K is the degeneracy of the Kth level. Since each rotational level is $(2K + 1)$-fold degenerate the result follows. By differentiating we find that $R(K)$ has a maximum at $K = [k_B T/(2B)]^{1/2} - 1/2$. The intensity of an absorption line is proportional to the number of molecules in the *lower* level and hence is proportional to $R(K)$.

11.2 Unlike rotational levels, vibrational levels are not degenerate so that $R(v) = \exp[(E_0 - E_v)/(k_B T)]$. For $v = 1$ we have $R(1) = \exp[-\hbar\omega_0/(k_B T)]$.

11.3 $\tilde{B} = 1.93$ cm^{-1}; $R_0 = 1.12$ Å; $k_s = 1.9 \times 10^3$ N m^{-1}.

11.4 (a) The Deslandres table is constructed from (11.25)

(b) The wave number of the $v' = 0$ to $v = 0$ transition is $\tilde{\nu} = \tilde{\nu}_{s's} - 801.54\ \text{cm}^{-1}$ $= 102\ 877.3\ \text{cm}^{-1}$

(c) To draw the Fortrat parabola and calculate the positions of the lines of each branch we use (11.26) with

$$(\tilde{\nu}^{\text{P}} - \tilde{\nu}) = 1.145K(K-1) - 2.010K(K+1)$$
$$= 1.145m(m-1) - 2.010m(m+1)$$

with $m = 0, 1, 2, 3, \ldots$ and $(\tilde{\nu}^{\text{R}} - \tilde{\nu})$ is given by the same formula with $m = -1, -2, -3, \ldots$.
The $R(1)$ line forms the band head at $(\tilde{\nu} + 2.29\ \text{cm}^{-1}) = 102\ 879.6\ \text{cm}^{-1}$ and the band is shaded to the red.

Chapter 12

12.3 For large r, solutions of the radial equation (12.41) having the form

$$u_l(k, r) = F_l(k, r) \exp(\pm ikr)$$

are assumed. Substituting this into (12.41) yields

$$\frac{F_l''(k, r)}{F_l(k, r)} \pm 2ik\frac{F_l'(k, r)}{F_l(k, r)} = U(r) + \frac{l(l+1)}{r^2}$$

Since $F_l(k, r)$ is a slowly varying function, we may drop the term F_l''/F_l and write approximately for large r

$$\pm\frac{F_l'(k, r)}{F_l(k, r)} = \frac{1}{2ik}\left[U(r) + \frac{l(l+1)}{r^2}\right]$$

The solution of this differential equation is

$$F_l(k, r) = \exp\left\{\pm\frac{1}{2ik}\int^r \left[U(r') + \frac{l(l+1)}{r'^2}\right] \mathrm{d}r'\right\}$$

Provided $r|V(r)| \to 0$ as $r \to \infty$, the integral on the right of this equation, and hence the function F_l, are independent of r in the limit $r \to \infty$.

12.6 Writing

$$l'(l'+1) = l(l+1) + A \qquad \textbf{(1)}$$

the radial equation (12.39) becomes

$$\left[\frac{\mathrm{d}^2}{\mathrm{d}r^2} + \frac{2}{r}\frac{\mathrm{d}}{\mathrm{d}r} - \frac{l'(l'+1)}{r^2} + k^2\right]R_l(k, r) = 0 \qquad \textbf{(2)}$$

The solution of (2) that is regular at the origin is $R_l(k, r) = Cj_{l'}(k, r)$, where C is a constant.

(a) Solving (1) for l', we obtain

$$l'_{\pm} = -\frac{1}{2} \pm \left[\left(l + \frac{1}{2}\right)^2 + A\right]^{1/2}$$

and we keep only the solution l'_+ since one must have $l' = l$ when $A = 0$. Using the asymptotic expression for $j_{l'_+}(kr)$, we have

$$R_l(k, r) \underset{r\to\infty}{\rightarrow} \frac{C}{kr} \sin\left(kr - \frac{l'_+\pi}{2}\right)$$

The phase shift δ_l, determined by the equation $-l'_+\pi/2 = -l\pi/2 + \delta_l$, is therefore given by

$$\delta_l = -\frac{\pi}{2}\left\{\left[\left(l + \frac{1}{2}\right)^2 + A\right]^{1/2} - \left(l + \frac{1}{2}\right)\right\}$$

and is independent of k. We note that when $(l + 1/2)^2 \gg A$, we have

$$\delta_l \simeq -\pi A/[2(2l + 1)] \tag{3}$$

(b) Since δ_l does not depend on k, we see from (12.60) that the scattering amplitude and the differential cross-section can be written, respectively, as

$$f(k, \theta) = \frac{1}{k} F(\theta), \qquad \frac{d\sigma}{d\Omega} = \frac{1}{k^2}|F(\theta)|^2$$

where the function

$$F(\theta) = \sum_{l=0}^{\infty} (2l + 1) \exp(i\delta_l) \sin \delta_l P_l(\cos \theta)$$

is independent of k. It follows that the angular distribution is independent of the energy of the scattered particle. In the forward direction,

$$F(\theta = 0) = \sum_{l=0}^{\infty} (2l + 1) \exp(i\delta_l) \sin \delta_l$$

We can separate this expression into two contributions as

$$F(\theta = 0) = \sum_{l=0}^{L-1} (2l + 1) \exp(i\delta_l) \sin \delta_l + \sum_{l=L}^{\infty} (2l + 1) \exp(i\delta_l) \sin \delta_l$$

and we choose L to be some large value of l beyond which δ_l is small and given by the expression (3). We then have

$$F(\theta = 0) = \sum_{l=0}^{L-1} (2l + 1) \exp(i\delta_l) \sin \delta_l - \frac{\pi A}{2} \sum_{l=L}^{\infty} 1$$

The series in the second term diverges so that the scattering amplitude, and hence the differential cross-section, diverge in the forward direction. The total cross-section is given by (12.64). As before, we can separate the summation on l into two contributions as

$$\sigma_{\text{tot}} = \frac{4\pi}{k^2} \sum_{l=0}^{L-1} (2l + 1) \sin^2\delta_l + \sum_{l=L}^{\infty} (2l + 1) \sin^2\delta_l$$

where L is chosen in the way described above. Hence

$$\sigma_{\text{tot}} = \frac{4\pi}{k^2} \sum_{l=0}^{L-1} (2l + 1) \sin^2\delta_l + \frac{\pi^2 A^2}{4} \sum_{l=L}^{\infty} \frac{1}{2l + 1}$$

and the total cross-section is seen to diverge since the series in the second term diverges.

12.7 Taking $A < 0$ we find that

$$l'_{\pm} = -\frac{1}{2} \pm \left[\left(l + \frac{1}{2} \right)^2 - |A| \right]^{1/2}$$

and we keep only the solution l'_+ since we must have $l' = l$ when $A = 0$. In addition, we must have $(l + 1/2)^2 > |A|$ for all l. This condition is satisfied for $|A| < 1/4$.

12.8 Starting from (12.63) we have

$$\int_{-1}^{1} P_L(\cos\theta) \frac{d\sigma}{d\Omega} d(\cos\theta) = \frac{1}{k^2} \sum_{l=0}^{\infty} \sum_{l'=0}^{\infty} (2l + 1)(2l' + 1) \exp[i(\delta_l - \delta_{l'})]$$

$$\times \sin\delta_l \sin\delta_{l'} \int_{-1}^{+1} P_L(\cos\theta) P_l(\cos\theta) P_{l'}(\cos\theta)\, d(\cos\theta)$$

Evaluating this expression with $L = 0$, we find

$$A = \frac{\sigma_{\text{tot}}}{4\pi} = \frac{1}{k^2} \sum_{l=0}^{\infty} (2l + 1) \sin^2\delta_l$$

Similarly, with $L = 1$, we have

$$B = \frac{6}{k^2} \sum_{l=0}^{\infty} (l + 1) \sin\delta_l \sin\delta_{l+1} \cos(\delta_{l+1} - \delta_l)$$

and with $L = 2$, one obtains

$$C = \frac{5}{k^2} \sum_{l=0}^{\infty} \left[\frac{l(l+1)(2l+1)}{(2l-1)(2l+3)} \sin^2 \delta_l \right.$$

$$\left. + \frac{3(l+1)(l+2)}{2l+3} \sin \delta_l \sin \delta_{l+2} \cos(\delta_{l+2} - \delta_l) \right]$$

12.11 Proceeding as in the case of the square well (12.108), but with U_0 replaced by $-U_0$, we find that

$$\delta_0 = -ka + \tan^{-1}\left(\frac{k}{K} \tanh (Ka) \right), \qquad K = (U_0 - k^2)^{1/2}$$

Hence, for $k \to 0$, we have

$$\delta_0 = ka \left[\frac{\tanh (\lambda_0 a)}{\lambda_0 a} - 1 \right], \qquad \lambda_0 = \sqrt{U_0}$$

and therefore, at zero energy,

$$\sigma_{\text{tot}} = \sigma_0 = 4\pi a^2 \left[\frac{\tanh (\lambda_0 a)}{\lambda_0 a} - 1 \right]^2$$

As $U_0 \to \infty$, this zero-energy cross-section becomes $\sigma_{\text{tot}} = 4\pi a^2$, in agreement with the result (12.122) obtained for the 'hard sphere' potential.

12.12 (a) $$(\tan \delta_0)^{\text{B1}} = -\frac{1}{k} \int_0^{\infty} \sin^2 (kr) U(r) \, \mathrm{d}r$$

$$= -\frac{U_0}{k} \int_0^{\infty} \sin^2 (kr) \frac{\mathrm{e}^{-\alpha r}}{r} \, \mathrm{d}r = -\frac{U_0}{4k} \log \left(1 + \frac{4k^2}{\alpha^2} \right)$$

(b) $$(\tan \delta_0)^{\text{B1}} = -\frac{U_0}{k} \int_0^{\infty} \sin^2 (kr) \frac{1}{(r^2 + d^2)^2} \, \mathrm{d}r$$

$$= -\frac{\pi U_0}{8kd^3} [1 - (1 + 2kd) \exp(-2kd)]$$

12.19 (a) $$f^{\text{B1}} = -U_0 \frac{2\alpha}{(\alpha^2 + \Delta^2)^2}$$

$$\sigma_{\text{tot}}^{\text{B1}} = \frac{16\pi U_0^2}{3} \frac{16k^4 + 12\alpha^2 k^2 + 3\alpha^4}{\alpha^4 (\alpha^2 + 4k^2)^3}$$

$$\sigma_{\text{tot}}^{\text{B1}} \underset{E \to \infty}{\sim} A E^{-1}, \qquad A = \frac{2\pi \hbar^2 U_0^2}{3m\alpha^4}$$

(b) $f^{B1} = -U_0 \frac{\pi^{1/2}}{4\alpha^3} \exp[-\Delta^2/(4\alpha^2)]$

$$\sigma_{tot}^{B1} = \frac{\pi^2 U_0^2}{8\alpha^4 k^2}[1 - \exp(-2k^2/\alpha^2)]$$

$$\sigma_{tot}^{B1} \underset{E\to\infty}{\sim} AE^{-1}, \qquad A = \frac{\pi^2 \hbar^2 U_0^2}{16 m \alpha^4}$$

(c) $f^{B1} = -\frac{U_0}{\Delta^3}(\sin \Delta a - \Delta a \cos \Delta a)$

$$\sigma_{tot}^{B1} = 8\pi U_0^2 a^6 F(x)$$

where

$$F(x) = \frac{1}{4x^2}\left(1 - \frac{1}{x^2} + \frac{\sin 2x}{x^3} - \frac{\sin^2 x}{x^4}\right)$$

and $x = 2ka$

$$\sigma_{tot}^{B1} \underset{E\to\infty}{\sim} AE^{-1}, \qquad A = \frac{\pi \hbar^2 a^4 U_0^2}{4m}$$

(d) $f^{B1} = -\frac{\pi U_0}{4d} \exp(-\Delta d)$

$$\sigma_{tot}^{B1} = \frac{\pi^3 U_0^2}{32 d^4 k^2}[1 - (4kd + 1)\exp(-4kd)]$$

$$\sigma_{tot}^{B1} \underset{E\to\infty}{\sim} AE^{-1}, \qquad A = \frac{\pi^3 \hbar^2 U_0^2}{64 m d^4}$$

Chapter 13

13.3 In atomic units $\psi_1(r') = \pi^{-1/2}\exp(-r')$, so that

$$V_{11}(\mathbf{r}) = -\frac{1}{r} + \frac{1}{\pi}\int_0^\infty \mathrm{d}r' r'^2 \int_{-1}^{-1} \mathrm{d}(\cos\theta') \int_0^{2\pi} \mathrm{d}\phi' \exp(-2r') \frac{1}{|\mathbf{r} - \mathbf{r}'|}$$

Expanding $|\mathbf{r} - \mathbf{r}'|^{-1}$ in spherical harmonics using equation (A4.25), we find that

$$V_{11}(r) = -\frac{1}{r} + 4\int_0^\infty r'^2 \exp(-2r')\left(\frac{1}{r_>}\right) \mathrm{d}r'$$

where $r_>$ is the greater of r and r'. Thus

$$V_{11}(r) = -\frac{1}{r} + \frac{4}{r}\int_0^r r'^2 \exp(-2r')\,\mathrm{d}r' + 4\int_r^\infty r' \exp(-2r')\,\mathrm{d}r'$$

The integration is elementary and we find

$$V_{11}(r) = -\left(1 + \frac{1}{r}\right)\exp(-2r)$$

13.4 Using the approximate wave function for the ground state of helium

$$\phi(r_1, r_2) = \left(\frac{Z_e^3}{\pi}\right)\exp[-Z_e(r_1 + r_2)], \text{ we have}$$

$$V_{11}(r_0) = -\frac{2}{r_0} + \left(\frac{Z_e^3}{\pi}\right)^2 \int \mathrm{d}\mathbf{r}_1 \int \mathrm{d}\mathbf{r}_2 \exp[-2Z_e(r_1 + r_2)]\left(\frac{1}{|\mathbf{r}_0 - \mathbf{r}_1|} + \frac{1}{|\mathbf{r}_0 - \mathbf{r}_2|}\right)$$

Proceeding as in Problem 13.3, we find that

$$V_{11}(r_0) = -2\left(Z_e + \frac{1}{r_0}\right)\exp(-2Z_e r_0)$$

13.9 From (13.144), we have

$$f_{11}^{\mathrm{B1}} = -\frac{1}{2\pi}\int \exp(\mathrm{i}\boldsymbol{\Delta}\cdot\mathbf{r}_0)V_{11}(r_0)\,\mathrm{d}\mathbf{r}_0$$

where $V_{11}(r_0)$ is the static potential obtained in Problem 13.4. Performing the integral, we find that

$$f_{11}^{\mathrm{B1}} = 4\left[\frac{8Z_e^2 + \Delta^2}{(4Z_e^2 + \Delta^2)^2}\right]$$

The corresponding elastic first Born differential cross section is given by

$$\frac{\mathrm{d}\sigma_{11}^{\mathrm{B1}}}{\mathrm{d}\Omega} = \left|f_{11}^{\mathrm{B1}}\right|^2$$

It is plotted in Fig. 13.16(b) for the case of a Hartree-Fock helium ground state wave function and an incident electron energy of 500 eV (dashed line).

Chapter 14

14.1 Let ${}^1f(\theta)$ and ${}^3f(\theta)$ be the scattering amplitudes for the singlet and triplet electronic spin states respectively and take the spin function of the proton to be $\alpha(\mathrm{p})$ and that of the deuteron to be $\chi(\mathrm{d})$, where χ can be either α or β. The asymptotic wave function for large R in the singlet state is then

$${}^1\psi_{\mathrm{A}} = [\exp(\mathrm{i}\mathbf{k}\cdot\mathbf{R}) + {}^1f(\theta)R^{-1}\exp(\mathrm{i}kR)]\alpha(\mathrm{p})\chi(\mathrm{d})2^{-1/2}[\alpha(1)\beta(2) - \beta(1)\alpha(2)]$$

where $\alpha(i)$, $\beta(i)$, $i = 1, 2$, are the spin functions of the two electrons.

Correspondingly the asymptotic wave function in the triplet state (with $M_S = 0$) is

$$^3\psi_A = [\exp(i\mathbf{k}\cdot\mathbf{R}) + {}^3f(\theta)R^{-1}\exp(ikR)]\alpha(p)\chi(d)2^{-1/2}[\alpha(1)\beta(2) + \beta(1)\alpha(2)]$$

A wave function in which the incident plane wave state describes hydrogen atoms with $F = 1$ and $M_F = 1$ has the asymptotic form

$$2^{-1/2}[^1\psi_A + {}^3\psi_A] = \exp(i\mathbf{k}\cdot\mathbf{R})[\alpha(p)\alpha(1)\chi(d)\beta(2)] + R^{-1}\exp(ikR)H$$

where

$$\begin{aligned} H &= 2^{-1}[^1f(\theta) + {}^3f(\theta)]\alpha(p)\alpha(1)\chi(d)\beta(2) + 2^{-1}[-{}^1f(\theta) + {}^3f(\theta)]\alpha(p)\beta(1)\chi(d)\alpha(2) \\ &= 2^{-1}[^1f(\theta) + {}^3f(\theta)]\alpha(p)\alpha(1)\chi(d)\beta(2) \\ &\quad + 2^{-3/2}[-{}^1f(\theta) + {}^3f(\theta)]2^{-1/2}[\alpha(p)\beta(1) - \beta(p)\alpha(1)]\chi(d)\alpha(2) \\ &\quad + 2^{-3/2}[-{}^1f(\theta) + {}^3f(\theta)]2^{-1/2}[\alpha(p)\beta(1) + \beta(p)\alpha(1)]\chi(d)\alpha(2) \end{aligned}$$

The first and third terms in this expression correspond to states with $F = 1$, while the second corresponds to $F = 0$. Thus the scattering amplitude for the transition $F = 1 \rightarrow F = 0$ is

$$g(\theta) = 2^{-3/2}[-{}^1f(\theta) + {}^3f(\theta)]$$

and

$$\sigma_{tot}(F = 1 \rightarrow F = 0) = \int |g(\theta)|^2 \, d\Omega$$

Expanding $^1f(\theta)$ and $^3f(\theta)$ in terms of partial waves, the result follows.

14.2 Consider $I = [i\partial/\partial t - H_e(t)]X_k^A(\mathbf{r}, t)$. We have to show that $I \rightarrow 0$ when $|t| \rightarrow \infty$. In the limit $|t| \rightarrow \infty$, $R \rightarrow \infty$ and there are two cases: either the electron is close to A, $r_A \ll r_B$, $r_B \rightarrow \infty$ so that $V_B \rightarrow 0$ and $V_{AB} \rightarrow 0$, or $r_A \gg r_B$. In the latter case $\psi_k^A(\mathbf{r}_A) \rightarrow 0$ and $I \rightarrow 0$. In the former case, we have for $|t| \rightarrow \infty$

$$I \rightarrow \left[i\frac{\partial}{\partial t} + \frac{1}{2}\nabla_{\mathbf{r}}^2 - V_A(r_A)\right]X_k^A(\mathbf{r}, t)$$

Now from (14.55a) $i\partial/\partial t X_k^A(\mathbf{r}, t) = i\{\dot{\mathbf{r}}_A\cdot\nabla_{\mathbf{r}_A}\psi_k^A(\mathbf{r}_A) - i(E_k^A + v^2/8)\}\exp[-iE_k^A t + i\mathbf{v}\cdot\mathbf{r}/2 - iv^2t/8]$ with $\dot{\mathbf{r}}_A = -\frac{1}{2}\dot{\mathbf{R}} = -\frac{1}{2}\mathbf{v}$, and also

$$\frac{1}{2}\nabla_{\mathbf{r}}^2 X_k^A(\mathbf{r}, t) = \left\{\frac{1}{2}\nabla_{\mathbf{r}_A}^2\psi_k^A(\mathbf{r}_A) + \frac{1}{2}\mathbf{v}\cdot\nabla_{\mathbf{r}_A}\psi_k^A(\mathbf{r}_A) - (v^2/8)\psi_k^A(\mathbf{r}_A)\right\}\exp[-iE_k^A t + i\mathbf{v}\cdot\mathbf{r}/2 - iv^2t/8]$$

Hence as $|t| \to \infty$ and for $r_A \ll r_B$

$$I \to \left\{\left[\frac{1}{2}\nabla^2_{\mathbf{r}_A} - V_A(r_A) + E_k^A\right]\psi_k^A(\mathbf{r}_A)\right\}\exp[-iE_k^At + i\mathbf{v}\cdot\mathbf{r}/2 - iv^2t/8]$$

From (14.56a) the right-hand side vanishes and $I \to 0$, which completes the verification. A similar argument applies to $X_j^B(\mathbf{r}, t)$.

14.3 Let

$$I = \langle\Psi_a(\mathbf{r}, t)|\Psi_a(\mathbf{r}, t)\rangle$$

then

$$i\frac{dI}{dt} = i\left\langle\frac{\partial}{\partial t}\Psi_a(\mathbf{r}, t)\middle|\Psi_a(\mathbf{r}, t)\right\rangle + i\left\langle\Psi_a(\mathbf{r}, t)\middle|\frac{\partial}{\partial t}\Psi_a(\mathbf{r}, t)\right\rangle$$

From (14.53) and (14.57), we have

$$i\left\langle\Psi_a(\mathbf{r}, t)\middle|\frac{\partial}{\partial t}\Psi_a(\mathbf{r}, t)\right\rangle$$

$$= \sum_k c_k^{A*}\langle F_k^A(\mathbf{r}, t)|H_e|\Psi_a(\mathbf{r}, t)\rangle + \sum_j c_j^{B*}\langle F_j^B(\mathbf{r}, t)|H_e|\Psi_a(\mathbf{r}, t)\rangle$$

$$= \langle\Psi_a(\mathbf{r}, t)|H_e|\Psi_a(\mathbf{r}, t)\rangle$$

Similarly,

$$i\left\langle\frac{\partial}{\partial t}\Psi_a(\mathbf{r}, t)\middle|\Psi_a(\mathbf{r}, t)\right\rangle = \left\langle -i\frac{\partial}{\partial t}\Psi_a(\mathbf{r}, t)\middle|\Psi_a(\mathbf{r}, t)\right\rangle$$

$$= -\langle\Psi_a(\mathbf{r}, t)|H_e|\Psi_a(\mathbf{r}, t)\rangle$$

Hence $dI/dt = 0$ and if $\langle\Psi_a(\mathbf{r}, t)|\Psi_a(\mathbf{r}, t)\rangle = 1$ at $t = -\infty$ it remains normalised to unity for all t. Now

$$\langle\Psi_a(\mathbf{r}, t)|\Psi_a(\mathbf{r}, t)\rangle = \sum_{k,k'} c_k^{A*}c_{k'}^A(S_{AA})_{kk'} + \sum_{kj} c_k^{A*}c_j^B(S_{AB})_{kj} + \sum_{jk} c_j^{B*}c_k^A(S_{BA})_{jk}$$

$$+ \sum_{jj'} c_j^{B*}c_{j'}^B(S_{BB})_{jj'}$$

$$= 1$$

In the limit $t \to \infty$, the overlaps $\mathbf{S}_{AB}$ and $\mathbf{S}_{BA} \to 0$ while $(S_{AA})_{kk'} \to \delta_{kk'}$, $(S_{AB})_{jj'} \to \delta_{jj'}$ and the result follows.

14.4 From (14.70),

$$(\mathbf{M}^*_{\mathrm{BA}})_{jk} - (\mathbf{M}_{\mathrm{AB}})_{kj} = \langle F_j^{\mathrm{B}} | H_{\mathrm{e}} | F_k^{\mathrm{A}} \rangle^* - \langle F_k^{\mathrm{A}} | H_{\mathrm{e}} | F_j^{\mathrm{B}} \rangle + \mathrm{i} \left\langle F_j^{\mathrm{B}} \left| \frac{\partial}{\partial t} \right| F_k^{\mathrm{A}} \right\rangle^*$$

$$+ \mathrm{i} \left\langle F_k^{\mathrm{A}} \left| \frac{\partial}{\partial t} \right| F_j^{\mathrm{B}} \right\rangle$$

$$= \mathrm{i} \left[\left\langle F_j^{\mathrm{B}} \left| \frac{\partial}{\partial t} \right| F_k^{\mathrm{A}} \right\rangle^* + \left\langle F_k^{\mathrm{A}} \left| \frac{\partial}{\partial t} \right| F_j^{\mathrm{B}} \right\rangle \right]$$

since H_{e} is Hermitian. Hence,

$$(\mathbf{M}^*_{\mathrm{BA}})_{jk} - (\mathbf{M}_{\mathrm{AB}})_{kj} = \mathrm{i} \frac{\partial}{\partial t} \langle F_k^{\mathrm{A}} | F_j^{\mathrm{B}} \rangle = \mathrm{i} \frac{\partial}{\partial t} (\mathbf{S}_{\mathrm{AB}})_{kj}$$

14.5 From (14.67), in the two-state approximation

$$\begin{aligned} \mathrm{i}(\dot{c}^{\mathrm{A}} + S_{\mathrm{AB}}\dot{c}^{\mathrm{B}}) &= M_{\mathrm{AA}}c^{\mathrm{A}} + M_{\mathrm{AB}}c^{\mathrm{B}} \\ \mathrm{i}(\dot{c}^{\mathrm{B}} + S_{\mathrm{BA}}\dot{c}^{\mathrm{A}}) &= M_{\mathrm{BB}}c^{\mathrm{B}} + M_{\mathrm{BA}}c^{\mathrm{A}} \end{aligned} \qquad \textbf{(1)}$$

since $S_{\mathrm{AA}} = S_{\mathrm{BB}} = 1$. We also have that

$$S_{\mathrm{AB}} = \int [\psi_{1s}(r_{\mathrm{A}})\psi_{1s}(r_{\mathrm{B}}) \exp(-\mathrm{i}\mathbf{v}\cdot\mathbf{r})] \, \mathrm{d}\mathbf{r}$$

$$S_{\mathrm{BA}} = \int [\psi_{1s}(r_{\mathrm{B}})\psi_{1s}(r_{\mathrm{A}}) \exp(\mathrm{i}\mathbf{v}\cdot\mathbf{r})] \, \mathrm{d}\mathbf{r}$$

since ψ_{1s} is real. By changing $\mathbf{r}$ into $-\mathbf{r}$ and taking the complex conjugate, we see that S_{AB} and S_{BA} are real and $S_{\mathrm{AB}} = S_{\mathrm{BA}} \equiv S$. Similarly we have $M_{\mathrm{AA}} = M_{\mathrm{BB}}$ and $M_{\mathrm{AB}} = M_{\mathrm{BA}}$ and M_{AA}, M_{AB} are real. From (1), we can write

$$\begin{aligned} \mathrm{i}\dot{c}^{\mathrm{A}} &= L_{\mathrm{AA}}c^{\mathrm{A}} + L_{\mathrm{AB}}c^{\mathrm{B}} \\ \mathrm{i}\dot{c}^{\mathrm{B}} &= L_{\mathrm{BB}}c^{\mathrm{B}} + L_{\mathrm{BA}}c^{\mathrm{A}} \end{aligned}$$

where

$$\begin{aligned} L_{\mathrm{AA}} &= L_{\mathrm{BB}} = (1 - S^2)^{-1}(M_{\mathrm{AA}} - SM_{\mathrm{AB}}) \\ L_{\mathrm{AB}} &= L_{\mathrm{BA}} = (1 - S^2)^{-1}(M_{\mathrm{AB}} - SM_{\mathrm{AA}}) \end{aligned}$$

Introducing the phase transformation

$$C^{\mathrm{A}} = \exp(\mathrm{i}\delta)c^{\mathrm{A}}; \qquad C^{\mathrm{B}} = \exp(\mathrm{i}\delta)c^{\mathrm{B}}$$

with

$$\delta = \int_{-\infty}^{t} L_{AA}\, dt'$$

we have

$$i\dot{C}^{A} = L_{AB}C^{B}$$

$$i\dot{C}^{B} = L_{AB}C^{A}$$

Setting $A_{\pm} = C^{A} \pm C^{B}$, the equations uncouple and

$$i\dot{A}_{\pm} = \pm L_{AB}A_{\pm} \tag{2}$$

The initial conditions at $t = -\infty$ are $c^{A} = 0$, $c^{B} = 1$. Hence $C^{A} = 0$, $C^{B} = 1$ and $A_{\pm}(t = -\infty) = \pm 1$. The solution of (2) with these initial conditions is

$$A_{\pm}(t) = \pm \exp\left[\pm i \int_{-\infty}^{t} L_{AB}\, dt'\right]$$

Taking the limit $t \to \infty$, we have

$$C^{A}(\infty) = \frac{1}{2}[A_{+}(\infty) + A_{-}(\infty)]$$

$$= i \sin\left[\int_{-\infty}^{\infty} L_{AB}\, dt\right]$$

and the probability for charge exchange, $|c^{A}(\infty)|^{2}$, is

$$|c^{A}(\infty)|^{2} = \sin^{2}\left[\int_{-\infty}^{\infty} L_{AB}\, dt\right]$$

which oscillates between 0 and 1 as v or b is varied.

14.9 Consider

$$I = \left[-\frac{1}{2}\nabla_{\mathbf{r}}^{2} - \frac{Z_{A}}{R} + V_{B}(r_{B}) - i\frac{\partial}{\partial t}\right]\tilde{X}_{i}^{B(+)}(\mathbf{r}, t)$$

$$= \left[-\frac{1}{2}\nabla_{\mathbf{r}}^{2} - \frac{Z_{A}}{R} + V_{B}(r_{B}) - i\frac{\partial}{\partial t}\right]X_{i}^{B}(\mathbf{r}, t)\exp[-i\sigma(t)] \tag{1}$$

where

$$\sigma(t) = \frac{Z_{A}}{v}\log(Rv - v^{2}t)$$

We have

$$\frac{\mathrm{d}}{\mathrm{d}t}\exp[-\mathrm{i}\sigma(t)] = -\mathrm{i}\frac{\mathrm{d}\sigma}{\mathrm{d}t}\exp[-\mathrm{i}\sigma(t)]$$

and

$$\frac{\mathrm{d}\sigma(t)}{\mathrm{d}t} = \frac{Z_\mathrm{A}}{v}\frac{1}{Rv - v^2t}(\dot{R}v - v^2) \qquad \textbf{(2)}$$

Since $R = (b^2 + v^2t^2)^{1/2}$, we have

$$\dot{R} = v^2t/R \qquad \textbf{(3)}$$

and from (2) and (3)

$$\frac{\mathrm{d}\sigma}{\mathrm{d}t} = -\frac{Z_\mathrm{A}}{R} \qquad \textbf{(4)}$$

Using (1) and (4), we have

$$I = \exp[-\mathrm{i}\sigma(t)]\left[-\frac{1}{2}\nabla_\mathbf{r}^2 + V_\mathrm{B}(r_\mathrm{B}) - \mathrm{i}\frac{\partial}{\partial t}\right]X_i^\mathrm{B}(\mathbf{r}, t)$$

Repeating the argument used in Problem 14.2, we see that

$$\left[-\frac{1}{2}\nabla_\mathbf{r}^2 + V_\mathrm{B}(r_\mathrm{B}) - \mathrm{i}\frac{\partial}{\partial t}\right]X_i^\mathrm{B}(\mathbf{r}, t) = 0$$

so that $\tilde{X}_i^{\mathrm{B}(+)}$ satisfies (14.125).

In the same way, it can be proved that $\tilde{X}_k^{\mathrm{A}(-)}$ satisfies (14.129).

14 Fundamental constants, atomic units and conversion factors

The physical constants listed in Table A14.1 are adapted from 'CODATA recommended values of the fundamental physical constants: 1998', by P.J. Mohr and B.N. Taylor, in *Reviews of Modern Physics* **72**, 351 (2000).

We recall that the quantities ε_0 and μ_0 are related by the formula

$$\varepsilon_0 \mu_0 c^2 = \kappa^2 \qquad \textbf{(A14.1)}$$

where c is the velocity of light in vacuum and κ is a coefficient depending on the system of units. In the rationalised MKSA (SI) units used in this book one has

$$\kappa = 1, \qquad \mu_0 = 4\pi \times 10^{-7}\ \text{H m}^{-1},$$
$$\varepsilon_0 = \frac{1}{\mu_0 c^2} = 8.854\,19 \times 10^{-12}\ \text{F m}^{-1} \qquad \textbf{(A14.2)}$$

The Gaussian (mixed) system of units uses electric units of the electrostatic CGS system, and magnetic units of the electromagnetic CGS system. In the Gaussian system of units one therefore has

$$\kappa = c, \qquad \varepsilon_0 = \frac{1}{4\pi}, \qquad \mu_0 = 4\pi \qquad \textbf{(A14.3)}$$

The quantum mechanical equations of atomic and molecular physics are considerably simplified if Hartree's atomic units (a.u.) are used. These units are defined in Table A14.2.

We note that since $m = e = \hbar = a_0 = 1$ in a.u., while $\alpha = 1/137.036$ is dimensionless, one has in particular (with $\kappa = 1$)

$$c = \alpha^{-1}\ \text{a.u.} = 137.036\ \text{a.u.}$$
$$\varepsilon_0 = 1/(4\pi)\ \text{in a.u.} \qquad \textbf{(A14.4)}$$
$$\mu_0 = 4\pi/c^2 = 4\pi\alpha^2\ \text{in a.u.}$$

We also give in Table A14.3 a few important conversion factors.

Table A14.1 Fundamental constants.

Quantity	*Symbol*	*Value*
Planck's constant	h	$6.626\,07 \times 10^{-34}$ J s
	$\hbar = \dfrac{h}{2\pi}$	$1.054\,57 \times 10^{-34}$ J s
Velocity of light in vacuum	c	$2.997\,92 \times 10^{8}$ m s^{-1}
Elementary charge (absolute value of electron charge)	e	$1.602\,18 \times 10^{-19}$ C
Permeability of free space	μ_0	$4\pi \times 10^{-7}$ H m^{-1} = $1.256\,64 \times 10^{-6}$ H m^{-1}
Permittivity of free space	$\varepsilon_0 = \dfrac{1}{\mu_0 c^2}$	$8.854\,19 \times 10^{-12}$ F m^{-1}
Gravitational constant	G	6.673×10^{-11} N m^2 kg^{-2}
Fine structure constant	$\alpha = \dfrac{e^2}{4\pi\varepsilon_0\hbar c}$	$\dfrac{1}{137.036} = 7.297\,35 \times 10^{-3}$
Avogadro's number	N_A	$6.022\,14 \times 10^{23}$ mol^{-1}
Faraday's constant	$F = N_A e$	$9.648\,53 \times 10^{4}$ C mol^{-1}
Boltzmann's constant	k_B	$1.380\,65 \times 10^{-23}$ J K^{-1}
Gas constant	$R = N_A k_B$	$8.314\,47$ J mol^{-1} K^{-1}
Atomic mass unit	a.m.u. = $\dfrac{1}{12} M_{^{12}C}$	$1.660\,54 \times 10^{-27}$ kg
Electron mass	m or m_e	$9.109\,38 \times 10^{-31}$ kg = $5.485\,80 \times 10^{-4}$ a.m.u.
Proton mass	M_p	$1.672\,62 \times 10^{-27}$ kg = $1.007\,28$ a.m.u.
Neutron mass	M_n	$1.674\,93 \times 10^{-27}$ kg = $1.008\,66$ a.m.u.
Ratio of proton to electron mass	M_p/m	1836.15
Electron charge to mass ratio	$-e/m$	$-1.758\,82 \times 10^{11}$ C kg^{-1}
Compton wavelength of electron	$\lambda_c = \dfrac{h}{mc}$	$2.426\,31 \times 10^{-12}$ m
Classical radius of electron	$r_0 = \dfrac{e^2}{4\pi\varepsilon_0 mc^2}$	$2.817\,94 \times 10^{-15}$ m
Bohr radius for atomic hydrogen (with infinite nuclear mass)	$a_0 = \dfrac{4\pi\varepsilon_0\hbar^2}{me^2}$	$5.291\,77 \times 10^{-11}$ m
Non-relativistic ionisation potential of atomic hydrogen for infinite nuclear mass	$I_P^H(\infty) = \dfrac{e^2}{8\pi\varepsilon_0 a_0} = \dfrac{1}{2}\alpha^2 mc^2$	$2.179\,87 \times 10^{-18}$ J = 13.6057 eV
Rydberg's constant for infinite nuclear mass	$\tilde{R}(\infty) = \dfrac{me^4}{8\varepsilon_0^2 h^3 c} = \dfrac{\alpha}{4\pi a_0}$	$1.097\,37 \times 10^{7}$ m^{-1}
Rydberg's constant for atomic hydrogen	$\tilde{R}(M_p)$	$1.096\,78 \times 10^{7}$ m^{-1}
Bohr magneton	$\mu_B = \dfrac{e\hbar}{2m}$	$9.274\,01 \times 10^{-24}$ J T^{-1}
Nuclear magneton	$\mu_N = \dfrac{e\hbar}{2M_p}$	$5.050\,78 \times 10^{-27}$ J T^{-1}
Electron magnetic moment	$\mathcal{M}_e$	$9.284\,76 \times 10^{-24}$ J T^{-1} = $1.001\,16\ \mu_B$
Proton magnetic moment	$\mathcal{M}_p$	$1.410\,61 \times 10^{-26}$ J T^{-1} = $2.792\,85\ \mu_N$
Neutron magnetic moment	$\mathcal{M}_n$	$-0.966\,24 \times 10^{-26}$ J T^{-1} = $-1.913\,04\ \mu_N$

Table A14.2 Atomic units.

Quantity	*Unit*	*Physical significance*	*Value*
Mass	m or m_e	Electron mass	$9.109\ 38 \times 10^{-31}$ kg
Charge	e	Absolute value of electron charge	$1.602\ 18 \times 10^{-19}$ C
Angular momentum	$\hbar$	Planck's constant divided by (2π)	$1.054\ 57 \times 10^{-34}$ J s
Length	a_0	Bohr radius for atomic hydrogen (with infinite nuclear mass)	$5.291\ 77 \times 10^{-11}$ m
Velocity	$v_0 = \alpha c$	Magnitude of electron velocity in first Bohr orbit	$2.187\ 69 \times 10^{6}$ m s^{-1}
Momentum	$p_0 = mv_0$	Magnitude of electron momentum in first Bohr orbit	$1.992\ 85 \times 10^{-24}$ kg m s^{-1}
Time	$\dfrac{a_0}{v_0}$	Time required for electron in first Bohr orbit to travel one Bohr radius	$2.418\ 88 \times 10^{-17}$ s
Frequency	$\dfrac{v_0}{2\pi a_0}$	Angular frequency of electron in first Bohr orbit (v_0/a_0) divided by (2π)	$6.579\ 69 \times 10^{15}$ s^{-1}
Energy	$\dfrac{e^2}{4\pi\varepsilon_0 a_0} = \alpha^2 mc^2$	Twice the ionisation potential of atomic hydrogen (with infinite nuclear mass)	$4.359\ 74 \times 10^{-18}$ J = 27.2114 eV
Wave number	$\dfrac{\alpha}{2\pi a_0} = 2\tilde{R}(\infty)$	Twice the Rydberg constant, i.e. twice the wave number corresponding to the ionisation potential of atomic hydrogen (with infinite nuclear mass)	$2.194\ 75 \times 10^{7}$ m^{-1}
Electric field strength	$\dfrac{e}{(4\pi\varepsilon_0)a_0^2}$	Strength of the Coulomb field experienced by an electron in the first Bohr orbit of atomic hydrogen (with infinite nuclear mass)	$5.142\ 21 \times 10^{11}$ V m^{-1}
Magnetic field strength	$\dfrac{\hbar}{ea_0^2}$		$2.350\ 52 \times 10^{5}$ T

Table A14.3 Conversion factors.

1 Å (ångström) = 0.1 nm = 10^{-10} m = 10^{-8} cm

1 fm (femtometre or fermi) = 10^{-6} nm = 10^{-15} m

λ (in Å) $\times$ $\tilde{\nu}$ (in cm^{-1}) = 10^8 (from $\lambda\tilde{\nu} = 1$)

$a_0 = 5.291\ 77 \times 10^{-11}$ m = 0.529 177 Å

$a_0^2 = 2.800\ 28 \times 10^{-21}\ m^2$

$\pi a_0^2 = 8.797\ 35 \times 10^{-21}\ m^2$

1 Hz = 1 s^{-1}

1 electron mass (m_e) = 0.510 999 MeV/c^2

1 proton mass (M_p) = 938.272 MeV/c^2

1 a.m.u. = $\frac{1}{12}M_{^{12}C}$ = $1.660\ 54 \times 10^{-27}$ kg = 931.494 MeV/c^2

1 J = 10^7 erg = 0.239 cal = $6.241\ 51 \times 10^{18}$ eV

1 cal = 4.184 J = 2.611×10^{19} eV

1 eV = $1.602\ 18 \times 10^{-19}$ J = $1.602\ 18 \times 10^{-12}$ erg

1 MeV = $1.602\ 18 \times 10^{-13}$ J = $1.602\ 18 \times 10^{-6}$ erg

1 eV corresponds to:
- a frequency of $2.417\ 99 \times 10^{14}$ Hz (from $E = h\nu$)
- a wavelength of $1.239\ 84 \times 10^{-6}$ m = 12 398.4 Å (from $E = hc/\lambda$)
- a wave number of $8.065\ 44 \times 10^5\ m^{-1}$ = 8065.44 cm^{-1} (from $E = hc\tilde{\nu}$)
- a temperature of $1.160\ 45 \times 10^4$ K (from $E = k_BT$)

1 cm^{-1} corresponds to:
- an energy of $1.239\ 84 \times 10^{-4}$ eV
- a frequency of $2.997\ 92 \times 10^{10}$ Hz

1 atomic unit of energy = 27.2114 eV corresponds to:
- a frequency of $6.579\ 69 \times 10^{15}$ Hz
- a wavelength of $4.556\ 33 \times 10^{-8}$ m = 455.633 Å
- a wave number of $2.194\ 75 \times 10^7\ m^{-1}$ = 219 475 cm^{-1}
- a temperature of $3.157\ 75 \times 10^5$ K

1 a.m.u. corresponds to an energy of 931.494 MeV = $1.492\ 43 \times 10^{-10}$ J

$k_BT = 8.617\ 34 \times 10^{-5}$ eV at $T = 1$ K

$hc = 1.239\ 84 \times 10^{-6}$ eV $\times$ m = 12 398.4 eV $\times$ Å

$\hbar c = 1.973\ 27 \times 10^{-7}$ eV $\times$ m = 1973.27 eV $\times$ Å

ΔE (in eV) $\times$ Δt (in s) = $6.582\ 11 \times 10^{-16}$ eV $\times$ s (from $\Delta E \Delta t = \hbar$)

References

Adams, C.S., Carnal, O. and Mlynek, J. (1994) *Advances in Atomic, Molecular and Optical Physics* **34**, 1.

Adams, C.S. and Riis, E. (1997) *Progress in Quantum Electronics* **21**, 1.

Adams, C.S., Sigel, M. and Mlynek, J. (1994) *Physics Reports* **240**, 143.

Adhikari, S.K. (1998) *Variational Principles and the Numerical Solution of Scattering Problems*. Wiley-Interscience, New York.

Andersen, N. (1997) *Physics Reports* **279**, 251.

Bandrauk, A.D. (Ed.) (1993) *Molecules in Laser Fields*. M. Dekker, New York.

Bashkin, S. and Stoner, J.O. (1975) *Atomic Energy Levels and Grotrian Diagrams*. North-Holland, Amsterdam.

Batani, D., Joachain, C.J., Martelluci, S. and Chester, A.N. (Eds.) (2001) *Atoms, Solids and Plasmas in Super-Intense Laser Fields*. Kluwer Academic-Plenum Publishers, New York.

Baudon, J., Mathevet, R. and Robert, J. (1999) *Journal of Physics B: Atomic, Molecular and Optical Physics* **32**, R173.

Bergquist, J.C., Jefferts, S.R. and Wineland, D.J. (2001) *Physics Today*, March 2001, p. 37.

Berman, P.R. (Ed.) (1997) *Atom Interferometry*. Academic Press, New York.

Bethe, H.A. and Salpeter, E.E. (1957) *Quantum Mechanics of One- and Two-Electron Atoms*. Springer-Verlag, Berlin.

Blümel, R. and Reinhardt, W.P. (1997) *Chaos in Atomic Physics*. Cambridge University Press, Cambridge.

Born, M. and Wolf, E. (1975) *Principles of Optics* 5th Edn. Pergamon Press, Oxford.

Borse, H. and Motz, L. (Eds.) (1966) *The World of the Atom*. Basic Books, New York.

Bowers, R. and Deeming, T. (1984) *Astrophysics* (2 vols). Jones and Bartlett, Boston, Massachusetts.

Brabec, T. and Krausz, F. (2000) *Reviews of Modern Physics* **72**, 545.

Bransden, B.H. (1983) *Atomic Collision Theory* 2nd Edn. Benjamin, New York.

Bransden, B.H. and Joachain, C.J. (2000) *Quantum Mechanics* 2nd Edn. Prentice Hall, Harlow.

Bransden, B.H. and McDowell, M.R.C. (1992) *Charge Exchange and the Theory of Ion–Atom Collisions*. Clarendon Press, Oxford.

Burcham, W.E. and Jobes, M. (1995) *Nuclear and Particle Physics*. Longman, London.

Burke, P.G. and Berrington, K.A. (Eds.) (1993) *Atomic and Molecular Processes: an R-Matrix Approach*. Institute of Physics Publishing, Bristol and Philadelphia.

Burke, P.G. and Joachain, C.J. (1995) *Theory of Electron–Atom Collisions. Part 1: Potential Scattering*. Plenum Press, New York and London.

Burke, P.G. and Joachain, C.J. (Eds.) (1997) *Photon and Electron Collisions with Atoms and Molecules*. Plenum Press, New York and London.

Burke, P.G. and Robb, W.D. (1975) *Advances in Atomic and Molecular Physics* **11**, 143.

Burnett, K., Reed, V.C. and Knight, P.L. (1993) *Journal of Physics B: Atomic, Molecular and Optical Physics* **26**, 561.

Byron, F.W. and Fuller, R.W. (1969) *Mathematics of Classical and Quantum Physics*. Addison-Wesley, Reading, Massachusetts.

Byron, F.W. and Joachain, C.J. (1977) *Physics Reports* **34** C, 233.

Byron, F.W. and Joachain, C.J. (1989) *Physics Reports* **179**, 211.

Casimir, H.B.C. (1963) *On the Interaction Between Atomic Nuclei and Electrons*. Freeman, San Francisco.

Chu, S.I. (1985) *Advances in Atomic and Molecular Physics* **21**, 197.

Clementi, E. and Roetti, C. (1974) *Atomic Data and Nuclear Data Tables* Vol. **14**, Nos 3–4. Academic Press, New York.

Codling, K. and Frasinski, L.J. (1994) *Contemporary Physics* **35**, 243.

Condon, E.U. and Shortley, G.H. (1951) *Theory of Atomic Spectra*. Cambridge University Press, London.

Connerade, J.P. (1998) *Highly Excited Atoms*. Cambridge University Press, Cambridge.

Demtröder, W. (1996) *Laser Spectroscopy* 2nd Edn. Springer-Verlag, Berlin.

Dirac, P.A.M. (1958) *The Principles of Quantum Mechanics* 4th Edn. Oxford University Press, Oxford.

Drake, G.W.F. (Ed.) (1996) *Atomic, Molecular and Optical Physics Handbook*. American Institute of Physics, New York.

Dreizler, R.M. and Gross, E.K.U. (1990) *Density Functional Theory*. Springer-Verlag, Berlin.

Duffin, W.J. (1968) *Advanced Electricity and Magnetism for Undergraduates*. McGraw-Hill, London.

Dwight, H.B. (1961) *Tables of Integrals and Other Mathematical Data*. Macmillan, London.

Edmonds, A.R. (1957) *Angular Momentum in Quantum Mechanics*. Princeton University Press, Princeton, New Jersey.

Englert, B.-G. (1988) *Semiclassical Theory of Atoms*. Springer-Verlag, Berlin.

Faisal, F.H.M. (1987) *Theory of Multiphoton Processes*. Plenum Press, New York.

Friedrich, H. (1998) *Theoretical Atomic Physics* 2nd Edn. Springer-Verlag, Berlin.

Gallagher, T.F. (1994) *Rydberg Atoms*. Cambridge University Press, Cambridge.

Gavrila, M. (Ed.) (1992) *Advances in Atomic, Molecular and Optical Physics* Suppl. **1**, 435.

Giusti-Suzor, A., Mies, F.H., DiMauro, L.F., Charron, E. and Yan, B. (1995) *Journal of Physics B: Atomic, Molecular and Optical Physics* **28**, 309.

Goldstein, H. (1980) *Classical Mechanics* 2nd Edn. Addison-Wesley, Reading, Massachusetts.

Grimm, R., Weidemüller, M. and Ovchinnikov, Y.B. (2000) *Advances in Atomic, Molecular and Optical Physics* **42**, 95.

Gupta, R. (1993) *Laser Spectroscopy (selected reprints)*. American Association of Physics Teachers, College Park, Pennsylvania.

Hänsch, T.W. and Walther, H. (1999) *Reviews of Modern Physics* **71**, S 242.

Harrison, E. (2000) *Cosmology* 2nd Edn. Cambridge University Press. Cambridge.

Helmerson, K., Hutchinson, D., Burnett, K. and Phillips W.D. (1999) *Physics World*, August 1999, p. 31.

Herzberg, G. (1945) *Infrared and Raman Spectra of Polyatomic Molecules*. Van Nostrand, New York.

Herzberg, G. (1950) *Spectra of Diatomic Molecules* 2nd Edn. Van Nostrand, New York.

Jackson, J.D. (1998) *Classical Electrodynamics* 3rd Edn. Wiley, New York.

Jastrow, R. and Thompson, M.H. (1977) *Astronomy: Fundamentals and Frontiers* 3rd Edn. Wiley, New York.

Joachain, C.J. (1983) *Quantum Collision Theory* 3rd Edn. North-Holland, Amsterdam.

Joachain, C.J. (1989) *Electron and Photon Collisions with Atoms*, in *Collision Theory for Atoms and Molecules*, ed. F. Gianturco. Plenum Press, New York, p. 59.

Joachain, C.J. (1994) *Theory of Laser-Atom Interactions*, in *Laser Interactions with Atoms, Solids and Plasmas*, ed. R.M. More. Plenum Press, New York, p. 39.

Joachain, C.J., Dörr, M. and Kylstra, N.J. (2000) *Advances in Atomic, Molecular and Optical Physics* **42**, 225.

Kessler, J. (1985) *Polarized Electrons* 2nd Edn. Springer-Verlag, Berlin.

King, F.W. (1999) *Advances in Atomic, Molecular and Optical Physics* **40**, 57.

Kittel, C. (1996) *Introduction to Solid State Physics* 7th Edn. Wiley, New York.

Kittel, C. and Kroemer, H. (1980) *Thermal Physics*. Freeman, San Francisco.

Knight, D.M. (1967) *Atoms and Elements: a Study of Theories of Matter in England in the 19th Century*. Hutchinson, London.

Knight, P.L., Lauder, M.A. and Dalton, B.J. (1990) *Physics Reports* **190**, 1.

Kroto, H.W., Fischer, J.E. and Cox, D.E. (Eds.) (1993) *The Fullerenes*. Pergamon Press, Oxford.

Kuhn, H.G. (1970) *Atomic Spectra* 2nd Edn. Longman, London.

Lambropoulos, P. (1976) *Advances in Atomic and Molecular Physics* **12**, 87.

Lambropoulos, P. and Tang, X. (1992) in *Atoms in Intense Laser Fields*, ed. M. Gavrila. Academic Press, Boston, p. 335.

Landau, L.D. and Lifshitz, E.M. (1965) *Quantum Mechanics. Nonrelativistic Theory*. Pergamon Press, Oxford.

Lemonde, P. (2001) *Physics World*, January 2001, p. 39.

Lepp, S., Stancil, P.C. and Dalgarno, A. (2002) *Journal of Physics B: Atomic, Molecular and Optical Physics* **35**, R57.

Major, F.G. (1998) *The Quantum Beat: The Physical Principles of Atomic Clocks*. Springer-Verlag, New York.

Marion, J.B. and Heals, M.A. (1980) *Classical Electromagnetic Radiation* 2nd Edn. Academic Press, New York.

Mathews, J. and Walker, R.L. (1973) *Mathematical Methods of Physics*. World Student Series Edn. Benjamin, New York.

Merzbacher, E. (1998) *Quantum Mechanics* 3rd Edn. Wiley, New York.

Messiah, A. (1968) *Quantum Mechanics*. North-Holland, Amsterdam.

Meystre, P. (2001) *Atom Optics*. Springer-Verlag, New York.

Mittleman, M.H. (1993) *Introduction to the Theory of Laser–Atom Interactions* 2nd Edn. Plenum Press. New York.

Mizushima, M. (1975) *The Theory of Rotating Diatomic Molecules*. Wiley, New York.

Morse, P.M. (1966) *Thermal Physics*. Benjamin, New York.

Parr, R.G. and Yang, W. (1989) *Density Functional Theory of Atoms and Molecules*. Oxford University Press, New York.

Pauling, L. and Wilson, E.B. (1935) *Introduction to Quantum Mechanics*. McGraw-Hill, New York.

Peterkop, R.K. (1977) *Theory of Ionization of Atoms by Electron Impact*. Colorado University Press.

Pilar, F.L. (1968) *Elementary Quantum Chemistry*. McGraw-Hill, New York.

Potvliege, R.M. and Shakeshaft, R. (1992) in *Atoms in Intense Laser Fields*, ed. M. Gavrila. Academic Press, Boston, p. 373.

Protopapas, M., Keitel, C.H. and Knight, P.L. (1997) *Reports on Progress in Physics* **60**, 389.

Ramsey, N.F. (1953) *Nuclear Moments*. Wiley, New York.

Ramsey, N.F. (1985) *Molecular Beams* 2nd Edn. Oxford University Press, London.

Richtmyer, F.K., Kennard, E.H. and Cooper, J.N. (1969) *Introduction to Modern Physics* 6th Edn. McGraw-Hill, New York.

Rose, M.E. (1957) *Elementary Theory of Angular Momentum*. Wiley, New York.

Sakurai, J.J. (1967) *Advanced Quantum Mechanics*. Addison-Wesley, Reading, Massachusetts.

Schiff, L.I. (1968) *Quantum Mechanics* 3rd Edn. McGraw-Hill, New York.

Scully, M.O. and Zubairy, M.S. (1997) *Quantum Optics*. Cambridge University Press, Cambridge.

Seaton, M.J. (1983) *Reports on Progress in Physics* **46**, 167.

Sengstock, K. and Ertmer, W. (1995) *Advances in Atomic, Molecular and Optical Physics* **35**, 1.

Series, G.W. (Ed.) (1988) *The Spectrum of Atomic Hydrogen: Advances*. World Scientific, Singapore.

Shakeshaft, R. and Spruch, L. (1979) *Reviews of Modern Physics* **51**, 369.

Slevin, J. (1984) *Reports on Progress in Physics* **47**, 461.

Sobelman, I.I. (1979) *Atomic Spectra and Radiative Transitions*. Springer-Verlag, Berlin.

Spruch, L. (1962) in *Lectures in Theoretical Physics*. Interscience, New York, **4**, 161.

Svanberg, S. (2001) *Atomic and Molecular Spectroscopy* 3rd Edn. Springer-Verlag, Berlin.

Svelto, O. (1998) *Principles of Lasers* 4th Edn. Plenum Press, New York and London.

Taylor, E.F. and Wheeler, J.A. (1966) *Spacetime Physics*. Freeman, San Francisco.

Thompson, R.C. (1993) *Advances in Atomic, Molecular and Optical Physics*, **31**, 63.

Thorp, J.S. (1969) *Masers and Lasers*. Macmillan, London.

Tonomura, A. (1993) *Electron Holography*. Springer-Verlag, New York.

Townes, C.H. and Schawlow, A.L. (1975) *Microwave Spectroscopy*. Dover, New York.

Walther, H. (1993) *Advances in Atomic, Molecular and Optical Physics*, **31**, 137.

Walther, H. (1994) *Advances in Atomic, Molecular and Optical Physics*, **32**, 379.

Weiner, J., Bagnato, V.S., Zilco, S. and Julienne, P.S. (1999) *Reviews of Modern Physics* **71**, 1.

Wheeler, J.A. (1978) *Mathematical Foundations of Quantum Theory*, ed. A.R. Marlow. Academic Press, New York.

Whittaker, E.T. and Watson, G.N. (1935) *A Course of Modern Analysis* 4th Edn. Cambridge University Press, London.

Winter, H. and Aumayr, F. (1999) *Journal of Physics B: Atomic, Molecular and Optical Physics* **32**, R39.

Author index

Subject index